Table of Atomic Weights

Elements by Name, Symbol, Atomic Number, and Atomic Weight
(Atomic weights are given to four significant figures for elements below atomic number 104. See Appendix D.)

Name	Symbol	Atomic Number	Atomic Weight	Name	Symbol	Atomic Number	Atomic Weight
Actinium	Ac	89	227.0	Mercury	Hg	80	200.6
Aluminum	Al	13	26.98	Molybdenum	Mo	42	95.94
Americium	Am	95	243.1	Neodymium	Nd	60	144.2
Antimony	Sb	51	121.8	Neon	Ne	10	20.18
Argon	Ar	18	39.95	Neptunium	Np	93	237.0
Arsenic	As	33	74.92	Nickel	Ni	28	58.69
Astatine	At	85	210.0	Nielsbohrium	Ns	107	—
Barium	Ba	56	137.3	Niobium	Nb	41	92.91
Berkelium	Bk	97	247.1	Nitrogen	N	7	14.01
Beryllium	Be	4	9.012	Nobelium	No	102	259.1
Bismuth	Bi	83	209.0	Osmium	Os	76	190.2
Boron	B	5	10.81	Oxygen	O	8	16.00
Bromine	Br	35	79.90	Palladium	Pd	46	106.4
Cadmium	Cd	48	112.4	Phosphorus	P	15	30.97
Calcium	Ca	20	40.08	Platinum	Pt	78	195.1
Californium	Cf	98	252.1	Plutonium	Pu	94	239.1
Carbon	C	6	12.01	Polonium	Po	84	210.0
Cerium	Ce	58	140.1	Potassium	K	19	39.10
Cesium	Cs	55	132.9	Praseodymium	Pr	59	140.9
Chlorine	Cl	17	35.45	Promethium	Pm	61	144.9
Chromium	Cr	24	52.00	Protactinium	Pa	91	231.0
Cobalt	Co	27	58.93	Radium	Ra	88	226.0
Copper	Cu	29	63.55	Radon	Rn	86	222.0
Curium	Cm	96	247.1	Rhenium	Re	75	186.2
Dysprosium	Dy	66	162.5	Rhodium	Rh	45	102.9
Einsteinium	Es	99	252.1	Rubidium	Rb	37	85.47
Erbium	Er	68	167.3	Ruthenium	Ru	44	101.1
Europium	Eu	63	152.0	Rutherfordium	Rf	104	—
Fermium	Fm	100	257.1	Samarium	Sm	62	150.4
Fluorine	F	9	19.00	Scandium	Sc	21	44.96
Francium	Fr	87	223.0	Seaborgium[1]	Sg	106	—
Gadolinium	Gd	64	157.2	Selenium	Se	34	78.96
Gallium	Ga	31	69.72	Silicon	Si	14	28.09
Germanium	Ge	32	72.59	Silver	Ag	47	107.9
Gold	Au	79	197.0	Sodium	Na	11	22.99
Hafnium	Hf	72	178.5	Strontium	Sr	38	87.62
Hahnium	Ha	105	—	Sulfur	S	16	32.07
Hassium	Ha	108	—	Tantalum	Ta	73	180.9
Helium	He	2	4.003	Technetium	Tc	43	98.91
Holmium	Ho	67	164.9	Tellurium	Te	52	127.6
Hydrogen	H	1	1.008	Terbium	Tb	65	158.9
Indium	In	49	114.8	Thallium	Tl	81	204.4
Iodine	I	53	126.9	Thorium	Th	90	232.0
Iridium	Ir	77	192.2	Thulium	Tm	69	168.9
Iron	Fe	26	55.85	Tin	Sn	50	118.7
Krypton	Kr	36	83.80	Titanium	Ti	22	47.88
Lanthanum	La	57	138.9	Tungsten	W	74	183.8
Lawrencium	Lr	103	260.1	Uranium	U	92	238.0
Lead	Pb	82	207.2	Vanadium	V	23	50.94
Lithium	Li	3	6.941	Xenon	Xe	54	131.3
Lutetium	Lu	71	175.0	Ytterbium	Yb	70	173.0
Magnesium	Mg	12	24.30	Yttrium	Y	39	88.91
Manganese	Mn	25	54.94	Zinc	Zn	30	65.39
Meitnerium	Mt	109	—	Zirconium	Zr	40	91.22
Mendelevium	Md	101	256.1				

[a]Names of elements 104, 105, and 107 to 109 have been endorsed by a committee of the American Chemical Society. The IUPAC recommends different names for elements 104 to 108.
[1]Proposed symbol and name

THEORY AND PRACTICE OF WATER AND WASTEWATER TREATMENT

THEORY AND PRACTICE OF WATER AND WASTEWATER TREATMENT

RONALD L. DROSTE

John Wiley & Sons, Inc.
New York • Chichester • Brisbane • Toronto • Singapore

Acquisitions Editor	Cliff Robichaud
Production Editor	Ken Santor
Designer	Laura Nicholls
Manufacturing Manager	Mark Cirillo
Illustration	Jaime Perea

This book was set in 10/12 Times Roman by Bi-Comp, Inc. and printed and bound by Hamilton Printing Company. The cover was printed by Lehigh Press, Inc.

Recognizing the importance of preserving what has been written, it is a policy of John Wiley & Sons, Inc. to have books of enduring value published in the United States printed on acid-free paper, and we exert our best efforts to that end.

The paper in this book was manufactured by a mill whose forest management programs include sustained yield harvesting of its timberlands. Sustained yield harvesting principles ensure that the number of trees cut each year does not exceed the amount of new growth.

Library of Congress Cataloging-in-Publication Data

Droste, Ronald L.
Theory and practice of water and wastewater treatment/by Ronald L. Droste.
p. cm.
Includes index.
ISBN 0-471-12444-3 (cloth : alk. paper)
1. Water—Purification. 2. Sewage—Purification. I. Title.
TD430.D76 1997 96-15477
628.1′62—dc20 CIP

Printed in the United States of America

10 9 8 7 6 5

ACKNOWLEDGMENTS

This book is dedicated to my family, who have shown endless support. The author is deeply indebted to previous instructors, colleagues, and students who have contributed to this work in numerous ways. There are four among the many who have made useful suggestions and to whom I wish to express my gratitude for extraordinary efforts: Ms. N. M. Gaudet, Mr. Z. Lukic, Mr. R. Marshall, and Dr. R. M. Narbaitz. I also wish to acknowledge the professional work of the staff at Wiley.

PREFACE

This book has been written as an undergraduate text based on the typical background of a civil engineering student. Necessary background information for design and assessment of treatment processes is included in the first sections. These sections do include some review of basic concepts from other courses and should not be considered as replacing the need for these courses. However, these concepts have been focused upon the needs of the water quality environmental engineer.

Basic theory of most treatment processes is presented in later sections. The variety of treatment processes is increasing and the issues surrounding their application are more subtle. An introductory coverage of most of the processes is given with more emphasis on common applications. Processes are grouped according to theoretical principles rather than their occurrence in water or wastewater treatment operations. It has been the experience of the author that many books on the subject simply present formulae without any detailed development. This book addresses that problem. The reader is made aware of theoretical and empirical developments. In addition to text references, key references are provided for consultation by the student who wishes to examine a topic in more detail or consult another presentation on the topic.

This book has a more intense focus on the mechanics of the processes used to treat water or wastewater. For instance, design of water distribution and wastewater collection systems are applied hydraulics exercises that are left to other courses. The theory presented in this book is at the level of an undergraduate. In the past much of this basic theory was left for the graduate level. As need and room in the undergraduate curriculum for environmental engineering grows, the opportunity now exists to bring undergraduates to a level of theoretical comprehension equivalent to that gained in other disciplines.

The material in this book includes substantial portions of three undergraduate courses given at the University of Ottawa. The first course covers Chapters 1–9, Sections 15.1–15.4, Chapter 16, and Chapter 21. This course is accompanied by weekly laboratory exercises. The second course covers Chapters 10–15 and a few introductory sections from Chapter 17. The remaining material is covered in an elective course with additional material. The extent of topic coverage in each course has changed as the environmental engineering curriculum and support courses have evolved. The material in the book can be organized in numerous ways depending on the backgrounds of students and the breadth and depth of the environmental engineering curriculum.

As a note to the reader, the theory presented is the first principles for design and operation of treatment processes. These principles govern the major phenomena occurring in a process but, as with all models, they are idealizations. Field studies are always required to confirm or refine the basic theory results into processes that will meet the objectives. Typical design ranges are given for most processes. There is a growing body of design handbooks available that note many practical considerations and experiences for consideration in design and process analysis. At the end of each chapter key references are supplied for the readers who wish to expand their coverage of the topics in the chapter.

Advanced study will yield understanding of missing components in some models and modify results obtained from basic models to results closer to those used in practice. But understanding the basic theory has often been the reason for significant advancements in the field. The development of tube and lamella clarifiers and new configurations for anaerobic biological wastewater treatment are two instances in which understanding and application of basic concepts for the former have resulted in detention time reductions of 50% or more and, in the latter, reductions by a factor of 10 or more.

Abbreviations and Acronyms Used in the Text

ABES–Associação Brasileria de Engenharia Sanitária e Ambiental
Acy–acidity
acu–apparent color unit
ADP–adenosine diphosphate
AIDS–acquired immunodeficiency syndrome
Alk–alkalinity
A/O–anaerobic/oxic
AMP–adenosine monophosphate
APHA–American Public Health Association
aq–aqueous
ASCE–American Society of Civil Engineers
ASME–American Society of Mechanical Engineers
asu–areal standard unit
ATA–anaerobic toxicity assay
atm–atmosphere
ATP–adenosine triphosphate
AWWA–American Water Works Association
BAT–best available technology
BDST–bed-depth-service time
BMP–biomethane potential
BOD–biochemical oxygen demand
BOD_u–ultimate BOD
bp–boiling point
Bq–Becquerel
Btu–British thermal unit
cap–capita
c/c–center to center
CCME–Canadian Council of Ministers of the Environment
CFU–colony forming unit
CH–carbonate hardness
Ci–curie
CM–complete mixed
COD–chemical oxygen demand
CSCE–Canadian Society of Civil Engineers
CSTR–completely stirred tank reactor
1-D–one-dimensional
DC–direct current
DDT–dichloro-diphenyl-trichloroethane
deg–degree
DF–dilution factor
DNA–deoxyribonucleic acid
DO–dissolved oxygen
DPD–*N,N*-diethyl-*p*-phenylenediamine
DSFF–downflow stationary fixed film
DST–defined substrate technology
EBCT–empty bed contact time
EC_{50}–see LC_{50}
EDTA–ethylene-diamine-tetraacetic acid
EMF–electromotive force
EPA–Environmental Protection Agency
eq–equivalent
ES–effective size
FAO–Food and Agriculture Organization
FC–fecal coliforms
FD–filtration and disinfection
FE–free energy
F:M–food: microorganism ratio
Fr–Froude number
FS–fecal streptococci
FSS–fixed suspended solids
ft–foot (feet)
g–gaseous; gram
GAC–granular activated carbon
gal–gallon (U.S.)
GLUMRB–Great Lakes Upper Mississippi River Board
GTP–guanosine triphosphate
HIV–human immunodeficiency viruses
h–hour
ha–hectare
hp–horsepower
HPC–heterotrophic plate count
HRT–hydraulic retention time
IAWPRC–International Association on Water Pollution Research and Control
IAWQ–International Association on Water Quality
i.d.–internal diameter
IJC–International Joint Commission
IMAC–interim maximum acceptable concentration
in.–inch
IRC–International Reference Centre
ISS–inert suspended solids
IUPAC–International Union of Pure and Applied Chemistry
L–length; liter
1–liquid
lb–pound
LC_{50}–lethal concentration for 50% of the organisms
LHS–left-hand side
LI–Langelier index
ln–natural (base *e*) or Naperian logarithm
log–base 10 logarithm
m–molal; meter
M–mass
M–molar
MAC–maximum acceptable concentration

Abbreviations and Acronyms Used in the Text (continued)

MAF–Ministry of Agriculture and Food
MCL–maximum contaminant level
MCLG–maximum contaminant level goal
MF–membrane filter
Mgal/d–million gallons (U.S.) per day
min–minute
ML–mixed liquor
MLVSS–mixed liquor volatile suspended solids
MLTSS–mixed liquor total suspended solids
mo.–month
MOE–Ministry of the Environment
mp–melting point
MPN–most probable number
MUG–methylumbelliferyl-β-*d*-glucuronide
MW–molecular weight
N–Newton
N–normal
NAD–nicotinamide adenine dinucleotide
NADP–nicotinamide adenine dinucleotide phosphate
NCH–noncarbonate hardness
ND–nondetectable
NOD–nitrogenous oxygen demand
NOEC–no observed effect concentration
NTA–nitriloacetic acid
NTU–nephelometric turbidity unit
o.d.–outside diameter
OF–overland flow land treatment
OI–odor index
ONPG–*o*-nitrophenol-β-*d*-galactopyranoside
ORP–oxidation–reduction potential
OUR–oxygen uptake rate
P/A–presence–absence
PAC–powdered activated carbon
PACT–powdered activated carbon activated sludge process
PF–plug flow
PFU–plaque forming unit
PHB–poly-β-hydroxybutyric acid
psi–pounds per square inch
PVC–polyvinylchloride
rad–roengten absorption dose
Re–Reynold's number
redox–oxidation–reduction
rem–roengten-equivalent-man
rev–revolution
RHS–right-hand side
RI–rapid infiltration land treatment
rms–root mean square
RO–reverse osmosis
rpm–revolutions per minute
s–solid; second(s)
SBR–sequencing batch reactor
SDWA–Safe Drinking Water Act
s.g.–specific gravity
SHE–standard hydrogen electrode
SI–Système International; saturation index
SOC–synthetic organic chemical
SR–slow rate land treatment
SRT–solids retention time
SS–suspended solids
STP–standard temperature and pressure
Sv–sievert
SVI–sludge volume index
SWTR–Surface Water Treatment Rule
T–time
TBOD–total biochemical oxygen demand
TC–total coliforms
TCA–tricarboxylic acid
TDS–total dissolved solids
TH–total hardness
THM–trihalomethane
TKN–total Kjeldahl nitrogen
TNTC–too numerous to count
TOC–total organic carbon
TON–threshold odor number
TOX–total organic halides
TSS–total suspended solids
TU–turbidity unit
TVA–Tennessee Valley Authority
UASB–upflow anaerobic sludge blanket
U.K.–United Kingdom
UNICEF–United Nations International Children's Emergency Fund
U.S.–United States
USEPA–United States Environmental Protection Agency
UV–ultraviolet
VOC–volatile organic chemical
VS–volatile solids
VSS–volatile suspended solids
W–watt
WAS–waste activated sludge
WEF–Water Environment Federation
WHO–World Health Organization
wk–week
WPCF–Water Pollution Control Federation
yr–year

CONTENTS

SECTION I

CHEMISTRY

CHAPTER 1

BASIC CHEMISTRY

This chapter gives a review of chemistry definitions and fundamentals to refresh and supplement background obtained in first courses. Our view is toward the applied chemistry of water and wastewater and their treatment.

1.1 DEFINITIONS

A fundamental substance that cannot be further decomposed by ordinary chemical means is an element. The smallest unit of an element is an atom. The names, symbols, atomic numbers, and atomic weights of the elements are provided for convenience in a periodic chart and a table located inside the front cover of the book. The symbols are not always the first letter(s) of the element names because they are derived from the Latin, Greek, or German names of the elements.

The atomic number is the number of protons in the atom, whereas the atomic weight reflects the number of protons and neutrons contained in the atom. Atoms with the same number of protons but different atomic weights are isotopes of an element. Atomic weights of the elements are referred to the weight of the ^{12}C isotope of carbon, which was assigned a value of exactly 12 by the International Union of Pure and Applied Chemistry (IUPAC).[1] The atomic weight also reflects the relative occurrence of isotopes of a given element. Isotopes contain the same number of protons, and therefore electrons, but differ in the number of neutrons. The precision of the atomic weight depends upon the natural variation in occurrence of the isotopes. The atomic weights are all constant within three significant digits, which is suitable for engineering calculations.

The gram atomic weight refers to the quantity of an element in grams equal to the atomic weight of the element. One gram atomic weight of an element or compound contains one Avogadro number (6.023×10^{23}) of each of the atoms contained in the chemical formula. The gram molecular weight refers to the molecular weight (MW) of a compound in grams.

1.2 THE EXPRESSION OF CONCENTRATION

Concentration of ions in a solution is expressed on the following bases: molar, molal, mole fraction, or mass concentration.

Molar. Molar concentration is the number of moles per volume of solution indicated by M. A 1 M solution contains one mole per liter.

[1]Abbreviations and acronyms used in the text are defined after the preface.

Molal. Molal concentration expresses the number of moles per 1 000 g of solvent (water). A molal concentration is indicated by m. Density and volume of a solution are influenced by the characteristics and quantities of dissolved substances and the temperature of the solution; thus the volume of 1 000 g of water containing dissolved or suspended substances will not exactly occupy 1 L. (Also there was a slight error in the density determination of pure water.) However, these differences are insignificant and molar and molal concentrations are practically equal for water solutions of interest to environmental engineers.

Mole Fraction. The equation defining the mole fraction x_i of the ith substance is

$$x_i = \frac{n_i}{\Sigma n_j}$$

where

n is the number of moles

the denominator is the total number of moles of all substances, including the solvent, in the solution (for k substances, $j = 1$ to k)

Expression of concentration on a mole fraction basis is most useful for thermodynamic analysis.

Mass Concentration. Concentration is often expressed in terms of parts per million parts (ppm) or mg/L. Sometimes parts per thousand (ppt) or parts per billion (ppb) are also used. The concentration of solute X in solvent Y in ppm is

$$\text{ppm} = \frac{\text{mass of substance}}{\text{mass of solution}} \qquad \text{e.g.,} \qquad \frac{x \text{ g of X}}{10^6 \text{ g of (Y + X)}} = x \text{ ppm}$$

Because 1 kg of solution with water as a solvent has a volume of approximately 1 liter,

$$1 \text{ ppm} \approx 1 \text{ mg/L}$$

Square brackets, [ ], around a species indicate concentration of the species (e.g., [Cl_2] indicates concentration of Cl_2). The concentration may be expressed in any of the above forms in an equation. The reader is cautioned to take note of the definition of concentration used in each equation.

Tables 1.1, 1.2, and 1.3, given later in this chapter, list various substances found in water. Many important elements exist in various combinations with other elements as discussed in the following sections. The concentrations of many compounds are often reported in terms of the key element in the compound. This is indicated by writing the chemical formula for the compound hyphen (-) the key element. For instance, the concentration of ammonia (NH_3) can be written as [NH_3] or [NH_3-N]. The former indicates the concentration of ammonia as ammonia and the latter indicates the concentration of ammonia as nitrogen. The -N is also written "as N," i.e., [NH_3 as N]. The transformation between the two concentration expressions is based on the gram molecular weight of the entity and the gram atomic weight of the element. For ammonia the gram molecular weight is 17 g and the gram atomic weight of nitrogen is 14 g. Therefore, 1 mg/L of NH_3 is equivalent to 14/17 mg/L or 0.82 mg/L of NH_3-N and vice versa. If only the chemical formula is indicated then the concentration is taken as being of the entity itself.

Sometimes the concentration of one species is expressed in terms of another chemical entity based on chemical reactions. Expressing the concentration in terms of one element or entity is convenient when the element is undergoing transformation from one species to another or when many substances have the same chemical effects.

■ Example 1.1 Expression of Concentration

A water contains 3.10 mg/L of H_2S, 1.40 mg/L of HS^-, and 26.50 mg/L of SO_4^{2-}. Express the concentrations of each species as S and as SO_4^{2-}. Base the SO_4^{2-} equivalency on the number of sulfur atoms in the entity. The elemental weight[2] of sulfur is 32.1 and the gram molecular weight of SO_4^{2-} is 96.1. The table below expresses the concentrations in the various forms.

		As S			As SO_4^{2-}		
Substance	[ ], mg/L	Factor[a]		[ ], mg/L	Factor[a]		[ ], mg/L
H_2S	3.10	32.1/34.1	H_2S-S	2.92	96.1/34.1	H_2S-SO_4^{2-}	8.74
HS^-	1.40	32.1/33.1	HS^--S	1.36	96.1/33.1	HS^--SO_4^{2-}	4.06
SO_4^{2-}	26.50	32.1/96.1	SO_4^{2-}-S	8.85	96.1/96.1	SO_4^{2-}-SO_4^{2-}	26.50
			Total	13.13		Total	39.30

[a] The factors are related to the actual entity.

As indicated in the table the total concentrations of all species as S or SO_4^{2-} are 13.13 and 39.30 mg/L, respectively.

1.3 IONS, MOLECULES, AND BONDING

Molecules are groups of atoms bonded together. Some atoms or groups of atoms (radicals) are able to completely give up or acquire one or more electrons. This results in a net charge on the species and allows it to bond electrostatically to an oppositely charged species. Charged species are called ions. The valence of an ion is equal to the number of electrons that the atom or group of atoms has lost or gained (see Table 1.1 for examples). In a solution, an ionic compound separates into its component ions. As more compound is added this continues to occur until the solution becomes saturated with the ions and precipitation (formation of a distinct solid phase) of the ionic compound occurs.

Whether a radical or compound actually exists is often a topic for advanced chemistry. Refer to Tables 1.1 and 1.2 and other tables throughout this book, which list information on common elements, radicals, and compounds of concern to environmental engineers.

The other major type of bond is a covalent bond, in which electrons are shared among atoms as opposed to being transferred between them. Covalent bonds are generally stronger than ionic bonds. Atoms forming a radical, which exhibits ionic behavior in solution (has a charge on the entity), are covalently bonded. In covalent bonds between like atoms such as Cl_2, and O_2, the electrons are equally shared. But when different atoms are covalently bonded, one atom may more strongly attract the electrons than the other, causing a negative charge distribution that is different from

[2]The atomic weights of all elements are given inside the front cover of the book. The periodic table of the elements is given inside the front cover of the book.

the positive charge distribution between the atoms. The bond is polar. The limiting case of a polar bond is an ionic bond. Polar bonds lead to formation of secondary bonds among compounds. These secondary bonds influence volatility, viscosity, melting point, and other characteristics of a solution or the pure substance. Ammonia (NH_3) is an example of a polar compound.

Resonance is another bonding phenomenon, wherein electrons are shared among atoms bonded together. Lewis structures are based on a theory stating, in its most elementary form, that atoms in a compound or ion share, give up, or take up electrons to attain the configuration of the nearest inert element. For many substances it is possible to write more than one Lewis structure. In these instances the compound is more stable (it takes more energy to break the bonds) than if only one structure could be written. Examples are given in Chapter 4.

Another significant secondary bond is the hydrogen bond, which is largely electrostatic in nature. When hydrogen is bonded to an atom that is highly electronegative (exerts a strong attraction for electrons) such as O, N, or F, it acts like a bare proton and is attracted to negative ions and the negative polar ends of other molecules. The energy of a hydrogen bond is generally around 5–10% of the energy of a typical covalent bond. Formation of hydrogen bonds significantly alters properties of solutions.

Overall stability of an ion or compound depends on a combination of many factors (Table 1.1), some of which have been touched on above, and others such as steric constraints that have not been mentioned.

The polarity and associated H-bond properties of the water molecule (Table 1.2) dictate to a large extent its chemical behavior. Water has a higher melting point, boiling point, heat of vaporization, and heat of fusion than other comparable hydrides (a hydride is a compound containing H and another element) such as H_2S, NH_3, and other common liquids. Because of its polarity, water is able to dissolve ionic and polar compounds. These properties are very important for life processes and environmental phenomena.

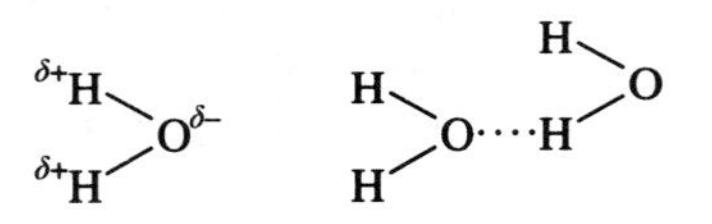

1.3.1 Oxidation Number

The oxidation number of an atom in a compound depends on the number of electrons that are associated with it. The oxidation number for an atom can vary depending on the compound. In compounds the electrons are associated with the atoms according to their electronegativity; electrons are assigned to the more electronegative element. The rules for assigning oxidation numbers are as follows.

1. The algebraic sum of the oxidation numbers of all atoms in a neutral compound is zero; otherwise, the sum must equal the charge on the ionic species.
2. The oxidation number of all elements in the free state is zero. When an atom is bonded to another chemically identical element, this bond makes no contribution to the oxidation state of either atom.
3. Alkali metals (those in the first column of the periodic table) assume a +1 oxidation number in their compounds; the alkaline earth elements (those in the second column) assume an oxidation number of +2.

TABLE 1.1 Basic Information on Common Elements

Element	Symbol	Atomic weight	Common valence[a]	Equivalent weight[b]	Other valence states
Aluminum	Al	27.0	3^+	9.0	
Arsenic	As	74.9	3^+	25.0	
Barium	Ba	137.3	2^+	68.7	
Boron	B	10.8	3^+	3.6	
Bromine	Br	79.9	1^-	79.9	
Cadmium	Cd	112.4	2^+	56.2	
Calcium	Ca	40.1	2^+	20.0	
Carbon	C	12.0	4^-, 4^+	3.0, 3.0	
Chlorine	Cl	35.5	1^-	35.5	
Chromium	Cr	52.0	3^+, 6^+	17.3, 8.7	2^+
Copper	Cu	63.5	2^+, 1^+	31.8, 63.5	
Fluorine	F	19.0	1^-	19.0	
Gold	Au	197.0	3^+	65.7	1^+
Hydrogen	H	1.0	1^+	1.0	
Iodine	I	126.9	1^-	126.9	
Iron	Fe	55.8	2^+, 3^+	27.9, 18.6	
Lead	Pb	207.2	2^+	103.6	4^+
Magnesium	Mg	24.3	2^+	12.2	
Manganese	Mn	54.9	2^+ 4^+, 7^+	27.5	
Mercury	Hg	200.6	2^+	100.3	1^+
Nickel	Ni	58.7	2^+	29.4	
Nitrogen	N	14.0	3^-, 5^+	4.7, 2.8	1^+, 2^+, 3^+, 4^+
Oxygen	O	16.0	2^-	8.0	
Phosphorus	P	31.0	5^+	6.0	3^-, 3^+
Potassium	K	39.1	1^+	39.1	
Selenium	Se	79.0	6^+	13.1	
Silicon	Si	28.1	4^+	6.5	4^-
Silver	Ag	107.9	1^+	107.9	
Sodium	Na	23.0	1^+	23.0	
Sulfur	S	32.1	2^-, 6^+	16.0, 5.4	4^+
Tin	Sn	118.7	2^+	59.4	4^+
Zinc	Zn	65.4	2^+	32.7	

[a]Valence is the charge on ionic substances such as Al^{3+} or Cl^- or the charge associated with the atom in a covalent grouping such as N(5^+) in NO_3^- or N(3^-) in NH_3. The single N^{5+} or N^{3-} ions do not exist naturally. Also see Section 1.3.1.

[b]Equivalent weight (combining weight) equals atomic weight divided by valence. The equivalent weights given in this column are based on the most common valence. Depending on the element and the circumstances of the reaction, equivalent weight may be for an acid–base or an oxidation–reduction reaction.

4. The oxidation state of oxygen is -2 except in peroxide, in which case it is -1.
5. Hydrogen has an oxidation number of $+1$ except when bonded to metallic hydrides such as NaH, in which case it has an oxidation number of -1.

From the general rule of assigning electrons to the more electronegative element, N in NCl_3 would have an oxidation number of $+3$ and each Cl atom would have an oxidation number of -1. However in NH_3, because of the general rule expressed through rule 5, N would have an oxidation number of -3 and each H has an oxidation

TABLE 1.2 Common Radicals Encountered in Water and Water Analyses

Name	Formula	Molecular weight	Valence	Equivalent weight[a]
Acetate	$C_2H_3O_2^-$	59.0	1^-	59.0
Ammonium	NH_4^+	18.0	1^+	18.0
Bicarbonate	HCO_3^-	61.0	1^-	61.0
Borate	BO_3^{3-}	58.8	3^-	19.6
Carbonate	CO_3^{2-}	60.0	2^-	30.0
Chlorate	ClO_3^-	83.5	1^-	83.5
Chlorite	ClO_2^-	67.5	1^-	67.5
Hypochlorite	OCl^-	51.5	1^-	51.5
Chromate	CrO_4^{2-}	116.0	2^-	58.0
Dichromate	$Cr_2O_7^{2-}$	216.0	2^-	108.0
Hydroxide	OH^-	17.0	1^-	17.0
Nitrite	NO_2^-	46.0	1^-	46.0
Nitrate	NO_3^-	62.0	1^-	62.0
Orthophosphate	PO_4^{3-}	95.0	3^-	31.7
Orthophosphate, monohydrogen	HPO_4^{2-}	96.0	2^-	48.0
Orthophosphate, dihydrogen	$H_2PO_4^-$	97.0	1^-	97.0
Permanganate	MNO_4^-	118.9	1^-	118.9
Bisulfate	HSO_4^-	97.0	1^-	97.0
Sulfate	SO_4^{2-}	96.0	2^-	48.0
Bisulfide	HS^-	33.1	1^-	33.1
Sulfite	SO_3^{2-}	80.0	2^-	40.0
Bisulfite	HSO_3^-	81.0	1^-	81.0

[a]Equivalent weights are given as in Table 1.1 based on weight of the ion divided by its valence. In redox reactions the equivalent weight will depend on the element being oxidized or reduced.

number of +1. There are some other rules that apply in special circumstances that will generally not be of concern in this text.

■ **Example 1.2 Oxidation Numbers**

What are the oxidation numbers of each atom in the following compounds?

(a) Na_2HPO_4

(b)
$$\begin{array}{l} H-C=C-CH_3 \\ \quad\;\; | \quad\; | \\ \quad\;\; H \quad H \end{array}$$

For Na_2HPO_4, by rule 3 each Na has an oxidation number of +1; by rule 5 each H has an oxidation number of +1; and, by rule 4 each O has an oxidation number of −2. The molecule has no net charge; therefore, from rule 1 the oxidation number on P is +5.

For the second compound, propene, the oxidation number on each H is +1 from rule 5. By rule 1, the 3 carbon atoms must have oxidation numbers that sum to −6. From rule 2, the carbon–carbon bonds make no contributions to the oxidation numbers. There are 2 Hs associated with the left C atom, 1 with the center C atom, and

3 with the right C atom and the oxidation numbers of the C atoms are, from left to right, −2, −1, and −3.

1.4 BALANCING REACTIONS

Two fundamental rules must be followed when balancing chemical reactions:

1. Mass must be conserved.
2. Charge must be conserved.

The first law states that the number of atoms of an element on the left-hand side of the equation must be equal to the number of atoms of the element on the right-hand side of the equation for each substance. This is the law of mass action. The atoms may have changed the atoms with which they are associated but they cannot be created or destroyed. Electrons do not have any significant mass associated with them and they may appear on one side of the equation and not on the other. They are given up or taken up by an ion, radical, or compound.

The second rule states that there cannot be a charge difference between either side of the reaction after adding up all charged species on each side of the reaction. Electrons, of course, do have a charge associated with them and they must be incorporated into the charge balance.

Consider the reaction:

$$aA + bB + \cdots \rightleftharpoons cC + dD + \cdots$$

Arrows going in either direction indicate that the reaction is reversible. All reactions are reversible but, for practical purposes, a significant number of reactions only go in one direction or conditions are provided that force the reaction to proceed in one direction. If the reaction is going to the right then the substances on the left-hand side of the equation are the reactants and those on the right-hand side are the products.

The lower case letters *a, b, c,* and *d* represent the stoichiometric combination values needed to satisfy the equation in accordance with the rules stated above. Do these numbers represent valence or oxidation numbers? Or are they related to valence or oxidation numbers? The stoichiometric numbers can be algebraically manipulated in any fashion considering the arrows to be equivalent to an equal sign. The following example demonstrates the additivity of chemical reactions:

$$\begin{array}{l} R_1 + R_2 \rightleftharpoons P_1 + P_2 \\ \underline{P_1 + R_3 \rightleftharpoons R_2 + P_3} \\ R_1 + R_3 \rightleftharpoons P_2 + P_3 \end{array}$$

Reactions must be written to describe processes that are actually feasible, i.e., the ions or compounds participating in the reaction must be true substances.

■ Example 1.3 Balancing Reactions

Balance the following reactions:

1. $MnO_2 + HCl \rightleftharpoons MnCl_2 + Cl_2 + H_2O$
2. $MnO_4^- + H_2O + e^- \rightleftharpoons MnO_2 + OH^-$

Both reactions represent true phenomena.

For reaction 1 it is observed that at least 4 Cl atoms must appear on the right-hand side. Multiplying HCl by 4 results in 4 H atoms on the left-hand side, which forces us to multiply H_2O by 2. A check on the number of atoms of each element on either side of the equation will show that the equation is balanced; thus the total masses on either side of the equation will be equal. No charged species participate in the reaction as written. The final balanced reaction is

$$MnO_2 + 4HCl \rightleftharpoons MnCl_2 + Cl_2 + 2H_2O$$

For reaction 2 there is an excess of oxygen and hydrogen on the left-hand side. Fixing the number of moles of MnO_4^- involved at 1 determines the number of moles of MnO_2 to be 1. Trial and error will show that OH^- must be multiplied by 4 and H_2O must be multiplied by 2 to balance the equation with respect to mass.

$$MnO_4^- + 2H_2O + e^- \rightleftharpoons MnO_2 + 4OH^-$$

Checking the charge balance, it is found that there is an excess of two negative charges on the right-hand side of the equation. The electrons (which do not influence the mass) must be increased to 3 to balance the equation.

$$MnO_4^- + 2H_2O + 3e^- \rightleftharpoons MnO_2 + 4OH^-$$

Electrons do not explicitly appear on the right-hand side of the equation; they have been incorporated into Mn in MnO_2. Check the oxidation numbers on Mn on either side of the equation.

1.5 OXIDATION–REDUCTION REACTIONS

Oxidation–reduction (redox) reactions involve transfer of electrons. A simple redox reaction has been examined in Example 1.3. In a redox reaction, one compound is giving up electrons and another compound is taking them up. The overall reaction can be formulated from two half-reactions, one producing electrons and the other receiving them; electrons will not explicitly appear in the overall balanced reaction. The substance giving up electrons is being oxidized, so it is the reducing agent; the substance receiving electrons is being reduced, so it is the oxidizing agent. A simple mnemonic device to discern which substance is the oxidizing agent is to note that reduction means a reduction in oxidation number; i.e., molecular oxygen is an oxidizing agent that receives electrons and has its oxidation number reduced from 0 to -1 or, more commonly, -2 in typical reactions.

Half-reactions are normally written as a reduction, a convention that will be followed in this book. Therefore when combining two half-reactions, one of them must be reversed. Table 1.3 lists common half-reactions of significance for water and wastewater treatment. These reactions are balanced and normalized to transfer of one electron for convenience in combining them. Simply subtracting one from another will produce a balanced reaction. Multiplying a half-reaction in this table by any number and combining it with the appropriate sign with any other reaction will produce an overall reaction that is balanced with respect to mass. Will the overall reaction be balanced with respect to charge?

As seen from Table 1.3, there is more than one possible product (oxidation state) for some substances (e.g., Mn). The particular species formed in a given reaction depend on the reaction conditions such as $[H^+]$ (pH, defined below) and other factors.

TABLE 1.3 Half-Reactions

Reaction no.	Half-reaction	$\Delta G°$, kcal/mole	$E°$, volts
1	$\frac{1}{2}Br_2(aq) + e^- = Br^-$	−25.2	1.09
2	$\frac{1}{2}BrCl + e^- = \frac{1}{2}Br^- + \frac{1}{2}Cl^-$	−31.1	1.35
3	$Ce^{4+} + e^- = Ce^{3+}$	−33.2	1.44
4	$\frac{1}{4}CO_3^{2-} + \frac{7}{8}H^+ + e^- = \frac{1}{8}CH_3COO^- + \frac{1}{4}H_2O$	−1.73	0.075
5	$\frac{1}{4}CO_3^{2-} + H^+ + e^- = \frac{1}{24}C_6H_{12}O_6 + \frac{1}{4}H_2O$	0.35	−0.001 5
6	$\frac{1}{2}Cl_2(aq) + e^- = Cl^-$	−32.1	1.391
7	$ClO_2 + e^- = ClO_2^-$	−26.6	1.15
8	$ClO_3^- + 2H^+ + e^- = ClO_2 + H_2O$	−26.6	1.15
9	$\frac{1}{2}OCl^- + H^+ + e^- = \frac{1}{2}Cl^- + \frac{1}{2}H_2O$	−39.9	1.728
10	$\frac{1}{8}ClO_4^- + H^+ + e^- = \frac{1}{8}Cl^- + \frac{1}{2}H_2O$	−31.6	1.37
11	$\frac{1}{6}Cr_2O_7^{2-} + \frac{7}{3}H^+ + e^- = \frac{1}{3}Cr^{3+} + \frac{7}{6}H_2O$	−30.7	1.33
12	$\frac{1}{2}Cu^{2+} + e^- = \frac{1}{2}Cu$	−7.78	0.337
13	$\frac{1}{2}Fe^{2+} + e^- = \frac{1}{2}Fe$	9.45	−0.409
14	$Fe^{3+} + e^- = Fe^{2+}$	−17.78	0.770
15	$\frac{1}{3}Fe^{3+} + e^- = \frac{1}{3}Fe$	0.84	−0.003 6
16	$H^+ + e^- = \frac{1}{2}H_2(g)$	0.00	0.00
17	$\frac{1}{2}H_2O_2 + H^+ + e^- = H_2O$	−40.8	1.77
18	$\frac{1}{2}Hg^{2+} + e^- = \frac{1}{2}Hg$	−19.7	0.851
19	$\frac{1}{2}I_2(aq) + e^- = I^-$	−14.3	0.62
20	$\frac{1}{5}IO_4^- + \frac{6}{5}H^+ + e^- = \frac{1}{10}I_2(g) + \frac{3}{5}H_2O$	−27.6	1.197
21	$\frac{1}{2}MnO_2 + 2H^+ + e^- = \frac{1}{2}Mn^{2+} + H_2O$	−27.9	1.208
22	$\frac{1}{5}MnO_4^- + \frac{8}{5}H^+ + e^- = \frac{1}{5}Mn^{2+} + \frac{4}{5}H_2O$	−34.4	1.491
23	$\frac{1}{3}MnO_4^- + \frac{4}{3}H^+ + e^- = \frac{1}{3}MnO_2 + \frac{2}{3}H_2O$	−39.2	1.695
24	$\frac{1}{6}NO_2^- + \frac{4}{3}H^+ + e^- = \frac{1}{6}NH_4^+ + \frac{1}{3}H_2O$	−20.75	0.898
25	$\frac{1}{8}NO_3^- + \frac{5}{4}H^+ + e^- = \frac{1}{8}NH_4^+ + \frac{3}{8}H_2O$	−20.33	0.880
26	$\frac{1}{3}NO_2^- + \frac{4}{3}H^+ + e^- = \frac{1}{6}N_2(g) + \frac{2}{3}H_2O$	−35.16	1.519
27	$\frac{1}{5}NO_3^- + \frac{6}{5}H^+ + e^- = \frac{1}{10}N_2(g) + \frac{3}{5}H_2O$	−28.73	1.244
28	$\frac{1}{4}O_2(aq) + H^+ + e^- = \frac{1}{2}H_2O$	−29.32	1.23
29	$\frac{1}{2}O_3(g) + H^+ + e^- = \frac{1}{2}O_2(g) + \frac{1}{2}H_2O$	−47.8	2.07
30	$\frac{1}{6}SO_4^{2-} + \frac{1}{3}H^+ + e^- = \frac{1}{6}S + \frac{2}{3}H_2O$	−8.24	0.357
31	$\frac{1}{8}SO_4^{2-} + \frac{5}{4}H^+ + e^- = \frac{1}{8}H_2S(aq) + \frac{1}{2}H_2O$	−7.00	0.303
32	$\frac{1}{4}SO_4^{2-} + \frac{5}{4}H^+ + e^- = \frac{1}{8}S_2O_3^{2-} + \frac{5}{8}H_2O$	−7.00	0.303
33	$\frac{1}{2}SO_4^{2-} + H^+ + e^- = \frac{1}{2}SO_3^{2-} + \frac{1}{2}H_2O$	0.93	−0.039
34	$\frac{1}{2}Zn^{2+} + e^- = \frac{1}{2}Zn$	17.6	−0.763

In many cases it is a matter of experience to know which product dominates. A variety of oxidation states may exist at any time and change as time proceeds.

Normalizing the half-reactions to one electron also conveniently gives the electron equivalence and equivalent weight of the species for a redox reaction. For instance, for reaction 14 in Table 1.3, the equivalence of Fe^{3+} or Fe^{2+} is 1 mole and the equivalent weight is 55.85 g. However, if the Fe^{3+} is reacting according to reaction 15, the electron equivalence is 0.33 mole and the equivalent weight is 18.6 g.

■ Example 1.4 Balancing Redox Reactions

Chlorine is to be used to oxidize sulfide to sulfate. What is the overall balanced reaction?

From Table 1.3, select the appropriate half-reactions. For Cl_2, the half-reaction is

$$\tfrac{1}{2}Cl_2 + e^- = Cl^-$$

For sulfide–sulfate, the half-reaction is

$$\tfrac{1}{8}SO_4^{2-} + \tfrac{5}{4}H^+ + e^- = \tfrac{1}{8}H_2S + \tfrac{1}{2}H_2O$$

One of these reactions must be reversed upon combining the two half-reactions. It has been specified that Cl_2 is the oxidant (E^o values, discussed in Section 2.2, indicate this will be the case); therefore, this reaction remains as written:

$$\tfrac{1}{2}Cl_2 + \tfrac{1}{8}H_2S + \tfrac{1}{2}H_2O = \tfrac{1}{8}SO_4^{2-} + \tfrac{5}{4}H^+ + Cl^-$$

Electrons do not appear in the balanced overall reaction. Chlorine has taken up electrons and sulfur has given up electrons; sulfur's oxidation number has changed from −2 to +6.

It can be observed from Table 1.3 that many redox reactions are influenced by pH ($[H^+]$) which is discussed in Chapter 3. The definition of pH is

$$pH = -\log[H^+] \tag{1.1}$$

From Eq. (1.1), $[H^+] = 10^{-pH}$.

To determine whether a low or high pH is desirable to promote formation of the products, the overall reaction is formulated. If H^+ ions appear on the left-hand side, a low pH (high $[H^+]$) will promote the formation of the products; if H^+ appears on the right-hand side, a high pH is necessary. Side reactions may also influence the pH and they must be considered when making the overall assessment of the effect of $[H^+]$.

For the case in which a half-reaction is not available, the core reaction participants are written and the following steps are performed to balance the reaction.

1. Write the core reactants and products on opposite sides of the equal sign. Balance the core reaction with respect to the atom that is changing oxidation state.
2. Balance the reaction with respect to oxygen by adding H_2O where needed.
3. Balance H by adding H^+ ions.
4. Balance the charge by adjusting the number of electrons.
5. Add any extraneous ions that do not participate in the redox reaction.

■ Example 1.5 Redox Half-Reaction

Write the balanced half-reaction for reduction of formate ($HCOO^-$) to organic matter, which will be represented as CH_2O.

Step 1
$HCOO^- + e^- = CH_2O$

Carbon is changing oxidation state (what is its change in oxidation number?) in this half-reaction. The stoichiometry remains 1:1 because there is only one carbon atom on either side of the equation. If a carbon compound that contained two carbon atoms was present on the right-hand side of the equation, then it would have to be multiplied by $\tfrac{1}{2}$.

Step 2
$HCOO^- + e^- = CH_2O + H_2O$

Step 3
$HCOO^- + 3H^+ + e^- = CH_2O + H_2O$

Step 4
$HCOO^- + 3H^+ + 2e^- = CH_2O + H_2O$

Step 5

If sodium formate has been used, the Na^+ may be incorporated on each side of the equation.

$Na^+ + HCOO^- + 3H^+ + 2e^- = CH_2O + H_2O + Na^+$

1.6 EQUILIBRIUM

Elementary reactions may be monomolecular, bimolecular, or rarely trimolecular. Complex reactions are composed of elementary reactions. Equilibrium is the state where the concentrations of all species are constant. In fact, equilibrium is a dynamic state where the rate of product formation is exactly equal to the rate of reactant formation. This is a statement of the principle of microscopic reversibility (Laidler, 1965). Although some chemical reactions practically go to completion, there is always at least an infinitesimal amount of both reactants and products present. In the chemical reaction

$$aA + bB + \cdots \rightleftharpoons cC + dD + \cdots \tag{1.2}$$

the arrows going in either direction indicate equilibrium.

The equilibrium expression for reaction (1.2) is

$$K = \frac{[C_C]^c[C_D]^d \cdots}{[C_A]^a[C_B]^b \cdots} \tag{1.3}$$

where

K is the equilibrium constant
$[C_A]$ is the concentration of substance A

The equilibrium constant corresponding to a chemical equation is always formed by placing the species on the right-hand side of the equation in the numerator and the species on the left-hand side in the denominator. The equilibrium expression corresponds to the manner in which the equation has been written. Normally the starting species are placed on the left-hand side and are referred to as reactants. The right-hand species are the products. This does not necessarily mean that the species on the right-hand side will have the highest concentrations at equilibrium.

In analytical methods or treatment situations it is normally desired to have a substance completely or nearly completely transformed to a particular product. The principle of microscopic reversibility allows this to be achieved. From the equilibrium expression it is seen that increasing or decreasing the concentration of any species changes the ratio of the products at equilibrium. This fact is used to displace the equilibrium concentrations in the desired direction. For instance in Eq. (1.2) if substance A is being transformed to substance C, to promote the transformation of A, substance B will be added to the water. This will decrease the concentration of A at equilibrium. Any species except A on the left-hand side of the reaction can be added to achieve enhanced transformation of A. Alternately decreasing the concentration of a species on the right-hand side of Eq. (1.2) by chemical reaction or other means is equivalent to increasing the concentration of a species on the left-hand side. The equilibrium relation is used to determine the amount of agent required to achieve the transformation.

The time to reach equilibrium varies with the type of reaction (e.g., acid–base or redox) the species involved, and environmental conditions such as the concentration of all ions and molecules, temperature, and pH. The equilibrium constant is a function of temperature (see Section 2.1.3).

The units for expressing concentration in an equilibrium expression vary depending on the phase of the substance. Ions or molecules dissolved in water are expressed as moles per liter. Gases in equilibrium with a solution have their concentrations expressed in atmospheres (atm). The concentration of the solvent, water, is assumed to be 1.0 regardless of whether water is being produced or consumed by the reaction within the water solution. A pure solid or liquid in equilibrium with a water solution is assumed to have a concentration of 1.0. Later sections and chapters discuss some of these special cases.

The equilibrium constant can be determined by carefully measuring the concentrations of all species with time. When the concentrations are stable, their values are put into the equilibrium expression to determine the equilibrium constant. Section 2.1 gives the thermodynamic basis for the equilibrium constant.

1.7 CONDUCTIVITY AND IONIC STRENGTH

Equilibrium relations depend on concentrations of participants in a reaction. The effective concentration of a substance depends on its concentration, its charge, and the electrical characteristics of the solution. The major influence on the electrical characteristics of the solution is the sum total of all charged species.

1.7.1. Conductance

Concentrations and charges of ions in a solution determine the ability of the solution to carry a current or the conductivity of the solution. Conductivity is measured by placing two electrodes in a solution and impressing a voltage drop across the electrodes. In accordance with Ohm's law, the current that flows depends on the resistance of the solution.

$$E = iR \tag{1.4}$$

where
- i is the current in amperes
- R is the resistance in ohms
- E is the voltage drop across the electrodes

The resistance is directly proportional to the separation of the electrodes and inversely proportional to the area of the electrodes.

$$R = \rho \frac{l}{A} \tag{1.5}$$

where
ρ is the specific resistance of the solution (ohms-cm)
l is the distance between electrodes (cm)
A is the area of the electrodes (cm^2)

Specific conductance, κ, which is the inverse of the specific resistance, is normally reported.

$$\kappa = \frac{1}{\rho} \tag{1.6}$$

where
κ is in mhos/cm (mho is ohm spelled backwards)

Specific conductance is related to the sum of charge carriers, which are the positive and negative ions in solution. The equivalent weight of an ion related to its charge carrying ability is simply its formula weight divided by the charge on the species. Ideally one equivalent weight of any substance has the same conductance. A 1 N (normal) solution contains 1 eq. wt. per liter and κ is related to equivalent conductance, λ, of a compound by

$$\kappa = k \frac{N}{1\,000} \lambda \tag{1.7}$$

where
N is the normality (no. of equivalents per liter)
k is an adjustment factor

If the solution is ideal, k equals 1. The equivalent conductances of all compounds or ions are not the same because of size and other characteristics of the ions involved. Mobility of an ion will have a primary influence on its ability to conduct current. Also, the equivalent conductance of a salt decreases as its concentration increases because of charge interaction effects.

Specific conductance is a useful tool to gauge the concentration of total dissolved solids (TDS) and related effects as discussed in the following section. *Standard Methods for the Examination of Water and Wastewater* (1992) notes the following approximate relation between TDS and specific conductance.

$$\text{TDS} = k\kappa \tag{1.8}$$

where
k varies from 0.55 to 0.90
κ is in μmhos/cm (mhos/cm $\times 10^{-6}$)
TDS is in mg/L

1.7.2 Ionic Strength

The ionic strength, μ, of a solution is related to the sum total of all charged species in the solution.

$$\mu = \tfrac{1}{2}\Sigma C_i z_i^2 \tag{1.9}$$

where
C_i is molar concentration of the ith ion
z_i is the charge on the ith ion

The following correlation was determined by Langelier (1936) for a number of waters:

$$\mu = 2.5 \times 10^{-5}(\text{TDS}) \tag{1.10}$$

where
TDS is in mg/L

Another correlation between specific conductance (κ) and ionic strength was derived by Russell (1976).

$$\mu = 1.6 \times 10^{-5}\,\kappa \tag{1.11}$$

where
κ is in μmhos/cm

The above correlations are useful for making quick estimates of the total dissolved solids content of a water. It must be kept in mind that the correlations are only approximate because nonionic species do not contribute to ionic strength and individual ionic species have different weights.

Also, the effective concentration or activity of a substance with respect to equilibrium calculations is related to the ionic strength of the solution. The effective concentration is determined by multiplying the actual concentration by an activity coefficient (Section 2.1.2). For most of the calculations in this book the activity coefficient will be assumed to be 1.0.

1.8 CHEMICAL KINETICS

Rate of a reaction is generally dependent on the concentration of one or more species involved in the reaction. Other influences, discussed in Sections 1.8.1 and 1.8.2, such as temperature and catalysis, can accelerate a reaction. When all intermediate steps of a reaction are known it is often possible to formulate theoretically the rate model exactly and verify it with experimental data. Otherwise the rate model is simply the empirical model that best correlates the experimental data.

A general reaction rate model is

$$\frac{dC_\text{A}}{dt} = \pm k C_\text{A}^a C_\text{B}^b \cdot \cdot \cdot C_\text{N}^n \tag{1.12}$$

where
k is the rate constant
C_A is the concentration of substance A, etc.
t is time

The exponents in the above equation may have any value (not necessarily an integer). The sum of the exponents gives the order of the reaction. The reaction is of order a with respect to substance A, order b with respect to B and so on. The rate constant applies at the experimental conditions. If k is positive, the reaction describes production; a negative k describes removal.

When all the intermediate steps are not known or when experimental data prove that some intermediates are not important influences on the overall rate of reaction, the following simplified expressions are often used.

1. Zero-Order Formulation

$$\frac{dC_A}{dt} = \pm k \tag{1.13}$$

In this case, the rate of change of the concentration of substance A is not significantly influenced by the amounts of any other substances present.

2. First-Order Formulation

$$\frac{dC_A}{dt} = \pm kC_A \tag{1.14}$$

This equation is one of the most often used formulations in environmental engineering to fit data from complex reactions. In complex reactions C_A is often a nonspecific measure of many substances. For instance, organic compounds can be measured by their carbon content. A logarithmic formulation for concentration variation with time indicates that the reaction rate depends on more readily reacted species followed by increasingly difficult to react substances that implicitly partake in the overall reaction.

3. Second-Order Formulation

Any of the following expressions describe second-order reactions.

$$\frac{dC_A}{dt} = \pm kC_A^2 \tag{1.15a}$$

$$\frac{dC_A}{dt} = \pm kC_A^a C_B^b \tag{1.15b}$$

$$\frac{dC_A}{dt} = \pm kC_B^2 \tag{1.15c}$$

In Eq. (1.15b), exponents on C_A and C_B must sum to 2.

1.8.1 Other Formulations

Other formulations have been empirically fit to data. Some of these are given below.

Retardant

The rate of reaction of some reactions is observed to increase or decrease with time beyond the change in reaction rate as a result of change in concentration of the reactant. This is known as a retardant reaction. The rate constant is adjusted in the following manner:

$$\frac{dC_A}{dt} = \pm \frac{k}{1 + \alpha t} C_A \tag{1.16}$$

where

α is a characteristic reaction constant

Effectively the reaction rate constant is decreased with time.

Autocatalytic

Some reactions accelerate spontaneously as time proceeds. The velocity of the reaction is dependent upon a catalytic effect of the products.

Consider the reaction

$$A + P \rightarrow P + P \tag{1.17}$$

where

A is the substance being converted

P is the product

For an elementary reaction the kinetic formulation is

$$\frac{dC_A}{dt} = -k_1 C_A - k_2 C_A C_P \tag{1.18}$$

At any time the total concentration of reactant and product, C_T is

$$C_T = C_A + C_P = C_{A0} + C_{P0} = \text{constant} \tag{1.19}$$

where

C_{A0} is the initial concentration of A

C_{P0} is the initial concentration of P

Substituting for C_P in Eq. (1.18) and using Eq. (1.19),

$$\frac{dC_A}{dt} = -k_1 C_A - k_2 C_A (C_T - C_A) \tag{1.20}$$

If $C_A = C$ and $C_{P0} = 0$,

$$\frac{dC}{dt} = -k_1 C - k_2 C(C_T - \mathrm{C})$$

There are other empirical models that can be used in any particular situation. The choice of a rate model depends on the model that most generally applies, giving the best statistical fit to the data.

When the stoichiometry of a reaction is known, the rate of change of one substance may be easily related to the rate of change of another. Consider the general reaction:

$$a\mathrm{A} + b\mathrm{B} + \cdots \rightleftharpoons c\mathrm{C} + d\mathrm{D} + \cdots \tag{1.21}$$

Dividing the equation by a,

$$\mathrm{A} + \frac{b}{a}\mathrm{B} + \cdots \rightleftharpoons \frac{c}{a}\mathrm{C} + \frac{d}{a}\mathrm{D} + \cdots \tag{1.22}$$

The equation states that 1 mole of A reacts with b/a moles of B to produce c/a moles of C and d/a moles of D.

The change in the number of moles of A, ΔN_A is

$$\Delta N_A = N_A - N_{A0} \tag{1.23}$$

where

the subscript 0 indicates the initial value

If $\Delta N_A = -1$ then $\Delta N_B = -b/a$, $\Delta N_C = c/a$, and $\Delta N_D = d/a$. The above equations may then be rearranged to

$$-\frac{\Delta N_A}{a} = -\frac{\Delta N_B}{b} = \frac{\Delta N_C}{c} = \frac{\Delta N_D}{d} \tag{1.24}$$

Differentiating Eq. (1.24) with respect to time yields

$$-\frac{1}{a}\frac{dN_A}{dt} = -\frac{1}{b}\frac{dN_B}{dt} = \frac{1}{c}\frac{dN_C}{dt} = \frac{1}{d}\frac{dN_D}{dt} \tag{1.25}$$

At equilibrium or steady state conditions the above derivatives are equal to 0.

■ Example 1.6 Change in Concentration in a Reaction

A decay reaction has been found to proceed according to a retardant model with $\alpha = 0.052\ h^{-1}$ and rate constant $k = 0.095\ h^{-1}$. What is the percentage decrease in concentration of the reactant after 5 h?

$$\frac{dC}{dt} = -\frac{k}{1+\alpha t}C \qquad \int_{C_0}^{C}\frac{dC}{C} = -k\int_0^t \frac{dt}{1+\alpha t} \qquad \ln C\Big|_{C_0}^{C} = -\frac{k}{\alpha}\ln(1+\alpha t)\Big|_0^t$$

$$\ln\frac{C}{C_0} = -\frac{k}{\alpha}\ln(1+\alpha t) \qquad \frac{C}{C_0} = (1+\alpha t)^{-\frac{k}{\alpha}}$$

Substituting for k, α, and t:

$$\frac{C}{C_0} = [1 + (0.052\ h^{-1})(5\ h)]^{-\frac{0.095\ h^{-1}}{0.052\ h^{-1}}} = 0.656 \text{ or } 65.6\%$$

The decrease in concentration is 34.4% after 5 h.

Catalysis

A catalyst is an agent that accelerates a reaction but it is not changed in composition as a result of the reaction. Dissolved metals can often serve as catalysts. Enzymes are biological molecules that speed reactions many hundreds or even thousands of times. Catalysts do not cause reactions that are thermodynamically impossible to happen; they only bring about the final equilibrium state more rapidly.

1.8.2 The Effect of Temperature on Rate of Reaction

An increase in temperature raises the energy level of the molecules and also produces a slight increase in the rate of collisions of molecules. Because of the elevated energy levels of the molecules the number of collisions in which the energy threshold for the reaction is exceeded is increased. From empirical observation, Arrhenius found that the rate constant for a reaction depends on temperature as follows:

$$\ln k = a - \frac{b}{T}$$

where

a and b are constants
T is temperature in °K

This equation can be related to the Boltzman equation with the result:

$$k = A \exp\left(-\frac{E_a}{RT}\right) \tag{1.26}$$

where

A is a constant related to the frequency of collisions
R is the universal gas constant
E_a is the activation energy for reaction

Equation (1.26) can be rearranged to

$$k_{T_2} = k_{T_1}\theta^{(T_2-T_1)} \tag{1.27}$$

where

k_{T_i} is the rate constant at temperature T_i (temperature is usually expressed in °C)
θ is a constant

Note that θ in Eq. (1.27) is not truly a constant when compared with Eq. (1.26), but over small temperature ranges it is approximately constant.

1.9 GAS LAWS

The gaseous state is one of the three fundamental states of matter. Gases do not have any surface and tend to completely fill any available space. The behavior of gases can be determined from an equation of state that incorporates two laws: Boyle's law and Charles's law.

Boyle's Law

In 1662, Boyle discovered that at a constant temperature the volume of a fixed amount of gas is inversely proportional to pressure. The law is simply stated in equation form as

$$pV = k \tag{1.28}$$

where

p is pressure
V is volume
k is a constant

Charles's Law

In 1787, J.A. Charles found that the volume of a fixed amount of gas at constant pressure is directly proportional to the absolute temperature. Stated in equation form,

$$\frac{V}{T} = k \tag{1.29}$$

where

T is temperature in °K

The Ideal Gas Law

Use of Avogadro's law allows Boyle's and Charles's laws to be incorporated into a general equation of state. Avogadro's law states that equal numbers of molecules of different gases occupy the same volume at a given temperature and pressure. The resulting equation of state is

$$pV = nRT \tag{1.30}$$

where

n is the number of moles of gas
R is the universal gas constant

Real gases deviate from the ideal gas law. For real gases the law is a good approximation at relatively low pressures and moderately high temperatures; however, many applications in environmental engineering occur in these ranges.

Dalton's Law

Dalton, in 1801, determined that the total pressure of a gas mixture is equal to the sum of the partial pressures exerted by each individual component gas. The partial pressure of a component gas in a mixture is the pressure the gas would exert if it alone occupied the whole volume of the mixture at the same temperature. For n components, the total pressure, p_T, is

$$p_T = p_1 + p_2 + \cdot \cdot \cdot + p_n \tag{1.31}$$

where

p_i is the partial pressure of the ith gas

Using the ideal gas law,

$$p_1V = n_1RT$$
$$p_2V = n_2RT$$
$$\vdots$$
$$p_nV = n_nRT$$

Combining the above,

$$p_TV = (n_1 + n_2 + \cdot \cdot \cdot + n_n)RT \tag{1.32}$$

It can also be determined for the ith component that

$$p_i = \frac{n_i}{n}p_T \qquad \text{or} \qquad p_i = \frac{V_i}{V}$$

where

n is the total number of moles of gas in the mixture
V_i is the portion of the total volume occupied by one component

All components are randomly dispersed in a gas.

One mole of an ideal gas, at a temperature of 0°C and pressure of 1 atm, occupies a volume of 22.414 L. Dalton's law and the ideal gas equation of state can be used to determine the concentration of a gas in a gas mixture when its partial pressure is known.

1.10 GAS SOLUBILITY: HENRY'S LAW

All gases are soluble to some degree in liquids. Henry's law describes the solubility relationship. It states that the weight of any gas that can dissolve in a given volume of liquid is directly proportional to the pressure that the gas exerts above the liquid.

TABLE 1.4 Henry's Law Constants[a]

Compound	K_H,[b] mg/L/atm	
	0°C	25°C
CH_4	3 968	2 152
Cl_2	1 460	980[c]
CO	43.8	27.6
CO_2	3 480	1 450
H_2	1.91	1.56
N_2	29.1	18.0
O_2	69.6	39.3
O_3	1 375	584
H_2S	6 640	3 505
SO_2	228 000	87 600

[a]Compiled from Lide and Frederikse (1994), Perry and Green (1984), and Weast (1984).
[b]K_H is highly variable with temperature.
[c]At 20°C.

The chemical equation describing the equilibrium relation is

$$G(g) \rightleftharpoons G(aq) \tag{1.33}$$

where
G is concentration of a gas
(aq) refers to aqueous (dissolved) and (g) refers to gaseous

The concentration of gas in the liquid is normally expressed in moles or milligrams per liter; the concentration of gas in a gas mixture is normally expressed in terms of partial pressure in atmospheres. Henry's law describes the equilibrium constant that applies to this situation.

$$[G(aq)] = K_H[G(g)] \quad \text{or} \quad K_H = \frac{[G(aq)]}{[G(g)]} \tag{1.34}$$

where
K_H is Henry's constant

Henry's law is a special case of an equilibrium expression (Eq. 1.3). Henry's constants for various gases are given in Table 1.4. As noted below the table, Henry's law constants are highly variable with temperature.

The units on Henry's constant vary from source to source. The concentration of a gas dissolved in the liquid may be expressed in milligrams per liter, milliliters of gas dissolved per liter, or moles per liter. Likewise, the partial pressure may be used to describe the gas phase concentration, or moles per liter of gas may be used. Sometimes Henry's constant is defined as the inverse of the definition in Eq. (1.34). Take care to note the definitions used in the reference. The universal gas law (Eq. 1.30) can be used to relate partial pressure to concentration in the gas phase.

The concentration of gas that exists in solution at equilibrium is referred to as the saturation concentration of the gas. The concentration of dissolved gas follows Henry's law regardless of chemical reactions that the gas may undergo with the liquid or components in the liquid. If the dissolved gas is below the saturation concentration, gas will dissolve in the liquid; if the dissolved gas concentration is above the saturation concentration, known as a supersaturated solution, the gas will normally leave the solution until the equilibrium condition is reached.

Natural waters are exposed to the atmosphere. Oxygen and nitrogen are the two major components of atmospheric air at concentrations of 21 and 79% by volume, respectively. The normal partial pressure of CO_2 in the atmosphere is $10^{-3.5}$ atm. Other gases of concern to environmental engineers are present in negligible amounts.

■ Example 1.7 Application of Henry's Law Equilibrium

For air with an oxygen content of 21% by volume, what is the saturation concentration, C_s, of oxygen in water at 20°C and 1 atm total pressure, given that Henry's constant is 43.3 mg/L/atm at that temperature?

Use Dalton's law to find the partial pressure of O_2.

$$p = 0.21 p_T = 0.21 \text{ atm } O_2$$
$$C_s = (43.3 \text{ mg/L/atm})(0.21 \text{ atm}) = 9.09 \text{ mg/L at } 20°\text{C}$$

1.11 SOLUBILITY PRODUCT

Like gases, all solids are also soluble to some degree in a liquid. A general chemical equation expressing this fact is

$$A_aB_b(s) \rightleftharpoons aA^{b+} + bB^{a-} \tag{1.35}$$

The equilibrium constant pertaining to this equation is known as the solubility product, K_{sp}.

$$K'_{sp} = \frac{[A^{b+}]^a[B^{a-}]^b}{[A_aB_b(s)]} \quad (1.36a) \qquad \text{or} \qquad K_{sp} = [A^{b+}]^a[B^{a-}]^b \quad (1.36b)$$

Equation (1.36a) is not used because the concentration (density) of the solid substance $A_aB_b(s)$ does not normally change significantly regardless of how much of it is present. Even though $A_aB_b(s)$ may be dispersed in a volume of liquid by mixing or for other reasons, the actual concentration of the solid material does not change in its own phase. The thermodynamic relations that are the basis for the equilibrium constant values are defined to be consistent with the latter definition of K_{sp} [Eq. (1.36b)].

The low solubility of substances is used both in chemical treatments to remove certain components and in analyses. The practical limits of solubility may be significantly different from the limits dictated by the K_{sp} because the time to reach the equilibrium condition is long. Also the rate of reaction may slow down significantly as equilibrium is approached. The practical limits will be influenced by the time allowed for the treatment. Table 1.5 gives K_{sp} constants for reactions of interest to environmental engineers.

■ Example 1.8 Application of Solubility-Product Equilibrium

How many milligrams of CaF_2 can be dissolved in 1 L of water that contains 1 mg/L of F^- at 25°C? Also, find the final amounts (in mg) of Ca^{2+} and F^- in the water. The

TABLE 1.5 Typical Solubility-Product Constants[a,b]

Equilibrium equation	K_{sp} at 25°C	Significance in environmental engineering
$AgCl \rightleftharpoons Ag^+ + Cl^-$	3×10^{-10}	Chloride analysis
$Al(OH)_3 \rightleftharpoons Al^{3+} + 3OH^-$	1×10^{-32}	Coagulation
$Fe(OH)_3 \rightleftharpoons Fe^{3+} + 3OH^-$	6×10^{-38}	Coagulation, iron removal, corrosion
$Fe(OH)_2 \rightleftharpoons Fe^{2+} + 2OH^-$	5×10^{-15}	Coagulation, iron removal, corrosion
$CaSO_4 \rightleftharpoons Ca^{2+} + SO_4^{2-}$	2×10^{-5}	Flue gas desulfurization
$CaF_2 \rightleftharpoons Ca^{2+} + 2F^-$	3×10^{-11}	Fluoridation
$CaCO_3 \rightleftharpoons Ca^{2+} + CO_3^{2-}$	5×10^{-9}	Hardness removal, scaling
$Ca(OH)_2 \rightleftharpoons Ca^{2+} + 2OH^-$	8×10^{-6}	Hardness removal
$MgCO_3 \rightleftharpoons Mg^{2+} + CO_3^{2-}$	4×10^{-5}	Hardness removal, scaling
$Mg(OH)_2 \rightleftharpoons Mg^{2+} + 2OH^-$	9×10^{-12}	Hardness removal, scaling
$SrCO_3 \rightleftharpoons Sr^{2+} + CO_3^{2-}$	5×10^{-9}	Hardness removal, scaling
$Cd(OH)_2 \rightleftharpoons Cd^{2+} + 2OH^-$	1×10^{-14}	Heavy metal removal
$Cr(OH)_3 \rightleftharpoons Cr^{3+} + 3OH^-$	6×10^{-31}	Heavy metal removal
$Cu(OH)_2 \rightleftharpoons Cu^{2+} + 2OH^-$	2×10^{-19}	Heavy metal removal, algae control
$Ni(OH)_2 \rightleftharpoons Ni^{2+} + 2OH^-$	2×10^{-16}	Heavy metal removal
$Pb(OH)_2 \rightleftharpoons Pb^{2+} + 2OH^-$	1×10^{-14}	Heavy metal removal
$Zn(OH)_2 \rightleftharpoons Zn^{2+} + 2OH^-$	3×10^{-17}	Heavy metal removal
$Mn(OH)_3 \rightleftharpoons Mn^{3+} + 3OH^-$	1×10^{-36}	Manganese removal
$Mn(OH)_2 \rightleftharpoons Mn^{2+} + 2OH^-$	8×10^{-14}	Manganese removal
$Ca_3(PO_4)_2 \rightleftharpoons 3Ca^{2+} + 2PO_4^{3-}$	1×10^{-27}	Phosphate removal
$CaHPO_4 \rightleftharpoons Ca^{2+} + HPO_4^{2-}$	3×10^{-7}	Phosphate removal
$BaSO_4 \rightleftharpoons Ba^{2+} + SO_4^{2-}$	1×10^{-10}	Sulfate analysis

[a] From various sources.
[b] Complexation also controls solubility (see Section 1.12).

water does not contain any Ca^{2+}. From Table 1.5 the K_{sp} for CaF_2 is 3.0×10^{-11} at this temperature.

According to the tables inside the front cover, the molecular weights (MWs) of Ca and F are 40 and 19, respectively.

The MW of CaF_2 is $40 + 2 \times 19 = 78$ g.

The initial concentration of F^- is

$$[F^-]_i = \frac{1\ \text{mg/L}}{19\ \text{g/mole}}\left(\frac{1\ \text{g}}{1\,000\ \text{mg}}\right) = 5.26 \times 10^{-5}\ M$$

The chemical equation for dissolution of CaF_2 is

$$\underset{x}{CaF_2(s)} \rightleftharpoons \underset{x}{Ca^{2+}} + \underset{2x}{2F^-}$$

Upon dissolution, x moles of CaF_2 will produce x moles of Ca^{2+} and $2x$ moles of F^- from the stoichiometry of the equation. The final concentration of F^- is

$$[F^-]_f = 2x + 5.26 \times 10^{-5}\ M$$

The corresponding K_{sp} expression is

$$[Ca^{2+}][F^-]^2 = K_{sp} \qquad x(2x + 5.26 \times 10^{-5})^2 = 3.0 \times 10^{-11}$$

$$4x^3 + (1.05 \times 10^{-4})x^2 + (2.77 \times 10^{-9})x = 3.0 \times 10^{-11}$$

$$x^3 + (2.63 \times 10^{-5})x^2 + (6.93 \times 10^{-10})x - 7.50 \times 10^{-12} = 0$$

Solving the cubic equation,

$$x = 1.86 \times 10^{-4}\ M$$

The amount of CaF_2 dissolved is

$$(1.86 \times 10^{-4}\ \text{moles})\left(\frac{78\ \text{g}}{\text{mole}}\right)\left(\frac{1\,000\ \text{mg}}{\text{g}}\right) = 14.5\ \text{mg}$$

The amount of Ca^{2+} dissolved in 1 L of the final solution is

$$(1.86 \times 10^{-4}\ \text{moles})\left(\frac{40\ \text{g}}{\text{mole}}\right)\left(\frac{1\,000\ \text{mg}}{\text{g}}\right) = 7.2\ \text{mg}$$

The amount of F^- dissolved in 1 L of the final solution is

$$(3.72 \times 10^{-4}\ \text{moles} + 5.26 \times 10^{-5}\ \text{moles})\left(\frac{19\ \text{g}}{\text{mole}}\right)\left(\frac{1\,000\ \text{mg}}{\text{g}}\right) = 8.1\ \text{mg}$$

1.12 COMPLEXES

Complexes consist of a central atom or ion surrounded by a set (usually two to nine) of other atoms, ions, or small molecules that are called ligands. These compounds are also known as coordination compounds. Complexes may be charged or uncharged. They are systems in which a definite number of groups (ligands) is arranged in a preferred geometric pattern around the central atom, wherein it is attempted to distribute the electrons uniformly. It is possible to have more than one central atom. Some common inorganic ligands are given in Table 1.6. The hydroxyl ion is another important ligand.

Complexes are usually present in small amounts but they may have a significant influence on the overall amount and distribution of a particular species in solution. This is particularly true for metals that are sparingly soluble. Equilibrium expressions describe complex formation and they are handled routinely in the same manner as other equilibria.

For Cu^{2+} complexes with NH_3 the following species are formed:

$$Cu^{2+} + NH_3 \rightleftharpoons CuNH_3^{2+} \qquad K_1$$
$$CuNH_3^{2+} + NH_3 \rightleftharpoons Cu(NH_3)_2^{2+} \qquad K_2$$
$$Cu(NH_3)_2^{2+} + NH_3 \rightleftharpoons Cu(NH_3)_3^{2+} \qquad K_3$$
$$Cu(NH_3)_3^{2+} + NH_3 \rightleftharpoons Cu(NH_3)_4^{2+} \qquad K_4$$

The K_i are equilibrium constants defined in the standard manner:

$$K_i = \frac{[Cu(NH_3)_i^{2+}]}{[Cu(NH_3)_{(i-1)}^{2+}][NH_3]} \qquad (i\text{: 1 to 4}) \qquad (1.37)$$

The K_i are also known as stepwise formation constants. It is also possible to define overall formation or stability constants, β_i as

$$\beta_i = \frac{[Cu(NH_3)_i^{2+}]}{[Cu^{2+}][NH_3]^i} \qquad (1.38)$$

TABLE 1.6 Stepwise Formation Constants for Various Ligand and Metal Complexes[a]

Ligand ion	Metal	Logarithm (base 10) of constants[b] K_1	K_2	K_3	K_4	Remark
Cl^-	Ag^+	3.30	1.30	0.36		
	Cd^{2+}	2.00	0.70	−0.59		
	Fe^{3+}	1.48	0.65	−1.0		
	Pb^{2+}	1.60	0.18	−0.1	−0.3	
	Zn^{2+}	−0.5	−0.5	1.00	−1.00	
F^-	Al^{3+}	6.16	5.05	3.91	2.71	In 0.53 *M* KNO_3
	Fe^{3+}	5.25	4.00	3.00		In 0.5 *M* $NaClO_4$
	Mg^{2+}	1.82				
	Zn^{2+}	1.26				
NH_3	Ag^+	3.37	3.84			
	Cd^{2+}	2.51	1.96	1.30	0.79	
	Cu^{2+}	3.99	3.34	2.73	1.97	
	Hg^{2+}	8.8	8.7	1.0	0.78	In 2 *M* NH_4NO_3
	Ni^{2+}	2.67	2.12	1.61	1.07	
	Zn^{2+}	2.18	2.25	2.31	1.96	

[a]From Sillén and Martel (1964). Permission granted by The Royal Society of Chemistry.
[b]Generally for temperatures near 25°C.

Stepwise formation constants for some complexes of interest to environmental engineers are given in Table 1.6. An example using these constants to determine the distribution of a metal among its complexes follows.

Note that uncharged entities such as AgCl are in solution as predicted by the equilibrium relations. There may also be precipitation of these substances depending on the concentrations of the free ions. The solubility product relation must be considered in association with the complex equilibria.

■ Example 1.9 Complex Formation

If concentrations of free Cd(II) and NH_3 are 20 μg/L and 1.8 mg/L, respectively, what are the concentrations of the ammonia complexes of cadmium and the total concentration of complexed cadmium in solution?

The MWs of Cd and NH_3 are 112.4 and 17, respectively, which result in molar concentrations of 1.78×10^{-7} and 1.06×10^{-3} *M*, respectively, of the uncomplexed ions. Using Eq. (1.37),

$$[CdNH_3^{2+}] = K_1[Cd^{2+}][NH_3] = (10^{2.51})(1.78 \times 10^{-7})(1.06 \times 10^{-3})$$
$$= 6.11 \times 10^{-8}\ M$$
$$[Cd(NH_3)_2^{2+}] = K_2[CdNH_3^{2+}][NH_3]$$
$$= (10^{1.96})(6.11 \times 10^{-8})(1.06 \times 10^{-3})$$
$$= 5.91 \times 10^{-9}\ M$$

In a similar fashion the concentrations of $Cd(NH_3)_3^{2+}$ and $Cd(NH_3)_4^{2+}$ are found to be 1.25×10^{-10} and 8.17×10^{-13} *M*, respectively. The total amount of complexed

Cd(II) in solution is

$$\Sigma_1^4 \mathrm{Cd(NH_3)_i^{2+}} = 6.71 \times 10^{-8}\,M = 7.5\ \mu\mathrm{g/L}$$

The complexed Cd(II) is 37.5% of the concentration of free Cd(II).

1.13 NUCLEAR CHEMISTRY

Most of the elements are stable but some elements have nuclei with unfavorable proton to neutron ratios and they spontaneously break down to more stable forms (nuclei with lower energy levels). In the transformation process energy is released and the unstable nuclei emit particles or radiation. The most typical decay processes are alpha decay, beta decay, and gamma radiation. Neutron capture is another process in the decay transformation. Alpha decay is the emission of a twice-ionized helium nucleus. Alpha particles have a large mass but small penetrating power. Beta decay is the emission of an electron or a positron, which is a positively charged particle with the mass of an electron. A gamma ray is true electromagnetic radiation similar to X rays; it is a photon traveling at the speed of light. All of these processes are referred to as radioactive decay.

All elements with atomic numbers greater than 83 or atomic weights greater than 200 are unstable. Some elements exist both as stable and as radioactive isotopes. Uranium (^{238}U, ^{235}U), thorium (^{232}Th) and potassium (^{40}K) make the greatest contributions to the natural radioactivity of the Earth's crust. Unstable nuclei can be created by bombardment of stable nuclei with neutrons or radiation. Industrial production of radioactive isotopes for research, medicinal, and military purposes has increased tremendously since World War II. Cosmic radiation from outer space converts some stable elements into radioactive isotopes.

A nuclide is a species of atom characterized by the number of protons, number of neutrons, and energy content of the nucleus. The decay process of an unstable isotope passes through a number of intermediate nuclides until the final stable isotope is produced. Table 1.7 shows the decay pathway for ^{238}U, which is of primary concern for health effects caused by natural radionuclides in drinking water. The decay process (alpha or beta) is indicated at each transformation.

1.13.1 Radioactivity Units

The decay process is described by a first-order reaction:

$$\frac{dN}{dt} = -kN \tag{1.39}$$

where

N is the number or mass of undecomposed nuclei

The radioactivity of a substance is usually described by its half-life ($t_{1/2}$), which is the time for 50% of the substance to decay.

$$t_{1/2} = \frac{0.693}{k} \tag{1.40}$$

TABLE 1.7 Uranium-238 Decay Series[a]

Nuclide	Historical name	Half-life
$^{238}_{92}U$	Uranium I	4.51×10^9 yr
↓ α		
$^{234}_{90}Th$	Uranium X	24.1 d
↓ β		
$^{234}_{91}Pa^m$	Uranium X	1.17 mo
99.87% (β) ↓ 0.13%		
$^{234}_{91}Pa$	Uranium Z	6.75 h
↓ (β)		
$^{234}_{92}U$	Uranium II	2.47×10^5 yr
↓ α		
$^{230}_{90}Th$	Ionium	8.0×10^4 yr
↓ α		
$^{226}_{88}Ra$	Radium	1 602 yr
↓ α		
$^{222}_{86}Rn$	Emanation radon	3.823 d
↓ α		
$^{218}_{84}Po$	Radium A	3.05 mo
99.98% (α) ↓ 0.02% (β)		
$^{214}_{82}Pb$	Radium B	26.8 mo
$^{218}_{85}At$	Astatine	≈2 s
(β from $^{214}_{82}Pb$; α from $^{218}_{85}At$) ↓		
$^{214}_{83}Bi$	Radium C	19.7 mo
99.98% (β) ↓ 0.02% (α)		
$^{214}_{82}Po$	Radium C′	164 μs
$^{210}_{81}Ti$	Radium C″	1.3 mo
(α from $^{214}_{82}Po$; β from $^{210}_{81}Ti$) ↓		
$^{210}_{82}Pb$	Radium D	21 yr
↓ β		
$^{210}_{83}Bi$	Radium E	5.01 d
≈100% (β) ↓ 0.000 13% (α)		
$^{210}_{84}Po$	Radium F′	138.4 d
$^{206}_{81}Ti$	Radium E″	4.19 mo
(α from $^{210}_{84}Po$) ↓		
$^{206}_{82}Pb$	Radium G	Stable

[a]From Lowry and Lowry (1988). Reprinted from *Journal of the American Water Works Association,* by permission. Copyright © 1988, American Water Works Association.

TABLE 1.8 Characteristics of Important Isotopes

Isotope	Radiation	Half-life	Specific activity, Ci/g
Bromine-78	β, γ	6.4 min	—
Carbon-14	β	5 800 yr	4.6
Cesium-137	β, γ	30 yr	87
Cobalt-60	β, γ	5.3 yr	1.1×10^3
Iodine-129	β, γ	1.7×10^7 yr	1.6×10^{-4}
Iodine-131	β, γ	8.1 d	1.23×10^5
Iron-59	β	45 d	4.9×10^4
Krypton-85	β, γ	11 yr	—
Phosphorous-32	β	14 d	2.8×10^5
Plutonium-239	α	2.4×10^4 yr	0.06
Potassium-40	β, γ	1.3×10^9 yr	6.9×10^{-6}
Radium-226	α, γ	1 600 yr	1.0
Radon-222	α, γ	3.8 d	1.6×10^5
Sodium-24	β, γ	15.0 h	—
Strontium-90	β	29 yr	140
Sulfur-35	β	88 d	—
Tellurium-132	β, γ	3.3 d	3.0×10^5
Thorium-232	α, γ	1.4×10^{10} yr	1.1×10^{-7}
Tritium	β	12 yr	9 700
Uranium-233	α, γ	1.6×10^5 yr	9.6×10^{-2}
Uranium-235	α, γ	7.1×10^8 yr	2.1×10^{-6}
Uranium-238	α, γ	4.5×10^9 yr	3.3×10^{-7}
Zinc-65	β, γ	245 d	8.2×10^3

As shown in Table 1.7, the decay of one radioactive nuclide does not usually terminate radioactivity because the daughter nuclide is an unstable isotope unless the transformation is the final step.

Pierre and Marie Curie were among the pioneers in research on radioactive elements and they introduced the term radioactivity. The unit of radioactivity is a curie, named in honor of them. The standard curie (Ci) is 3.7×10^{10} disintegrations per second. One curie of an alpha, beta, or gamma emitter is that quantity of material that emits 3.7×10^{10} alpha particles, beta particles, or photons per second, respectively. This is a large quantity of radiation and milli- down to picocuries are more commonly used.

The Becquerel (Bq) is the SI unit of radioactivity and is defined as 1 nuclear transformation per second. It is equivalent to approximately 27 picocuries.

Table 1.8 summarizes half-lives and radiation types of important natural and human-generated radioactive isotopes. Isotopes with intermediate half-lives are considered the most dangerous. An isotope with a short half-life decays rapidly before moving any significant distance; an isotope with a long half-life does not emit a significant quantity of radiation.

■ Example 1.10 Radioactive Decay

This example illustrates principles in Section 1.8 as well as Section 1.13. What quantities of ionium and radium exist after 10 000 years, if the initial deposit only contained 10 g of ionium?

From Table 1.7, ionium decays to radium, which in turn decays to radon. Define the masses of ionium and radium as m_I and m_{ra}, respectively and their respective decay rates by k_I and k_{ra}. The decay process consists of a series of consecutive reactions. The rate of decay of ionium is described by Eq. (1.39).

$$\frac{dm_I}{dt} = -k_I m_I \qquad \int_{m_{I0}}^{m_I} \frac{dm_I}{m_I} = -k_I \int_0^t dt \Rightarrow m_I = m_{I0} e^{-k_I t}$$

The production of radium is proportional to the decay of ionium and radium also decays by a first-order reaction.

$$\frac{dm_{ra}}{dt} = f\left(\frac{dm_{ra}}{dt}\right)_{\text{production}} - \left(\frac{dm_{ra}}{dt}\right)_{\text{decay}} = fk_I m_I - k_{ra} m_{ra} = fk_I m_{I0} e^{-k_I t} - k_{ra} m_{ra}$$

The factor f in the above equation accounts for the mass difference between radium and ionium. Rearranging the above equation, the following differential equation is obtained.

$$\frac{dm_{ra}}{dt} + k_{ra} m_{ra} = fk_I m_{I0} e^{-k_I t}$$

This equation can be solved by use of an integrating factor, $e^{k_{ra}t}$ (see the Appendix) which is used to multiply each side of the equation.

$$e^{k_{ra}t}\left(\frac{dm_{ra}}{dt} + k_{ra} m_{ra}\right) = \frac{d}{dt}(m_{ra} e^{k_{ra}t})$$

$$\int d(m_{ra} e^{k_{ra}t}) = m_{ra} e^{k_{ra}t} = fk_I m_{I0} \int e^{(k_{ra}-k_I)t}\, dt = \frac{fk_I m_{I0}}{k_{ra} - k_I} e^{(k_{ra}-k_I)t} + C$$

$$m_{ra} = \frac{fk_I m_{I0}}{k_{ra} - k_I} e^{-k_I t} + Ce^{-k_{ra}t}$$

There was no radium at the initial time, and this is used to solve for the integration constant, C.

$$0 = \frac{fk_I m_{I0}}{k_{ra} - k_I} e^{-k_I(0)} + Ce^{-k_{ra}(0)} \Rightarrow C = -\frac{fk_I m_{I0}}{k_{ra} - k_I}$$

$$m_{ra} = \frac{fk_I m_{I0}}{k_{ra} - k_I}(e^{-k_I t} - e^{-k_{ra}t})$$

Now f = (226 g radium produced)/(230 g ionium decayed) = 0.983. From Table 1.7 and Eq. (1.40) the rate constants are

$$\text{Ionium:} \quad k_I = \frac{0.693}{t_{1/2}} = \frac{0.693}{8.0 \times 10^4 \text{ yr}} = 8.66 \times 10^{-6} \text{ yr}^{-1}$$

$$\text{Radium:} \quad k_{ra} = \frac{0.693}{t_{1/2}} = \frac{0.693}{1\,602 \text{ yr}} = 4.33 \times 10^{-4} \text{ yr}^{-1}$$

Substituting for the factors in the equations for m_I and m_{ra},

$$m_I = (10 \text{ g})e^{-(8.66 \times 10^{-6} \text{yr}^{-1})(10\,000 \text{ yr})} = 9.17 \text{ g}$$

$$m_{ra} = \frac{0.983(8.66 \times 10^{-6} \text{ yr}^{-1})(10 \text{ g})}{(4.33 \times 10^{-4} - 8.66 \times 10^{-6})\text{yr}^{-1}} \{e^{[-(8.66 \times 10^{-6} \text{yr}^{-1})(10\,000 \text{ yr})]} - e^{[-(4.33 \times 10^{-4} \text{yr}^{-1})(10\,000 \text{ yr})]}\} = 0.18 \text{ g}$$

This example is rather hypothetical because other species would be present.

QUESTIONS AND PROBLEMS

1. Express the following concentrations (given for water solutions) in ppm, ppb, molarity, and molality.
   (a) 4.2 mg N per liter;
   (b) 12 μg H_2S per liter;
   (c) 1.36×10^{-3} moles $NaHCO_3$ per liter.
2. What are the mole fractions of each of the constituents and water in a solution that contains 75 mg/L of NaCl, 120 mg/L of $C_6H_{12}O_6$, 8 mg/L of O_2, 150 mg/L of $Ca(HCO_3)_2$, 45 mg/L of $MgSO_4$, and 15 mg/L of KNO_3?
3. Express the concentrations of 0.40 mg/L of NO_2^-, 1.90 mg/L of NO_3^-, 0.70 mg/L of NH_3, and 8.90 mg/L of NH_4^+ as N to find the total concentration of all species as N.
4. The density of dry air at 0°C and 1 atm is 1.292 9 g/L. If the concentration of carbon monoxide (CO) is 2.0 ppm in dry air at 0°C and 1 atm, what is its concentration in $\mu g/m^3$?
5. What are the oxidation numbers on each of the atoms in the following compounds: (a) $Fe(NH_4)_2(SO_4)_2$; (b) $H_3C{-}C{=}O$.

   $$\begin{array}{c} H_3C-C=O \\ \quad\ \ | \\ \quad\ \ O-H \end{array}$$

6. What is the average oxidation number on the C atoms in $C_6H_{12}O_6$?
7. What are the oxidation numbers of chlorine in each of the following compounds (all atoms except chlorine have their common oxidation numbers): sodium perchlorate ($NaClO_4$); sodium chlorate ($NaClO_3$); chlorine dioxide (ClO_2); sodium hypochlorite (NaOCl); hypochlorous acid (HOCl); hydrochloric acid (HCl); monochloramine (NH_2Cl); dichloramine ($NHCl_2$); and nitrogen trichloride (NCl_3)?
8. Balance the following reactions:
   (a) $H_2O + CO_2 \rightleftharpoons C_6H_{12}O_6 + O_2$
   (b) $Al_2(SO_4)_3 + Ca(HCO_3)_2 \rightleftharpoons Al(OH)_3 + CO_2 + CaSO_4$
9. Nitrate is produced from the oxidation of ammonia in aerobic (oxygen is utilized) biological wastewater treatment. (a) Determine which of the following core reactions is feasible and balance it.

   $$NH_3 + O_2 \rightleftharpoons NO_3^-$$
   $$NH_3 + O_2 \rightleftharpoons NO_3^- + H_2O$$
   $$NH_3 + O_2 \rightleftharpoons NO_3^- + H_2O + H^+$$
   $$NH_3 + O_2 \rightleftharpoons NO_3^- + H_2O + OH^-$$
   $$NH_3 + O_2 \rightleftharpoons NO_3^- + H_2$$

   Actually in natural waters most of the ammonia exists as ammonium ion, NH_4^+. (b) Write the correct expression for the oxidation of ammonium ion with nitrate as a product.
10. What is the equivalent weight of O (oxygen) for electron transfer? Is the equivalent weight of O_2 the same or different?
11. What is the half-reaction for the oxidation of NH_4^+ to N_2? What is the equivalent weight of ammonium ion in this reaction?
12. What is the chemical equation for oxidation of NO_2^- to NO_3^- by hypochlorite ion (OCl^-)?
13. (a) Would oxidation of sulfide to sulfate, using chlorine as an oxidant, be favored by a high or by a low concentration of hydrogen ions? (b) If the oxidant was changed to permanganate (MnO_4^-) and MnO_2 was a product, answer the same question.
14. Give some reasons that the specific conductance of a solution is not simply the ideal equivalent conductance multiplied by the number of equivalents in solution.
15. Determine an approximate relation between specific conductance and total dissolved solids for tap water from Eqs. (1.10) and (1.11) and compare it with Eq. (1.8).

16. Derive Eq. (1.27) from Eq. (1.26).

17. What is the concentration of oxygen in moles/L and mg/L in 1 L of air when the temperature is 25°C and the total pressure is 1 atm? The air contains 21% oxygen by volume.

18. What is the density of air at a pressure of 1 atm that only contains nitrogen and oxygen at 79 and 21% by volume, respectively?

19. What is the saturation concentration of dissolved nitrogen in water in mg/L at a total atmospheric pressure of 1 atm and temperatures of (a) 0°C and (b) 25°C? Air contains 79% nitrogen by volume.

20. In a deep shaft biological treatment system, air is injected into wastewater moving down one shaft and up through an adjacent shaft. Assuming that wastewater is similar to tap water, what is the saturation concentration of oxygen in the wastewater at the bottom of a 100 m deep shaft? Use a temperature of 25°C for convenience.

21. Henry's constant is dimensionless if the concentrations in the gas and liquid are each defined as moles per liter. What are the values of Henry's constant for CH_4 and Cl_2 at 0°C if the concentrations in Eq. (1.34) are given as molar concentrations?

22. (a) After 2 h what is the concentration of a substance that decays according to a first-order reaction when the initial concentration is 2.6×10^{-4} *M* and the rate constant at 10°C is 0.063 h^{-1}? If the temperature is raised to 30°C and the value of θ is 1.062, what is the percentage decrease in concentration of this substance after 2 h?

(b) Perform the same exercise as in (a) except assume a second-order reaction applies with a rate constant of 106.8 L/mole/h and the other conditions are as given in (a).

23. What is the value of θ that corresponds to a doubling of the rate of reaction for a 10°C increase in temperature? The reaction is an elementary zero-, first-, or second-order reaction.

24. Plot the percentage of reactant remaining as a function of time for a retardant decay reaction with $k = 0.22$ d^{-1} and $\alpha = 0.008\,5$ d^{-1}. Let time vary from 0 to 10 days.

25. From each of the following values for water solubility, calculate the corresponding solubility-product constant: (a) $Mg_3(PO_4)_2$, 6.1×10^{-5} *M*; (b) FeS, 6.3×10^{-9} *M*; (c) CuF_2, 7.4×10^{-3} *M*.

26. Forty milligrams of calcium sulfate, $CaSO_4$, and 100 mg of sodium carbonate, Na_2CO_3 are added to 1 L of water. The solubility product of $CaCO_3$ is 5.00×10^{-9}. Assuming that $[H^+]$ is less than 10^{-10} (i.e., the CO_3^{2-} does not form bicarbonate), what is the final concentration of each dissolved species at equilibrium?

27. A water treatment plant operator wishes to fluoridate the water at a concentration of 1.0 mg/L F^-. The operator is concerned because the treated water contains calcium concentrations of up to 150 mg/L and precipitation of CaF_2 may occur. Assuming that CaF_2 is the only possible species to be formed, what are the limits on Ca^{2+} concentration for F^- at 1.0 mg/L and what are the limits for F^- concentration for Ca^{2+} at 150 mg/L? The K_{sp} for CaF_2 is 3.0×10^{-11}.

28. What is a ligand?

29. What are the concentrations of the chloride complexes of Hg^{2+} if the free Hg^{2+} concentration is 0.10 mg/L and the free Cl^- concentration is 0.5 mg/L?

30. What is the stability constant, β_3, for $AgCl_3^{2-}$?

31. How many grams of each of the following species are required to produce 0.1 Bq of activity [0.1 Bq/L is the World Health Organization (WHO) guideline for α activity in water]: radium-226, potassium-40, cobalt-60, cesium-137, and radon-222?

32. Why does radioactive decay of an unstable isotope generally not terminate radiation?

33. Starting with 1 g of uranium I, how many grams of $^{234}_{91}Pa^{m}$ would be formed after 1 000 years?

KEY REFERENCES

BENEFIELD, L. D., J. F. JUDKINS, AND B. L. WEAND (1982), *Process Chemistry for Water and Wastewater Treatment,* Prentice-Hall, Englewood Cliffs, NJ.

BUTLER, J. N. (1964), *Ionic Equilibrium: A Mathematical Approach,* Addison-Wesley, Don Mills, ON.

DENBIGH, K. (1981), *The Principles of Chemical Equilibrium,* 4th ed., Cambridge University Press, London.

SAWYER, C. N., P. L. MCCARTY, AND G .F. PARKIN (1994), *Chemistry for Environmental Engineering,* 4th ed., McGraw-Hill, Toronto.

SNOEYINK, V. L. AND D. JENKINS (1980), *Water Chemistry,* John Wiley & Sons, Toronto.

REFERENCES

LAIDLER, K. J. (1965), *Chemical Kinetics,* 2nd ed., McGraw-Hill, Toronto.

LANGELIER, W. F. (1936), "The Analytical Control of Anti-Corrosion Water Treatment," *J. American Water Works Association,* 28, pp. 1500–1521.

LIDE, D. R. AND H. P. R. FREDERIKSE (1994), *Handbook of Chemistry and Physics,* 75th ed., CRC Press, Boca Raton, FL.

LOWRY, J. D. AND S. B. LOWRY (1988), "Radionuclides in Drinking Water," *J. American Water Works Association,* 80, 7, pp. 50–64.

PERRY, R. H. AND D. W. GREEN, eds. (1984), *Perry's Chemical Engineers' Handbook,* 6th ed., McGraw-Hill, Toronto.

RUSSELL, L. L. (1976), *Chemical Aspects of Groundwater Recharge with Wastewaters,* Ph.D. Thesis, University of California, Berkeley.

SILLÉN, L. G. AND A. E. MARTEL (1964), *Stability Constants of Metal-Ion Complexes,* Special Publication No. 17, The Chemical Society, London.

Standard Methods for the Examination of Water and Wastewater (1992), 18th ed., American Public Health Association, Washington, DC.

WEAST, R. C., ed. (1984), *Handbook of Chemistry and Physics,* 65th ed., CRC Press, Boca Raton, FL.

CHAPTER 2

THE THERMODYNAMIC BASIS FOR EQUILIBRIUM

The driving force for a chemical reaction is attainment of a state in which the energy of the system and its surroundings is minimized. The study of energy transformations in chemical systems is thermodynamics, which provides us with very powerful relations to predict changes in systems. There are many valuable textbooks on this subject that can be consulted to gain insight into this discipline.

2.1 THERMODYNAMIC RELATIONS

Some of the more important thermodynamic relations are presented here with a brief explanation that is more utilitarian than theoretical.

2.1.1 The First and Second Laws of Thermodynamics

The first law of thermodynamics is logical and straightforward. It simply states:

Energy is conserved.

All the energy transformations in a process must sum to zero. The form of the energy may change, for instance, from chemical to heat energy or from potential to kinetic energy, but the total amount of all forms of energy remains constant before, during, and after the process. Energy may be input into process 2 from process 1 but the total energy state of the two processes (the system) remains constant.

The second law is a bit more difficult to formulate and it has been stated in a number of different forms. Some examples of the second law will be presented first. It is observed that crowding does not naturally occur. There is a net transformation of high-grade energy to low-grade energy in a system and its surroundings. The following statement of the second law is equivalent to the preceding statements.

The amount of useful work that can be obtained from an energy transformation is less than the amount of energy transformed.

Friction accompanies all mechanical processes and its analog is present in chemical or other energy transformations. An increase in overall randomness in a system and its surroundings accompanies an increase in organization in part of the system. Therefore, attaining a given level of high-grade energy will require a greater amount of energy input. Beyond the energy lost by convection and other processes, the sun is the ultimate source of the excess energy required to maintain the level of organization on Earth because life processes continuously degrade the quality of energy on Earth.

Ideal processes are frictionless and therefore allow energy transformations without dissipation of waste energy. It is impossible to make an ideal process but a close approximation to an ideal process can be achieved by advancing a process slowly in small incremental steps. In this manner it is possible to move from one state to another and back again with very little loss of useful energy. An ideal process is therefore a reversible process. Real processes are irreversible and require useful energy input to return to their former state. The amount of waste energy produced depends on how slowly or rapidly the change occurs. An explosion produces much more waste energy than a controlled heating.

The second law has been established by repeated observations. Mathematical formulations proceed from this tenet.

2.1.2 Free Energy

The driving force for all chemical reactions may also be stated as the aim to obtain a state of randomness or stability in which there is no production of entropy, which is a measure of disorder. For reactions, the useful concept describing energy transformations is the change in free energy, ΔG, which is related to entropy, internal energy, and work. Free energy is given the symbol G in honor of J. W. Gibbs, a pioneer in elucidating free energy relations.

For any reaction to occur there must be a net gain in entropy for the system and its surroundings. The ΔG function actually describes the amount of useful work that is theoretically available from a reaction or, conversely, the amount of work required to make the reaction proceed. Free energy changes alone are necessary to characterize the relation between initial and final states of a biochemical or chemical reaction. Because it is a state function, it is independent of path and gives no indication of the rate at which processes occur.

Each chemical species in any state has a partial molal (or molar) free energy or chemical potential defined by

$$\mu = \mu^o + RT \ln a \tag{2.1}$$

where

μ is chemical potential of the species
μ^o is the chemical potential of the species in its standard state
R is the universal gas constant (1.987 cal/K/mole)
T is temperature in °K
a is activity of the species

The standard state value is measured at a temperature of 298°K (25°C) and pressure of 1 atm. The activity of a species is dependent on the concentration of the species. Curly brackets, { }, are used to indicate the activity of species A at concentration [A].

$$a_A = \{A\} = \gamma_A[A] \tag{2.2}$$

where

γ_A is an activity coefficient

The activity coefficient is a function of the ionic strength (Section 1.7.2) of the solution and usually less than 1. However in dilute solutions the value of γ is sufficiently close to 1 for dissolved species and can be ignored. Therefore for dilute solutions, for a substance A, Eq. (2.1) becomes

$$\mu_A = \mu_A^o + RT \ln [A] \tag{2.3}$$

where

$[A]$ is the molar concentration of species A

Note that there is a hidden dimensional constant with a value of 1 L/mole multiplying $[A]$ in the logarithmic term.

The concentration of gases is expressed in terms of partial pressure in atmospheres.

$$\mu_g = \mu_g^o + RT \ln p_g \tag{2.4}$$

where

p_g is the partial pressure of the gas

[There is also a hidden dimensional constant with a value of 1 atm^{-1} multiplying p_g in the logarithmic term of Eq. (2.4).] The partial pressure of a gas may also have to be adjusted by an activity coefficient with an expression similar to Eq. (2.2) but it will generally be ignored throughout this text.

For the reaction

$$aA + bB + \cdots \rightleftharpoons cC + dD + \cdots \tag{2.5}$$

for given concentrations of reactants and products, the free energy of the reactants is

$$a\mu_A + b\mu_B + \cdots \tag{2.6a}$$

and the free energy of the products is

$$c\mu_C + d\mu_D + \cdots \tag{2.6b}$$

The change in free energy for the reaction as written is calculated by subtracting the free energy of the reactants from the free energy of the products:

$$\begin{aligned} \Delta G &= (c\mu_C + d\mu_D + \cdots) - (a\mu_A + b\mu_B + \cdots) \\ &= [c(\mu_C^o + RT \ln a_C) + d(\mu_D^o + RT \ln a_D) + \cdots] \\ &\quad - [a(\mu_A^o + RT \ln a_A) + b(\mu_B^o + RT \ln a_B) + \cdots] \end{aligned} \tag{2.7}$$

Equation (2.7) can be rearranged to

$$\Delta G = \Delta G^o + RT \ln \left(\frac{a_C^c \times a_D^d \times \cdots}{a_A^a \times a_B^b \times \cdots} \right) \tag{2.8}$$

where

ΔG^o is the standard free energy change for the reaction

$\Delta G^o = c\mu_C^o + d\mu_D^o + \cdots - (a\mu_A^o + b\mu_B^o + \cdots)$

a_I^i indicates the activity of substance I raised to the power i, where i is the stoichiometric number for the species in the balanced reaction

Equation (2.8) is the basic working equation. It is observed that the free energy change depends upon the standard free energy change for the reaction and the concentrations (activities) of the participating species. In this text, activities are generally assumed to be equal to concentrations. In some instances this assumption can lead to highly erroneous results, particularly when high concentrations are involved. For water solutions, even a highly concentrated wastewater, concentrations of substances are small relative to the concentration of water and the assumption is usually justified.

Thermodynamic relations only provide information on the ultimate equilibrium state of a system. It is generally true that as the change in free energy between the products and reactants increases, the more quickly the reaction proceeds.

It is impossible to calculate the absolute free energy (chemical potential) of any species but as seen from Eq. (2.7) only relative changes are important. A few definitions make it possible to compute relative free energy changes. The benchmark for computing relative free energy changes is the reaction between H^+ and H_2 gas when both species are at unit activity.

$$2H^+ + 2e^- \rightarrow H_2(g) \qquad \Delta G^\circ = 0 \tag{2.9}$$

(The letter in parentheses defines the state of the substance: aq, aqueous or in solution; g, gas; l, liquid; s, solid.) The free energy change of this reaction is defined to be zero at standard conditions. The standard free energy or chemical potential of any species is defined by measuring free energy for a system when hydrogen species in Eq. (2.9) are at unit activity and the species of interest are at unit activity.

The standard state free energy of any element in its native state is defined to be zero at any temperature, if it is at unit activity. As well, by convention, the standard state free energy of the electron at unit activity is defined to be equal to zero. Later this will be useful to describe half-reactions for redox processes.

The standard free energies of formation of many compounds and ions are available in numerous references. Table 2.1 lists some of interest. Table 1.3 also lists standard free energies for some half-reactions.

One of the more obvious ways to calculate ΔG° for a reaction is to set $\Delta G = 0$ in Eq. (2.8). If this is the case, equilibrium exists (products and reactants are at the same chemical potential) and an equilibrium constant, K, describes the relation among the concentrations of species involved.

For the reaction described by Eq. (2.5) and the free energy relation given by Eq. (2.8),

$$K = \left(\frac{\{C\}^c\{D\}^d \cdots}{\{A\}^a\{B\}^b \cdots}\right)_{eq} = \left(\frac{[\gamma_C C]^c[\gamma_D D]^d \cdots}{[\gamma_A A]^a[\gamma_B B]^b \cdots}\right)_{eq} \tag{2.10}$$

where

the subscript eq means the activities or concentrations are measured at equilibrium

In dilute solutions where γ_i's are close to 1 the equation for the equilibrium constant is given by

$$K = \left(\frac{[C]^c[D]^d \cdots}{[A]^a[B]^b \cdots}\right)_{eq} \tag{2.11}$$

which will be the working equation throughout most of the text. (The parentheses and eq subscript will be dropped.)

Equations (2.10) and (2.11) are statements of the law of mass action. Henry's law (Eq. 1.34) and the solubility product expression (Eqs. 1.36a and b) are also expressions of the law of mass action. The dimensions on K vary with the exponents and concentration dimensions in Eqs. (2.10) and (2.11).

Substituting Eq. (2.11) into Eq. (2.8) and setting $\Delta G = 0$,

$$\Delta G^\circ = -RT \ln K \qquad \text{or} \qquad K = e^{-\frac{\Delta G^o}{RT}} \tag{2.12}$$

Expression of Concentration in Equilibrium Expressions

As noted in Section 1.6 the units for expressing concentration in an equilibrium expression vary depending on the phase of the substance. Ions or molecules dissolved

TABLE 2.1 Standard Enthalpies and Free Energies of Formation at 25°C[a]

Substance	ΔH°_{298} kcal/mole	ΔG°_{298} kcal/mole	Substance	ΔH°_{298} kcal/mole	ΔG°_{298} kcal/mole
$AgCl$ (s)	−30.36	−26.22	H^+ (aq)	0	0
Ca^{2+} (aq)	−129.77	−132.18	H_2 (g)	0	0
$CaCO_3$(s), calcite	−288.45	−269.78	HCO_3^-	−165.18	−140.26
CaF_2 (s)	−290.3	−277.7	H_2CO_3 (aq)	−167.0	−148.94
$Ca(OH)_2$(s)	−235.80	−214.33	H_2O (l)	−68.32	−56.69
CaO (s)	−151.9	−144.4	H_2O (g)	−57.80	−54.64
$CaSO_4 \cdot 2H_2O$ (s)	−483.06	−429.19	H_2S (g)	−4.815	−7.892
C (s), graphite	0	0	H_2S (aq)	−9.4	−6.54
CH_4 (g)	−17.89	−12.13	HS^- (aq)	−4.22	3.01
CH_3CH_3 (g)	−20.24	−7.86	H_2SO_4 (l)	−193.91	−164.9
CH_3COOH (aq)	−116.74	−95.51	HgO (s) (red)	−21.68	−13.99
CH_3COO^- (aq)	−116.84	−89.0	Mg^{2+} (aq)	−110.41	−108.99
$C_2H_5COO^-$ (aq)	—	−86.3	$Mg(OH)_2$ (s)	−221.00	−199.27
C_2H_5COOH (l)	−122.1	−91.65	Mn^{2+} (aq)	−53.3	−54.4
CH_3OH (l)	−57.02	−41.92	MnO_2 (s)	−124.2	−111.1
C_2H_5OH (l)	−66.36	−43.44	Na^+	−57.28	−61.59
n-C_3H_7OH (l)	—	−42.02	$NaCl$ (s)	−98.232	−91.785
$C_6H_{12}O_6$ (aq)	—	−219.22	NH_3 (g)	−11.04	−3.98
$C_{12}H_{22}O_{11}$ (aq)	—	−370.90	NH_3 (aq)	−19.32	−6.35
C_6H_6 (l)	11.72	29.76	NH_4^+ (aq)	−31.74	−18.97
C_6H_5OH (l)	−37.80	−11.38	NO_2^- (aq)	−25.4	−8.9
Cl^- (aq)	−39.9	−31.35	NO_3^- (aq)	−49.37	−26.61
ClO_2^- (aq)	−15.9	5.1	NO (g)	21.600	20.719
ClO_3^- (aq)	−24.85	−0.8	O_2 (g)	0	0
CO (g)	−26.42	−32.78	O_2 (aq)	−3.9	3.93
CO_2 (g)	−94.05	−94.25	OH^- (aq)	−54.957	−37.594
CO_2 (aq)	−93.69	−92.31	PbO (s) (red)	−52.40	−45.25
CO_3^{2-} (aq)	−161.63	−126.17	S^{2-} (aq)	10	20
F^- (aq)	−78.66	−66.08	SO_2 (g)	−70.96	−71.79
Fe^{2+} (aq)	−21.0	−20.30	SO_4^{2-} (aq)	−216.90	−177.34
Fe^{3+} (aq)	−11.4	−2.52	Zn^{2+} (aq)	−36.43	−35.18
$Fe(OH)_3$ (s)	−197.0	−166.0	ZnS (s)	−48.50	−47.40

[a]From Dean (1985), Pedley et al. (1986), Thauer et al. (1977), and other sources.

in water are expressed in moles per liter. A pure solid or liquid in equilibrium with a water solution is assumed to have a concentration of 1.0. Gases in equilibrium with a solution have their concentrations expressed in atmospheres (atm). The concentration of the solvent water is assumed to be 1.0 regardless of whether water is being produced or consumed by reactions within the water solution. The actual basis for expressing the concentration of water or other liquid or solid phases in equilibrium with water is a mole fraction basis.

Equilibrium constants are based on the above assumptions, which will be applied throughout this text. The assumption that the concentrations of solids in equilibrium with water have a concentration of unity (or a constant density) is not always correct because of changes in the crystalline state of the solid. The assumption that water itself has a concentration of unity is quite reasonable for most dilute solutions of concern to environmental engineers because the change in moles of water present resulting from any reaction is negligible. However, in the case of reaction of hydrogen and oxygen gases to form water, the change in the concentration of water could not be assumed to be negligible and the actual concentration of water would have to be considered.

A few examples will be given to illustrate use of the equations and definitions.

■ Example 2.1 Standard Free Energy Change

For the formation of water vapor from $H_2(g)$ and $O_2(g)$, K is known to be 1.55×10^7 at a temperature of 2 000°K (1 727°C).

The reaction is

$$2H_2(g) + O_2(g) \rightleftharpoons 2H_2O(g)$$

$$\begin{aligned}\Delta G^\circ &= -RT \ln K \\ &= -(1.987)(2\,000) \ln(1.55 \times 10^7) \\ &= -65.8 \text{ kcal/mole}\end{aligned}$$

This gives the standard free energy of formation of water vapor at 2 000°K from hydrogen and oxygen gas because the latter two substances have standard free energies of zero. The negative sign indicates that free energy is released and the reaction would proceed spontaneously in the direction as written.

There are numerous other methods for measuring and calculating ΔG° for a reaction. These methods employ heat capacities of substances, electromotive forces (EMF) of electrochemical cells, and heat released under specified conditions. These measured values have been tabulated as noted above (see Table 2.2) and can be used to find ΔG° for many other reactions solely by calculation (see Section 2.2).

The following example illustrates the use of component reactions that are combined to form a new master reaction. The component reactions are stoichiometrically balanced and added together. If the activities of all components are set equal to unity, the ΔG° of the master reaction can be shown to be equal to the sum of the constituent reactions where each constituent is multiplied by its stoichiometric coefficient.

■ Example 2.2 Standard Free Energy Change

It is known that at 298°K = 25°C,

$$2H_2(g) + O_2(g) \rightleftharpoons 2H_2O(g) \qquad \Delta G^\circ_{298} = -109.28 \text{ kcal/mole} \qquad \text{(i)}$$

$$CO(g) + H_2O(g) \rightleftharpoons CO_2(g) + H_2(g) \qquad \Delta G^\circ_{298} = -6.82 \text{ kcal/mole} \qquad \text{(ii)}$$

Multiplying Eq. (ii) by 2 and adding it to Eq. (i) results in the new master reaction (Eq. iii).

$$2CO(g) + O_2(g) \rightleftharpoons 2CO_2(g) \qquad \text{(iii)}$$

The ΔG°_{298} for reaction (iii) is

$$\Delta G^\circ = (2 \times -6.81) - 109.28 = -122.92 \text{ kcal/mole}$$

Careful examination of Eq. (2.7) will show the above relation to be true. From Eqs. (2.7) or (2.8) it is noted that $\Delta G = \Delta G^\circ$ if all species are at unit activity.

The final Eq. (iii) states that (free energy is abbreviated to FE)

$$\begin{aligned}\Delta G^\circ_{iii} &= 2\text{FE } CO_2(g) - 2\text{FE } CO(g) - \text{FE } O_2(g) \\ &= 2\text{FE } H_2O - 2\text{FE } H_2(g) - \text{FE } O_2(g) \\ &\quad + 2[\text{FE } CO_2(g) + \text{FE } H_2(g) - \text{FE } CO(g) - \text{FE } H_2O(g)] \\ &= \Delta G^\circ_i + 2\Delta G^\circ_{ii}\end{aligned}$$

Example 2.2 has assumed that species are at unit activity. It is a simple matter to calculate the free energy change when substances are not at unit activity. For example, calculate the free energy change when the concentrations of the substances involved in reaction (iii) in Example 2.2 are changed from unity as shown in the following example.

■ Example 2.3 ΔG for Concentrations Different from Unity

For reaction (iii) in Example 2.2, the gaseous concentrations of CO_2, CO, and O_2 are 0.50, 0.20, and 0.40 atm, respectively. Assuming that concentrations are equal to activities, from Eq. (2.8):

$$\begin{aligned}\Delta G &= \Delta G^\circ + RT \ln \frac{[CO_2]^2}{[CO]^2[O_2]} \\ &= -122.92 + (1.987 \times 10^{-3})(298) \ln \left[\frac{(0.5)^2}{(0.20)^2(0.4)}\right] \\ &= -121.29 \text{ kcal}\end{aligned}$$

In biochemical systems it is redox reactions that release energy: the removal of electrons from an electron donor is accompanied by a release of energy. Conversely, the fabrication of new cellular components (complex molecules from simple ones) involves energy and electron input. Synthesis does not violate any rules of thermodynamics in that it is achieved at the expense of more entropy production in the surroundings.

As noted in Chapter 1, any redox reaction can be considered to be made from two half-reactions: an electron-producing half-reaction and an electron-receiving half-reaction. The free energy change of a half-reaction may be calculated. Consider the reduction of sulfate to sulfur.

$$SO_4^{2-} + 8H^+ + 6e^- \rightarrow S + 4H_2O$$

Any reaction in which oxygen is part of an ion or compound being oxidized or reduced will involve water. By dividing the above equation by 6, the equation is normalized to involve one electron.

$$\tfrac{1}{6}SO_4^{2-} + \tfrac{4}{3}H^+ + e^- \rightarrow \tfrac{1}{6}S + \tfrac{2}{3}H_2O$$

■ Example 2.4 Standard Free Energy Change of a Half-Reaction

This example is concerned with calculating the standard free energy change for the above reaction, given that the ΔG° values for S, H^+, H_2O, and SO_4^{2-} are 0, 0, −56.69, and −177.34 kcal, respectively, at 298°K.

$$\Delta G^\circ = \tfrac{1}{6}(\text{FE S}) + \tfrac{2}{3}(\text{FE } H_2O) - \tfrac{1}{6}(\text{FE } SO_4^{2-}) - \tfrac{4}{3}(\text{FE } H^+) - 1(\text{FE } e^-)$$
$$= 0 + \tfrac{2}{3}(-56.69) - \tfrac{1}{6}(-177.34) - 0 - 0$$
$$= -8.24 \text{ kcal/mole}$$

2.1.3 Temperature Effects on the Equilibrium Constant

Enthalpy is the heat that is released or taken up in a reaction. The standard enthalpy change for a reaction is calculated in the same manner as the standard free energy change (see Example 2.2). The standard enthalpy change ΔH° for the reaction in Eq. (2.5) is

$$\Delta H^\circ = cH_c^\circ + dH_D^\circ + \cdots - (aH_A^\circ + bH_B^\circ + \cdots) \tag{2.13}$$

where
H_i° is the standard enthalpy change for species i

Standard enthalpies for a number of substances are given in Table 2.1.

The standard enthalpy change is most useful to calculate changes in the equilibrium constant for a reaction. The van't Hoff equation, derived from thermodynamic principles, provides the relation:

$$\frac{d(\ln K)}{dT} = \frac{\Delta H^\circ}{RT^2}$$

where
K is the equilibrium constant
T is temperature in °K

Integrating the above expression,

$$\ln \frac{K_{T_1}}{K_{T_2}} = \frac{\Delta H^\circ}{R}\left(\frac{1}{T_2} - \frac{1}{T_1}\right) \tag{2.14}$$

This equation provides the basis for calculating the change in the equilibrium constant as a function of temperature.

■ **Example 2.5 Temperature Variation of the Equilibrium Constant**

Determine the equilibrium constant for the reaction below at temperatures of 25 and 50°C.

$$S^{2-} + 2H^+ \rightleftharpoons H_2S$$

The standard free energy and enthalpy for each substance (at 25°C) are

S^{2-}:	$\Delta G^\circ = 20$ kcal/mole	$\Delta H^\circ = -10$ kcal/mole
H^+:	$\Delta G^\circ = 0$	$\Delta H^\circ = 0$
H_2S:	$\Delta G^\circ = -7.892$ kcal/mole	$\Delta H^\circ = -4.815$ kcal/mole

The free energy and enthalpy changes for the reaction are

$$\Delta G^\circ = -7.892 - [20 + 2(0)] = 12.1 \text{ kcal/mole}$$
$$\Delta H^\circ = -4.815 - [-10 + 2(0)] = 5.2 \text{ kcal/mole}$$

$$T = 25°C = 25 + 273.2 = 298.2°K$$
$$R = 1.987\,2 \times 10^{-3}\ \text{kcal/°K/mole}$$

The equilibrium constant at 25°C is

$$K = e^{-\frac{\Delta G^o}{RT}} = \exp\left[-\frac{12.1\ \text{kcal/mole}}{(1.987 \times 10^{-3}\ \text{kcal/°K/mole})(298.2°K)}\right] = 1.35 \times 10^{-9} = K_{25}$$

$$T = 50°C = 50 + 273.2 = 323.2°K$$

$$\ln\frac{K_{T_1}}{K_{T_2}} = \frac{\Delta H^o}{R}\left(\frac{1}{T_2} - \frac{1}{T_1}\right)$$

$$\frac{K_{T_1}}{K_{T_2}} = \exp\left[\frac{\Delta H^o}{R}\left(\frac{1}{T_2} - \frac{1}{T_2}\right)\right]$$

$$= \exp\left[\frac{5.2\ \text{kcal/mole}}{1.987 \times 10^{-3}\ \text{kcal/°K/mole}}\left(\frac{1}{323.2°K} - \frac{1}{298.2°K}\right)\right] = 0.507$$

$$K_{50} = K_{25}(0.507) = (1.35 \times 10^{-9})(0.507) = 6.85 \times 10^{-10}$$

2.2 Redox Potentials

Chemical energy can be used to generate electrical energy and vice versa. This is the concept applied in batteries; it also applies to corrosion and is useful in analyses. The driving force for a reaction is the free energy change, which can be directly related to an electrical potential.

An electrochemical cell (Fig. 2.1) is a special device for measuring the potential of half-reactions. In an electrochemical cell the concentrations of the substances are usually adjusted to be at unit activity for convenience. The cell pictured in Fig. 2.1 contains a solid Zn electrode and Zn^{+} and SO_4^{2-} ions in the right half-cell and a solid Cu electrode and Cu^{2+} and SO_4^{2-} ions in the left half-cell. The half-reactions involved are

$$Zn^{2+} + 2e^- = Zn \qquad Cu^{2+} + 2e^- = Cu$$

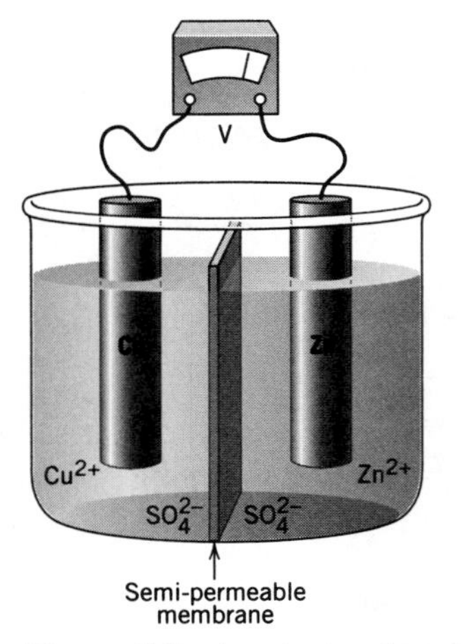

Figure 2.1 An electrochemical cell.

Observation over a period of time will confirm that the Zn bar becomes pitted and a deposit forms on the Cu bar. This shows that Zn has a greater tendency to give up electrons than copper. The characteristics of the cell allow an electrical potential to be measured and useful work can be obtained from the setup.

Also, an examination of free energy change with each of the ions at unit activity would show that Cu^{2+} oxidizes Zn according to the following reaction.

$$Zn + Cu^{2+} = Zn^{2+} + Cu$$

For concentrations of Zn and Cu ions each at unit activity, a potential of 1.107 V would be measured.

The barrier between the two half-cells is a semipermeable membrane that allows SO_4^{2-}, in this circumstance, to migrate between the two cells. The migration of SO_4^{2-} ions is necessary because as the reaction proceeds there will be a buildup of positive charge in the right cell and a decrease of positive charge in the left cell. Sulfate migration preserves neutrality. The complete circuit consists of an external and an internal circuit. The external circuit is the wires connecting the two electrodes. The internal circuit is the solution itself, containing the ions that are able to move freely.

The other important feature of the membrane is prevention of Zn^{2+} and Cu^{2+} from passing between the cells. This forces electrons (current) to flow through the external circuit and allows the potential to be measured. If Cu^{2+} and Zn^{2+} ions were allowed to move through the membrane, oxidation and reduction would proceed directly in solution and the measured potential would not reflect the total amount of electrons being transferred. In the absence of the external circuit no electron flow would be possible and no reaction would occur when the semipermeable membrane is present.

A potentiometer is a device used to measure potential of reactions. In an electrochemical cell as described above, reaction would proceed and concentrations and the voltage would continuously change. In a potentiometer a voltage opposing the voltage in the electrochemical cell is imposed to prevent the flow of current. The concentrations remain constant and a stable reading may be obtained.

The electrodes are named according to the redox reaction occurring at the electrode. The cathode is the electrode at which reduction takes place. Cations receive electrons at this electrode. The anode is the electrode at which oxidation takes place. Electrons are removed from anions at this electrode. Electrons always flow from the anode to the cathode in the external circuit.

Electrochemical cells can be constructed for a large number of substances. The substance does not need to be in a solid state to be used in a cell. An inert electrode such as platinum, which conducts current but does not participate in the reaction of interest, may be used. The potential for any oxidation–reduction reaction is measured against a standard hydrogen electrode (SHE), which consists of hydrogen gas continuously bubbled into a solution to maintain a pressure of 1 atm and $[H^+]$ in solution at unit activity (Fig. 2.2). Platinum is immersed in the solution and wired to the external circuit. The semipermeable membrane in this configuration is replaced with a U-tube containing KCl gel, which makes electrical contact with each solution and has an insignificant effect on potentials. Potassium or chloride ions migrate from the KCl gel to preserve charge neutrality depending on the reactions in each cell. The reaction for the SHE is

$$H_2(g) = 2H^+ + 2e^- \qquad E^\circ = 0\ V$$

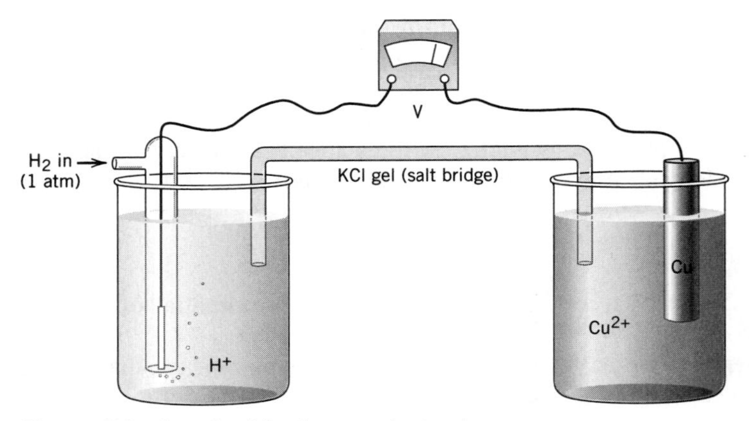

Figure 2.2 Standard hydrogen electrode.

where
 E° is the standard potential

The standard potential of this electrode is assigned a value of 0, which is in accord with the assignment of a free energy change of 0 (Eq. 2.9) for this reaction. The half-reactions in each cell of an electrochemical cell form a couple. The standard potential for any substance at standard conditions of unit activity and temperature of 298°K coupled to a SHE is measured directly.

For the generalized reaction

$$aA + bB + \cdots \rightleftharpoons cC + dD + \cdots \tag{2.5}$$

The equation describing the potential, E, for the reaction is the Nernst equation, which is closely related to the equation for free energy change.

$$E = E^\circ - \frac{RT}{nF} \ln \left[\frac{a_C^c \times a_D^d \times \cdots}{a_A^a \times a_B^b \times \cdots} \right] \tag{2.15}$$

where
 F is the Faraday constant (96 493 coulombs/equivalent)
 n is the number of electrons transferred (no. of equivalents/mole)

E and E° are readily related to ΔG and ΔG° by comparing Eqs. (2.8) and (2.15).

$$E = -\frac{\Delta G}{nF} \quad (2.16a) \qquad E^\circ = -\frac{\Delta G^\circ}{nF} \quad (2.16b)$$

The number of electrons transferred is not explicitly contained in Eq. (2.15) but it can be assessed from the half-reactions involved. As before, concentrations are used as an approximation of activities.

Using the definition of R as 8.314 V-coulombs/°K/mole, assuming a temperature of 25°C, substituting concentrations for activity, and changing to a base 10 logarithm,

$$E = E^\circ - \frac{0.059}{n} \log \left(\frac{[C]^c[D]^d \cdots}{[A]^a[B]^b \cdots} \right) \tag{2.17}$$

Equation (2.17) may be written for a half-reaction as well as a complete reaction.

In the case of the copper–zinc electrochemical cell, the Nernst equations for the two half-reactions are:

$$E_{Zn} = E^{o}_{Zn} - \frac{0.059}{2}\log\frac{[Zn]}{[Zn^{2+}]} = E^{o}_{Zn} - \frac{0.059}{2}\log\frac{1}{[Zn^{2+}]}$$

$$E_{Cu} = E^{o}_{Cu} - \frac{0.059}{2}\log\frac{[Cu]}{[Cu^{2+}]} = E^{o}_{Cu} - \frac{0.059}{2}\log\frac{1}{[Cu^{2+}]}$$

$[Zn] = [Cu] = 1$ because these are solid phases.

Significance of E^o

E^o is the standard potential, which as G^o (or ΔG^o), is characteristic of the species. Consider the two half-reactions,

$$A + e^- \rightarrow A^- \qquad E^o_A$$

$$B + e^- \rightarrow B^- \qquad E^o_B \qquad E^o_A > E^o_B$$

If the concentration of each species is unity, then the species with the higher E^o will be the oxidant (substance A for the above two reactions).

Cell or Couple Potential

In the overall reaction, the half-reaction with the higher potential will go as written, the other half-reaction will be reversed. The Nernst equations for the above two half-reactions are

$$E_A = E^o_A - 0.059\log\frac{[A^-]}{[A]}$$

$$E_B = E^o_B - 0.059\log\frac{[B^-]}{[B]}$$

If $E_A > E_B$, the cell or couple potential (E_c) is the difference between E_A and E_B. It is the driving force for the redox reaction. However, it does not provide any information on the rate of reaction, only on the direction in which the reaction will naturally proceed. For the A and B couples, the overall reaction is

$$A + e^- \rightarrow A^- \qquad E^o_A$$

$$\underline{B^- - e^- \rightarrow B} \qquad -E^o_B$$

$$A + B^- \rightarrow A^- + B \qquad E^o_A - E^o_B$$

$$E_c = E_A - E_B = E^o_A - E^o_B - 0.059\log\frac{[A^-][B]}{[A][B^-]}$$

If the number of electrons involved in a half-reaction is not equal to 1, this will influence the stoichiometry of the overall reaction but not the potential associated with a half-reaction. Consider the reaction

$$C + 4e^- \rightarrow 2D^{2-} + X \qquad E^o_{CD}$$

The Nernst equation for this half-reaction is

$$E_{CD} = E^o_{CD} - \frac{0.059}{4}\log\frac{[D^{2-}]^2[X]}{[C]}$$

This equation would not change if the reaction were written as

$$100C + 400e^- \rightarrow 200D^{2-} + 100X \qquad E^\circ_{CD}$$

The Nernst equation for this reaction is

$$E_{CD} = E^\circ_{CD} - \frac{0.059}{400}\log\frac{[D^{2-}]^{200}[X]^{100}}{[C]^{100}} = E^\circ_{CD} - \frac{0.059}{4}\log\frac{[D^{2-}]^2[X]}{[C]}$$

Combining the C–D^{2-} half-reaction with the A–A^- half-reaction (assuming $E_A > E_{CD}$),

$$4A + 2D^{2-} + X \rightarrow 4A^- + C$$

$$E_c = E_A - E_{CD} = E^\circ_A - E^\circ_{CD} - \frac{0.059}{4}\log\frac{[A^-][C]}{[A]^4[D^{2-}]^2[X]}$$

from which it can be shown that the individual Nernst equations that make up E_c have not changed in the overall equation.

The cathode (electrode where reduction occurs) will have a higher potential than the anode (where oxidation occurs). The cell potential for a reaction to proceed as written will be positive. Therefore,

$$E_c = E_{cathode} - E_{anode} \tag{2.18}$$

Note that in the case of an electrochemical cell, E_c refers to the potential difference between the couples that exist in separate well-defined cells. In general as different waters mix or substances are added to a water, half-reaction couples are set up and there will be potential differences between the couples. A potential difference between couples is E_c regardless of whether these couples occur in separate cells or not. An electrochemical cell allows the potential difference to be measured accurately. This information is useful for defining potential differences in an actual solution when concentrations are known and determining where the reaction will proceed.

Oxidation–Reduction Potential and System Potential

When equilibrium exists, $E_c = 0$. In the ideal case, the potentials of all possible half-reactions are equal. This potential is known as the system potential (E_{sys}).

$$E_{sys} = E_1 = E_2 = \cdots = E_i = \cdots = E_n \qquad (E_c = 0) \tag{2.19}$$

Note to use Eq. (2.19) with the Nernst equation, each reaction must be written as a reduction, which is consistent with $E_c = 0$.

When equilibrium does not exist, the system potential is not well defined but it will be somewhere in between the potentials of the couples that exist in the system.

The oxidation–reduction potential (ORP) is the system potential of a solution. It is measured by an inert electrode that transmits potential. The electrode is usually a platinum wire. As noted before, platinum is a good choice for the electrode because it does not significantly influence the potential that it transmits and it is sensitive to many species.

To complete the external circuit, a reference electrode is wired to the voltmeter and inserted into the solution. Reference electrodes maintain a constant potential and do not significantly participate in reactions of interest in the solution. Their only purpose is to make electrical contact with the solution. The choice and makeup of a reference electrode varies depending on the situation in which it will be used. Effectively the reference electrode can be considered to be one half-cell of an electrochemical cell. The other half-cell of the electrochemical cell is the solution with the other electrode immersed in it. The half-cells are separated and make electrical contact by a membrane or other equivalent means in the reference electrode.

For a solution at equilibrium with its potential measured by an ideal electrode, Eq. (2.19) will apply. The electrode will transmit E_{sys}. Defining the reference electrode potential as E_{ref} and considering it to be the anode, the potential reading on the voltmeter will be

$$E_{meter} = E_c = E_{sys} - E_{ref} \tag{2.20}$$

The ORP is usually reported with respect to the hydrogen scale, i.e., the result is normalized with respect to using the SHE as the reference electrode.

Because natural waters are unlikely to be in a state of complete equilibrium, the ORP is a gross indicator of the state of oxidation–reduction in the system. Besides this, the potential reading is influenced by how well the electrode "sees" species in solution among other factors. But this measurement is still a useful tool in many circumstances.

The general form of the Nernst equation for a half-reaction is

$$\text{ORP} = E_{sys} = E^\circ - \frac{0.059}{n} \log \frac{[\text{reduced species}]}{[\text{oxidized species}]} \tag{2.21}$$

The charge balance always sums to neutrality, but it is seen from Eq. (2.21) that as the concentrations of species able to donate electrons (reduced species) increase, the ORP decreases and vice versa. Thus as oxygen or other oxidants are added to a solution, the ORP rises. Water with a pH of 7.00 and temperature of 25°C in equilibrium with oxygen at a partial pressure of 0.21 atm has an ORP of 800 mV on the hydrogen scale. At high ORP there will be a predominance of ions such as SO_4^{2-}, NO_3^-, and Fe^{3+} as opposed to their reduced forms of S^{2-}, NH_4^+, and Fe^{2+}. As free oxygen is removed from a water by chemical or biological reactions, the state of the water changes from an aerobic to an anoxic state, wherein the concentration of oxidized species is predominant compared to their reduced counterparts but there is no significant oxygen concentration. Many microorganisms are able to use SO_4^{2-}, NO_3^-, or other oxidized species as electron acceptors. As reaction progresses and oxidized species are consumed and the reduced species predominate, the water reaches an anaerobic state. In anaerobic reactors, ORPs in the range of −300 to −400 mV exist.

■ Example 2.6 Use of the Nernst Equation

A platinum electrode and reference electrode were used to measure the ORP of a water. The meter reading was 0.01 V. The reference electrode was known to have a potential of 0.25 V and it can be considered to be the anode. The pH of the water was 8.21. If the water was in equilibrium, what is the ratio of $[NH_4^+]$ to $[NO_3^-]$? Was there any free dissolved oxygen in the water?

The platinum electrode senses the potential of the solution. The potential reading on the meter is E_c. From Eq. (2.20),

$$E_{sys} = E_{meter} + E_{ref} = 0.01 \text{ V} + 0.25 \text{ V} = 0.26 \text{ V}$$

From Table 1.3, the half-reaction for NH_4^+–NO_3^- is

$$\tfrac{1}{8}NO_3^- + \tfrac{5}{4}H^+ + e^- = \tfrac{1}{8}NH_4^+ + \tfrac{3}{8}H_2O \qquad E^\circ = 0.880$$

and the corresponding Nernst equation set equal to E_{sys} is

$$E_{sys} = 0.26 = 0.880 - \frac{0.059}{8} \log \frac{[NH_4^+]}{[NO_3^-][H^+]^{10}}$$

The ratio of ammonium to nitrate is

$$\log \frac{[NH_4^+]}{[NO_3^-]} = -\frac{8(0.26 - 0.88)}{0.059} - 10pH = 84.1 - 82.1 = 2.0$$

The ratio is found by taking the antilog of both sides.

$$\frac{[NH_4^+]}{[NO_3^-]} = 100$$

To find the concentration of dissolved oxygen that would be in equilibrium at this potential, use the half-reaction for dissolved oxygen.

$$\tfrac{1}{4}O_2(aq) + H^+ + e^- = \tfrac{1}{2}H_2O \qquad E^\circ = 1.27$$

$$0.26 = 1.27 - \frac{0.059}{4} \log \frac{1}{[O_2][H^+]^4}$$

Solving the above equation for $[O_2]$, the equilibrium $[O_2]$ is found to be $10^{-35.6}$ or effectively 0. Anaerobic conditions exist.

2.3 CORROSION

The principles of electrochemistry are the underlying reasons for corrosion. When a metal has a potential that is anodic with respect to its environment, it will tend to corrode or dissolve. Corrosion is a very complex phenomenon as will be seen from this brief discussion. Many factors mediate to accelerate and retard the rate of corrosion and the final determination of remedial measures may be largely a matter of trial and error.

Two dissimilar metals in contact with each other and exposed to a conductive solution form an electrochemical cell. The metals themselves form the external circuit and the ions in solution form an internal circuit. A potential difference will exist between the metals promoting the flow of electrons. The less resistant metal (M_1) becomes the anode and corrodes. The other metal (M_2) becomes the cathode.

$$M_1 \rightarrow M_1^{n+} + ne^-$$

$$M_2^{n+} + ne^- \rightarrow M_2$$

An example of this type of situation is shown in Fig. 2.3. It is not necessary to have Cu^{2+} ions in solution; any electron acceptor is suitable. This type of corrosion is

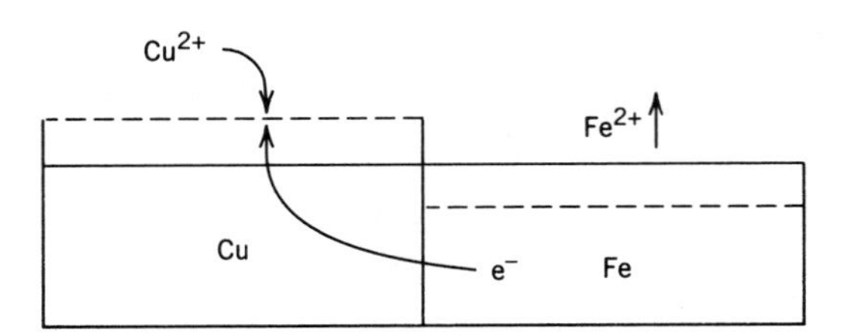

Figure 2.3 Galvanic corrosion.

TABLE 2.2 EMF Series[a] for Metals (25°C)

Element	Electrode reaction	Standard potential, V
(Noble end)		
Gold	$Au^{3+} + 3e^- \rightarrow Au$	1.42
Platinum	$Pt^{2+} + 2e^- \rightarrow Pt$	1.2
Silver	$Ag^+ + e^- \rightarrow Ag$	0.800
Copper	$Cu^+ + e^- \rightarrow Cu$	0.522
Copper	$Cu^{2+} + 2e^- \rightarrow Cu$	0.345
Hydrogen	$2H^+ + 2e^- \rightarrow H_2(g)$	0.000 (reference)
Lead	$Pb^{2+} + 2e^- \rightarrow Pb$	−0.126
Tin	$Sn^{2+} + 2e^- \rightarrow Sn$	−0.136
Nickel	$Ni^{2+} + 2e^- \rightarrow Ni$	−0.250
Cobalt	$Co^{2+} + 2e^- \rightarrow Co$	−0.277
Cadmium	$Cd^{2+} + 2e^- \rightarrow Cd$	−0.402
Iron	$Fe^{2+} + 2e^- \rightarrow Fe$	−0.44
Chromium	$Cr^{3+} + 2e^- \rightarrow Cr$	−0.71
Zinc	$Zn^{2+} + 2e^- \rightarrow Zn$	−0.762
Manganese	$Mn^{2+} + 2e^- \rightarrow Mn$	−1.05
Aluminum	$Al^{3+} + 3e^- \rightarrow Al$	−1.67
Beryllium	$Be^{2+} + 2e^- \rightarrow Be$	−1.70
Magnesium	$Mg^{2+} + 2e^- \rightarrow Mg$	−2.34
Sodium	$Na^+ + e^- \rightarrow Na$	−2.712
(Active end)		

[a]After Bosich (1970).

called galvanic corrosion. The potential difference between the two metals accelerates the removal of electrons from the more active metal.

Metals have been organized by their electrochemical activity. Table 2.2 lists the standard electromotive force (EMF) values (E° values) for various metals. The values in Table 2.2 are not necessarily equal to those in Table 1.3 because the EMF values are more reflective of the effective potential of a metal. The more active the metal (the lower its position in the table), the more likely it is to be oxidized (lose electrons) or corrode. The noble elements at the top of the table tend to be cathodic with respect to the surrounding environment, and thus not readily oxidizing. From examination of the Nernst equation, it is seen that a lower ORP promotes the dissolution of metals; i.e., the ionized metal concentration increases.

In general for corrosion to occur, dissimilar metals are not required. Direct attack by oxidizing agents will cause corrosion. Furthermore, differences in the composition of a metal or imperfections in the metal will cause potential differences within it. Likewise, differences in concentrations of species in solution at two different locations will set up a potential difference.

Oxygen is present in most waters and it causes corrosion because of its oxidizing power. The complexity of the corrosion process is shown by the following equations, which illustrate how oxygen can be involved in both the cathodic and anodic reactions for the corrosion of iron.

$$4Fe + 2O_2 + 8H_2O \rightarrow 4Fe(OH)_3 + 4H^+ + 4e^- \qquad \text{(anodic reaction)}$$

$$O_2 + 4H^+ + 4e^- \rightarrow 2H_2O \qquad \text{(cathodic reaction)}$$

Another route of corrosion is through high hydrogen ion concentrations, which cause direct acid attack as illustrated with the following reaction.

$$Fe + 2H^+ \rightarrow Fe^{2+} + H_2$$

The H_2 may react with O_2 to form water.

Pipes are subject to corrosion from both the inside and the outside. On the outside, external direct current (DC) enters and travels along a pipe (the cathodic area) and leaves the pipe at another area (the anodic area). Sources of DC can be electrified transit systems, cathodic protection systems for other structures, DC welding equipment, and crane operations. This is electrolytic corrosion, which is similar to galvanic corrosion.

Soil characteristics influence the rate of external corrosion in a number of ways. Electrochemical differences in the soil along a pipe can create a galvanic cell. The resistance of the soil affects the current carrying ability of the soil. The soil can also influence the buildup or removal of corrosion products on the structure being corroded. These products will produce an insulating barrier to the further corrosion of the structure, which is one example of polarization phenomena. Polarization is the reduction in potential difference from the ideal potential difference. This lowers the rate of corrosion. O'Day (1989) has discussed other factors causing external corrosion of pipes.

Grounding of electrical systems to household water supply plumbing systems increases corrosion and therefore should not be done. Not only will the stray current–induced corrosion increase the system deterioration but lead and other toxic metals in solder and pipes will also appear at elevated levels in the drinking water (Lee et al., 1989).

2.3.1 Microbial Corrosion

Many bacteria, algae, and fungi cause corrosion directly or indirectly. At locations of bacteria colonies, pH and dissolved oxygen concentrations are different from the bulk phase concentrations and concentrations at other locations on a structure or pipe. Anaerobic colonies also cause differences in O_2 concentration from the bulk phase because they manufacture a protective biofilm that shields them from oxygen. Thus, concentration cells are set up that cause electron flow and corrosion.

Bacteria that derive their energy from the oxidation of inorganic compounds are primary agents of corrosion. Sulfate-reducing bacteria aid in the corrosion of sewers. These anaerobes derive energy from the following reaction:

$$SO_4^{2-} + 8H_{ads} \rightarrow S^{2-} + 4H_2O$$

where

H_{ads} is hydrogen adsorbed to the metal surface

The removal of elemental hydrogen that has adsorbed to a metal surface accelerates the corrosion of the surface. The adsorbed hydrogen retards the transfer of electrons to an oxidizing agent by acting as an insulator. Furthermore, in steel pipes, the sulfide produced in this reaction reacts with Fe^{2+} to produce FeS. FeS is more efficient than steel for promotion of the reaction:

$$2H^+ + 2e^- \rightarrow H_2$$

This further accelerates corrosion.

In sewers the evolution of hydrogen sulfide gas produced by the sulfate-reducing bacteria is taken to advantage by aerobic sulfide-oxidizing bacteria adhering to the crown of the pipes. The reaction is

$$H_2S + 2O_2 \rightarrow SO_4^{2-} + 2H^+$$

The H_2SO_4 is a much stronger acid than H_2S, which enhances acid attack. These bacteria can also oxidize elemental sulfur. Other bacteria produce strong organic acids as byproducts.

Another group of bacteria that is significant in pipe corrosion is the iron-oxidizing bacteria. The energy-producing reaction for these bacteria is

$$2Fe + 1\tfrac{1}{2}O_2 \rightarrow Fe_2O_3$$

The colonies form iron crust deposits known as "tubercles." Corrosion is accelerated at the location of these tubercles.

2.3.2 Corrosion Prevention

Various constituents in water influence the rate of corrosion; the more significant chemical factors that influence the rate of corrosion are given in Table 2.3. Water

TABLE 2.3 Effects of Water Constituents on Corrosion[a]

Factor	Effect
Alkalinity (buffering capacity; see Chapter 3)	Alkalinities may help form protective coating; helps control pH change. Low to moderate alkalinity reduces corrosion of most materials. High alkalinities increase corrosion of copper and lead.
Ammonia	Ammonia may increase the solubility of some metals such as copper and lead through the formation of complexes.
Ca and Mg (hardness)	Ca may precipitate as $CaCO_3$ and thus provide protection and reduced corrosion rates. Ca and Mg may enhance the buffering effect of alkalinity and pH.
Chloride and sulfate	High levels of chloride or sulfate increase corrosion of iron, copper, and lead.
Chlorine residual	Chlorine residual increases metallic corrosion, particularly for copper, iron, and steel.
Copper	Copper causes pitting in galvanized pipe.
Dissolved oxygen	Dissolved oxygen increases the rate of many corrosion reactions.
Hydrogen sulfide	Hydrogen sulfide increases corrosion rates.
Magnesium and other trace metals	Trace metals may inhibit the formation of the more stable crystalline form of $CaCO_3$ (calcite) and favor the less stable (more soluble) crystalline form (aragonite). See the text for discussion of inhibitors.
pH	Low pH may increase corrosion rate: high pH may protect pipes and decrease corrosion rates or could cause dezincification of brasses.
Natural color, organic matter	Organic matter may decrease corrosion by coating pipe surfaces. Some organics can complex metals and accelerate corrosion.
Total dissolved solids (TDS)	High TDS increases conductivity and corrosion rate.

[a]From M. R. Schock (1990), "Internal Corrosion and Deposition Control," in *Water Quality and Treatment*, 4th ed., F. W. Pontius, ed., McGraw-Hill, Toronto, reproduced with permission of McGraw-Hill, Inc.

treatment plant operators should consider these factors in controlling the treatment process to provide safe water that will be less corrosive in the distribution system.

Beyond the adjustment of chemical factors there are a number of measures that can be taken to retard the rate of corrosion.

Protective Coatings. Coating materials subject to corrosion provides a barrier to electron transfer. All metal surfaces should be coated, including the noble metals. The noble metals may corrode themselves. The surfaces of metals that will be cathodic should be coated to minimize opportunity for the circuit to be completed.

Corrosion Resistant Alloys. These hybrid metals have been developed by metallurgists and chemists to strongly hold metal ions within the crystalline structure of the metal. Thus they are very stable and resistant to oxidation.

Inhibitors. Inhibitors are substances that react in water to form a compound that lays down a protective barrier to electron transfer. Anodic inhibitors migrate toward anodic sites and are oxidized, forming a film that is relatively impenetrable. An example is chromate ion, which has an overall reaction as follows:

$$2CrO_4^{2-} - 6e^- \rightarrow Cr_2O_3$$

The intermediate reactions involved are not definitely known but the oxide does form a layer of insulation.

Cathodic inhibitors stop the flow of H^+ or O_2 to the cathodic site. Calcium carbonate is an example of a cathodic inhibitor and one of the most common means used to control corrosion in water distribution system pipes (see Section 15.3).

Aluminum is fairly low on the EMF scale (Table 2.2) yet it does not readily corrode. This is because it reacts with oxygen to form a thin film of aluminum oxide that protects the aluminum from further oxidation.

Pure Noble Metals. The noble metals are high in the EMF series and tend to be cathodic with respect to the surrounding environment. However, they are expensive and most of them are not rigid enough for structural support. In any case, they will have some imperfections in molecular structure that will tend to cause corrosion.

Uniform Environmental Conditions. Uniform environmental conditions cause uniform potentials to exist, minimizing the existence of concentration cells. Therefore, mixing a solution is desirable if possible. Pipe laid underground or a structure below ground is exposed to a soil solution that has a potential relative to the structure. Uniform backfill will spread the attack over a larger surface area instead of concentrating it, which would result in earlier failure.

Cathodic Protection. This method consists of rendering the structure to be protected a cathode in relation to its surrounding environment. The potential of the structure is raised such that it becomes less likely to donate electrons. This occurs at the expense of either electrical or chemical energy. Another substance will be sacrificed. There are two methods of cathodic protection: galvanic and electrolytic, which are discussed next.

Galvanic Cathodic Protection

In this method a sacrificial anode, made of a metal lower in standard potential than the metal to be protected, is connected to the structure (Fig. 2.4). The anode is consumed and must be periodically replaced.

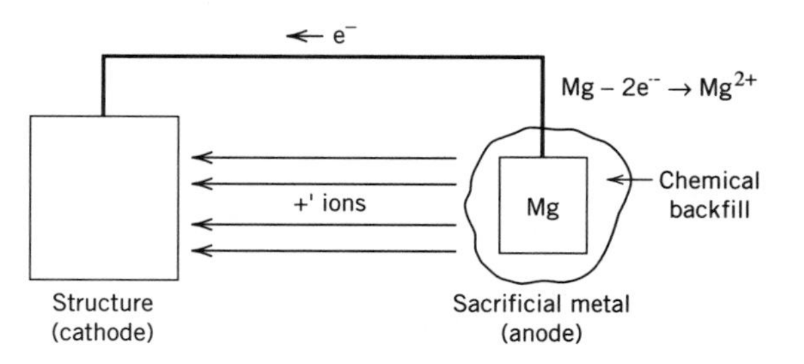

Figure 2.4 Galvanic cathodic protection.

The sacrificial anodes are usually made of Zn or Mg. The lifetime of a sacrificial anode depends on the current that flows from it. For a current, i, flowing through an electrochemical cell, the mass deposited in the cathodic cell or mass consumed in the anodic cell is proportional to time, current, and equivalent weight of the metal. In equation form:

$$m \propto it(\text{eq. wt.})$$

where

m is the mass deposited or consumed
i is the current in coulombs/s
t is time

The proportionality constant is the inverse of the Faraday constant, F (96 500 coulombs/eq).

$$m = \frac{it(\text{eq. wt.})}{F} \tag{2.22}$$

Solving Eq. (2.22) for time and incorporating an efficiency factor, E, to account for extraneous losses,

$$t = \frac{EFm}{i(\text{eq. wt.})} \tag{2.23}$$

The efficiency factors are approximately 0.90 and 0.50 for Zn and Mg, respectively (Bosich, 1970).

TABLE 2.4 Soil Corrosivity Rating Based on Resistivity[a]

Soil resistivity, Ω-cm	Corrosivity
0–2 000	Very corrosive
2 000–5 000	Corrosive
5 000–10 000	Moderately corrosive
10 000–25 000	Mildly corrosive
25 000–50 000	Less corrosive
50 000–100 000	Progressively noncorrosive

[a]After O'Day (1989). Reprinted from *Journal of the American Water Works Association,* by permission. Copyright © 1989, American Water Works Association.

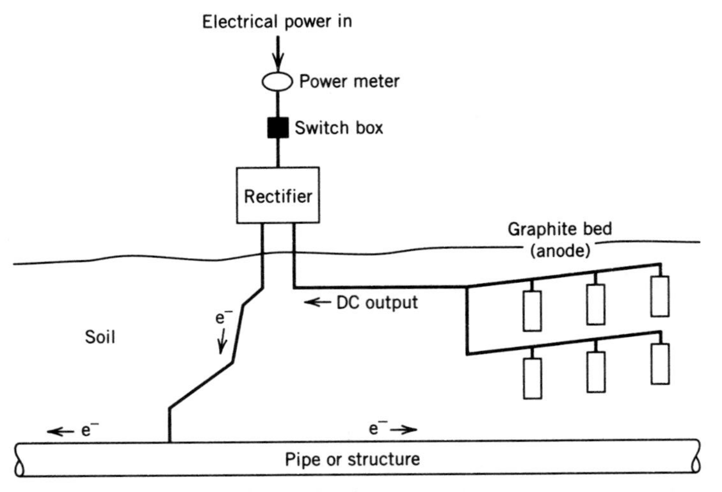

Figure 2.5 Electrolytic cathodic protection.

Studies have shown that current in the range of 10–215 mamp/m^2 of pipe surface is needed to protect a pipe. For normal conditions, a range of 10–30 mamp/m^2 is typical (Bosich, 1970). The amount of current that can flow from the anode depends on the soil resistance, which can be measured. Qualitative estimates of soil corrosivity as a function of resistance are given in Table 2.4. There are also special electrodes to measure soil potential.

Electrolytic Cathodic Protection

Electrolytic cathodic protection uses a rectifier, which is essentially an electron pump, to bring electrons to the structure (Fig. 2.5). The anode may or may not be reactive. Nonreactive graphite (carbon), which is an excellent conductor, is commonly used for the anode. In this case, ions surrounding the anode will be oxidized. The driving force for the reaction is an external power supply.

QUESTIONS AND PROBLEMS

1. What is the release or gain of free energy at standard conditions for each mole of water that is evaporated?
2. Using standard free energies, what is the equilibrium constant for the dissociation of acetic acid ($CH_3COOH \rightleftharpoons H^+ + CH_3COO^-$) at standard conditions?
3. Using the half-reactions in Table 1.3, what are the equations of the overall balanced reaction, the standard free energy change associated with the reaction, and the equilibrium constant for the oxidation of ferrous iron to ferric iron with chlorine?
4. Using information in Tables 1.3 and 2.1, determine the standard free energy of formation for the perchlorate ion (ClO_4^-).
5. What is the ratio of the equilibrium constant (K_{sp}) for the formation of calcite ($CaCO_3$) at temperatures of 20 and 80°C?

6. Consult the nomograph for free carbon dioxide determination in the Appendix. This nomograph solves the equation

$$K = \frac{[H^+][HCO_3^-]}{[H_2CO_3^*]}$$

where $H_2CO_3^*$ is free carbon dioxide and bicarbonate alkalinity gives the concentration of HCO_3^- (and pH is $\gamma_H[H^+]$). The total dissolved solids is used to calculate the activity coefficient corrections of the appropriate species. (At TDS = 0 the activity coefficients are 1 and there is no effect of this parameter on the equilibrium constant.) Does the equilibrium constant for this reaction follow the van't Hoff equation (Eq. 2.14)?

7. Explain the function of a semipermeable membrane in an electrochemical cell.

8. What is E° for the reaction $ZnS \rightleftharpoons Zn^{2+} + S^{2-}$?

9. Considering only the half-reaction involving MnO_4^- and Mn^{2+} given below, at what system potential will the concentration of Mn^{2+} ions begin to exceed the concentration of MnO_4^- ions if (a) the pH is 5.0 and the temperature is 25°C and (b) the pH is 4.5 and the temperature is 5°C?

$$MnO_4^- + 8H^+ + 5e^- = Mn^{2+} + 4H_2O \qquad E^\circ = 1.491 \text{ V}$$

10. (a) The following substances are added to 1 L of water: 1.0 g of solid copper, 1.0 g of solid zinc, 100 mg of $ZnSO_4$, and 100 mg of $CuSO_4$. The temperature is 25°C. Assuming that only zinc and copper will react and that all dissolved copper is in Cu(II) form, what are the concentrations of dissolved zinc and copper at equilibrium? What is E_{sys} at equilibrium?
(b) What is the initial potential difference between the couples?

11. Determine the equilibrium constant for the formation of $Mg(OH)_2(s)$ from standard free energy changes at 25°C.

12. Would it be possible to measure ORP of a sample with the ORP electrode immersed in the sample and the reference electrode (a) placed in the air or (b) placed in a separate beaker containing distilled water? Explain your answers.

13. In Example 2.6, what is the ORP with respect to the hydrogen scale?

14. What is the potential (ORP) at which the concentration of NH_4^+ is equal to the concentration of NO_3^- ions? The pH of the solution is 6.58 and its temperature is 25°C. What would be the reading on a voltmeter for a water at these conditions if the reference electrode had a potential of 0.24 V and it was the anode?

15. (a) Explain why water containing a higher concentration of dissolved ions promotes corrosion reactions more readily than water with a low concentration of dissolved ions. (b) Does the same reasoning apply to a water that contains a high concentration of glucose, which is a nonionizing compound?

16. Based on potential considerations, which metal would be more likely to corrode if tin were in contact with steel?

17. List and discuss all the factors that would influence the corrosion of a water distribution pipe laid underground.

18. How do microorganisms influence corrosion?

19. How does a uniform backfill retard the rate of corrosion?

20. Does galvanic or electrolytic cathodic protection protect the inside, the outside, or both surfaces of a pipe against corrosion?

21. What is the cost break-even ratio in the price per kilogram of Mg and Zn, respectively, for the same current flow and design lifetime of a sacrificial anode to be used in a galvanic cathodic protection system?

22. What is the annual rate of consumption of a sacrificial zinc anode used to protect a 100 m length of pipe with an o.d. of 315 mm in a soil where the average current is 17 $mamp/m^2$?

KEY REFERENCES

BENEFIELD, L. D., J. F. JUDKINS, AND B. L. WEAND (1982), *Process Chemistry for Water and Wastewater Treatment,* Prentice-Hall, Englewood Cliffs, NJ.

DENBIGH, K. (1981), *The Principles of Chemical Equilibrium,* 4th ed., Cambridge University Press, London.

SAWYER, C. N., P. L. MCCARTY, AND G. F. PARKIN (1994), *Chemistry for Environmental Engineering,* 4th ed., McGraw-Hill, Toronto.

SNOEYINK, V. L. AND D. JENKINS (1980), *Water Chemistry,* John Wiley & Sons, Toronto.

STUMM, W. AND J. J. MORGAN (1981), *Aquatic Chemistry: An Introduction Emphasizing Equilibria in Natural Waters,* 2nd ed., John Wiley & Sons, Toronto.

REFERENCES

BOSICH, J. F. (1970), *Corrosion Prevention for Practicing Engineers,* Barnes & Noble, New York.

DEAN, J. A., ed., (1985), *Lange's Handbook of Chemistry,* 13th ed., McGraw-Hill, Toronto.

LEE, R. G., W. C. BECKER, AND D. W. COLLINS (1989), "Lead at the Tap: Sources and Control," *J. American Water Works Association,* 81, 7, pp. 52–62.

O'DAY, D. K. (1989), "External Corrosion in Distribution Systems," *J. American Water Works Association,* 81, 10, pp. 45–52.

PEDLEY, J. B., R. D. NAYLOR, AND S. P. KIRBY (1986), *Thermochemical Data of Organic Compounds,* Chapman and Hall, New York.

SCHOCK, M. R. (1990), "Internal Corrosion and Deposition Control," in *Water Quality and Treatment,* 4th ed., F. W. Pontius, ed., McGraw-Hill, Toronto, pp. 997–1111.

THAUER, R. K., K. JUNGERMANN, AND K. DECKER (1977), "Energy Conservation in Chemotrophic Bacteria," *Bacteriological Reviews,* 41, 1, pp. 100–180.

CHAPTER 3

ACID–BASE CHEMISTRY

The concentration of free hydrogen ions is one of the major controlling factors of the state of a water because H^+ ions are directly or indirectly involved in a large number of reactions. In most situations the concentration of H^+ or OH^- is small or insignificant compared to the concentrations of other species but this does not mitigate the influence of these ions as controlling variables of the state of the water.

3.1 pH

Some water molecules will dissociate according to the following reaction:[1]

$$H_2O \rightleftharpoons H^+ + OH^- \tag{3.1}$$

The expression for the equilibrium constant for this reaction is

$$K_{eq} = K'_w = \frac{[H^+][OH^-]}{[H_2O]} \tag{3.2}$$

However, the concentration of water does not change significantly because of this reaction or by the participation of H^+ and OH^- in other reactions in usual circumstances. As pointed out in the previous chapter, the concentration of water is taken to be unity and thermodynamic equations such as equilibrium expressions incorporate this definition. Therefore the equilibrium expression for the dissociation of water is

$$K_w = [H^+][OH^-] \tag{3.3}$$

The values of K_w at different temperatures are given in Table 3.1. The value at 25°C will be used in many examples for convenience.

The magnitude of $[H^+]$ can vary over a wide range and a logarithmic scale is most convenient for describing its concentration. The "p" notation defined as

$$pX = -\log_{10} X \qquad (p = -\log_{10}) \tag{3.4}$$

is used. The p is a mathematical operator that can be applied to any coefficient or expression.

At 25°C,

$$pH + pOH = 14 \tag{3.5}$$

[1]A more accurate description of this dissociation is $2H_2O \rightleftharpoons H_3O^+ + OH^-$. The H^+ ion does not exist alone; the proton is hydrated.

TABLE 3.1 K_w at Different Temperatures[a]

Temperature, °C	K_w
0	1.2×10^{-15}
10	2.9×10^{-15}
20	6.8×10^{-15}
25	1.01×10^{-14}
30	1.47×10^{-14}
50	5.48×10^{-14}

[a]From Harned and Owen (1958).

An equation describing the variation of the equilibrium constant for water with temperature is (Hanson and Cleasby, 1990):

$$pK_w = -6.0875 + 4\,470.99/T + 0.017\,06T$$

where
T is in °K

3.2 ACIDS AND BASES

Acids and bases are substances that influence the pH of a solution. One of the earlier definitions of an acid is the Arrhenius theory of ionization.

Acid: $HA \rightleftharpoons H^+ + A^-$

Base: $BOH \rightleftharpoons B^+ + OH^-$

Salt: $HA + BOH \rightleftharpoons BA + H_2O$

An acid is a substance that ionizes producing H^+ and an anion; a base produces OH^- and a cation upon ionization. Salts are produced by the reaction of an acid and a base; water is necessarily a byproduct.

This concept is limited to ionic solutions only but it will suffice for many applications in this book. A more comprehensive definition is the Brönsted and Lowry concept, developed in 1923.

An acid is a proton donor.
A base is a proton acceptor.

This concept extends to substances that do not explicitly contain hydrogen ions. The dissociation (equilibrium) constants of a number of acids and bases that are important in environmental engineering are given in Table 3.2.

Strong acids (e.g., HCl) dissociate completely. There are essentially no HCl entities in solution; only the component ions exist. The same is true for strong bases such as NaOH. Only when the solution has an extremely low pH in the case of an acid, or an extremely high pH in the case of a base, would the assumption of complete dissociation be negated when the strongly dissociating substance is added to solution. Strong acids have pK values less than 3.

Weak acids and bases are only partly dissociated in solution. The equilibrium expression provides the ratio of dissociated ions to undissociated compound. The H^+ (or OH^-) concentration is the governing factor in dictating the ratio of the salt ion

TABLE 3.2 Dissociation Constants at 25°C

Substance	Equilibrium equation	K	pK	Significance
Acetic acid	$CH_3COOH \rightleftharpoons H^+ + CH_3COO^-$	1.8×10^{-5}	4.7	Organic wastes, anaerobic digestion
Ammonia	$NH_3 + H_2O \rightleftharpoons NH_4^+ + OH^-$	1.8×10^{-5}	4.7	Disinfection
Boric acid	$H_3BO_3 \rightleftharpoons H^+ + H_2BO_3^-$	5.8×10^{-10}	9.2	Nitrogen analysis
Butyric acid	$C_3H_7COOH \rightleftharpoons H^+ + C_3H_7COO^-$	1.5×10^{-5}	4.8	Anaerobic digestion
Carbonic acid	$H_2CO_3^* \rightleftharpoons H^+ + HCO_3^-$	4.3×10^{-7}	6.4	Numerous applications
	$HCO_3^- \rightleftharpoons H^+ + CO_3^{2-}$	4.7×10^{-11}	10.3	
Hydrochloric acid	$HCl \rightleftharpoons H^+ + Cl^-$	Strong	≈ -3	Analyses
Hydrocyanic acid	$HCN \rightleftharpoons H^+ + CN^-$	7.2×10^{-10}	9.1	Toxicity
Hydrofluoric acid	$HF \rightleftharpoons H^+ + F^-$	6.75×10^{-4}	3.2	Fluoridation
Hydrosulfuric acid (hydrogen sulfide)	$H_2S \rightleftharpoons H^+ + HS^-$	9.1×10^{-8}	7.0	Odor, corrosion, anaerobic digestion, toxicity
	$HS^- \rightleftharpoons H^+ + S^{2-}$	1.3×10^{-13}	12.9	
Hypochlorous acid	$HOCl \rightleftharpoons H^+ + OCl^-$	2.9×10^{-8}	7.5	Disinfection
Nitric acid	$HNO_3 \rightleftharpoons H^+ + NO_3^-$	0.10	−1.0	Nitrification, analyses
Nitrous acid	$HNO_2 \rightleftharpoons H^+ + NO_2^-$	5.1×10^{-4}	3.29	Nitrification
Perchloric acid	$HClO_4 \rightleftharpoons H^+ + ClO_4^-$	Strong	≈ -7	Analyses
Phenol	$C_6H_5OH \rightleftharpoons H^+ + C_6H_5O^-$	1.2×10^{-10}	9.9	Tastes, odors
Phosphoric acid	$H_3PO_4 \rightleftharpoons H^+ + H_2PO_4^-$	7.5×10^{-3}	2.1	Buffer, nutrient
	$H_2PO_4^- \rightleftharpoons H^+ + HPO_4^{2-}$	6.2×10^{-8}	7.2	
	$HPO_4^{2-} \rightleftharpoons H^+ + PO_4^{3-}$	4.8×10^{-13}	12.3	
Potassium hydroxide	$KOH \rightleftharpoons K^+ + OH^-$	Strong (base)		Analyses
Propionic acid	$C_2H_5COOH \rightleftharpoons H^+ + C_2H_5COO^-$	1.3×10^{-5}	4.9	Anaerobic digestion
Sodium hydroxide	$NaOH \rightleftharpoons Na^+ + OH^-$	Strong (base)		Analyses, neutralization
Sulfuric acid	$H_2SO_4 \rightleftharpoons H^+ + HSO_4^-$	Strong	≈ -3	Coagulation, pH
	$HSO_4^- \rightleftharpoons H^+ + SO_4^{2-}$	1.2×10^{-2}	1.9	Control, analyses
Sulfurous acid	$H_2SO_3 \rightleftharpoons H^+ + HSO_3^-$	1.7×10^{-2}	1.8	Dechlorination
	$HSO_3^- \rightleftharpoons H^+ + SO_3^{2-}$	6.3×10^{-8}	7.2	

concentration (the salt ion is the ion or molecule associated with H^+ or OH^-) to the undissociated compound concentration, if equilibrium exists. Consider acetic acid, which is a weak acid:

$$CH_3COOH \rightleftharpoons CH_3COO^- + H^+ \qquad K = 1.8 \times 10^{-5} \qquad (pK = 4.74) \quad (3.6)$$

The equilibrium expression is

$$K = \frac{[CH_3COO^-][H^+]}{[CH_3COOH]} \quad \text{or} \quad \frac{[CH_3COO^-]}{[CH_3COOH]} = \frac{K}{[H^+]} \quad (3.7)$$

At pH = 4.74 there will be equal amounts (or concentrations) of CH_3COOH and CH_3COO^-. The amount of CH_3COO^- in solution relative to the amount of CH_3COOH becomes very large as $[H^+]$ decreases or vice versa. The tendency for acetic acid to give up a hydrogen ion is much greater as the concentration of hydrogen ions is decreased. For a base the situation is reversed.

A salt is formed by the reaction of an acid and a base according to the following reaction:

$$HA + BOH \rightleftharpoons H_2O + BA \rightarrow H_2O + B^+ + A^-$$

When a salt is added to water, it ionizes completely and the product ions hydrolyze to the extent predicted by the equilibrium expressions associated with the ions. Some salt may precipitate depending on the solubility product discussed in Section 1.11.

■ Example 3.1 Acid–Base Equilibrium

An example problem using equilibrium expressions from Table 3.2 is as follows. How much sodium propionate, C_2H_5COONa (NaPr), must be added to a liter of water containing 0.800×10^{-3} moles of HCl to attain a pH of 4.75?

From Table 3.2, HCl is a strong acid and HPr is a weak acid.

Because HCl is a strong acid, it completely dissociates and the initial pH and pOH of the water are

$$[H^+] = 8.00 \times 10^{-4}\,M \qquad \text{and} \qquad pH = -\log[H^+] = 3.10$$
$$pOH = -\log[OH^-] = 14.00 - 3.10 = 10.90 \qquad \text{and} \qquad [OH^-] = 1.25 \times 10^{-11}\,M$$

The final pH is 4.75, which corresponds to $[H^+]$ and $[OH^-]$ of 1.78×10^{-5} and 5.62×10^{-10} *M*, respectively.

Now consider the amount of NaPr that must be added. NaPr is a salt and HPr is an acid. The equilibrium expression that applies is

$$K = 1.3 \times 10^{-5} = \frac{[H^+][Pr^-]}{[HPr]} \tag{i}$$

The chemical equations that apply are

$$NaPr \rightarrow Na^+ + Pr^- \tag{ii}$$
$$H^+ + Pr^- \rightleftharpoons HPr \tag{iii}$$
$$H^+ + OH^- \rightleftharpoons H_2O \tag{iv}$$

The difference in the initial and final $[H^+]$ is

$$\Delta[H^+] = 8.00 \times 10^{-4} - 1.78 \times 10^{-5} = 7.82 \times 10^{-4}\,M$$

The difference between the final and initial $[OH^-]$ is

$$\Delta[OH^-] = 5.62 \times 10^{-10} - 1.25 \times 10^{-11} = 5.49 \times 10^{-10}\,M$$

The H^+ can be removed only by forming H_2O or HPr, but from the initial and final OH^- concentrations, OH^- must be produced. The only source of OH^- ions is from the dissociation of additional water molecules according to Eq. (iv); therefore this equation actually proceeds from right to left. The OH^- production is less than 10^{-9} *M* and it is accompanied by production of an equal amount of H^+ ions. The amount of H^+ removed is equal to the initial amount of H^+ plus the production of H^+ from dissociation minus the final amount of H^+. The contribution of H^+ from the dissociation of water is negligible.

Using Eq. (iii), the amount of HPr formed is 7.82×10^{-4} *M*. The amount of H^+ produced from H_2O dissociation is negligible compared to H^+ used to form the weak acid HPr.

From the equilibrium expression (i), the ratio of $[Pr^-]$ to [HPr] is

$$\frac{[Pr^-]}{[HPr]} = \frac{1.3 \times 10^{-5}}{1.78 \times 10^{-5}} = 0.73$$

Using this ratio the concentration of Pr^- is

$$[Pr^-] = 0.73[HPr] = 0.73(7.82 \times 10^{-4}) = 5.71 \times 10^{-4}\,M$$

The total amount of NaPr to be added consists of the NaPr consumed by H^+ to produce HPr and the NaPr dissociated to produce Pr^-.

$$[\text{NaPr}] \text{ added} = 7.82 \times 10^{-4} + 5.71 \times 10^{-4} = 1.35 \times 10^{-3}\ M$$

3.2.1 Conjugate Acids and Bases

The loss of H^+ by an acid HA, gives rise to a potential H^+ acceptor, A^-, i.e., A^- is a conjugate base of HA. Similarly, the loss of OH^- by the base BOH gives rise to a potential OH^- acceptor, B^+, which is indirectly a proton donor and therefore is the conjugate acid of BOH.

$$\underset{\text{Acid}}{HA} \rightleftharpoons H^+ + \underset{\text{Conjugate base}}{A^-} \qquad K_A$$

$$\underset{\text{Base}}{BOH} \rightleftharpoons \underset{\text{Conjugate acid}}{B^+} + OH^- \qquad K_B$$

For a conjugate acid–base pair it is readily shown that

$$K_A K_B = K_w \tag{3.8}$$

3.3 EQUIVALENTS AND NORMALITY

In an acid or base context, the equivalent weight of a compound is that weight that contains one gram atom of available H^+ or its chemical equivalent. Equivalent weights are based on gram atoms; milliequivalent weights (meq) are conveniently based on milligram atoms. The equivalent weight is often equal to the molecular weight of a compound or ion divided by its valence number. A one normal (1 *N*) solution contains one equivalent weight of compound.

■ Example 3.2 Equivalent Weight Determination

What are the equivalent weights of (a) $Ca(HCO_3)_2$ (b) NaH_2PO_4?

(a) $Ca(HCO_3)_2$ dissociates as follows:

$$Ca(HCO_3)_2 \rightarrow Ca^{2+} + 2HCO_3^-$$

Each HCO_3^- has the potential to give up 1 H^+ or, alternatively, Ca^{2+} has the potential to accept 2 OH^-s. The equivalent weight of this compound is its MW (162 g) divided by 2, which makes its equivalent weight 81 g.

(b) NaH_2PO_4 dissociates as follows:

$$NaH_2PO_4 \rightarrow Na^+ + H_2PO_4^-$$

The definition of normality for NaH_2PO_4 is ambiguous. Following the definition of normality as a measure of the available H^+, each $H_2PO_4^-$ has 2 H^+s and the equivalent weight of the compound would be its MW (120 g) divided by 2. However, each $H_2PO_4^-$ could take up one H^+, which would yield an equivalent weight of 120 g. Using molarity avoids confusion in labeling bottles. The normality of the compound depends on whether it is participating in an acid or a base reaction.

Some compounds do not donate or take up H^+ or OH^- ions directly. The equivalent weight of these compounds is determined after their reaction with water as shown in the following example.

■ Example 3.3 Equivalent Weight Determination

What is the equivalent weight of CO_2?

The hydrolysis reaction for CO_2 is

$$CO_2 + H_2O \rightarrow H_2CO_3$$

Each H_2CO_3 molecule has the potential to donate 2 H^+ ions:

$$H_2CO_3 \rightarrow 2H^+ + CO_3^{2-}$$

Therefore the equivalent weight of CO_2 is its MW (44 g) divided by 2, which results in an equivalent weight of 22 g.

The equivalent weight of a compound or ion depends on the circumstances of the reaction of interest. For instance, for a multiprotic acid H_3X that is participating in a reaction (e.g., a titration) where the endpoint is fixed when the acid has only given up two hydrogen ions, the equivalent weight of H_3X is its gram molecular weight divided by 2. However, when equivalent weights are tabulated, they are normally based on the total potential of the substance to gain or produce H^+ ions.

3.4 SOLUTION OF MULTIEQUILIBRIA SYSTEMS

The solution of multi-equilibria problems follows this procedure:

1. Identify all important species participating in the reaction.
2. Write all known equilibrium relations involving the species identified in step 1. These include acid–base and redox equilibria as well as equilibria between phases such as Henry's law and K_{sp} relations. Do not forget the equilibrium relation for the dissociation of water.
3. Write the mass balance relations that apply.
4. Write the charge balance equation for the system.
5. Make any simplifying assumptions.
6. Ensure that the number of equations is equal to the number of unknowns. If more equations are needed, return to steps 2 or 3.
7. Solve the system of equations algebraically or with computer methods for systems of equations. If computer methods are used, check that a realistic solution has been found (e.g., the pH is between 0 and 14).
8. Check the assumptions made in step 5.

Regarding step 3, if the total amount of sulfur ($[S_T]$) in a water is known, then including species listed in Tables 1.2, 1.3, and 3.2 in a mass balance:

$$[S_T] = [S^{2-}] + [HS^-] + [H_2S] + [SO_4^{2-}] + [HSO_4^-] + [H_2SO_4] + [S] + 2[S_2O_3^{2-}] + [SO_3^{2-}] + [HSO_3^-] + [H_2SO_3]$$

This mass balance does not include possible complexes of sulfur that could occur among other possible sulfur species. The solution to a multiequilibria problem can become quite involved if all possible entities are considered. The key to solving multiequilibria problems is step 5, where simplifying assumptions are made. Many of the species in the mass balance can be ignored because they are present in minor amounts under normal conditions. For instance, the concentrations of HSO_4^- and H_2SO_4 are negligible in all but highly acidic solutions. Typically these substances would not be included in the mass balance expression.

The assumption that concentrations of HSO_4^- and H_2SO_4 are negligible is a routine assumption. Similarly, the assumption that strong bases are completely dissociated is normal. Other assumptions that are reasonable starting points are as follows. Weak acids are not dissociated to any significant extent in acidic solutions. Weak bases are not dissociated to any extent in basic solutions. A judgment is made on the final pH and the commensurate assumptions are made. Of course, species at low concentrations are included in the appropriate equilibrium expressions. The starting assumptions should be adequate to reduce the solution to solving one or two equations. The information obtained from this solution is systematically used to solve the remaining equations, which ultimately provides verification of the assumptions or disproves them.

If the assumptions have been incorrect, the first solution of the equations will indicate the direction in which adjustments must be made and an iterative approach can be used to solve the problem. The alternative to simplifying assumptions is to write and solve all of the equations algebraically. This process is lengthy and involved but it will lead to the correct answer.

Experience is often required to eliminate certain entities from consideration. In this text, problems will focus on the major possible species as shown in tables in the text or specified in the problems.

3.5 BUFFERS

Buffers are important in analyses and in natural waters. Buffers resist pH change. This has obvious importance for life forms that usually can survive only within a narrow pH range. In analyses, desired reactions can be forced to a specified degree of completion or the rate of reaction may be increased by governing the pH in an appropriate range.

Buffers rely upon the combined effects from a weak acid or base and the salt of the weak acid or base. This is called the common ion effect. Consider a solution of acetic acid, which is a weak acid, and its salt, sodium acetate. The reactions in water are

$$CH_3COOH \rightleftharpoons CH_3COO^- + H^+ \qquad K_A \tag{3.9}$$

$$CH_3COONa \rightarrow CH_3COO^- + Na^+ \tag{3.10}$$

The salt is highly soluble and dissociates completely. The equilibrium expression is

$$K_A = \frac{[CH_3COO^-][H^+]}{[CH_3COOH]} \tag{3.11}$$

which can be rearranged to

$$pH = pK_A + \log\frac{[CH_3COO^-]}{[CH_3COOH]} \tag{3.12}$$

or in general,

$$pH = pK_A + \log\frac{[\text{salt}]}{[\text{acid}]} \qquad \text{or} \qquad pH = pK_A - p\frac{[\text{salt}]}{[\text{acid}]} \tag{3.13}$$

Equation (3.13) is known as the Henderson–Hasselbach equation. If all species are in equilibrium, the ratio of [salt] : [acid] for any acid or base is readily determined from this equation at a given pH.

When acid is added to the above solution, the acetate ion will combine with H^+ to form weakly dissociated acetic acid until essentially all of the acetate ion is exhausted. When base is added to the solution, the OH^- consumes H^+ ions that are replaced by H^+ ions donated by dissociation of the acetic acid until essentially all of the acetic acid is dissociated.

■ Example 3.4 Buffer

An analyst needs a solution with a pH that is fairly constant around a value of 4.74. Absorption of CO_2 and lab fumes and the reagents and impurities in the reagents added to the buffer solution will cause deviation from the desired pH. Noting that the pK_A of acetic acid (HAc) is 4.74, the analyst decides to make a solution containing acetate and acetic acid. Using Eq. (3.13), the ratio of salt to acid in the solution should be 1.0 to obtain the desired pH.

The analyst chooses to make the solution by adding 0.1 mole of HAc and 0.1 mole of NaAc to 1 L of water. The pH is calculated as follows.

$$\underset{0.1 - x}{CH_3COOH} \rightarrow \underset{x}{CH_3COO^-} + \underset{x}{H^+}$$

$$\underset{0.1 - 0.1 = 0}{CH_3COONa} \rightarrow \underset{0.1}{CH_3COO^-} + \underset{0.1}{Na^+}$$

CH_3COO^- is produced by both reactions. The equilibrium expression is

$$K_A = \frac{[CH_3COO^-][H^+]}{[CH_3COOH]} = \frac{(0.1 + x)x}{0.1 - x}$$

To solve the above expression, the analyst recognizes that HAc is a weak acid and assumes that x is small compared to 0.1. The equation reduces to

$$K_A = \frac{0.1x}{0.1} = 1.8 \times 10^{-5}$$

and x is readily determined to be 1.8×10^{-5} (i.e., the pH is 4.74). The assumption that x was small compared to 0.1 was correct.

Is the solution a buffer? Addition of 0.05 mole of a strong acid or strong base to 1 L of an unbuffered water would result in a pH of 1.30 or 12.70, respectively. Let us find the pH of our buffered solution after addition of the same quantities of either acid or base.

When 0.05 mole of HCl is added, the H^+ ions will react with Ac^- to form weakly dissociated HAc.

$$HCl + Ac^- \rightarrow HAc + Cl^-$$

Again using the assumption that the contribution of Ac^- and H^+ from dissociation of HAc is small, the resulting concentrations of species involved are

$$[Ac^-] = 0.1 - 0.05 = 0.05\ M \qquad [HAc] = 0.1 + 0.05 = 0.15\ M$$

The pH is found from Eq. (3.13).

$$pH = pK_A + \log\frac{[\text{salt}]}{[\text{acid}]} \quad \text{or} \quad pH = 4.74 + \log\frac{[0.05]}{[0.15]} = 4.26$$

A check will prove the assumption to be correct.

When 0.05 mole of NaOH is added, the OH^- ions will react with the HAc.

$$NaOH + HAc \rightarrow Ac^- + Na^+ + H_2O$$

Using the assumption that the contributions of Ac^- and H^+ from dissociation of HAc are small, the resulting concentrations of species involved are

$$[Ac^-] = 0.1 + 0.05 = 0.15\ M$$
$$[HAc] = 0.1 - 0.05 = 0.05\ M$$

The pH is

$$pH = pK_A + \log\frac{[\text{salt}]}{[\text{acid}]} \quad \text{or} \quad pH = 4.74 + \log\frac{[0.15]}{[0.05]} = 5.22$$

A check will again prove the assumption to be correct.

The pH change of the buffered solution is much less dramatic than the pH change of an unbuffered solution. The effective buffering capacity of the solution is approximately equal to the amounts of HAc and Ac^- in the initial solution.

If a weak acid or base with a pK value exactly at the desired pH cannot be found, the ratios of the buffering agent acid and salt are adjusted to provide the desired pH in the buffered solution.

3.5.1 Dilution of a Buffered Solution

Addition of water has no effect on the pH of a buffered solution unless the dilution is large. This can be proven by using the Henderson–Hasselbach equation.

Consider a solution that contains m moles of salt and n moles of acid in a volume V. The pH is

$$pH = pK_A + \log\frac{[\text{salt}]}{[\text{acid}]} = pK_A + \log\frac{[m/V]}{[n/V]} = pK_A + \log\left(\frac{m}{n}\right)$$

This equation proves that dilution has no effect on the pH of a buffered solution until the dilution is large enough to cause the contribution of H^+ from the dissociation of water to be significant. At this point the buffering capacity is negligible and the resulting equations are of little practical significance.

3.5.2 The Most Effective pH for a Buffer

The buffering capacity of a solution is quantitatively determined by measuring the pH change upon addition of a given amount of strong acid or base.

In general terms, consider a buffer solution made from the addition of a strong base, BOH, to a solution containing a weak acid, HA. The equations to be used are

$$BOH + HA \rightarrow B^+ + A^- + H_2O \tag{3.14}$$

$$HA \rightarrow H^+ + A^- \tag{3.15}$$

The amounts of HA and A^- are determined assuming that the contribution of A^- from the dissociation of HA is negligible as demonstrated in Example 3.4. In this case the salt ion (A^-) is formed by reaction (3.14); there is no direct addition of a salt containing A^-. If the starting concentration of HA in terms of equivalents in a volume, V, is n/V and the equivalents concentration of BOH added is m/V, the equilibrium expression is calculated as follows:

$$K_A = \frac{[H^+][A^-]}{[HA]} = \frac{[H^+]m}{(n-m)} \tag{3.16}$$

$$[H^+] = K_A \frac{(n-m)}{m} \quad \text{or} \quad pH = pK_A + \log\left(\frac{m}{n-m}\right) \tag{3.17}$$

The rate of change of pH with respect to addition of BOH is

$$2.303 \frac{dpH}{dm} = \frac{1}{n-m} + \frac{1}{m} \tag{3.18}$$

The question is, when is the above expression a minimum, which is found by taking the derivative and setting it equal to zero?

$$\frac{d^2pH}{dm^2} = \frac{n(2m-n)}{m^2(n-m)^2} = 0 \tag{3.19}$$

This implies that $m = n/2$. It is not possible for n to be zero because there was HA in the starting solution.

Substituting this into the pH expression (Eq. 3.17), it is found that the solution is most highly buffered when

$$pH = pK_A \quad \text{and} \quad [\text{salt}] = [\text{acid}]$$

3.6 ACID–BASE TITRATIONS

A titration is a volumetric form of analysis in which a measured amount of solution of known concentration (the titrant) is added to a sample. (See Chapter 5 for more discussion of titrations.) When an acid or a base is added to a solution, all species with a capability of donating or accepting H^+ or OH^- ions react to attain a new state of equilibrium. Acid–base reactions are usually rapid, occurring in a matter of seconds. The amount of pH change after the addition of titrant depends on the buffering compounds in solution.

3.6.1 Titration of Strong Acids and Bases

The presence of significant amounts of a strong acid in a water will cause the pH to be low. To examine the change in pH with addition of a strong base consider the following example.

TABLE 3.3 pH Variation in a Strong Acid–Strong Base Titration

Volume of NaOH added, mL	$[H^+]$ *M*	pH
0.00	1.00×10^{-1}	1.00
10.00	4.27×10^{-2}	1.37
20.00	1.11×10^{-2}	1.95
24.00	2.05×10^{-2}	2.69
24.90	2.00×10^{-4}	3.70
24.99	2.00×10^{-5}	4.70
25.00	1.00×10^{-7}	7.00
25.01	5.0×10^{-10}	9.30
25.10	5.0×10^{-11}	10.30
26.00	5.0×10^{-12}	11.30
30.00	1.0×10^{-12}	12.00
40.00	4.3×10^{-13}	12.37

Consider a 25-mL volume of a 0.10 *M* HCl solution that is titrated with 0.10 *M* NaOH. Because the reactants are strong, they are completely dissociated and the titration reaction simply is

$$H^+ + OH^- \rightarrow H_2O$$

The variation of pH after additions of various amounts of titrant is tabulated in Table 3.3. The data in Table 3.3 are plotted in Fig. 3.1. The curve for the titration of a strong base with a strong acid is also shown on the figure. It is the reverse mirror image of the first curve.

It is observed that strong acids and bases effectively buffer the pH at low and high values, respectively. The pH fluctuation around a value of 7 is dramatic. The equivalence point is the pH at which the equivalents of titrant added are equal to the initial equivalents in solution. For this titration the equivalence point (endpoint for stopping the titration) is a pH of 7. The curves in Fig. 3.1 are typical for all titrations involving strong acids and strong bases.

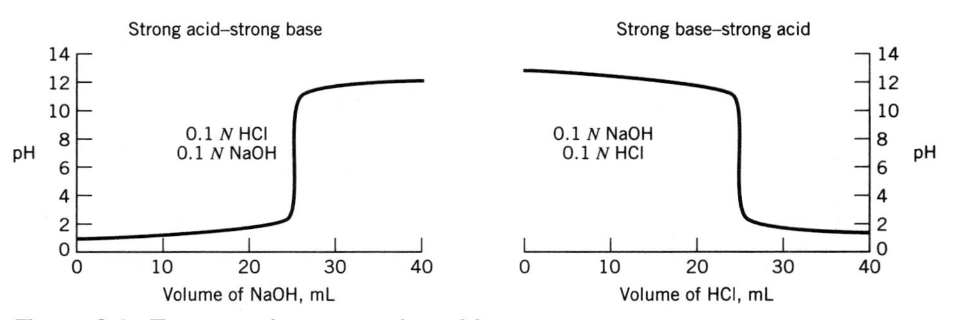

Figure 3.1 Titration of strong acids and bases.

TABLE 3.4 pH Variation in a Weak Acid–Strong Base Titration

Volume of NaOH added, mL	$[H^+]$ *M*	pH
0.00	2.00×10^{-3}	2.70
10.00	2.69×10^{-5}	4.57
20.00	5.01×10^{-6}	5.35
24.00	7.50×10^{-7}	6.12
24.90	7.9×10^{-8}	7.10
25.00	2.0×10^{-9}	8.70
25.10	5.0×10^{-11}	10.30
26.00	5.1×10^{-12}	11.29
30.00	1.1×10^{-12}	11.96
40.00	4.3×10^{-13}	12.37

3.6.2 Titration of Weak Acids and Bases

To illustrate the variation of pH in a solution that contains a weak acid when titrated with a strong base, consider a 25 mL volume of 0.10 *M* acetic acid (HAc) titrated with 0.10 *M* NaOH. The reaction is

$$HAc + NaOH \rightarrow H_2O + Na^+ + Ac^-$$

In addition, until the HAc has been consumed, the dissociation of the remaining HAc must be considered.

$$HAc \rightleftharpoons H^+ + Ac^- \qquad K_{eq} = 1.8 \times 10^{-5}$$

The data for this titration are given in Table 3.4. See Example 3.5 for the calculation of the solution pH after addition of titrant. The data in Table 3.4 are plotted in Fig. 3.2 along with the data for a 0.10 *M* solution of NH_3 (a weak base) titrated with 0.10 *M* HCl.

The following observations are made from Fig. 3.2 compared to Fig. 3.1. The equivalence points for these titrations have shifted from neutrality. Also the range of pH variation near the equivalence point has decreased compared to the strong acid–strong base situation.

For the HAc titration with NaOH, the titration curve is the nearest to being flat

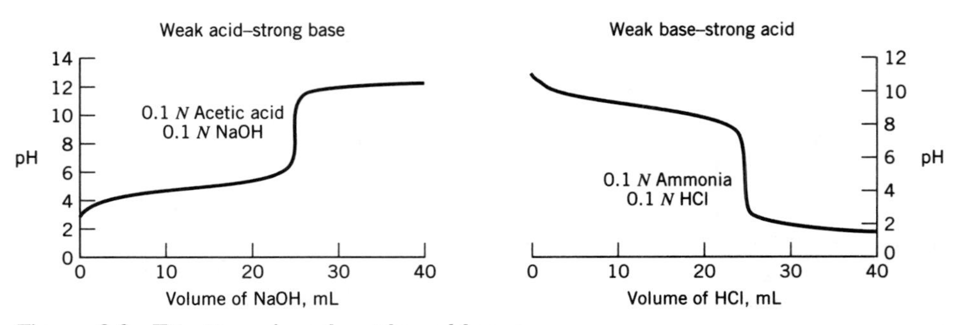

Figure 3.2 Titration of weak acids and bases.

at a pH of 4.74, which is the pH of maximum buffering capacity as previously illustrated in Section 3.5.

The correct solution of any system can be made by using the equilibria, mass balance, and charge balance equations without any simplifying assumptions. However, in many cases it is possible to make simplifying assumptions and reduce the effort in solving the problem. The important point is to check whether the assumptions have been valid. This approach is used in the example below for pH variation in a titration.

■ Example 3.5 pH Determination in a Titration

Find the pH of the solution, when (a) 20 and (b) 25 mL of 0.10 N NaOH have been added to 25 mL of 0.10 N acetic acid. Note that the latter pH is the equivalence (endpoint) pH for the titration. The temperature is 25°C.

(a) NaOH is a strong base and dissociates completely. The OH^- ion then reacts with HAc to form Ac^- as follows.

$$NaOH + HAc \rightarrow Na^+ + Ac^- + H_2O$$

The volume of the solution is 20 + 25 = 45 mL. The equivalents of acid originally present are

$$25 \text{ mL} \times 0.10\,N \times \frac{1 \text{ eq/L}}{1\,N} \times \frac{1 \text{ L}}{1\,000 \text{ mL}} = 2.50 \times 10^{-3} \text{ eq}$$

The equivalents of base added are

$$20 \text{ mL} \times 0.10\,N \times \frac{1 \text{ eq/L}}{1\,N} \times \frac{1 \text{ L}}{1\,000 \text{ mL}} = 2.00 \times 10^{-3} \text{ eq}$$

The reaction and equilibrium relation to be satisfied are:

$$HAc \rightleftharpoons H^+ + Ac^- \qquad K_{HAc} = \frac{[H^+][Ac^-]}{[HAc]} = 1.8 \times 10^{-5}$$

Because HAc is a weak acid, there are $2.50 \times 10^{-3} - 2.00 \times 10^{-3} = 5.00 \times 10^{-4}$ eqs of HAc and 2.00×10^{-3} eqs of Ac^- in the 45-mL solution. The concentrations of these substances are

$$[HAc] = \frac{5.00 \times 10^{-4} \text{ eq}}{45 \text{ mL}} \times \frac{1 \text{ mole}}{\text{eq}} \times \frac{1\,000 \text{ mL}}{1 \text{ L}} = 1.11 \times 10^{-2}\,M$$

$$[Ac^-] = \frac{2.00 \times 10^{-3} \text{ eq}}{45 \text{ mL}} \times \frac{1 \text{ mole}}{\text{eq}} \times \frac{1\,000 \text{ mL}}{1 \text{ L}} = 4.44 \times 10^{-2}\,M$$

Substituting these concentrations into the Henderson–Hasselbach (equilibrium) expression the solution pH is

$$pH = pK_{HAc} + \log\frac{[Ac^-]}{[HAc]} = 4.75 + \log\frac{4.44 \times 10^{-2}}{1.11 \times 10^{-2}} = 5.35$$

The assumption on the weakly acidic behavior of HAc can be verified.

(b) The equations noted for part (a) apply. In this case the volume of the solution is 25 + 25 = 50 mL. The equivalents of base added is the same as the starting amount of HAc and equal to 2.5×10^{-3} eqs. It is true that the amount of Ac^- is nearly equal to 2.5×10^{-3} eqs but the amount of HAc in solution is not zero.

Another equation is required to solve the system. Recall the charge conservation principle. The statement of charge conservation for this system is

$$[H^+] + [Na^+] = [OH^-] + [Ac^-]$$

$[Na^+]$ is known. Assume

$$[Ac^-] = \frac{2.5 \times 10^{-3}\text{ eq}}{50\text{ mL}} \times \frac{1\text{ mole}}{\text{eq}} \times \frac{1\,000\text{ mL}}{\text{L}} = 0.050\ M$$

$[OH^-]$ can be put in terms of $[H^+]$ by use of the equilibrium relation for the dissociation of water:

$$[H^+] + [OH^-] \rightleftharpoons H_2O \qquad K_w = 1 \times 10^{-14}$$

Substituting this into the charge balance equation and rearranging,

$$[H^+] + [Na^+] = \frac{K_w}{[H^+]} + [Ac^-] \Rightarrow [H^+]^2 + [H^+]([Na^+] - [Ac^-]) - K_w = 0$$

Because $[Ac^-] = [Na^+] = 0.050\ M$, the quadratic reduces to $[H^+]^2 = K_w = 10^{-14}$ with a solution of $[H^+] = 1 \times 10^{-7}$ or pH = 7.00. To check the assumption, use the equilibrium relation for HAc.

$$[HAc] = \frac{[H^+][Ac^-]}{K_{HAc}} = \frac{(10^{-7})(0.050)}{1.8 \times 10^{-5}} = 2.78 \times 10^{-4}$$

The assumption that $[Ac^-] = 0.050\ M$ (or that there were 2.5×10^{-3} eq of Ac^- in solution) is reasonable because $0.050 - 2.78 \times 10^{-4} = 0.049\,7$, which is an error of approximately 0.6%. However, a further check will show that the error with respect to the hydrogen ion concentration is much larger.

The amount of HAc consumed is

$$\begin{aligned} HA_i - HA_f &= (25\text{ mL})\left(\frac{0.10\text{ eq}}{\text{L}}\right)\left(\frac{1\text{ mole}}{\text{eq}}\right)\left(\frac{1\text{ L}}{1\,000\text{ mL}}\right) \\ &\quad - (50\text{ mL})\left(2.78 \times 10^{-4}\frac{\text{mole}}{\text{L}}\right)\left(\frac{1\text{ L}}{1\,000\text{ mL}}\right) \\ &= 2.486 \times 10^{-3}\text{ moles} \end{aligned}$$

If 2.50×10^{-3} moles of NaOH have been added, then the excess OH^- is 1.39×10^{-5} moles that are present in the final 50 mL volume, which is equivalent to $[OH^-] = 2.78 \times 10^{-4}\ M$ or a pOH of 3.56. This corresponds to a pH of 10.44, which is significantly different from 7.00.

Our assumption has effectively dictated that the only source of H^+ or OH^- ions is from the dissociation of water itself, which results in the small concentration of $10^{-7}\ M$ for H^+. Small errors in the amount of HAc that dissociates can dramatically affect the H^+ concentration in this circumstance.

A pOH of 3.56 or pH of 10.44 is not the final answer. For this special situation we must solve the system without any simplifying assumptions. The total amount of acetate, Ac_T is

$$Ac_T = [HAc] + [Ac^-] = 0.050\ M$$

Substituting this into the equilibrium relation for HAc and solving for $[Ac^-]$,

$$K_{HAc} = \frac{[H^+][Ac^-]}{Ac_T - [Ac^-]} \qquad [Ac^-] = \frac{Ac_T K_{HAc}}{[H^+] + K_{HAc}}$$

Replacing $[Ac^-]$ in the charge balance with this relation:

$$[H^+] + [Na^+] = \frac{K_w}{[H^+]} + \frac{Ac_T K_{HAc}}{[H^+] + K_{HAc}}$$
$$\Rightarrow [H^+]^3 + [H^+]^2([Na^+] + K_{HAc})$$
$$+ [H^+]([Na^+]K_{HAc} - Ac_T K_{HAc} - K_w) - K_w K_{HAc} = 0$$

Substituting the known values into the above equation and solving it for the only feasible $[H^+]$ yields a pH of 8.72. From the equilibrium expression and mass balance for acetate, $[Ac^-] = 0.050\,0$ M and $[HAc] = 5.54 \times 10^{-6}$ M. The amount of undissociated HAc is significant (and therefore the unreacted OH^- from the NaOH is also significant) with respect to the dissociation of water.

Part (a) should also be checked to ensure that the charge balance relation is satisfied.

Figure 3.3 is a comparison of titration curves for acids with varying dissociation constants. The pH change near the equivalence point decreases as the dissociation constant for the acid decreases. For a dissociation constant below a value of about 10^{-9}, an acid–base titration is not a useful analytical tool to measure the acid concentration.

3.7 NATURAL BUFFERING OF WATERS FROM CARBON DIOXIDE AND RELATED COMPOUNDS

The most significant buffering system in natural waters and wastewaters is due to carbon dioxide and its related species: carbonic acid (H_2CO_3), bicarbonate (HCO_3^-), and carbonate (CO_3^{2-}). Carbon dioxide and its related species exhibit weak acid behavior, which provides buffering capacity. These species are present in relatively large amounts in waters because of biological activity that produces CO_2 and because of

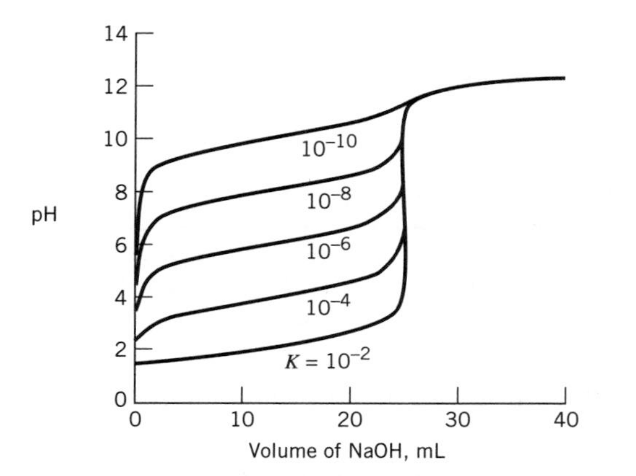

Figure 3.3 Titration curves for acids of varying strength.

the high degree of solubility of CO_2 as a result of its hydrolysis. The equilibria among these are described by the following equations.

$$CO_2(g) \rightleftharpoons CO_2(aq) \qquad K_H \qquad (3.20)$$

$$CO_2 + H_2O \rightleftharpoons H_2CO_3 \qquad K_0 \qquad (3.21)$$

$$H_2CO_3 \rightleftharpoons HCO_3^- + H^+ \qquad K_1' \qquad (3.22a)$$

$$HCO_3^- \rightleftharpoons CO_3^{2-} + H^+ \qquad K_2 \qquad (3.23)$$

The reaction in Eq. (3.21) is quite sensitive to the ionic strength and temperature of the water. Normally the concentrations of CO_2 and H_2CO_3 are considered together with the following mass balance,

$$[H_2CO_3^*] = [CO_2] + [H_2CO_3] \qquad (3.24)$$

and the following equation is used to describe the first dissociation of carbonic acid.

$$H_2CO_3^* \rightleftharpoons HCO_3^- + H^+ \qquad (3.22b)$$

Note that the equilibrium constant for carbonic acid in Table 3.2 is reported for $H_2CO_3^*$. Example 3.6 gives the derivation of the equilibrium constant for the dissociation of $H_2CO_3^*$. The $H_2CO_3^*$ concentration is the free carbon dioxide concentration of a water. A nomograph to determine free carbon dioxide concentration as a function of pH, temperature, total dissolved solids concentration (ionic strength), and total alkalinity is given in the Appendix. The nomograph implements the solution of the appropriate equilibrium expressions.

■ Example 3.6 Equilibrium Constant for Dissociation of $H_2CO_3^*$

Determine the dissociation constant for $H_2CO_3^*$ in terms of the equilibrium constants for CO_2 and H_2CO_3.

The equilibrium expressions for Eqs. (3.21), (3.22a), and (3.22b) are given in Eqs. (i), (ii), and (iii), respectively.

$$K_0 = \frac{[H_2CO_3]}{[CO_2]} \quad \text{(i)} \qquad K_1' = \frac{[HCO_3^-][H^+]}{[H_2CO_3]} \quad \text{(ii)} \qquad K_1 = \frac{[HCO_3^-][H^+]}{[H_2CO_3^*]} \quad \text{(iii)}$$

We can use Eq. (3.24) to relate $[H_2CO_3^*]$ to $[H_2CO_3]$ and $[CO_2]$ and then use the equilibrium expressions.

$$K_1 = \frac{[HCO_3^-][H^+]}{[H_2CO_3] + [CO_2]}$$

Eliminating $[CO_2]$ in the above equation with Eq. (i),

$$K_1 = \frac{[HCO_3^-][H^+]}{[H_2CO_3] + [CO_2]} = \frac{[HCO_3^-][H^+]}{[H_2CO_3] + \dfrac{[H_2CO_3]}{K_0}} = \frac{[HCO_3^-][H^+]}{[H_2CO_3]\left(1 + \dfrac{1}{K_0}\right)} = \frac{K_1'}{1 + \dfrac{1}{K_0}} = \frac{K_0 K_1'}{1 + K_0}$$

Note that Henry's law for CO_2 dissolution can also be reformulated in terms of $H_2CO_3^*$ such that

$$CO_2(g) \rightleftharpoons H_2CO_3^* \qquad \text{and} \qquad [H_2CO_3^*] = K_H'[CO_2(g)]$$

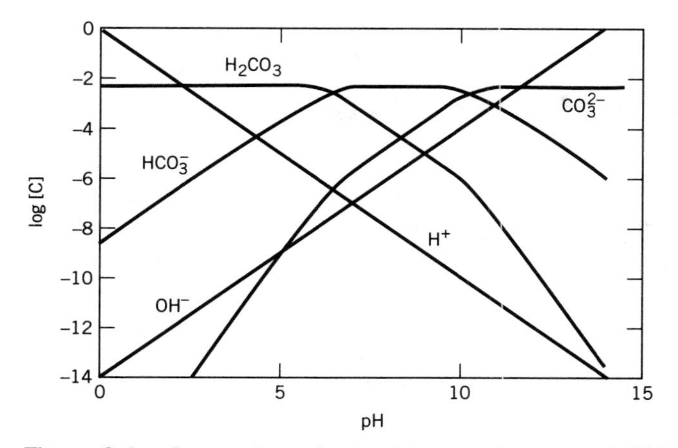

Figure 3.4a Inorganic carbon species as a function of pH (total concentration of inorganic carbon = 0.005 *M*).

3.7.1 Acidity and Alkalinity

Alkalinity (Alk) is defined as the ability of a water to neutralize acids; it is a measure of buffering capacity against a pH drop. In addition to inorganic carbon forms, $[OH^-]$ may be significant. The definition of alkalinity, assuming that only inorganic carbon is significant, is

$$[\text{Alk}] = [HCO_3^-] + [CO_3^{2-}] + [OH^-] \qquad (3.25)$$

In this equation all concentrations are expressed in eq/L.

If other ions that can take up H^+ ions are present in significant concentrations, they must be included on the right-hand side of the equation. Examples of other ions that can contribute to alkalinity are: $SiO(OH)_3^-$, PO_4^{3-}, HPO_4^{2-}, and $H_2PO_4^-$.

Figure 3.4a gives the distribution of inorganic carbon among its three forms as a function of pH. The curves for the carbon species maintain their relative spacing regardless of the total concentration of inorganic carbon in solution. Figure 3.4b, which is partly derived from Fig. 3.4a, shows the forms of alkalinity that are significant in various pH ranges. These figures are based on the equilibrium constants for the dissociations of carbonic acid. If the pH of a water is above 10, the presence of strong base is indicated and $[OH^-]$ or hydroxyl alkalinity becomes significant.

Acidity (Acy) is the opposite of alkalinity: it is the capacity to neutralize bases. Hydrogen ion itself can neutralize bases, and assuming that only CO_2-related species

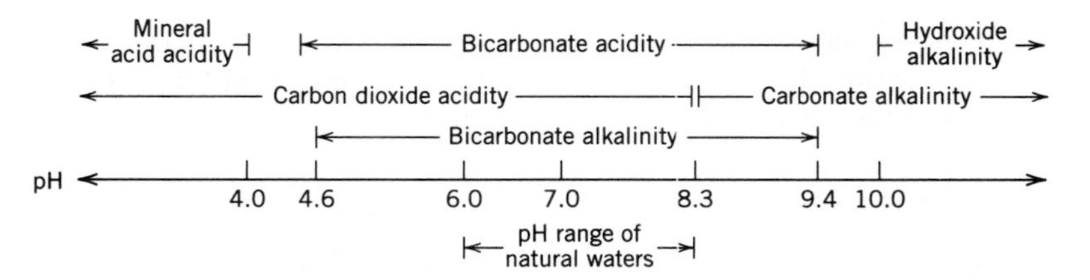

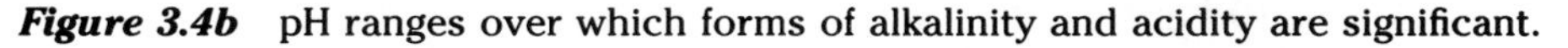

Figure 3.4b pH ranges over which forms of alkalinity and acidity are significant.

are significant, it is defined as

$$[Acy] = [H_2CO_3^*] + [HCO_3^-] + [H^+] \tag{3.26}$$

In this equation all concentrations are expressed in eq/L.

If the presence of other ions that can take up OH^- is significant, they must be included on the right-hand side of the equation. Examples of other possible acidity species are: HOCl, HPO_4^{2-}, $H_2PO_4^-$, and H_3PO_4. It is observed from Fig. 3.4b that below a pH of 4.6 the concentration of H^+ ions is significant, which indicates a significant concentration of mineral or strong acids. Bicarbonate ion appears in both alkalinity and acidity expressions because it can neutralize acids or bases.

Regardless of the components of alkalinity and acidity that are present in a water, the carbon dioxide titration endpoints of pH 4.6 (carbonic acid) and 8.3 (bicarbonate) are often used in the determination of alkalinity and acidity in situations where carbon dioxide species are significant and even in circumstances where they are present in small amounts compared to other alkalinity- or acidity-causing agents. Functional rules based on the results of the titration are developed in these circumstances. For instance, the wine industry uses the bicarbonate endpoint to determine acidity even though the primary acids present are tartaric, malic, and citric acids, which have dissociation constants much different from CO_2 species.

Nomographs to determine the concentrations of hydroxyl, bicarbonate, and carbonate alkalinities as functions of pH, temperature, total dissolved solids, and total alkalinity are given in the Appendix. These charts incorporate the effects of ionic strength and temperature on the equilibrium constant.

■ Example 3.7 Alkalinity Determination

A water with a pH of 8.00 contains HPO_4^{2-}, $H_2PO_4^-$, HCO_3^-, and NH_3 at 7.6, 2.4, 52, and 1.5 mg/L, respectively. What is its alkalinity in eq/L and as $CaCO_3$?

The concentrations of each of these species are

$$[HPO_4^{2-}] = \frac{7.6 \text{ mg/L}}{96\,000 \text{ mg/mole}} = 7.92 \times 10^{-5}\, M$$

$$[H_2PO_4^-] = \frac{2.4 \text{ mg/L}}{97\,000 \text{ mg/mole}} = 2.47 \times 10^{-5}\, M$$

$$[HCO_3^-] = \frac{52 \text{ mg/L}}{61\,000 \text{ mg/mole}} = 8.52 \times 10^{-4}\, M$$

$$[NH_3] = \frac{1.5 \text{ mg/L}}{17\,000 \text{ mg/mole}} = 8.82 \times 10^{-5}\, M$$

A check of equilibrium relations will show that the concentrations of the other possible alkalinity ions are insignificant. Except for HPO_4^{2-} all species have the capability of taking up 1 H^+ ion. The total alkalinity of the water is

$$\begin{aligned}[Alk] &= (2 \text{ eq/mole})(7.92 \times 10^{-5}\, M) \\ &\quad + (1 \text{ eq/mole})(2.47 \times 10^{-5}\, M + 8.52 \times 10^{-4}\, M + 8.82 \times 10^{-5}\, M) \\ &= 1.12 \times 10^{-3} \text{ eq/L}\end{aligned}$$

$CaCO_3$ has 50 g/eq. The alkalinity expressed as $CaCO_3$ is

$$[\text{Alk}] = (1.12 \times 10^{-3}\ \text{eq/L})\left(\frac{50\ \text{g}}{\text{eq}}\right)\left(\frac{1\ 000\ \text{mg}}{\text{g}}\right) = 56\ \text{mg/L as } CaCO_3$$

The contributions of ions other than HCO_3^- is small, which is typical.

QUESTIONS AND PROBLEMS

1. What is the molar concentration of water in pure water? Calculate the equilibrium constant at 25°C for the dissociation of water when the concentration of water is incorporated into the denominator of the equilibrium expression.
2. Show by actual calculations whether the pH of pure water will increase or decrease with an increase in temperature using the data for K_w at 0, 25, and 50°C given in the text.
3. Is a negative pH theoretically possible?
4. The pH of a solution is 5.92. What are the pOH, the hydrogen ion concentration, and the hydroxyl ion concentration at 25°C?
5. What is the equilibrium constant and pK_{eq} for the following reaction?

$$NH_4^+ \rightleftharpoons NH_3 + H^+$$

6. At a pH of 7.79 what are the ratios of the following species: H_3PO_4 and $H_2PO_4^-$; NH_3 and NH_4^+; and HOCl and OCl^-?
7. The hydrogen ion concentration in a dilute solution of sulfuric acid is 2×10^{-5} *M*. What is the pH value? What is the pOH value at 25°C?
8. What is the hydrogen ion concentration and pH of a 0.200 *M* solution of acetic acid at 25°C?
9. (a) How many mL of a 0.10 *N* solution of NaOH must be added to a 25 mL solution of 0.10 *N* HAc to obtain a pH of 9.75? What is the concentration of HAc at this pH?
   (b) Answer the same question with the following changes: [HAc] = 0.010 *N* in 25 mL and the normality of the titrant is 0.020 *N*. The final pH is 11.00.
10. What are the conjugate acids or bases of the following: HCl, SO_4^{2-}, Na^+, NH_3, KOH, HOCl, and HF?
11. The dissociation constants for a monoacid, its conjugate base, and water are K_a, K_b, and K_w, respectively. Prove that $K_A K_B = K_w$ for a conjugate acid–base pair.
12. What is the equivalent weight of NH_3 for an acid–base reaction?
13. What is the equivalent weight of Na_2HPO_4 in an acid reaction (i.e., H^+ is being added)?; in a basic reaction?
14. (a) Explain why a weak acid or base without its salt is not a good buffer.
   (b) Explain why a strong acid or strong base buffers a solution at a low or high pH, respectively.
15. In the buffer example (Example 3.4), what is the contribution of H^+ from the dissociation of water in the initial solution containing only HAc and NaAc?
16. (a) What is the hydrogen ion concentration in 500 mL of 0.100 *M* solution of acetic acid at 25°C, if the solution contains additional 2.00 g of acetate ions added in the form of sodium acetate (i.e., 2.78 g NaAc were added).
   (b) What will be the hydrogen ion concentration if 4 mmoles of NaOH are introduced into this buffered solution?
   (c) What is the pH in each case?
17. In what ratio must H_3PO_4 and NaOH be added to a liter of water to buffer it at a pH of 4.00? (Hint: $[HPO_4^{2-}]$ and $[PO_4^{3-}]$ will be negligible.)

18. What is the ratio of propionic acid and sodium propionate that must be added to a water to initially buffer it at a pH of 5.63?

19. Show by calculations that the equivalence pH for the conditions given in Table 3.3 is 7.0.

20. Prepare a table similar to Table 3.4 for the titration of 25 mL of 0.10 M NH_3 with 0.10 M HCl.

21. Would Cl^-, H_2S, HS^-, and S^{2-} contribute to the alkalinity or acidity of water?

22. Write the balanced reaction for the oxidation of NH_4^+ to NO_3^-. How much alkalinity as $CaCO_3$ is consumed by this reaction?

23. Four liters of water are equilibrated with a gas mixture containing carbon dioxide at a partial pressure of 0.3 atm. Henry's law constant for $H_2CO_3^*$ solubility is 2.0 g/L-atm.
    (a) How many grams of CO_2 are dissolved in the water? What is the pH of this water? (Neglect the secondary dissociation of carbonic acid.)
    (b) Assume that $H_2CO_3^*$ dissolves first without dissociating, achieving equilibrium with the atmosphere, and then the solution is closed to the atmosphere. $H_2CO_3^*$ then dissociates (neglect the secondary dissociation). Find the pH of this solution and compare it to the result obtained in (a).

24. What is the equilibrium constant for the reaction below in terms of K_0, K_1', and K_2 as defined by Eqs. (3.21), (3.22a), and (3.23), respectively?

$$H_2CO_3^* \rightleftharpoons CO_3^{2-} + 2H^+$$

25. Can alkalinity be expressed as HCl?

26. Why is bicarbonate ion a component of both alkalinity and acidity?

27. What are the values of the following ratios at pHs of 4.6 and 8.3: $H_2CO_3^*:[HCO_3^-]$ and $[HCO_3^-]:[CO_3^{2-}]$? Do these results agree with Fig. 3.4a?

28. What are the alkalinity and acidity of a water that contains 79 mg/L of HCO_3^- and has a pH of 7.3? Assume a temperature of 25°C.

KEY REFERENCES

BENEFIELD, L. D., J. F. JUDKINS, AND B. L. WEAND (1982), *Process Chemistry for Water and Wastewater Treatment,* Prentice-Hall, Englewood Cliffs, NJ.

BUTLER, J. N. (1964), *Ionic Equilibrium: A Mathematical Approach,* Addison-Wesley, Don Mills, ON.

SAWYER, C. N., P. L. MCCARTY, AND G. F. PARKIN (1994), *Chemistry for Environmental Engineering,* 4th ed., McGraw-Hill, Toronto.

SNOEYINK, V. L. AND D. JENKINS (1980), *Water Chemistry,* John Wiley & Sons, Toronto.

REFERENCES

HANSON, A. T. AND J. L. CLEASBY (1990), "The Effects of Temperature on Turbulent Flocculation: Fluid Dynamics and Chemistry," *J. American Water Works Association,* 82, 11, pp. 56–73.

HARNED, H. S. AND B. B. OWEN (1958), *The Physical Chemistry of Electrolyte Solutions,* 3rd ed., van Nostrand Reinhold, New York.

CHAPTER 4

ORGANIC AND BIOCHEMISTRY

Organic chemistry is the study of compounds of carbon. Living systems form organic molecules but there is also an abundance of synthetic organics. Modern industry manufactures something on the order of 10 000 new compounds per year, most of which are organic. More than 1 100 000 organic compounds have been prepared. Determining their behavior in the environment and methods for treatment of these compounds presents a formidable task for the chemist, biologist, ecologist, and environmental engineer.

In this chapter the major classes of organic compounds are described. A brief overview of metabolism is also presented to gain some appreciation of the means of processing foodstuffs and the complexity of metabolism.

4.1 CARBON

Carbon is a unique atom, with properties that account for the limitless compounds it can form. The bonds between carbon atoms are stable, which allows chains to be formed. Atoms near carbon in the periodic table can also bond to carbon to form chains. Each carbon atom can form four bonds. The bonds are covalent in nature; that is, electrons are shared between the participating atoms. If the atoms bonded to carbon have an electronegativity that is significantly different from carbon, then the bond will have a polar character, with the electrons being shared unequally between the atoms. The degree of polarity in an organic molecule determines its solubility in the polar solvent, water. Because of the four bonds and chaining ability of carbon, many different configurations are possible for the same group of atoms, which is known as isomerism.

The principal atoms that form compounds with carbon are hydrogen and oxygen. Nitrogen, phosphorus, and sulfur are the most common minor elements found in naturally occurring organic molecules. Carbon dioxide and its related species are inorganic ions or compounds.

4.2 PROPERTIES OF ORGANIC COMPOUNDS

There are more known compounds of carbon than any other element except hydrogen. Sawyer et al. (1994) have summarized the major characteristics of organic compounds compared to inorganic compounds. A brief explanation is added to most of their points.

1. Organic compounds are usually combustible.
   A large amount of energy is released as an organic compound is oxidized to CO_2, H_2O, and other oxides. The energy release propagates the reaction.
2. Organic compounds usually have lower melting and boiling points.

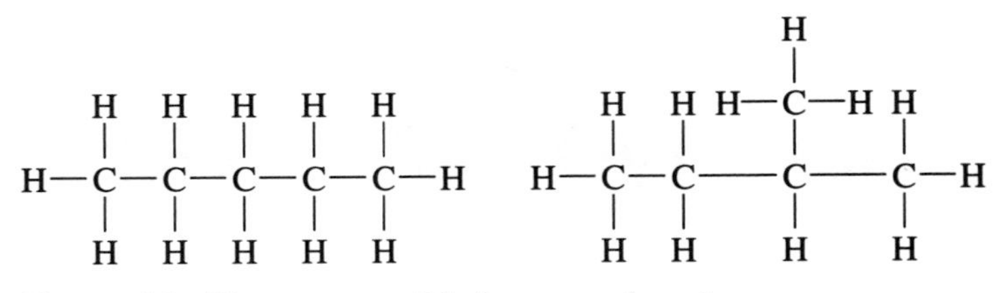

Figure 4.1 The two possible isomers of pentane.

Covalent bonding within organic molecules reduces their bonding or attraction to identical molecules or other molecules.

3. Organic compounds are usually less soluble in water.
   The nonpolar nature of carbon–carbon bonds reduces solubility of these compounds in the polar solvent, water.
4. Several isomers exist for a given formula.
   The isomers are a result of different geometries for a compound or result from changing the bonding location of atoms or groups of atoms.
5. Reactions of organic compounds are usually molecular rather than ionic, which slows their rate of reaction compared to ionic compounds.
6. The valency of carbon allows it to form multiple bonds and carbon has the ability to form strong bonds with itself. This can result in high molecular weight compounds, often well over 1 000.
7. Most organic compounds can serve as a source of food for bacteria and other microorganisms.
   Organic compounds are energy rich (carbon is in a reduced state) and contain the raw materials for cell fabrication.

In general as the size of an organic molecule increases, the melting and boiling points begin to rise (see Table 4.1). Hydrocarbons are organic compounds that only contain hydrogen and carbon. There are three major types of organic compounds: aliphatic, aromatic, and heterocyclic groups. Aliphatic compounds are chained and branched compounds. Examples of isomers for a five-carbon compound are illustrated in Fig. 4.1. Because carbon atoms can be rotated about their bonds, a bend in the compound does not produce a new isomer. Only changes in association of atoms produces an isomer. As atoms are substituted for H, more isomers are possible, as illustrated in Fig. 4.2.

4.3 FUNCTIONAL GROUPS

Functional groups strongly influence the behavior of organic compounds. Three major functional groups are the alcohol, $-OH$; carboxyl, $-COOH$; and, carbonyl, $>C=O$,

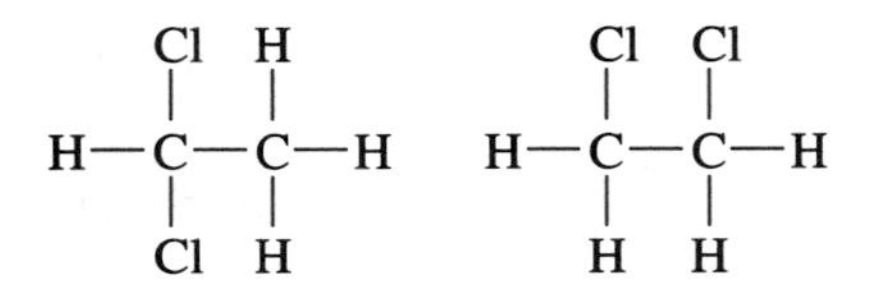

Figure 4.2 Isomers of dichloroethane.

groups. The alcohol group is similar in behavior to water; that is, it is neither strongly acidic nor basic but the bond is polar. The carboxyl group is acidic in nature. The

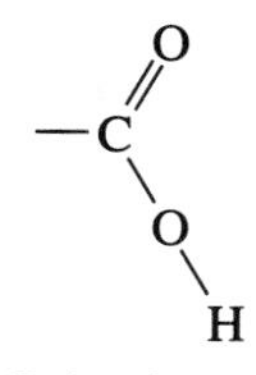

Carboxyl group

electronegativity of the two oxygen atoms pulls electrons away from the hydrogen atom, which then dissociates as H^+. The carbonyl group is an extremely important group in modern synthetic organic chemistry. Because of the high electronegativity difference between carbon and oxygen, the bond in the carbonyl group provides a highly reactive site.

4.4 ALIPHATIC COMPOUNDS

The aliphatic compounds can be broken down into alkanes, which contain only single carbon–carbon bonds; alkenes, which contain at least one double bond between two carbon atoms; and alkynes, which contain at least one triple bond between a carbon pair. Compounds that contain double or triple carbon bonds are unsaturated compounds. Double or triple carbon bonds are major reaction sites. Many synthetic reactions break these multiple bonds and add halogens or other atoms to the carbon.

TABLE 4.1 Physical Properties of *n*-Alkanes, $CH_3(CH_2)_{n-2}CH_3$[a]

n	name	bp, °C at 760 mm Hg	mp °C	density, g/cm³ at 20°C
1	Methane	−161.5	−183	0.424[b]
2	Ethane	−88.6	−172	0.546[b]
3	Propane	−42.1	−188	0.501[c]
4	Butane	−0.5	−135	0.579[c]
5	Pentane	36.1	−130	0.626
6	Hexane	68.7	−95	0.659
7	Heptane	98.4	−91	0.684
8	Octane	125.7	−57	0.703
9	Nonane	150.8	−54	0.718
10	Decane	174.1	−30	0.730
11	Undecane	195.9	−26	0.740
12	Dodecane	216.3	−10	0.749
15	Pentadecane	270.6	10	0.769
20	Eicosane	342.7	37	0.786[d]
30	Triacontane	446.4	66	0.810[d]

[a]From Roberts et al. (1971).
[b]At the boiling point.
[c]Under pressure.
[d]For the supercooled liquid.

The chained compounds that can occur in this group have a fairly predictable gradation in physical properties as an additional carbon atom is added to the structural backbone. This concept is known as homology and the compounds form a homologous series. The names that use Greek prefixes and a few physical properties of some chained alkanes are given in Table 4.1. Greek prefixes are used in naming complex organic molecules. For instance, diethyl in the name of a long compound would indicate two two-carbon groups.

Aldehydes and Ketones

All aldehydes contain the carbonyl group with a hydrogen atom bonded to the carbon atom in the carbonyl group. Their general formula is

$$R{-}C{\overset{\displaystyle =O}{\underset{\displaystyle \diagdown H}{}}}$$

Ketones have the general formula

$$R{-}\overset{\overset{\displaystyle O}{\|}}{C}{-}R'$$

The carbonyl group is the most important functional group in synthetic organic chemistry.

The structural formulas for formaldehyde, a preservative and a toxic compound, and acetone (dimethyl ketone), a common household solvent, are shown below:

$$\underset{\text{Formaldehyde}}{H{-}\overset{\overset{\displaystyle O}{\|}}{C}{-}H} \qquad \underset{\text{Acetone}}{CH_3{-}\overset{\overset{\displaystyle O}{\|}}{C}{-}CH_3}$$

Esters

Esters are compounds formed by the reaction of acids and alcohols and they correspond to salts in inorganic chemistry. The general formula of an ester is R—COO—R′. A typical reaction for the hydrolysis formation of an ester is

$$RCO{-}OH + H{-}OR' \rightarrow H_2O + RCOOR'$$

Alcohols and Ethers

Alcohols have the functional group, —OH, and general formula ROH; ethers are compounds that have the characteristic formula R—O—R′. They can be regarded as substitution products of water. Alcohols tend to be very weakly acidic, giving up the H^+ ion from the —OH group because of the electronegativity of oxygen.

One way of preparing ethers is to dehydrate two molecules of alcohol. Ethers are relatively low in chemical reactivity except in the presence of free molecular oxygen, which causes the formation of unstable, explosive-prone peroxides. Ethers are widely used industrial solvents. Low molecular weight ethers are highly flammable.

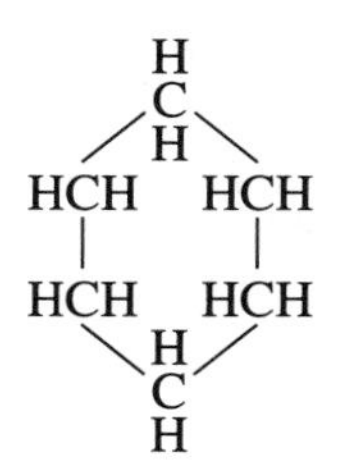

Figure 4.3 Cyclohexane.

Cyclic Aliphatic Compounds

Cyclic aliphatic compounds, as the name implies, consist of carbon atoms singly bonded together to form a ring compound. They may contain functional groups. There is no resonance in these compounds and their behavior is similar to the chained aliphatics. Cyclohexane is shown in Fig. 4.3.

Reactions of Aliphatic Compounds

In general, aliphatic compounds can be oxidized by oxygen or other oxidizing agents. Halogens will substitute themselves for hydrogen, forming haloorganics, many of which are toxic or carcinogenic. Aerobic bacteria have enzymes that catalyze the oxidation of hydrocarbons to alcohols and then further to carbon dioxide and water. The bacteria work on the terminal carbon atom and carry this on until the whole molecule is oxidized.

4.5 NITROGEN-CONTAINING COMPOUNDS

Amines are derivatives of ammonia, NH_3. Primary, secondary, and tertiary amines have one, two, or three of the hydrogen atoms, respectively, substituted with alkyl groups. Tertiary amines combine with alkyl halides to form quarternary ammonium salts. These salts have bactericidal properties that make them useful disinfecting agents.

An amide is another functional group consisting of a primary amine–carbonyl combination: $-(C{=}O)-NH_2$. It is a reactive group of interest to synthetic organic chemists.

Amino acids have at least one amine ($-NH_2$) and one carboxylic group per molecule. Because of the presence of the amino group and the carboxylic acid group these molecules have both acidic and basic properties. The general structural formula for an amino acid is shown at the margin.

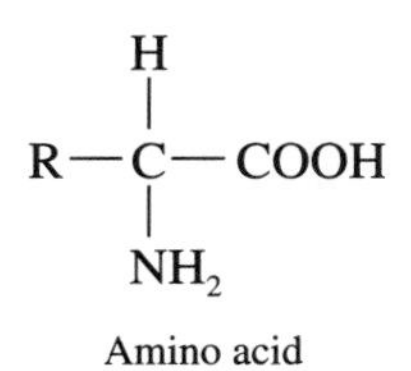

Amino acid

The hydrolysis of two amino acids, which involves the reaction of the carboxylic group on one acid with the amine group on the other with a molecule of water leaving, results in the formation of a peptide bond. Amino acids can thus be chained together

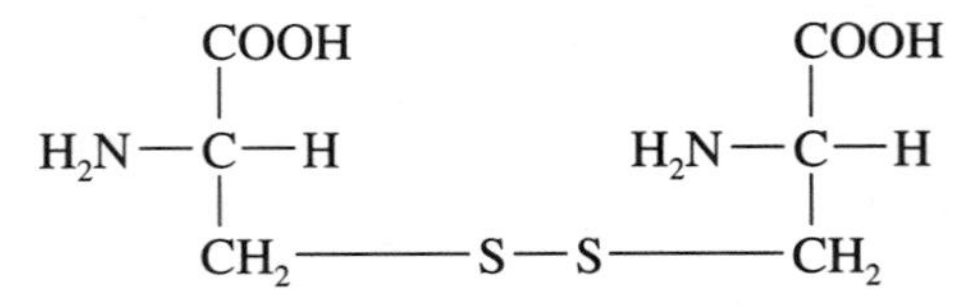

Figure 4.4 An isomer of cysteine.

in an infinite number of ways to form proteins. The 20 or so amino acids that are essential to life have both carboxylic and amino groups bonded to the same carbon atom. The formula for glutamic acid, which is an amino acid, is

$$HO_2C(CH_2)_2\underset{\displaystyle NH_2}{\underset{|}{C}}HCO_2H$$

Some amino acids, such as cysteine and methionine, contain sulfur. One of the isomers of cysteine is shown in Fig. 4.4.

Amino acids are degraded to pyruvate, acetyl-coenzyme A (acetyl-CoA), or other intermediates of the tricarboxylic acid (TCA) cycle (Section 4.11), which is a primary metabolic pathway.

4.6 AROMATIC COMPOUNDS

Aromatic compounds contain the benzene ring (Fig. 4.5). Many of the compounds containing the benzene ring exhibit a characteristic odor, which is why this class of compounds is termed aromatic.

Benzene is a compound that exhibits resonance, which is a very stable configuration because of the sharing of the electrons in the bond among the six participating carbon atoms. These compounds are difficult to degrade because of this property. Resonance in the benzene ring is often indicated by drawing a circle within the ring

A carbon atom in the ring may have an —OH group attached to it, in which case the compound is referred to as a phenol; an —OOH group, which makes the ring an acid; or a —CHO or a —(CO)—R group, which makes the compound an aldehyde or a ketone, respectively. It is possible to attach carbon chains to benzene rings, and

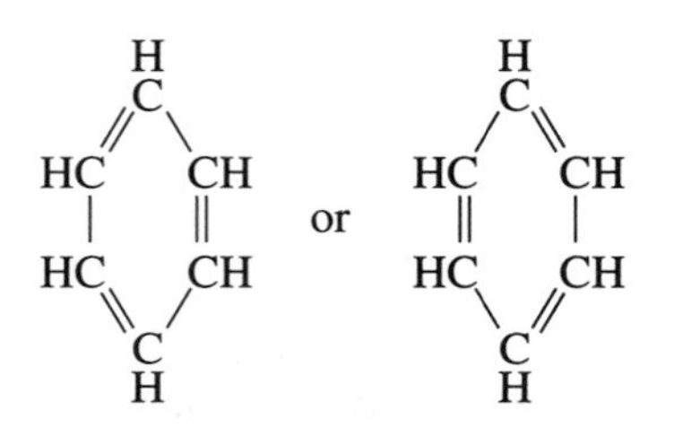

Figure 4.5 Benzene.

polyring aromatic hydrocarbons exist. There are many naturally occurring aromatics as well as synthetic aromatics.

OH
Cl Cl
Cl Cl
Cl

PCP

Chlorinated benzene derivatives are often quite toxic. Pentachlorophenol (PCP), shown above, is one of the most widely used pesticides and is one of the most toxic compounds because of its high number of chlorine atoms.

4.7 COMPOUNDS OF SULFUR

Similarly to ammonia, sulfuric acid may have either one or both H or OH atoms substituted with organic chains to form various compounds. Two examples of these compounds are diphenyl sulfone and methane sulfonic acid, with the following structures.

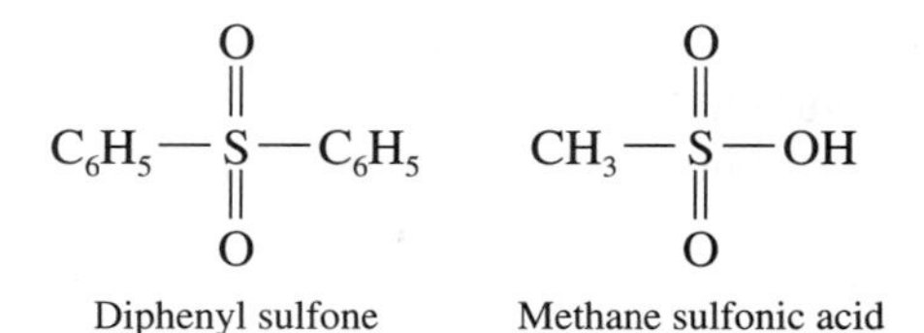

Diphenyl sulfone   Methane sulfonic acid

A more interesting group of compounds is the derivatives of H_2S. These compounds are also known as mercaptans. Like hydrogen sulfide, the mercaptans have disagreeable odors. The odors from many of these compounds are noticeable at very low concentrations. An example of a mercaptan is diethyl sulfide, which has the formula $C_2H_5SC_2H_5$.

4.8 NATURALLY OCCURRING ORGANIC COMPOUNDS

Organisms from the smallest to the largest require certain compounds to exist. These compounds are found in all organisms to varying degrees. There are three major classes of naturally occurring organic compounds: carbohydrates, proteins, and fats and oils.

Carbohydrates

Carbohydrates contain C, H, and O exclusively. Their general formula is $(CH_2O)_n$. The simplest carbohydrates or sugars are monosaccharides of which glucose, $C_6H_{12}O_6$,

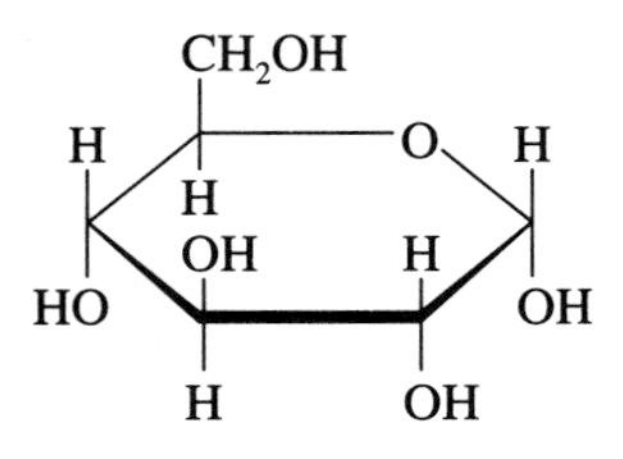

Figure 4.6 Glucose.

(Fig. 4.6) is a typical example. Disaccharides contain two monosaccharides bonded together (common household sugar, sucrose, is a disaccharide of glucose and fructose) and polysaccharides are chains of monosaccharides. Two main classes of polysaccharides are energy storage compounds and structural compounds. Energy storage compounds are readily degraded. They are used by the cell in energy-demanding situations. Starch is an example of an energy storage carbohydrate.

Structural compounds serve to maintain the integrity of the cell and as such they are resistant to degradation. Cellulose is a good example of a structural carbohydrate.

Proteins

Proteins are complex molecules with no general formula that contain C, H, O, N, P, and S along with trace occurrences of other elements. N, P, and S are contained in lesser amounts than C, H, and O on a molar basis. The building blocks of proteins are amino acids. There are 20 amino acids that form the building blocks of proteins. There are other amino acids that perform other functions. Amino acids have acidic and basic properties and may be polar or nonpolar, among other characteristics. The properties of amino acids and their sequence in a protein give the protein four levels of structure, which is critical to the function of the protein.

The biological functions of proteins are multivariate. The main classes of proteins are enzymes, structural compounds, and transport compounds. Enzymes catalyze reactions, accelerating them 10 to 100 000 or more times. Structural elements keep the cell intact. Transport proteins serve to move compounds within the cell and between the cell and its environment.

In metabolism, proteins are broken down into amino acids and amino acids are resynthesized into proteins, but the pathways for each process are not the reverse of the other.

Fats and Oils

Fats and oils have the general formula $C_nH_{2n+1}COOH$. Lightweight compounds are oily and high weight compounds are waxlike. Many of these compounds are sparingly soluble in water, which can be a treatment problem because microorganisms lack the ability to secrete emulsifying agents to keep the fatty acids in suspension. Many are odoriferous. Formulas for two typical fatty acids are given in Fig. 4.7.

$CH_3(CH_2)_{16}COOH$ Stearic acid

$CH_3(CH_2)_7CH{=}CH(CH_2)_7COOH$ Oleic acid (an unsaturated acid)

Figure 4.7 Typical fatty acids.

Fatty acids play an important role in metabolism as energy-rich fuels with more energy per unit weight than carbohydrates.

4.9 BIOCHEMISTRY

Given the diversity of species, there is a remarkable similarity among the metabolic machineries of all life forms, from the smallest to the largest. Metabolism is the sum total of all chemical reactions within the organism. These reactions proceed in accordance with thermodynamic principles. All heterotrophic organisms ultimately obtain their energy from oxidation–reduction reactions.

Metabolism is broken down into processes to obtain energy and processes that use this energy to synthesize new cell components and perform other functions. Dissimilation or catabolism is the breakdown of energy-rich (reduced) compounds to obtain energy. Assimilation, anabolism, or synthesis is the fabrication of complex molecules from simpler ones.

Aerobic organisms use oxygen as the ultimate electron acceptor in catabolism. Respiration is the oxidation of fuels with molecular oxygen. Anaerobic organisms do not use oxygen and essentially break organic molecules into a component that is oxidized and another component that becomes reduced while the organism derives some energy from the process. The extraction of chemical energy from substrates in the absence of molecular oxygen is known as fermentation.

Metabolic processes occur through a series of steps depending on the presence of enzymes and the substance to be degraded or fabricated. The series of steps is known as a metabolic pathway. The study of metabolism is involved. To obtain an appreciation of metabolic pathways two major pathways are presented summarily here. There are many other pathways. Before the pathways are examined, two important energy carriers are discussed.

The energy obtained from oxidation is stored in energy carriers that release energy in synthesis reactions. Adenosine triphosphate (ATP) is the primary energy carrier in a cell. The structure of the molecule is shown in Fig. 4.8. As adenosine monophosphate

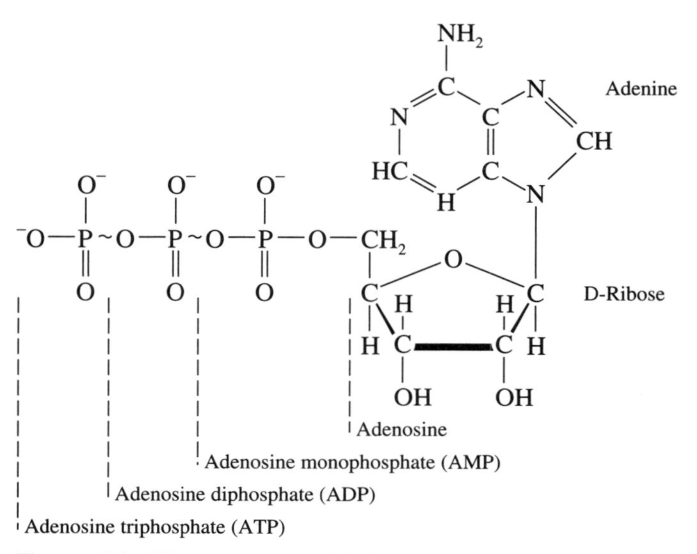

Figure 4.8 The structure of ATP.

(AMP) adds successively one (forming adenosine diphosphate, ADP) and another inorganic phosphate group, the bonds holding these phosphate groups become more energy rich. These high-energy bonds are indicated by "~" in Fig. 4.8. ADP–ATP transformations are involved in oxidation–reduction reactions, the energy-rich ATP being formed in oxidation reactions (catabolism) and the lower energy ADP being formed with the release of inorganic phosphorus in reduction (synthesis) reactions. This compound is representative of a wide variety of energy carriers. There is no electron transfer involved in the formation of ATP, ADP, or AMP.

The ATP content of microorganisms is variable (Levin et al., 1975) but for both aerobic and anaerobic microorganisms that are found in sewage treatment processes, ATP content of viable cells is in the range of 1–2 μg ATP/mg cell (Weddle and Jenkins, 1971; Chung and Neethling, 1990).

Another important class of compounds involved in metabolism is dehydrogenase enzymes. These compounds participate in reactions by being reversibly reduced and oxidized in transferring electrons from the substrate to oxygen or the ultimate electron acceptor in a generally long series of steps. Nicotinamide adenine dinucleotide (NAD) (its structure is shown in Fig. 4.9) is a common electron carrier representative of those that are associated with enzymes of this class.

The oxidized form of NAD is symbolized as NAD^+ and the reduced form is symbolized as NADH. In fact, two hydrogen atoms are involved in the transformation from NAD^+ to NADH. One H atom is released as H^+ and the other H atom is transferred along with two electrons to NAD^+. The transformation of NADH to NAD^+ directly or indirectly involves the formation of an energy carrier such as ATP in aerobic respiration. In an anaerobic fermentation, the H atom and the associated

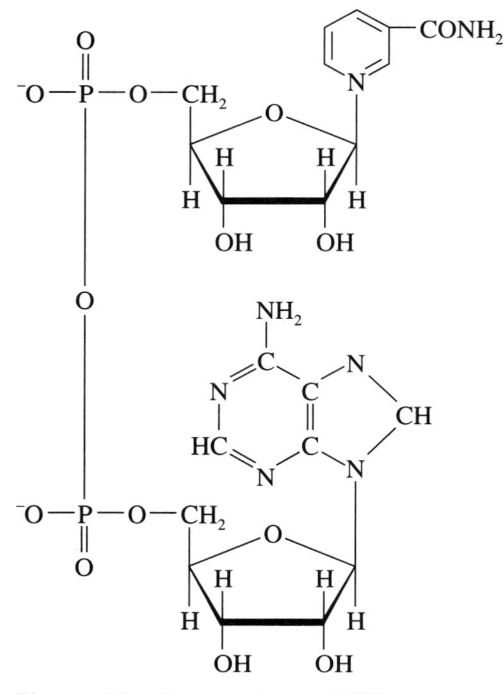

Figure 4.9 The structure of NAD.

electrons are transferred to another organic molecule. ATP is also formed in anaerobes as a result of these electron transfers.

The concentrations of ATP and dehydrogenase enzymes can be used as a measure of activity of cultures.

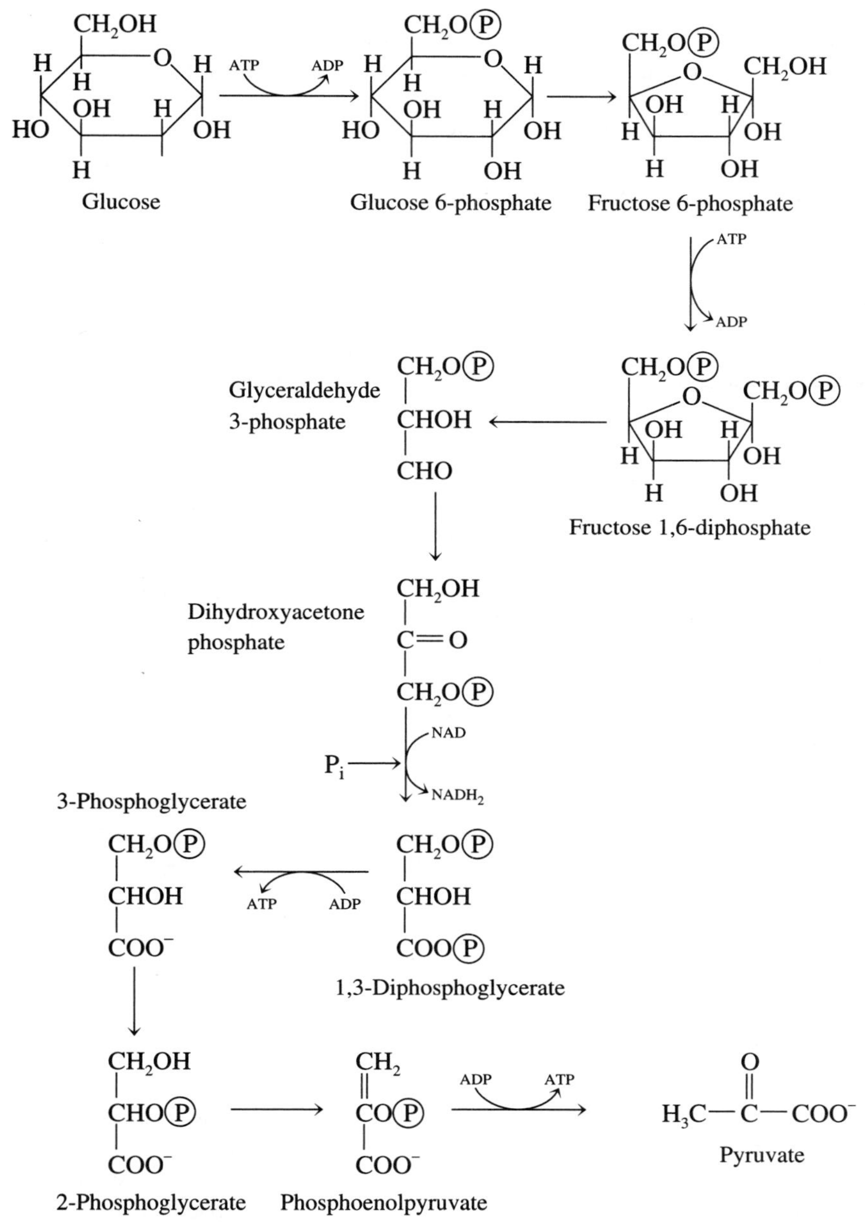

Figure 4.10 Glycolysis.

4.10 GLYCOLYSIS

Glycolysis is a fermentation process whereby glucose is broken down to lactic acid. The ability to use this pathway (also known as the Embden–Meyerhof–Parnas pathway) exists in, of course, anaerobic organisms, and many aerobic organisms, which may use this pathway in preparation for aerobic completion of the oxidation of glucose. The pathway is shown in Fig. 4.10.

Glycolysis is catalyzed by the action of 11 enzymes. It is seen from the figure that the first step, in which phosphate is added to glucose (phosphorylation), requires an initial expenditure of energy by the use of one ATP molecule. The net release of ATP from the process is 2 ATP molecules per molecule of glucose. As noted on the figure, other monosaccharides can also enter the cycle. Other fermentations proceed in a similar fashion, although the final products vary. For instance, ethanol is a major product of organisms used in the brewery industry. The change occurs in the last step, where pyruvate is converted to acetaldehyde, which is then converted to ethanol.

The free energy change associated with the conversion of glucose to 2 lactate molecules is −47.0 kcal/mol.

4.11 THE TRICARBOXYLIC ACID CYCLE

Glycolysis only releases a small amount of the chemical energy contained within glucose. Complete oxidation of glucose to CO_2 and H_2O magnifies the yield of energy considerably. The theoretical free energy change for complete oxidation of one molecule of glucose to CO_2 and H_2O is −686.0 kcal/mol.

The TCA cycle (Fig. 4.11) is sometimes referred to as the Krebs cycle after the man who first postulated its existence. It is almost universally present in aerobic organisms and is a primary pathway. In catabolism the cycle proceeds clockwise as it is represented in Fig. 4.11.

In an aerobic organism the NADH formed in glycolysis can ultimately transfer its electrons to oxygen and form six molecules of ATP in addition to the two molecules of ATP formed from the original glucose molecule. Thus glycolysis will yield a net gain of eight ATP molecules in aerobic organisms. The pyruvate formed in glycolysis enters the TCA cycle. As shown on the figure, GTP (guanosine triphosphate) is formed in the TCA cycle. This compound is similar to ATP. Also, $NADPH_2$ (nicotinamide adenine dinucleotide phosphate) and $FADH_2$ (flavin adenine dinucleotide) are formed. Their roles are essentially the same as NADH.

Because 1 mole of glucose yields 2 moles of pyruvate and 6 moles of ATP and 1 mole of pyruvate yields 15 moles of ATP, the aerobic biooxidation of 1 mole of glucose to carbon dioxide and water yields 36 moles of ATP. About 73% of the theoretical energy yield of oxidation of glucose is captured in ATP. There are 10 enzymes involved in the TCA cycle and other enzymes are involved in the ultimate transport of electrons to oxygen.

4.12 ENZYME KINETICS

Enzymes catalyze reactions by forming an intermediate compound with the reactants (substrate). The intermediate compound is metastable and normally breaks down to yield the unchanged enzyme and the products of the overall reaction. As with all chemical reactions, each step is reversible (the principle of microscopic reversibility). The reaction sequence can be represented as follows.

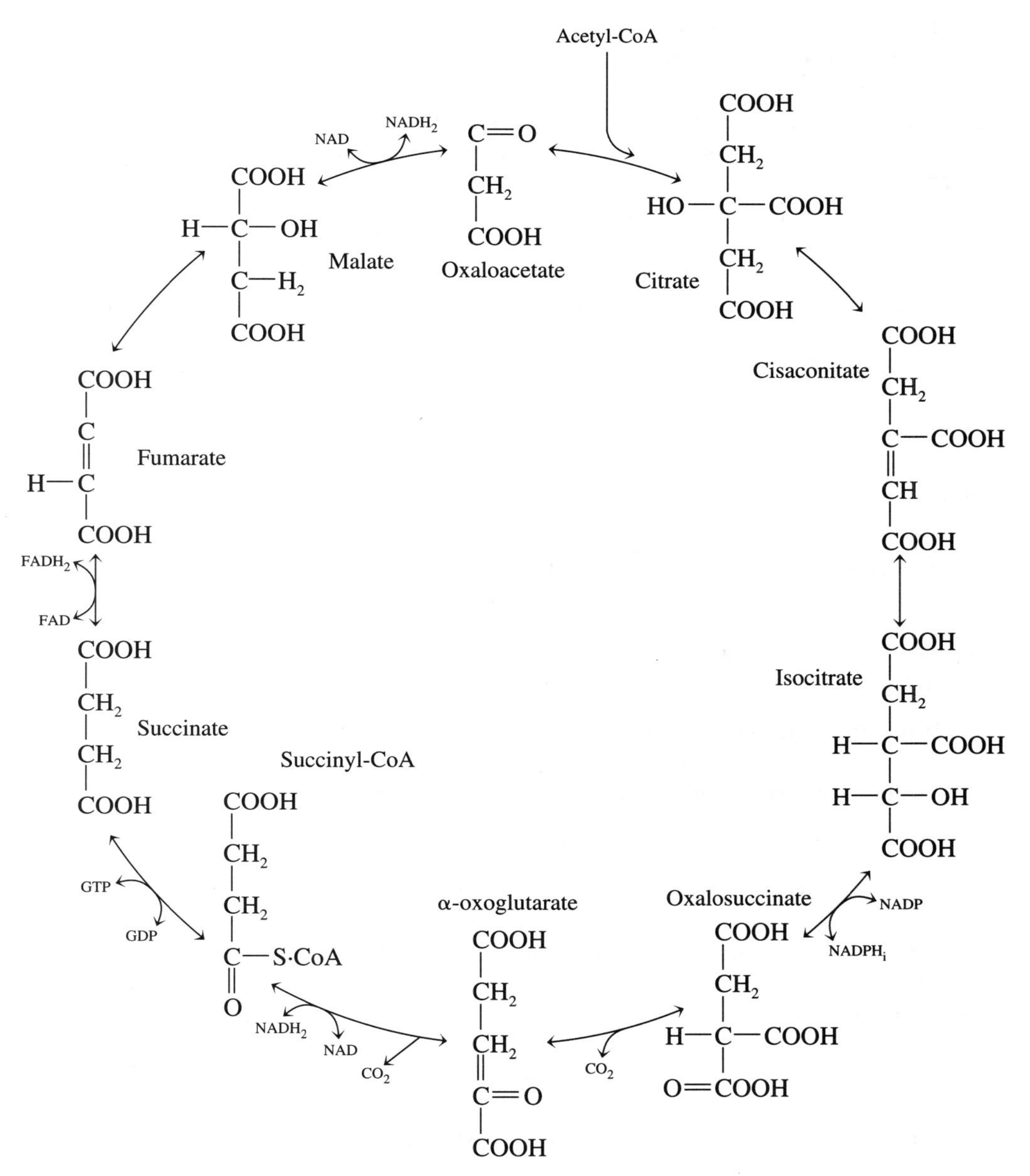

Figure 4.11 The tricarboxylic acid cycle.

$$E + S \underset{k_2}{\overset{k_1}{\rightleftarrows}} ES \underset{k_4}{\overset{k_3}{\rightleftarrows}} P + E \tag{4.1}$$

where

E is enzyme
S is substrate
P is product
ES is the intermediate enzyme-substrate complex
k_i are rate constants

If the reaction is proceeding at a uniform rate, the rates of formation and destruction of the intermediate are equal, although the concentrations of reactants and products are continuously changing. This situation often occurs, particularly when the enzyme concentration is low compared to the concentrations of the reactants. Assuming normal first-order reactions with respect to each entity apply, the following equation describes the balance for [ES].

$$\frac{d[\mathrm{ES}]}{dt} = k_2[\mathrm{E}][\mathrm{S}] - k_1[\mathrm{ES}] + k_3[\mathrm{E}][\mathrm{P}] - k_4[\mathrm{ES}] = 0 \tag{4.2}$$

The rate of formation of ES from product and enzyme is normally negligible, i.e., $k_3 \approx 0$.

The overall velocity of the reaction depends on the rate of product formation:

$$v = \frac{d[\mathrm{P}]}{dt} = k_4[\mathrm{ES}] \tag{4.3}$$

where
v is the overall velocity of the reaction

The following mass balance also applies at any time.

$$[\mathrm{E}]_0 = [\mathrm{E}] + [\mathrm{ES}] \tag{4.4}$$

where
$[\mathrm{E}]_0$ is the initial concentration of enzyme

It is difficult to measure the amount of ES present for use in Eq. (4.4); therefore, Eqs. (4.2) and (4.4) are used to develop a formula in terms of more readily assessed parameters. The following expression for [ES] will be found.

$$[\mathrm{ES}] = \frac{k_2[\mathrm{E}]_0[\mathrm{S}]}{k_2[\mathrm{S}] + k_1 + k_4} = \frac{[\mathrm{E}]_0[\mathrm{S}]}{[\mathrm{S}] + \dfrac{k_1 + k_4}{k_2}} \tag{4.5}$$

The determination of the individual rate constants involved is difficult and an overall constant, K, known as the Michaelis–Menten (they made the original development) constant or half-velocity constant (see Problem 15), is defined as

$$K = \frac{k_1 + k_4}{k_2} \tag{4.6}$$

Substituting Eqs. (4.5) and (4.6) into Eq. (4.3) yields

$$v = \frac{k_4[\mathrm{E}]_0[\mathrm{S}]}{K + [\mathrm{S}]} \tag{4.7}$$

The maximum rate of reaction (v_{max}) will occur when $[\mathrm{ES}] = [\mathrm{E}]_0$. Therefore,

$$v = k_4[\mathrm{E}]_0 = v_{max} \tag{4.8}$$

Substituting Eq. (4.8) into Eq. (4.7) results in

$$v = \frac{v_{max}[\mathrm{S}]}{K + [\mathrm{S}]} \tag{4.9}$$

Equation (4.9) is known as the Michaelis–Menten equation. It was originally developed by them to describe enzyme kinetics but has been found to apply to many

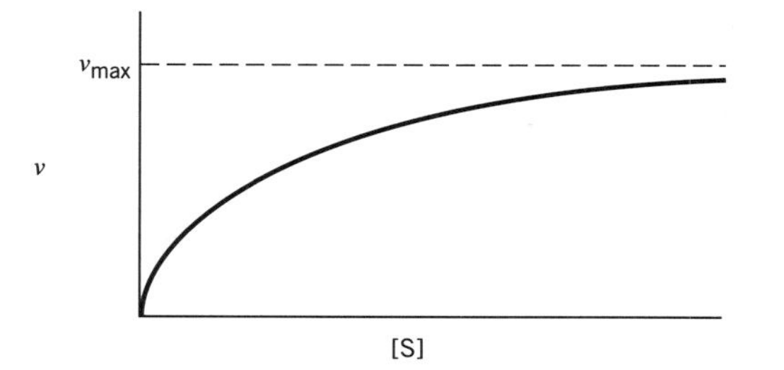

Figure 4.12 The Michaelis–Menten equation.

other types of reactions. The equation is plotted in Fig. 4.12. It is observed from the plot that when substrate concentration is high, the rate of reaction is zero order; when it is low the velocity is first order. In between, the reaction velocity has an order between 0 and 1.

The rate of product formation is related to the rate of substrate removal and the stoichiometry of the overall reaction. The constant relating moles (or mass) of substrate is absorbed into v_{max}. If mass m of substrate is required to form 1 mass unit of product,

$$\frac{d[P]}{dt} = \frac{v'_{max}[S]}{K + [S]} = -m\frac{d[S]}{dt} \quad \text{or} \quad \frac{d[S]}{dt} = -\frac{(v'_{max}/m)[S]}{K + [S]} = -\frac{v_{max}[S]}{K + [S]} \tag{4.10}$$

where

[P] and [S] are in mass or molar concentration units

■ Example 4.1 Application of the Michaelis–Menten Equation

What is the time for 90% completion of a reaction described by the Michaelis–Menten equation when the initial substrate concentration is 32 mg/L, v_{max} = 4.3 mg/L-h, and K = 1.5 mg/L?

The concentration at 90% completion will be 0.10(32 mg/L) = 3.2 mg/L. Integrating Eq. (4.10),

$$K\int_{[S]_0}^{[S]} \frac{d[S]}{[S]} + \int_{[S]_0}^{[S]} d[S] = -v_{max}\int_0^t dt \Rightarrow K\ln\frac{[S]}{[S]_0} + [S] - [S]_0 = -v_{max}t$$

$$t = -\frac{K\ln([S]/[S]_0) + [S] - [S]_0}{v_{max}}$$

$$= -\frac{(1.5\text{ mg/L})\ln(0.10) + 3.2\text{ mg/L} - 32\text{ mg/L}}{4.3\text{ mg/L-h}} = 7.5\text{ h}$$

The two constants in the equation, K and v_{max}, may be readily determined when reaction velocity and substrate concentration data are available. A common method for determining the constants is to invert the equation and plot $1/v$ against 1/[S]. This plot is known as a Lineweaver–Burke plot.

$$\frac{1}{v} = \frac{1}{v_{max}} + \frac{K}{v_{max}}\frac{1}{[S]} \tag{4.11}$$

The slope and intercept define the constants. A Lineweaver–Burke plot weights data at lower substrate concentrations more than data taken at high substrate concen-

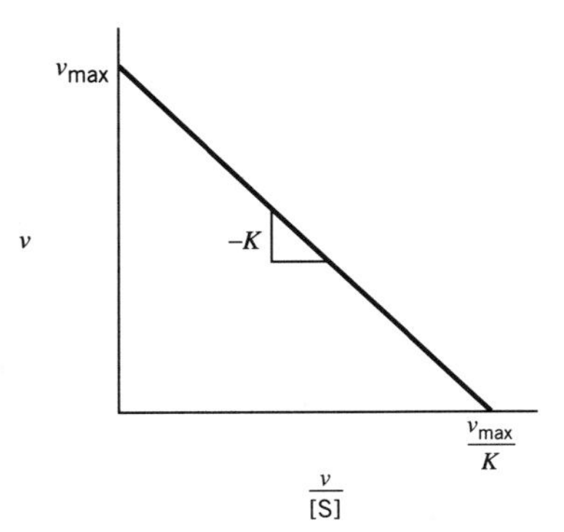

Figure 4.13 Eadie–Hofstee plot.

trations. Furthermore, lower substrate concentrations are determined with less accuracy than high substrate concentrations. There are other methods of algebraically manipulating Eq. (4.9) to obtain v_{max} and K from the measured data. Rearranging Eq. (4.9) into the form of Eq. (4.12) and plotting v against $v/[S]$ results in an Eadie–Hoftsee plot (Fig. 4.13). The Eadie-Hoftsee plot magnifies departures from linearity that may not be apparent in a double reciprocal (Lineweaver–Burke) plot.

$$v = -K\frac{v}{[S]} + v_{max} \tag{4.12}$$

Inhibitory substances limit the action of enzymes by binding with a substrate or the enzyme. It can be demonstrated that inhibitory substances effectively increase the half-velocity constant. Toxic substances destroy the enzyme. It can be seen from Eqs. (4.7) and (4.8) that a decrease in $[E]_0$ decreases the maximum velocity of the reaction.

QUESTIONS AND PROBLEMS

1. Compare the amount of oxygen required to oxidize methane and ethane completely in terms of g O_2/g of compound and g O_2/g C.
2. Define isomerism and illustrate with structural formulas for the compounds with molecular formula C_6H_{14}.
3. What are the major functional groups and how do they affect organic molecules? What are the general chemical formulas for alcohols, aldehydes, ketones, esters, and ethers?
4. In general, how do organic compounds differ from inorganic compounds?
5. Why does the carboxyl group behave as an acid?
6. Write the chemical reaction for the formation of an ether from ethanol (C_2H_5OH) and methanol (CH_3OH).
7. What are the three major classes of naturally occurring organics and general characteristics of compounds in each group?
8. What is necessary to saturate oleic acid (Fig. 4.6)? Write the formula of the saturated compound.
9. The generic chemical formulas for carbohydrates and fatty acids are $(CH_2O)_n$ and $C_nH_{2n+1}COOH$, respectively. Consult a reference on organic chemistry and list three or four carbohydrates and three or four fatty acids (not found in this text) and their chemical formulas to prove that these formulas are generally correct.

10. Why can aerobic microorganisms extract more energy than anaerobic microorganisms from metabolism of a given compound?
11. Write the chemical equations for the oxidation of glucose to pyruvate ($C_3H_3O_3^-$) and for the oxidation of pyruvate to CO_2 and H_2O with oxygen as the oxidant. Combine these two equations to find the overall equations for the oxidation of glucose with oxygen.
12. What are the characteristics of an aromatic compound?
13. Write the structural formula for phenol. Why is it difficult to degrade? Is the compound acidic or basic?
14. What are roles of ATP and NAD in metabolism?
15. In the Michaelis–Menten equation, what is the reaction velocity in terms of v_{max} when $K = [S]$?
16. How long will it take for 95% conversion of a substrate with an initial concentration of 280 mg/L, if $v_{max} = 1.45$ mg/L/h and $K = 165$ mg/L?
17. For $v_{max} = 2.08$ mg/L/h and $K = 45$ mg/L plot the reaction velocity as a function of substrate concentration up to $[S] = 500$ mg/L. At what substrate concentration does the Michaelis–Menten reaction velocity differ by 10% from the first-order and zero-order reaction velocities that apply when the substrate concentrations are low and high, respectively?
18. Graphically compare reaction velocity as a function of substrate concentration for a v_{max} of 5.0 mg/L/d and K values of 10, 100, and 1 000 mg/L. Make the plots for substrate concentrations up to 5 000 mg/L. At what substrate concentrations does the reaction become essentially first order (90% of the first-order velocity) for each K value?
19. What are v_{max} and K for the following data? Use both Lineweaver–Burke and Eadie–Hofstee methods to determine the constants.

$S \times 10^2$ (mol/L)	1.0	0.80	0.60	0.40	0.20
$v \times 10^4$ (mol/L/min)	6.4	5.8	4.79	3.80	2.19

20. Could an Eadie–Hofstee or Lineweaver–Burke approach be used to evaluate the coefficients for a reaction model of

$$v = \frac{kS^2}{K + S^2}$$

21. Would you expect v_{max} only or both v_{max} and K to be affected by a temperature change?

KEY REFERENCES

LEHNINGER, A. L. (1975), *Biochemistry,* 2nd ed., Worth Publishers, New York.

ROBERTS, J. D., R. STEWART, AND M. C. CASERIO (1971), *Organic Chemistry: Methane to Macromolecules,* W. A. Benjamin, Menlo Park, CA.

SAWYER, C. N., P. L. MCCARTY, AND G. F. PARKIN (1994), *Chemistry for Environmental Engineering,* 4th ed., McGraw-Hill, Toronto.

REFERENCES

CHUNG, Y. C. AND J. B. NEETHLING (1990), "Viability of Anaerobic Digester Sludge," *J. of Environmental Engineering, ASCE,* 116, 2, pp. 330–342.

LEVIN, G. V., J. R. SCHROT, AND W. C. HESS (1975), "Methodology for Application of Adenosine Triphosphate Determination in Wastewater Treatment," *Environmental Science and Technology,* 9, 10, pp. 961–965.

WEDDLE, C. L. AND D. JENKINS (1971), "The Viability and Activity of Activated Sludge," *Water Research,* 5, 8, pp. 621–640.

CHAPTER 5

ANALYSIS AND CONSTITUENTS IN WATER

Analysis of a water for its constituents provides an opportunity to apply and exploit chemical phenomena to expand our understanding of them and arrive at concrete results to evaluate the state of the water vis-à-vis its health aspects, use as a resource, and role in the ecosystem. This chapter covers additional theory and techniques that have not been discussed in previous chapters.

The cookbook for environmental engineers is *Standard Methods for the Examination of Water and Wastewater* (1992), hereafter referred to as *Standard Methods.* Detailed procedures for most of the analytical procedures (physical, chemical, biological, and others) of interest to the environmental engineer are contained in this reference, which is compiled in consultation with experts around the world. This book is revised at least every 5 years or sooner to remain current. Brief background information and qualifying remarks are given with each procedure. As well, the precision and accuracy of the procedures are given based on the results from many qualified laboratories. Use of this reference ensures that consistent techniques are used and the analyst is aware of the validity and shortcomings of results reported by anyone using the techniques prescribed in it.

5.1 TITRATION

A titration is a volumetric form of analysis that can be applied to many chemical phenomena. The procedure consists of adding a measured volume of reagent (the titrant) to a sample until a point is reached where the amount of reagent added is

equivalent to the amount of the constituent being analyzed in the sample. The sample may have undergone preliminary treatment (e.g., digestion) to release or transform the substance of interest into a form that is amenable to titration. The concentration of reagent in the titrant ultimately is related to a gravimetric determination (based on a weight measurement).

The concentration of the titrant is expressed in terms of normality, *N*. A 1 *N* solution contains 1 equivalent weight of reactant per liter as defined earlier. It is worth mentioning again that the equivalent weight depends on the context of the reaction or the reaction products that exist at the endpoint of the titration. The total capacity of the titrant to give up H^+ ions or take up electrons, for example, may not be exhausted. The normality of a substance usually differs if the substance is participating in an acid–base reaction compared to a redox reaction.

5.2 PRIMARY STANDARDS

Primary standards are chosen and maintained to preserve the highest amount of accuracy. Titrants are calibrated against primary standards or, in fact, may be primary standards themselves. The following considerations apply to the choice of a primary standard.

The primary standard should

1. be readily obtainable in a pure condition
2. be capable of being dried at temperatures slightly above 100°C without decomposition
3. not be hygroscopic
4. not be oxidized by air or adulterated by CO_2
5. be capable of being tested for impurities
6. have a high equivalent weight; this minimizes weighing errors and relative errors dependent on the sensitivity of the balance
7. participate in well-defined reactions that attain equilibrium rapidly
8. be inexpensive

The shelf life of primary standards must be determined and new reagents prepared whenever the shelf life is approached or contamination has occurred. Biological growth and sunlight often cause deterioration of reagents.

5.3 ACID–BASE TITRATIONS

The variation in pH of a solution that contains acid or base and titrated with base or acid, respectively, has been discussed in Section 3.5. The only additional comments to be made are on methods to determine the equivalence point pH. The solution should be stirred during the titration. The pH can be directly monitored during the titration, which provides the most accurate point for stopping the titration.

An entirely suitable method for determining the stopping point is the use of an indicator. Indicators are convenient and are most commonly utilized in routine work. An indicator produces a visible change in color as the pH changes; therefore it is sensitive to the presence of H^+ ions. An equation describing the reaction of an indicator with H^+ ions is

$$\underset{\text{acid color}}{\mathrm{HIn}} \rightleftharpoons \mathrm{H}^+ + \underset{\text{base color}}{\mathrm{In}^-} \tag{5.1}$$

The predominance of either the acidic or basic species determines the color of the solution. The equilibrium expression for the above reaction is

$$K_{\mathrm{In}} = \frac{[\mathrm{H}^+][\mathrm{In}^-]}{[\mathrm{HIn}]} \tag{5.2}$$

Typically a five- to tenfold ratio of concentration of one form to the other is required for the color change to be apparent to visual observation.

$$\frac{[\mathrm{In}^-]}{[\mathrm{HIn}]} \leq \frac{1}{10} \qquad \text{acid color} \tag{5.3a}$$

$$\frac{[\mathrm{In}^-]}{[\mathrm{HIn}]} \geq 10 \qquad \text{base color} \tag{5.3b}$$

Applying the Henderson–Hasselbach equation (Eq. 3.13) to Eqs. (5.3a) and (5.3b) determines the effective pH range for color change of the indicator.

$$\mathrm{pH} \leq -1 - \log K_{\mathrm{In}} \qquad \text{acid color}$$

$$\mathrm{pH} \geq 1 - \log K_{\mathrm{In}} \qquad \text{base color}$$

and the effective pH range for the indicator is

$$\text{pH range} = \mathrm{p}K_{\mathrm{In}} \pm 1 \tag{5.4}$$

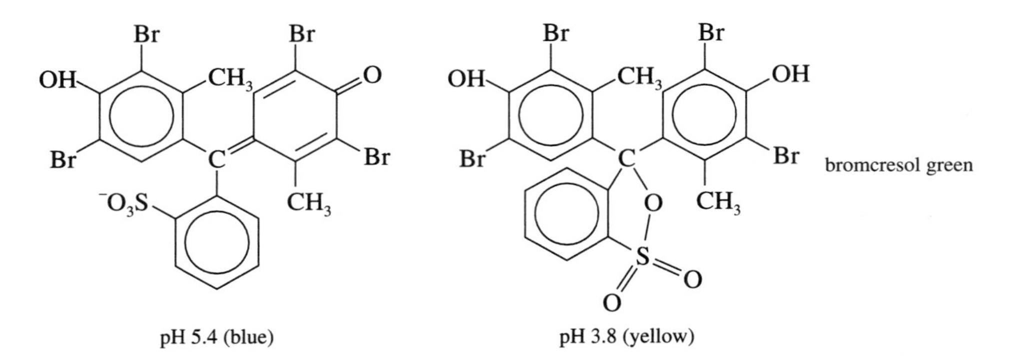

pH 5.4 (blue) pH 3.8 (yellow)

It is seen that choice of an indicator depends on its equilibrium constant and the pH inflection range of the species being titrated. Many indicators are able to be used when strong acids or bases are being titrated; acids or bases with smaller dissociation constants are limited to a smaller number of indicators with K_{In} in the appropriate range. The indicator itself participates in the equilibria in the solution. Therefore only a small amount of indicator is used to render its effects on the solution pH negligible. Acid–base indicators are generally weak organic acids or bases; commonly used acid–base indicators in water analyses are given in Table 5.1.

■ Example 5.1 Influence of an Indicator on a Titration

Indicator solutions are typically made using 0.5–1 g of indicator per liter. Determine the equivalent $CaCO_3$ concentration of an acid–base indicator in a solution to be

TABLE 5.1 Acid–Base Indicators[a]

Indicator	pH transition range	pK[b]
Bromphenol blue	3.0 (yellow)–4.6 (purple)	4.10
Bromcresol green	3.8 (yellow)–5.4 (blue)	4.90
Bromthymol blue	6.0 (yellow)–7.6 (blue)	7.30
Chlorophenol red	4.8 (yellow)–6.4 (purple)	6.25
Cresol red	0.2 (red)–1.8 (yellow)	—
	7.0 (yellow)–8.8 (purple)	8.46
2,6-Dinitrophenol	1.7 (colorless)–4.7 (yellow)	3.70
Metacresol purple	1.2 (red)–2.8 (yellow)	1.51[c]
	7.4 (yellow)–9.0 (purple)	8.32
Methyl orange	3.1 (red)–4.4 (yellow)	3.46
Methyl red	4.4 (red)–6.2 (yellow)	5.00
Phenolphthalein	8.2 (colorless)–9.8 (purple)	—
Phenol red	6.4 (yellow)–8.2 (red)	8.00
Thymol blue	1.2 (red)–2.8 (yellow)	1.65
	8.0 (yellow)–9.6 (blue)	9.20

[a]From Bányai (1972).
[b]All values at 20°C and ionic strength of 0.
[c]At an ionic strength of 0.1.

titrated. The indicator has a molecular weight of 250 g and 1 drop of indicator solution is to be added to a sample volume of 25 mL for the titration. A drop contains approximately 0.05 mL. The indicator has an equivalent weight of 250 g.

The concentration of indicator in the indicator solution ($[I]_I$) is

$$[I]_I = \left(1\,\frac{g}{L}\right)\left(\frac{1\text{ mole}}{250\text{ g}}\right) = 4.00 \times 10^{-3}\,M \text{ or } 4.00 \times 10^{-3}\text{ eq/L}$$

The concentration of indicator equivalents in the sample solution ($[I]_s$) is

$$[I]_s = \frac{(0.05\text{ mL})(4.00 \times 10^{-3}\text{ eq/L})}{25\text{ mL}} = 8.00 \times 10^{-6}\text{ eq/L}$$

The equivalent weight of $CaCO_3$ is 100 g/2 = 50 g.

$$[CaCO_3] = \left(8.00 \times 10^{-6}\,\frac{eq}{L}\right)\left(\frac{50\text{ g}}{eq}\right)\left(\frac{10^3\text{ mg}}{g}\right) = 0.40\text{ mg/L}$$

This amount of indicator is equivalent to only a very small amount of $CaCO_3$ in the sample and can be ignored in usual circumstances.

5.4 COMPLEX AND PRECIPITATE FORMATION TITRATIONS

The formation of precipitates or complexes removes the free ion from solution. An indicator sensitive to the free ion signals its presence or absence and allows for easy determination of the species. The agent causing the formation of the complex or precipitate must have a stronger affinity for the species of interest than other competing complexing agents.

An examination of solubility product constants (Section 1.11) shows which reactions may be feasible for this type of analysis. A common analysis for chloride (argentometric method) uses silver nitrate as a titration agent. White silver chloride precipitates. In this case the indicator used is the chromate ion, which forms a reddish-brown precipitate (Ag_2CrO_4). Silver chloride is less soluble than silver chromate (see Problem 2); therefore the chloride is precipitated first. After the chloride is removed, further addition of silver forms the highly visible silver chromate precipitate. Because there is some consumption of titrant by the indicator, a blank containing only the indicator is titrated and the volume of titrant used for the blank is subtracted from the volume of titrant used for the sample.

Complex formation titrations can be used to determine either the concentration of a metal or a ligand such as chloride or cyanide. The most common complex formation reagent is ethylenediaminetetraacetic acid (EDTA) shown below. This organic ligand is able to form up to six bonds with a metal ion. Ligands that form multiple bonds with a metal are known as chelating agents (from the Greek *chela,* meaning claw). EDTA combines with metal ions on a 1:1 ratio regardless of the charge on the metal ion.

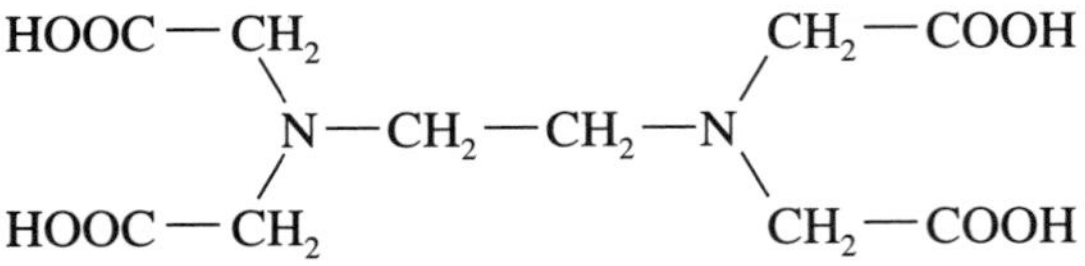

```
HOOC—CH2                    CH2—COOH
       \                   /
        N—CH2—CH2—N                   EDTA
       /                   \
HOOC—CH2                    CH2—COOH
```

The volumetric calcium and total hardness procedures in *Standard Methods* (1992) use EDTA as the titrant. The indicators used in these titrations are metallochromic indicators which are themselves chelating agents that form a colored complex with a metal ion. EDTA is a stronger complexing agent than the indicator. The first additions of EDTA will sequester metal ions that are not complexed with the indicator. As the concentration of uncomplexed metals decreases with further addition of EDTA, the indicator begins to release metal ions. When EDTA has sequestered the metal ions from the indicator, the indicator changes color, signaling the endpoint of the titration.

5.5 REDOX TITRATIONS AND POTENTIOMETRIC ANALYSES

The principles for a redox titration are similar to an acid–base titration except that the potential of the solution is changing with the addition of titrant, which is an oxidant or reductant. The solution potential changes after the addition of titrant and reaction of it with the species in solution. A plot of system potential versus amount of titrant added will produce curves similar to those in Figs. 3.1–3.3. There will be a large change in system potential around the equivalence point.

■ Example 5.2 Potential Variation During Titration and Equivalence Potential

Consider the titration of Ce^{4+} with Fe^{2+}. The initial concentrations in a volume of 50 mL are $[Ce^{4+}] = 2.70 \times 10^{-3}$ *M* and $[Ce^{3+}] = 1.50 \times 10^{-4}$ *M*. The solution is being titrated with 0.020 *N* Fe^{2+}. The half reactions are

$$Ce^{4+} + e^- = Ce^{3+} \qquad E^\circ = 1.44 \qquad \text{(i)}$$

$$Fe^{3+} + e^- = Fe^{2+} \qquad E^\circ = 0.771 \qquad \text{(ii)}$$

The Nernst equations for these reactions are, respectively,

$$E_{Ce} = E^{o}_{Ce} - 0.059 \log \frac{[Ce^{3+}]}{[Ce^{4+}]} \quad \text{(iii)}$$

$$E_{Fe} = E^{o}_{Fe} - 0.059 \log \frac{[Fe^{2+}]}{[Fe^{3+}]} \quad \text{(iv)}$$

The overall chemical equation is

$$Fe^{2+} + Ce^{4+} \rightarrow Fe^{3+} + Ce^{3+} \quad \text{(v)}$$

This reaction reaches equilibrium almost immediately upon addition of titrant. As Fe^{2+} is added to a solution containing Ce^{4+}, the system potential decreases. Before the equivalence point is reached, the concentration of Fe^{2+} in solution becomes exceedingly small. However, the concentrations of Ce^{4+} and Ce^{3+} are significant. If the initial amount of Ce^{4+} in solution is known, the system potential can be determined using Eqs. (iii) and (v). After the equivalence point is passed, the concentration of Fe^{2+} becomes significant and Eqs. (iv) and (v) can be used to determine the system potential.

For instance, after 5.0 mL of titrant has been added (do not forget the total volume of the solution):

$$[Ce^{4+}] = 6.36 \times 10^{-4}\,M \qquad [Fe^{3+}] = 1.82 \times 10^{-3}\,M$$
$$[Ce^{3+}] = (1.50 \times 10^{-4})(0.050/0.055) + 1.82 \times 10^{-3} = 1.96 \times 10^{-3}\,M$$

From Eq. (iii): $E_{Ce} = 1.44 - 0.059 \log \dfrac{1.96 \times 10^{-3}}{6.36 \times 10^{-4}} = 1.41$ V

Because $E_{Ce} = E_{Fe} = E_{sys}$, Eq. (iv) can be used to find $[Fe^{2+}]$.

$$[Fe^{2+}] = [Fe^{3+}]10^{\frac{E^{o}_{Fe} - E_{sys}}{0.059}} = (1.82 \times 10^{-3})10^{\frac{0.771-1.41}{0.059}} = 2.68 \times 10^{-14}\,M$$

The assumption that $[Fe^{2+}]$ is small is correct.

At equivalence the concentration of Fe^{2+} is not exceedingly small because there is not an excess of Ce^{4+} ions. To avoid iterative solution of the mass balance and potential equations a slightly different approach is taken. The system potential can be determined by adding Eqs. (iii) and (iv) and considering the mass relations that exist at the equivalence point.

$$2E_{eq} = E^{o}_{Ce} + E^{o}_{Fe} - 0.059 \log \frac{[Ce^{3+}][Fe^{2+}]}{[Ce^{4+}][Fe^{3+}]} \quad \text{(vi)}$$

where

E_{eq} is the system potential at the equivalence point

The volume of titrant added at equivalence is determined from

$$N_t V_t = N_{Ce} V_0$$

where

N_t is the normality of the titrant; N_{Ce} is the normality of the solution

V_t and V_0 are volume of titrant and initial volume of the solution, respectively

$$V_t = N_{Ce}V_0/N_t = (2.70 \times 10^{-3}\text{ eq/L})(0.050\text{ L})/(0.020\text{ eq/L}) = 0.006\,75\text{ L} = 6.75\text{ mL}$$

From the stoichiometry of the reaction and at the equivalence point, the amount of Fe^{2+} added must be equal to the amount of Ce^{4+} that was initially present.

$$([Fe^{2+}] + [Fe^{3+}])V_{eq} = [Ce^{4+}]_0 V_0$$

where

V_{eq} is the volume of the solution at equivalence

Also at equivalence, the number of equivalents of Fe^{2+} that are present will be equal to the number of unreacted equivalents of Ce^{4+}.

$$[Fe^{2+}]V_{eq} = [Ce^{4+}]V_{eq} \qquad \text{(vii)}$$

From the overall reaction, the equivalents of Ce^{3+} produced are equal to the equivalents of Fe^{3+} that are present. $[Ce^{3+}]$ is referred to the volume of the solution at equivalence.

$$[Fe^{3+}]V_{eq} = \Delta[Ce^{3+}]V_{eq} \qquad \text{(viii)}$$

The total concentration of Ce^{3+} at the equivalence point depends on the initial concentration and the amount of Ce^{3+} produced.

$$[Ce^{3+}]V_{eq} = [Ce^{3+}]_0 V_0 + \Delta[Ce^{3+}]V_{eq} \qquad \text{(ix)}$$

Substituting the above relations into Eq. (vi) and reducing the equation,

$$\begin{aligned} 2E_{eq} &= E^{\circ}_{Ce} + E^{\circ}_{Fe} - 0.059 \log \frac{[Ce^{3+}][Fe^{2+}]}{[Ce^{4+}][Fe^{3+}]} \\ &= E^{\circ}_{Ce} + E^{\circ}_{Fe} - 0.059 \log \left(\frac{\dfrac{[Ce^{3+}]_0 V_0 + [Fe^{3+}]V_{eq}}{V_{eq}}}{[Fe^{3+}]} \right) \end{aligned} \qquad \text{(x)}$$

It is also the case that $[Ce^{4+}]$ and $[Fe^{2+}]$ are very small and $[Ce^{3+}]$ and $[Fe^{3+}]$ comprise most of the Ce and Fe species present at the equivalence point. Therefore it will be assumed that

$$[Fe^{3+}] = (0.02 \text{ mole/L})(0.006\ 75 \text{ L})/(0.056\ 75 \text{ L}) = 0.002\ 379\ M$$

Substituting the values into the above equation:

$$2E_{eq} = 1.44 + 0.771 - 0.059 \log \left[\frac{\dfrac{(1.50 \times 10^{-4})(0.050) + (0.002\ 379)(0.056\ 75)}{0.056\ 75}}{0.002\ 379} \right]$$

$$= 2.21$$

$$E_{eq} = 2.21/2 = 1.11 \text{ V}$$

Check to see that the assumptions are correct. Use Eqs. (iii) and (iv) to verify that $[Ce^{4+}]$ and $[Fe^{2+}]$ are negligible compared to their counterparts.

5.5.1 Indicators for Potentiometric Analysis

The potential of the solution being titrated can be monitored directly or redox indicators may be used to indicate the endpoint of a redox titration. A redox indicator will exhibit a color change in the system potential inflection range for the titration.

A general redox equation for the indicator is

$$\underset{\substack{\text{oxidized}\\\text{color}}}{\text{In}} + ne^- \rightleftharpoons \underset{\substack{\text{reduced}\\\text{color}}}{\text{In}^{n-}} \tag{5.5}$$

with a corresponding Nernst equation of

$$E_{\text{In}} = E^{\circ}_{\text{In}} - \frac{0.059}{n} \log \frac{[\text{In}^{n-}]}{[\text{In}]} \tag{5.6}$$

As in the case of an acid–base indicator, typically a ratio of 1 : 10 for the oxidized and reduced species is required to produce a visible color change. Following the same approach for an acid–base indicator (Section 5.3), the E_{In} range can be determined to be

$$E_{\text{In}}\ \text{range} = E^{\circ}_{\text{In}} \pm \frac{0.059}{n} \tag{5.7}$$

Bishop (1972) and Ottaway (1972) give thorough discussions of redox indicators.

5.5.2 Specific Ion Electrodes

A convenient analytical tool is an electrode that is sensitive to one ion or a group of ions. The electrode is a specific ion electrode that relies on the principles of electrochemistry. The most common application is the pH electrode. Essentially the electrode contains a membrane that is sensitive to the substance of interest and transmits the potential of this substance in solution to an external circuit. An electrode contains a solution of a compound(s) that reacts only with the ion or species of interest. A significant amount of research has gone into the development of these compounds to make them specific to only one substance in the presence of many possible interfering agents.

The pH electrode (Fig. 5.1) will be examined. The pH electrode's membrane is made of glass with a special composition that is sensitive to H^+ ions only. The body of the electrode may also be made of glass of a different composition to provide rigidity; more commonly the electrode body is made of plastic. The electrode contains an HCl solution that has H^+ ($[H^+]_1$) at a constant activity. It achieves electron transfer

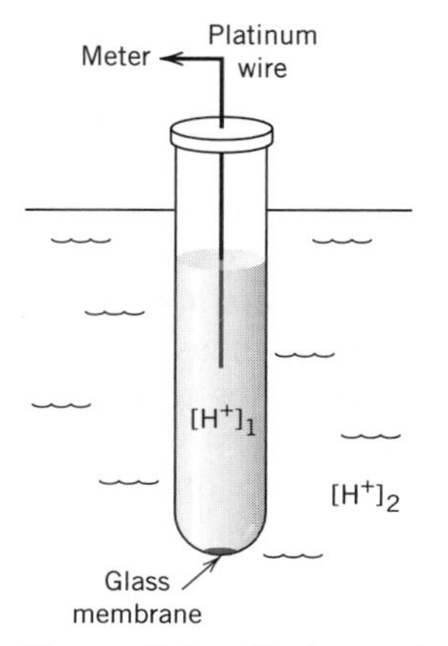

Figure 5.1 pH electrode.

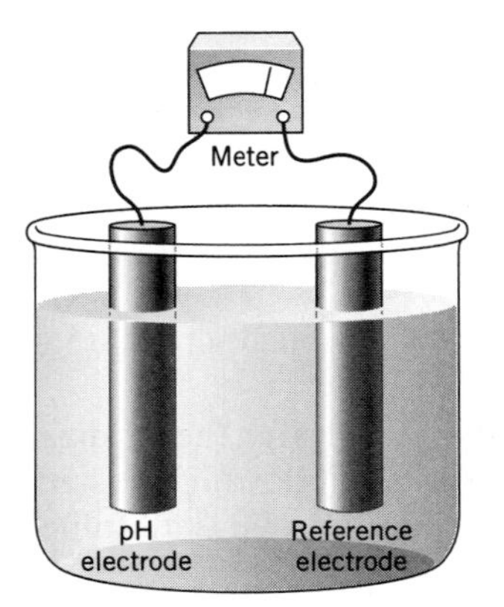

Figure 5.2 pH meter and electrodes.

through the external circuit by means of a wire (usually platinum), which is immersed in the electrode solution and connected to a meter.

The glass membrane does not allow H^+ ions to move from the solution to the electrode or vice versa and the electrode is sealed to prevent the evaporation of water from the electrode solution (which would change the activity of $[H^+]_1$). The concentration of H^+ ions in solution ($[H^+]_2$) sets up a potential difference across the membrane. The Nernst equation for potential for this electrode can be written as

$$E_{pH} = E^{\circ}_{pH} - 0.059 \log \frac{[H^+]_1}{[H^+]_2} \tag{5.8}$$

where

E°_{pH} is the standard potential of the pH electrode

To complete the circuit through the meter and the solution, a reference electrode is provided (Fig. 5.2). The reference electrode makes electrical contact with the solution and has a nonvarying potential, E_{ref}. The meter measures the difference between E_{pH} and E_{ref}. Considering the reference electrode as the anode,

$$E_{cell} = E_{pH} - E_{ref} = E^{\circ}_{pH} - E_{ref} - 0.059 \log \frac{[H^+]_1}{[H^+]_2} \tag{5.9}$$

The only variable in the equation is the concentration of H^+ ions in solution. Because slight changes in the response characteristics of either electrode can occur over a period of time, the meter is calibrated using a solution of known pH. The voltage difference sensed by the meter is output in terms of volts or a pH reading.

The above concept is readily applied to other electrodes that contain solutions that are specific to one substance only. The electrode is calibrated by measuring the potentials of a series of known solutions and plotting the potentials versus concentration on semilog graph paper. A straight line will be obtained according to an equation similar to Eq. (5.9). The slope of the line will be influenced by the number of electrons involved in the half-reactions as shown in the following general equation.

$$E_{cell} = E_{el} - E_{ref} = E^{\circ}_{el} - E_{ref} - \frac{0.059}{n} \log \frac{[X]_1}{[X]_2}$$

where

E_{el}, E^{o}_{el} are the electrode and standard electrode potentials, respectively
X is the target species

5.6 COLORIMETRIC ANALYSES

Colorimetric analysis depends on the formation of a colored species by the substance in question. Many different substances can be analyzed by this method. Principles of light transmittance need to be understood to relate light intensity to the amount of color-producing substance.

5.6.1 The Beer–Lambert Laws for Light Transmittance

Two basic laws govern the phenomenon of light transmittance.

Lambert's Law

The intensity of light is reduced by absorption by the medium. Each layer of medium reduces the passage of light proportionately, which results in a first-order law.

$$T = \frac{I}{I_0} = 10^{-kl} \tag{5.10}$$

where

T is transmittance
I is intensity of light at any distance from the surface
I_0 is intensity of light at the surface
k is absorptivity
l is length of absorbing medium

Beer's Law

The intensity of light is also reduced by the absorbing species in the medium. Similar to Lambert's law, each layer of absorbing species reduces the passage of light proportionately.

$$T = \frac{I}{I_0} = 10^{-k'C} \tag{5.11}$$

where

C is concentration
k' is molar absorptivity

Beer–Lambert Law

The Beer and Lambert laws may be combined to give

$$T = \frac{I}{I_0} = 10^{-k''lC} \tag{5.12}$$

where

k'' is an absorption coefficient

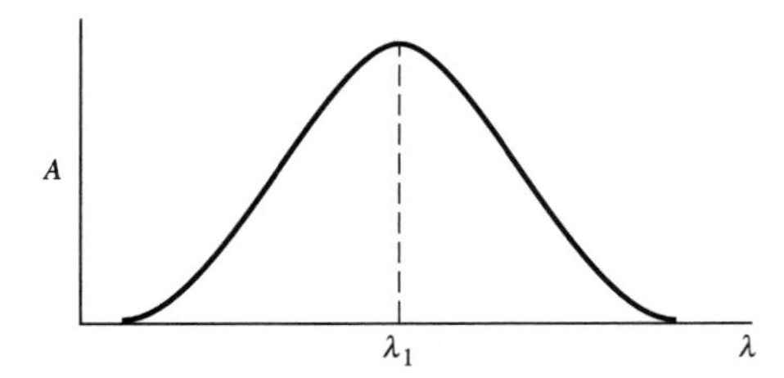

Figure 5.3 Absorbance as a function of wavelength.

The converse of transmittance is absorbance, which is defined in the following manner.

$$A = \log \frac{I_0}{I} = k''lC = -\log T \tag{5.13}$$

The incident light on the sample contained in a cell is difficult to measure accurately because of reflection of light at the glass interfaces and scattering of light in the solvent. Therefore all comparisons are made against a blank prepared in exactly the same manner as the standard solutions and unknowns.

The Beer–Lambert law is subject to some limitations that cause deviation from the linear relation between absorbance, A, shown in Section 5.6.2, and concentration. These deviations are generally not present in solutions in which the concentration of colored species is less than 0.01 M. The Beer–Lambert law is also applicable to the attenuation of light in a natural water.

Because of its molecular structure, etc., each species absorbs light best at one wavelength (λ_1 in Fig. 5.3). The absorption curve is determined by measuring the absorption at different wavelengths.

Colorimetric data are interpreted in two ways:

a. Color Comparison Tubes. A series of color comparison tubes is made from standards of known concentration. These knowns are treated with the procedure to render the target species into the colored compound. The unknowns are treated in the same manner and their color is compared visually against the color comparison standards.

b. Spectrophotometer. A spectrophotometer generates monochromatic light that is directed through the cell containing the sample to a photoelectric cell, which measures the intensity of light (Fig. 5.4). A series of standards at known concentrations is tested according to the procedure and their transmittances are measured

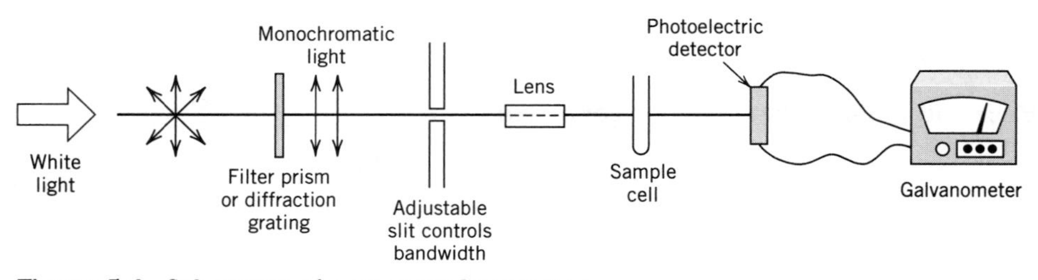

Figure 5.4 Schematic of a spectrophotometer.

with the spectrophotometer. These data are plotted on semilog paper against concentration. The transmittance of an unknown treated according to the procedure is measured and its concentration is determined from this calibration curve.

Atomic absorption spectroscopy is a more advanced technique for metal analysis related to colorimetric analysis. The method requires very small sample volumes and can detect very low concentrations of a metal. Light generated by a metal in an excited state is directed at a sample in an excited state. The amount of light absorbed is proportional to the element's concentration in the sample.

5.6.2 Use of a Blank and Other Corrective Procedures in Colorimetric Analysis

The basic relations describing absorbance and transmittance have been derived for the ideal situation. In laboratory analysis, however, matters become more complicated because of impurities in the reagents and variation in equipment function. One of the most common techniques employed in overcoming these variations is to incorporate a blank into the procedure and make all measurements relative to it. In the case of light transmission, the total absorbance, A, is due to the concentration of the species of interest and other species not of interest, which are interferences. Furthermore, reflections, r, are caused by the surfaces of the vessel and by suspended particles. These are effectively absorbances because the photodetector does not see the reflected light.

The total absorbance is

$$A = k_i l C_i + k_2 l C_2 + \cdot \cdot \cdot + k_n l C_n + r \tag{5.14}$$

where

C_i is concentration of the ith absorbing species
r is total reflected light (expressed as absorbance)

Actually,

$$r = r_v + r_s \tag{5.15}$$

where

r_v is reflection from the vessel surfaces
r_s is reflection from suspended matter

Reflection by suspended matter (r_s) is reduced to a minimum (insignificant) value by filtering the samples. The sample and blank containers are all made of the same quality glass so that r_v is nearly the same for all of them.

The reagent used may contain slight amounts of interfering agents. Therefore, the same amounts of reagent to be added to the sample are added to distilled water in the blank. For either a blank or sample, the modified form of Eq. (5.14) is

$$A = k_0 l C_0 + k_i l C_i + r_v \tag{5.16}$$

In this equation, C_0, is the concentration of the species of interest; $k_i l C_i$ is the absorbance of all interfering agents lumped together and expressed as an effective concentration C_i; and, r_v is as defined previously. Note that the blank and samples contain approximately the same amount of interfering agents from the reagents but the samples may contain additional interference from other species present in them. These latter interferences may in some cases be chemically removed but commonly they are not significant. The influence of interferences is minimized by using a procedure that is specific only to the ion of interest.

In the blank, $C_0 = 0$, and its absorbance, A_b, is

$$A_b = \log\frac{I_0}{I_b} = k_i l C_i + r_v \tag{5.17}$$

where

I_0 is the intensity of the light source

I_b is the intensity of light after passing through the blank

For a sample,

$$A_s = \log\frac{I_0}{I_s} = k_0 l C_0 + k_i l C_i + r_v \tag{5.18}$$

where

subscript s refers to sample

From Eq. (5.17),

$$I_b = I_0 e^{-2.3(k_i l C_i + r_v)} \tag{5.19}$$

and from Eq. (5.18),

$$I_s = I_0 e^{-2.3(k_0 l C_0 + k_i l C_i + r_v)} \tag{5.20}$$

Combining these equations results in,

$$\frac{I_b}{I_s} = e^{2.3 k_0 l C_0} \tag{5.21}$$

Equation (5.21) demonstrates that making all measurements relative to the blank eliminates absorbance caused by reagent impurities and vessel reflections. Other errors such as voltage fluctuations, differences in glass composition, and other interfering dissolved species or suspended particles in the samples are not eliminated.

5.7 PHYSICAL ANALYSES

The characterization of solids is one of the most common assessments of water quality. Turbidity is associated with suspended solids concentrations.

5.7.1 Solids

Solids in a water fall into one of the following categories.

1. Dissolved
2. Colloidal
3. Suspended

Dissolved solids are truly in solution and pass through a filter. The solution consisting of the dissolved components and water is homogeneous, forming a single phase. Colloidal solids are uniformly dispersed in solution but they form a solid phase that is distinct from the water phase. Colloidal solutions are termed sols.

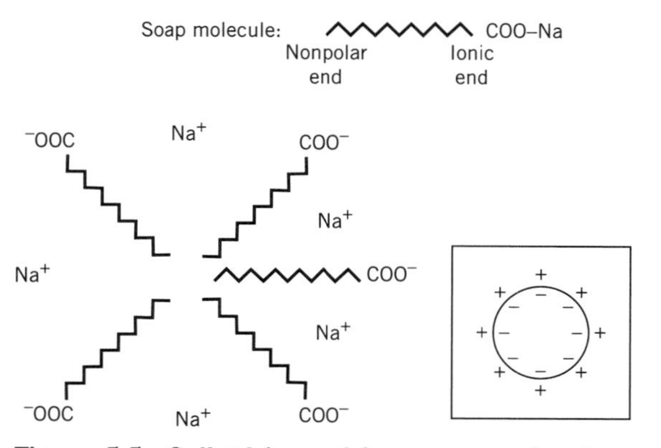

Figure 5.5 Colloid formed from soap molecules.

Colloidal particles are charged entities that derive their stability primarily from their charge characteristics. An example is the soap molecule shown in Fig. 5.5. Soap is the sodium salt of a fatty acid derived from animal fat. One end of the molecule is ionic and the other end of the hydrocarbon chain is nonpolar. In a water solution the soap molecules dissociate. The nonpolar ends of the molecule group together with the polar ends extending into the polar water solution. A negatively charged particle surrounded by positively charged ions results (see the insert in Fig. 5.5). This type of colloid is known as a micelle.

Suspended solids are also a separate phase from the solution. Some suspended solids are classified as settleable solids. Settleable solids are determined by placing a sample in a cylinder and measuring the amount of solids that have settled after a set amount of time. The time is usually set at 30 min or 1 h. The size of solids increases going from dissolved solids to suspended solids (Fig. 5.6).

Water is removed from a sample by subjecting the sample to a temperature near 103°C, which is slightly above the boiling point of water and low enough to prevent organic compounds from burning. Volatile solids are primarily organic solids that will burn in the presence of oxygen. High temperatures in the range of 550–600°C will cause complete combustion of the organic matter but some volatile organics may be lost. Some inorganic solids may also be lost at these temperatures, e.g.,

$$Ca(HCO_3)_2 \xrightarrow{\Delta} CaO + 2CO_2\,(g) + H_2O\,(g)$$

$$(NH_4)_2CO_3 \xrightarrow{\Delta} 2NH_3\,(g) + CO_2\,(g) + H_2O(g)$$

The amount of inorganic solids lost through these reactions is not normally significant. The solids that remain after combustion will be inert or fixed inorganic solids.

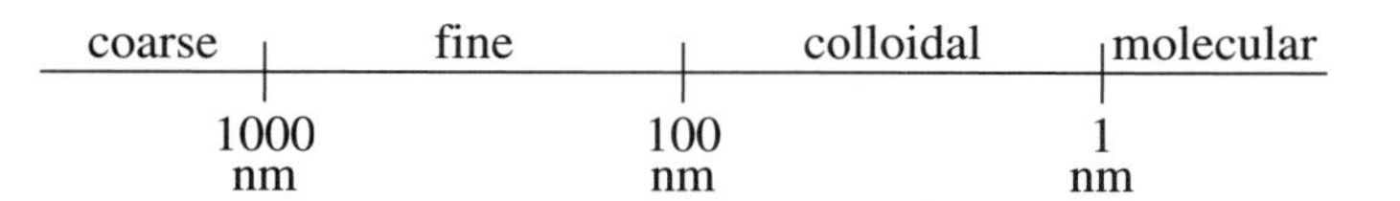

Figure 5.6 Size ranges of solids. Adapted from Sawyer et al. (1994).

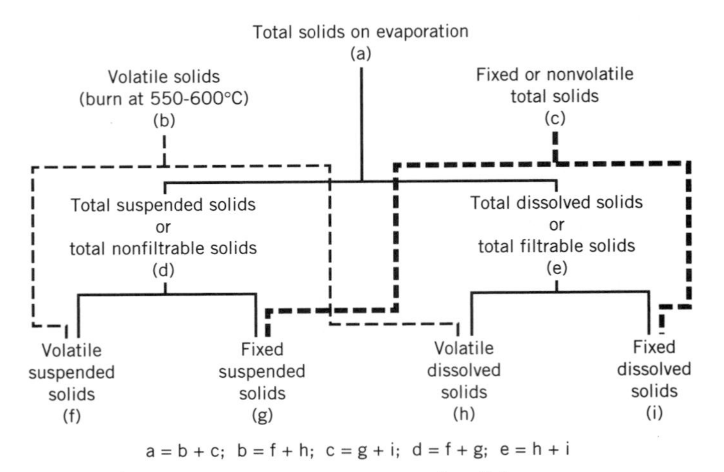

Figure 5.7 Relations among types of solids.

Figure 5.7 characterizes the makeup of solids in a water sample. Earlier editions of *Standard Methods* referred to solids as residue.

5.7.2 Color and Turbidity

Color and turbidity are important aesthetic parameters for drinking waters. They also influence the biotic community in ecosystems.

Turbidity

Turbidity is a result of the scattering and absorption of light by suspended solids. A rough correlation exists between suspended solids concentration and turbidity. Raleigh's law describes scattering of white light by suspended particles.

$$I_s \propto \frac{V^2}{\lambda^4} n \tag{5.22}$$

where
- I_s is the intensity of scattered light
- V is the volume of particles
- n is the number of particles
- λ is the wavelength of light

From Eq. (5.22) it is observed that size and concentration of particles influence the measurement of turbidity. A natural water or wastewater will contain many different sized particles at different concentrations; the relation between suspended solids concentration and turbidity can be highly variable. However, particularly in water treatment operations where suspended solids concentrations are low, turbidity can be a very useful parameter.

Turbidity of a drinking water is always an aesthetic concern. The nature of the solids causing the turbidity may have other health ramifications. Turbidity in natural waters reduces light transmittance and affects the species that may survive in the waters.

Color

Color is imparted to a water by dissolved constituents that absorb white light and emit light at specific wavelengths. Color of a water is also influenced by its turbidity. Humic and fulvic acids impart color to natural waters. Color of a natural water approximates the copper-orange color of a platinum–cobalt solution produced by dissolving specified amounts of potassium chloroplatinate, K_2PtCl_6, and cobaltous chloride, $CoCl_2$, in water. The sample of water is visually compared against standards prepared with these agents or against colored disks in a comparator. The disks have the same color as the platinum–cobalt solution when the disk is viewed over a tube of distilled water. A tube containing the sample is compared to the tube containing distilled water with the disks placed over it.

Industrial wastes may contain significant concentrations of compounds that produce any color in the rainbow. In these cases a series of standards of any compound causing the color may be prepared or a spectrophotometer may be used to assess the amount of color.

5.8 CHROMATOGRAPHIC ANALYSES

Chromatography is a technique that allows the analyst to separate, identify, and quantify components of a solution. It is a powerful technique that allows for determination of minute quantities of a substance. Chromatographic separations involve a stationary phase and a mobile phase. The sample being analyzed is passed by the stationary phase. The stationary phase has a variable attraction for each of the components in the sample; thus the rate of movement of each component by the stationary phase differs. A detector sensitive to the substances of interest will respond as the substances exit from the stationary phase.

Liquid and gas chromatography are the two most common methods. Liquid chromatography uses liquid as a carrier (the mobile phase) and gas is used as the carrier in gas chromatography. The stationary phase is packed inside a tubular column. The packing is porous to allow movement of the mobile phase. The sample is introduced at the head of the column and moved through the column by the mobile phase. The components separate into bands that travel through the column.

The detector response is proportional to the concentration of the substance. There are a wide variety of detectors such as thermal conductivity or infrared detectors that may be used to monitor the effluent from the column. A typical output from a chromatographic separation is shown in Fig. 5.8. The response curve is composed of a baseline and series of peaks. The baseline is characteristic of the carrier. The peaks are due to the presence of one or more components in the exiting carrier. If the proper column and conditions of analysis are used the desired components exit alone at different times. The height of a peak or the area under a peak can be used to calculate the concentration of the substance. Using the area is more accurate.

The rate of movement of a substance through a column depends on its characteristics such as size of molecule and its polarity, packing characteristics, the type and rate of flow of the carrier, and temperature.

It can be a lengthy and tedious task to determine the correct packing, carrier, and its flow rate, and other conditions to obtain good separations for a particular sample. Calibration is performed by injecting a sample with a known concentration of a substance to be analyzed for and noting the time of its exit and the area of its peak.

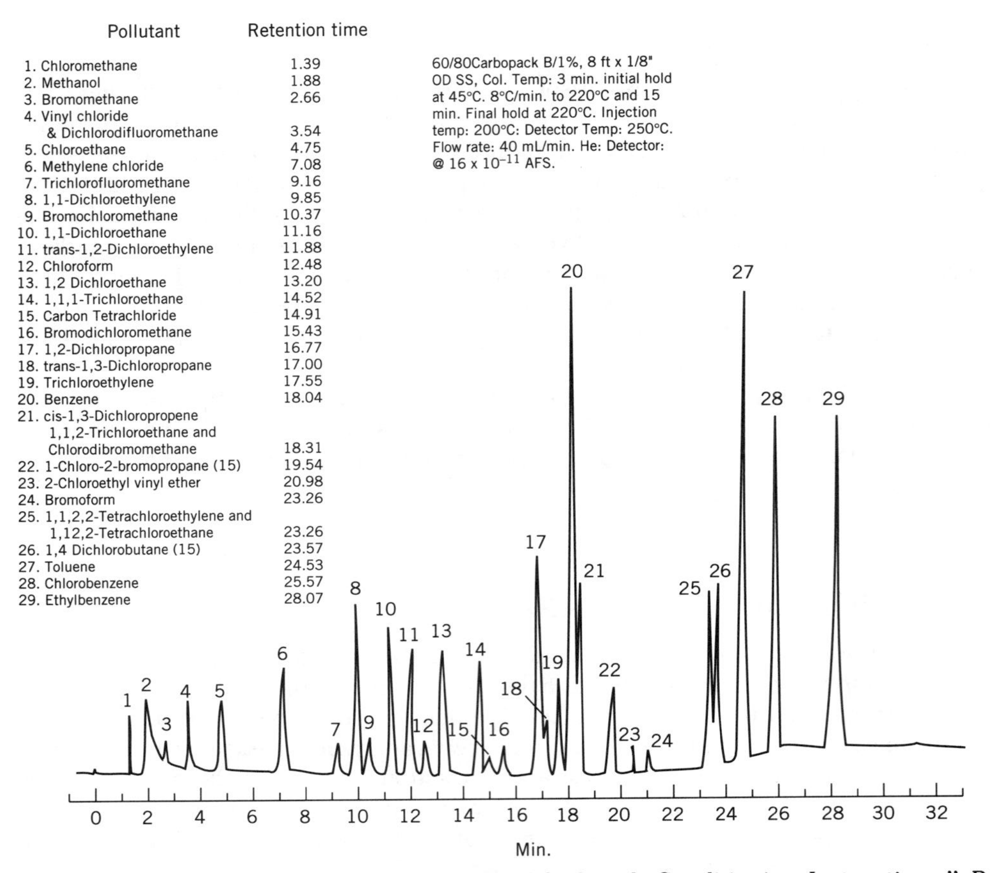

Figure 5.8 A typical chromatogram. From "Carbpack Conditioning Instructions." Reprinted with permission of Supelco, Inc., Bellefonte, PA 16823.

5.9 DETERMINATION OF ORGANIC MATTER

There are a number of measures of organic matter that are of use in various situations. Volatile solids is a crude measure of organic matter. Organic matter is most often assessed in terms of the oxygen required to completely oxidize the organic matter to CO_2, H_2O, and other oxidized species. Besides expressing the organics in terms of a common denominator, this is practical because an important consequence of the presence of organic matter is consumption of oxygen. The dissolved oxygen concentration in waters is vital for the maintenance of higher life forms. In natural bodies of water dissolved oxygen is depleted by the aerobic biological decomposition of organic matter.

The theoretical oxygen demand of a compound is calculated by writing the balanced reaction for the compound with oxygen to produce CO_2, H_2O, and oxidized inorganic components. If the organic compounds and their concentrations are known, the theoretical oxygen demand can be accurately calculated, but this is an impossible task for most natural waters and wastewaters because of the number of substances that are present. However, analytical techniques are evaluated in laboratory situations where the amounts and types of organic matter are known. Known amounts of specific compounds that span various classes of compounds are tested with the analytical procedure to establish the limitations of the procedure.

Another important method of expressing organic matter is in terms of its carbon content. Carbon is the primary constituent of organic matter.

The reactions to determine the amount of oxygen or carbon are based on the average formula weight of all organic compounds considering their respective concentrations in the water. This results in a hybrid organic compound of formula $C_nH_aO_bN_c$ where n is the number of moles of organic carbon, and a, b, and c are, respectively, the number of moles of organically bound hydrogen, oxygen, and nitrogen in a liter of water. The coefficients, n, a, b, and c are usually not whole numbers.

■ Example 5.3 Oxygen Demand and Total Organic Carbon Concentrations

What are the theoretical oxygen demand and organic carbon concentration of a water that contains the following components:

glucose ($C_6H_{12}O_6$), 150 mg/L   benzene (C_6H_6), 15 mg/L

The equations for total oxidation of the two constituents are

$$\underset{180}{C_6H_{12}O_6} + \underset{192}{6O_2} \rightarrow \underset{264}{6CO_2} + \underset{108}{6H_2O}$$

$$\underset{78}{C_6H_6} + \underset{240}{7.5O_2} \rightarrow \underset{264}{6CO_2} + \underset{64}{3H_2O}$$

From the equations it is determined that the theoretical oxygen demand for glucose is (192/180) = 1.07 mg O_2/mg of glucose and the theoretical oxygen demand of benzene is (240/78) = 3.08 mg O_2/mg of benzene.

The total theoretical oxygen demand of the solution is

$$1.07 \times 150 \text{ mg/L} + 3.08 \times 15 \text{ mg/L} = 206.7 \text{ mg } O_2/\text{L}$$

The carbon content of glucose is 72 mg C/180 mg glucose = 0.40 mg C/mg glucose, and the carbon content of benzene is 72 mg C/78 mg benzene = 0.92 mg C/mg benzene.

The total organic carbon concentration of this solution is

$$0.40 \times 150 \text{ mg/L} + 0.92 \times 15 \text{ mg/L} = 73.8 \text{ mg C/L}$$

What is the formula weight of organic matter in this solution?

The concentrations in mg/L of each atom in the organic matter are

$$\text{C:} \quad 150 \times \frac{72}{180} + 15 \times \frac{72}{78} = 73.8 \text{ mg/L}$$

$$\text{H:} \quad 150 \times \frac{12}{180} + 15 \times \frac{6}{78} = 11.2 \text{ mg/L}$$

$$\text{O:} \quad 150 \times \frac{96}{180} = 80.0 \text{ mg/L}$$

Converting the above into molar concentrations:

$$\text{C:} \quad 73.8 \times \frac{1}{12\,000} = 6.15 \times 10^{-3}\ M$$

$$\text{H:} \quad 11.2 \times \frac{1}{1\,000} = 1.12 \times 10^{-2}\ M$$

$$\text{O:} \quad 80.0 \times \frac{1}{16\,000} = 5.00 \times 10^{-3}\ M$$

The formula for the organic matter is $C_{6.15}H_{11.2}O_{5.0}$ at a concentration of 1.00 mM. Obviously this formula does not represent the structure of the organic matter but only the relative amounts of the elements contained in it.

5.9.1 Chemical Oxygen Demand

Chemical oxygen demand (COD) is the amount of oxygen required to stabilize organic matter determined by using a strong oxidant. Theoretically any strong oxidant could be used. Ideally the oxidant should be able to oxidize any organic compound and the oxidant should be inexpensive. The first criterion is the most important. Considering these criteria, the oxidant of choice is dichromate ($Cr_2O_7^{2-}$), which also meets other criteria in Section 5.2.

The analysis is conducted by adding strong acid to the sample along with silver, which is a catalyst, and a known amount of dichromate. The strong acid helps to digest (break down) complex molecules. Mercuric sulfate is also added to complex chlorides that interfere with the test. To accelerate the reaction and achieve the maximum effectiveness of the oxidant, the sample with added reagents is heated in a flask, which is connected to a condensing tube to minimize loss of volatile organics.

General Reaction for COD

In the COD reaction, dichromate ($Cr_2O_7^{2-}$) oxidizes organic matter to end products of CO_2 and H_2O while becoming reduced to Cr^{3+}. The oxidant contains oxygen; also H^+ ions will be consumed in the oxidation (see the half-reaction for $Cr_2O_7^{2-}$). From Table 1.3, the equivalent weight of dichromate is 36.0 g and the equivalent weight of oxygen is 8 g, which are used to calculate the oxygen demand.

The general reaction takes the following form:

$$C_nH_aO_b + cCr_2O_7^{2-} + fH^+ \rightarrow dCr^{3+} + gCO_2 + eH_2O \qquad (5.23)$$

where

c, d, f, g, and e are the stoichiometric coefficients required to balance the reaction

From examination of Eq. (5.23) it is obvious that $g = n$ on the right-hand side (RHS). Using this and making mass balance relations between the left-hand (LHS) and RHS components, the number of unknowns can be further reduced.

	LHS	RHS
C:	n	n
H:	$a + f$	$2e$
O:	$b + 7c$	$2n + e$
Cr:	$2c$	d

From the information in the mass balance table, the equation can be reduced to

$$C_nH_aO_b + cCr_2O_7^{2-} + (2b + 14c - 4n - a)H^+ \rightarrow 2cCr^{3+} + nCO_2 + (b + 7c - 2n)H_2O \quad (5.24)$$

A further reduction in unknowns can be made by considering the charge balance equation.

$$-2c + 2b + 14c - 4n - a = (2)(3)c \qquad \text{or} \qquad 6c = -2b + 4n + a$$

This equation is readily substituted into the coefficients for H^+ and H_2O. The resulting equation is

$$C_nH_aO_b + cCr_2O_7^{2-} + 8cH^+ \rightarrow 2cCr^{3+} + nCO_2 + \frac{a + 8c}{2}H_2O \quad (5.25)$$

The same final result in Eq. (5.24) could have been obtained from an electron balance.

The oxidation numbers on C in $C_nH_aO_b$ and CO_2 are

$$C_nH_aO_b: \quad -\frac{a - 2b}{n}; \quad CO_2: \quad 4$$

The total number of electrons lost by carbon in the oxidation is

$$4n + a - 2b$$

The number of electrons gained by each mole of dichromate is 6. Therefore the total number of electrons gained by dichromate is $6c$. The number of electrons gained is equal to the number of electrons lost. Therefore,

$$6c = 4n + a - 2b$$

This equation can be used to modify the coefficients for H^+ and H_2O in Eq. (5.24) in the same manner as above. Equation (5.25) will ultimately be obtained.

The charge balance equation and electron balance equation are not independent relations; therefore only one of them provides new information.

■ Example 5.4 COD Calculation

A 10-mL sample diluted to 25 mL with distilled water was digested according to the standard procedure. It was determined that 3.12×10^{-4} moles of dichromate (DC) were consumed by the sample. What is the COD of the sample?

Dilution with distilled water does not affect the amount of organic matter present. From the half-reactions (Table 1.3) for DC and oxygen, 1 mole of DC has 6 electron equivalents and 1 mole of O_2 has 4 equivalents. The equivalent weight of oxygen is 32.0 g/4 eq = 8.0 g/eq. The COD of the sample is

$$\text{COD} = \frac{3.12 \times 10^{-4}\ \text{moles DC}}{10\ \text{mL}}\left(\frac{6\ \text{equiv}}{\text{mole DC}}\right)\left(\frac{8\ \text{g O}_2}{\text{equiv}}\right)\left(\frac{1\,000\ \text{mL}}{\text{L}}\right)\left(\frac{1\,000\ \text{mg}}{\text{g}}\right)$$

$$= 1\,498\ \text{mg O}_2/\text{L}$$

COD is normally reported in terms of mass O_2/L, not in terms of moles DC/L or eq/L.

Interferences with the COD Test

Ammonia can interfere to a small extent with the COD determination if significant amounts of chloride are present and high concentrations of dichromate are used (Kim, 1989; *Standard Methods,* 1992). Kim (1989) found that a false COD equivalent to about 4% of the ammonia COD was exerted using a 0.25 *N* dichromate solution when equimolar amounts of ammonium and chloride were present, even when a large excess of mercuric sulfate was present to complex the chloride. When a 0.025 *N* solution of dichromate was used, the ammonia was not oxidized. If nitrogen is contained in significant amounts in the organic matter it will be released as NH_4^+ and subject to minor oxidation if high concentrations of dichromate are used and chloride concentrations are significant.

Hydrogen peroxide (H_2O_2) may be present in industrial wastewaters and interfere with COD determinations by reducing the amount of dichromate consumption. The measured COD must be corrected by subtracting $0.25[H_2O_2]$ (Talinli and Anderson, 1992). $[H_2O_2]$ is in mg/L and determined independently.

Functionally, the results of a COD determination include whatever substances are oxidized by dichromate under the conditions of the test. The results may be greater or less than the true theoretical oxygen demand of all organics present in the sample. Not all organic compounds are subject to oxidation by dichromate under the conditions of this test. Aromatic compounds and pyridine (a common biomolecule) are refractory to this analysis. Also, inorganic reducing agents will result in an inorganic chemical oxygen demand (or false positive COD). Some inorganic reducing agents, Cl^- and NO_2^-, can be removed or prevented from oxidation by dichromate (*Standard Methods,* 1992). Furthermore, the test does not distinguish between biodegradable and nonbiodegradable organics. The test is relatively rapid; normally 2–3 h of digestion suffice to complete the oxidation.

In theory any strong oxidant will suffice. The limitations of the test regarding inorganic reducing agents and biodegradability of the organic matter will apply to any oxidant. The cost and broad-spectrum oxidizing capabilities of dichromate have favored it over other oxidants. The test is the best practical measure of the oxygen demand of a water.

5.9.2 Biochemical Oxygen Demand

Biochemical oxygen demand (BOD) is defined as the amount of oxygen required for the biological decomposition of organic matter under aerobic conditions at a standardized temperature and time of incubation. The amount of oxygen required in various periods of time will depend upon the concentration of organic matter, temperature, concentration of bacteria, nature of the organic matter, and the type of bacteria. See Chapter 18 for a description of an anaerobic "BOD" procedure.

The BOD test has many limitations but it is a well-established procedure that has not changed substantially through the years. The first laboratory procedure for determination of BOD is believed to have been performed in 1870 (Logan and Wagenseller, 1993).

Assuming that the rate of oxidation of organic matter at any instant of time is proportional to the amount of oxidizable matter present, i.e., a first-order reaction,

$$\frac{dL}{dt} \propto -L \qquad \text{or} \qquad \frac{dL}{dt} = -k_1 L \tag{5.26}$$

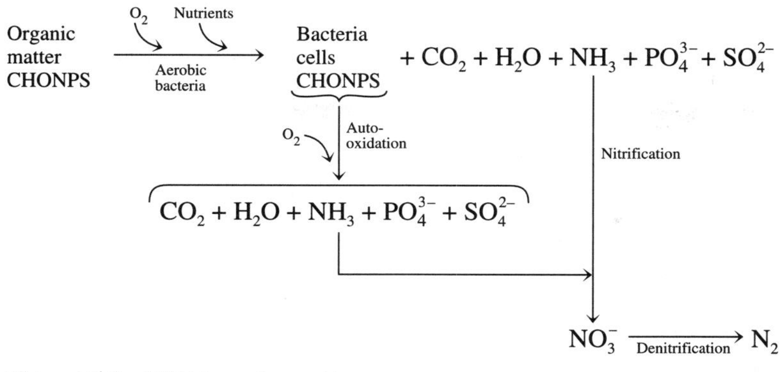

Figure 5.9 BOD transformations.

where

L is the concentration of organic matter expressed as O_2

k_1 is a rate constant

The BOD test is performed in the laboratory by diluting the wastewater sample with water containing sufficient amounts of dissolved oxygen and nutrients, and measuring the depletion in dissolved oxygen after a fixed time of incubation at a fixed temperature. Figure 5.9 illustrates transformations in BOD exertion.

The amount of BOD exerted, from Eq. (5.26), is dependent on time and the amount of degradable organic matter. The ultimate BOD (sometimes written as BOD_u), L_a, is for most practical purposes a constant, although it has been found by some investigators that L_a increases with temperature; i.e., increasing temperature promotes the degradation of more difficult to degrade substances. The ultimate BOD is usually taken to be equal to the COD in the absence of a measurement of the ultimate BOD.

A BOD progression curve is shown in Fig. 5.10. The amount of BOD exerted at any time is equal to the difference between the BOD existing at the initial time and the BOD remaining at any time, L.

$$x = L_a - L = L_a(1 - 10^{-k_1' t}) = L_a(1 - e^{-k_1 t}) \tag{5.27}$$

where

x is BOD exerted at any time, t

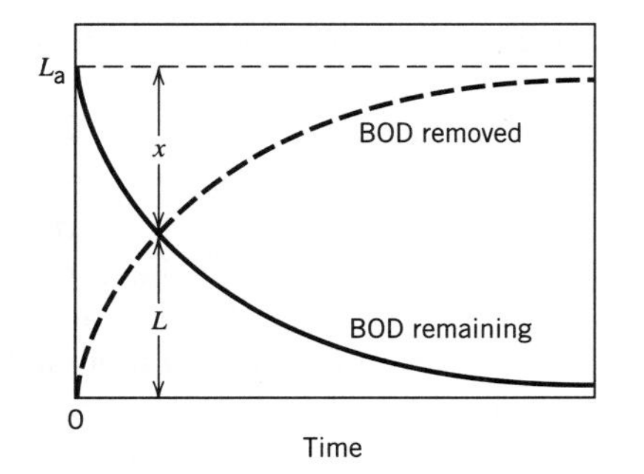

Figure 5.10 BOD progression.

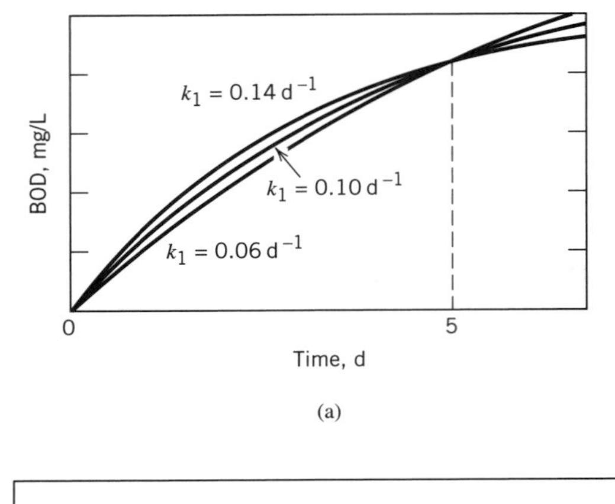

(a)

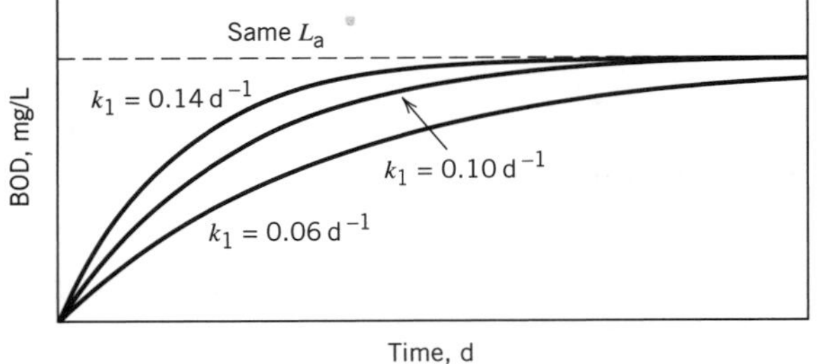

(b)

Figure 5.11 Comparison of ultimate and 5-day BODs for different rate constants.

The BOD determination is normally made for a time period of 5 d. Figures 5.11a and b compare the progression curves and ultimate BOD for samples that have the same 5-d BOD and the same ultimate BOD, respectively.

BOD exertion is a very complex process. It is dependent on the nature of the seed microorganisms and their acclimation to the organic materials in the waste. Environmental conditions such as pH, temperature, or the presence of nutrients and other metabolites all influence the rate of BOD exertion. In the BOD procedure measures are taken to provide controlled environmental conditions to minimize the effects of adverse environmental factors and standardize test conditions. The presence of inhibitory substances can highly influence the results of a BOD determination. Toxic or inhibitory substances can delay or retard the rate of BOD exertion to totally preventing any BOD exertion.

BOD exertion is the sum total of a number of processes qualitatively depicted in Fig. 5.12. Organic matter present in the sample is metabolized primarily by bacteria in the early stages. Some of the organics are oxidized, which causes oxygen uptake, and the remainder are transformed into new bacteria cells. As the supply of external substrate becomes scarce, the major source of organics is the bacteria. Some bacteria die through starvation and other bacteria predate on the living and dead bacteria present. If protozoans are present in the seed, they can also contribute to the removal of organic matter initially present in the sample but bacteria are much more efficient in competing for the substrate. However, protozoans will increase because of predation on the increasing population of bacteria. As the food source for the bacteria diminishes to small amounts the protozoans will eventually become dominant. At this stage

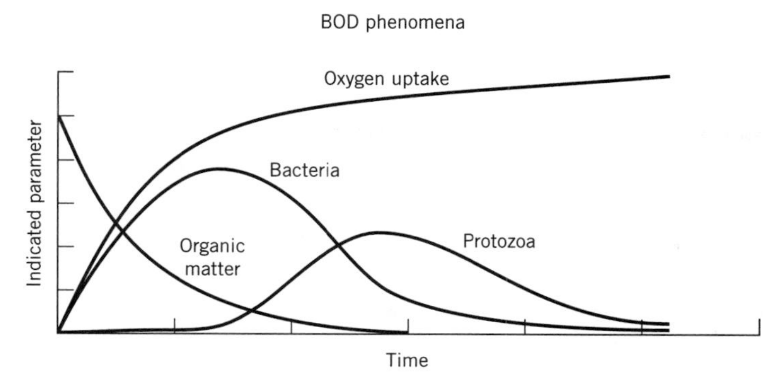

Figure 5.12 Phenomena in a BOD bottle. Adapted from A. F. Gaudy, "Biochemical Oxygen Demand," in *Water Pollution Microbiology,* vol. 1, R. Mitchell, ed., copyright © 1972 by John Wiley & Sons, Inc. Reprinted by permission of John Wiley & Sons, Inc.

bacteria and protozoans continue to feed on living and dead cells. The original organic matter is recycled through a number of organisms, with oxygen being consumed as the organic matter is metabolized for each biomass transformation. There are a succession of microorganisms and overall decrease in the amount of organic matter in the bottle as the cycle progresses. Oxygen will be utilized for metabolism until it is exhausted or the amount of organic matter decreases to negligible amounts and the last aerobic microorganism dies.

The overall rate of BOD exertion or oxygen uptake exhibits two phases. The readily metabolized substrate components in the sample will be removed rather rapidly with a concomitant consumption of oxygen. After organic matter is synthesized into bacteria and protozoan cells the process of organic matter metabolism decreases because the transformed organic matter is not as easily metabolized. The first-order BOD model described above fits a single rate constant over the whole oxygen uptake curve. The rate constant is a weighted average of the rate of oxygen uptake throughout the cycle. Different mixtures of organic components and seed populations can shift the duration and maximum values of individual microbial groups. However, it is generally true that oxygen uptake in the initial stages will be significantly higher than in the latter stages not only because of the higher amounts of organic matter that are present but also because the initial degradability of the organic matter is higher than after it has been transformed into microbial cells.

The procedure for laboratory determination of BOD is described in the "Laboratory Determination of BOD" section. It usually involves dilution of the wastewater sample, which may reduce the concentrations of toxic or inhibitory substances below the threshold at which they significantly affect microbial activity. This can cause complications in a BOD progression exercise where different dilutions are used.

The removal of degradable organics generally does not proceed uniformly. Readily metabolized substrates are rapidly removed, followed by a slower degradation of the biomass with different species becoming dominant as the nature of organic matter changes. Biomass itself degrades slowly because cells are resilient to attack to maintain their integrity. Because of the complexities involved, BOD determinations are sometimes difficult to reproduce. BOD determinations should be interpreted with care. Comparisons of BOD among different wastes are difficult to make because of these reasons.

A BOD determination for a period of t days (typically BODs are determined for a 5-d incubation period) is not meaningful as a measure of the total biodegradable strength of a sample unless the rate constant is known. A BOD progression analysis as discussed in Section 5.9.4 must be performed to determine k_1. A commonly assumed value for k_1 is 0.1 d^{-1} (base 10 basis) at 20°C, which results in a BOD_5: ultimate BOD ratio ($x_5 : L_a$) of 2:3.

Effects of Temperature on BOD Exertion

Temperature exerts an influence on the rate of oxidation of organic matter. The rate constant is influenced by temperature in the following manner:

$$k_T = k_{20}\theta^{(T-20)} \tag{5.28}$$

where
 k_T is the rate constant at a temperature T (°C)
 θ is a constant

The value of θ is commonly taken to be 1.047, although its value depends on the waste. Laboratory studies must be performed to determine the value of θ that applies.

■ Example 5.5 BOD Exertion

An experimenter set a sample for a BOD determination in an incubator at 20°C. The sample had an ultimate BOD of 330 mg/L and the rate constant for this waste was 0.13 d^{-1} (base 10) at 20°C. On the beginning of day 3 another person adjusted the temperature of the incubator to 25°C for another analyses. The incubator temperature changed to the new temperature within 20 minutes. What were the true 5-d BOD of the sample and the 5-d BOD that was determined? Assume $\theta = 1.047$.

The true x_5 of the sample was

$$x_5 = L_a(1 - 10^{-5k}) = (330\ \text{mg/L})[1 - 10^{-(5\ \text{d})(0.13\ \text{d}^{-1})}] = 0.776(330\ \text{mg/L}) = 256\ \text{mg/L}$$

At $T = 25$°C,

$$k_{25} = k_{20}\theta^{(25-20)} = (0.13\ \text{d}^{-1})(1.047)^5 = (0.13\ \text{d}^{-1})(1.26) = 0.16\ \text{d}^{-1}$$

After 2 d the concentration of organic matter in the sample is

$$L_2 = L_a 10^{-2k_{20}} = (330\ \text{mg/L})[10^{-(2)(0.13)}] = 181\ \text{mg/L}$$

The amount of BOD exerted was 330 − 181 = 149 mg/L

The time for the temperature change is small compared to the time scale of the determination and can be ignored. The BOD exerted over days 3–5 is

$$x_{3\text{-}5} = L_2(1 - 10^{-3k_{25}}) = (181\ \text{mg/L})[1 - 10^{-(3\ \text{d})(0.16\ \text{d}^{-1})}] = 0.669(181\ \text{mg/L}) = 121\ \text{mg/L}$$

The 5-d BOD determined was

$$x_2 + x_{3\text{-}5} = 149 + 121 = 270\ \text{mg/L}$$

Laboratory Determination of BOD

The BOD of a sample is measured by adding an aliquot of sample to a bottle and filling the bottle with dilution water. Dilution water is distilled water that has been supplemented with buffering agents, nutrients, and trace metals and aerated for a significant period of time to raise the concentration of dissolved oxygen (DO) to saturation. The chemicals provide a controlled optimum environment to an extent. The bottle is manufactured to contain a fixed volume (normally 300 mL). Because of the variability of the test, analyses are performed in duplicate.

The initial dissolved oxygen concentration (DO_i) is measured and the sample is incubated in the dark at a constant temperature, usually 20°C, for a fixed period of time. At the end of the time period the final dissolved oxygen (DO_f) is measured. BOD is calculated from the equation

$$BOD_t = (DO_i - DO_f)\frac{V_b}{V_s} = \Delta DO(DF) \tag{5.29}$$

where

BOD_t is the BOD after t days
V_b is the volume of the bottle
V_s is the volume of sample added to the bottle
DF is a dilution factor $= V_b/V_s$

An acclimated seed may need to be added to the bottle. If the sample is from a lake, stream, or sewage treatment plant effluent, there is no need to add a seed because the water will contain a significant number of acclimatized aerobic bacteria. However, some industrial wastes may need to have a seed developed by adding some soil to a portion of waste and aerating it. Soil is a medium that supports a wide variety of microorganisms and after aeration with the waste a small volume of this water will contain a sufficient number of microorganisms adapted to the waste. Alternately, sewage treatment plant effluent may be used to seed a sample. The volume of seed added to the sample must be measured and a determination of the seed BOD must be made. The calculation for BOD_t is

$$BOD_t = \left[(DO_i - DO_f) - (DO_{isc} - DO_{fsc})\frac{V_{ss}}{V_{sc}}\right](DF) \tag{5.30}$$

where

DO_{isc} is the initial DO of the seed control
DO_{fsc} is the final DO of the seed control
V_{ss} is the volume of seed added to the sample
V_{sc} is the volume of seed in the seed control

It is imperative to have an acclimated seed present in the sample. This will ensure that the maximum rate of BOD exertion occurs.

BOD determinations are performed in the dark to prevent the photosynthetic production of oxygen from algae. Most natural water samples will contain algae in significant concentrations. The algae themselves are part of the organic matter in the sample but their effects on the BOD determination will be variable. Algae obviously do not die immediately when placed in a dark environment; in fact they remain viable for some time even under the severe conditions of anoxic or anaerobic conditions (Golueke et al., 1957). In a dark environment algae respire and the respiration rate diminishes with time (Oswald and Ramani, 1976). These researchers found that algae respiration rates were 0.15–0.25 mg O_2/mg algae/d immediately after the cells were placed in the dark. Rates declined to 0.02–0.05 mg O_2/mg algae/d after 24 h of darkness. They found that over a 5-d period the cumulative amount of oxygen consumed per milligram of algae was around 0.25 mg. Therefore algae respiration will contribute to the observed BOD.

The different environment in the BOD bottle compared to the environment from which the sample was taken may contribute to accelerating the dieoff of algae. In any case, in BOD determinations made over long periods of time, dieoff of algae cells and degradation of this organic matter will contribute to the BOD determination.

■ Example 5.6 BOD Determination Example No. 1

This type of computation will be used on those samples that contain raw or settled sewage, other types of wastes or stream, lake, or similar samples where there is no requirement for seed.

Given:

Bottle capacity	300. mL
mL of waste added to bottle	5.00 mL
Initial dissolved oxygen in bottle containing waste and dilution water	7.80 mg/L
Dissolved oxygen in bottle containing waste and dilution water after 5 days incubation	3.40 mg/L

Computations:

$$\text{Depletion of oxygen resulting from waste} = (7.80 - 3.40)\ \text{mg/L} = 4.40\ \text{mg/L}$$

(Note: 4.40 mg/L = 56% depletion)

$$\text{Dilution factor} = 5.00\ \text{mL in } 300.\ \text{mL} = \frac{300}{5.00} = 60.0$$

$$\text{BOD} = \text{dilution factor} \times \text{depletion} = 60.0 \times 4.40\ \text{mg/L} = 264.\ \text{mg/L}$$

■ Example 5.7 BOD Determination Example No. 2

This type of computation will be necessary for waste samples that require the addition of seed.

Given:

Bottle capacity	300. mL

<u>Seed control</u>

Milliliters of seed added to bottle	5.00 mL
Initial dissolved oxygen in bottle containing seed and dilution water	8.90 mg/L
Dissolved oxygen in bottle containing seed and dilution water after 5-d incubation	5.90 mg/L

<u>Sample</u>

Milliliters of seed added to bottle	1.00 mL
Milliliters of sample added to bottle	40.0 mL
Initial dissolved oxygen in bottle containing seed, sample, and dilution water	8.80 mg/L

Dissolved oxygen in bottle containing seed, sample, and dilution water after 5 d incubation 3.75 mg/L

Computations:

Depletion of dissolved oxygen in the bottle containing seed, sample, and dilution water:

$$= (8.80 - 3.75) \text{ mg/L}$$
$$= 5.05 \text{ mg/L}$$

Depletion of dissolved oxygen in bottle containing seed and dilution water (seed control):

$$= (8.90 - 5.90) \text{ mg/L}$$
$$= 4.00 \text{ mg/L}$$

$$\text{Ratio of seed in sample to seed in seed control} = \frac{1.00}{5.00} = 0.200$$

$$\text{Dilution factor for sample} = \frac{300.}{40.0} = 7.50$$

$$\text{BOD of sample} = 7.50 \times [5.05 - 4.00(0.200)]\text{mg/L} = 31.9 \text{ mg/L}$$

Although it is not necessary to calculate the BOD of the seed control in order to find the BOD of the sample, the BOD of the seed control is:

$$\text{Dilution factor for seed control} = \frac{300.}{5.00} = 60.0$$

$$\text{BOD of seed control} = 60.0 \times 4.00 = 240. \text{ mg/L}$$

Note: The dissolved oxygen uptake of the sample is calculated by correcting the total dissolved oxygen uptake in the bottle for oxygen uptake of the seed.

BOD exertion can be measured continuously with commercially available instruments or a Warburg apparatus. The Warburg apparatus is a simple device in which a manometer is connected to a vessel to which the sample and seed culture (if required) are added. The flask is agitated in a controlled temperature water bath. A concentrated potassium hydroxide solution is placed within a separate container within the reaction vessel to absorb CO_2. The gas pressure change in the reaction vessel will be solely a result of the consumption of oxygen and it is readily measured with the manometer.

Commercial respirometers are similar in some respects to a Warburg apparatus but they offer better performance. In these devices a large reaction vessel (usually 1 L) is placed in a controlled temperature water bath and mixing is applied to the sample. A concentrated potassium hydroxide solution in a reservoir above the sample is used to absorb CO_2. Oxygen is generated by an electrolysis reaction in response to pressure decrease in the sample vessel. The larger sample volume and measurement of current consumed provide a more accurate determination of BOD. The electronic output is convenient for monitoring oxygen consumption. Mixing in a Warburg apparatus or commercial apparatus is necessary to ensure efficient mass transfer of oxygen from the gas to the liquid phase and mixing is also beneficial for reaction conditions in the sample. Mixing is usually not provided in a standard BOD determination.

A novel approach to BOD analysis has been presented by Logan and Wagenseller (1993). The procedure is termed "headspace" BOD analysis. An undiluted sample is placed in a test tube or small bottle with sufficient volume remaining in the container for air. The initial DO of the sample is measured and the container is sealed immediately after addition of the sample. Samples are incubated at the desired temperature and mixed (e.g., on a shaker table) to maintain equilibrium between the liquid and gas phases in the container. At the end of the incubation period the DO of the sample is measured. Knowing the volumes of sample and headspace in the container and atmospheric pressure when the samples were sealed, the BOD may be calculated by utilizing Henry's law. The production of CO_2 and other gases during BOD exertion does not interfere with the results because only the partial pressure of oxygen is related to its concentration in the liquid.

Excellent comparisons were achieved between this procedure and the standard procedure. The procedure is simple and various ranges of BOD can be analyzed by varying the volume of sample compared to the total volume of the container. Dilution errors are minimized in this procedure and the highest substrate concentrations attainable, which promote the most rapid BOD exertion, exist in every situation. Seed and nutrients must be added if the waste is deficient. Toxicity would have to be assessed by serial dilution of the sample.

Significant Figures

In the computation of BOD values above, the results were 264. and 31.9 mg/L for Examples 5.6 and 5.7, respectively. These computations indicate an accuracy greater than that which exists, regardless of the accuracy of volumetric and dissolved oxygen measurements. It will be found, even when using the same dilutions, that there will be differences in BOD values among duplicate and triplicate samples as great as 50 mg/L, and quite frequently as great as 20 mg/L for samples containing raw and settled sewage. Likewise, on samples containing wastes that have received biological treatment, it will be found that there are frequently differences as great as 10 mg/L. Reasons for this variability have been commented on earlier.

Obviously an accuracy much greater than actually exists is indicated, if the above two example values are reported as 264. mg/L and 31.9 mg/L. For practical purposes, all values that have dissolved oxygen depletions between 30 and 90% should be averaged and this average reported to the nearest 10 for values greater than 100, and to the nearest 5 for values less than 100. For values less than 10, it will be necessary to report the BOD to the nearest unit. Therefore, the above two BOD values would be reported as 260 mg/L and 30 mg/L, respectively.

Immediate Oxygen Demand

The presence of reducing agents such as S^{2-} or Fe^{2+} may cause a significant amount of oxygen to be consumed by direct chemical oxidation. These reactions will be essentially complete in a very short period of time. The time for this immediate chemical oxygen demand (not to be confused with COD described in the previous section) is arbitrarily taken to be 15 min. The DO depletion in 15 min is simply subtracted from the total DO depletion to calculate the BOD. The immediate oxygen demand, often called the immediate BOD (BOD_I), is calculated by

$$BOD_I = \Delta DO_{15\text{ min}} \times DF \tag{5.31}$$

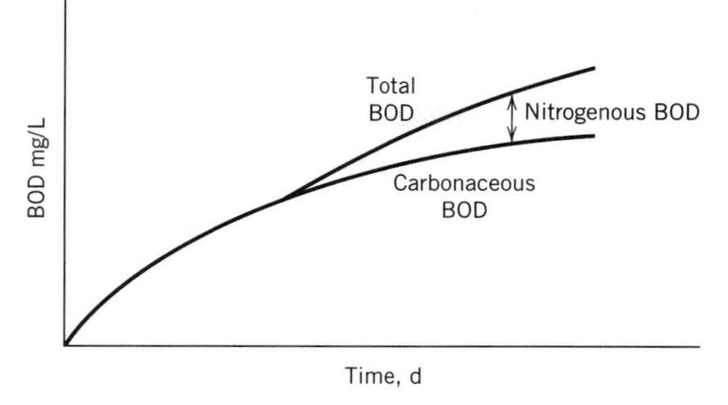

Figure 5.13 Nitrogenous BOD.

5.9.3 Carbonaceous and Nitrogenous BOD

The primary products of oxidation of organic matter will be CO_2, H_2O, and NH_3. The first stage or carbonaceous BOD consists of oxidizing organic matter to these products. Oxidation of ammonia to nitrate (nitrification) constitutes a second stage BOD that occurs simultaneously with the oxidation of carbonaceous BOD (Fig. 5.13). Nitrification can be modeled by a first-order reaction with the appropriate lag time. The reactions of interest are

$$NH_3 + \tfrac{3}{2}O_2 \xrightarrow{\textit{Nitrosomonas}} NO_2^- + H^+ + H_2O \qquad (5.32a)$$

$$NO_2^- + \tfrac{1}{2}O_2 \xrightarrow{\textit{Nitrobacter}} NO_3^- \qquad (5.32b)$$

$$NH_3 + 2O_2 \rightarrow NO_3^- + H^+ + H_2O \qquad (5.32c)$$

Two separate genera of bacteria (*Nitrosomonas* and *Nitrobacter*) are involved in the conversion of ammonia to nitrate as indicated in Eqs. (5.32a) and (5.32b). The reaction in Eq. (5.32b) depends on the rate at which the reaction in Eq. (5.32a) proceeds. Nitrite is generally not thermodynamically stable in natural water environments and it is readily transformed into nitrate.

The organisms responsible for nitrification do not oxidize carbon. They obtain energy from reactions (5.32a) and (5.32b) and use carbon dioxide as a carbon source for cell synthesis. The second stage or nitrogenous oxygen demand (NOD) is seen to be 3.76 mg O_2/mg NH_3. The oxidation of ammonia normally does not become significant until after 7 or 8 d have elapsed, which is one reason for limiting a BOD determination to 5 d. It takes some time for breakdown of the organic matter and production of NH_3. After a suitable amount of NH_3 is present, the nitrifier population begins to increase to significant numbers and nitrification causes a significant amount of oxygen to be consumed. Nitrification can be suppressed by the addition of inhibitors such as thiourea to a sample or pasteurization of the sample followed by addition of a seed that does not contain nitrifying bacteria. Young (1973) found that 2-chloro-6-(trichloromethyl) pyridine (TCMP) was a very effective suppressant of nitrification.

If all of the DO in the BOD bottle is consumed then some microorganisms will be able to use nitrate, if it is present, as a terminal electron acceptor, converting the nitrate into nitrogen gas as indicated on Fig. 5.9. Some organic matter is consumed by denitrifying bacteria. The BOD determination will not be valid because the DO has been totally depleted.

5.9.4 Determination of the Rate of BOD Exertion and Ultimate BOD

Although variation in BOD_5 reflects variation in the ultimate BOD, the relation of short-term (e.g., 5-d) tests to the ultimate BOD should be determined to calculate the total strength of a water from a BOD_5 determination.

Limitations of the BOD Test for Biological Wastewater Treatment Process Design

The BOD test is a suitable indicator of biologically removable organic matter in a water sample but it is subject to a number of limitations. It is highly dependent on the care taken to maintain an optimal environment with acclimatized microorganisms and a progression should have been run to obtain meaningful results. The test will not reflect rates of organic matter removal in a biological wastewater treatment process, where environmental conditions are significantly different from conditions in a BOD bottle. Wastewater treatment reactors have much higher concentrations of microorganisms, which is the main difference.

Analysis of a BOD Progression

The first-order equation is only an approximation to the overall rate of organic matter removal in the BOD bottle as discussed above. The two parameters that must be assessed in a BOD determination are the rate constant and the ultimate BOD. A BOD progression is required to estimate both of these parameters. In a BOD progression analysis, a wastewater sample is separated into a number of aliquots for which BOD determinations will be performed for times ranging from near 0 to 20 or more days. The upper time limit depends on the time for the ultimate BOD to be exerted. More dilution will be applied to aliquots incubated for longer periods of time. Various statistical approaches are applied to analyze the data according to Eqs. (5.26) or (5.27). Two of the methods are discussed next. The method of moments described by Moore et al. (1950) is another valid approach.

Thomas's Graphical Method. In finding k and L_a, from x–t data, difficulty arises because of the exponential term in Eq. (5.27). There are various ways to handle this problem, one of which is to replace the exponential term with a nonexponential function that closely approximates the original function.

Thomas (1950) recognized that the $(1 - 10^{-kt})$ term is similar to the function[1]

$$2.3kt\left(1 + \frac{2.3}{6}kt\right)^{-3}$$

The similarity is seen by series expansion of each expression.

$$(1 - 10^{-kt}) = 2.3kt\left[1 - \frac{1}{2}(2.3kt) + \frac{1}{6}(2.3kt)^2 - \frac{1}{24}(2.3kt)^3 + \cdots\right] \quad (5.33a)$$

$$2.3kt\left(1 + \frac{2.3}{6}kt\right)^{-3} = 2.3kt\left[1 - \frac{1}{2}(2.3kt) + \frac{1}{6}(2.3kt)^2 - \frac{1}{21.6}(2.3kt)^3 + \cdots\right] \quad (5.33b)$$

The expansions of Eqs. (5.33a) and (5.33b) are identical for the first three terms and the difference in the fourth term is small. Therefore, x can be approxi-

[1]The first step in the regression of nonlinear equations is often to find a linear equation that has a series expansion similar to the original equation. This equation is then solved to estimate the parameters.

mated by

$$x = 2.3L_a kt\left(1 + \frac{2.3}{6}kt\right)^{-3} \tag{5.34}$$

This equation can be rearranged into a linear equation by solving for t/x,

$$\frac{t}{x} = \frac{1}{2.3kL_a}\left(1 + \frac{2.3}{6}kt\right)^{3}$$

Taking the cube root and rearranging,

$$\left(\frac{t}{x}\right)^{1/3} = \left(\frac{1}{2.3kL_a}\right)^{1/3} + \left[\frac{(2.3k)^{2/3}}{6L_a^{1/3}}\right]t \tag{5.35}$$

This equation has a slope, S, and intercept, I, that depend only on the two unknowns.

$$S = \frac{(2.3k)^{2/3}}{6L_a^{1/3}} \tag{5.36a}$$

$$I = \left(\frac{1}{2.3kL_a}\right)^{1/3} \tag{5.36b}$$

Solving these two equations, the unknowns are found to be:

$$k = \frac{6S}{2.3I} = 2.61\frac{S}{I} \qquad (k \text{ in base } 10) \tag{5.37a}$$

$$L_a = \frac{1}{2.3kI^3} \tag{5.37b}$$

The procedure for the Thomas method is to:

1. Construct a table with columns for t_i, x_i, t_i/x_i, and $(t_i/x_i)^{1/3}$.
2. Plot $(t/x)^{1/3}$ versus t and draw the best fitting curve.
3. Obtain S and I from the curve and use Eqs. (5.37a) and (5.37b) to solve for k and L_a.

The method is approximate. Experimental values of $x > 0.9L_a$ should not be used in fitting the line because deviations from the exact equation become significant when about 90% of the BOD has been exerted.

Least Squares Regression. Regression analyses can be applied to x–t data by differentiating Eq. (5.27) and applying a logarithmic transformation to produce a linear equation.

$$\frac{dx}{dt} = k_1 L_a e^{-k_1 t} \tag{5.38}$$

The logarithmic transformation of Eq. (5.38) is

$$\ln\left(\frac{dx}{dt}\right) = \ln(k_1 L_a) - k_1 t \tag{5.39}$$

which has the general form: $y = mt + b$.

The standard least squares technique[2] (see any elementary statistics book, e.g., Kennedy and Neville, 1986) is applied to minimize the sum of the squares of the

[2]The equation to be regressed is nonlinear. Taking the logarithm of the equation may change the variance structure of the data. This may distort the weights given to portions of the data and thus lend some error to parameter estimation. However, in many cases the procedure gives good estimates of the parameters.

residuals, which results when the parameters k_1 and L_a are chosen and applied in Eq. (5.39) with the measured data.

$$R = \ln\left(\left.\frac{dx}{dt}\right|_i\right) - \ln(k_1 L_a) + k_1 t_i \tag{5.40}$$

where

$\left.\frac{dx}{dt}\right|_i$ is the estimate of the derivative at time t_i

R is the residual

The central difference approximation is used to find the derivatives:

$$\left.\frac{dx}{dt}\right|_i \approx \frac{x_{i+1} - x_{i-1}}{t_{i+1} - t_{i-1}}$$

For the last derivative in (x–t) data pairs, use

$$\left.\frac{dx}{dt}\right|_n = \frac{x_n - x_{n-1}}{t_n - t_{n-1}} \qquad \text{which applies at } t = \frac{t_n + t_{n-1}}{2}$$

The slope and the intercept of the best fit line determine k_1 and L_a.

5.9.5 Total Organic Carbon

Total organic carbon (TOC) is the organic carbon content of a sample. It is usually determined with an instrument. Inorganic carbon (CO_2, HCO_3^-, and CO_3^{2-}) must first be purged from the sample under acidic conditions (why?) or alternatively it may be assessed through an acidity analysis. A sample is injected into the instrument that uses a catalyst and heat while supplying oxygen to convert organic C into CO_2. The amount of CO_2 produced is measured for the known volume of sample. See Example 5.3 for the calculation of TOC when the sample composition is known.

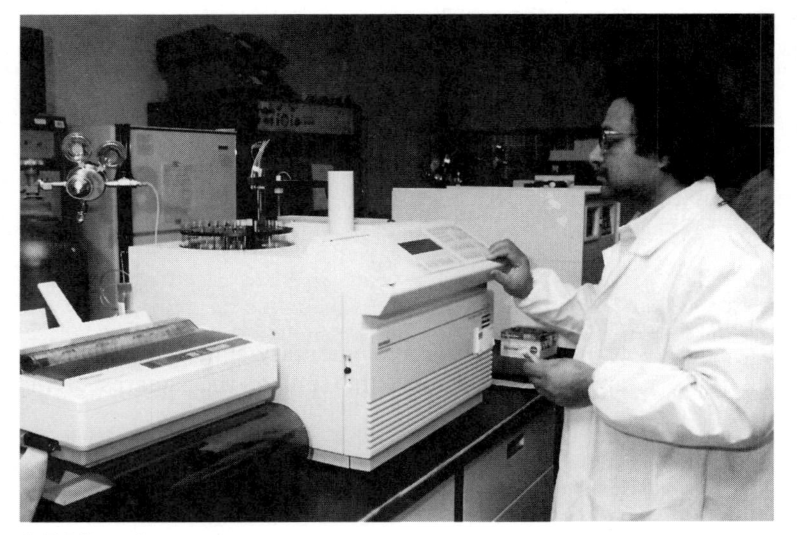

A TOC analyzer

Drawbacks associated with this analysis are the expense of the instrument and the average oxidation state of the organic carbon is not determined. The oxidation number on carbon may range from -4 to $+4$ (CH_4 to CO_2). Also the technique does not distinguish between biologically degradable and nondegradable substances. These limitations make TOC the lesser reported parameter of TOC, BOD, and COD. On the positive side, almost all organic carbon is measured by this technique, which is rapid and reliable.

The ratios of BOD_5 and COD to each other and the ratio of either of these parameters to TOC are not constant for wastewaters. The relative biodegradability and state of oxidation of organics in the waste influence these ratios. Ratios of BOD_5 : TOC and COD : TOC ranged from 1.28 to 2.53 and from 1.75 to 6.65, respectively, for a number of chemical wastes (Ford and Eckenfelder, 1970). In a biological wastewater treatment process, these ratios change as the waste becomes degraded by the microorganisms. BOD_5 becomes very low in an efficient process. COD will decrease because of oxidation of organics but it will be higher than BOD because of the production of some substances that are difficult to degrade. TOC will decrease but not by as much as the other two parameters. Therefore the ratios of BOD_5 or COD to TOC will decrease with biological treatment (see Section 17.3.1).

QUESTIONS AND PROBLEMS

1. Plot the data from Table 3.3 for the range of volumes of NaOH added between 24.0 and 26.0 mL. Use the full page of normal graph paper for the plot. Draw the line of best fit through the points. How much error would occur in the titration if an indicator that changed color at a pH of 8.3 was used as the stopping point?
2. The solubility products for Ag_2CrO_4 and AgCl are 2×10^{-12} and 3×10^{-10}, respectively. Chromate ion (CrO_4^{2-}) is added at about 300 mg/L in the standard procedure for Cl^- determination. For this concentration of chromate how much Cl^- can be in solution? (Hint: find the amount of Ag^+ in solution.) If 0.7 mg of Ag_2CrO_4 must form in a 100-mL sample to give a noticeable brownish-red color to the analyst, how much $AgNO_3$ must be added to a 100-mL sample of water containing a Cl^- concentration of 28 mg/L?
3. Identify the components of the EDTA molecule that are responsible for its name.
4. (a) What is the equivalence potential of 100 mL of a solution containing 1.00×10^{-3} moles of Ce^{4+} titrated with a 0.010 *N* solution of Fe^{2+}? The initial concentration of Ce^{3+} is negligible.
   (b) What is the equivalence potential of 100 mL of a solution containing 1.00×10^{-3} moles of Ce^{4+} and 1.00×10^{-4} moles of Ce^{3+} titrated with a 0.010 *N* solution of Fe^{2+}?
   (c) Will the difference in equivalence potentials determined in (a) and (b) have a significant influence on the results of the titration if the same indicator is used? (A few calculations may help prove your point.)
5. In Example 5.2, does an equation similar to Eq. (vi) apply at any time during the titration?
6. Derive Eq. (5.7).
7. Dr. Drastic wishes to give the impression that he can change water into wine (or at least show that he has the potential for doing this). He has prepared a solution with resorufin (a derivative of resazurin), an indicator that has an $E°$ of -0.051 V at a pH of 7.00. Resorufin is an indicator that is colorless in its reduced state and pink in its oxidized state. The reduced form gives up 2 electrons when it is oxidized. The redox equation for resorufin (R) at a constant pH can be represented as

   $$\underset{\text{Pink}}{R} + 2e^- = \underset{\text{Colorless}}{R^{2-}} \qquad E° = -0.051 \text{ V (at pH} = 7.00)$$

   He has prepared a 1-L solution from distilled water and added a buffering agent to maintain the pH at 7.00. Initially the buffered solution contained 9.00 mg/L of oxygen and no sulfate. Sodium sulfite was added to the solution at 800 mg/L to remove the oxygen. A small amount of resorufin (insignificant compared to other species) was added to the sulfited solution. A 1-L solution of perchloric acid ($HClO_4$) will be mixed with 1 L of sulfite solution to produce a pink solution. $HClO_4$ is a very strong acid that ionizes to produce ClO_4^- (perchlorate) and the perchloric acid solution is also buffered at a pH of 7.00. Perchlorate is a strong oxidizing agent. There was no chloride ion initially in the solution to which the perchlorate was added. Ignoring any addition of oxygen, what is the minimum concentration of perchlorate in the second solution required to obtain a definite pink color (a ratio of 10 is required)? The temperature will be 25°C the night of the show.

8. Explain how a specific ion electrode functions.

9. Light at a certain wavelength was beamed through a sample of distilled water and its intensity decreased by 10%. In a sample of water in a similar cell, the intensity of light transmitted was 82% of the intensity generated by the light source. What is the absorbance resulting from suspended solids and other matter in the sample?

10. An analyst found that the transmittance of a sample that contained 3.00 mg/L of an absorbent substance is 0.847 in a cell with a width of 2.00 cm. What is the transmittance of the same substance at a concentration of 2.00 mg/L in a cell with a width of 1.00 cm?

11. How would a spectrophotometric analysis be affected by (a) a voltage fluctuation and (b) suspended solids in the sample?

12. Characterize the different forms of solids that can occur in a water sample.

13. How can some inorganic solids be lost when a total suspended solids–volatile suspended solids analysis is performed? Write the chemical equations given in the text illustrating this phenomenon. Can you write any other chemical equations for loss of inorganic components at high temperatures besides those given in this chapter?

14. (a) In the COD determination a solution containing dichromate is titrated with ferrous ion. Prepare a table listing the concentrations of Fe^{2+}, Fe^{3+}, $Cr_2O_7^{2-}$, and Cr^{3+} with the following additions of 3.0 *M* $Fe(NH_4)_2(SO_4)_2$ to the solution described below: 10.0, 30.0, 50.0, 80.0, 95.0, 100.0, and 105.0 mL. The solution initially contains 0.05 moles of potassium dichromate in 100 mL of water. The pH of the solution is initially 1.0 and may be assumed to remain at this value throughout the titration. Include in your table a column for the system potential after each addition of titrant and prepare a plot of system potential versus amount of titrant added. Use appropriate assumptions about the concentrations of the various ions before, after, and at equivalence. If the concentration of an ion or ions is negligible do not iterate to find its exact concentration. At equivalence use the fact that the amount of ferrous iron added must be equivalent to the initial amount of dichromate present. Also the dichromate has almost totally been converted to Cr^{3+} at this point.

(b) Calculate the concentration of Fe^{2+} after the addition of 10.0 mL of 3.0 *M* $Fe(NH_4)_2(SO_4)_2$ and equilibrium is established. If a second aliquot of 20.0 mL of titrant is added instantaneously to this solution, calculate the cell potential ($E_{DC} - E_{Fe}$) that exists after addition of titrant but before any reaction has occurred.

(c) The reduction of $Cr_2O_7^{2-}$ consumes a significant amount of H^+ ions that could have a significant effect on the potential. Consult the latest edition of *Standard Methods* to find out how the analysis is conducted and explain how this problem is handled.

15. In the COD procedure in *Standard Methods,* 25.0 mL of 0.041 7 M $K_2Cr_2O_7$ and 75 mL of concentrated H_2SO_4 are added to a 50 mL sample. Conc. H_2SO_4 has a specific gravity of 1.835 and contains 97% H_2SO_4. If the sample is distilled water, how much hydrogen ion will be consumed if the solution is titrated with ferrous iron? Do you expect a significant pH change when a dichromate-sample solution is titrated with ferrous ion?

16. Does a pH change influence the COD content of a sample?

17. If a water contained an agent that could complex significant amounts of Fe^{2+}, would a lower or higher COD be reported than the actual value?

18. What is the COD of a water that contains 100 mg/L of phenol (C_6H_5OH)?

19. General formulas for carbohydrate, protein, and fats are CH_2O, $C_{16}H_{24}O_5N_4$, and $C_8H_{16}O$, respectively. If a wastewater contains carbohydrates, proteins, and fats in ratios of 50 : 40 : 10 on a COD basis (N does not contribute to the COD), what are the concentrations of carbohydrates, proteins, and fats if the total COD concentration is 1 000 mg/L?

20. (a) What is the COD of a water in terms of *n, a,* and *b* for a water that contains organic matter that is represented by the formula $C_nH_aO_b$? Assume that there are no other substances that consume oxygen.

(b) Determine *c* in terms of *n, a, b,* and *z* for the following reaction:

$$C_nH_aO_bN_z + cCr_2O_7^{2-} + fH^+ \rightarrow dCr^{3+} + gCO_2 + eH_2O + hNH_3$$

21. Quadrivalent cerium [Ce(IV)] can be used in the COD test although results are generally not as satisfactory as with dichromate. Find the general reaction (equivalent to Eq. 5.25) for oxidation of an organic compound using Ce(IV) which is reduced to Ce(III) in the reaction.

22. What result would be obtained for COD of a sample if it contained organic matter with a COD of 568 mg/L resulting from organic carbon and an organic and ammonia nitrogen content of 28 mg/L as N. The digestion was carried out with 0.25 *N* dichromate and the chloride concentration of the sample was known to be between 200 and 300 mg/L.

23. What effects do inorganic reducing agents have on COD and COD determination?

24. Comment on the statement: "Some wastes have a nonbiodegradable BOD."

25. Describe and discuss some of the reasons for variability in a BOD determination.

26. Once the stopper is placed in a BOD bottle, the system is closed. Does the total amount of organic matter (microorganisms and nonliving organic matter) remain constant?

27. If a rate constant has a value of 0.092 d^{-1} on a base 10 basis, what is its value on a base *e* basis with units of min^{-1}?

28. The 5-d BOD of a sewage sample is 200 mg/L. What is its first stage ultimate BOD at 20°C? k_{20} (base 10) = 0.16 d^{-1}.

29. Construct a BOD progression curve for a sample that has a first stage rate constant 0.16 d^{-1} (base 10) and a nitrogenous first-order rate constant of 0.08 d^{-1} (base 10). Nitrification is not significant until after a period of 7.5 d has elapsed. The ultimate first stage BOD is 335 mg/L and the ultimate nitrogenous oxygen demand is 90 mg/L. What is the total BOD exerted at 10 days?

Assume θ is 1.047 for Problems 30–32.

30. The BOD of a sewage sample incubated for 2 d at 30°C has been found to be 140 mg/L. What will be the 5-d 20°C BOD? k_{20} (base 10) = 0.16 d^{-1}.

31. If the 3-d, 20°C BOD of a sample is 200 mg/L, what will be its 7-d, 25°C BOD? k_{20} (base 10) = 0.16 d^{-1}.

32. If the BOD of a wastewater sample shows a first stage BOD value of 188 mg/L and a reaction velocity constant, k_1 (base *e*), of 0.52 d^{-1} at 15°C, what is its expected 5-d BOD at 20°C?

33. What are the ultimate BODs of 100 mg of each of the following substances: (a) nitrite, (b) nitrate, (c) ammonia, and (d) ammonium ion? What are the oxygen demands of ammonia and ammonium ion, respectively, when they are expressed as N? Is it more practical to express ammonia as ammonia and ammonium as ammonium or to express both as N when referring to their oxygen demand (explain)?

34. Assuming that the cumulative respiration of algae over 5 d is 0.25 mg O_2/mg algae, what are the algae concentrations in the sample that would result in a DO decrease of (a) 2.5 mg/L and (b) 2.75 mg/L resulting from algal respiration in a sample with a BOD_5 of 20 mg/L resulting from other organic matter? The dilution factor is 10 in each case.

35. While conducting BOD tests on a wastewater sample, the following measurements were taken. Calculate the BOD of the sample.

Sample no.	1	2	3
Dilution, %	1	2	3
Dissolved oxygen decrease, mg/L	2.7	4.9	7.2

36. What are the permissible dilution limits to attain a DO decrease between 2.0 and 7.0 mg/L for a sample with a BOD_5 of 1 500 mg/L?

37. Develop the equations for BOD calculation for the "headspace" BOD procedure.

38. A BOD determination was made on an industrial waste that required addition of a seed culture. The seed culture had a BOD of 320 mg/L and 1.00 mL of seed was added to the standard 300 mL BOD bottle, which contained 8 mL of the wastewater. A DO decrease of 6.25 mg/L was measured in the BOD bottle after 5 d. What was the BOD_5 of the wastewater?

39. A BOD determination was performed using 10 mL of sample in a 300-mL bottle. It was discovered that the DO meter was improperly calibrated for the initial DO determination; therefore, a DO determination was made 24 h after the sample was set. The DO was 7.20 mg/L. The DO after 5 d of incubation at 20°C was measured at 3.05 mg/L. The rate constant for this waste is known to be 0.11 d^{-1} (base 10). What was the BOD_5 of this sample?

40. (a) Is it possible to have a sample with an ultimate BOD greater than the COD? (b) Is it possible for BOD to decrease without a decrease in COD? (c) Is it possible for COD to decrease without a reduction in BOD?

41. Discuss the practicality of accelerating the BOD test by adding a large amount of seed organisms.

42. The BOD data in the following table were obtained for a wastewater sample. Estimate the rate constant and ultimate BOD of the waste using (a) the Thomas method and (b) the regression method.

time, d	1	2	3	4	5	6	7	8	9	10
BOD, mg/L	190	306	360	438	500	490	510	500	524	542

43. The energy yield from the oxidation of ammonia to nitrite yields more energy than the oxidation of nitrite to nitrate. (a) Why would you expect this to be true? (b) Would you expect a higher concentration of *Nitrosomonas* or of *Nitrobacter* in a reactor where nitrification is proceeding?

44. What is the relative improvement in accuracy of a BOD determination as the number of sample aliquots increases from one to two to three?

45. (a) What is the 5-d BOD of a sample that has a concentration of 159 mg/L of organic matter that can be represented as $C_{4.2}H_{6.1}N_{0.8}O_2$? The rate constant is 0.14 d^{-1} (base 10) and nitrification was not significant until day 6.
(b) What are the ultimate carbonaceous and nitrogenous BODs, COD, and TOC of this waste?

KEY REFERENCES

SAWYER, C. N., P. L. MCCARTY, AND G. F. PARKIN (1994), *Chemistry for Environmental Engineering*, 4th ed., McGraw-Hill, Toronto.

SKOOG, D. A. AND D. M. WEST (1974), *Analytical Chemistry: An Introduction*, 2nd ed., Holt, Rinehart, Winston, Toronto.

SNOEYINK, V. L. AND D. JENKINS (1980), *Water Chemistry,* John Wiley & Sons, Toronto.

Standard Methods for the Examination of Water and Wastewater, (1992), 18th ed. American Public Health Association, Washington, DC.

VOGEL, A. (1961), *A Textbook of Quantitative Inorganic Analysis,* 3rd ed., Longman, London.

REFERENCES

BÁNYAI, É. (1972), "Acid–Base Indicators," in *Indicators, E.* Bishop, ed., Pergamon Press, Toronto, pp. 65–176.

BISHOP, E. (1972), "Oxidation–Reduction Indicators of High Formal Potential," in *Indicators,* E. Bishop, ed., Pergamon Press, Toronto, pp. 531–684.

FORD, D. L. AND W. W. ECKENFELDER, JR. (1970), *Water Pollution Control: Experimental Procedures for Process Design,* Pemberton Press, New York.

GAUDY, A. F., JR. (1972), "Biochemical Oxygen Demand," in *Water Pollution Microbiology,* vol. 1, R. Mitchell, ed., John Wiley & Sons, Toronto, pp. 305–332.

GOLUEKE, C. G., W. J. OSWALD, AND H. B. GOTAAS (1957), "Anaerobic Fermentation of Algae," *Applied Microbiology,* 5, 1, pp. 47–55.

KENNEDY, J. B. AND A. M. NEVILLE (1986), *Basic Statistical Methods for Engineers and Scientists,* 3rd ed., Harper & Row, New York.

KIM, B. R. (1989), "Effect of Ammonia on COD Analysis," *J. Water Pollution Control Federation,* 61, 5, pp. 614–617.

LOGAN, B. E. AND G. A. WAGENSELLER (1993), "The HBOD Test: A New Method for Determining Biochemical Oxygen Demand," *Water Environment Research,* 65, 7, pp. 862–868.

MOORE, E. W., H. A. THOMAS, JR., AND W. B. SNOW (1950), "Simplified Method for Analysis of BOD Data," *Sewage and Industrial Wastes,* 22, 10, pp. 1343–1353.

OSWALD, W. J. AND R. RAMANI (1976), "The Fate of Algae in Receiving Waters," in *Ponds as a Waste Treatment Alternative, Water Resources Symposium No. 9,* E. F. Gloyna, J. F. Malina, Jr., and E. M. Davis, eds., University of Texas, Austin, TX, pp. 111–121.

OTTAWAY, J. M. (1972), "Oxidation–Reduction Indicators of $E^{\circ} < 0.76$ Volt," in *Indicators,* E. Bishop, ed., Pergamon Press, Toronto, pp. 469–530.

TALINLI, I. AND G. K. ANDERSON (1992), "Interference of Hydrogen Peroxide on the Standard COD Test," *Water Research,* 26, 1, pp. 107–110.

THOMAS, H. A., JR. (1950), "Graphical Determination of BOD Curve Constants," *Water and Sewage Works,* 97, pp. 123–124.

YOUNG, J. C. (1973), "Chemical Methods for Nitrification Control," *J. Water Pollution Control Federation,* 45, 4, pp. 637–646.

SECTION II

MICROORGANISMS IN WATER AND WATER QUALITY

CHAPTER 6

MICROBIOLOGY

Microorganisms are significant in water and wastewater because of their roles in disease transmission and they are the primary agents of biological treatment. They are the most diverse group of living organisms on earth and occupy important niches in the ecosystem. Within the microbial realm, there are species able to attack iron and able to survive and proliferate in extreme environments ranging from, for instance, low to high pH or temperature. Their simplicity and minimal survival requirements allow them to exist in diverse situations.

6.1 BACTERIA

The simplest wholly contained life systems are bacteria or prokaryotes, which are the most diverse group of microorganisms. They are characterized by the lack of a nuclear membrane and their machinery of metabolism is not contained in organelles. They reproduce by simple fission. A typical bacterial cell is shown in Fig. 6.1.

The bacterial cell is enclosed within a cell wall that is a semirigid polymeric membrane on the order of 100 nm thick. The cell wall maintains the integrity of the cell, holding it together against osmotic pressure gradients that occur because of concentration differences between the cell contents and the surrounding liquid. The cytoplasmic membrane is immediately inside the cell wall and it controls passage of nutrients and other compounds into and out of the cell.

The cytoplasm is a solution that contains the biomolecules essential for metabolism. The fundamental controlling biomolecules, deoxyribonucleic acid (DNA), and others are contained within the nuclear material. The primary component of prokaryotic cells, as for all living systems, is water, which is approximately 75% of the cell by weight.

The following features are not common to all bacteria. A flagellum is a means for cell motility. It is moved in a whiplike fashion to propel the cell. A cell may have one, two, or many all around the cell. Pili, singular pilus (also called fimbriae, singular fimbria), are smaller appendages unrelated to motility. One type of pilus is involved in bacterial conjugation but the functions of other types of pili are obscure.

Granules, aggregates, or other inclusions are observed in many bacteria. The nature of these inclusions differs in different microorganisms. Poly-β-hydroxybutyric acid (PHB) is a carbon or energy containing compound that is a long polymeric linkage of β-hydroxybutyric acid molecules (Fig. 6.2). These polymers aggregate into granules. Starch and glycogen, which are also carbon- and energy-containing polymers, are stored in granules. Inorganic phosphate is stored as polyphosphates in granules called volutin and sulfur may be stored in sulfur droplets.

The cell may be surrounded by a slime layer which is usually composed of polysaccharides, polypeptides, or polysaccharide–protein complexes. When the layer is com-

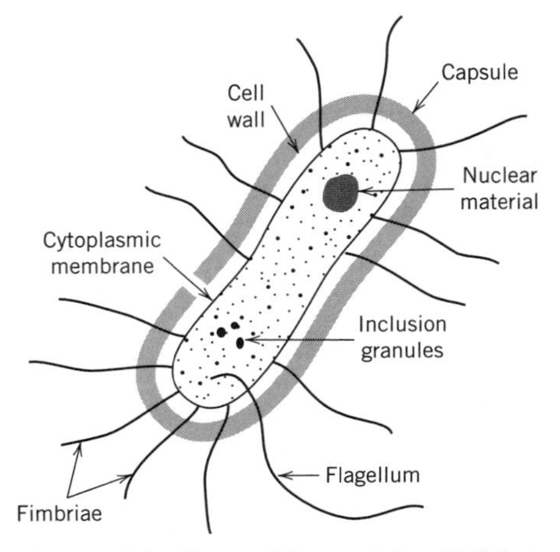

Figure 6.1 Typical bacterial cell (Mara, 1974). Used with permission of Churchill Livingstone.

pact it is known as a capsule; when it is diffuse it is known as a slime layer. The functions of the capsule are not clear.

Some bacteria are able to form spores (endospores or free spores) in adverse circumstances. A spore is a nonvegetative state that bears little resemblance to the living cell. The protoplasm is covered by many coatings, making it heat resistant and resistant to lack of moisture. Upon finding itself in an agreeable environment, the spore reverts to a vegetative cell. This trait is common in two genera of bacteria, *Bacillus* and *Clostridium.* Spores are remarkably heat resistant, which has implications for sterilization. The structure of a spore is much more complex than the vegetative cell.

Fundamental cell shapes are circular (coccus), rod (bacillus), bacteria which are curved (vibrio), and spiral (spirillum) (Fig. 6.3). Bacteria of a species may be singular, chained, or clumped together. Individual cells range in size from 0.125 μm diameter to 1.35 $\times$ 30 μm.

6.1.1 Classification of Bacteria

There are many means of classifying bacteria, ranging from purely physical characteristics to physiological responses.

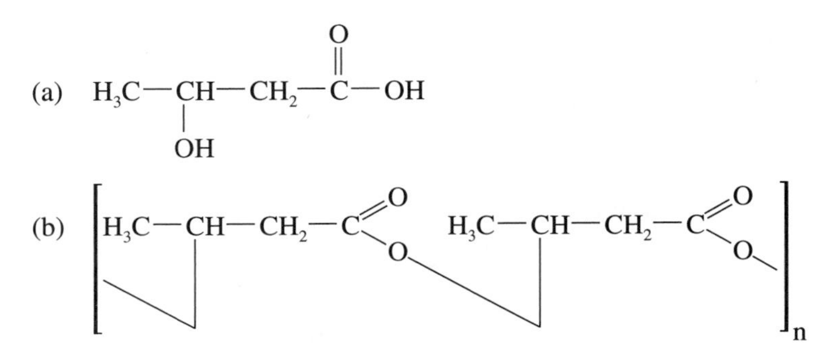

Figure 6.2 (*a*) β-Hydroxybutyric acid (*b*) poly-β-hydroxybutyric acid.

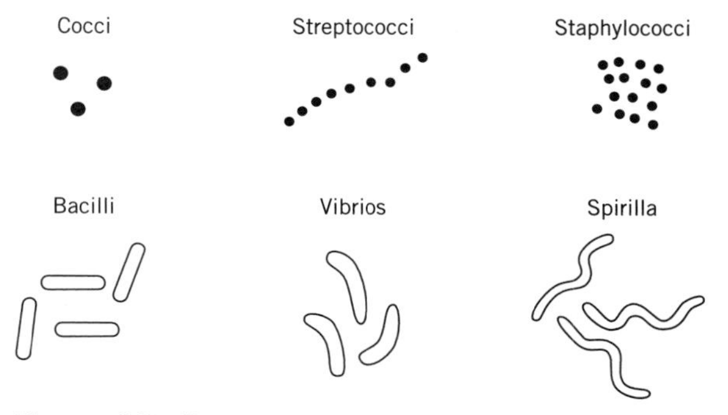

Figure 6.3 Bacteria shapes.

Nutritional Types

Bacteria and all life forms require an energy source and a carbon source. The same substrate may serve both functions. Energy may be obtained from sunlight (phototrophy), oxidation of reduced organic matter (heterotrophy), or oxidation of inorganic compounds (autotrophy). Organisms that do not require a reduced organic (carbon) source for energy are referred to as autotrophs (from the Greek *auto-* self; *troph-* feed). Heterotrophs (Greek *hetero-* many) obtain energy from reduced carbon-containing compounds and these compounds also serve as a carbon source.

Organisms that obtain energy from chemical oxidation of organic or inorganic matter are chemosynthesizers as opposed to those that obtain energy from the sun, known as photosynthesizers. Photosynthesizers need inorganic or organic sources of carbon. Bacteria are compared to plants and animals in Fig. 6.4. Plants obtain their energy by photosynthesis and use inorganic carbon as a carbon source. Animals obtain their energy from oxidation of organic matter, which also serves as a carbon source. Bacteria fall into these categories but there are also some bacteria that fall into the other energy and food combinations as indicated in the figure. The majority of bacteria are heterotrophs. Organisms that live on dead or decaying organic matter are sometimes called saprophytes or saprobes.

Some bacteria are able to live on a basic inorganic salts medium; the nutrients may need to be in a reduced form; for instance, nitrogen must be supplied as ammonia. Others require certain prefabricated growth factors ranging from amino acids to vitamins.

Oxygen Requirements (ORP)

The requirement for oxygen, or more precisely the oxidation–reduction potential (ORP), for survival and optimum growth of bacteria spans a wide range. Aerobic bacteria use oxygen for metabolism; strict or obligate aerobes are killed by lack of

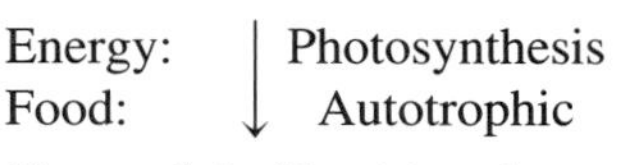

Figure 6.4 Nutritional types of bacteria.

oxygen. Anaerobic bacteria do not use oxygen for metabolism; strict anaerobes are killed by the presence of free oxygen.

There is a third group of bacteria that is able to switch modes from aerobic to anaerobic metabolism depending on the availability of free oxygen. These bacteria are termed facultative. Aerobic metabolism is preferred because of the greater yields of energy per unit of foodstuff processed. The availability of oxygen influences the ORP and state of oxidation–reduction of compounds. For instance, as ORP drops nitrates become converted to ammonia. The range of fluctuation of ORP able to be tolerated varies for different species of facultative bacteria.

Besides nutritional and oxygen needs, bacteria are classified according to other criteria discussed later.

6.2 EUKARYOTES

Eukaryotes are the next higher stage of living organisms from prokaryotes. The major characteristic of this group compared to prokaryotes is the existence of a nucleus enclosed within a nuclear membrane (Fig 6.5). The area outside the nucleus is the cytoplasm. Aerobic cells contain a mitochondrion, which is an organelle that contains enzymes associated with metabolism. Photosynthesizers (algae) contain a chloroplast, which is fundamental to conversion of sunlight into usable energy. Other organelles that may be found in eukaryotic cells are Golgi bodies, which store and transport material. Vacuoles that contain a variety of storage products may be found.

Eukaryotic cells are about an order of magnitude larger than bacteria, although some algae such as seaweed are macroscopic in size, growing to over 30 m in length. Few anaerobic eukaryotes are found because of their higher levels of complexity and size.

6.2.1 Algae

Algae are photosynthesizers that occur in all natural waters. They can be unicellular or multicellular. There are three large groups characterized by their color: green,

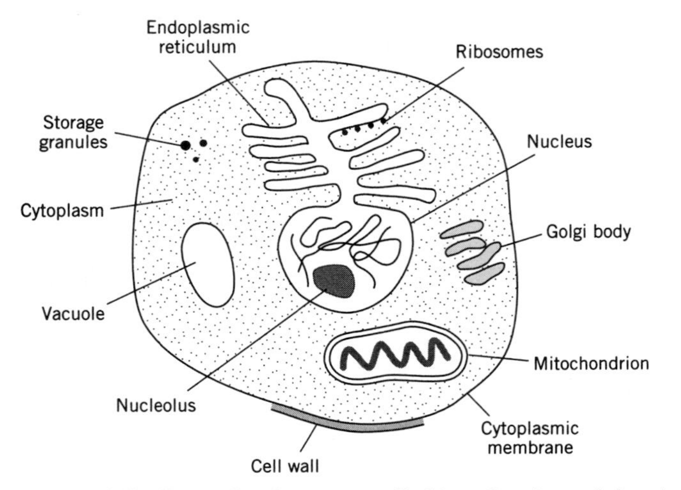

Figure 6.5 Typical eukaryotic cell. After Gaudy and Gaudy (1980).

brown, and red. The green color of most algae is due to chlorophyll *a*, which is essential for capture of light and photosynthesis. All algae contain chlorophyll *a* but the green color of chlorophyll *a* can be masked by other pigmented chlorophylls that absorb light at different wavelengths. In a study of Florida waters, the chlorophyll *a* content of algae ranged from 0.001 to 9.1% of the algae biomass, with a median value of 0.24% (Canfield et al., 1985). The data were log-normally distributed. In the same study, median nitrogen and phosphorus contents of the algae were 23 and 0.80%, respectively.

Algae play a role in some wastewater treatment processes, particularly stabilization ponds. Many algae species are harmless, but in water treatment algae are a nuisance. Large amounts of certain algae species can lead to taste and odor problems from either byproducts of their metabolism or decay of the cells. Some species produce toxins.

Excessive algal blooms are a characteristic of eutrophied inland waters resulting from natural aging or anthropogenic pollution, particularly nutrient enrichment. Some algae can produce toxins that are retained by the shellfish that consume the algae. Consumption of the shellfish by humans may lead to death, depending on quantity consumed and other variables.

Bluegreen algae or cyanobacteria (these microorganisms are at the very bottom of taxonomic classification and their distinction between bacteria and algae is not clear) include several species that are able to produce hepatotoxins and neurotoxins. *Microcystis aeruginosa, Oscillatoria agardhii,* and *Anabaena flos-aquae* are species able to produce these toxins and are commonly found in blooms.

6.2.2 Fungi

Fungi are generally filamentous and have a true cell wall. Individual filaments are known as hyphae that may have no crosswalls or be divided at irregular intervals by crosswalls. Yeasts are nonfilamentous fungi that reproduce by a process known as budding. A small bubble is produced on the mother cell that grows to about the same size as the mother cell, then a crosswall is formed and the new cell separates.

Most fungi are aerobic. Fungi can tolerate a lower pH than bacteria and their nitrogen and phosphorous requirements are lower than requirements for bacteria. These characteristics make them valuable for treating some industrial wastewaters. However, filamentous forms are difficult to settle and significant growth of fungi in wastewater treatment plants can lead to poor effluent quality.

6.2.3 Protozoa

Protozoa are single-celled organisms. Many of these are motile because of flagella or cilia or they move by means of pseudopodia (i.e., amoeboid protozoa). Many protozoa feed on prokaryotes and other eukaryotes. There are also protozoans that are saprobes; however, bacteria degrade organic matter more efficiently than protozoans.

Protozoa have a minor role in most wastewater treatment processes but there have been many studies associating performance of aerobic biological treatment processes (activated sludge) with protozoan communities. Protozoans can be identified directly with a microscope by a skilled observer. Protozoan identification and enumeration provides a rapid, inexpensive tool for monitoring process performance. Curds and Cockburn (1970) carried out a comprehensive study of activated sludge plants. Simplifying the complex taxonomy required to identify protozoa to a level usable by wastewater treatment operators is a significant problem (Kinner and Curds, 1987).

There are a number of protozoans that cause disease. Sporozoa is a major group of protozoa, all of which are parasites and cause disease. This group forms spores (these are not true spores as for bacteria but they are similar) at one stage in their life cycle. *Plasmodium* is a member of this group and this species causes malaria.

Some protozoa are able to form cysts. A cyst, similar to a bacterial spore, is a nonvegetative state resistant to inhospitable environments. The organism returns to an active state upon reaching favorable conditions.

6.3 OTHER MICROORGANISMS

Other microorganisms that fall outside of the above classes are also found in natural waters and wastewaters.

Viruses

Viruses are noncellular entities that contain protein and nucleic acids. A protein coat (capsid) surrounds the nucleic acid molecule (genome). The longest dimension of the largest virus particle (the smallpox virus) is 200 nm. Viruses are unable to reproduce or metabolize on their own. They are obligate parasites that invade a cell and direct the cell metabolism to manufacturing new viruses. The host cell ultimately dies. Viruses are host specific, which make the analysis of viruses difficult.

Rotifers

Rotifers are simple multicellular organisms at the first stage above single-celled organisms. They have cilia, used for locomotion and food currents, located around their mouth. These aerobic microorganisms are naturally found in marine and fresh waters in relatively high numbers.

Worms

Worms have an elongated body and move with an undulating motion. These simple animals are sometimes found in sewage treatment processes but their greater significance is the diseases that a number of them cause. There are many worms classified into a number of different phyla. Roundworms or nematodes are estimated to be the second most numerous group of organisms after insects. Flatworms are distinguished by a flat body. The most familiar free-living flatworms are the planaria commonly studied in high school biology courses and found in quiescent waters throughout the world.

6.4 GROWTH OF MICROORGANISMS

Microorganisms are able to grow at very high rates under suitable conditions because of their relatively simple structures and growth requirements. A particular environment will favor some species over others.

Metabolism

Metabolism is the process of chemical and energy transformations that keeps the organism alive and reproductive. It is broken down into synthesis (anabolism or assimilation) in which foodstuffs (substrate) are transformed into complex entities and catabolism (dissimilation) in which substances are broken down into simpler ones to provide energy. Some of the biochemical pathways are discussed in Chapter 4.

The rate of growth is influenced by chemical and physical variables. Many of these have been commented on in Section 6.1.1. The presence or absence of oxygen is a primary growth requirement. Temperature and pH are primary control variables. The availability of substrate, nutrients, and other minerals and compounds all influence the rate of metabolism and growth. Toxins inhibit metabolism and growth.

The influence of pH, temperature, and other environmental conditions on the rate of metabolism and growth for a single species or a group of microorganisms can be systematically evaluated by holding conditions constant and varying one of the parameters. Generally a bell-shaped response curve will be obtained.

Metabolism can considerably accelerate the rate of chemical reactions. For instance, Buisman et al. (1990) found biological sulfide oxidation was 6 to 75 times higher than the noncatalyzed chemical oxidation of sulfide.

Temperature

Microorganisms that grow well at low temperatures (0–15°C) are classified as psychrophilic. Microorganisms that grow well in the temperature range of 15–45°C are mesophilic. Thermophilic microorganisms grow best above a temperature of 45°C. There are some thermophiles that can live in nature at temperatures as high as 85°C but no organism has been cultured that has a temperature optimum greater that 75°C. No single organism can grow over the whole temperature range.

Salt and Sugar Concentration

Dissolved substances produce an osmotic pressure on cells as well as contributing to other effects. Organisms that grow best in the presence of high concentrations of solutes are osmophilic. Microorganisms that tolerate high concentrations of salt are known as halophiles or halotolerant. Extremely halophilic (halo-obligatory) microorganisms require salt concentrations over 12% for growth. If a microorganism is tolerant of a high concentration of sugar as opposed to a salt, then it is termed saccharophylic.

6.4.1 Growth of Pure Cultures

Most of the prokaryotes and unicellular algae, fission yeasts, and most protozoa reproduce by binary fission. This leads to exponential growth of a pure culture under ideal conditions. Ideal conditions consist of an excess availability of substrate and other growth requirements as well as lack of inhibitory substances. Consider a single microorganism in a vessel containing an ideal medium for the species. The generation time, g, is the time for a cell to divide into two daughter cells. The number of cells would increase as follows:

$$1 \xrightarrow{g} 2 \xrightarrow{g} 4 \xrightarrow{g} 8 \xrightarrow{g} \cdots$$

The number of microorganisms present at any time is

$$N = N_0 2^{t/g} = N_0 e^{kt} \tag{6.1}$$

where
N is number of microorganisms
N_0 is the number of microorganisms at $t = 0$
k is a rate constant and $k = 1/g \ln 2$

The generation time may not be constant for each cycle or individual cells; therefore, g represents the average generation time for the culture. The rate of increase of microorganisms is

$$\frac{dN}{dt} = kN \tag{6.2}$$

The rate of substrate removal is proportional to the rate of microorganism growth.

$$\frac{dS}{dt} \propto -\frac{dN}{dt} \quad \text{or} \quad \frac{dS}{dt} = -\frac{1}{Y}\frac{dN}{dt} \tag{6.3}$$

where
S is substrate concentration
Y is a yield factor (mass of microorganisms formed/mass of substrate removed)

Eventually substrate, a nutrient, or a trace element will reach a short supply for the number of microorganisms present. At this time the growth rate will begin to decrease. It is also possible that a buildup of toxic byproducts of metabolism is responsible for retardation of growth.

Assuming that substrate is the limiting factor, the rate of microbial growth becomes dependent on the amount of substrate present as well as the number of microorganisms in the vessel.

$$\frac{dN}{dt} = k'SN \tag{6.4}$$

where
k' is a rate constant

After this stage any combination of limiting substance or toxic byproduct buildup will cause the number of viable microorganisms to decrease.

A complete growth cycle for a pure culture of microorganisms started in a vessel containing an ideal medium is shown in Fig. 6.6. The growth phases are indicated on the diagram. The first phase is an acclimation phase during which the culture fabricates the metabolic machinery to handle the substrate. During this phase the growth rate may be quite low or even nonexistent, depending on the history of the seed.

This phase is followed by the log or exponential growth phase. The only factor limiting growth during this phase is the number of microorganisms present. At the point where a substance becomes limiting or concentration of a toxic byproduct becomes significant, the declining growth phase begins. There is still a net increase in the number of viable microorganisms. After a period of time, a stationary phase is reached. There is no change in the number of viable microorganisms during this phase.

Following the stationary phase, the supply of substrate dwindles past the critical amount required to sustain the numbers present or toxins or a combination of toxins and lack of substrate cause numbers to decline. This decline often accelerates into a log death phase.

All phases may be variable in length or even nonexistent depending on the species and medium. Monod (1949) has given a comprehensive discussion of microbial growth. Monod, from experience, formulated an empirical expression that describes growth and substrate removal during the log growth and substrate limiting phases.

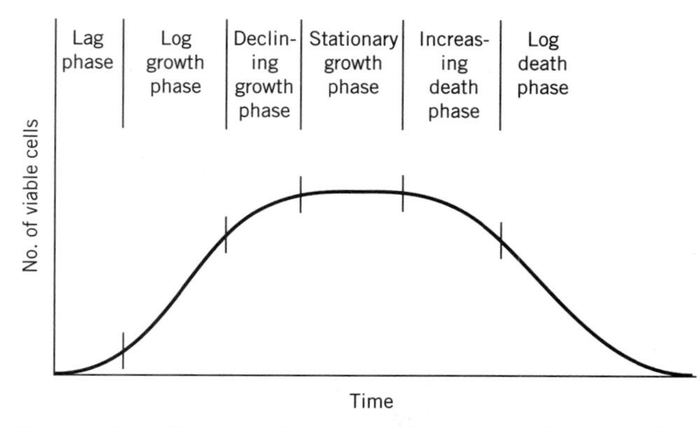

Figure 6.6 Growth of a pure culture in an ideal medium.

$$\frac{dS}{dt} = -\frac{kXS}{K + S} \tag{6.5}$$

where

k is the maximum specific rate of substrate removal (mass of substrate removed/ mass of microorganisms/time)

X is concentration of microorganisms (mass/volume)

K is a constant

Equation (6.5) is formulated in terms of concentrations. The concentration of microorganisms may be expressed in terms of numbers or mass per unit volume. The mass concentration is more commonly used because of the ease of its determination. It is recognized that individual cells within the culture vary in mass and that mass of the cells may change with time. The constant K is also called the half-velocity constant. This hyperbolic equation is of the same form as the Michaelis–Menten equation [Eq. (4.9), also see Fig. 4.12] developed from theory for enzyme kinetics. The analogy between enzyme kinetics and microbial growth is reasonable because microorganisms possess a collection of enzymes.

It can be shown that Eq. (6.5) reduces to Eqs. (6.3) and (6.4) at high and low values of substrate concentration, respectively, using the yield factor Y, given in Eq. (6.3). The yield factor is precisely defined as the average yield of microorganisms

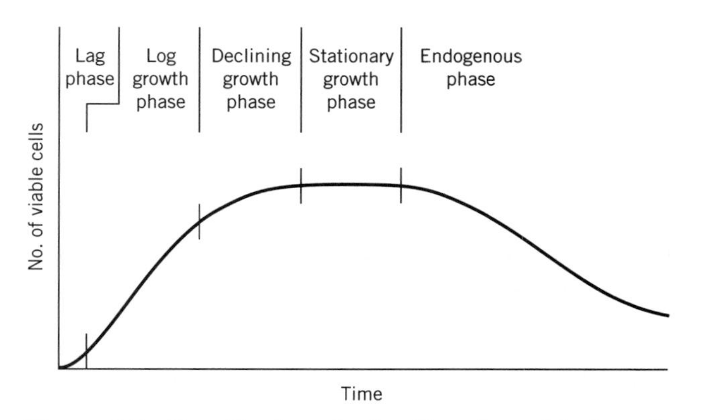

Figure 6.7 Growth of a mixed culture in an ideal medium.

produced from removal of a unit of substrate under ideal growth conditions. Dieoff of microorganisms will be negligible in these circumstances and Y is known as the true yield constant.

■ Example 6.1 Growth of a Pure Culture

The data in the table below were obtained for the growth of a pure culture of *Escherichia coli* in nutrient broth at a temperature of 37°C. Determine the generation time of *E. coli* and the duration of the lag and log growth phases. Comment on the suitability of turbidity as a measure of viable microorganisms.

The bacteria count data were plotted on a semilog scale and the turbidity data were plotted on an arithmetic scale in the figure below.

Time (h)	0	1	2	3	4	5	6	7	12	24
Bacteria No./mL	4.30×10^4	3.50×10^4	8.90×10^4	5.20×10^5	2.60×10^6	1.60×10^7	7.10×10^7	1.20×10^8	2.40×10^8	5.90×10^8
Turbidity (TU)	0	0	0	0	0.4	7.5	19.2	22.4	45.6	71.1

From the plot, the lag phase lasted approximately 1.5 h. The log growth phase occurred between 1.5 h and 6 h, for a duration nearly 4.5 h. Regression of the count data between 2 and 6 h using Eq. (6.1) yielded a slope of 1.68 h^{-1}.

The generation time was

$$g = \frac{\ln 2}{k} = \frac{0.693}{1.68\ \text{h}^{-1}} = 0.413\ \text{h} = 24.8\ \text{min}$$

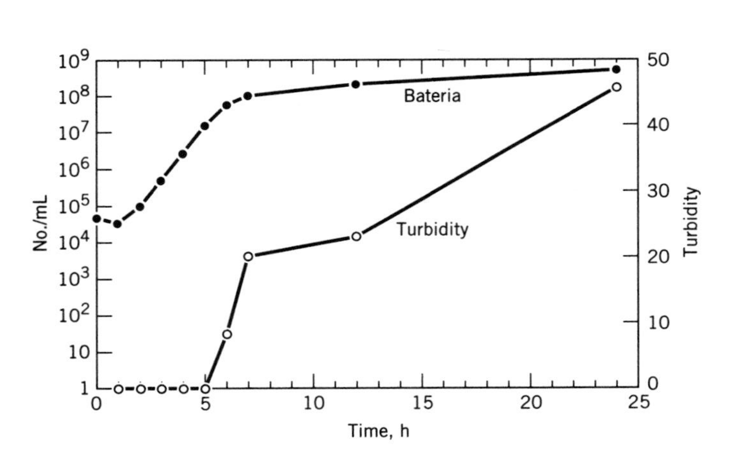

When the turbidity curve is visually compared to the bacterial numbers curve, turbidity is a reasonable measure of viable mass during the growth phase but turbidity is not an accurate measure of bacterial numbers when suspended debris accumulates in the culture after death begins to occur to a significant extent. The curves indicate that there was a significant amount of dieoff between 12 and 24 h, although the culture was primarily in a stationary phase over this period.

The yield factor may also be expressed with units of COD of biomass formed per COD of substrate transformed or removed. The yield factor with the above units is simply multiplied by the COD per unit biomass. The yield factor expressed on a COD or BOD basis is always less than 1. For every unit of substrate removed a portion of the substrate will be synthesized into biomass, whereas the remainder will be used (oxidized) to provide energy for cell synthesis and other life functions.

6.4.2 Growth of Mixed Cultures

The growth of mixed cultures is quite similar to the growth of pure cultures. Mixed cultures, because of the variety of microbial species contained within them, are better able to survive and grow in any given situation. Figure 6.7 (cf. Fig. 6.6) shows a typical growth curve for a mixed culture placed in a suitable medium.

The following features are different for growth of a mixed culture compared to a pure culture. The lag or acclimation phase is generally of shorter duration for a mixed culture. The probability of a few of the species in the mixed culture being adapted to the medium at inoculation is higher because of the diversity of organisms present. Therefore, log growth is more quickly initiated. The rate of increase in the log growth phase may accelerate as more species acclimate to the medium.

There will likely be a continuous succession of dominant species in a mixed culture as substrate availability and environmental conditions change within the medium. The duration of the log growth, declining growth, stationary, and endogenous phases are all more likely to be longer than the respective phases for a pure culture because of the diversity of species in a mixed culture. The viability of the culture as a whole is extended in the log death phase because the waste products of species and dead cells from starved species become substrate for other species. This phase is more typically known as the endogenous phase for mixed cultures. Predation can play a significant role in all phases.

Equations (6.2)–(6.5) apply to the growth of mixed cultures as well as to pure cultures.

6.4.3 Viability and Mass in Growing Cultures

The equations in the previous sections have been formulated in terms of numbers or mass of viable microorganisms. Mass of suspended matter is often used as an approximation of mass of viable cells. The distinction between the two measures becomes more extreme as the declining growth phase occurs and carries on to the following phases.

In the log growth phase, when there is an abundance of substrate, there will only be a small to negligible amount of death of microorganisms. Average cell weights may change somewhat as time progresses but the relation between viable cells and solid, suspended microbial mass will be more or less constant. Once dieoff becomes significant, the relation between viable cells and suspended matter weakens. The accumulation of dead cells, even though cell lysis is significant, will contribute to measurements of suspended solids. In the declining growth phases and thereafter, viable cells may represent only a small portion of the total cells or suspended mass.

Typically VSS (volatile suspended solids; see Fig. 5.7 in Sec. 5.7.1) are used as a measure of microorganisms. Volatile solids are those that will burn in the presence of oxygen and are a better measure of organics than total suspended solids (TSS) which includes inert solids. This is not discounting the fact that all cells contain some inorganic substances. VSS is convenient to determine. However, for many cultures involved in wastewater treatment where substrate concentrations are low relative to the microbial mass, VSS will not be indicative of the numbers of viable microorganisms, which are the active mass responsible for treatment. Thus, correlations and rate constants developed using VSS should only be expected to apply under conditions that are similar to those for which the correlation has been developed.

It is not practical in terms of day-to-day operation to measure numbers or mass of individual species in a mixed culture to which any adventitious microorganism has entry. VSS is the measure of choice.

6.5 IDENTIFICATION AND ENUMERATION OF MICROORGANISMS

The identification and enumeration of microorganisms is generally a tedious task involving a series of tests, although for a significant number of microorganisms, relatively rapid procedures have been developed. This section gives a brief overview of the procedures. A good manual on laboratory procedures for microbiology (including *Standard Methods,* 1992) should be consulted for details.

6.5.1 Tests and Procedures

The identification procedures consist of physical examination and measurement of response of the microorganism to physiological conditions and reaction with various compounds. For bacteria, *Bergey's Manual of Systematic Bacteriology* (1984) provides the definitive classification of these organisms based on their characteristics. The tests outlined below are conducted systematically in accordance with this manual to suggest choices of the next level of tests to finally arrive at a definitive identification of the species. It must be cautioned that responses to any of the tests can be atypical and final identification is based on responses to a large number of tests.

In a mixed culture, the species of interest must be separated from the other species to obtain a pure culture for examination. The mixed culture is dispersed in a liquid medium containing agar that solidifies and the isolated colonies that grow are sampled and subjected to tests to confirm their identification. Commonly the mixed culture is grown in an enrichment medium that favors the growth of certain species over others. Microorganisms that grow on the enrichment medium are then subjected to further examination for known characteristics to confirm the presence or absence of suspected species.

Sterilization and Aseptic Techniques

All microbiological analyses are conducted using sterile apparatus and aseptic techniques. Sterilization of apparatus is accomplished with heat, chemical agents, filtration, or radiation. The most effective form of sterilization is with an autoclave, which is a pressure cooker that subjects the apparatus or medium to steam at 121°C. These conditions are maintained for at least 15 min. No organism has been known to survive these conditions. The presence of water is essential to the efficacy of autoclaving.

Heating glassware to 160°C and holding it at that temperature for at least 1 h is another technique.

Some apparatus cannot physically withstand autoclaving or heating and alternative sterilization techniques are used. Ethylene dioxide is a gas that kills microorganisms and may be employed. Ultraviolet and gamma radiation can also be utilized to kill microorganisms. Media that will decompose upon heating can be sterilized by passing it through filters that contain pores that are too small for microorganisms to pass.

Aseptic techniques consist of common sense measures to minimize the potential for contamination. Flaming forceps and picks to red hot conditions before using them is standard practice. Clean clothing and countertops disinfected with a chemical solution are other standard measures. More extreme measures may be dictated in certain circumstances. Lab manuals (*Standard Methods,* 1992) and other references (Mara, 1974) provide information on aseptic practices and sterilization techniques.

Selective and Differential Enrichment Media

A medium that is specific to one group of microorganisms allows them to outgrow other species. The medium is tailored to meet the requirements of the group of interest and other environmental conditions such as temperature are controlled to the optimum value for this group. This gives the group a competitive advantage over other species; i.e., the medium is selective.

It is unlikely that a medium can be made selective to the point where only the desired group will grow and all other species will grow at an insignificant rate. Therefore the medium usually has a component that reacts in a specific manner for the specified group and causes a visible indication of the presence of the desired group. This trait makes the medium differential. It may still be possible for other species to cause the desired differential reaction and further tests will therefore be required, but lack of a positive reaction is normally sufficient evidence that the group of interest is not present.

Physical Examination

Standard physical examinations include microscopic and naked eye observations of the morphology of the cells and colonies on agar slants and plates. The color and texture of the colonies and other visible features all provide clues for the identification of the culture. Some species grow on or near the surface of media, others grow deeper into the media.

The size of the cells and other features such as growth in clumps and chains are recorded.

Staining Cells

Staining techniques are important to identify cell components. Various dyes are used with sometimes more than one used for one test.

The most important staining technique is the Gram stain. The technique involves two dyes, iodine, and a solvent. The primary dye is crystal violet (purplish color) and the secondary dye is safranin (red color). The stain that remains is related to the cell wall characteristics. Those cells that are blue after the procedure is finished are termed gram positive and those that are red upon completion are termed gram negative. Most cells of living organisms are gram positive. Yeast, some bacteria, and a few molds are gram negative. As for most of the tests, results are not always consistent for the same species. This test normally eliminates a large number of microorganisms from consideration and is one of the first tests performed in an identification sequence.

Another important staining technique is the acid-fast stain. An organism is called acid fast if it resists destaining with acids or alcohol, following staining with hot carbol–fuchsin. Other stains exist to differentiate cell components and inclusions.

Physiological Tests

The response of the microorganism to various concentrations of salts, sugars, hydrogen ions (pH), and other substances such as milk constitute important differential tests. Varying the temperature of incubation is also included in this group of tests. The intensity of exposure of the microbe to the stimulus is varied systematically with careful observation and measurement of the response to characterize the culture.

Growth Factors and Metabolic Tests

Microorganisms need certain growth factors such as trace minerals, prefabricated amino acids, or proteins to survive and grow. The tests referred to above are often conducted with a general medium that contains adequate amounts of growth factors but is altered only with respect to the substance of interest. To determine the need for growth factors, media are made without vitamins, amino acids, or other substances and the survival of the species is observed. Furthermore, not only the ability to utilize a specific substrate but also the production of specific products from a given substrate provides information on the biochemical traits of the organism.

Antigens are foreign macromolecules that result in the formation of antibodies. Serology is the study of antigen–antibody reactions, which are used to classify microorganisms into serotypes.

6.5.2 Enumeration of Microorganisms

Enumeration techniques must also effectively identify the microorganisms of interest. Because of the microscopic size of microorganisms, it is necessary and convenient to develop measurement methods that produce more readily observed results. There is no chemical analysis that can definitively identify a complex entity such as a microorganism. However, chemical agents can be associated with a culture either through their production or through requirements by the bacteria and thus be an important tool in devising enumeration techniques. Most techniques rely on the rapid growth of microorganisms, where one microorganism can produce a visible colony or cause identifiable chemical reactions within a short period of time.

Heterotrophic Plate Count

The heterotrophic plate count (HPC, formerly known as the standard plate count) is a direct plate count method designed to measure the total bacterial population of a water. With this objective it is necessary to design a medium that will support the growth of all bacteria in a reasonable period of time. This requirement is a particularly severe constraint because of the diversity of species and their individual growth requirements (temperature, nutrient, growth factors, and others) that have been discussed previously. Therefore, no medium and other environmental conditions will suffice for all microorganisms. However, there are several general purpose media that do satisfy the metabolic needs of the majority of microorganisms and one of these is employed in the method. Temperatures of incubation are either 20 or 35°C.

The procedure for the HPC is to shake the sample and prepare dilutions as necessary. A small volume, on the order of 0.1–1 mL, is withdrawn from the sample and added to a petri dish with the medium that contains agar. The sample and the medium are thoroughly mixed and the covered petri dish is placed in an incubator. The agar will harden and prevent the microorganisms from moving. Depending on the temperature, most of the microorganisms that can grow will produce a visible colony in 24 to 48 h, although in some instances incubation times as long as 4 d may be required to achieve the best results. The colonies are counted to compute the concentration based on the volume of sample that has been plated.

The HPC is used as a monitor of water quality in water treatment operations. The test can provide a general measure of microorganism activity in natural waters and wastewater treatment operations.

Most Probable Number Techniques

Most probable number (MPN) techniques (Greenwood and Yule, 1917) consist of taking a series of aliquots of different volume from a sample containing the culture and placing the aliquots in separate test tubes containing an enrichment medium. If an aliquot contains one or more of the microorganisms for which the test is designed, a positive reaction will occur in the test tube. The number of positive tubes is related to the density of microorganisms in the original sample. The most probable numbers are based on standard statistical analysis.

It is common to examine three sets of sample aliquots. The first set consists of five 10 mL aliquots from the sample; the second consists of five 1 mL aliquots; and, the third set consists of five 0.1 mL aliquots. The aliquots are placed in separate tubes containing the enrichment medium. If an aliquot contains one or more bacteria of the species that will grow and cause the desired reaction at a visible or measurable level, a positive result will be obtained. The Poisson distribution describes the probability of r occurrences of an event when the mean frequency of occurrence is μ. The expression (Kennedy and Neville, 1986) is

$$p_r = \frac{e^{-\mu}\mu^r}{r!} = p \tag{6.6}$$

where

p_r (or p) is the probability of r occurrences of the event (finding bacteria in a sample)

A binomial distribution (Kennedy and Neville, 1986) describes the probability of an event succeeding z times in n trials. The expression is

$$p_z = \frac{n!}{z!(n-z)!} p^z (1-p)^{(n-z)} \tag{6.7}$$

If the concentration of bacteria in the sample is C (No./mL) and the volume of a sample aliquot is V_i (mL), the mean number of bacteria in the volume is

$$\mu = V_i C \tag{6.8}$$

As noted above, only one bacterium is needed to provide a positive tube; therefore, Eq. (6.8) is substituted into Eq. (6.6) with r set equal to 0, resulting in Eq. (6.9a), which gives the probability of a tube not being positive. Because all probabilities sum to 1, the probability of finding one or more bacteria in a sample and thus a positive tube when the mean number is μ is given by Eq. (6.9b).

$$p = e^{-V_iC} \quad (6.9a) \qquad 1 - p = 1 - e^{-V_iC} \quad (6.9b)$$

Using the above equations, the probability of finding a, b, and c positive tubes in three sets of tubes with n_1, n_2, and n_3 tubes in each set where sample volumes of V_1, V_2, and V_3 are used, respectively, is

$$p_{a,b,c} = \frac{n_1!n_2!n_3!}{a!(n_1 - a)!b!(n_2 - b)!c!(n_3 - c)!}(1 - e^{-V_1C})^a(e^{-V_1C})^{(n_1-a)} \times (1 - e^{-V_2C})^b(e^{-V_2C})^{(n_2-b)}(1 - e^{-V_3C})^c(e^{-V_3C})^{(n_3-c)} \quad (6.10)$$

where

$p_{a,b,c}$ is the probability of finding a, b, and c positive tubes

From an analysis, a, b, and c will be determined for volumes V_1, V_2, and V_3 with n_1, n_2, and n_3 sets of tubes, respectively. The problem is to estimate the bacterial density that is most likely to result in these numbers of positive tubes. A given bacteria concentration results in a given probability. Statistical techniques of maximum likelihood estimation are used to transform the data, weight it, and assume an underlying distribution of the small number of samples. McCrady (1915) put forward the basic technique, which has been refined by many authors (see, for example, Finney, 1988).

The MPN is the bacterial density that produces the highest probability with a given set of constants. A procedure that gives an approximate answer without considering weighting factors and other corrective measures is to find the value of C that produces the maximum value of Eq. (6.10) by solving the following equation:

$$\frac{dp_{a,b,c}}{dC} = 0 \quad (6.11)$$

Standard deviations, confidence limits, and other measures of spread of the MPN have been tabulated. See Table A.1 in the Appendix (or *Standard Methods,* 1992) for MPN values for selected aliquot combinations.

It may be necessary to subject the positive tubes from the preliminary analysis to one or more tests or enrichment media. This additional information finally confirms the presence of the suspected species or group in the positive tubes. The MPN technique is the method against which other bacteriological analyses are compared to assess their accuracy.

Membrane Filter Techniques

Over the past 30 years, membrane filter (MF) analyses have developed into a formidable tool for the identification and enumeration of microorganisms. The procedure consists of passing a known volume of sample through a filter that contains openings that are smaller than the species or group being analyzed. A vacuum is required to accomplish the filtration. The filter is then placed in a petri dish containing a selective and differential medium. The dish is incubated at the appropriate temperature. Single bacteria will give rise to visible colonies during incubation. Colonies exhibiting the characteristic reaction are counted (Fig. 6.8). This direct count is assumed to be the number of microorganisms from this group that were present in the volume of sample that was filtered.

The technique assumes that each colony has derived from a single cell and it is essential that the sample be well shaken to break up clumps and distribute the microorganisms on the filter. The analyses are usually performed in duplicate to improve the accuracy of the result and check the consistency of the analyses. It is

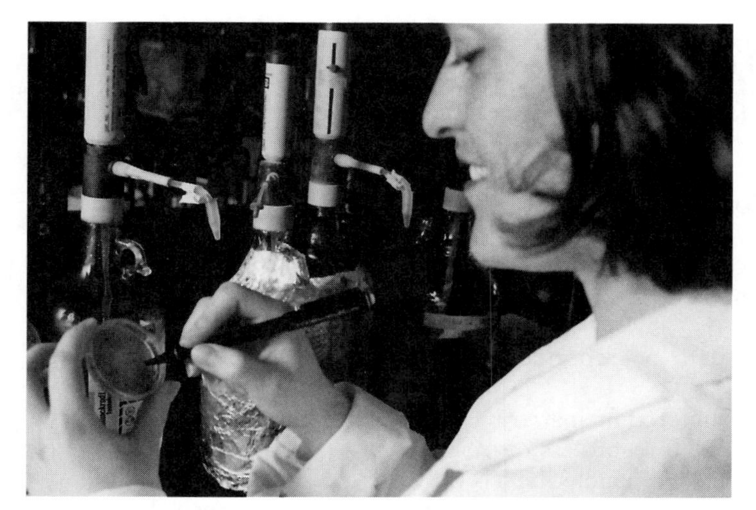

Figure 6.8 Counting colonies on a membrane filter.

necessary to perform serial dilutions of the sample in instances when no firm estimate of the final count is available.

Effective selective and differential media have been developed for a number of microorganisms of importance to water supply and sanitation engineers. The MF techniques have been tested against MPN techniques to ensure their reliability.

The MF technique is rapid and convenient. If a high concentration of suspended solids exists, filtration time may be too lengthy and the MPN technique is the only alternative.

Practical Considerations in Determining Mean Values

Arithmetic and geometric means are used in reporting coliform count values. The geometric mean is commonly used because counts are often log-normally distributed. The median is also useful in describing data. These measures are defined as follows:

$$\text{Arithmetic mean} = \frac{\Sigma y_i}{n}$$

$$\text{Geometric mean} = \sqrt[n]{y_1 \times y_2 \times \cdot \cdot \cdot \times y_n}$$

where

y_i is an individual observation

n is the total number of observations

Median: if n is odd, use the middle value

if n is even, use the average of $\frac{n}{2}$ and $\left[\frac{n}{2} + 1\right]$ values

The example below discusses some problems in describing bacteria concentration data.

■ Example 6.2 Interpretation of Microorganism Assay Data

Fecal coliform counts taken during a 1-month period of time are reported below in ascending order. (The counts are per 100-mL sample.)

Count: 0, <25, 60, 85, 168, 234, 330, TNTC

where
TNTC means too numerous to count

How should these data be reported for median, arithmetic mean, and geometric mean?

Some problems encountered in answering these questions are: If 0 is considered, the geometric mean will be 0 regardless of other data. What value does one use for the <25 count? How does one handle the TNTC value?

Using the zero value would give an unacceptable geometric mean. Eliminating the zero value also biases the results. A practical solution is to change it to 1.

Because no information is available for the <25 observation, the best estimate for the true value would be 12 or 13, although any value from 0 to 24 could be correct. Using 13 would be the more conservative choice.

If the TNTC value is deleted, serious bias is again introduced. If a value of 330 can be counted then it is reasonably assumed that the TNTC represents a number of 331 or larger. This is the number that should be used without any further information even though it probably underestimates the actual value.

Results:

Median	127
Geometric mean	64
Arithmetic mean	153

6.5.3 Detection of Viruses in Water

Viruses are also significant agents of waterborne disease. Identification and enumeration of viruses are much more difficult than they are for bacteria or protozoans. Viruses consist of a nucleic acid core (genome) that is surrounded by a shell of proteins known as a capsid. In some types of viruses the capsid is also enveloped by a lipid bilayer membrane. Viruses are unable to reproduce by themselves but must invade a host cell to replicate.

Techniques to detect viruses involve two stages: a concentration stage followed by an assay to determine the number and type of viruses. Five techniques may be used to concentrate viruses: passive adsorption, directed adsorption, ultrafiltration, direct physicochemical flocculation and phase separation, and affinity chromatography. The directed adsorption technique is most commonly used (Hurst et al., 1989). Directed adsorption is essentially the filtration of a volume of water. The viruses are accumulated on a filter or medium. There are a variety of chemical agents and media that can be used to improve the capture of viruses.

TABLE 6.1 Effect of pH on Simultaneous Adsorption of Three Virus Types onto Different Materials[a]

Filter type[b]	pH	Extent of viral adsorption (%)			
		f2[c]	ϕX174[c]	P1[d]	Average
A	3.5	81	84	76	80
	7.0	59	63	48	57
	8.0	ND[e]	ND	ND	
	8.5	40	74	54	56
B	3.5	99	80	84	88
	7.0	100	55	41	65
	8.0	99	66	42	69
	8.5	93	69	48	70
C	3.5	100	99	99	99
	7.0	88	61	98	82
	8.0	100	55	96	84
	8.5	100	56	95	84

[a]Reprinted from Hurst et al. (1989) *Journal of the American Water Works Association,* by permission.
[b]Three filters with different characteristics were tested.
[c]Virus types F2 and ϕX174 are bacteriophages that infect *E. coli.*
[d]Virus type P1 is a human polio virus type 1, a human enteric virus.
[e]ND = nondetectable.

The viruses must then be eluted (washed) from the medium. Again, different eluant agents can be used with varying efficacy. The overall efficiency of viral recovery can be quite low depending on the choice of media and eluant (Table 6.1).

Assay techniques include direct observation of virus particles by electron microscopy, observation of living host cells that have been infected with viruses, or adding tagged antibodies or nucleic acids that bind with biomolecules in the viruses. Some techniques such as electron microscopy are unable to distinguish whether a virus particle is infectious.

Cytopathogenicity and plaque assays are two commonly used assays, although Hurst et al. (1989) note that in situ nucleic acid hybridization is superior because of shorter times for the assay and greater sensitivity. Cytopathogenicity assays are based on the appearance of visibly observable (by microscope) changes in the morphology of living host cells when infected by viruses. Plaque assays detect the production of focal areas of death within an infected culture of cells. A plaque is an area of a viral colony on a plate culture formed by lysis of a host cell. A stain may facilitate detection. Plaques can normally be detected without the aid of a microscope.

6.5.4 Techniques for Enumerating Other Microorganisms

Larger organisms present in natural water and sewage samples can be counted directly. An overview of various techniques is given by Fox et al. (1981). This reference also contains color plates of a number of larger microorganisms found in sewage.

When the concentrations of the microorganisms are low, a concentration approach is required, otherwise a sample aliquot is drawn and diluted. For raw sewage, Fox et al. recommend concentrating the solids from samples ranging from 4 to 40 L to obtain a representative sample. Large microorganisms will settle and sedimentation procedures will concentrate ova, cysts, arthropod (insects, mites, centipedes, etc.)

larvae, worms, and macroinvertebrates. Besides plain sedimentation, centrifugation can be used to concentrate sewage microorganisms if the microorganism can withstand the force. The converse of sedimentation is flotation. A flotation medium is a concentrated solution of a salt or sucrose. Ova or cysts will rise to the top.

Smaller microorganisms can be counted with a hemacytometer. A hemacytometer consists of a glass chamber with a grid etched on it. The cell is covered with a coverslip and a sample is allowed to flow under the coverslip. Some time is given for the microorganisms to settle and then the count is made under microscope. The count is translated into a concentration with a factor dependent on grid size and depth of the cell.

There are other types of chambers for direct enumeration. The Whipple disk is often used for plankton enumeration. The Whipple disk is an ocular gridded disk that is placed in the eyepiece of a microscope to perform a grid count. Various grid sizes are present on the disk for use with varying densities of microorganisms in the aliquot.

Staining is often essential for the identification of microorganisms, particularly protozoa. Some stains are fixatives that will kill the species; others will allow observation of the viable forms for some time. The stain is added to the dispersed pellet obtained from centrifugation or the concentrated sludge from the sedimentation procedure.

Adult forms of pathogenic parasites (those which invade host organs and tissues) are never found in processed sewage. Only the potentially transmissible forms such as ova may be observed (Fox et al., 1981).

QUESTIONS AND PROBLEMS

1. How does a chemosynthetic autotroph obtain its energy and what is its carbon source?
2. Consult *Bergey's Manual* to find whether *Escherichia coli* has simple or complex growth requirements. Can members of the genus *Escherichia* ferment acetate or citrate as a sole carbon source?
3. On the taxonomic scale, why are there few anaerobic organisms higher than prokaryotes?
4. A medium containing an abundance of substrate was seeded with 20 mg/L VSS of an active mixed culture. After 8.0 h the VSS concentration was measured at 590 mg/L. Data indicate that there was a lag time of 0.5 h. Assuming log growth, what is the average generation time of this culture?
5. Prove that Eq. (6.5) reduces to Eqs. (6.3) and (6.4) as substrate concentration varies from a high value to a low value.
6. What is the order of the substrate removal expression (Eq. 6.5) with respect to substrate when the following conditions exist: $k = 1.15\ d^{-1}$, $K = 100$ mg/L, $X = 2\,200$ mg/L, and $S = 250$ mg/L? What is the rate of substrate removal? (Recall that the Michaelis–Menten equation has an order from 0 to 1 with respect to substrate.)
7. Why is a mixed culture hardier than a pure culture for any given medium? Compare and contrast the growth phases of a pure and mixed culture for any given medium.
8. Would turbidity be a reasonable measure of microorganisms throughout the growth cycle of a pure or mixed culture placed in a medium containing only dissolved components? Explain your answer.
9. What are the advantages and disadvantages of using VSS as a measure of viable microorganisms?
10. (a) What is a facultative microorganism? (b) Would a facultative microorganism choose aerobic or anaerobic metabolism if it had a choice?

11. If ideal conditions existed for 30 h, how many organisms would exist in a culture that started with one organism that had a doubling time of 25 min, after 10, 20, and 30 h?

12. If 10% of a species of microorganisms was being removed from a culture every hour, what must be the doubling time of this species to sustain it in the culture?

13. If the yield factor is 0.60 mg biomass COD/mg substrate COD removed, calculate the amount of substrate removed after 6 h for the data in Example 6.1. Assume that the mass of an *E. coli* cell is 10^{-12} g.

14. Three species of microorganisms have generation times of 20, 75, and 125 min, respectively. The average weight of cells is the same for all species. At the initial time of an experiment, a culture was seeded with equal amounts of each species. (a) Plot the mass or numerical ratios of species 2 and 3 to species 1 as a function of time over a 24-h period of time for 1 h intervals assuming that generation times are constant. (b) If the weight of cells of species 2 and 3 are 110 and 130%, respectively, of the weight of species 1, perform the same exercise as in (a) if equal masses of each species were seeded into the medium.

15. Describe the features of a medium that is selective and differential.

16. Why is the MPN procedure the standard against which other enumeration techniques are compared?

17. A sample was analyzed with three sets containing five tubes each that contained 10-, 1-, and 0.1-mL aliquots, respectively. What is the most probable number of bacteria in the sample if three of the 10-mL tubes, two of the 1-mL tubes, and one of the 0.1-mL tubes were positive? (a) Use Table A-2 in the Appendix. (b) Use Eqs. (6.10) and (6.11). The result from these two equations will not exactly agree with the result from the table as discussed in the text.

18. It is a rule of thumb that the rate of a chemical reaction doubles with each 10°C increase in temperature. Discuss the effects of a temperature increase on the rate of growth of pure and mixed cultures of microorganisms placed in an ideal medium.

19. One liter of sterile medium containing a COD of 8 500 mg/L is seeded with 1 mg of microorganisms as COD from the same species. The medium is highly buffered and at the ambient temperature this species can double every 2.1 h. (a) If 1 mg of substrate COD produces 0.66 mg (as COD) of biomass (the other 0.34 mg of COD is oxidized to provide energy for growth), assuming log growth, how much biomass as COD and substrate exist after 24 h? (b) If the microorganisms grow in the log phase until the supply of original substrate is exhausted, how much time will elapse? Why should the medium be buffered?

20. For three sets of five samples containing 10-, 1-, and 0.1-mL aliquots, respectively, what is the probability of finding 3, 3, and 1 positive tubes in the respective sample sets when the sample contains a bacteria concentration of 36 microorganisms/100 mL?

KEY REFERENCES

Bergey's Manual of Systematic Bacteriology (1984), The Williams & Wilkins Co., Baltimore.

GAUDY, A. F. JR., AND E. T. GAUDY (1980), *Microbiology for Environmental Scientists and Engineers*, McGraw-Hill, Toronto.

MARA, D. D. (1974), *Bacteriology for Sanitary Engineers*, Churchill Livingstone, London.

REFERENCES

BUISMAN, C., P. IJSPEERT, A. JANSSEN, AND G. LETTINGA (1990), "Kinetics of Chemical and Biological Sulphide Oxidation in Aqueous Solutions," *Water Research*, 24, 5, pp. 667–671.

CANFIELD, D. F., S. B. LINDA, AND L. M. HODGSON (1985), "Chlorophyll-Biomass-Nutrient

Relationships for Natural Assemblages of Florida Phytoplankton," *Water Research Bulletin*, 21, 3, pp. 381–391.

CURDS, C. R. AND A. COCKBURN (1970), "Protozoa in Biological Sewage Treatment Processes-II. Protozoa as Indicators in the Activated Sludge Process," *Water Research*, 4, 3, pp. 225–236.

FINNEY, D. J. (1988), *Statistical Methods in Biological Assay*, 3rd ed., Oxford University Press, New York.

FOX, J. C., P. R. FITZGERALD, AND C. LUE-HING (1981), *Sewage Microorganisms: A Color Atlas*, Lewis Publishers, Chelsea, MI.

GREENWOOD, M. AND G. U. YULE (1917), "The Statistical Interpretation of Some Bacteriological Methods Employed in Water Analysis," *J. Hygiene,* 16, pp. 36–54.

HURST, C. J., W. H. BENTON, AND R. E. STETLER (1989), "Detecting Viruses in Water," *J. American Water Works Association*, 81, 9, pp. 71–80.

KENNEDY, J. B. AND A. M. NEVILLE (1986), *Basic Statistical Methods for Engineers and Scientists*, 3rd ed., Harper & Row, New York.

KINNER, N. E. AND C. R. CURDS (1987), "Development of Protozoan and Metazoan Communities in Rotating Biocontactor Biofilms," *Water Research,* 4, 4, pp. 481–490.

MCCRADY, M. H. (1915), "The Numerical Interpretation of Fermentation-Tube Results," *J. Infectious Disease*, 17, pp. 183–212.

MONOD, J. (1949), "The Growth of Bacterial Cultures," *Annual Reviews of Microbiology*, 3, pp. 371–394.

Standard Methods for the Examination of Water and Wastewater (1992), 18th ed., American Public Health Association, Washington, DC.

CHAPTER 7

WATER, WASTES, AND DISEASE

Because water is used as a carriage vehicle for excreta, many pathogens (disease-causing microorganisms) enter waters via this route. Also, a number of diseases are indirectly associated with water and waste disposal. Stormwater runoff is the natural route for surface wash and transport of fecal material to waters.

A number of considerations apply to a pathogenic agent. Factors that may be important are the following. Some microorganisms have a latency period between excretion from a carrier or infected individual and becoming infective. The infective stage of some pathogens lasts more than 1 year, whereas for others it is only a few days or less. In the tables in the following sections, the existence of major reservoirs other than humans will be indicated. Some pathogens are able to multiply outside their human hosts. Vaccines exist for some of the diseases; sometimes the vaccines produce side effects. Infective doses vary from organism to organism and the state of health and immunity of the exposed individual. Almost all pathogens can be spread by humans in a symptomless carrier state (Feachem et al., 1981). See Feachem et al. (1981) and Benenson (1985) for more details on these aspects.

It is the very young and elderly who suffer the most severe consequences of the diseases, although increased occurrence of the diseases takes its toll by decreasing life spans of those who survive infancy.

7.1 AGENTS OF DISEASE

Microorganisms are responsible for a wide spectrum of diseases. Except for helminths and parasitic diseases spread by vectors, most of the pathogenic agents discussed in the following sections have to be orally ingested to cause the disease. Therefore hand-to-mouth, water, and food are the most common routes for disease transmission.

7.1.1 Bacterial Pathogens

Common bacterial agents of disease are given in Table 7.1; the table is not a complete survey. Many of these agents are associated with diarrhea or dysentery (gastroenteritis), which is a more severe upset of the digestive system involving inflammation of the stomach and intestinal linings. Blood is excreted with dysentery. The organism causing typhoid fever is a notable exception that does not cause diarrhea. Agents found in excreta are obvious candidates for waterborne transmission, particularly because water is used as a carriage vehicle for wastes.

Perhaps the most spectacular disease is cholera. In this disease the microbial agent, *Vibrio cholerae,* produces a substance that upsets the isotonic balance in the fluids in the digestive system causing excessive fluid loss. Death can occur in 3 to 12 h. There

TABLE 7.1 Bacterial Pathogens

Agent	Disease	Reservoir
Campylobacter spp.	Diarrhea	Animals
Escherichia coli	Diarrhea	Humans
Legionella spp.	Pneumonia	Aquatic habitats
Leptospira spp.	Leptospirosis	Humans and animals
Mycobacterium spp.	Tuberculosis	Humans and cattle
Pasturella tularensis	Tularemia	Humans and animals
Salmonella spp.	Food poisoning, salmonellosis	Humans and animals
Salmonella paratyphi	Paratyphoid fever	Humans
Salmonella typhi	Typhoid fever	Humans
Shigella spp.	Bacillary dysentery	Humans
Vibrio cholerae	Cholera	Humans
Vibrio spp.	Diarrhea	Humans and animals
Yersinia spp.	Yersiniosis (acute enteric disease)	Humans and animals

is no vaccine that imparts any lasting immunity. The disease can be counteracted by replacing fluids either orally or intravenously.

Some of the agents are found in numerous habitats and are quite hardy. For example, *V. cholerae* and other vibrios are part of the indigenous microflora of most estuaries. They are able to persist in demanding environments. The harsh conditions of anaerobic digestion of rum distillery effluent did not decrease survival of *V. cholerae* compared to its survival in untreated effluent (Rojas and Hazen, 1989).

Some of the agents must be viable when ingested to cause the disease. This is not the case for food poisoning caused by *Salmonella* species. Food that is inoculated with the bacteria provides a suitable environment for the bacteria to multiply and produce their toxin. Then when the food is ingested the effects of the toxin are manifest. Salmonella food poisoning is often fatal.

There are waterborne disease-causing agents that do not normally reside in humans or animals. *Legionella* is an example. *Legionella* species have only been identified recently because of their specific media and culturing requirements. The major illness caused by these microorganisms is pneumonia, which is often accompanied by severe gastrointestinal symptoms (Muraca et al., 1988). Healthy individuals are rarely susceptible to the disease. *Legionella* have a natural habitat of lakes and rivers and they can readily colonize water distribution systems or home heating units (Muraca et al., 1988; Witherell et al., 1988). The mode of infection appears to be inhalation of contaminated aerosols. Aerosol dissemination from ventilation–dehumidification systems has often been implicated as the transmission route; however, showering has not been implicated (Muraca et al., 1988).

Many of these diseases are essentially unknown in developed countries, although there are instances of these diseases because of breakdowns in preventative measures and travel of people to countries where the diseases are found. Proper water treatment, food handling, and home hygiene practice are effective prophylactic measures.

7.1.2 Viral Pathogens

Viruses were first recognized as agents of disease through observation of the harmful effects they had on plants and animals in the late 1800s. Since then viruses have been identified that cause almost every type of disorder, including death, known to medical science. Viruses invade cells and take over their machinery, disrupting cell function or causing death of the cell. Viruses infect all living organisms from bacteria to plants, animals, and humans. Unlike bacteria, of which the largest number are beneficial to their host or the environment, viruses are responsible for damaging cells.

Virology is a more complex and costly endeavor than bacteriology or other disciplines of biology. Primarily for this reason, the state of knowledge on viruses, in general, is not as refined as knowledge in other areas (Katzenelson, 1978). The transmission of viruses through water can be magnified by a person becoming infected by this route and then spreading the disease by coughing or sneezing aerosols, which contaminate others through inhalation or ingestion of viruses spread on foods. Table 7.2 lists some viruses that have been implicated as waterborne.

Many of the viruses listed in Table 7.2 are members of the family of enteroviruses, which replicate in the intestinal tract. However, except for hepatitis A, low-level transmission of enteroviruses by water did not appear to contribute appreciably to the incidence of infection by enteroviruses [International Association on Water Pollution Research and Control (IAWPRC) Study Group, 1991]. Reoviruses are also considered of minor epidemiological importance in water. Other viruses such as adenoviruses are mainly respiratory tract viruses but they have been found in feces. The "numerous conditions" for any given group include most of the following: fever, diarrhea, respiratory illness, meningitis, and paralysis. Certain adenoviruses cause eye infections contracted in swimming pools.

The rotaviruses are considered as important agents of childhood and possibly adult gastroenteritis (IAWPRC Study Group, 1991). There is strong evidence that water transmission is significant for a variety of ill-characterized viruses that cause gastroenteritis.

The acquired immunodeficiency syndrome (AIDS) is caused by a group of viruses known as human immunodeficiency viruses (HIV). When a person is infected with these viruses, death usually results although the time for this to occur may be extended. This group of viruses is very fragile outside the human body and the possibility of waterborne transmission is essentially zero (Riggs, 1989). Similar conclusions were arrived at by Johnson et al. (1994), who evaluated the risk of HIV infection to wastewater treatment plant personnel. HIV was found to be fairly stable in wastewater and sterile water for up to 12 hours but both are a much more hostile environment than cell growth media (Casson et al., 1992). HIV survival was significantly less than that of poliovirus.

Viruses causing the common cold are also unable to survive outside the human body for any significant time and their transmission through water is not of concern.

TABLE 7.2 Viral Pathogens[a]

Agent	Disease	Reservoir
Adenoviruses	Numerous conditions	Humans
Coxsackie viruses	Numerous conditions	Humans
Echoviruses	Numerous conditions	Humans
Hepatitis A	Infectious hepatitis	Humans
Polioviruses	Poliomyelitis	Humans
Reoviruses	Numerous conditions	?
Rotaviruses	Gastroenteritis	?

[a]Data from Feachem et al. (1981) and IAWPRC Study Group, (1991).

TABLE 7.3 Protozoan Pathogens

Agent	Disease	Reservoir
Entamoeba histolytica	Colonic ulceration, dysentery	Humans
Giardia lamblia	Diarrhea	Humans
Balantidium coli	Mild diarrhea, colonic ulceration	Humans and animals
Cryptosporidium	Diarrhea	Animals and humans

7.1.3 Protozoan Pathogens

Pathogenic protozoa are listed in Table 7.3; there are many others that may be infective. On a worldwide basis, *Entamoeba histolytica* infections are most common. In an adverse environment many protozoa may form cysts, which are a nonvegetative hardy form that reverts to a viable cell when conditions are suitable. All of the protozoans given in Table 7.3 are able to form cysts.

Cryptosporidium is a more recent addition to the list; it was first identified as a human pathogen in 1976. It is now being researched for its potential for waterborne transmission; it has been documented as the causative agent in six waterborne disease outbreaks worldwide (Rose, 1990). *Cryptosporidium parvum* is the major species responsible for clinical illness in humans and animals (Rose, 1988). It has been identified as causing a severe disease in individuals with compromised immune systems (Navin and Juranek, 1984).

7.1.4 Helminths

Helminths are parasitic worms many of which require one or more intermediate hosts in their life cycle. Roundworms or hookworms are nematodes, flatworms or flukes are trematodes, and tapeworms are cestodes. Some of these cause serious illness but many cause few symptoms. In any case, they all debilitate their host deriving their sustenance at the host's expense. They may damage organs. Most of the helminths with the exception of *Schistosoma haematobium* and guinea worms (dracunculiasis disease) are passed in feces. *Schistosoma haematobium* is passed in urine. Guinea worms erupt from blisters on the skin when the infected person is near water.

Except for *Stronglyoides*, helminths do not multiply within the human host. The life of many of these worms within the body is several years. For many worms, drugs exist that will rid them from the body. Table 7.4 lists common helminthic pathogens.

A number of helminths are quite hardy, yet the northern climates prove to be too severe for the majority of them. *Ascaris lumbricoides*, which is one of the hardiest of any pathogen, has been known to survive through harsh wastewater treatment processes and in northern climates.

The life cycle of the schistosome is pictured in Fig. 7.1. Schistosomiasis (commonly known as bilharzia) is one of the most prevalent diseases in the world. The intermediate host in this instance is a snail that is about the size of the tip of an adult's little finger. The snail lives in slow moving waters; reservoirs provide ideal breeding conditions for the snail in this regard. The second larval stage is able to penetrate the skin of any human being whether they are healthy or not.

TABLE 7.4 Helminthic Pathogens[a]

Agent	Common name of pathogen	Pathogen	Transmission	Distribution
Ascariasis	Roundworm	*Ascaris lumbricoides*	Human–soil–human	Worldwide
Clonorchiasis	Chinese liver fluke	*Clonorchis sinensis*	Animal or human–aquatic snail–fish–human	S.E. Asia
Diphyllobothriasis	Fish tapeworm	*Diphyllobothrium latum*	Human or animal–copepod–fish–human	Widely distributed
Dracunculiasis	Guinea worm disease	*Dracunculus medinensis*	Human–cyclops–human	Africa, India, Middle East
Enterobiasis	Pinworm	*Enterobius vermicularis*	Human–human	Worldwide
Fascioliasis	Sheep liver fluke	*Fasciola hepatica*	Sheep–aquatic snail–aquatic vegetation–human	Worldwide
Fasciolopsiasis	Giant intestinal fluke	*Fasciolopsis buski*	Human or pig–aquatic snail–aquatic vegetation–human	S.E. Asia, mainly China
Gastrodiscoidiasis	—	*Gastrodiscoides hominis*	Pig–aquatic snail–aquatic vegetation–human	India, Vietnam, Philippines, Bangladesh
Heterophyiasis	—	*Heterophyes heterophyes*	Dog or cat–brackish water snail–brackish water fish–human	Middle East, southern Europe, Asia
Hookworm	Hookworm	*Ancylostoma duodenale* *Necator americanus*	Human–soil–human	Mainly in warm, wet climates
Hymenolepiasis	Dwarf tapeworm	*Hymenolepsis* spp	Human or rodent–human	Worldwide
Metagonimiasis	—	*Metagonimus yokogawai*	Dog or cat–aquatic snail–freshwater fish–human	China, Japan, Korea, Siberia, Taiwan
Opisthorchiasis	Cat liver fluke	*Opisthorchis felineus*	Animal–aquatic snail–fish–human	Thailand, U.S.S.R.
Paragonimiasis	Lung fluke	*Paragonimus westermani*	Pig, human, dog, cat or other animal–aquatic snail–crab or crayfish–human	S.E. Asia; scattered foci in Africa and S. America
Schistosomiasis	Bilharzia	*Schistosoma haematobium*	Human–aquatic snail–human	Africa, India, Middle East
		S. mansoni	Human–aquatic snail–human	Africa, Arabia, Latin America
		S. japonicum	Animals and human–snail–human	S.E. Asia
Strongyloidiasis	Threadworm	*Stronglyoides stercoralis*	Human–human; (dog–human?)	Mainly in warm climates
Taeniasis	Beef tapeworm	*Taenia saginata*	Human–cow–human	Worldwide
	Pork tapeworm	*T. solium*	Human–pig–human or human–human	Worldwide
Trichuriasis	Whipworm	*Trichuris trichiura*	Human–soil–human	Worldwide

[a]From Feachem et al. (1981).

7.1.5 Vectors of Disease

There are many diseases that are indirectly associated with the provision of water or the proper disposal of excreta. An adequate amount of water used for personal hygiene limits the occurrence of many vectors of disease on the body, thus minimizing the risk of biting by the agent and subsequent infection. Water supply and wastewater removal schemes must take into account the breeding habitats of vectors associated with diseases that are endemic to an area. Again the northern climates prove to be too severe for the survival of most of the agents listed in Table 7.5, which lists some of the more significant vectors and associated diseases.

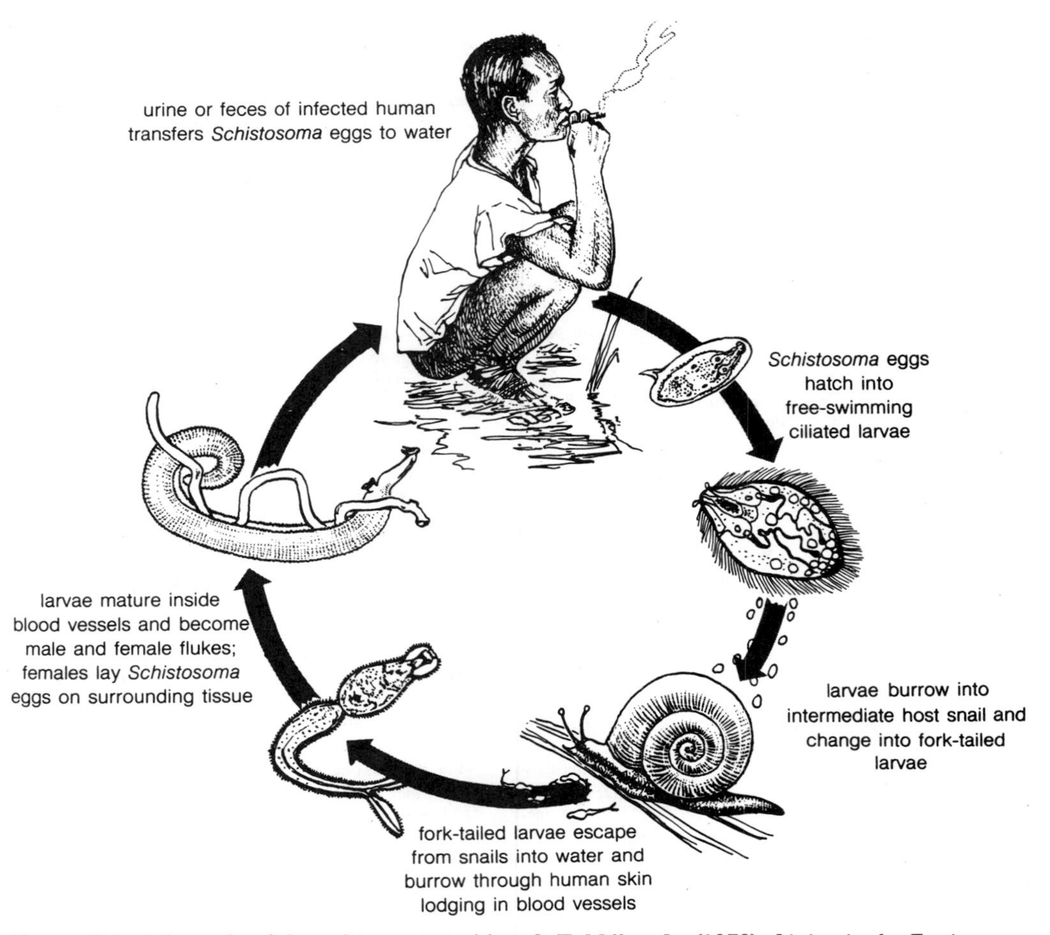

Figure 7.1 Life cycle of the schistosome. After G. T. Miller, Jr. (1979), *Living in the Environment,* 2nd ed., copyright Wadsworth Publishing Company, Belmont, CA. Reprinted with permission.

There are many other prophylactic measures beyond water–excreta management that may be exercised. These range from vaccination to chemical control of breeding grounds and use of nets.

7.2 INDICATORS OF FECAL CONTAMINATION

To identify and enumerate the pathogens in a water would require an army of microbiologists and excessive time, and be prohibitively expensive for analyses on any routine basis. Therefore, there is a need to find a suitable indicator that would signal the potential presence of pathogenic microorganisms. The golden rule for an indicator microorganism is: The indicator must be present when pathogens are present and absent when pathogens are absent. Given the association of pathogens with human and animal feces and taking into account other practical considerations, the traits of

TABLE 7.5 Vectors of Disease

Agent	Disease	Reservoir
Flies	Numerous	Organic matter
Black flies (*Simulium* spp.)	River blindness	
Mangrove fly (*Chrysops* spp.)	Loiasis	
Tsetse fly (*Glossina* spp.)	Sleeping sickness	
Louse	Relapsing fever, typhus	
Mosquitoes		
Aedes aegypti	Dengue fever	
Aedes aegypti *Haemagogus* spp.	Yellow fever	
Anopheles	Malaria	Stagnant water
Culex pipens *Mansonia* spp. *Anopheles* *Aedes* spp.	Filariasis, elephantiasis and encephalitis	
Rats	Numerous	Organic matter

an ideal indicator microorganism are as follows:

1. The microorganism should only originate in the digestion tract of humans and warm-blooded animals.
2. The microorganism should be easily, rapidly, and reliably identified and enumerated.
3. The analysis should be inexpensive.
4. The indicator should survive longer than pathogens in the extraintestinal environment.
5. The indicator should occur in high numbers.
6. The indicator should not be pathogenic itself.

The first characteristic is based on the fecal origin of pathogens. An indicator microorganism that has a natural habitat in the gut of mammals and in soils or on plants would be less desirable. The presence of a microorganism originating from soil or on a plant in a water may lead to unnecessary precaution or alarm when in fact only stormwater runoff has contributed to its occurrence. Because animals carry smaller loads of pathogens that are able to infect humans, two indicators, one for humans and one for animals, would be ideal. More accurate assessment of health hazards could be made and the source of pollution would be more readily identifiable.

The second and third characteristics are fairly obvious. Indicator microorganism tests are performed on a regular basis (many times a day for large cities) for almost every public water supply in the developed world. A small cost saving translates into a large amount of money on a global basis. However, economy cannot be allowed to compromise reliability of the analyses.

The ability of the indicator to outsurvive pathogens when the indicator and pathogens are excreted into the extraintestinal environment is a critical factor. There have been instances, particularly in the tropics, where pathogens have been able to multiply for periods of time in the warm waters found there. An ideal indicator should exhibit similar behavior.

The fifth point is related to characteristics 2 and 4. The higher the number of indicator organisms the less difficult it is to obtain accurate counts, and in their absence there is more assurance that pathogens have been removed. Even in the worst circumstances, the number of healthy persons far exceeds the number of ill persons and pathogens comprise only a small percentage of the flora of sewage.

The last characteristic concerns the health of the analysts. Identifying and enumerating microorganisms involves culturing the organisms, which increases their numbers and poses additional risk to the personnel performing the tests.

An organism possessing all of these characteristics allows one to confidently assume that a reduction in the number of indicators implies an equal or greater reduction in pathogens of fecal origin. Meeting all characteristics listed above is a difficult task but significant advances have been made in the past 10 years. The flora of feces are briefly examined, followed by a discussion of the current indicator microorganisms of choice.

7.2.1 The Bacterial Flora of Feces

The chemical and water composition of feces is variable and complex. Different diets and climates also influence the composition of feces and thus the flora present in it. Likewise dietary practices and other factors affect the volume and weight of feces. From a survey of studies around the world the average wet weight of feces produced on a daily basis by individuals ranged from 91 to 489 g (Feachem et al., 1981). As water content increases, the volume of fecal material increases. Few organisms are found in urine because of its high ammonia content and low pH.

Bacteria dominate the living organisms found in feces, which makes some of them candidates for an indicator organism. The total bacterial population in 1 g of feces has been estimated to be equivalent to 1.8×10^{12} *Escherichia coli* cells (Lamanna and Mallette, 1965).

Except for pathogens, fecal bacteria are commensal and necessary for digestion. As one would expect, more bacteria in feces are facultative or anaerobic microorganisms. The environment in the intestine is rich in organic material, which would require high amounts of oxygen to maintain aerobic bacteria. Maintaining an aerobic environ-

TABLE 7.6 Fecal Microflora of Different Human Populations[a]

		Mean $\log_{10}$ number of bacteria per gram of feces						
Diet	Country	*Entero-bacteria*[b]	*Streptococci*	*Lacto-bacilli*	*Clostridia*	*Bacter-oides*	*Bifido-bacteria*	*Eubacteria*
Largely carbohydrate	India	7.9	7.3	7.6	5.7	9.2	9.6	9.5
	Japan	9.4	8.1	7.4	5.6	9.4	9.7	9.6
	Uganda	8.0	7.0	7.2	5.1	8.2	9.4	9.3
	Hong Kong	7.0	5.8	6.1	4.7	9.8	9.1	8.5
Mixed Western	England	7.9	5.8	6.5	5.7	9.8	9.9	9.3
	Scotland	7.6	5.3	7.7	5.6	9.8	9.9	9.3
	U.S.A.	7.4	5.9	6.5	5.4	9.7	9.9	9.3
	Denmark	7.0	6.8	6.4	6.3	9.8	9.9	9.3
	Finland	7.0	7.8	8.0	6.2	9.7	9.7	9.5

[a]From Feachem et al. (1980).
[b]Fecal coliform bacteria are in the family of *Enterobacteriaceae*.

TABLE 7.7 Microbial Flora of Animal Feces[a]

	Average density × 10^{-3} per gram[b]				
Animal group	Fecal coliforms	Fecal streptococci	*Clostridium perfringens*	*Bacteroides*	*Lactobacilli*
Farm animals					
Cow	230	1 300	0.2	[c]	0.25
Pig	3 300	84 000	3.98	500	251 000
Sheep	16 000	38 000	199	[c]	79
Horse	12.6	6 300	[c]	[c]	10 000
Duck	33 000	54 000	—	—	—
Chicken	1 300	3 400	0.25	[c]	316 000
Turkey	290	2 800	—	—	—
Animal pets					
Cat	7 900	27 000	25 100	795 000	63 000
Dog	23 000	980 000	251 000	500 000	39.6
Wild animals					
Mice	330	7 700	[c]	795 000	1 260 000
Rabbits	0.020	47	[c]	396 000	[c]
Chipmunk	148	6 000	—	—	—
Human	13 000	3 000	1.58	5 000 000	630 000

[a]From Geldreich (1978).

[b] In this text, when a table heading indicates that the number is multiplied by some factor, the recorded number must be divided by this factor to obtain the correct result. For instance, the fecal coliform concentration per gram in cow feces is $z = 230 \times 10^3$. The number recorded is $z \times (10^{-3}) = 230 \times 10^3 \times 10^{-3} = 230$.

[c]Present in negligible amounts.

ment in the intestine would be counterproductive for the host because aerobic microorganisms would oxidize the organic matter and reduce its value for the host. Table 7.6 lists the average concentrations of major groups of bacteria found in human feces. It is observed that its bacteriological composition is relatively uniform around the world. However, the presence of pathogens is highly variable depending on occurrence of diseases and carriers of pathogens as well as other variables.

Because animals are a reservoir for some diseases that afflict humans, the flora of animal feces is also pertinent. Geldreich (1978) has compiled information (Table 7.7) on the occurrence in animals of microorganisms that are candidates for indicator microorganisms (discussed next).

7.3 INDICATOR MICROORGANISMS

The use of bacteria as indicators of the sanitary quality of water dates back to the late 1800s when von Fritsch described *Klebsiella pneumoniae* and *K. rhinoscleromatis* as microorganisms characteristically found in feces (Geldreich, 1978), which is the major source of microbiological health hazards in natural waters.

7.3.1 Coliforms

The coliform groups, fecal and total, are the most widely used indicator organisms now and historically. The total coliform group consists of members whose normal habitat is the intestines of humans and warm- and cold-blooded animals; some members are naturally found in the soil and vegetation as well as feces. The technical definition of this group is all of the aerobic and facultative anaerobic gram-negative, non-spore-forming, rod-shaped bacteria that ferment lactose with gas formation within 48 h at 35°C.

The total coliform group is widely used as the indicator organism of choice for drinking water. The major criticism of this group as an indicator is that some of its members (such as *Enterobacter aerogenes*) are widely distributed in the environment. However, when it is assumed that all of its members originated in feces, a safety factor is provided. This approach is reasonable when applied to the provision of safe drinking water but suffers when fecal contamination of a water is suspected.

The fecal coliform subgroup of total coliforms is a much more specific indicator of fecal contamination. Among the coliforms in human feces, 96.4% are fecal coliforms; in the excreta of warm-blooded animals, 93–98% of the total coliforms are fecal coliforms (Geldreich, 1978). *Escherichia coli* is the predominant member of the fecal coliform group. Members of this subgroup are much less able to multiply in the extraintestinal environment than other members of the total coliform group. Data from numerous stream pollution investigations indicate that fecal coliforms will not persist in receiving waters with a BOD less than 30 mg/L (Geldreich, 1978). Fecal coliforms and *E. coli,* in particular, were originally chosen as fecal pollution indicators because of their relation to the typhoid–paratyphoid group and because they occur in large numbers.

Gannon and Busse (1989) found ratios of *E. coli* to fecal coliforms to be in the range of 0.82–1.34 for storm drains and streams during and after runoff. In tests of water samples, each test for an indicator is performed independently and separately. *E. coli* is a subgroup of fecal coliforms but variation in testing techniques can lead to ratios greater than 1.

Salmonella species are most often correlated with the occurrence of indicators. From a survey of studies, findings demonstrate that in estuarine waters with <200 fecal coliforms per 100 mL, *Salmonella* occurrences ranged from 6.5 to 31%; at fecal coliform densities >1 000 per 100 mL, the frequency of *Salmonella* occurrences at least doubles (Geldreich, 1978). Similar correlation trends have been found for enteroviruses in seawater and fecal coliform–*Salmonella* in fresh waters (Geldreich, 1978). In general there is wide variation in the data.

Assays for Total and Fecal Coliforms

The classic test for total coliforms is a most probable number (MPN) procedure (Sec. 6.5.2) involving a series of steps to establish beyond reasonable doubt their presence and numbers. The first stage involves incubation of the sample aliquot in lactose or lauryl tryptose containing broth at 35°C for approximately 24 or 48 h and noting the production of gas. Gas production constitutes a positive presumptive test. A small aliquot from a positive presumptive test is then subjected to further incubation in brilliant green lactose bile broth at 35°C for 48 h or on Endo or eosin methylene blue agar at 35°C for 24 h. Gas production in the broth or formation of typical coliform colonies on the agar is considered a positive confirmed test.

The final completed test is made from small aliquots taken from typical colonies formed on Endo or eosin methylene blue agar. These aliquots are transferred to lactose- or lauryl tryptose broth-containing tubes and to corresponding nutrient agar slants. Gram-stained preparations are made from colonies on agar slants corresponding to broth tubes that showed gas production after 24 or 48 h incubation at 35°C. A gram-negative reaction is a positive completed test.

The whole series of tests may take 5 d. When the later stages are not performed (which is often the case) it is always assumed that the results would have been positive. Serial dilutions of a sample are made to establish MPN estimations of total coliform microorganisms in the sample.

To determine the presence of fecal coliforms, small samples are taken from positive presumptive test tubes cultured as described for the first stage total coliform test and inoculated into EC (tryptose, lactose, bile salts) broth medium and incubated at 44.5°C for about 24 h. Alternatively, a small sample from a positive presumptive tube may be incubated in boric acid lactose broth at 43°C for about 24 h. Gas production in either broth and the completed test with a gram-negative reaction is a positive result.

Membrane procedures have been developed for each group. Different media and different incubation temperatures (35°C for total coliforms and 44.5°C for fecal coliforms) are used for each group. It is essential that temperature control at the required temperatures be maintained during culturing of the bacteria or erroneous results will be obtained. A modified membrane filter technique for fecal coliforms has been developed for chlorinated effluents for which the standard membrane filter (MF) technique is inadequate (Hendricks, 1978).

The current fecal coliform membrane procedure requires careful attention to detail and it still produces false positive and negative results (Katamay, 1990). Up to 10% of *E. coli* are not detected by the fecal coliform test. About 15% of *Klebsiella pneumoniae* organisms are thermotolerant and produce a positive fecal coliform reaction. This organism can be found in pristine sites and not be of fecal origin. For instance, pulp and paper mill wastewater may contain very high concentrations of these organisms (Clark and Allan, 1994). On the other hand, *E. coli* is exclusively of fecal origin. There is movement to base standards on *E. coli* as opposed to fecal coliforms determined by the MF analysis.

A new defined-substrate technology (DST) that directs metabolism of target bacteria to specific indicator nutrients has been developed to detect total coliforms and *E. coli* simultaneously (Katamay, 1990). The method (commercially known as the Colilert test) is simple, inexpensive, and reliable. The technique involves adding a water sample to the powdered medium and incubating it. The hydrated medium is initially colorless.

The medium contains two sugar–dye complexes that are specific to coliform and *E. coli* organisms. The compound specific to the coliform group is *o*-nitrophenol β-D-galactopyranoside (ONPG). When this compound is metabolized by a member of the coliform group, *o*-nitrophenol (the nonfood part of the ONPG complex) is released and produces a yellow color. The other compound is methylumbelliferyl-β-D-glucuronide (MUG) which can be metabolized by *E. coli* organisms because *E. coli* (and few other microorganisms) produces the enzyme β-glucuronidase. Upon metabolism, 4-methylumbelliferone (the nonfood part of MUG) is released. This dye exhibits intense fluorescence at 388 nm.

There are no other organic substances in the medium. If only one coliform or *E. coli* organism is present initially, about 24 h is required to produce results. For quantitative results, an MPN procedure (multiple tubes with different dilutions) can be used.

The minimal media ONPG–MUG (MMO–MUG) test was approved by the USEPA for detection of *E. coli*. Tests using EC medium plus MUG and nutrient agar plus MUG have also been approved (Pontius, 1993). The rapidity of this procedure compared to the conventional MPN confirmed procedure has prompted a number of agencies to consider rewriting some standards in terms of *E. coli* as opposed to fecal coliforms.

7.3.2 Fecal Streptococci

Fecal streptococci are also a primary indicator of fecal contamination. This group consists mainly of strains of *Streptococcus bovis, S. durans, S. equinus, S. faecalis,* and *S. faecium*. Analysis is for the group as a whole. Many varieties and biotypes of these species have been reported (Kenner, 1978). *Streptococcus bovis* and *S. equinus* only originate from warm-blooded animals, which is taken advantage of in tracing pollution sources. *Streptococcus faecalis* var. *liquifaciens* is ubiquitous and may affect the precision of this indicator system at low counts.

In human feces the fecal coliform to fecal streptococci ratio is greater than 4.0, whereas for animals the ratio is less than 1.0 (Geldreich, 1978; Kenner, 1978). These ratios are only valid for relatively recent pollution.

Members of this group are more halotolerant than coliforms; indeed, elevated NaCl concentrations are incorporated into selective media for them. In general, fecal streptococci are more fastidious in their nutritional requirements (Kenner, 1978). Membrane filter techniques have been developed for the analysis of fecal streptococci.

All of the major strains of fecal streptococci (*S. faecium*, *S. faecalis*, *S. durans,* and *S. hirae*) produce the enzyme β-glucosidase and a DST assay using MPN procedures appears to be feasible for routine fecal streptococci examination (Hernandez et al., 1993). The fluorogenic substrate was 4-methylumbelliferyl-β-glucoside. The enzyme releases 4-methylumbelliferyl, which is fluorescent and measured at 366 nm. Minor members of the fecal streptococci group and other nonfecal cocci that can be easily mistaken for fecal streptococci in the routine membrane procedure are unable to grow in a medium containing the fluorogenic substrate and thallous carbonate nalidixic acid with incubation at 44°C. Some nonfecal streptococci species rarely reported in water were able to grow in the medium. The test has not yet been approved.

7.3.3 *Clostridium perfringens*

Clostridium perfringens is a prominent member of the intestinal flora of humans and animals (Tables 7.6 and 7.7), although present at significantly smaller numbers than the above microorganisms. It is a major cause of wound infection and can cause food poisoning; however, no data suggest that waterborne transmission is significant in the epidemiology of diseases caused by this organism (Cabelli, 1978). Sorensen et al. (1989) found that concentrations of *C. perfringens* spores in samples of cattle, horse, and sheep feces were lower than those reported for human feces. *C. perfringens* is exclusively of fecal origin (Mara, 1974).

The characteristic of this microorganism that makes it useful as an indicator is its ability to form a spore in adverse circumstances. Its persistence is therefore much longer than members of the two groups discussed in Sections 7.3.1 and 7.3.2. A major criticism of the above groups is their disappearance before most of the enteroviruses; because *C. perfringens* survives environmental stress better it is a viable indicator. Sand filtration and chlorination decrease concentrations of *C. perfringens* spores by a factor of 10 or more, respectively (Sorensen et al., 1989).

The above indicators are cultured aerobically. *C. perfringens* is an anaerobe, which makes its analysis more difficult; however, *C. perfringens* can tolerate small amounts of oxygen without a loss in recovery (Geldreich, 1978).

Presence–Absence Test for Multiple Indicators

Quantitative assessment of specific microorganisms provides the best information on the quality of a water. Not all indicator microorganisms are necessarily present in a contaminated water; therefore, a comprehensive analysis would monitor multiple indicators. *Standard Methods* (1992) provides a medium (MacConkey-PA) and procedure for simple presence–absence testing of water for total coliforms. A medium containing lactose, lauryl tryptose, and tryptone (STM-PA) was suggested by Clark et al. (1982) as a suitable medium for the detection of total and fecal coliforms and fecal streptococci. The test is a simplification of the multiple tube fermentation technique and only gives qualitative results. A sample is inoculated into the medium and, after incubation, a color change indicates the presence of one or more of the indicators. Martins et al. (1991) found that STM-PA medium was more efficacious in responding to total and fecal coliforms than MacConkey-PA medium for incubation periods of 24–48 h (48 h is recommended). Fecal streptococci were not detected in MacConkey-PA but incubation periods of up to 5 d may be necessary for detection of low densities of fecal streptococci with STM-PA.

The presence–absence test advantages are its simplicity and multi-indicator sensitivity with a single culture bottle. Its qualitative result is a drawback but it can be used to augment system monitoring. A large number of samples can be processed compared to individual analyses for the indicators.

7.3.4 Other Indicator Bacteria

One of the two predominant members of feces is *Bacteroides*. It is a strict anaerobe, which causes analytical difficulties; furthermore, tests for these organisms are involved (Geldreich, 1978). It will probably not make any inroad into the indicator systems until the assay problems are overcome.

Bifidobacteria are nonsporulating anaerobic microorganisms that occur in large numbers in feces. They are the first microorganisms to colonize the gastrointestinal tract of the newborn. They are later equaled by other bacteria in predominance. Three factors limit the usefulness of *Bifidobacteria* as an indicator: significant extrafecal sources, poor survival in the aquatic environment, and multiplication in water when nutrient is present (Cabelli, 1978). Cabelli (1978) suggests that the one characteristic that may provide a role for these organisms in an indicator system is their specific association with the fecal wastes of mammals as opposed to lower animals and insects. Rapid, inexpensive isolation techniques have not yet been developed for this group.

Pseudomonas aeruginosa is an opportunistic pathogen that can cause disease in debilitated persons. It is also responsible for eye and ear infections, which has made it a candidate for an indicator of swimming pool and recreational water quality. However *P. aeruginosa* appears to be ubiquitous in the environment and may multiply under natural conditions (Cabelli et al., 1976). An MF technique has been developed for enumeration of *P. aeruginosa* (*Standard Methods*, 1992).

7.3.5 Coliphages

Phages are viruses. Enteric viruses are excreted in human feces in concentrations as high as 10^6 viruses per gram of feces (Olivieri, 1982). Viruses are more stable than the indicator bacteria discussed previously and they also survive longer than bacterial pathogens. An increasing amount of research is being devoted to coliphages as indicators of fecal contamination. As their name indicates, these phages are specific to the host bacteria, *E. coli*. Evidence suggests that coliphages and the most common pathogenic viruses are more resistant to chlorination than *E. coli* (Grabow, 1968). Coliphages are directly correlated with their host (Borrego et al., 1987). The presence of fecal coliforms was never observed in the absence of coliphages (Borrego et al., 1987; Ratto et al., 1989). Coliphages are more resistant to dieoff than coliforms. The *E. coli* C bacteriophage was concluded to be the best indicator among several bacteriophage groups of remote pollution in marine waters (Cornax et al., 1991).

Analysis for coliphages is more rapid than techniques for bacterial indicators. A plaque is a viral colony and the concentration of viruses is reported in terms of plaque-forming units (PFU). *Standard Methods* (1992) includes a tentative procedure for coliphage detection. The procedure consists of adding sample to medium containing a host *E. coli* strain and adding an agent to enhance PFU visibility. Plaques can be observed within 4–6 h; 12–14 h are needed for full plaque development. Membrane filter techniques, where the water is filtered through 0.45-μm filters that are then placed on a medium containing host cells for plaque development, are being developed (Sobsey et al., 1990).

Different groups of bacteriophages are detected depending on the host strain of *E. coli*. Somatic coliphages are phages that adsorb to receptors in the cell wall of F^- *E. coli* hosts (F^+ strains have a male sex pilon). F-specific bacteriophages adsorb to F- or sex-pili strains of various bacteria. F-specific bacteriophages contain single-stranded RNA or DNA nucleic acid. F-specific RNA bacteriophages are relatively resistant to conditions applied in water treatment (IAWPRC Study Group, 1991).

Borrego et al. (1987) argue for the adoption of coliphages as a standard indicator and others (Ratto et al., 1989) argue that coliphages should be included as part of any indicator scheme because of the occurrence of coliphages in the absence of other indicators. Based on studies correlating coliphages to enteric viruses, Palmateer et al. (1991) and Moriñigo et al. (1992) support coliphage and bacteriophage enumeration as part of health risk assessment of recreational waters. Dutka et al. (1987) suggested a recreational water standard of 20 coliphages per 100 mL. An IAWPRC study group (1991) contends that further study is required to correlate phage survival and viruses. Some phages are known to be able to multiply in unpolluted waters. The problems of phages as indicators do not appear to be different from the abnormalities of bacterial indicators. The major barrier against adoption of phages as indicators is the small number of studies. More experimental data will provide the needed information on the behavior of phages and define their role as indicators.

7.3.6 Problems of Indicator Systems

The reduced incidence of waterborne microbiological disease in countries where indicators are routinely monitored attests to the validity of the systems. The most reliable assessment will be obtained when all three indicators are examined. Waterborne disease still occurs in countries where indicator tests are routine. Some of these incidents are due to momentary breakdown in treatment processes and escape of pollutant slugs from too infrequent monitoring. However, there are many instances where absence of one or more indicators has been indicated from examination but the water has carried fecal originated pathogens that have caused disease (Bosch et al., 1991; Geldreich, 1978; Hendricks, 1978; Kenner, 1978).

The microorganism traits that a medium cultivates define the species that will grow on it. Microbiologists are continuously examining characteristics of fecal and pathogenic microorganisms in an effort to refine tests for them. Natural variability in microorganisms and variability in their response to different environments dictate that aberrations from assumptions on the relation of indicators to pathogens will exist. Improper lab techniques will compound the problems. Coliform tests on waters other than fresh waters give rise to the most exceptions (Hendricks, 1978). Therefore there is more need in these circumstances to consult the literature for possible modifications to standard procedures.

In general, a massive amount of evidence supports the indicators discussed in this chapter as the most suitable choices at present. The confidence that is placed in the indicator concept decreases as the number of individuals who contribute feces decreases. There will always be a small component of a population that does not contribute the indicators for various reasons. Fluctuations are always wider for smaller groups.

7.3.7 Survival of Indicator Microorganisms

Many factors influence the survival of microorganisms within and outside their natural habitat. The availability of food and nutrient sources and the competition for them are primary determinants of the rate of multiplication or dieoff. When microorganisms are taken from their natural habitat adverse factors override supportive factors and net dieoff begins, although, as noted before, short instances of multiplication have occurred.

Besides actual dieoff or decay, other factors affect the presence of bacteria in a sample. Dilution will reduce concentrations in an easily quantifiable manner. Coagulation, flocculation, adsorption, and sedimentation are physical–chemical phenomena that influence the concentration of bacteria in suspension.

As noted previously, the availability of food and nutrients are major controllers of dieoff. Solar radiation increases the rate of dieoff of indicator microorganisms. The ultraviolet component of sunlight has relatively small effect on dieoff because of its large attenuation coefficient in water (Carlucci and Pramer, 1959; Mitchell and Chamberlin, 1978). Gameson and Saxon (1967) and Bellair et al. (1977) have conclusively demonstrated the beneficial effects of sunlight on the decay of indicator microorganisms. The dieoff rate is directly related to radiation intensity in the water.

$$k_r = \alpha I \tag{7.1}$$

where

k_r is the dieoff coefficient resulting from radiation
α is a constant (cm^2/cal)
I is the intensity of radiation ($cal/cm^2/d$)

The value of α for fecal coliforms in a freshwater lake was found to be 0.008 2 cm^2/cal by Auer and Niehaus (1993).

Predation and bacteriophages play a role in reduction of indicators. Algae can produce antibacterial toxins. Chemical toxins and other physicochemical factors such as osmotic or pH effects may produce significant detrimental physiological effects enhancing dieoff. The high concentration of salt in brackish or seawaters is a significant factor. Temperature is also an important factor.

The isolation and quantification of each individual mechanism of disappearance or death is exceedingly difficult. In situ studies are necessary to determine rate constants. The influence of all factors is normally lumped into one constant, k, which is used in a first-order decay model:

$$\frac{dC}{dt} = -kC \tag{7.2}$$

where
C = concentration of bacteria
k = overall disappearance or dieoff rate constant
t = time

$$k = (k_d + k_r + k_s + k_p)\theta^{(T-20)}$$

where
k_d is the dieoff coefficient in the dark
k_s is the settling coefficient
k_p is a predation coefficient
θ is an Arrhenius constant
T is temperature in °C

Other factors as just discussed may be included in the above equation if data are available.

TABLE 7.8 Freshwater Decay Rates of Coliform Bacteria[a]

System	Temperature indication	t_{90} h	k h^{-1}
Cumberland River	Summer	10	0.23
Glatt River	—	2.1	1.1
Groundwater stream	10°C	110	0.021
Leaf River (Mississippi)	—	135	0.017
Lower Illinois River	June–September	27	0.085
	October and May	63	0.037
	December–March	90	0.026
	April and November	80	0.029
Missouri River	Winter	115	0.020
Ohio River	Summer (20°C)	47	0.049
	Winter (5°C)	51	0.045
Sacramento River	Summer	32	0.072
"Shallow turbulent stream"	—	3.6	0.63
Tennessee River (Chattanooga)	Summer	42	0.055
Tennessee River (Knoxville)	Summer	53	0.043
Upper Illinois River	June–September	27	0.085
	October and May	22	0.105
	December–March	95	0.024
	April and November	53	0.043
Maturation ponds	—	28	0.083
	19°	33	0.07
Oxidation ponds	20°C	21.3	0.108
Wastewater lagoon	7.9–25.5°C	79–276	0.029–0.008 3

[a]From Mitchell and Chamberlin (1978).

TABLE 7.9 Seawater Decay Rates of Coliform Bacteria[a]

Location	Previous sewage treatment	t_{90} h	k h^{-1}
Denmark	None	2.0	1.15
England	None	0.78–3.50	0.66–2.90
Gentoffe, Denmark	None	1.16	1.98
Istanbul, Turkey	None	0.80–3.00	0.77–2.88
Manila Bay,	None	1.78–3.45	0.67–1.30
Philippines[b]	None	2.16–2.84	0.81–1.06
Nice, France	None	1.5	1.54
Rio de Janeiro, Brazil	None	<1.0	>2.3
Santa Barbara, California	Primary	0.37–5.47	0.42–6.01
Santa Monica, California	Secondary	6.5	0.354
Seaside Heights, New Jersey	Primary	1.05	2.2
Sidmouth and Bridport, England	—	$0.57 - \gg 4$	$\ll 0.56 - 4.04$
Titahi Bay, New Zealand	None	0.65	3.54
Tema, Ghana	None	1.33	1.73

[a]From Mitchell and Chamberlin (1978).
[b]Data are from two separate studies conducted at different times.

Results are often reported in terms of t_{90}, which is the time required for a 90% decrease (one log reduction) in concentration.

$$t_{90} = \frac{\ln 10}{k} \tag{7.3}$$

Mitchell and Chamberlin (1978) have summarized decay rates of coliform bacteria in freshwater (Table 7.8) and seawater (Table 7.9). Seawater was found to be much more bactericidal for coliforms. Cooler temperatures tend to prolong the survival of coliforms. Typical reductions in indicator microorganisms and pathogens for various wastewater treatment processes are given in Table 7.10.

Katzenelson (1978) has commented on the wide variability of data on virus survival but noted that the outstanding fact is the impressive resistance of viruses to inactivation

TABLE 7.10 Removal of Indicators and Pathogens in Sewage Treatment Processes[a]

Type of sewage treatment	Removal range for various organisms, %
Septic tanks	25–75
Primary	5–40
Activated sludge	25–99
Trickling filters	18–99
Anaerobic digestion	25–92
Waste stabilization ponds	60–99
Tertiary (flocculation, sand filtration, etc.)	93–99.99

[a]From Geldreich (1978).

in natural waters compared to indicator bacteria. Viruses cannot increase in numbers outside living cells.

■ Example 7.1 Rate of Inactivation of Indicator Microorganisms

Studies on stormwater retention ponds in Nepean, Ontario have shown that indicator microorganisms in runoff from a storm event tend to maintain relatively constant numbers for approximately 24 h. After this time, numbers decrease to 10% of their initial number after 3 d. What is the model for dieoff and the rate constant?

The 1-d period in which there is little change should be considered a lag time, after which dieoff ensues. The t_{90} time is 3 d. The appropriate model for this situation is

$$C = C_0 e^{-k(t-t_l)}$$

where

t_l is a lag time (1 day in this case)

$$t_{90} = \frac{\ln 10}{k}$$

$$k = \frac{\ln 10}{t_{90}} = \frac{2.303}{3\ \text{d}} = 0.77\ \text{d}^{-1}$$

Removal of Pathogens in Water and Wastewater Processes

Except for disinfection, wastewater treatment operations (see Chapter 9) are more focused on reduction of the organic matter and chemical agents in a wastewater. But pathogens and indicators will be removed to some extent in most operations. Studies mentioned here provide some additional information to that contained in the tables and, also, chapters discussing the individual treatments should be consulted. Removals of vegetative indicator organisms, bacteriophages, and enteroviruses ranged from 93.7 to 98.4% (1.2–1.8 $\log_{10}$) in a standard activated sludge process (Nieuwstad et al., 1988). Spores of sulfite-reducing clostridia were reduced by 75–80% (0.6–0.7 $\log_{10}$). Addition of iron(III) chloride to the aeration basin to enhance precipitation of phosphate in the secondary clarifier did not enhance removals of these organisms. Filtration of effluent from the activated sludge process with iron(III) chloride added resulted in further removals of the above groups from 84 to 99.2% (0.8–2.1 $\log_{10}$). Addition of iron(III) chloride to the effluent from the conventional activated sludge process followed by flocculation and sedimentation resulted in removals similar to the filtered effluent.

In a 10-year study, average enteric virus removal was 99.8% in activated sludge processes (Yanko, 1993). Filtration and chlorination of clarified effluent from the activated sludge processes resulted in only one virus being detected in 590 samples.

Very harsh environments exist in anaerobic sludge digesters. A study of fecal coliform, fecal streptococci, *E. coli*, and virus survival in a conventional digestion process operated at 35°C (mesophilic) with 10 d or longer detention times, found one to two log reductions for these indicators and enterovirus (Lee et al., 1989). Operation of the process at a temperature of 53°C (thermophilic) greatly enhanced removals of these organisms to three to four log reductions. Thermophilic digestion reduced densities to essentially nondetectable levels.

TABLE 7.11 Occurrence of Parasites and Free-living Organisms in Sewage Solids[a]

	Mean number of organisms/100 g of solids[b]				
Organisms	Lemont	West–Southwest	Hanover Park	Calumet	North Side
Raw Sewage					
Ascaris lumbricoides eggs	0	0	78	0	29
Toxocara sp. eggs	246	0	90	79	232
Toxocaris leonina eggs	0	0	15	0	0
Trichuris sp. eggs	0	154	0	0	93
Enterobius vermicularis eggs	109	0	0	0	0
Cestode eggs	0	0	0	0	0
Coccidian cysts	4	0	0	0	20
Entamoeba coli cysts	0	0	0	0	0
Nematode eggs	4 889	81 744	1 316	1 935	3 236
Nematode adults	4 735	3 139	3 983	2 473	4 415
Nematode larvae	5 516	4 540	5 322	3 296	7 712
Mite eggs	6 059	2 825	2 408	1 644	3 574
Mites	908	198	202	269	315
Dipteran eggs	122	0	31	35	20
Dipteran larvae	0	22	29	6	12
Rotifers	23 439	725	1 927	1 171	791
Cladocera	0	15	369	569	63
Copepods	0	22	0	98	0
Tardigrada	15	11	0	0	0
Unchlorinated Effluent					
Ascaris lumbricoides eggs	0	29	0	0	0
Toxocara sp. eggs	0	86	0	0	0
Toxocaris leonina eggs	0	0	0	0	0
Trichuris sp. eggs	0	21	0	0	0
Enterobius vermicularis eggs	0	0	0	0	0
Cestode eggs	0	22	0	0	0
Coccidian cysts	0	68	0	0	220
Entamoeba coli cysts	0	5	0	0	0
Nematode eggs	705	1 531	5 752	1 087	11 826
Nematode adults	1 212	3 139	15 402	10 060	67 152
Nematode larvae	1 422	4 540	23 391	9 945	82 733
Mite eggs	611	2 825	2 612	111	1 533
Mites	597	198	426	784	1 098
Dipteran eggs	0	0	216	0	0
Dipteran larvae	1 024	22	38	132	212
Rotifers	116 057	725	4 384	2 217	10 528
Cladocera	83	15	0	148	0
Copepods	461	22	294	6 142	0
Tardigrada	0	11	82	0	119
Anaerobically Digested Sludge					
Ascaris lumbricoides eggs	48	144	232	14	
Toxocara sp. eggs	440	153	283	157	
Toxocaris leonina eggs	0	18	0	13	
Trichuris sp. eggs	6	28	9	11	
Enterobius vermicularis eggs	0	0	0	0	
Cestode eggs	0	53	0	0	
Coccidian cysts	94	268	17	21	
Entamoeba coli cysts	0	14	0	0	
Nematode eggs	4 311	996	2 363	2 476	
Nematode adults	1 134	74	159	425	
Nematode larvae	1 013	111	314	408	
Mite eggs	26 558	4 154	6 087	4 345	
Mites	420	28	2 133	55	
Dipteran eggs	109	24	32	50	
Dipteran larvae	0	7	0	8	
Rotifers	1 355	229	434	396	
Cladocera	11	44	23	152	
Copepods	0	0	7	13	
Tardigrada	15	0	0	0	

[a]Reprinted with permission from J. C. Fox, P. R. Fitzgerald, and C. Lue-Hing (1981), *Sewage Microorganisms: A Color Atlas,* Lewis Publishers, Chelsea, MI. Copyright Lewis Publishers, an imprint of CRC Press, Boca Raton, FL.

[b]Samples were taken biweekly during Feb.–Mar., 1976 from five metropolitan sewage treatment plants in the Chicago area. The number of samples ranged from 27 to 30 at each location.

Although anaerobic digestion appears to be a more stressful environment than aerobic digestion, Scheuerman et al. (1991) found that poliovirus 1, echovirus 1, and rotavirus SA-11 had mean daily changes near $-0.50 \log_{10}$ for both aerobic digestion at a temperature close to 28°C and anaerobic digestion at a temperature close to 33°C. Only temperature had a significant effect on virus inactivation; solids concentration and pH were not correlated with virus survival.

Numbers of parasites and other large microorganisms found in sewage, in unchlorinated effluent from sewage treatment plants, and in anaerobically digested sludge for plants in the Chicago area are compared in Table 7.11. The viability of the organisms or eggs was not determined. Many microorganisms are concentrated in sludge and sludge digestion devitalizes a good number of them. Note that the numbers reported in this table are based on sewage solids. The solids content of effluent is very low and the concentrations of microorganisms per liter of effluent would generally be very small. The occurrence of microorganisms at various stages of treatment is variable depending on conditions in and preceding the location. A few microorganisms are able to multiply under conditions of regular wastewater treatment operations but generally pathogenic microorganisms are reduced in numbers and concentration.

Sewage is sometimes treated by discharge on land and sludge is often disposed on land. Total and fecal coliform decay rates were monitored for a loamy fine sand mixed with sludge and exposed to different rainfall rates (Zyman and Sorber, 1988). Decay rates were apparently independent of rainfall rates as long as moisture was present. Dieoff rate constants (base *e*) of 0.25 and 0.35 d^{-1} were observed for total and fecal coliforms, respectively, when moisture was present. Under lack of moisture, decay rate constants increased to 0.44 and 0.56 d^{-1}, respectively.

The primary objective of water treatment operations is eradication of pathogens. However many of the individual processes are concerned with aesthetic improvement of the water. Chapter 9 gives general descriptions of water treatment operations. Some comments on studies that have examined pathogen or indicator removals with various process modifications are given here to provide an indication of results that may be achieved. Using clay as an adsorbent with alum as a coagulant (a small amount of polyacrylic coagulant aid was also added to improve floc strength) followed by sedimentation resulted in mean removals of 96.8% (1.5 $\log_{10}$) and 98.4% (1.8 $\log_{10}$) for a poliovirus strain and phage, respectively (Gersberg et al., 1988). Sand filtration of the clarified effluent resulted in additional 99% (2.0 $\log_{10}$) and 84% (0.8 $\log_{10}$) removals, respectively. With chlorine disinfection that produced free chlorine residuals from 6.6 to 9.6 mg/L, overall log 10 removals greater than 6.5 were achieved.

QUESTIONS AND PROBLEMS

1. List three pathogens and their associated diseases from the bacteria, protozoan, and virus groups.
2. Describe the life cycle of the helminth responsible for bilharzia.
3. Consult local authorities to find incidents of waterborne diseases in your community over the past 10 years. What were the agents and reasons for the outbreaks?
4. What are the characteristics of an ideal indicator microorganism for fecal contamination? Discuss each group of indicator microorganisms against these characteristics.
5. Discuss the pros and cons of potential alternative indicator bacteria.
6. What is the mean ratio of fecal streptococci to fecal coliforms for animals and to *Enterobacter* for humans using the data in Tables 7.6 and 7.7? How do these ratios change for excreta deposited in a water?

7. Explain the Colilert test.

8. What is a coliphage?

9. Why is salt incorporated at higher than normal concentrations in media for fecal streptococci? What are the major strains of bacteria that make up this group?

10. If the indicator count decreased from 1.12×10^6/100 mL to 4.83×10^4/100 mL in 36 h, what is the t_{90} in this situation?

11. In the summer, how many days would it take coliform bacteria in the Ohio river to decrease 90%, 99%, 99.9%, and 99.99% from the initial concentration (see Table 7.8)?

12. On a day with average cloud cover, the maximum sunlight intensity at a location was 79.7 cal/cm^2/d. The sun was above the horizon for 14.4 h. Sunlight intensity follows a sinusoidal pattern when the sun is above the horizon. What is the average daily value of the solar radiation dieoff coefficient at the surface of a lake if $\alpha = 0.008\ 2$ cm^2/cal?

13. If a rapid test for *Salmonella* species enumeration were developed, would it be a suitable replacement for fecal coliforms as an indicator microorganism?

14. Would fecal coliforms or *E. coli* be a reliable indicator for ascarids?

15. Samples from a stream had an initial ratio of fecal streptococci to fecal coliforms of 9.0 that decreased to a ratio of 1.0 after 3.20 d travel time in the stream. The t_{90} for fecal coliforms was 58 h. What was the dieoff coefficient for fecal streptococci?

16. If a sample from a stream contained significant numbers of *Clostridium perfringens* but none or few fecal coliforms or fecal streptococci, would the source of contamination be near the sampling point or far away?

17. Based on information in Table 7.10 what is the range of removal for indicator microorganisms for wastewater that passes through primary treatment, a trickling filter, an activated sludge process, and tertiary treatment? Give the minimum and maximum values after each stage of treatment.

KEY REFERENCES

BENENSON, A. S., ed. (1985), *Control of Communicable Diseases in Man,* 14th ed., American Public Health Association, Washington, DC.

BERG, G., ed. (1978), *Indicators of Viruses in Water and Food,* Ann Arbor Science Publishers, Ann Arbor, MI.

FEACHEM, R. G., D. J. BRADLEY, H. GARELICK, AND D. D. MARA (1981), *Health Aspects of Excreta and Sullage Management—A State-of-the-Art Review,* The World Bank, Washington, DC.

PIPES, W. O., ed. (1982), *Bacterial Indicators of Pollution,* CRC Press, Boca Raton, FL.

Standard Methods for the Examination of Water and Wastewater (1992), 18th ed., American Public Health Association, Washington, DC.

REFERENCES

AUER, M. T. AND S. L. NIEHAUS (1993), "Modeling Fecal Coliform Bacteria-I. Field and Laboratory Determination of Loss Kinetics," *Water Research,* 27, 4, pp. 693–701.

BELLAIR, J. T., G. A. PARR-SMITH, AND I. G. WALLIS (1977), "Significance of Diurnal Variations in Faecal Coliform Dieoff Rates in the Design of Ocean Outfalls," *J. Water Pollution Control Federation,* 49, 9, pp. 2022–2030.

BORREGO, J. J., M. A. MORIÑIGO, A. DE VICENTE, R. CORNAX, AND P. Romero (1987), "Coliphages as an Indicator of Faecal Pollution in Water. Its Relationship with Indicator and Pathogenic Microorganisms," *Water Research,* 21, 12, pp. 1473–1480.

Bosch, A., F. Lucena, J. M. Diez, R. Gajardo, M. Blasi, and J. Jofre (1991), "Waterborne Viruses Associated with Hepatitis Outbreak," *J. American Water Works Association,* 83, 3, pp. 80–83.

Cabelli, V. J. (1978), "Obligate Anaerobic Bacterial Indicators," in *Indicators of Viruses in Water and Food,* G. Berg, ed., Ann Arbor Science Publishers, Ann Arbor, MI, pp. 171–200.

Cabelli, V. J., H. Kennedy, and M. A. Levin (1976), "*Pseudomonas aeruginosa*-Fecal Coliform Relationships in Estuarine and Fresh Recreational Waters," *J. Water Pollution Control Federation,* 48, 2, pp. 367–376.

Carlucci, A. F. and D. Pramer (1959), "Factors Affecting the Survival of Bacteria in Sea Water," *Applied Microbiology,* 7, 6, pp. 388–392.

Casson, L. W., C. A. Sorber, R. H. Palmer, A. Enrico, and P. Gupta (1992), "HIV Survivability in Wastewater," *Water Environment Research,* 64, 3, pp. 213–215.

Clark, J. A., C. A. Burger, and L. E. Sabatinos (1982), "Characterization of Indicator Bacteria in Municipal Raw Water, Drinking Water, and New Main Water Samples," *Canadian J. Microbiology,* 28, 9, pp. 1002–1013.

Clark, T. A. and P. Allan (1994), "Bacteriological Water Quality of Pulp and Paper Mill Discharges: an Emerging Compliance Problem," *Appita,* 47, 6, pp. 467–476.

Cornax, R., M. A. Moriñigo, M. C. Balebona, D. Castro, and J. J. Borrego (1991), "Significance of Several Bacteriophage Groups as Indicators of Sewage Pollution in Marine Waters," *Water Research,* 25, 6, pp. 673–678.

Dutka, B. J., A. El Shaarawi, M. T. Martins, and P. S. Sanchez (1987), "North and South American Studies on the Potential of Coliphage as a Water Quality Indicator," *Water Research,* 21, 9, pp. 1127–1134.

Fox, J. C., P. R. Fitzgerald, and C. Lue-Hing (1981), *Sewage Microorganisms: A Color Atlas*, Lewis Publishers, Chelsea, MI.

Gameson, A. L. H. and J. R. Saxon (1967), "Field Studies on Effect of Daylight on Mortality of Coliform Bacteria," *Water Research,* 1, 4, pp. 279–295.

Gannon, J. J. and M. K. Busse (1989), "E. coli and Enterococci Levels in Urban Stormwater, River Water and Chlorinated Treatment Plant Effluent," *Water Research,* 23, 9, pp. 1167–1176.

Geldreich, E. E. (1978), "Bacterial Populations and Indicator Concepts in Feces, Sewage, Stormwater and Solid Wastes," in *Indicators of Viruses in Water and Food,* G. Berg. ed., Ann Arbor Science Publishers, Ann Arbor, MI, pp. 51–97.

Gersberg, R. M., S. R. Lyon, R. Brenner, and B. V. Elkins (1988), "Performance of a Clay-Alum Flocculation (CCBA) Process for Virus Removal from Municipal Wastewater," *Water Research,* 22, 11, pp. 1449–1454.

Grabow, W. O. K. (1968), "The Virology of Wastewater Treatment," *Water Research,* 2, 10, pp. 675–701.

Hendricks, C. W. (1978), "Exceptions to the Coliform and the Fecal Coliform Tests," in *Indicators of Viruses in Water and Food,* G. Berg, ed., Ann Arbor Science Publishers, Ann Arbor, MI, pp. 99–145.

Hernandez, J. F., A. M. Pourcher, J. M. Delattre, C. Oger, and J. L. Loeuillard (1993), "MPN Miniaturized Procedure for the Enumeration of Fecal Enterococci in Fresh and Marine Waters: The MUST Procedure," *Water Research,* 27, 4, pp. 597–606.

IAWPRC Study Group (1991), "Bacteriophages as Model Viruses in Water Quality Control," *Water Research,* 25, 5, pp. 529–545.

Johnson, R. W., E. R. Blatchley, III, and D. R. Mason (1994), "HIV and the Bloodborne Pathogen Regulation: Implications for the Wastewater Industry," *Water Environment Research,* 66, 5, pp. 684–691.

Katamay, M. M. (1990), "Assessing Defined-Substrate Technology for Meeting Monitoring

Requirements of the Total Coliform Rule," *J. American Water Works Association,* 82, 9, pp. 83–87.

KATZENELSON, E. (1978), "Survival of Viruses," in *Indicators of Viruses in Water and Food,* G. Berg, ed., Ann Arbor Science Publishers, Ann Arbor, MI, pp. 39–50.

KENNER, B. A. (1978), "Fecal Streptococcal Indicators," in *Indicators of Viruses in Water and Food,* G. Berg, ed., Ann Arbor Science Publishers, Ann Arbor, MI, pp. 147–170.

LAMANNA, C. AND M. F. MALLETTE (1965), *Basic Bacteriology, Its Biological and Chemical Background,* William & Wilkins, Baltimore.

LEE, K. M., C. A. BRUNNER, J. B. FARRELL, AND A. E. ERLAP (1989), "Destruction of Enteric Bacteria and Viruses during Two-Phase Digestion," *J. Water Pollution Control Federation,* 61, 8, pp. 1421–1429.

MARA, D. D. (1974), *Bacteriology for Sanitary Engineers,* Churchill Livingstone, London.

MARTINS, M. T., D. M. MYAKI, V. H. PELLIZARI, C. ADAMS, AND N. R. S. BOSSOLAN (1991), "Comparison of the Presence–Absence (P–A) Test and Conventional Methods for Detection of Bacteriological Water Quality Indicators," *Water Research,* 25, 10, pp. 1279–1283.

MILLER, G. T., JR. (1979), *Living in the Environment,* 2nd ed., Wadsworth Publishing Company, Belmont, CA.

MITCHELL, R. AND C. CHAMBERLIN (1978), "Survival of Indicator Microorganisms," in *Indicators of Viruses in Water and Food,* G. Berg, ed., Ann Arbor Science Publishers, Ann Arbor, MI, pp. 15–37.

MORIÑIGO, M. A., D. WHEELER, C. BERRY, C. JONES, M. A. MUÑOZ, R. CORNAX, AND J. J. BORREGO (1992), "Evaluation of Different Bacteriophage Groups as Faecal Indicators in Contaminated Natural Waters in Southern England," *Water Research,* 26, 3, pp. 267–271.

MURACA, P. W., V. L. YU, AND J. E. STOUT (1988), "Environmental Aspects of Legionnaires' Disease," *J. American Water Works Association,* 80, 2, pp. 78–86.

NAVIN, T. R. AND D. D. JURANEK (1984), "Cryptospordiosis: Clinical Epidemiologic and Parasitologic Review," *Reviews of Infectious Diseases,* 6, 3, pp. 313–327.

NIEUWSTAD, T. J., E. P. MULDER, A. H. HAVELAAR, AND M. V. OLPHEN (1988), "Elimination of Micro-Organisms from Wastewater by Tertiary Precipitation and Simultaneous Precipitation Followed by Filtration," *Water Research,* 22, 11, pp. 1389–1397.

OLIVIERI, V. P. (1982), "Bacterial Indicators of Pollution," in *Bacterial Indicators of Pollution,* W. O. Pipes, ed., CRC Press, Boca Raton, FL.

PALMATEER, G. A., B. J. DUTKA, E. M. JANZEN, S. M. MEISSNER, AND M. G. SAKELLARIS (1991), "Coliphage and Bacteriophage as Indicators of Recreational Water Quality," *Water Research,* 25, 3, pp. 355–357.

PONTIUS, F. W. (1993), "Federal Drinking Water Update," *J. American Water Works Association,* 85, 2, pp. 42–51.

RATTO, A., B. J. DUTKA, C. VEGA, C. LOPEZ, AND A. EL-SHAARAWI (1989), "Potable Water Safety Assessed by Coliphage and Bacterial Tests," *Water Research,* 23, 2, pp. 253–255.

RIGGS, J. L. (1989), "AIDS Transmission in Drinking Water: No Threat," *J. American Water Works Association,* 81, 9, pp. 69–70.

ROJAS, Y. A. AND T. C. HAZEN (1989), "Survival of *Vibrio Cholerae* in Treated and Untreated Rum Distillery Effluents," *Water Research,* 23, 1, pp. 102–113.

ROSE, J. (1990), "Emerging Issues for the Microbiology of Drinking Water," *Water Engineering and Management,* 137, 7, pp. 23–29.

ROSE, J. B. (1988), "Occurrence and Significance of Cryptosporidium in Water," *J. American Water Works Association,* 80, 2, pp. 53–58.

SCHEUERMAN, P. R., S. R. FARRAH, AND G. BITTON (1991), "Laboratory Studies of Virus Survival During Aerobic and Anaerobic Digestion of Sewage Sludge," *Water Research,* 25, 3, pp. 241–245.

Sobsey, M. D., K. J. Schwab, and T. R. Handzel (1990), "A Simple Membrane Filter Method to Concentrate and Enumerate Male-Specific RNA Coliphages," *J. American Water Works Association,* 82, 9, pp. 52–59.

Sorensen, D. L., S. G. Eberl, and R. A. Dicksa (1989), "*Clostridium Perfringens* as a Point Source Indicator in Non-Point Polluted Streams," *Water Research,* 23, 2, pp. 191–197.

Witherell, L. E., R. W. Duncan, K. M. Stone, L. J. Stratton, L. Orciari, S. Kappel, and D. A. Jilson (1988), "Investigation of Legionella pneumophila in Drinking Water," *J. American Water Works Association,* 80, 2, pp. 87–93.

Yanko, W. A. (1993), "Analysis of 10 Years of Virus Monitoring Data from Los Angeles County Treatment Plants Meeting California Wastewater Reclamation Criteria," *Water Environment Research,* 65, 3, pp. 221–225.

Zyman, J. and C. A. Sorber (1988), "Influence of Simulated Rainfall on the Transport and Survival of Selected Organisms in Sludge-Amended Soils," *J. Water Pollution Control Federation,* 60, 12, pp. 2105–2114.

CHAPTER 8

WATER COMPONENTS AND QUALITY STANDARDS

Out of the approximately 100 elements that are naturally found on the earth, the molecular basis for all living systems is remarkably similar (Lehninger, 1975). Sixteen elements have been found to be essential to living systems and 22 at most are found in living systems; therefore, there has been selectivity in the evolution of living systems. Besides the laws of chemistry and thermodynamics, which govern biochemical phenomena, evolution is a process of inheritance as well as change. The elements that form the backbone of organic biomolecules are present in the highest amounts. Other elements are present in varying amounts. Table 8.1 lists elements found in living systems.

The last six elements in Table 8.1 are generally not essential to living systems except to some algae and bacteria.

8.1 TOXICITY OF ELEMENTS AND COMPOUNDS

The other elements not listed in Table 8.1 have varying effects on living systems, ranging from innocuous to extremely toxic. Natural defense mechanisms have evolved in organisms allowing them to cope to an extent with harmful substances. Some elements that are harmless or even essential to living systems cause harm to ecosystems when their concentrations rise beyond natural levels. A pollutant is often defined as a substance that is "out of place" in the environment. Imbalances in a few elements or compounds can result in the total disruption of an ecosystem. Anthropogenic sources disperse substances into the environment, threatening ecosystems that have evolved over long periods of time.

Mercury is an element that illustrates the variability in toxicity response of some substances. Mercury has been used for centuries in various applications such as in tooth amalgams, as a fungicide, and in many industrial uses. Elemental mercury is relatively harmless unless it is vaporized and inhaled directly into the lungs. The toxic forms of mercury for ingestion are methyl mercury (CH_3Hg^+ and $CH_3 \cdot Hg \cdot CH_3$) and inorganic salts, particularly mercuric chloride ($HgCl_2$). Methyl mercury can be formed by bacteria in sediment and acidic waters. Inorganic ionized mercury is acutely toxic. Elemental mercury has a relatively short residence time in the body but methyl mercury compounds stay in the body 10 times longer than metallic mercury and they cause malfunction of brain, nervous system, kidneys, and liver as well as birth defects. Methyl mercury is cumulative in the food chain. Phenyl mercury compounds ($C_6H_5Hg^+$ and

TABLE 8.1 Elements Found in Living Systems

Groups of elements	Remark
C, H, O, N, P, S	The major constituents of organic molecules
Ca, Cl, K, Mg, Na	Elements found above trace levels
Co, Cu, Fe, Mn, Zn	Trace elements
Al, B, I, Mo, Si, V	Generally not essential

$C_6H_5 \cdot Hg \cdot C_6H_5$) are moderately toxic with short retention times in the body but these compounds are rapidly transformed in the environment to release inorganic mercury.

A survey of harmful effects of some elements and compounds is presented in Table 8.2. The survey is by no means exhaustive.

Other heavy metals besides mercury can be methylated biologically or abiotically. Formation of organometal compounds affects their accumulation and toxicity to humans and other life forms. Ferguson (1990) has discussed chemistry and environmental impacts of heavy metals.

The variety of synthetic organic compounds is enormous. The bank of data describing their toxicity and environmental effects is growing continuously and only a few of the classes or specific compounds are commented on in this book. A good reference on environmental data for organic chemicals has been compiled by Verschueren (1983).

8.2 CONTAMINANTS IN WATER

Contaminants enter a water supply from natural and anthropogenic sources. Direct discharge of effluents from population centers and industries are point sources and stormwater runoff washes residues of products applied to the land and transported by air into receiving waters. Contamination of water by sewage is the major source of waterborne disease.

The application of certain agents such as chlorine and aluminum in water treatment processes is associated with low-risk contamination. As an example, alum is the most common coagulant used in water treatment operations. High intake of aluminum has been associated with neuropathological disorders, one of which is Alzheimer's disease (Crapper et al., 1973; Kopeloff et al., 1942), although the link between aluminum and Alzheimer's disease is tenuous. The American Water Works Association (AWWA) goal for Al in drinking water is 50 μg/L.

Corrosion of materials in the distribution system or household plumbing and appurtenances may be a significant source of contaminants in drinking water. Schock and Neff (1988) have shown that new chrome-plated brass faucets were a significant source of Pb, Zn, and Cu contamination of drinking water, particularly when water had been standing in them for some time. Lead concentrations in standing (8-h standing period) water samples exceeded United States maximum contaminant levels (MCL) of 0.05 mg/L for more than 50% of the samples from some sites. As time went on, leaching of metals decreased, presumably because of formation of protective metal oxide coatings. Obviously, standards for metals should be cognizant of heavy metal intake from these extraneous sources.

Addition of nutrients to fresh waters from agricultural runoff and treated or untreated domestic sewage is making blooms of algae more common. Cyanobacteria (bluegreen algae), particularly *Anabaena* species, which are neurotoxic, and *Microcystis,* which is hepatotoxic (toxic to the liver), are more common in drinking waters (Falconer et al., 1989). Besides being toxic, these organisms also cause distinct tastes and odors. Copper (usually applied as copper sulfate) is a control measure for these microorganisms and algae in general, but they pass through the normal treatment operations of flocculation, sedimentation, and rapid sand filtration. Specialized activated carbon treatment is required to remove the toxicity associated with these bacteria (Falconer et al., 1989).

Agricultural runoff is also the source of many pesticides used in modern agricultural practice. The water-soluble pesticides, alachlor, metachlor atrazine, linuron, cyanazine, metribuzin, carbofuran, and simazine were found in concentrations up to 22 μg/L in midwestern rivers in the United States (Miltner et al., 1989). Pesticide occurrence was somewhat predictable following springtime thaw, tillage, and agrichemical application. Concentrations in the rivers increased significantly during periods of runoff and these pesticides were present in agricultural watersheds for periods of 2–6 months or longer following application.

Runoff from solid waste disposal sites is another significant source of water pollution. Runoff leaches soluble substances and suspends particulates at these sites and transports them to streams.

Emissions to air cause air pollution and many substances are returned to land and water by rainfall. Acid rain is one of the most familiar of these phenomena. The oxides of sulfur and nitrogen, produced by combustion of fuels, hydrolyze to cause a reduction in pH of the rain.

8.2.1 Common Contaminants

Additional information on a few contaminants is provided in the following sections.

Nitrate

The application of fertilizers also results in elevated concentrations of nitrate, which, as noted in Table 8.2, can be toxic to infants. The disease caused by nitrate and nitrite is methemoglobinemia. Nitrates ingested through food or water are converted to nitrites by bacteria in the digestive system of infants. Nitrite reacts with hemoglobin, preventing it from carrying oxygen, thus the common name "blue baby disease" for white-skinned persons. This condition is generally confined to infants up to the age of 6 months.

The standard for nitrate is based on a United States survey in 1951 in which no observed cases of methemoglobinemia occurred when nitrate-N was less than 10 mg/L. The ubiquitous use of fertilizers commonly causes groundwaters in agricultural areas to exceed the nitrate standard (Bouchard et al., 1992). Certain regions have geologic deposits of nitrate salts that can significantly affect groundwater supplies. European drinking waters also commonly exceed the nitrate standard.

Fluoride

Fluoride is a naturally occurring element in many water supplies around the world. Concentrations range from insignificant to over 50 mg/L. Surface waters generally do not contain fluoride in excess of 0.3 mg/L. Low concentrations of fluoride are beneficial to the formation of teeth that resist dental caries. The fluoride ion replaces hydroxide

TABLE 8.2 A Sampler of Toxic Elements and Compounds[a]

Substance	Remarks
Antimony, Sb	Little research on Sb. Accumulates in liver and is detrimental to the heart in humans. Can be accumulated by marine organisms.
Arsenic, As	Acute or chronic toxicity to humans. Toxic to all life. Byproduct of smelting ores and used in other industries.
Barium, Ba	Ingested Ba salts are highly toxic to humans. Usually found in trace amounts in natural waters but surface water concentrations are sometimes as high as 0.340 mg/L. May be toxic to plants if present above trace amounts.
Beryllium, Be	Extremely toxic to all life. Usually naturally present in concentrations less than 0.000 1 mg/L in surface waters. Oxides and hydroxides are insoluble within normal pH ranges.
Boron, B	No evidence of accumulation in humans. Large amounts may produce digestive difficulties and nerve disorders.
Bromine, Br	Free bromine, Br_2, is a strong oxidant not found naturally. Bromine salts are harmless.
Cadmium, Cd	Cumulative, highly toxic in humans and livestock. Affects all life. Protects other metals against oxidation; also used in other industries.
Chlorine, Cl	Same as bromine.
Chromium, Cr	Natural Cr is rare. Cr(VI) is the toxic form to humans. Cr(III) is slowly oxidized to Cr(VI) in waters. Toxic to plants. Varying tolerance to Cr salts in aquatic life.
Cobalt, Co	Low toxicity to humans; essential in trace amounts.
Copper, Cu	Essential to humans in small daily amounts (2.0 mg). Upper limits not determined but water is very distasteful at 1–5 mg/L Cu. Essential to all life but is toxic at differing levels to plants and aquatic life.
Cyanide, CN^-	Cyanide renders tissues incapable of oxygen exchange. It is not cumulative and it is biodegradable in streams. CN^- behaves like halides.
Fluorine, F	Fluoride has been shown to reduce dental caries. Above concentration guidelines there is no further reduction in caries but mottling increases. Natural F^- concentrations are generally low but wide fluctuations occur.
Lead, Pb	Cumulative in humans and livestock. Human absorption of ingested lead is small; single large doses are not a problem.
Lithium, Li	Higher concentrations cause phytotoxicity.
Mercury, Hg	Toxic to all forms of life. Mercury is very slowly excreted from the human body. Methyl mercury is 50 times more toxic than inorganic mercury.
Molybdenum, Mo	Essential micronutrient. Does not accumulate and humans can tolerate large quantities. Plants accumulate Mo in their foliage but at normal concentrations in water it is not harmful. Some grazing animals, e.g., cattle, sheep, and swine, exhibit sensitivity to this metal.
Nickel, Ni	Low oral toxicity to humans. Toxic to plants and marine life.
Nitrogen, N	
Ammonia, NH_3	Nontoxic to humans at natural levels. Fish can not tolerate large quantities of ammonia.
Nitrate, NO_3^-	Toxic to infants at high concentrations.
Nitrite, NO_2^-	More toxic than nitrate but less chemically stable than nitrate and generally found in low concentrations.

TABLE 8.2 Continued

Substance	Remarks
Organic N	No health effects per se.
Phenol	Taste and odor from these compounds are more significant than their toxicity. They exhibit direct toxicity to fish.
Selenium, Se	Cumulative poison in humans and animals. Moderately toxic to plants.
Silver, Ag	Cumulative in human tissue resulting in blue-gray discoloration of skin (argyria). Toxic to aquatic organisms.
Sodium, Na	Harmful to some persons with cardiac problems. Destructive to soils.
Strontium	The stable isotope is a widespread minor component of igneous rocks. It is similar to calcium and of no health significance. Industrial sources of radioactive strontium are insignificant.
Sulfide, S^{2-}	Undissociated hydrogen sulfide (H_2S) is the toxic entity. Also toxic to aquatic life.
Thallium, Tl	Cumulative poison with sublethal effects such as hair loss and hypertension. Thallium has been shown to inhibit photosynthesis and acts as a neuropoison to fish and aquatic invertebrates.
Tungsten	Highly insoluble in water and little information on health effects.
Uranium, U	Uranium and its salts are quite toxic to humans and have also been reported to be toxic to aquatic organisms.
Vanadium, V	Low concentrations are not toxic to humans. Toxicity has been demonstrated in plants.
Zinc, Zn	Relatively nontoxic to humans and animals. Essential nutrient for life. Only at high concentrations has it been found toxic to plants. However, zinc is acutely and chronically toxic to aquatic organisms.

[a]Adapted from Inland Waters Directorate (1979).

ion in the apatite that forms the enamel in teeth. High concentrations lead to mottling (brownish discoloration) of teeth or dental fluorosis. Fluoride concentrations above 1.5 mg/L must be reduced to prevent fluorosis.

Fluoride is added to deficient waters to produce a concentration near 1 mg/L in the finished water. Studies have shown that the level of fluoride ingestion from water with this concentration provides optimal dental protection with no risk of mottling. Opposition to fluoridation remains and there have been some cogent arguments made against fluoridation (Camp and Meserve, 1974). Modern dentistry and home dental hygiene reduce the need for publicly enforced fluoridation. But the benefits of fluoridation are still substantial. There have been numerous studies on the physiological effects of ingestion of fluoride. The overwhelming research evidence indicates that fluoride near 1 mg/L with normal water consumption is safe with a good margin of safety. Dental associations in Canada and the United States widely encourage the addition of fluoride to water when it is lacking. As of 1986, 39.2% of the Canadian population was receiving naturally or artificially fluoridated water at or near recommended levels (Droste, 1987). The decision to fluoridate a water is made at the local level in Canada.

Detergents

Historically, soaps were made from animal fats. These soaps are precipitated by multivalent cations in water. Industry developed synthetic detergents that have under-

TABLE 8.3 Composition of PO_4^{3-} and CO_3^{2-} Built Detergents[a]

Component	Percentage by weight in detergents	
	PO_4^{3-}	CO_3^{2-}
Surfactants	18	22
Linear akylbenzene sulfonates	7.0	14
Tallow alcohol sulfates	5.5	—
Ethyloctyl sulfosuccinates	5.5	6.0
Sodium sulfosuccinates	—	2.0
Builders		
Sodium tripolyphosphates	24	—
Sodium carbonates	—	20
Corrosion inhibitors		
Sodium silicates ($SiO_2:Na_2O_2$)	12 (1.6)	20 (2.4)
Suspending agents		
Carboxylmethylcellulose	0.30	—
Whitening (fluorescent) agents		
Optical brightners	0.13	0.17
Coloring matter and fragrance		
Perfume	0.15	0.20
Dye	0.14	0.80
Fillers		
Sodium sulfates	37	32
Water	6.0	4.0
Polyethylene glycol	0.90	0.90

[a]From B. J. Alhajjar, J. M. Harkin, and G. Chesters (1989), "Detergent Formula and Characteristics of Wastewater in Septic Tanks," *J. Water Pollution Control Federation,* 61, 5, pp. 605–613, © WEF, 1989.

gone a continuous evolution in the past half-century to maintain the performance of a laundering agent that is unaffected by the mineral content of the water and does not cause water quality problems. At least, any such agents should be readily treatable at a wastewater treatment facility. Alkylbenzenesulfonates and phosphate builders were major constituents of detergents up to the 1960s. The alkylbenzenesulfonates were not readily biodegradable and rivers were known to become covered with foam caused by this agent. The phosphorus accelerated eutrophication of lakes. To remedy the problem with refractory alkylbenzenesulfonates, a biodegradable linear alkyklbenzenesulfonate was substituted for the nonlinear alkylbenzenesulfonates.

Nitrilotriacetic acid (NTA) is one agent that was used to replace phosphate builders. NTA is a chelating agent incorporated into detergents to complex the water hardness cations (Section 15.1) and prevent them from reacting with the detergent molecule.

Detergents remain a major source of phosphorus in wastewaters and receiving waters. Polyphosphate is used as a complexing agent in some detergents and carbonate is used to remove Ca^{2+} by precipitation in others. The characteristics of PO_4^{3-} and CO_3^{2-} built detergents are given in Table 8.3. Regulation of phosphate in detergents can result in significant reductions of phosphorus in streams. A statewide ban on phosphorus-containing detergents in Virginia reduced the total contribution of phosphorus from Richmond, Virginia to the James River by a factor greater than 5 (Hoffman and Bishop, 1994).

8.2.2 Carcinogens

Cancer-causing agents are a major concern of society. Increasing life spans and increased production of cancer-causing agents, among other factors, contribute to increased mortality rates from cancer. There are many factors that contribute to the potency of an agent causing cancer. Naturally occurring carcinogens exist in foods that are staples in diets throughout the world. There do not appear to be any threshold levels for carcinogenic agent concentrations. The risk of contracting the disease is directly proportional to the exposure to the agent; however, there is wide variability in the potency of agents. The World Health Organization (WHO) sets guideline values for carcinogens in drinking water at a level where the lifetime exposure will increase the risk by one additional cancer per 100 000 (a risk factor of 10^{-5}) for the population ingesting the drinking water (WHO, 1993).

Determination of Carcinogenicity

Assessing the carcinogenicity of a substance is a difficult task. The disease develops over long periods of time, which is a major complicating factor. There are a number of techniques employed to determine the cancer causing potential of a substance. The assessment of the risk is made using statistical analysis of a control population and an experimental population. Many studies have been devoted to comparing cancer occurrence among segments of the human population where significant differences exist between one group and the other (e.g., smokers versus nonsmokers; miners versus the general population). There are often many confounding variables in these studies.

Lab studies utilize rats or other animals (that exhibit metabolic traits similar to humans) in well-controlled conditions where only one substance is examined. These studies are costly. It takes a large number of subjects to prove with confidence that an agent poses a risk at a level of 1 per 100 persons (1%), which is a very high risk level. New techniques have somewhat overcome this difficulty.

One of the newest techniques for genotoxicity (mutagenicity, carcinogenicity, or teratogenicity) is the Ames test (Ames et al., 1975). The test is a measure of the ability of an agent to cause mutation. A mutation is an unexpected change in genetic material in a cell. Cancerous cells are mutated cells that grow at an uncontrolled rate.

This test uses a strain of bacteria (*Salmonella typhimurium*) that has been genetically engineered to remove the ability to produce the essential amino acid histidine. In addition, the bacteria are modified in other ways to make them more susceptible to mutagenic agents. A DNA repair mechanism has been removed from the cells. The cell wall has been removed to allow substrates to more easily enter the cell for metabolism.

Exposing the cells to a mutagenic agent increases the likelihood that the cells will recover the ability to produce histidine. The cells are mixed with an extract of rat liver to subject them to mammalian metabolic processes and separated into two groups: a control and tester group. The media for both groups also contains a small amount of histidine, which will support bacterial growth for a short period of time.

The chemical agent being examined is added to the tester bacteria plate. There will be some cell growth in the control plate because of natural reversion of some cells to a histidine production state. The number of cells that grow in the tester plate compared to the control depends on whether the chemical agent-induced mutation allows the bacteria to produce histidine. This test and others like it only measure one specific feature and their quantitative relation to long-term rodent bioassay results is not clear. The test depends on the correlation between mutagenicity and carcinogenicity and it has performed well (Meier and Daniel, 1990) in this regard. Therefore it will continue to be a valuable tool for analysis of water for health effects (Meier and Daniel, 1990).

8.2.3 Toxicity Determination

Toxicity is a general term that refers to the poisonous effect of an agent to a living organism. Species exhibit varying responses to toxic agents. Some agents are essential in small amounts and toxic at higher concentrations. Toxicity is also complicated by some agents that are cumulative and others that are transformed chemically or biochemically into harmless compounds. Furthermore, toxic agents can have synergistic interactions wherein the total toxicity of a sample is greater than the sum of toxicities of individual components. Antagonistic interactions may also occur wherein the total toxicity of a sample is less than the sum of toxicities of individual components. In the past toxicity was narrowly confined to agents that were potentially toxic to humans. Current regulations are being broadened to include agents that exhibit aquatic toxicity.

Toxicity determinations are generally divided into two ranges of response: subchronic and acute. Subchronic tests are designed to determine the concentration level

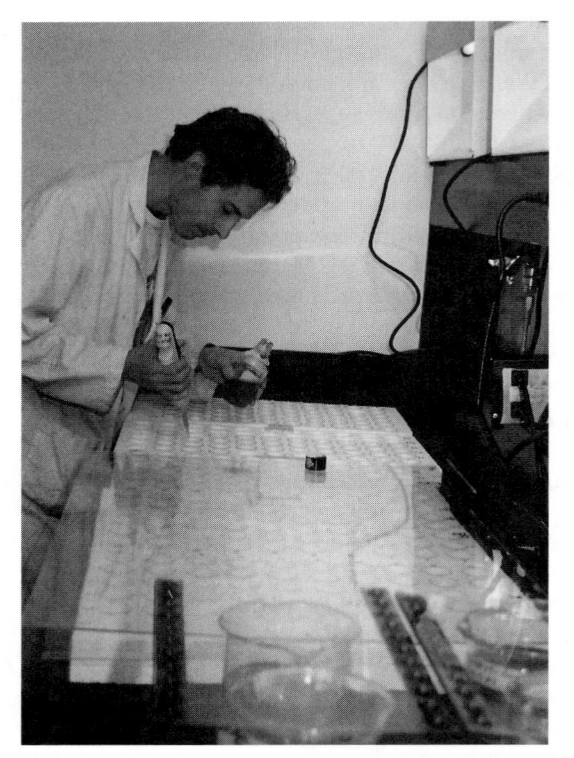

Figure 8.1 Subchronic toxicity testing using *Ceriodaphnia dubia* (water fleas). Varying concentrations of the substance or wastewater to be evaluated are added to the test chambers. Feed for the *Ceriodaphnia* is added (shown in the figure). Then a young adult organism is added to each vessel and reproduction is monitored in all vessels over several days, usually about 7 d, which will allow reproduction of three generations.

TABLE 8.4 Test Organisms for Toxicity Bioassays[a]

Type of test	Freshwater organism	Marine organism
Subchronic	Algae (*Selnasatrum capricorutum*), *Ceriodaphnia dubia*	Sheepshead minnow, inland silverside *Mysidopsis bahia,* sea urchin, *Champia parvula*
Acute	Bluegill (*Lepomis macrochirus*), *Ceriodaphnia dubia,* zooplankton, *Daphnia magna, Daphnia pulex,* fathead minnow (*Pimephales promelas*), rainbow trout (*Salmo gairdneri*)	Atlantic silverside (*Menidia menidia*), grass shrimp (*Palaemonetes pugio*), inland silverside (*Menidia beryllina*), mysids (*Mysidopsis bahia*), sheepshead minnow (*Cyprindon variegatus*)

[a]After Isom (1990).

where a substance has an effect on the survival, growth, and reproduction (Fig. 8.1) of a test species. Subchronic tests last a few days and chronic tests have a duration over a number of life cycles. Acute tests are short-term survival assays conducted over 24- to 96-h periods of a test organism exposed to different concentrations of an agent. The categorization of responses from the various tests ranges from the "no-observed-effect concentration" (NOEC) to the concentration that is lethal to 50% of the test organisms (LC_{50} or EC_{50}).

There have been a multitude of tests devised to measure toxicity. Standard protocols have been developed to minimize variability in response; however, there is still considerable variation in results reported from different laboratories performing replicate tests (Isom, 1990). The choice of species and their health and developmental stage as well as the number of species involved influence the results. Toxicity variation among species undermines the rational basis for design of toxicity measurement but the consistent application of a toxicity protocol can provide useful information on changes in the amount of toxicity. Furthermore there is a growing database of information relating responses from various procedures.

The expense of tests can be high. Subchronic tests are generally more costly than acute tests because of the longer times involved. Test organisms range from bacteria to fish from both marine and freshwater environments. Table 8.4 lists commonly used test organisms.

The costs and time involved in conventional toxicity assays has led to the development of a bacterial assay (the Microtox® assay), which is a rapid and inexpensive test for acute toxicity assessment (Chang et al., 1983). The test utilizes a luminescent marine bacterium, *Photobacterium phosphoreum.* Test conditions are maintained to cause a culture of these bacteria to produce light. Exposure of a test culture aliquot to a toxicant that affects these bacteria results in the inhibition of activity of the culture or the death of cells. The luminescence of the test culture is compared against a control by measuring the light output with a photometer. The bacteria used in this test are reported to be quite sensitive to toxicants (e.g., Blum and Speece, 1991) and there have been a significant number of studies relating results from this assay to more conventional assays.

8.2.4 Radioactive Constituents

The major health effects of radiation are an increase in the occurrence of cancer and genetic defects. Radiation induces changes in the cell that lead to cancer. Some radioactive elements such as uranium can have a chemically toxic effect on organs. High doses of radiation can cause death in a short period of time.

Radiation units in terms of disintegrations per second have been defined in Section 1.13. The biological effects of radiation depend on the ionizing effects of radiation in cells. The roentgen (named after W. Roentgen, who discovered X rays) is the unit of gamma or X-radiation intensity. The roentgen is defined as the quantity of gamma or X radiation that will produce ions carrying one electrostatic unit of electricity of either sign in 0.001 293 g (1 cm^3) of dry air at 0°C and 760 mm pressure. This is equivalent to 1.61×10^{12} ion pairs per gram of air and corresponds to the absorption of 83.8 ergs of energy. The sievert (Sv) is another measure of radiation dose equal to approximately 8.38 roentgens.

The rad (roentgen-absorption-dose) is another measure of the dose of radiation delivered by radioactivity. One rad equals 100 ergs of energy deposited per unit mass of any absorbing material. A rad is not necessarily a measure of the damage resulting to humans. Rads are modified for humans by multiplication with a factor called the relative biological effectiveness factor that gauges how effective the deposited material is in damaging human tissue. The resulting measure is the roentgen equivalent—man (rem). Recommended maximum exposures for the general public and for workers in the radiation industry are 0.5 and 5 rem/yr, respectively.

The issue of biological effectiveness of a given dose of radiation is complicated by the accumulation of some isotopes in the body; for instance, iodine is accumulated in the thyroid gland. The penetrating power of radiation is also an important factor.

Alpha particles are relatively massive and travel only a short distance in air and cannot generally penetrate the dead layer of skin of a human. Thus alpha emitters are only dangerous if they are ingested. Then a particle is able to do great damage inside the body over the short pathway in which its energy is deposited. Beta particles are lighter than alpha particles and are generally considered to have about 100 times the penetrating power of alpha particles (although there is considerable variation from this value). The most penetrating forms of radiation are neutrons, which are rare, and gamma rays. Gamma rays are 10 000 times more penetrating than alpha particles (Lowry and Lowry, 1988).

Table 8.5 lists United States data on the annual effective dose equivalents from radiation sources. Radon-222 and its progeny are the major radiation hazards. Radon is ubiquitous, with outdoor levels generally measuring less than 0.5 pCi/L of air (Horgan, 1994). Radon-222 is volatile and its presence in drinking water is therefore not significant (Lowry and Lowry, 1988). But it is released into the home through normal water use and presents a health threat, especially to the lungs. Elevated levels of ^{222}Rn are almost exclusively confined to groundwater sources.

8.3 Taste and Odor

There are many natural sources of taste and odor causing compounds in addition to synthetic chemicals. Taste and odor problems with a water supply elicit rapid response from consumers. Odor emanation from wastewater treatment works is also a cause of complaints.

TABLE 8.5 Annual Effective Dose Equivalent to People in the United States from Various Sources[a]

Source	Dose equivalent mSv/yr	Relative contributions %
Radon (airborne)	2.0	56
Other natural sources	1.0	28
Diagnostic X rays	0.39	11
Nuclear medicine	0.14	4
Occupational	0.009	<1
Miscellaneous environmental exposure	0.006	<1
Consumer products	0.005	<1
Total	3.6	100

[a]From Lowry and Lowry (1988). Reprinted from *Journal of the American Water Works Association,* by permission. Copyright © 1988, American Water Works Association.

The most well-known natural odoriferous compound is H_2S, commonly found in well water. Anaerobic conditions generate H_2S, which is readily soluble in water. H_2S is one of the few taste and odor causing compounds addressed in water quality standards. Even though H_2S is toxic, taste and odor problems occur at much lower concentrations. Many other taste and odor causing compounds have been identified but they are generally not addressed beyond the general requirement that water be inoffensive to consumers. The problem is complicated by many factors. First, there is wide variability in the sensitivity of individuals to different compounds and the threshold concentrations for detecting the compounds. Mineral content and synergisms among water components produce different taste and odor reactions. Many taste and odor compounds are noticeable at concentrations less than 1 ppb. For instance, geosmin, an earthy smelling byproduct of bluegreen algae and *Actinomycetes,* is detectable at concentrations of 0.2 ppb.

The sources of taste and odor compounds are many. Decaying organic matter releases taste and odor causing compounds into waters. Phenols are found naturally in fossil fuels but they are also produced in many industries. Phenols are detectable at low concentrations and they are also mildly toxic. Chlorine substitution on phenols increases both taste and odor and toxicity effects. The earliest microorganisms identified with tastes and odors were the algae. Some taste and odor causing algae are shown in Fig. 8.2.

Actinomycetes are moldlike bacteria, widely distributed in the environment, that produce many offensive tastes and odors. Tastes and odors associated with them have been described as woody, haylike, marshy, musty, manurial, potato bin, and bitter. Taste- and odor-causing microbes are generally not harmful to the health of persons drinking water containing viable cells of these microbes. Therefore, their allowable concentrations are not directly prescribed.

Table 8.6 lists some chemicals with high odor recognition. The odor potential of a substance depends not only on the concentration at which the odor is noticed but also on the ability of the agent to volatilize. Thus Verschueren (1983) used the odor index (OI) proposed by Hellman and Small (1974) as well as the 100% recognition

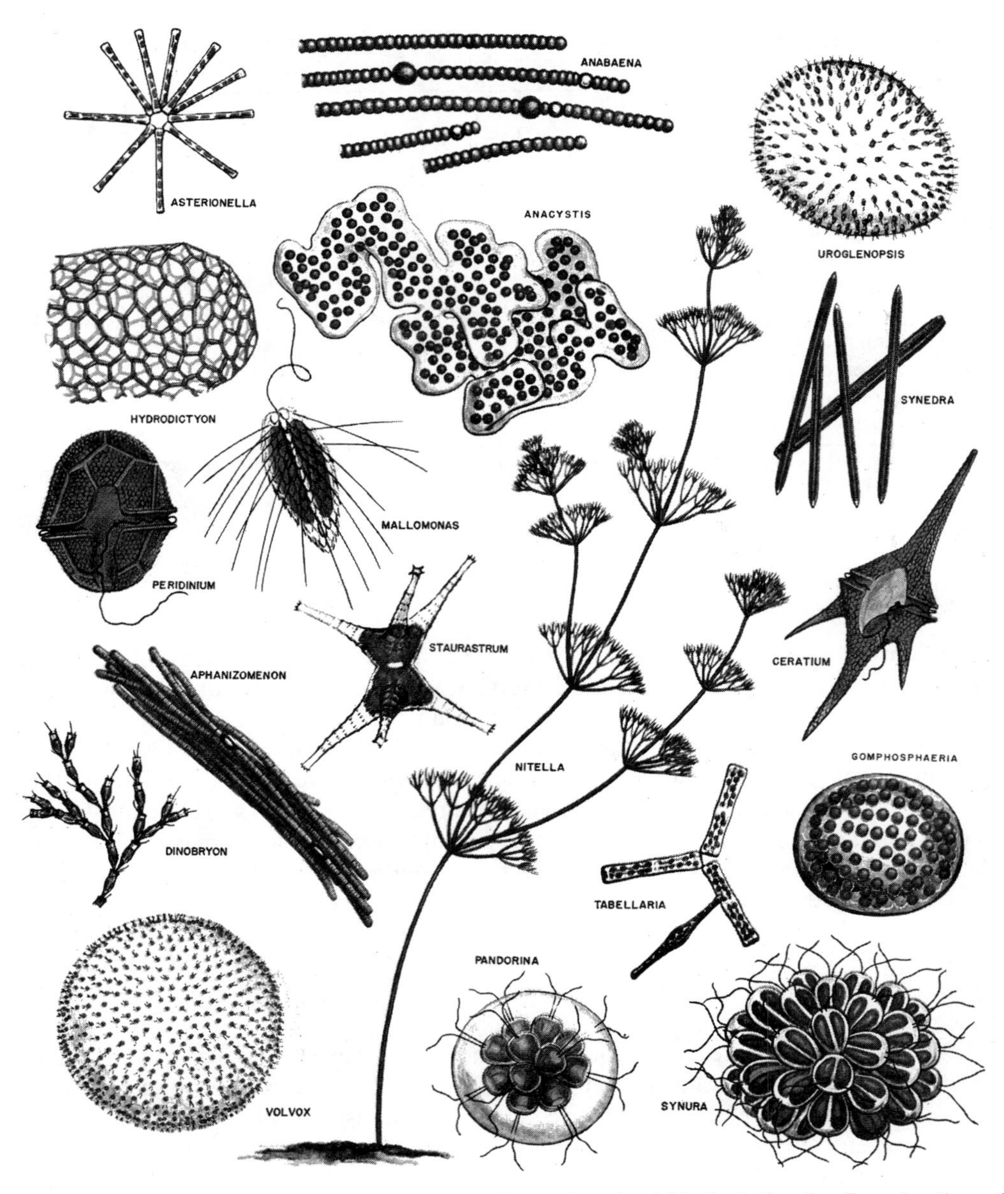

Figure 8.2 Taste and odor causing algae. From *Standard Methods for the Examination of Water and Wastewater,* 18th ed. Copyright 1992 by the American Public Health Association, the American Water Works Association, and the Water Environment Federation. Reprinted with permission.

level to characterize odor potential. The equation for OI is

$$\mathrm{OI} = \frac{\text{vapor pressure (ppm)}}{\text{100\% odor recognition threshold (ppm)}} \tag{8.1}$$

where

ppm is on a volume basis and 1 atm = 1 000 000 ppm

TABLE 8.6 Odor Potential of Selected Chemical Compounds[a]

Chemical	Formula	Molecular weight	Odor index[b]	100% recognition
Ethylesters				
Ethylbutyrate	$C_3H_7COOC_2H_5$	116	1 982 000	7 ppb
Ethyldecanate	$C_9H_{19}COOC_2H_5$	200	—	0.17 ppb
Mercaptans				
Methylmercaptan	CH_3SH	48	53 300 000	35 ppb
Ethylmercaptan	CH_3CH_2SH	62	289 500 000	2 ppb
Propylmercaptan	$CH_3CH_2CH_2SH$	76	263 000 000	0.7 ppb
Butylmercaptan	$CH_3(CH_2)_3SH$	90	49 000 000	0.8 ppb
Phenylmercaptan	C_6H_5SH	110	940 000	0.2 ppb
Sulfides				
Hydrogen sulfide	H_2S	34	17 000 000	1 ppm
Methylsulfide	$(CH_3)_2S$	62	2 760 000	0.1 ppm
Ethylsulfide	$(CH_3-CH_2)_2S$	90	14 400 000	4 ppb
Isoamylsulfide	$[(CH_3)_2CH(CH_2)_2]_2S$	174	1 640 000	0.4 ppb
Butyrates				
Methylbutyrate	$CH_3-CH_2-CH_2-COOCH_3$	102	11 000 000	3 ppb
Ethylbutyrate	$CH_3-CH_2-CH_2-COOC_2H_5$	116	1 982 000	7 ppb
Amines				
Ammonia	NH_3	17	167 300	55 ppm
Methylamine	CH_3NH_2	31	940 000	3 ppm
Ethylamine	$CH_3CH_2NH_2$	45	1 445 000	0.8 ppm
Dimethylamine	$(CH_3)_2NH$	45	280 000	6 ppm
Trimethylamine	$(CH_3)_3N$	59	493 500	4 ppm
Alkenes				
1-Butene	$CH_3CH_2CH{=}CH_2$	56	43 480 000	0.07 ppm
Isobutene	$(CH_3)_2C{=}CH_2$	56	4 640 000	0.6 ppm
1-Pentene	$CH_3CH_2CH_2CH{=}CH_2$	70	376 000 000	2 ppb
Ethers				
Ethylether	$CH_3CH_2OCH_2CH_3$	74	1 939 000	0.3 ppm
Isopropylether	$(CH_3)_2CHOCH(CH_3)_2$	100	3 227 000	0.06 ppm
Aldehydes				
Formaldehyde	$HCHO$	30	5 000 000	1 ppm
Acetylaldehyde	CH_3CHO	44	4 300 000	0.3 ppm
Propionaldehyde	CH_3CH_2CHO	58	3 865 000	0.08 ppm
Butyraldehyde	$CH_3CH_2CH_2CHO$	72	2 395 000	40 ppb
Acids				
Propionic acid	CH_3CH_2COOH	74	112 300	40 ppb
Valeric acid	$CH_3(CH_2)_3COOH$	102	256 300	0.8 ppb

[a]Adapted from K. Verschueren (1983), *Handbook of Environmental Data on Organic Chemicals,* Van Nostrand Reinhold, New York, NY. See this reference for a more comprehensive listing.
[b]See text for definition of OI. OIs were calculated at 20°C.

The 100% odor recognition level is the concentration at which 100% of the odor panel members perceive the odor. Compounds listed in Table 8.6 are those that have an odor potential in the medium (OI = 100 000 to 1 000 000) to high range (OI > 1 000 000).

8.4 BASES FOR STANDARDS

Standards drive the need and degree of water and wastewater treatment. The promulgation and enforcement of standards that ensure reasonable protection of public health and environmental quality are required. In many countries water quality criteria are incorporated into laws that specify substantial penalties for their violation.
Standards for drinking water quality are generally established on the following bases:

1. **Established or Ongoing Practices.** Practices that have been used for many years without noticeable harmful effects are used as a basis to set water quality criteria. Hidden dangers can exist in merely adopting historical practice without scientific analysis. Aluminum and chlorine are two examples of substances used without any noticeable harmful effects until careful analysis has shown that the former is associated with Alzheimer's disease and the latter forms carcinogenic compounds.
2. **Experimentation with Animals.** Historically this has been the primary method for assessing chemicals that are toxic to humans. It is most desirable to use animals with a physiological response as similar as possible to humans, e.g., rats, to find exposure levels that produce harmful effects. These studies tend to be expensive. Recent tests, such as the Ames test (Ames et al., 1975) described in Section 8.2.2 which indicates the carcinogenicity of a compound, use bacteria to assess whether a chemical agent causes mutations. These tests are increasingly being used as an alternative to animal testing.
3. **Human Exposure.** Epidemiology or studies that show direct causal evidence between exposure to an agent in water and disease or toxic effects in humans are the most reliable source of information. Experimenting with humans directly is undertaken only in unusual circumstances; however, there are numerous occurrences of long-term or sudden contamination of water supplies that have led to mild or disastrous results. Some benefit can be derived from these unfortunate situations by careful assessment of exposure levels and resulting health deterioration in the affected population.
4. **Statistical Comparisons.** Statistical analysis of data from two populations can confirm differences in the occurrence of disease and in conjunction with other studies, the possible causes of the disease. The interpretation of data from these studies is often very difficult because of the impossibility of isolating two naturally occurring populations that differ only in one aspect of their behavior or exposure to a substance.

Once a threshold level for a causative agent is established from the best available information, this level is divided by a factor of 100 or more to provide a suitable safety factor to arrive at a standard. Risk analysis is increasingly being applied. The objective is to regulate all harmful substances at the same level of risk and other substances at levels that are inoffensive or aesthetically pleasing to the majority of consumers. Drinking water is not the major route for exposure to many chemical substances. Period and degree of exposure of a population from all sources (air, water, and food) are taken into account. Technological and economic constraints may mitigate the immediate attainment of a given standard.

8.4.1 Environmental Water Quality Standards

Establishing standards for general environmental quality is based on approaches similar to those outlined above. The focus of many regulations is based on the health threat to humans as opposed to threats to the ecosystem, which are extremely difficult to specify.

Water quality criteria do not offer the same degree of protection for all forms of life in an ecosystem. Large degrees of response variation occur in different life forms and among different individuals of a species. Therefore testing must include a wide range of biological processes. Values of society as well as scientific information influence standards. Often there is conflict between the desire for a more comfortable existence for humans and the preservation of part of an ecosystem. Complex interrelationships among species and the environment make the task of establishing environmental standards very difficult.

The approach taken in Canada for development of guidelines for aquatic life adopts the following philosophy of the International Joint Commission (IJC) and the Ontario Ministry of the Environment [Canadian Council of Ministers of the Environment (CCME), 1972]. It states that guidelines "are set at such values as to protect all forms of aquatic life and all aspects of the aquatic life cycles. The clear intention is to protect all life stages during indefinite exposure to water."

8.5 STANDARDS FOR DRINKING WATER

Drinking water standards around the world are in a continuous state of evolution as more information becomes available and is evaluated. No single standard for drinking water quality suffices for all countries but there is a considerable degree of agreement on contaminants and their allowable concentrations (Sayre, 1988). Yet different approaches to regulation and different conditions in countries will maintain differences in standards currently enforced. Although standards and monitoring programs are in place for most public water supplies around the world, bottled water, which is increasingly popular, is often not regulated.

The first priority of water suppliers in all countries is to ensure that drinking water is bacteriologically safe. In the United States, reporting of waterborne disease outbreaks has been and continues to be voluntary. Based on the available data the incidence of waterborne diseases had declined from 8 cases per 100 000 person-years during 1920–1940 to 4 cases during 1971–1980 (Craun, 1986).

Over the last few decades, the number of chemicals appearing in the standards has increased and will continue to increase as more data become available.

8.5.1 International Drinking Water Standards

The WHO is an international body and, using experts from around the world, has developed **guidelines** (WHO, 1984) to be used as a basis for developing standards in all countries, particularly those countries that lack the resources to perform the basic information gathering and assessment tasks involved. WHO notes that the guidelines are to be considered in the environmental, social, economic, and cultural milieu of the country. The guidelines have undergone various revisions through the years.

8.5.2 U.S. Safe Drinking Water Act

The U.S. Safe Drinking Water Act (SDWA) was signed into **law** in 1974 and mandated the establishment of drinking water regulations that were the first to apply to all public water systems in the United States. The U.S. Environmental Protection Agency

(USEPA) was to establish maximum contaminant level goals (MCLGs) and maximum contaminant levels (MCLs) for 83 contaminants. MCLs were to be set as close to MCLGs as feasibly possible, taking cost into consideration (Ohanian, 1992).

MCLs have been set for some contaminants and a timetable has been set to establish the others. Treatment technologies must be included in the standards for each chemical or group of chemicals. The EPA must also specify the analytical method best suited to detect the amount of contaminant in drinking water. Work is continuing to refine analytical techniques and MCLs for all contaminants.

An MCL is an enforceable standard. The SDWA authorizes the USEPA to seek injunctive relief (i.e., a court order requiring a water system to undertake certain tasks to achieve compliance) and civil penalties of up to $25 000 per day for violations of an MCL or a USEPA administrative order (Koorse, 1990). The adverse health effect of a contaminant need not be proven conclusively prior to regulation. Furthermore, the EPA was required to develop a list of contaminants known or anticipated to occur in public water systems that might require regulation under the SDWA. These contaminants were placed on the drinking water priority list (DWPL) as of 1989. Regulations for many contaminants on the original DWPL have now been developed.

Monitoring and reporting requirements are specified in the regulations. The frequency of monitoring depends on the constituent, water source, population served, and other factors. The risk of waterborne disease is immediate and high; therefore monitoring indicator organisms is frequently performed (Table 8.7). The monitoring frequency of other constituents will in general be lower.

The SDWA and regulations developed under it are extensive. Requirements for treatment are specified in certain circumstances. The public must be notified when drinking water standards are violated; the notification procedures depend on the acuteness of the violation. MCLGs are less than MCLs based on reasonable attainability with the best available technology (BAT) or limits of analytical methods for a

TABLE 8.7 SDWA Routine Sampling Requirements for Total Coliforms

Population served	Minimum no. of routine samples per month	Population served	Minimum no. of routine samples per month
25–1 000	1	59 001–70 000	70
1 001–2 500	2	70 001–83 000	80
2 501–3 300	3	83 001–96 000	90
3 301–4 100	4	96 001–130 000	100
4 101–4 900	5	130 001–220 000	120
4 901–5 800	6	220 001–320 000	150
5 801–6 700	7	320 001–450 000	180
6 701–7 700	8	450 001–600 000	210
7 701–8 500	9	600 001–780 000	240
8 501–12 900	10	780 001–970 000	270
12 901–17 200	15	970 001–1 230 000	300
17 201–21 500	20	1 230 001–1 520 000	330
21 501–25 000	25	1 520 001–1 850 000	360
25 001–33 000	30	1 850 001–2 270 000	390
33 001–41 000	40	2 270 001–3 020 000	420
41 001–50 000	50	3 020 001–3 960 000	450
50 001–59 000	60	3 960 001 or more	480

substance. Whenever a substance cannot be precisely determined at the MCLG level, the MCL is set at a factor of 5 to 10 times the method detection limit (Koorse, 1990). As treatment and analytical technologies progress, the goals will be implemented. Not all rules were in effect in 1990, but a time schedule has been set for testing and enacting outstanding rules.

Complying with standards does not guarantee the same level of protection for all because of differences in monitoring requirements. A large number of samples is required to minimize confidence limits on quality variation and the resources needed to obtain the data become excessive for smaller communities. There will always be compromise involved for monitoring requirements because of economies of scale for large communities compared with smaller communities.

There are other problems with some of the regulations. The presence/absence (P/A) rule for coliforms (Section 8.6.1) is an example of a rule based on flawed statistical assumptions. The issue is discussed by Borup (1992). One of the flaws arises from the requirement that no more than 5% of all samples from a distribution system may show the presence of coliforms. If a portion of the distribution system is compromised, the number of samples from this section may be much less than 5% of the total samples and no problem would be identified. In Chapter 16 the Surface Water Treatment Rule (SWTR) with respect to disinfection is discussed. In formulating this rule there were improper assumptions about the performance of treatment units.

Despite its shortcomings, the SDWA is the most advanced water standard in the world. Continuing research will resolve some of the inequities and contradictions in current regulations to further improve them.

8.5.3 Canadian Water Quality Guidelines

In Canada, water is considered a natural resource, which under the constitution of the country gives the provinces wide jurisdiction over it. The federal Department of Health and Welfare develops unenforceable **guidelines** and the provinces are free to determine whether the guidelines are adopted. Currently only the provinces of Alberta and Quebec have incorporated the federal guidelines into legislation (Decker and Long, 1992). Other provinces have only informally adopted the guidelines, which are not legally enforceable. Canadian guidelines specify maximum acceptable concentration (MAC) or interim MAC (IMAC).

8.6 DRINKING WATER STANDARDS

The guidelines and standards set by the WHO, United States, and Canada are given together for comparison in the sections that follow. The latest WHO guidelines were set in 1993 and are taken from WHO (1993). The Canadian guidelines are as of November 1993 and were compiled from information in Environmental Health Directorate (1993) and Cotruvo and Vogt (1990).

The United States standards are as of May 1994 (USEPA, 1994). There are a number of substances for which the USEPA has set tentative MCLs that are not included in the tables below. Regulations for many of these will likely be in place in the near future.

8.6.1 Microbiological Parameters

All three sets of standards agree that the desirable state of a drinking water has no total coliforms in it.

TABLE 8.8 United States Microbiological Parameters[a]

	MCLG	MCL
Giardia	0	FD[b]
Heterotrophic plate count	—	FD
Legionella	0	FD
Total coliforms	0	P/A[c]
Viruses	0	FD

[a]See notes at beginning of Section 8.6
[b]Filtration and disinfection (see text).
[c]Presence/absence of coliforms in samples (see text).

WHO Guidelines for Microbiological Quality

Any water intended for drinking should contain fecal[1] and total coliform counts of 0, respectively, in any 100 mL sample. When either of these groups of bacteria are encountered in a sample, immediate investigative action should be taken. Repeat sampling is the minimum investigative action and if either count is again positive the cause should be investigated. For systems where a sufficient number of samples are taken throughout any year, for water sampled within the distribution system, 95% of samples should not contain any coliform organisms in a 100 mL sample. As a footnote to the bacteriological guidelines, WHO (1993) notes that rural supplies in developing countries will often contain fecal and total coliform organisms and the authority in the country should set interim standards for progressive improvement of these supplies. Total coliforms are not an acceptable indicator of the hygienic state of a rural water supply, particularly in tropical areas where many bacteria with no health significance will occur in untreated water supplies.

Virological examination of a water at a site is generally not feasible and no guideline value has been set. No guideline values have been set for pathogenic protozoa, helminths, and other free-living organisms such as algae that are commonly found in water supplies.

United States Standards for Microbiological Quality

Table 8.8 lists MCLGs and MCLs for microbiological parameters in the United States.

In the P/A criterion, small systems cannot have more than one positive coliform sample per month; for large systems, not more than 5% of total coliform samples per month may be positive.

Various filtration and disinfection (FD) schemes are specified in the United States regulations for different water quality conditions.

Canadian Guidelines for Microbiological Quality

The MAC for total coliforms is 0/100 mL. However water that fulfills the following conditions is considered to be in compliance with the MAC.

[1]WHO (1993) uses the designation of thermotolerant coliforms instead of fecal coliforms noting that the assay for fecal coliforms will culture some thermotolerant coliforms, as noted in Section 7.3.1, that originate from decaying plant materials, soils and industrial effluents when these sources exist. *E. coli* and related coliform species of fecal origin are also cultivated but the fecal coliform designation for the assay is incorrect on a strict basis. This text will continue to use the common term fecal coliform with the understanding that it is subject to inaccuracies when the MF technique is used.

1. No sample should contain more than 10 total coliforms per 100 mL, none of which should be fecal coliforms.
2. No consecutive samples from the same site should show the presence of coliform organisms.
3. For community drinking water supplies:
   a. Not more than 10% of the samples based on a minimum of 10 samples should show the presence of coliform organisms
   b. Not more than one sample from a set of samples taken from a community on a given day should show the presence of coliform organisms.

8.6.2 Chemical and Physical Quality

Tables 8.9 and 8.10 list inorganic and organic chemical agents that affect health.

It is generally accepted that only the zero-, di-, tri-, and hexavalent oxidation states of chromium have biological importance. The trivalent form is essential in microamounts (50–200 μg/d) for humans (Anon., 1989). Hexavalent chromium is the toxic form of chromium and causes a variety of ill effects.

TABLE 8.9 Inorganic Constituents of Health Significance[a]

	Concentration, mg/L			
	Guideline		U.S. standards	
Constituent	WHO[b]	Canada[c]	MCLG	MCL
Antimony	0.005		0.006	0.006
Arsenic	0.01	0.025[d]	0	0.05
Asbestos			7×10^6 [e]	7×10^6 [e]
Barium	0.7	1.0	2	2
Beryllium			0.004	0.004
Boron	0.3	5[d]		
Cadmium	0.003	0.005	0.005	0.005
Chromium	0.05	0.05	0.1	0.1
Cyanide (as CN)	0.07	0.02	0.2[f]	0.2[f]
Fluoride	1.5	1.5	4	4
Lead	0.01	0.01	0	TT[g]
Mercury, total	0.001	0.001	0.002	0.002
Molybdenum	0.07			
Nickel	0.02		0.1	0.1
Nitrate (as N)	11.3	10	10	10
Nitrite (as N)	0.91		1	1
Selenium	0.01	0.01	0.05	0.05
Thallium			0.000 5	0.002

[a]See notes at beginning of Section 8.6.
[b]Reproduced by permission from WHO (1993), *Guidelines for Drinking Water Quality,* vol. 1: *Recommendations,* 2nd ed., World Health Organization, Geneva.
[c]MAC.
[d]Interim value (IMAC).
[e]Based on fibers >10 μm.
[f]Proposed.
[g]Treatment technology specified.

TABLE 8.10 Organic Constituents of Health Significance[a]

Constituent	Concentration, mg/L			
	Guideline		U.S. standards	
	WHO[b]	Canada[c]	MCLG	MCL
Acrylamide	0.000 5		0	TT[d]
Alachlor	0.02		0	0.002
Aldicarb	0.01	0.009	0.007[e]	0.007[e]
Aldicarb sulfone			0.007[e]	0.007[e]
Aldicarb sulfoxide			0.01	0.01
Adipate			0.4	0.4
Aldrin and dieldrin	0.000 03	0.000 7		
Atrazine	0.002	0.005[f]	0.003	0.003
Ainphos-methyl		0.02		
Bendiocarb		0.04		
Benyazone	0.03			
Benz[*a*]anthracene			0[g]	0.000 1[g]
Benzene	0.1	0.005	0	0.005
Benzo[*a*]pyrene	0.000 7	0.000 01	0	0.000 2
Benzo[*b*]fluoranthene			0[g]	0.000 2[g]
Benzo[*k*]fluoranthene			0[g]	0.000 2[g]
Bromoxynil		0.005[f]		
Butyl benzyl phthlate fluoranthene			0[g]	0.1[g]
Carbaryl (sevin)		0.09		
Carbofuran	0.005	0.09	0.04	0.04
Carbon tetrachloride	0.002	0.005	0	0.005
Chlordane	0.000 2	0.007	0	0.002
Chlorodibromomethane				0.08[h]
Chlorobenzenes	0.2-1			
Chloroform	0.03			0.08[h]
Chlorophenols	[i]			
Chlorotoluron	0.030			
Chlorpyrifos		0.09		
Chrysene			0[g]	0.000 2[g]
Cyanazine		0.01[f]		
2,4,D (2,4-dichlorophenoxyacetic acid)	0.03	0.1[f]	0.07	0.07
DDT	0.002	0.03		
Dalapon			0.2	0.2
Diazinon		0.02		
Dibenz[*a,h*]anthracene			0[g]	0.000 3[g]
Dicamba		0.12		
Dichlorobenzene			0.075	0.075
1,2-Dichlorobenzene	1.0	0.2	0.6	0.6
1,4-Dichlorobenzene	0.30	0.005	0.6	0.6
1,2-Dibromo-3-chloropropane (DBCP)	0.001		0	0.000 2
1,2-Dichloroethane	0.03	0.05	0	0.005
1,1-Dichloroethylene	0.03		0.007	0.007
cis-1,2-Dichloroethylene	0.05		0.07	0.07
trans-1,2-Dichloroethylene	0.05		0.1	0.1
Dichloromethane	0.02	0.05	0	0.005
2,4-Dichlorophenol		0.9		
1,2-Dichloropropane	0.02		0	0.005
Diclofop-methyl		0.009		
Di[2-ethyl]adipate			0.4	0.4
Diethylhexyl phthlate			0	0.006
Dimethoate		0.02[f]		
Dinoseb		0.01	0.007	0.007
Diquat		0.07	0.02	0.02
Diuron		0.15		
EDTA	0.2			
Endothall			0.1	0.1
Endrin			0.002	0.002
Epichlorohydrin	0.000 4		0	TT[d]
Ethylenedibromide (EDB)			0	0.000 05
Ethylbenzene	0.3	0.002 4[j]	0.7	0.7
Fenoprop	0.009			
Glyphosate		0.28[f]	0.7	0.7

TABLE 8.10 Continued

Constituent	Concentration, mg/L			
	Guideline		U.S. standards	
	WHO[b]	Canada[c]	MCLG	MCL
Heptachlor	0.000 03[k]	0.003[k]	0	0.000 4
Heptachlor epoxide	0.000 03[k]	0.003[k]	0	0.000 2
Hexachlorobenzene	0.001		0	0.001
Hexachlorobutadiene	0.000 6			
Hexachlorocyclopentadiene			0.05	0.05
γ-HCH (lindane)	0.002	0.004	0.000 2	0.000 2
Indeno[1,2,3-*c,d*]pyrene			0[g]	0.000 4[g]
Isoproturon	0.009			
Malathion		0.19		
Mecoprop	0.01			
Methoxychlor	0.02	0.9	0.04	0.04
Metolachlor	0.01	0.05[f]		
Metribuzin		0.08		
Molinate	0.006			
Monochlorobenzene		0.08	0.1	0.1
NTA	0.2	0.40		
Oxamyl			0.2	0.2
Paraquat		0.01[f]		
Parathion		0.05		
Pendimethalin	0.02			
Pentachlorophenol	0.009	0.06	0	0.001
Permethrin	0.02			
Phorate		0.002[f]		
Picloram		0.19[f,g]		
Polychlorinated biphenyls			0	0.000 5
Propanil	0.02			
Pyridate	0.1			
Simazine	0.002	0.01[f]	0.004	0.004
Styrene	0.02		0.1	0.1
2,4,5-T (2,4,5-Trichlorophenoxyacetic acid)	0.009	0.28		
2,3,7,8-TCDD (dioxin)			0	3×10^{-8}
2,4,5-TP [2(2,4,5-trichlorophenoxy) propionic acid]			0.05	0.05
Temephos		0.28[f]		
Terbufos		0.001[f]		
Tetrachloroethylene	0.04		0	0.005
2,3,4,6-Tetrachlorophenol		0.1		
Toluene	0.7	0.024[j]	1	1
Toxaphene			0	0.003
Triallate		0.23		
Tributyltin oxide	0.002			
1,2,4-Trichlorobenzene			0.07	0.07
1,1,1-Trichloroethane	2		0.2	0.2
1,1,2-Trichloroethane			0.003	0.005
Trichloroethylene	0.07	0.05	0	0.005
2,4,6-Trichlorophenol	0.01	0.005		
Trifluralin	0.02	0.045[f]		
Trihalomethanes		0.1		0.08[i]
Vinyl chloride	0.005	0.002	0	0.002
Xylene	0.5	0.3[j]	10	10

[a]See notes at beginning of Section 8.6.
[b]Reproduced by permission from WHO (1984), *Guidelines for Drinking Water Quality,* vol. 1., *Recommendations,* 2nd ed., World Health Organization, Geneva.
[c]MAC.
[d]Treatment technology specified.
[e]Draft.
[f]IMAC.
[g]Proposed.
[h]The total of THM can not exceed 0.08 mg/L.
[i]Chlorophenols have no health-related guideline value set by WHO but odor thresholds have been set. The odor thresholds vary between 0.1 and 300 μg/L for different chlorophenols.
[j]Aesthetic objective.
[k]The sum of heptachlor and heptaclor epoxide must not exceed the value.

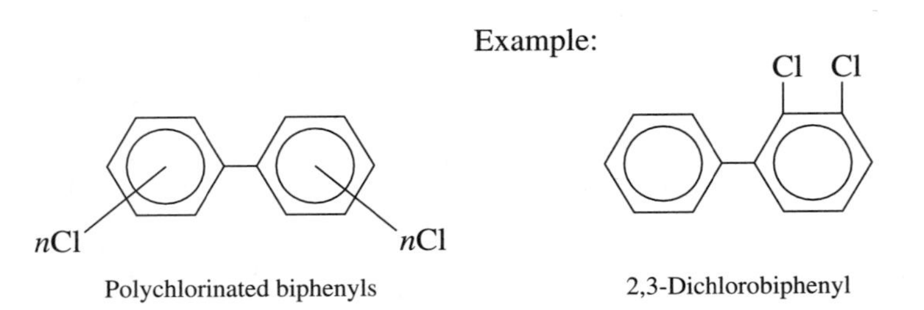

Polychlorinated biphenyls

2,3-Dichlorobiphenyl

8.6.3 Aesthetic Quality

Some of the substances in Table 8.11 also have health significance but undesirable organoleptic properties are generally produced at lower concentrations.

8.6.4 Radiological Constituents

In the United States and Canada, formerly only gross activities of radiological constituents were specified. Now standards have been set for specific substances considering

TABLE 8.11 Aesthetic Parameters[a]

	Concentration, mg/L		
	Guideline		U.S.[b]
Constituent	WHO[c]	Canada	MCL
Aluminum	0.2		0.05 to 0.2
Ammonia	1.5		
Chloride	250	250	250
Color	15 (units)	15 (units)	15 (units)
Copper	1.0	1	TT[d]
Corrosivity			Noncorrosive
Foaming agents			0.5
Hardness (as $CaCO_3$)	500		
Hydrogen sulfide	0.05	0.05	
Iron	0.3	0.3	0.3
Manganese	0.1	0.05	0.05
Taste and odor	Inoffensive	Inoffensive	3 TON[e]
pH	preferably <8.0	6.5–8.5	6.5–8.5
Silver			0.1
Sodium	200	200	
Solids, total dissolved	1 000	500	500
Sulfate	250	500	250
Temperature	No value set	≤15°C	
Turbidity (NTU)	5	1	0.5–1.0
Zinc	3.0	5	5

[a]See notes at beginning of Section 8.6.
[b]Except for copper, these are secondary contaminants.
[c]Reproduced by permission from WHO (1984), *Guidelines for Drinking Water Quality,* vol. 1., *Recommendations,* World Health Organization, Geneva.
[d]Treatment technology specified.
[e]Only odor is specified as 3 threshold odor numbers (TON).

TABLE 8.12 Radiological Parameters[a]

	Concentration, Bq/L			
	Guideline		U.S. standards	
Constituent	WHO[b]	Canada	MCLG	MCL
Cesium-137		50		
Iodine-131		10		
Radium-226		1	0	20 pCi/L
Radium-228			0	20 pCi/L
Radon			0	300 pCi/L
Strontium-90		10		
Tritium		40 000		
Uranium		0.1 mg/L	0	20 μg/L
Gross alpha activity	0.1		0	15 pCi/L
Gross beta activity	1		0	4 mrem

[a]See notes at beginning of Section 8.6
[b]Reproduced by permission from WHO (1993), *Guidelines for Drinking Water Quality,* vol. 1, *Recommendations,* 2nd ed., World Health Organization, Geneva.

their propensity to stay within the body and their biological effects. Table 8.12 lists current standards.

In Canada the basis of the radiological guidelines is a dose of 0.1 mSv from 1 year's consumption of drinking water which is assumed to be 730 L.

Tables 8.8–8.12 not only illustrate the different stages of regulation development but also different philosophies of regulation. Some constituents that are controlled by one country have been eliminated as a hazard by the other. Differences in guidelines also reflect differing approaches to risk assessment.

8.6.5 Other Water Use Standards

In addition to drinking water standards, guidelines have been developed for many other uses of water. For example, in Canada, federal guidelines for the following water uses can be found in CCME (1992).

Raw Water for Drinking Water Supply
Recreational Water Quality and Aesthetics
Freshwater Aquatic Life
Industrial Water Supplies
Agricultural Uses

8.7 CANADIAN INLAND WATER QUALITY

Table 8.13 presents a survey of raw water quality for inland waters in Canada. These data have been compiled from a number of sources. Some of the information is based on limited studies. These data are the best representative data for Canada but typical values and ranges for contaminants in any raw water supply can be significantly different from those in Table 8.13. Therefore local information must always be obtained.

TABLE 8.13 Canadian Raw Water Quality[a]

Constituent	Concentration[b]		Comment
	Range	Typical	
Inorganic contaminants			
Aluminum		<1 mg/L	
Ammonia	ND[c]–2 mg/L		
Antimony		<0.01 mg/L	Limited data indicate ND
Arsenic		<0.05 mg/L	Total dissolved arsenic
Asbestos	$<10^6$–10^7 fibers/L	10^6 fibers/L	
Barium		<0.01 mg/L	Limited data
Boron	ND–2 mg/L	<0.5 mg/L	Median values <0.05 mg/L
Cadmium		<0.01 μg/L	Value for drinking water supplies
Chloride		<10 mg/L	
Chromium	0.01–0.023 mg/L		Total dissolved chromium
Copper		<5 μg/L	
Cyanide (total)	Limited data		Western Canada sites have yielded values as high as 0.06 mg/L
Fluoride	0.01–4.5 mg/L		
Iron		<0.5 mg/L	Surface waters only
Lead		<0.05 mg/L	
Manganese	—	—	
Mercury		<0.001 mg/L	Outside of known contaminated areas
Nitrate	—	—	
Silver			Silver contamination is seldom reported
Sodium		9 mg/L	Mean value for N. American rivers
Sulfate	10–80 mg/L		For surface waters; groundwaters can be higher
Organic contaminants			
Atrazine	ND–26.9 μg/L		
Azinphos-methyl	<0.01–<0.05 μg/L		Limited data
Chlordane		0.05 μg/L	
Dicamba	ND–trace		
DDT	0.005–3.24 μg/L		
2,4-D	2–3 μg/L		
Dieldrin		0.038 μg/L	
Heptachlor epoxide	0.005–0.007 μg/L	0.005 μg/L	
Lindane		0.003 μg/L	
Malathion	13 ng/L–<0.05 μg/L		
Methoxychlor	0.012–0.03 μg/L		
Nitriloacetic acid (NTA)		<12 μg/L	78% of samples between 1972 and 1975 were below 12 μg/L
Parathion	ND		Current detection level is <0.01 μg/L
Toxaphene	0.1–108.3 μg/L	1 μg/L	
Radionuclides			
Cesium-137	0.68 nBq/L–2.10 mBq/L		Range is for the Great Lakes
Radium-226		3.66 mBq/L	
Strontium-90	16.6–37.8 mBq/L		Range is for the Great Lakes

[a]Adapted from Department of National Health and Welfare (1993).
[b]Most of the data are for surface waters.
[c]ND, nondetectable.

8.8 WATER CONSUMPTION

The quantity of water consumed is variable depending on water supply, climate, and cultural habits of the population. Herschel (1973) analyzed the data of Frontinus, water commissioner of Rome in AD 97 to arrive at an average consumption of 144 L/cap/d for its one million or so inhabitants. The availability of water and societal habits have considerably increased the average use of water from Roman times.

Rural households tend to use less water than urban residences because city residents have more water-using appliances and practices than do rural residents. Water demands can vary from 180 to 1 500 L/cap/d (MOE, 1985a). In Ontario, average daily water demands in the range of 270–450 L/cap/d are recommended. [McLellon (1991) reported that students in Florida used water at the average rate, 337 L/cap/d.] Table 8.14 gives unit water demands for commercial and institutional establishments.

Northern communities in Canada are remote with varying accessibility to water supplies. Design water consumption for various types of water delivery and collection systems are given in Table 8.15 for these communities. In piped, pressurized water supplies, water bleeder systems to prevent the pipes from freezing can significantly increase the rate of water consumption.

Table 8.16 gives a typical breakdown of domestic water use for various purposes in the United States or Canada. Flushing the toilet is the major use of water. Older toilets consume 18–25 L/flush. In an effort to reduce water consumption, in 1989 the state of Massachusetts required all new and replacement toilet installations to use no more than 6.05 L (1.6 gal) per flush (Vickers, 1989). This regulation replaced an existing standard that required toilets to use no more than 13.2 L/flush. The old standard was more progressive than standards in most states and provinces. Savings

TABLE 8.14 Commercial and Institutional Water Demand[a]

Source	Average daily water use[b]
Shopping centers	2.5–5.0 L/m^2
	60–120 gal/1 000 ft^2
	(based on total floor area)
Hospitals	900–1 800 L/bed
	240–480 gal/bed
Schools	70–140 L/student
	18–36 gal/student
Travel trailer parks	
Without individual hookups	340 L/site
	90 gal/site
With individual hookups	800 L/site
	210 gal/site
Campgrounds	225–570 L/campsite
	60–150 gal/campsite
Mobile home parks	1 000 L/unit
	265 gal/unit
Motels	150–200 L/bed
	40–53 gal/bed
Hotels	225 L/bed
	60 gal/bed

[a]From MOE (1985a).
[b]These are design values for Ontario.

TABLE 8.15 Residential Water Use for Northern Canadian Communities[a]

	Water use		Range	
Source	L/cap/d[b]	gal/cap/d[b]	L/cap/d	gal/cap/d
Nonpressure water system with bucket	10	2.6	5–25	1.3–6.6
Trucked water delivery				
Nonpressure system with bucket toilets	10–25	2.6–6.6	5–50	1.3–13
Nonpressure system, holding tank with wastewater pumpout collection	40	10.6	20–70	5.3–18
Pressure water system, normal flush toilet and wastewater pumpout collection	90	24	40–250	11–66
Piped water supply and vacuum or pressure sewers	145	38	60–250	16–66
Piped water supply and gravity sewers	225	60	100–400	26–106

[a]From Smith and Knoll (1986).
[b]Design values.

in water consumption up to 70% could be realized if a household was to replace 25 L flush toilets with 6 L flush toilets.

Low-volume flush toilets are as efficient or more efficient than the larger volume units that they replace. Ultra-low-volume toilets using 3 L/flush have been designed and tested in residential districts (Anderson and Siegrist, 1989). They performed as satisfactorily as conventional toilets (using more than 13 L/flush).

The quantity of water actually drunk is a very small percentage of the total amount of water used. The total amount of tap water consumed by drinking either tap water itself or in tapwater–based beverages such as coffee, tea, or rehydrated soup or milk in Canada averaged 1.34 L/cap/day (0.35 gal/cap/d) (Health and Welfare Canada, 1981). There was little variation in consumption with sex, but Fig. 8.3 shows consumption variation with age which reflects differences in body weight. This information is used in formulating water quality guidelines.

As might be expected, consumers who were less satisfied with the taste of their water tended to drink a greater percentage of water-based beverages, probably to disguise the taste of the water. There was not a significant change in total water consumption as satisfaction with the water taste changed.

The price of water and the rate schedule are important factors influencing water consumption. A survey of Canadian water pricing practice in 1987 found that 37% of

TABLE 8.16 Typical Domestic Water Use (Various Sources)

Water use	Percent of total use
Toilet flushing	40
Bathing	30
Laundry	15
Kitchen	10
Other	5

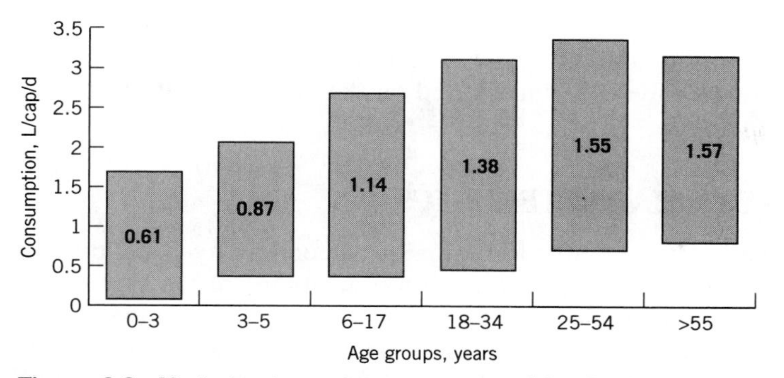

Figure 8.3 Variation in tap water consumed by drinking. From *Tap Water Consumption in Canada,* Report No. 82-EHD-80, Health Canada, 1981. Reproduced with permission of the Minister of Supply and Services Canada, 1995.

the population was paying a flat rate for water, i.e., a fixed rate was charged regardless of the amount of water consumed (Environment Canada, 1989). The benefits of this policy are facile administration and its easy comprehension by the public. Obviously this pricing structure does not encourage conservation of water. An increasing block rate structure where the price of water increases in successive consumption blocks is the strongest conservation measure. Two percent of Canada's population had this type of rate schedule. A declining block rate (the reverse of the increasing block rate) and a constant unit rate apply to 34 and 27% of the population, respectively. The marginal price of water to consumers ranged from 14 to 81 ¢/m^3 ($0.53 to 3.07/1 000 gal), with an average value of 38 ¢/m^3 ($1.44/1 000 gal) for the whole country.

8.9 CANADIAN FEDERAL WASTEWATER QUALITY GUIDELINES

Canadian wastewater effluent guidelines for federal establishments are given in Table 8.17. Provinces often adopt these guidelines in their own regulations or guidelines.

TABLE 8.17 Canadian Federal Wastewater Quality Guidelines[a]

Substance	Limit
BOD_5	20 mg/L
TSS	25 mg/L
Fecal coliforms	400/100 mL (after disinfection)
Cl_2 residual	0.5 mg/L min (after 30 min contact time)
	1.0 mg/L max. (may require dechlorination before discharge)
pH	6–9
Phenols	20 μg/L
Total P	1.0 mg/L
Oils and grease	15 mg/L
Temperature	May not alter temperature of receiving water by more than 1°C
Dilution is not an acceptable treatment.	

[a]From Environment Canada (1976). Reproduced with permission of the Minister of Supply and Services Canada, 1995.

Industry specific guidelines also exist for a number of industries. Effluent concentrations are often based on the amount of raw material processed, although stream assimilative capacity may override guidelines based on the amount of production.

8.10 WASTEWATER CHARACTERISTICS

Domestic wastewater varies widely from community to community depending on water quality, use and conservation practices, cultural attributes of the population, industries present and treatment applied at industry locations, as well as other factors such as infiltration in the sewer system. The major source of organics in domestic wastewater is human excreta. Feachem et al. (1981) arrive at typical daily BOD_5 contributions to sewage per adult from excreta ranging from 39 to 42 g (urine, 10.3 g; fecal material, 24.7–30.6 g; and anal cleansing materials, 2.0–3.5 g). Most adults produce 1 to 1.3 kg of urine per day, which is a rich source of nitrogen. The ammonia and urea (which is readily hydrolyzed to ammonia) content of urine is typically around 550 and 24 000 mg/L, respectively. Sullage and industrial wastes will alter the per capita contributions of organics and other substances. Table 8.18 provides information on typical composition of medium strength domestic wastewater. Nemerow and Dasgupta (1991) provide information on quality of wastewater produced from various industries.

Size ranges of organic particulate matter in wastewater are given in Table 8.19. The size distribution of particulates is site specific but a significant fraction of the organic matter is in the colloidal and supracolloidal size ranges. As wastewater passes through treatment operations, degradation and other removal processes shift the distribution of organic matter toward smaller sizes.

Raunkjær et al. (1994) analyzed the protein, carbohydrate, and lipid fractions in domestic wastewater and reviewed results from other studies. The ranges of average results for these components expressed as a percentage of total COD from all studies were 8–28, 6–18, and 12–31 for protein, carbohydrates, and lipids, respectively.

TABLE 8.18 Composition of Medium Strength Untreated Domestic Wastewater[a]

Constituent	Concentration, mg/L
Bacteria[b]	10^7–2×10^8
Total solids	450
Total volatile solids	300
Suspended solids	250
Volatile suspended solids	200
Total dissolved solids	200
BOD_5	150–250
Nitrate and nitrite nitrogen as N	<0.6
Organic nitrogen as N	25–85
Ammonia nitrogen as N	15–50
Total phosphorus	6–12
Soluble phosphorus	4–6

[a]From MOE (1978).

[b]Bacteria are given as No./100 mL.

TABLE 8.19 Size Distribution of Organic Matter in Treated and Untreated Domestic Sewage[a]

	Percent of organic matter contained in indicated size range, μm							
	<0.001		0.001–1		1–100		>100	
Wastewater sample	Ave.	Range	Ave.	Range	Ave.	Range	Ave.	Range
Untreated wastewater	31	12–50	14	9–16	25	10–30	37	15–43
Primary effluent	43[b]	35–51[b]	10[b]	2–19[b]	24[b]	13–34[b]	28[b]	5–60[b]
Activated sludge effluent	52	74–79	3	2–5	22	16–31	2	1–3
	33[b]	26–46[b]	5[b]	2–9[b]	35[b]	20–49[b]	27[b]	13–49[b]
Trickling filter effluent[c]	40	—	60	—	—	—	—	—
Wastewater sludges[c]								
Primary	5	—	1	—	4	—	90	—
Secondary	3	—	0.1	—	1	—	96	—
Anaerobic	5	—	3	—	19	—	72	—

[a]Reprinted from A. D. Levine, G. Tchobanoglous, and T. Asano, "Size Distribution of Particulate Contaminants and Their Impact on Treatability," *Water Research,* 25, pp. 911–922, Copyright 1991, with kind permission from Elsevier Science Ltd., The Boulevard, Langford Lane, Kidlington 0X5 1GB UK.
[b]Size ranges of <0.1 μm; 0.1–1 μm; 1–12 μm, and >12 μm apply to these data.
[c]Results are for only one study.

Graywater

Graywater is defined as all wastewater produced from a household excluding toilet wastes. There may be potential for reuse of graywater, reducing demands for domestic water consumption; however, fecal coliforms and other indicator microorganisms can be found in graywater in significant numbers. Rose et al. (1991) found fecal coliforms ranging from 10^4 to 10^7/100 mL. Other data on the quality of graywater are given in Table 8.20.

TABLE 8.20 Characteristics of Graywater[a]

Constituent	Units	Range	Average	Average in tap water
pH		5–7	6.54	6.6
Alkalinity	mg/L	149–198	158	131
Ammonia nitrogen	mg/L	0.15–3.2	0.74	0
Nitrate	mg/L	0–4.9	0.98	1.0
Total nitrogen	mg/L	0.6–5.2	1.7	1.0
Chloride	mg/L	3.1–12	9.0	10
Hardness	mg/L	112–152	144	142
Phosphate	mg/L	4–35	9.3	3.1
Sulfate	mg/L	12–40	22.9	0
Turbidity	NTU	20–140	76.3	0.8

[a]Reprinted from J. B. Rose, G. S. Sun, C. P. Gerba, and N. A. Sinclair, "Microbial Quality and Persistence of Enteric Pathogens in Graywater from Various Sources," *Water Research,* 25, pp. 37–42, Copyright 1991, with kind permisison from Elsevier Science Ltd., The Boulevard, Langford Lane, Kidlington 0X5 1GB UK.

TABLE 8.21 Design Sewage Flows[a]

Source	Average daily flow[b]
Domestic sewage	225–450 L/cap/d (60–120 gal/cap/d)
Shopping centers	2 500–5 000 L/1 000 m^2 (60–120 gal/1 000 ft^2) (based on total floor area)
Hospitals	900–1 800 L/bed (240–480 gal/bed)
Schools	70–140 L/student (18–36 gal/student)
Travel trailer parks	
Without individual hookups	340 L/site (90 gal/site)
With individual hookups	800 L/site (210 gal/site)
Campgrounds	225–570 L/campsite (60–150 gal/campsite)
Mobile home parks	1 000 L/unit (265 gal/unit)
Motels	150–200 L/bed (40–53 gal/bed)
Hotels	225 L/bed (60 gal/bed)
Industrial areas	
Light industrial area	35 m^3/ha/d (3 750 gal/acre/d)
Heavy industry	50 m^3/ha/d (5 350 gal/acre/d)

[a]From MOE (1985b).
[b]These are design values for Ontario.

8.11 WASTEWATER PRODUCTION

Flows for sewage treatment plants are based on the design population and commercial and industrial activity. Historical data should be gathered to find existing flow information. The plant must be able to handle all flows anticipated in the design period. In Ontario, design periods for sewage treatment plants have generally been for 10–20 years as opposed to sewer systems which are designed for 20- to 40-year periods (MOE, 1985b). Accepted practice has been to add an extraneous flow allowance of 90 L/cap/d (24 gal/cap/d) to the average flow and 227 L/cap/d (60 gal/cap/d) to the peak flow. These are approximately equivalent to 3.7 m^3/ha/d (400 gal/acre/d) (average) and 9.2 m^3/ha/d (1 000 gal/acre/d) (peak). Table 8.21 gives typical design sewage flows from various sources. Actual flows and design criteria will vary from region to region. Small communities generate lower flows than larger cities. In northern Canadian communities, nearly all of the water consumed is returned as wastewater and the design values given in Table 8.15 apply for wastewater flows. A typical rule of thumb for estimating domestic wastewater flows is 380 L/cap/d (100 gal/cap/d).

Peaking factors are multiplication factors used to estimate high and low flows with respect to the average flow. Design peaking factors are usually specified by the authority. Gaines (1989) has studied the distribution of sewage flows and the city of Denver used the following technique to determine peaking factors.

The distribution of wastewater flows approximately follows a normal distribution. Minimum, maximum, and peak flows are related to the average flow with an equation of the following form.

$$Q_x = a(Q_{ave})^b \tag{8.2}$$

or

$$\ln Q_x = \ln a + b \ln Q_{ave}$$

where

Q_x is the instantaneous or average peak flow, maximum 1- or 6-h flow, or other flow specification

a and b are regression coefficients

Q_{ave} is the average flow

The coefficients in Eq. (8.2) are developed from linear regression of historical data for the flow specification (e.g., 1 h peak flow). The coefficients a and b in Eqs. (8.3a) and (8.3b) depend on the confidence level and the flow factor defined below. From linear regression, the coefficients a and b are dependent on the desired confidence level given by the following relations:

$$\ln a = A \pm Z \times S_1 \tag{8.3a}$$

$$b = B \pm Z \times S_2 \tag{8.3b}$$

where

A is the linear regression intercept

B is the linear regression slope

S_1 and S_2 are the standard deviations of the intercept and slope, respectively

Z is a factor related to the desired confidence level

The confidence level is the probability that an event will not be exceeded.

$$q = 1 - p \tag{8.4}$$

where

q is the confidence level

p is the probability of the event being exceeded

A plot of the data on normal probability paper (which should yield an approximately straight line) readily shows the event that exceeded p percent of time. For instance, data on the maximum 1-h flows for each day are gathered over any time period. The percent of these flows that are less than specified values of flow are plotted on probability paper and the best fitting straight line is drawn through them, which provides a cumulative frequency distribution. From the plot the maximum 1-h flow that is exceeded 1% (or any other percent) of the time is readily determined.

The values of Z are the distance from the mean in terms of standard deviations for various confidence levels. Values of Z at selected confidence levels are given in Table 8.22.

TABLE 8.22 *Z* Factor at Various Confidence Levels

Confidence level %	Z
99.9	3.08
99	2.33
95	1.64
90	1.28
75	0.67
67	0.44
50	0.00

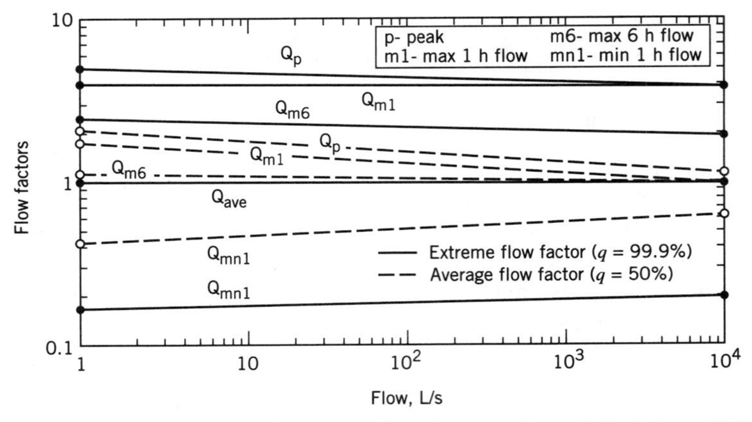

Figure 8.4 Average and extreme flow factors. From J. B. Gaines (1989), "Peak Sewage Flow Rate: Prediction and Probability," *J. Water Pollution Control Federation,* 61, 7, pp. 1241–1248, © WEF, 1989.

The instantaneous peak flow (Q_p) is defined as the maximum flow during a 24-h period. The maximum (Q_{m1}) and minimum (Q_{mn1}) 1-h flows are defined as the flows that are exceeded for 1 and 23 h, respectively. Similarly the flow that is exceeded for 6 h is Q_{m6}. Flow factors are used to design sewers and wastewater treatment facilities. The flow factor for a flow specification, Q_x is given by

$$F_x = \frac{Q_x}{Q_{ave}} = e^{(A \pm ZS_1)} \times Q^{(B-1 \pm ZS_2)} \tag{8.5}$$

where

F_x is the flow factor (e.g., F_{mn1} is the 1-h minimum flow factor corresponding to Q_{mn1}/Q_{ave}.)

In the exponents, use $A + ZS_1$ and $B - 1 + ZS_2$ when peak or maximum flows are being considered; use $A - ZS_1$ and $B - 1 - ZS_2$ when minimum flows are being considered.

The flow factors are a function of the confidence level or exceedance probability that the designer chooses.

Fig. 8.4 is a typical plot of peaking, maximum, and minimum flow factors constructed using Eq. (8.5). Note that the level of confidence (solid lines in the figure are the 99.9% confidence level and dashed lines are the 50% confidence level) significantly influences the comparisons. The average ($q = 50\%$) instantaneous peak flow is actually lower than the 99.9% confidence level estimate of the maximum 6-h flow rate. Which confidence level to use depends on the severity of the situation and the regulations in effect.

Ideally measurements of inflow and infiltration (I/I) should be made when collecting flow data and the wastewater flows should be corrected for I/I but this is not usually done. Therefore the regression equations reflect I/I as well as wastewater production and applying them to a future condition assumes that the amount of I/I is approximately the same. If I/I is a significant portion of the wastewater flow and measures are definitely going to be taken to correct the situation, best estimates of current I/I should be used to modify the current data and arrive at the equations. Then anticipated I/I can be added after the projected wastewater flow is estimated from the equation.

QUESTIONS AND PROBLEMS

1. What are the federal and provincial (state) drinking water guidelines?
2. Would breaking a mercury thermometer in your mouth and ingesting some of the mercury (but no glass) pose a serious threat to your health?
3. What is methemoglobinemia?
4. Distinguish between a carcinogen and a mutagen.
5. (a) Describe the features of the Ames test. (b) Search the literature for information on other similar tests and write a detailed description of the analyses.
6. What is the cost of water in your community? Is it broken down into drinking water, sewage, and sewer costs?
7. What is the purpose of the Microtox® assay? Describe the test.
8. It is inevitable that research on the health effects of various agents will suggest standards that are below detectability limits of current instruments when safety factors are applied or risk analysis is performed. The defensibility of regulations when standards are below detectability limits is difficult if not impossible in court. Discuss your opinions on the regulation of substances in this situation.
9. At a public meeting on fluoridation of drinking water in Ottawa, ON, a disputant of the practice argued that hydrofluoric acid, which is so strong that it is used to etch glass, is commonly added to water as the fluoride agent (and related effects would occur in humans). Discuss the merits of this argument.
10. Research and summarize the health effects of one of the following: (a) aluminum; (b) fluoride; (c) nitrates; (d) trihalomethanes.
11. What are the design guidelines for water and wastewater treatment works in your community?
12. Keep track of water use and wastewater production by yourself and residents of your dwelling for 1 week and report the results. Note the amount of water used for each activity (drinking, laundry, lawn watering, food preparation/cooking, toilet flushing, bathing/personal hygiene, and car washing are suggested categories).
13. Find the detectability limits of aluminum for various analytical techniques.
14. Gaines (1989) reports the following regression results (for Eq. 8.2) for the city of Denver. (a) For an average flow of 150 L/s find the peak flow that will only be exceeded 10% of the time and the corresponding flow factor. (b) For the same average flow as in (a) find the 1 h minimum flow that will be exceeded 10% of the time.

    Peak flow: $A = 0.778\ 0$, $S_1 = 0.280\ 0$; $B = 0.935\ 8$, $S_2 = 0.001\ 1$

    Minimum 1h flow: $A = -0.790\ 4$, $S_1 = 0.393\ 1$; $B = 1.044\ 3$, $S_2 = 0.002\ 9$

KEY REFERENCES

Calabrese, E. J., C. E. Gilbert, and H. Pastides, eds. (1989), *Safe Drinking Water Act,* Lewis Publishers, Chelsea, MI.

Department of National Health and Welfare (1993), *Water Treatment Principles and Applications, Guidelines for Canadian Drinking Water Quality,* M. Sheffer, ed., Canadian Water and Wastewater Association, Ottawa, ON.

USEPA (U.S. Environmental Protection Agency) (1976), *Quality Criteria for Water,* USEPA, Washington, DC.

Verschueren, K. (1983), *Handbook of Environmental Data on Organic Chemicals,* Van Nostrand Reinhold, New York.

REFERENCES

ALHAJJAR, B. J., J. M. HARKIN, AND G. CHESTERS (1989), "Detergent Formula and Characteristics of Wastewater in Septic Tanks," *J. Water Pollution Control Federation,* 61, 5, pp. 605–613.

AMES, B. N., J. MCCANN, AND E. YAMASAKI (1975), "Methods for Detecting Carcinogens and Mutagens with the Salmonella/Mammalian-Microsome Test," *Mutation Research,* 31, 6, pp. 347–363.

ANDERSON, D. L. AND R. L. SIEGRIST (1989), "The Performance of Ultra-Low-Volume Flush Toilets in Phoenix," *J. American Water Works Association,* 81, 3, pp. 52–57.

Anon. (1989), *J. of the International Register of Potentially Toxic Chemicals Devoted to Information on Hazardous Chemicals,* United Nations Environment Programme, 9, 2, pp. 19–21.

BLUM, D. J. W. AND R. E. SPEECE (1991), "A Database of Chemical Toxicity to Environmental Bacteria and Its Use in Interspecies Comparisons and Correlations," *J. Water Pollution Control Federation,* 63, 3, pp. 198–207.

BORUP, M. B. (1992), "Presence–Absence Coliform Monitoring Has Statistical Limitations," *J. American Water Works Association,* 84, 3, pp. 66–71.

BOUCHARD, D. C., M. K. WILLIAMS, AND R. Y. SURAMPALLI (1992), "Nitrate Contamination of Groundwater: Sources and Potential Health Effects," *J. American Water Works Association,* 84, 9, pp. 85–90.

CAMP, T. R. AND R. L. MESERVE (1974), *Water and Its Impurities,* 2nd ed., Dowden, Hutchinson & Ross, Stroudsburg, PA.

CCME (Canadian Council of Ministers of the Environment) (1992), *Canadian Water Quality Guidelines,* CCME, Inland Waters Directorate, Ottawa, ON.

CHANG, J. C., P. B. TAYLOR, AND F. R. LEACH (1981), "Use of the Microtox Assay System for Environmental Samples," *Bulletin of Environmental Contaminant Toxicology,* 26, pp. 150–156.

COTRUVO, J. A. AND C. D. VOGT (1990), "Rationale for Water Quality Standards and Goals," in *Water Quality and Treatment,* 4th ed., F.W. Pontius, ed., McGraw-Hill, Toronto, pp. 1–62.

CRAPPER, D. R., S. S. KRISHNAN, AND A. J. DALTON (1973), "Brain Aluminum in Alzheimer's Disease and Experimental Neurofibrillary Degeneration," *Science,* 180, pp. 511–513.

CRAUN, G. F. (1986), "Statistics of Waterborne Outbreaks in the US (1920–80)," in *Waterborne Disease in the United States,* G. F. Craun, ed., CRC Press, Boca Raton, FL.

DECKER, K. C. AND B. W. LONG (1992), "Canada's Cooperative Approach to Drinking Water Regulations," *J. American Water Works Association,* 84, 4, pp. 120–128.

DROSTE, R. L. (1987), *Fluoridation in Canada as of 1986,* report for Health and Welfare Canada, Ottawa, Canada.

Environment Canada (1976), *Guidelines for Effluent Quality and Wastewater Treatment (Federal Establishments)*, Environment Canada, Ottawa.

Environment Canada (1989), "Municipal Water Rates in Canada: Current Practices and Prices," No. En 37-84/1989E, Ministry of Supply and Services, Ottawa, Canada.

Environmental Health Directorate (1993), *Summary of Guidelines for Canadian Drinking Water Quality (11/93)*, Health Canada, Ottawa, Canada.

FALCONER, I. R., M. T. C. RUNNEGAR, T. BUCKLEY, V. L. HUYN, AND P. BRADSHAW (1989), "Using Activated Carbon to Remove Toxicity from Drinking Water Containing Cyanobacterial Blooms," *J. American Water Works Association*, 81, 2, pp. 102–105.

FEACHEM, R. G., D. J. BRADLEY, H. GARELICK, AND D. D. MARA (1981), *Health Aspects of Excreta and Sullage Management—A State-of-the-Art Review*, The World Bank, Washington, DC.

FERGUSON, J. E. (1990), *The Heavy Elements: Chemistry, Environmental Impact and Health Effects,* Pergamon Press, Toronto.

GAINES, J. B. (1989), "Peak Sewage Flow Rate: Prediction and Probability," *J. Water Pollution Control Federation,* 61, 7, pp. 1241–1248.

Health and Welfare Canada (1981), *Tap Water Consumption in Canada,* Report No. 82-EHD-80, Department of National Health and Welfare, Ottawa, Canada.

HELLMAN, T. M. AND F. H. SMALL (1974), "Characterization of Odor Properties of 101 Petrochemical Using Sensory Methods," *J. Air Pollution Control Association,* 24, 10, pp. 979–982.

HERSCHEL, F., Transl. (1973), *The Water Supply of the City of Rome, A.D. 97,* by S.J. Frontinus, New England Water works Association, Boston.

HOFFMAN, F. A. AND J. W. BISHOP (1994), "Impacts of a Phosphate Detergent Ban on concentrations of Phosphorus in the James River, Virginia," *Water Research,* 28, 5, pp. 1239–1240.

HORGAN, J. (1994), "Radon's Risks," *Scientific American,* 271, 2, pp. 14–16.

Inland Waters Directorate (1979), *Water Quality Sourcebook, A Guide to Water Quality Parameters,* Environment Canada, Inland Waters Directorate, Ottawa, Canada.

ISOM, B. G. (1990), "Aquatic Toxicity Testing," in *Toxicity Reduction in Industrial Effluents,* P. W. Lankford and W. W. Eckenfelder, Jr., eds., Van Nostrand Reinhold, New York, pp. 18–34.

KOORSE, S. J. (1990), "MCL Noncompliance: Is the Laboratory at Fault," *J. American Water Works Association,* 82, 2, pp. 53–58.

KOPELOFF, L. M., S. E. BARRERA, AND N. KOPELOFF (1942), "Recurrent Conclusive Seizures in Animals Produced by Immunologic and Chemical Means," *American J. of Psychiatry,* 98, pp. 881–902.

LEHNINGER, A. L. (1975), *Biochemistry,* 2nd ed., Worth Publishers, New York.

LEVINE, A. D., G. TCHOBANOGLOUS, AND T. ASANO (1991), "Size Distribution of Particulate Contaminants and Their Impact on Treatability," *Water Research,* 25, 8, pp. 911–922.

LOWRY, J. D. AND S. B. LOWRY (1988), "Radionuclides in Drinking Water," *J. American Water Works Association,* 80, 7, pp. 50–64.

MCLELLON, W. M. (1991), "Student Water Use," *J. American Water Works Association,* 83, 4, pp. 132–133.

MEIER, J. R. AND F. B. DANIEL (1990), "The Role of Short-Term Tests in Evaluating Health Effects Associated with Drinking Water," *J. American Water Works Association,* 82, 11, pp. 48–56.

MILTNER, R. J., D. B. BAKER, T. F. SPETH, AND C. A. FRANK (1989), "Treatment of Seasonal Pesticides in Surface Waters," *J. American Water Works Association,* 81, 1, pp. 43–52.

MOE (1978), *Basic Sewage Treatment Operation,* Ontario Ministry of the Environment, Toronto.

MOE (1985a), *Guidelines for the Design of Water Distribution Systems,* Ontario Ministry of the Environment, Toronto.

MOE (1985b), *Guidelines for the Design of Sanitary Sewage Systems,* Ontario Ministry of the Environment, Toronto.

NEMEROW, N. L. AND A. DASGUPTA (1991), *Industrial and Hazardous Waste Treatment,* Van Nostrand Reinhold, New York.

OHANIAN, E. V. (1992), "New Approaches in Setting Drinking Water Standards," *J. of the American College of Toxicology,* 11, 3, pp. 321–324.

RAUNKJÆR, K., T. HVITVED-JACOBSEN, AND P. H. NIELSEN (1994), "Measurement of Pools of Protein, Carbohydrate and Lipid in Domestic Wastewater," *Water Research,* 28, 2, pp. 251–262.

ROSE, J. B., G. S. SUN, C. P. GERBA, AND N. A. SINCLAIR (1991), "Microbial Quality and Persistence of Enteric Pathogens in Graywater from Various Sources," *Water Research,* 25, 1, pp. 37–42.

SAYRE, I. M. (1988), "International Standards for Drinking Water," *J. American Water Works Association,* 80, 1, pp. 53–60.

Schock, M. R. and C. H. Neff (1988), "Trace Metal Contamination from Brass Fittings," *J. American Water Works Association,* 80, 11, pp. 47–56.

Smith, D. W. and H. Knoll, eds. (1986), *Cold Climate Utilities Manual,* Canadian Society for Civil Engineering, Montreal, PQ.

Standard Methods for the Examination of Water and Wastewater (1992), 18th ed., APHA, AWWA, WPCF, American Public Health Association, Washington, DC.

USEPA (U.S. Environmental Protection Agency) (1994), "Drinking Water Regulations and Health Advisories," Office of Water, USEPA, Washington, DC.

Vickers, A. (1989), "New Massachusetts Toilet Standard Sets Water Conservation Precedent," *J. American Water Works Association,* 81, 3, pp. 48–51.

WHO (World Health Organization) (1993), *Guidelines for Drinking-Water Quality,* vol. 1, Recommendations, 2nd ed., Geneva.

SECTION III

WATER AND WASTEWATER TREATMENT

CHAPTER 9

WATER AND WASTEWATER TREATMENT OPERATIONS

Water and wastewater treatment plants are designed on a unit operations concept in which one operation is optimized to accomplish one task, although more than one problem substance may be remedied in the task. Each operation generally has ramifications on other downstream treatment processes, and tradeoffs between increasing the efficiency of one process or another depend on water characteristics and costs of each operation. A brief description of the processes used for water and wastewater treatment is given here with an explanation of where the process may occur in the treatment stream.

There are a number of manufacturers that supply units designed for one process to prefabricated package plants that incorporate a number of individual unit operations. Package plants may be suitable for smaller installations; larger installations are usually custom designed.

Basins in a treatment operation may be built in parallel or in series. Either option may be dictated by the efficiency of the process and other considerations. Multiple units also provide the capability to treat water or wastewater while one unit is out of service because of breakdown or routine maintenance. In smaller plants, where the design of multiple units is not economical or feasible, bypasses are designed to accommodate shutdown of a unit.

9.1 WATER TREATMENT OPERATIONS

The choice of treatment operations depends on the quality and variability of the raw water source and the treatment objectives, which may vary for industrial as opposed to municipal needs. A thorough survey of the quality and quantity of all possible sources is the first and most important step for designing a water supply process. Compromising the water survey can prove to be very costly in the long run through payment for more complex and expensive treatment operations. Water treatment operations must be designed to handle the extremes in raw water quality variation to provide an acceptable product water at all times.

The unit operations that may be incorporated into a water treatment plant are listed alphabetically below with some comments on their location in the treatment train.

Activated Carbon. Activated carbon is a broad-scale adsorbent of dissolved substances. Dissolved, colloidal, and particulate substances are attracted and attached to the surface of the carbon particles. It is used to remove taste and odor causing compounds as well as toxic organic chemicals. Precipitation and other chemical reactions also occur on the carbon surface. Activated carbon may be used for **dechlorination**.

A variety of carbon adsorbers can be designed, including batch and continuous flow units. The adsorption capacity of the carbon is eventually exhausted. The carbon is regenerated by heating the carbon, which burns and volatilizes the substances accumulated on it. Instead of heat, strong acids or bases, or other solvent solutions can be used to regenerate the carbon. Smaller operations normally do not regenerate the carbon on site.

Activated carbon can also be used as one of the media in a **filtration** unit.

Aeration. Aerators expose water to air to remove volatile dissolved components that are in excess of their saturation concentration. Some toxic organics are volatile. Taste- and odor-causing compounds may be removed to satisfactory levels. Groundwaters may have a high CO_2 content that will be stripped to tolerable levels in aeration.

The transfer of gases from the atmosphere to the water will also be effected. The addition of dissolved oxygen will enhance the oxidation of iron, manganese, and other metals to higher and more insoluble oxidation states. These precipitates will be removed in sedimentation basins and filtration units.

Aeration is one of the first treatment operations applied to a water. It can be designed as an aesthetically pleasing spray aerator open to public view.

Air Flotation. Air flotation is used to separate suspended matter from a water and sometimes used as an alternative to sedimentation. It is primarily used to thicken chemical sludge suspensions. Under pressurized conditions air is injected into the water. The water is released to a vessel at atmospheric pressure that causes the release of the dissolved air as tiny bubbles, which form nuclei to buoy up suspended particulates that are skimmed from the surface.

Biological Treatment. Biological treatment is used for ammonia nitrogen conversion to nitrate, nitrogen removal, and removal of organics in water. It can be accomplished in **filters, fluidized beds,** or **packed beds** containing granular activated carbon. Biological treatment of water is an emerging treatment.

Ammonia nitrogen makes a water biologically unstable and high concentrations may be converted to nitrate. Biological removal of nitrogen is accomplished by converting nitrates to nitrogen gas (denitrification). The nitrogen gas is stripped from the water. It is an alternative to **ion exchange**.

Biological removal of organic matter lessens the amount of disinfection byproducts formed when disinfection agents are added at the end of the treatment plant and renders a water that is less hospitable for growth or regrowth of microorganisms. The water is more biologically stable.

Chemical Feed Mixers. Many processes rely on the addition of chemical agents.[1] Mixers are designed to disperse the chemicals rapidly and thoroughly throughout the water.

Coagulant Recovery. The sludge generated with alum and iron coagulant salts may be regenerated by addition of acid. **Recalcination** is used to recover calcium oxide (CaO) from calcium carbonate sludge by heating the sludge to drive off carbon dioxide. Calcium oxide is then **slaked** by adding water to convert it to lime [$Ca(OH)_2$].

Coagulation. Coagulation is the process of adding chemical reagents in a mixing device to destabilize colloidal particles and allow them to agglomerate or flocculate with

[1]Some chemical agents used in water and wastewater treatment are given in the Appendix along with their bulk densities.

other suspended particles to form larger more readily settled particles. Coagulation reactions are fast and occur in the rapid mixing device. It is essential that the coagulant be dispersed throughout the water to contact and react with the target substances before the coagulant is consumed in side reactions with water itself.

Disinfection. Disinfection is the removal or inactivation of pathogenic microorganisms (not necessarily sterilization). Chemical agents, commonly chlorine or its derivatives, may be used or the water may be exposed to **ultraviolet** (UV) light or radiation. **Ozone** is becoming more widely used as a disinfectant. The disinfection tank or device (such as a UV chamber) maintains the water in contact with the dose of disinfectant for a time long enough to assure the required log reductions in indicator bacteria. It is exceedingly rare to find a raw water that would not require disinfection.

North American practice advocates the addition of a small amount of chlorine (and possibly ammonia) to form chloramines, which maintain a small disinfectant residual in the distribution system when other disinfectants are used as the primary disinfectant.

Disinfection is the last treatment applied to a water.

Filtration. Filtration accomplishes polishing of a water and is required for almost every water. Filtration follows sedimentation if it is provided. Water moves through tanks that contain sand and other types of media. Fine solids that did not settle out in a sedimentation basin will be entrapped in the filter. There also will be significant removal of bacteria in a filter but not enough (log reduction of bacteria is usually required) to provide a safe water. Larger microorganisms such as protozoans are completely removed in a properly operated filter.

There are two filtration alternatives in common use. **Slow sand filters** have only sand media. They are cleaned by scraping off the top layer of media on a periodic basis as the filter clogs. **Rapid filters** are sand filters or multimedia filters that have anthracite, sand, and possibly other media in them. Loading rates of rapid filters are much higher than slow sand filters. Rapid filters are cleaned by backwashing—reversing the flow of water through the media and pumping at a rate sufficient to expand the media. Backwashing is necessary every 1–4 d depending on influent water quality.

The influent to rapid filters generally must have a coagulating agent added to it at some upstream location.

Flow through rapid and slow sand filters is due to gravity. **Pressure filters,** where water is forced through the filter by applied pressure in a completely enclosed unit, are used in some smaller installations.

Roughing filters that contain coarse media may be used to prefilter water with a very high suspended solids content.

Raw water that is of high quality may require filtration only to remove the small quantities of suspended solids that are present. Otherwise rapid filters are preceded by coagulation, flocculation, and sedimentation.

Flocculation. Flocculators provide gentle agitation of a water that has been coagulated to promote particle contact and formation of larger particles. Hydraulic or mechanically driven flocculators may be designed. Flocculators follow the rapid mixing coagulation tank and precede sedimentation and filtration units.

Fluoridation. Fluoride is added to waters to reduce the incidence of dental caries in the population. The fluoride is added by a chemical feeder. Pumping and flow through the distribution pipes will ensure that the fluoride ion is thoroughly dispersed in the water. Fluoride is added at the end of the treatment train, where its concentration will not be affected by coagulation, precipitation, sedimentation, and adsorption processes.

When fluoride concentrations in the raw water exceed recommended limits, ion exchange with activated alumina is generally used to remove the excess fluoride.

Ion Exchange. Ion exchange can be used to remove hardness ions, nitrates, or other inorganic constituents. Ion exchangers contain large molecular weight organic substances (resins) that have the ability to exchange one ion for another. For instance there are resins that will exchange hydrogen ions for calcium ions or other positively charged ions. Naturally occurring ion exchangers (crystalline aluminosilicates) are known as zeolites. Eventually the ion exchanger's capacity is exhausted and concentrated solutions of acid, base, or salt are used to regenerate the ion exchanger.

Pipes, Channels, and Other Conduits. Flow in pipes and channels should have a minimum velocity of 0.3 m/s (1 ft/s) to avoid deposition of solids. Recommended velocities at various locations in the plant depend on the state of the water.

Prechlorination or Pre-oxidation. Instead of using aeration, high concentrations of nuisance metals such as Fe and Mn and other metals that may be toxic can be oxidized to more insoluble states by chlorine or other oxidants. The precipitates will be removed by following processes of coagulation, flocculation, sedimentation, and filtration. Sulfides are also oxidized to sulfates. This treatment is also used to retard microbial growth throughout the plant.

This operation is, as the name implies, at the front end of the treatment plant. Elemental chlorine should not be used at this point because it will have a greater tendency to form chloro-organics, some of which are carcinogenic. Other chlorine derivatives such as chloramines or other oxidation agents such as permanganate can be used.

Recarbonation. In water softening plants, it is often necessary to add an excess of lime to facilitate removal of hardness ions. After sedimentation of the $CaCO_3$ and $Mg(OH)_2$ precipitates, the pH of the water is high because of the excess lime. Carbon dioxide is added to neutralize the excess OH^-.

Reverse Osmosis and Electrodialysis. These technologies are used for the removal of high concentrations of dissolved solids. Reverse osmosis essentially "filters" dissolved solids from the water by forcing the water through a membrane by applying pressure in excess of the osmotic pressure of the dissolved components in the solution. Electrodialysis uses electric current that induces ions to migrate through a membrane. These technologies are primarily used to desalinate brackish waters. They can also be used for softening.

Suspended solids must be removed to a low level before water is subjected to reverse osmosis to prevent fouling of the membrane.

Screens and Bar Racks. Screens and bar racks are used to remove coarse debris from a raw water source. This debris may damage pumps or deposit in channels, causing clogging. They are located at the intake for the water source and at the beginning of the treatment plant. Screenings are collected and disposed at a **landfill** site.

Sedimentation. Exposing the water to relatively quiescent conditions will allow settleable solids to be removed by the action of the force of gravity. The sludge accumulated in these tanks may be disposed in landfills or the water source downstream of the withdrawal point for the water supply. Sedimentation that has not been preceded by coagulation and flocculation is known as plain sedimentation. Raw waters that contain a high sediment load may be settled in a plain sedimentation basin to remove the readily settled particulates. Then a chemical assist may be provided through addition of coagulant followed by flocculation and another sedimentation basin to remove slower settling particulates.

Waters, particularly groundwaters that have a low concentration of suspended solids, may not require sedimentation. They can be directly filtered.

Softening. Softening is the removal of multivalent cations (hardness) in the water. As noted above, **ion exchange** can be used to accomplish this task. Lime–soda ($Ca(OH)_2$–Na_2CO_3) softening involves the addition of these agents in a mixer, followed by flocculation, sedimentation, and filtration. The alkalinity added by lime and soda ash causes $CaCO_3$ and $Mg(OH)_2$ to precipitate and be removed in the sedimentation and filtration units. Softening will be practiced in conjunction with coagulation. It is possible to recover the lime.

Super-chlorination. Chlorine is a strong oxidizing agent that can destroy some taste and odor compounds. Also, in plants where filtration is not present and there is a significant risk of protozoan pathogens, elevated levels of chlorination beyond those normally required to disinfect a water will be required. After the chlorine is added, the water is held in a contact basin for the required contact time. Super-chlorination will have to be followed by **dechlorination** to reduce residual chlorine and the tastes and odors arising from chlorine itself. The dechlorinating agent is usually sulfite, although activated carbon can also be used for dechlorination.

Ultrafiltration. In ultrafiltration or **nanofiltration,** pressure is used to drive a liquid through a membrane that is permeable to some components. It is similar to reverse osmosis. The process is used to separate higher molecular weight compounds from a water. It is not commonly used for water treatment. **Microfiltration** uses membranes with larger sized openings to filter small particulates, bacteria, and protozoans from a water.

Water Stabilization. Water stabilization refers to the adjustment of the pH, alkalinity, and calcium content of a water such that there is a slight tendency for the finished water to precipitate $CaCO_3$. A thin deposit of $CaCO_3$ in the distribution system pipes will form a protective barrier to retard corrosion of the pipes by agents in the finished water. If softening is practiced, the process must be adjusted to produce the desired concentrations. It may be necessary to add acid or alkaline agents to achieve the desired balance among the three components.

Not all of the above treatment processes will be required for a raw water. There is usually more than one process or a combination of processes that can be used for the same purpose. Also the efficiency of a single unit operation may be significantly influenced by upstream operations. It is even possible for treatment to produce negative results. For example, Himberg et al. (1989) found in a few instances that a conventional treatment process as depicted in Fig. 9.1 produced negative removals of bluegreen algae toxins presumably because of cell lysis and release of toxins to solution in an otherwise proper functioning treatment train. Typical flow charts for water treatment operations are shown in Figs. 9.1, 9.2, and 9.3. There are many possible substitutions and configuration changes possible. Water treatment processes are classified according to their degree of contaminant removal in Table 9.1.

Reservoirs

Reservoirs are often required to provide a steady supply of water over periods of drought. Some improvement in the quality of a river water can also be obtained by

TABLE 9.1 General Effectiveness of Water Treatment Processes for Contaminant Removal[a,b]

Contaminant categories	Aeration and stripping	Coagulation processes, sedimentation, filtration	Lime softening	Ion exchanges		Membrane processes			Chemical oxidation, disinfection	SH Adsorption		
				Anion	Cation	Reverse osmosis	Ultra-filtration	Electro-dialysis		Granular activated carbon	Powdered activated carbon	Activated alumina
A. Primary Contaminants												
1. Microbial and turbidity												
Giardia lamblia	P	G–E	G–E	P	P	E	E	—	E	F	P	P–F
Legionella	P	G–E	G–E	P	P	E	E	—	E	F	P	P–F
Total coliforms	P	G–E	G–E	P	P	E	E	—	E	F	P	P–F
Turbidity	P	E	G	F	F	E	E	—	P	F	P	P–F
Viruses	P	G–E	G–E	P	P	E	E	—	E	P	P	P–F
2. Inorganics												
Arsenic (+3)	P	F–G	F–G	G–E	P	F–G	—	F–G	P	F–G	P–F	G–E
Arsenic (+5)	P	G–E	G–E	G–E	P	G–E	—	G–E	P	F–G	P–F	E
Asbestos	P	G–E	—	—	—	—	—	—	P	—	—	—
Barium	P	P–F	G–E	P	E	E	—	G–E	P	P	P	P
Cadmium	P	G–E	E	P	E	E	—	E	P	P–F	P	P
Chromium (+3)	P	G–E	G–E	P	E	E	—	E	F	F–G	F	P
Chromium (+6)	P	P	P	E	P	G–E	—	G–E	P	F–G	F	P
Cyanide	P	—	—	—	—	G	—	G	E	—	—	—
Fluoride	P	F–G	P–F	P–F	P	E	—	E	P	G–E	P	E
Lead	P	E	E	P	F–G	E	—	E	P	F–G	P–F	P
Mercury (inorganic)	P	F–G	F–G	P	F–G	F–G	—	F–G	P	F–G	P–F	P
Nickel	P	F–G	E	P	E	E	—	E	P	F–G	P–F	P
Nitrate	P	P	P	G–E	P	G	—	G	P	P	P	P
Nitrite	F	P	P	G–E	P	G	—	G	G–E	P	P	P
Radium-226, -228	P	P–F	G–E	P	E	E	—	G–E	P	P–F	P	P–F
Selenium (+4)	P	F–G	F	G–E	P	E	—	E	P	P	P	G–E
Selenium (+6)	P	P	P	G–E	P	E	—	E	P	P	P	G–E
3. Organics												
VOCs	G–E	P	P–F	P	P	F–E	F–E	F–E	P–G	F–E	P–G	P
SOCs	P–F	P–G	P–F	P	P	F–E	F–E	F–E	P–G	F–E	P–E	P–G
Pesticides	P–F	P–G	P–F	P	P	F–E	F–E	F–E	P–G	G–E	G–E	P–G
THMs	G–E	P	P	P	P	F–G	F–G	F–G	P–G	F–E	P–F	P
THM precursors	P	F–G	P–F	F–G	—	G–E	F–E	G–E	F–G	F–E	P–F	P–F
Hepatotoxins and neurotoxins from bluegreen algae[c]	P	P	P	—	—	—	—	—	P[d]	E	P	—

B. Secondary Contaminants												
Carbon dioxide	G–E	P–F	E	P	P	P	P	P	P	P	P	P
Chloride	P	P	P	F–G	P	G–E	P	G–E	P	P	P	—
Color	P	F–G	F–G	P–G	—	—	—	—	F–E	E	G–E	G
Copper	P	G	G–E	P	F–G	E	—	E	P–F	F–G	P	—
Hardness	P	P	E	P	E	E	G–E	E	P	P	P	P
Hydrogen sulfide	F–E	P	F–G	P	P	P	P	P	F–E	F–G	P	P
Iron	F–G	F–E	E	P	G–E	G–E	G	G–E	G–E	P	P	P
Manganese	P–F	F–E	E	P	G–E	G–E	G	G–E	F–E	P	P	P
Methane	G–E	P–E	P	P	P	P	P	P	P	F–E	P–G	P
Sulfate	P	P	P	G–E	P	E	P	E	P	P	P	G–E
Taste and odor	F–E	P–F	P–F	P–G	—	—	—	—	F–E	G–E	G–E	P–F
TOC	F	P–F	G	—	G–E	E	—	E	P	—	—	—
Total dissolved solids	P	P	P–F	P	P	G–E	P–F	G–E	P	P	P	P
Zinc	P	F–G	G–E	P	G–E	E	—	E	P	—	—	—
C. Proposed Contaminants												
Aluminum	P	F	F–G	P	G–E	E	—	E	P	—	—	—
Disinfection byproducts	—	P–E	P–F	P–F	—	P	F–G	F–G	F–G	F–E	P–G	—
Radon	G–E	P	P	P	P	P	P	P	P	E	P–F	P
Silver	F–G	G–E	P	G	—	—	—	P	F–G	P–F	—	—
Uranium	P	G–E	G–E	E	G–E	E	—	E	P	F	P–F	G–E

[a]From C. L. Hamann, J. B. McEwen, and A. G. Myers (1990), "Guide to Selection of Water Treatment Processes," in *Water Quality and Treatment,* 4th ed., F. W. Pontius, ed., American Water Works Association, McGraw-Hill, pp. 157–187. Reproduced with permission of McGraw-Hill, Inc.

[b]P, poor (0–20% removal); F, fair (20–60% removal); G, good (60–90% removal); E, excellent (90–100% removal); "—", not applicable/insufficient data.

[c]After Himberg et al. (1989).

[d]With ozone as the oxidant.

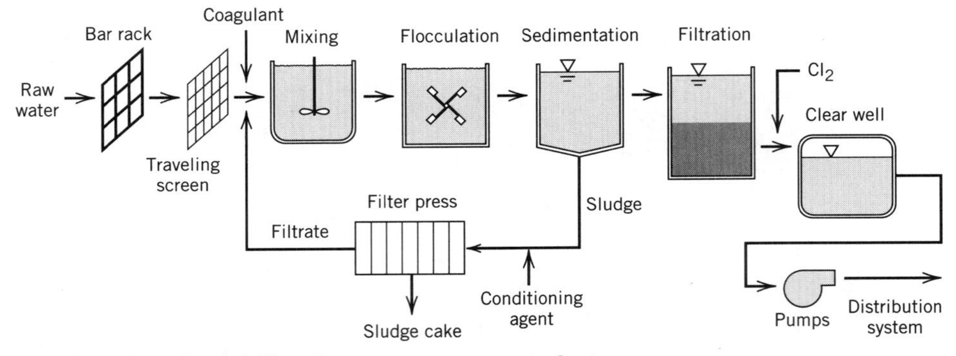

Figure 9.1 Rapid sand filtration water treatment plant.

building an impoundment from which the water will be drawn. Quiescent conditions in a reservoir will allow suspended solids to settle. Turbidity and some organic matter will be removed. The holding time in a reservoir will cause the devitalization and death of waterborne pathogens. Sunlight will photo-oxidize some color-causing compounds. Some equalization of quantity and quality fluctuations will also be achieved.

Softening may occur naturally in small lakes or reservoirs. Hot summer days promote high growth rates of algae, which remove carbon dioxide from the water. As a consequence, the pH of the water rises and the remaining dissolved inorganic carbon species are converted into carbonates. The low solubility product of Ca^{2+} and CO_3^{2-} will precipitate $CaCO_3$. However this advantage is not permanent because when the pH decreases, the precipitated $CaCO_3$ can be redissolved if the reservoir water is below saturation. Reservoir impoundment may cause changes in water quality because of taste and odor compound production. Vegetative growth will occur in shallow bodies of water, causing changes in water quality. Growth of bacteria on the reservoir bottom may cause taste and odor problems. Also, algal growth may cause taste and odor problems.

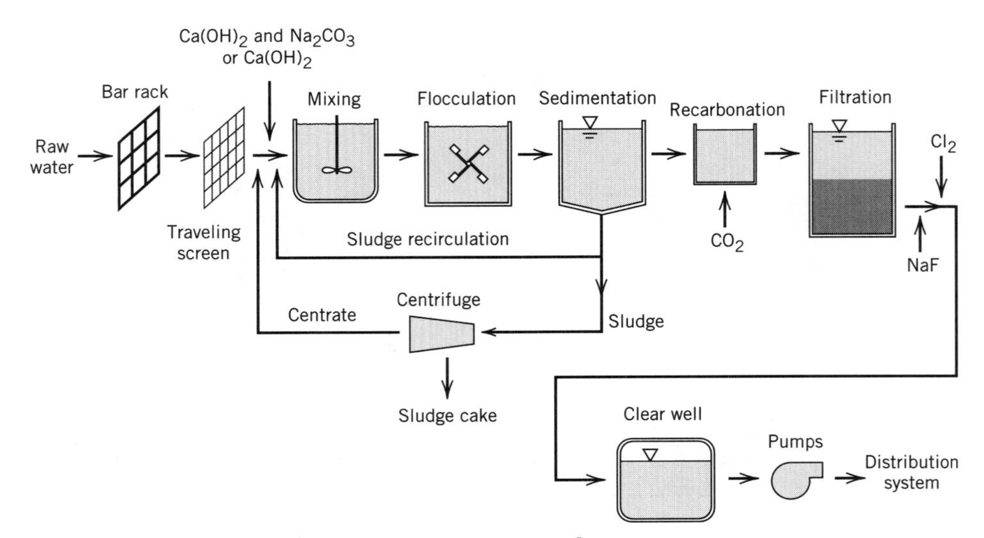

Figure 9.2 Lime–soda softening water treatment plant.

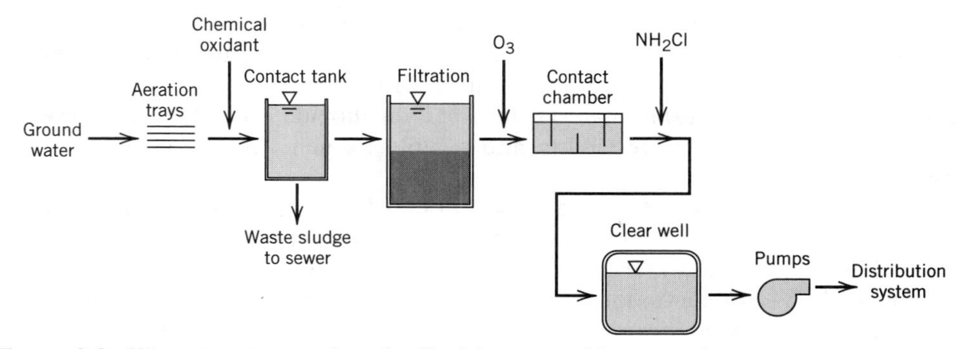

Figure 9.3 Water treatment plant for Fe–Mn removal in groundwater.

The benefits of building an impoundment in a river only for quality improvement must be weighed against the disadvantages. The costs of impoundment are generally high. Preliminary treatment gained in the reservoir is a permanent gain.

9.1.1 Home Water Treatment Units

A number of point-of-use (applied at the water tap) or point-of-entry (applied at water supply point to the residence) treatment units, including water distillation units, activated carbon filters, reverse osmosis systems, and ion exchangers, have appeared on the market in recent years. Myths about water quality coupled with exaggerated claims from many manufacturers of these devices can lead to erroneous conclusions by the consumer.

Distillation units produce water of high purity. The other units treat water as described above depending on the treatment process incorporated into the unit. The onus is on the user to maintain and operate these units properly according to manufacturer's instructions. Chlorine or its derivatives will be removed in carbon filters and the filter medium provides support surfaces for bacteria to colonize. Bacteria can populate filters, particularly those containing activated carbon, to high levels resulting in slug discharges (Geldreich et al., 1985). There is a risk that adventitious, opportunistic pathogens may proliferate. Many activated carbon filters are impregnated with silver, which is toxic to microorganisms, to retard microbial growth. When these devices are properly designed and used they can supplement a central treatment system (Regunathan et al., 1983).

9.2 WASTEWATER TREATMENT UNIT OPERATIONS

Wastewaters are normally treated by a combination of physical–chemical and biological operations. However, it is possible to treat wastewaters solely with physical–chemical methods. Again, operations are listed alphabetically for easy reference.

Activated Carbon Treatment. Activated carbon may be added to biological treatment processes or used as a treatment by itself. Activated carbon is used to adsorb toxic substances such as metals and pesticides that may be present in a wastewater.

Air Flotation. Air flotation is used to separate suspended matter from a wastewater. It is primarily used to thicken biological or chemical sludge suspensions. See the description of the process given above.

Aerobic Biological Treatment. Many types of processes are possible for the aerobic biological removal of dissolved and suspended organics. In the process, air (oxygen) is supplied to microorganisms that are in contact with the wastewater. The microorganisms metabolize the organic material into carbon dioxide and other end products and new biomass. The putrescibility and soluble oxygen demand is reduced to a small amount.

Some processes such as **trickling filters** or **biodisks (rotating biological contactors)** use a solid support media to provide surfaces on which bacteria grow and accumulate. In a trickling filter wastewater is sprayed over a rock or synthetic media bed. Wastewater is applied on a continuous or an intermittent basis. Bacteria attach and grow on the media. Natural air currents supply oxygen. Biodisks work on a similar principle. A series of disks are partially submerged in a tank of wastewater. The disks are rotated to expose the biomass-covered disks to the atmosphere, where oxygen transfer to the biomass occurs. Providing immobilized media for microorganism attachment and growth is known as a fixed film process.

The common **activated sludge** process is a suspended growth process where the microorganisms are mixed with the wastewater. Air is pumped into the basin to supply oxygen and mixing. Pure oxygen may be supplied as an alternative to air but supplemental mixing is required. There are many variations of the activated sludge process.

Processes may also be designed to oxidize nitrogen or remove nitrogen from the wastewater. It is also possible to design a process for increased removal of phosphorus.

All aerobic biological treatment processes convert some of the dissolved organics into biological solids that must normally be removed in a sedimentation basin following the biological treatment unit. Aerobic biological treatment is often preceded by sedimentation to physically remove suspended organics that would be degraded in the biological treatment process at more expense.

Air Strippers. Volatile compounds or gases present at supersaturated concentrations are amenable to air stripping. Air stripping units are devices that enhance the mass transfer between liquid and the atmosphere. This is accomplished by increasing the surface area of the liquid to maximize its exposure to the atmosphere. There are many types of stripping devices, ranging from highly efficient countercurrent flow to diffused aeration systems. In countercurrent flow devices, the wastewater is injected at the top of a column that contains media such as ceramic berl saddles and air is injected at the bottom of the column.

Diffused aerators inject air into the liquid through porous plates located near the bottom of a tank.

Ammonia Stripping. Ammonia is a volatile component that may be removed by exposing the wastewater to the atmosphere. A high pH favors the formation of ammonia as opposed to ammonium ion. Air strippers are used to remove the unionized ammonia.

Anaerobic Biological Treatment. High-strength wastewaters are amenable to treatment by an anaerobic process. Industrial and agricultural wastewaters are often highly concentrated. Anaerobic treatment occurs in enclosed reactors to prevent access of oxygen. The anaerobic microorganisms that are in contact with the wastewater convert the dissolved and suspended organics into biomass and methane. Significant quantities of methane are produced and it may be used as fuel. As in aerobic biological treatment operations, some processes incorporate solid support media for the microorganisms (fixed film processes), whereas others keep the microorganisms in suspension.

The biological solids produced in an anaerobic process must be settled in a clarifier that may follow the process or be incorporated into the anaerobic reactor. Fixed film and suspended growth processes among other options are used in anaerobic treatment.

Anaerobic digesters are often used to treat solids produced in an aerobic treatment process and in the primary clarifier, considerably reducing the volume of solids finally to be removed from the plant.

Fixed film processes include anaerobic filters and fluidized beds. The suspended growth process is also known as an anaerobic contact process. A unique process in anaerobic treatment is the upflow anaerobic sludge blanket reactor, wherein dense granules of microorganisms are retained in the reactor by inertia.

It is not necessary to presettle the influent to an anaerobic reactor.

Chemical Feed Mixers. Physical–chemical plants rely heavily on the addition of chemical agents.[2] Chemical agents are also used in biological treatment plants. Mixers are designed to disperse the chemicals rapidly and thoroughly throughout the water.

Coagulant Recovery. Physical–chemical plants use methods for coagulant recovery described under water treatment.

Coagulation. This process has the same objectives as noted for water treatment and it is used in physical–chemical wastewater treatment processes. Also, chemicals may be added to precipitate a particular constituent such as phosphorus. Alum is commonly applied to effluent from an aeration basin for biological treatment to ensure that phosphorus is precipitated in the final clarifier.

Comminutors. Comminutors are devices that macerate rags, sticks, paper, and other gross solid debris in the wastewater that can clog channels or otherwise obstruct flow through the plant. These devices are placed downstream of the grit chamber. The cuttings from the comminutor are returned directly to the wastewater.

Disinfection. Clarified effluent from the plant may be disinfected to reduce pathogen concentrations prior to discharge to the receiving body of water. Chlorine is usually the disinfectant applied, although UV units are being used at greater frequency. A plug flow contact chamber is used to maintain the wastewater in contact with the disinfectant for the specified time.

Equalization. When flow quantity and quality variations are significant, holding basins are incorporated near the front of the treatment train to allow wastewater to be input to the plant at a more uniform rate and quality. More uniform flows and concentrations reduce the variability of treatment and allow a more compact wastewater treatment plant to be designed with higher utilization of all units. The equalization basin is mixed and aeration may be applied to avoid septic conditions. The amount of biological treatment provided by aerating the wastewater will not be significant. Some stripping of volatile compounds will occur.

Filtration. Filtration is often used in physical–chemical wastewater treatment processes to remove fine colloidal or solid particles that cannot be settled in a sedimentation basin. Filtration is also used finally to polish effluent from clarifiers that have settled effluent from a biological treatment process. It is also used to remove algae in effluent from a stabilization pond process.

Flocculation. This process has the same objectives as noted for water treatment. It is not commonly used in treatment plants that use biological processes but it is an important operation in physical–chemical treatment plants.

[2]Some chemical agents used in water and wastewater treatment are given in the Appendix along with their bulk densities.

Grit Chambers. After the wastewater is screened, grit is removed. Grit chambers are sedimentation basins designed to remove nonputrescible matter such as silt and sand that cause abrasive wear on channels and pumps and can accumulate and clog channels. Because the materials removed in a grit chamber are essentially nonbiodegradable, they are collected and disposed in a **landfill** site.

Modern grit chambers are usually aerated chambers with an offset aerator that induces a small degree of water circulation to keep lighter organic particulates in suspension while the heavier grit materials settle.

Incineration. Incineration is used for final stabilization of sludge. An incinerator is a furnace for high-temperature combustion of the sludge. Without emission control, incinerators produce a significant amount of atmospheric contamination.

Lagoons. Lagoons are mechanically aerated ponds that provide aerobic biological treatment.

Membrane Technologies. The membrane technologies of **reverse osmosis** and **electrodialysis** described under water treatment can be applied to wastewaters for selective removal and recovery of species.

Neutralization. Industrial wastes with extremely high or low pH are neutralized by acid or base addition. If biological treatment is to be used, a pH near neutrality is required.

Oxidation Ditches. An oxidation ditch is an oval channel with a mechanical aeration device installed to provide aerobic biological treatment.

Pipes, Channels, and Other Conduits. Flow in pipes and channels should have a minimum velocity of 0.6–0.9 m/s (2–3 ft/s) to avoid deposition of solids (WEF and ASCE, 1992). Some facilities use minimum velocities as low as 0.3 m/s (1 ft/s) but the higher minimum is preferred by more operators.

Recarbonation. Carbon dioxide is added to neutralize excess OH^- ion that was added earlier in the process for coagulation–flocculation reactions.

Screens and Bar Racks. Screens and bar racks are used to remove coarse debris from wastewater for the same reasons as in a water treatment plant. This debris may damage pumps or deposit in channels, causing clogging. Bar racks are located at the intakes to wet wells in pumping stations in the sewer system as well as at the intake to the wastewater treatment plant. At the wastewater treatment plant screens will also be used after the bar racks. Materials removed on the screens and racks are relatively nonbiodegradable. They are collected and disposed in a **landfill** site.

Sedimentation. Primary clarifiers are designed to remove settleable solids from a wastewater before biologically treating it for dissolved organics. Removal of biodegradable solids by sedimentation is much less expensive than treating them aerobically in an aeration basin.

Secondary clarifiers follow biological treatment units and are designed to remove biomass formed during biological treatment and other solids present in the influent to the biological treatment unit. In addition to clarification of the wastewater, a secondary clarifier for a biological treatment process is designed to perform some thickening of the biological sludge that accumulates in it.

Sedimentation basins in physical–chemical treatment systems will follow coagulation and flocculation units.

Sludge Concentration and Dewatering. The purpose of sludge concentration units is to reduce the volume of sludge for ultimate disposal in a **landfill,** incinerator, or on land. **Vacuum dewatering, filter presses,** and **centrifugation** are commonly used to separate the solids from the water. Chemical agents are added to the sludge going into the dewatering unit to improve the efficiency of the device. Sludges are also dewatered in drying beds.

Sludge Digestion. Sludge generated in aerobic biological treatment and removed from primary clarifiers may be reduced in quantity and volume by microbial action in aerobic or anaerobic sludge digesters. **Aerobic digesters** are often used at smaller installations. Larger installations use **anaerobic digesters.**

Sludge Thickening. Thickening is a settling process where the sludge concentrates in a basin by the action of gravity. The supernatant from a thickener contains significant concentrations of colloidal and suspended matter and it is returned to the primary clarifier to pass through the treatment process.

Stabilization Ponds. Stabilization ponds are a series of ponds that settle and biologically treat a wastewater. They are not normally used with other treatment processes except possibly a final filter or screen to remove effluent suspended solids. Mechanical aeration is normally not provided in stabilization ponds. In the final ponds in a series of stabilization ponds, fish may be grown for harvest, which is referred to as **aquaculture.** Aquaculture is not common in North America but it is practiced in other parts of the world.

Ultrafiltration. Ultrafiltration or **nanofiltration** and the related process of **microfiltration** have been described under water treatment operations. They can be used for filtration and the recovery of compounds in a wastewater.

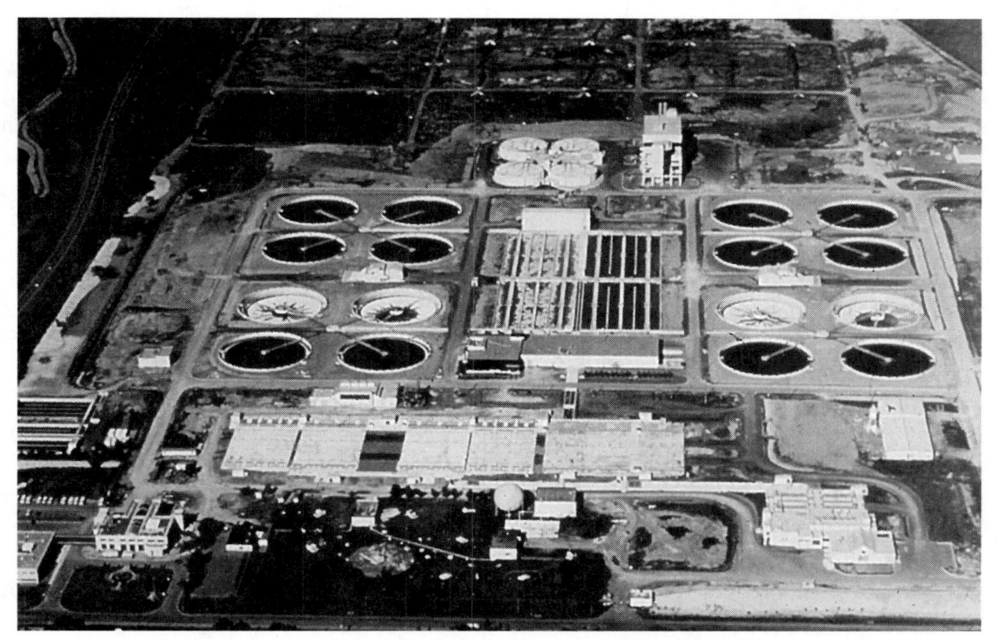

Secondary biological wastewater treatment plant at Ottawa, ON. (Photo courtesy of Walter Horvath.)

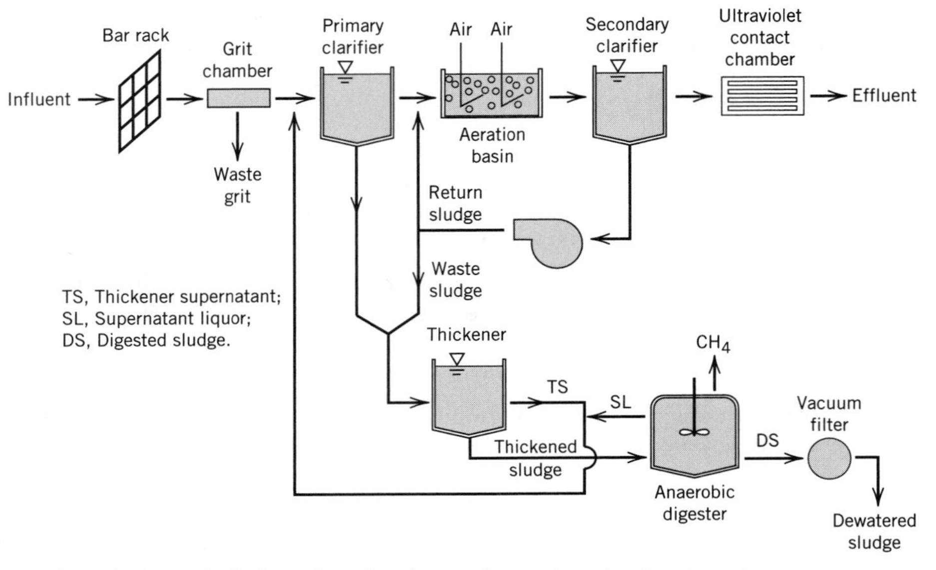

Figure 9.4 Activated sludge plant for domestic wastewater treatment.

A typical wastewater treatment plant incorporating biological treatment is shown in Fig. 9.4; a physical–chemical treatment plant for wastewater is shown in Fig. 9.5. The major difference between physical–chemical treatment plants and biological treatment plants is the processes used to remove dissolved organic matter. The quantities of sludge produced by a physical–chemical treatment will be higher than for a plant incorporating biological treatment because of the coagulation agents added and the absence of biological oxidation of organics. However, not all wastes are amenable to biological treatment. A multitude of changes could be made to the treatment plant schemes shown in Figs. 9.4 and 9.5. As in water treatment the treatment plant is an integrated works where the performance of any individual process is linked to the processes that precede it.

The effectiveness of various treatment processes in removing classes of contaminants is given in Table 9.2.

A wastewater treatment plant that only incorporates sedimentation as the major treatment operation is often referred to as a primary treatment plant. Provision of biological treatment is termed secondary treatment. Physical–chemical treatment plants are often referred to as advanced wastewater treatment or tertiary treatment operations. Biological treatment processes designed for nutrient removal are advanced treatment processes. The physical–chemical processes of filtration and carbon adsorption can be added to the biological treatment process depicted in Fig. 9.4 to produce a very highly purified wastewater. In the past, in extreme circumstances, wastewater has been purified to be used directly as a water supply.

Land application of raw sewage is another alternative that has been used throughout the world but is generally not accepted in North America because of concern for pathogens. Heavy metals will be problematic for land application of industrial wastes. Irrigation, groundwater recharge, and fertilization (nutrient recycle) are some benefits of land application of sewage. Land application of primary or secondary treated sewage

TABLE 9.2 Contaminant Removal in Wastewater Treatment Processes

		Sedimentation[a]				Biological treatment[a]						
Contaminant category	Air stripping[a]	Coagulation, flocculation, sedimentation	Primary	Following biological treatment	Following biol. treatment with chemical addition to influent	Conventional aerobic treatment	Biological denitrification	Low-loading trickling filter	High-loading trickling filter	Anaerobic treatment	Disinfection[a]	Carbon adsorption[a]
Suspended organic matter	—	G–E	F–G	G–E	G–E	—	—	—	—	F–G	—	G–E
Dissolved organic matter	P	P–F	—	P	P	G–E	G–E	G–E	G–E	G	—	G–E
Ammonia nitrogen	G–E	—	—	—	—	P	G–E	G–E	P	P	F–G	F–G
Inorganic nitrogen	—	—	—	—	—	P	G–E	P	P	P	—	P
Phosphorus	—	G–E	—	—	G–E	P–F	P–F	P–F	P–F	P	—	—
Sulfides	G–E	—	—	—	—	E	E	E	G–E	—	G–E	G–E
VOC	G–E	—	—	—	—	G–E	G–E	G–E	G–E	G–E	—	—
SOC	—	P–F	—	—	—	P–E	P–E	P–E	P–E	P–E	P	G–E
Pesticides	—	—	—	—	—	P–E	P–E	P–E	P–E	P–E	P	G–E
Heavy metals	—	G–E	P–F	G–E	G–E	—	—	—	—	—	—	G–E
Pathogens	—	P	P	P	P	P–F	P–F	P–F	P–F	P–F	G–E	—

[a]P, poor (0–20% removal); F, fair (20–60% removal); G, good (60–90% removal); E, excellent (90–100% removal); "—", not applicable.

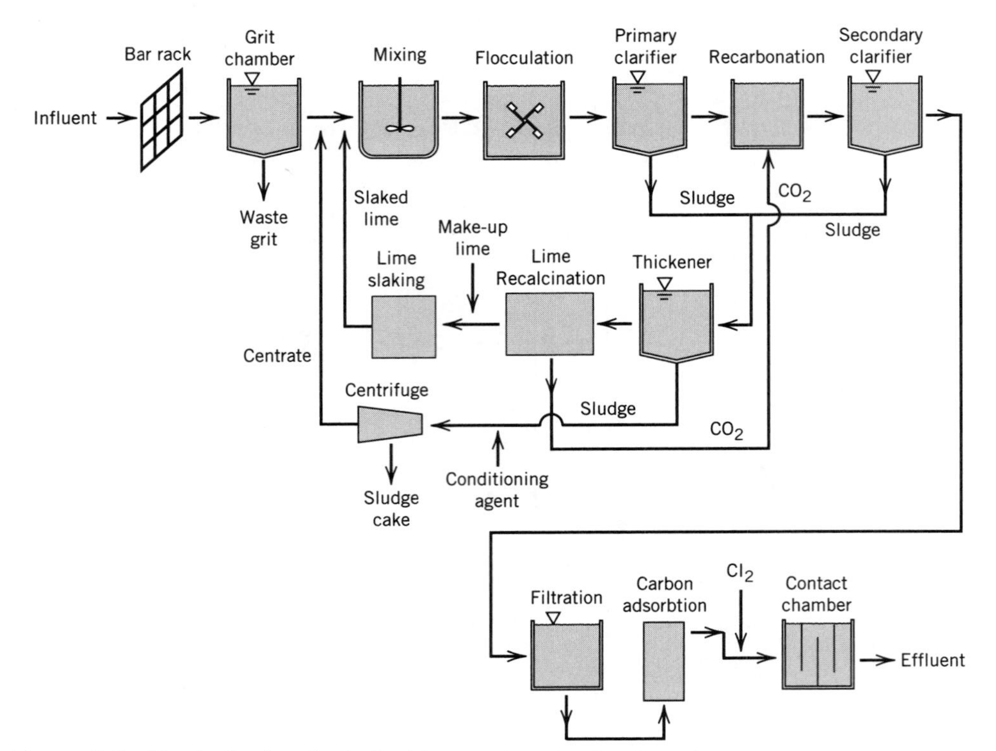

Figure 9.5 Physical–chemical plant for wastewater treatment.

is also practiced and encouraged where feasible. Raw or partially treated sewage may also be discharged to wetlands.

Landfills are the most common means of disposal of sludge produced from treatment operations but a significant quantity of sludges generated from domestic wastes can be applied to soils.

9.3 HYDRAULIC DESIGN OF WATER AND WASTEWATER TREATMENT PLANTS

The hydraulic design of water and wastewater treatment plants is a challenging exercise. Plants are designed to flow as much as possible by gravity, minimizing the number of pumps in the plant. Near the front end of the treatment works enough head is supplied by pumping if necessary to drive the water through the treatment operations that follow. Efficient hydraulic design provides treatment over a wide range of flows at a minimum energy expenditure. Inefficient hydraulic design results in higher energy consumption and costs for the plant throughout its service time. The designer must be cognizant of possible future modifications and expansions and their hydraulic requirements. To provide the necessary flexibility to handle the flow variation within treatment and physical constraints and other requirements, a number of alternatives should be examined.

The plant must be designed to function under all hydraulic conditions from low to high flows. Because two or more units are usually designed for any treatment operation, the extreme high flow condition for a treatment will be when one of the basins is out of service and the remaining basins must accommodate the flow. The hydraulic conditions should always be checked for the low, average, and high flows and also for when one basin is out of service. Backwater conditions are generally avoided; the plant must not flood at any time and surcharged sewers cannot be tolerated for wastewater treatment plants.

The critical design flow depends on the treatment unit. Practical considerations are applied. At extreme conditions guidelines are often violated. For instance at low flow, velocities in channels may not be high enough to prevent deposition of solids in channels. Designing to achieve a scouring velocity at all flows is impractical because higher energy losses will be incurred and abrasive scour may occur at higher flows. As long as flow velocities that will scour channels or pipes occur for a significant time during a day, then low velocity conditions can be tolerated. Extreme conditions and suboptimal treatment conditions that accompany them occur with lower frequency. Minimal treatment objectives must be established. Then flow and load variation throughout the design period of the plant must be examined to find the plant capacity. Additional units that are idle at times may be required to provide the flexibility to handle flow and load variation at the minimum operating cost within the treatment requirements.

Another restriction that applies to hydraulic design of plants is the general necessity to use standard sizes. Devices are supplied in standard sizes to minimize fabrication costs, improve quality control, and facilitate construction. Custom sizes are used infrequently because of the additional expense. Therefore, hydraulic computations are adjusted to the nearest standardized size that provides a more conservative condition.

Basic hydraulic formulas commonly used in treatment plants are given in this section. Hydraulic design information on specific operations is also found in later chapters. In addition there are sources such as Benefield et al. (1984), Kawamura (1991), Qakim (1985), Sanks (1978), and Schulz and Okun (1984) that provide more detailed information on hydraulic design of treatment units. Handbooks on hydraulics and manufacturers also provide information on energy losses and hydraulic performance for various devices.

Flow in Pressurized Pipes

Flow in pressurized conduits is described by the Bernoulli equation:

$$\frac{v_1^2}{2g} + z_1 + \frac{p_1}{\rho g} = \frac{v_2^2}{2g} + z_2 + \frac{p_2}{\rho g} + h_L \tag{9.1}$$

where

subscripts 1 and 2 refer to the upstream and downstream locations
v is velocity
g is the acceleration of gravity
ρ is the density of the fluid
p is pressure
z is elevation from a fixed datum
h_L is the headloss from location 1 to location 2

Headloss in pipe flow is calculated with the Darcy–Weisbach equation:

$$h_L = f\frac{L}{D}\frac{v^2}{2g} \tag{9.2}$$

where
f is a friction factor
L is length of pipe
D is pipe diameter

The headloss is a function of the velocity head. When irregularly shaped conduits are used, the hydraulic radius (Eq. 9.3) is substituted for D in Eq. (9.2).

$$R_h = \frac{A_f}{P} \tag{9.3}$$

where
R_h is the hydraulic radius
A_f is the cross-sectional area of flow
P is the wetted perimeter

The friction factor is a function of the Reynold's number.

$$\text{Re} = \frac{\rho v D}{\mu} \tag{9.4}$$

where
Re is the Reynold's number
μ is the viscosity

The Moody diagram (given in the Appendix) describes the variation of f with Re.

Flow in Open Channels

The energy equation for flow in channels exposed to the atmosphere is usually written with the channel bottom taken as the datum. The total head is the depth of flow plus the velocity head, which is referred to as the specific energy.

$$y_1 + \frac{v^2}{2g} = y_2 + \frac{v_2^2}{2g} + h_L \tag{9.5}$$

where
y is the depth of flow

The Darcy–Weisbach equation can be used to find headloss in open channel flow situations but there are a number of other relations that have been developed. In uniform flow the water surface is parallel to the channel bottom. Manning's equation (Eq. 9.6) is the equation most commonly applied to calculate energy losses in open channel flow.

$$v = \frac{1}{n} R_h^{2/3} S^{1/2} \tag{9.6}$$

where
n is a roughness coefficient
S is the slope of the energy line (and the slope of the channel), $S = h_L/L$

Channels are often designed with flat slopes which are easier to construct and install. Varied flows may require solution of gradually or rapidly varied flow equations, which are covered in some of the references given earlier and in texts on open channel hydraulics. Manning's equation is applied discretely in these solutions.

Several flow control points will exist in a treatment plant that control the hydraulic profile upstream or downstream from the control point. A known depth (stage)–discharge condition exists at a flow control point. The Froude (Fr) number (Eq. 9.7) dictates whether gravity flow is sub- or supercritical. At $\text{Fr} < 1$ flow is subcritical and when $\text{Fr} > 1$ flow is supercritical. Subcritical flows and depths are controlled by downstream control points; supercritical flows and depths are controlled by upstream control points. Most flows through treatment plants are subcritical. Any transition between sub- and supercritical flow or vice versa produces a control point at which critical flow occurs and $\text{Fr} = 1$. The critical flow condition provides the depth–discharge relation.

$$\text{Fr} = \frac{v}{\sqrt{gy}} \tag{9.7}$$

Weirs, discussed below, with free fall over them are the most common control devices. Valves can also be used to achieve flow control.

Other Losses

Besides friction losses along the walls of a conduit, any change in direction or velocity causes increased headloss. Obstructions such as valves for flow control or orifices for flow measurement result in headloss. An obstruction may be installed for the sole purpose of causing headloss to equalize flows. Losses from these appurtenances are termed minor losses in fluid mechanics texts but losses associated with these devices are often the major losses in the short conduits installed in treatment works.

The losses to be considered for any unit are:

1. Friction losses through the unit
2. Entrance losses
3. Exit losses
4. Losses through appurtenances

The losses are usually proportional to the velocity head. Table 9.3 lists the headloss factors for both pressurized and open channel conduits. The higher values are used for design purposes because of fouling through microbial growth, deposition, or corrosion. These losses are summed for all conduits and basins to find the total head requirement for the plant. Other miscellaneous losses such as the additional depth required to ensure free fall over a weir must also be included. Additional head allowance is also provided for future expansions. Wall friction losses in treatment basins are usually negligible because of the relatively low flow velocities through basins, and the water surface in a basin without a significant number of baffles is assumed to be horizontal.

Other flow devices that are regularly used in treatment works are weirs, orifices, and baffles. Weirs are used to control the flow from basins and to measure flow. Maintenance of free flow conditions from a weir will require that the water surface in the receiving channel be lowered. The water surface is typically lowered 5–10 cm below the crest of the weir. Weirs have the disadvantage that solids may settle ahead of the weir and a higher headloss is associated with maintenance of free fall conditions.

The equations for weirs are based on idealized flow over the weir and an empirically determined discharge coefficient that accounts for aeration of the nappe, contraction, and surface tension effects. For a sharp-crested weir with free fall that spans the entire

TABLE 9.3 Headlosses in Appurtenances[a]

Appurtenance (alphabetically)	K[b]
Butterfly valves	
Fully open	0.3
Angle closed, $\theta = 10°$	0.46
$\theta = 20°$	1.38
$\theta = 30°$	3.6
$\theta = 40°$	10
$\theta = 50°$	31
$\theta = 60°$	94
Check (reflux) valves	
Ball type (fully open)	2.5–3.5
Horizontal lift type	8–12
Swing check	0.6–2.3
Swing check (fully open)	2.5
Contraction—sudden	
4:1 (in terms of velocity at small end)	0.42
2:1	0.33
4:3	0.19
Also see Reducers	
Diaphragm valve	
Fully open	2.3
$\frac{3}{4}$ open	2.6
$\frac{1}{2}$ open	4.3
$\frac{1}{4}$ open	21.0
Elbow—90°	
Flanged—regular	0.21–0.30
Flanged—long radius	0.18–0.20
Intersection of two cylinders (welded pipe—not rounded)	1.25–1.8
Screwed—short radius	0.9
Screwed—medium radius	0.75
Screwed—long radius	0.60
Elbow—45°	
Flanged—regular	0.20–0.30
Flanged—long radius	0.18–0.20
Screwed—regular	0.30–0.42
Elbow—22.5° (flanged)	0.10–0.12
Enlargement—sudden	
1:4 (in terms of velocity at small end)	0.92
1:2	0.56
3:4	0.19
Also see Increasers	
Entrance losses	
Bell-mouthed	0.04
Pipe flush with tank	0.5
Pipe projecting into tank (Borda entrance)	0.83–1.0

Appurtenance (alphabetically)	K[b]
Obstruction in pipes	
(in terms of pipe velocities)	
pipe to obstruction area ratio	
1.1	0.21
1.4	1.15
1.6	2.40
2.0	5.55
3.0	15.0
4.0	27.3
5.0	42.0
6.0	57.0
7.0	72.5
10.0	121.0
Orifice meters (in terms of pipe velocities)	
orifice to pipe diameter ratio	
0.25 (1:4)	4.8
0.33 (1:3)	2.5
0.50 (1:2)	1.0
0.67 (2:3)	0.4
0.75 (3:4)	0.24
Outlet losses	
Bell-mouthed outlet	$0.1(v_1^2/2g - v_2^2/2g)$
Sharp-cornered outlet	$(v_1^2/2g - v_2^2/2g)$
Pipe into still water or air (free discharge)	1.0
Plug globe or stop valve	
Fully open	4.0
$\frac{3}{4}$ open	4.6
$\frac{1}{2}$ open	6.4
$\frac{1}{4}$ open	780
Reducers	
Ordinary (in terms of velocity at small end)	0.25
Bell mouthed	0.10
Standard	0.04
Bushing or coupling	0.05–2.0
Return bend (2 Nos. 90°)	
Flanged—regular	0.38
Flanged—long radius	0.25
Screwed ends	2.2
Siphon	2.8
Sluice gates	
Contraction in conduit	0.5
Same as conduit width without top submergence	0.2
Submerged port in 30 cm wall	0.8
Tees	
Standard—bifurcating	1.5–1.8
Standard 90° turn	1.80

TABLE 9.3 Continued

Appurtenance (alphabetically)	K^b	Appurtenance (alphabetically)	K^b
Slightly rounded	0.23	Standard—run of tee	0.60
Strainer and foot valve	2.50	Reducing—run of tee	
Gate valves		2:1 (based on velocity at smaller end)	0.90
Fully open	0.19		
$\frac{3}{4}$ closed	1.15	4:1	0.75
$\frac{1}{2}$ closed	5.6	Venturi meters	
$\frac{1}{4}$ closed	24.0	Headloss occurs mostly in and downstream of throat, but losses are given in terms of velocity at inlet end to assist in design.	
Also see Sluice gates			
Increasers			
$0.25(v_1^2/2g - v_2^2/2g)$ where v_1 is velocity at small end		Long tube type—throat-to-inlet diameter ratio	
		0.33 (1:3)	1.0–1.2
Miter bends			
Deflection angle, θ		0.50 (1:2)	0.44–0.52
5°	0.016–0.024	0.67 (2:3)	0.25–0.30
10°	0.034–0.044	0.75 (3:4)	0.20–0.23
15°	0.042–0.062	Long tube type–throat-to-inlet diameter ratio	
22.5°	0.066–0.154	0.33 (1:3)	2.43
30°	0.130–0.165		
45°	0.236–0.320	0.50 (1:2)	0.72
60°	0.471–0.684	0.67 (2:3)	0.32
90°	1.129–1.265	0.75 (3:4)	0.24
		Y branches (regular)	0.50–0.75

[a]From Amirtharajah (1978) and Kawamura (1991).
[b]The headloss is given by $h_L = Kv^2/2g$ unless otherwise indicated.

width of a channel (known as a suppressed weir),

$$Q = C_d \left(\frac{2}{3}\right) L\sqrt{2g}H^{3/2} \tag{9.8}$$

where

- Q is the volumetric flow rate
- C_d is the discharge coefficient
- L is the length of the weir
- H is the depth of water above the weir

Evett and Liu (1987) give the following equation for C_d that is applicable for SI or U.S. units.

$$C_d = 0.611 + 0.075\left(\frac{H}{H_w}\right) \tag{9.9}$$

where

- H_w is the height of the weir

A contracted weir is a weir that does not span the entire length of the channel. It is also referred to as a notched weir. For a rectangular, sharp-crested notched weir (Evett and Liu, 1987):

$$Q = C_d L^{1.02}\sqrt{2g}H^{1.47} \tag{9.10}$$

where

C_d is 1.69 (SI) with Q in m^3/s for L and H in m; 3.10 (U.S.) with Q in ft^3/s for L and H in ft

V-notched weirs are also used, particularly for sedimentation basins. They are discussed in Section 11.6. The above equations do not apply to submerged weirs.

Slight improvement in accuracy of headloss calculations for weirs can be gained by using the formulations in American Society of Mechanical Engineers (ASME, 1971) but Eqs. (9.8) through (9.10) are simpler and suitable for water or wastewater treatment plant problems.

Application of Bernoulli's equation results in the equation for a freely discharging orifice.

$$Q = C_d a \sqrt{2gh} \tag{9.11}$$

where

a is the area of the orifice
h is the piezometric head above the center of the orifice

The discharge coefficient varies with the hydraulic smoothness of the orifice. For a sharp-edged orifice, $C_d = 0.62$.

When the water level drops below the top of the orifice, the orifice behaves as a weir. When the orifice is submerged, the discharge through it is governed by the difference in piezometric head across the orifice.

$$Q = C_d a \sqrt{2g\Delta h} \tag{9.12}$$

where

$\Delta h = h_1 - h_2$

The discharge coefficient for sharp-edged submerged orifices is also 0.62.

Baffles are often used in basins to dissipate energy of incoming flows and disperse the jet. For a rectangular baffle (reaction baffle) the drag force from the baffle is

$$F_D = \tfrac{1}{2}\rho C_D A_b v^2$$

where

F_D is the drag force
C_D is the drag coefficient
A_b is the area of the baffle
v is the approach velocity to the baffle

Converting the above equation to energy loss per unit weight of water,

$$h_L = C_D \frac{v^2}{2g}\left(\frac{A_b}{A}\right) \tag{9.13}$$

where

A is the cross-sectional area of the channel

Rouse (1959) gives the value of C_D as 1.2.

QUESTIONS AND PROBLEMS

1. What unit operations are incorporated into the local water and wastewater treatment plants?
2. Describe the changes in water quality after each unit operation in Figs. 9.1 and 9.2.

3. Which unit operations would be most likely required for treating a groundwater as opposed to a surface water?
4. What is the risk associated with point-of-use water treatment devices?
5. Would you expect water produced from a water treatment plant to contain more dissolved inorganic or more dissolved organic components?
6. Describe the general features of each unit operation in the treatment schemes in Figs. 9.4 and 9.5.
7. List all wastewater treatment operations that would produce or remove suspended solids. Describe the removal or solids transformation in each process.
8. (a) Describe the purposes of (i) equalization basins, (ii) grit chambers, and (iii) stabilization ponds in wastewater treatment.
   (b) Describe the purposes of (i) flocculation, (ii) softening, and (iii) disinfection in water treatment.
9. You have been assigned the task of making a preliminary choice on unit operations to be incorporated into a water treatment plant. The data for the raw water are given below. Draw the flow diagram of the units that you will incorporate into your design. In brief statements, give reasons for choosing these units. Use the WHO standards in Chapter 8 as the basis for determining whether a substance requires treatment.

River data		
pH		7.7
Alkalinity		180 mg/L as $CaCO_3$
Total hardness		300 mg/L as $CaCO_3$
Calcium hardness		250 mg/L as $CaCO_3$
Iron		0.3 mg/L
Manganese		0.3 mg/L
Turbidity		300 NTU
Water temperature	Mean	18°C
	Low	10°C
	High	25°C
Coliform count		3 000 MPN/100 mL

   Assume other substances are within acceptable ranges.
10. Would you consider a biological or a physical–chemical treatment process to be better for treating the wastewater produced from a potato processing plant? Why?
11. What are the losses in the following system at a temperature of 15°C? A 6.0 cm diameter pipe flush with a tank and below the surface of water in the tank conveys water horizontally 2.0 m, then turns 90° in a flanged regular elbow and travels vertically down for 0.5 m. Then the pipe turns horizontally through another 90° flanged regular elbow. The horizontal length is 1.5 m and the pipe enters another large basin where the water velocity may be neglected. The pipe entrance to the tank is flush with the tank wall. There is a fully open gate valve in the line and the flow velocity is 1.2 m/s. Assume that the roughness height of the pipe is 0.000 08 m.

KEY REFERENCE

HAMANN, C. L., J. B. MCEWEN, AND A. G. MYERS (1990), "Guide to Selection of Water Treatment Processes," in *Water Quality and Treatment,* 4th ed., F. W. Pontius, ed., American Water Works Association, McGraw-Hill, Toronto, pp. 157–187.

REFERENCES

AMIRTHARAJAH, A. (1978), "Design of Granular-Media Filter Units," in *Water Treatment Plant Design,* R. L. Sanks, ed., Ann Arbor Science, Ann Arbor, MI, pp. 675–737.

American Society of Mechanical Engineers (1971), *Fluid Meters, ASME,* New York.

Benefield, L. D., J. F. Judkins, Jr., and A. D. Parr (1984), *Treatment Plant Hydraulics for Environmental Engineers,* Prentice-Hall, Englewood Cliffs, NJ.

Evett, J. B. and C. Liu (1987), *Fundamentals of Fluid Mechanics,* McGraw-Hill, Toronto.

Geldreich, E. E., R. H. Taylor, J. C. Blannon, and D. J. Reasoner (1985), "Bacterial Colonization of Point-of-Use Water Treatment Devices," *J. American Water Works Association,* 77, 2, pp. 72–80.

Himberg, K., A. M. Keijola, L. Hiisvirta, H. Pyysalo, and K. Sivonen (1989), "The Effect of Water Treatment Processes on the Removal of Hepatotoxins from *Microcystis* and *Oscillatoria* Cyanobacteria: A Laboratory Study," *Water Research,* 23, 8, pp. 979–984.

Kawamura, S. (1991), *Integrated Design of Water Treatment Facilities,* Wiley-Interscience, Toronto.

Qakim, S. R. (1985), *Wastewater Treatment Plants: Planning, Design and Operation,* Holt, Rinehart and Winston, Toronto.

Regunathan, P., W. H. Beauman, and E. G. Kreusch (1983), "Efficiency of Point-of-Use Treatment Devices," *J. American Water Works Association,* 75, 1, pp. 42–50.

Rouse, H., ed. (1959), *Engineering Hydraulics,* John Wiley & Sons, Toronto.

Sanks, R. L., ed. (1978), *Water Treatment Plant Design,* Ann Arbor Science, Ann Arbor, MI.

Schulz, C. R. and D. A. Okun (1984), *Surface Water Treatment for Communities in Developing Countries,* John Wiley & Sons, Toronto.

WEF and ASCE (1992), *Design of Wastewater Treatment Plants,* vol. I, WEF, Washington, DC.

CHAPTER 10

MASS BALANCES AND HYDRAULIC FLOW REGIMES

The concepts in this chapter apply to any water or wastewater unit operation. Mass balances relate influent flow rate and concentrations to effluent flow rates and concentrations by accounting for removal or transformation phenomena. The pattern of flow in a reactor is influenced by reactor geometry and mixing devices, which in turn influence the kinetics of a unit operation.

10.1 SETUP OF MASS BALANCES

There are a number of different notations used in textbooks and the literature for describing mass balances. Each method, of course, ultimately leads to the same differential equation. A few different approaches are discussed here along with other considerations in formulating the correct equation. In particular, the assumptions that underlie the development of an equation are important. Assumptions are normally made to simplify a complex situation; the assumptions must be examined to determine whether they are reasonable for the circumstances under consideration.

Mass balances are applied to rivers, lakes, or treatment basins, where the problem is to find the concentration of a substance at a location or its rate of change in a section. Varying degrees of complexity are reached as more parameters are introduced and when the parameters vary with time (i.e., when non-steady state conditions exist). Often it is not necessary to consider variation at all times. Only the worst case or perhaps the worst and best case scenarios need to be considered to arrive at the design constraints (costs and material requirements). The steady state condition in the majority of circumstances will render an intractable equation analytically solvable, greatly reducing the costs (normally associated with computer runs) and time involved in analyzing the problem.

If a reactor is complete mixed (CM) [sections below discuss CM and plug flow (PF) hydraulic regimes in detail], i.e., uniform conditions exist throughout the reactor, the whole volume is most conveniently considered. If PF conditions exist (water flows through the reactor as a series of plugs) within the reactor, an elemental volume must be considered, because each elemental volume is different with respect to the mass and concentration of reacting substances contained in it. In the latter case the average contribution of the whole reactor volume to change in a substance's concentration is obtained by integrating the elemental mass balance over the whole volume.

A substance is transported into a given volume of a reactor (which may be a vessel or a section of a river or lake) by bulk flow (convection or advection) or other hydraulic phenomena such as dispersion. Bulk flow is merely the displacement of

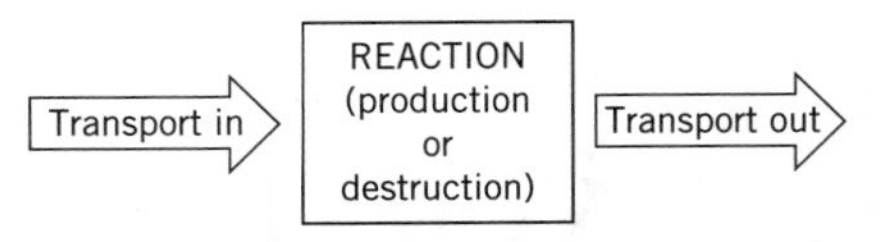

Figure 10.1 Qualitative process description.

liquid in a volume with incoming fluid. Dispersion is a function of turbulence where random fluctuations in the movement of fluid transport dissolved and suspended matter in random directions. Diffusion, which is transport through the random motion of molecules, may also be important. Transport out of the volume is normally due to the same mechanisms as transport into the volume. A case in which in and out transport mechanisms differ would occur if the substance is transformed into a gas in the reactor, for instance, partial oxidation of large molecules into smaller volatile compounds.

The general situation is qualitatively described by Fig. 10.1.

Reaction of the substance in question may be production or destruction caused by chemical, biochemical, or physical (such as temperature) phenomena. During the time in which the substance resides in the volume, it is exposed to the reaction mechanism, producing an alteration in its concentration. The general word statement describing the situation in Fig. 10.1 is

$$\text{IN} - \text{OUT} + \text{GENERATION} = \text{ACCUMULATION} \tag{10.1}$$

In Eq. (10.1), IN − OUT refers to the net transport of a substance into the reactor, GENERATION refers to the net amount of production or destruction by reaction or physical processes, and ACCUMULATION is the amount left over. Physical removal is a transport process but it is usually viewed as a generation term.

Mention should be made here of time and its meaning in steady state, which is commonly assumed to exist in the first analysis, and time of reaction, which refers to rate of change of a substance with time. The assumption of steady state means that all quantities are constant in value at each point in the system. This does not imply that they are equal throughout the system, only that a value at a given point does not change with time. On the other hand, two points in a PF system are separated by a distance of finite travel time, and the substance will be decayed (or produced) in the period of time it takes to travel from one point to another.

Once Eq. (10.1) is written, the important processes must be identified and mathematically formulated. In environmental engineering, the rate of reaction, r, is often formulated as shown below. The Monod or Michaelis–Menten formulation (Eq. 4.9) is also commonly used.

$$r = \pm kC^n \tag{10.2}$$

where

k is a rate coefficient
C is concentration (usually mg/L)
n is a constant (not necessarily with an integer value)

The units on r are always mass/volume/time regardless of the value of n. The units on k are adjusted to make the equation dimensionally consistent. Many reactions in environmental engineering are reasonably described with a first-order model, i.e.,

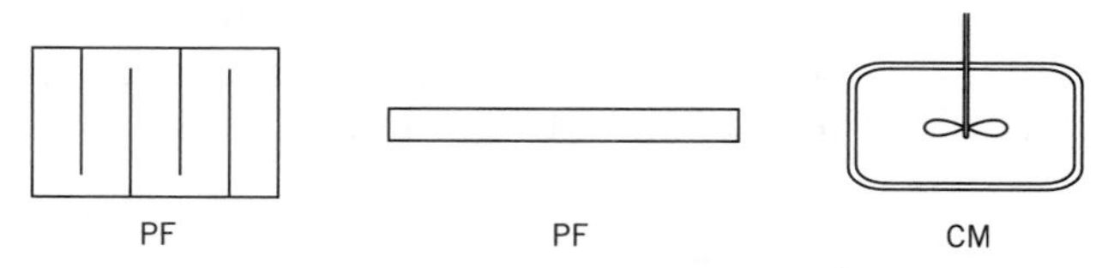

Figure 10.2 Basin characteristics for CM and PF basins.

n is equal to 1. A positive sign on the right-hand side corresponds to production; a negative sign refers to destruction.

10.1.1 Mixing Characteristics of Basins

The degree of mixing that exists in a basin affects reaction kinetics in a fundamental manner. The addition of a reagent to a water often requires intense mixing to achieve contact between the reagent and the substance in the water before the reagent is dissipated. On the other hand, some processes such as sedimentation rely on the existence of quiescent conditions for maximum efficiency. The two extremes in hydraulic mixing regime are CM (sometimes referred to as a completely stirred tank reactor, CSTR) and PF.

An influent particle of water to a CM basin is instantaneously dispersed throughout the entire contents of the basin. Uniform conditions exist throughout the basin; i.e., there is no spatial variation in any component. The concentrations of components in the effluent are the same as their concentrations in the basin. In an ideal basin it would not make any difference where the inlet and outlet points were located. Impellers, pumping, and gas or liquid recirculation are means of increasing the degree of mixing (Fig. 10.2).

The opposite end of the spectrum is PF. In a PF hydraulic regime, the water flows through the reactor as a series of plugs and there is no transfer of contents between plugs. Vertical and lateral (into the paper) mixing are consistent with this definition but longitudinal (left to right) mixing is not. Inlet and outlet locations should be at opposite ends of a PF basin. PF conditions are achieved by designing long narrow reactors or by placing baffles in a reactor (Fig. 10.2). Baffles provide a physical barrier to mixing.

10.1.2 Mass Balances for PF Reactors

In a PF situation the mass balance must be taken over an elemental volume. The typical situation is shown in Fig. 10.3 where Q is the volumetric flow rate, A is the cross-sectional area, and Δx is the length of the element.

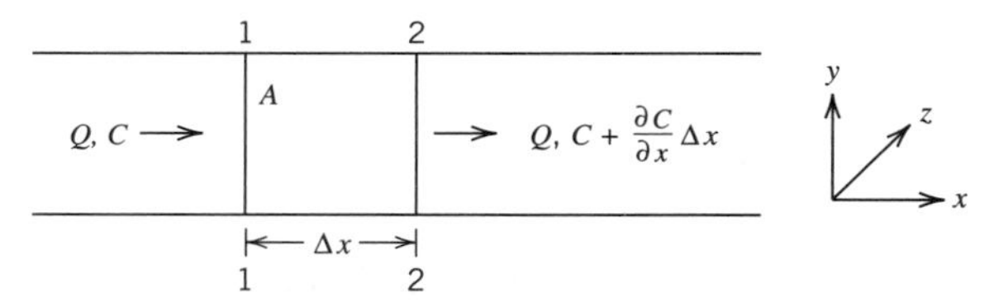

Figure 10.3 Elemental volume for PF.

For the basic model, area and volumetric flow rate will be assumed to be constant and change in the y and z directions is assumed to be zero or insignificant, i.e., one-dimensional change exists. The following two relations are used in all derivations:

$$\Delta V = A\,\Delta x \tag{10.3a}$$

$$v = \frac{Q}{A} \tag{10.3b}$$

where
V is volume
Q is volumetric flow rate
A is cross-sectional area of the element
v is velocity

Method I

Assuming a steady volumetric flow rate, the following mass balance applies to the element over a time period, $\Delta t = (t_2 - t_1)$:

$$Q\left[\frac{C_1(t_1) + C_1(t_2)}{2}\right]\Delta t - Q\left[\frac{C_2(t_1) + C_2(t_2)}{2}\right]\Delta t + r\,\Delta V\,\Delta t = \left[\frac{C_1(t_2) + C_2(t_2)}{2} - \frac{C_1(t_1) + C_2(t_1)}{2}\right]\Delta V \tag{10.4a}$$

In the above C_i refers to the concentration at interface i. If a decay reaction is assumed, the expression for r is

$$r = -k\left[\frac{C_1(t_1) + C_1(t_2) + C_2(t_1) + C_2(t_2)}{4}\right]^n$$

Dividing by Δt and letting $\Delta t \to 0$, changes terms in the equation in the following manner.

$$\frac{C_i(t_1) + C_i(t_2)}{2} \to C_i$$

$$\left[\frac{\frac{C_1(t_2) + C_2(t_2)}{2} - \frac{C_1(t_1) + C_2(t_1)}{2}}{\Delta t}\right]\Delta V \to \frac{\partial \overline{C}}{\partial t}\partial V$$

where

$$\overline{C} = \frac{C_1 + C_2}{2}$$

Making these substitutions into the reaction expression and Eq. (10.4a),

$$QC_1 - QC_2 + r\,\Delta V = \frac{\partial \overline{C}}{\partial t}\Delta V$$

or

$$QC_1 - QC_2 - k\overline{C}^n\,\Delta V = \frac{\partial \overline{C}}{\partial t}\Delta V \tag{10.4b}$$

The left-hand side of the equation describes the spatial variation and the right-hand side describes the time variation.

Equation (10.4b) is commonly written as the immediate starting equation for the mass balance.

Assuming steady state and a first-order decay ($r = -kC$), substituting for these quantities and ΔV, and rearranging Eq. (10.4b):

$$Q(C_1 - C_2) - k\overline{C}A\,\Delta x = 0 \tag{10.4c}$$

Now,

$$C_1 - C_2 = -\Delta C$$

Substituting the above into Eq. (10.4c) and dividing by Δx,

$$-Q\frac{\Delta C}{\Delta x} = k\overline{C}A \tag{10.4d}$$

Letting $\Delta x \to 0$ and dividing by A,

$$\frac{Q}{A}\frac{\partial C}{\partial x} = -kC \qquad \text{or} \qquad v\frac{\partial C}{\partial x} = -kC \tag{10.4e}$$

In the above two equations $\partial C/\partial x$ can be changed to dC/dx since C is only a function of x.

Because $v = dx/dt$, by application of the chain rule Eq. (10.4e) can be changed to

$$\frac{dx}{dt}\frac{dC}{dx} = \frac{dC}{dt} = -kC \tag{10.4f}$$

In the above equation time is synonymous with distance. Steady state conditions exist.

In the literature there are other equivalent methods to arrive at the governing differential equation. A couple of the methods will be commented on here. All procedures when correctly applied must arrive at the same equation. The choice of the approach is arbitrary; however, sometimes one is more convenient than the others.

Method II

In this method the concentrations at interfaces 1 and 2 are formulated as C and $C + (\partial C/\partial x)\,\Delta x$, respectively. Placing these in the mass balance:

$$QC - Q\left(C + \frac{\partial C}{\partial x}\Delta x\right) + r\,\Delta V = \frac{\partial \overline{C}}{\partial t}\Delta V \tag{10.5a}$$

(Why is C, not $\overline{C}$, used in the partial derivative term in the expression for concentration at interface 2?)

Assuming steady state, substituting for r and ΔV, expanding the equation, and simplifying:

$$-Q\frac{\partial C}{\partial x}\Delta x - kCA\,\Delta x = 0 \tag{10.5b}$$

Following the procedures outlined above, Eqs. (10.4e) and (10.4f) will be derived.

Method III

In the final method the notation to describe the value of a substance z at $x + \Delta x$ is

$$z|_{x+\Delta x}$$

Formulating the mass balance,

$$QC|_x - QC|_{x+\Delta x} + r\,\Delta V = \frac{\partial \overline{C}}{\partial t}\Delta V \tag{10.6a}$$

Applying the steady state condition and substituting for ΔV,

$$QC|_x - QC|_{x+\Delta x} + rA\,\Delta x = 0 \tag{10.6b}$$

Substituting for r, noting that Q is constant, and dividing by Δx:

$$Q\left[\frac{C|_x - C|_{x+\Delta x}}{\Delta x}\right] = k\overline{C}A \tag{10.6c}$$

As $\Delta x \rightarrow dx$, the expression within the brackets becomes $(-\partial C/\partial x)$. In this case C is only a function of x and it becomes $(-dC/dx)$ as before. After this step the procedure is the same as outlined from Eqs. (10.4d) to (10.4f).

Note that if Q were variable over the distance, the left-hand side of Eq. (10.6c) would become:

$$\left[\frac{QC|_x - QC|_{x+\Delta x}}{\Delta x}\right]_{\Delta x \rightarrow dx} \rightarrow -\frac{\partial(QC)}{\partial x} = -C\frac{\partial Q}{\partial x} - Q\frac{\partial C}{\partial x}$$

Now consider variation in area, which would be more realistic for a river. The definition sketch in Fig. 10.3 applies except that the area of interface 1 is different from the area of interface 2. If Q is constant and input of the substance to the system is constant, do steady state conditions exist? Using the notation in Method II, the formulation of the mass balance is the same as Eq. (10.5a).

$$QC - Q\left(C + \frac{\partial C}{\partial x}\Delta x\right) + r\,\Delta V = \frac{\partial \overline{C}}{\partial t}\Delta V \tag{10.7}$$

Because A_1 is different from A_2 the following expression must be substituted for the elemental volume.

$$\Delta V = \overline{A}\,\Delta x = \left(\frac{A_1 + A_2}{2}\right)\Delta x$$

The assumption of steady state is valid for the conditions just described; velocity varies from point to point but it (along with the other parameters) is constant at a given point. Applying this condition and substituting for ΔV and r:

$$-Q\frac{\partial C}{\partial x}\Delta x - k\overline{CA}\,\Delta x = 0 \tag{10.8}$$

Letting $\Delta x \rightarrow dx$, $\overline{CA}$ becomes CA, the concentration in the plane multiplied by the area of the plane.

$$Q\frac{\partial C}{\partial x} = -kCA \tag{10.9}$$

Dividing by A and noting that area is a function of x, $A(x)$:

$$\frac{1}{A(x)}Q\frac{dC}{dx} = -kC \tag{10.10}$$

This equation can be integrated as follows.

$$\int_{C_0}^{C} \frac{dC}{C} = -\frac{k}{Q}\int_0^x A(x)\,dx \tag{10.11}$$

But,

$$\frac{Q}{A(x)} = v(x) = \frac{dx}{dt}$$

This equation can be solved for dt,

$$dt = \frac{A(x)}{Q}\,dx = \frac{dx}{v(x)}$$

Substituting this relation into Eq. (10.11),

$$\int_{C_0}^{C} \frac{dC}{C} = -k\int_0^t dt \tag{10.12}$$

In this case time and distance are not linearly related.

For an example, examine a channel that has a constantly increasing area, $A(x)$, given by

$$A(x) = A_0(1 + ix)$$

where

A_0 and i are constants

Using Eq. (10.11),

$$\int_{C_0}^{C} \frac{dC}{C} = -\frac{k}{Q}\int_0^x A_0(1 + ix)\,dx \tag{10.13}$$

which integrates to:

$$\ln\frac{C}{C_0} = -\frac{kA_0}{Q}\left(x + \frac{ix^2}{2}\right) \tag{10.14}$$

Also, from Eq. (10.12),

$$\ln\frac{C}{C_0} = -kt \tag{10.15}$$

and

$$v(x) = \frac{Q}{A(x)} = \frac{Q}{A_0(1 + ix)} = \frac{dx}{dt} \tag{10.16}$$

To find the x–t relation,

$$\int_0^t dt = \frac{A_0}{Q}\int_0^x (1 + ix)\,dx \tag{10.17}$$

$$t = \frac{A_0}{Q}\left(x + i\frac{x^2}{2}\right) \tag{10.18}$$

Substituting Eq. (10.18) into Eq. (10.14),

$$\ln \frac{C}{C_0} = -kt \qquad (10.19)$$

which is the same equation (Eq. 10.15) derived before.

■ Example 10.1 Plug Flow Reactor Volume

Determine the volume of a PF reactor required to give a treatment efficiency of 95% for a substance that decays according to half-order kinetics with a rate constant of 0.05 $(\text{mg/L})^{1/2}/\text{h}$. The flow rate is steady at 300 L/h (79.2 gal/h) and the influent concentration is 150 mg/L.

The rate expression is

$$r = -kC^{1/2}$$

The required effluent concentration is

$$C = 0.05\ (150\ \text{mg/L}) = 7.5\ \text{mg/L}$$

Substituting the relation for r into Eq. (10.5a), the mass balance for steady state conditions over an elemental volume becomes

$$QC - Q\left[C + \frac{\partial C}{\partial x}\Delta x\right] - kC^{1/2}\,\Delta V = 0$$

This reduces to the integral

$$\int_{C_0}^{C} \frac{dC}{C^{1/2}} = -\frac{k}{Q}\int_0^V dV \rightarrow V = 2\frac{Q}{k}(C_0^{1/2} - C^{1/2})$$

$$V = 2\frac{300\ \text{L/h}}{0.05\ (\text{mg/L})^{1/2}/\text{h}}[(150\ \text{mg/L})^{1/2} - (7.5\ \text{mg/L})^{1/2}]$$
$$= 114 \times 10^3\ \text{L} = 114\ \text{m}^3\ (3.23\ \text{ft}^3)$$

10.1.3 Mass Balances and Reaction for CM Basins

The only difference in making a mass balance for a CM reactor (Fig. 10.4) compared to a PF reactor is to make the mass balance over the entire reactor volume (replace ΔV with V in the equations). The concentration of a substance in the reactor is the same as its concentration in the effluent as discussed in a later section.

Setting up a mass balance:

$$\text{In} - \text{Out} + \text{Generation} = \text{Accumulation}$$

$$QC_0 - QC + rV = V\frac{dC}{dt} \qquad (10.20)$$

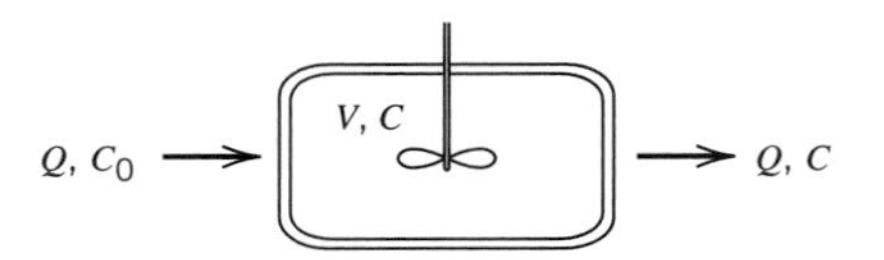

Figure 10.4 CM reactor.

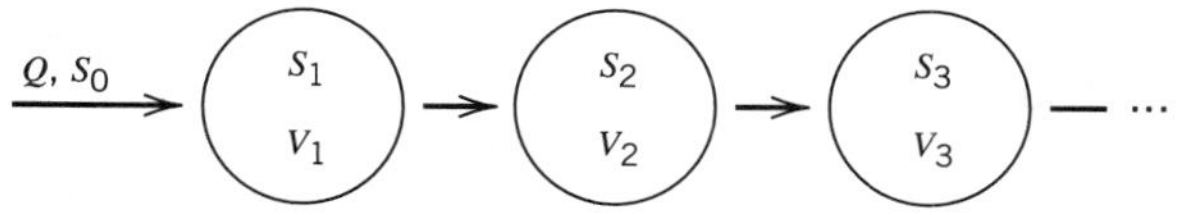

Q, S_0 → S_1 V_1 → S_2 V_2 → S_3 V_3 — ...

Figure 10.5 CM reactors in series.

Assuming steady state conditions and substituting a first-order decay reaction for r,

$$QC_0 - QC - kCV = 0$$

Solving for the effluent concentration,

$$C = \frac{C_0}{1 + kt_d} \qquad \left(t_d = \frac{V}{Q}\right) \tag{10.21}$$

where
t_d is the liquid detention time in the vessel

The dilution rate (D) is the inverse of detention time.

$$D = \frac{1}{t_d} = \frac{Q}{V}$$

A series of CM basins are used in some situations as shown in Fig. 10.5. In this case the effluent from one reactor becomes the influent for the next reactor. Using the symbols in Fig. 10.5, the equations describing concentration of S for the first two reactors at steady state conditions are:

$$QS_0 - QS_1 - k_1S_1V_1 = 0 \tag{10.22a}$$
$$QS_1 - QS_2 - k_2S_2V_2 = 0 \tag{10.22b}$$

where
k_i is the rate coefficient for the ith reactor

Similar equations can be written for the other reactors. For the ith reactor:

$$S_i = \frac{S_{i-1}}{1 + k_i t_{di}} \tag{10.23}$$

The influent (S_0) and effluent (S_e) concentrations for an n reactor system are related by:

$$\frac{S_e}{S_0} = \frac{S_n}{S_0} = \frac{S_1}{S_0} \times \frac{S_2}{S_1} \times \cdots \times \frac{S_n}{S_{n-1}} = \frac{1}{1 + k_1 t_{d1}} \times \frac{1}{1 + k_2 t_{d2}} \times \cdots \times \frac{1}{1 + k_n t_{dn}} \tag{10.24}$$

If the reactors are identical ($V_i/Q = t_d$) and the rate coefficient does not change throughout the system ($k_i = k$),

$$S_e = \frac{S_0}{(1 + kt_d)^n} = \frac{S_0}{\left(1 + k\dfrac{V}{nQ}\right)^n} \tag{10.25}$$

where
V is the total volume in the system

As the number of reactors in the system increases, the system response approaches the response obtained from a PF system. The total reactor volume required to obtain a given degree of treatment will dramatically decrease for a system containing a series of CM reactors when the reaction kinetic expression is of order greater than zero.

■ Example 10.2 Complete Mixed Reactor Volume

(a) Determine the volume of a CM reactor required to give a treatment efficiency of 95% for a substance that decays according to half-order kinetics with a rate constant of 0.05 (mg/L)$^{1/2}$/h. The flow rate is steady at 300 L/h (79.2 gal/h) and the influent concentration is 150 mg/L.

(b) Determine the volumes of two identical CM reactors in series to provide the same degree of treatment for the conditions given in (a).

Compare results with those of Example 10.1.

For part (a): The required effluent concentration is 7.5 mg/L as given in Example 10.1 and the rate expression is

$$r = -kC^{1/2}$$

Substituting the relation for r into Eq. (10.20), the mass balance for steady state conditions over the volume is

$$QC_0 - QC - kC^{1/2}V = 0$$

Solving this for V and substituting the given information,

$$V = \frac{QC_0 - QC}{kC^{1/2}} = \frac{\left(300\,\dfrac{\text{L}}{\text{h}}\right)\left(150\,\dfrac{\text{mg}}{\text{L}} - 7.5\,\dfrac{\text{mg}}{\text{L}}\right)}{\left[0.05\,\dfrac{(\text{mg/L})^{1/2}}{\text{h}}\right]\left(7.5\,\dfrac{\text{mg}}{\text{L}}\right)^{1/2}} = 3.12 \times 10^5\ \text{L} = 312\ \text{m}^3$$

In U.S. units:

$$V = \frac{\left(79.2\,\dfrac{\text{gal}}{\text{h}}\right)\left(\dfrac{3.79\ \text{L}}{\text{gal}}\right)\left(150\,\dfrac{\text{mg}}{\text{L}} - 7.5\,\dfrac{\text{mg}}{\text{L}}\right)}{\left[0.05\,\dfrac{(\text{mg/L})^{1/2}}{\text{h}}\right]\left(7.5\,\dfrac{\text{mg}}{\text{L}}\right)^{1/2}} = 3.12 \times 10^5\ \text{L}$$

$$= (3.12 \times 10^5\ \text{L})\left(\frac{1\ \text{m}^3}{10^3\ \text{L}}\right)\left(\frac{3.28\ \text{ft}}{\text{m}}\right)^3 = 11\,010\ \text{ft}^3$$

For part (b): Using the approach outlined above, the mass balance expressions are

$$QC_0 - QC_1 - kC_1^{1/2}V_1 = 0$$
$$QC_1 - QC_2 - kC_2^{1/2}V_2 = 0$$
$$V_1 = V_2$$

Solving the second mass balance for C_1,

$$C_1 = C_2 + kC_2^{1/2}\frac{V}{Q}$$

and substituting this into the first mass balance:

$$QC_0 - QC_2 + kC_2^{1/2}V - k\left(C_2 + kC_2^{1/2}\frac{V}{Q}\right)^{1/2} V = 0$$

$$\left(300\frac{\text{L}}{\text{h}}\right)\left(150\frac{\text{mg}}{\text{L}} - 7.5\frac{\text{mg}}{\text{L}}\right) + \left[0.05\frac{(\text{mg/L})^{1/2}}{\text{h}}\right]\left(7.5\frac{\text{mg}}{\text{L}}\right)^{1/2} V$$

$$- \left[0.05\frac{(\text{mg/L})^{1/2}}{\text{h}}\right]\left\{\left(7.5\frac{\text{mg}}{\text{L}} + \frac{\left[0.05\frac{(\text{mg/L})^{1/2}}{\text{h}}\right]\left(7.5\frac{\text{mg}}{\text{L}}\right)^{1/2} V}{300\frac{\text{L}}{\text{h}}}\right)^{1/2}\right\} V = 0$$

Reducing the above equation and eliminating the units,

$$42\,750 + 0.136\,9V - 0.05(7.5 + 0.000\,456\,4V)^{1/2}V = 0$$

where

V is in L

This equation can be solved by a nonlinear equation numerical method or by trial and error. V for a single reactor is found to be

$$V = 146 \times 10^3 \text{ L} = 146 \text{ m}^3 \text{ (5 155 ft}^3\text{)}$$

The total volume in this system is $2(146 \text{ m}^3) = 292 \text{ m}^3$ (10 311 ft^3).

A comparison table for the results is

Volume (m^3) Comparison for the Three Systems

PF	Single CM reactor	Two CM reactors in series
114	312	292

Examples 10.1 and 10–2 illustrate the kinetic behavior of CM compared to intermediate mixed and PF systems. When the same reaction model (except for zero-order reactions) applies regardless of the mixing regime, a PF system is always the most efficient. As the total reactor volume in a CM system is divided into more compartments, less total volume is required. Fundamentally, when a reaction process is in any way dependent on the concentration of a reactant, there will be a concentration gradient in a PF reactor but a complete or partially mixed reactor will dilute higher concentration elemental volumes with lower concentration elemental volumes. Thus, $\pm kC$ (or similar expressions) will be lower in portions of the complete or partially mixed reactor compared to a PF reactor.

10.1.4 Batch Processes

In a batch process, a reactor is filled and any required reagents are added to the reactor. If the reagents are added along with the influent, the reaction begins as the reactor is filling. Once the reactor is full, addition of influent is terminated. After this time the reactor contents are held for a specified time and then the reactor is drained. The process is repeated on a cyclic basis. To handle a continuous flow of water more than one batch reactor is needed and the series of reactors is staggered throughout the cycle phases.

Regardless of the degree of mixing when the reactor is full, all of the reactor contents are subject to the reaction conditions for the holding time until the reactor is drained. A batch process is similar to a PF reactor in that each volume of water resides in the reactor for the reaction time. The total time for processing a water in a batch reactor consists of the fill time, the reaction time, and the drain time. Depending on the reactor operation there may be reaction occurring during either or both of the fill and drain periods.

10.2 FLOW ANALYSIS OF CM AND PF REACTORS

The actual flow characteristics through a basin rarely achieve the desired ideal regime but basins are designed to approach one regime or the other depending on the process. A basin with an intermediate degree of mixing has dispersed flow (nonideal flow). As illustrated in Examples 10.1 and 10.2 the degree of mixing exerts a fundamental influence on the volume or time required for a given amount of reaction.

Tracers (dyes, electrolytes, and radioactive isotopes) are used to characterize the degree of mixing. The tracer must be conservative. A conservative tracer is soluble and does not participate in any reaction and it is not adsorbed or absorbed by the reactor or its contents. The tracer molecules are assumed to be moved about in the same manner as the water molecules and therefore their flow pattern will mimic the liquid flow pattern. One could consider that the tracer coats the water molecules to which it is added. If a conservative tracer is added at the inlet of a basin as one short impulse, the tracer will appear at the outlet over a period of time following a curve that reflects the degree of mixing as shown in Fig. 10.6. The tracer curves for CM and PF flows are derived in the following sections.

10.2.1 Analysis of Complete Mixed Reactors

Consider a reactor of volume, V, receiving a steady flow, Q, and assume ideal CM conditions exist within the reactor (see Fig. 10.4). At time, $t = 0$, the concentration of tracer in the vessel is 0 and immediately a known amount of conservative tracer is introduced into the reactor in a single slug with no tracer in the influent after this time. Will this be a steady state situation?

Setting up a mass balance for the tracer:

$$\text{In} - \text{Out} + \text{Generation} = \text{Accumulation}$$

$$QC_{\mathrm{I}} - QC + 0 = V\frac{dC}{dt}$$

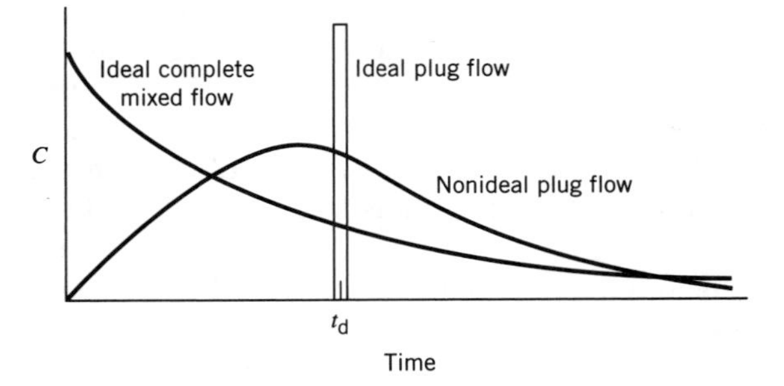

Figure 10.6 Effluent tracer concentrations for varying degrees of mixing.

where
C_I is influent tracer concentration
C is concentration of tracer in the reactor

Because the tracer is a conservative substance, the generation term is zero. The above equation can be arranged to

$$\frac{dC}{dt} + \frac{1}{t_d} C = \frac{1}{t_d} C_I \tag{10.26}$$

In Eq. (10.26) C_I is a forcing function that is variable in time and also the exit of tracer from the reactor will be variable in time. The equation is a linear first-order differential equation, readily solved by LaPlace transforms or other basic techniques. The LaPlace transforms (L[ ]) for a function and its derivative are

$$\mathrm{L}[C] = \int_0^\infty Ce^{-st}\,dt$$

$$\mathrm{L}[C'] = s\mathrm{L}[C] - C(0)$$

where
C' indicates the derivative
s is a constant

Transforming Eq. (10.26) yields

$$s\mathrm{L}[C] - C(0) + \frac{1}{t_d}\mathrm{L}[C] = \frac{1}{t_d}\mathrm{L}[C_I] \tag{10.27}$$

The initial condition is

$$C(0) = 0 \tag{10.28}$$

Substituting this into the above equation gives

$$s\mathrm{L}[C] + \frac{1}{t_d}\mathrm{L}[C] = \frac{1}{t_d}\mathrm{L}[C_I] \tag{10.29}$$

Solving Eq. (10.29) for L[C],

$$\mathrm{L}[C] = \frac{\frac{1}{t_d}\mathrm{L}[C_I]}{s + \frac{1}{t_d}} \tag{10.30}$$

The inverse transform of Eq. (10.30) gives $C(t)$.

$$C(t) = \mathrm{L}^{-1}\left[\frac{\frac{1}{t_d}\mathrm{L}[C_I]}{s + \frac{1}{t_d}}\right] \tag{10.31}$$

Before solving Eq. (10.31), the form of the forcing function, $C_I(t)$ needs to be known. Adding a slug of tracer is equivalent to imposing an impulse (at time t_1) defined by the Dirac delta function, $\delta(t - t_1)$. The proof of this follows.

In the time from t_1^- to t_1^+ the tracer concentration in the tank changes from 0 to C_0. Therefore,

$$Q \int_{t_1^-}^{t_1^+} C_I(t)\, dt = \int_{t_1^-}^{t_1^+} \frac{d[VC(t)]}{dt}\, dt = VC(t_1^+) - VC(t_1^-) \tag{10.32a}$$

V and Q are constant in this equation. Simplifying the equation,

$$\int_{t_1^-}^{t_1^+} C_I(t)\, dt = t_d[C(t_1^+) - C(t_1^-)] \tag{10.32b}$$

Differentiating each side of Eq. (10.32b),

$$C_I(t) = \frac{t_d[C(t_1^+) - C(t_1^-)]}{dt} = t_d \frac{C(t_1^+) - C(t_1^-)}{t_1^+ - t_1^-} = t_d \frac{dC}{dt}$$

Substituting for the concentrations at times t_1^+ and t_1^- results in the following expression for the derivative.

$$\frac{dC}{dt} = \frac{C_0 - 0}{t_1^+ - t_1^-} = \frac{C_0}{t_1^+ - t_1^-}$$

The Dirac delta is defined as

$$\int_{-\infty}^{\infty} \delta(t_1^+ - t_1^-)\, dt = \int_{t_1^-}^{t_1^+} \delta(t_1^+ - t_1^-)\, dt = 1$$

or

$$\delta(t - t_1) = \frac{1}{t_1^+ - t_1^-} \tag{10.33}$$

It is an ideal spike function with no width, infinite height, and an area of 1.

Using these definitions and letting the tracer be injected at time $t_1 = 0$,

$$\frac{dC}{dt} = C_0\delta(t) \qquad \text{and} \qquad C_I(t) = t_d C_0 \delta(t)$$

The LaPlace transform of $\delta(t - t_1)$ is

$$\text{L}[\delta(t - t_1)] = e^{-st_1}$$

The LaPlace transform for $C_I(t)$ is

$$\text{L}[t_d C_0 \delta(t)] = t_d C_0$$

Substituting this into Eq. (10.31) results in

$$C(t) = \text{L}^{-1}\left[\frac{C_0}{s + \dfrac{1}{t_d}}\right]$$

and

$$C(t) = C_0 e^{-t/t_d} \tag{10.34}$$

10.2.2 Simplified Analysis of Complete Mixed Reactors

The development given in Section 10.2.1 is the more rigorous analysis of a CM reactor. The same result can be obtained from a much more simplified approach. Equation (10.20) applies at all times. If the equation is applied at the instant after the slug input of tracer has entered the basin, the influent concentration, C_I is unchanging with a value of 0. The initial concentration of tracer in the basin can be determined from the known mass, M, of tracer input into the basin of volume, V.

$$C_0 = \frac{M}{V}$$

Therefore, taking the initial time at the instant immediately after the tracer has entered the basin, Eq. (10.20) becomes

$$-QC = V\frac{dC}{dt} \tag{10.35}$$

which is integrated between the following limits to determine the equation for the reactor and exit tracer concentration:

$$\int_{C_0}^{C} \frac{dC}{C} = -\frac{Q}{V}\int_0^t dt \tag{10.36}$$

The resulting equation is the same as Eq. (10.34).

$$C(t) = C_0 e^{-t/t_d}$$

■ Example 10.3 Tracer Exit Curve

What is the exit tracer concentration curve for a CM reactor of volume V that receives a steady flow containing a concentration of tracer, C_I, beginning at time 0? The initial concentrations of tracer in the reactor and influent were 0.

The mass balance expression is

$$QC_I - QC = V\frac{dC}{dt}$$

Rearranging the equation, noting $Q/V = 1/t_d$ and integrating,

$$\int_0^C \frac{dC}{C_I - C} = \frac{Q}{V}\int_0^t dt \Rightarrow \ln\left(\frac{C_I - C}{C_I}\right) = -\frac{t}{t_d}, \quad C = C_I(1 - e^{-t/t_d})$$

The tracer exit curve is shown in the accompanying figure. Ultimately the effluent tracer concentration equals the influent concentration because there is no transformation of the tracer.

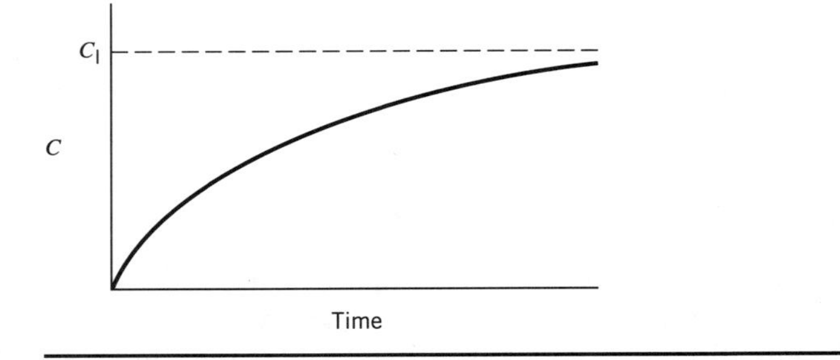

10.2.3 Fundamental Analysis of Plug Flow

The definition sketch used for the CM reactor in Fig. 10.4 will apply to the fundamental analysis of PF except that there is no mixer in a PF basin. A steady flow situation exists. Because there is no mixing allowed between adjacent planes in a PF reactor, it is expected that a slug of tracer will appear in the effluent at the time of exit of the plane containing the slug of tracer. A rigorous proof of this is offered next.

A slug of tracer is injected into the incoming flow at time $t = 0$. It is assumed that the tracer is completely distributed throughout the elemental volume (plane) that receives the tracer. It is also assumed that the plug of influent receiving the tracer also fully expands across the cross-section when it enters the reactor.

For a PF situation, the mass balance on tracer cannot be made over the whole reactor volume because the reactor is not homogeneous. Consider an elemental volume for any plug in the reactor (Fig. 10.7).

Setting up the mass balance for this elemental volume:

$$\text{In} - \text{Out} + \text{Generation} = \text{Accumulation} \tag{10.37}$$

$$QC - Q(C + dC) + 0 = dV \frac{d\left[C + \dfrac{dC}{2}\right]}{dt}$$

where

$C + \dfrac{dC}{2}$ is the average concentration in dV

Substituting $A\,dx$ for dV into the mass balance expression and expanding it,

$$-Q\,dC = A\,dx\left[\frac{dC}{dt} + \frac{1}{2}d\left(\frac{dC}{dt}\right)\right] = A\,dx\frac{dC}{dt}$$

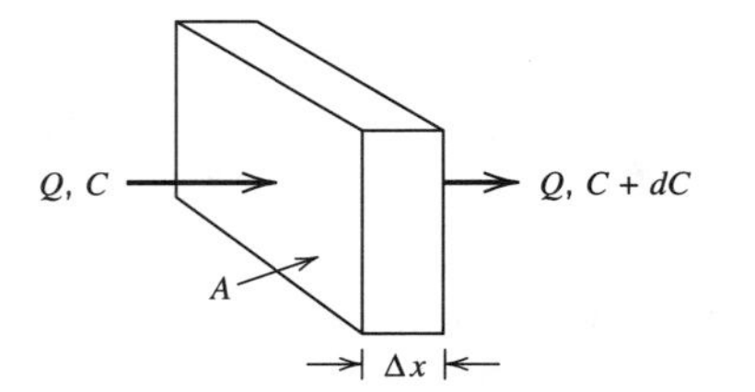

Figure 10.7 Elemental volume of PF reactor.

which reduces to

$$-\frac{Q}{A}\frac{dC}{dx} = \frac{dC}{dt} \tag{10.38}$$

because $d(dC/dt)$ is a second-order differential.

But

$$\frac{Q}{A} = v = \frac{dx}{dt}$$

Using this relation, the following relation is found.

$$-\frac{dC}{dt} = \frac{dC}{dt} \tag{10.39}$$

The only way for this equation to hold is for dC/dt to be zero. The PF situation can be solved rigorously using LaPlace transforms. Equation (10.38) is a statement of the continuity equation.

$$-v\frac{\partial C}{\partial x} = \frac{\partial C}{\partial t} \tag{10.40}$$

where

$C = C(x,t)$

Transforming Eq. (10.40) for the time domain:

$$\mathrm{L}[C(x,t)] = \int_0^\infty C(x,t)e^{-st}\,dt = Y(x,s)$$

where

Y is the transformed concentration.

Tracer is added at $t = t_1$ to water just entering the basin, but the initial condition is

$$C(x,t) = 0 \qquad \text{at } t = 0 \tag{10.41}$$

Let $C_t(x,t)$ be the derivative of C with respect to t.

$$\mathrm{L}[C_t(x,t)] = sY - C(x,0) = sY$$

Substituting this into the partial differential equation (Eq. 10.40):

$$sY = -v\frac{\partial Y}{\partial x} \tag{10.42}$$

Because Y is the transform with respect to t, the derivative with respect to x is not affected. Equation (10.42) is an ordinary first-order differential equation that is readily solved.

$$\begin{aligned}\frac{dY}{Y} &= -\frac{s}{v}dx \\ Y &= K_1 e^{-\frac{s}{v}x}\end{aligned} \tag{10.43}$$

To find K_1, the initial condition must be transformed because the differential equation has been transformed. The initial condition is

$$C(x,t) = C_0(t - t_1) \qquad \text{at } x = 0 \tag{10.44}$$

Proof that the initial condition constant is C_0 [with dimensions of (mass-time)/volume] and not t_d as for the CM case, is offered below. Transforming Eq. (10.44),

$$Y(0,s) = \mathrm{L}[C(0,t)] = C_0 e^{-st_1} \tag{10.45}$$

Using Eqs. (10.43) and (10.45),

$$\begin{aligned} Y &= C_0 e^{-st_1} = K_1 e^{-\frac{s}{v}(0)} \\ K_1 &= C_0 e^{-st_1} \end{aligned} \tag{10.46}$$

Substituting Eq. (10.46) into Eq. (10.43) yields

$$Y = C_0 \exp\left[-s\left(t_1 + \frac{x}{v}\right)\right] \tag{10.47}$$

Taking the inverse of Eq. (10.47) gives

$$C(x,t) = C_0 \delta\left(t - t_1 - \frac{x}{v}\right) \tag{10.48}$$

This equation has values

$$C(x,t) = C_0 \qquad \text{when } t = t_1 + \frac{x}{v}$$

$$C(x,t) = 0 \qquad \text{otherwise}$$

If $t_1 = 0$, Eq. (10.48) states that the concentration is zero everywhere except at the plane $x = vt$; that is, it describes a plane moving with velocity, v, through the vessel, which is ideal PF.

There is no kinetic advantage gained by providing a series of PF reactors as for CM reactors because of the nature of flow in a PF reactor. However, PF reactors in series can allow the staged introduction of chemicals or changing the physical conditions in each reactor to optimize treatment.

Derivation of the Forcing Function for PF Conditions

Similar to the CM case,

$$Q \int_{t_1^-}^{t_1^+} C_I(0,t)\, dt = \int_{t_1^-}^{t_1^+} \frac{d[dV(C)]}{dt} dt = dV[C(t_1^+) - C(t_1^-)] \tag{10.49}$$

In the case of PF, the concentration rises from 0 to C_0 in the plane entering at $t = t_1$. The volume of the plane is dV or $A\ dx$.

Therefore, Eq. (10.49) can be rearranged to

$$\int_{t_1^-}^{t_1^+} C_I(0,t)\, dt = \frac{dV}{Q}[C(0,t_1^+) - C(0,t_1^-)] = \frac{A\ dx}{Q}(C_0 - 0) \tag{10.50}$$

Differentiating Eq. (10.50) and proceeding as before,

$$C_I(0,t) = \frac{A\ dx}{Q} C_0 \delta(t - t_1)$$

Noting that

$$Q = \frac{A\,dx}{dt}$$

and substituting it into the above expression,

$$C_I(0,t) = C_0\,dt\,\delta(t - t_1) \tag{10.51}$$

Equation (10.51) is similar to the expression found for the CM vessel. If the spike concentration (in the plane) is to be C_0, dt must be set equal to 1 in the above expression. Alternatively, let M be the mass of tracer introduced in the time, dt.

$$Q \int_{t_1^-}^{t_1^+} C_I(0,t)\,dt = M \tag{10.52}$$

Dividing by Q and differentiating each side,

$$C_I(0,t) = \frac{M}{Q}\left(\frac{1}{t_1^+ - t_1^-}\right) = \frac{M}{Q}\,\delta(t - t_1) \tag{10.53}$$

but $Q = dV/dt$ and $C_0 = M/dV$; therefore,

$$C_I(0,t) = \frac{M}{dV}\,dt\,\delta(t - t_1) = C_0 dt\,\delta(t - t_1)$$

The quantity $C_0 dt$ is constant and has units of (mass-time)/volume. The actual observed concentration will depend on the length of time, dt, it takes to introduce the tracer, which will determine the length of the elemental volume. In an instant of time (1.0dt), the mass, M, must be introduced into the plane of width dx to achieve a concentration of C_0 in this plug.

10.2.4 Other Flow Irregularities: Dead Volume and Short Circuiting

Flow that passes through the reactor in a very short time relative to the average detention time is short circuited. Naturally in a CM reactor, a large quantity of the flow passes through the reactor in a time less than the average detention time. Flow that leaves the reactor before the detention time because of the hydraulic mixing regime is not short circuited. However, true short circuiting can exist in a CM reactor. Short-circuited flow is the excess flow that leaves the reactor in a relatively short period of time beyond the flow that leaves dictated by the hydraulic mixing. The tracer curve for a CM basin with some short circuiting is illustrated in Fig. 10.8.

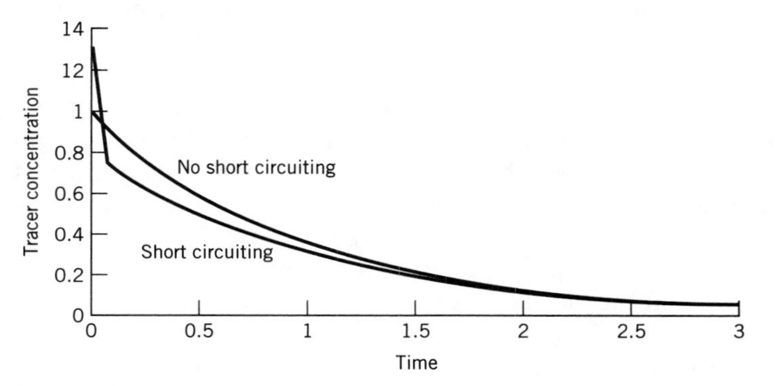

Figure 10.8 Short-circuited flow in a CM basin.

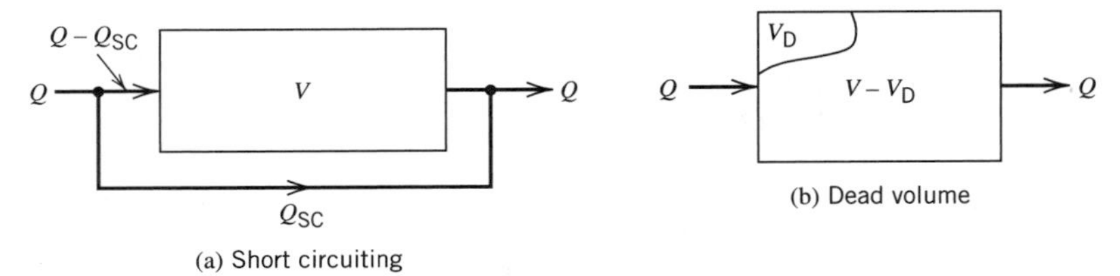

Figure 10.9 Short circuiting and dead volume.

Short circuiting can be caused by improper baffling, improper location of influent and effluent pipes, or more often by a density difference between the influent and reactor contents. (If ideal CM conditions existed, could the influent and effluent points be located next to each other?)

A conceptualization of short circuiting (Q_{SC}) is illustrated in Fig. 10.9a. The short-circuited flow occurs within the reactor and the viewer, of course, is only able to observe the total flow going into or exiting the reactor. However, short-circuited flow manifests its behavior according to the conceptualization.

Time dependent phenomena will be significantly affected by short-circuited flow with essentially no treatment of the portion of influent that is short circuited. The average detention time of the total flow in the basin remains the same and the remainder of the flow (the portion that is not short circuited) is treated for a longer period of time than the average detention time.

Short-circuited flow can be analyzed by only considering the effective flow $Q' = Q - Q_{SC}$ into the basin and mixing the untreated flow, Q_{SC}, with the treated portion.

Dead volume is a portion of the reactor volume that is unused. Reactor geometry, obstructions, inadequate mixing, and density currents can cause flow to bypass a portion of the reactor. The effective volume of the reactor is decreased from its physical volume, resulting in a decrease in the treatment time. Dead volume (V_D) is illustrated in Fig. 10.9b. Use a volume $V' = V - V_D$ as the reactor volume in this case.

10.2.5 Typical Flow Characteristics of Basins

The degree of mixing in a basin is assessed by the use of tracers as noted above. In the first analysis, tracer tests determine the flow regime and the appropriate model (CM or PF). When intermediate degrees of mixing occur, assuming that the basin is CM will always provide the most conservative estimate of basin performance (see Example 10.2), although when the flow regime approaches PF the factor of safety may be excessive. Conceptualizing the basin as a series of CM compartments can be used for a more accurate assessment of intermediate mixing situations. The analysis is similar to that presented in Section 10.1.3. This common approach implies certain assumptions about how mixing occurs but the approach is reasonable when water is the liquid (Lawler and Singer, 1993). Levenspiel (1972) gives a more thorough analysis and discussion of these issues. Lawler and Singer (1993) give an excellent application of the concepts that are beyond the scope of this presentation.

The following indexes based on the tracer exit curve for a slug input of tracer give a gross characterization of the flow situation. Flow that is truly short circuited and flow that leaves early because of mixing cannot be distinguished by these measures. Therefore short-circuited flow will refer to both mechanisms below.

The following definitions apply:

t_i time of first appearance of the tracer
t_d detention time
t_P time of the mode
t_g time of the mean
t_{10} time of passage of 10% of the tracer
t_{90} time of passage of 90% of the tracer

1. t_i/t_d
This is a measure of the degree of mixing. Smaller values indicate more mixing: 0 for ideal mixing and 1.0 for ideal PF. For a typical sedimentation basin it usually ranges between 0.2 and 0.3.
2. t_P/t_d
This indicates the average short circuiting, dead space, and effective tank volume. For a typical sedimentation basin it ranges from 0 to 0.8.
3. t_{90}/t_{10}
This provides a measure of the dispersion index. It is 1.0 for an ideal PF basin and 21.9 for an ideal mixed (CM) basin.
4. $1 - t_P/t_g$
This is referred to as the index of short circuiting. It is 0 for an ideal PF basin and 1.0 for an ideal CM basin.
5. $(t_{90} - t_P)/(t_P - t_{10})$
If this index is equal to 1.0 then the dispersion curve is symmetrical about the time t_P/t_d.

The spread of the concentration curve depends on the reactor dispersion number, which is given by D/vL, where D is the longitudinal or axial dispersion coefficient, v is the mean displacement velocity, and L is the length of the reactor.

10.2.6 Measurement of Dispersion

A better description of the mixing regime and other flow irregularities is obtained by considering the whole tracer curve. See Levenspiel (1972) for a more extensive discussion. It is essential that the tracer used to assess mixing, dead space, and short circuiting be conservative and not influence the flow pattern in a significant way by causing, for example, density currents. Common dyes that have been used are Rhodamine B or fluorescein. Radioactive tracers such as tritium are another option. Sodium chloride can also be employed but using high concentrations of this salt can result in a significant density difference between the water in the vessel and sodium chloride–containing influent. Sodium chloride has the advantage of being readily measured with a conductivity meter. Dextran blue (Jiminez et al., 1988) was found to be better than a number of tracers for submerged biological filters because of its high molecular weight, which minimized adsorption and diffusion in the biomass and packing.

Lithium is commonly used as a tracer in anaerobic reactors. At high concentrations lithium can inhibit activity in an anaerobic reactor (Anderson et al., 1991). Maximum lithium concentrations of 10–20 mg/L in a CM reactor are recommended; when PF conditions are suspected, lithium concentration delivered from a pulse input should not exceed 1.0 g/L to avoid inhibition.

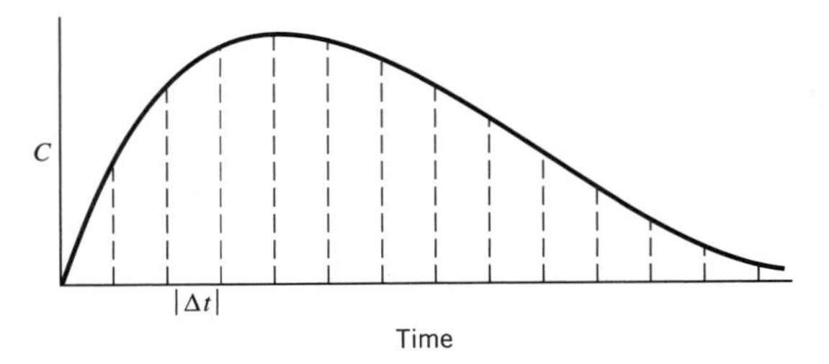

Figure 10.10 General tracer exit curve.

10.3 DETENTION TIME IN VESSELS

Perhaps surprisingly, the mixing patterns in vessels do not affect the average retention time of a particle of fluid in a vessel, as the developments in the following sections will demonstrate.

10.3.1 Average Detention Time

The average retention time of a particle of tracer in a vessel is determined by weighting the exiting tracer masses by the time each elemental mass has spent in the reactor. For the general tracer exit curve shown in Fig. 10.10, each elemental mass, m_i, stayed a time of t_i.

The average detention time is

$$\bar{t} = \sum \frac{m_i t_i}{\Sigma m_i} \tag{10.54}$$

$$m_i = QC_i \, \Delta t$$

$$\bar{t} = \frac{\Sigma t_i QC_i \, \Delta t_i}{\Sigma QC_i \, \Delta t_i} = \frac{Q}{M} \int tC(t) \, dt \tag{10.55}$$

where

M is the total mass of tracer injected

By applying the above equation to the tracer exit curves for CM and PF conditions, the average retention time of a particle of tracer in each case is seen to be V/Q. If an actual tracer exit curve does not produce this result, then dead space exists in the basin. The average retention of a particle of water is the same as the average retention time of a particle of conservative tracer.

10.3.2 The Effects of Flow Recycle on Detention Time

Many treatment operations incorporate the recycle of a portion of effluent flow into the basin to improve treatment. The activated sludge process, for example, returns a portion of the underflow from a clarifier that follows the reactor. This situation

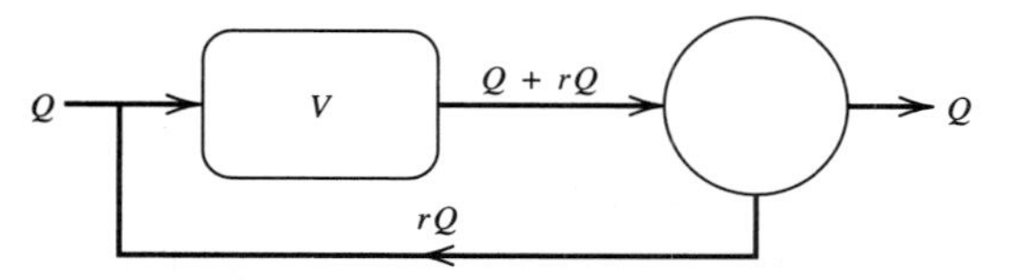

Figure 10.11 System with recycled flow.

will be examined for its effects on the average detention time of a particle of water in the reactor.

The sketch in Fig. 10.11 shows the flow pattern of the system. Water exits the primary basin and passes through a second basin, from which a portion of the water is recycled to the influent pipe of the first basin. The recycled flow and the influent are well mixed at their confluence. Although the recycled flow was taken from the second basin, as shown in this figure, the same analysis would apply if the recycled flow had been taken directly from the effluent from the first basin. For an ideal CM basin, the recycled flow can be returned at any point to the basin without consequence. Practically, addition and withdrawal points should be separated by as great a distance as possible to prevent short circuiting. For a basin with a PF hydraulic flow regime the recycled flow would normally be mixed with the fresh influent before the combined flow entered the reactor.

In Fig. 10.11, the ratio of recycled flow, Q_r, to influent flow is r. The volume of the primary basin is V.

$$r = \frac{Q_r}{Q} \tag{10.56}$$

To calculate the average residence time of water in the first basin, an elemental volume of water, Δq will be followed. Regardless of the mixing regime in the basin, the average residence time of the elemental volume during one pass, t_p, will be

$$t_\mathrm{p} = \frac{V}{Q + rQ} \tag{10.57}$$

A portion of the effluent from the primary basin is recycled. The recycled fraction of the elemental volume is

$$\frac{rQ}{Q + rQ}\Delta q \tag{10.58}$$

This recycled water again stays in the basin for an average time of t_p. A fraction of this fraction will be recycled:

$$\frac{rQ}{Q + rQ}\left(\frac{rQ}{Q + rQ}\Delta q\right) = \left(\frac{rQ}{Q + rQ}\right)^2 \Delta q \tag{10.59}$$

This process is repeated an infinite number of times, with the exponent increasing with each pass. The total retention time for a unit elemental volume (t_T) can now be formulated as

$$
\begin{aligned}
t_T &= \frac{V}{Q + rQ} + \left(\frac{rQ}{Q + rQ}\right)\left(\frac{V}{Q + rQ}\right) + \left(\frac{rQ}{Q + rQ}\right)^2\left(\frac{V}{Q + rQ}\right) + \cdots \\
&\quad + \left(\frac{rQ}{Q + rQ}\right)^i\left(\frac{V}{Q + rQ}\right) + \cdots \\
&= \frac{V}{Q}\left(\frac{1}{1 + r}\right)\left[1 + \frac{r}{1 + r} + \left(\frac{r}{1 + r}\right)^2 + \cdots \left(\frac{r}{1 + r}\right)^i + \cdots\right]
\end{aligned}
\tag{10.60}
$$

This series is recognized to be a geometric series with a general term,

$$
\begin{aligned}
x &= \frac{r}{1 + r} \\
t_T &= \frac{V}{Q}\left(\frac{1}{1 + r}\right)\sum_{n=1}^{\infty} x^{n-1} = \frac{V}{Q}\left(\frac{1}{1 + r}\right)\left(\frac{1}{1 - x}\right) = \frac{V}{Q}\left(\frac{1}{1 + r}\right)\left(\frac{r + 1}{1 + r - r}\right) = \frac{V}{Q} = t_d
\end{aligned}
\tag{10.61}
$$

This proves that the average retention in the basin is not affected by the amount of flow recycled (theoretically r may range from 0 to infinity). Recycle is an internal phenomenon within the system. In this case, the system consists of the two basins. If an enclosed line is drawn around the two basins, the input rate to the system is Q which is equal to the output rate. The volume of the first basin is V and the hydraulic retention time in the basin is simply V/Q. The same is true for the second basin. A volume of influent to the system stays in the second basin an average time calculated by dividing the basin's volume by the volumetric flow rate into the system.

This analysis should avoid confusion over the retention time for an individual pass (t_p) through a basin and the overall average retention time (t_d) in the basin. This distinction will have ramifications on the kinetics of treatment processes to be discussed in later chapters.

Recycling flow into a CM basin for the purpose of increasing *fluid* contact time is a fallacy. There are other reasons for recycling flow from a following basin into a preceding basin. One reason may be the possibility of short circuited flow in the primary basin. Recycling flow from the secondary basin would increase exposure of the fluid to reaction in the first basin.

10.3.3 The Effects of Recycle on Mixing

In a CM basin, recycling some of the effluent does not influence the degree of mixing because mixing is already at the extreme upper limit. It is sometimes not appreciated that recycle will increase the overall amount of mixing in a PF basin even though the flow regime remains PF during an individual pass of a plug through the basin. Recycle causes a portion of the effluent to be mixed with the influent.

For example, if a slug input of tracer produces a concentration of C_0 over a small time dt at a time t_d (or V/Q) from a PF reactor without recycle, recycle from the effluent with an r value of 1 would produce the tracer exit curve shown in Fig. 10.12 for the same input. In this case, one-half of the effluent flow is returned to the influent to the basin and the remaining portion exits along with the corresponding amount of tracer. Therefore, tracer appears at time intervals of $0.5t_d$ at concentrations that are 50% of the previous value. The tracer exit curve takes on characteristics of the curve expected from a CM basin. An infinite amount of recycle would produce a CM exit curve; a CM basin, in fact, has an infinite amount of internal recycle.

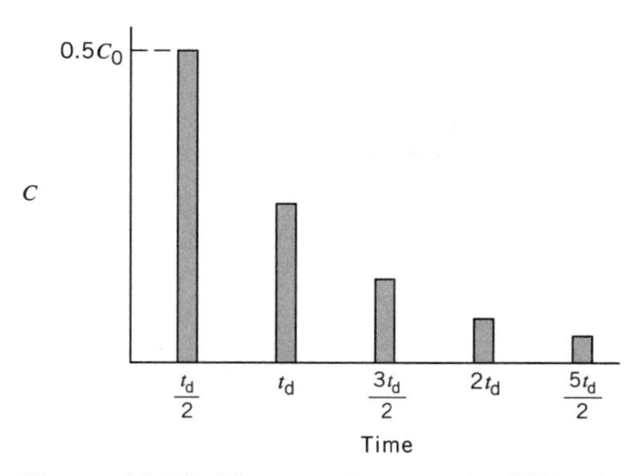

Figure 10.12 Tracer exit curve for PF basin with recycle.

10.4 FLOW AND QUALITY EQUALIZATION

Variation in flow rate and concentration adds variability to the treatment operation. More operator skill and careful monitoring and control are required to produce an acceptable treatment. Some of the treatment basins will have to be designed to accommodate the larger flows or additional basins will have to be provided, even though these flow rates will only exist for a part of the day, unless the excess flow is to be bypassed. Wastewater treatment operations are subject to variable flows and quality and providing an equalization basin will smooth variation in flow rate and quality, with resultant steady operation of the treatment operation. Water treatment plants are normally run at a more constant flow rate because of the storage reservoirs provided in the system.

For wastewater treatment plants it may be possible to use excess volume in large interceptor sewers entering the plant for equalization storage. USEPA (1974) recommends nightly or weekly drawdown of the interceptor system to flush out solids deposited during previous storage periods.

In wastewater treatment operations, when organic loadings vary by more than 4:1 based on 4 hour composite samples, equalization is recommended (Eckenfelder and Ford, 1970). An equalization basin affords the opportunity to pre-aerate the sewage before primary sedimentation, which results in improved settling of suspended solids (SS). To maintain aerobic conditions in a sewage equalization basin, air should be supplied in the range of 9–15 L air/m^3 of basin volume (USEPA, 1974).

A time-varying wastewater flow rate is shown in Fig. 10.13. The total volume, V, of water delivered over a 24-h period is

$$V = \int_0^{24} Q(t)\, dt \tag{10.62}$$

and the average flow rate, $\overline{Q}$, over this period is

$$\overline{Q} = \frac{1}{24}\int_0^{24} Q(t)\, dt \tag{10.63}$$

The volume of a basin required to store water to allow water to be delivered at a constant flow rate is computed by considering the volume of water delivered when the flow rate is higher than the average flow rate. In Fig. 10.13 the area above the average flow rate line and the corresponding areas below this line have been labeled

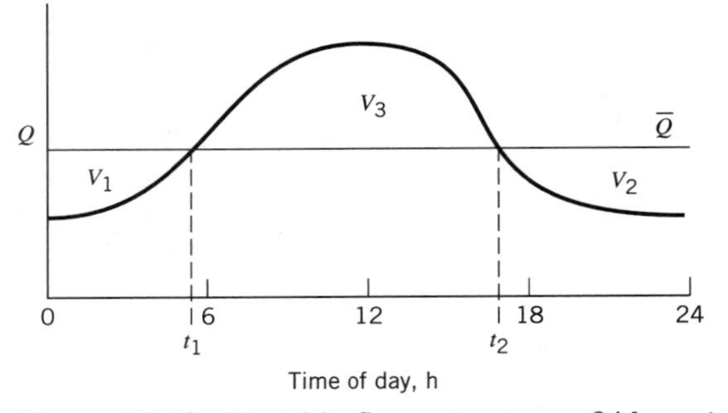

Figure 10.13 Variable flow rate over a 24-h period.

V_3, V_1, and V_2, respectively. It is clear that

$$V_3 = V_1 + V_2$$

The volume required for an equalization basin is V_3. Using a volume balance on the equalization basin, starting with an empty basin at time t_1,

$$\text{IN} - \text{OUT} = \text{ACCUMULATION}$$

$$\int_{t_1}^{t_2} Q\,dt - \overline{Q}(t_2 - t_1) = V_3$$

From time t_2 to 24 hours,

$$\int_{t_2}^{24} Q\,dt - \overline{Q}(24 - t_2) = -V_2$$

and for time 0 to t_1

$$\int_{0}^{t_1} Q\,dt - \overline{Q}(t_1 - 0) = -V_1$$

A more typical sewage flow variation has the shape of the curve shown in Fig. 10.14a. There are no differences to the approach outlined above if the double humped portion of the curve lies above the average flow rate line. However, if the valley extends below the average flow rate line, as shown in Fig. 10.14b, the basin is designed for a volume of

$$V_5 \qquad \text{if } V_4 < V_2$$

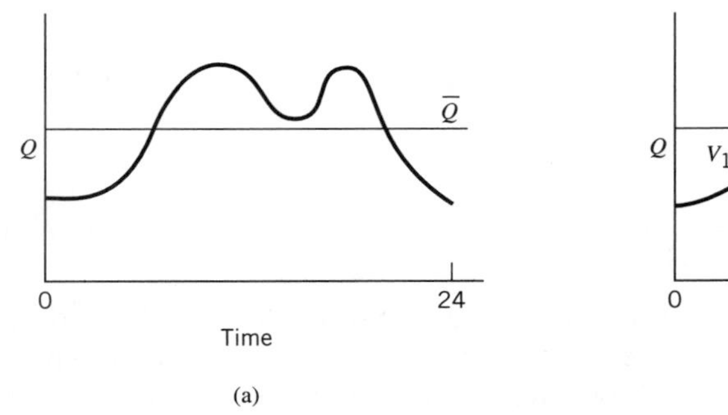

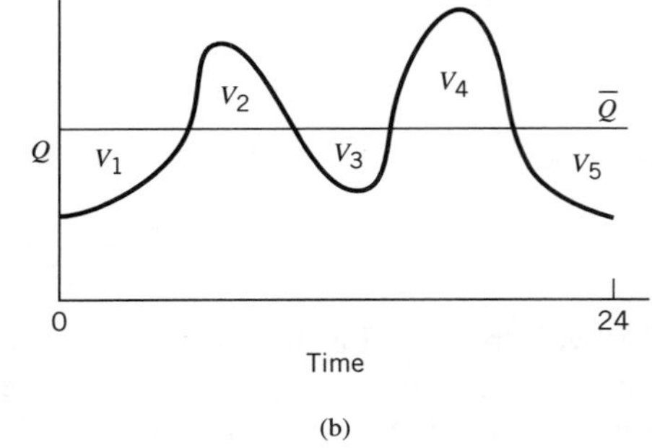

Figure 10.14 Diurnal variation of flow rate.

or

$$V_5 + V_4 - V_2 \qquad \text{if } V_4 > V_2$$

The quantity variation will be rectified by the approach outlined above but the quality variation of flow exiting from the equalization basin will not be totally smoothed. Quality or loading depends on both quantity and concentration. The following equation would be used to calculate the average loading rate.

$$\overline{QC} = \frac{1}{24}\int_0^{24} Q(t)C(t)\,dt \tag{10.64}$$

To design a basin to produce a constant loading would require continuous monitoring of the constituent's concentration in the influent. Mixing characteristics in the equalization basin would have to be known and a variable flow rate from the basin would be used to produce a constant loading. Variable concentrations of different components will make it impossible to achieve a uniform loading for all components. However there will be some mixing in the basin and loading variation is normally dampened considerably by pumping at a constant flow rate.

There will also be some equalization of flow rate and quality as the water passes through basins in the treatment plant because of mixing and the hydraulic characteristics of the channels. Most of the basins in a treatment plant discharge flow over a weir. This will have a considerable effect on reducing the flow variation (see Problem 30).

10.5 SYSTEM MATERIAL BALANCES

Individual operations are assembled together to form the complete treatment train. Chemical addition, sludge withdrawal, flow recycle, and chemical and biological transformations cause the liquid and solids flows to differ at various locations within the treatment plant. Sometimes split treatment schemes are used for one part of the treatment. In split treatment, a portion of the water is treated to a degree higher than the desired level, while the remainder bypasses the treatment. The two streams are combined to form the product water from the unit operation.

To characterize the system, the performance of each individual operation, including each of their inputs and outputs, must be known. The principles of mass balances discussed earlier apply to each basin. There is usually more than one basin or reactor for each unit operation. Some operations receive input from more than one other unit operation. For instance, a solids digester in a wastewater treatment operation typically has influent from both the primary and secondary clarifiers. The influent is a function of the sludge concentrations in each of the clarifiers. The number of unit operations and the number of cross-connections among them, as well as the variability of the influent quality and flow, increase the challenge of finally sizing the number of basins and characterizing the overall system inputs and outputs.

The approach to making the system balances is to lay out the complete plant in a logical sequence as done in Figs. 9.1–9.5. To facilitate making mass balances, a color scheme can be used or liquid flow lines can be made solid and solids flow lines can be dotted. Chemical feed lines can be dashed. The process performance characteristics of each unit operation (e.g., chemical concentration in the process, quantity of solids generated or destroyed) are indicated along with other a priori known characteristics such as the concentrations of commercially available agents. The influent flow and its quality characteristics are listed at the front of the plant. Then mass balances are made for each individual operation. All inputs and outputs are isolated. The inputs and transformations in an operation determine its outputs. This procedure is applied to each unit operation of the plant. The process is not necessarily straightforward because of return flows from later treatment operations to operations near the front of the treatment train. A trial and error approach may be required to achieve the correct balance.

For each individual unit operation, and for the system as a whole, the sum of liquid inputs must equal the sum of liquid outputs. Each basin or device is isolated and the symbolic representation of Eq. (10.1) is written for the unit. A box is drawn around the unit and all lines that intersect the box are included in the materials balance expression. Lines going into the box appear on the input side of the materials balance expression and those going out of the box appear on the output side. In addition any generation or destruction of a constituent by chemical or other means must be included in the materials balance. Mass balances for any constituent must be fully accounted for in the input–generation–output balances. When flow enters a Tee joint, the concentration of dissolved or SS does not change through the joint.

To perform a materials balance exercise it is not necessary to know the detention times or other design features of the units, which will be covered in later chapters. The process is best illustrated by example. In the example, the necessary performance characteristics of various processes will be supplied. However, it is assumed that principles covered in preceding chapters (particularly on chemical reaction stoichiometry) are to be applied as necessary. As the first exercise in designing a plant, typical performance of individual operations is assumed and modified by any data that are available. Later chapters cover the design of individual operations, after which this exercise can be repeated taking into account the specific design details to achieve the expected performance of each operation.

■ Example 10.4 Mass Balance

The problem is to find the flow rates and solids concentrations at all locations within an activated sludge wastewater treatment plant as illustrated in Fig. 9.4. A centrifuge will be used to dewater the digested sludge from the anaerobic digester instead of a vacuum filter. Influent data and unit operation performance information are given below. The process has been redrawn in Fig. 10.15 with symbols for flow and solids concentrations throughout the plant.

The definition of symbols in Fig. 10.15 is as follows. The list is arranged according to the sequence through the treatment train.

- S_0 influent BOD_5 concentration
- Q_0 influent flow rate
- X_0 influent SS concentration (excluding screenings and grit)
- Q_{Sc} flow rate after the screens
- Q_{Scw} volumetric removal rate of screenings
- X_{Scw} concentration (density) of screenings
- Q_g flow rate after the grit chamber
- Q_{gw} volumetric removal rate of grit

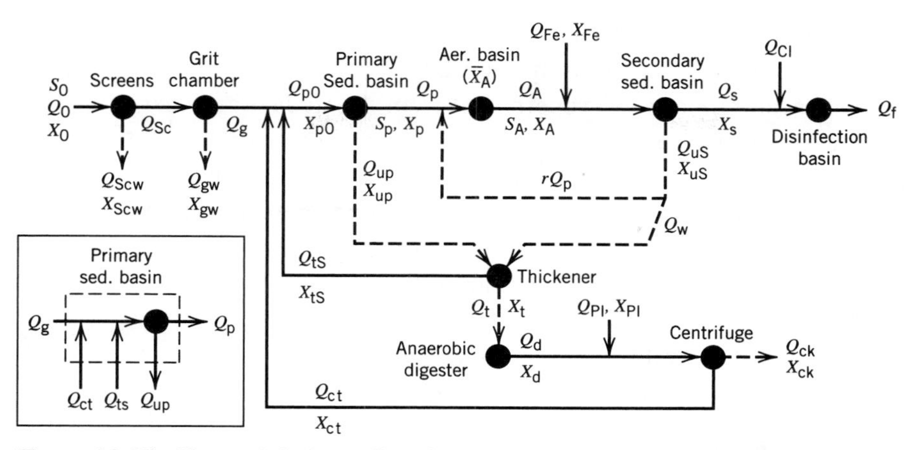

Figure 10.15 Materials balance flow diagram for an activated sludge wastewater treatment plant. The symbols are defined in Example 10.4.

X_{gw}	concentration (density) of grit collected in the grit chamber
Q_{p0}	total flow rate into the primary sedimentation basin
X_{p0}	SS concentration of influent to the primary sedimentation basin
Q_p	flow rate of clarified supernatant from the primary sedimentation basin
X_p	SS concentration of clarified supernatant from the primary sedimentation basin
S_p	BOD_5 concentration in the effluent from the primary sedimentation basin
Q_{up}	flow rate of underflow from the primary sedimentation basin
X_{up}	SS concentration in the underflow from the primary sedimentation basin
Q_A	flow rate of effluent from the aeration (activated sludge) basin
X_A	SS concentration in the effluent from the aeration basin
S_A	BOD_5 concentration in the effluent from the aeration basin
Q_S	flow rate of clarified supernatant from the secondary sedimentation basin
X_S	SS concentration of clarified supernatant from the secondary sedimentation basin
Q_{uS}	flow rate of underflow from the secondary sedimentation basin
X_{uS}	SS concentration in the underflow from the secondary sedimentation basin
r	ratio of recycled flow from underflow of secondary sedimentation basin to flow from primary sedimentation basin
Q_{Fe}	flow rate of chemical feed for ferric chloride
X_{Fe}	SS concentration in the ferric chloride chemical feed
Q_{Cl}	gas flow rate of chemical feed for chlorine
Q_f	flow rate from the disinfection basin
Q_w	flow rate of waste activated sludge from the underflow of the secondary sedimentation basin
Q_t	flow rate of underflow from the sludge thickener

X_t SS concentration in the underflow from the sludge thickener
Q_{tS} flow rate of supernatant from the sludge thickener
X_{tS} SS concentration in the supernatant from the sludge thickener
Q_d flow rate from the underflow of the anaerobic digester
X_d SS concentration in the underflow from the anaerobic digester
Q_{pl} flow rate of chemical feed for conditioning polymer for the centrifuge
X_{pl} SS concentration in the polymer chemical feed
Q_{ck} volumetric removal rate of cake from the centrifuge
X_{ck} SS concentration in the cake from the centrifuge
Q_{ct} flow rate of centrate from the centrifuge
X_{ct} SS concentration in the centrate from the centrifuge

The following information is given:

$Q_0 = 437$ ML/d (115.5 Mgal/d) $\quad S_0 = 320$ mg/L $\quad X_0 = 330$ mg/L

S_0 and X_0 do not include material that will be removed by the screens or in the grit chamber.

Screens: The quantity of screenings collected is 0.005 m³/1 000 m³ (67 ft³/100 Mgal)

Grit chamber: The quantity of grit collected is 0.008 m³/1 000 m³ (107 ft³/100 Mgal)

Primary sedimentation basin: The SS removal is 55%.
The BOD_5 removal is 35%.
The underflow concentration of SS is 3.8%.

Activated sludge aeration basin: Effluent soluble BOD_5 is 5 mg/L.
The net SS yield based on influent BOD_5 and soluble effluent BOD_5 is 0.65 mg SS produced/mg BOD_5 removed.
The concentration of SS in the aeration basin is 2 000 mg/L.

Secondary sedimentation basin: Ferric iron is being added to the influent to the clarifier at 5 mg/L as Fe. The coagulant is $FeCl_3$. The feed solution of $FeCl_3$ contains $FeCl_3$ at 20% by weight. The clarified overflow (effluent) from this basin contains 10 mg/L of SS.
The underflow SS concentration is 0.75%.

Thickener: The thickener captures 85% of the solids (i.e., 85% of influent solids appears in the underflow).
The underflow contains SS at 4.5%.

Disinfection: Chlorine gas is being added at a dose of 1.5 mg/L.

Anaerobic digester: SS reduction of 55% is achieved.

Centrifuge: 9 kg of polymer are being added to each tonne of solids. The feed solution of polymer contains polymer at 80 g/L.
The unit captures 97.5% of the solids.
The cake SS content is 32%.

For convenience it will be assumed that the specific gravity (s.g.) of SS is 1.00. The error associated with this assumption is small. (Chapter 20 discusses the equations necessary to correct this problem.) It is also assumed that there is no soluble BOD_5 generation in the anaerobic digester. Likewise, it is assumed that the thickener supernatant and centrifuge centrate do not contain soluble BOD. The only points where BOD_5 transformations occur are after the primary clarifier and the aeration basin. After consultation of later chapters, corrections for these assumptions and other refinements on the BOD and SS transformations can be made and incorporated into the material balance exercise but the procedure for determining flows and solids concentrations is as presented here.

The solids concentration in the aeration basin is assumed to be uniform (i.e., CM conditions exist).

Liquid, solids, and substrate balances can be written at each node in the system. For example, the box in Fig. 10.15 isolates the liquid flow streams for the primary clarifier and will be used below to construct the flow balance for this unit.

Proceeding first with the balances for liquid flows:

Screens

$$Q_0 = Q_{Sc} + Q_{Scw} \tag{i}$$

Grit Chamber

$$Q_{Sc} = Q_g + Q_{gw} \tag{ii}$$

Primary Sedimentation Basin

$$Q_g + Q_{ct} + Q_{tS} = Q_p + Q_{up} \tag{iii}$$

$$Q_g + Q_{ct} + Q_{tS} = Q_{p0} \tag{iv}$$

Activated Sludge Aeration Basin

$$Q_p + rQ_p = Q_A \tag{v}$$

Secondary Sedimentation Basin

$$Q_A + Q_{Fe} = Q_S + Q_{uS} \tag{vi}$$

$$Q_{uS} = Q_w + rQ_p \tag{vii}$$

Disinfection Basin

$$Q_S + Q_{Cl} = Q_f$$

Q_{Cl} is a gas flow rate and will not affect the liquid flow rate. Therefore, this equation simplifies to:

$$Q_S = Q_f \tag{viii}$$

Thickener

$$Q_{up} + Q_w = Q_{tS} + Q_t \tag{ix}$$

Anaerobic Digester

$$Q_t = Q_d \tag{x}$$

Centrifuge

$$Q_d + Q_{pl} = Q_{ck} + Q_{ct} \tag{xi}$$

In addition to these liquid balances for each individual unit, an overall liquid balance can be made for the process. This equation is not an independent relation from the above equations; it can be constructed from the individual unit balances.

There are 22 variables in the above equations. Q_{Cl} has been eliminated and only one of the variables, Q_0, has been specified. We need some more equations!

The solids and substrate balance equations are now formulated.

Screens

The volumetric collection rate of screenings has been given; therefore, screenings collected will be calculated on a volumetric basis with $X_{Scw} = 0.005\ m^3/1\,000\ m^3$. Note that screenings are not included in the influent solids concentration.

$$Q_0 X_{Scw} = Q_{Scw} \tag{xii}$$

Grit Chamber

As for screenings, grit is not included in the influent solids concentration and the concentration of grit removed is $X_{gw} = 0.008\ m^3/1\,000\ m^3$ on a volumetric basis.

$$Q_{Sc} X_{gw} = Q_{gw} \tag{xiii}$$

Primary Sedimentation Basin

Solids Balances

There has been no change in X_0 in the first two operations.

$$Q_g X_0 + Q_{ct} X_{ct} + Q_{tS} X_{tS} = Q_p X_p + Q_{up} X_{up} \tag{xiv}$$

$$Q_g X_0 + Q_{ct} X_{ct} + Q_{tS} X_{tS} = Q_{p0} X_{p0} \tag{xv}$$

The removal ratio for the primary clarifier has been specified; it will be represented by R_p.

$$\frac{Q_p X_p}{Q_{p0} X_{p0}} = R_p \tag{xvi}$$

Substrate Balance

There has been no change in the BOD_5 concentration in the first two operations. Also by assumption there is no BOD_5 in Q_{ct} or Q_{tS}. In the information supplied, the reduction factor for BOD_5 in the primary clarifier is $f_{pBOD} = 0.35$. The BOD_5 in the underflow from the primary clarifier is also ignored. Therefore the substrate balance is

$$Q_g S_0 = Q_p S_p + f_{pBOD} Q_g S_0 \tag{xvii}$$

The last term in Eq. (xvii) describes the removal of BOD_5 in the primary clarifier.

Activated Sludge Aeration Basin

Solids Balance

There is generation of solids in the aeration basin which will be termed ΔX in the mass balance.

$$Q_p X_p + r Q_p X_{uS} + \Delta X = Q_A X_A \tag{xviii}$$

Substrate Balance

The substrate balance for the aeration basin includes a term (ΔS) for removal of BOD_5.

$$Q_p S_p + rQ_p S_A = Q_A S_A + \Delta S \tag{xix}$$

To determine ΔX in Eq. (xviii), the net yield (designated Y) was supplied in the given information. The net yield is based on the removal of BOD_5 in the aeration basin.

$$\Delta X = Y \Delta S \tag{xx}$$

Secondary Sedimentation Basin

The dosing rate of $FeCl_3$ is defined as D_{Fe}.

$$(Q_{Fe} X_{Fe})/Q_A = D_{Fe} \tag{xxi}$$

The Fe^{3+} will be transformed by chemical reaction to $Fe(OH)_3$. X'_{Fe} is used to designate $Fe(OH)_3$.

$$Q_A X_A + Q_{Fe} X'_{Fe} = Q_S X_S + Q_{uS} X_{uS} \tag{xxii}$$

There is no change in solids concentration where Q_{uS} splits into Q_p and Q_w.

Disinfection Basin

There is no change in SS concentration in the disinfection basin because chlorine is a soluble gas. A very small amount of BOD will be oxidized by chlorine but it will assumed to be negligible.

Thickener

The capture ratio for the thickener will be defined as C_t.

$$Q_{up} X_{up} + Q_w X_{uS} = Q_{tS} X_{tS} + Q_t X_t \tag{xxiii}$$

$$\frac{Q_t X_t}{Q_w X_{uS} + Q_{up} X_{up}} = C_t \tag{xxiv}$$

Anaerobic Digester

Solids reduction in the anaerobic digester has been specified. A factor, f_{AD}, will be used to describe the amount of solids reduction in the digester.

$$Q_t X_t = Q_d X_d + f_{AD} Q_t X_t \tag{xxv}$$

Centrifuge

The dosing rate for polymer is defined as D_{pl}. The capture ratio for the centrifuge is defined as C_c.

$$(Q_{pl} X_{pl})/(Q_d X_d) = D_{pl} \tag{xxvi}$$

$$Q_d X_d + Q_{pl} X_{pl} = Q_{ck} X_{ck} + Q_{ct} X_{ct} \tag{xxvii}$$

$$\frac{Q_{ck} X_{ck}}{Q_d X_d + Q_{pl} X_{pl}} = C_c \tag{xxviii}$$

There are 21 new variables (including ΔX and ΔS) in the above 17 equations (Eqs. xii–xxviii). However, 13 of these variables are specified in the information supplied. There are a total of 28 unknowns and an equal number of equations; therefore the system is determined. As the equations are solved, keeping units with the parameters will provide a check to ensure that the calculations are correct. Also, it may be helpful to isolate two or more units in the system to solve for some of the variables.

The solids removals in the screens and grit chamber are straightforward. From Eq. (xii):

$$Q_{Scw} = (437 \times 10^3 \text{ m}^3/\text{d}) \times (0.005 \text{ m}^3/1\,000 \text{ m}^3) = 2.19 \text{ m}^3/\text{d}$$

In U.S. units:

$$Q_{Scw} = (115.5 \text{ Mgal/d}) \times (67 \text{ ft}^3/100 \text{ Mgal}) = 77 \text{ ft}^3/\text{d}$$

Using Eq. (i),

$$Q_{Sc} = 437 \times 10^3 \text{ m}^3/\text{d} - 2.19 \text{ m}^3/\text{d} = 437 \times 10^3 \text{ m}^3/\text{d}$$

From Eq. (xiii),

$$Q_{gw} = (437 \times 10^3 \text{ m}^3/\text{d}) \times (0.008 \text{ m}^3/1\,000 \text{ m}^3) = 3.50 \text{ m}^3/\text{d} \ (124 \text{ ft}^3/\text{d})$$

Using Eq. (ii),

$$Q_g = 437 \times 10^3 \text{ m}^3/\text{d} - 2.19 \text{ m}^3/\text{d} = 437 \times 10^3 \text{ m}^3/\text{d}$$

The equations can be solved algebraically or by making a few reasonable assumptions and iteratively solving the equations. The latter approach is chosen here.

The main liquid stream has its solids removed in the primary and secondary clarifiers. The substrate in the influent is removed in the aeration basin. These phenomena will be used to derive the first estimates of unknown flows. In the primary clarifier, 55% of SS are removed. The first estimate of the underflow from this clarifier can be based on the influent solids concentration. From Eqs. (xv) and (xvi) and ignoring Q_{ct} and Q_{tS} in Eq. (xv),

$$\text{SS removed in primary clarifier} = 0.55\left(437 \times 10^3 \frac{\text{m}^3}{\text{d}}\right)\left(330 \frac{\text{mg}}{\text{L}}\right)\left(\frac{1\,000 \text{ L}}{\text{m}^3}\right)\left(\frac{1 \text{ kg}}{10^6 \text{ mg}}\right)$$

$$= 79\,315 \text{ kg/d}$$

$$\text{In U.S. units: } = 0.55(115.5 \text{ Mgal/d})\left(330 \frac{\text{mg}}{\text{L}}\right)\left(\frac{8.34 \text{ L} - \text{lb}}{\text{Mgal} - \text{mg}}\right) = 174\,800 \text{ lb/d}$$

The solids content of the underflow from the clarifier is 3.8% or 38 g/L based on a s.g. of 1.00 for the solids. Q_{up} can be directly calculated by

$$Q_{up} = \frac{79\,315 \dfrac{\text{kg}}{\text{d}}}{0.038 \dfrac{\text{kg}}{\text{L}}}\left(\frac{1 \text{ m}^3}{1\,000 \text{ L}}\right) = 2\,087 \text{ m}^3/\text{d}$$

$$\text{in U.S. units: } Q_{up} = \frac{174\,800 \dfrac{\text{lb}}{\text{d}}}{0.038 \dfrac{\text{kg}}{\text{L}}}\left(\frac{0.454 \text{ kg}}{\text{lb}}\right)\left(\frac{1 \text{ gal}}{3.79 \text{ L}}\right)\left(\frac{1 \text{ Mgal}}{10^6 \text{ gal}}\right) = 0.551 \text{ Mgal/d}$$

Now Q_p can be calculated from Eq. (iii):

$$Q_p = Q_g - Q_{up} = 437 \times 10^3 \text{ m}^3/\text{d} - 2\,087 \text{ m}^3/\text{d} = 434.9 \times 10^3 \text{ m}^3/\text{d}$$

The effluent solids content in Q_p from the primary clarifier is

$$X_p = (1 - R_p)X_0 = (1 - 0.55)(330 \text{ mg/L}) = 149 \text{ mg/L}$$

The effluent substrate content in Q_p from the primary clarifier is found from Eq. (xvii),

$$S_p = \frac{(1 - f_p)Q_g S_0}{Q_p} = \frac{(1 - 0.35)(437 \times 10^3 \text{ m}^3/\text{d})(320 \text{ mg/L})}{434.9 \times 10^3 \text{ m}^3/\text{d}} = 209 \text{ mg/L}$$

In the equations for the aeration basin, the unknowns are r, ΔX, Q_A,, and ΔS. The only source of BOD_5 is the influent to the aeration basin and the final BOD_5 concentration from the aeration basin is given as 5 mg/L. Therefore, using Eqs. (v) and (xix),

$$\Delta S = Q_p(S_p - S_A) = \left(434.9 \times 10^3 \frac{\text{m}^3}{\text{d}}\right)\left(209 \frac{\text{mg}}{\text{L}} - 5 \frac{\text{mg}}{\text{L}}\right)\left(\frac{10^3 \text{ L}}{\text{m}^3}\right)\left(\frac{1 \text{ kg}}{10^6 \text{ mg}}\right)$$
$$= 88\,719 \text{ kg/d}$$

Now Eq. (xx) can be used to find ΔX. Then Eqs. (xviii) and (v) can be solved together to find r and Q_A. For ΔX,

$$\Delta X = Y\Delta S = 0.65(88\,719 \text{ kg/d}) = 57\,667 \text{ kg/d}$$

Combining Eqs. (xviii) and (v),

$$Q_p X_p + rQ_p X_{uS} + \Delta X = (1 + r)Q_p X_A$$

The underflow from the secondary clarifier, X_{uS} has been given as 7 500 mg/L (0.75%) and the aeration basin SS concentration (X_A) has been given as 2 000 mg/L.

$$r = \frac{Q_p(X_p - X_A) + \Delta X}{Q_p(X_A - X_{uS})}$$

$$= \frac{\left(434.9 \times 10^3 \frac{\text{m}^3}{\text{d}}\right)\left(149 \frac{\text{mg}}{\text{L}} - 2\,000 \frac{\text{mg}}{\text{L}}\right)\left(\frac{1\,000 \text{ L}}{\text{m}^3}\right)\left(\frac{1 \text{ kg}}{10^6 \text{ mg}}\right) + 57\,667 \frac{\text{kg}}{\text{d}}}{\left(434.9 \times 10^3 \frac{\text{m}^3}{\text{d}}\right)\left(2\,000 \frac{\text{mg}}{\text{L}} - 7\,500 \frac{\text{mg}}{\text{L}}\right)\left(\frac{1\,000 \text{ L}}{\text{m}^3}\right)\left(\frac{1 \text{ kg}}{10^6 \text{ mg}}\right)} = 0.31$$

From Eq. (v),

$$Q_A = (1 + \text{r})Q_p = (1 + 0.31)(434.9 \times 10^3 \text{ m}^3/\text{d}) = 570.8 \times 10^3 \text{ m}^3/\text{d}$$

The concentration of Fe^{3+} in the influent to the secondary clarifier is 5 mg/L. The amount of $FeCl_3$ added will be 14.5 mg/L. The concentration of $FeCl_3$ in the feed stream is 0.20 kg/L. Q_{Fe} can be determined from Eq. (xxi).

$$Q_{Fe} = \frac{D_{Fe}Q_A}{X_{Fe}} = \frac{\left(14.5 \frac{\text{mg}}{\text{L}}\right)\left(570.8 \times 10^3 \frac{\text{m}^3}{\text{d}}\right)}{0.20 \frac{\text{kg}}{\text{L}}\left(\frac{10^6 \text{ mg}}{\text{kg}}\right)} = 41.4 \text{ m}^3/\text{d}$$

$FeCl_3$ will react with alkalinity to form $Fe(OH)_3$. The $Fe(OH)_3$ formed is X'_{Fe} in Eq. (xxii), which can be considered to be part of the influent solids to the secondary clarifier. For every 14.5 mg/L of $FeCl_3$, 9.6 mg/L of $Fe(OH)_3$ will be formed.

Equations (vi) and (xxii) can be combined to solve for Q_S and Q_{uS}. X_S has been specified as 10 mg/L.

$$Q_A X_A + Q_{Fe} X'_{Fe} = (Q_A + Q_{Fe} - Q_{uS}) X_S + Q_{uS} X_{uS}$$

which can be rearranged to solve for Q_{uS}.

$$Q_{uS} = \frac{Q_A(X_A - X_S) + Q_{Fe}(X'_{Fe} - X_S)}{X_{uS} - X_S}$$

$$= \frac{\left(570.8 \times 10^3 \frac{m^3}{d}\right)\left(2\,000 \frac{mg}{L} - 10 \frac{mg}{L}\right) + \left(41.4 \frac{m^3}{d}\right)\left(9.6 \frac{mg}{L} - 10 \frac{mg}{L}\right)}{\left(7\,500 \frac{mg}{L} - 10 \frac{mg}{L}\right)}$$

$$= 151.7 \times 10^3 \text{ m}^3/\text{d}$$

From Eqs. (vi) and (vii),

$$Q_S = Q_A + Q_{Fe} - Q_{uS} = (570.8 \times 10^3 + 41.4 - 151.7 \times 10^3) \text{ m}^3/\text{d} = 419.1 \times 10^3 \text{ m}^3/\text{d}$$

$$Q_w = Q_{uS} - rQ_p = [151.7 \times 10^3 - (0.31)(434.9 \times 10^3)] \text{ m}^3/\text{d} = 16.9 \times 10^3 \text{ m}^3/\text{d}$$

The effluent flow from the treatment plant, Q_f is equal to Q_S from Eq. (viii) and is 419.1×10^3 m^3/d.

Because the dose of Cl_2 is 1.5 mg/L, the flow rate of Cl_2 gas can be calculated.

Applying Eq. (xxiv), $Q_t X_t$ may be found.

$$Q_t X_t = C_t(Q_w X_{uS} + Q_{up} X_{up})$$

$$= 0.85\left[\left(16.9 \times 10^3 \frac{m^3}{d}\right)\left(7\,500 \frac{mg}{L}\right) + \left(2\,087 \frac{m^3}{d}\right)\left(38\,000 \frac{mg}{L}\right)\right]\left(\frac{10^3 \text{ L}}{1 \text{ m}^3}\right)\left(\frac{1 \text{ kg}}{10^6 \text{ mg}}\right)$$

$$= 175.1 \times 10^3 \text{ kg/d}$$

The thickener underflow is determined using X_t = 0.045 kg/L (4.5%).

$$Q_t = \frac{Q_t X_t}{X_t} = \frac{\left(175.1 \times 10^3 \frac{kg}{d}\right)\left(\frac{1 \text{ m}^3}{1\,000 \text{ L}}\right)}{0.045 \frac{kg}{L}} = 3\,891 \text{ m}^3/\text{d}$$

Q_t is also the flow rate from the anaerobic digester according to Eq. (x).

The supernatant flow from the thickener is (Eq. ix),

$$Q_{tS} = Q_{up} + Q_w - Q_t = (2\,087 + 16.9 \times 10^3 - 3\,891) \text{ m}^3/\text{d} = 15.1 \times 10^3 \text{ m}^3/\text{d}$$

The concentration of solids in the supernatant is (Eq. xxiii):

$$X_{tS} = \frac{Q_{up} X_{up} + Q_w X_{uS} - Q_t X_t}{Q_{tS}}$$

$$= \frac{\left(2\,087 \frac{m^3}{d}\right)\left(38\,000 \frac{mg}{L}\right) + \left(16.9 \times 10^3 \frac{m^3}{d}\right)\left(7\,500 \frac{mg}{L}\right) - \left(3\,891 \frac{m^3}{d}\right)\left(45\,000 \frac{mg}{L}\right)}{15.1 \times 10^3 \frac{m^3}{d}}$$

$$= 2\,050 \text{ mg/L}$$

The anaerobic digester's output loading is (Eq. xxv):

$$Q_d X_d = (1 - f_{AD}) Q_t X_t = (1 - 0.55)(175.1 \times 10^3 \text{ kg/d}) = 78\,795 \text{ kg/d}$$

The concentration of solids in the effluent from the digester is

$$X_d = Q_d X_d / Q_d = (78\,795 \text{ kg/d})/(3\,891 \text{ m}^3\text{/d}) = 20.3 \text{ kg/m}^3 = 20.3 \text{ g/L}$$

The polymer is applied at 9 kg/tonne of solids (D_p). The application rate of polymer (A_{pl}) to be added to the flow rate of 392 m³/d from the digester containing 79 380 kg of solids is

$$A_{pl} = \left(78\,795 \frac{\text{kg}}{\text{d}}\right)\left(\frac{1 \text{ tonne}}{1\,000 \text{ kg}}\right)\left(\frac{9 \text{ kg}}{\text{tonne}}\right) = 709 \text{ kg/d}$$

The polymer feed flow rate is (Eq. xxvi),

$$Q_{pl} = \frac{D_{pl} Q_d X_d}{X_{pl}} = \frac{\left(9 \frac{\text{kg}}{\text{tonne}}\right)\left(3\,891 \frac{\text{m}^3}{\text{d}}\right)\left(20.3 \frac{\text{kg}}{\text{m}^3}\right)\left(\frac{1 \text{ tonne}}{1\,000 \text{ kg}}\right)}{80 \frac{\text{g}}{\text{L}}\left(\frac{1\,000 \text{ L}}{\text{m}^3}\right)\left(\frac{1 \text{ kg}}{1\,000 \text{ g}}\right)} = 8.9 \text{ m}^3\text{/d}$$

Equation (xxviii) is used to find the centrifuge cake output loading rate, $Q_{ck} X_{ck}$.

$$\begin{aligned} Q_{ck} X_{ck} &= C_c (Q_d X_d + Q_{pl} X_{pl}) \\ &= 0.975 \left[\left(3\,891 \frac{\text{m}^3}{\text{d}}\right)\left(20.3 \frac{\text{g}}{\text{L}}\right) + \left(8.9 \frac{\text{m}^3}{\text{d}}\right)\left(80 \frac{\text{g}}{\text{L}}\right)\right]\left(\frac{1\,000 \text{ L}}{\text{m}^3}\right)\left(\frac{1 \text{ kg}}{1\,000 \text{ g}}\right) \\ &= 77\,707 \text{ kg/d} \end{aligned}$$

The volumetric rate of cake production is

$$Q_{ck} = \frac{Q_{ck} X_{ck}}{X_{ck}} = \frac{77\,707 \frac{\text{kg}}{\text{d}}}{0.32 \frac{\text{kg}}{\text{L}}}\left(\frac{1 \text{ m}^3}{1\,000 \text{ L}}\right) = 243 \text{ m}^3\text{/d}$$

Using Eq. (xi), Q_{ct} can be determined as

$$Q_{ct} = Q_d + Q_{pl} - Q_{ck} = (3\,891 + 8.9 - 243) \text{ m}^3\text{/d} = 3\,657 \text{ m}^3\text{/d}$$

Equation (xxvii) can be used to find the concentration of solids in Q_{ct}.

$$\begin{aligned} X_{ct} &= \frac{Q_d X_d + Q_{pl} X_{pl} - Q_{ck} X_{ck}}{Q_{ct}} \\ &= \frac{\left(3\,891 \frac{\text{m}^3}{\text{d}}\right)\left(20.3 \frac{\text{kg}}{\text{m}^3}\right) + \left(8.9 \frac{\text{m}^3}{\text{d}}\right)\left(80 \frac{\text{kg}}{\text{m}^3}\right) - \left(243 \frac{\text{m}^3}{\text{d}}\right)\left(320 \frac{\text{kg}}{\text{m}^3}\right)}{3\,657 \frac{\text{m}^3}{\text{d}}} \\ &= 0.53 \text{ kg/m}^3 = 530 \text{ mg/L} \end{aligned}$$

The first estimates of all quantities are now complete. At least another iteration should be performed using the values obtained for thickener and centrifuge supernatant return flows and their respective SS concentrations to obtain better accuracy. This is left as an exercise for the student (see Problem 33).

Flow m^3/d	SS mg/L	Flow m^3/d	SS mg/L	Flow m^3/d	SS mg/L
Q_0 437×10^3	X_0 330	Q_p 435×10^3	X_p 149	Q_S 419×10^3	X_S 10
Q_{Scw} 2.19	X_{Scw}[a] 0.005	Q_{ts} 15.3×10^3	X_{tS} 2 036	Q_w 16.9×10^3	
Q_{Sc} 437×10^3		Q_{ct} 3 679	X_{ct} 580	Q_t 3 891	X_t 45[b]
Q_{gw} 3.50	X_{gw}[a] 0.008	Q_{up} 2 087	X_{up} 38[b]	Q_d 3 891	X_d 20.3[b]
Q_g 437×10^3		Q_A 571×10^3	X_A 2 000	Q_{ck} 243	X_{ck} 320[b]
Q_{p0} 437×10^3	X_{p0} 330	Q_{uS} 152×10^3	X_{uS} 7 500	Q_{Fe} 41.4	X_{Fe} 200[b]

[a]In $m^3/1\,000\ m^3$.
[b]In g/L.

Examine the flow rates and solids concentrations around the system that are given in the table above. It is seen that the underflows from the clarifiers are small compared to the main liquid stream. In the first iteration, it is often a good assumption to ignore the adjustments in the main liquid flow resulting from these reductions.

QUESTIONS AND PROBLEMS

1. Would mixing in only the vertical and lateral directions negate PF conditions? Would it negate CM conditions? Explain.
2. Why or why not can mass balances be made over the whole basin volume for PF and CM basins?
3. What are the differences among bulk flow, advection, diffusion, and dispersion? Would suspended solids be subject to all phenomena?
4. What volume of a PF reactor is required to remove 90% of a substance that decays according to first-order kinetics with a rate constant of 0.05 d^{-1}? The flow rate is 395 m^3/d (104 350 gal/d).
5. Compare the total reactor volume required for 90% removal of a substance that decays according to first-order kinetics with a rate constant of 0.05 d^{-1} in a CM system that contains a single reactor against a system that contains three CM reactors of equal volume in series.
6. (a) What are the relative volume requirements for a CM reactor compared to a PF reactor if first-order kinetics apply and treatment efficiencies of 50, 75, 90, 95, and 99% are required? The same rate constant and flow rate apply to each system.
   (b) Perform the same exercise if second-order kinetics apply.
7. What are the characteristics of an ideal tracer?
8. Using the results from the exit concentration curve for a slug input of dye into a CM reactor, prove that the average residence (detention) time of a particle of dye, and thus of a particle of water, in a CM unit is V/Q for a reactor with a volume, V, receiving a steady input flow of Q.
9. An experimenter wishes to conduct steady state studies on a continuous flow CM reactor. It is generally thought that data can be safely taken after at least 95% of the initial material that has been in the reactor at startup is washed out. How many detention times must this experimenter wait until data should be taken?
10. A reactor is operated in the following manner: (i) the reactor fills with influent; (ii) a chemical agent is added and mixing is applied to the basin; and (iii) after the reaction is completed, mixing is terminated and the basin is emptied to begin another cycle. Would this reaction be modeled with CM or PF kinetics?
11. What is the expression describing the concentration of tracer in the effluent from a CM

reactor that receives a steady concentration, C_I, of conservative tracer in a flow, Q for a finite period of time t_0 to t_1? The reactor initially had no tracer in it.

12. Why can recycle flow be delivered to any point in a CM basin? Would the same principle apply to a PF basin?

13. Prove that the ratio t_{90}/t_{10} should be 21.9 for an ideal CM basin receiving a slug input of tracer.

14. A CM reactor receives influent containing 10.0 mg/L of tracer for 2 h. Then tracer addition is terminated but the flow remains steady. The volume of the reactor is 10 L and the flow rate is 2 L/h. What is the concentration of tracer in the reactor 1 h after tracer addition is terminated? The reactor had an initial concentration of tracer of 1.0 mg/L when tracer addition commenced.

15. In Example 10.3 it is shown that the curve describing concentration of tracer in the effluent from a CM reactor receiving a constant concentration of tracer in the influent starts at a concentration of zero. However, when a slug input of tracer is added to a CM reactor, the initial concentration of tracer in the reactor and effluent is C_0 (see Eq. 10.34). In both instances the concentration of tracer in the reactor before tracer addition was zero. Explain this seemingly contradictory situation.

16. Stormwater runoff is often treated in flow-through ponds that approach CM conditions when the runoff event is occurring. Some treatment occurs during the runoff event but the bulk of treatment occurs by retention of the stormwater in the pond until the next event occurs. A pond with a volume of 15 000 m^3 (530 000 ft^3) receives a runoff volume of 10 000 m^3 (2.64×10^6 gal) delivered uniformly over the 4 h duration of the event. What fraction of the influent 10 000 m^3 (2.64×10^6 gal) volume exits the pond during the event and what is the fraction that remains until the next event occurs? The pond was full at the beginning and end of the event.

17. If you have studied convolution in hydrology (or elsewhere), apply it to determine the effluent tracer concentration change with time from a CM basin receiving pulse inputs with a concentration of 100 μg/L of tracer lasting 1 min and followed by a 4 min interval until the next pulse. The flow is 550 m^3/d and the detention time in the basin is 6 h. Assume the pulses are impulses occurring over 5 min. (Is this reasonable?) Use discrete convolution to find the effluent tracer concentration immediately after the sixth pulse.

18. If short-circuited flow is defined as the flow that stays in the basin for less than 5% of the average detention time, how much flow is "naturally" short circuited in an ideal CM basin?

19. What portion of a CM reactor is occupied by water that has stayed in the reactor for more than 4 detention times?

20. What is the average retention time of a particle of water in a PF reactor with a portion of the reactor effluent directly returned to the front of the basin? Draw the exit tracer curves for a PF basin with a volume of V and an influent flow of Q that has effluent recycle with r values of 0.5 and 2. Plot mass of tracer exiting versus t/t_d.

21. (a) Compare the effluent tracer concentration from the final reactor of two CM reactors each with volume $V/2$ in series with the effluent tracer concentration from one CM reactor of volume V. A slug input of tracer was used and the same mass of tracer was injected into each system. For both systems plot C/C_0 against t/t_d, where C_0 and t_d apply to the single-reactor system. Travel time between reactors is negligible.
(b) Perform the same exercise for three reactors in series, each with volume $V/3$.

22. As more CM reactors in series are provided, the system becomes more "PF." For a substance that decays according to first-order kinetics, what is the equivalent PF rate constant that would be used to describe treatment in a series of three CM reactors in terms of Q, V (the total volume in the system), and the rate constant observed for the CM system?

23. What is the effluent concentration of a substance that enters the first reactor in a series of three CM reactors at a concentration of 560 mg/L? Each reactor has a volume of 80

m^3 and the flow rate is 150 m^3/h. The substance decays according to second-order kinetics with a rate constant of 0.063 L/mg/h.

24. An industry uses a special cleaning compound to wash some equipment every 8 h. The cleaning compound contains a water-soluble substance that is inhibitory to their biological treatment process above a threshold concentration. The biological treatment process has a CM reactor without recycle. The reactor volume is 50 m^3 (13 210 gal) and the influent flow rate is steady at 10 m^3/h (2 641 gal/h). The average amount of the inhibitory compound used in each washing and discharged to the waste stream is 75 g. Assuming that the inhibitory compound is delivered in a slug, what are the minimum and maximum concentrations of this compound in the biological reactor after a large number of cycles?

25. In Eq. 10.4f, how is time related to distance?

26. A CM basin with a volume of 475 m^3 receives a flow of 2 000 m^3/d. The basin has a dead volume of 42 m^3. Find the effluent tracer curve if a mass of 5 g of tracer is added instantaneously to the influent. Also plot the effluent tracer curve for the reactor without dead volume.

27. Does moving the location of dead volume change its effect on the performance of (a) a CM reactor (b) a PF reactor?

28. (a) For the flow rate data given in the table below, what is the minimum volume required for an equalization basin to produce a constant flow rate? What is the average hydraulic residence time in the equalization basin?

Time, h	0000	0100	0200	0300	0400	0500	0600	0700	0800
Q, m^3/s	0.11	0.15	0.20	0.26	0.30	0.34	0.36	0.37	0.37
Time, h	0900	1000	1100	1200	1300	1400	1500	1600	1700
Q, m^3/s	0.35	0.31	0.26	0.25	0.26	0.29	0.34	0.40	0.46
Time, h	1800	1900	2000	2100	2200	2300	2400		
Q, m^3/s	0.49	0.49	0.46	0.31	0.26	0.20	0.11		

(b) If the concentration of a constituent in the influent to the equalization basin is constant over the 24-h period, will the load of the constituent from the basin be constant? If the concentration of the constituent in the basin influent varies over the 24-h period, will the basin produce a constant loading rate?

29. A basin of volume V receives a steady flow Q but the concentration of a nonremovable component in the influent varies according to

$$C = C_1 + C_2 \sin \omega t \qquad (C_1,\ C_2,\ \text{and } \omega \text{ are constants})$$

What is the expression for effluent concentration of this component if the basin is (a) CM; (b) PF?

30. Flow into a basin follows the equation

$$Q = 7\,500 + 3\,900 \sin 0.2618t$$

where

Q is in m^3/d and t is in h

(a) Find the discharge curve over a 24 h period using the above equation if the inflow is into a rectangular basin with a length:width ratio of 3:1 and overflow straight edge weir that has a length that is twice the width of the basin. The plan area of the basin is 166 m^2 and the discharge coefficient for the weir is 0.62. Compare the ratio of the maximum and minimum values of the inflow to the ratio of the maximum and minimum values of the outflow. (Hint: Both parts a and b will require a numerical solution.)

(b) Perform the same exercise for a circular basin with the same plan area and an overflow straight edge weir extending around the periphery of the basin.

31. An engineer wishes to evaluate the mean detention time of stormwater in a stormwater collection pond. The engineer has written a versatile computer program to convert a rainfall record over time into runoff, which is routed into a pond. The treatment algorithm for the pond allows the engineer to simulate varying degrees of mixing in the pond and any type of reaction. The pond to be evaluated is always full and water exits the pond by means of an overflow weir. Flow into the pond is highly variable over any period of time because of the intermittent, variable nature of rainfall. To evaluate the residence time distribution (and mean detention time) of stormwater runoff in the pond the engineer chooses to run the program with a 20-year historical rainfall record that will provide ample runoff flow variation. The engineer will input into the runoff a constant concentration of a substance that will be decayed by a first-order reaction in the pond. The average effluent concentration, C_i, of the substance in each 1 000 m^3 of runoff will be calculated by the program. Then using the equation

$$C_i = C_0 e^{-kt}$$

(where k is known and constant, C_0 is the influent concentration, and C_i is the concentration of the ith pond effluent 1 000 m^3 volume), the time that each volume has stayed in the pond will be calculated by

$$t = -\frac{1}{k} \ln \frac{C_i}{C_0}$$

Statistical analyses of the times will provide the mean, standard deviation, and cumulative residence time distribution of runoff volume. What is the error in the engineer's reasoning? What could be done to obtain the correct result? It is assumed that the pond is not mixed between events. Does it matter if the pond is CM or PF during a runoff event?

32. A system with three treatment units has an influent flow, Q, of 3 800 m^3/d. The volumes of the first and second basins are 950 and 500 m^3, respectively. Flow from the second basin is recycled at a rate of $0.50Q$ to the first basin, Also, 20% of the flow coming into the second basin is sent to the third basin. Seventy percent of the flow from the third basin is returned to the first basin. What are the detention times in the first and second basins based on the total flow entering them? What volume is required in the third basin to provide a detention time of 15 d based on the total flow entering it?

33. Perform another iteration in Example 10.4 to find all the quantities asked for in the example.

34. Assuming that the solids in the underflow from the secondary sedimentation basin split in proportion to the flow, what is the concentration of Fe in the aeration basin for Example 10.4? Use the flows in the table at the end of the example. Assume that the Fe concentrations in the supernatant from the thickener and centrate from the centrifuge are insignificant and that all Fe forms precipitate and settles in the secondary clarifier. All Fe in the influent to the process is removed in the primary clarifier.

35. Perform a mass balance for a water treatment plant with the treatment sequence indicated in Fig. 9.1. Ignore screenings removed from the bar rack and traveling screen. Use the following information.
influent flow—235 ML/d influent SS—5.0 mg/L
Assume that the s.g. of all chemical and other solids is 1.00.
Alum is the coagulant added to achieve a concentration of 22 mg/L $Al_2(SO_4)_3$. The concentration of the alum feed is 10% $Al_2(SO_4)_3 \cdot 14H_2O$. There is sufficient alkalinity in the water to react with the alum and form the precipitate $Al(OH)_3$, which does not hydrate.
The sedimentation basin removes 80% of the SS.
The underflow from the sedimentation basin contains SS at 0.55%.
The remainder of the suspended particles are removed in the rapid sand filter.
The filter backwash uses 2% of the product water from the filter. The filter backwash water is returned ahead of the mixing unit. Assume that the backwash flow rate is continuous.
Chlorine is added at 1.0 mg/L to the effluent before discharge to the clear well.

Fluoride is added as NaF from a dry feeder to achieve a concentration of 1.0 mg F^-/L in the effluent before discharge to the clear well.
The vacuum filter achieves a cake content of 35% solids by weight. The vacuum captures 95% of the influent solids. A polymer is added to the influent sludge at a dose rate of 5 kg polymer/tonne of solids. The concentration of polymer in the polymer feed is 90 g/L.

Determine the flows and suspended solids concentrations before and after each unit. Also determine the mass feed rates of chlorine and sodium fluoride to be added on a daily basis.

36. What is the error in development of Eq. (10.39)?

KEY REFERENCE

LEVENSPIEL, O. (1972), *Chemical Reaction Engineering,* John Wiley & Sons, Toronto.

REFERENCES

ANDERSON, G. K., C. M. M. CAMPOS, C. A. L. CHERNICHARO, AND L. C. SMITH (1991), "Evaluation of the Inhibitory Effects of Lithium When Used as a Tracer for Anaerobic Digesters," *Water Research,* 25, 7, pp. 755–760.

ECKENFELDER, W. W., JR. AND D. L. FORD (1970), *Water Pollution Control,* Pemberton Press, New York.

JIMINEZ, B., A. NOYOLA, B. CAPDEVILLE, M. ROUSTAN, AND G. FAUP (1988), "Dextran Blue Colorant as a Reliable Tracer in Submerged Filters," *Water Research,* 22, 10, pp. 1253–1257.

LAWLER, D. F. AND P. C. SINGER (1993), "Analyzing Disinfection Kinetics and Reactor Design: A Conceptual Approach Versus the SWTR," *J. American Water Works Association,* 85, 11, pp. 67–76.

USEPA (1974), *Process Design Manual for Upgrading Existing Wastewater Treatment Plants,* Center for Environmental Research Information, Cincinnati, OH.

SECTION IV

PHYSICAL–CHEMICAL TREATMENT PROCESSES

CHAPTER 11

SCREENING AND SEDIMENTATION

Screening and sedimentation are inexpensive physical processes that are widely incorporated into treatment operations for water, wastewater, and stormwater runoff. The basic laws of physics and fluid mechanics govern the processes.

11.1 SCREENS AND BAR RACKS

Screens (Fig. 11.1) and bar racks are located at intakes from rivers, lakes, and reservoirs for water treatment plants or at the wet well into which the main trunk sewer discharges for a wastewater treatment plant. They are also located before pumps in stormwater and wastewater pumping stations. They are almost always provided at these locations. These devices remove coarse debris (such as rags, solids, and sticks), which may damage pumps or clog downstream pipes and channels. The spacing of the bars may be coarse, with 50–150 mm (2–6 in.) openings; medium, with 20–50 mm (0.8–2 in.) openings; or fine screens, with openings of 10 mm (0.4 in.) or less (ASCE and WPCF, 1977; Degrémont, 1973).

To prevent the settling of coarse matter, the velocity in the approach channel to the screens should not be less than 0.6 m/s (2 ft/s). The ratio of the depth to width in the approach channel ranges from 1 to 2.

Coarse screens may be installed on an incline to facilitate the removal of debris. The headloss through the screens is a function of the flow velocity and the openings in the screens. A sketch of the water profile through a screen is given in Fig. 11.2. Bernoulli's equation is used to analyze the headloss.

$$h_1 + \frac{v^2}{2g} = h_2 + \frac{v_{sc}^2}{2g} + \text{losses} \tag{11.1}$$

where

h_1 is the upstream depth of flow
h_2 is the downstream depth of flow
g is the acceleration of gravity
v is the upstream velocity
v_{sc} is the velocity of flow through the screen

The losses are usually incorporated into a coefficient.

$$\Delta h = h_1 - h_2 = \frac{1}{2gC_d^2}(v_{sc}^2 - v^2) \tag{11.2}$$

where

C_d is a discharge coefficient (typical value = 0.84)

Figure 11.1 Screens at a wastewater treatment plant.

Alternatively, an orifice equation is often applied to the velocity through the screen.

$$\Delta h = \frac{v_{sc}^2}{2gC_d^2} = \frac{1}{2g}\left(\frac{Q}{C_d A}\right)^2 \tag{11.3}$$

where
  Q is the volumetric flow rate
  A is the area of the openings

The value of C_d in Eq. (11.3) will be supplied by the manufacturer or can be obtained from experimentation. The design value for the area is the open area of the screens for mechanically cleaned screens. The value of the discharge coefficient supplied by a manufacturer may take into account the open area of the screens. The open area of a screen may be considerably reduced by the space taken by the mesh. If the screens are to be manually cleaned, the open area should be taken as 50% of the open area (the half-clogged condition). The headloss is estimated at the maximum flow condition.

Screens or racks may be cleaned by hand or automatically. Screenings are generally nonputrescible. They are collected and hauled away to an incinerator or landfill disposal site.

Screens for Water Treatment Plants

When raw water is being withdrawn from the surface of a river, coarse screens (75 mm or 3 in. or larger) are installed to prevent the intake of small logs or other floating

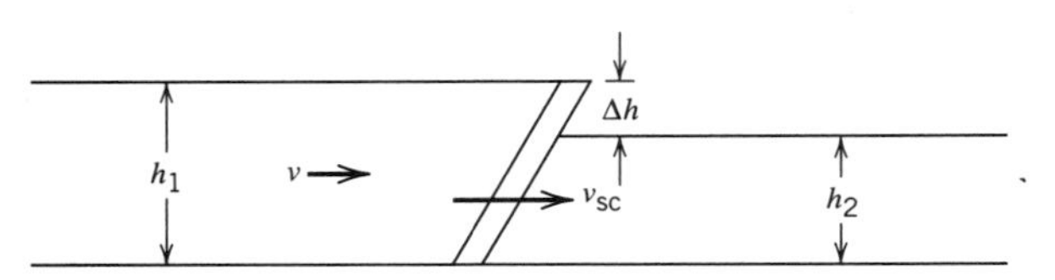

Figure 11.2 Water profile through a screen.

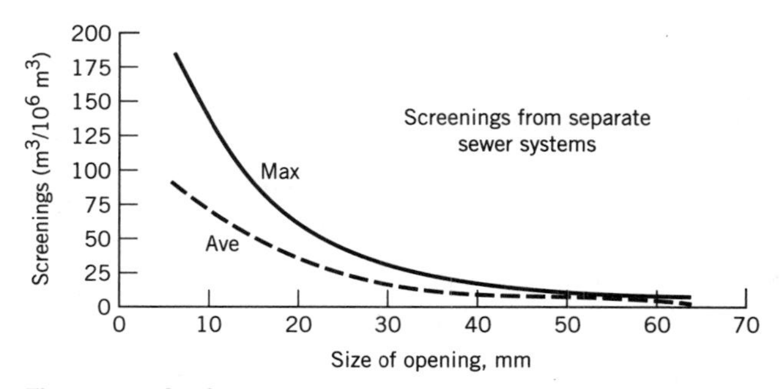

Figure 11.3 Screenings volume variation with bar opening for separate sewer systems. From WEF and ASCE (1992), *Design of Municipal Wastewater Treatment Plants,* vol. 1, WEF, © WEF 1992.

debris. For a submerged intake from a reservoir or lake, smaller coarse screens can be used. Screens at these intakes are not usually mechanically cleaned. These coarse screens are usually followed by screens with smaller openings at the treatment plant. The screens at the treatment plant may be mechanically or manually cleaned depending on the size of the operation.

Quantities of screenings collected at water treatment installations are highly variable depending on the opening of the bars and screens and the raw water source. Screenings may be washed back into the water source. The primary water treatment installation for the city of Ottawa, Ontario, which draws water from the Ottawa River, has mechanically cleaned traveling screens with openings of 1.3 mm and the screenings are washed back into the Ottawa River. The secondary water treatment facility has manually cleaned 1.3-mm (0.05-in.) opening screens that collect an average volume of 0.29 L/1 000 m^3 (0.039 ft^3/Mgal) on an annual basis. Quantities collected during the spring freshet are much higher than at other times during the year.

Screens at Wastewater Treatment Plants

Coarse screens with openings from 50 to 150 mm (2 to 6 in.) are used ahead of raw wastewater pumps (ASCE and WPCF, 1977). Screens with smaller openings (25 mm) are suitable for most other devices or processes (ASCE and WPCF, 1977). A screen with smaller openings would be installed at the beginning of the treatment plant after the water is pumped from the trunk sewer or influent wet well, which are protected by coarse bar racks. At medium to large installations, mechanically cleaned screens are used to reduce labor costs, provide better flow conditions, and improve capture.

Figure 11.3 is a design chart for estimating the maximum and average quantities of coarse screenings collected from separate sewer systems as a function of the bar opening size. It is based on data collected from 133 installations in the United States. Table 11.1 provides other information on screenings collected from separate and combined sewer systems. Combined sewer areas can produce several times the amount of screenings collected from separate sewered areas during storm flows. The peak daily collection can vary as high as 20:1 on an hourly basis.

Microstrainers

Microstrainers (Fig. 11.4) have been used to reduce suspended solids for raw waters that contain high concentrations of algae or to further reduce suspended solids in

TABLE 11.1 Coarse Screenings Characteristics[a]

Item	Range
Quantities	
Separate sewer system	
Average	3.5–35 L/1 000 m^3 (0.47–4.7 ft^3/Mgal/d)
Peaking factor (hourly flows)	1:1–5:1
Combined sewer system	
Average	3.5–84 L/1 000 m^3 (0.47–11 ft^3/Mgal/d)
Peaking factor (hourly flows)	2:1–>20
Solids content	10–20%
Bulk density	640–1 100 kg/m^3 (40–70 lb/ft^3)
Volatile content of solids	70–95%
Fuel value	12 600 kJ/kg (5 400 Btu/lb)

[a]From WEF and ASCE (1992), *Design of Municipal Wastewater Treatment Plants,* vol. 1, WEF, © WEF 1992.

effluent from secondary clarifiers following biological wastewater treatment. The microstrainer is made of a very fine fabric or screen wound around a drum. The drum is typically 75% submerged and rotated with water commonly flowing from the inside to the outside of the drum. In some strainers flow moves from the perimeter to the center. The solids deposit is removed by water jets that can be activated by a pressure differential across the screen. The water jets are directed at the exposed drum surface and collected in a channel under the top of the drum.

Openings in microstrainers vary from 20 to 60 μm. Suspended solids will be removed but bacteria will not be removed to any significant extent. To minimize slime growth that will cause high headloss, ultraviolet light may be applied to the strainer.

The headloss performance of a microstrainer is evaluated by a semi-empirical equation. It is observed that the headloss is directly proportional to the flow rate,

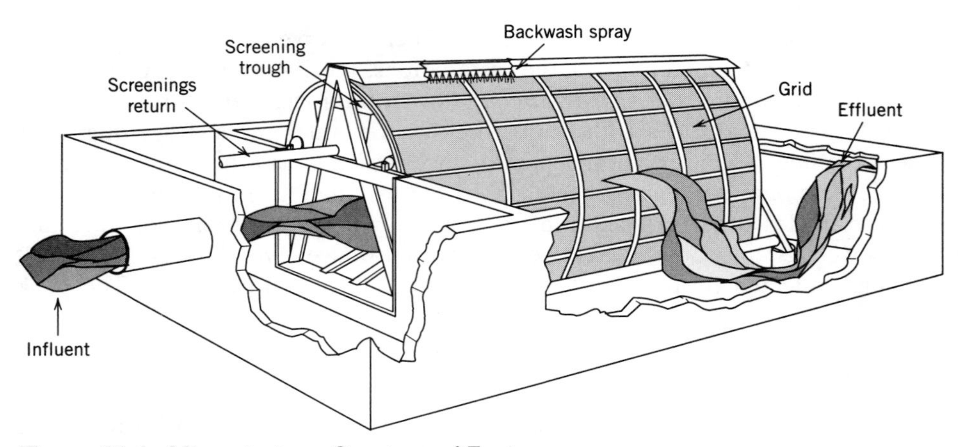

Figure 11.4 Microstrainer. Courtesy of Envirex.

TABLE 11.2 Microstrainer Design Parameters[a]

Item	Typical value
Screen mesh	20–25 μm
Submergence	75% of height (66% of area)
Hydraulic loading	12–24 $m^3/m^2/h$ (300–600 $gal/ft^2/d$) of submerged drum surface area
Headloss through screen (h_L)	7.5–15 cm (3–6 in.)
Max. h_L	30–45 cm[b] (12–18 in.)
Peripheral drum speed	4.5 m/min at 7.5-cm h_L (15 ft/min at 3-in. h_L) 40–45 m/min at 15-cm h_L (130–150 ft/min at 6-in. h_L)
Typical drum diameter	3m (10 ft)
Washwater flow	2% of throughput at 345 kN/m^2 (50 psi) 5% of throughput at 100 kN/m^2 (14.5 psi)

[a]After USEPA (1975).
[b]Typical designs provide an overflow to bypass part of the flow when h_L exceeds 15–20 cm (6–8 in.).

degree of clogging, and time, and inversely proportional to the surface area (A) of the strainer. These parameters are incorporated into a first-order relation:

$$\frac{dh_L}{dt} = k\frac{Q}{A}h_L \tag{11.4}$$

where
k is a characteristic loss coefficient

The above equation integrates to:

$$h_L = h_0 e^{k\frac{Q}{A}t} \tag{11.5}$$

where
h_0 is the headloss of the clean strainer

The loss coefficient for the strainer should be experimentally determined.

Typical design parameters for solids removal from secondary effluents are given in Table 11.2. The USEPA (1975) surveyed a number of microstrainers treating secondary effluent with solids concentrations in the range of 6–65 mg/L and found average removals from 43 to 85%. Microstrainers also find application in the treatment of stormwater runoff and the polishing of effluents (removal of algae) from stabilization pond systems.

11.2 SEDIMENTATION

Sedimentation is the physical separation of suspended material from a water by the action of gravity. It is a common operation for water treatment and found in almost all wastewater treatment plants. It is less costly than many other treatment operations.

11.2.1 Particle Settling Velocity

Before a basin to settle particles is designed, the settling velocities of the particles must be known. The physical characteristics of a particle determine its settling velocity.

Consider a particle falling in a body of fluid with the following assumptions:

1. The particle is discrete and its size and shape do not change.
2. Infinite size vessel.
3. Viscous fluid.
4. Single particle.
5. Quiescent fluid.

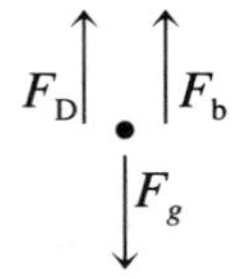

The forces acting on the particle are the effective gravitational force, F_1, and the drag force, F_D, caused by fluid resistance. The effective gravitational force (downward) is the difference between the gravitational force, F_g, and the buoyant force, F_b.

$$F_n = F_g - F_b = (\rho_p - \rho)gV_p \tag{11.6}$$

where
F_n is the net downward force
ρ_p is the density of the particle
ρ is the density of water
V_p is the volume of the particle

The drag force (F_D) can be found from dimensional analysis to be

$$F_D = \tfrac{1}{2}\rho C_D A_p v^2 \tag{11.7}$$

where
C_D is the drag coefficient
A_p is the cross-sectional area of the particle
v is the settling velocity of the particle

The force balance applying to the particle while it is accelerating is

$$m\vec{a} = \vec{F}_g + \vec{F}_b + \vec{F}_D$$

where
a is the rate of acceleration of the particle
m is the mass of the particle
the arrows represent vector quantities

Removing the vector notation from this equation and substituting for the forces in the vertical direction results in the governing differential equation.

$$-\rho_p V_p a = -\rho_p V_p \frac{dv}{dt} = -(\rho_p - \rho)gV_p + \tfrac{1}{2}\rho C_D A_p v^2$$

The settling velocity increases in a very short time from 0 to a constant ultimate settling velocity (see Problem 8). Taking a force balance after the ultimate settling velocity has been reached ($a = 0$):

$$(\rho_p - \rho)gV_p = \tfrac{1}{2}\rho C_D A_p v^2 \tag{11.8}$$

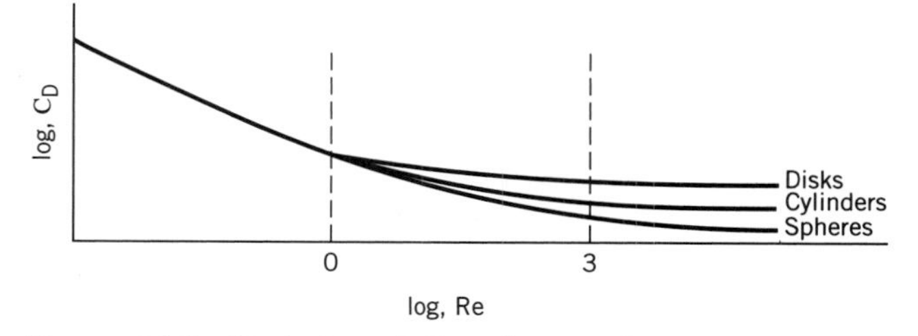

Figure 11.5 Variation of C_D with particle geometry.

Solving for v:

$$v = \sqrt{\frac{2g}{C_D}\frac{V_P}{A_p}\left(\frac{\rho_P - \rho}{\rho}\right)} \tag{11.9}$$

For a spherical particle of diameter, d,

$$v = \sqrt{\frac{4}{3}\frac{gd}{C_D}\left(\frac{\rho_P - \rho}{\rho}\right)} \tag{11.10}$$

The drag coefficient, C_D, is not constant but varies with the Reynold's number, Re, and with the shape of the particle.

$$\text{Re} = \frac{\rho v d}{\mu} \tag{11.11}$$

For spherical particles the following relations apply:

$$\text{Re} < 1: \qquad C_D = \frac{24}{\text{Re}} \tag{11.12a}$$

Re <1 is the laminar range also known as the Stoke's range. The next range is the transition between laminar and turbulent settling.

$$1 < \text{Re} < 10^3: \qquad C_D = \frac{24}{\text{Re}} + \frac{3}{\text{Re}^{0.5}} + 0.34 \qquad \text{or} \qquad C_D = \frac{18.5}{\text{Re}^{0.6}} \tag{11.12b}$$

For fully developed turbulent settling:

$$\text{Re} > 10^3: \qquad C_D = 0.34 \text{ to } 0.40 \tag{11.12c}$$

C_D varies with the effective resistance area per unit volume of the particle as shown in Fig. 11.5.

11.3 TYPE I SEDIMENTATION

The design of an ideal settling basin is based on the removal of all particles that have a settling velocity greater than a specified settling velocity. The work of Hazen (1904) and Camp (1945) provides the basis of sedimentation theory and basin design. Type I sedimentation refers to discrete particle settling.

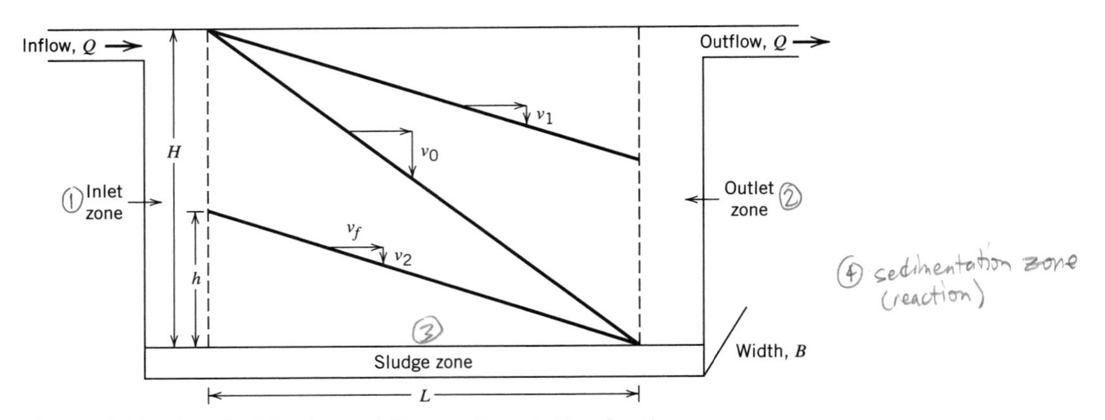

Figure 11.6 An ideal horizontal flow sedimentation basin.

11.3.1 Theory

A definition sketch for an ideal, horizontal flow, rectangular section basin is given in Fig. 11.6. In Fig. 11.6, H is the effective depth of the settling zone and v_f is the longitudinal velocity of the water. The width of the basin is B. The settling velocities v_1 and v_0 apply to two different particles entering at the top of the basin. The settling velocity v_2 applies to a particle entering the settling zone at a height, h, above the sludge zone.

There will be dissipation of energy (turbulence) near the entrance as the flow profile through the basin is established. There is assumed to be no settling in the inlet zone. A similar phenomenon occurs at the exit side as the flow streamlines turn upward, and no settling is assumed to occur in the outlet zone. Sludge accumulates in the sludge zone, which is not part of the effective settling zone.

Other important assumptions are as follows:

1. There is uniform dispersion of water and suspended particles in the inlet zone. Therefore, the suspended solids concentration is the same at all depths in the inlet zone.
2. Continuous flow at a constant rate (steady flow) exists.
3. Once a particle enters the sludge zone, it remains there (i.e., there is no resuspension of settled particles).
4. The flow-through period is equal to the detention time, i.e., there is no dead space or short circuiting in the volume above the sludge zone.
5. PF conditions exist.
6. Settling is ideal discrete particle sedimentation.
7. Particles move forward with the same velocity as the liquid.
8. There is no liquid movement in the sludge zone.

The design volume must be related to the influent flow rate and the particle settling velocity. The particle that takes the longest time to remove will be one that enters at the top of the effective settling zone. The design settling velocity is v_0, which is the settling velocity of the particle that settles through the total effective depth of the tank in the theoretical detention time. The flow-through velocity is v_f.

$$t_d = \frac{V}{Q} \tag{11.13}$$

$$v_f = \frac{Q}{BH} \tag{11.14}$$

Because the particle must travel the length and depth of the basin in the time t_d,

$$v_0 t_d = H \tag{11.15a}$$

$$v_f t_d = L \tag{11.15b}$$

Substituting Eq. (11.15b) into Eq. (11.15a) and using Eq. (11.14):

$$\frac{L}{v_f} = \frac{H}{v_0} \quad \text{and} \quad v_0 = v_f \frac{H}{L} \quad \text{or} \quad v_0 = \frac{QH}{BHL} = \frac{Q}{BL} \tag{11.16}$$

The surface overflow rate, Q/A_s (A_s is the surface area of the basin) is defined by

$$v_0 = \frac{Q}{BL} = \frac{Q}{A_s} \tag{11.17}$$

This proves that the sedimentation basin design is independent of the depth and depends only on the surface overflow or loading rate (Q/A_s) for particles with a specified settling velocity v_0. From this it also follows that the sedimentation efficiency is also theoretically independent of detention time in the basin. This fact is not a mathematical curiosity. Consider a basin with the flow at the bottom of the basin and uniformly distributed introduced across the plan (surface) area, resulting in an upflow velocity v_0. Any particle with a settling velocity greater than v_0 will be removed (settled) after being introduced into the basin regardless of the residence time of the water in the basin. Likewise, any particle with a settling velocity less than v_0 will eventually exit with the effluent overflowing from the basin.

Horizontal (or radial) flow and upflow are the two possible operational modes of a settling basin. In either case all particles with settling velocities greater than v_0 will be removed. In the horizontal flow mode some particles with settling velocities less than v_0 will also be removed if they enter the basin at a depth less than H. Assumption 1 above is critical to analysis of the total removal. Assume that a particle with settling velocity, v, which is less than v_0, will travel a vertical distance h in the time t_d.

$$h = v t_d \tag{11.18}$$

All particles with settling velocity v that enter at a depth h or lower will be removed. The criterion for removal of particles with this settling velocity is

$$\frac{h}{H} \leq \frac{v}{v_0} \tag{11.19}$$

Because all particles with settling velocity v are uniformly distributed throughout the inlet depth (assumption 1), the fractional removal, r, of particles with this settling velocity is

$$r = \frac{h}{H} = \frac{v}{v_0} \tag{11.20}$$

The settling velocity distribution for a suspension can be determined from a column settling test as described in Section 11.4. The results of the test provide data to construct

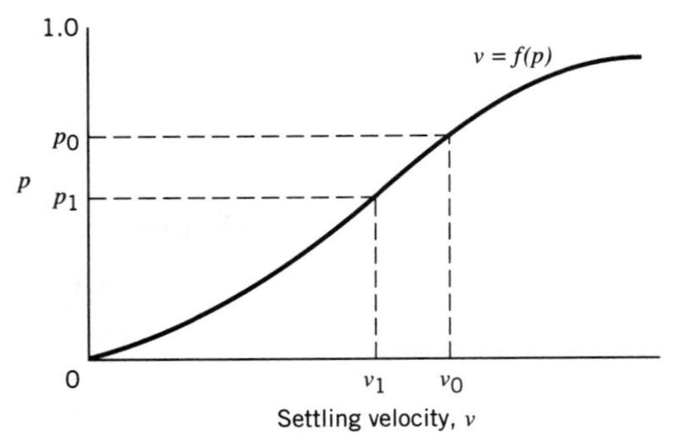

Figure 11.7 Settling velocity curve for a suspension, where p is the weight fraction of particles with settling velocity less than stated velocity.

a plot as shown in Fig. 11.7, which is a cumulative settling velocity frequency distribution.

The total removal R (fraction by weight) of all particles is the sum of the fractional removals of each fraction of particles. Applying Eq. (11.20) to each fraction Δp,

$$R = 1 - p_0 + \sum r_i$$

$$R = 1 - p_0 + \frac{v_0 + v_1}{2v_0}(p_0 - p_1) + \frac{v_1 + v_2}{2v_0}(p_1 - p_2) + \cdots + \frac{v_i + v_{i+1}}{2v_0}(p_i - p_{i+1}) + \cdots \tag{11.21a}$$

which in the limit is

$$R = (1 - p_0) + \frac{1}{v_0}\int_0^{p_0} v\, dp \tag{11.21b}$$

where

p_0 is the fraction of particles by weight with a settling velocity equal to or less than v_0

A polynomial fit can be applied to data used to construct the curve in Fig. 11.7 to use Eq. (11.21b).

For circular tanks with an inlet in the middle and radial flow under the same assumptions as above, it can also be shown that

$$v_0 = \frac{Q}{A_s} \qquad r = \frac{h}{H} = \frac{v}{v_0}$$

and the same overall removal expressions (Eqs. 11.21a and 11.21b) apply (see Problem 5).

Equation (11.17) also holds for vertical flow tanks. Particles with a settling velocity less than the upflow velocity are entrained in the upward flow and washed out of the system.

11.3.2 Overall Removals and Overflow Rates

When the settling velocity distribution of the suspension is known (such as given in Fig. 11.7), it is a routine matter to construct a plot of overall removals versus the

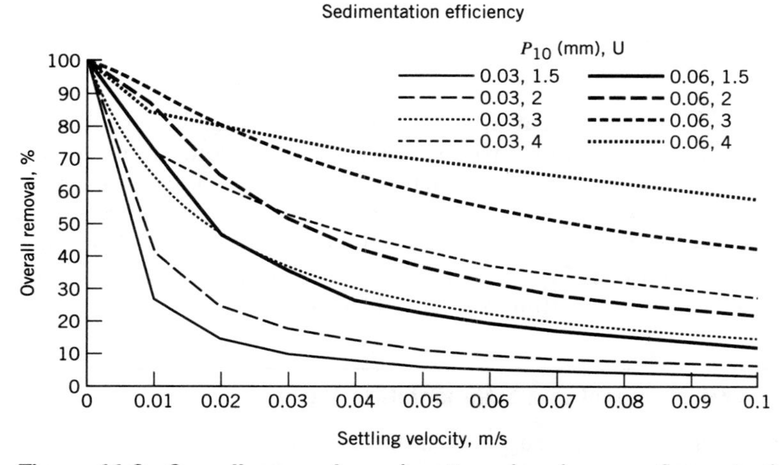

Figure 11.8 Overall removal as a function of surface overflow rate. See text for definition of terms.

surface overflow rate using Eqs. (11.21a) or (11.21b). Figure 11.8 shows typical curves. The general shape of these curves suggests that a parametric equation can be used to describe them. A solids size distribution is characterized by the effective size, P_{10} (the size through which 10% by weight of the particles pass) and the uniformity coefficient, U (the ratio of sizes through which 60 and 10% by weight of the particles, respectively, pass). These parameters and size distributions are discussed in Section 14.2.1. Bhargava and Rajagopal (1989) have determined a correlation that readily provides the overall removal curve when the particle size distribution and specific gravity (s.g.) of a suspension are known. The correlation should be used for suspensions that contain a large quantity of inorganic matter such as surface waters or raw sewage.

The equation for removal determined by Bhargava and Rajagopal (1989) for a temperature of 30°C and particle s.g. of 2.65 is

$$\frac{1}{R} = \left[\frac{U}{177.88 + 44.71U}\right] + \left\{\frac{1}{\exp\left[(3.186 \times 10^{-3}U + 2.036)\ln P_{10} + \exp\left(\frac{\ln U + 4.627}{2.809}\right)\right]}\right\} v_0 \qquad (11.22)$$

where

P_{10} is expressed in mm
v_0 is expressed in $m^3/m^2/s$

For temperatures different from 30°C or particle s.g.s different from 2.65, the settling velocity in the second term of Eq. (11.22) must be adjusted to the equivalent settling velocity at 30°C and s.g. of 2.65 using Eq. (11.10). For a particle with a terminal settling velocity in the Stoke's range, the temperature and s.g. correction factors (C_T and C_{sg}, respectively) to multiply v_0 are

$$C_T = \frac{\nu_T}{\nu_{30}} \qquad (11.23a) \qquad\qquad C_{sg} = \frac{2.65 - 1}{S_s - 1} \qquad (11.23b)$$

where

S_s is the s.g. of the suspension

ν is kinematic viscosity

The s.g. of inorganic particles is near 2.65 and the s.g.s of organic particles are normally in the range of 1.001–1.01. For mixed suspensions, the overall removal will be the weighted sum of the removals of each fraction. Under conditions of a typical settling test it may be assumed that the organic matter removal is about 5%. The TSS–VSS analysis may be used to discriminate between inorganic and organic matter. When inorganic solids predominate, Eq. (11.10) can be used to calculate the nominal diameters of the fractions from the settling velocity distribution.

■ Example 11.1 Overall Removal in a Sedimentation Basin

What is the overall removal of a sewage suspension that contains 76% inorganic matter with a s.g. of 2.557 and 24% organic matter with a s.g. of 1.096 at a temperature of 28.5°C and surface overflow rate of 0.008 $m^3/m^2/s$? The P_{10} and U values at 30°C for the inorganic particles were found to be 0.057 5 mm and 1.315, respectively, based on a settling test.

From Eq. (11.22), the overall removal of inorganic solids with specified P_{10} and U values and a s.g. of 2.65 is calculated to be 69.3%. For the mixed suspension, the overall solids removal at 30°C and $v_0 = 0.008\ m^3/m^2/s$ will be

$$R = 69.3 \times 0.76 + 5.0 \times 0.24 = 53.9\%$$

The removal of the inorganic solids (R_I) is 52.7%.

Correcting the overflow rate to 28.5°C and a s.g. of 2.557,

$$R_{28.5,2.557} = \frac{1}{5.556 \times 10^{-3} + \left(\frac{0.828 \times 10^{-6}}{0.800 \times 10^{-6}}\right)\left(\frac{2.65 - 1}{2.557 - 1}\right)\left(\frac{0.008}{0.902\ 5}\right)} = 65.5$$

$$R_I = 65.5 \times 0.76 = 49.8\%$$

and the overall removal for the suspension is 51.0%.

Theory dictates that increasing the surface area of a settling basin will improve its performance. Lamella and tube clarifiers, discussed in Section 11.4.2, exploit the concepts from the basic theory resulting in the design of clarifiers with very high loading rates.

11.4 TYPE II SEDIMENTATION

Under quiescent conditions suspended particles in many waters exhibit a natural tendency to agglomerate or the addition of chemical agents promotes this tendency. This phenomenon is known as flocculent or type II sedimentation. Analysis of type II sedimentation proceeds from the principles of type I sedimentation.

As particles settle and coalesce with other particles, the sizes of particles and their settling velocities will increase. The trajectory traced by a settling particle will be

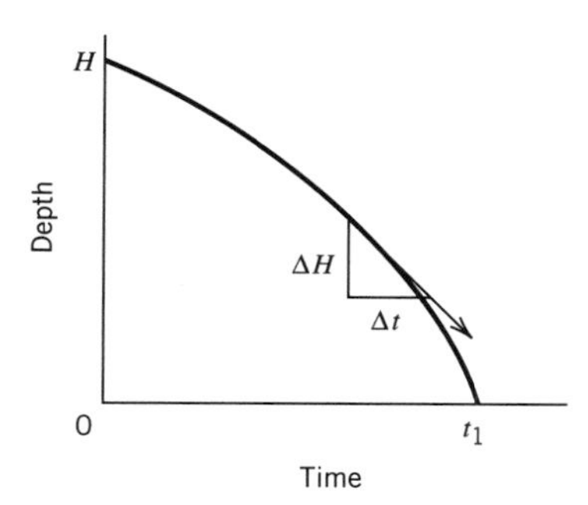

Figure 11.9 Settling trajectory in type II sedimentation.

curvilinear (Fig. 11.9) because of the increase in its settling velocity as other particles attach to it.

The instantaneous settling velocity is the tangent to the curve. The average settling velocity for the particle in Fig. 11.9 is

$$\bar{v} = \frac{H}{t_1} \tag{11.24}$$

The average settling velocity distribution for the suspension is continually changing with time as shown in Fig. 11.10.

To design a basin for flocculent settling, the average settling velocity distribution variation with time must be found to calculate the total removal as time (or volume of the basin) increases. At some point an incremental increase in the volume of the basin (which increases the detention time) will not produce a significant increase in the amount of solids removed. There is no theoretical means of predicting the amount of flocculation and settling velocity distribution variation for a suspension. A laboratory analysis as described in the following section is required.

Laboratory Determination of Settling Velocity Distribution

The water to be analyzed must have the same coagulants and other agents added that will be applied in the field situation. The suspension is mixed and added to a column that has approximately the same depth as the anticipated settling basin. Because type II sedimentation is time–depth dependent, more representative settling curves are obtained when the column depth is near the prototype basin depth.

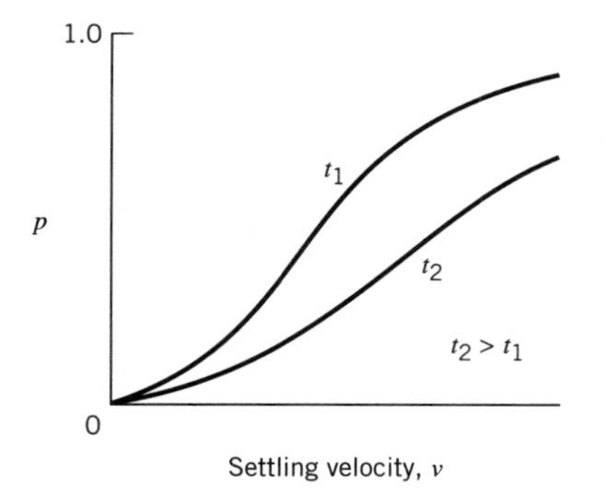

Figure 11.10 Settling velocity distribution at various times.

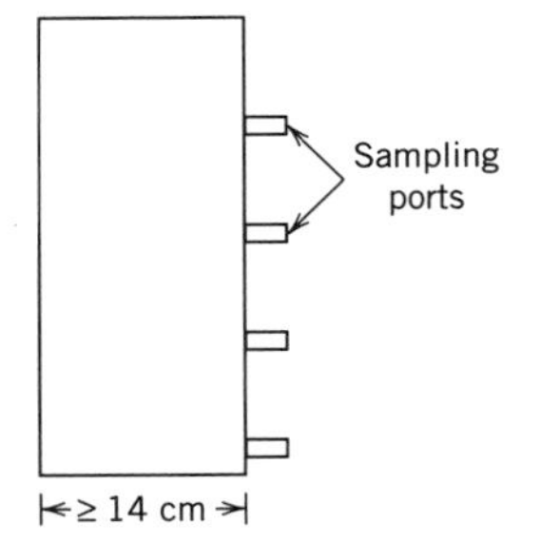

Figure 11.11 Settling column.

The column (Fig. 11.11) is normally made of clear plastic so that one may visually observe the process. Sampling ports are uniformly spaced along the length of the column. The bottom port will provide samples that are indicative of the compaction of the settled sludge. The effective settling depth is the depth above the bottom port. The column internal diameter should be at least 14 cm (5.5 in.) to avoid bridging of the suspension and other wall effects. After the initial sample is taken, samples are taken from each port at uniform time intervals, noting the time and port number.

The volume of samples removed from the column causes the water surface elevation to descend, which should be accounted for in processing the data.

11.4.1 Type II Sedimentation Data Analysis

The analysis of data gathered as outlined in the previous section is best presented by example. The object of the analysis is to obtain a plot of the settling trajectories for various fractions of the suspended solids. Then the total removals at any time may be estimated. For a column with a total depth of 240 cm (7.87 ft) and sampling ports spaced at 60-cm (1.97-ft) intervals, the data in Table 11.3 have been obtained. The effective depth of the sedimentation basin under consideration is 1.8 m (5.91 ft). The initial concentration of SS was 430 mg/L.

TABLE 11.3 Raw Data

	Concentration, mg/L		
Time min	60 cm (1.97 ft)	120 cm (3.94 ft)	180 cm (5.91 ft)
5	357	387	396
10	310	346	366
20	252	299	316
30	198	254	288
40	163	230	252
50	144	196	232
60	116	179	204
75	108	143	181

TABLE 11.4 Percentage Solids Removed

Time min	Solids removed, % 60 cm (1.97 ft)	120 cm (3.94 ft)	180 cm (5.91 ft)
5	17.0	10.0	7.9
10	28.0	19.5	14.9
20	41.5	30.5	26.0
30	54.0	40.9	33.0
40	62.0	46.5	41.4
50	66.5	54.4	46.0
60	73.0	58.6	52.5
75	75.0	66.7	57.9

The first step is to convert the concentrations to percentage removals at each depth.

$$r_{\text{ti,d}}(\%) = \frac{C_0 - C_{\text{ti,d}}}{C_0} \times 100$$

where
the subscript ti,d indicates time i and depth, d, respectively

The results of these calculations are listed in Table 11.4.

The desired plot of the settling trajectories of various fractions of the suspended solids (see Fig. 11.13) can be obtained by constructing a depth–time plot with percentage solids removed as a parameter. An intermediate step improves the interpolation that is required (Ramalho, 1977).

A plot of the percentage solids removed at each depth versus time is constructed using the data in Table 11.4. This is done in Fig. 11.12. A smooth curve is drawn between the data points for each depth.

Figure 11.12 can be used to easily estimate the time required to attain a specified removal at a given depth. The times to attain a given removal at each depth are found by extending a horizontal line from the removal to the curves and dropping vertical

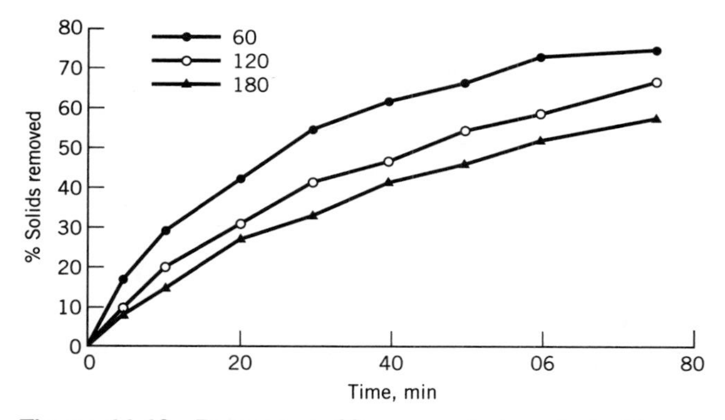

Figure 11.12 Percentage SS removed at each depth.

TABLE 11.5 Interpolated Percentage Solids Removed

% SS removed	t, min 60 cm (1.97 ft)	120 cm (3.94 ft)	180 cm (5.91 ft)
5	1.2	2.5	3.7
10	2.5	5.0	6.5
20	6.7	11.0	14.5
30	11.7	19.0	25.0
40	18.0	30.0	39.0
50	27.0	44.0	56.5
60	38.5	61.5	77.5
70	55.0	87.5	—
75	75.0	—	—

lines at the intersections. These times are then tabulated for the removals at each depth (Table 11.5).

Using the data in Table 11.5, isoconcentration (or isoremoval) lines are now constructed on a depth versus time plot (Fig. 11.13).

The curves on Fig. 11.13 trace the settling trajectories of particulate fractions of the suspension. The actual particle makeup of each fraction is continuously changing as the particles coalesce and these curves represent the gross phenomena.

Now for any time, a p–v plot similar to Fig. 11.7 can be made and the overall removal for the suspension can be determined in the same manner outlined for discrete particle settling. The data from Fig. 11.13 can be used directly to estimate the total removal. To find the total removal at any chosen time, a vertical line is projected upward. It is most convenient to choose times that are at the end of an isoconcentration line. (Why?) When a time of 39 min is chosen for the data, it is seen that 40% of the particles are completely removed; i.e., 40% of the particles had an average settling

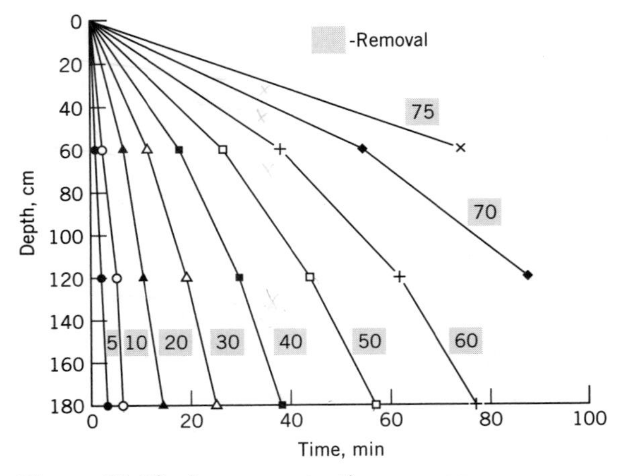

Figure 11.13 Isoconcentration curves.

velocity greater than or equal to

$$\frac{180 \text{ cm}}{39 \text{ min}} = 4.62 \text{ cm/min} \qquad \left(\frac{5.9 \text{ ft}}{39 \text{ min}} = 0.152 \text{ ft/min}\right)$$

in the first 39 min of settling.

The other fractions are partially removed. To estimate the removal of these fractions, median lines should be drawn between the isoconcentration curves. Where possible the median lines should be located based on the average depth between two isoconcentration lines at a given time. For example, at a time of 57 min the 50 and 60% isoconcentration lines intersect the vertical at depths of 180 and 106 cm (5.90 and 3.48 ft), respectively. The median line between these isoconcentration lines should pass through the point (57 min, 143 cm) (57 min, 4.69 ft).

The fractional removal of the 40–50% fraction (10% of the particles) is calculated by reading the depth at the intersection of the vertical and the median isoconcentration lines for this fraction (130 cm; 4.27 ft). This is the average depth that this fraction reached in 39 min. In a manner analogous to the discrete particle settling analysis, the average settling velocity of a fraction compared to the design settling velocity will dictate the percentage removal of the fraction.

$$\frac{v_i}{v_0} = \frac{\dfrac{d_i}{t_d}}{\dfrac{D}{t_d}} = \frac{d_i}{D} \tag{11.25}$$

where

d_i is average depth reached by the ith fraction in the time t_d
D is the total effective settling depth
v_i is the average settling velocity of the ith fraction

The fractional removal, r_i, of the fraction, Δp_i, is

$$r_i = \frac{d_i}{D} \Delta p_i \tag{11.26}$$

The fractional removals for the data above in a time of 39 min are

$$\frac{130}{180}(50 - 40) = 7.2$$

$$\frac{78}{180}(60 - 50) = 4.3$$

$$\frac{48}{180}(70 - 60) = 2.7$$

$$\frac{30}{180}(75 - 70) = 0.8$$

The same fractional removals would be obtained using U.S. units.

The removal of the fraction between 75 and 100% is small and is ignored. The upper value of 100% removal would probably exist only at the surface plane of the volume. A lower upper limit would be dictated by the presence of colloidal particles that are practically unsettleable.

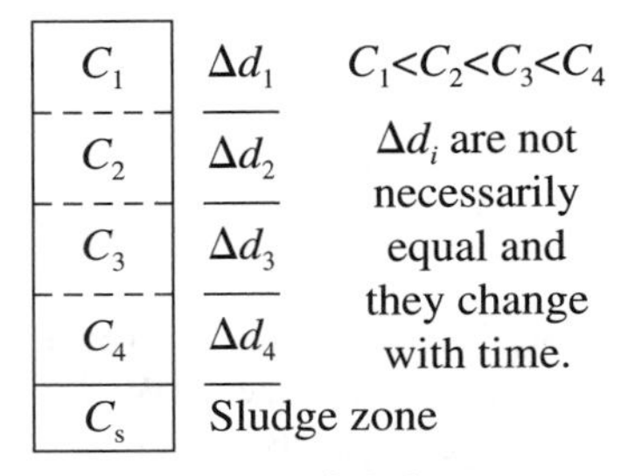

Figure 11.14 Solids concentrations in the column.

The total removal, R, is in this case

$$R = r_0 + \sum r_i = 40 + 7.2 + 4.3 + 2.7 + 0.8 = 55.0\% \tag{11.27}$$

This procedure is repeated for different times and the overall removals at each selected time are tabulated to construct a graph as shown later in Fig. 11.15.

Alternative Method for Calculating Total Removal

As noted, the method just given is analogous to the procedure for discrete particle settling. An equivalent method to find the total removal is to examine the amount of removal in each section of the column (Fig. 11.14) and sum them.

The initial suspended mass, M_0, in the column is

$$M_0 = C_0 A D \tag{11.28}$$

where

A is the cross-sectional area of the column
C_0 is the initial concentration of suspended particles

Referring to Fig. 11.14, the suspended mass, M_t, at any time is

$$\begin{aligned} M_t &= C_1V_1 + C_2V_2 + C_3V_3 + C_4V_4 \\ &= A(C_1\Delta d_1 + C_2\Delta d_2 + C_3\Delta d_3 + C_4\Delta d_4) \end{aligned} \tag{11.29}$$

The percentage removal on any isoconcentration line in Fig. 11.13 is

$$r_i = \frac{C_0 - C_i}{C_0} \times 100$$

A vertical line projected upward from a chosen time in Fig. 11.13 intersects isoconcentration lines that determine each column section length, Δd_i. The average removal in a section of column is the average of the isoconcentrations that define Δd_i. Assuming the section numbering to start from the lower depth, the total removal R is

$$R = \frac{\Delta d_1}{D}\left(\frac{r_1 + r_2}{2}\right) + \frac{\Delta d_2}{D}\left(\frac{r_2 + r_3}{2}\right) + \cdots \frac{\Delta d_i}{D}\left(\frac{r_i + r_{i+1}}{2}\right) + \cdots \tag{11.30}$$

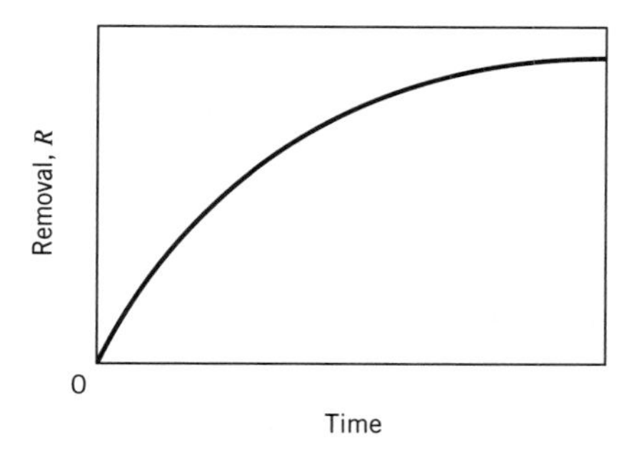

Figure 11.15 Overall removal versus detention time.

Equation (11.30) is applied for different times and the results for overall removal at various times can be tabulated and plotted as shown in Fig. 11.15.

Sizing the Basin

A graph of total removal (R) versus time will provide a design curve. A typical graph is shown in Fig. 11.15. The removal curve will eventually become nearly horizontal as time increases. The design point is in the region where the marginal increase in removal is less than the marginal increase in time, which is equivalent to the size of the clarifier. Costs of the clarifier compared to costs associated with other solids removal processes will determine the optimum design point.

The design detention time from the laboratory column is equivalent to a design settling velocity of D/t_d, which is also equal to the design surface loading rate (Q/A_s). To translate the laboratory data to the field, where nonideal flow conditions exist and sedimentation does not occur under completely quiescent conditions, safety factors of 1.25–1.75 are applied to the detention time and the surface overflow rate.

1. Multiply the design t_d based on the column performance by 1.25–1.75.
2. Divide the design Q/A_s based on the column performance by 1.25–1.75.

Application of the safety factors will result in an increase in the surface area of the settling basin.

11.4.2 Tube and Lamella Clarifiers

Theoretically the efficiency of a clarifier is independent of depth as discussed previously. Fundamentally, this is because the liquid upflow velocity in the basin must be less than the velocity of the slowest settling particle that is to be removed. The pioneer environmental engineer, T. R. Camp (1945), attempted to exploit this concept by inserting a number of subfloors into horizontal sedimentation basins to increase the surface area. Practically, sludge removal (each floor would need a scraping device) was a problem and the idea became dormant until the 1960s, with the development of tube or lamella settlers, which are an interesting and effective application of the concept.

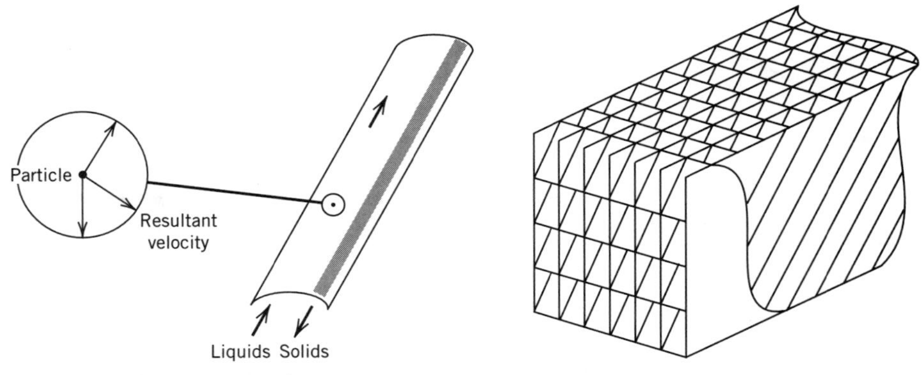

Figure 11.16 Tube clarifier.

Tube settlers are plastic (PVC) modules with uniformly spaced, inclined channels (Fig. 11.16). Lamella settlers have uniformly spaced, inclined panels (Fig. 11.17). Lamella settlers can be made with plastic, rawhide, or other available resilient materials. Both types of clarifiers solve the problem of sludge removal. The resultant velocity on a particle from the upward flow of water and the vertically downward settling velocity of the particle directs the particle to the bottom wall of a tube or toward the lamella. The particle then slides down the surface and exits at the bottom to be collected in a sludge chamber (see the insert in Fig. 11.16).

The theory of these clarifiers is discussed by Yao (1970) but is beyond the scope of this textbook. The increased surface area available in tube or lamella clarifiers allows surface loading rates based on plan area that are two or more times higher than loading rates for conventional clarifiers with the same or better performance (Richter, 1987). The Reynold's number in the tubes or between lamella plates is very low compared to the Reynold's number for a conventional clarifier, which will be in the turbulent range. Turbulence effectively decreases the settling velocity of particles and causes resuspension of settled particles by scour. Existing clarifiers can be upgraded

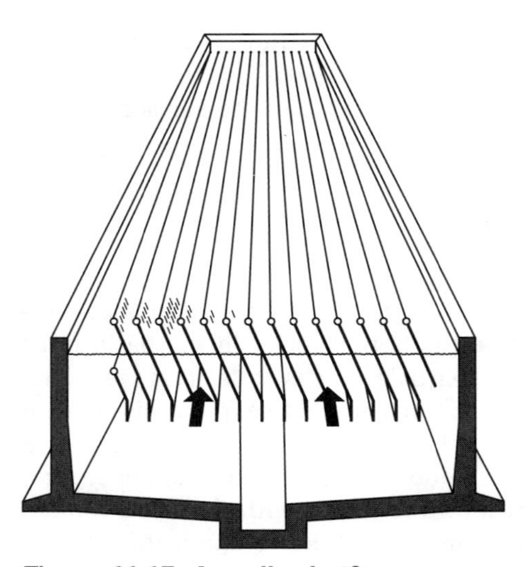

Figure 11.17 Lamella clarifier.

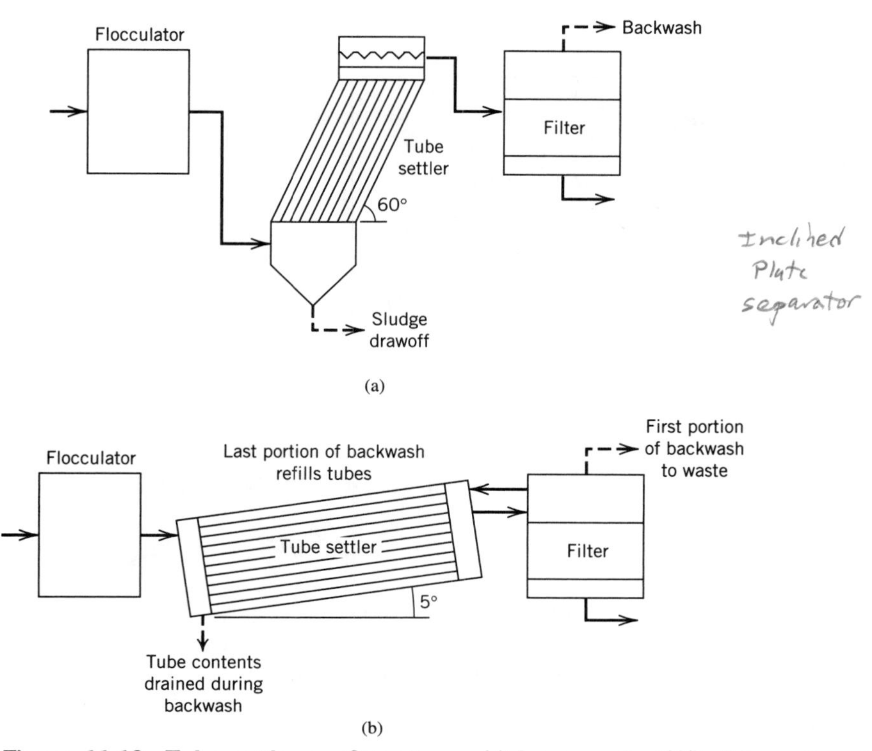

Figure 11.18 Tube settler configurations. (a) Large angle (60°); (b) small angle (5°).

to higher loading rates by the installation of a tube module or lamella. Lamellae can be installed in a concentric array in circular clarifiers. Both tubes and lamellae can be installed in horizontal flow sedimentation basins and upflow solids contact clarifiers. There are a number of companies that supply ready-made tube modules.

The settlers are installed in two different ways. Steeply inclined settlers at angles of 60° or more are stand alone units that are essentially self-cleansing (Fig. 11.18a). From time to time a high-pressure wash may need to be applied to the module to remove particles and biological growth that have accumulated on the settler walls.

The other alternative is to reduce the angle of inclination to a small value, but the angle must be high enough that solids are not washed out with the clarified effluent (Fig. 11.18b). Angles as small as 5° have been used. Solids will not travel significantly downward in a clarifier at this angle of inclination. But these clarifiers are designed in conjunction with rapid sand filters (Chapter 14) such that the discharge of water from backwashing the filter is directed through the clarifier to clean the accumulated solids from the walls. Normal backwash quantities and velocities are adequate to scour the tubes. The first portion of the backwash is not directed into the tube settler because it is laden with solids removed from the filter; the intermediate portion is used to wash the tube module; and the latter portion of the backwash water is used to fill the tube settler.

Conventional inlets can be used for lamella clarifiers when they are configured as indicated in Fig. 11.17. However, in tube settlers, front end inlets produce flow patterns that do not distribute the flow uniformly to the clarifier module (Fig. 11.19). If lamellae are oriented perpendicular to the inlet flow the same problem will occur.

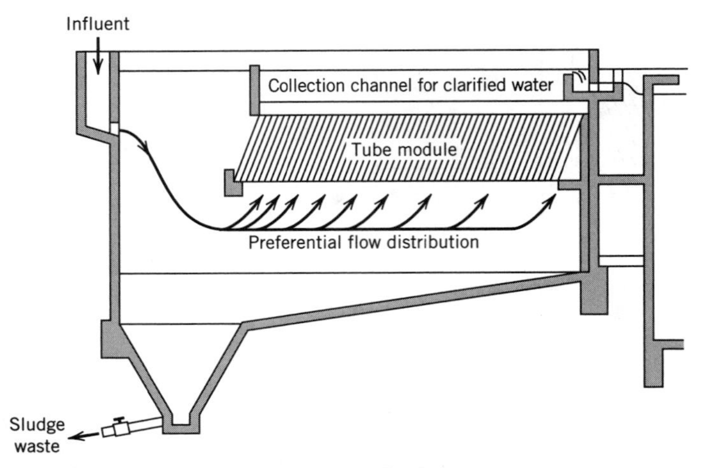

Figure 11.19 Flow distribution from a conventional inlet in a tube settler. From Di Bernardo, L. and ABES, ed. (1993).

An improved inlet arrangement used in Brazil is shown in Fig. 11.20 (Di Bernardo, 1993). Flow is evenly discharged along the length of the clarifier by a series of ports. Sludge is withdrawn through tubes (minimum 38-mm or 1.5-in. diameter) equidistantly spaced along the bottom of the clarifier. Di Bernardo (1993) has also given other sludge removal designs for tube settlers.

11.5 TYPE III SEDIMENTATION: ZONE SETTLING

When solids concentrations become high, forces between the particles become significant and settling is hindered by the additional resistance to movement provided by

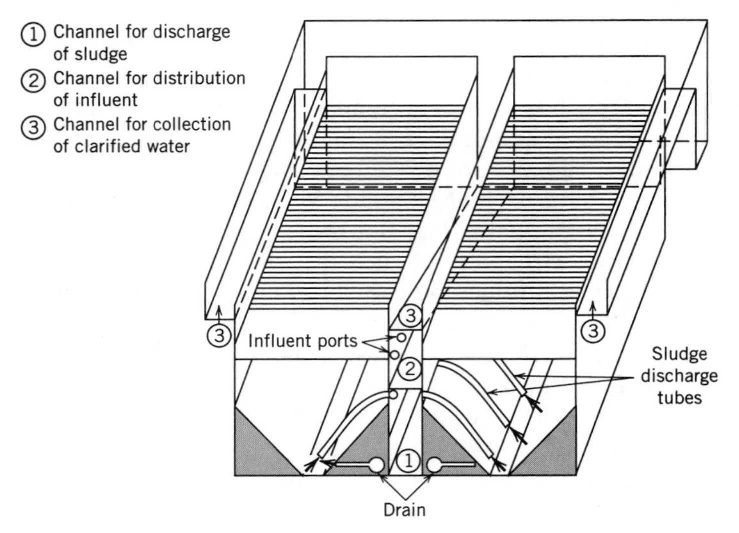

Figure 11.20 Influent distribution for tube settler. From Di Bernardo, L. and ABES, ed. (1993).

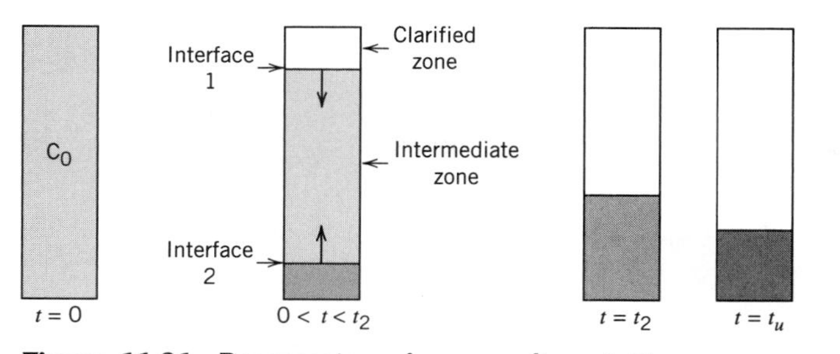

Figure 11.21 Progression of zone sedimentation.

other particles. The suspension tends to settle *en masse.* A clarified zone exists in the upper portion of the clarifier, followed by a zone in which the suspension is moving down and concentrating toward a bottom layer, where slowly compacting sludge exists. This behavior is exhibited by flocculent suspensions with solids concentrations above 500 mg/L. Effluents from biological treatment units such as activated sludge processes or trickling filters exhibit zone settling. Properly designed and operated zone clarifiers can produce a clarified effluent with a very low concentration of suspended solids. The only further treatment that normally may be applied to the clarified effluent from a settler receiving effluent from a biological treatment process is disinfection.

The progression of zone sedimentation in a batch process is shown in Fig. 11.21. Starting from a uniform concentration of C_0, the formation of two interfaces becomes apparent as the suspension settles. A relatively clear water exists above the top interface and a concentrated sludge exists below the bottom interface. These two interfaces are propagated downward and upward, respectively, as time goes on. At time t_2 the interfaces meet. After this time, compaction of the sludge occurs relatively slowly until its ultimate compaction limit is reached.

The data for zone sedimentation are gathered from a laboratory column settling test using a graduated cylinder. A 1- or 2-L cylinder may be used, although the latter is recommended to prevent bridging and minimize wall effects. To further prevent these two phenomena from influencing the results, gentle stirring at around 1 rpm is recommended. The initial concentration of the suspension is measured and the height of the top interface is monitored with time.

Continuous flow clarifiers that receive these suspensions are normally designed to accomplish sedimentation and some thickening of the settled sludge. These clarifiers are usually circular in plan, which facilitates collection of sludge by scrapers that travel around the floor of the clarifier. A schematic of a clarifier, identifying the zones that will occur in a clarifier receiving a continuous flow, is shown in Fig. 11.22 and schematic views of a circular secondary clarifier are shown in Fig. 11.23. In a continuous flow situation with relatively constant influent flow and solids concentration, a dynamic steady state condition is established as solids continuously move through the sludge blanket into the concentrated sludge zone. In zone I there is an upflow velocity caused by fluid movement; in zone II, it is assumed that the net movement of fluid is downward.

It is the settling velocity of the **suspension** that governs the design of the clarifier in zone I (the sludge blanket zone). The drawoff of the thickened sludge at the bottom of the tank influences settling in zone I. However, the thickening of sludge that occurs in zone II (concentrated sludge zone) is not influenced by the drawoff of clarified effluent at the top of the tank. The detention time of sludge in this zone is independent of the detention time in zone I.

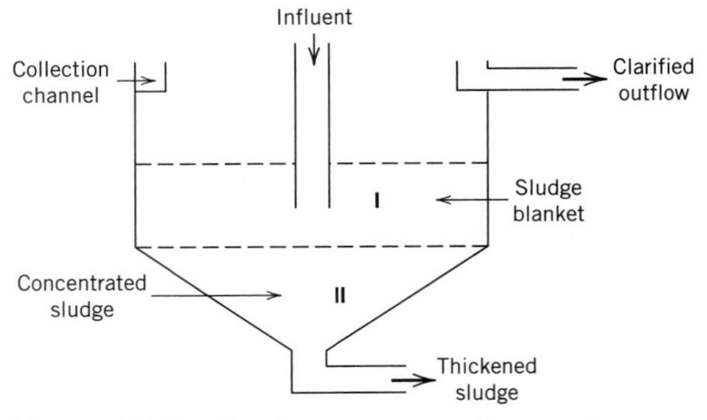

Figure 11.22 Clarifier for zone sedimentation.

11.5.1 Analysis of Zone Sedimentation

The governing principle for design of a sedimentation basin is $v_s \leq Q/A_s$. The limiting settling velocity of the suspension will be the design settling velocity, v_s, which will dictate the surface area of the basin. A plot of the top interface height with time will result in a curve as shown in Fig. 11.24.

The slope of the line at any time gives the settling velocity of the interface. Because compaction occurs in zone II the problem is to find the limiting settling velocity of the suspension before compaction begins. Kynch (1952) analyzed type III behavior. The important assumption in his analysis was that settling velocity of a layer is solely a function of the concentration of solids in the layer. The solids flux, N (mass/area/time), is a function of concentration and velocity.

$$v = v(C) \qquad N = vC$$

Examine the elemental volume of liquid in the cylinder in Fig. 11.25. If N is taken to be positive downward then

$$N = -Cv \tag{11.31}$$

because v will be a negative number.

The mass balance for the volume is

$$\begin{gathered} \text{IN} - \text{OUT} + \text{GENERATION} = \text{ACCUMULATION} \\ \left(N + \frac{\partial N}{\partial y}dy\right)A - NA + 0 = \frac{\partial C}{\partial t}A\,dy \end{gathered} \tag{11.32}$$

where

y is the vertical distance

There is no production or destruction of solids. Expanding and simplifying the above equation, the governing differential equation is

$$\frac{\partial N}{\partial y} = \frac{\partial C}{\partial t} \tag{11.33}$$

Differentiating Eq. (11.31) with respect to depth,

$$\frac{\partial N}{\partial y} = -\frac{\partial (Cv)}{\partial y} = -v\frac{\partial C}{\partial y} - C\frac{\partial v}{\partial y} \tag{11.34}$$

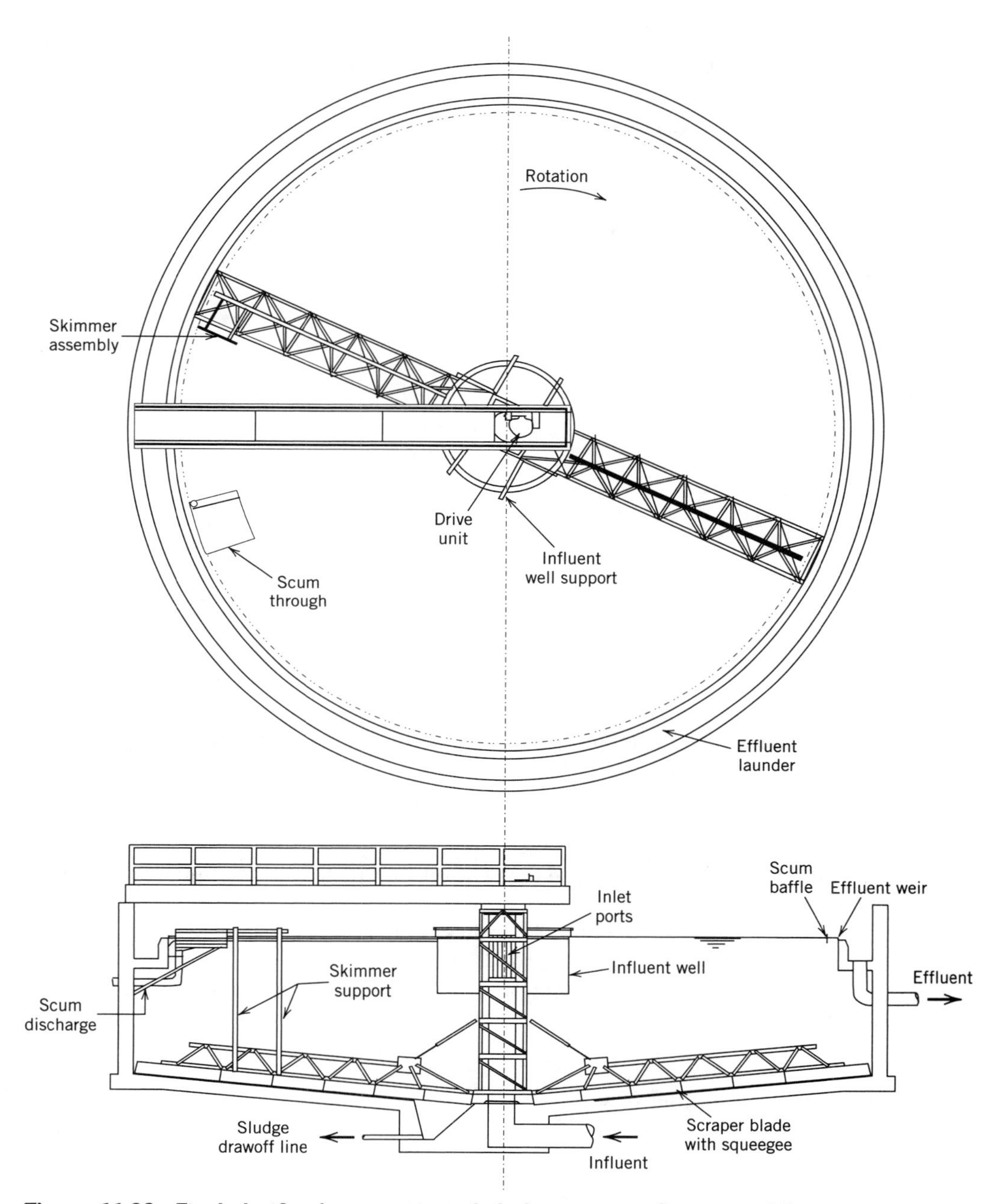

Figure 11.23 Final clarifier for an activated sludge process. Courtesy of Envirex.

Because $v = v(C)$, N is only a function of concentration and, using the chain rule,

$$\frac{\partial N}{\partial y} = \frac{\partial N}{\partial C}\frac{\partial C}{\partial y} \qquad (11.35a) \qquad \text{also,} \qquad \frac{\partial v}{\partial y} = \frac{\partial v}{\partial C}\frac{\partial C}{\partial y} \qquad (11.35b)$$

From Eq. (11.33),

$$\frac{\partial C}{\partial t} - \frac{\partial N}{\partial y} = 0 \quad \text{or using Eq. (11.35a)} \quad \frac{\partial C}{\partial t} - \frac{\partial N}{\partial C}\frac{\partial C}{\partial y} = 0 \qquad (11.36)$$

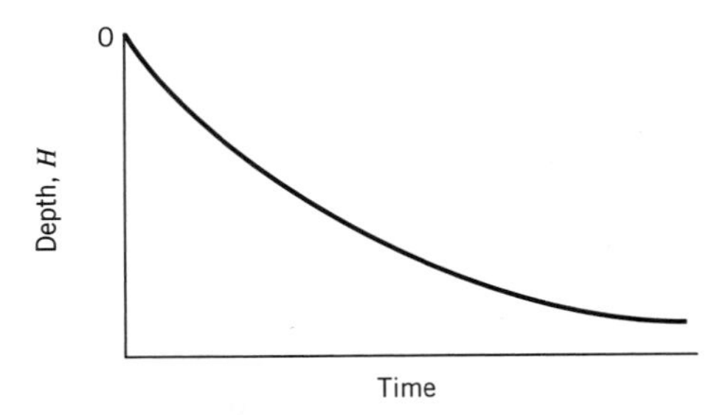

Figure 11.24 Interface settling rate in type III sedimentation.

Using Eq. (11.34),

$$\frac{\partial C}{\partial t} + v\frac{\partial C}{\partial y} + C\frac{\partial v}{\partial y} = 0$$

Substituting Eq. (11.35b) into this equation,

$$\frac{\partial C}{\partial t} + v\frac{\partial C}{\partial y} + C\frac{\partial v}{\partial C}\frac{\partial C}{\partial y} = 0 \qquad \text{and} \qquad \frac{\partial C}{\partial t} + \left(v + C\frac{\partial v}{\partial C}\right)\frac{\partial C}{\partial y} = 0 \quad (11.37)$$

The expression in parentheses in Eq. (11.37) can be obtained by differentiating Eq. (11.31) with respect to C.

$$\frac{\partial N}{\partial C} = -\frac{\partial (Cv)}{\partial C} = -v - C\frac{\partial v}{\partial C} = -V \tag{11.38}$$

Equation (11.38) is the definition of V, which is discussed later. From Eqs. (11.36) and (11.38) it is also found that

$$\frac{\partial C}{\partial t} - \frac{\partial N}{\partial C}\frac{\partial C}{\partial y} = 0 = \frac{\partial C}{\partial t} + V\frac{\partial C}{\partial y} \tag{11.39}$$

From this development, it is seen that V is only a function of concentration because $v = f_1(C)$ and $\partial v/\partial C = f_2(C)$.

The interface between the clarified water and suspended solids zones is moving down at a velocity, $v = -dy/dt$. What is V?

Considering Eq. (11.39) and letting $V = K_1$, a constant, implies that concentration C is constant. Therefore,

$$C(y, t) = K_2 \qquad \text{if } V = K_1$$

Taking the differential of C,

$$dC = \frac{\partial C}{\partial y}dy + \frac{\partial C}{\partial t}dt = 0 \text{ along } C = K_2 \tag{11.40}$$

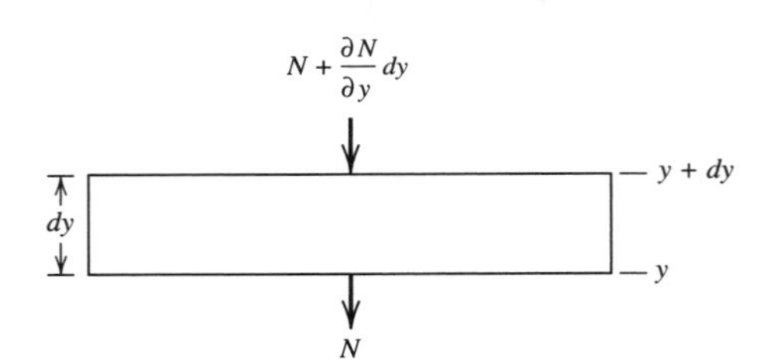

Figure 11.25 Elemental volume of liquid in the cylinder.

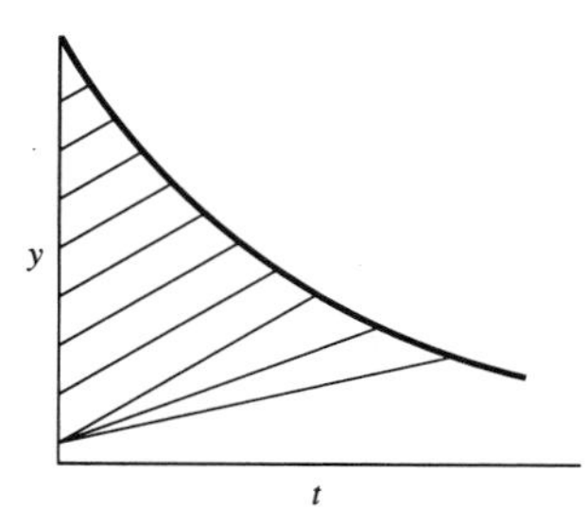

Figure 11.26 Rate at which layers of different concentration are propagated to the surface.

This results in

$$\frac{\partial C}{\partial y}\frac{dy}{dt} + \frac{\partial C}{\partial t} = 0 \tag{11.41}$$

When Eq. (11.41) is compared with Eq. (11.39) it is seen that $V = dy/dt$ along lines of constant C. Elementary calculus shows this to be true for Eq. (11.39). These lines are described by

$$\int_{y_0}^{y} dy = V\int_{0}^{t} dt$$

which integrates to

$$y = y_0 + Vt \tag{11.42}$$

Equation (11.42) holds along each line where C is a constant. On a plot they appear as shown in Fig. 11.26. The lines have a constant slope V and terminate when they meet the interface curve. The lines actually describe the rate at which a layer of a given concentration C is propagated to the surface of the suspension. Because the initial concentration of the suspension was uniform, V is constant for all lines (i.e., they are parallel) until the upper and lower interfaces meet and more highly concentrated layers are propagated to the surface. To design a clarifier properly, the velocity of the slowest moving layer in the clarifier zone (as opposed to the sludge thickening zone) must be assessed. Solving Eqs. (11.39) and (11.42) together gives the interface point.

Practically it is difficult to apply the preceding theory to continuous flow situations and real sludges exhibit behavior that deviates to some extent from the ideal Kynch curve (Vesilind and Jones, 1990). The design of a clarifier is based on the tangent to the settling curve determined at the point where the critical (rate controlling) concentration reaches the surface. Dick and Ewing (1967) have given a discussion of zone sedimentation theory as applied to biological suspensions. Dick and co-workers have developed the theory most commonly applied to continuous flow type III sedimentation.

11.5.2 Design of a Basin for Type III Sedimentation

The data from the progression of batch sedimentation depicted in Figs. 11.21 and 11.24 must be translated to the continuous flow situation. The clarifier must be designed with the minimum surface area to allow for maximum thickening and settling of the suspension. The analysis and design technique for type III continuous flow clarifiers

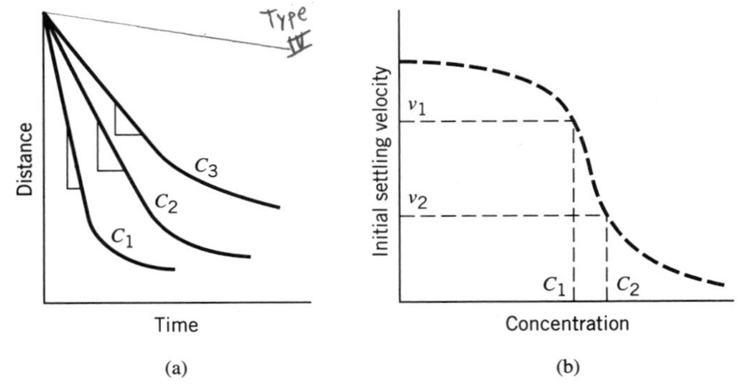

Figure 11.27 Preliminary plots for gravity flux determination.

was developed by Dick (1970). The analysis applies to zone I, where the limiting conditions for clarification exist.

Gravity Flux

Solids in the clarified zone are transported downward by gravity and by bulk transport because of solids removal by the underflow. The hindered settling velocity, v_h, of a suspension is the settling velocity of the interface during the initial straight line portion of the interface settling curve. The gravity flux, N_g, is

$$N_g = C_i v_h \tag{11.43}$$

where

C_i is the initial concentration of the suspension

Different dilutions of the suspension are settled in the laboratory to obtain data for a plot as shown in Fig. 11.27a. The variation of the initial or hindered settling with concentration can be plotted as shown in Fig. 11.27b.

From the curve in Fig. 11.27b, the gravity mass flux can be calculated using Eq. (11.43). The curve can be plotted as shown in Fig. 11.28. At low concentrations of the suspension, the flux is low because of the small amounts of solids; at high solids concentrations the flux is low because of extremely hindered settling.

Vesilind (1979) observed that a general equation with two adjustable parameters, a and b, of the form

$$v_h = ae^{-bC} \tag{11.44a}$$

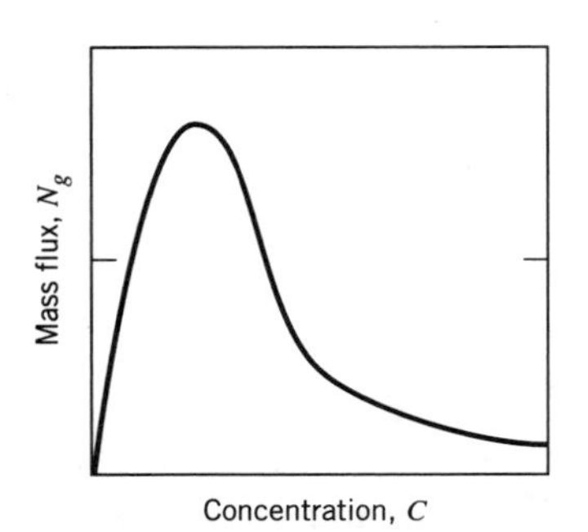

Figure 11.28 Mass flux resulting from gravity.

could be used to describe the settling velocity of a suspension at any concentration, C. Regression analysis can be used to establish the values of a and b. After an extensive investigation of many nonchemically amended activated sludges, Wahlberg and Keinath (1988) found the following regression equation:

$$v_h = [25.3 - 0.061(\text{SVI})]e^{[-0.426+(0.003\,84)(\text{SVI})-(0.000\,054\,3)(\text{SVI})^2]C} \quad (11.44b)$$

where

SVI is the stirred sludge volume index measured in a 1-L cylinder as prescribed by *Standard Methods* (1992). SVI is discussed in Section 17.6.1.

$35 \text{ mL/g} \leq \text{SVI} \leq 220 \text{ mL/g}$

The relation in Eq. (11.44b) does not necessarily hold for plants where chemical addition for phosphorus removal or improved settling is practiced or for other sludges. But for appropriate sludges the equation can be readily used to establish the gravity flux curve. Keinath (1990) has used this relation to develop design charts.

Wilson and Lee (1982) have used the following equation to describe initial settling velocities of the suspension.

$$v_h = aC^b \quad (11.44c)$$

Experimental verification of Eqs. (11.44a)–(11.44c) should be performed. A variety of mathematical formulations of the settling velocity for the suspension have been reported. Once the appropriate equation is established it can be substituted into the flux equations to facilitate a mathematical analysis. Otherwise the discrete v_h data can be used to construct the N_g curve and perform the design as outlined next.

Underflow Flux

In addition to gravity induced settling, the underflow withdrawal increases the downward movement of solids. The underflow velocity, U_b, is

$$U_b = \frac{Q_u}{A_s} \quad (11.45)$$

where

Q_u is the volumetric underflow flow rate

A_s is the surface area of the clarifier

The underflow flux in the clarified zone is

$$N_u = CU_b \quad (11.46)$$

where

N_u is the underflow flux

Note that the underflow flux varies with the local concentration.

Total Flux

The total flux, N, is the sum of the above two fluxes and is plotted in Fig. 11.29 along with its component fluxes.

$$N = N_g + N_u = Cv_h + CU_b \quad (11.47)$$

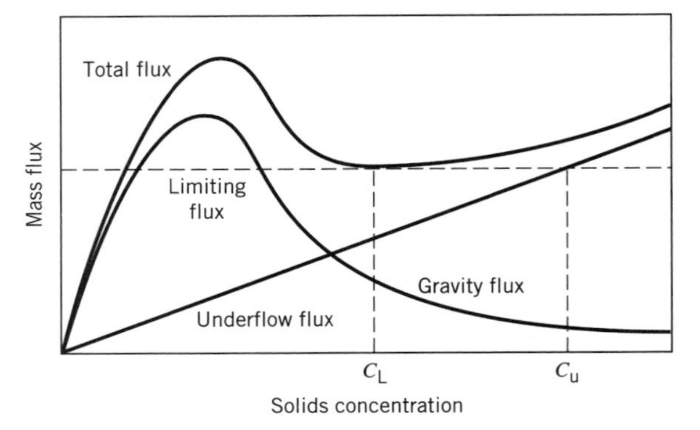

Figure 11.29 Total mass flux as a function of concentration.

The solids transmitting capacity of each layer varies with the concentration of the layer. The overall flux usually has a maximum and minimum, as shown in Fig. 11.29, and the minimum point on the curve determines the design area of the clarifier. The minimum solids handling capacity of the suspension is the limiting flux, N_L. The concentration at which the limiting flux occurs is C_L. The clarifier must be able to handle the mass flow rate of solids coming into it; therefore,

$$A_s N_L \geq QC_0 \tag{11.48}$$

where
 Q is the influent volumetric flow rate
 C_0 is the concentration of solids in the influent

The design area of the clarifier is then

$$A_s \geq \frac{QC_0}{N_L} \tag{11.49}$$

Effectively the area of the clarifier is being sized to maintain the upward velocity (flux) of water at or less than the minimum settling velocity that will occur in the solids suspension as the concentration in the suspension increases to higher values. It is assumed that the influent jet is dispersed and distributed over the whole plan area of the clarifier.

At the withdrawal point for the thickened solids at the bottom of the clarifier, there is no gravity flux. All solids are removed by bulk flow (U_b).

$$C_u Q_u = QC_0 = A_s N_L \tag{11.50}$$

By extending the limiting flux line to its intersection with the underflow flux line and dropping a perpendicular, the concentration of solids in the underflow, C_u, can be determined. Choosing a lower value for $U_b(Q_u)$ will result in a higher value of C_u and the total flux curve will shift down. Of course, the value of C_u must be physically attainable. The values of U_b for biological sludges are typically in the range of 25–50 cm/h (0.84–1.7 ft/h).

There is no upward flux of water in the lower compaction zone section in the clarifier. The influent jet is dissipated and distributed above the bottom layers of the compaction zone. Theoretically, the compaction zone can be designed to hold the sludge as long as desired. However, on a practical basis, biological sludges cannot be held in the clarifier for excessive time or denitrification and anaerobic decomposition become significant. Any oxygen present in the influent to the clarifier will be rapidly consumed in the clarifier. Denitrification occurs if a significant amount of nitrate is present, with the production of nitrogen gas. In time the microorganisms will acclimate to the anaerobic conditions and begin to anaerobically metabolize residual substrate present in the wastewater and, also, the microorganisms will begin to digest themselves. Anaerobic decay results in the production of gases such as methane that have low solubilities. Carbon dioxide will also be produced in significant quantities. Gases will leave solution and form bubbles, which will attach to some settled particles and buoy them up and out of the clarifier, significantly deteriorating the quality of the clarified effluent.

Alternate Method for Determining the Minimum Area of a Clarifier for Type III Sedimentation

There is a more convenient method to size the area of a clarifier considering different underflow rates. The underflow concentration, C_u, is chosen based on the most reasonable lab or field data. There are five unknowns needed to solve for the minimum area of the clarifier: N_L, C_L, v_L (the velocity of the suspension resulting from gravity at the limiting flux), U_b, and C_u.

At the limiting flux in the clarifier, the following relations hold:

$$N_L = C_L v_L + C_L U_b = N_{gL} + C_L U_b \tag{11.51}$$

where

N_{gL} is the gravity flux component of the limiting flux

In the underflow:

$$N_L = C_u U_b \tag{11.52}$$

For a given set of conditions, N_L is constant; Eq. (11.51) can be rearranged to

$$N_{gL} = N_L - C_L U_b \tag{11.53}$$

If C_u and U_b are chosen, Eqs. (11.52) and (11.53) along with the gravity flux curve define the system.

The procedure to use the equations is as follows. Consider the equation,

$$N_g = N_L - CU_b \tag{11.54}$$

which is similar to Eq. (11.53). The following conditions apply.

$$C = 0 \qquad N_g = N_L$$
$$C = C_u \qquad N_g = 0$$

Also consider the equation

$$N_g = Cv_L \tag{11.55}$$

If v_L is constant, this curve produces a straight line beginning at the origin. At the limiting flux, $C = C_L$ and $N_{gL} = C_L v_L$. Equations (11.54) and (11.55) and the N_g curve must intersect at the same point. The point of intersection occurs after the rise

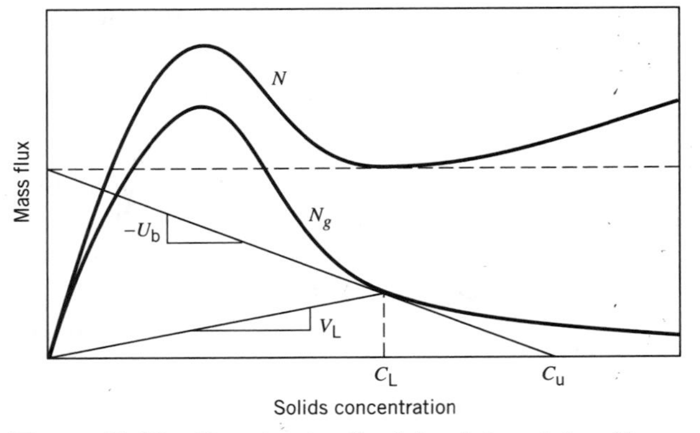

Figure 11.30 Alternate method for determining N_L.

in the N_g curve. The situation is shown in Fig. 11.30. The total flux curve (N) is also drawn for reference. It is not necessary to draw the line for Eq. (11.55). Once U_b is selected, which determines the slope of Eq. (11.54), the line is simply moved to its point of tangency with the N_g curve and N_L and C_u are read at the intersection of the line with the y and x axes, respectively.

This method conveniently allows the designer to evaluate the effects of different underflow velocities on the design. Different values of U_b should be checked, corresponding to the expected maximum and minimum concentrations of thickened sludge in the underflow. In a biological treatment unit the concentration of sludge in the return line to the biological unit in turn influences the volume of flow that is recycled and consequently the influent volumetric flow rate to the clarifier.

■ Example 11.2 Design of a Clarifier Using the Solids Flux Method

The data for initial hindered settling velocities of a suspension are given in the following table. Size the area of a clarifier to handle a flow of 2 300 m^3/d containing a TSS concentration of 2 100 mg/L. The minimum underflow SS concentration is 10 000 mg/L. What is the rate of underflow?

C, mg/L	1 000	2 000	4 000	6 000	8 000	10 000	15 000
v_h, m/h	3.74	2.82	2.26	1.04	0.49	0.25	0.072

The N_g values were obtained from $N_g = Cv_h$ and are given in the following table.

C, mg/L	1 000	2 000	4 000	6 000	8 000	10 000	15 000
N_g, $g/m^2/s$	1.04	1.57	2.51	1.73	1.09	0.69	0.30

They were plotted in the following figure.

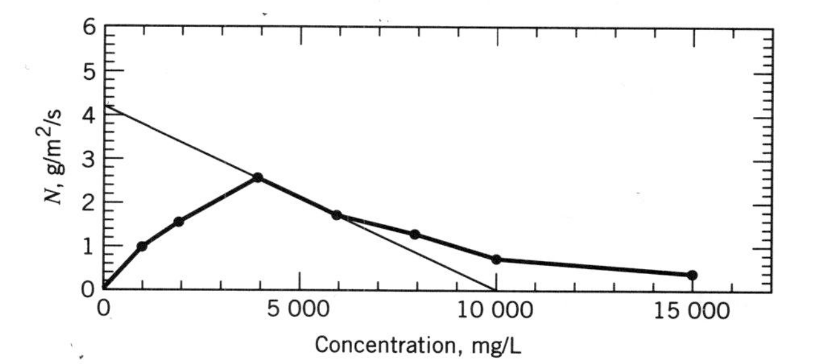

A tangent was drawn to the N_g curve passing intercepting the x axis at the desired underflow concentration, 10 000 mg/L. From the intersection of the tangent line and the y axis, the limiting flux is 4.2 g/m²/s.

From Eq. (11.49) the minimum surface area of the clarifier is

$$A_s = \frac{QC_0}{N_L} = \frac{(2\,300\ \text{m}^3/\text{d})(2\,100\ \text{mg/L})}{4.2\ \text{g/m}^2/\text{s}}\left(\frac{1}{86\,400\ \text{s}}\right)\left(\frac{1\,000\ \text{L}}{1\ \text{m}^3}\right)\left(\frac{1\ \text{g}}{1\,000\ \text{mg}}\right) = 13.3\ \text{m}^2$$

$$QC_0 = Q_u C_u$$

$$Q_u = \frac{C_0}{C_u}Q = \frac{2\,100\ \text{mg/L}}{10\,000\ \text{mg/L}}(2\,300\ \text{m}^3/\text{d}) = 483\ \text{m}^3/\text{d}$$

11.6 WEIR–LAUNDER DESIGN

Rising water in a sedimentation basin flows over a weir into a channel or launder that conveys the collected water to the exit channel or pipe. Weirs are located as far as possible from the basin inlet. Weir loading rates are specified to prevent strong updrafts that would carry solids out of the basin. Weir loading rates are specified later on for clarifiers in water and wastewater treatment. As the depth of the basin increases higher weir loading rates have less influence on the performance of the clarifier.

For basins that are not covered, the weir is frequently of the V-notch type (Fig. 11.31) to minimize wind effects. Also, straight edge weirs that are not perfectly level do not have uniform flow over the entire weir. This will cause uneven flow patterns in the sedimentation basin and deterioration of its performance. Submerged orifices are also sometimes designed to discharge effluent from a clarifier.

V-Notch weirs must have a depth that permits discharge of the peak flow through the basin. Spacing of the notches is in the range of 150–300 mm (6–12 in.) center to center.

The discharge through a V-notch weir is given by (Vennard and Street, 1982),

$$Q = \frac{8}{15}C_d\sqrt{2g}\tan\frac{\theta}{2}H_w^{5/2} \tag{11.56}$$

Figure 11.31 V-notch weir.

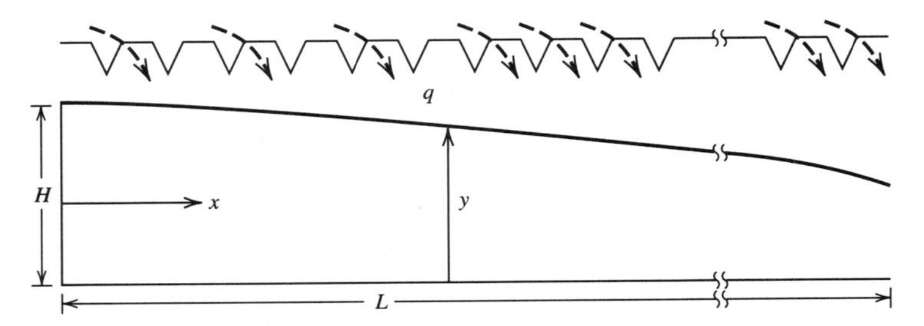

Figure 11.32 Definition sketch for flow in a launder.

where
θ is the angle of the V-notch
C_d is the discharge coefficient
H_w is the depth of water over the weir

The discharge coefficient is around 0.62. The depth–width relation for a V-notch (triangular) weir is

$$\frac{w}{2H_w} = \tan\frac{\theta}{2} \tag{11.57}$$

where
w is the width of the weir at any height H_w

Flow in the launder is spatially varied flow; refer to the definition sketch in Fig. 11.32. These flumes are usually built with no slope. The momentum principle is used for analysis.

The following symbols are used in the development:

A cross sectional area
b launder width
F force
g acceleration of gravity
H water depth at upstream end of launder
L launder length
P pressure
q discharge per unit length
Q total volumetric discharge rate
x distance from upstream end of launder
y depth of flow in the launder at any x
v flow velocity
ρ density of water

Friction (F_f in Fig. 11.33) will be ignored in the development of the governing equation. It will be accounted for after the equation is derived. This approach is an approximate but reasonable solution of the problem. Steady state conditions are also assumed. Benefield et al. (1984) give a program to perform the numerical integration of the flow equations for this lateral spillway channel and arrive at a more exact solution.

In a circular basin, the flow into the launder is uniform around the circumference. Furthermore, the flow in the launder is symmetrical for the two halves of the launder. At the point opposite from the launder discharge outlet, the flow splits and travels in opposite directions. Therefore the development is for half of the channel with a length of $L/2$. To develop the equation governing flow in either half of the channel, examine an elemental volume (Fig. 11.33) indicating parameters to be used in a momentum analysis.

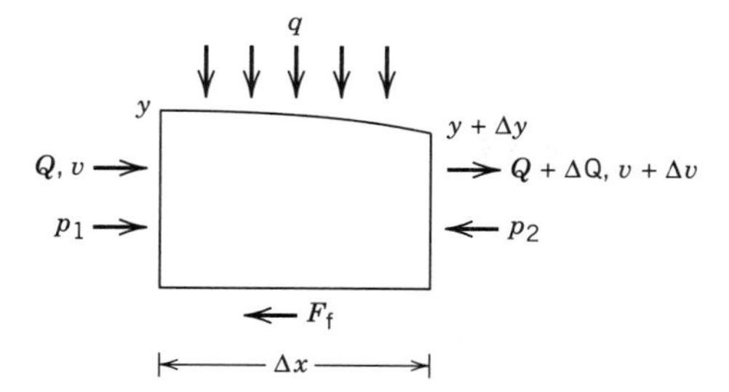

Figure 11.33 Elemental volume in a launder.

The momentum principle is

$$\Sigma \vec{F} = \frac{d(m\vec{v})}{dt} \tag{11.59}$$

and it is applied in the x direction. The following relations are used:

$$Q = qx \qquad dQ = q\,dx \qquad A = by \qquad dA = b\,dy$$

$$v = \frac{Q}{A} = \frac{Q}{by} \qquad dv = \frac{dQ}{A} - \frac{Q}{A^2}dA$$

The resultant pressure force is

$$\begin{aligned} F = p_1A_1 - p_2A_2 &= \rho gby\frac{y}{2} - \rho gb(y+dy)\frac{(y+dy)}{2} \\ &= \tfrac{1}{2}\rho gby^2 - \tfrac{1}{2}\rho gb[y^2 + 2y\,dy + (dy)^2] \\ &= -\rho gby\,dy \end{aligned} \tag{11.60}$$

Evaluating the right-hand side of the momentum equation for the horizontal direction:

$$\begin{aligned} \frac{\Delta(mv)}{\Delta t} &= \rho(Q+dQ)(v+dv) - \rho Qv \\ &= \rho Qv + \rho Q\,dv + \rho v\,dQ + \rho\,dQdv - \rho Qv \\ &= \rho Q\,dv + \rho v\,dQ \end{aligned} \tag{11.61}$$

Equating Eqs. (11.60) and (11.61),

$$\rho Q\,dv + \rho v\,dQ = -\rho gby\,dy$$

Substituting for dv in the above equation,

$$\frac{Q}{A}dQ - \frac{Q^2}{A^2}dA + \frac{Q}{A}dQ = -gb\,dy$$

Substituting for A, dA, Q, and dQ results in

$$\frac{2qx}{by}q\,dx - \frac{q^2x^2}{by^2}dy = -gby\,dy$$

The above equation can be simplified to

$$2x\frac{dx}{dy} - \frac{x^2}{y} = -g\frac{b^2y^2}{q^2} \quad (11.62a) \qquad \text{or} \qquad \frac{d(x^2)}{dy} - \frac{x^2}{y} = -g\frac{b^2y^2}{q^2} \quad (11.62b)$$

This equation can be integrated to

$$x^2 = Cy - \frac{gb^2y^3}{2q^2} \tag{11.63}$$

The boundary condition is

$$x = 0, \qquad y = H$$

The boundary condition is used to find the integration constant, C.

$$C = \frac{gb^2H^2}{2q^2}$$

Substituting for C, the equation for H becomes

$$H = \left(y^2 + \frac{2q^2x^2}{gb^2y}\right)^{0.5} \tag{11.64}$$

To determine H, it must be recognized that a free fall condition exists at the end of the launder, which means that critical flow exists there. For critical flow, the depth is y_c and the Froude number (Fr) is equal to 1.

$$\mathrm{Fr} = 1 = \frac{v}{\sqrt{gy_c}} \qquad \text{or} \qquad v = \sqrt{gy_c} \tag{11.65}$$

The velocity at the end of the launder is

$$v = \frac{Q}{A} = \frac{qL}{2by_c}$$

Substituting this into Eq. (11.65) and solving for y_c yields

$$y_c = \left[\frac{(qL)^2}{4b^2g}\right]^{1/3} \tag{11.66}$$

Equation (11.66) is used to find y_c, which occurs at $x = L/2$. This information may be used in Eq. (11.64) to find H.

Headloss can be estimated from the Chezy or Manning equations. Using Manning's equation,

$$v = \frac{1}{n} R^{2/3} S^{1/2} \tag{11.67}$$

where

R is the hydraulic radius $= \dfrac{by}{b + 2y}$
S is the slope of the energy line
n is a roughness coefficient

The value of n is 0.011 to 0.015 for smooth concrete; n values for other materials can be obtained from standard references (Chow, 1959). The value of n should be slightly increased to account for curvature in the lateral and turbulence from the falling water. For smooth concrete use $n = 0.014$. At any distance x, $v = qx/by$. Substituting this into the above equation and solving for S,

$$S = \left[\left(\frac{qxn}{by}\right)\left(\frac{b + 2y}{yb}\right)^{2/3}\right]^2 = \frac{(qx)^2n^2}{(by)^{10/3}}(b + 2y)^{4/3} \tag{11.68}$$

Equation (11.64) is used to determine the value of y at any x. Equation (11.68) is applied to find the energy slope at the midpoint of the flume, S_{MP}, and at the end (depth of flow is y_c), S_c. The energy slope at the beginning of the flume is 0. The headloss (h_L) in each half of the flume is estimated from

$$h_L = \left(\frac{0 + S_{MP}}{2}\right)\frac{L}{4} + \left(\frac{S_{MP} + S_c}{2}\right)\frac{L}{4} \tag{11.69}$$

The headloss is added to the value of H determined from Eq. (11.64). The depth of the launder is increased by 5–10 cm (2–4 in.) to ensure free fall from the weir. Freeboard depth is also added to prevent splashing and to account for nonuniform flow variation over the weir because of the wind or flow distribution in the clarifier and uncertainty in the estimation of the flow. A freeboard of 10 to 20% of the depth calculated from the headloss and H is used. The total depth of the launder is the sum of freeboard, H, headloss, and free fall allowance.

11.7 CLARIFIER DESIGN FOR WATER AND WASTEWATER TREATMENT

The design ranges for clarification basins used in water treatment plant are highly variable depending on the quality of raw water and type of floc formed, which is dependent on the coagulant used and the operation of the flocculation process (see Chapter 13). There are also a variety of designs used ranging from rectangular horizontal flow basins to circular and lamella clarifiers. Solids contact clarifiers incorporate coagulation, flocculation, and clarification into a single unit and are usually circular. Ranges for design variables for different configurations are given in Table 11.6. Handbooks or other references should be consulted to find the narrower ranges for different

TABLE 11.6 Clarifiers in Water Treatment[a]

Item	Value
Rectangular and Circular Clarifiers	
Depth, m (ft)	2.4–4.9 (8–16)
Overflow rate, $m^3/m^2/d$ ($gal/ft^2/d$)	20–70 (490–1 720)
Weir loading rate, $m^3/m/d$ (gal/ft/d)	Less than 1 250 (100 000)
Maximum length of rectangular basin, m (ft)	70–75 (230–250)
Circular basin maximum diameter, m (ft)	38 (125)
Upflow Solids Contact Clarifiers	
Depth, m (ft)	2.5–3 (8–10)
Overflow rate, $m^3/m^2/d$ ($gal/ft^2/d$)	24–550 (590–13 500)
Inclined Tube or Lamella Clarifiers	
Inclined length, m (ft)	1–2 (3.3–6.6)
Angle of inclination (°)	7–60
Tube diameter or plate spacing, cm (in)	Near 5 (2)
Overflow rates based on plan area, $m^3/m^2/d$ ($gal/ft^2/d$)	2–8 times rate for conventional clarifiers, 88–178 (2 160–4 370)
Depth, m (ft)	6–7 (20–23)

[a]Compiled from Gregory and Zabel (1990), ASCE and AWWA (1990), AWWA (1971), Culp and Culp (1974).

TABLE 11.7 Clarifiers in Wastewater Treatment[a]

Item	Value
Primary Clarifiers[b]	
Overflow rate, $m^3/m^2/d$ ($gal/ft^2/d$)	
For average dry weather flow rate	32–49 (785–1 200)
For peak flow condition	49–122 (1 200–3 000)
Sidewater depth, m (ft)	2.1–5 (6.9–16.4)
Weir loading rate[c], $m^3/m/d$ (gal/ft/d)	125–500 (10 000–40 000)
Secondary Clarifiers	
Overflow rate[d], $m^3/m^2/d$ ($gal/ft^2/d$)	
For average dry weather flow rate	16–29 (393–712)
For peak flow condition	41–65 (1 006–1 595)
Sidewater depth, m (ft)	3.0–5.5 (9.8–18)
Floor slope	Nearly flat to 1 : 12
Maximum diameter, m (ft)	46 (150)

[a]From WEF and ASCE (1992), *Design of Municipal Wastewater Treatment Plants,* vol. 1, WEF, © WEF 1992.
[b]Criteria are based on the maximum ranges specified by a number of firms and agencies reported in WEF and ASCE (1992).
[c]Generally for average flow conditions.
[d]For circular or rectangular tanks.

types of sediment and coagulants. The majority of suspended solids removal will occur in the sedimentation basin in a water treatment process.

Dissolved air flotation is also used for clarification of flocculated waters in water treatment plants. See Chapter 19 for a description of this process.

Clarifiers used in wastewater treatment may be rectangular, square, or circular. Circular clarifiers are most commonly used for both primary and secondary clarifiers. Design guidelines for primary and secondary clarifiers are given in Table 11.7. Variation among design codes is large. Primary clarifiers are designed more conservatively if sedimentation is the only treatment and if activated sludge is being returned to the primary clarifiers. Rectangular tanks are generally designed with the same criteria as circular tanks. Length to width ratios employed for design of rectangular primary clarifiers range from 3 : 1 to 5 : 1, although values for existing tanks range from 1.5 : 1 to 15 : 1 (WEF and ASCE, 1992).

Properly designed and operated primary clarifiers should remove 50–65% of the influent TSS.

■ Example 11.3a Clarifier Design

Determine the number of clarifiers and surface area of a primary clarifier system for a wastewater treatment plant. Use circular clarifiers that meet criteria given in Table 11.7. The minimum, average, and maximum flows have been determined to be 0.174, 0.347, and 0.868 m^3/s (6.18, 12.4, and 30.9 ft^3/s), respectively. These flows correspond to 15 000, 30 000, and 75 000 m^3/d (3.96, 7.93, and 19.8 Mgal/d), respectively.

A minimum of two sedimentation basins should be installed. For two basins, with one basin out of service, the maximum flow into the basin will be 0.868 m^3/s. At low flow conditions, only one basin will be in service. The required surface area of the basins at various flow conditions was calculated using the criteria in Table 11.7. The results are given in the following table.

Flow m^3/d	Number of basins	Surface overflow rate m^3/m^2/d	Surface area m^2
30 000	2	32	469
30 000	2	49	306
75 000	2	49	765
75 000	2	122	307
75 000	1	122	615

From the results in this table it appears that a surface area of 450 m^2 for each clarifier will suffice. The overflow rates for various flow conditions for the design surface area are given in the following table.

Flow, m^3/m^2/d	Number of basins	Q/A, m^3/m^2/d
15 000	1	33.3
30 000	2	33.3
75 000	2	83.3
75 000	1	166.6

The peak flow condition with only one basin in service is above the recommended value but this condition is rare. Furthermore, in this system activated sludge–secondary clarification will follow primary treatment and provide further opportunity to remove the suspended solids.

The radius of each clarifier will be

$$\pi r^2 = A \qquad r = \sqrt{A/\pi} = \sqrt{(450 \text{ m}^2)/\pi} = 12.0 \text{ m (39.4 ft)}$$

■ Example 11.3b Weir-Launder Design

Design the overflow weirs and collection launders for the clarifiers in Example 11.3a. The launders will be made from smooth concrete; take n to be 0.014.

The weir length around the perimeter of a basin is

$$L = 2\pi r = 2\pi(12.0 \text{ m}) = 75.4 \text{ m}$$

The weir loading rates at peak flow with one and both basins in service are

$$\text{Both basins: } q = Q/L = (75\,000 \text{ m}^3/\text{d})/2(75.4 \text{ m}) = 497 \text{ m}^3/\text{m/d} = 0.005\,75 \text{ m}^3/\text{m/s}$$

$$\text{Single basin: } q = (75\,000 \text{ m}^3/\text{d})/(75.4 \text{ m}) = 995 \text{ m}^3/\text{m/d} = 0.011\,5 \text{ m}^3/\text{m/s}$$

The weir loading rate is satisfactory at average flow, which is less than the peak flow with two basins in service.

The weir and collection launder must be designed for the peak flow condition with only one basin in service. A V-notch weir will be used with an angle of 90° and spacing of 25.0 cm center to center (c/c). The number of V-notches is

$$n = (75.4 \text{ m})/(0.250 \text{ m}) = 301.6$$

Use 302 notches.

The flow per notch, q_n is

$$q_n = Q/n = (0.868\ m^3/s)/302 = 0.002\ 87\ m^3/s\ (0.101\ ft^3/s)$$

For a discharge coefficient of 0.62 and peak flow through one basin, the depth of water over the crest of a V-notch is (Eq. 11.56)

$$H_w = \left(\frac{15Q}{8C_d\sqrt{2g}\tan\frac{\theta}{2}}\right)^{2/5} = \left[\frac{15(0.002\ 87\ m^3/s)}{8(0.62)\sqrt{2(9.81\ m/s^2)}}\right]^{2/5} = 0.082\ 6\ m = 8.26\ cm\ (3.25\ in.)$$

The elevation of water over the crest of the weir should be increased above this value by a safety factor. A safety factor of 15% will be used. The total depth of the weir is

$$H_w = 1.15(8.26\ cm) = 9.50\ cm\ (3.74\ in.)$$

The width of a V-notch at the top is

$$w = 2H_w = 19.0\ cm\ (7.48\ in.)$$

The launder width is somewhat arbitrary; after a few trial calculations a launder width of 0.70 m (2.30 ft) was chosen. The peak flow with only one basin in service is used. To establish the depth of flow in the launder Eq. (11.66) is used to find the critical depth at the discharge point from the launder.

$$y_c = \left[\frac{(qL)^2}{4b^2g}\right]^{1/3} = \left\{\frac{[(0.011\ 5\ m^3/m/s)(75.4\ m)]^2}{4(0.70\ m)^2(9.81\ m/s^2)}\right\}^{1/3} = 0.339\ m\ (1.11\ ft)$$

Now apply Eq. (11.64) to find the upstream depth.

$$H = \left(y^2 + \frac{2q^2x^2}{gb^2y}\right)^{0.5} = \left[(0.339\ m)^2 + \frac{2(0.011\ 5\ m^3/m/s)^2\left(\frac{75.4\ m}{2}\right)^2}{(9.81\ m/s^2)(0.70\ m)^2(0.339\ m)}\right]^{0.5}$$

$$= 0.588\ m\ (1.93\ ft)$$

The headlosses are found by applying Eq. (11.67) at the midpoint ($x = L/4$) found from Eq. (11.63).

$$H^2 = y^2 + \frac{2q^2x^2}{gb^2y}$$

$$(0.588\ m)^2 = y_{mp}^2 + \frac{2(0.011\ 5\ m^3/m/d)^2\left(\frac{75.4\ m}{4}\right)^2}{(9.81\ m/s^2)(0.70\ m)^2 y_{mp}} \Rightarrow y_{mp} = 0.56\ m\ (1.84\ ft)$$

$$S_{MP} = \frac{\left(q\frac{L}{4}\right)^2 n^2}{(by_{mp})^{10/3}}(b + 2y_{mp})^{4/3}$$

$$= \frac{\left[(0.011\ 5\ m^3/m/d)\left(\frac{75.4\ m}{4}\right)(0.014)\right]^2}{[(0.70\ m)(0.56\ m)]^{10/3}}[0.70\ m + 2(0.56\ m)]^{4/3} = 0.000\ 464$$

At the discharge end of the launder,

$$S_c = \frac{\left[(0.011\,5\ \mathrm{m^3/m/d})\left(\frac{75.4\ \mathrm{m}}{2}\right)(0.014)\right]^2}{[(0.70\ \mathrm{m})(0.339\ \mathrm{m})]^{10/3}}[0.70\ \mathrm{m} + 2(0.339\ \mathrm{m})]^{4/3} = 0.006\,83$$

The total headloss is approximately

$$h_L = \left(\frac{0 + S_{MP}}{2}\right)\frac{L}{4} + \left(\frac{S_{MP} + S_c}{2}\right)\frac{L}{4} = \left(\frac{0.000\,464}{2}\right)\left(\frac{75.4\ \mathrm{m}}{4}\right)$$
$$+ \left(\frac{0.000\,464 + 0.006\,83}{2}\right)\left(\frac{75.4\ \mathrm{m}}{4}\right)$$
$$= 0.073\ \mathrm{m}\ (0.24\ \mathrm{ft})$$

The total depth of flow in the basin is 0.588 m + 0.073 m = 0.661 m (2.16 ft)

Using a factor of 10% for freeboard and adding 5 cm to ensure free fall, the total depth provided in the launder is

$$H_t = 1.10(0.661\ \mathrm{m}) + 0.05\ \mathrm{m} = 0.777\ \mathrm{m},\ \text{say}\ 0.78\ \mathrm{m}\ (2.56\ \mathrm{ft})$$

Depth in Sedimentation Basins

Because sedimentation is theoretically independent of depth, the question may be asked why any significant depth at all is provided. All models assume that the full cross-sectional or surface area of the clarifier is utilized equally by the incoming flow and that suspensions are uniformly distributed. Nonideal flow patterns that result in dead volume, in particular, reduce the effective area as well as the effective volume of the clarifier. Providing more depth allows flow patterns to develop at the expense of the additional volume but with improvement in performance. Also, providing more depth minimizes scour of settled solids. Figure 11.34 shows a circular clarifier design in which the location of the effluent weir promotes a flow pattern that utilizes more of the volume of the clarifier.

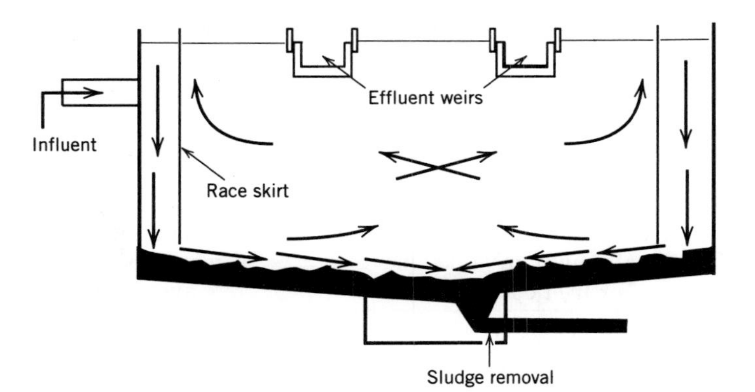

Figure 11.34 Locating the collection weirs near the center of a circular clarifier improves the hydraulic performance of the clarifier. Courtesy of Lakeside Equipment Corp.

11.8 GRIT CHAMBERS

Grit chambers are sedimentation basins placed at the front of wastewater treatment plants to remove sand, egg shells, coffee grounds, and other nonputrescible materials that may clog channels or cause abrasive wear of pumps and other devices. Because the grit material is nonputrescible no further treatment is required before disposal at the ultimate site. The grit is collected in containers or directly in truck beds and hauled away at required intervals.

Grit is defined as sand, gravel, or other mineral matter that has a nominal diameter of 0.15–0.20 mm or larger. Actually, grit will also include smaller mineral particles that may settle as well as nonputrescible organic matter such as rags, coffee grounds, vegetable cuttings, ash, clinker, wood pieces, and tea leaves. Even though some of the grit components such as coffee grounds are organic, they are essentially nonbiodegradable over time spans for grit collection and disposal. The quality and quantity of grit in the sewage determine the design factors and choice of grit removal method.

The amount of grit collected is a function of the removal device, its operation, and the quantity of grit in the sewage and therefore varies over a wide range. Table 11.8 gives typical values for grit quantities. Grit solids content varies from 35 to 80% and volatile content from 1 to 55% (USEPA, 1979). Grit that is washed should achieve a solids content of 70–80%, with a minimum of putrescible matter. The bulk density of grit is from 1 450 to 1 750 kg/m^3 (90–110 lb/ft^3).

Generally grit chambers are designed to remove all particles with a nominal diameter of 0.20 mm (particles retained on a 65-mesh screen) or larger and with a s.g. of 2.65 (sand) (Camp, 1942). The settling velocity of these particles at 10°C is usually taken to be 2.3 cm/s (4.5 ft/min) based on curves of sewage grit settling velocities given by Camp (1942). Using Eq. (11.10) the settling velocity of a particle with these characteristics can be calculated to be near 2.3 cm/s (4.5 ft/min) but the angularity of typical grit particles causes a small deviation from the calculated value. Sometimes grit removal devices are designed to remove 0.15-mm sand particles (retained on a 100-mesh screen) with a settling velocity of 1.30 cm/s (2.6 ft/min) taken from Camp's curve.

It is not desirable to remove any organic matter in grit chambers because no further treatment of the grit is necessary or provided. The chamber must be designed to scour the lighter organic particles while the heavier grit particles remain settled. Different types of devices can accomplish this: (1) constant velocity horizontal flow channels, (2) rectangular grit chambers with a grit washing device, and (3) aerated grit chambers.

TABLE 11.8 Estimated Grit Quantities[a]

Type of system	Average quantity of grit (typical range), m^3/1 000 m^3 or ft^3/1 000 ft^3	Ratios of maximum day to average day
Separate	0.004–0.037	1.5 to 3.0:1
Combined	0.004–0.18	3.0 to 15.0:1

[a]From WEF and ASCE (1992), *Design of Municipal Wastewater Treatment Plants,* vol. 1, WEF, © WEF 1992.

11.8.1 Horizontal Flow Grit Chambers

Horizontal flow grit chambers are open channels with detention times sufficient to allow design particles to settle and designed to maintain a constant velocity sufficient to scour organics. A weir at the end of the channel or channel geometry provides velocity control. These types of grit removal devices are more commonly found in older installations. They may be manually cleaned or have mechanical sludge (grit) scrapers installed in them.

The Camp–Shields equation (Camp, 1942) is used to estimate the velocity of scour necessary to resuspend settled organics.

$$v_s = \sqrt{\frac{8kgd}{f}\left(\frac{\rho_p - \rho}{\rho}\right)} \tag{11.70}$$

where

v_s is the velocity of scour
d is nominal diameter of the particle
k is an empirically determined constant
f is the Darcy–Weisbach friction factor

The above equation is based on a tractive force analysis. The constant k is related to the "stickiness" of the organic material. The values of f and k were found by Camp (1942) to be 0.02 and 0.04–0.06, respectively, for domestic sewage.

It is common practice to maintain a horizontal flow velocity between 15 and 30 cm/s (0.50–1.0 ft/s) to keep organic particles in suspension. A velocity of 38 cm/s (1.25 ft/s) will scour 0.2-mm diameter quartz particles or organic particles up to a nominal diameter of 1.5 mm.

The problem in designing a horizontal flow chamber is to maintain the constant velocity in the channel, although varying quantities of flow are passed through the chamber over a 24-h period. If the channel is rectangular and discharges over a rectangular weir, the discharge relation based on Bernoulli's equation is

$$Q = C_d A\sqrt{2gH} = CwH^{3/2} \tag{11.71}$$

where

w is the width of the channel
A is the cross-sectional area of the channel
C_d is the discharge coefficient
C is equal to $C_d\sqrt{2g}$
H is depth of flow in the channel

The horizontal velocity, v_h, is related to the discharge rate and channel geometry by

$$v_h = \frac{Q}{A} = \frac{Q}{wH} = CH^{1/2} = C\left(\frac{Q}{Cw}\right)^{1/3} \tag{11.72}$$

If the ratio of the maximum to the minimum volumetric flow ($Q_{max} : Q_{min}$) is 5 : 1, which is common, the ratio of the corresponding maximum and minimum horizontal

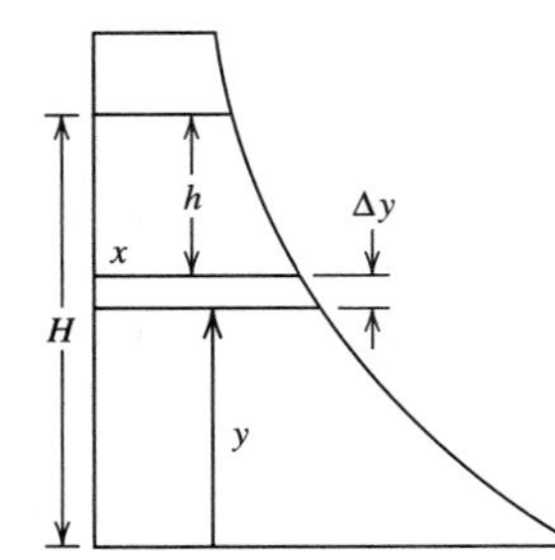

Figure 11.35 Definition sketch for grit chamber weir.

velocities ($v_{hmax} : v_{hmin}$) is

$$\frac{v_{hmax}}{v_{hmin}} = \frac{\dfrac{Q_{max}}{wH_{max}}}{\dfrac{Q_{min}}{wH_{min}}} = \frac{Q_{max}}{Q_{min}}\frac{H_{min}}{H_{max}}$$

$$\frac{v_{hmax}}{v_{hmin}} = \frac{Q_{max}}{Q_{min}}\frac{\left(\dfrac{Q_{min}}{Cw}\right)^{2/3}}{\left(\dfrac{Q_{max}}{Cw}\right)^{2/3}} = \left(\frac{Q_{max}}{Q_{min}}\right)^{1/3} = 5^{1/3} = 1.71$$

where

H_{max} and H_{min} correspond to Q_{max} and Q_{min}, respectively

Therefore, the shape of either the channel or the weir has to be modified to maintain a satisfactory flow-through velocity.

Velocity Control with a Weir

A flow control weir at the end of the channel is needed that satisfies the requirement that

$$v_h = \frac{Q}{wH} = \text{constant} \qquad \text{i.e.,} \qquad wv_h = \frac{Q}{H} = \text{constant} \tag{11.73}$$

because the width of the channel is constant. The weir is known as a Sutro or proportional weir.

What is the shape of the weir? Figure 11.35 is a definition sketch used to solve the problem. As shown in the figure, the weir is an opening through which the fluid moves as opposed to a plate over which the water flows. In Fig. 11.35 the following definitions apply:

y is the distance above the weir crest

x is the width of the weir at any point y

h is the depth of water above the elevation y

The head, h, over any elemental area $x\,dy$ is given by

$$h = H - y$$

The flow through an elemental area dA is

$$dQ = C_d\sqrt{2gh}\,dA = C_d\sqrt{2g(H - y)}\,x\,dy \tag{11.74}$$

The total flow is

$$Q = \int_0^Q dQ = C\int_0^H \sqrt{H - y}\,x\,dy \tag{11.75}$$

where

$C = C_d\sqrt{2g}$

The problem is to find a function for x such that Eq. (11.75) satisfies Eq. (11.73). The section below has the procedure to arrive at the required function that is given in Eq. (11.76).

$$xy^{1/2} = k \qquad \text{or} \qquad x = \frac{k}{y^{1/2}} \tag{11.76}$$

where

k is a constant

The function in Eq. (11.76) should be checked to see that it does satisfy the constraint. Substituting Eq. (11.76) into Eq. (11.75),

$$Q = C\int_0^H k\sqrt{\frac{H - y}{y}}\,dy \tag{11.77}$$

To integrate the above equation, let $z = y/H$, then $dy = H\,dz$.

Also, at

$$y = 0,\ z = 0; \qquad y = H,\ z = 1$$

Substituting these into the integral

$$Q = CkH\int_0^1 \sqrt{\frac{1 - z}{z}}\,dz = \frac{CkH\pi}{2} \tag{11.78a}$$

The function in the integral in Eq. (11.78a) may be integrated by trigonometric substitution or found from integral tables.

Therefore, $\dfrac{Q}{H} = Ck\dfrac{\pi}{2}$, which is a constant.

Derivation of an x–y Relation for a Sutro Weir

The x–y relation must be such that $Q = KH$, where K is a constant. When Eq. (11.76) is examined it is seen that H is contained in the integral limit and the function. To make the solution convenient, H is removed from the integral limit. Define

$$z = y/H$$
$$y = H,\ z = 1; \qquad y = 0,\ z = 0$$

and

$$dz = \frac{1}{H}dy \qquad \text{or} \qquad dy = H\,dz$$

Substituting the above into Eq. (11.76),

$$Q = CH \int_0^1 \sqrt{H - zH}\, x\, dz \tag{11.78b}$$

Factoring out H,

$$Q = CH^{3/2} \int_0^1 \sqrt{1 - z}\, x\, dz \tag{11.78c}$$

When Eq. (11.78c) is integrated, a function for x is needed such that the result of the integration is $H^{-1/2}$. The integration will produce a number, say N. For the relation $x = y^m$, the result of the integration is

$$Q = NCH^{(3/2+m)}$$

From this it is seen that $m = -\frac{1}{2}$ satisfies the requirement. The general relation between x and y is therefore

$$x = ky^{-1/2}$$

For the design, wv_h must be chosen. $C = C_d\sqrt{2g}$ is known.
Substituting,

$$\frac{Q}{H} = wv_h = CK\frac{\pi}{2}$$

Therefore,

$$k = \frac{2wv_h}{C\pi} \qquad \text{and} \qquad x = \frac{2wv_h}{C\pi y^{1/2}} \tag{11.79}$$

But it is impossible to have this weir in reality because at $y = 0, x = \infty$. Therefore, the bottom portion of the weir is slightly modified as shown in Fig. 11.36. The weir is cut off, forming a lower rectangular section. The height of the rectangular section is a. The width of the weir is b at its crest.

Similar to Eq. (11.75), the discharge through the rectangular section, Q_r, is

$$Q_r = C_d b\sqrt{2g} \int_0^a \sqrt{H - y}\, dy = \frac{2}{3} bC_d\sqrt{2g}[H^{3/2} - (H - a)^{3/2}] \tag{11.80}$$

Equation (11.80) was developed ignoring the approach velocity in the channel, which is found to be reasonable in practice.

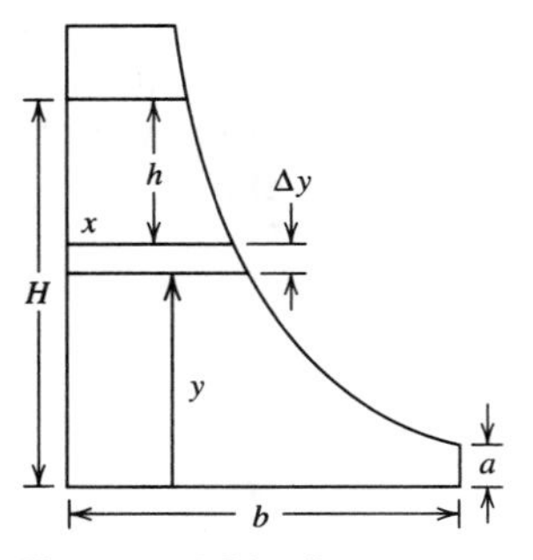

Figure 11.36 Sutro weir.

Similarly, the discharge through the curved section, Q_c, is

$$Q_c = C_d\sqrt{2g}\int_a^H \sqrt{H - y}\, x\, dy \quad (11.81)$$

The total discharge, Q_t, is

$$Q_t = Q_r + Q_c = C_d\sqrt{2g}\left\{\frac{2}{3}b[H^{3/2} - (H - a)^{3/2}] + \int_a^h \sqrt{H - y}\, x\, dy\right\} \quad (11.82)$$

The following describes how the slightly modified relation between x and y is derived.

The x–y relation given in Eq. (11.79) can be applied but Pratt (1914) has derived a more suitable relation for the modified weir. The total discharge must be proportional to the depth, H, above an arbitrary reference level. The reference level is chosen as $a/3$ because this choice facilitates computations. Then,

$$Q_t = K\left(H - \frac{a}{3}\right) \quad (11.83)$$

where
K is a constant

The constant in Eq. (11.83) may be found by noting that at $H = a$, $Q_c = 0$. When Eqs. (11.82) and (11.83) are equated it is found that

$$K\left(a - \frac{a}{3}\right) = \frac{2}{3}C_d b a^{3/2}\sqrt{2g} = \frac{2}{3}aK$$

and

$$K = C_d b\sqrt{2ga} \quad (11.84)$$

This may be used with Eq. (11.83) to find

$$\frac{2}{3}b[H^{3/2} - (H - a)^{3/2}] + \int_a^H \sqrt{H - y}\, x\, dy = b\left(H - \frac{a}{3}\right)a^{1/2} \quad (11.85)$$

Equation (11.85) is easier to solve with the following definitions:

$$z = y - a,\ dz = dy$$
$$h = H - a$$
$$y = a,\ z = 0; \qquad y = H,\ z = H - a = h$$

Substituting these definitions into Eq. (11.85) and rearranging it:

$$b\left(H - \frac{a}{3}\right)a^{1/2} - \frac{2}{3}b[(a + h)^{3/2} - h^{3/2}] = \int_0^h \sqrt{h - z}\, x\, dz \quad (11.86)$$

The term $(a + h)^{3/2}$ on the left-hand side may be expanded in series. If it is assumed that x has the following form:

$$x = M_0 + M_1 z^{1/2} + M_2 z^{3/2} + M_3 z^{5/2} + \cdots$$

This may be substituted into the integral on the right-hand side of Eq. (11.86) and then the integration can be performed. After like coefficients on each side are equated, it will be found that

$$x = b - \frac{2b}{\pi}\left[\frac{\left(\frac{z}{a}\right)^{1/2}}{1} - \frac{\left(\frac{z}{a}\right)^{3/2}}{3} + \frac{\left(\frac{z}{a}\right)^{5/2}}{5} - \frac{\left(\frac{z}{a}\right)^{7/2}}{7} + \cdots\right]$$

This series is recognized to be the power series for

$$\tan^{-1}(z/a)^{1/2} \qquad z \geq a$$

In addition to Eq. (11.80), the other equations for a Sutro weir are

$$x = b\left(1 - \frac{2}{\pi}\tan^{-1}\sqrt{\frac{y-a}{a}}\right) \tag{11.87}$$

$$Q_t = C_d b\sqrt{2ag}\left(h + \frac{2a}{3}\right) \tag{11.88}$$

where

h is the depth of water above the rectangular section of the weir

Also note that y is measured from the weir crest

The discharge coefficient is approximately 0.62 (ASCE and WPCF, 1977). A stilling well with a depth recorder installed can be used to calculate the discharge according to Eq. (11.88).

A proportional weir is two Sutro weirs mirrored together as shown in Fig. 11.37. The formulas for total discharge (Q_{tp}) and discharge through the rectangular section (Q_{rp}) of this weir are

$$Q_{tp} = 2Q_t \quad (11.89a) \qquad Q_{rp} = 2Q_r \quad (11.89b)$$

Design Notes for a Sutro Weir Grit Chamber

There should always be two channels or one channel with a bypass installed to allow for a channel being taken out of service for repair and maintenance. A rectangular overflow weir should be designed above the Sutro or proportional weir to accommodate unexpected large flows. Also a freeboard of 30–45 cm (1–1.5 ft) should be designed for the channel.

The recommended minimum value for a is 25 mm (1 in.) for domestic wastewater to prevent fouling of the weir (ASCE and WPCF, 1977). The weir crest should be located at a minimum of 100 mm (4 in.) but preferably 300 mm (12 in.) above the chamber bottom.

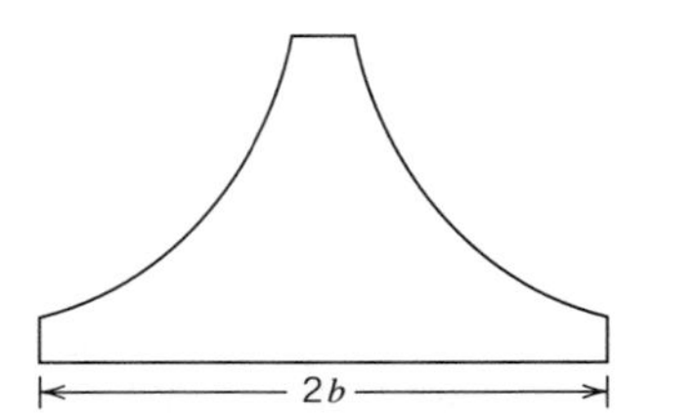

Figure 11.37 Proportional weir.

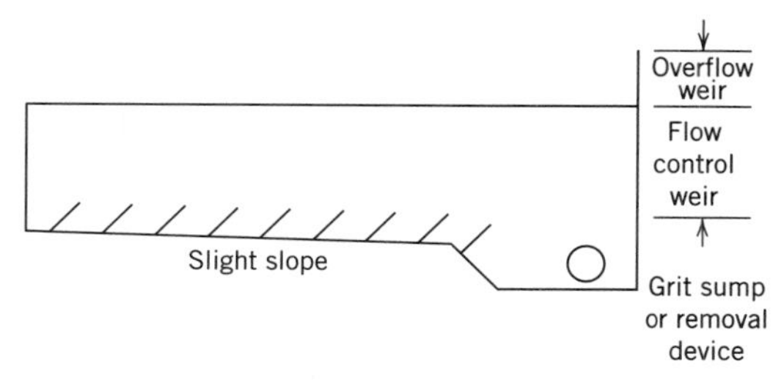

Figure 11.38 Grit chamber.

A grit well should be provided as shown in Fig. 11.38 and the channel should have a slight slope toward the grit well. The volume provided for grit storage depends on the frequency of cleaning and quantity of grit in the wastewater. A mechanical grit collection device may be used, which relaxes the 100–300 mm (4–12 in.) weir crest elevation requirement. Grit removal will be more efficient in this case.

The width of the channel is not directly specified by the equations. The width of the channel is governed by the size of the trash rakes (either manual or mechanical). The practical minimum width is 30 cm (12 in.).

For best flow distribution, the inlet pipe to the chamber should be located on the centerline. The flow velocity will not be constant across the channel cross-section; the velocity will be greater at the bottom of the channel.

Once the channel cross-section is fixed, the horizontal velocity in the channel should be checked at $Q_{\max}$, Q_{ave}, and $Q_{\min}$. According to Eq. (11.83), the depth–discharge relation is not exactly constant. However, deviation from the desired horizontal flow velocity should not be large in a properly designed channel.

The length of the chamber is determined by the horizontal velocity in the chamber and the settling velocity of the design particle, which are specified.

$$v_{\text{h}} = \frac{Q}{wH} \qquad v_{\text{s}} = \frac{Q}{wL}$$

where

L is the length of the chamber

v_{s} is the settling velocity of the design grit particle

The maximum depth of flow in the channel is determined by $Q_{\max}$ and the channel geometry. The length of the channel is related to the maximum depth by

$$\frac{v_{\text{h}}}{v_{\text{s}}} = \frac{\dfrac{Q}{wH}}{\dfrac{Q}{wL}} = \frac{L}{H} \tag{11.90}$$

Water depths in horizontal flow chambers are typically in the range of 0.6–1.5 m (2–5 ft) (WEF and ASCE, 1992). Lengths range from 3 to 25 m (10–80 ft). The length of the chamber is increased beyond the theoretical value by a factor of 10–50% to account for nonidealities in the flow and settling of the particles.

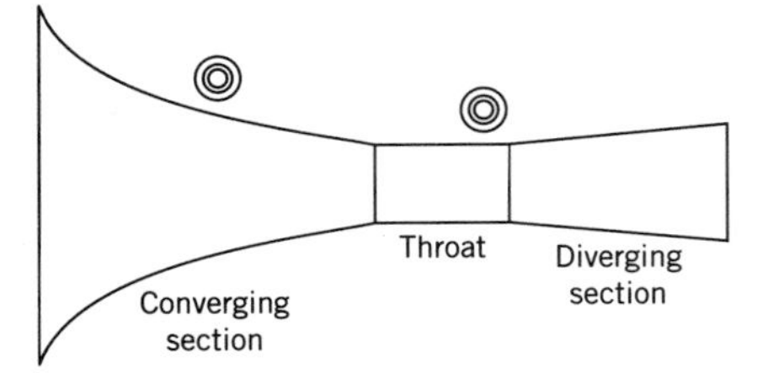

Figure 11.39 Parshall flume.

Channel with Varying Cross-Section

A channel with a varying cross-section geometry is an alternative to modifying the weir. The channel may discharge into a Parshall flume (Fig. 11.39) to measure the discharge accurately. In this case there is a rectangular control section at the end of the channel but the channel width varies as a function of depth. Figure 11.40 is a definition sketch.

Let w_t be the width of the rectangular throat (control) section in the Parshall flume at the downstream end of the channel; w_t is constant. The flow through the throat is

$$Q = C_d w_t H_t \sqrt{2gH_t} \tag{11.91}$$

where

H_t is the depth in throat

Again Eq. (11.73) must be satisfied. From Eq. (11.73), which applies to discharge in the channel,

$$dQ = v_h\, dA = v_h w\, dy \tag{11.92}$$

Differentiating Eq. (11.91),

$$dQ = C_d \frac{3}{2} \sqrt{2g} w_t y^{1/2}\, dy \tag{11.93}$$

A function for w must be found using these relations. Setting Eq. (11.92) equal to Eq. (11.93),

$$v_h w\, dy = C_d \frac{3}{2} \sqrt{2g} w_t y^{1/2}\, dy \tag{11.94}$$

and solving for w,

$$w = \frac{3C_d\sqrt{2g}}{2v_h} w_t y^{1/2} \tag{11.95}$$

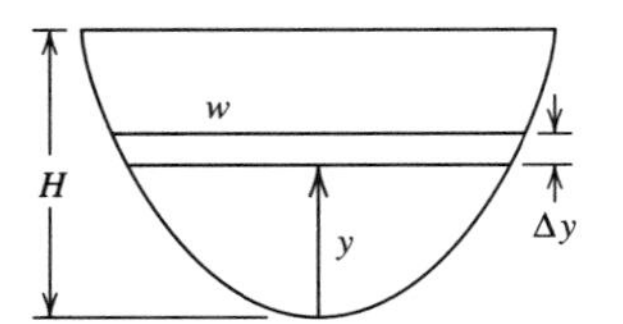

Figure 11.40 Grit chamber with varying geometry.

The equation for the channel geometry describes a parabolic section. In this case, because y is synonymous with H, it is found that $\frac{dQ}{dy} = \frac{dQ}{dH} = v_h w$ from Eqs. (11.93) and (11.94). In the case of the Sutro weir, it is not possible to obtain the required equation for dQ/dH directly from Eq. (11.75).

Design Notes for a Parabolic Grit Chamber

In actual practice the parabolic shape is approximated with a trapezoid to reduce construction costs. As above, one channel and a bypass or two or more channels should be installed. When the number of channels is determined, the maximum, average, and minimum flows in a channel can be calculated. When one channel is out of service, the flow to this channel will be diverted to the other channels, resulting in an emergency flow, Q_{emerg}. The emergency flow in a channel should be based on the maximum flow into the set of grit chambers with one channel out of service. These flows, Q_{emerg}, Q_{max}, Q_{ave}, and Q_{min}, are used to design the shape and length of a channel.

The procedure to determine the trapezoidal shape of the channel is first to choose the horizontal velocity in the chamber. The horizontal velocity is usually 0.30 m/s (1 ft/s) as noted above. Using Eqs. (11.91) and (11.95) the discharge is related to the width, x, of the channel at a depth y by Eq. (11.96a). The equation is then rearranged to find the depth as a function of the other parameters (Eq. 11.96b).

$$Q = \frac{3}{2} x v_h y \qquad (11.96a) \qquad\qquad y = \frac{3}{2}\frac{Q}{x v_h} \qquad (11.96b)$$

because

$$Q = v_h A$$

where

A is the cross-sectional area of flow

For a parabola, from Eq. (11.96a): $A = \frac{2}{3}xy$

Q_{max} is used for Q and a value for x_{max} is chosen to solve Eq. (11.96b) for y_{max}. The width, x_{max}, should be in the range of $1\text{–}1.25 y_{max}$ to keep the slope of the trapezoid sidewalls sufficiently steep to minimize debris accumulation on them.

The depth and width of the downstream control section can now be determined. Critical flow conditions exist in the throat of the Parshall flume. The specific energy relation that applies is

$$y + \frac{v_h^2}{2g} = d_c + \frac{v_c^2}{2g} + 0.1\frac{v_c^2}{2g} \qquad (11.97)$$

where

d_c is the depth of flow in the throat (equivalent to H_t)

v_c is the velocity of flow in the throat

The factor 0.1 in Eq. (11.97) is a reasonable value (10%) to gauge the headloss for a well-rounded smooth approach to the throat.

At the critical depth, the Froude number (Fr) is equal to 1. The critical depth occurs at the throat entrance.

$$\text{Fr} = \frac{v_c}{\sqrt{g d_c}} = 1 \qquad (11.98a) \qquad \text{and} \qquad d_c = \frac{v_c^2}{g} \qquad (11.98b)$$

Substituting for d_c in Eq. (11.97),

$$y + \frac{v_h^2}{2g} = \frac{v_c^2}{g} + \frac{v_c^2}{2g} + 0.1\frac{v_c^2}{2g} = 3.1\frac{v_c^2}{2g} \tag{11.99}$$

and this equation may be solved for v_c.

$$v_c = \left[\frac{2g}{3.1}\left(y + \frac{v_h^2}{2g}\right)\right]^{1/2} \tag{11.100}$$

The depth, d_c, may now be found from Eq. (11.98a) and the following equations are used to find the width of the throat, w_t.

$$A_t = \frac{Q}{v_c} = w_t d_c \tag{11.101a}$$

$$w_t = \frac{Q}{v_c d_c} \tag{11.101b}$$

where
A_t is the cross-sectional area of flow in the throat

Now that the throat width is determined, finding the width of the channel at the other discharge rates is facilitated. The area of the throat is related to v_c by the following equations.

$$w_t d_c = A_t = \frac{v_c^2 w_t}{g} \tag{11.102a}$$

and

$$v_c = \frac{Q}{A_t} \tag{11.102b}$$

and

$$A_t = \frac{\left(\frac{Q}{A_t}\right)^2 w_t}{g} \tag{11.103a}$$

and

$$A_t = \sqrt[3]{\frac{Q^2 w_t}{g}} \tag{11.103b}$$

Equation (11.103b) is used to find the throat area at the other flows. Equations (11.102a) and (11.102b) are used to find d_c and v_c, respectively, and y is then found from Eq. (11.99). Equation (11.96a) is finally used to find the width of the trapezoid at Q_{ave} and Q_{min}.

Lengths and depths of parabolic grit chambers are in the ranges given for rectangular chambers (see the section entitled Design Notes for a Sutro Weir Grit Chamber).

■ Example 11.4 Parabolic Grit Chamber Design

Design a horizontal flow grit chamber with varying cross-sectional area for a plant with an average flow of 5.00×10^4 m^3/d (13.2 Mgal/d), minimum flow of 2.50×10^4 m^3/d (6.60 Mgal/d) and maximum flow of 12.0×10^4 m^3/d (31.7 Mgal/d). Use 3 channels in the grit chamber that should remove 65 mesh particles. Each channel should handle a flow of 6.00×10^4 m^3/d (15.8 Mgal/d) to allow maintenance of one channel. The flow-through velocity (v_h) will be 0.30 m/s (0.98 ft/s). Size the grit storage area for an accumulation of 35 m^3/10^6 m^3 (4.7 ft^3/Mgal) sewage to be removed manually once per week.

$$Q_{max} = \frac{12.0 \times 10^4 \text{ m}^3/\text{d}}{3} = 4.00 \times 10^4 \text{ m}^3/\text{d} = 0.463 \text{ m}^3/\text{s}$$

In U.S. units:

$$Q_{max} = \frac{31.7\ \text{Mgal/d}}{3} = 10.6\ \text{Mgal/d} = \left(10.6 \times 10^6 \frac{\text{gal}}{\text{d}}\right)\left(\frac{1\ \text{ft}^3}{7.48\ \text{gal}}\right)\left(\frac{1\ \text{d}}{86\,400\ \text{s}}\right)$$

$$= 16.4\ \text{ft}^3/\text{s}$$

$$Q_{ave} = (5.00 \times 10^4\ \text{m}^3/\text{d})/3 = 1.67 \times 10^4\ \text{m}^3/\text{d} = 0.193\ \text{m}^3/\text{s}\ (6.81\ \text{ft}^3/\text{s})$$

$$Q_{min} = (2.50 \times 10^4\ \text{m}^3/\text{d})/3 = 8.33 \times 10^3\ \text{m}^3/\text{d} = 0.096\,5\ \text{m}^3/\text{s}\ (3.41\ \text{ft}^3/\text{s})$$

$$Q_{emerg} = 6.00 \times 10^4\ \text{m}^3/\text{d} = 0.694\ \text{m}^3/\text{s}\ (24.5\ \text{ft}^3/\text{s})$$

The design settling velocity, v_s, of a 65-mesh particle is 1.13 m/min (3.71 ft/min) or 0.019 m/s (0.061 8 ft/s).

Use a fixed width control section with a smooth approach. Assume losses = 10% of velocity head in the throat.

Governing Equations

In the channel, $Q = vA$

In the rectangular throat control section, $Q = Cw_t H^{3/2}$

From Eq. (11.95) the width in the channel varies according to

$$w = \frac{3}{2}\frac{Cw_t}{v} y^{1/2} = \frac{3}{2}\frac{Cw_t}{v} H^{1/2}$$

$$Q = Cw_t H^{3/2} \qquad w = T\ \text{(top width)}$$

$$Q = \frac{2}{3} vwH = \frac{2}{3} vTH$$

Q_{max}

The top width T should be equal to $1\text{–}1.25H_{max}$. Also the width will be affected by the Parshall flume geometry that is chosen. Using a value of $1.20H_{max}$ (at Q_{max}), H_{max} is determined from

$$Q = \frac{2}{3} vTH \qquad Q = \frac{2}{3} v(1.20H)H$$

$$H = \sqrt{\frac{3Q}{2.40v}} = \sqrt{\frac{3(0.463\ \text{m}^3/\text{s})}{2.40(0.30\ \text{m/s})}} = 1.39\ \text{m; in U.S. units: } H = \sqrt{\frac{3(16.4\ \text{ft}^3/\text{s})}{2.40(0.98\ \text{ft/s})}} = 4.57\ \text{ft}$$

$T = (1.20)(0.95\ \text{m}) = 1.67\ \text{m}$; in U.S. units: $T = (1.20)(4.57\ \text{ft}) = 5.48\ \text{ft}$

Upstream Energy = Downstream Energy + Losses

$$H + \frac{v^2}{2g} = d_c + \frac{v_c^2}{2g} + 0.1\frac{v_c^2}{2g} \qquad d_c = \frac{v_c^2}{g}$$

$$H + \frac{v^2}{2g} = 3.1\frac{v_c^2}{2g} \qquad v_c = \sqrt{\frac{2g}{3.1}\left(H + \frac{v^2}{2g}\right)}$$

$$v_c = \left\{\frac{2(9.81\ \text{m/s}^2)}{3.1}\left[1.39\ \text{m} + \frac{(0.30\ \text{m/s})^2}{2(9.81\ \text{m/s}^2)}\right]\right\}^{1/2} = 2.97\ \text{m/s}$$

$$\text{In U.S. units: } v_c = \left\{\frac{2(32.2\ \text{ft/s}^2)}{3.1}\left[4.57\ \text{ft} + \frac{(0.98\ \text{ft/s})^2}{2(32.2\ \text{ft/s})^2)}\right]\right\}^{1/2} = 9.76\ \text{ft/s}$$

$$d_c = \frac{v_c^2}{g} = \frac{(2.97 \text{ m/s})^2}{9.81 \text{ m/s}^2} = 0.90 \text{ m; in U.S. units: } d_c = \frac{(9.76 \text{ ft/s})^2}{32.2 \text{ ft/s}^2} = 2.96 \text{ ft}$$

$$A_c = \frac{Q}{v_c} = \frac{(0.463 \text{ m}^3\text{/s})}{2.97 \text{ m/s}} = 0.156 \text{ m}^2\text{; in U.S. units: } A_c = \frac{(16.4 \text{ ft}^3\text{/s})}{9.76 \text{ ft/s}} = 1.68 \text{ ft}^2$$

$$w_t = \frac{A_c}{d_c} = \frac{0.156 \text{ m}^2}{0.90 \text{ m}} = 0.17 \text{ m; in U.S. units: } w_t = \frac{1.68 \text{ ft}^2}{2.96 \text{ ft}} = 0.57 \text{ ft}$$

***Q*ave**

$$v_c = \frac{Q}{A_c} \qquad A_c = d_c w_t \qquad d_c = \frac{v_c^2}{g} \qquad A_c = \sqrt[3]{\frac{Q^2 w_t}{g}}$$

$$A_c = \sqrt[3]{\frac{(0.193)^2(0.17)}{9.81}} = 0.086 \text{ m}^2$$

$$d_c = \frac{A_c}{w_t} = \frac{0.086}{0.17} = 0.51 \text{ m} \qquad v_c = \frac{0.193}{0.086} = 2.24 \text{ m/s}$$

$$H = 3.1\frac{v_c^2}{2g} - \frac{v^2}{2g} = 3.1\frac{(2.24)^2}{2(9.81)} - \frac{(0.30)^2}{2(9.81)} = 0.79 \text{ m (2.59 ft)}$$

$$T = \frac{3}{2}\frac{(0.193)}{(0.79)(0.30)} = 1.22 \text{ m (4.00 ft)}$$

***Q*min**

$$A_c = \sqrt[3]{\frac{(0.096\,5)^2(0.17)}{9.81}} = 0.054 \text{ m}^2$$

$$d_c = \frac{0.054}{0.17} = 0.32 \text{ m} \qquad v_c = \frac{0.096\,5}{0.054} = 1.79 \text{ m/s}$$

$$H = 3.1\frac{(1.79)^2}{2(9.81)} - \frac{(0.30)^2}{2(9.81)} = 0.50 \text{ m (1.65 ft)}$$

$$T = \frac{3}{2}\frac{(0.096\,5)}{(0.30)(0.501)} = 0.96 \text{ m (3.16 ft)}$$

***Q*emerg**

$$A_c = \sqrt[3]{\frac{(0.694)^2(0.17)}{9.81}} = 0.21 \text{ m}^2$$

$$d_c = \frac{0.21}{0.17} = 1.23 \text{ m} \qquad v_c = \frac{0.694}{0.21} = 3.31 \text{ m/s}$$

$$H = 3.1\frac{(3.31)^2}{2(9.81)} - \frac{(0.30)^2}{2(9.81)} = 1.73 \text{ m (5.66 ft)}$$

$$T = \frac{3}{2}\frac{(0.694)}{(0.30)(1.73)} = 2.01 \text{ m (6.58 ft)}$$

Because T is larger than T assumed (1.65 m), either T must be adjusted or the velocity will vary from 0.30 m/s. Because emergency flow situations occur rarely, allow the velocity to vary. To find the velocity,

$$v = \frac{3}{2}\frac{Q}{HT} \qquad H + \frac{v^2}{2g} = 3.1\frac{v_c^2}{2g} \Rightarrow v = \sqrt{3.1v_c^2 - 2gH}$$

$$\sqrt{3.1v_c^2 - 2gH} = \frac{3}{2}\frac{Q}{HT}$$

$$3.1v_c^2 - 2gH = \left(\frac{3}{2}\frac{Q}{T}\right)^2\left(\frac{1}{H}\right)^2 \qquad 3.1v_c^2H^2 - 2gH^3 = \left(\frac{3}{2}\frac{Q}{T}\right)^2$$

$$(3.1)(3.31)^2H^2 - 2(9.81)H^3 = \left[\frac{3}{2}\frac{(0.694)}{(1.65)}\right]^2$$

$$34.0H^2 - 19.6H^3 = 0.398 \Rightarrow H = 1.73 \text{ m } (5.67 \text{ ft})$$

$$v = \frac{3}{2}\frac{(0.694)}{(1.73)(1.65)} = 0.36 \text{ m/s}$$

The velocity variation (20%) is tolerable.

Length of the Tank

At Q_{max}, $H = 0.95$ m

$$L = \frac{H}{v_s}v_h = \frac{(1.39 \text{ m})}{(0.019 \text{ m/s})} \times 0.30 \text{ m/s} = 22.0 \text{ m};$$

$$\text{in U.S. units: } L = \frac{(4.56 \text{ ft})}{(0.062 \text{ ft/s})} \times 0.98 \text{ ft/s} = 72.1 \text{ ft}$$

For Q_{emerg},

$$L = \frac{(1.73 \text{ m})}{(0.019 \text{ m/s})} \times 0.36 \text{ m/s} = 32.8 \text{ m } (108 \text{ ft})$$

Use $L = 33.0$ m (108 ft) to provide a 50% safety factor. Field studies should be performed to establish the required safety factor.

Grit Accumulation

Base the design on the average flow for one channel.

For 1 week: $Qt = (1.67 \times 10^4 \text{ m}^3/\text{d})(7 \text{ d}) = 1.17 \times 10^5 \text{ m}^3$

In U.S. units: $Qt = (4.41 \times 10^6 \text{ gal/d})(7 \text{ d}) = 3.09 \times 10^7 \text{ gal}$

The accumulation is: $\frac{35 \text{ m}^3}{10^6 \text{ m}^3}(1.17 \times 10^5 \text{ m}^3) = 4.10 \text{ m}^3$

In U.S. units: $\frac{4.7 \text{ ft}^3}{10^6 \text{ gal}}(3.09 \times 10^7 \text{ gal}) = 145 \text{ ft}^3$

The area required for grit storage is $\frac{4.10 \text{ m}^3}{33.0 \text{ m}} = 0.124 \text{ m}^2 \ (1.34 \text{ ft}^2)$

Check the width at $H = 0$ for the required depth–width relation to accommodate the minimum and average flows.

Q_{ave}: $H = 0.79$ m	$w = 1.22$ m	$0.5w = 0.61$ m
Q_{min}: $H = 0.50$ m	$w = 0.96$ m	$0.5w = 0.48$ m

The equation of the line describing the slope of the lower portion of the grit chamber is

$$H = 2.23(0.5)w - 0.57$$
$$\text{At } H = 0, w = 0.51 \text{ m } (1.68 \text{ ft})$$

Using this width, the depth required for grit storage is

$$\frac{0.124 \text{ m}^2}{0.51 \text{ m}} = 0.24 \text{ m } (0.80 \text{ ft})$$

If in the future a mechanical rake is to be installed (an expensive option for this small installation), manufacturer's specifications should be consulted. It may be necessary to adjust the grit accumulation dimensions to fit standard equipment. Mechanical grit removal devices will result in lower accumulations of grit.

Small changes in the channel geometry will not have a significant effect on the flow velocity. In any case the channel width should be maintained as near as possible to the widths calculated for Q_{min}, Q_{ave}, and Q_{max}. The flows will be in this range most of the time. The velocity in the grit storage area is assumed to be zero because this zone has a wall at either end.

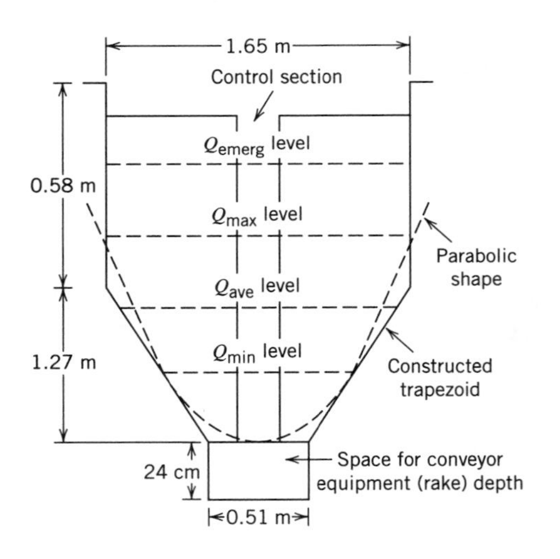

See the sketch for the final design. A freeboard of 15 cm (0.49 ft) was added to the depth of the channel.

11.8.2 Aerated Grit Chambers

A common approach currently used in medium to large plants for grit removal is aerated grit chambers (Fig. 11.41). In aerated grit chambers a spiral flow pattern is induced in the sewage as it moves through the tank by supplying air from a diffuser located on one side of the tank. The tank inlet and outlet are positioned to direct the flow perpendicular to the spiral roll pattern. The roll velocity is sufficient to maintain lighter organic particles in suspension while allowing the heavier grit particles to settle. The air supply must be adjustable to provide the "optimum" roll velocity for different

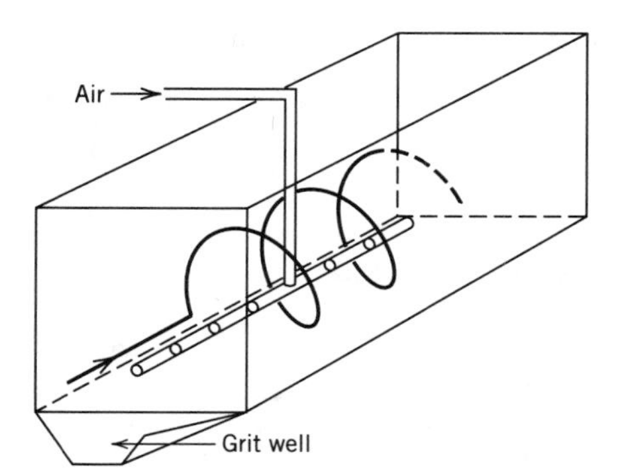

Figure 11.41 Aerated grit chamber.

conditions. Adjustment of the air supply is easily performed, which makes the device flexible.

The WEF and ASCE (1992) state that a minimum hydraulic detention time of 3 min at maximum instantaneous flow rates will capture 95% of 0.21-mm diameter grit. Baffle and diffuser locations and air flow rate are more important design parameters than detention time. Other design values for an aerated grit chamber are given in Table 11.9.

11.8.3 Square Tank Degritter

An early design for grit removal was a square, constant level, short–detention time tank. These tanks are also known as detritus tanks. The corner zones are sloped into the central circular grit collection zone. A horizontal arm with flow deflectors rotates above the circular grit collection zone to maintain circulation within the tank. The influent enters through baffles along one side of the tank and the effluent exits over a weir on the opposite side. Because flow control is not required, the operation of these tanks is simple.

These basins are designed to maintain a detention time of 1 min or less at the design flow rate (WEF and ASCE, 1992). The maximum day flow rate may be used as the design flow rate. The area of the tank is based on the grit particle size desired

TABLE 11.9 Aerated Grit Chamber Design Information[a]

Item	Range	Comment
Dimensions		
Depth, m (ft)	2.1–4.9 (7–16)	Varies widely
L : W ratio	3 : 1–5 : 1	
W : D ratio	1 : 1–5 : 1	2 : 1 typical
Minimum detention time at peak flow, min	2–5	3 typical
Air supply (medium to coarse bubble diffusers)		
m^3/min/m (ft^3/min/ft)	0.27–0.74 (3–8)	0.45 (4.8) typical
Distance from bottom, m (ft)	0.6–1.0 (2–3.3)	
Traverse roll velocity, m/s (ft/s)	0.60–0.75 (2–2.5)	

[a]Partially from WEF and ASCE (1992), *Design of Municipal Wastewater Treatment Plants,* vol. 1, WEF, © WEF 1992; and Metcalf & Eddy, Inc. (1991).

TABLE 11.10 Theoretical Maximum Overflow Rate for Detritus Tanks[a]

Particle size		Settling velocity[b]		Theoretical overflow rate[c]	
Diameter, mm	Approx. mesh	cm/min	ft/min	$m^3/m^2/d$	gal/1 000 ft^2/d
0.83	20	494	16.2	7 120	175
0.59	28	363	11.9	5 200	128
0.46	35	247	8.10	3 550	87
0.33	48	186	6.10	2 670	66
0.25	60	165	5.41	2 370	58
0.21	65	131	4.30	1 890	46
0.18	80	116	3.81	1 670	41
0.15	100	91	2.99	1 320	32

[a]Source: WEF and ASCE (1992), *Design of Municipal Wastewater Treatment Plants,* vol. 1, WEF, © WEF 1992.
[b]At a liquid temperature of 15.5°C and s.g. of 2.65.
[c]A safety factor of 2.0 is applied to the rates in this table to account for hydraulic inefficiency (see text).

to be removed (Table 11.10). A safety factor of 2 is generally applied to the theoretical surface overflow rate to allow for inlet and outlet turbulence in addition to the short circuiting that will occur in the basin. Metcalf and Eddy (1991) note that a common criterion for design is to remove 95% of the 100-mesh particles at peak flow. Given the surface overflow rates and detention time criteria, the tanks will be shallow. An additional depth of 150–250 mm (6–10 in.) is added to accommodate the raking mechanism.

Hydraulic retention of the liquid within these tanks varies directly with the flow rate. This is a disadvantage because, particularly at low flow rates, significant quantities of organic material will be removed with the grit. It is necessary to use a grit washing device on the material collected from the detritus tank. The typical design criteria ensure that most of the grit is removed at all flow rates.

Grit Washing

In some treatment facilities a grit chamber may not be installed and the grit is accumulated with sludge removed in the primary clarifiers. A cyclone degritter is used to separate grit from the sludge in this instance and when grit is collected in detritus tanks. Sometimes it is necessary to wash grit collected in other types of grit chambers.

A cyclone degritter uses centrifugal force to separate the heavier particles of grit from the lighter particles. The cyclone degritter feed enters the unit tangentially and creates a vortex wherein the heavier particles are transported to the walls of the device. The unit is inclined and the solids travel downward while the liquid and lighter particles exit at the top. Manufacturers supply design information.

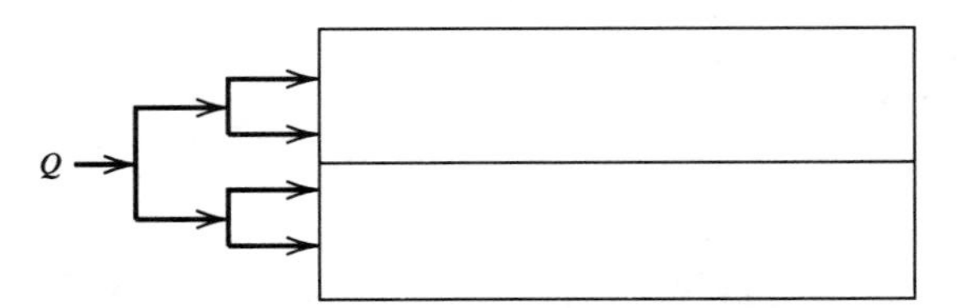

Figure 11.42 Inlet arrangement for horizontal flow sedimentation tanks.

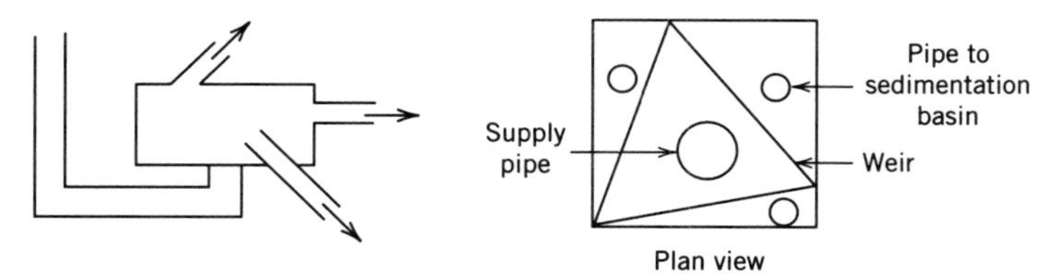

Figure 11.43 Distribution box for uneven number of pipes.

Other types of grit washing devices are available from manufacturers. Washed grit is nonputrescible and does not have a significant insect or rodent attraction.

11.9 INLET HYDRAULICS FOR SEDIMENTATION BASINS

Inlets to horizontal sedimentation tanks and grit chambers must be designed to distribute the flow as uniformly as possible over the tank cross-section and dissipate the kinetic energy of the influent jet to lessen turbulence in the tank. The task may be relatively simple if the tank arrangement is as shown in Fig. 11.42. When the tank elevations are the same, headlosses and the discharges through each inlet pipe are equally subject to random variation of solids deposition, for example, in the influent lines. When the number of tanks is uneven, a distribution box may be used to deliver the flow to the tanks (Fig. 11.43).

It may not be feasible to have either of the above situations. A common header may be necessary to distribute flow to a number of tanks as shown in Fig. 11.44. In this case the orifices can be designed to have a large, flow controlling headloss relative to headloss in the supply channel or pipe. Then velocity head will be converted to pressure head as flow moves along the supply manifold. It is desired to have each lateral discharging nearly the same flow. Hudson (1981) and Hudson et al. (1979) present a suitable approach to design a manifold (Fig. 11.45). The coefficients in the development apply to square-edged laterals that discharge below the water surface in the basin.

The headloss from a point in the manifold upstream of a lateral to the discharge from the lateral consists of (1) friction losses in the manifold and the lateral, (2) entrance losses to the lateral, and (3) exit losses from the lateral. Friction losses have been ignored because they are normally insignificant; however, they can be incorporated into the development by using the Darcy–Weisbach equation.

Figure 11.44 Single distribution pipe inlet arrangement.

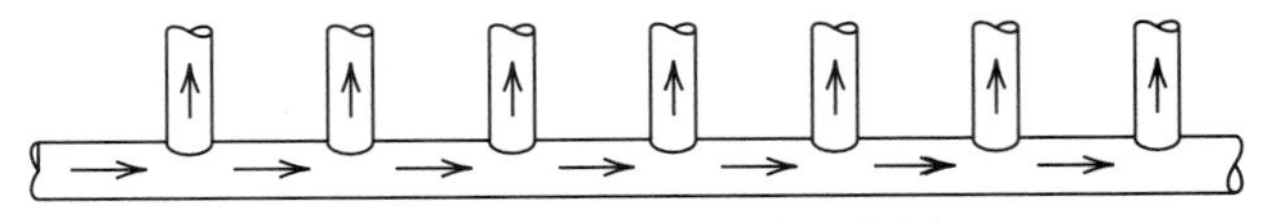

Figure 11.45 Flow distribution manifold.

The sum of the losses between points 1 and 2 (Fig. 11.46) is

$$h_L = h_{entry} + h_{exit} = h_e + \frac{v_l^2}{2g} \tag{11.104}$$

where

h_{entry} or h_e is the entrance loss to the lateral
h_{exit} is the exit loss from the lateral
v_l is the velocity in the lateral

The entrance loss is different from a submerged orifice and is a function of the velocity in the manifold as given by Eq. (11.105).

$$h_e = \left[\phi\left(\frac{v_m}{v_l}\right) + \theta\right]\frac{v_l^2}{2g} \tag{11.105}$$

where

ϕ and θ are constants
v_m is the velocity in the manifold

The equation for headloss through a lateral becomes

$$h_L = \left[\phi\left(\frac{v_m}{v_l}\right)^2 + \theta\right]\frac{v_l^2}{2g} + \frac{v_l^2}{2g} = \beta\frac{v_l^2}{2g} \tag{11.106}$$

where

$$\beta = \phi\left(\frac{v_m}{v_l}\right)^2 + \theta + 1$$

The constants ϕ and θ were determined from experimental measurements and two distinct correlations were found for long and short laterals (Table 11.11).

In ideal flow distribution, the flow from each lateral is the same and the headloss through each lateral is the same. It is assumed that when the manifold is delivering flow to more than one basin, the water surface elevation is the same in each basin.

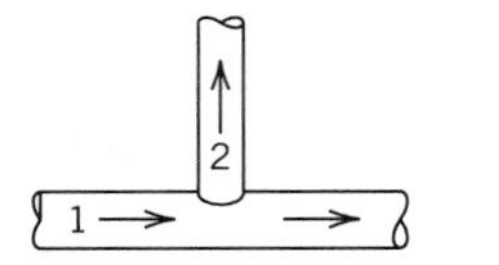

Figure 11.46 A lateral.

TABLE 11.11 Entrance Headloss Parameters[a]

Lateral length[b]	ϕ	θ
Short	1.67	0.7
Long	0.90	0.4

[a]From Hudson (1981).
[b]Short laterals have a length less than three pipe diameters. Large laterals have a length substantially larger than three pipe diameters.

This assumption is not exactly correct because the flow distribution will not be ideal; however, this approach is the best method to design manifolds.

$$\frac{\beta_1 v_{l1}^2}{2g} = \frac{\beta_2 v_{l2}^2}{2g} = \cdot \cdot \cdot = \frac{\beta_i v_{li}^2}{2g} = \text{constant} \tag{11.107}$$

from which

$$v_{li} = v_{l1}\sqrt{\frac{\beta_1}{\beta_i}} \tag{11.108}$$

Now for the same size laterals,

$$\sum q_i = Q \quad (11.109a) \qquad\qquad q_i = a v_{li} \quad (11.109b)$$

where
q_i is the discharge from the ith lateral
Q is the flow into the manifold
a is the cross-sectional area of a lateral

Substituting Eqs. (11.109b) and (11.107) into Eq. (11.109a):

$$Q = a v_{l1}\left(1 + \sqrt{\frac{\beta_1}{\beta_2}} + \sqrt{\frac{\beta_1}{\beta_3}} + \cdot \cdot \cdot + \sqrt{\frac{\beta_1}{\beta_n}}\right)$$

where
n is the number of laterals

$$v_{li} = \frac{Q}{a\sqrt{\beta_1}}\left(\sum_n \frac{1}{\beta_i}\right)^{-1} \tag{11.110}$$

The above equations are applied iteratively to determine the actual flow distribution. The procedure is to choose manifold and lateral sizes. The number of laterals dictates the flow and velocity through each lateral. The flow and velocity in the manifold is calculated after each lateral and β_i is calculated. Then Eq. (11.108) is applied for v_{l1} and the procedure is repeated until the process converges. An example follows.

Final selection of manifold and lateral diameters is a matter of trial and error until suitable flow distribution is achieved over the design flow range. Perfect flow distribution can be achieved by tapering the manifold as recommended by Hudson (1981).

■ Example 11.5 Flow Distribution From a Manifold

Find the flow distribution for a flow of 0.35 m³/s that is to be distributed to two sedimentation basins. There will be 4 laterals discharging to each basin. Assume that the laterals are short. The diameters of the manifold and lateral are 40.6 cm and 13 cm, respectively. Ignore friction losses in the manifold and laterals.

The following tables outline the calculations. For the first iteration the flow is equally distributed through the laterals and the velocity in each lateral is calculated. The flow in the manifold (col. 4) is the flow immediately above a lateral, from which the velocity is calculated. β_i is calculated using the definition in Eq. (11.106).

(1)	(2)	(3)	(4)	(5)	(6)	(7)	(8)
Lateral No.	q_i, m³/s	v_{li}, m/s	Q_{mi} m³/s	v_{mi}, m/s	$\left(\frac{v_m}{v_l}\right)^2$	β_i	$\sqrt{\frac{1}{\beta_i}}$
First iteration							
1	0.043 8	3.296	0.350	2.703	0.673	2.824	0.595
2	0.043 8	3.296	0.306	2.366	0.515	2.560	0.625
3	0.043 8	3.296	0.263	2.028	0.378	2.332	0.655
4	0.043 8	3.296	0.219	1.690	0.263	2.139	0.684
5	0.043 8	3.296	0.175	1.352	0.168	1.981	0.711
6	0.043 8	3.296	0.131	1.014	0.095	1.858	0.734
7	0.043 8	3.296	0.088	0.676	0.042	1.770	0.752
8	0.043 8	3.296	0.044	0.338	0.011	1.718	0.763
							Σ 5.518
Second iteration							
1	0.037 8	2.844	0.350	2.703	0.904	3.209	0.558
2	0.039 6	2.987	0.310	2.397	0.644	2.776	0.600
3	0.041 5	3.130	0.269	2.076	0.440	2.435	0.641
·	·	·	·	·	·	·	·
8	0.048 4	3.647	0.037 8	0.292	0.006	1.711	0.765
							Σ 5.443

After five iterations the values for q_i are

i	1	2	3	4	5	6	7	8
q_i	0.034 8	0.038 2	0.041 2	0.043 9	0.046 0	0.047 7	0.048 8	0.049 4

In the second iteration, Eq. (11.110) is used to calculate v_{l1} using $\Sigma\sqrt{1/\beta_i}$ from the first iteration. Then Eq. (11.108) is used to calculate the other lateral velocities using β_i from the first iteration. The other columns are then calculated.

The variation in flow from the first to the last lateral is (0.049 4/0.034 8) = 1.42, which is generally acceptable.

Inlet Baffling

The influent jet to a horizontal sedimentation tank has a high amount of kinetic energy that must be dissipated. Also, the influent must be distributed throughout the depth as well as across the width of the tank to set up a horizontal flow pattern through the tank. Figure 11.47 shows various baffled inlet arrangements that have been used for energy dissipation and flow distribution in sedimentation tanks. When these inlet devices are viewed toward the cross-sectional area perpendicular to the influent, there

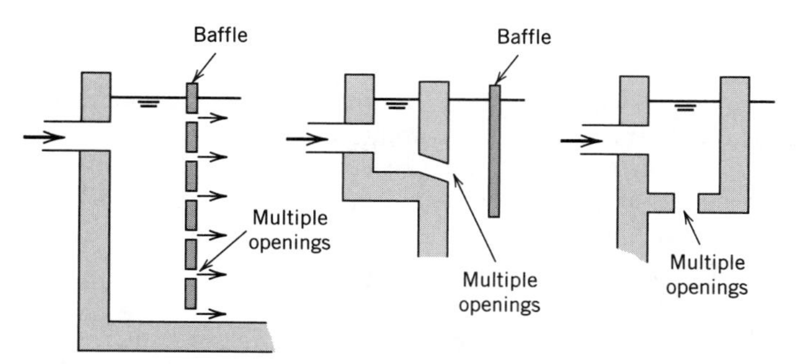

Figure 11.47 Inlets for sedimentation tanks. From T. J. McGhee (1991), *Water Supply and Sewerage,* 6th ed., McGraw-Hill, Toronto, used with permission of McGraw-Hill, Inc.

is a series of uniformly spaced openings across the width of the tank. Figure 13.14 gives such a view for the first type of inlet shown in Fig. 11.47.

QUESTIONS AND PROBLEMS

1. A mechanically cleaned wastewater bar screen is constructed using 6.5-mm-wide bars spaced 5.0 cm apart center to center. The wastewater flow velocity in the channel immediately upstream of the screen will vary from 0.4 to 0.9 m/s. What is the design headloss for the screens at the two extremes of flow? Assume that the friction coefficient is 0.84.
2. Assume that the water temperature is 10°C and the depth of the settling tank is 3.0 m (9.80 ft). Calculate the theoretical settling velocity (in cm/s or ft/h), tank loading (m^3/m^2/d or gal/ft^2/d) and detention time (h) required for the removal of the following particles in a settling tank. Assume discrete particle settling.
   (a) Sand particles with a relative density of 2.65 and diameter of 0.001 cm to be removed in a presettling tank.
   (b) Alum floc with a relative density of 1.002 and diameter of 1.0 mm to be removed in a settling tank after coagulation.
   (c) Calcium carbonate precipitates with a relative density of 1.20 and diameter of 0.10 mm to be removed in a settling tank after softening treatment.
   (d) What will be the corresponding design loading rate and detention time for the above three cases if the efficiency of the settling basins is 50%, assuming that removal is linearly related to efficiency?
3. How will the values of parameters requested in Problem 2 change for the alum floc in part (b) if the water temperature is 25°C instead of 10°C?
4. For the particles described in Problem 2(c), compare the effects of a 10% increase in viscosity, particle diameter, and s.g. on the ultimate settling velocities of the particles. Vary only one parameter at a time. What is the temperature change required to produce a 10% increase in viscosity? What is the temperature change required to produce a 10% decrease in settling velocities of these particles?
5. Prove that the design settling velocity for a circular sedimentation basin is Q/A_s, where A_s is the effective surface area in which sedimentation occurs. Sedimentation occurs after the inlet zone that extends to a radius r_i and the outlet zone begins at radius, r_o. Flow proceeds from the cylindrical inlet zone radially outward and the flow is uniformly distributed throughout the depth of the clarifier. The radial velocity is dr/dt. Plug flow and other assumptions given for a rectangular sedimentation basin apply.
6. English units were used in older textbooks and regulations were specified in these units. For clarifiers the units used for surface loading rates were gal/d/ft^2 and weir loading rates

were gal/d/ft of weir length. Convert 0.010 m/s ($m^3/s/m^2$) to gal/d/ft^2 and 200 m^3/m/d to gal/d/ft.

7. A rectangular clarifier with a length: width ratio of 3 : 1 receives a flow of 850 m^3/d. The clarifier's depth is 4.0 m and the detention time of the water in the clarifier is 2.4 h. What are the surface overflow rate and horizontal flow-through velocity if the flow is distributed uniformly across the cross-sectional area of the tank?

8. For a water temperature of 10°C and alum floc with a s.g. of 1.002 and a diameter of 1.00 mm, what are the times for a floc particle starting from rest to attain velocities of 90 and 99% of its ultimate settling velocity?

9. What is the overall removal for a suspension that contains 90% inorganic matter with a s.g. of 1.20 and P_{10} of 0.060 mm at a temperature of 15°C in a basin with a surface overflow rate of 0.002 $m^3/m^2/s$? The uniformity coefficient of the particles is 1.52. Assume that the removal of organic matter is 5%.

10. A settling column is 2.5 m high with sampling ports located at 0.5 m intervals. A suspension with a concentration of 345 mg/L is placed in the column and uniformly mixed at the beginning of the analysis. For the velocity distribution in the following table, what are the concentrations of suspended solids at each sampling port (except the bottom port) for 15-min time intervals over 3 h? The settling velocity of each fraction remains constant over the time period.

Fraction of particles	0.16	0.12	0.18	0.20	0.15	0.08	0.11
Settling velocity, m/h	0.11	0.32	0.57	0.85	1.11	1.39	1.85

11. Assume that the settling velocity distribution in Problem 10 is the initial settling velocity for a flocculent suspension and that the settling velocity of each fraction increases linearly at a rate of $0.06v_{\text{initial}}$ per hour.
(a) Solve Problem 10 for this variable settling velocity situation.
(b) What is the total removal at a time of 1.5 h?
(c) Problems 10 and 11(a) assume that the sample volumes removed from the column are insignificant. If the settling column in Problem 10 had an i.d. of 15.0 cm and 50-mL sample aliquots were withdrawn at each port at 30-min intervals (no aliquots were withdrawn at time 0), determine the suspended solids concentrations at each port at 15-min intervals for the flocculent (variable velocity) suspension. Perform the calculations for a total time period of 3 h. The samples were taken in order from the top port to the lowest port (no samples were removed from the bottom port). Assume that sample withdrawal does not disturb the solids distribution.

12. A settling column analysis is performed on a dilute suspension of particles from a water. The suspension exhibited discrete particle settling characteristics. Data collected from samples taken at a depth of 1.5 m are as follows:

Time required to settle 1.5 m, min	0.6	1.2	2.2	4.5	6.0	8.8
Fraction of particles settled	0.40	0.50	0.65	0.82	0.91	0.96

(a) Plot the velocity distribution for the suspension.
(b) Find the overall removal if the surface loading rate of a clarifier is designed to be 2 400 m^3/m^2/d.
(c) For a lab settling column with a height of 200 cm, find the concentration of solids at 25-cm intervals in the column at 1-min intervals using the above data. Report only concentrations between 0 and 175 cm. Assume the suspension is uniformly distributed throughout the column at the initial time. The initial concentration of suspended solids is 160 mg/L.

13. Problems associated with flow (such as short circuiting) through continuous flow clarifiers diminish their performance. An engineer proposes to change existing clarifiers to sequencing batch clarifiers to provide ideal, quiescent settling conditions. In the operation of sequencing batch clarifiers only one clarifier at a time is receiving influent and after the fill period of the current clarifier is finished, influent is directed to another clarifier. While one clarifier is filling, quiescent conditions in other clarifiers allow the solids to settle. Also, clarified effluent must be withdrawn from one of the clarifiers not receiving influent. Each clarifier proceeds through a sequence of fill, settle, and drain. In a two-clarifier system, settle and drain periods occur in one clarifier while the other is filling. The process is repeated in a cyclical manner.

The existing clarifiers were designed for a flow of 8 000 m^3/d and an overflow rate of 35 $m^3/m^2/d$ and have a depth of 3.0 m and a length : width ratio of 4 : 1.

(a) What are the dimensions of the continuous flow clarifiers if the system contains (i) two clarifiers and (ii) three clarifiers?

(b) In the operation of the clarifiers as a sequencing batch system the engineer proposes to displace only the volume in the top 2.5-m depth of a clarifier to minimize disturbance of the settled sludge. During the fill period some settled solids are resuspended and it is assumed that no solids settle during the turbulent conditions that exist in this period. The total depth of the sequencing batch clarifiers is 3.0 m. What are the volumes of the clarifiers when the system is operated as a sequencing batch system for (i) the two-clarifier and (ii) the three-clarifier systems? Assume that the time to drain a clarifier is 20 min in both cases. Also assume that sedimentation is 20% more efficient in the sequencing batch system.

(c) What improvement in sedimentation efficiency (or what is the effective settling velocity) is required for the two tank sequencing batch systems to have the same total volume as the continuous flow system?

14. A laboratory settling analysis has been performed on a suspension exhibiting flocculent (type II) sedimentation with the results given in the following table. The column had a diameter of 14 cm and total depth of 2.40 m, with sampling ports spaced every 0.60 m. The initial concentration of the suspension was 480 mg/L.

	Suspended solids concentration, mg/L		
Time min	Tap 1 0.60 m	Tap 2 1.20 m	Tap 3 1.80 m
5	385	420	438
10	334	379	405
20	265	330	355
30	216	279	315
40	180	250	280
50	154	213	251
60	123	182	217
75	106	148	192

(a) Prepare the following graphs (list data used in preparing the graphs in tables):
  (i) Percentage SS removed versus time
  (ii) Isoremoval curves
  (iii) Total % removal versus detention time
  (iv) Total % removal versus surface overflow rate

(b) Size a primary clarifier for 50% removal based on the data in (a). The design flow rate is 4 000 m^3/d. What will the underflow flow rate from the clarifier be if solids are compacted to a 1.5% concentration? Use safety factors of 1.5 for the detention time and surface overflow rate of the clarifier, respectively.

(c) If the volumetric flow rate doubles but the concentration and settling characteristics of the suspended solids do not change, what is the expected total removal? What is the expected removal if the flow rate increases by 10%?

15. What considerations apply to the sizing the depths of the clarified and sludge collection zones of a circular secondary clarifier in a biological treatment process? What considerations would apply to locating the depth of the outlet of the inlet pipe? Describe tracer tests that may be conducted to assure reliable performance of the clarifier in the field.

16. Assume a temperature of 15°C for the following problems.

(a) Calculate the Reynold's number for a horizontal flow sedimentation basin that is 2.0 m deep and will remove particles with a settling velocity of 1.2 mm/s. The design influent flow rate is 1 600 m^3/d. The length : width ratio of the clarifier is 3 : 1.

(b) Calculate the Reynold's number for a tube settler that receives the same flow as in (a) but the volume of the tubes is only one-half of the volume of the clarifier in (a) (see Fig. 11.18a). The tubes are square and have a side of 5.0 cm and length of 2.0 m.

(c) Calculate the Reynold's number for the clarifier in (a) with lamella placed in the clarifier as indicated in Fig. 11.17, except that the lamella are uniformly inclined without any vertical section at the bottom end. The lamella are inclined at an angle of 60° with 10 cm of free surface over the lamella and 0.50 m depth below the lamella. The lamella are separated by a distance of 15 cm.

17. Is it desirable to set lamella or tubes vertically in a clarifier? What are the effects of changing the angle of inclination of lamella or tubes?

18. Assuming that a particle has a constant settling velocity, what is the settling velocity of a particle that has taken 6 h to settle a depth of 3.0 m (9.80 ft) in an ideal circular clarifier with Q/A = 35 $m^3/m^2/d$ (860 $gal/d/ft^2$). The influent flow is uniformly distributed across the plan area of the clarifier. Underflow (Q_u/A) is being removed from the clarifier at a rate of 7.5 $m^3/m^2/d$ (184 $gal/d/ft^2$)

19. Size the area of a secondary clarifier using the solids flux method. The anticipated critical conditions in the activated sludge treatment process are as follows: the influent flow ranges from 1 200 to 2 000 m^3/d and solids concentration is 3 500 mg/L. The clarifier must perform under all of these conditions. The ultimate compacted sludge concentration is expected to vary from 10 000 to 20 000 mg/L. Use the initial settling velocity data in the following table for your analysis. The solids exiting in the clarified overflow can be assumed to be negligible.

C, mg/L	1 000	2 000	3 000	5 000	7 000	10 000	15 000
v, m/h	8.50	3.10	1.68	0.74	0.44	0.29	0.12

20. Perform a regression analysis to determine the parameters in Eqs. (11.44a) and (11.44c) for the settling data in Problem 19. Which equation is best and what are its parameters?

21. In the alternate method for determining the area of a clarifier for type III sedimentation, the unknowns are N_L, C_L, v_L, U_b, C_u, N_{gL}, C, and N_g which totals 8. The solution requires choice of U_b and the use of Eqs. (11.51)–(11.55). Noting that Eq. (11.51) is actually two equations, the total number of equations is 6. What is the missing equation to solve the system?

22. A new regional water treatment facility is being built to supply water at minimum, average, and maximum flow rates of 40 000, 90 000, and 210 000 m^3/d, respectively to the surrounding communities. Rectangular lamella clarifiers will be used. Studies have indicated that the clarifiers perform adequately at maximum surface overflow rates between 95 and 140 $m^3/m^2/d$ but the upper limit should not be exceeded. All clarifiers will have the same dimensions and their length ratio will be 4 : 1. Size the basin surface areas and determine whether 2, 3, or 4 basins should be used in the plant.

23. Calculate the dimensions of a launder around the periphery of a 10.0 m radius sedimentation basin. A critical flow condition exists at the outlet end of the launder. The design flow for this basin is 5 000 m^3/d. Assume that the channel width is 300 mm and that the flow will divide equally around the two halves of the basin. The launder is made with smooth concrete.

24. Integrate $Q = CkH \int_0^1 \sqrt{\frac{1-z}{z}}\, dz$. Show your work.

25. The average daily wastewater flow rate for a city is 17.6 L/s, the peak hourly rate is 28.4 L/s, and the minimum hourly rate is 7.1 L/s. Design a grit chamber with a Sutro weir flow control device. Assume a bypass channel has been provided for emergency situations. Provide a grit sump in your design. Assume the grit particle to be removed has a settling velocity of 2.3 cm/s (65-mesh sand) and that the chamber flow through velocity is to be maintained at 0.30 m/s.

26. Design an aerated grit removal chamber for a maximum flow of 0.075 m^3/s and removal of 65-mesh sand grit particles. Specify dimensions of the chamber and air requirements.

27. What are the area and depth for a square grit removal chamber for the flow conditions in Problem 25? Use the Metcalf and Eddy design criteria given in the text.

28. Design a horizontal flow trapezoidal grit chamber with a varying cross-sectional area for a plant with an average flow of 1.0×10^4 m^3/d (2.64 Mgal/d) a minimum flow of 0.50×10^4 m^3/d (1.32 Mgal/d), and a maximum flow of 3.0×10^4 m^3/d (7.93 Mgal/d). The grit chamber should remove 65-mesh particles, have 2 channels, and handle a flow of 2.00×10^4 m^3/d (5.28 Mgal/d) with one channel out of service. The horizontal flow-through velocity will be 0.30 m/s (1 ft/s).
   (a) Provide a sketch indicating the dimensions of a channel.
   (b) How much storage will need to be provided for a grit accumulation of 40 m^3 grit/10^6 m^3 (5.35 ft^3/Mgal) wastewater, if grit is to be removed once a week?

29. Find the flow distribution in a manifold–lateral system when the influent flow is 0.66 m^3/s. The system is supplying 3 rectangular sedimentation basins in parallel with 4 evenly spaced laterals for each basin. The diameter of the manifold is 46 cm and the diameter of each lateral is 10 cm. Ignore friction losses in the system and assume the laterals are "short."

KEY REFERENCE

WEF and ASCE (1992), *Design of Municipal Wastewater Treatment Plants,* vol. I, Washington, DC.

REFERENCES

ASCE and AWWA (1990), *Water Treatment Plant Design,* 2nd ed., McGraw-Hill, Toronto.

ASCE and WPCF (1977), *Wastewater Treatment Plant Design,* ASCE, New York.

AWWA (1971), *Water Quality and Treatment,* 3rd ed., McGraw-Hill, Toronto.

Benefield, L. D., J. F. Judkins, Jr., and A. D. Parr (1984), *Treatment Plant Hydraulics for Environmental Engineers,* Prentice-Hall, Englewood Cliffs, NJ.

Bhargava, D. S. and K. Rajagopal (1989), "Modeling for Class-I Sedimentation," *J. Environmental Engineering Division, ASCE,* 115, EE6, pp. 1191–1198.

Camp, T. R. (1942), "Grit Chamber Design," *Sewage Works J.,* 14, pp. 368–381.

Camp, T. R. (1945), "Sedimentation and the Design of Settling Tanks," *ASCE Transactions,* 111, pp. 895–936.

Chow, V. T. (1959), *Open Channel Hydraulics,* McGraw-Hill, Toronto.

Culp, G. L. and R. L. Culp (1974), *New Concepts in Water Purification,* Van Nostrand Reinhold, Toronto.

Degrémont (1973), *Water Treatment Handbook,* Taylor and Carlisle, New York.

Di Bernardo, L. (1993), *Métodos e Téchnicas de Tratamento de Água,* vol. I, Associação Brasileria de Engenharia Sanitária e Ambiental, Rio de Janeiro, Brazil.

DICK, R. I. (1970), "Role of Activated Sludge Final Settling Tanks," *J. Sanitary Engineering Division, ASCE,* 96, SA2, pp. 423–426.

DICK, R. I. AND B. EWING (1967), "Evaluation of Activated Sludge Thickening Theories," *J. Sanitary Engineering Division, ASCE,* 93, SA4, pp. 9–29.

GREGORY, R. AND T. F. ZABEL (1990), "Sedimentation and Flotation," in *Water Quality and Treatment,* 4th ed., F. W. Pontius, ed., McGraw-Hill, Toronto, pp. 367–453.

HAZEN, A. (1904), "On Sedimentation," *Transactions of the American Society of Civil Engineers,* 53, pp. 46–88.

HUDSON, H. E., JR. (1981), *Water Clarification Processes: Practical Design and Evaluation,* Van Nostrand Reinhold, Toronto.

HUDSON, H. E., JR., R. B. UHLER, AND R. W. BAILEY (1979), "Dividing-Flow Manifolds with Square-Edged Laterals," *J. Environmental Engineering Division, ASCE,* 104, EE4, pp. 745–755.

KEINATH, T. M. (1990), "Diagram for Designing and Operating Secondary Clarifiers According to the Thickening Criterion," *J. Water Pollution Control Federation,* 62, 3, pp. 254–258.

KYNCH, G. J. (1952), "A Theory of Sedimentation," *Transactions, Faraday Society,* 48, pp. 166–176.

McGHEE, T. J. (1991), *Water Supply and Sewerage,* 6th ed., McGraw-Hill, Toronto.

METCALF AND EDDY, INC. (1991), *Wastewater Engineering: Treatment, Disposal, and Reuse,* 3rd ed., G. Tchobanoglous and F. L. Burton, eds., McGraw-Hill, Toronto.

PRATT, E. A. (1914), "Another Proportional-Flow Weir; Sutro Weir," *Engineering News,* 72, 9, pp. 462–463.

RAMALHO, R. S. (1977), *Introduction to Wastewater Treatment Processes,* Academic Press, Toronto.

RICHTER, C. J. (1987), Personal communication, SANEPAR, Curitiba, Brazil.

Standard Methods for the Examination of Water and Wastewater (1992), 18th ed., APHA, Washington, DC.

USEPA (1975), *Process Design Manual for Suspended Solids Removal,* No. EPA 25/1–75-003a, USEPA, Washington, DC.

USEPA (1979), *Process Design Manual for Sludge Treatment and Disposal,* No. EPA 625/1–79-011, USEPA, Washington, DC.

VENNARD, J. K. AND R. L. STREET (1982), *Elementary Fluid Mechanics,* 6th ed., John Wiley & Sons, Toronto.

VESILIND, P. A. (1979), *Treatment and Disposal of Wastewater Sludges,* Ann Arbor Science, Ann Arbor, MI.

VESILIND, P. A. AND G. N. JONES (1990), "A Reexamination of the Batch-Thickening Curve," *Water Environment Research,* 62, 7, pp. 887–893.

WAHLBERG, E. J. AND T. M. KEINATH (1988), "Development of Settling Flux Curves Using SVI," *J. Water Pollution Control Federation,* 60, 12, pp. 2095–2100.

WILSON, T. E. AND J. S. LEE (1982), "Comparison of Final Clarifier Design Techniques," *J. Water Pollution Control Federation,* 54, 10, pp. 1376–1381.

YAO, K. M. (1970), "Theoretical Study of High-Rate Sedimentation," *J. Water Pollution Control Federation,* 42, 2, pp. 218–228.

CHAPTER 12

MASS TRANSFER AND AERATION

Mass transfer phenomena govern removal of volatile substances from a water and dissolution of gases in a water. Besides bulk flow, diffusion and its hydraulic analog, dispersion, are also mass transfer operations. Membrane technologies such as reverse osmosis depend on diffusion of substances through a porous media.

12.1 FICK'S LAW

Nature strives to maintain equilibrium. If a quantity of dye is introduced into a volume of water, random movement of the dye molecules because of their thermal energy ultimately results in a uniform concentration of the dye in the fluid. Fick developed the theory describing this phenomenon, known as diffusion. Consider a vessel partitioned in the middle with the liquid on the left side containing dye or any other soluble substance at a uniform concentration and the liquid in the other compartment containing pure water. If the partition is removed without causing disturbance of the fluid in either compartment, net movement of the dye and other substances will be from left to right until a uniform concentration throughout the entire volume is achieved. The situation for an elemental volume is shown in Fig. 12.1.

It is experimentally verified that the movement of mass per unit time in the positive x direction is directly proportional to the area of the element and the concentration gradient across the element. Mass is transferred in the direction of decreasing concentration (i.e., $C + \Delta C$ is less than C), which necessitates a negative sign in the equation.

$$\frac{\Delta m}{\Delta t} \propto -A\frac{\Delta C}{\Delta x} \qquad \text{or} \qquad \frac{\Delta m}{\Delta t} = -DA\frac{\Delta C}{\Delta x}$$

where

m is mass
t is time
A is area of the element
C is concentration
x is distance
D is a diffusion coefficient

The above equation is usually formulated in terms of a mass flux, N, which is the rate per unit area (velocity is the volumetric flux). In the limit, the above equation becomes

$$N = \frac{1}{A}\frac{dm}{dt} = -D\frac{dC}{dx} \tag{12.1}$$

A more complete development would also consider the movement of the medium into the diffusing substance, but this phenomenon is normally insignificant when the

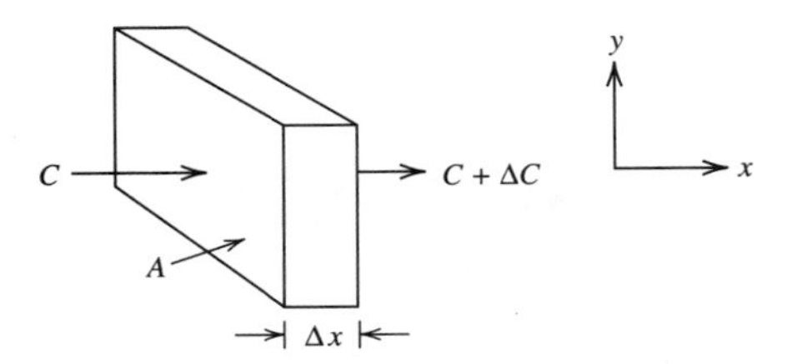

Figure 12.1 Mass transfer across an elemental volume.

amount of diffusing substance is small compared to the amount of medium. Also, in the preceding development mass transfer occurs only in the longitudinal direction; in other instances diffusion may be significant in all three directions (for instance, if the dye has been initially deposited at one location in the fluid). This is the subject of more advanced presentations.

The diffusion coefficient depends on a number of factors: (1) the size and other characteristics of the diffusing substance, (2) the diffusing medium, (3) temperature, and (4) mixing.

Smaller molecules diffuse more rapidly than larger molecules. Attractive forces between the substance and the medium also influence the mobility of the diffusing substance. The density of the medium is a primary influence. A gas will diffuse much more rapidly through other gases as opposed to a liquid. Temperature affects the kinetic energy and random movement of molecules (Brownian motion).

Equation (12.1) was developed for diffusion alone. The hydraulic mixing and turbulence induced in a medium, if present to any significant degree, are much more important than Brownian motion in mass transport. An equation similar to Eq. (12.1) is used to describe these phenomena, although the "diffusion" coefficient reflects the amount of mixing. Diffusion coefficients for a few substances in water at 25°C with no induced mixing are listed in Table 12.1.

TABLE 12.1 Diffusion Coefficients of Substances in Water at 25°C[a]

Solute	Concentration M/L	$D \times 10^5$ cm^2/s
Strong Electrolytes		
$CaCl_2$	0.1	1.11
HBr	0.1	3.16
HCl	0.1	3.05
KBr	0.1	1.87
KCl	0.1	1.84
KI	0.1	1.87
NaCl	0.1	1.52
Weak or Nonelectrolytes		
Citric acid	0.1	0.66
Glucose	0.022	0.67
Glycine	0	1.06
Sucrose	0.011	0.52

[a]From Weast (1984).

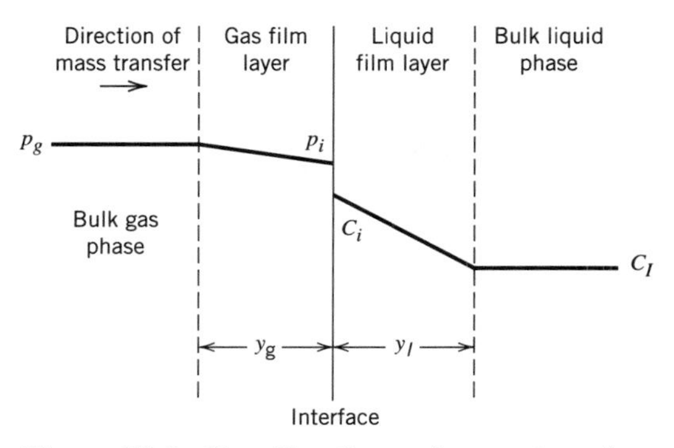

Figure 12.2 Two-film theory for gas transfer.

12.2 GAS TRANSFER

Figure 12.2 shows a definition sketch for the two-film theory for gas transfer (Lewis and Whitman, 1924). The gas–liquid interface is hypothesized to consist of a gas film and a liquid film through which gas is transferred by molecular diffusion. The same amount of mass is being transferred per unit time through each film. Applying Fick's law to each film, the rate of diffusion or mass transfer, dm/dt, is given by:

$$\frac{dm}{dt} = -K_c \frac{D_g}{y_g} A(p_i - p_g) = -\frac{D_l}{y_l} A(C_l - C_i) \tag{12.2a}$$

or

$$\frac{dm}{dt} = k_g A(p_g - p_i) = k_l A(C_i - C_l) \tag{12.2b}$$

where

dm/dt = time rate of mass transfer by diffusion
D_l = diffusion coefficient for the gas in water
D_g = diffusion coefficient for the gas in the gas phase
y_l = hypothetical liquid film thickness
y_g = hypothetical gas film thickness
A = interfacial area
C_i = gas concentration in liquid at the interface
C_l = gas concentration in the liquid bulk phase
p_i = partial pressure of gas in the gas phase at the interface
p_g = partial pressure of gas in the bulk gas phase
K_c = a mass transfer constant
k_l = liquid film coefficient = D_l/y_l
k_g = gas film coefficient = $K_c D_g/y_g$

Gas phase concentrations are most commonly expressed as partial pressures. K is a dimensional constant relating partial pressure of a gas to its concentration in the gas phase. The value of K_c is determined from the gas equation of state

$$pV = nRT$$

$$[C_g] = \frac{n}{V} = \frac{p}{RT}$$

where

$[C_g]$ is on a moles/L basis

Using the molecular weight (MW) of the compound,

$$C_g = \text{MW}\frac{n}{V} = \text{MW}[C_g] = \frac{(\text{MW})p}{RT}$$

where

C_g is on a mass/L basis

Thus K_c has the value of

$$K_c = \frac{\text{MW}}{RT}$$

Mass transfer may be limited by either the gas or the liquid phase, although it is usually diffusion in the liquid phase that controls the overall rate of mass transfer because of the much lower diffusivity in that phase. Because the mobility of molecules in the gas phase is much greater than the mobility of molecules in the liquid phase, the gas film thickness is much smaller than the liquid film thickness. This means that $k_l \ll k_g$. Henry's law (Eq. 1.34) is assumed to apply at the interface. If this is the case and $k_l \ll k_g$, the concentration of diffusing substance in the liquid at the surface of the interface will be the saturation concentration, C_s, determined using the partial pressure of the diffusing substance in the bulk gas phase.

$$C_i = C_s = K_H p_g \tag{12.3}$$

Substituting this into the liquid film expression,

$$\frac{dm}{dt} = k_l A(C_s - C_l)$$

It is common to divide each side of the equation above by V, the unit volume, to find

$$\frac{d(m/V)}{dt} = \frac{dC_l}{dt} = k_l a(C_s - C_l) \tag{12.4}$$

where

$a = A/V$, the specific area

If k_l is not significantly less than k_g but Henry's law still applies at the interface, it can be shown that an expression of the same form as Eq. (12.4) still applies.

$$\frac{dC_l}{dt} = K_l a(C_s - C_l) \tag{12.5}$$

where

K_l is an overall mass transfer coefficient

The overall mass transfer coefficient is a function of the resistances (k_g and k_l) in each phase. It is not always possible to calculate a. If this is the case, the mass transfer equation is

$$\frac{dC_l}{dt} = K(C_s - C_l) \tag{12.6}$$

where
$K = K_l a$

Note that in Eqs. (12.5) and (12.6), if C_l is less than C_s, mass is transferred into the liquid and if C_l is greater than C_s, mass is transferred out of the liquid. Mass is transferred to attain the gas and liquid concentrations that will satisfy Henry's law.

The preceding developments have implicitly assumed a batch or plug flow situation. The rate model is

$$r_g = K_l a(C_s - C_l) \qquad \text{or} \qquad r_g = K(C_s - C_l) \tag{12.7}$$

where
r_g is the mass transfer rate (mass/volume/time)

Mass balances in a batch or plug flow reactor lead to the differential equations (12.4) and (12.5).

To attain the highest transfer efficiency:

1. The renewal rate of the interface should be high; i.e., in both the liquid and gas phases, the elements of fluid in the bulk phase should be brought to the interface as rapidly as possible.
2. The interface thickness should be small.
3. More contact area should be available.
4. Conditions 1–3 should be attained with a minimum expenditure of energy.

Some clarification of the first point is necessary. Consider the liquid phase with a concentration C_l that is less than C_i. Point 1 indicates that the contact (or residence) time of an elemental volume at the interface should be small. While an elemental volume resides at the interface, the concentration of the substance being transferred into the element begins to approach the saturation concentration, thus decreasing the concentration difference term $C_i - C_l$ to $C_i - C_l'$ (C_l' is greater than C_l) in the element. Therefore, the rate of mass transfer into the element also decreases. It is desirable to move a new element from the bulk phase to the interface because it will have a lower concentration than the element at the interface, which has begun to approach saturation. Therefore, rapid circulation of elements to the interface results in higher mass transfer rates.

A longer overall contact time between gas and liquid results in the highest amount of mass being transferred. The overall contact time is the average time for which the total liquid volume is in contact with the gas. The first point is concerned with elemental volumes that should only reside at the surface for a minimal amount of time to accomplish some mass transfer and be rapidly replaced with another elemental volume. This will maintain the concentration gradient at the surface at its maximum value.

12.2.1 Calculating the Mass Transfer Coefficient

The mass transfer coefficient is calculated experimentally by measuring the concentration of the diffusing substance in the liquid at various times. It is important that the experimental apparatus reproduce the levels of turbulence that will exist in the field unit and will also cause the fluid to break into smaller water particles in the same manner as the field unit. These constraints can make model studies difficult for some types of aeration systems. Integrating Eq. (12.5), which applies to a batch reactor, results in C_l as a function of time.

$$\int_{C_0}^{C} \frac{dC_l}{C_s - C_l} = K_l a \int_0^t dt$$
$$C = C_s - (C_s - C_0)e^{-K_l at} \quad (12.8)$$

The saturation concentration of the diffusing substance may be found by applying Henry's law at the temperature of the experiment or by simply letting the experiment run for a significant period of time until the concentration in the liquid remains constant. A plot of $C_s - C$ versus time on semilog paper yields $K_l a$ as the slope. If the surface area of the gas in contact with the liquid is well defined, K_l may be separated from the slope term. Once the mass transfer coefficient is known, Eq. (12.8) can be used to determine the contact times necessary to provide the desired amount of treatment for various influent concentrations.

■ Example 12.1 Determination of $K_l a$

Dissolved oxygen (DO) concentrations measured at various times for a submerged turbine aerator are listed in the following tabulation. Find the mass transfer coefficient ($K_l a$) for this aerator.

Time, min	0	2	4	6	8	10	12
DO, mg/L	0	2.75	4.80	6.15	7.00	7.45	8.05
Time, min	15	20	30	40	50	60	
DO, mg/L	8.30	8.55	8.75	8.70	8.75	8.75	

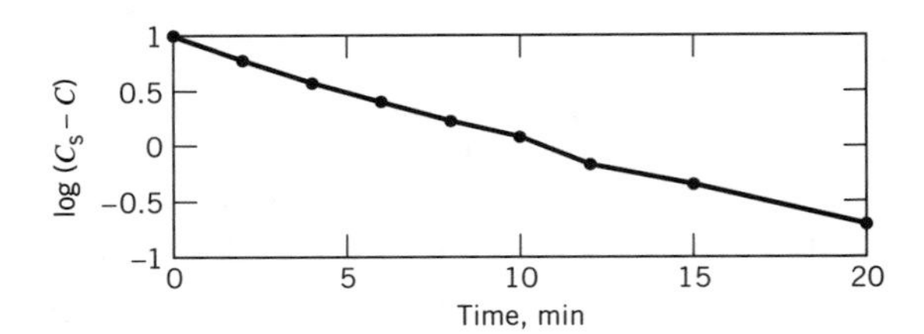

The DO concentration stabilizes at 8.75 mg/L after aeration for a lengthy period of time. This is taken as the saturation concentration of oxygen under the experimental conditions. The C–t data are plotted as shown in the sketch on semilog paper according to Eq. (12.8).

The slope of the line is 2.3 $K_l a$. $K_l a$ is calculated to be 0.087 min^{-1} = 5.2 h^{-1}.

In the two-film model, the mass transfer coefficient is proportional to the diffusivity (see Eqs. 12.2a and 12.4). Other developments, beyond the scope of this text, have shown that the overall mass transfer coefficient may be proportional to the diffusivity raised to a power between 1/2 and 1 (Danckwerts, 1951; Higbie, 1935). Munz and Roberts (1989) have shown that when the dimensionless Henry's constant (C_g/C_l with both expressed in mass/volume) is greater than 0.55, the mass transfer coefficient of volatile organics is directly proportional to that of oxygen for surface aerators.

12.2.2 The Effects of pH on Mass Transfer

The mass transfer rate coefficient, $K_L a$ is independent of pH but the rate of mass transfer (Eq. 12.7) is subject to the influence of pH when the gas participates in acid–base reactions. Many gases such as CO_2, NH_3, and H_2S undergo hydrolysis reactions and are speciated among various forms. The ionized forms are not volatile. The concentration of undissociated species of volatile acids such as acetic acid ($HC_2H_3O_2$) is also a function of pH.

To apply Eq. (12.7) correctly, C_s is determined from Henry's law based on the concentration of the gas in the gaseous phase. C_l is the concentration of the undissociated form of the species in solution. As mass transfer proceeds the solution pH changes because of the addition or removal of species that influence the pH (e.g., CO_2). Acid–base reactions occur rapidly and the assumption that equilibrium conditions with respect to pH are instantaneously re-established as the species is transferred has been found to be reasonable (Howe and Lawler, 1989). The solution of the mass transfer problem is performed by applying mass transfer and pH transformations in parallel. Howe and Lawler (1989) give more discussion of the procedure, which is generally beyond the scope of this presentation.

The important point to be drawn is that acidic substances should be stripped at low pH and basic substances should be stripped at high pH, where the undissociated species will predominate. Another factor that may influence the rate of mass transfer is the system oxidation–reduction potential (ORP). For instance, oxygen transferred into a groundwater containing sulfides will raise the ORP and promote the oxidation of sulfides to sulfate. Oxidation–reduction reactions are not as rapid as acid–base reactions and the assumption that redox equilibrium is maintained does not hold. It becomes more difficult to separate the overall rate of change between mass transfer and conversion of a species to a different form. If the rate of oxidation is not determined, the mass transfer rate coefficient is an apparent value that strictly applies only to similar conditions and may need to be modified as conditions change significantly.

12.3 AERATION IN WATER AND WASTEWATER TREATMENT

Aeration serves a number of useful purposes in water treatment. Many of the problem substances that aeration removes are most generally found in groundwaters. These substances must be present at levels that pose a threat to health or cause excessive consumption of chemicals in the water treatment process before a separate unit operation of aeration is incorporated into the treatment train. Using aeration to achieve mixing in flocculation basins (flocculation and coagulation are discussed in Chapter 13) can be considered in some cases, although the degree of aeration treatment in flocculators that have diffused air mixing is not likely to be as great as in a standard aeration unit. Also the degree of removal of some compounds will have been changed by the addition of coagulants in the coagulation process that precedes flocculation.

The primary application of aeration in wastewater is to supply oxygen to aerobic biological treatment processes. Air stripping to remove toxic volatile organics is another wastewater application.

The theory of aeration is explained above and applied to various methods in the following sections but it is always necessary to conduct laboratory studies to confirm rate constants and removals. Note that because of the hydrolysis reactions of many gases, pH is an important factor influencing the removal of a number of compounds.

The primary purposes and effects of aeration in water treatment are as follows.

Algal Growth. Open air aeration systems expose the water to sunlight, which promotes the growth of algae. These algae may produce taste and odor causing compounds or their decomposition in the distribution system may lead to similar problems. Aeration is generally not an efficient method for removing the taste and odor compounds produced by algae because the algal oils causing taste and odor are not volatile to a significant extent.

Carbon Dioxide. The high solubility of CO_2 reduces the pH of the water, which causes excessive consumption of lime or other neutralization agents in coagulation and softening processes. The corrosiveness of the water is also higher at lower pH values.

Exposure of water droplets to air for 2 s will lower the concentration of CO_2 by 70 to 90%. Carbon dioxide can also be removed (converted into bicarbonates) by the addition of lime. It is recommended that aeration be used if CO_2 concentrations are greater than 10 mg/L; otherwise lime addition should be used to neutralize the CO_2 (AWWA, 1971). The economics of each process must be carefully compared at expected concentrations to select the appropriate process.

Corrosiveness. The addition of oxygen to water makes the water more corrosive. As noted previously, the removal of CO_2 will have the beneficial effect of raising the pH.

Hydrogen Sulfide. The solubility of H_2S is also very high because of its ability to react with water and form the ions HS^- and S^{2-}. Sulfides cause unpleasant tastes and noxious odors in a water. At high concentrations, H_2S can be lethal. Removal of H_2S is not as rapid as removal of CO_2, and as CO_2 is removed the pH rises, which causes the increased formation of the ionized forms of sulfides. These ionized forms are not removed by aeration. However, the atmospheric concentration of H_2S is essentially zero, which enhances its rate of mass transfer. The standard (in Canada) for total sulfide concentration in drinking water is 0.05 mg/L.

Iron and Manganese. Iron and manganese are dissolved by groundwaters that have a high dissolved CO_2 content and have a low pH. The more soluble forms of iron and manganese are their lower oxidation states, Fe(II) and Mn(II). The addition of oxygen by aeration promotes the oxidation of iron and manganese to their relatively insoluble forms, Fe(III) and Mn(IV). The precipitates formed will be removed in coagulation–sedimentation and filtration operations later on in the treatment train.

Methane. Methane, or marsh gas as it is commonly known, may be present in some waters at significant concentrations. It is very insoluble in water and its presence creates an explosion hazard.

Organic Matter. The oxidation of organic matter and other inorganic substances causing oxygen demand will be accomplished to only a very limited extent by the addition of oxygen through aeration. This is not a reason to design an aeration system.

Oxygen. Oxygen will be transferred to waters below saturation concentration with effects described under various other categories.

Taste and Odor. Taste and odor causing compounds are often volatile and these are removed to a significant extent by aeration. Also, the addition of oxygen will improve the taste of a water to a limited extent.

Volatile Organics. Trace volatile organics that may be present in surface and groundwaters are stripped from the water by aeration. The degree of aeration required depends on the concentration and solubility of the compounds of interest. Stripping volatile toxic substances from a water or wastewater transfers the problem to the atmosphere. There is movement to require passing the exhaust gases from the stripper through activated carbon or other treatments to remove the contaminants. Using an air stripper with activated carbon treatment of the exhaust gas may be more economical than treating the water directly with activated carbon, which more rapidly exhausts

TABLE 12.2 Hazardous Oxygen Concentration Levels[a]

Concentration, %	Effect[b]
21	Normal percentage of O_2 in air
16	8 h exposure is acceptable
14	Slight difficulty in breathing, ringing in ears
12	Thought is cloudy, drowsiness
10	Unconsciousness
8	Death

[a]From *WPCF Highlights* (1980), 17, 6, © WEF 1980.
[b]An oxygen deficiency is most likely to be found (a) down low, (b) where ventilation is poor, (c) in deep manholes, (d) in wet pits, and (e) near siphons.

its capacity by removing many other compounds that are not problematic in addition to the desired substance(s).

There are also disadvantages associated with aeration. It is not possible to achieve complete removal of some compounds, for instance, CO_2.

Hazards Associated with Oxygen and Hydrogen Sulfide

When hydrogen sulfide and other gases are being removed indoors, one must ensure that adequate ventilation exists to prevent accumulation of high concentrations of the toxic gas or conditions that may cause asphyxiation through the displacement of oxygen with another gas. Dangerous levels for oxygen and hydrogen sulfide are given in Tables 12.2 and 12.3, respectively. Environmental engineers should always be aware of these hazards and ensure that work crews take appropriate measures to protect themselves. Anaerobic conditions in sewers can result in accumulation of high concentrations of hydrogen sulfide. Hydrogen sulfide poisoning is one of the leading causes of accidents in the field. It has the particularly malicious property of paralyzing the olfactory organs at higher concentrations, where it is most toxic.

TABLE 12.3 Hazardous Concentration Levels for Sulfides[a]

Concentration ppm	Effect[b]
Nil	Normal concentration of H_2S in air
5	Moderate odor, readily detectable
10	Eye irritation begins
30	Strong, unpleasant odor of rotten eggs
100	Coughing, loss of smell in 2–16 min
200–300	Red eyes, rapid loss of smell, breathing irritation
300–700	Unconsciousness and possibly death in 30–60 min
700–1 000	Rapid unconsciousness, death in a few minutes
1 000–2 000	Instant unconsciousness, death in a few minutes
4 300	Lower explosive limit

[a]From *WPCF Highlights* (1980), 17, 6, © WEF 1980.
[b]Hydrogen sulfide is most likely to be found (a) down low, (b) where flow is blocked, (c) where ventilation is inadequate, (d) in large sewers flowing slowly and nearly full, (e) in wet pits, and (f) at the end of a long force main.

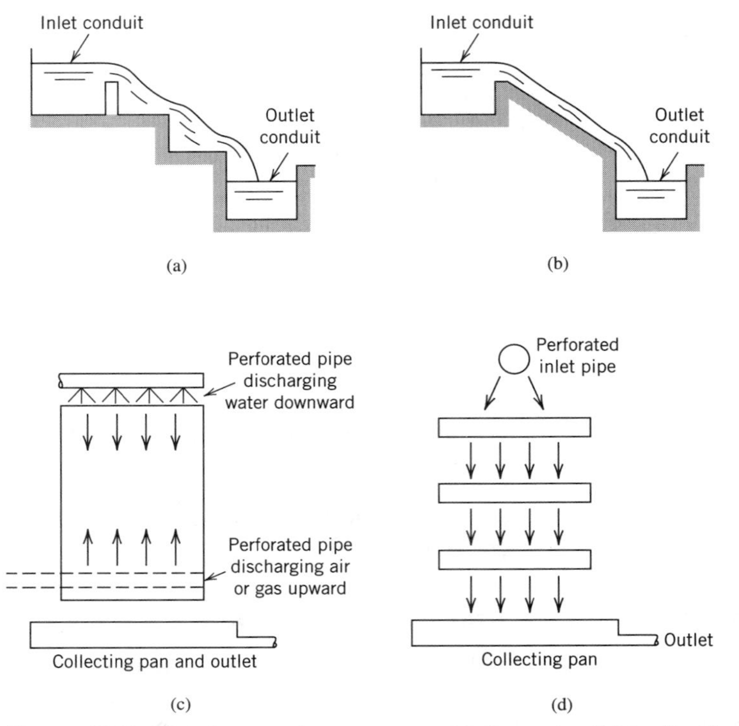

Figure 12.3 Gravity aeration systems. (a) Cascade; (b) inclined apron, possibly studded with riffle plates; (c) tower with countercurrent flow air (gas) and water; and (d) stack of perforated plates, possibly containing perforated media. From G. M. Fair, J. C. Geyer, and D. A. Okun, *Water Purification and Wastewater Treatment and Disposal,* vol. 2, copyright © 1968 by John Wiley & Sons, Inc. Reprinted by permission of John Wiley & Sons, Inc.

12.4 DESIGN OF AERATION SYSTEMS

Aeration systems are designed to promote turbulence and break the water into smaller volumes or droplets, increasing the surface area for mass transfer. Gravity or pressurized flow systems may be used.

12.4.1 Gravity Aerators

Some gravity flow systems are illustrated in Fig. 12.3. These systems are used for water treatment. In a cascade aerator, water falls down a series of steps. The splashing of the water creates turbulence and water droplets, although not too efficiently compared to other aeration methods. An inclined apron, which may contain studs, works on similar principles to the cascade aerator. Another technique is to discharge the water onto a stack of perforated plates. The plates may contain media that promote oxidation of iron and manganese.

Cascade aerators are analyzed from fundamental mechanics principles. Free fall occurs over each step and gravity is the major force.

$$a = \frac{dv}{dt} = -g \qquad \text{and} \qquad v = v_0 - gt \tag{12.9}$$

where

a is acceleration
v is vertical velocity
t is time
g is the acceleration of gravity

The initial vertical velocity is zero. The time, t_h, to fall the height of one step, h, is calculated from

$$v = \frac{dh}{dt} = -gt \qquad \text{and} \qquad h = \tfrac{1}{2}gt_h^2 \tag{12.10}$$

$$t_h = \sqrt{\frac{2h}{g}} \tag{12.11}$$

In a total height, H, which is divided into n individual steps, the total contact time between water and air, t_c, is

$$t_c = n\sqrt{\frac{2H/n}{g}} = \sqrt{\frac{2Hn}{g}} \tag{12.12}$$

Head requirements vary from 1 to 3 m (3–10 ft). The ASCE and AWWA (1990) report that carbon dioxide reduction often ranges from 20 to 45% in cascade aerators. Space requirements are in the range of 9.8–12.3 $cm^2/m^3/d$ (40–50 ft^2/Mgal/d) of flow.

Multiple tray aerators are similar to cascade aerators. Media that may be used on the trays to enhance the oxidation of iron and manganese are coke, stone, or ceramic balls ranging in size from 50 to 150 mm diameter. The deposited oxides on the medium will have a catalytic effect on the oxidation of iron and manganese in the water. Spacing between the trays ranges from 30 to 75 cm (12–30 in.) and water application rates are generally in the range of 50–75 m^3/h per m^2 (20–30 gal/min per ft^2) of tray area (ASCE and AWWA, 1990). Three to nine trays are used.

12.4.2 Spray Aerators

Spray aerators are efficient but they require a relatively large land area to collect the water. They are also aesthetic and are often designed for this reason at water treatment plants. In fixed spray aerators water is discharged from a pipe with a number of orifices inclined from the horizontal to achieve a waterfall effect. For n orifices discharging at approximately the same rate, the total flow, Q, is

$$Q = nq = nC_d a\sqrt{2gh} \tag{12.13}$$

where

C_d is the discharge coefficient for an orifice
q is the discharge through an individual orifice
a is the area of an orifice
h is the head on an orifice

TABLE 12.4 Discharge Coefficients for Different Shapes of Openings

Opening	Sharp edged	Rounded	Streamline designed
C_d	0.6	0.8	0.85–0.92

The discharge coefficient depends on the hydraulic characteristics of the orifice. Table 12.4 lists values of C_d for various orifice shapes. Figure 12.4 shows a layout for an aerator and various streamline designed nozzle types.

The discharge, q, and velocity, v_0, through an individual orifice are

$$q = \frac{Q}{n} = C_d a \sqrt{2gh} \tag{12.14}$$

$$v_0 = \frac{q}{a} = C_v \sqrt{2gh} \tag{12.15}$$

where

C_v is the velocity coefficient of the orifice

The value of C_v is typically 0.95.

Section 11.9 should be reviewed for the development describing varying flow and headloss in a multiple-orifice distribution pipe. A movable spray aerator with arms rotating in a horizontal plane over a circular basin that collects the water could also be designed but is not as common. Section 17.14.2 gives design details of the distributor.

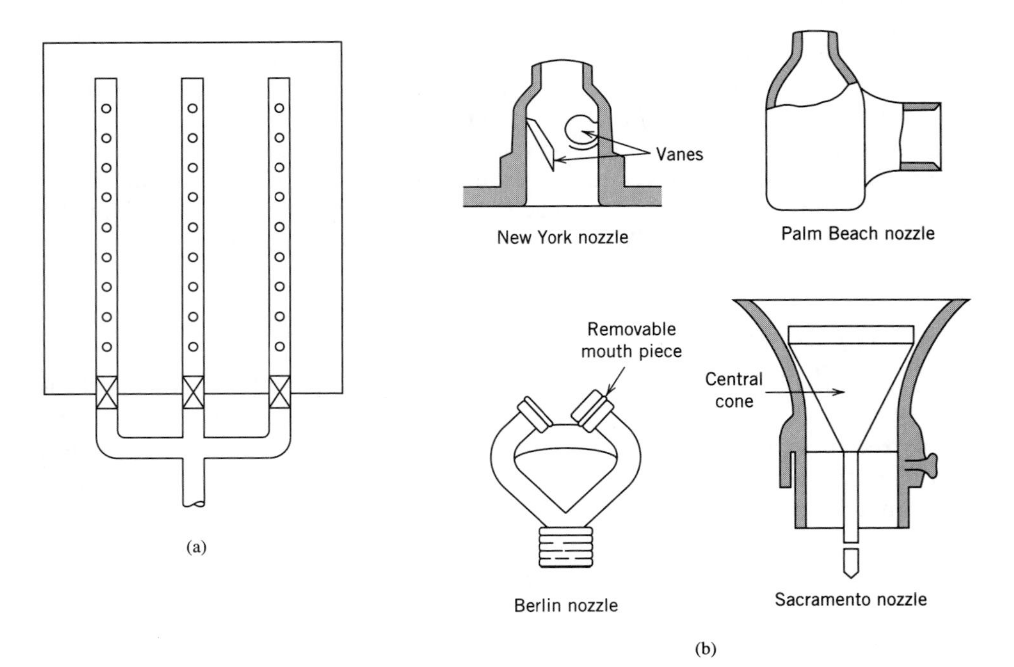

Figure 12.4 Spray aerator and nozzles. (a) Nozzled aerator; (b) aerator nozzles. From G. M. Fair, J. C. Geyer, and D. A. Okun, *Water Purification and Wastewater Treatment and Disposal,* vol. 2, copyright © 1968 by John Wiley & Sons, Inc. Reprinted by permission of John Wiley & Sons, Inc.

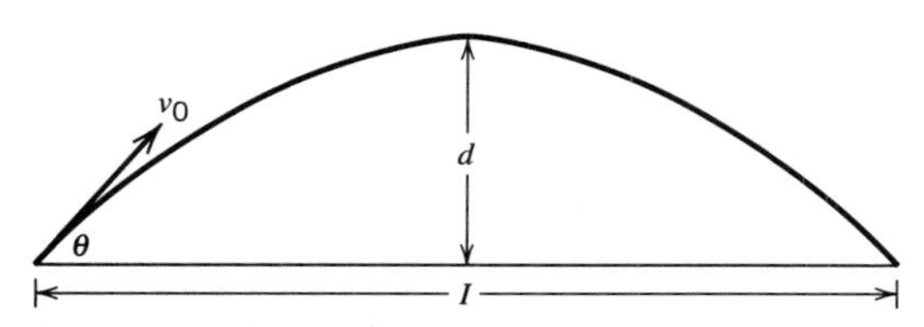

Figure 12.5 Spray aerator water trajectory.

For an orifice inclined at an angle θ from the horizontal, the water issuing from the orifice follows a parabolic trajectory (Fig. 12.5).

Again the situation is analyzed using basic mechanics principles. The following definitions are used in the developments.

t_r time of rise of a droplet

T total contact time

v_0 initial velocity of the jet

v_y vertical component of velocity

v_x horizontal component of velocity

d height of rise of the jet

l horizontal distance of travel of the jet

Wind effects are ignored in the initial development. They will be accounted for later. Now,

$$v_x = v_0 \cos \theta$$

at all times and

$$v_{y0} = v_0 \sin \theta.$$

At the top of the trajectory, the rise velocity is 0 and the time to reach the maximum height is t_r.

$$a = \frac{dv_y}{dt} = -g$$

and

$$\int_{v_{y0}}^{0} dv_y = -g \int_0^{t_r} dt \Rightarrow v_{y0} = gt_r$$

The time of rise is one-half of the total contact time.

$$t_r = \frac{v_{y0}}{g} = \frac{v_0 \sin \theta}{g} = \frac{T}{2} \tag{12.16}$$

Using Eq. (12.15) to solve for the driving head,

$$T = \frac{2C_v \sin \theta \sqrt{2gh}}{g} \quad \text{and} \quad h = \frac{gT^2}{8C_v^2 \sin^2 \theta} \tag{12.17}$$

The vertical distance of travel is

$$\frac{dy}{dt} = v_y = v_{y0} - gt \qquad \int_0^d dy = v_{y0}\int_0^{t_r} dt - g\int_0^{t_r} t\,dt$$
$$d = v_{y0}t_r - \tfrac{1}{2}gt_r^2 = \frac{v_0^2 \sin^2\theta}{g} - \frac{v_0^2 \sin^2\theta}{2g} = \frac{v_0^2 \sin^2\theta}{2g} \tag{12.18}$$

The horizontal distance of travel is

$$\frac{dx}{dt} = v_0 \cos\theta \qquad \int_0^l dx = \int_0^T dt = v_0 T\cos\theta = 2v_0 t_r \cos\theta \tag{12.19}$$

Substituting for t_r from Eq. (12.16) and using $2\cos\theta\sin\theta = \sin 2\theta$,

$$l = \frac{v_0^2 \sin 2\theta}{g} \tag{12.20}$$

Equations (12.18) and (12.20) can be expressed in terms of the driving head by using Eq. (12.15).

$$d = hC_v^2 \sin^2\theta \quad (12.21) \qquad l = 2hC_v^2 \sin 2\theta \tag{12.22}$$

If wind velocity is significant, the wind's drag force on the water droplets carries droplets further in the direction in which the wind is blowing. The wind velocity, v_w, results in an additional horizontal velocity component on a water droplet, v_{xw}.

$$v_{xw} = C_w v_w \tag{12.23}$$

where
C_w is a drag coefficient of the wind

This additional velocity is additive with the horizontal velocity component at the orifice and the additional distance, l_w, traveled by a water particle is

$$l_w = C_w v_w T \tag{12.24}$$

The drag coefficient, C_w, is found to be approximately 0.6.

A significant wind velocity increases the mass transfer coefficient and less exposure time is required when the wind is blowing. Design wind conditions and exposure time must be used to size the collection basin to collect the water that has traveled the maximum horizontal distance.

The ASCE and AWWA (1990) report that nozzles for existing aerators are usually from 25 to 40 mm (1–1.5 in.) in diameter with discharge ratings from 0.28 to 0.56 m^3/min (75–150 gal/min) at a pressure rating of 69 kPa (10 psi). Nozzle spacing varies from 0.6–3.7 m (2–12 ft). General removals of carbon dioxide are 70% or higher.

12.4.3 Diffused Aerators

Submerged diffused aerators are a common device for transferring oxygen in aerobic biological treatment systems. They are also used for air stripping volatile organics from contaminated waters, although they are not quite as efficient as countercurrent flow packed media aeration towers for this purpose. Air strippers are specialized treatment devices beyond the scope of this text. Montgomery (1985) gives a discussion of their design and operation.

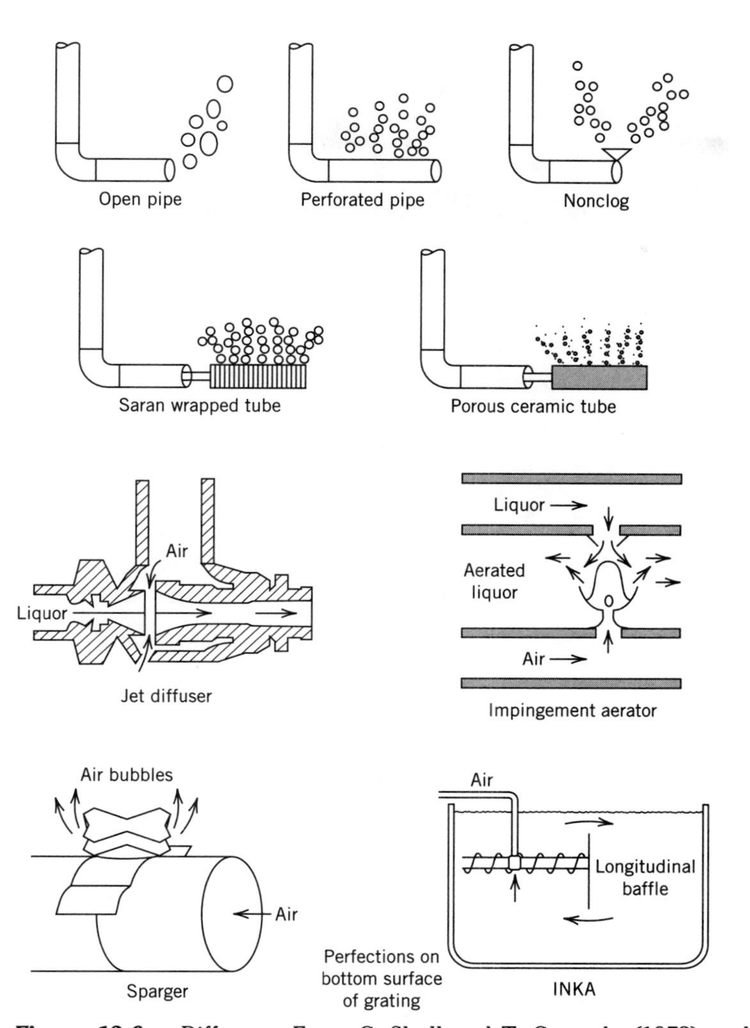

Figure 12.6a Diffusers. From G. Shell and T. Cassady (1973) and A. L. Downing (1960), *J. of the Institution of Public Health Engineers,* 59, p. 87. Used with permission.

The rate of gas transfer depends on the number and size of bubbles in the system. Smaller bubbles have a greater surface area to volume ratio and are more efficient than larger sized bubbles for mass transfer. The improvement in mass transfer efficiency must be evaluated against the increased headlosses associated with the smaller orifices required to create the smaller bubbles. Smaller orifices are also more likely to become clogged with biological growth. Diffusers used in biological wastewater treatment are shown in Figs. 12.6a and 12.6b.

A sketch of a diffuser in a basin is shown in Fig. 12.7. To calculate the rate of gas transfer, the contact time and mass transfer coefficient are required. The following symbol definitions apply to Fig. 12.7:

Q_a volumetric airflow rate

h depth of fluid over the diffusers

v_b rise velocity of the bubbles relative to the fluid

Figure 12.6b Diffusers (a) aeration panel (Parkson Aeration Panel™), (b) fine bubble, (c) flexible tube (Wyss® Flex-A-Tube®), (d) coarse bubble. Courtesy of Parkson Corp.

V_b volume of an air bubble at the surface of the liquid

v_l average upward velocity of the liquid

The surface area for gas transfer depends on the total number of bubbles in the basin, n.

$$n = \frac{\text{volume of air in the tank}}{\text{volume of a single bubble}} = \left(\frac{Q_a}{V_b}\right)\frac{kh}{v_b + v_l}$$

where

k is a factor to account for the increased depth resulting from the entrainment of bubbles

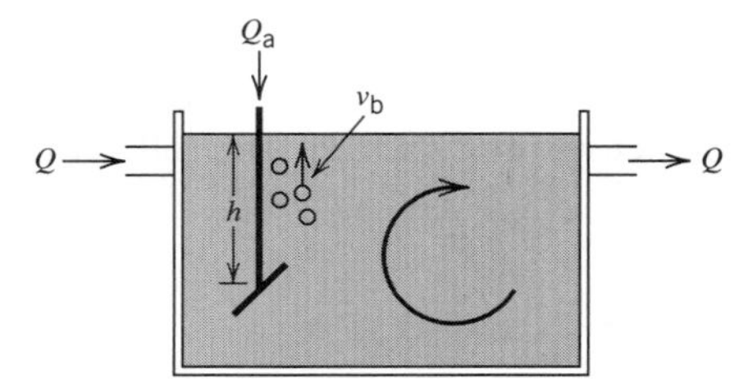

Figure 12.7 Diffused aerator.

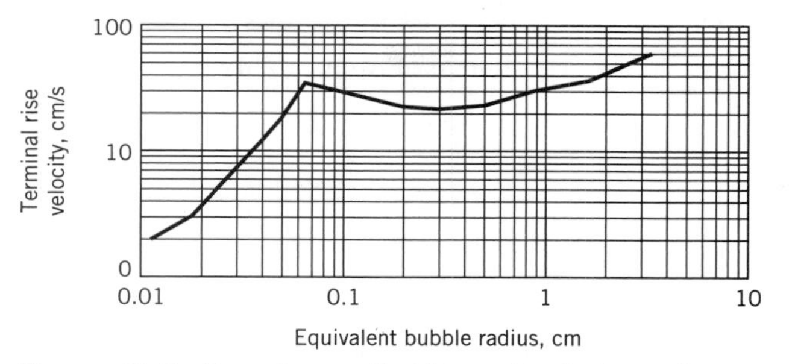

Figure 12.8 Experimentally determined rise velocities of bubbles in distilled water. Adapted from Haberman and Morton (1953).

From experimentation (Bewtra and Nicholas, 1964), it is found that

$$\frac{kv_b}{v_b + v_l} \approx 1$$

Therefore

$$n = \left(\frac{Q_a}{V_b}\right)\frac{h}{v_b} \tag{12.25}$$

If there is no coalescence of bubbles, the number of bubbles in the basin does not change; however, the volume of a bubble will change as the bubble rises and pressure (p) decreases. The volume change does not need to be considered if both Q_a and V_b are calculated at atmospheric pressure (see Section 13.3.2 for a discussion of bubble volume variation with pressure).

The terminal rise velocity of a bubble is achieved almost immediately after the bubble exits the orifice. (The equations for rise velocity of a bubble are the same as those for settling velocity of a particle, Eqs. 11.6–11.8.) The nominal diameter of a bubble is close to the diameter of the opening in the diffuser. Drag forces, surface tension effects, and the shape of the bubble change as a function of the bubble size and they influence the rise velocity of the bubble. Wallis (1969) gives a discussion of these factors. An experimental determination of the rise velocity of bubbles of various sizes in distilled water is given in Fig. 12.8.

Peebles and Garber (1953) developed equations to describe the terminal rise velocities of single bubbles as a function of three dimensionless groups (Re_b, the Reynold's number for the bubble, G_1, and G_2).

$$Re_b = \frac{2\rho_l v_b r}{\mu_l} \tag{12.26}$$

$$G_1 = \frac{g\mu_l^4}{\rho_l \sigma^3} \tag{12.27}$$

$$G_2 = \frac{gr^4 v_\infty^4 \rho_l^3}{\sigma^3} \tag{12.28}$$

TABLE 12.5 Terminal Rise Velocity of Single Bubbles in Liquids[a]

	Terminal velocity expression[b]	Range of Applicability	Eq. No.
Region 1	$v_b = \dfrac{2r^2(\rho_l - \rho_g)g}{9\,\mu_l}$	$\mathrm{Re}_b < 2$	(12.30a)
Region 2	$v_b = 0.33g^{0.76}\left(\dfrac{\rho_l}{\mu_l}\right)^{0.52} r^{1.28}$	$2 < \mathrm{Re}_b < 4.02G_1^{-2.214}$	(12.30b)
Region 3	$v_b = 1.35\left(\dfrac{\sigma}{\rho_l r}\right)^{0.50}$	$4.02G_1^{-2.214} < \mathrm{Re}_b < 3.10G_1^{-0.25}$ or $16.32G_1^{-0.144} < G_2 < 5.75$	(12.30c)
Region 4[c]	$v_b = 1.53\left(\dfrac{g\sigma}{\rho_l}\right)^{0.25}$	$3.10G_1^{-0.25} < \mathrm{Re}_b < 5.75 < G_2$	(12.30d)

[a]After Peebles and Garber (1953).
[b]ρ_g is the density of the gas phase.
[c]The constant 1.53 is suggested by Harmathy (1960) as a better value than 1.18 reported by Peebles and Garber.

where
- ρ_l is the density of the liquid
- r is the nominal radius of the bubble
- v_∞ is the terminal rise velocity of the bubble
- μ_l is the dynamic viscosity of the liquid
- σ is the surface tension of the liquid

The equations for the rise velocity assume that the bubble density is negligible compared to the liquid density. Four regions were defined and the equations for each region are given in Table 12.5.

To calculate the surface area of a single bubble, the radius, r, of the bubble at mid-depth in the tank should be used. The surface area of a bubble (A_b) is

$$A_b = 4\pi r^2 \tag{12.29a}$$

and their total surface area is

$$A = n4\pi r^2 \tag{12.29b}$$

The equation for A is used to calculate a (A/V) in Eq. (12.4) or (12.7) and related equations. Strictly, the surface area of the tank should also be considered but it is usually ignored because it is small in comparison to the total bubble surface area. V is the liquid volume of the tank.

In water treatment processes, diffused aerators are commonly 2.7–4.6 m (9–15 ft) deep and 3–9 m (10–30 ft) wide (ASCE and AWWA, 1990). Ratios of width to depth should not exceed 2 to ensure effective mixing. The detention time depends on the desired removals at the limiting mass transfer rate. The diffusers are usually placed along one side of the tank to induce a spiral flow through the tank. The diffusers may be located near mid-depth to reduce energy requirements to compress the air. The amount of air required ranges from 0.075 to 1.12 m^3 air per m^3 (0.01–0.15 ft^3 per gal) of water.

12.5 AERATION IN BIOLOGICAL WASTEWATER TREATMENT

Oxygen supply is vital to all forms of aerobic biological treatment and is one of the more expensive operations in the process. Because gas transfer causes turbulence and mixing, aeration devices play a significant role in mixing and they are normally designed to supply the required degree of mixing besides providing enough oxygen.

The fundamental equation of gas transfer (Eq. 12.7) applies. The rate constant and saturation concentration are temperature dependent. Saturation concentrations of oxygen for the common temperature range are given in the Appendix. The mass transfer coefficient varies with temperature according to the Arrhenius equation:

$$K_T = K_{20}\theta^{(T-20)} \tag{12.31}$$

where

T is temperature, °C

θ is a constant

The common value of θ reported in the literature is 1.024, although many different values have been found. Furthermore, it must be remembered that θ actually varies slightly with temperature, as well as with other environmental conditions and fluid characteristics. Therefore, different values of θ should be used for different temperature ranges.

Aeration efficiency is normally tested using tap water in a basin that is different from the prototype in a number of aspects. Model studies must be carefully performed to ensure that a meaningful comparison of various aeration devices is obtained for the prototype. There are also a number of correction factors that must be considered to determine in situ rates of oxygen transfer when wastewater is the medium.

The saturation concentration of oxygen in wastewater will be different from that in tap water.

$$C_{s,ww} = \beta C_{s,Tw} \tag{12.32}$$

where

ww refers to wastewater

Tw refers to tap water

β is a correction factor → $0.80 \le \beta < 1.0$ typically 0.93–0.96

The saturation concentration of oxygen in water or tap water is calculated from Henry's law, is found from lab studies, or is available in tables for different temperatures. Besides being corrected for temperature, C_s must also be corrected for altitude because solubility of a gas varies with pressure (according to Henry's law). The equation describing this effect is

$$C_{s,alt} = C_{s,sl}\left[1 - \frac{\text{altitude (m)}}{9\,450}\right] \tag{12.33}$$

where
alt refers to altitude
sl refers to sea level

Equation (12.33) incorporates a polynomial fit to pressure variation with altitude.

The rate transfer coefficient in wastewater will also be different from that in tap water.

$$K_{ww} = \alpha K_{Tw} \tag{12.34}$$

where
α is a correction factor → $0.5 \le \alpha \le 0.95$ typically $0.80 - 0.85$

The factors α and β are variable from waste to waste and should be measured from laboratory or field tests in each instance. However, basin geometry effects, the other major factor responsible for variation in K, are difficult to evaluate unless the prototype is used in the tests. Common values for α and β are 0.80–0.85 and 1.0, respectively.

Formulating Eq. (12.7) for a wastewater,

$$r'_{ww,T} = \alpha K_{20,Tw}(1.024)^{(T-20)}(\beta C_s - C_l) \tag{12.35}$$

where
$r'_{ww,T}$ is rate of O_2 transfer for wastewater

Manufacturers commonly report their data for aeration equipment in tap water at 20°C. When they conduct their tests the tap water is deoxygenated such that $C_{l,Tw} = 0$. Sulfite is added to water in excess of the stoichiometric requirement to remove oxygen (Eq. 12.36). Cobalt is added in the form of cobalt chloride to catalyze the deoxygenation reaction, which is quite fast. A cobalt concentration of no higher than 0.05 mg/L is recommended; however, if the tap water contains high concentrations of chloramines (greater than 0.3 mg/L) then the cobalt dose should be increased (Terry and Thiem, 1989). Excess cobalt ion does not interfere with DO measurement by probe, but it can cause interference with the Winkler iodometric method for DO determination.

$$2Na_2SO_3 + O_2 \xrightarrow{Co} 2Na_2SO_4 \tag{12.36}$$

The deoxygenating agents are mixed with the water and the aeration device is activated. The change in DO concentration is measured over time.

Taking the ratio of Eqs. (12.35) and (12.7), where Eq. (12.7) is applied at these specified testing conditions results in

$$\frac{r'_{ww,T}}{r'_{Tw,20}} = \frac{(\beta C_s - C_l)\alpha(1.024)^{(T-20)}}{9.09} \tag{12.37}$$

where
9.09 mg/L is the saturation concentration of O_2 in tap water at 20°C

For activated sludge and aerobic digestion processes the DO content needed is usually between 0.2 and 2.0 mg/L to prevent oxygen diffusion limitations from hindering the rate of substrate removal by the suspended microorganisms. A value of 1.0 mg/L is safe for most processes except under unusual (e.g., bulked) conditions. The higher, more conservative value (2.0 mg/L) should be used for design purposes. The Great Lakes Upper Mississippi River Board (GLUMRB) has 10 member states: Illinois, Indiana, Iowa, Michigan, Minnesota, Missouri, Ohio, Pennsylvania, and Wisconsin. GLUMRB (1976) has set standards for water works developed from observations of many treatment processes. GLUMRB specifies that 150% of air requirements should be available.

Manufacturers do not usually report their ratings in terms of mg/L/time (r') but in terms of oxygen transferred per unit of energy (r).

$$r = \frac{\text{kg O}_2 \text{ transferred}}{\text{MJ}} \quad \text{(SI)} \qquad \text{or} \qquad r = \frac{\text{lb O}_2 \text{ transferred}}{\text{hp-h}} \quad \text{(U.S.)} \tag{12.38}$$

where

MJ = 10^6 J

Aeration efficiency is also described by the oxygen transfer efficiency, E.

$$E = \frac{\text{mass of oxygen utilized}}{\text{mass of oxygen supplied}} \times 100 \tag{12.39}$$

The former measure is most useful because it is used for estimating energy costs for supplying oxygen. The r value relates oxygen transfer to gross energy input to the aeration device; i.e., the efficiency of the aerator is included in the r rating. Incorporating mass transfer factors, the field rate of oxygen transfer is

$$r_{\text{ww,T}} = \frac{r_{\text{Tw,20}}(\beta C_s - C_l)\alpha(1.024)^{(T-20)}}{9.09} \tag{12.40}$$

Using Eq. (12.40) and knowing the amount of oxygen required in the biological treatment process, the total power requirements for the aerators may be estimated. The oxygen supplied to the wastewater is the oxygen consumed by the microorganisms. Excess oxygen supplied increases the DO content of the water in the basin and is transported out of the basin with the effluent. The total oxygen required per unit time is estimated from the anticipated process operating conditions, the flow rate, and amount of oxygen demand in the sewage. The aerators must be sized for the extreme conditions considering all of the above factors. At higher temperatures the saturation concentration of oxygen is lower but mass transfer coefficients are higher. Extreme values of sewage oxygen demand satisfied in treatment will be higher at high temperatures.

12.5.1 Aeration Devices in Wastewater Treatment

Surface and submerged aerators are employed in biological wastewater treatment units or pond systems. Mechanical surface aerators draw water up and throw it up and radially outward. Two types of mechanical aerators are shown in Figs. 12.9 and 12.10. Draft tubes may be installed with surface aerators to ensure that water is drawn up from the bottom of the basin, which promotes better mixing.

Submerged aerators are porous or nonporous diffusers with a variety of designs, some of which are illustrated in Figs. 12.6(a) and (b). Porous diffusers are made from rigid materials such as ceramic, which have small openings that control the size of the gas bubble emerging from them. Perforated membranes made of flexible rubber or plastic can also be used. Porous diffusers are available in plates, domes (Fig. 12.11),

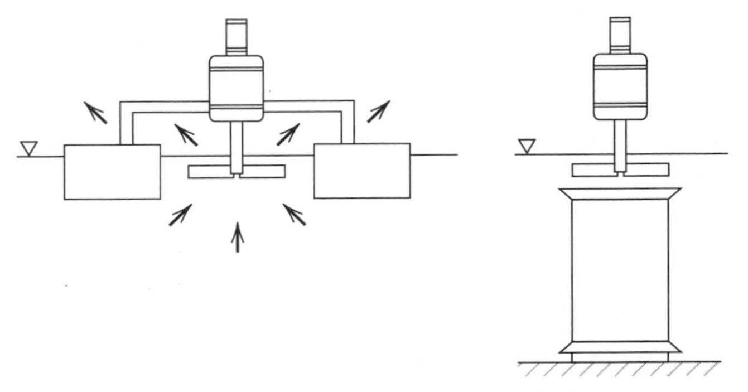

Figure 12.9 Mechanical surface aerators. Left, surface aerator; right, surface aerator with draft tube. Courtesy of Degrémont Infilco.

discs, and tubes. The USEPA (1989) has provided a thorough discussion of porous diffusers. Nonporous diffusers have orifices with larger openings than porous diffusers. Perforated piping, spargers, and slotted tubes are typical nonporous diffuser designs. The in situ oxygen transfer efficiencies of all of these devices are a function of many variables and data must be interpreted with caution. WEF and ASCE (1992) have given more detailed information on the various types of aeration systems and their performance characteristics. All types of systems are widely used.

Figure 12.10 Surface aerators in an activated sludge process.

Figure 12.11 Ceramic dome diffusers in an activated sludge basin. Courtesy of Regional Municipality of Ottawa-Carleton, Ontario.

There are many factors affecting the efficiency of oxygen transfer. Groves et al. (1992) and the USEPA (1989) have conducted a number of field evaluations of diffused aeration systems and found the general trends given in Table 12.6.

A recent concept in aeration for oxidation ditches is the total barrier oxidation ditch (Fig. 12.12). A vertical barrier wall is installed across the entire cross-section of a ditch and all circulating flow is intercepted and passed through draft tube turbine aerators. In the aerators, compressed air supplied by blowers is introduced into a turbine assembly through a sparge ring located beneath the turbine blades. The turbine assembly forces the aerated wastewater through a J-tube extension and imparts sufficient energy to circulate it at the desired velocity. Oxygen transfer efficiency depends on blower air supply rates and velocities of circulation. Clean water test results for oxygen transfer at two installations were found to average 1.04 kg O_2/kWh (based on energy consumption by the blower and turbine); α values were in the range of 0.76–0.87 for process wastewater (Boyle et al., 1989). K_{Tw} was found to be 2.07 h^{-1}.

12.5.2 Power Requirements for Mixing

The other function of aeration noted earlier is mixing. The only effective method to evaluate mixing is by placing the aerators in the prototype, which is impossible at the design stage. For the initial estimates of the mixing requirement, one must rely on

TABLE 12.6 Factors Affecting Oxygen Transfer in Submerged Diffused Aeration Systems[a]

Factor	Effect on oxygen transfer
Equipment Factors	
Diffuser type	Fine bubble diffusers have higher oxygen transfer than coarse bubble diffusers.
Diffuser density	A larger number of fine pore diffusers produces higher oxygen transfer.
Diffuser submergence	As submergence increases the percentage of oxygen transfer increases.
Diffuser layout	Grid patterns produce higher oxygen transfer than diffusers placed along one side (spiral roll) or in the center.
Diffuser age	Changes in membrane materials may cause a decrease in oxygen transfer.
Flow regime	Plug flow systems have higher oxygen transfer efficiency than step feed basins.
Basin geometry	Tanks that are more square have less variation in oxygen transfer throughout the tank than long, narrow tanks.
Operation Factors	
Solids retention time	Systems with higher solids retention times have higher oxygen transfer.
Nitrification	Systems that are nitrifying have higher oxygen transfer efficiencies than those that are not.
Food : microorganism ratio	Increases in food : microorganism ratio cause a decrease in oxygen transfer.
Airflow rate per diffuser	As airflow rate per diffuser increases, the oxygen transfer efficiency of most fine bubble devices decreases. For other devices, the opposite may be true.
Mixed liquor DO concentration	The percentage oxygen transfer efficiency decreases as DO increases.
Diffuser fouling	Fouling decreases oxygen transfer.
Wastewater Characteristics	
Temperature	An increase in temperature increases oxygen transfer.
Wastewater constituents	An increase in interfering agents such as surfactants causes a decrease in oxygen transfer.

[a]After Groves et al. (1992) and USEPA (1989).

field experience gained from similarly designed units. It is evident that power should be dissipated uniformly throughout the basin volume, resulting in no dead (unmixed) zones. The power input for mixing normally ranges from 13 to 26 kW/1 000 m^3 (0.37–0.74 kW/1 000 ft^3) of basin volume. This is effective power input into the basin, which excludes mechanical and electrical losses in the aerator. The efficiency of power transfer to the liquid for a typical aerator is approximately 75–80% of the input power to the aerator. Using the previously calculated value of gross power required for oxygen input and reducing it by the efficiency factor enables one to calculate the power delivered to the basin. Model studies can be helpful. If the mixing power requirement is not met, aerators will have to be operated at higher rates than required to supply oxygen; otherwise supplemental mixers must be added.

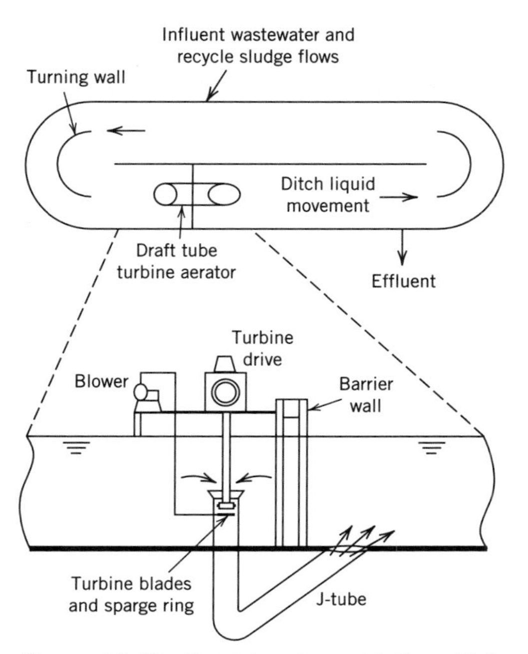

Figure 12.12 Total barrier oxidation ditch concept and draft tube aerator. From W. C. Boyle, M. K. Stenstrom, H. J. Campbell, Jr., and R. C. Brenner (1989), "Oxygen Transfer in Clean and Process Water for Draft Tube Turbine Aerators in Total Barrier Oxidation Ditches," *J. Water Pollution Control Federation,* 61, 8, pp. 1449–1463, © WEF 1989.

■ Example 12.2 Aeration System

An activated sludge basin with dimensions of 3.5 m × 12 m × 3.2 m deep is treating wastewater at a flow rate of 650 m^3/d. The COD reduction through oxidation is 120 mg/L. DO in the basin is kept at 1.5 mg/L. A diffused aeration system is used to supply oxygen with an oxygen transfer efficiency of 12% and an oxygen transfer rate of 0.35 kg O_2/MJ in tap water at 20°C. The efficiency of the compressors for the diffused aeration system is 80%. The α and β factors for the wastewater are 0.85 and 0.95, respectively.

Determine the power consumption of the diffused aeration system, volume of air supplied per day, and power delivered per 1 000 m^3 of aeration basin volume at an operating temperature of 10°C.

The temperature of the wastewater is 273 + 10 = 283°K

The saturation concentration of oxygen at 10°C is 11.3 mg/L.

The rate of oxygen utilization is

$$L = Q(\Delta\text{COD}) = \left(650\,\frac{\text{m}^3}{\text{d}}\right)\left(120\,\frac{\text{mg}}{\text{L}}\right)\left(\frac{1\,000\text{ L}}{1\text{ m}^3}\right)\left(\frac{1\text{ kg}}{10^6\text{ mg}}\right) = 78\text{ kg/d}$$

The volumetric rate of air supply is

$$Q_a = \frac{L}{E} = \frac{78\text{ kg/d}}{\dfrac{12\text{ kg}}{100\text{ kg}}}\left(\frac{1\text{ mole}}{32\text{ g}}\right)\left(\frac{1\,000\text{ g}}{\text{kg}}\right)\left(\frac{22.4\text{ L}}{\text{mole}}\right)\left(\frac{1\text{ m}^3}{1\,000\text{ L}}\right)\left(\frac{1\text{ m}^3\text{ air}}{0.21\text{ m}^3\text{ oxygen}}\right)\left(\frac{283°\text{K}}{273°\text{K}}\right)$$

$$= 2\,246\text{ m}^3/\text{d}$$

The rate of transfer in the aeration basin is

$$r_{\text{ww,T}} = \frac{r_{\text{Tw,20}}(\beta C_s - C_1)\alpha(1.024)^{(T-20)}}{9.09} = \frac{(0.35\text{ kg/MJ})[0.95(11.3) - 1.5](0.85)(1.024)^{(10-20)}}{9.09}$$

$$= 0.238\text{ kg/MJ}$$

The power consumption is

$$P = \frac{78\text{ kg/d}}{0.238\text{ kg/MJ}} = 328\text{ MJ/d} = (328\text{ MJ/d})\left(\frac{1\text{ d}}{86\,400\text{ s}}\right)\left(\frac{1\text{ W}}{1\text{ J/s}}\right)\left(\frac{1\text{ kW}}{1\,000\text{ W}}\right) = 3.80\text{ kW}$$

The volume of the aeration basin is $V = (3.5\text{ m})(12\text{ m})(3.2\text{ m}) = 134.4\text{ m}^3$
The power dissipated in the aeration basin is

$$P_{\text{d}} = \frac{(0.80)3.80\text{ kW}}{134.4\text{ m}^3}\left(\frac{1\,000\text{ m}^3}{1\,000\text{ m}^3}\right) = 22.6\text{ kW/1 000 m}^3$$

QUESTIONS AND PROBLEMS

1. In the gas phase, if D for oxygen is 0.178 $\text{cm}^2\text{/s}$, y is 1.00×10^{-4} cm, and the partial pressures of oxygen in the bulk phase and at the interface are 0.20 and 0.21 atm, respectively, what is the rate of mass transfer of oxygen across an interface area of 1 m^2? The temperature is 20°C.
2. This problem uses hypothetical values of parameters in Eq. (12.2a) to illustrate its application and relation to subsequent equations.

   MW of diffusing substance = 84

$D_g = 260\text{ cm}^2\text{/s}$	$D_L = 8\text{ cm}^2\text{/s}$	$A = 50\text{ cm}^2$	$T = 20°\text{C}$
$p_g = 0.09$ atm	$C_L = 1.2$ mg/L	$K_H = 0.233$ mg/L-atm	

   (a) What are the values of p_i, C_i, and dm/dt (i) if y_g and y_L are each 1.00×10^{-2} cm and (ii) if y_g and y_L are 1.00×10^{-3} and 1.00×10^{-1} cm, respectively? Is mass being transferred into or out of the solution?
   (b) What is the overall mass transfer coefficient, K_l for (i) and (ii) in part (a) if the volume of the solution is 1 m^3?
3. Why will a shorter contact time for elemental volumes at the surface of an aeration basin result in an overall increase in the rate of mass transfer for an aeration device?
4. What is the most important reason that a unit mass of a smaller log burns more rapidly than a unit mass of a larger log?
5. Qualitatively analyze the perception of odor from a local wastewater treatment plant for those exposed to the odor by an increase in wind velocity in their direction. Assume that the production of odoriferous compounds at the plant is constant but the wind speed in one direction has increased. All other factors (who knows what they may be) such as humidity remain constant. What equation applies and what are the changes in each factor?
6. Describe the effects of aeration on possible constituents in a water.
7. What are the hazards associated with hydrogen sulfide?
8. What is the time of exposure of water falling through a distance of 3.0 m in a gravity cascade aerator in (a) a single descent and (b) four descents?
9. Calculate the total contact time for descending through a series of perforated plates as a function of the separation distance of the plates, d, and the number of plates, n.
10. (a) Estimate the driving head that will expose water for 2.00 s from a fixed spray aerator,

assuming that the velocity coefficient of the nozzle is 0.95. Make the calculations for (i) a vertical jet and (ii) for a jet inclined 60° from the horizontal. Also find the vertical length of rise and the horizontal length of travel.

(b) Assuming that the coefficient of drag for wind on the water droplets is 0.6, what is the distance that a wind velocity of 20 km/h will carry the water for the above situation?

Solve the problem with SI or U.S. units at the discretion of the instructor.

11. Would a high or a low pH favor the removal of (a) oxygen, (b) carbon dioxide, (c) methane, and (d) hydrogen sulfide in an aeration process?

12. How much sodium sulfite is required to deoxygenate a 5 m^3 volume of water with a DO content of 9.0 mg/L? How much cobalt chloride must be added to this volume to attain a concentration of 0.05 mg/L Co?

13. What is the number of bubbles in an aeration basin that is supplied 0.60 m^3 of air per m^3 (0.080 ft^3 per gal) of water? The bubbles have an average diameter of 0.25 cm and rise at a uniform velocity of 18 cm/s (0.59 ft/s). The flow rate of water is 30 000 m^3/d (7.93 Mgal/d) and the detention time of water in the basin is 15 min. The depth of liquid in the basin is 2.8 m (9.18 ft). What is the gas surface area per unit volume of liquid?

14. If the concentration of DO is uniform in a deep shaft aerator, what is the ratio of the rate of oxygen transfer at the surface and bottom of an 80-m-deep shaft for the following conditions. Each m^3 of water in the shaft contains 10^7 air bubbles and the bubbles do not coalesce in the shaft. The bubbles can be assumed to be spherical, with a diameter of 3 mm at atmospheric pressure. The concentration of DO in the shaft is 4.5 mg/L and $\beta = 1.0$. Depletion of oxygen in the gas bubbles can be ignored. The temperature in the shaft is 10°C.

15. (a) Determine K_la for the DO data given in the following table. The data were gathered for tap water at a temperature of 22°C with an excess of sodium sulfite added for deoxygenation.

Time, min	0.0	1.0	2.0	3.0	4.0	5.0	6.0
C_l, mg/L	0.00	2.20	3.85	5.05	6.00	6.65	7.10
Time, min	8.0	10.0	12.0	15.0	20.0	25.0	30.0
C_l, mg/L	7.85	8.25	8.55	8.60	8.70	8.75	8.75

(b) If the test were conducted on deoxygenated wastewater with $\alpha = 0.90$ and $\beta = 0.82$ at the same temperature, what would be the DO values at the times given in part (a)?

16. Would increasing the DO concentration in an activated sludge aeration basin increase the energy efficiency of an aerator (r-kg O_2/MJ) if the rate of oxygen consumption does not change? Answer the same question if the rate of oxygen consumption increased with an increase in DO concentration.

17. At a temperature of 10°C, the DO concentration in a complete mixed aeration basin of a biological treatment unit is 2.40 mg/L and steady state conditions exist. $\beta = 1.0$ and $\theta = 1.020$. The mass of oxygen transferred to raise the influent DO to the effluent DO is negligible compared to the mass of oxygen transferred for microorganism consumption.

(a) If the rate of oxygen consumption by the microorganisms did not change but the temperature increased to 17°C and all other conditions remained the same, determine the DO concentration in the aeration basin at steady state.

(b) What is the minimum percentage increase in the rate of oxygen consumption at 17°C compared to the rate at 10°C that would reduce DO in the aeration basin to zero? All conditions except rate of oxygen consumption are the same at each temperature.

18. A 225 m^3 (59 440 gal) aeration basin in a biological treatment process is receiving a flow of 1 300 m^3/d (0.343 Mgal/d). The influent to the basin has a DO of 0 and the effluent from the basin contains a DO of 2.05 mg/L. The flow regime in the basin was complete mixed. COD balances around the basin show that the rate of oxygen consumption was

105 kg/d (231 lb/d). The temperature of wastewater in the basin was 16°C. The β factor was evaluated at 0.92 for the wastewater. What was the average $K_l a$ for the aerators in this basin?

19. A wastewater flow of 3 800 m^3/d is being treated in a complete mixed activated sludge aeration basin with a volume of 875 m^3. At a temperature of 10°C with influent and effluent DO concentrations of 0 and 2.40 mg/L, respectively, the oxygen consumption is 120 mg/L. β has been determined to be 0.95 for this wastewater; $\theta = 1.024$ for this system. What is the effluent DO at equilibrium conditions if the wastewater flow increases to 4 900 m^3/d, temperature is 8°C, and oxygen consumption decreases to 100 mg/L? Influent DO is 0 and there is no change in operation of the aerators.

20. In selecting aeration equipment for an activated sludge basin, units from two manufacturers are being evaluated. Manufacturer A's aerators are rated at 0.50 kg O_2/MJ, whereas manufacturer B's aerators are rated at 0.45 kg O_2/MJ. The effective power input of aerator A is estimated to be 77% of the power input to the compressor to supply oxygen; for B, it is estimated to be 79%. The elevation of the plant is approximately 750 m above sea level. The DO in the basin is to be designed for 2.0 mg/L and the maximum anticipated oxygen requirement is 0.49 kg $O_2/m^3/d$. The minimum and maximum temperatures in the aeration basin are anticipated to be 8 and 18°C, respectively. Use $\alpha = 0.80$ and $\beta = 1.0$. Assuming the mixing power requirement is 22 kW/1 000 m^3, which aerator is the preliminary choice?

KEY REFERENCE

Cornwell, D. A. (1990), "Air Stripping and Aeration," in *Water Quality and Treatment,* 4th ed., F. W. Pontius, ed., McGraw-Hill, Toronto, pp. 229–268.

REFERENCES

ASCE and AWWA (1990), *Water Treatment Plant Design*, 2nd ed., McGraw-Hill, Toronto.

AWWA (1971), *Water Treatment Plant Design*, AWWA, Denver, CO.

Bewtra, J. K. and W. R. Nicholas (1964), "Oxygenation from Diffused Air in Aeration Tanks," *J. Water Pollution Control Federation,* 36, 10, pp. 1195–1224.

Boyle, W. C., M. K. Stenstrom, H. J. Campbell, Jr., and R. C. Brenner (1989), "Oxygen Transfer in Clean and Process Water for Draft Tube Turbine Aerators in Total Barrier Oxidation Ditches," *J. Water Pollution Control Federation,* 61, 8, pp. 1449–1463.

Danckwerts, P. V. (1951), "Significance of Liquid-Film Coefficients in Gas Absorption," *Industrial and Engineering Chemistry,* 46, 6, pp. 1460–1467.

Downing, A. L. (1960), "Aeration in the Activated Sludge Process," *J. of the Institution of Public Health Engineers,* 59, pp. 80–118.

Fair, G. M, J. C. Geyer, and D. A. Okun (1968), *Water Purification and Wastewater Treatment and Disposal,* vol. 2, John Wiley & Sons, Toronto.

GLUMRB (1976), *Recommended Standards for Water Works,* Great Lakes-Upper Mississippi River Board of State Sanitary Engineers, Health Education Service, Albany, NY.

Groves, K. P., G. T. Daigger, T. J. Simpkin, D. T. Redmon, and L. Ewing (1992), "Evaluation of Oxygen Transfer Efficiency and Alpha-Factor on a Variety of Diffused Aeration Systems", *Water Environment Research,* 64, 5, pp. 691–698.

Haberman, W. L. and R. K. Morton (1953), David W. Taylor Model Basin Report 802, U.S. Navy Dept., Washington, DC.

Harmathy, T. Z. (1960), "Velocity of Large Drops and Bubbles in Media of Infinite or Restricted Extent," *J. American Institute of Chemical Engineers,* 6, pp. 281–288.

HIGBIE, R. (1935), "The Rate of Absorption of Pure Gas into a Still Liquid during Short Periods of Exposure," *Transactions American Institute of Chemical Engineering,* 31, pp. 365–389.

HOWE, K. J. AND D. F. LAWLER (1989), "Acid–Base Reactions in Gas Transfer: A Mathematical Approach," *J. American Water Works Association,* 81, 1, p. 61–66.

LEWIS, W. K. AND W. E. WHITMAN (1924), "Principles of Gas Absorption," *Industrial and Engineering Chemistry,* 26, pp. 382–385.

MONTGOMERY, JAMES, M., Consulting Engineers, Inc. (1985), *Water Treatment Principles and Design,* John Wiley & Sons, Toronto.

MUNZ, C. AND P. V. ROBERTS (1989), "Gas- and Liquid-Phase Mass Transfer Resistances of Organic Compounds during Mechanical Surface Aeration," *Water Research,* 23, 5, pp. 589–601.

PEEBLES, F. N. AND H. J. GARBER (1953), "The Motion of Gas Bubbles in Liquids," *Chemical Engineering Progress,* 49, pp. 88–97.

SHELL, G. AND T. CASSADY (1973), "Selecting Mechanical Aerators," *Industrial Water Engineering*, Jul/Aug, pp. 21–25.

USEPA (1989), *Fine Pore Aeration Systems,* No. EPA-623/189023, USEPA, Washington, DC.

TERRY, D. W. AND L. T. THIEM (1989), "Potential Interferences in Catalysis of the Unsteady-State Reaeration Technique," *J. Water Pollution Control Federation,* 61, 8, pp. 1464–1470.

WALLIS, G. B. (1969), *One-Dimensional Two-Phase Flow,* McGraw-Hill, Toronto.

WEAST, R. C. (1984), *CRC Handbook of Physics and Chemistry,* 67th ed., CRC Press, Boca Raton, FL.

WEF and ASCE (1992), *Design of Municipal Wastewater Treatment Plants,* vol. 1, WEF, Alexandria, VA.

CHAPTER 13

COAGULATION AND FLOCCULATION

13.1 COAGULATION

Coagulation is the destabilization of colloidal particles. The particles are essentially coated with a chemically sticky layer that allows them to flocculate (agglomerate) and settle in a reasonable period of time. Coagulation of waters to aid their clarification has been practiced since ancient times (Baker, 1981). Many naturally occurring compounds from starch to iron and aluminum salts can accomplish coagulation. In addition, synthetic cationic, anionic, and nonionic polymers are very effective coagulants but are usually more costly than natural compounds.

Coagulation and flocculation are used in both water and wastewater treatment processes. In water treatment it is usually cost effective to apply coagulation and flocculation to remove colloidal and small particles that settle slowly. Coagulation–flocculation can also be applied to enhance the removal of solids in highly concentrated natural waters that contain significant amounts of settleable solids (Guibai and Gregory, 1991). Commonly, presedimentation without coagulant addition or a roughing filter is used to remove high concentrations of settleable solids before coagulation–flocculation–sedimentation.

Biological wastewater treatment processes produce microorganisms that naturally flocculate themselves and other suspended matter, although it may be necessary to add coagulating agents to assist their flocculation in times of poor performance. Coagulants may also be added continuously for nutrient removal. Physical–chemical wastewater treatment processes rely on coagulation–flocculation for removal of suspended matter.

The ability of an agent to coagulate a water is related to its charge. The size of synthetic polymers is also a factor. Table 13.1 lists the relative coagulating power of several common salts. There is more than an order of magnitude increase in the effectiveness of an ion as its charge increases by one. This is a statement of the Schultze–Hardy rule based on the work of these two researchers in 1882 and 1900, respectively.

The most common coagulants are alum (aluminum sulfate) and iron salts, with alum being the most extensively used agent. The multivalent characteristic of these cations strongly attracts them to charged colloidal particles and their relative insolubility ensures their removal to a high degree.

Typical reactions of aluminum and iron salts in water are shown in Table 13.2. These salts consume alkalinity, which may necessitate the addition of an alkaline agent. Lime is usually the least expensive source of alkalinity. The pH of coagulation is critical. Ferric salts work best in a pH range of 4.5–5.5, whereas aluminum salts are most effective around a pH range of 5.5–6.3. These pH values should be attained

TABLE 13.1 Coagulating Power of Inorganic Electrolytes

	Relative power of coagulation	
Electrolyte	Positive colloids	Negative colloids
$NaCl$	1	1
Na_2SO_4	30	1
Na_3PO_4	1 000	1
$BaCl_2$	1	30
$MgSO_4$	30	30
$AlCl_3$	1	1 000
$Al_2(SO_4)_3$[a]	30	>1 000
$FeCl_3$	1	1 000
$Fe_2(SO_4)_3$[a]	30	>1 000

[a]The most common coagulating agents.

after the coagulant is added. If necessary, the pH may be adjusted with acid or alkalinity. Note that alum and some iron salts are supplied in hydrated states. Dry alum is hydrated with 14.3–18 water molecules [$Al_2(SO_4)_3 \cdot 14.3H_2O$ or $Al_2(SO_4)_3 \cdot 18H_2O$]. Table A.3 in the Appendix gives concentrations of commercially available coagulating agents.

Temperature exerts an effect on the efficiency of coagulation and flocculation. Metal salts given in Table 13.2 form hydroxide precipitates, which logically leads to the supposition that pOH is an important factor in the chemistry of the process. Temperature affects the equilibrium constant of water, as discussed in Section 3.1;

TABLE 13.2 Chemistry of Aluminum, Iron Salts, and Lime Coagulation

1. Alum

$Al_2(SO_4)_3 \cdot 18H_2O + 6H_2O \rightarrow 2Al(OH)_3(s) + 6H^+ + 3SO_4^{2-} + 18H_2O$

With an increase in H^+, pH is depressed and no more $Al(OH)_3$ is formed. If natural alkalinity is present, then

$HCO_3^- + H^+ \rightleftharpoons H_2CO_3 \rightleftharpoons CO_2 + H_2O$

$Al_2(SO_4)_3 \cdot 18H_2O + 3Ca(HCO_3)_2 \rightarrow 2Al(OH)_3(s) + 3CaSO_4 + 6CO_2 + 18H_2O$

If natural alkalinity is insufficient, then lime or caustic soda can be added.

$AL_2(SO_4)_3 \cdot 18H_2O + 3Ca(OH)_2 \rightarrow 2Al(OH)_3(s) + 3CaSO_4 + 18H_2O$

2. Sodium Aluminate

$Na_2Al_2O_4 + Ca(HCO_3)_2 + 2H_2O \rightarrow Al(OH)_3(s) + CaCO_3(s) + Na_2CO_3$

3. Ferrous Sulfate

$FeSO_4 \cdot 7H_2O + Ca(OH)_2 \rightarrow Fe(OH)_2 + CaSO_4 + 7H_2O$

$4Fe(OH)_2 + O_2 + 2H_2O \rightarrow 4Fe(OH)_3(s)$

4. Chlorinated Copperas

$3FeSO_4 \cdot 7H_2O + 1.5Cl_2 \rightarrow Fe_2(SO_4)_3 + FeCl_3 + 21H_2O$

$2FeCl_3 + 3Ca(HCO_3)_2 \rightarrow 2Fe(OH)_3(s) + 3CaCl_2 + 7CO_2$

This reaction takes place if natural alkalinity is present in a sufficient amount. Otherwise, if lime is added,

$2FeCl_3 + 3Ca(OH)_2 \rightarrow 2Fe(OH)_3(s) + 3CaCl_2$

5. Ferric Sulfate

$Fe_2(SO_4)_3 + 3Ca(HCO_3)_2 \rightarrow 2Fe(OH)_3(s) + 3CaSO_4 + 6CO_2$

$Fe_2(SO_4)_3 + 3Ca(OH)_2 \rightarrow 2Fe(OH)_3(s) + 3CaSO_4$

therefore, maintaining the same pH at all temperatures will result in varying pOHs. For both iron and alum salts, Hanson and Cleasby (1990) found that a constant pOH over the temperature range of 5–20°C produced the best coagulation–flocculation results.

The association of aluminum with Alzheimer's disease was discussed in Chapter 8. Use of aluminum salts as coagulants may increase the concentration of Al in product water. Letterman and Driscoll (1988) surveyed 91 plants in North America that use alum and found that 75% of the plants produced water with 210 μg/L or less total Al; the 50th percentile value was 90 μg/L. The results of their survey were in good agreement with an earlier survey (Miller et al., 1984). High raw water Al concentrations were associated with higher residual Al concentrations. Effective removal of particulate matter by filtration minimizes residual Al concentrations. Letterman and Driscoll's study found that lime used for pH adjustment after filtration may be an important source of residual Al in product water.

For in-line filtration of cold waters (temperature <3°C) with turbidities less than 2 nephlometric turbidity units (NTU), ferric chloride removed turbidity more effectively than alum on a mole to mole of metal ion basis (Haarhoff and Cleasby, 1988) but alum caused slower head loss development.

Metal precipitates are often colloidal. Conventional treatment of metal finishing wastes involves the addition of base to precipitate metals as hydroxides. Coagulants such as iron salts are often added to improve solids/liquid separation, with further benefits of contaminant adsorption onto iron hydroxide. The benefits of iron hydroxide are directly related to its concentration (Edwards and Benjamin, 1989). In general, Edwards and Benjamin (1989) found that iron hydroxide removed an equal or greater percentage of soluble Cu, Cd, Zn, Cr(III), Ni, and Pb from a waste at all pHs than base addition to form metal hydroxide precipitates. Cr(VI) was not removed in the presence or absence of iron. A pH range of 8–12.5 resulted in removals $>98\%$ of all metals [except Cr(VI)] in the presence of iron hydroxide.

Synthetic coagulating agents are widely available. Cationic, anionic, and nonionic polymers have all been found to provide excellent results in different situations. These agents are usually more costly than alum or iron salts but much smaller dosages are required. Polymers do not produce voluminous, gelatinous flocs as their inorganic counterparts do.

Figure 13.1 gives general formulas of synthetic polymers. Nonionic polymers are almost exclusively polyacrylamides with a molar mass between 1 and 30 million. Anionic polyelectrolytes have masses of several million. They contain groups that permit absorption and negatively ionized groups (carboxyl or sulfuric groups), which extend the polymer. Soda ash (Na_2CO_3) has been used to partially hydrolyze the polyacrylamide groups in the figure. Cationic polymers have molar masses less than one million. Their chains have amine, imine, or quaternary ammonium groups that produce the positive charge.

Activated silica is another commonly used coagulation agent. Sodium silicate (Na_2SiO_3) is "activated" by acidification. The chemistry of activated silica is fairly complex but Stumm et al. (1967) have provided the basic explanation of the process. When a concentrated solution of sodium silicate is acidified it becomes over saturated with respect to the precipitation of SiO_2. The precipitation process begins with the formation of polysilicate polymers that contain $-Si-O-Si-$ linkages. These polysilicates react further by cross-linking and aggregation to form negatively charged silica sols. The solution contains anionic polysilicates and other silicate species, which are effective coagulants. Over time, SiO_2 will precipitate and the solution will lose much

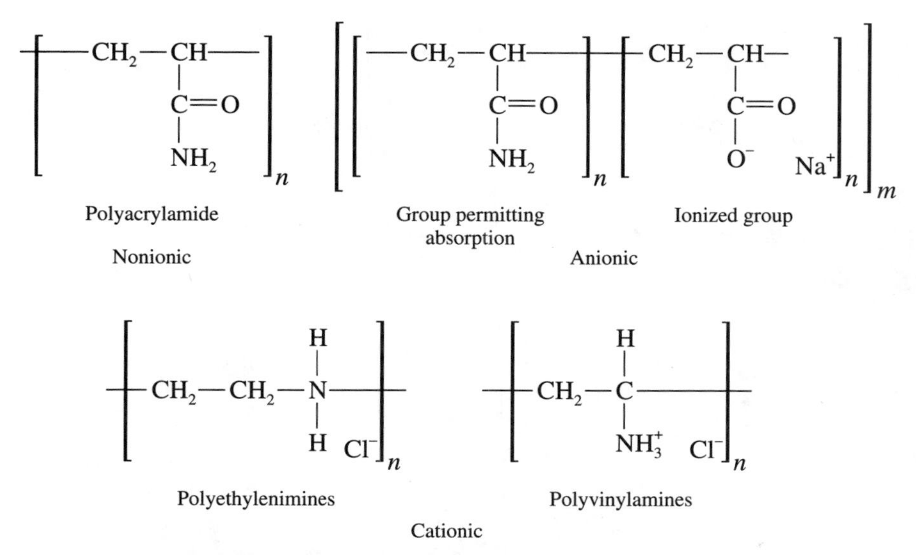

Figure 13.1 Typical formulas of coagulant polymers. After Degrémont Infilco (1979).

of its coagulating strength. Although the acidified solution is thermodynamically unstable, it can remain active for periods up to a few weeks.

Ozone has been found to be an effective coagulant aid (Singer, 1990). Low dosages (<3 mg/L and often 0.5–1.5 mg/L) are most effective and overdosing can actually deteriorate coagulation. Ozone does not improve ultimate particle removal but it facilitates the removal of readily coagulatable material with greater economy of coagulant (Singer, 1990).

There are also natural polyelectrolytes that are effective coagulant aids. Extracts from seeds, leaves, and roots have been used for centuries in the far east to clarify waters. Chitosan is a cationic polymer (molecular weight approximately 10^6) made by acidifying chitin, which is the organic skeletal substance of shells of crustaceans such as lobsters, shrimp, and crabs. Chitosan is biodegradable and nontoxic. Feed solutions of chitosan must have a pH below 6.5 (Kawamura, 1991a). Doses of 0.2 mg/L improved flocculation with alum significantly and reduced the alum dose.

Sodium alginate is an extract from certain brown seaweeds (kelp) and is widely used in foods. Sodium alginate in doses less than 1.0 mg/L also improved coagulation with alum (Kawamura, 1991a). The preparation and application of these and other natural agents are described by Schulz and Okun (1984).

The effectiveness and required doses of coagulants for a water are evaluated by use of a jar testing machine (Fig. 13.2). The apparatus is operated to simulate a mixing, flocculation, and settling cycle. Varying amounts of coagulants, alkalinity agents, and coagulant aids are added at the same time to the jars that contain the water to be treated. To ensure that the coagulants are added at the same time to all of the beakers, an arm with six test tubes attached and spaced at the separation distance of the beakers should be used to add the coagulants. The jars are then mixed at high speed for a short period of time, around 1 min. This rapid mixing phase is followed by a 20- to 40-min period of gentle mixing to promote formation of flocs. Mixing is then terminated and the impellers are withdrawn from the jars. The suspension is allowed to settle for a period of 15–60 min. Comparisons of initial and final turbidities determine the

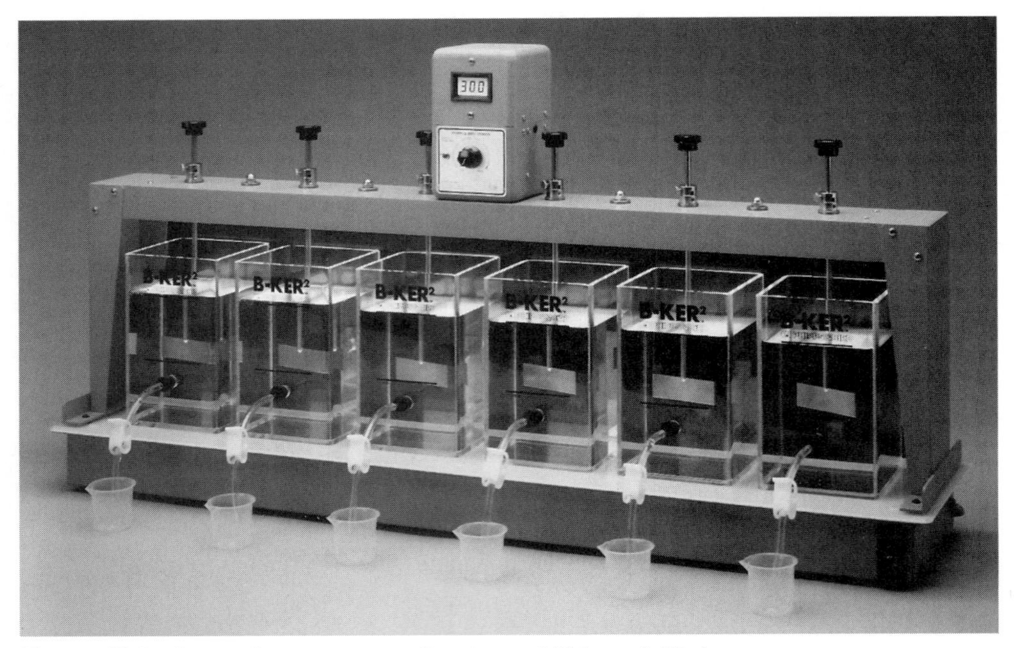

Figure 13.2 Jar testing apparatus. Courtesy of Phipps & Bird.

choice and effective doses of reagents and other conditions such as alkalinity and pH requirements. Cost and other considerations such as delivery systems determine the coagulant to be used.

Coagulants may be fed in dry or liquid form (see Table A.3 in the Appendix). Metering pumps used to deliver liquid agents can usually be varied over a 1 : 10 flow rate to accommodate variable doses. Manufacturers specify other requirements of chemical feed systems.

Recovery of Alum and Iron Coagulants

Alum and iron salts may be recovered from sludge by adding acid. If sulfuric acid is used,

$$\mathrm{Me(OH)}_x + x\mathrm{H}^+ + \frac{x}{2}\mathrm{SO}_4^{2-} \rightarrow \mathrm{Me(SO_4)}_{x/2} + x\mathrm{H_2O} \tag{13.1}$$

Other metals will be solubilized along with the coagulant metal. Reuse of the recovered coagulant will recycle metals that should again be precipitated along with the coagulant. The increased risk associated with acid coagulant recovery has decreased the frequency of this practice. Careful monitoring is required to ensure that product water does not exhibit elevated metals concentrations when coagulant recovery and reuse is practiced.

Iron hydroxide metals containing sludge was able to be regenerated at least 50 times by addition of acid regeneration solution with a pH of 3.5 or 4.5 (Edwards and Benjamin, 1989). After the sludge was exposed to the regeneration solution for 10 to 30 min, the remaining iron solids were separated and reused to treat waste. Addition of acid released the removed metals from the iron hydroxide sludge.

13.2 MIXING AND POWER DISSIPATION

Coagulants and coagulant aids must be rapidly dispersed throughout a water to ensure maximum contact between the reagent and suspended particles; otherwise, the coagulant will react with water and dissipate some of its coagulating power. Mixers are also required to disperse other chemical agents in waters or wastewaters. Flocculators operate at lower degrees of mixing to promote particle contact but prevent breakup of the large floc particles formed.

Three phenomena contribute to mixing: molecular diffusion (perikinetic motion), eddy diffusion, and nonuniform flow. Molecular diffusion is due to thermally induced Brownian motion and is not significant compared to the other two phenomena. The latter two phenomena are functions of the degree of turbulence in the basin. Mixing is caused by two layers of water traveling at different velocities. Hydraulically induced motion is also referred to as orthokinetic motion. The velocity gradient (dv/dy) is proportional to the amount of energy dissipated in the fluid. Camp and Stein (1943) developed the basic theory of power input for mixing that is most commonly applied today.

To relate dv/dy to the power input into a basin, a force balance and power balance are needed. Newton's law of viscosity will be useful for calculating energy dissipation. For one-dimensional (1-D) flow it is

$$\tau_{yx} = -\mu \frac{dv_x}{dy} \tag{13.2}$$

where

τ_{yx} is the shear force per unit area in the x direction or the momentum transferred in the y direction resulting from motion in the x direction
μ is the dynamic viscosity
v_x is the velocity in the x direction
y is the vertical direction

A force (F) balance relates shear and pressure (p) forces (Fig. 13.3). Assuming 1-D laminar flow conditions for the elemental volume in Fig. 13.3,

$$\Sigma F_x = 0 = p\Delta y\,\Delta z - \left(p + \frac{\partial p}{\partial x}\Delta x\right)\Delta y\,\Delta z + \tau\Delta x\,\Delta z - \left(\tau + \frac{\partial \tau}{\partial y}\Delta y\right)\Delta x\,\Delta z \tag{13.3}$$

Expanding and simplifying Eq. (13.3),

$$\frac{\partial p}{\partial x} = -\frac{\partial \tau}{\partial y} \tag{13.4}$$

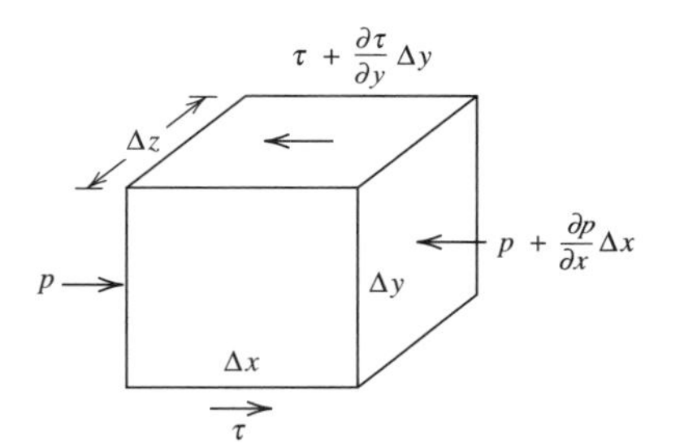

Figure 13.3 Force balance over an elemental volume of fluid.

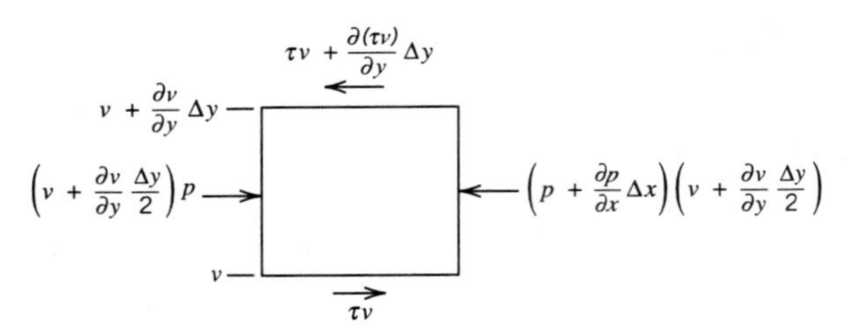

Figure 13.4 Power balance on an elemental volume.

The force balance is performed in a similar manner for the other two directions, although the gravity force is incorporated into the balance for the y direction.

A schematic for the power balance for an element (1-D flow) is shown in Fig. 13.4. Power is expressed by $P = \vec{F} \cdot \vec{v}$. Setting up the balance in the x direction for power into and out of the element,

$$P_{in} - P_{out} = \Delta P = \left(v + \frac{\partial v}{\partial y}\frac{\Delta y}{2}\right)p\Delta y\,\Delta z - \left(p + \frac{\partial p}{\partial x}\Delta x\right)\left(v + \frac{\partial v}{\partial y}\frac{\Delta y}{2}\right)\Delta y\,\Delta z + \tau v\,\Delta x\,\Delta z - \left(\tau v + \frac{\partial(\tau v)}{\partial y}\Delta y\right)\Delta x\,\Delta z \tag{13.5a}$$

Expanding and simplifying Eq. (13.5a),

$$-v\frac{\partial p}{\partial x}\Delta x\,\Delta y\,\Delta z - v\frac{\partial \tau}{\partial y}\Delta x\,\Delta y\,\Delta z - \tau\frac{\partial v}{\partial y}\Delta x\,\Delta y\,\Delta z = \Delta P \tag{13.5b}$$

Substituting the result of the force balance (Eq. 13.4) into the above equation and noting that $\Delta x \Delta y\, \Delta z = \Delta V$,

$$-\tau\frac{\partial v}{\partial y} = \frac{\Delta P}{\Delta V} \tag{13.5c}$$

Applying Newton's law of viscosity (Eq. 13.2),

$$\mu\left(\frac{\partial v}{\partial y}\right)^2 = \frac{\Delta P}{\Delta V}$$

Assuming all elemental volumes are the same in this well-mixed basin,

$$\frac{\Delta P}{\Delta V} = \frac{P}{V}$$

and defining G as the velocity gradient, dv/dy,

$$\mu G^2 = \frac{P}{V} \qquad \text{and} \qquad G = \left(\frac{P}{\mu V}\right)^{1/2} \tag{13.6}$$

P is the net power dissipated in the element. In fact, power will be dissipated in all three directions and G is the root-mean-square (rms) velocity gradient with dimensions of T^{-1}. Turbulent flow conditions normally exist in mixers and flocculators but Eq. (13.6) has been found to provide suitable results to relate power to mixing characteristics in basins regardless of the flow regime. The dimensionless number, Gt_d, has been correlated to performance of mixers and flocculators. For mixers, power dissipation should be high and time of reaction relatively short.

Calculation of the rate of energy dissipation depends upon the device used for mixing. Devices used for mixing range from power driven impellers to baffled basins. Power dissipation in hydraulic devices is due to the viscosity of the liquid. The equation for power dissipation in a hydraulic device is

$$P = \rho g Q h_L \tag{13.7}$$

where

h_L is the headloss in the device
ρ is the density of the liquid.
g is the acceleration of gravity
Q is the volumetric flow rate

Once P is calculated it is inserted into Eq. (13.6) to find the velocity gradient. Expressions for other power dissipating devices are developed as they are discussed in the following sections.

13.3 MIXERS

Because coagulation reactions are rapid, a short detention time is all that is necessary, but a high degree of turbulence is required. The AWWA (1969) recommends design values in Table 13.3 for mixers in water treatment.

G values up to 5 000 s^{-1} have been used. Mixing units should be designed for the maximum day flow. Letterman et al. (1973) found the following empirical correlation that relates the key parameters of G, t_d, and concentration of coagulant for an impeller rapid mixing device using alum as a coagulant.

$$G t_{dopt} C^{1.46} = 5.9 \times 10^6$$

where

t_{dopt} is the optimum detention time
C is the concentration of alum in mg/L

Conditioning chemicals are added to sludges to aid dewatering. High-shear conditions in the mixer can destroy the structure of the sludge. Kawamura (1991b) recommends G values of 100–150 s^{-1} for mixers that apply conditioning agents to sludge.

There are two means of coagulation: sweep coagulation and charge neutralization. During sweep coagulation, voluminous amounts of iron or aluminum hydroxide precipitates are formed to interact with the colloids in the raw water. In typical water treatment practice, the water is supersaturated three to four orders of magnitude beyond the solubility of the metal (iron or aluminum) so that the metal hydroxide precipitates quickly in 1–7 s (AWWA Committee, 1989). Transport interaction between colloids and coagulants during initial coagulant addition and rapid mixing are less important than the subsequent flocculation step.

TABLE 13.3 G–t_d Values for Mixers[a]

t_d, s	20	30	40	>40
G, s^{-1}	1 000	900	790	700

[a]After AWWA (1969).

For charge neutralization to occur effectively, the coagulants must be dispersed rapidly (<0.1 s is desirable) and high-intensity mixers are required (AWWA Committee, 1989). When the metal coagulant is added to the water, within microseconds up to a second, charged metal hydrolysis products are formed. These charged species must be transported to the colloids before the ultimate hydroxide precipitation product is formed.

13.3.1 Mechanical Mixers

Impeller driven mixers are the most efficient devices to rapidly disperse coagulants. Mixing in these devices is a function of the geometry of the basin and impeller, fluid characteristics, and power expenditure (Uhl and Gray, 1966). Gravity effects are generally not important. Correlations for power input and the Reynold's number (Re) have been developed for many types of mixing devices.

The following relations exist in an impeller mixer:

$$v \propto ND \tag{13.8a}$$

$$A \propto D^2 \tag{13.8b}$$

$$P \propto F_D v = F_D N D \tag{13.8c}$$

where

v is velocity of the impeller
N is the rate of revolution
D is the impeller diameter
F_D is the drag force
P is the power expenditure

Using relations in Eqs. (13.8a)–(13.8c) a power number analogous to a drag coefficient or friction factor is formulated. A drag coefficient is the ratio of drag force to kinetic energy of the impeller and a characteristic area.

$$C_D = \frac{2F_D}{\rho A v^2} = \frac{2F_D v}{\rho A v^3} \tag{13.9a}$$

$$\phi = \frac{P}{\rho N^3 D^5} \tag{13.9b}$$

where

C_D is a drag coefficient
ϕ is the power number

The Reynold's number for the impeller is

$$\mathrm{Re} = \frac{\rho N D^2}{\mu} \tag{13.10}$$

From experiment it is found that

$$\phi = K\mathrm{Re}^p \tag{13.11}$$

where

K is a characteristic constant of an impeller and tank geometry
$p = -1$ (laminar)
$= 0$ (turbulent)

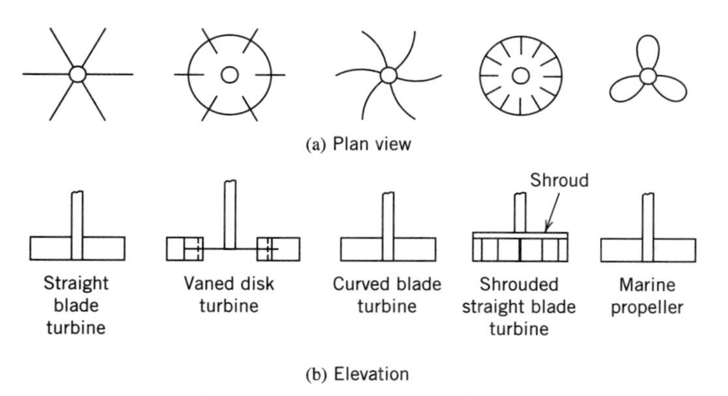

Figure 13.5 Types of impellers.

The laminar or viscous range exists when Re $\lesssim$ 10. For fully developed turbulent flow the power number is constant.

Types of impellers are shown in Fig. 13.5. The following definitions apply to impellers.

Propeller	an impeller that is curved around its axis in a screwlike manner.
Pitch	the advance per revolution considering the propeller as a screw. A square pitch means the pitch is equal to the diameter.
Turbine (Fig. 13.6)	the blade may be flat or curved.
Shroud	a full or partial plate added to the top or bottom planes of a radial flow turbine.

A typical plot of power number variation with the Reynold's number is given in Fig. 13.7 (note its similarity to a Moody diagram). Similar plots should be developed for each configuration.

Table 13.4 lists experimentally determined values of K for various types of impellers in a circular tank where vortex conditions have been eliminated by having four baffles at the tank wall, each with a length of 10% of the tank diameter. Separate correlations should be developed for each tank–impeller configuration. Equation (13.11), with coefficients given in Table 13.4, applies only to the straight portions of the curve in Fig. 13.7.

Under turbulent flow conditions, power requirements in a baffled vertical circular tank are the same as power requirements for a baffled vertical square tank when the diameter of the circular tank is equal to the width of a square tank (Reynolds, 1982).

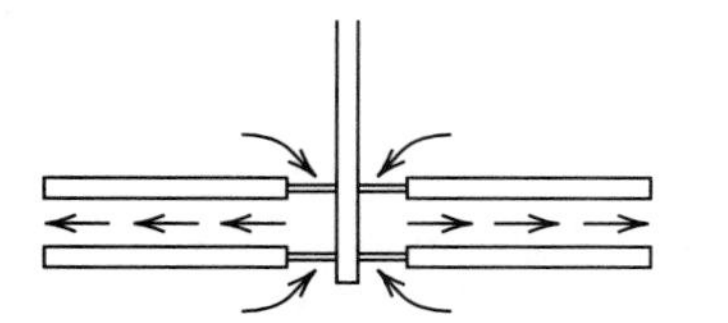

Figure 13.6 Radial flow turbine.

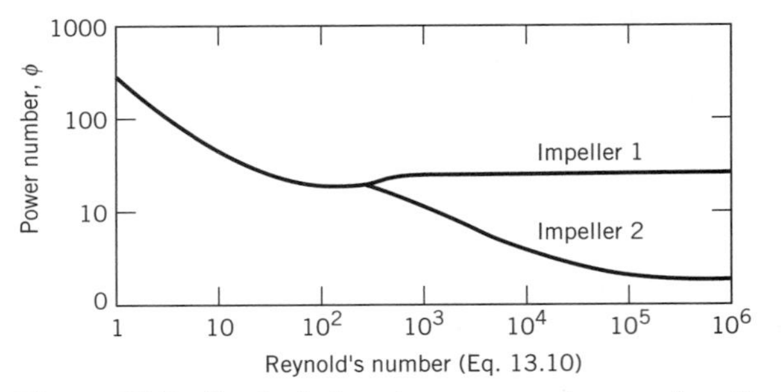

Figure 13.7 Typical plot of power number as a function of the Reynold's number. Reprinted with permission from J. H. Rushton (1952), "Mixing of Liquids in Chemical Processing," *Industrial and Engineering Chemistry,* 44, 12, pp. 2931–2936, copyright 1952, American Chemical Society.

In an unbaffled square tank the power imparted is about 75% of the power imparted in a baffled square or a baffled circular tank (Reynolds, 1982).

TABLE 13.4 Power Coefficients for Various Impellers[a,b]

Impeller	Viscous range	Turbulent range
Propeller, square pitch, 3 blades	41.0	0.32
Propeller, pitch of 2, 3 blades	43.5	1.00
Turbine, 4 flat blades, vaned disk	71.0	6.30
Turbine, 6 flat blades, vaned disk	71.0	6.30
Turbine, 6 curved blades	70.0	4.80
Turbine, 6 arrowhead blades	71.0	4.00
Fan turbine, 6 blades at 45°	70.0	1.65
Flat paddle, 2 blades	36.5	1.70
Shrouded turbine, 2 curved blades	97.5	1.08
Shrouded turbine, 6 curved blades	97.5	1.08
Shrouded turbine with stator, no baffles	172.5	1.12
Flat paddles, 2 blades, $D/w^c = 6$	36.5	1.60
Flat paddles, 2 blades, $D/w^c = 8$	33.0	1.15
Flat paddles, 4 blades, $D/w^c = 6$	49.0	2.75
Flat paddles, 6 blades, $D/w^c = 6$	71.0	3.82

[a]Reprinted with permission from J. H. Rushton (1952), "Mixing of Liquids in Chemical Processing," *Industrial and Engineering Chemistry,* 44, 12, pp. 2931–2936, copyright 1952, American Institute of Chemical Engineers; and J. H. Rushton and J. Y. Oldshue (1953), "Mixing—Present Theory and Practice," *Chemical Engineering Progress,* 49, 4, pp. 161–168, used with permission of the American Institute of Chemical Engineers.

[b]For baffled cylindrical tanks with four baffles at the tank wall, baffle width is 10% of the tank diameter. The impeller width is one third of tank diameter, D.

[c]w is the impeller width.

13.3.2 Pneumatic Mixers

Diffused aerators may be used to provide mixing. The rising gas bubbles produced by the diffuser cause circulation of the liquid. The calculation of the effective power input into the basin depends on the following variables (Bewtra and Nicholas, 1964):

Q_a = volumetric airflow rate
γ = weight density of the fluid
γ_a = weight density of air
ρ = density of the fluid-air mixture
ρ_a = density of air
h = depth of fluid over the diffusers
v_b = rise velocity of the bubbles relative to the fluid
V_{b0} = volume of an air bubble at the surface of the liquid
V_b = volume of an air bubble at any depth
v_l = average upward velocity of the liquid

The rate of work (P_b) done on the fluid by a single rising bubble is

$$P_b = Fv_b \tag{13.12}$$

where
F is the net weight of fluid displaced by the bubble

For an individual bubble,

$$F = (\gamma - \gamma_a)V_b \tag{13.13}$$

The total rate of work done depends on the total number of bubbles in the basin, n, which has been shown to be (Section 12.4.3):

$$n = \left(\frac{Q_a}{V_b}\right)\frac{h}{v_b} \tag{13.14}$$

The average upward liquid velocity depends on the depth of expansion of liquid in the basin during its detention time in the basin. Using the last two relations and Eq. (13.12), the total power expenditure, P, is

$$P = nP_b = (\gamma - \gamma_a)V_b\left(\frac{Q_a}{V_b}\right)\frac{kh}{v_b + v_l}v_b \tag{13.15}$$

It was also noted in Section 12.4.3 that

$$\frac{kv_b}{v_b + v_l} \approx 1 \qquad \text{also} \qquad (\gamma - \gamma_a) \approx \gamma$$

The volume of a bubble will change as the bubble rises and pressure (p) decreases. Boyle's law describes the pressure–volume relation for a gas.

$$pV_b = C$$

where
C is a constant

The volumetric flow rate of air is related to standard pressure and temperature. If a bubble has a volume V_{b0} at atmospheric pressure (10.3 m or 33.9 ft of H_2O), which occurs at the surface of the liquid, its volume at a depth h (V_{bh}) will be

$$V_{bh} = \frac{10.3}{10.3 + h} \times V_{b0}$$

The average volume of an air bubble as it moves up the column of water is

$$\overline{V_b} = \frac{1}{h}\int_0^h V_b\,dh = \frac{10.3\,V_{b0}}{h}\int_0^h \frac{dh}{10.3 + h} = \frac{10.3\,V_{b0}}{h}\ln\left(\frac{10.3 + h}{10.3}\right)$$

Substituting the above relations into Eq. (13.15) and simplifying,

$$P = (10.3)\gamma Q_a \ln\left(\frac{10.3 + h}{10.3}\right) = (10.3)\rho g Q_a \ln\left(\frac{10.3 + h}{10.3}\right) = p_a Q_a \ln\left(\frac{10.3 + h}{10.3}\right) \qquad (13.16)$$

where

p_a is atmospheric pressure (kN/m^2), h is in m, and Q_a is in m^3/s

$$\text{In U.S. units: } P = p_a Q_a \ln\left(\frac{33.9 + h}{33.9}\right)$$

where

p_a is in lb$_f$/ft^2, h in ft, Q_a in ft^3/s

The same equation for power dissipation can be obtained from considering the isothermal expansion of the bubbles as they rise. For an isothermal process,

$$p_a V_{b0} = p V_b$$

where

p is pressure at any depth

The work, W, of expansion of the bubble is equal to the work of compression.

$$W = \int p\,dV_b$$

Noting that $W/t = P$ (where t is time) and $V_{b0}/t = Q_a$, the above expression can be used to find the relation given by Eq. (13.16).

13.3.3 Hydraulic Mixers

Coagulants may be applied at any point where turbulence is high. Examples of suitable locations are below weirs and at the suction side of pumps. Pumps will be required for most installations. Weirs may be used for flow measurement. Baffled chambers can also be used to promote turbulence for mixing. Huisman et al. (1981) and Schulz and Okun (1984) describe a variety of hydraulic devices with design details. Kawamura (1991b) recommends that headlosses for hydraulic devices should be at least 0.6 m (2 ft) to ensure good mixing; however, there is a range in values reported by others.

Headloss may be calculated for weirs and used for calculation of pumping energy requirements but using these values to calculate G values for mixing will only provide a rough estimate of the effective power for mixing. Monk and Trussell (1991) suggest that the minimum headloss at a weir should be 10 cm (4 in.) for mixing. Losses of electrical energy in a pump motor do not impart mixing energy. Equation (13.7) describes the rate of energy dissipation in hydraulic devices. To calculate G, the volume in which the energy is dissipated must be known. The ASCE and AWWA (1990) recommend that a volume equivalent to 2 s detention time be used when mixing takes

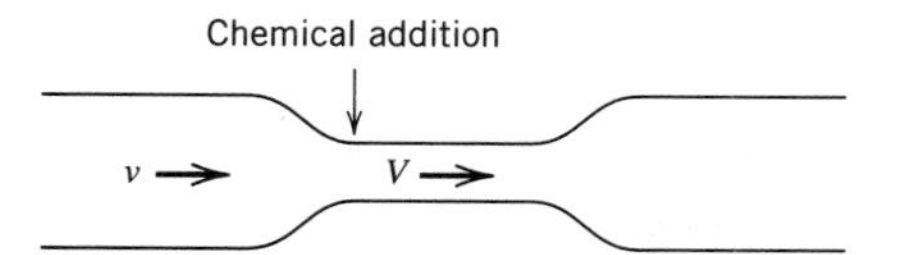

Figure 13.8 Venturi section for chemical feeding.

place in a pipe or open channel with a velocity greater than 0.5 m/s (1.6 ft/s). If the velocity is lower than this value or the possibility of backflow eddies exists, the entire volume of the conduit should be used.

When coagulant is to be added in a channel such as below a weir or at a hydraulic jump, the dispensing pipe for the coagulant should extend across the channel with uniformly spaced ports to achieve the most effective distribution of the coagulant in the flow. Schulz and Okun (1984) recommend that the height of the dispensing pipe over a weir should be at least 0.3 m (1 ft) to give the coagulant stream a sufficient velocity to penetrate the nappe thickness. This recommendation would apply to other hydraulic mixing devices.

Venturi Sections and Hydraulic Jumps

Venturi sections and hydraulic jumps may be designed for dispersion of coagulants. A Venturi section is shown in Fig. 13.8. The reduced pressure in the throat of the section aspirates the chemical feed solution into the flow. Turbulence generated in the throat and as the flow jet expands upon exiting the throat causes mixing.

Application of Bernoulli's equation to a Venturi flume will result in an orifice equation.

$$h_L = C_d \frac{V^2}{2g} \tag{13.17}$$

where

C_d is a discharge coefficient

The discharge coefficient depends on the entrance pipe or channel area and the flow area in the throat as well as frictional losses in the section. Parshall flumes, which are commonly used for flow measurement, are Venturi flumes designed for a critical flow condition.

Hydraulic jumps may be designed for the dispersion of coagulants (Huisman et al., 1981). The turbulence generated in the jump will provide suitable mixing. A jump is schematically shown in Fig. 13.9.

Chow (1959) or any text on open channel hydraulics gives a thorough analysis of hydraulic jumps. For a flat channel and ignoring friction effects along the walls, the momentum principle is used for analysis.

$$\frac{\Delta(m\vec{v})}{\Delta t} = \Sigma \vec{F}$$

Applying the equation in the x direction:

$$(\rho Q\vec{v} \cdot \vec{n})|_1 + (\rho Q\vec{v} \cdot \vec{n})|_2 = p_1A_1 - p_2A_2 = \tfrac{1}{2}\rho g b y_1^2 - \tfrac{1}{2}\rho g b y_2^2 \tag{13.18}$$

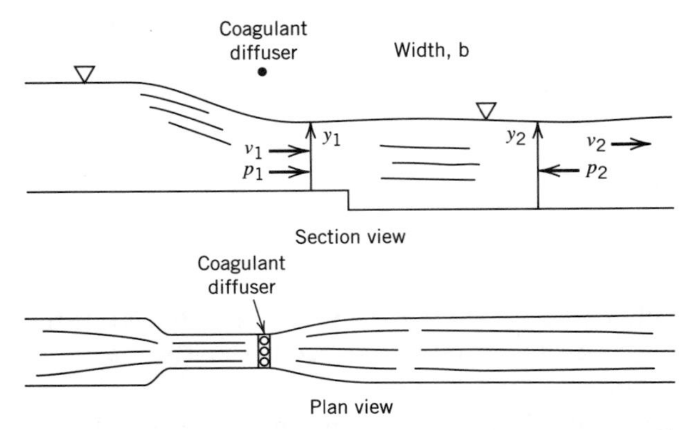

Figure 13.9 Definition sketch for hydraulic jump. From L. Huisman, J. M. De Azevedo Netto, B. B. Sundaresan, J. N. Lanoix, and E. H. Hofkes (1981), *Small Community Water Supplies: Technology of Small Water Supply Systems in Developing Countries,* Technical Paper No. 18, International Reference Center, The Netherlands. Reprinted by permission of Intermediate Technology Publications.

where

$\vec{n}$ is the unit vector normal to the surface
p is the average pressure on a face
A is cross-sectional area
y is depth of water
b is width of the jump
subscripts 1 and 2 refer to locations upstream and downstream of the jump, respectively

After the vector multiplications are performed,

$$\rho Q v_2 - \rho Q v_1 = \tfrac{1}{2}\rho g b y_1^2 - \tfrac{1}{2}\rho g b y_2^2$$

The velocity is related to the cross-sectional area of the flow by

$$v = \frac{Q}{A} = \frac{Q}{by}$$

Substituting this into the above expression and simplifying it gives the relation between the upstream and downstream depths.

$$\frac{y_2}{y_1} = \frac{1}{2}(\sqrt{8\mathrm{Fr}_1^2 + 1} - 1) \tag{13.19}$$

where

Fr_1 is the Froude number $= v_1/\sqrt{g y_1}$

To assure the location of the jump, the floor should be dropped and the drop is generally placed at the end of an expansion of supercritical flow (Schulz and Okun, 1984). For a good jump, $\mathrm{Fr}_1 \geq 2$ and the depth ratio $y_2/y_1 > 2.38$. There is appreciable energy loss and, although not all of the energy is dissipated usefully, there is sufficient turbulence to provide good mixing.

The headloss is calculated from Bernoulli's equation applied at the upstream and downstream depths. Typical headlosses are 0.3 m (1 ft) or greater, although 0.6 m (2 ft) is recommended as noted earlier.

$$h_L = y_1 + \frac{v_1^2}{2g} - \left(y_2 + \frac{v_2^2}{2g}\right) \tag{13.20}$$

■ Example 13.1 Weir Mixer

An engineer must perform calculations to design a weir for mixing at various flow conditions. G values of 1 000 s^{-1} must be achieved. The downstream channel must have a minimum velocity, v, of 0.5 m/s. One flow condition to be examined is $Q = 28\,000$ m^3/d. The downstream channel will be rectangular. Size the weir at a temperature of 20°C.

The velocity heads of the upstream and downstream waters will be ignored.

From the Appendix, at 20°C, $\rho = 998.2$ kg/m^3; $\mu = 1.002 \times 10^{-3}$ N-s/m

The datum will be the elevation of the downstream channel. The best hydraulic section for an open channel satisfies the condition that

$$w = 2d$$

where

w and d are the channel width and depth, respectively

The width and depth are determined by

$$v = \frac{Q}{wd} = \frac{Q}{2d^2} \qquad d = \sqrt{\frac{Q}{2v}} = \sqrt{\frac{\left(28\,000\,\frac{\text{m}^3}{\text{d}}\right)\left(\frac{1\text{ d}}{86\,400\text{ s}}\right)}{2(0.50\text{ m/s})}} = 0.57\text{ m}$$

$$w = 2d = 2(0.57\text{ m}) = 0.64\text{ m}$$

From Eqs. (13.6) and (13.7),

$$G = \sqrt{\frac{\rho g Q h_L}{\mu V}} \qquad \text{or} \qquad h_L = \frac{\mu G^2 V}{\rho g Q}$$

$$h_L = \frac{\left(1.002 \times 10^{-3}\,\frac{\text{kg}}{\text{m-s}}\right)(1\,000\text{ s}^{-1})^2}{\left(998.2\,\frac{\text{kg}}{\text{m}^3}\right)\left(9.81\,\frac{\text{m}}{\text{s}^2}\right)}(2\text{ s}) = 0.20\text{ m}$$

In this equation $V/Q = 2$ s when the velocity is 0.5 m/s or larger.

The depth of the upstream water must be 0.20 m + 0.57 m = 0.77 m.

The height of the water over the weir, h, can be calculated from a weir formula:

$$Q = \frac{2}{3} C_d w \sqrt{2gh} \qquad h = \frac{1}{2g}\left(\frac{3Q}{2C_d w}\right)^2$$

where

C_d is a discharge coefficient

$$h = \frac{1}{2\left(9.81\,\frac{\text{m}}{\text{s}^2}\right)}\left[\frac{3\left(28\,000\,\frac{\text{m}^3}{\text{d}}\right)\left(\frac{1\text{ d}}{86\,400\text{ s}}\right)}{2C_d\,(1.14\text{ m})}\right]^2 = \frac{0.009\,3\text{ m}}{C_d^2}$$

It is apparent that the depth of water over the weir will be less than 50% of the height of the weir. The discharge coefficient can be found from handbooks or fluid mechanics texts to be approximately 0.62 (see Shames, 1982, for instance).

$$h = (0.009\,3\text{ m})/(0.62)^2 = 0.024\text{ m}$$

The height of the weir, h_w, above the datum must be

$$h_w = 0.52\text{ m} - 0.076\text{ m} = 0.44\text{ m}$$

The design must be compatible with hydraulic grades in the upstream and downstream sections. It will not be possible to maintain constant G or optimum hydraulic sections at all flow conditions that will occur throughout the year. Similar calculations must be made for other flow conditions and overall evaluation must be performed to determine the final design.

13.4 FLOCCULATORS

There are a wide variety of devices that can be used to accomplish the more gentle mixing required for flocculation. Basins with mechanically driven paddles are common. Also, pneumatic flocculators designed on the principles given in Section 13.3.2 can be used. Devices in the hydraulic category are pipes, baffled channels, pebble bed flocculators, and spiral flow tanks. Headloss calculations are made from basic fluid mechanics principles.

The work of Smoluchowski (1916) relates particle contact to the velocity gradient or G. Many studies have shown that flocculation efficiency is related to the G value. As noted, Eq. (13.6) was developed for laminar flow and assumed to apply to turbulent flow. Recently Cleasby (1984) has made a thorough theoretical analysis of flocculation and power dissipation in turbulent flow. He has given convincing arguments that G is related to $(P/V)^{2/3}$ under some conditions. Note that viscosity is absent in this expression; thus, temperature does not influence the phenomenon. The development is beyond the scope of this text but it does indicate that flocculation efficiency should be compared against $(P/\mu V)^{1/2}$ and $(P/V)^{2/3}$ to determine the better correlation for design and operation. In most cases the former relation is applicable.

Flocculators are commonly designed to have Gt_d values in the range of 10^4 to 10^5. G values may range from 10 to 60 s^{-1} and detention times are typically in the range of 15–45 min. Mixing in an individual flocculator basin causes the hydraulic flow regime to approach complete mixed conditions. Therefore, it is desirable to have two or more basins in series to ensure that all particles are exposed to the mixing for a significant amount of the total detention time and to promote plug flow through the system as a whole. Furthermore, providing basins in series allows the G value to be decreased from one compartment to the next as the average floc size increases. This is known as tapered flocculation, which produces better results. The total detention time for all compartments should be in the range suggested above. Units are also designed in parallel or with bypasses to allow for a unit being taken out of service.

Paddle Flocculators

Paddle flocculators have compartments that each have one or more sets of paddles mounted in them. The paddles may be oriented in any direction; Figs. 13.10 and 13.11 show two orientations. The useful power input imparted by a paddle to the fluid

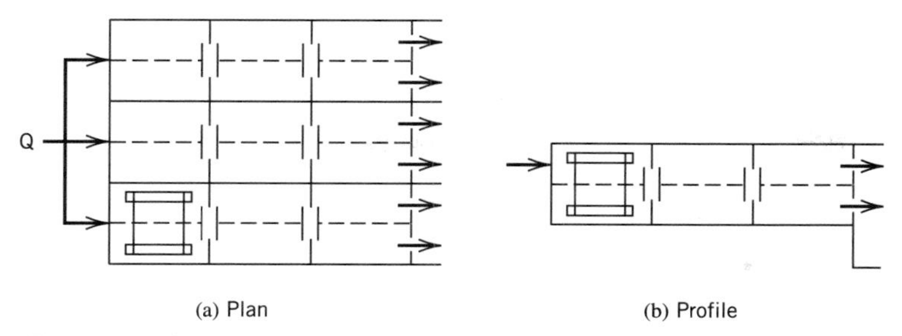

Figure 13.10 Horizontal flow paddle flocculator with blades parallel to flow. Adapted from AWWA (1969). Reprinted from *Water Treatment Plant Design,* by permission. Copyright © 1969, American Water Works Association.

depends on the drag force and the relative velocity of the fluid with respect to the paddle. A schematic of a paddle blade is shown in Fig. 13.12. The following symbol definitions apply to the sketch.

v_f = fluid velocity

v_p = paddle velocity

r_i = distance from shaft to inner edge of the paddle

r_o = distance from shaft to outer edge of the paddle

b = length of the paddle

N = rate of revolution of the shaft (rpm)

The velocity of the fluid will be less than the velocity of the paddle by a factor, k, because of slip. If baffles (called stators) are placed along the walls in a direction

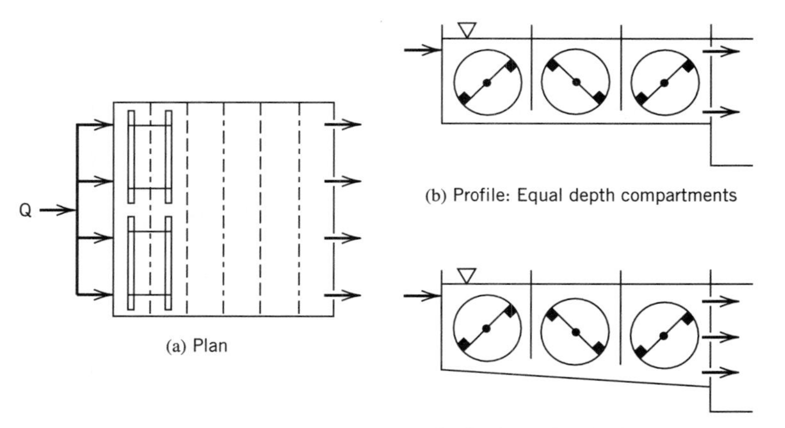

Figure 13.11 Horizontal flow paddle flocculators with blades perpendicular to flow. Adapted from AWWA (1969). Reprinted from *Water Treatment Plant Design,* by permission. Copyright © 1969, American Water Works Association.

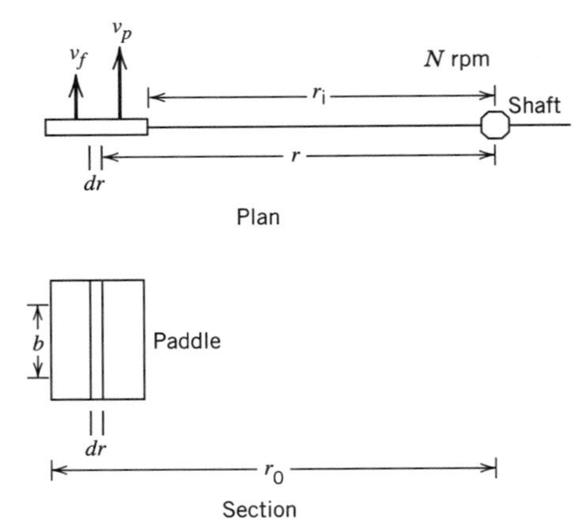

Figure 13.12 Paddle blade in a flocculator.

perpendicular to the fluid movement, the value of k decreases because these baffles obstruct the movement of fluid. The frictional dissipation of energy depends on the relative velocity, v.

$$v = v_p - v_f = v_p - kv_p = v_p(1 - k) \tag{13.21}$$

The velocity of the paddle at a distance r from the shaft is

$$v_p = \frac{2\pi N}{60} r \tag{13.22}$$

The equation for drag force of the paddle on the fluid is the usual equation,

$$F_D = \tfrac{1}{2}\rho C_D A v^2$$

where
- A is the area of the paddle
- F_D is the drag force
- C_D is a drag coefficient

The power input imparted to the fluid by an elemental area of the paddle is

$$dP = dF_D \cdot v = \tfrac{1}{2}\rho C_D v^3 \, dA \tag{13.23}$$

Noting that $dA = b\,dr$ and substituting Eqs. (13.21) and (13.22) into Eq. (13.23) and integrating,

$$\begin{aligned} P &= \frac{1}{2}\rho C_D b \left[\frac{2\pi N}{60}(1 - k)\right]^3 \int_{r_i}^{r_o} r^3\,dr \\ &= (1.44 \times 10^{-4}) C_D \rho b [N(1 - k)]^3 (r_o^4 - r_i^4) \end{aligned} \tag{13.24}$$

(SI: ρ, kg/m^3; b, m; r, m; P, watts)
(U.S.: ρ, slug/ft^3; b, ft; r, ft; P, ft-lb/s)

There is usually more than one set of paddles on each arm of the shaft. Equation (13.24) should be applied individually to each paddle and the results summed to obtain

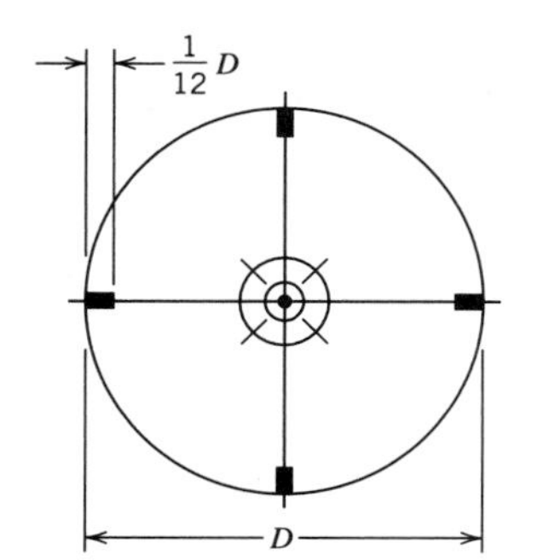

Figure 13.13 Paddle with stators.

the total power dissipation. Recommended design parameters for a paddle flocculator are as follows:

Peripheral speed of the paddles, 0.09–0.9 m/s (0.3–3 ft/s)

k = 0.25 without stators, 0–0.15 with stators

$C_D \approx 1.8$ for flat blades

Area of the blades is 15–20% of the tank cross-sectional area (if the paddles are oriented parallel to the flow, use 15–20% of the tank side area).

There are no specific guidelines for depth–area ratios in a paddle flocculator but the maximum depth should not exceed 5 m (16.4 ft) or unstable flows will result (Montgomery, 1985). Ground levels and plant layout dictate feasible depths. A freeboard of 0.5 m (1.6 ft) should be provided. If flocculators are followed by rectangular, horizontal flow sedimentation basins, the widths in both basins should be the same.

A section with a paddle arm and stators is shown in Fig. 13.13. The spacing, size, and number of paddles on a paddle arm can be varied to provide different G values. Variable speed motors should be provided to change the power input as required with changes in temperature, flow rate, and water quality. There must be space for shaft supports between adjacent paddle sets.

The compartments in a paddle flocculator are often separated with a baffle wall (Fig. 13.14) to equalize flow distribution. When sedimentation tanks directly follow

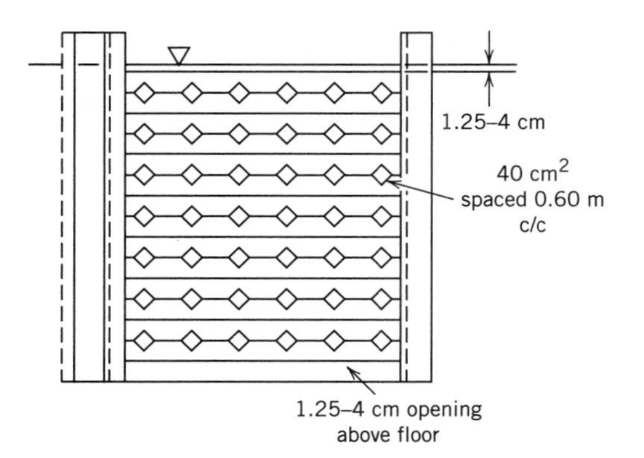

Figure 13.14 Typical baffle wall. Adapted from J. M. Montgomery, Consulting Engineers, Inc. (1985), *Water Treatment Principles and Design,* copyright © 1985 by John Wiley & Sons, Inc. Reprinted by permission of John Wiley & Sons, Inc.

flocculation basins, baffle walls (also known as diffuser walls) are the preferred inlet devices. Design guidelines for baffle walls vary among various sources (ASCE and AWWA, 1990; Kawamura, 1973, 1991b). The area of the baffle orifices is approximately 3–6% of the wall area or provides a velocity of 0.3 m/s (1 ft/s) under maximum flow conditions (ASCE and AWWA, 1990; Kawamura, 1973). The size of an orifice should be between 40 and 175 cm^2 (6–27 in.^{2}). The baffle wall is raised 1.25–4 cm (0.5–1.5 in.) above the floor to allow easy cleaning of sludge deposits. A water clearance of 1.25–4 cm (0.5–1.5 in.) over the baffle wall is provided to pass scum through the flocculator. Orifice size and spacing is somewhat variable to meet the flow-through velocity constraint. Uniform flow through the baffle wall is assumed and the area above and below the baffle wall is ignored.

Turbines or propellers may be used instead of paddles. It will be difficult to install a paddle system in a small installation. Propeller area will not be 15–20% of the cross-sectional area. Equations (13.9b), (13.10), and (13.11) would be used to calculate power and rotational speed requirements. Monk and Trussell (1991) provide equations and the design procedure for vertical shaft flocculators and Amirtharajah (1978) provides a design example for an axial flow propeller flocculator.

■ Example 13.2 Paddle Flocculator Design

Design a three-compartment flocculator with the configuration of Fig. 13.11b to treat a design flow of 15 000 m^3/d (3.96 Mgal/d). The design G values are 45, 20, and 10 s^{-1} in each successive compartment and the total detention time in the flocculator will be 30 min. The maximum length of an individual paddle should not exceed 3.5 m (11.5 ft) and the width of a paddle should be between 10 and 20 cm (0.33–0.66 ft). The minimum clear space between the outer paddle and the floor and water surface should be 20 cm (8 in.). Shaft supports require a 30-cm (1-ft) spacing. The distance of a paddle from a wall should be at least 20 cm (8 in.). The lowest temperature expected is 10°C, which is the design temperature. Use 1:4 variable speed motors. The maximum depth (D) of the flocculator is 4.3 m (14.1 ft). Assume C_D and k are 1.8 and 0.25, respectively.

The total volume (V) of the flocculator is

$$V = Qt_d = \left(15\,000\,\frac{\text{m}^3}{\text{d}}\right)(30\text{ min})\left(\frac{1\text{ h}}{60\text{ min}}\right)\left(\frac{1\text{ d}}{24\text{ h}}\right) = 313\text{ m}^3$$

$$\text{In U.S. units: } V = \left(3.96\times10^6\,\frac{\text{gal}}{\text{d}}\right)(30\text{ min})\left(\frac{1\text{ h}}{60\text{ min}}\right)\left(\frac{1\text{ d}}{24\text{ h}}\right)\left(\frac{1\text{ ft}^3}{7.48\text{ gal}}\right) = 11\,029\text{ ft}^3$$

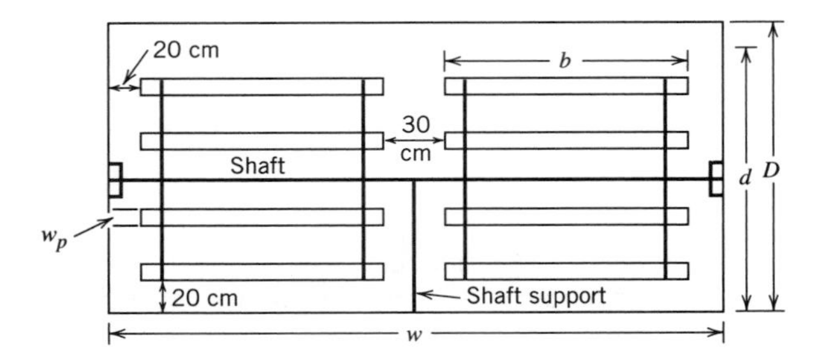

For a single compartment use a volume of 104 m^3 ($\approx$ 313 $m^3/3$) (similarly, the volume of a chamber in U.S. units is 3 676 ft^3). The effective depth (d) is 3.80 m (12.5 ft) with a freeboard of 0.50 m (1.64 ft or 20 in.).

With shafts located in the center of the compartment, the maximum length of a paddle arm (l_a) is determined from the surface and floor clearances and the depth.

$$l_a = \tfrac{1}{2}(3.80 \text{ m} - 2 \times 0.20 \text{ m}) = 1.70 \text{ m}$$

$$\text{In U.S. units: } l_a = \tfrac{1}{2}(12.5 \text{ ft} - 2 \times 1.64 \text{ ft}) = 4.60 \text{ ft}$$

(A single paddle arm would exceed the maximum length specification.) The minimum length of a compartment (l_c), considering wall spacings, is

$$l_c = 2l_a + 2 \times 0.20 \text{ m} = 2(1.70 \text{ m}) + 0.40 \text{ m} = 3.80 \text{ m}$$

$$\text{In U.S. units: } l_c = 2l_a + 2 \times 0.66 \text{ ft} = 2(4.60 \text{ ft}) + 1.32 \text{ ft} = 10.5 \text{ ft}$$

The cross-sectional area (A) and width (w) of a compartment (and the basin) are

$$A = (104 \text{ m}^3)/(3.80 \text{ m}) = 27.4 \text{ m}^2 \qquad w = (27.4 \text{ m}^2)/(3.80 \text{ m}) = 7.20 \text{ m}$$

$$\text{In U.S. units: } A = (3\,676 \text{ ft}^3)/(12.5 \text{ ft}) = 294 \text{ ft}^2 \qquad w = (294 \text{ ft}^2)/(12.5 \text{ ft}) = 23.5 \text{ ft}$$

Two adjacent sets of paddles can be installed in this width. A sketch of the front view of the flocculator is shown earlier. The available width (w_a) is determined by subtracting the wall clearance and shaft support widths from the total width.

$$w_a = 7.20 \text{ m} - 0.30 \text{ m} - (2 \times 0.20 \text{ m}) = 6.50 \text{ m}$$

$$\text{In U.S. units: } w_a = 23.5 \text{ ft} - 1.0 \text{ ft} - (2 \times 0.66 \text{ ft}) = 21.2 \text{ ft}$$

The length of a paddle (b) will be 3.25 m (10.6 ft). Choose the area of the paddles (A_p) to be 17% of the cross-sectional area

$$A_p = 0.17\ (27.4 \text{ m}^2) = 4.66 \text{ m}^2 \qquad \text{in U.S. units: } A_p = (0.17)(294 \text{ ft}^2) = 50.0 \text{ ft}^2$$

If each paddle set has four paddles, the width of a paddle (w_p) is

$$w_p = (4.66 \text{ m}^2)/8/(3.25 \text{ m}) = 0.18 \text{ m} = 18 \text{ cm}$$

$$\text{In U.S. units: } w_p = (50.0 \text{ ft}^2)/8/(10.6 \text{ ft}) = 0.59 \text{ ft} = 7.1 \text{ in.}$$

This is a suitable paddle width. Therefore each paddle set will include four paddles. The outer paddle will be located at the end of the paddle arm. The inner paddle will be located midway between the shaft and the outer paddle.

The power requirement for the first chamber is

$$P = \mu V G^2 = (1.307 \times 10^{-3} \text{ kg/s-m})(104 \text{ m}^3)(45 \text{ s}^{-1})^2 = 275.3 \text{ W}$$

$$\text{In U.S. units: } P = \left(2.735 \times 10^{-5} \frac{\text{lb-s}}{\text{ft}^2}\right)(3\,676 \text{ ft}^3)(45 \text{ s}^{-1})^2 \left(\frac{1 \text{ hp}}{550 \frac{\text{ft-lb}}{\text{s}}}\right) = 0.370 \text{ hp} \tag{i}$$

The power expenditure for a paddle (P_p) is

$$P_p = (1.44 \times 10^{-4}) C_D \rho b [N(1 - k)]^3 (r_o^4 - r_i^4)$$

$$P_p = (1.44 \times 10^{-4})(1.8)(999.7)(3.25)(0.75)^3 N^3 (r_o^4 - r_i^4) = 0.355 N^3 (r_o^4 - r_i^4)$$

$$\text{In U.S. units: } P_p = (1.44 \times 10^{-4})(1.8)(1.94)(10.6)(0.75)^3 N^3 (r_o^4 - r_i^4) = 0.002\,25 N^3 (r_o^4 - r_i^4)$$

For an outer paddle with inner and outer radii of 1.52 and 1.70 m (5.00 and 5.59 ft), respectively, the power expenditure (P_o) is

$$P_o = 0.355N^3[(1.70)^4 - (1.52)^4] = 1.071N^3 \text{ W}$$
$$\text{In U.S. units: } P_o = 0.002\,25N^3[(5.59)^4 - (5.00)^4] = 0.791N^3 \text{ ft-lb/s}$$

For an inner paddle with inner and outer radii of 0.76 and 0.94 m (2.49 and 3.08 ft), respectively, the power expenditure (P_i) is

$$P_i = 0.355N^3[(0.94)^4 - (0.76)^4] = 0.159N^3 \text{ W}$$
$$\text{In U.S. units: } P_i = 0.002\,25N^3[(3.08)^4 - (2.49)^4] = 0.116N^3 \text{ ft-lb/s}$$

The total power expenditure in the compartment as a function of N is

$$P = 4(P_o + P_i) = 4(1.071 + 0.159)N^3 = 4.92N^3 \text{ W}$$
$$\text{In U.S. units: } P = 4(0.791 + 0.116)N^3 = 3.63N^3 \text{ ft-lb/s} \qquad \text{(ii)}$$

By definition in Section 13.4, the units on N are rpm. N is found by equating Eqs. (i) and (ii).

$$N = \left(\frac{275.3}{4.92}\right)^{1/3} = 3.82 \text{ rpm}$$

$$N = \left[\frac{(0.370 \text{ hp})\left(550\dfrac{\text{ft-lb/s}}{\text{hp}}\right)}{3.63}\right]^{1/3} = 3.83 \text{ rpm}$$

The peripheral velocity of the outer paddle (Eq. 13.22) is

$$v_p = \frac{2\pi N}{60}r = \frac{2\pi(3.82 \text{ rev/min})}{60 \text{ s/min}}(1.70 \text{ m}) = 0.68 \text{ m/s}$$

$$\text{In U.S. units: } v_p = \frac{2\pi(3.82 \text{ rev/min})}{60 \text{ s/min}}(5.59 \text{ ft}) = 2.24 \text{ ft/s}$$

which is within the acceptable range of 0.09–0.9 m/s (0.3–3 ft/s).

Choosing the design N value to be in the middle of the range for a 1:4 variable speed motor, the following relations exist between the minimum, maximum, and middle rotational speeds (N_{min}, N_{max}, and N_{mid}, respectively).

$$\frac{N_{min} + N_{max}}{2} = N_{mid} \quad N_{max} = 4N_{min} \quad N_{min} = 0.4N_{mid}$$

Therefore, $N_{min} = 1.53$ rpm and $N_{max} = 6.12$ rpm.

The required G values in the second and third compartments are obtained by reducing the rotational speed of the paddles in these compartments. For the second compartment, the G value is 20 s^{-1}. Using Eq. (i), the power dissipation in the compartment is

$$P = \mu VG^2 = (1.307 \times 10^{-3} \text{ kg/s-m})(104 \text{ m}^3)(20 \text{ s}^{-1})^2$$
$$= 54.4 \text{ W}$$

Using Eq. (ii), N is found to be

$$N = \left(\frac{54.5}{4.92}\right)^{1/3} = 2.23 \text{ rpm}$$

The peripheral velocity of the outer paddle is

$$v_p = \frac{2.23}{3.82}(0.68) = 0.40 \text{ m/s}$$

The speed range of the motor is from 0.89 to 3.57 rpm.
Similar calculations for the third compartment yield

$$P = 13.6 \text{ W} \qquad N = 1.40 \text{ rpm} \qquad v_p = 0.25 \text{ m/s}$$

The speed range of the motor is from 0.56 to 2.24 rpm.

■ Example 13.3 Baffle Wall Design

Design baffle walls to be placed between compartments and at the exit end of the flocculator in the previous example. Calculate the headloss through the baffle wall. Use a design similar to Fig. 13.14 with sharp-edged, diamond-shaped orifices.

The width of the basin is 7.20 m (23.5 ft) and its effective depth is 3.80 m (12.5 ft). The design flow is

$$Q = 15\,000 \text{ m}^3/\text{d} = \left(15\,000 \frac{\text{m}^3}{\text{d}}\right)\left(\frac{1 \text{ d}}{86\,400 \text{ s}}\right) = 0.174 \text{ m}^3/\text{s} \ (6.13 \text{ ft}^3/\text{s})$$

The velocity through the baffle wall should not exceed 0.3 m/s (1 ft/s). The total area of the orifices is

$$v = Q/\Sigma a \qquad \Sigma a = Q/v = (0.174 \text{ m}^3/\text{s})/(0.3 \text{ m/s}) = 0.580 \text{ m}^2 \ (6.24 \text{ ft}^2)$$

If an orifice has sides of 7 cm and an opening of 49 cm² (0.004 9 m²), the number of orifices is

$$N = (0.580 \text{ m}^2)/(0.004\,9 \text{ m}^2) = 118$$

Define the number of orifices per unit height and unit width as n_h and n_w, respectively. A clearance of 1.5 cm will be provided above and below the baffle wall. The height of the baffle wall will be 3.77 m (12.4 ft). Then for uniform spacing of the orifices,

$$n_h n_w = N \qquad \frac{n_h}{3.77} = \frac{n_w}{7.20} \qquad \frac{7.20}{3.77} n_h = n_w \qquad 1.91 n_h = n_w$$

$$1.91 n_h^2 = N \qquad n_h = \sqrt{\frac{118}{1.91}} = 7.9 \qquad n_w = 1.91(7.9) = 15.1$$

Choose n_h and n_w to be 8 and 15, respectively. The total number of orifices is $8 \times 15 = 120$. Space the orifices (7.20 m)/15 = 0.48 m (1.57 ft) c/c horizontally and (3.77 m)/8 = 0.47 m (1.54 ft) c/c vertically.

The flow through an orifice is

$$q = Q/N = (0.174 \text{ m}^3/\text{s})/120 = 0.001\,45 \text{ m}^3/\text{s} \ (0.051\,2 \text{ ft}^3/\text{s})$$

The velocity through an orifice will be less than 0.3 m/s (1 ft/s) because the number of orifices has been slightly increased and there will be some flow over and under the baffle wall.

The headloss (Δh) through the baffle wall is calculated with the equation for a submerged orifice. The discharge coefficient is 0.62 for sharp-edged orifices.

$$q = C_d a \sqrt{2g\Delta h}$$

$$\Delta h = \frac{(q/a)^2}{2gC_d^2} = \frac{(0.001\,45\ \text{m}^3/\text{s})^2}{2(9.81\ \text{m/s}^2)(0.62)^2(0.004\,9\ \text{m}^2)^2} = 0.011\,6\ \text{m or } 1.2\ \text{cm}$$

In U.S. units the headloss is 0.5 in.

Pipes

Coagulating agents can be added to pipes where mixing and flocculation will occur. If water is being pumped over a significant distance, for example from a reservoir, it may be feasible to use the pipe as a flocculator. The Darcy–Weisbach equation (Eq. 13.25) is suitable for estimating headloss in pipes to be used in Eqs. (13.7) and (13.6) for calculating the velocity gradient.

$$h_L = f\frac{Lv^2}{D2g} \tag{13.25}$$

where
- f is the friction factor taken from a Moody diagram
- L is the length of the pipe
- v is the average velocity in the pipe
- D is the diameter of the pipe

Because the losses are a function of the Reynold's number, the amount of power dissipation varies with the quantity or velocity of flow in the pipe.

Baffled Channels

Baffled channels are another hydraulic alternative to achieve mixing or flocculation. Baffled channels may be the around the end type or the over and under flow type. The sketch in Fig. 13.15 would be viewed in plan for the former and from the side for the latter. Velocities in the channels are recommended to be in the range of 0.1–0.4 m/s (0.3–1.3 ft/s) and the detention time is 15 to 20 min (Huisman et al., 1981). The headloss primarily depends on the losses incurred in the flow around the ends of the baffles.

$$h_L = K\frac{v^2}{2g} \tag{13.26}$$

where
- K is a loss coefficient

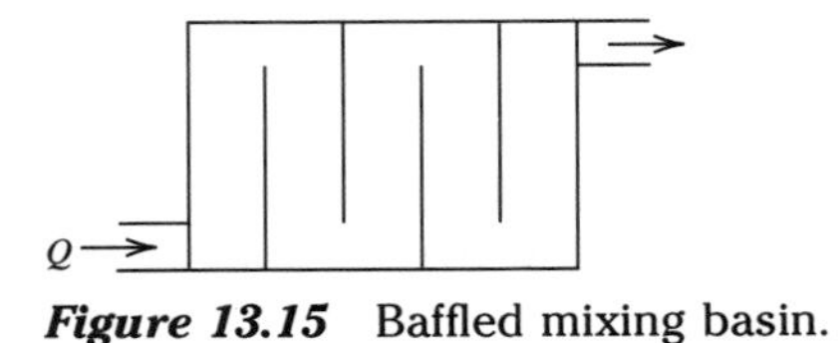

Figure 13.15 Baffled mixing basin.

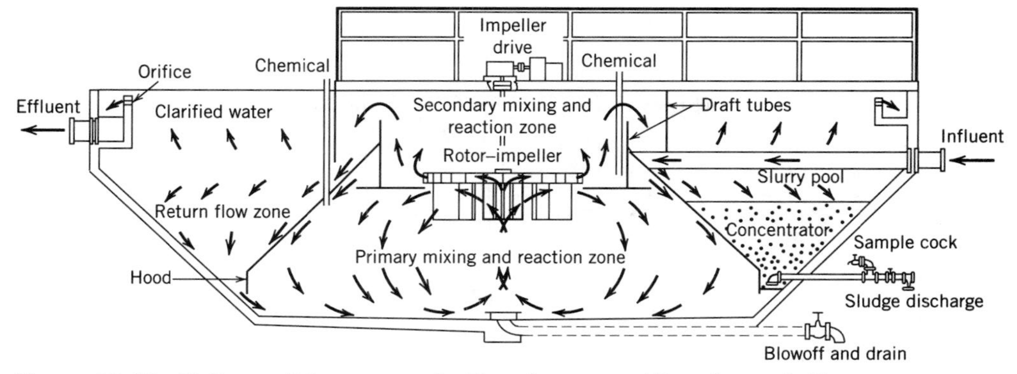

Figure 13.16 Upflow solids contact clarifier. Courtesy of Degrémont Infilco.

The value of K is typically assumed to be in the range of 2–3.5. Bhargava and Ojha (1993) found that the lower value of 2 was best from a pilot unit study.

In horizontal flow (around the end) flocculators, Schulz and Okun (1984) recommend that the distance between baffles should not be less than 45 cm (18 in.) to permit cleaning. The clear distance between the end of each baffle and the wall should be about $1\frac{1}{2}$ times the distance between baffles but not less than 60 cm (2 ft) and the depth of the water in the tank should not be less than 1.0 m (3.3 ft).

For vertical flow (over and under) flocculators Schulz and Okun (1984) recommend baffle spacing of at least 45 cm (1.5 ft) and clear space between the baffle and tank bottom or between the baffle and water surface of $1\frac{1}{2}$ times the baffle spacing. Tank depth should be two to three times the spacing between baffles.

Tapered flocculation can be achieved in either type of baffled flocculator by varying the spacing of the baffles. The spacing requirements of the baffles limit these designs to plants with larger flows. Typical flow variations will cause velocity gradients to be outside the desired range during flow extremes.

Upflow Solids Contact Clarifier

Upflow solids contact clarifiers (Fig. 13.16) provide mixing, flocculation, and clarification in a single unit. A significant reduction in total space for the operations is attained. More skill is required to operate these units. The coagulating agent is added and mixed with the water in the central zone of the unit, where a significant amount of flocculation occurs. Floc formed in the central zone exits into the clarification zone. A dense sludge blanket forms in the clarification zone, which aids in floc entrapment and clarification. The sludge blanket improves flocculation–clarification efficiency, but it takes time to establish the sludge blanket after it is drained for routine maintenance or lost through operational upset. Design guidelines for upflow solids contact clarifiers are given in Table 11.6.

Alabama Flocculator

Another type of hydraulic flocculator, known as the "Alabama" flocculator, is shown in Fig. 13.17 (Huisman et al., 1981). Water flows through a series of chambers, entering each chamber through an inlet turned upward. The water then travels down to the outlet (inlet for the next chamber). These units were first developed and used in Alabama and they have been used successfully in Latin America.

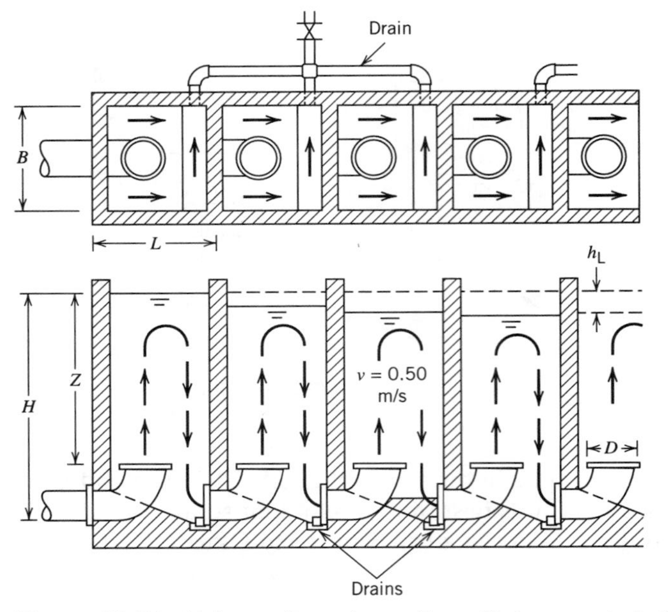

Figure 13.17 Alabama flocculator. From Huisman, et al. (1981), *Small Community Water Supplies,* IRC.

For effective flocculation the outlets should be placed about 2.50 m (8.2 ft) below the water level. The headloss is normally about 0.35–0.50 m (1.1–1.6 ft) for the entire unit. The resulting velocity gradient is in the range of 40–50 s^{-1}.

Common design criteria are given in Table 13.5. The diameter (D) of the inlet is chosen to maintain the inlet velocity between 0.4 and 0.55 m/s (1.3–1.8 ft/s). Other guidelines for design of an Alabama flocculator are given in Table 13.6.

Huisman et al. (1981) also describe other hydraulic flocculators and mixers, which are often appropriate for developing countries because they have no mechanical components.

Spiral Flow Tanks

In these basins influent is inlet near the bottom of the basin and directed tangentially toward the side of the basin. A spiral flow pattern ensues. Influent and effluent locations

TABLE 13.5 Alabama Flocculator Design Criteria[a,b]

	Range	
Item	SI units	U.S. units
Rated capacity per unit chamber	25–50 L/s/m^2	0.6–1.2 gal/s/ft^2
Velocity at turns	0.40–0.60 m/s	1.3–2.0 ft/s
Length of unit chamber (L)	0.75–1.50 m	2.5–5.0 ft
Width (B)	0.50–1.25 m	1.6–4.1 ft
Depth (Z)	2.50–3.50 m	8.2–11.4 ft
Total detention time in flocculator	15–25 min	15–25 min

[a]From Huisman et al. (1981).
[b]Refer to Fig. 13.17 for symbol definitions.

TABLE 13.6 Guidelines for Design of an Alabama Flocculator[a]

Q		Width (*B*)		Length (*L*)		Diameter (*D*)		Unit chamber area		Unit chamber volume	
L/s	gal/s	m	ft	m	ft	mm	in.	m^s	ft^2	m^3	ft^3
20	5.3	0.60	2.0	0.75	2.5	250	9.8	0.45	4.8	1.3	46
40	10.6	0.80	2.6	1.00	3.2	350	13.8	0.80	8.6	2.4	85
60	15.9	1.00	3.2	1.20	4.0	400	15.7	1.20	12.9	3.6	127
80	21.1	1.20	4.0	1.40	4.6	450	17.7	1.60	17.3	4.8	170
100	26.4	1.40	4.6	1.60	5.2	500	19.7	2.00	21.5	6.0	212

[a]From Huisman et al. (1981), *Small Community Water Supplies,* IRC.

are on opposite sides of the tank. The headloss in these basins must be empirically determined; it is a function of the tank geometry and velocity of the influent jet.

Pebble Bed Flocculators

For any flocculation device, decreasing the detention time in an individual compartment but increasing the number of compartments increases the efficiency of flocculation. This is because the flow regime through the flocculation system approaches plug flow conditions. Directing flow through gravel media in a vertical tank provides a large number of void openings in which flocculation will occur. Larger gravel media are used to provide larger void openings for mixing and to prevent medium clogging. This type of flocculator is known as a pebble bed flocculator. These devices are relatively insensitive to flow variation and their operation is simple, which makes them appropriate for small communities. Clogging is not a problem unless there is silt or sand in the influent to the flocculator (Richter, 1986). The units can be flushed with water at a high velocity in the upflow direction when necessary.

Headloss in pebble bed flocculators can be calculated by the Carman–Kozeny equation (Eq. 13.27); this equation is fully developed in the next chapter. Detention times are based on the void volume of the flocculator. Studies in Brazil (Richter, 1977, 1986) and elsewhere (Kuczynski, 1985) found that media with nominal diameters in the range of 10–20 mm (0.4–0.8 in.) gave the best results. Gt_d values in the range of 1.4×10^4 to 1.6×10^4 were optimal. High G values, in the range of 50–80 s^{-1}, provided the best flocculation, although satisfactory results were obtained at G values as low as 30 s^{-1}. At G values in the optimal range, very short detention times, on the order of 5 min, are required to produce a well-flocculated effluent. The porosity (void volume/total volume) of gravel in the size range of 1–50 mm will be near 0.40.

$$\frac{h_L}{L} = \frac{150\mu}{\rho g} \frac{(1-e)^2}{e^3} \frac{v_s}{(\psi d)^2} + 1.75 \frac{1-e}{e^3} \frac{v_s^2}{\psi d g} \tag{13.27}$$

where

h_L is headloss through a depth, L
e is porosity of the medium
ψ is sphericity of the medium
d is nominal diameter of the medium
v_s is the superficial velocity (Q/A_s where A_s is the surface area of the unit)

Similar studies with a horizontal flow, gravel packed, baffled flocculator demonstrated improvement in performance over a baffled channel flocculator without gravel (Ayoub and Nazzal, 1988). Gravel sizes in the range of 5.1–7.6 cm (2.0–3.0 in.) diameter gave performance that was 6% better than gravel in the size range of 1.3–2.5 cm (0.5–1.0 in.). Performance only marginally improved above Gt_d values of 1.5×10^4. Contact times ranged from 1.3 to 9.5 min.

QUESTIONS AND PROBLEMS

1. How does charge on an ion generally affect its ability to coagulate colloids?
2. How much alkalinity (as $CaCO_3$) is consumed by a dose of 40 mg/L of each of the following coagulants: $Al_2(SO_4)_3 \cdot 14.2H_2O$, $FeSO_4 \cdot 7H_2O$, $FeCl_3$, and $Fe_2(SO_4)_3$?
3. Write the chemical equation for the reaction of alum with sodium carbonate.
4. If a pH of 6.0 is effective for flocculation at a temperature of 20°C, what is the pH at 5°C that provides the same pOH?
5. If a pOH of 8.3 produces the best results for coagulation–settling of a water, how much lime should be added to a water with an alkalinity of 135 mg/L as $CaCO_3$ and pH of 7.5 that will be dosed with alum [$Al_2(SO_4)_3 \cdot 18H_2O$] at 55 mg/L when the water temperature is (a) 22°C and (b) 8°C?
6. Would it be desirable to use ozone as the sole coagulant?
7. For a flow of 13 500 m^3/d containing 55 mg/L of suspended solids, ferric sulfate is used as a coagulant at a dose of 50 mg/L. (a) Assuming that there is little alkalinity in the water, what is the daily lime dose? (b) If the sedimentation basin removes 90% of the solids entering it, what is the daily solids production from the sedimentation basin?
8. Design a square rapid mixing basin for a water treatment plant with a design flow of 8.0×10^3 m^3/d (2.11 Mgal/d). The basin depth is equal to 1.25 times the width. The design velocity gradient will be 1 000 s^{-1} at 15°C and the detention time is to be 30 s. A vane disk impeller with six flat blades is to be used and the tank is not baffled. Determine the basin dimensions, and the input power required. Find the impeller speed if the impeller diameter is 50% of the basin width.
9. Why is rapid mixing required for coagulant addition? Discuss the intensity of mixing required in a flocculation basin.
10. A sludge contains 155 mg/L of Fe as $Fe(OH)_3$. What is the minimum concentration of H_2SO_4 that must be added to the sludge to recover the iron?
11. Distinguish between sweep coagulation and charge neutralization coagulation.
12. What volumetric airflow rate is required to maintain a G value of 500 s^{-1} in a basin that is 2.75 m deep and provides a liquid detention time of 5 min? Perform the exercise for water temperatures of 5 and 18°C for a liquid flow rate of 0.21 m^3/s.
13. Derive the equation: $\frac{y_2}{y_1} = \frac{1}{2}(\sqrt{8Fr_1^2 + 1} - 1)$ for a hydraulic jump.
14. A hydraulic flocculator with around the end baffles is designed to treat a flow of 10 000 m^3/d (2.64 Mgal/d). The flocculator has width, length, and depth dimensions of 12, 12, and 1.45 m (39.4, 39.4, and 4.76 ft) respectively. There are 29 evenly spaced baffles in the flocculator and the flow velocity around the ends is the same as the flow velocity in the channels. The loss coefficient around the baffles has been found to be 3.1. The temperature is 10°C.
    (a) Calculate the headloss in the channels using the Darcy–Weisbach equation. Assume smooth walls.
    (b) Calculate the headloss around the bends and compare it to the headloss in the channels.

(c) What are G, t_d, and Gt_d for this basin?
(d) Answer (c) if Q decreases by one-half and if Q doubles into this basin.

15. Provide the development that proves that the work of isothermal expansion of gas bubbles as they rise in a tank results in the same expression for power input as given by Eq. (13.16).

16. Discuss the hydraulic flow regime in a flocculation system. Is it plug flow or complete mixed?

17. Why does tapered flocculation improve the performance of a flocculator?

18. A water treatment plant is processing a flow of 42 000 m^3/d. The flocculation system consists of three identical parallel units. A side view of one of the units is shown in the accompanying sketch. The dimensions of each flocculation unit are 4.5 m wide by 4.5 m deep by 18.8 m long. The inlet baffle is located 0.3 m from the front wall. Each unit has three sets of paddles mounted on a horizontal shaft. There are three sets of paddles attached to the shaft. The first set of paddles has four paddles on each arm (total of eight paddles) with centers located at 1.9, 1.7, 1.5, and 1.3 m from the shaft. The second set has three paddles on each arm with centers located at 1.9, 1.7, and 1.5 m from the shaft, and the third set has two paddles per arm with centers located at 1.9 and 1.5 m from the shaft. Each paddle is 0.1 m wide and 4.5 m long. The first paddle set is 0.8 m from the inlet baffle and the paddle set are separated by a distance of 0.8 m.

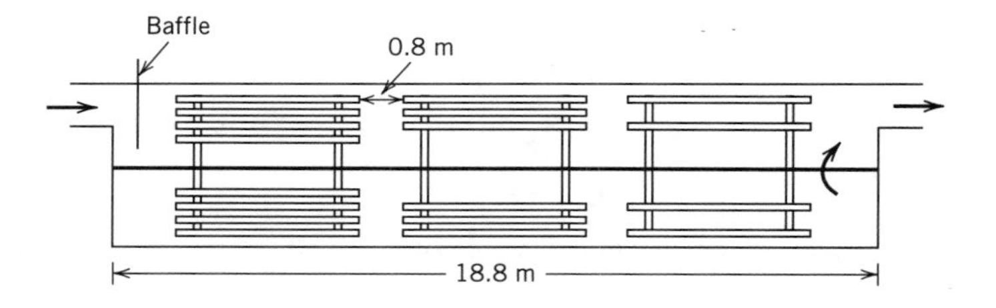

At temperatures of 2 and 25°C what is the rotational speed of the shaft to achieve a mean velocity gradient of 20 s^{-1} in a unit? What is the local velocity gradient for each paddle set? $C_D = 1.8$, $k = 0.25$.

19. Design a flocculation basin for a flow of 30×10^3 m^3/d (7.93 Mgal/d). The basin will have horizontal shafts with paddle wheels as indicated in Fig. 13.11b and three compartments of equal depth. The detention time will be 36 min. Provide tapered flocculation with G values at 10°C in the three compartments of 50, 25, and 10 s^{-1}, respectively. The paddles will have a width of 15 cm (5.9 in.) and a length of 2.50 m (8.20 ft). The outside blades should clear the floor by 0.30 m (1.0 ft) and be 0.30 m (1.0 ft) below the surface of the water. Other design specifications are: four blades per paddle wheel; minimum clear spacing of the blades, 30 cm (1 ft); clear space between adjacent paddle wheels, in the range of 60–90 cm (2–3 ft); wall clearance, between 30 and 45 cm (1 and 1.5 ft). The compartments should be square in profile. Determine the number of paddle sets and design the unit to meet other criteria given in the text. Find:
(a) The basin dimensions
(b) The paddle wheel design
(c) The power requirements for each compartment and the total power required
(d) The rotational speed range if 1:4 variable speed motors are used

20. How much error is introduced in calculating the rate of power dissipation for a diffuser system if the variation in bubble size with depth is ignored? Perform your analysis for diffuser depths of 1, 2, and 3 m.

21. What are the basin and chamber dimensions for an Alabama flocculator designed to treat a flow of 6 100 m^3/d? If G is 45 s^{-1}, what is the headloss through each chamber and the entire unit at a temperature of 25°C?

22. Design a cylindrical pebble bed flocculator to treat an average flow of 4 800 m^3/d for Gt_d

values in the range of 1.4×10^4 to 1.6×10^4 with G in the range of 40–80 s^{-1}. Gravel with a specific gravity of 2.65 and nominal diameter of 15 mm will be used. The porosity of the gravel is 0.42 and its sphericity is 0.90. The design temperature is 10°C. Find the diameter and height of the bed.

KEY REFERENCES

American Water Works Association Research Foundation (1991), *Mixing in Coagulation and Flocculation,* A. Amirtharajah, M. M. Clark, and R. R. Trussell, eds., American Water Works Association, Denver, CO.

AMIRTHARAJAH, A. AND C. R. O'MELIA (1990), "Coagulation Processes: Destabilization, Mixing and Flocculation," in *Water Quality and Treatment,* 4th ed., F. W. Pontius, ed., McGraw-Hill, Toronto, pp. 269–365.

HUDSON, H. E., JR. (1981), *Water Clarification Processes: Practical Design and Evaluation,* Van Nostrand Reinhold, Toronto.

MONTGOMERY, J. M., Consulting Engineers, Inc. (1985), *Water Treatment Principles and Design,* John Wiley & Sons, Toronto.

REFERENCES

AMIRTHARAJAH, A. (1978), "Design of Flocculation Systems," in *Water Treatment Plant Design for the Practicing Engineer,* R. L. Sanks, ed., Ann Arbor Science, Ann Arbor, MI, pp. 195–229.

ASCE and AWWA (1990), *Water Treatment Plant Design,* 2nd ed., McGraw-Hill, Toronto.

AWWA (1969), *Water Treatment Plant Design,* Denver, CO.

AWWA Committee (1989), "Committee Report: Coagulation as an Integrated Water Treatment Process," *J. American Water Works Association,* 81, 10, pp. 72–78.

AYOUB, G. A. AND F. F. NAZZAL (1988), "Gravel Packed Baffled Channel Flocculator," *J. of Environmental Engineering, ASCE,* 114, 6, pp. 1448–1463.

BAKER, M. N. (1981), *The Quest for Pure Water,* vol. I, American Water Works Association, Denver, CO.

BEWTRA, J. K. AND W. R. NICHOLAS (1964), "Oxygenation from Diffused Air in Aeration Tanks," *J. Water Pollution Control Federation,* 36, 10, pp. 1195–1224.

BHARGAVA, D. S. AND C. S. P. OJHA (1993), "Models for Design of Flocculating Baffled Channels," *Water Research,* 27, 3, pp. 465–475.

CAMP, T. R. and P. C. STEIN (1943), "Velocity Gradients and Internal Work in Fluid Motion," *J. Boston Society of Civil Engineers,* 30, 4, pp. 219–227.

CLEASBY, J. L. (1984), "Is Velocity Gradient a Valid Turbulent Flocculation Parameter," *J. of Environmental Engineering, ASCE,* 114, 5, pp. 875–897.

CHOW, V. T. (1959), *Open Channel Hydraulics,* McGraw-Hill, Toronto.

Degrémont Infilco (1979), *Water Treatment Handbook,* Taylor and Carlisle, New York.

EDWARDS, M. AND M. M. BENJAMIN (1989), "Regeneration and Reuse of Iron Hydroxide Adsorbents in Treatment of Metal-Bearing Wastes," *J. Water Pollution Control Federation,* 61, 4, pp. 481–490.

GUIBAI, L. AND J. GREGORY (1991), "Flocculation and Sedimentation of High-Turbidity Waters," *Water Research,* 25, 9, pp. 1137–1143.

HAARHOFF, J. AND J. L. CLEASBY (1988), "Comparing Aluminum and Iron Coagulants for In-Line Filtration of Cold Water," *J. American Water Works Association,* 80, 4, pp. 168–175.

HANSON, A. T. AND J. L. CLEASBY (1990), "The Effects of Temperature on Turbulent Flocculation: Fluid Dynamics and Chemistry," *J. American Water Works Association,* 82, 11, pp. 56–73.

HUISMAN, L., J. M. DE AZEVEDO NETTO, B. B. SUNDARESAN, J. N. LANOIX, AND E. H. HOFKES (1981), *Small Community Water Supplies: Technology of Small Water Supply Systems in Developing Countries,* Technical Paper No. 18, International Reference Center, The Netherlands.

KAWAMURA, S. (1973), "Coagulation Considerations," *J. American Water Works Association,* 65, 6, pp. 417–423.

KAWAMURA, S. (1991a), "Effectiveness of Natural Polyelectrolytes in Water Treatment," *J. American Water Works Association,* 83, 10, pp. 88–91.

KAWAMURA, S. (1991b), *Integrated Design of Water Treatment Facilities,* John Wiley & Sons, Toronto.

KUCZYNSKI, L. (1985), *Evaluation of the Flocculation Process in Granular Media,* M.A.Sc. Thesis, University of Ottawa, Ottawa, Canada.

LETTERMAN, R. D. AND C. T. DRISCOLL (1988), "Survey of Residual Aluminum in Filtered Water," *J. American Water Works Association,* 80, 4, pp. 154–158.

LETTERMAN, R. D., J. E. QUON, AND R. S. GEMMELL (1973), "Influence of Rapid-Mix Parameters on Flocculation," *J. American Water Works Association,* 65, 11, pp. 716–722.

MILLER, R. G., F. C. KOPFLER, K. C. KELTY, J. A. STOBER, AND N. S. ULMER (1984), "The Occurrence of Aluminum in Drinking Water," *J. American Water Works Association,* 76, 1, pp. 84–91.

MONK, R. D. G. AND R. R. TRUSSELL (1991), "Design of Mixers for Water Treatment Plants: Rapid Mixing and Flocculators," in *Mixing in Coagulation and Flocculation,* American Water Works Association, Denver, CO, pp. 380–419.

REYNOLDS, T. D. (1982), *Unit Operations and Processes in Environmental Engineering,* Wadsworth, Belmont, CA.

RICHTER, C. A. (1977), *Water Treatment Plant for Small Communities,* Report for SANEPAR, Curitiba, Brazil.

RICHTER, C. A. (1986), Personal communication, SANEPAR, Curitiba, Brazil.

RUSHTON, J. H. (1952), "Mixing of Liquids in Chemical Processing," *Industrial and Engineering Chemistry,* 44, 12, pp. 2931–2936.

RUSHTON, J. H. AND J. Y. OLDSHUE (1953), "Mixing—Present Theory and Practice," *Chemical Engineering Progress,* 49, 4, pp. 161–168.

SHAMES, I. H. (1982), *Mechanics of Fluids,* McGraw-Hill, Toronto.

SCHULZ, C. R. AND D. A. OKUN (1984), *Surface Water Treatment for Communities in Developing Countries,* John Wiley & Sons, Toronto.

SINGER, P. C. (1990), "Assessing Ozone Research Needs in Water Treatment," *J. American Water Works Association,* 82, 10, pp. 78–88.

SMOLUCHOWSKI, M. v. (1916), "Drei Vortäge über Diffusion, Brownsche Molekularbewegung und Koagulation von Kolloidteilchen (Three Lectures on Diffusion, Brownian Motion, and Coagulation of Colloidal Particles)," *Zeitschrift fuer Physik,* 17, pp. 585–599.

STUMM, W., H. HUPER, AND R. L. CHAMPLIN (1967), "Formulation of Polysilicates as Determined by Coagulation Effects," *Environmental Science and Technology,* 1, 3, pp. 221–227.

UHL, V. W. AND J. B. GRAY (1966), *Mixing Theory and Practice,* vol. 1, Academic Press, New York.

CHAPTER 14

FILTRATION

Sand filtration was thought for some time to be the treatment for rendering seawater drinkable (Baker, 1981). Although this theory was debunked, filtration is the bulwark of water treatment. This is reflected by water treatment plants being commonly called filtration plants or simply "the filters." Filtration, especially when joined with chemical coagulation, produces clear water very low in turbidity. Significant removal of bacteria and other microbes also occurs in filtration. Craun (1988) concludes that in all but exceptional situations, effective filtration of surface waters must be provided to minimize waterborne disease outbreak. Another application of filtration in water treatment is preliminary treatment of a raw water with high suspended solids content. Filters with very coarse media, known as roughing filters, are used. Filtration is also used for polishing wastewaters, particularly effluents from stabilization pond systems.

Removal in a filter is accomplished by a number of mechanisms (Tchobanoglous and Eliassen, 1970). Straining, sedimentation, flocculation, and nine other chemical and physical mechanisms have been identified; some are indicated in Fig. 14.1. It is generally accepted that under the conditions of water filtration the dominant mechanisms are diffusion and sedimentation (Amirtharajah, 1988). Biological growth in a filter can significantly affect its performance and influence the predominant removal mechanisms.

Sand is the most common medium; however, other media such as crushed anthracite (hard coals), crushed magnetite, and garnet, besides inert synthetic media are used. The medium size and the pore openings to which it gives rise are important characteristics influencing removal. These characteristics also determine to a large degree the hydraulic performance of the filter.

Removal in a filter is highly dependent on the surface area of the media particles. The surface area of media available in a given volume of filter is large. Consider a 1 m^3 volume of filter with a medium that has a typical porosity of 0.40. Sand particles in filters often have a nominal diameter near 0.50 mm. Using this information and assuming that the medium particles are spheres, the number of particles per cubic meter of filter is 9.17×10^9 and the gross surface area of the particles is 7.20×10^3 m^2/m^3. The effective surface area is less than this value because particles are shielded by each other. Even assuming that only 1% of the surface area is effective yields a substantial increase in the available surface area compared to the same volume devoid of media.

14.1 SLOW SAND FILTERS AND RAPID FILTERS

The first filtration operations (dating from 1829 in England) were designed simply to pass water through a bed of sand without any chemical or mechanical assists to the process. The process is similar to withdrawing water from an infiltration gallery placed in the sand bed of a river. The flow rates per unit surface area in these filters are low

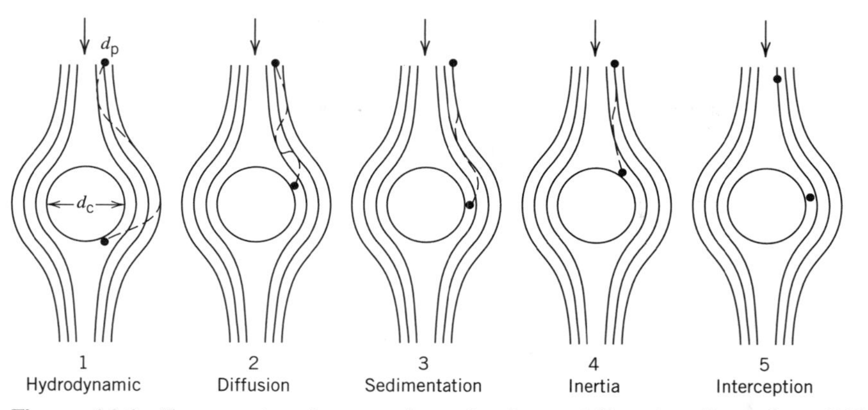

Figure 14.1 Transport and removal mechanisms of filtration. From Ives (1982).

compared to later developments, and they have become known as "slow sand filters." Many cities in Europe incorporated filtration designed along these lines into their water treatment processes and they are still in use today. Slow sand filters are still designed — they are a viable alternative to more recent rapid filters in many circumstances.

Slow sand filters go through a ripening phase of a few weeks after their startup. During this phase a dense microbial zoogleal or gelatinous growth establishes itself in the upper layers of the filter. It is in this layer that most of the removal of suspended and colloidal particles occurs. After a period of time, headloss increases to the cutoff point and a small layer of medium is scraped off the top of the filter. The biological growth extends below the layer that was removed and filter performance is not impaired. This cycle is repeated until a minimum depth of medium remains in the filter. At this time the discarded medium is washed and returned to the filter.

Rapid filters (Fig. 14.2) were conceived in North America as an alternative to slow sand filters. Slow sand filters concentrate removals in the upper layers of the filter, and the rapid filter was designed to utilize the entire depth of a filter bed more fully to attain a higher throughput of water for a given surface area. A higher loading rate produces more rapid headloss development. This and the deeper penetration of solids into the bed limits the only feasible means of rejuvenating the filter media to backwashing the filter. During backwash, water is forced through the filter in the upward direction at a velocity sufficient to expand the media. Cleansing occurs by scour caused by hydraulic shear forces on the media and by abrasive scour resulting from particles rubbing against each other. The former is the more important cleansing mechanism (Amirtharajah, 1978a).

As the expanded media settles after backwashing is terminated, larger particles tend to settle towards the bottom of the filter. Larger void spaces are associated with larger particles of media and the filter becomes a reverse graded sieve, with smaller openings at the top and larger openings at the bottom. The fine media particles will accumulate at the top of the media resulting in clogging at the top layer and little use of the whole filter depth. This leads to a requirement for more uniform media in a rapid filter.

The higher throughput of water and the lack of significant biological growth necessitates the addition of chemical coagulating agents to the influent to a rapid filter.

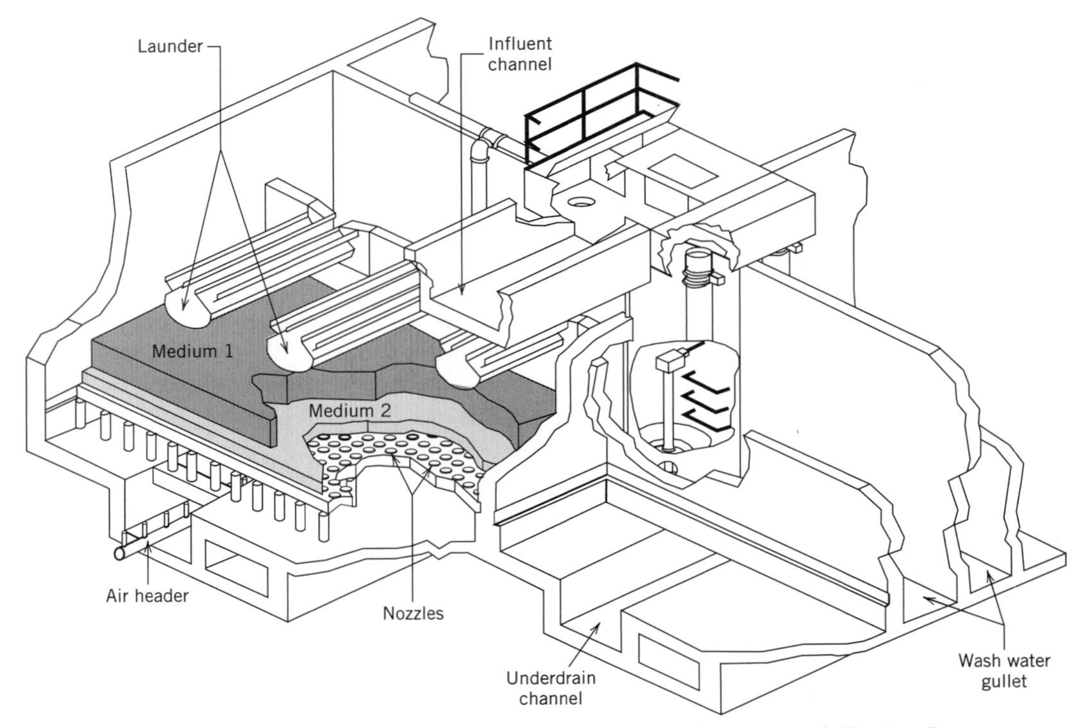

Figure 14.2 Typical rapid multimedia filter waterway. Courtesy of Eimco Process Equipment.

The above characteristics and other features of slow and rapid sand filters are compared in Table 14.1.

14.2 FILTERING MATERIALS

The porosity, e, of the filter depends on how well the particles fit together. As the particles become less spherical, the porosity of a given volume increases. Also, as particles become less spherical, their surface area increases, which has a beneficial effect on removal mechanisms that depend on surface area. The measure of shape is sphericity, ψ, defined as the ratio of the surface area of the equivalent volume sphere to the actual surface area of the particle.

$$\psi = \frac{(\text{surface area of a sphere})/V_{\text{sphere}}}{(\text{surface area of a particle})/V_{\text{particle}}} \qquad V_{\text{sphere}} = V_{\text{particle}}$$

The volume (V_s) and surface area (A_s) of a sphere are calculated from

$$V_s = \frac{\pi d^3}{6} \qquad A_s = \pi d^2$$

where

d is the diameter of the sphere

TABLE 14.1 General Features of Construction and Operation of Slow and Rapid Sand Filters[a]

Item	Slow sand filters[b]	Rapid sand filters[b]
Rate of filtration	1 to 4 to 8 $m^3/m^2/d$ (25 to 100 to 200 $gal/ft^2/d$)	100–475 $m^3/m^2/d$ (2 500–11 650 $gal/ft^2/d$)
Depth of bed	0.3 m (1 ft) of gravel; 1.0–1.5 m (3.3–5 ft) of sand	0.5 m (1.6 ft) of gravel; 0.75 m (2.5 ft) of sand
Size of sand	Effective size: 0.15 to 0.3 to 0.35 mm; Uniformity coefficient: 2 to 2.5 to 3 (unstratified)	Effective size: 0.45 mm and higher; Uniformity coefficient: 1.5 and lower (stratified)
Length of run	20 to 30 to 120 d	12 to 24 to 72 h
Penetration of suspended matter	Superficial (only the top layer is cleaned)	Deep (whole bed is washed)
Preparatory treatment of water	Generally aeration, but flocculation and sedimentation can be included.	Flocculation and sedimentation are essential
Method of cleaning	(1) Scraping off surface layer of sand and washing removed sand (2) Washing surface sand in place by traveling washer	Scour by mechanical rakes, air or water and removal of dislodged material by upward backwash flow
Costs		
Construction	Higher	Lower
Operation	Lower	Higher
Depreciation	Lower	Higher
Amount of wash water	0.2–0.6% of water filtered	1–6% of the water filtered

[a]Adapted in part from G. M. Fair, J. C. Geyer, and D. A. Okun (1968), *Water Purification and Wastewater Treatment and Disposal,* vol. 2, copyright © 1968 by John Wiley & Sons, Inc. Reprinted by permission of John Wiley & Sons, Inc.
[b]Average values are underlined.

Denoting the surface area and volume of an irregularly shaped particle as A_p and V_p, respectively, ψ is

$$\psi = \frac{6\pi d^2}{\pi d^3}\frac{V_p}{A_p} \quad (V_p = V_s) \qquad \text{and} \qquad \frac{A_p}{V_p} = \frac{6}{\psi d} \tag{14.1}$$

(Many authors define $A_p/V_p = S$, which appears throughout their equations). The surface area of a sphere is smaller than for any other particle geometry; therefore ψ is always less than 1.

Table 14.2 gives a classification of media shapes and porosities. Media commonly used for filtration are shown in Table 14.3 along with their properties.

14.2.1 Grain Size and Distribution

The effective sizes (defined in Eq. 14.2) of available media may be too coarse or too fine and they may not be of the required uniformity. Grain size distribution in the medium is determined from a sieve analysis. Size openings of the United States sieve

TABLE 14.2 Particle Sphericity and Porosity

Description	Sphericity (ψ)	Typical porosity (e)
Spherical	1.00	0.38
Rounded	0.98	0.38
Worn	0.94	0.39
Sharp	0.81	0.40
Angular	0.78	0.43
Crushed	0.70	0.48

series are given in Table 14.4. The cumulative percent weight of medium passing a given sieve size is plotted on either log-normal or arithmetic probability paper, whichever gives the best straight line relation. The mean size and standard deviation of the medium can be read or calculated from the 50th and 16th percentile values, respectively.

The size–frequency parameters that are used to characterize the media for filtration are the effective size, ES, which is the 10th percentile value, P_{10}, and the uniformity coefficient, U, which is the ratio of the P_{60} to P_{10} value.

$$\text{Effective size} = \text{ES} = P_{10} \tag{14.2}$$

$$\text{Uniformity coefficient} = U = P_{60}/P_{10} \tag{14.3}$$

where

P represents the percent by weight equal to or less than the size

Note that as the uniformity coefficient increases, the medium is less uniform.

The 10th percentile value is chosen for the ES because it has been found by Hazen (1892) that the hydraulic resistance of sand beds is relatively unaffected by size variation (up to a uniformity coefficient of 5.0) as long as the 10th percentile sand size remains unchanged. The uniformity coefficient describes 50% of the sand relative to the ES.

The d_{10}, d_{60}, and d_{90} values are used in various equations describing filter behavior. These are the diameters of the 10th, 60th, and 90th percentile sand sizes, respectively. If the media sizes are assumed to have a log-normal distribution, the following equation describes the relation between the d_{90} and d_{10} sizes:

$$d_{90} = d_{10}U^{1.67} \tag{14.4}$$

TABLE 14.3 Filter Media Characteristics

Material	Shape	Sphericity	Relative density	Porosity %	Effective size mm
Silica sand	Rounded	0.82	2.65	42	0.4–1.0
Silica sand	Angular	0.73	2.65	53	0.4–1.0
Ottawa sand	Spherical	0.95	2.65	40	0.4–1.0
Silica gravel	Rounded		2.65	40	1.0–50
Garnet			3.1–4.3		0.2–0.4
Crushed anthracite	Angular	0.72	1.50–1.75	55	0.4–1.4
Plastic		Any characteristics of choice			

TABLE 14.4 United States Standard Sieve Size Openings

Sieve designation number[a]	Size of opening mm	Sieve designation number[a]	Size of opening mm
200	0.074	30	0.59
140	0.105	25	0.71
100	0.149	20	0.84
80	0.178	18	1.00
70	0.210	16	1.19
60	0.249	14	1.41
50	0.297	12	1.68
45	0.350	8	2.36
40	0.419	6	3.36
35	0.500	4	4.76

[a]Approximately the number of meshes per inch.

The sand or other media available may not meet ES and uniformity coefficient specifications and so will have to be tailored to requirements. A uniformity coefficient below 1.3 is generally not achievable by manufacturers and a value of 1.5 will have a cost premium associated with it. For less expensive, less uniform media, the fines may be washed out and the coarse particles can be screened out. An outline of the procedure (Fair, et al., 1968) to determine the sizes above and below which media should be discarded, is given next.

From a medium that does not meet size specifications there is a usable portion, P_{use}, a portion that is too fine, P_f, and a portion that is too coarse, P_c. Sand is used as the stock medium here.

Therefore,

$$P_{use} + P_f + P_c = 100 \tag{14.5}$$

All of the sand that lies between the specified sizes of the 10th percentile value and the 60th percentile value is usable.

$$d_{60} = Ud_{10}$$

For the stock sand, P_{st10} and P_{st60} are defined as the percentages of stock sand that are less than the specified P_{10} and P_{60} sizes, respectively.

The amount of sand that lies between the P_{10} and P_{60} sizes comprises 50% of the specified sand. The total usable sand is

$$P_{use} = 2(P_{st60} - P_{st10}) \tag{14.6}$$

Ten percent of the usable sand can be below the specified P_{10} size. For the portion of the stock sand that is smaller than the specified P_{10} size, it would be most desirable to remove the smaller sizes and retain the larger sizes in the amounts required. The percentage of usable stock sand below the P_{10} size is equal to $0.1P_{use}$. The percentage of stock sand that is too fine is

$$P_f = P_{st10} - 0.1P_{use} = P_{st10} - 0.2(P_{st60} - P_{st10}) \tag{14.7}$$

The percentage of stock sand that is too coarse is the remaining portion determined from Eq. (14.5).

$$\begin{aligned} P_c &= 100 - P_f - P_{use} \\ &= 100 - P_{st10} + 0.2(P_{st60} - P_{st10}) - 2(P_{st60} - P_{st10}) \\ &= 100 - P_{st10} - 1.8(P_{st60} - P_{st10}) \end{aligned} \tag{14.8}$$

Again it is desirable to retain sand as close as possible to the P_{st60} size. The coarse portion P_c, to be removed, will consist of the largest grain sizes.

■ Example 14.1 Determination of Usable Sand from a Stock Sand

Figure 14.3 and the following table give the size distribution by weight of a local sand that has an effective size of 0.031 cm and a uniformity coefficient of 2.3. A log-normal distribution satisfactorily describes the medium's size variation as observed from the plot. The filter sand specifications are an ES (d_{10}) of 0.050 cm and a uniformity coefficient of 1.4.

The d_{60} size is

$$d_{60} = Ud_{10} = 1.4(0.05 \text{ cm}) = 0.70 \text{ cm}$$

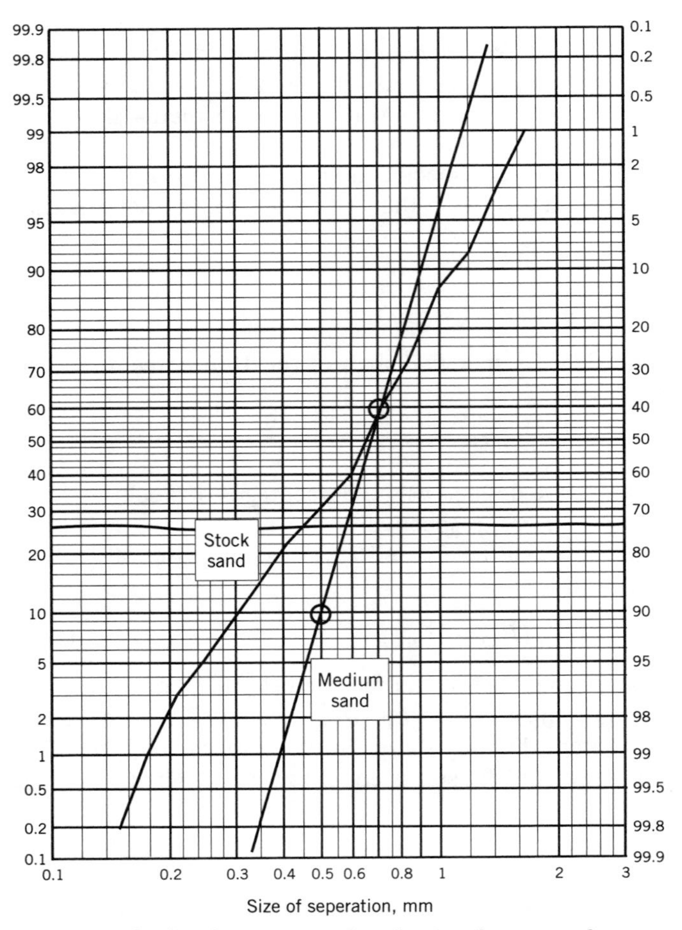

Figure 14.3 Sand grain size distribution for example.

Size of opening mm	Cumulative weight %	Size of opening mm	Cumulative weight %
0.149	0.2	0.59	40
0.178	1.0	0.71	60
0.210	3.0	0.84	72
0.249	5.1	1.00	85
0.297	8.9	1.19	92
0.350	15	1.41	97
0.419	22	1.68	99
0.500	30		

From Eq. (14.6), the proportion of usable stock sand is

$$2(60\% - 30\%) = 60\%$$

The percentage of sand that is too fine from Eq. (14.7) is

$$30\% - 0.2(60\% - 30.0\%) = 24\%$$

Therefore, it is desired to remove the smallest 24% of the stock sand, which is all sand below the 0.044 cm size. From Eq. (14.8) the percentage of the local sand that is too coarse is

$$100\% - 24\% - 60\% = 16\%$$

The largest 16% of stock sand or sand with a diameter above 0.085 cm is to be discarded.

Media that are too coarse must be removed by sieving. Fines may be removed from a medium by passing water in an upflow direction through the medium. The water velocity required to transport a particle will be just beyond the settling velocity of the particle. A particle moving upward at a constant velocity, v_u, will not be accelerating and the drag force on the particle will depend on the relative velocity of the fluid with respect to the particle. The force balance is as given by Eq. (11.8) using the relative velocity in the drag force term.

$$(\rho_p - \rho)gV_p = \tfrac{1}{2}\rho C_D A_p v_r^2$$

where

v_r is the relative velocity of the fluid with respect to the particle. $v_r = v_f - v_u$ (v_f is the fluid velocity).

Other terms are as defined for Eq. (11.8).

This equation is solved using the drag coefficient relations, Eqs. (11.12a)–(11.12c); however, the Reynold's number should be calculated with the relative velocity of the particle with respect to the fluid.

14.3 HEADLOSS IN FILTERS

Headloss in a filter is a complex function of flow rate, pressure, influent suspended solids concentration, and characteristics of the suspended solids and filter media.

It continuously varies with time and position in the bed. A classical equation was developed by Carman (1937) to describe overall headloss in porous media. The derivation can be started from the Darcy–Weisbach equation for pressurized flow in a closed conduit because flow through a filter is pressurized flow.

$$h_{\mathrm{L}} = f\frac{L}{D}\frac{v^2}{2g} \tag{14.9}$$

where

h_{L} = headloss
L = length of travel (bed depth)
f = friction factor
D = pipe diameter
v = velocity of flow

Laminar flow conditions are assumed. The problem is to adapt this equation to a filter (porous media), incorporating relevant filter characteristics that are easily determined.

1a. The diameter, D, is replaced by the hydraulic radius, R, of a noncircular section.

$$R = \frac{\text{area}}{\text{wetted perimeter}}$$

For a circular section,

$$R = \frac{\pi D^2}{4\pi D} = \frac{D}{4}$$

Substituting the hydraulic radius into the equation results in:

$$h_{\mathrm{L}} = f\frac{Lv^2}{8gR}$$

1b. The hydraulic radius of a filter is not well defined but a reasonable working definition of the hydraulic radius can be obtained by considering the definition of the hydraulic radius. The hydraulic radius relates the area of flow to the wetted perimeter providing resistance to flow. If the numerator and denominator are multiplied by suitable length parameters, dimensionality of the hydraulic radius is maintained.

$$R = \frac{\text{area}}{\text{wetted perimeter}} \approx \frac{\text{volume (available for flow)}}{\text{total surface area of the particles}}$$

The hydraulic radius can now be related to media characteristics. Defining the number of particles as N and the volume and surface area of a single particle as V_{p} and A_{p}, respectively, the total volume, V_{Tp} and total surface area, A_{Tp} of the particles are

$$V_{\mathrm{Tp}} = NV_{\mathrm{p}} \qquad A_{\mathrm{Tp}} = NA_{\mathrm{p}}$$

The volume available for flow can be calculated using the porosity, e, of the filter bed.

$$e = \frac{\text{void volume}}{\text{filter volume}} \qquad 1 - e = \frac{\text{volume of particles}}{\text{filter volume}}$$

Using the above two relations,

$$V = \frac{NV_p}{1-e} \quad (14.10a) \qquad V_v = \frac{eNV_p}{1-e} \quad (14.10b)$$

where
V is the volume of the filter bed
V_v is the void volume

Substituting the appropriate relations into the modified definition of the hydraulic radius,

$$R = \frac{\dfrac{eNV_p}{1-e}}{NA_p} = \left(\frac{e}{1-e}\right)\frac{V_p}{A_p} \quad (14.11)$$

2. The volume and surface area of the particles are related by their sphericity. From Eq. (14.1),

For a spherical particle $\dfrac{V_p}{A_p} = \dfrac{d}{6}$ For an irregularly shaped particle $\dfrac{V_p}{A_p} = \psi\dfrac{d}{6}$

3. Substituting for the hydraulic radius and V_p/A_p the equation for headloss becomes

$$\frac{h_L}{L} = f\frac{3v^2(1-e)}{4g\psi de}$$

The velocity, v, in the above equation is the average velocity in the pores. The superficial velocity, v_s, or velocity related to the surface area of the filter is normally used in the final equation. The superficial velocity is also referred to as the surface loading rate.

$$v_s = \frac{Q}{A_s} \qquad \text{and} \qquad v_s = ev$$

where
Q is the volumetric flow rate
A_s is the surface area of the filter

The resulting equation, known as the Carman–Kozeny equation, is

$$\frac{h_L}{L} = f\frac{3(1-e)v_s^2}{4g\psi de^3} = f_f\left(\frac{1-e}{e^3}\right)\frac{v_s^2}{\psi dg} \quad (14.12)$$

where
the numerical constants are incorporated into the filter friction factor, f_f

The friction factor, f_f, is a function of the Reynold's number.

$$\text{Re} = \frac{\rho v_s d}{\mu} \quad (14.13a) \qquad \text{Re} = \frac{\rho v_s \psi d}{\mu} \quad (14.13b)$$

The definition used for Re may vary from author to author. Using the definition in Eq. (14.13b), a commonly used relation for f_f is (Ergun, 1952),

$$f_f = 150\frac{1-e}{\text{Re}} + k \quad (14.14)$$

where

k is a constant

Ergun originally reported the value of k to be 1.75 and this value has been commonly accepted. Substituting Eq. (14.14) into Eq. (14.12):

$$\frac{h_L}{L} = \frac{150\mu}{\rho g}\frac{(1-e)^2}{e^3}\frac{v_s}{(\psi d)^2} + k\frac{1-e}{e^3}\frac{v_s^2}{\psi d g} \tag{14.15}$$

The first term in Eq. (14.15) is due to losses under laminar flow conditions and the second term applies to losses caused by turbulent conditions. Camp (1964) found that laminar flow conditions applied up to a Re number (based on Eq. 14.13a) of 6. When Re is less than 6, the second term of Eq. (14.15) can be ignored.

Equation (14.15) applies to the clean bed. As time passes, the porosity in the filter decreases because of the accumulation of solids and, concomitantly, the friction factor increases because of the restricted paths available for flow. It is difficult to model the change in the rate of headloss as the run progresses. Practically, filters are operated until a terminal headloss of 1.5 to 2 m (5 to 6.5 ft) is reached.

Metcalf and Eddy (1991) present a semi-empirical approach based on the work of Tchobanoglous and Eliassen (1970) for modeling headloss development in wastewater filters.

14.3.1 Grain Size Distribution and Headloss

A more refined estimate of the initial headloss can be made by taking into account the size distribution of the medium. From a sieve analysis the percentage by weight of each size of particles will be known. The medium can be separated into fractions by weight, x_i, of particles of mean nominal diameter, d_i.

$$\Sigma x_i = 1.0$$

Assuming that the porosity is the same throughout the entire bed, the length, l_i, associated with particles of size d_i is

$$l_i = x_i L$$

The headloss in each depth l_i will vary because d_i changes; the friction factor (Eq. 14.14) also changes because of the variation in d_i.

$$h_{Li} = \left(\frac{1-e}{e^3}\right)\frac{v_s^2}{\psi g} f_{fi}\frac{l_i}{d_i}$$

where

h_{Li} is the headloss of the ith layer

$$h_L = \sum h_{Li} = \left(\frac{1-e}{e^3}\right)\frac{v_s^2}{\psi g} L \sum f_{fi}\frac{x_i}{d_i} \tag{14.16}$$

If a functional relation is developed between x_i and d_i and f_f is relatively constant:

$$h_L = \left(\frac{1-e}{e^3}\right)\frac{v_s^2}{\psi g} L f_{fi} \int_0^1 \frac{dx}{d} \tag{14.17}$$

In a multimedia filter the above equations are applied separately to each medium if porosity varies from one medium to another.

■ Example 14.2 Initial Headloss in a Dual-Media Filter

Calculate the initial headloss in a dual-media filter containing anthracite and Ottawa sand with depths of 0.45 and 0.30 m (1.5 and 1 ft), respectively. The effective size and uniformity coefficient of the anthracite are 0.85 mm and 1.5, respectively. The corresponding characteristics for the Ottawa sand are 0.55 mm and 1.35, respectively. The sphericities of the sand and anthracite are 0.95 and 0.72, respectively. Use a s.g. of 1.5 for anthracite and information in Table 14.3 for other media characteristics. Perform the calculation for a temperature of 10°C and surface velocity of 175 $m^3/m^2/d$ (4 290 $gal/ft^2/d$).

The P_{60} sizes for the media are

anthracite: $d_{60} = Ud_{10} = 1.5(0.85\text{ mm}) = 1.27\text{ mm}$

sand: $d_{60} = Ud_{10} = 1.35(0.55\text{ mm}) = 0.74\text{ mm}$

The media size distributions are obtained by plotting the P_{10} and P_{60} sizes for each medium on probability paper and drawing a straight line through them. The media size distribution data obtained from these plots are tabulated below.

Size Distribution of Media

Percentiles (by weight) of media	d_1 mm	d_2 mm	Mean size[a] mm
Anthracite			
5–20[b]	0.72	1.00	0.85
20–40	1.00	1.18	1.09
40–60	1.18	1.27	1.22
60–80	1.27	1.53	1.39
80–95[b]	1.53	1.81	1.66
Sand			
5–20[b]	0.51	0.61	0.56
20–40	0.61	0.68	0.64
40–60	0.68	0.74	0.71
60–80	0.74	0.82	0.74
80–95[b]	0.82	0.93	0.87

[a]The mean size is the geometric mean size because a probability plot is used. $d = \sqrt{d_1 d_2}$.

[b]The 5th and 95th percentile sizes were chosen to represent the extreme sizes.

The headloss calculations are performed by calculating f_{fi} (Eq. 14.14) and then calculating the term after the summation sign in Eq. (14.16) to each layer. The results are shown for each layer in the table of media sizes. Then Eq. (14.16) is applied to each medium.

Headloss Calculations

Percentiles (by weight) of media	Mean size mm	Re[a]	f_{ti}	x_i/d_i[b] mm^{-1}	$f_{ti}\frac{x_i}{d_i}$ mm^{-1}
Anthracite					
5–20	0.85	1.02	67.6	0.236	15.9
20–40	1.09	1.31	53.2	0.184	9.8
40–60	1.22	1.48	47.4	0.163	7.7
60–80	1.39	1.68	41.8	0.143	6.0
80–95	1.66	2.01	35.3	0.120	4.2
				Total	43.7
Sand					
5–20	0.56	0.82	111.5	0.359	40.0
20–40	0.64	0.95	96.8	0.311	30.0
40–60	0.71	1.04	88.0	0.282	24.8
60–80	0.74	1.15	80.3	0.257	20.6
80–95	0.87	1.28	71.8	0.229	16.4
				Total	131.9

[a]The porosities of anthracite and sand are 0.55 and 0.40, respectively, from Table 14.3.

[b]The fractions, x_i, were taken to be 0.20 for each mean size.

$$h_L = h_{La} + h_{Ls} = \left(\frac{1-e_a}{e_a^3}\right)\frac{v_s^2}{\psi_a g}L_a\sum\left(f_{fai}\frac{x_{ai}}{d_{ai}}\right) + \left(\frac{1-e_s}{e_s^3}\right)\frac{v_s^2}{\psi_s g}L_s\sum\left(f_{fsi}\frac{x_{si}}{d_{si}}\right)$$

$$h_L = \frac{(1-0.55)\left[175\,\frac{\text{m}}{\text{d}}\left(\frac{1\text{ d}}{86\,400\text{ s}}\right)\right]^2(0.45\text{ m})}{(0.55)^3(0.72)(9.81\text{ m/s}^2)}(43.7\times10^3\text{ m}^{-1})$$

$$+\frac{(1-0.40)\left[175\,\frac{\text{m}}{\text{d}}\left(\frac{1\text{ d}}{86\,400\text{ s}}\right)\right]^2(0.30\text{ m})}{(0.40)^3(0.95)(9.81\text{ m/s}^2)}(131.9\times10^3\text{ m}^{-1})$$

$$= 0.032\text{ m} + 0.163\text{ m} = 0.195\text{ m}$$

In U.S. units:

$$h_L = \frac{(1-0.55)\left[4\,290\frac{\text{gal}}{\text{ft}^2\text{-d}}\left(\frac{1\text{ ft}^3}{7.48\text{ gal}}\right)\left(\frac{1\text{ d}}{86\,400\text{ s}}\right)\right]^2(1.5\text{ ft})}{(0.55)^3(0.72)(32.2\text{ ft/s}^2)}(43.7\times10^3\text{ m}^{-1})\left(\frac{1\text{ m}}{3.28\text{ ft}}\right)$$

$$+\frac{(1-0.40)\left[4\,290\frac{\text{gal}}{\text{ft}^2\text{-d}}\left(\frac{1\text{ ft}^3}{7.48\text{ gal}}\right)\left(\frac{1\text{ d}}{86\,400\text{ s}}\right)\right]^2(1\text{ ft})}{(0.40)^3(0.95)(32.2\text{ ft/s}^2)}(131.9\times10^3\text{ m}^{-1})\left(\frac{1\text{ m}}{3.28\text{ ft}}\right)$$

$$= 0.102\text{ ft} + 0.543\text{ ft} = 0.645\text{ ft}$$

14.4 BACKWASHING FILTERS

When headloss through the filter reaches a set value, a rapid filter is backwashed to remove the accumulated solid matter (Fig. 14.4). At many plants water sent to a filter

Figure 14.4 Formation and removal of clouds of particulate debris during the initial stage of backwash.

is either prechlorinated or has had another disinfection agent added to minimize microbial growth and associated headloss development. Filtered water is used for backwashing, which consumes from 1 to 5% of the product water. The backwash water is supplied to the filter through a system of underdrains (Section 14.5). As the rate of backwashing increases, the frictional dissipation of energy increases with the velocity squared according to the Carman-Kozeny equation. The media remains at rest until the drag forces of the water on the media are sufficient to suspend the particles. When the particles are no longer resting on each other but are supported by the drag forces from the water, the bed is fluidized. After this point further increases in the backwash velocity will cause an increase in the local water velocity and the media particles will further expand or separate until the drag force (proportional to the square of the local velocity) decreases to a value sufficient to support the particles.

As backwash velocity is further increased, either the expansion becomes sufficient to cause media washout or the particles acquire an upward velocity and media washout occurs. Bed expansions from 15 to 35% are typically achieved in backwashing (Cleasby, 1990).

14.4.1 Total Head Requirements for Backwashing

A portion of the filtered water may be pumped to a head tank to supply the water for backwashing or water may be simply pumped from the supply reservoir. The backwash water is input through the collection underdrain system of the filter. The sediment-laden wash water is collected in launders (Fig. 14.2) and wasted. The collection launders are designed in the same manner as collection launders in sedimentation basins (see Section 11.6).

Figure 14.5 is a schematic of a bed during backwashing. The total head required equals the depth of water in the filter plus losses through the system. The losses can be individually quantified. Losses in the supply pipes can be computed using the Darcy–Weisbach or equivalent formulations. Losses in the underdrains are estimated

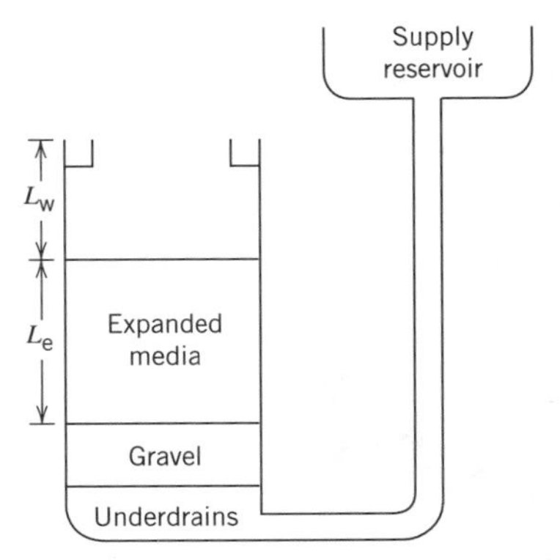

Figure 14.5 Filter arrangement for backwashing.

using pipe flow and orifice equations. The gravel support layers do not expand during backwashing and the Ergun equation (Eq. 14.15) is suitable to calculate headlosses through these layers.

Losses in the Expanded Media

For a single-medium bed the drag force, F_D, on the particles must equal their effective weight. The effective weight of the particles per unit volume of expanded medium is

$$\frac{\text{effective weight of media}}{\text{volume of media}} = (\rho_s - \rho)g(1 - e_e)$$

where
e_e is the porosity of the expanded medium
ρ_s is the density of the medium particles

Equating this to the drag force per volume of expanded medium,

$$(\rho_s - \rho)g(1 - e_e) = \frac{F_D}{A_s L_e}$$

where
L_e is the depth of the expanded media

Now each side of this equation may be multiplied by L_e and equated to the headloss through the expanded medium.

$$(\rho_s - \rho)(1 - e_e) g L_e = \frac{F_D}{A_s} = \rho g h_{Le} \tag{14.18}$$

where
h_{Le} is the headloss of the expanded medium

The effective weight of the particles at rest is the same as the effective weight of the particles when they are suspended.

$$(\rho_s - \rho)g(1 - e_e)L_e = (\rho_s - \rho)g(1 - e)L \tag{14.19}$$

$$h_{Le} = \frac{(\rho_s - \rho)}{\rho}(1 - e_e)L_e = \frac{(\rho_s - \rho)}{\rho}(1 - e)L \tag{14.20}$$

From Eq. (14.20) the relation between the expanded bed depth and the depth of the bed at rest can be found.

$$\frac{L_e}{L} = \frac{1-e}{1-e_e} \tag{14.21}$$

14.4.2 Backwash Velocity

As shown in the previous section, headloss determination during backwash is based on sound physics principles and easily measured parameters. Bed expansion, on the other hand, is a complex function of particle sphericity and other medium and fluid characteristics. There are a number of approaches to calculating backwash velocity and the resulting expansion. Cleasby and co-workers have been instrumental in developing and evaluating the theory and empirical expressions.

The velocity required to initiate fluidization is the minimum fluidization velocity, v_{mf}. At the point of incipient fluidization the bed is still a fixed bed. The Ergun equation (Eq. 14.15) can be equated to the equation for headloss across the bed (Eq. 14.20) to arrive at an equation for v_{mf}.

$$\frac{h_{Le}}{L_e} = \frac{150\mu}{\rho g}\frac{(1-e)^2}{e^3}\frac{v_{mf}}{(\psi d)^2} + 1.75\frac{1-e}{e^3}\frac{v_{mf}^2}{\psi d g} = \frac{(\rho_s - \rho)}{\rho}(1-e) \tag{14.22a}$$

This equation can be simplified to

$$\frac{150\mu}{d^2}\left(\frac{1-e}{\psi^2 e^3}\right)v_{mf} + 1.75\left(\frac{1}{\psi e^3}\right)\frac{\rho}{d}v_{mf}^2 = (\rho_s - \rho)g \tag{14.22b}$$

Equation (14.22b) can be rearranged into a quadratic and solved for v_{mf}, which is a function of the porosity, shape factor, and densities of the media and the fluid. Wen and Yu (1966) determined empirical correlations that described the two terms in parentheses on the left-hand side of Eq. (14.22b). The correlations were based on data for the porosities at minimum fluidization for particles with different sphericities.

$$\frac{1-e}{\psi^2 e^3} = 11 \quad (14.23a) \qquad \frac{1}{\psi e^3} = 14 \tag{14.23b}$$

These correlations facilitate the solution of Eq. (14.22b) and eliminate the necessity of accurately determining ψ. Making the substitutions for these two terms,

$$\frac{1\,650\mu}{d^2}v_{mf} + 24.5\frac{\rho}{d}v_{mf}^2 = (\rho_s - \rho)g$$

This equation can be rearranged in terms of a Reynold's number and another nondimensional parameter. The appropriate definition of the Reynold's number for minimum fluidization (Re_{mf}) is

$$\mathrm{Re}_{mf} = \frac{\rho v_{mf} d}{\mu} \tag{14.24}$$

Rearranging the equation,

$$24.5\left(\frac{\rho v_{mf} d}{\mu}\right)^2 + 1\,650\frac{\rho v_{mf} d}{\mu} - \frac{d^3\rho(\rho_s - \rho)g}{\mu^2} = 0$$

The last term on the left-hand side of the equation is the dimensionless Galileo number, Ga.

$$\mathrm{Ga} = \frac{d^3\rho(\rho_s - \rho)g}{\mu^2} \quad (14.25)$$

Substituting for the Reynold's and Galileo numbers and solving the quadratic:

$$\mathrm{Re}_{mf} = [(33.7)^2 + (0.040\,8)\mathrm{Ga}]^{1/2} - 33.7 \quad (14.26)$$

Cleasby and Fan (1981) report that this is a suitable relation to determine the minimum fluidization velocity and present data to support this contention. In a graded medium, the velocity to expand the particles will vary as a function of particle size. The d_{90} size of the largest medium in a multimedia filter is a suitable size to calculate the minimum fluidization velocity for the filter (Cleasby and Fan, 1981, 1982). The minimum fluidization velocity will not be sufficient to expand the coarse media in the bed. Cleasby and Fan (1981) recommend a backwash rate equal to $1.3v_{mf}$ to ensure adequate movement of the grains.

In North American practice, typical backwash velocities expand the media 15–35% as noted in Section 14.4. Equation (14.26) does not provide an estimation of the expansion as accurately as the techniques developed next.

Previous attempts to relate bed expansion and backwash velocity equations were based on the Richardson–Zaki equation (Eq. 14.27; Richardson and Zaki, 1954) which these researchers found to empirically correlate data for two-phase flow systems.

$$e_e = \left(\frac{v_b}{v_i}\right)^n \quad (14.27)$$

where

v_b is the superficial backwash velocity
v_i is an intercept velocity
n depends on the medium

The intercept velocity is found by plotting e_e versus v_b on a log–log plot and extrapolating the line to $e_e = 1$, where v_i is read. The slope of the line is n. Although n has been commonly reported as having a value near 0.22 for sand, Amirtharajah (1978b) reports that an n value in the range of 0.40–0.50 is more correct for sand in a filter.

The settling velocity of a representative medium particle was initially used as the intercept velocity. This does not provide a good estimate of the expansion as discussed by Cleasby and Fan (1981) and Dharmarajah and Cleasby (1986). Cleasby and Fan (1981) developed relations based on the Richardson–Zaki equation that provided reasonable estimations of the amount of expansion in filters in many cases; however, there remained significant deviations from the predictions of the equations. They determined correlations to find n and v_i using a fairly involved procedure. Later, the work of Dharmarajah and Cleasby (1986) improved on the work of Cleasby and Fan (1981) to find a more universally accurate and easily applied correlation.

The correlation was based on a Reynold's number and voidage function for pressure drop correlation. The Reynold's number for a fluidized bed incorporates the interstitial velocity through the bed.

$$v_{bi} = v_b/e \quad (14.28)$$

where

v_{bi} is the interstitial velocity during backwash

The characteristic length for the Reynold's number is the hydraulic radius of the filter defined previously by Eq. (14.11). Therefore for an expanded bed, the Reynold's number (Re_f) definition is

$$Re_f = \frac{\rho\, v_b}{\mu\, e_e}\left(\frac{e_e}{(1-e_e)}\frac{V_p}{A_p}\right) = \frac{\rho v_b}{\mu(1-e_e)}\left(\frac{V_p}{A_p}\right) = \frac{\rho v_b \psi d}{6\mu(1-e_e)} \tag{14.29}$$

The pressure drop–voidage correlation was based on the Carman–Kozeny equation, Eq. (14.12) and the headloss equation (Eq. 14.20) for a fluidized bed. In the Carman–Kozeny equation, f_f is a nondimensional parameter that relates headloss to flow conditions and medium characteristics. Formulating Eq. (14.12) for f_f in terms of backwash velocity and expanded bed porosity:

$$f_f = \frac{h_L/L_e}{\dfrac{(1-e_e)}{e_e^3}\dfrac{v_b^2}{\psi dg}}$$

Now Eq. (14.20) is substituted for h_L/L_e.

$$f_f = \frac{\dfrac{(\rho_s-\rho)}{\rho}(1-e_e)}{\dfrac{(1-e_e)}{e_e^3}\dfrac{v_b^2}{\psi dg}} = \frac{(\rho_s-\rho)}{\rho}\frac{e_e^3\psi dg}{v_b^2} \tag{14.30}$$

Equations (14.29) and (14.30) were combined into a new nondimensional parameter, Θ, which is not a function of the backwash velocity, v_b.

$$\Theta = f_f \times Re_f^2 = \frac{(\rho_s-\rho)}{\rho}\frac{e_e^3\psi dg}{v_b^2} \times \left(\frac{\rho v_b \psi d}{6\mu(1-e_e)}\right)^2 = \frac{e_e^3}{(1-e_e)^2}\left[\frac{\rho(\rho_s-\rho)g(\psi d)^3}{36\mu^2}\right] \tag{14.31}$$

A stepwise regression was used to determine the correlation between Θ and Re_f (which is a function of v_b) and ψ.

For $Re_f < 0.2$,

$$\Theta = 18.1\ Re_f \tag{14.32a}$$

For $Re_f > 0.2$,

$$\log(\Theta/6) = 0.565\,43 + 1.093\,48 \log Re_f + 0.179\,79 (\log Re_f)^2 - 0.003\,92 (\log Re_f)^4 - 1.5 (\log \psi)^2 \tag{14.32b}$$

Excellent agreement was achieved with Eqs. (14.32a) and (14.32b) for a wide variety of media at different expanded porosities above approximately 10% expansion of the bed (Cleasby, 1995). Dharmarajah and Cleasby (1986) report that the correlations are good up to $e_e = 0.85$ and for media with a s.g. < 4.3, which is above the s.g. of all media commonly used in filters. The procedure of Dharmarajah and Cleasby cannot be used to back-calculate the minimum fluidization velocity for a medium because the correlation does not apply at expansions less than approximately 10%.

The procedure to apply Eq. (14.32b) is to select a backwash velocity, v_b. Then by trial and error find the value of e_e that satisfies Eq. (14.32b) when Θ is calculated with Eq. (14.31) and Re_f is calculated by Eq. (14.29). The fluid characteristics of density and viscosity are specified at the desired temperature. The sphericity characteristic of

TABLE 14.5 Fluidization Velocities for Uniformly Sized Media[a]

For U.S. standard sieves			Mean size	Flow rate to achieve 10% expansion at 25°C m/h (gal/min/ft²)		
Passing	mm	Retained	mm	Anthracite[a]	Sand[b]	Garnet[b]
7	2.830	8	2.59	90.4 (37.0)		
8	2.380	10	2.18	73.3 (30.0)		
10	2.000	12	1.84	58.7 (24.0)	100.2 (41.0)	
12	1.680	14	1.54	48.9 (20.0)	80.7 (33.0)	
14	1.410	16	1.30	38.4 (15.7)	66.0 (27.0)	119.8 (49.0)
16	1.190	18	1.09	30.6 (12.5)	51.3 (21.0)	97.8 (40.0)
18	1.000	20	0.92	24.2 (9.9)	40.1 (16.4)	78.2 (32.0)
20	0.841	25	0.78	20.5 (8.4)	30.8 (12.6)	66.0 (27.0)
25	0.707	30	0.65	17.1 (7.0)	22.0 (9.0)	53.8 (22.0)
30	0.595	35	0.55		15.4 (6.3)	44.0 (18.0)
35	0.500	40	0.46		13.2 (5.4)	33.5 (13.7)
40	0.420	45	0.38		9.8 (4.0)	27.6 (11.3)
50	0.297	60 (0.25 mm)	0.27			15.4 (6.3)

[a]From USEPA (1977).
[b]s.g.s of anthracite, sand, and garnet = 1.7, 2.65, and 4.1, respectively.

the medium is used. Cleasby (1990) uses the d_{50} diameter of the medium as the representative particle diameter. A simple computer program (in Basic) is given by Dharmarajah and Cleasby (1986) to avoid iterative manual solution of the problem. Once e_e is found, Eq. (14.21) can be used to find the depth of the expanded bed.

Empirical observed data on the velocities required to expand common media are given in Table 14.5. The results obtained from the preceding procedure should be near the appropriate fluidization velocity in Table 14.5 with some allowance for different degrees of expansion. Temperature correction factors to be applied to the fluidization velocities in Table 14.5 for temperatures other than 25°C are given in Table 14.6.

Headloss and Expansion in a Stratified Bed

For a stratified bed the expansions and headlosses of each layer are calculated in the following manner. Each layer has a length l_i when the bed is at rest.

TABLE 14.6 Temperature Correction Factors for Fluidization Velocities[a]

Temperature, °C	Correction factor[b]
30	1.09
25	1.00
20	0.91
15	0.83
10	0.75
5	0.68

[a]From USEPA (1977).
[b]Multiply the velocities in Table 14.5 by the correction factor.

Recall,

$$l_i = x_i L \qquad \Sigma l_i = L \qquad \Sigma l_{ei} = L_e$$

where

l_{ei} is the expanded length of the ith layer

The headloss through each layer from Eq. (14.18) is

$$h_{li} = \frac{(\rho_{si} - \rho)}{\rho}(1 - e_i)l_i \tag{14.33}$$

and

$$h_{Le} = \Sigma h_{li} \tag{14.34}$$

Each layer will expand to a different degree. Once v_b is established, the expanded length of each layer can be calculated from Eq. (14.33) after applying the procedure given in Section 14.4.2 to each layer. The particle diameters, sphericities, and densities must be known for each layer. For a single-medium filter, it is recommended to divide the bed into five zones of equal weight. For a multimedia filter, each medium should be divided into three to five zones of equal weight. Now,

$$1 - e_i = \frac{V_{Tpi}}{A_s l_i} \qquad \text{and} \qquad 1 - e_{ei} = \frac{V_{Tpi}}{A_s l_{ei}}$$

where

V_{Tpi} is the volume of particles in the ith layer

From these equations it can be determined that

$$(1 - e_i)l_i = (1 - e_{ei})l_{ei} \tag{14.35}$$

In multimedia beds, the effective sizes and uniformity coefficients of each medium will influence the degree of intermixing among the different types of media. As more intermixing occurs, the advantage of a stratified filter with porosities decreasing from the top to the bottom of the filter decreases. However, the medium with the lower density will be predominant in the upper layers of the bed and the medium with the highest density will be predominant in the lower layers of the bed. There is disagreement on how much intermixing should be allowed. Pilot studies are definitely in order to assess optimal media characteristics.

■ Example 14.3 Backwash Velocity and Bed Expansion

(a) Find the minimum fluidization velocity at 10°C for the rapid filter described in Example 14.2.

(b) Also find the backwash velocity to achieve an average bed expansion of 10% at 10°C.

Solution to Part (a)

At 10°C:

$\rho = 999.7\ \text{kg/m}^3 \qquad \mu = 1.307 \times 10^{-3}\ \text{N-s/m}^2 = 1.307 \times 10^{-3}\ \text{kg/m-s}$

$\rho = 1.94\ \text{slug/ft}^3 \qquad \mu = 2.735 \times 10^{-5}\ \text{lb-s/ft}^2$

Equation (14.26) is used to find v_{mf}.

The recommended particle size to be used in Eq. (14.24) is the d_{90} size of the largest medium. From Eq. (14.4) for the anthracite,

$$d_{90} = 0.85 \text{ mm}[10^{1.67(\log 1.5)}] = 1.67 \text{ mm}$$

The Reynold's number for minimum fluidization is

$$\text{Re}_{mf} = \frac{\rho d_{90}}{\mu} v_{mf} = \frac{\left(999.7 \dfrac{\text{kg}}{\text{m}^3}\right)(1.67 \text{ mm})\left(\dfrac{1 \text{ m}}{1\,000 \text{ mm}}\right)}{1.307 \times 10^{-3} \dfrac{\text{kg}}{\text{m-s}}} v_{mf} = 1\,277\, v_{mf}$$

where

v_{mf} is in m/s

$$\text{In U.S. units: Re}_{mf} = \frac{\left(1.94 \dfrac{\text{slug}}{\text{ft}^3}\right)(1.67 \text{ mm})\left(\dfrac{1 \text{ in.}}{25.4 \text{ mm}}\right)\left(\dfrac{1 \text{ ft}}{12 \text{ in.}}\right)\left(\dfrac{1 \text{ lb-s}^2/\text{ft}}{\text{slug}}\right)}{2.735 \times 10^{-5} \dfrac{\text{lb-s}}{\text{ft}^2}} v_{mf} = 389 v_{mf}$$

where

v_{mf} is in ft/s

The other parameter required in Eq. (14.26) is the Galileo number (Eq. 14.25).

$$\text{Ga} = \frac{d_{90}^3 \rho(\rho_s - \rho)g}{\mu^2}$$

$$= \frac{(1.67 \text{ mm})^3 \left(\dfrac{1 \text{ m}}{1\,000 \text{ mm}}\right)^3 \left(999.7 \dfrac{\text{kg}}{\text{m}^3}\right)(1\,500 - 999.7)\left(\dfrac{\text{kg}}{\text{m}^3}\right)\left(9.81 \dfrac{\text{m}}{\text{s}^2}\right)}{\left(1.307 \times 10^{-3} \dfrac{\text{kg}}{\text{m-s}}\right)^2}$$

$$= 13\,365$$

In U.S. units:

$$\text{Ga} = \frac{(1.67 \text{ mm})^3 \left(\dfrac{1 \text{ in.}}{25.4 \text{ mm}}\right)^3 \left(\dfrac{1 \text{ ft}}{12 \text{ in.}}\right)^3 \left(1.94 \dfrac{\text{slug}}{\text{ft}^3}\right) \times \left[(1.5 - 1.0)\left(1.94 \dfrac{\text{slug}}{\text{ft}^3}\right)\right]\left(32.2 \dfrac{\text{ft}}{\text{s}^2}\right)\left(\dfrac{1 \text{ lb-s}^2/\text{ft}}{\text{slug}}\right)^2}{\left(2.735 \times 10^{-5} \dfrac{\text{lb-s}}{\text{ft}^2}\right)^2}$$

$$= 13\,323$$

Substituting the values into Eq. (14.26):

$1\,277\, v_{mf} = [(33.7)^2 + 0.040\,8(13\,365)]^{1/2} - 33.7 = 7.3$

In U.S. units: $389\, v_{mf} = [(33.7)^2 + 0.040\,8(13\,323)]^{1/2} - 33.7 = 7.3$

$v_{mf} = 7.3/1\,277 = 5.72 \times 10^{-3}$ m/s = 20.6 m/h

In U.S. units: $v_{mf} = 7.3/389 = 0.018\,8$ ft/s = 67.6 ft/h

Using the recommended value of $1.3v_{mf}$, the backwash velocity required to achieve movement of all grains in the bed would be 26.8 m/h (87.8 ft/h).

Applying the same procedure to the sand medium, $d_{90} = 0.91$ mm, $\text{Re}_{mf} = 696v_{mf}$, Ga = 7 136, and

$$v_{mf} = 5.85 \times 10^{-3} \text{ m/s} = 21.1 \text{ m/h (69.2 ft/h)}$$

Applying the adjustment, $1.3v_{mf} = 27.4$ m/h (90.0 ft/h). Because this value is higher than the value calculated for the anthracite, it would be a better choice.

Solution to Part (b)

To achieve a bed expansion of 10%, the length of the expanded bed will be

$$L_e = 1.10L = 1.10\,(0.45 \text{ m} + 0.30 \text{ m}) = 0.825 \text{ m}$$

In U.S. units: $L_e = 1.10L = 1.10\,(1.50 \text{ ft} + 1.00 \text{ ft}) = 2.75$ ft

From Example 14.2 the porosities of the anthracite and sand at rest are 0.55 and 0.40, respectively. The expansions of the sand and anthracite layers will be calculated separately, with e_{se} and e_{ae} representing their expanded porosities, respectively. From the size distribution table in Example 14.2 the d_{50} sizes for the sand and anthracite are 0.71 and 1.22 mm, respectively.

Equations (14.31) and (14.32) must be solved by trial and error for the correct solution. Incorporating the media and fluid characteristics into the right-hand side of Eq. (14.31):

$$\Theta = \frac{e_e^3}{(1 - e_e)^2}\left[\frac{\rho(\rho_s - \rho)g(\psi d)^3}{36\mu^2}\right]$$

For the sand:

$$\Theta_s = \frac{e_{se}^3}{(1 - e_{se})^2} \times \left\{\frac{\left(999.7 \dfrac{\text{kg}}{\text{m}^3}\right)(2\,650 - 999.7)\left(\dfrac{\text{kg}}{\text{m}^3}\right)\left(9.81 \dfrac{\text{m}}{\text{s}^2}\right)[0.95(0.71 \text{ mm})]^3\left(\dfrac{1 \text{ m}}{10^3 \text{ mm}}\right)^3}{36\left(1.307 \times 10^{-3} \dfrac{\text{kg}}{\text{m-s}}\right)^2}\right\}$$

$$= 80.72 \frac{e_{se}^3}{(1 - e_{se})^2}$$

In U.S. units:

$$\Theta_s = \frac{e_{se}^3}{(1 - e_{se})^2}\left\{\frac{\left(1.94 \dfrac{\text{slug}}{\text{ft}^3}\right)(2.65 - 1.00)\left(1.94 \dfrac{\text{slug}}{\text{ft}^3}\right)\left(32.2 \dfrac{\text{ft}}{\text{s}^2}\right) \times [0.95(0.71 \text{ mm})]^3\left(\dfrac{1 \text{ ft}}{304.8 \text{ mm}}\right)^3\left(\dfrac{1 \text{ lb-s}^2/\text{ft}}{\text{slug}}\right)^2}{36\left(2.735 \times 10^{-5} \dfrac{\text{lb-s}}{\text{ft}}\right)^2}\right\}$$

$$= 80.47 \frac{e_{se}^3}{(1 - e_{se})^2}$$

The Reynold's number for the sand, Re_{fs}, is

$$Re_{fs} = \frac{\rho v_b \psi d_s}{6\mu(1-e_{se})} = \frac{\left(999.7\,\frac{kg}{m^3}\right)(0.95)(0.71\text{ mm})\left(\frac{1\text{ m}}{10^3\text{ mm}}\right)}{6\left(1.307\times10^{-3}\,\frac{kg}{m\text{-}s}\right)}\frac{v_b}{(1-e_{se})} = 86.0\frac{v_b}{(1-e_{se})}$$

In U.S. units:

$$Re_{fs} = \frac{\left(1.94\,\frac{slug}{ft^3}\right)(0.95)(0.71\text{ mm})\left(\frac{1\text{ ft}}{304.8\text{ mm}}\right)\left(\frac{1\text{ lb-s}^2/\text{ft}}{slug}\right)}{6\left(2.735\times10^{-5}\,\frac{lb\text{-}s}{ft^2}\right)}\frac{v_b}{(1-e_{se})} = 26.2\frac{v_b}{(1-e_{se})}$$

For the anthracite:

$$\Theta_a = \frac{e_{ae}^3}{(1-e_{ae})^2}\times \left\{\frac{\left(999.7\,\frac{kg}{m^3}\right)(1\,500-999.7)\left(\frac{kg}{m^3}\right)\left(9.81\,\frac{m}{s^2}\right)[0.72(1.22\text{ mm})]^3\left(\frac{1\text{ m}}{10^3\text{ mm}}\right)^3}{36\left(1.307\times10^{-3}\,\frac{kg}{m\text{-}s}\right)^2}\right\}$$

$$= 54.02\frac{e_{ae}^3}{(1-e_{ae})^2}$$

The Reynold's number for the anthracite, Re_{fa}, is

$$Re_{fa} = \frac{\left(999.7\,\frac{kg}{m^3}\right)(0.72)(1.22\text{ mm})\left(\frac{1\text{ m}}{10^3\text{ mm}}\right)}{6\left(1.307\times10^{-3}\,\frac{kg}{m\text{-}s}\right)}\frac{v_b}{(1-e_{ae})} = 112.0\frac{v_b}{(1-e_{ae})}$$

The expansions were first calculated for a minimum fluidization velocity of 20.6 m/h (67.6 ft/h). Then the backwash velocity was increased incrementally. A spreadsheet was used to facilitate the calculations. Once e_e is found that satisfies Eq. (14.32b), Eq. (14.35) is used to calculate the expanded depth. The results are shown in the following table.

Bed Expansion as a Function of Backwash Velocity

	Sand				Anthracite				
v_b m/h	e_e	Re_f[a]	Θ	l_e m	e_e	Re_f[a]	Θ	l_e m	L_e[b] m
20.6	0.424	0.85	18.6	0.31	0.501	1.28	27.3	0.41[c]	—
24	0.443	1.03	22.6	0.32	0.523	1.57	34.0	0.42[c]	—
28	0.464	1.25	28.1	0.34	0.548	1.93	43.6	0.45	0.79
32	0.484	1.48	34.4	0.35	0.571	2.32	54.7	0.47	0.82
36	0.501	1.72	40.8	0.36	0.592	2.75	67.4	0.50	0.86
38	0.510	1.85	44.6	0.37	0.601	2.96	73.7	0.51	0.88

[a]Because Re_f was always >0.2, only Eq. (14.32b) was used to calculate Θ.
[b]L_e is the sum of the expanded depths of the two layers.
[c]The expanded porosities are less than the rest porosity and the depth of the layer should be adjusted to its unexpanded depth.

A backwash velocity of 32 m/h (105 ft/h) will achieve an overall bed expansion of 10%. Note that the two layers are expanded to different degrees. The minimum fluidization velocity does not achieve any expansion of the anthracite layer; however, the backwash rate of 26.8 m/h ($1.3v_{mf}$ for the anthracite; 88 ft/h in U.S. units) is near the velocity predicted by Eqs. (14.31) and (14.32) to achieve expansion of this layer. On the other hand, the minimum fluidization velocity for the sand layer is near the minimum fluidization velocity calculated with Dharmarajah and Cleasby's approach. With the 1.3 adjustment factor (resulting in a velocity of 27.3 m/h), the expansion of the sand layer is about 10% which should be sufficient to expand the largest particles.

The results in the table are reasonably close to the observations recorded in Table 14.5. Bed fluidization is a complex hydrodynamic phenomenon dependent on a number of fluid and medium characteristics as well as the manner in which the flow is introduced into the bed. A more accurate estimate of the expansion could be obtained by applying the equations to individual layers within each medium.

14.5 SUPPORT MEDIA AND UNDERDRAINS IN SAND FILTERS

The media in a filter are supported by graded gravel layers that prevent the media from reaching and clogging the water collection underdrains. Figure 14.6 shows two commonly used gravel layer gradations. Headloss through the gravel layers is strictly a function of the filtration velocity. The gravel layers do not clog as filtration progresses. The Ergun equation (Eq. 14.15) can be used to calculate the headloss through each layer in a manner similar to the procedure in Example 14.2. Porosities in gravel vary from 0.18 to 0.35 for coarse to fine gravel (Reed et al., 1988).

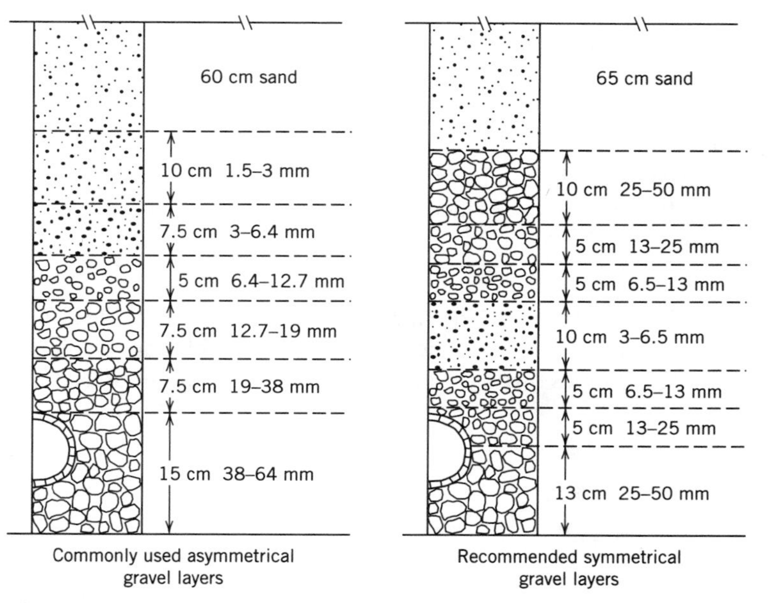

Figure 14.6 Supporting gravel layers for sand filters. From G. M. Fair, J. C. Geyer, and D. A. Okun (1968), *Water Purification and Wastewater Treatment and Disposal,* vol. 2, copyright © 1968 by John Wiley & Sons, Inc. Reprinted by permission of John Wiley & Sons, Inc.

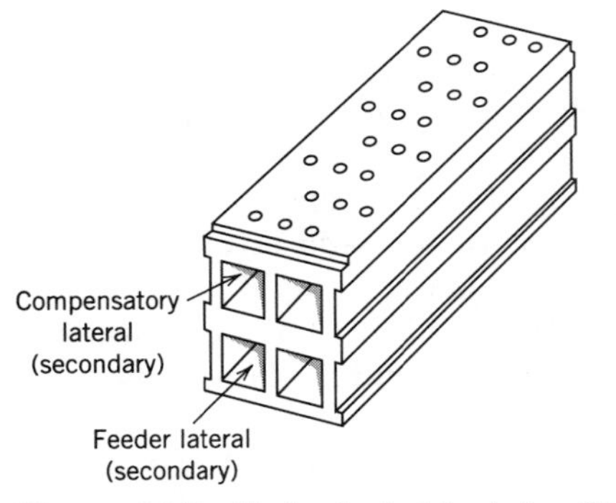

Figure 14.7 Underdrain block for filters.

Backwash velocities will not be sufficient to expand the supporting gravel layers. The backwash velocity should be slowly increased over a period of at least 30 s to avoid disturbing the supporting gravel layers. Headloss through the gravel layers during backwash is also calculated with the Ergun equation.

Underdrain systems serve both collection of the filtered water and uniform distribution of the backwash water. There are a variety of underdrain designs. Figure 14.7 shows one type of underdrain that may be made of plastic or vitrified clay. These blocks are laid across the filter bottom to form continuous channels that feed into larger collection pipes. Some companies supply strainer underdrains, as shown in Fig. 14.8, that do not require a supporting gravel layer.

Headloss through the underdrain openings during filtration or backwash is calculated with an orifice equation. The flow is distributed through all of the orifices to calculate the velocity through each orifice.

$$h_L = \Sigma h_l = nC_d \frac{v^2}{2g} \tag{14.36}$$

where

h_L is the total headloss through the orifices
h_l is the headloss through an individual orifice
C_d is the discharge coefficient for the orifice
n is the number of orifices

The orifices drain into channels that feed into larger collection channels. Flow in all underdrains is pressurized and the Darcy-Weisbach equation is suitable for calculating headloss in these channels. Headlosses through valves, bends, and other appurtenances

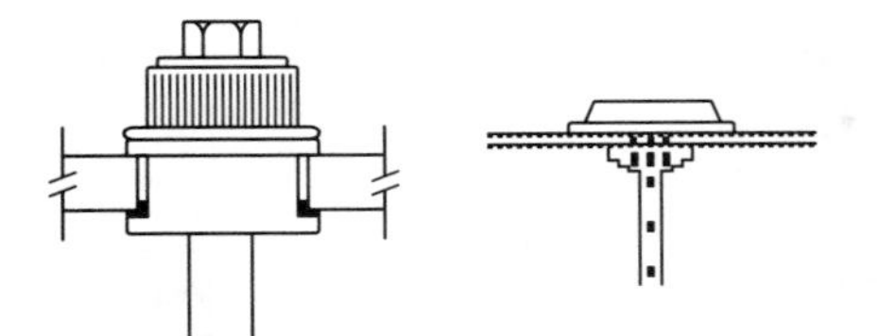

Figure 14.8 Strainers used in false-bottom underdrains without gravel.

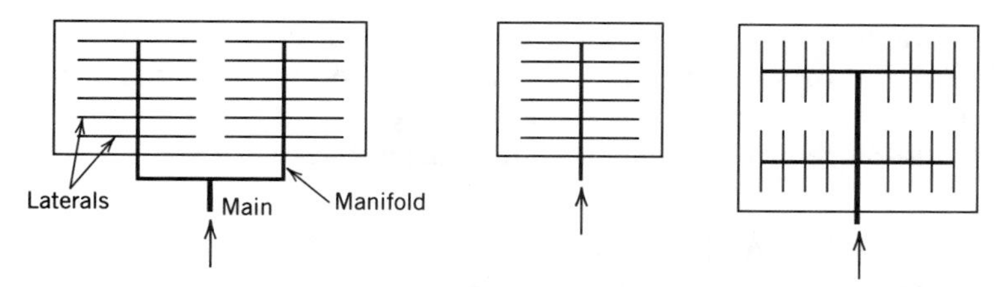

Figure 14.9 Filter underdrain systems. From G. M. Fair, J. C. Geyer, and D. A. Okun (1968), *Water Purification and Wastewater Treatment and Disposal,* vol. 2, copyright © 1968 by John Wiley & Sons, Inc. Reprinted by permission of John Wiley & Sons, Inc.

in the channels and pipes from the filter are proportional to the velocity head through the device.

Pipe lateral systems can also be used for underdrains as shown in Fig. 14.9. Laterals are perforated with orifices placed directly downward or in two rows at 45° to the vertical. Orifice diameters range between 6 and 20 mm (0.2–0.8 in.) and are spaced at 6.5–30 cm (2.5–12 in.). Spacing of laterals is equal to the spacing of orifices. High headloss and poor distribution of backwash have been encountered with these systems which has led to a decline in their use in recent times (ASCE and AWWA, 1990).

Other Design Features of Filters

Conduits in filters are designed for velocities near the following ranges:

Influent pipe to filters	0.6–1.8 m/s (2–6.0 ft/s)
Effluent pipe carrying filtered water	0.9–1.8 m/s (3.0–6.0 ft/s)
Drains carrying spent backwash water	1.2–2.4 m/s (4–8 ft/s)
Wash water line (influent)	2.4–3.7 m/s (8–12 ft/s)
Filter to waste drain	3.7–4.8 m/s (12–15.5 ft/s)

Auxiliary Wash Systems

The primary scouring mechanism to cleanse the media during backwashing is hydrodynamic shear (Amirtharajah, 1978a); abrasive scour caused by particle collisions is not significant. Auxiliary wash systems are incorporated into filter systems to promote particle collisions and improve backwashing performance. Two types of auxiliary wash systems are used: surface wash and air scour. The latter is commonly used in Europe.

A surface wash system supplies jets of water from nozzles located 2.5–5 cm (1–2 in.) above the fixed bed surface. The nozzles are directed 15 to 45° below the horizontal. Operating pressures are typically in the range of 350–520 kPa (50–75 psi) (Cleasby, 1990). Either a fixed pipe grid or rotating pipes are used. The nozzle orifice sizes are 2 to 3 mm (0.08–0.12 in.). The surface wash is initiated 1 or 2 min before the backwash flow is started and continued until 2 or 3 min before the backwash flow is terminated. Surface wash systems affect only the upper layers of the expanded bed.

Air scour systems deliver air across the entire area of the filter. The air is introduced at the bottom of the filter and causes particle contact to occur throughout the entire depth of media. These systems are more effective than surface wash systems.

There are two alternatives for applying air scour: air scour applied before the fluid backwash and simultaneous air scour and fluid backwash. In the former case air is supplied for 2 to 5 min before the fluid backwash. The water level in the filter is lowered because the entrained air adds volume to the filter. The expanded volume must not be allowed to rise above the overflow weirs and cause loss of media. After the air scour is terminated, the bed is backwashed at a velocity sufficient to fluidize the bed. Entrained air will be removed from the filter during backwash.

14.6 FILTER BEDS FOR WATER AND WASTEWATER TREATMENT

The design of a rapid filter system for water treatment depends on the treatment objectives and the pre-treatment that has been applied to the filter influent. Design information for filter beds for various applications is given in Table 14.7. The minimum number of individual filter beds is two. When only two beds are installed, a single bed must be capable of meeting water demands during periods of shutdown of either filter for maintenance and backwashing. In medium to large installations (flow greater than 35 000 m^3/d or 10 Mgal/d) at least four beds should be installed (ASCE and AWWA, 1990). The practical maximum area of an individual bed is approximately 150 m^2 (1 600 ft^2) (Kawamura, 1991).

Filtration of wastewater is becoming a more common practice to enhance suspended solids removal. More stringent wastewater treatment standards promote the practice. Typical design information for wastewater treatment filter beds is given in Table 14.8. For wastewater, monomedium filters are more common than dual-media filters. The maximum area of an individual filter for wastewater treatment is the same as for a water treatment filter (Culp et al., 1978).

For intermittent filtration of effluent from stabilization ponds, there are two basic configurations: single-stage intermittent filters or intermittent filters in series USEPA (1983). Single-stage intermittent filters use sand medium with a small effective size in the range of 0.20–0.30 mm. Uniformity coefficients are high, ranging from 5 to 10 for

TABLE 14.7 Design Features of Filter Beds for Water Treatment[a]

	Effective size	Total depth	
	mm	m	ft
A. Common U.S. practice after coagulation and settling			
1. Sand alone	0.45–0.55	0.6–0.7	2–2.3
2. Dual media	0.9–1.1	0.6–0.9	2–3
Add anthracite (0.1 to 0.7 of bed)			
3. Triple media	0.2–0.3	0.7–1.0	2.3–3.3
Add 0.1 m (0.3 ft) garnet			
B. U.S. practice for direct filtration			
Practice not well established. With seasonal diatom blooms, use coarser top size. Dual media coal, 1.5 mm effective size			
C. U.S. Practice for Fe and Mn filtration			
1. Dual media similar to A-2			
2. Single medium	<0.8	0.6–0.9	2–3
D. Coarse single-medium filters washed with air and water simultaneously			
1. For coagulated and settled water	0.9–1.0	0.9–1.2	3–3.9
2. For direct filtration	1.4–1.6	1–2	3.3–6.6
3. For Fe and Mn removal	1–2	1.5–3	4.9–9.8

[a]From J. L. Cleasby, (1990), "Filtration," in *Water Quality and Treatment,* 4th ed., F. W. Pontius, ed., McGraw-Hill, Toronto, reproduced with permission of McGraw-Hill, Inc.

TABLE 14.8 Design Features of Monomedium Filter Beds for Wastewater Treatment[a]

Characteristic	Value	
	Range	Typical
Shallow bed (stratified)		
Sand		
Depth, cm (in.)	25–30 (10–12)	28 (11)
Effective size, mm	0.35–0.6	0.45
Uniformity coefficient	1.2–1.6	1.5
Filtration rate, m/h (gal/ft^2/min)	5–15 (2–6)	7 (3)
Anthracite		
Depth, cm (in.)	30–50 (12–20)	40 (16)
Effective size, mm	0.8–1.5	1.3
Uniformity coefficient	1.3–1.8	1.6
Filtration rate, m/h (gal/ft^2/min)	5–15 (2–6)	7 (3)
Conventional (stratified)		
Sand		
Depth, cm (in.)	50–76 (20–30)	60 (24)
Effective size, mm	0.4–0.8	0.65
Uniformity coefficient	1.2–1.6	1.5
Filtration rate, m/h (gal/ft^2/min)	5–15 (2–6)	7 (3)
Anthracite		
Depth, cm (in.)	60–90 (24–36)	76 (30)
Effective size, mm	0.8–2.0	1.3
Uniformity coefficient	1.3–1.8	1.6
Filtration rate, m/h (gal/ft^2/min)	5–20 (2–8)	10 (4)
Deep bed (unstratified)		
Sand		
Depth, cm (in.)	90–180 (36–72)	120 (48)
Effective size, mm	2–3	2.5
Uniformity coefficient	1.2–1.6	1.5
Filtration rate, m/h (gal/ft^2/min)	5–24 (2–10)	12 (5)
Anthracite		
Depth, cm (in.)	90–215 (36–84)	150 (60)
Effective size, mm	2–4	2.75
Uniformity coefficient	1.3–1.8	1.6
Filtration rate, m/h (gal/ft^2/min)	5–24 (2–10)	12 (5)

[a]Metcalf and Eddy (1991), *Wastewater Engineering: Treatment, Disposal, Reuse,* 3rd ed., G. Tchobanoglous and F. L. Burton, eds., McGraw-Hill, Toronto, reproduced with permission of McGraw-Hill, Inc.

a number of installations. The other alternative is to employ two or more intermittent filters in series. Coarser media with an effective size of 0.60–0.70 mm are used in the first filter whereas the subsequent filters use media with smaller effective sizes in the range of 0.15–0.40 mm.

14.7 AIR BINDING OF FILTERS

A major problem of filter operation is air binding or the formation of gas bubbles in the filter. The release of dissolved gases can dislodge accumulated solids from the media, driving them deeper into the media and increasing the possibility of their escape

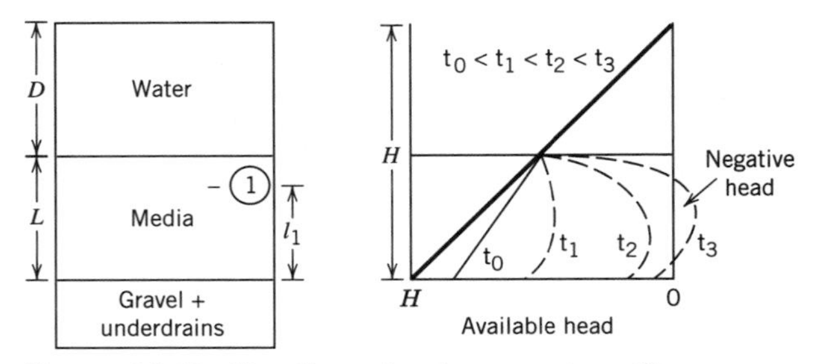

Figure 14.10 Headloss development in a filter.

in the filter effluent. Also, the entrapped gases take a portion of the media out of service, which leads to more rapid deterioration of the effluent quality and shortens run time. Air binding of a filter may result from:

1. Reduction of pressure in the filter to a value less than atmospheric pressure (negative head)
2. Increase of water temperature in the filter
3. Release of oxygen by algae growing in the filter

Measures can be taken to eliminate or reduce the possibility of occurrence of pressures less than atmospheric pressure in the filter. An examination of headloss and pressure variation within the filter is required; calculating the total headloss through the filter will not illustrate the problem. Figure 14.10 shows the progression of headloss development over time in a filter measured by inserting piezometer taps at various depths in the filter. The depth of water over the filter may be constant or vary with time but the headloss progression is similar in any case. The effluent for the filter depicted in Fig. 14.10 is discharged over a weir located at the same elevation as the bottom of the media in the filter. The pressure distribution for a filter with no water flowing simply follows a 45° line.

The available head for frictional dissipation of energy is the piezometric head minus the elevation head when the effluent weir from the filter is located at the same elevation as the bottom of the filter. As water flows downward for a distance it gains available head because of the additional depth of water above it; however, available head is decreased because of losses incurred in traveling the distance.

Once water begins to flow through a clean filter, the initial headloss follows the Carman–Kozeny equation and headloss is a linear function of depth (time t_0). Even in a multimedia filter, solids removal will be greatest in the upper layers and headloss will also be largest in these layers. Therefore, as time goes on, the available head curve becomes skewed to the right in the upper layers. As the water descends from the layer of maximum headloss, it recovers head. Flow is in accord with Darcy's law (see Problem 18). The available head at the bottom of the media must always be equal to the sum of the headlosses through the gravel, underdrains, and other appurtenances in line before the effluent overflow weir.

Extended operation of the filter before backwashing can produce a situation where the available head drops to a negative value in the upper layers (time t_3 in Fig. 14.10). This may cause air binding of the filter. This situation can occur when the overflow weir is located at the same elevation as the bottom of media in the filter and will be examined first. Consider a location (1) within the media (Fig. 14.10). Choosing a datum at the bottom of the media and applying Bernoulli's equation, the absolute pressure at this point may be determined. The energy at location 1 will be related to the energy at the water surface above the media (location 0). The velocity head is insignificant above and in the filter.

$$p_{atm} + \rho g H = p_1 + \rho g l_1 + \rho g h_{L,0-1} \tag{14.37}$$

where

$h_{L,0-1}$ is the headloss between the water surface and location 1
p_{atm} is atmospheric pressure
l_1 is the elevation of point 1 above the datum

Solving for p_1, the pressure at point 1 above the datum,

$$p_1 = p_{atm} + \rho g H - \rho g l_1 - \rho g h_{L,0-1} \tag{14.38}$$

It is seen that if $l_1 + h_{L,0-1}$ is greater than H, then the pressure decreases below atmospheric pressure. It is possible for this to occur in the upper layers given the higher removal of solids and the associated greater headlosses there. Similar results can be obtained by writing Bernoulli's equation between location 1 and the effluent weir located at the same level as the bottom of the filter media.

This situation may be remedied by increasing the elevation of the effluent weir (Fig. 14.11). If the effluent weir is located at the same elevation as the top of the media, the possibility of negative head never arises. Again, considering point 1 within the media and relating the total energy at this location to the total energy at the weir (now located at a height, L, above the datum) through Bernoulli's equation:

$$p_1 + \rho g l_1 = p_{atm} + \rho g L + \rho g h_{L,1\text{-}w} \tag{14.39}$$

where

$h_{L,1\text{-}w}$ is the headloss from location 1 to the top of the weir

Solving for p_1,

$$p_1 = p_{atm} + \rho g L - \rho g l_1 + \rho g h_{L,1\text{-}w} \tag{14.40}$$

Because $L \geq l_1$, Eq. (14.40) shows that $p_1 > p_{atm}$. When the effluent weir is located at the height of the surface of the media, available head within the media is a monotonically decreasing function of depth below the media surface (Fig. 14.11b). The head at the bottom of the media is equal to the headlosses in the gravel, underdrain system, and other appurtenances in the filter effluent line. No available head is gained below the surface of the media.

This solution solves the problem, but as Monk (1984) advises, the safety factor involved is excessive. Increasing the elevation of the weir will require an increase in

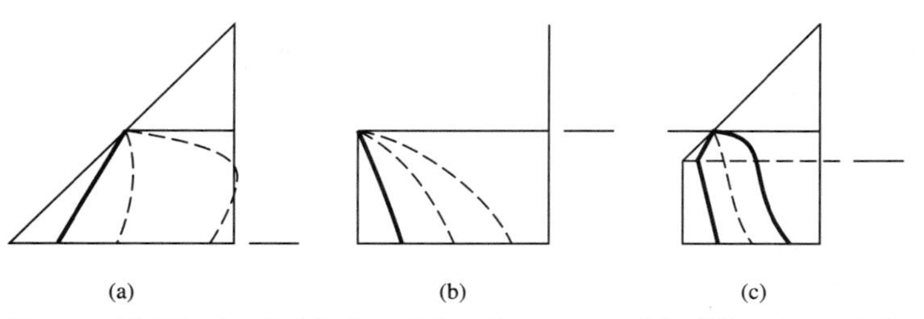

Figure 14.11 Available head development with different weir locations. Weir located at (a) bottom of filter; (b) top of media; (c) an intermediate depth.

depth of water over the media to achieve the same throughput of water. Some economies can be gained by lab studies on the headloss performance of the filter under probable operating conditions and the weir can be lowered and still provide a reasonable safety factor against the development of negative head. The available head curve for this situation is shown in Fig. 14.11(c). It is a combination of curves applying to the cases where the weir is located at the bottom (Fig. 14.11a) or top (Fig. 14.11b) of the filter.

14.8 RAPID FILTRATION ALTERNATIVES

There are some features that apply to all rapid filtration operations. It is essential that the influent to a rapid filter be coagulated. There will be a slight deterioration in filtrate quality after backwashing. Sudden increases in the rate of filtration also deteriorate the filter effluent quality. The bed must not be allowed to dewater.

Single-Medium and Multimedia Filters

Most of the solids removal in single-medium (sand) filters occurs in the top layers of the filter. The full depth of the medium is not effectively used and headloss increases rapidly. Replacing the upper depth of sand with coarser anthracite medium definitely retards the rate of headloss development and increases the length of a filter run. A dual-media filter does not necessarily improve the quality of the filtrate (Cleasby, 1990) but there is no deterioration of the filtrate over the longer run times.

The extension of the dual-media filter is the triple-media (multimedia) filter, which uses garnet or ilmenite media of finer size than sand in the bottom layer of the filter. As a result of a finer grain size being included in the filter, the initial clean bed headloss is larger for triple-media filters compared to single- or dual-media filters. But triple-media filters outperform single-medium filters for headloss development (Cleasby, 1990). The incorporation of finer sized media is expected to improve effluent quality. Cleasby (1990), in a review of multimedia filter studies, was unable to conclude that triple-media filters resulted in superior effluent quality compared to dual-media filters. Some studies were poorly designed and there was a lack of studies comparing dual- and triple-media filters. A comparison of dual-media and mixed media (anthracite, sand, and two size ranges of garnet) performance at a full-scale plant installation showed no significant differences in performance (Barnett et al., 1992).

Granular activated carbon (GAC) can be used as the top medium for taste and odor control and to adsorb organic compounds. GAC has a lower density than anthracite, which affects backwashing requirements.

Constant- and Declining-Rate Filtration

The rate of water throughput in a filter is a function of the headloss through the filter system and the driving head of water over the filter.

$$Q = f(h_D - h_c - h_u - h_f - h_v) \tag{14.41}$$

where

h_D is the driving head (depth of water over the filter)
h_c is the clean bed headloss
h_u is the headloss through the gravel and underdrain system
h_f is the friction headloss, resulting from solids accumulation in the filter

h_v is the headloss through valves and other appurtenances in the effluent pipe from the filter

A constant-rate filter produces water at a rate that is more or less constant. The flow rate may be influent or effluent controlled. If the supply of water to the filter is constant the headloss in the filter (h_f) will increase as the run progresses to cause the water depth over the filter to increase to maintain the throughput.

If effluent control is used, a constant water level over the filter can be maintained by progressively opening a valve on the effluent line as h_f increases during the run to maintain $h_f + h_v$ constant. Either alternative can be used, or a constant feed rate with some variation in depth of the water over the filter and some valve control can also be used.

For a filter operated in the declining-rate mode, the water level above the filter is held more constant throughout the run, resulting in a declining rate of water throughput. Therefore, declining-rate filtration has a lower total head requirement than constant-rate filtration that does not rely on valve control of the effluent. Declining-rate filters can be designed in a bank of filters in which a filter is backwashed by the effluent from the other filters without any requirement for a pump. For this design a minimum of four filters is required. The filters are fed by a common header and the same water level exists over all filters. The underdrain systems from each filter feed into a common collection pipe. The filter most recently backwashed has the highest filtration rate, whereas the dirtiest filter has the slowest filtration rate. Amirtharajah (1978b) has given a detailed design example of a declining rate filtration system.

Effluent quality is better from a declining-rate filter compared to a constant-rate filter, although there is some debate over this issue. Neither mode of operation results in rapid changes in filtration rate when the filters are properly controlled. Constant-rate filters will be operated until the desired terminal headloss is reached without any special instrumentation. In a bank of declining-rate filters, instrumentation must be provided to determine the flow rate through each filter. The headloss is the same for each filter at all times.

14.8.1 Direct Filtration

Direct filtration refers to omitting sedimentation from the treatment train. Coagulation is still essential for the filter influent in this circumstance. Less coagulant is generally used for water to be directly filtered to promote the formation of smaller, filterable floc rather than large, readily settled floc. A separate flocculation unit may precede the filter or flocculation may occur in the filter.

Direct filtration is used for good quality surface water because of the cost savings associated with elimination of sedimentation. The quality of water defined as the "perfect candidate" for direct filtration is as follows (AWWA Committee, 1980):

Color	$<$ 40 color units
Turbidity	$<$ 5 NTU
Algae	$<$ 2 000 asu/mL (asu- areal standard unit)
Iron	$<$ 0.3 mg/L
Manganese	$<$ 0.05 mg/L

Direct filtration was successful in treating high-turbidity water (20–30 NTU to 40–60 NTU) with deep bed (2.0 m or 6.6 ft) dual-media filters with the application

TABLE 14.9 Filtration Rates in Pressure Sand Filters[a]

Effective size, mm	0.35	0.55	0.75	0.95
Filtration rate, m/h	25–35	40–50	55–70	70–90
($gal/ft^2/h$)	(615–860)	(980–1 230)	(1 350–1 720)	(1 720–2 210)

[a]From Degrémont (1979).

of the correct filter aids (Lodgson et al., 1993). Bench scale tests are required to definitely determine if a water is suitable for direct filtration.

14.9 PRESSURE FILTERS

Pressure filters (Fig. 14.12) are enclosed in usually metal containers and can be operated in an upflow or downflow mode. The filters can be mono- or multimedia and they are cleansed by backwashing. Air scour may be used with the backwash. Maximum headloss in pressure filters is 2 to 20 m (6–65 ft) (Degrémont, 1979). Filtration rates are a function of media size as given in Table 14.9. The filtration rate may range up to 25–50 m/h (615–1 230 $gal/ft^2/h$) in dual-media pressure filters.

14.10 SLOW SAND FILTERS

The design and operation of slow sand filters are relatively simple compared to rapid filters. These features make them primary alternatives for applications in developing countries. The costs of cleaning the filter and the land area required have discouraged their use in North America except for smaller installations, where the advantage of simpler operation is also important (Leland and Damewood, 1990). Cleaning the filter can be labor intensive, which is usually a benefit in developing countries given the large supply of labor.

As noted in Section 14.1, the filters rely on the formation of a dense microbial growth in the upper layers of the filter. This layer is known as the schmutzedecke (dirty layer). Bacteria, protozoa, and other large microorganisms such as helminths and suspended solids are effectively removed in this layer. Removal of bacteria is not complete, but *E. coli* removals of 10^2–10^3 can be expected (Hofkes, 1981). *Giardia* and *Crytosporidium* cysts were removed to levels greater than 99.9% in a properly operated slow sand filter (Schuler et al., 1991).

Countries in the tropical climates are best suited for slow sand filters because the warm temperatures promote the establishment of a very active schmutzedecke. In northern climates, the filter will have to be enclosed to maintain a reasonable growth during the winter season. The large surface area of the filter makes it less practical to enclose it.

Influent turbidity should be less than 50 turbidity units for trouble-free operation but higher turbidities can be tolerated for short periods of time (Hofkes, 1981; Huisman and Wood, 1974; van Dijk and Oomen, 1978). If raw water turbidities exceed these values then some form of pretreatment such as plain sedimentation or a roughing filter will be required.

Commissioning a filter should be performed by filling the filter from the bottom with clean water. This will ensure that air entrained in the pores of the media will be driven out. After the media is submerged, operation of the filter is commenced. It will

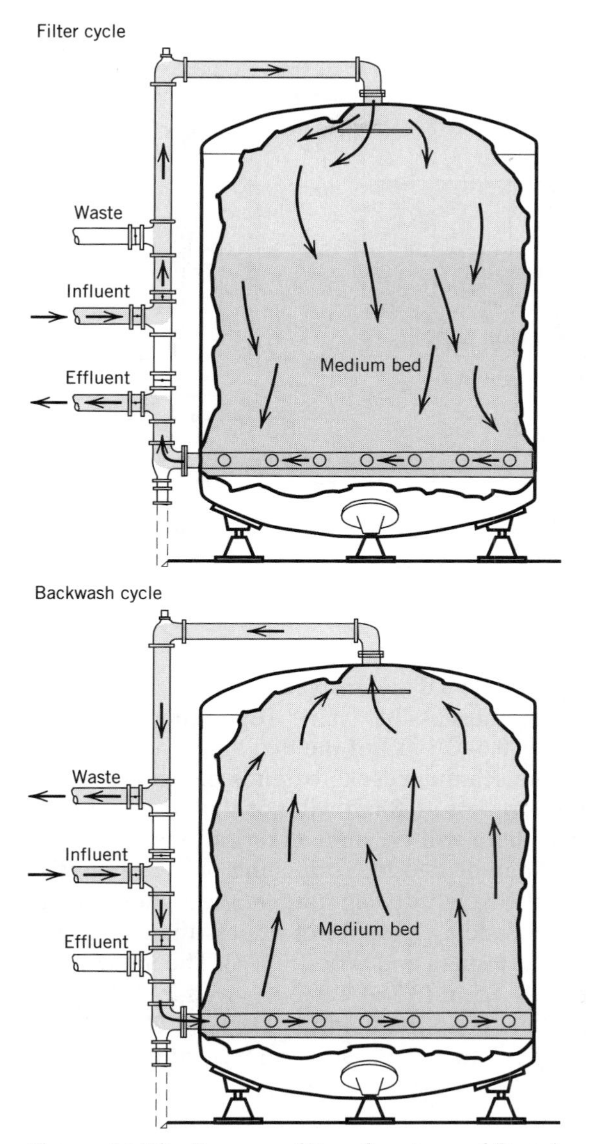

Figure 14.12 Pressure filter. Courtesy of Degrémont Infilco.

take up to a few weeks to establish the schmutzedecke and produce effluent of acceptable quality. It is not necessary to operate a slow sand filter continuously; however, when the filter is out of operation the water level must not be allowed to fall below the media. This will cause the schmutzedecke to dry out and lose activity.

A filter exposed to sunlight may experience excessive algae growth, which leads to rapid clogging of the filter, and the decay of the algae may impart undesirable tastes and odors to the product water. The problem of algae growth is solved by covering the filters.

A schematic of a slow sand filter is given in Fig. 14.13. Media characteristics for a slow sand filter are given in Table 14.1. Flow through the filter may be controlled

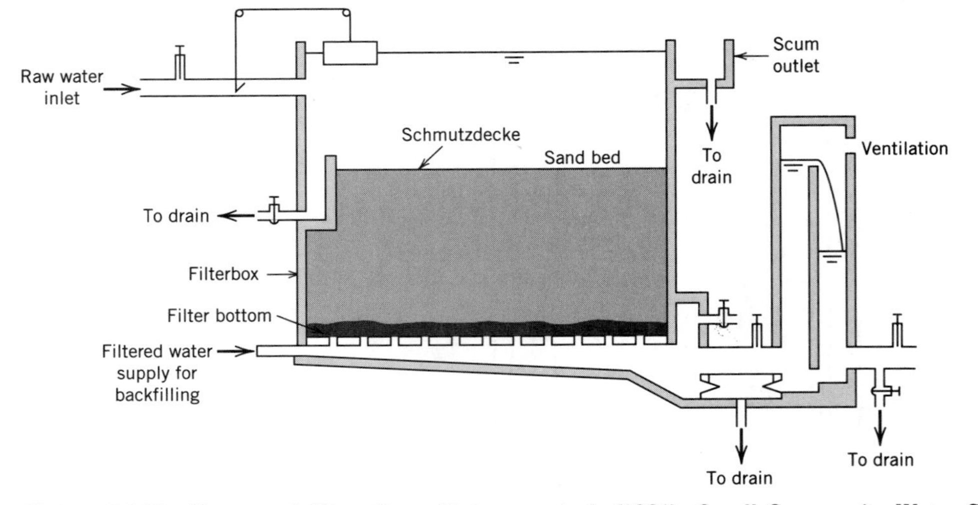

Figure 14.13 Slow sand filter. From Huisman et al. (1981), *Small Community Water Supplies,* IRC.

by either an outlet or an inlet valve, adjusted on a daily or so basis. When the headloss through the filter reaches the maximum permissible value (operating heads are 1.0–1.5 m or 3.3–4.9 ft), the top 1.5–2 cm (0.6–0.8 in.) of the bed are scraped off and operation is resumed (Hofkes, 1981). The schmutzedecke extends below a depth of 2 cm (0.8 in.); therefore, removal of the top 2 cm (0.8 in.) will not significantly impair the performance of the filter. However, there will be some deterioration of effluent quality after scraping and it may take a day or two for reripening of the filter. Two filters will allow for an uninterrupted operation, producing water at the highest quality.

The final depth of the filter before resanding should not be less than 0.5–0.8 m (1.6–2.6 ft) to ensure good water quality (Huisman and Wood, 1974). The initial depth of the media is normally in the range of 1–1.5 m (3.3–4.9 ft). As given in Table 14.1, the surface loading rate of slow sand filters is normally in the range of 2–8 $m^3/m^2/d$ (50–200 $gal/ft^2/d$). Kawamura (1991) reports that the area of an individual filter bed should be less than 3 000 m^2 (32 500 ft^2) but the average size is in the range of 100–200 m^2 (1 100–2 150 ft^2).

14.11 BIOLOGICAL FILTRATION FOR WATER TREATMENT

Biological filtration is a standard treatment for wastewaters but biological filtration for water purification is a recent development. Biological filtration, as the name states, encourages microbial growth in filters to enhance their performance beyond solely filtration. Microorganisms consume organic matter in the water and render a water that is biologically stable—the potential for growth or regrowth of microorganisms in the finished water is diminished. The reduction of organic matter removes precursors of disinfection byproducts (trihalomethanes and others), reducing the potential for their formation when chlorine is applied as the final treatment.

Biological filtration can also be used for nitrification and removal of nitrates (denitrification; see Section 17.10.1). Ammonia in drinking water is a cause of biologically unstable water, promoting bacterial growth in the distribution system. Besides filters, fluidized beds and GAC beds are alternatives for biological filtration (Rittman and Snoeyink, 1984). Iron and manganese can also be removed with biological filtration (see Section 15.4).

Biological treatment of wastewater is discussed in Chapters 17, 18, and 19 and many of the principles developed in those chapters apply to biological filtration. The media in filters provide a surface to support microbial growth, which applies principles well known in biological wastewater treatment. There are obvious significant differences, however, between wastewater and water used to produce drinking water. The concentration of organic matter is much lower in the partially treated water being delivered to a biological filter for drinking water compared to a wastewater. The degradability of natural organic matter, consisting of humic and fulvic acids, is also significantly less than many of the organics in sewage. Measures must be taken to enhance the biological activity in water treatment plant filters.

For organic matter to be removed, it must be biodegradable. To improve the biodegradability of organic matter, pre-oxidation of the organic matter in the water is applied. The oxidants break down the large molecular weight organics, making them more amenable to biological uptake, but the complete oxidation of the organics to carbon dioxide and other mineral products will not be achieved. An inordinate amount of oxidant would be required to provide complete oxidation. The judicious application of oxidant allows the biological treatment to become effective, minimizing the overall costs of the additional treatment.

Ozone (see Chapter 16) is by far the most common oxidation agent used. Other advanced oxidation processes can be applied to enhance the biodegradability of natural organic matter. Pre-oxidation must not leave any residual biocidal activity in the water being sent to the filter or microbial growth will be retarded. Ozone is an ideal agent because of its rapid decomposition. Upstream chlorination is not beneficial for biological filtration because of the residual bactericidal properties of chlorine derivatives (see Chapter 16). Biological growth in the filter decreases time intervals between backwashing (Golgrabe et al., 1993).

Slow sand filters rely on biological activity for their performance and thus provide biological filtration to an extent. However, their effectiveness in removing natural dissolved organic matter is enhanced by pre-oxidizing the filter influent with ozone (Malley et al., 1993). Rapid filters may also have some biological activity depending on the upstream treatments but their biological performance is affected to a much larger extent by preparation of the water for biological treatment. Using GAC as one of the media in a filter also generally improves biological performance. The additional surface area of the carbon can adsorb organic matter and expose it for long contact times to microorganisms, even though empty bed contact times of the water in the filter are only a few minutes. Dissolved and colloidal carbon removed by biological activity ranges from 0.2 to 1.5 mg/L in typical installations (Rittman and Snoeyink, 1984).

The reader is encouraged to study concepts in later chapters on biological treatment and consider their application in biological filters. There are many intricacies of the process that are beyond the scope of this presentation. There are other alternatives to filters such as biological fluidized bed treatment. Biological filtration is used in a number of European installations but the process is used sparingly in North America. The quality of raw waters in Europe demands more advanced measures to produce a wholesome water but there is increasing interest in biological filtration in North America for the improvements in water quality that it provides.

QUESTIONS AND PROBLEMS

1. Calculate the gross specific surface area (m^2/m^3) available in a sand filter that has particles with a nominal diameter of (a) 0.40 mm and (b) 1.0 mm. Assume that the porosity is 0.40.
2. What is the purpose of increasing the number of media types in a rapid filter?
3. Is a uniformity coefficient less than 1.0 possible?
4. Prove that the interstitial velocity is equal to the superficial velocity divided by the porosity.
5. The available head remaining at the bottom of the medium in a filter is required to drive the water through the gravel, underdrains, and other devices on the way to the effluent weir. Under what conditions will the head available at this location remain constant? When will it vary?
6. (a) Derive the formula $d_{90} = d_{10}U^{1.67}$ for a log-normal distribution. The Appendix has equations for a normal distribution. The following values apply to the equations: $z = 1.28$, $F(z) = 0.90$; $z = -1.28$, $F(z) = 0.10$; and $z = 0.25$, $F(z) = 0.60$.
   (b) Derive an equivalent relation between d_{90} and d_{10} if a normal distribution describes media variation.
   (c) Which medium would be more uniform: a log-normally distributed medium or a normally distributed medium? Both media have the same uniformity coefficient.
7. Sand medium with an effective size of 0.45 mm and uniformity coefficient of 1.35 is required for a rapid filter. Stock sand has an effective size of 0.55 mm and uniformity coefficient of 2.55. Determine the percentage of usable sand and the lower and upper size limits for discarding unusable stock sand.
8. What is the minimum upflow velocity required to wash out sand particles with a nominal diameter of 4.4×10^{-2} cm or smaller at a temperature of 20°C? (Solve the problem in SI or U.S. units at the discretion of the instructor.)
9. What is the velocity head expressed in atmospheres for a rapid filter operated at a surface loading rate of 475 m/d when the local porosity is 0.40?
10. Compare and discuss the headloss with depth for a slow sand filter and a rapid sand filter. Construct a qualitative graph to illustrate the differences.
11. Use log probability and arithmetic probability paper to assess the sizes in 10% increments of sand medium with an effective size of 0.45 mm and a uniformity coefficient of 1.4. The porosity and sphericity are 0.45 and 0.85, respectively. The depth of the bed is 0.75 m. Calculate the initial headloss through each medium at a loading rate of 175 $m^3/m^2/d$ and temperature of 15°C. Is the difference significant?
12. A rapid sand filter with a bed depth of 60 cm contains only sand that has a sieve analysis given in the accompanying table. The filter is to be operated at a loading rate of 120 $m^3/m^2/d$ and temperature of 15°C. The porosity for the filter is 0.39 and the sphericity of the particles is 0.92.

Sieve Analysis

Sieve number	Percentage (by weight) of sand retained	Mean size mm
16–18	1.6	1.09
18–20	5.3	0.92
20–25	12.5	0.77
25–30	25.1	0.65
30–35	23.4	0.54
35–40	15.1	0.46
40–45	13.8	0.38
45–50	2.2	0.32
50–100	0.8	0.27

(a) What are the effective size and uniformity coefficient for the medium?
(b) What is the initial headloss through this medium?
(c) What are the minimum fluidization velocity and headloss through the medium at this velocity?
(d) Using the d_{50} diameter calculate the backwash velocity to expand the bed by 15%. Also calculate the headloss in the filter at this backwash velocity.
(e) Divide the sand into five equal layers by weight and find the expanded depth of each layer at the velocity calculated in (d).

13. A dual-media rapid filter contains anthracite with an effective size of 0.9 mm, sphericity of 0.75, porosity of 0.48, and depth of 0.40 m and sand with an effective size of 0.48 mm, sphericity of 0.90, porosity of 0.40, and depth of 0.40 m. The uniformity coefficients are 1.5 for each of the media. Perform all calculations for a temperature of 10°C. Divide each medium into three layers on a weight basis.
(a) Calculate the initial headloss through the bed if the loading rate is 200 $m^3/m^2/d$.
(b) Calculate the minimum fluidization velocity for the bed.
(c) Calculate the backwash velocity that will expand the largest layer of sand by 15% using the procedure of Dharmarajah and Cleasby. Find the total expansion of the bed at this backwash velocity.
(d) What is the total headloss through the expanded bed at the backwash velocity calculated in (c)?

14. Using the information in Tables 14.5 and 14.6, determine the backwash velocities to expand the d_{90} particles in each of the layers of the filter in Problem 13 by 10% at a temperature of 10°C.

15. Design a trimedia filter for coagulated and flocculated water using the typical values given in Table 14.7 and other tables in the chapter for a flow in the range of 16 000–25 000 m^3/d (4.23–6.60 Mgal/d). Use uniformity coefficients of 1.5, 1.4, and 1.4 for the anthracite, sand, and garnet, respectively. What are the size ranges of the media (d_{10}–d_{90}), area and depth of the filter, and volumes of each medium?

16. Calculate the headloss though the asymmetrical gravel support layer pictured in Fig. 14.6 for a filtration surface loading rate of (a) 275 m/d and (b) a backwash velocity of 38 m/h. Base your analysis on the mean size particle in each layer and use a temperature of 8°C. Use a porosity of 0.35 and a sphericity of 0.90.

17. Prove that available head is a monotonically decreasing function of depth in a filter with the effluent weir located at the elevation of the surface of the media.

18. Headloss data for a filter of 0.6 m depth are tabulated below. Assume that headloss from the top of the water to the top of the media is negligible and ignore the velocity head. What are the piezometric and gauge pressure heads at each elevation in the media for the following situations?

Depth below top of media, m	0.10	0.20	0.30	0.40	0.50	0.60
Headloss, m	0.45	0.35	0.13	0.05	0.04	0.03
Σ(headloss), m	0.45	0.80	0.93	0.98	1.02	1.05

(a) water depth of 0.6 m above the media and effluent weir located at the same elevation as the bottom of the media
(b) water depth of 1.2 m and the weir located at the same elevation as the bottom of the media
(c) and water depth of 1.2 m and effluent weir located at the same elevation as the top of the media

What is the headloss through the gravel, underdrains, and piping to the effluent weir in each case? Would the flow being handled by the filter be greater, equal, or lower in cases (b) and (c) compared to (a)? What is the minimum height of the effluent weir above the bottom of the media and the corresponding depth of water over the media to guarantee that negative head will not occur?

19. Prove with equations that a temperature rise in the filter could lead to air binding in the filter if the local pressure is constant.

20. In a bank of filters operated in a declining-rate mode, why is the headloss the same at any time for each filter during a run? (Note that the headloss varies with time.)

21. (a) Would it be reasonable to have a multimedia slow sand filter? Explain.
(b) Would it be worthwhile to use media in a slow sand filter with lower uniformity coefficients (approaching those used in rapid filters)?

22. (a) Design a slow sand filter to treat a flow for 5 000 people with a maximum water demand of 300 L/cap/d (79 gal/cap/d). Use a depth of 1.25 m (4.10 ft) and other typical design values given in the text. What is the size range of the medium and what volume of medium is required?
(b) What is the expected time period for the bed to be replaced, assuming that the average time of a run is 30 d and that 2 cm (0.80 in.) of medium will be removed at the end of each run? Calculate the time period for the bed to be replaced if the final bed depth is (i) 0.5 m (1.64 ft) and (ii) 0.8 m (2.62 ft) and assuming that (i) a run lasts an average of 20 days and 2 cm (0.80 in.) of medium are removed at each cleaning and (ii) a run lasts an average of 60 days and 1.5 cm (0.59 in.) of medium are removed at each cleaning.

23. What are the goals of biological filtration for water treatment? What process modifications are necessary to promote the process?

KEY REFERENCES

CLEASBY, J. L. (1990), "Filtration," in *Water Quality and Treatment,* 4th ed., F. W. Pontius, ed., McGraw-Hill, Toronto, pp. 455–560.

HUISMAN, L. AND W. E. WOOD, (1974), *Slow Sand Filtration,* WHO, Geneva.

MONTGOMERY, J. M., Consulting Engineers, Inc. (1985), *Water Treatment Principles Design,* John Wiley & Sons, Toronto.

SANKS, R. L., ed. (1978), *Water Treatment Plant Design,* Ann Arbor Science, Ann Arbor, MI.

REFERENCES

AMIRTHARAJAH, A. (1978a), "Optimum Backwashing of Sand Filters," *J. Environmental Engineering Division, ASCE,* 104, EE5, pp. 917–932.

AMIRTHARAJAH, A. (1978b), "Design of Granular Media Filter Units," in *Water Treatment Plant Design,* R. L. Sanks, ed., Ann Arbor Science, Ann Arbor, MI, pp. 675–737.

AMIRTHARAJAH, A. (1988), "Some Theoretical and Conceptual Views of Filtration," *J. American Water Works Association,* 80, 12, pp. 36–46.

ASCE and AWWA (1990), *Water Treatment Plant Design,* 2nd ed., McGraw-Hill, Toronto.

AWWA Committee (1980), "The Status of Direct Filtration," *J. American Water Works Association,* 72, 7, pp. 405–411.

BAKER, M. N. (1981), *The Quest for Pure Water,* vol. I, American Water Works Association, Denver, CO.

BARNETT, E., R. B. ROBINSON, D. W. LOVEDAY, AND J. SNYDER (1992), "Comparing Plant-Scale Dual- and Mixed-Media Filters," *J. American Water Works Association,* 84, 6, pp. 76–81.

CAMP, T. R. (1964), "Theory of Water Filtration," *J. Sanitary Engineering Division, ASCE,* 90, SA4, pp. 1–30.

Carman, P. C. (1937), "Fluid Flow through a Granular Bed," *Transactions of the Institution of Chemical Engineers,* 15, pp. 150–156.

Cleasby, J. L. (1995), personal communication.

Cleasby, J. L. and K. Fan (1981), "Predicting Fluidization and Expansion of Filter Media," *J. of the Environmental Engineering Division, ASCE,* 107, EE3, pp. 455–472.

Cleasby, J. L. and K. Fan (1982), Closure to "Predicting Fluidization and Expansion of Filter Media," *J. of the Environmental Engineering Division, ASCE,* 108, EE5, pp. 1083–1087.

Craun, G. F. (1988), "Surface Water Supplies and Health," *J. American Water Works Association,* 80, 2, pp. 40–52.

Culp, R. L., G. M. Wesner, and G. L. Culp (1978), *Handbook of Advanced Wastewater Treatment,* Van Nostrand Reinhold, Toronto.

Dharmarajah, A. H. and J. L. Cleasby (1986), "Predicting the Expansion of Filter Media," *J. American Water Works Association,* 78, 12, pp. 66–76.

Degremont (1979), *Water Treatment Handbook,* 5th ed., Halstead Press, Toronto.

Ergun, S. (1952), "Determination of Geometric Surface Area of Crushed Porous Solids," *Analytical Chemistry,* 24, pp. 388–393.

Fair, G. M., J. C. Geyer, and D. A. Okun (1968), *Water and Wastewater Engineering,* vol. 2, *Water Purification and Wastewater Treatment and Disposal,* John Wiley & Sons, Toronto.

Golgrabe, J. C., R. S. Summers, and R. J. Miltner (1993), "Particle Removal and Head Loss Development in Biological Filters," *J. American Water Works Association,* 85, 12, pp. 94–106.

Hazen, A. (1892), Annual Report of the Massachusetts State Board of Health, Boston, MA.

Hofkes, E. H., ed., (1981), *Small Community Water Supplies,* WHO International Reference Center, The Hague.

Ives, K. J. (1982), "Fundamentals of Filtration," Proc. of Symposium on Water Filtration, European Federation of Chemical Engineering, Antwerp, Belgium, pp. 1–11.

Kawamura, S. (1991), *Integrated Design of Water Treatment Facilities,* John Wiley & Sons, Toronto.

Leland, D. E. and M. Damewood III (1990), "Slow Sand Filtration in Small Systems in Oregon," *J. American Water Works Association,* 82, 6, pp. 50–59.

Lodgson, G. S., D. G. Neden, A. M. D. Ferguson, and S. D. LaBonde (1993), "Testing Direct Filtration for the Treatment of High-Turbidity Water," *J. American Water Works Association,* 85, 12, pp. 39–46.

Malley, J. P., Jr., T. T. Eighmy, M. R. Collins, J. A. Royce, and D. F. Morgan (1993), "The Performance and Microbiology of Ozone-Enhanced Biological Filtration," *J. American Water Works Association,* 85, 12, pp. 47–57.

Metcalf and Eddy, Inc. (1991), *Wastewater Engineering: Treatment, Disposal, Reuse,* 3rd ed., G. Tchobanoglous and F. L. Burton, eds., McGraw-Hill, Toronto.

Monk, R. D. G. (1984), "Improved Methods of Designing Filter Boxes," *J. American Water Works Association,* 76, 8, pp. 54–59.

Reed, S. C., E. J. Middlebrooks, and R. W. Crites (1988), *Natural Systems for Waste Management and Treatment,* McGraw-Hill, Toronto.

Richardson, J. F. and W. N. Zaki (1954), *Transactions of the Institution of Chemical Engineers,* 32, pp. 35–53.

Rittman, B. E. and V. L. Snoeyink (1984), "Achieving Biologically Stable Water," *J. American Water Works Association,* 76, 10, pp. 106–114.

Schuler, P. F., M. M Ghosh, and P. Gopalan (1991), "Slow Sand and Diatomaceous Earth Filtration of Cysts and Other Particulates," *Water Research,* 25, 8, pp. 995–1005.

Tchobanoglous, G. and R. Eliassen (1970), "Filtration of Treated Sewage Effluent," *J. Sanitary Engineering Division, ASCE,* 96, SA2, pp. 243–265.

USEPA (1977), *Wastewater Filtration Design Considerations,* No. EPA-625/4–74-007a, Center for Environmental Research Information, Cincinnati, OH.

USEPA (1983), *Municipal Wastewater Stabilization Ponds: Design Manual,* USEPA, Washington, DC.

van Dijk, J. C. and J. H. C. M. Oomen, (1978), *Slow Sand Filtration for Community Water Supply in Developing Countries: A Design and Construction Manual,* WHO International Reference Centre, The Hague.

Wen, C. Y. and Y. H. Yu (1966), "Mechanics of Fluidization," *Chemical Engineering Progress Symposium Series,* 62, pp. 100–111.

CHAPTER 15

PHYSICAL–CHEMICAL TREATMENT FOR DISSOLVED CONSTITUENTS

The removal of dissolved constituents in a water, particularly dissolved ions, is often a challenging problem. Chemical precipitation is used with flocculation, sedimentation, and filtration to remove many dissolved components. Adsorption and ion exchange are alternatives that do not require other unit operations.

Another significant physical–chemical treatment is oxidation, which is further discussed in Chapter 16.

15.1 WATER SOFTENING

Water softening is the removal of certain dissolved minerals in water that cause scaling in boilers, form deposits on pipes, and cause excessive consumption of soaps made from natural animal fats. The minerals responsible for these phenomena are referred to as hardness ions. The term hardness arose because of a colloquial reference in earlier times to the difficulty of laundering in waters containing large concentrations of mineral ions. Soaps made from animal fats readily precipitate with hardness ions. Therefore, when water is treated for the removal of hardness ions it is softened and there is a reduction in the consumption of soap as well as the mineral concentration (TDS). Modern synthetic detergents are not affected by the concentration of hardness ions.

Hardness concentration also influences the tendency of a water to protect or corrode distribution pipes. Industries are particularly concerned about the scale formation potential of a water. Hardness ions can also contribute color or influence the taste of products made from the water, which are important considerations for certain industries. Industries often treat water beyond standards for municipal water supplies.

There have been some studies that indicate a link between hardness concentration (particularly calcium and magnesium) and cardiovascular disease (Neri and Johansen, 1978). However, the link is tenuous and many confounding variables existed in studies (Comstock, 1979).

Hardness in water is due to cations such as calcium and magnesium (divalent cations) and to a lesser extent to aluminum, iron, and other divalent and trivalent cations. Calcium and magnesium are the two major cations responsible for hardness in natural waters. Therefore for most waters:

$$\text{TOTAL HARDNESS} = \text{CALCIUM HARDNESS} + \text{MAGNESIUM HARDNESS}$$

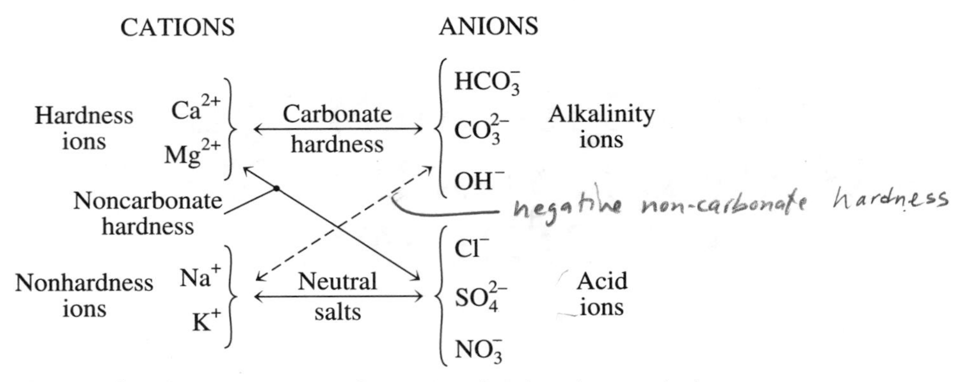

Figure 15.1 Major cations and associated anions in a typical water.

The cations are usually associated with the following anions: HCO_3^-, CO_3^{2-}, and OH^- (these are alkalinity ions), and SO_4^{2-}, NO_3^-, and Cl^- (these are acidity-associated ions). Carbonate hardness is the amount of hardness ions associated with alkalinity ions.

The hydroxyl ion concentration is normally insignificant compared to the hardness ions concentrations. The difference between the equivalents of the hardness ions, calcium and magnesium (and any other significant hardness ions), and the equivalents of bicarbonate and carbonate anions is noncarbonate hardness as indicated on Fig. 15.1. Noncarbonate hardness (also known as permanent hardness) is more costly to remove in precipitation processes as discussed below.

To remove the major ions causing hardness, an examination of solubility products is useful. Table 15.1 provides K_{sp}s and solubilities of various magnesium and calcium salts. It is observed that $CaCO_3$ and $Mg(OH)_2$ are relatively insoluble and hardness removal is commonly achieved by formation of these compounds. Furthermore, bicarbonate ion is usually naturally present in significant amounts in water. Because calcium is the most common ion causing hardness and calcium is commonly removed by formation of calcium carbonate with a convenient molecular weight of 100 (equivalent weight of 50), hardness is normally expressed in terms of $CaCO_3$.

TABLE 15.1 Solubilities of Possible Compounds Related to Water Hardness

Compound	Solubility, mg/L: Cold	Solubility, mg/L: Hot (25°C)	K_{sp}
$CaCO_3$	14	18	5×10^{-9}
$CaCl_2$		745 000	1 590 000
$Ca(OH)_2$	1 850	770	8×10^{-6}
$CaSO_4$	2 090	1 620	2×10^{-5}
$MgCO_3$	542 500	727 000	—
$Mg(OH)_2$	9	40	9×10^{-12}
$MgSO_4$	260 000	738 000	—
$Sr(OH)_2$	4 100	218 300	—
$SrCO_3$	11	650	—

TABLE 15.2 Hardness Ranges

Degree of hardness	Hardness concentration, mg/L as $CaCO_3$
Moderately hard	60–120
Hard	120–180
Very hard	180 and over

Water is considered soft when hardness is $\leq$60 mg/L as $CaCO_3$. Table 15.2 gives qualitative ranges of hardness.

15.2 LIME–SODA SOFTENING

Raising the pH of a water by addition of alkalinity will convert bicarbonates into carbonates and $CaCO_3$ formation and precipitation will ensue. Also, $Mg(OH)_2$ will be formed and precipitate. Slaked lime or lime [$Ca(OH)_2$], which is hydrated calcium oxide (CaO), is usually the least expensive source of alkalinity. Calcium oxide is also known as quicklime. Sodium hydroxide, NaOH (caustic soda), can also be used for alkalinity depending on economics and other factors such as availability and ease of handling. When insufficient bicarbonate concentrations exist, sodium carbonate (Na_2CO_3), known as soda ash, is added to supply both alkalinity and carbonate ions. The stoichiometry of the reactions is straightforward. The reactions involved in the lime–soda process are given below.

Any carbon dioxide present in the raw water consumes lime according to reaction (15.1). As lime is added, the pH rises and HCO_3^- is converted to CO_3^{2-} and the reactions in Eqs. (15.2) and (15.3a) occur. As more lime is added, the concentration of OH^- ions becomes significant and reactions (15.3b) and (15.5) proceed. Reaction (15.6) applies only when soda ash is added because some of the Ca^{2+} and Mg^{2+} ions are associated with noncarbonate anions. Soda ash is more expensive than lime.

$$Ca(OH)_2 + CO_2 \rightarrow CaCO_3(s) + H_2O \tag{15.1}$$

$$Ca(HCO_3)_2 + Ca(OH)_2 \rightarrow 2CaCO_3(s) + 2H_2O \tag{15.2}$$

$$Mg(HCO_3)_2 + Ca(OH)_2 \rightarrow CaCO_3(s) + MgCO_3 + 2H_2O \tag{15.3a}$$

$$MgCO_3 + Ca(OH)_2 \rightarrow Mg(OH)_2(s) + CaCO_3(s) \tag{15.3b}$$

Adding the above two reactions, the net lime requirement for magnesium carbonate hardness is

$$Mg(HCO_3)_2 + 2Ca(OH)_2 \rightarrow Mg(OH)_2(s) + 2CaCO_3(s) + 2H_2O \tag{15.4}$$

$$MgX + Ca(OH)_2 \rightarrow Mg(OH)_2(s) + CaX \tag{15.5}$$

$$CaX + Na_2CO_3 \rightarrow CaCO_3(s) + Na_2X \tag{15.6}$$

where

X is a noncarbonate anion such as SO_4^{2-}, NO_3^-, or Cl^-

Inorganic carbon alkalinity is first associated with Ca^{2+} because of the sequence of reactions that occurs on addition of lime. Bicarbonate is converted to carbonate by the addition of alkalinity and $CaCO_3$ forms and precipitates. $MgCO_3$ does not precipitate under normal circumstances but after HCO_3^- has been converted to CO_3^{2-}, additional OH^- does cause the precipitation of $Mg(OH)_2$. Therefore, when magnesium removal is desired, an amount of lime sufficient to convert alkalinity to carbonate must be added in addition to the amount of lime required to remove the magnesium.

Although Table 15.1 indicates that calcium concentrations below 20 mg/L as $CaCO_3$ can be achieved, the practical lower limit of removal is near 30 mg/L as $CaCO_3$ (0.6 meq/L). When the starting concentrations of Ca^{2+} and CO_3^{2-} are different, precipitation of $CaCO_3$ occurs until the ion with the lower initial concentration reaches 0.6 meq/L. This assumption is somewhat conservative (see Problem 3). Performance data are required to determine the actual residual calcium hardness. Environmental conditions and residence time in the reactor influence the extent to which equilibrium is achieved. For instance, Cole (1976) found that water treated with lime took 90 min at 1°C to reach the minimum hardness concentration, whereas at room temperature the same water took only 10 min to reach the same concentration.

The practical lower limit for removal of magnesium is 10 mg/L as $CaCO_3$ (0.2 meq/L). Excess lime at approximately 35 mg/L as $CaCO_3$ beyond stoichiometric requirements is necessary to raise the pH to levels to ensure $Mg(OH)_2$ formation. A pH above 10.5 must be achieved in order to precipitate $Mg(OH)_2$. The actual amount of excess alkalinity required is dependent on the buffering capacity of the water. Carbon dioxide can be added to neutralize the excess lime in the treated water and lower the pH to the desired value as shown by the following equations.

$$2OH^- + CO_2 \rightarrow CO_3^{2-} + H_2O \qquad (15.7)$$

$$CO_3^{2-} + H_2O + CO_2 \rightarrow 2HCO_3^- \qquad (15.8)$$

An alternative to remove the Ca^{2+} resulting from excess lime is to add Na_2CO_3. The Na_2CO_3 is added along with the lime to the water entering the basin, where precipitation of $CaCO_3$ and $Mg(OH)_2$ will occur. The excess OH^- in the effluent from this basin is neutralized by addition of acid to achieve the pH desired for a stable water (Section 15.3). This approach requires only a single basin for hardness removal, in contrast to split recarbonation processes discussed in the next section.

15.2.1 Treatment Methods for Hardness Removal

Softening can be accomplished in conjunction with coagulation, flocculation, and sedimentation of suspended solids as shown in Fig 9.2. There are also a variety of split treatment processes (Fig. 15.2). These processes are often used to control magnesium concentrations in the finished water. Magnesium silicate scale formed at high temperatures is problematic. A split recarbonation process is used to remove magnesium from waters that have high amounts of noncarbonate hardness (Montgomery, 1985). There is no bypass ($x = 0$ in Fig. 15.2). Excess lime is required to precipitate magnesium and recarbonation is required to neutralize the excess lime. If recarbonation is split into two stages where only enough CO_2 is applied in the first stage to react with the excess lime and produce carbonate, at the expense of an additional settling basin more calcium may be precipitated. The second stage recarbonation will be used for final pH adjustment. A single-stage process will produce water with more alkalinity and higher hardness concentrations.

In a split treatment process a portion of the water is treated for hardness (particularly magnesium) removal and recombined with a bypassed stream. The combined flow may be further treated for calcium reduction as shown in Fig. 15.2. Economies are gained by treating smaller volumes of water. Also magnesium concentrations in

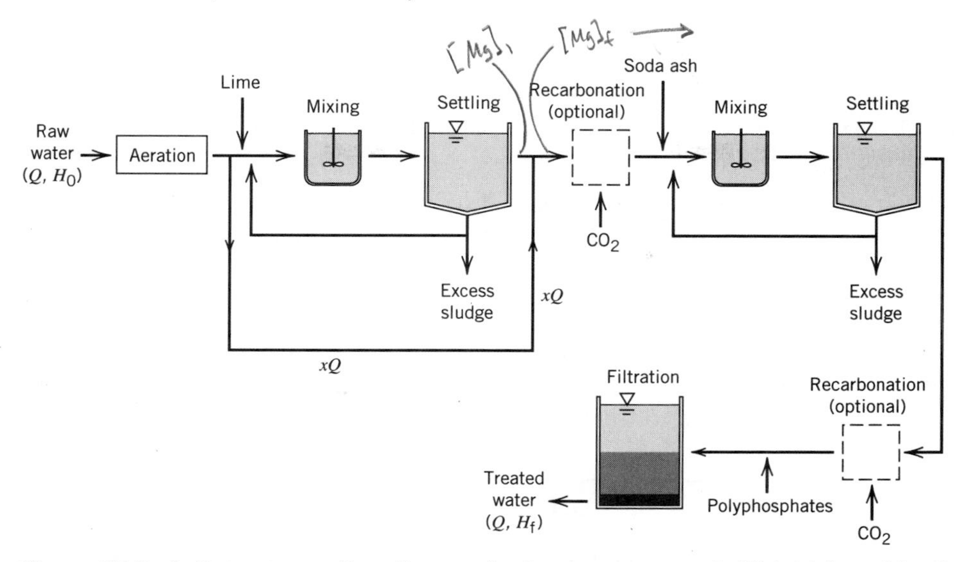

Figure 15.2 Split treatment flow diagram for hardness removal. (H_0 initial total hardness; H_f final total hardness). Adapted from J. L. Cleasby and J. H. Dillingham (1966), "Rational Aspects of Split Treatment," *J. of the Sanitary Engineering Division, ASCE,* 92, SA2, pp. 1–7. Reprinted with permission of ASCE.

the product water can be controlled in a split treatment process. Chemical costs are also reduced in a split treatment process. The bypassed flow can be treated with coagulation agents as required for removal of suspended solids in the settling basin.

A split treatment process is normally applied to a water with a low amount of noncarbonate hardness and magnesium hardness reduction is desired (Montgomery, 1985). In a split treatment process all lime based on the total flow is added to the stream entering the first stage to cause reactions (15.1)–(15.5). There is normally a large excess of lime in the first basin and magnesium concentrations below 0.2 meq/L down to practically zero can be attained in the settled effluent from this stage. Recarbonation of the effluent from the first stage is almost never required because there is sufficient alkalinity and carbon dioxide in the bypassed flow to react with the excess lime. There will be no precipitation of magnesium after the first stage because the pH will be lowered below solubility product values for $Mg(OH)_2$ precipitation.

Mixing the treated and bypassed streams results in more carbonate formation and subsequent precipitation of calcium carbonate. If further reduction of noncarbonate hardness is required the influent to the second stage can be supplemented with Na_2CO_3 to reduce the noncarbonate Ca^{2+} according to reaction (15.6). Another settling basin is required to settle the precipitate formed by these reactions. Finally, CO_2 is added to cause reaction (15.8) to bring the pH to the desired final value.

The final desired magnesium concentration dictates the fraction of flow that is bypassed in a split treatment process. Water heater fouling problems are avoided when magnesium concentrations are 40 mg/L as $CaCO_3$ (0.8 meq/L) or lower (Larson et al., 1959). Because no magnesium is removed in the second stage,

$$(1 - x)Q[\mathrm{Mg}]_1 + xQ[\mathrm{Mg}]_0 = Q[\mathrm{Mg}]_f \quad \text{or} \quad x = \frac{[\mathrm{Mg}]_f - [\mathrm{Mg}]_1}{[\mathrm{Mg}]_0 - [\mathrm{Mg}]_1} \tag{15.9}$$

where

x is bypassed fraction of influent flow, Q

$[Mg]_0$, $[Mg]_1$, and $[Mg]_f$ are the influent, first stage, and final magnesium concentrations, respectively

The maximum fraction that can be bypassed occurs when $[Mg]_1 = 0$.

Recycling a portion of the sludge from the clarifier promotes reaction and settling in the process. Doses higher than stoichiometric requirements may be required to achieve desired results.

Carbon dioxide can be supplied in bulk containers or generated by burning a hydrocarbon such as gas or oil or burning coal or coke. Polyphosphates are synthetic phosphate compounds that sequester hardness ions, keeping them in solution. Polyphosphates are often added to treated waters to prevent deposit buildup on pipes.

■ Example 15.1 Lime and Soda Determination for Hardness Removal

Using the river data in Problem 9 in Chapter 9, find the lime and soda requirements to treat the water to a final hardness of (a) 100 mg/L as $CaCO_3$ and (b) the practical limit in a single-stage process. In a single-stage process no CO_2 is added in the hardness precipitation reactor. Assume the excess lime requirement to remove magnesium is 0.75 meq/L. The pertinent data from the problem are as follows:

	mg/L as $CaCO_3$		
pH	Alkalinity	Total hardness	Calcium hardness
7.7	180	215	160

It is assumed that the alkalinity is due totally to inorganic carbon species and the concentrations of iron and manganese are ignored. The concentration of free CO_2 ($H_2CO_3^*$) will be checked to determine whether it is significant. The dissociation constant for $H_2CO_3^*$ (Table 3.2) is 4.3×10^{-7}. $[H^+] = 10^{-pH} = 10^{-7.7} = 2.00 \times 10^{-8}\ M$.

$$[HCO_3^-] = \left(180 \frac{\text{mg } CaCO_3}{\text{L}}\right)\left(\frac{122 \text{ mg } HCO_3^-}{100 \text{ mg } CaCO_3}\right)\left(\frac{1 \text{ mole}}{61\,000 \text{ mg}}\right) = 3.60 \times 10^{-3}\ M$$

From the equilibrium expression:

$$[H_2CO_3^*] = \frac{[H^+][HCO_3^-]}{K} = \frac{(2.00 \times 10^{-8})(3.60 \times 10^{-3})}{4.3 \times 10^{-7}} = 1.67 \times 10^{-4}\ M$$

The concentration of $H_2CO_3^*$ in meq/L is

$$[H_2CO_3^*] = (1.67 \times 10^{-4}\ M)(2\,000 \text{ meq/mole}) = 0.33 \text{ meq/L}$$

$$\text{Mg hardness} = \text{Total hardness} - \text{Ca hardness} = 215 \text{ mg/L} - 160 \text{ mg/L}$$
$$= 55 \text{ mg/L as } CaCO_3$$

$$[Mg^{2+}] = \left(55 \frac{\text{mg } CaCO_3}{\text{L}}\right)\left(\frac{1 \text{ meq}}{50 \text{ mg } CaCO_3}\right) = 1.10 \text{ meq/L}$$

$$[Ca^{2+}] = \left(160 \frac{\text{mg } CaCO_3}{\text{L}}\right)\left(\frac{1 \text{ meq}}{50 \text{ mg } CaCO_3}\right) = 3.20 \text{ meq/L}$$

Because there is 1 meq/mmole for HCO_3^-,

$$[HCO_3^-] = 3.60 \text{ meq/L}$$

The Ca carbonate hardness is 3.20 meq/L and the Mg carbonate hardness is 0.40 meq/L. There is 0.70 meq/L of Mg noncarbonate hardness. A residual hardness of 100 mg/L as $CaCO_3$ is equivalent to 2 meq/L.

(a) The amount of the total hardness to be removed (ΔTH) is

$$\Delta TH = 3.20 + 1.10 - 2.00 = 2.30 \text{ meq/L}$$

It will only be necessary to remove calcium hardness. The lime requirement is equal to the CO_2 content and the amount of calcium to be removed, which is increased by 0.6 meq/L to compensate for the residual CO_3^{2-}.

$$[Ca(OH)_2] = [CO_2] + [Ca(HCO_3)_2] + 0.60 \text{ meq/L} = 0.33 + 2.30 + 0.60 = 3.23 \text{ meq/L}$$

The final concentrations (meq/L) of ions in the water are: $[Ca^{2+}]$, 0.90; $[Mg^{2+}]$, 1.10; $[CO_3^{2-}]$, 0.60; $[HCO_3^-]$, 0.70

(b) In this case the lime requirement using Eqs. (15.1)–(15.6) is

$$\begin{aligned}[Ca(OH)_2] &= [CO_2] + [Ca(HCO_3)_2] + 2 \times [Mg(HCO_3)_2] + [MgX] + \text{excess} \\ &= 0.33 + 3.20 + 2 \times 0.40 + 0.70 + 0.75 = 5.78 \text{ meq/L}\end{aligned}$$

If only this amount of lime were added, the amounts of Ca^{2+} and CO_3^{2-} formed would be

$$\begin{aligned}[CO_3^{2-}] &= 0.33 + 6.40 + 0.80 = 7.53 \text{ meq/L} \\ [Ca^{2+}] &= 3.20 + 5.78 = 8.98 \text{ meq/L}\end{aligned}$$

The CO_3^{2-} is limiting and the residual Ca^{2+} and CO_3^{2-} concentrations would be

$$[CO_3^{2-}] = 0.60 \text{ meq/L} \qquad [Ca^{2+}] = 8.98 - 6.93 = 2.05 \text{ meq/L}$$

The residuals for Mg^{2+} and OH^- would be

$$\begin{aligned}[Mg^{2+}] &= 1.10 - (0.40 + 0.70) + 0.20 = 0.20 \text{ meq/L} \\ [OH^-] &= 5.78 - (0.33 + 3.20 + 0.40 + 0.90) = 0.95 \text{ meq/L}\end{aligned}$$

The soda ash requirement is

$$[Na_2CO_3] = [MgX] + \text{excess lime} = 0.70 + 0.75 = 1.45 \text{ meq/L}$$

The final composition (meq/L) of the water leaving the settling basin is: $[Ca^{2+}]$, 0.60; $[Mg^{2+}]$, 0.20; $[CO_3^{2-}]$, 0.60; $[OH^-]$, 0.95; $[Na^+]$, 1.45.

Acid addition will be required for OH^- neutralization and pH adjustment of the effluent.

■ Example 15.2 Hardness Removal in a Split Treatment Process

Determine the bypassed fraction and doses of lime and soda ash to be applied at each stage in a split treatment process to a groundwater that contains the following softening related constituents:

$[CO_2]$, 0.40 meq/L; $[Ca^{2+}]$, 3.60 meq/L; $[Mg^{2+}]$, 2.00 meq/L; [alkalinity], 5.80 meq/L

The product water should have a magnesium concentration of 50 mg/L as $CaCO_3$ (1 meq/L) and a total hardness concentration of 100 mg/L as $CaCO_3$ (2 meq/L) or lower. Perform the analysis when the concentration of magnesium from the first stage process is 0.2 meq/L. Assume the excess lime required to precipitate $Mg(OH)_2$ is 0.8 meq/L.

From Eq. (15.9),

$$x = \frac{[Mg]_f - [Mg]_1}{[Mg]_0 - [Mg]_1} = \frac{1.0 - 0.2}{2.0 - 0.2} = 0.444$$

Because the alkalinity exceeds the sum of the concentrations of the hardness ions, the raw water calcium carbonate hardness (CH) is 3.60 meq/L and the magnesium CH is 2.00 meq/L. There is no need to add Na_2CO_3 because there is no noncarbonate hardness.

The final concentration of calcium is 2.00 − 1.00 = 1.00 meq/L.

There are two constraints in this problem:

1. To precipitate $Mg(OH)_2$, lime must be added in excess of the carbon dioxide and alkalinity to result in free hydroxyl ions in the first stage.

2. There must be at least 0.8 meq/L excess OH^- to reduce Mg^{2+} to 0.2 meq/L.

The total equivalents of lime required is the amount required to remove CO_2, CaCH, and MgCH plus excess.

CO_2: from Eq. (15.1), 0.40 eq/L

CaCH: from Eq. (15.2), 3.60 − 1.00 = 2.60 meq/L

MgCH: from Eq. (15.4), 2(2.00 − 1.00) = 2.00 meq/L

Excess for Ca removal: The usual excess is 0.6 meq/L

The lime requirement is 0.40 + 2.60 + 2.00 + 0.6 = 5.60 meq/L based on the total flow. The fraction of flow entering the first stage is 1 − 0.444 = 0.556. The dose of lime (L) to be added to the flow entering the first stage is

$$0.556\ L = 5.60\ \text{meq/L} \qquad \text{or} \qquad L = (5.60\ \text{meq/L})/0.556 = 10.07\ \text{meq/L}$$

Because the CO_2 and HCO_3^- concentrations sum to 6.20 meq/L there is 4.07 meq/L of OH^-, which satisfies constraint 1.

After addition of the lime but before reaction,

$$[Ca^{2+}] = 10.07 + 3.60 = 13.67\ \text{meq/L} \qquad [OH^-] = 10.07\ \text{meq/L}$$

After reaction of OH^- with CO_2 and HCO_3^-,

$$[CO_3^{2-}] = 0.40 + 2(5.80) = 12.00\ \text{meq/L}$$

$CaCO_3$ is precipitated. CO_3^{2-} is limiting and its concentration is 0.60 meq/L after precipitate formation.

$$[Ca^{2+}] = 13.67 - (12.00 - 0.60) = 2.27\ \text{meq/L}$$

After reaction of Mg^{2+} with OH^-,

$$[Mg^{2+}]_1 = 0.20\ \text{meq/L}$$
$$[OH^-] = 10.07 - 0.40 - 5.80 - (2.00 - 0.20) = 2.07\ \text{meq/L}$$

This is considerably in excess of the 0.8 meq/L constraint. It is probable that more than 1.8 meq/L of Mg^{2+} will be removed but the calculations will be carried on assuming $[Mg^{2+}]_1$ = 0.20 meq/L.

After mixing the first stage and bypassed flows, the following concentrations result:

$$[OH^-] = \frac{0.556(2.07) + 0}{1} = 1.15 \text{ meq/L}$$

$$[Ca^{2-}] = \frac{0.556(2.27) + 0.444(3.60)}{1} = 2.86 \text{ meq/L}$$

$$[HCO_3^-] = 0.444(5.80) = 2.58 \text{ meq/L} \qquad [CO_2] = 0.444(0.40) = 0.18 \text{ meq/L}$$

$$[Mg^{2+}] = 0.556(0.2) + 0.444(2.00) = 1.00 \text{ meq/L}$$

$$[CO_3^{2-}] = 0.556(0.60) = 0.33 \text{ meq/L}$$

OH^- will react with CO_2 and HCO_3^- to produce

$$[CO_3^{2-}] = 0.33 + 0.18 + 2(1.15 - 0.18) = 2.45 \text{ meq/L}$$

CO_3^{2-} is limiting again. The final concentrations of all species (meq/L) in the water are $[Ca^{2+}]$, 1.01; $[CO_3^{2-}]$, 0.60; $[Mg^{2+}]$, 1.00; $[OH^-]$, 0.

Because of the operation of the process wherein all of the lime is added to the first stage, a sufficient excess for $Mg(OH)_2$ removal results and it is not necessary to add the $Mg(OH)_2$ excess into the lime dose.

If this water were to be treated in a conventional process the lime requirement would be

$$[Ca(OH)_2] = 0.40 + 5.80 + 2(2.00 - 1.00) + 0.80 = 9.00 \text{ meq/L}$$

to be added to the whole flow to meet the effluent magnesium objective. Calcium would be removed to 0.6 meq/L, which is better than the requirement, but constraint 1 necessitates this condition.

The lime savings is 3.4 meq/L. Also, the effluent from the conventional process would contain 0.8 meq/L of OH^-, requiring CO_2 or other neutralizing agent.

15.2.2 Bar Graphs

The solution of softening problems is conveniently accomplished using bar graphs showing the concentrations of the species involved at various stages during the process. On a bar graph, all concentrations are expressed in meq/L. The first step in a water softening analysis is to measure the concentrations of the major cations and anions. After the concentrations are converted to meq/L, the sums of the cation and anion concentrations should be approximately equal; otherwise there has been an error in analysis or computations, or a significant ion has not been measured.

The use of bar graphs is illustrated by an example. Consider the water with the concentrations of major ions listed in Table 15.3.

Usually only the ions given in Table 15.3 are significant in natural waters. The bar graph corresponding to the raw water is given in Fig. 15.3a. Carbon dioxide is given as the first component. It is not a cation or an anion. The cations are arranged in the order of Ca^{2+}, Sr^{2+}, Mg^{2+}, which is followed by other monovalent cations in any order. The anions are arranged with OH^-, CO_3^{2-}, HCO_3^-, followed by SO_4^{2-} and other monovalent anions in any order. The hypothetical concentrations for $Ca(HCO_3)_2$, $CaSO_4$, $MgSO_4$, and NaCl can be determined easily and are given on the graph for this water. The bar graph provides an easy check on the total cation and anion charge balance. The sum of the positive ions should be within 0.1–0.2 meq/L of the sum of the negative ions.

TABLE 15.3 Water Components

Constituent	Concentration mg/L	Equivalent weight	meq/L
CO_2	9.5	22	0.43
Ca^{2+}	98	20	4.90
Mg^{2+}	27	12.2	2.20
Na^+	6.5	23	0.28
HCO_3^-	281	61	4.60
SO_4^{2-}	120	48	2.50
Cl^-	6.1	35.5	0.18

It is desired to attain hardness removal to the practical limits of 30 mg/L of $CaCO_3$ (0.6 meq/L) and 10 mg/L of $MgSO_4$ (0.2 meq/L) using a split recarbonation method. The lime requirement is equal to the CO_2, $Ca(HCO_3)_2$, and Mg^{2+} concentrations plus the excess lime (only when Mg removal is desired). The lime requirement is determined from examination of Eqs. (15.1)–(15.6) and the composition of the water. In this case it is

$$\begin{aligned} Ca(OH)_2 &= CO_2 + Ca(HCO_3)_2 + Mg^{2+} + \text{excess} \\ &= 0.43 + 4.60 + 2.20 + 1.25 = 8.48 \text{ meq/L} \end{aligned}$$

The excess lime was chosen to be 1.25 meq/L, which is somewhat higher than the more typical value of 0.7 meq/L. The excess lime requirement will be dictated by trial and error for the water and treatment times in each situation.

This amount of lime will remove the CO_2, and Ca^{2+}, or CO_3^{2-} will be removed down to the level of 0.60 meq/L, depending on which of these ions is limiting. Likewise, Mg^{2+} is removed only to a level of 0.2 eq/L. The state of the water after addition of the lime is shown in Fig. 15.3b. Figure 15.3c shows the state of the water after reaction of OH^- with CO_2 and HCO_3^-. Note that the total alkalinity has not changed but the forms of alkalinity have changed.

After precipitation of the remaining $CaCO_3$ and $Mg(OH)_2$ the water is in the state shown in Fig. 15.3d. Note that the cation and anion orders given previously are maintained throughout all of the bar graphs. The water in Fig. 15.3d is the effluent from the first stage of the process.

Now CO_2 is added to convert the excess OH^- (1.45 meq/L from excess lime and Mg^{2+} that was not precipitated) to CO_3^{2-}, and Na_2CO_3 is added before the second mixing (reaction) vessel to precipitate the remaining Ca^{2+} as $CaCO_3$ in the second settling basin. The Na_2CO_3 requirement is

$$Na_2CO_3 = 4.35 - 1.45 - 0.60 = 2.30 \text{ meq/L}$$

Figure 15.3e shows the water after recarbonation and addition of Na_2CO_3. After final precipitation of $CaCO_3$ the water is in the state shown in Fig. 15.3f. Recarbonation or acid addition to achieve the desired pH for stabilization of the water may now be performed.

Lime Recovery and Sludge Reduction

Softening sludges, generated in basins that are separate from primary sedimentation basins, contain only $CaCO_3$ and $Mg(OH)_2$. Softening processes may be operated to

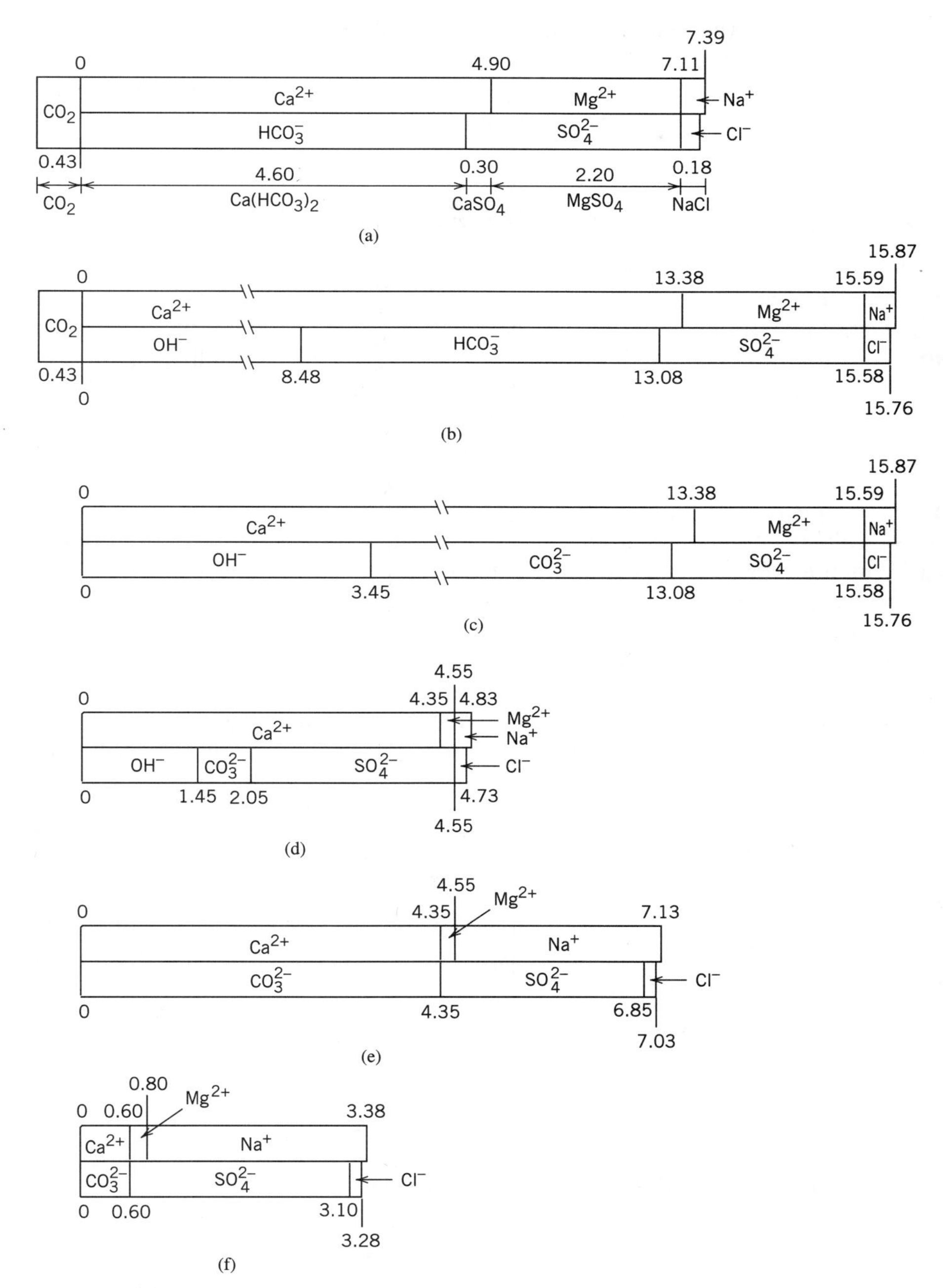

Figure 15.3 State of the water at stages during treatment.

exclusively remove $CaCO_3$ in one basin. Lime is readily recovered from calcium carbonate sludge by heating it to drive off CO_2.

$$CaCO_3 \xrightarrow{\Delta} CaO + CO_2 \tag{15.10}$$

Lime recovery reduces the amount of lime required as well as decreases the quantity of sludge. For waters with a high alkalinity content, each mole of calcium removed produces 2 moles of calcium carbonate as shown in reaction (15.2). Not all sludge will need to be treated for lime recovery to sustain the process.

15.3 CHEMICAL STABILIZATION OF WATER

Water leaving the treatment plant should not be corrosive to the pipes in the distribution system or in households. To protect pipes from corrosive agents in the water, the pH, $[Ca^{2+}]$, and alkalinity content of the water are adjusted to the calcium carbonate saturation equilibrium value at the temperature of the water. Normally a slight tendency to precipitate calcium carbonate is maintained. This preserves a film of calcium carbonate in distribution pipes and retards corrosion of the pipes (see Section 2.3.2). This process is referred to as chemical stabilization of the water.

Water is considered to be stable when it will neither dissolve nor deposit calcium carbonate, i.e., it is just saturated with calcium carbonate. A water that has a tendency to dissolve calcium carbonate is an aggressive water. An aggressive water is not necessarily innately more corrosive than a stable water. But because of its ability to remove the protective $CaCO_3$ barrier, an aggressive water will indirectly enhance the corrosion of the pipes.

The chemical reactions involved are the solubility product relation for calcium carbonate and a carbonate equilibrium expression. Bicarbonate alkalinity will be the predominant form of alkalinity at pHs of treated water. The two equations can be added to find a single equation relating $[Ca^{2+}]$, pH ($[H^+]$), and alkalinity as follows:

$$Ca^{2+} + CO_3^{2-} \rightleftharpoons CaCO_3(s) \qquad 1/K_{sp} \tag{15.11}$$

$$HCO_3^- \rightleftharpoons H^+ + CO_3^{2-} \qquad K_2 \tag{15.12}$$

$$Ca^{2+} + HCO_3^- \rightleftharpoons H^+ + CaCO_3(s) \tag{15.13}$$

To accurately determine the pH required to precipitate $CaCO_3$, activity coefficients must be incorporated into the equilibrium expressions (see Eq. 2.10). The equilibrium expressions for the above equations are

$$\text{For Eq. (15.11):} \qquad K_{sp} = \gamma_{Ca^{2+}}[Ca^{2+}]\gamma_{CO_3^{2-}}[CO_3^{2-}]$$

$$\text{For Eq. (15.12):} \qquad K_2 = \frac{\gamma_{CO_3^{2-}}[CO_3^{2-}]\gamma_{H^+}[H^+]}{\gamma_{HCO_3^-}[HCO_3^-]}$$

Therefore the equilibrium expression for Eq. (15.13) is

$$K = \frac{K_{sp}}{K_2} = \frac{\gamma_{Ca^{2+}}[Ca^{2+}]\gamma_{HCO_3^-}[HCO_3^-]}{\gamma_{H^+}[H^+]} \tag{15.14}$$

A pH meter actually measures the activity of the hydrogen ion.

$$pH_{meter} = -\log \gamma_{H^+}[H^+]$$

Defining the pH corresponding to the saturation condition as pH_s, from Eq. (15.14),

$$\begin{aligned} pH_s &= -\log \gamma_{H^+}[H^+] = \log K - \log \gamma_{Ca^{2+}}[Ca^{2+}] - \log \gamma_{HCO_3^-}[HCO_3^-] \\ &= \log K_{sp} - \log K_2 - \log [Ca^{2+}] - \log [HCO_3^-] - \log \gamma_{Ca^{2+}}\gamma_{HCO_3^-} \end{aligned}$$

Using the "p" notation, the definition of alkalinity given by Eq. (3.25), and assuming that the carbonate ion concentration is negligible,

$$pH_s = pK_2 - pK_{sp} - \log [Ca^{2+}] + S - \log [\text{Alkalinity}] \quad (15.15)$$

where

S is a salinity correction factor equal to $-\log \gamma_{Ca^{2+}}\gamma_{HCO_3^-}$

(Note that $[Ca^{2+}]$ and [Alkalinity] are in mole/L and alkalinity is assumed to be $[HCO_3^-]$.)

The factor S adjusts the equation for the true activities of the ions in the equilibrium expressions. Equation (15.15) is valid in the pH range of 6.0–8.5 in which most treated waters fall (AWWA Joint Task Group, 1990). This equation is based on the assumption that bicarbonate ion is the only significant alkalinity ion.

Equation (15.15) should be applied at the temperature of the treated water. The equilibrium constants, K_{sp} and K_2, and activity coefficients are functions of temperature. Table 15.4 gives values of S and the equilibrium constants over a broad temperature range. Note that pK_{sp} values in Table 15.4 are based on the formation of calcite, which is one crystalline form of $CaCO_3$. Other isomorphs of $CaCO_3$ that may form are aragonite and veterite. The solubility products for these other forms are slightly different from that for calcite and their formation changes pH_s. This is a complicating factor in any pH_s calculation. However, the AWWA Joint Task Group (1990) points out that the most common form of $CaCO_3$ in fresh waters is calcite.

The parameters involved are all readily assessed by simple lab procedures. It is also possible to determine pH_s for a water by keeping the water in contact with pure $CaCO_3$ overnight which is sufficient time to establish equilibrium conditions. Then the pH, which will be pH_s, should be measured.

TABLE 15.4 Equilibrium Constants and Salinity Factors for Saturation Index[a]

			S[d]				
			TDS, mg/L				
Temperature, °C	pK_2[b]	pK_{sp}[b,c]	50	150	400	1 000	1 500
5	10.55	8.39	0.082 5	0.137	0.210	0.300	0.345
10	10.49	8.41	0.083 2	0.138	0.211	0.303	0.348
15	10.43	8.43	0.083 8	0.139	0.213	0.305	0.351
20	10.38	8.45	0.084 5	0.140	0.215	0.308	0.354
25	10.33	8.48	0.085 4	0.142	0.217	0.311	0.358
30	10.29	8.51	0.086 1	0.143	0.219	0.314	0.362
35	10.25	8.54	0.086 9	0.144	0.221	0.318	0.366
40	10.22	8.58	0.087 9	0.146	0.224	0.322	0.370
45	10.20	8.62	0.088 8	0.148	0.226	0.325	0.375
50	10.17	8.66	0.089 8	0.149	0.229	0.329	0.379
60	10.14	8.76	0.091 9	0.153	0.235	0.337	0.389
70	10.13	8.87	0.094 1	0.157	0.241	0.346	0.400
80	10.13	8.99	0.096 5	0.161	0.247	0.356	0.411
90	10.14	9.12	0.099 0	0.165	0.254	0.366	0.423

[a]Adapted from AWWA Joint Task Group (1990).
[b]From Plummer and Busenberg (1982).
[c]For calcite as recommended by AWWA Joint Task Group (1990).
[d]Based on Eq. (1.10) for ionic strength and the activity expression given by AWWA Joint Task Group (1990).

If the pH of the water is less than pH_s, there will be no deposition of $CaCO_3$ and hence the possibility of corrosion is enhanced. If the pH of the water is greater than pH_s then $CaCO_3$ can be deposited in pipes.

Langelier (1936) performed the development just described and the value of $pH - pH_s$ (pH is the actual pH of the water being tested) is called the Langelier index or saturation index (SI).

$$SI = pH - pH_s \tag{15.16}$$

It is common practice to adjust waters to an SI value of 0.2 by addition of the appropriate alkalinity or acidity agent. Lime changes both calcium and alkalinity concentrations. Other agents that can be used to adjust the water to the desired conditions are Na_2CO_3, CO_2, and strong acids or bases such as HCl and NaOH. These agents will change the ratio of $H_2CO_3^*$ and HCO_3^- from which the new pH of the water can be calculated with Eq. (3.22b).

The SI, although widely used, is only a qualitative indication of the amount of potential $CaCO_3$ deposition. A larger value of the index does not necessarily mean that more $CaCO_3$ will deposit and, at the extreme case of a pH greater than pK_2 for carbonic acid, an undersaturated solution will yield a positive index value. Snoeyink and Jenkins (1980) and Rossum and Merrill (1983) offer a comprehensive discussion of the index, pointing out other deficiencies.

There are a number of other indexes of a similar nature to the SI (Rossum and Merrill, 1983; AWWA Joint Task Group, 1990). The modified Caldwell–Lawrence approach (Caldwell and Lawrence, 1953; Merrill, 1978), which is beyond the scope of this text, offers a relatively easy approach to quantitatively estimating the extent of potential precipitation. Other indexes or methods to quantify the deposition potential of $CaCO_3$ require computer code or extensive calculations. These indexes usually consider more chemical equilibria, particularly the formation of complexes, and so produce more accurate results. The programs for calculating the indexes are readily run on a personal computer with a relatively small amount of memory.

■ Example 15.3 Saturation Index

The calcium concentration of a water is 42 mg/L and the alkalinity concentration is 60 mg/L as $CaCO_3$. TDS have been measured at 120 mg/L. The pH of the water is 7.73 and its temperature is 12°C. Find the SI of this water.

Equation (15.15) is required to calculate pH_s. The values for pK_2, pK_{sp}, and S are obtained from Table 15.4. At the given values of temperature and TDS it will be necessary to interpolate for each parameter. The values for pK_2, pK_{sp}, and S are

$$pK_2 = 10.49 + \frac{(12-10)}{(15-10)}(10.43 - 10.49) = 10.47$$

$$pK_{sp} = 8.41 + \frac{(12-10)}{(15-10)}(8.43 - 8.41) = 8.42$$

$$S_{12,50} = 0.083\,2 + \frac{(12-10)}{(15-10)}(0.083\,8 - 0.083\,2) = 0.083\,44$$

$$S_{12,150} = \frac{(12-10)}{(15-10)}(0.139 - 0.138) = 0.138\,4$$

$$S_{12,120} = S_{12,50} + \frac{(120-50)}{(150-50)}(S_{12,150} - S_{12,50}) = 0.083\,44 + \frac{(70)}{(100)}(0.138\,4 - 0.083\,44) = 0.122$$

Converting the calcium and alkalinity concentrations to M:

$$[Ca^{2+}] = \left(42\,\frac{mg}{L}\right)\left(\frac{1\ mole}{40\ g}\right)\left(\frac{1\ g}{1\ 000\ mg}\right) = 1.05 \times 10^{-3}\ M$$

$$[Alkalinity] = \left(60\,\frac{mg\ CaCO_3}{L}\right)\left(\frac{61\ g\ HCO_3^-}{50\ g\ CaCO_3}\right)\left(\frac{1\ mole}{61\ g\ HCO_3^-}\right)\left(\frac{1\ g}{1\ 000\ mg}\right)$$
$$= 1.20 \times 10^{-3}\ M$$

Substituting these values into Eq. (15.15),

$$pH_s = 10.47 - 8.42 - \log(1.05 \times 10^{-3}) + 0.122 - \log(1.20 \times 10^{-3}) = 8.07$$

and

$$SI = 7.73 - 8.07 = -0.34$$

The water is undersaturated and it will be necessary to add strong base or Ca^{2+} [$Ca(OH)_2$ accomplishes both] to raise the SI.

Corrosion results in the addition of metals to the water and wastewater. The bulk of metals in a wastewater is removed with the sludge in a wastewater treatment process. Another benefit of water stabilization is reduction in the metal content of sludges, which improves the reusability of the sludge from the wastewater treatment plant (Kuchenrither et al., 1992).

15.4 IRON AND MANGANESE REMOVAL

Iron and manganese are discussed together because they commonly occur together in raw waters and the problems resulting from them and their methods of treatment are similar. Iron and manganese are minerals that cause staining of plumbing fixtures and laundered clothes as well as produce distinct tastes and odors in a drinking water. Iron and manganese also contribute to the hardness of a water. These are aesthetic problems; there is no health risk associated with excessive amounts of iron and manganese. The WHO standards for acceptable concentrations of iron and manganese are 0.3 and 0.1 mg/L, respectively.

The solubilities of iron and manganese are primarily controlled by their oxidation state. The redox reactions are

$$Fe^{3+} + e^- \rightarrow Fe^{2+} \tag{15.17a}$$

$$Mn^{4+} + 2e^- \rightarrow Mn^{2+} \tag{15.17b}$$

The higher oxidation states of both iron and manganese are soluble to an insignificant degree in waters in normal pH ranges. At these oxidation states, the precipitates of Fe_2O_3 and MnO_2 form. The iron oxide (rust) has a reddish-brown color and the manganese dioxide has a brown to brownish-black appearance.

In the presence of dissolved oxygen, iron and manganese are oxidized to their insoluble oxidation states of Fe(III) and Mn(IV) according to redox reactions (15.17a) and (15.17b). Reducing (anaerobic) conditions cause the dissolution of these minerals. Therefore, groundwaters are particularly susceptible to excess concentrations of these minerals when soil contains these minerals. Also, in the bottom sediments of streams and lakes, anaerobic conditions exist and precipitated organic matter releases iron and manganese to the water. Redox reactions are slower than acid–base reactions.

Environmental conditions such as pH, temperature, complexing agents, and other factors influence the rate of a redox reaction. Sometimes iron and manganese are found in aerated surface waters because the rate of conversion of soluble iron and manganese to their stable, insoluble forms is slower than their rate of production or input from the sources.

The treatments for iron and manganese removal are based on accelerating their rate of oxidation. Manganese is more difficult to remove than iron. Copper sulfate catalyzes the oxidation of manganese. Contact processes form MnO_2, which also catalyzes the oxidation of Mn(II). Iron and manganese precipitates are colloidal in nature; therefore, they settle slowly. Filtration will efficiently remove them but other processes remove the iron and manganese precipitates on contact.

Addition of oxidizing agents is a commonly used method for iron and manganese removal. Chlorine or permanganate are the chemical agents most often used. Other common oxidizing agents such as chlorine dioxide and ozone are also effective. Chloramines will not convert ferrous iron to ferric iron or manganese to the +4 state. The overall chemical redox reactions are obtained by combining the appropriate half-reactions from Table 1.3. The redox reactions using permanganate as the oxidant are given in Eqs. (15.18) and (15.19a).

$$3Mn^{2+} + 2MnO_4^- + 4OH^- \rightarrow 5MnO_2(s) + 2H_2O \quad (15.18)$$

$$3Fe^{2+} + MnO_4^- + 4H^+ \rightarrow 3Fe^{3+} + MnO_2 + 2H_2O \quad (15.19a)$$

The Fe^{3+} reacts with alkalinity in the water to form $Fe(OH)_3$.

$$Fe^{3+} + 3OH^- \rightarrow Fe(OH)_3(s) \quad (15.19b)$$

The overall reaction for iron removal is

$$3Fe^{2+} + MnO_4^- + 5OH^- + 2H_2O \rightarrow 3Fe(OH)_3(s) + MnO_2 \quad (15.20)$$

From the reactions it is seen that the oxidation of iron with permanganate is favored by a low pH and the oxidation of manganese is favored by a high pH. However, a high pH favors the formation of $Fe(OH)_3$ and the overall reaction for iron removal is favored by a high pH. When potassium permanganate is used as the oxidant the formation of pink water is possible if doses are too high. The amount of permanganate required may be less than the stoichiometric amount because of the catalytic action of MnO_2 on the oxidation of both soluble iron and manganese.

The reactions for oxidation of iron and manganese with oxygen, chlorine, and chlorine dioxide are as follows:

$$4Fe^{2+} + O_2 + 10H_2O \rightarrow 4Fe(OH)_3 + 8H^+ \quad (15.21)$$

$$2Fe^{2+} + Cl_2 + 6H_2O \rightarrow 2Fe(OH)_3 + 6H^+ + 2Cl- \quad (15.22)$$

$$Fe^{2+} + ClO_2 + 3H_2O \rightarrow Fe(OH)_3 + ClO_2^- + 3H^+ \quad (15.23)$$

$$2Mn^{2+} + O_2 + 2H_2O \rightarrow 2MnO_2 + 4H^+ \quad (15.24)$$

$$Mn^{2+} + Cl_2 + 2H_2O \rightarrow MnO_2 + 4H^+ + 2Cl^- \quad (15.25)$$

$$Mn^{2+} + 2ClO_2 + 2H_2O \rightarrow MnO_2 + 2ClO_2^- + 4H^+ \quad (15.26)$$

Chemical oxidizing agents are usually added near the beginning of a treatment process to allow them to react with iron and manganese in following units operations (Glaze, 1990). It may be necessary to provide a holding tank for reaction depending on which operations follow the application of the agent.

Coagulation, flocculation, and sedimentation will remove some of the iron and manganese precipitates formed from addition of oxygen or chemical agents. Filtration is almost always necessary to remove the fine iron and manganese precipitates to a satisfactory degree, regardless of whether sedimentation is used.

Greensand

Greensand (also known as glauconite) is a granular, naturally occurring mineral with a greenish color. This zeolite has a composition of $(K,Na,Ca)_{1.2\text{-}2}$–$(Fe^{3+},Al,Fe,Mg)_4$ $Si_{7\text{-}7.6}Al_{1\text{-}1.4}O_2(OH)_4 \cdot nH_2O$. Iron and manganese can be removed by greensand medium that has a manganese coating. Adsorbed manganese catalyzes the oxidation of Fe(II) and Mn(II). The mechanism of removal is a combination of sorption and oxidation. Greensand is regenerated with potassium permanganate. The regenerant removes the sorbed, oxidized iron and manganese and restores the removal capability of the greensand. The reactions for oxidation and regeneration are

$$Z - MnO_2 + \begin{cases} Fe^{2+} \\ Mn^{2+} \end{cases} \rightarrow Z - Mn_2O_3 + \begin{cases} Fe^{3+} \\ Mn^{3+} \\ Mn^{4+} \end{cases} \tag{15.27}$$

$$Z - Mn_2O_3 + KMnO_4 \rightarrow Z - MnO_2 \tag{15.28}$$

where

Z is the greensand

Calcium and magnesium are not removed in a greensand filter. However, in addition to iron and manganese, greensand can remove hydrogen sulfide, phenols, and radium-226 (Brinck et al., 1978; Sayell and Davis, 1975; Weber, 1972).

Manganese greensand beds are usually placed in pressure filters operated in the downflow mode. It is recommended that the pressure differential across the bed not exceed 83 kPa (12 psi) (Sayell and Davis, 1975). When iron or sulfide is the primary contaminant to be removed, potassium permanganate and other oxidants are continuously added to the influent to the greensand filter. Backwashing is required only infrequently. If influent manganese concentrations are significant, then backwashing will be required on a regular basis. The manganese dioxide coating of the greensand will grow when manganese concentrations are significant because some of the Mn(II) will be oxidized and retained by the greensand. Plugging of the greensand medium can be largely eliminated by placing a filter ahead of the greensand filter and feeding the potassium permanganate ahead of the rapid filter.

Waters containing 18 mg/L Fe(II) or more can be easily treated in a greensand filter. Proper process operation produces effluents with iron concentrations not exceeding 0.1 mg/L and manganese concentrations less than 0.01 mg/L (Sayell and Davis, 1975).

Aeration

Aeration is another alternative for the removal of iron and manganese. The aeration configuration depicted in Fig. 12.3d is effective for oxidation of Fe(II). The reaction is given by Eq. (15.21). Manganese is more difficult to oxidize than iron. Addition of coke coated with oxides to the trays will effectively catalyze the oxidation of Mn(II) to its insoluble forms. Sedimentation and filtration will remove the iron and manganese precipitates formed in the aeration device.

Sequestering Iron and Manganese

Instead of removing iron and manganese, sequestering agents may be added to a water that bind the iron and manganese in polymeric colloidal complexes which prevent them from forming color or turbidity within detention times in the distribution system. Sodium silicates, phosphates, or polyphosphates are the common sequestering agents used. Chlorine must be added simultaneously with sodium silicate to oxidize iron and manganese (Robinson et al., 1992). Iron is most effectively sequestered with silicate; manganese is less susceptible. Silicate doses in the range of 5–25 mg/L as SiO_2 are normal.

Biological Removal of Iron and Manganese

All of the processes previously discussed for iron and manganese removal are physical–chemical processes. A recent innovation has been the development of biological iron and manganese removal (Mouchet, 1992). The processes are complex and only a brief description is given here (see Mouchet, 1992, for more detail). The processes use filters that support autotrophic iron- and manganese-oxidizing bacteria. Although some bacteria can oxidize both species, others can oxidize only one of them and the optimal conditions for iron removal are different from those for manganese removal. Two separate filters are usually required if both iron and manganese must be removed.

The influent must be adjusted to ORP and pH conditions that promote the activity of the respective groups of bacteria. Dissolved oxygen content of the influent is critical. Conditions that cause physical–chemical removal of iron or manganese will decrease activity of iron and manganese bacteria and ultimately impair process performance. A water must be fully nitrified before biological manganese removal can take place. Higher filtration rates in the range of 25–40 m/h (80–130 ft/h) are used in iron removal filters; typical filtration rates in manganese filters are 10–40 m/h (33–130 ft/h). The higher filtration rates in iron or manganese filters are achieved with coarse sand media with an effective size (d_{10}) of 0.95–1.35 mm or higher. Backwashing frequency depends on the raw water quality but the quantity of backwash water should be well below 1% of the product water.

There are more than 100 bio-Fe or -Mn operations in France alone and other installations throughout Europe. The process greatly reduces the use of chemical agents and is claimed to reduce capital and operating costs (Mouchet, 1992). Sludge produced from the process does not pose treatability problems.

15.5 PHOSPHORUS REMOVAL FROM WASTEWATER BY CHEMICAL PRECIPITATION

Phosphorus is usually the limiting nutrient for eutrophication in inland receiving waters for wastewater treatment plant effluent; therefore, phosphorus concentrations in effluents are controlled. Biological wastewater treatment processes remove some phosphorus in the biological sludge that is formed in these processes but the amount of P removal is normally inadequate for typical wastewaters.

Agents used to precipitate dissolved phosphorus are salts of the metals calcium (Ca^{2+}), iron (either Fe^{2+} or Fe^{3+}), or aluminum [either alum, $Al_2(SO_4)_3 \cdot 18H_2O$ or sodium aluminate, $Na_2Al_2O_4$]. The chloride and sulfate salts of Fe^{2+} and Fe^{3+} can be used. Pickle liquor, which is a waste product of the steel industry that contains ferrous iron in either a sulfuric or hydrochloric acid solution, may also be used. All of these metals form sparingly soluble salts with dissolved forms of phosphate. Besides forming precipitates with dissolved phosphorus, these agents also aid the coagulation–flocculation of suspended solids, which enhances their settling. Any improvement in the removal of biological solids or other inorganic solids that contain phosphorus further contributes to total phosphorus reduction.

Orthophosphate (dissolved inorganic phosphate) equilibrium reactions are given in Table 3.2. In the pH range of wastewaters (around 7), the dominant forms of orthophosphate are $H_2PO_4^-$ and HPO_4^{2-}. As the pH increases the equilibrium moves toward formation of PO_4^{3-}. Condensed phosphates (polyphosphates) are also present in fresh wastewaters but microorganisms convert polyphosphates into orthophosphates in the sewers and biological treatment processes.

The chemistry of phosphate precipitate formation is complex because of complexes formed between phosphate and metals and between metals and other ligands in the wastewater. Side reactions of the metals with alkalinity to form hydroxide precipitates are another factor to be considered. The common precipitates formed by the metals given above are given in Table 15.5 along with the optimal pH range for phosphate precipitation.

Practically the amount of phosphorus removal at different conditions can be assessed in jar tests (Section 13.1). The dose of coagulant required is a function of pH. Typical curves for residual phosphorus versus dose and pH are given in Figs. 15.4a and 15.4b.

The amount of metal needed to remove phosphorus exceeds the stoichiometric amount at the low levels at which phosphorus is present in a wastewater. Typical phosphorus concentrations in raw wastewater are in the range of 4–8 mg P/L, which is low, but effluent concentrations are commonly specified to be 1 mg/L or less to minimize environmental effects. Considering the general relation for solubility product for species listed in Table 15.5,

$$[H_xPO_4^{3-x}] = \sqrt[n]{\frac{K_{sp}}{[Me^{a+}]^b}}$$

where

$[Me^{a+}]$ is the concentration of metal
n and b depend on the metal

TABLE 15.5 Phosphate Precipitates[a]

Metal	Precipitates	pH range	Comment
Ca^{2+}	Various calcium phosphates, e.g., $Ca_3(PO_4)_2$, $Ca_5(OH)(PO_4)_3$, $CaHPO_4$	≥10	Produces lowest residual P concentrations. The alkalinity of the water determines the dose because of the formation of $CaCO_3$.
	$CaCO_3$	≤9.5	Residual P in the range of 1–2 mg/L.
Fe^{2+}	$Fe_3(PO_4)_2$, $Fe_x(OH)_y(PO_4)_3$, $Fe(OH)_2$, $Fe(OH)_3$	6–8.5	There will be some oxidation of Fe^{2+} to Fe^{3+}.
Fe^{3+}	$Fe_x(OH)_y(PO_4)_z$, $Fe(OH)_3$	6–8.5	
Al^{3+}	$Al_x(OH)_y(PO_4)_3$, $Al(OH)_3$	6–8.5	

[a]Reprinted with permission from D. Jenkins and S. W. Hermanowicz (1991), "Principles of Chemical Phosphate Removal," in *Phosphorus and Nitrogen Removal from Municipal Wastewater,* 2nd ed., R. I. Sedlak, ed., Chelsea, MI, copyright Lewis Publishers, an imprint of CRC Press, Boca Raton, FL.

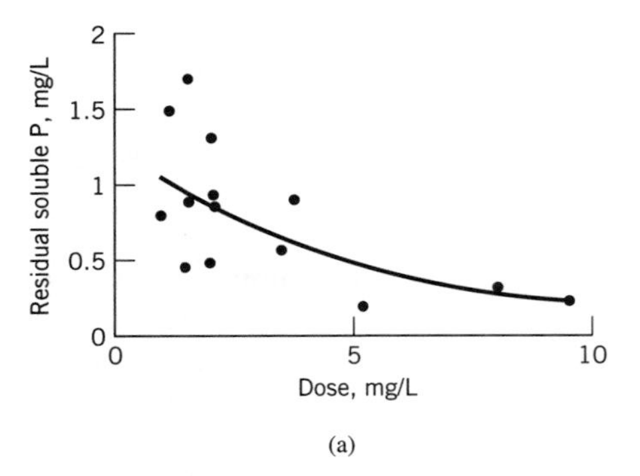

(a)

Figure 15.4a Residual soluble P as a function of dose. From USEPA (1987c)

The inverse relation between phosphate species concentration and metal ion concentration dictates larger doses of the metal agent as smaller residual concentrations of phosphorus are required. This partially explains the nature of the curve in Fig. 15.4a. There will also be a higher concentration of complexes formed at the higher metal doses.

Lime can achieve the lowest residual phosphate concentration because of the formation of apatites. Apatites (apatites are calcium-, hydroxide-, and phosphate-containing compounds) such as $Ca_5(OH)(PO_4)_3$ are highly insoluble; however, a high pH must be attained to realize their formation. Because calcium forms $CaCO_3$ near pH 9.5 and pHs higher than 9.5 are required for significant apatite formation, the alkalinity of the water is the governing factor for the required dose of lime. Very high doses of lime are required, which makes this choice costly compared to other metal agents, although the lowest residual P concentrations can be achieved.

For iron and aluminum salts, without jar tests, the general guideline for the metal dose to achieve total soluble residual P concentrations of from 1 to 2 mg P/L is 1 mole of metal per mole of phosphorus (Jenkins and Hermanowicz, 1991; USEPA 1987a). As the metal:P ratio is increased to 1.5 and 2, residual P concentrations decrease to 0.5 and 0.3 mg/L, respectively.

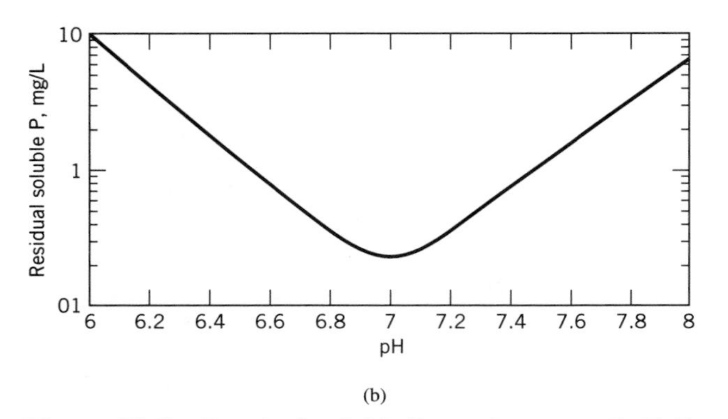

(b)

Figure 15.4b Residual soluble P as a function of pH. Reprinted with permission from D. Jenkins and S. W. Hermanowicz (1991), "Principles of Chemical Phosphate Removal," in *Phosphorus and Nitrogen Removal from Municipal Wastewater,* 2nd ed., R. I. Sedlak, ed., Chelsea, MI, copyright Lewis Publishers, an imprint of CRC Press, Boca Raton, FL.

Addition of metal coagulants to wastewater can have a significant impact on the amount of sludge produced. The common addition points for metals in biological treatment plants are before the primary clarifier or before the secondary clarifier (see Fig. 9.4). In the physical–chemical plant depicted in Fig. 9.5, P removal would occur in the primary clarifier. In a biological treatment plant, metal salt addition to the primary clarifier can result in sludge mass increases of from 50 to 100% from this clarifier and an overall increase of 60–70% in the sludge mass produced by the plant. If the metal agent is added before the secondary clarifier the increases in sludge mass are 35–45% from the clarifier and 10–25% for the whole plant (USEPA, 1987b). These salts will also have an impact on sludge dewaterability. Chemical costs and sludge handling costs must be considered together for optimization.

15.6 ION EXCHANGE

Ion exchange is the exchange of ions in solution for other ions on a medium. Ion exchange is primarily used for hardness removal in water treatment. In wastewater treatment it can be used for removal of toxic metals or recovery of precious metals. There are many natural substances that are able to exchange ions. Chemists have developed many synthetic media that are highly specific for target ions and are very efficient.

Synthetic resins are most commonly used because of their better performance. A resin is a three-dimensional hydrocarbon network to which functional groups (Section 4.3) are attached. These groups contain the exchangeable ions and are soluble in water. The exchange reaction is

$$nR^-B^+ + A^{n+} \rightarrow R_{n-}A^{n+} + nB^+ \qquad (15.29)$$

where

R_{n-} indicates that n functional groups are coordinated to A^{n+}

The equilibrium relation for the above equation is defined as a quotient, Q, or a selectivity quotient, Q_s, which is similar to an equilibrium constant.

$$Q = \frac{[B^+][R_n^- A^{n+}]}{[R^-B^+][A^{n+}]} \qquad (15.30a)$$

$$Q_s = \frac{[B^+]^n[R_n^- A^{n+}]}{[R^-B^+]^n[A^{n+}]} \qquad (15.30b)$$

where

$[R_n^- A^{n+}]$ and $[R^-B^+]$ are expressed either as (moles X)/(g of resin) or as (moles X)/(volume of resin)

It is essential that the effective concentrations of $R_n^- A^{n+}$ and R^-B^+ be used in the above expressions; i.e., activity coefficients for the operating conditions must be assessed.

Ion exchange resins are designed to remove ions of a certain class. In general there will be exchange of more than one ion species. The selectivity characteristics of resins that are available from manufacturers and the presence and concentration of dissolved solids in the water dictate the most favorable choice of a resin to achieve removal of the target species. Resins are subject to fouling. Iron and manganese can be problematic in some circumstances, coating resin particles irreversibly. Resins are good solid coagulants; therefore, suspended solids concentration of the influent should be low (Sanks, 1978). Influent turbidities less than 2 TU are recommended.

The total exchange capacities for a number of commercial resins ranged from 2.5 to 4.9 meq/g of resin on a dry basis (Flick, 1991). Resins have different swelling characteristics and the total exchange capacity on a volumetric basis ranged from 1.0 to 4.0 meq/mL.

The saturation capacity of a resin can be measured simply by means of a titration. The procedure is:

1. Regenerate the resin with acid (base).
2. Rinse the resin with distilled water or water that does not contain any ions that are exchangeable with the resin. This water may be prepared by passing it through an active resin. The rinse will remove any excess regenerating agent in the resin column.
3. Measure the resin volume.
4. Titrate the resin with base (acid).

The exchange capacity of the resin is calculated from

$$\text{Exchange capacity} = \frac{N_t V_t}{\text{volume of resin}} \tag{15.31}$$

where

N_t is the normality of the titrant
V_t is the volume of titrant used

The residence time of water in ion exchangers and other packed-bed reactors is often characterized in terms of the empty bed contact time (EBCT), which is simply the depth of media divided by the superficial velocity or the total bed volume divided by the volumetric flow rate (V/Q).

The theory and design of ion exchangers are the same as for adsorbers, which are described in the next section.

Activated Alumina

Activated alumina is a granular natural zeolite that is preferentially selective for fluoride, arsenic, selenium, phosphate, and silica (Clifford, 1990; Rubel and Woosely, 1979; Trussell et al., 1980). The most common application of activated alumina has been for defluoridation. Trussell et al. (1980) reported the order of preference of activated alumina at pH 6.5 for these substances as

$$OH^- > H_2PO_4^- > F^- > H_2AsO_4^- > HSeO_3^-$$

The pH is critical for the performance of activated alumina. A pH of 5.5 is generally reported as optimal for removal of fluoride, As(V), and Se(IV) (Rubel and Woosely, 1979; Trussell et al., 1980). If the pH is greater than 9.5, activated alumina acts as a cation exchanger. An EBCT of at least 5 min with activated alumina particle sizes between 28 and 48 mesh is most commonly employed (Rubel and Woosely, 1979). Sodium hydroxide at concentrations between 0.5 and 5% is used to regenerate activated alumina.

Ammonia Removal by Ion Exchange

Wastewaters high in ammonia can be treated with ion exchangers using the natural zeolite clinophlolite (also called clinoptilolite) which exhibits unusual selectivity for ammonium ion in the presence of calcium, magnesium, and sodium ions (Eckenfelder and Argaman, 1991; Sims and Hindin, 1978). Other forms of inorganic nitrogen are not removed by the zeolite. The capacity of the exchanger is approximately 2 meq/g of zeolite. Regeneration of the zeolite can be accomplished with neutral sodium chloride or alkaline reagents such as calcium or sodium hydroxide. The bed may become unstable with high pH regenerants and decompose, although high-pH regeneration may require as little as one third of the regenerant (Sims and Hindin, 1978). A 2% NaCl solution required 25–30 bed volumes for regeneration. Ammonia can be biologically removed or stripped from the regenerant.

15.7 FLUORIDATION AND DEFLUORIDATION

The fluoride objective for finished waters is around 1 mg F^-/L for most communities in North America (Section 8.6.2). For water deficient in fluoride, fluoride is added after filtration or after activated carbon treatment if this treatment is used. The common fluoridation agents are sodium fluoride, sodium silicofluoride (Na_2SiF_6), and fluorosilicic acid (H_2SiF_6).

Natural fluoride concentrations in water supplies (both surface and groundwaters) exceed desirable levels at various locations throughout the world. Groundwater concentrations are generally higher than surface water concentrations. In some cases the concentrations can be quite high; for instance, in India, some water supplies contain F^- as high as 19 mg/L (Bulusu et al., 1979); in Africa, concentrations over 50 mg/L have been observed. Fluoride content in natural drinking water sources in Canada ranges from less than 0.01 to 4.5 mg F^-/L (Dept. of National Health and Welfare, 1980).

Defluoridation is costly. Methods for fluoride removal include chemical precipitation, ion exchange, or ion exchange/absorption processes. Activated alumina discussed above has been the most common agent used for defluoridation. Bone char, activated carbon, tricalcium phosphate, lime, magnesium oxide, and other anion exchangers also have been used or studied (Bulusu et al., 1979).

Bone char, which is naturally high in calcium, is pulverized bone that has been carbonized by heating in the absence of air or it may be burnt in the presence of air. It has been commonly used but it is more expensive than activated alumina (ASCE and AWWA, 1990). Its capacity is around 1 000–1 500 mg F^-/L of bone char (Bulusu et al., 1979). It is regenerated with caustic solution. Variable results and generally poor performance of activated carbon have been reported.

Alum and lime are used to precipitate fluoride in a defluoridation technique developed in India and referred to as the "Nalgonda" technique (Nawlakhe et al., 1975). The amount of alum required increases with the amount of fluoride in the water. Alkalinity must be adjusted with lime (which is added before the alum) to ensure that the alum is precipitated or, when excess alkalinity is present, more alum must be added to achieve the desired removal. Table 15.6 provides the guidelines found from the study by Nawlakhe et al. (1975). The Nalgonda technique is simple, with significant cost savings compared to other techniques (Bulusu et al., 1979). Culp and Stoltenberg (1958) found that lime–alum addition for fluoride removal would be more expensive than activated alumina at one location but there were several advantages of the former compared to the latter considering the characteristics of the water to be treated. The simplicity of the lime–alum techniques was also noted as a primary advantage.

TABLE 15.6 Guidelines for Alum Dose (mg Al/L) to Obtain Excessive (E) and Permissive (P) Fluoride Levels[a,b]

Raw water F^-, mg/L	Water alkalinity, mg $CaCO_3$/L											
	80		125		200		400		600		1 070	
	E[c]	P[c]	E	P	E	P	E	P	E	P	E	P
2	—	8	—	11	—	17	—	24	—	31	—	40
3	6	d	9	17	12	23	20	31	27	40	33	59
4	11		15	d	16	31	23	36	31	46	43	72
6	d		d		31	d	35	55	46	72	59	93
8					d		51	d	57	86	72	110
10									64	d	89	130

[a]Reprinted with permission from W. G. Nawlakhe, D. N. Kulkarni, B. N. Pathak, and W. L. Bulusu (1975), "Defluoridation of Water by Nalgonda Technique," *Indian J. of Environmental Health,* 17, 1, p. 52.
[b]See Bulusu et al. (1979) for additional data.
[c]E, excessive limit of 2 mg/L; P, permissive limit of 1 mg/L.
[d]Not possible to reach the limit.

The Nalgonda technique can be applied for any size treatment operation but it is practical for small villages in developing countries, where treatment to international standards may not be economical; therefore, guidelines for higher levels of effluent fluoride are included in Table 15.6.

15.8 MEMBRANE PROCESSES

Membrane treatment processes are used to separate dissolved and colloidal constituents from a water. In membrane treatment, water or components in water are driven through a membrane under the driving force of a pressure, electrical potential, or concentration gradient. Necessarily, particulates are also trapped in the fine openings of membranes. Significant advances have been made in deign of membranes for selectivity and efficiency over the past two decades. The main application of membrane technology in water treatment is in the desalination of brackish waters, with over 4 000 land-based plants (AWWA Membrane Committee, 1992). However, membrane treatment has been used for filtration, removal of microorganisms, hardness, volatile organics and other soluble organics, and biological treatment (AWWA Membrane Committee, 1992; Castro and Zander, 1995; McLeaf and Schroeder, 1995; Yoo et al., 1995). Membrane technologies will be increasingly used in the future for water treatment to meet more stringent water quality regulations. Sengupta and Shi (1992) used a membrane to selectively recover aluminum from acidified clarifier sludge and to avoid the problems of heavy metals and manganese recovery in the recycled alum. Metals are also released with acidification of the sludge and are recycled with the aluminum, posing a health risk.

An important application in wastewater treatment is the recovery of precious metals in certain industries; filtration is another application. Membranes can be used to remove colloidal matter in both water and wastewater treatment. Specific ion probes are applications of membranes in which certain ions or molecules, such as oxygen in the case of dissolved oxygen probes, are selectively allowed to migrate through the membranes.

A semipermeable membrane is selective to the species it passes. The size of the openings in the membrane are a major determinant of species that can pass because the openings present a physical barrier to any substances that are larger than the openings. Other chemical characteristics of the membrane also influence the substances that it can pass. Charge or polarity, shape, and size are characteristics of species that influence their ability to migrate through a membrane. Membranes are usually made from organic polymers, including cellulose acetate, polysulfone, polyamide, polyurea, and polycarbonate.

Membrane separation is a mass transfer phenomenon. Underlying principles for mass transfer have been presented in Chapter 12 where a concentration gradient is the driving force. A fairly direct application of these principles is separation of metals or measurement of dissolved oxygen. In these instances, metals are removed from solution by plating out; for oxygen, it is consumed on one side of the membrane. A concentration gradient is created across the membrane. A selective membrane prevents or minimizes the passage of interfering substances to the side where consumption is occurring.

Mass transfer through a membrane proceeds in both directions. Membrane characteristics govern the direction in which the solutes and solvent are traveling. The ultimate state of any system is an equilibrium condition where the concentrations of all species are equal. With membrane selection technologies energy in one form or another must be applied to counteract and reverse natural tendencies. Reverse osmosis (RO) is a good illustration of the principles besides being the most common water treatment application.

When a semipermeable membrane separates two solutions with different concentrations of solute, the fluid or solvent passes through the membrane to equalize the concentrations of all species on either side of the membrane. The passage of solvent as opposed to solutes is osmosis. The chemical imbalance on either side of the membrane causes water to flow against a pressure gradient as illustrated in Fig. 15.5a. Ultimately the forces generated by the chemical imbalance are opposed by the physical pressure force and a static condition occurs. The pressure that exists in this state is the osmotic pressure, Δp_o.

If a pressure force is applied to the more highly concentrated solution, the flow of liquid can be reversed as shown in Fig. 15.5b, which is RO. As water is removed from the right-hand cell the concentration difference increases as well as the osmotic pressure. A supply of brackish water is fed to the cell to keep osmotic pressures at values near the lowest value.

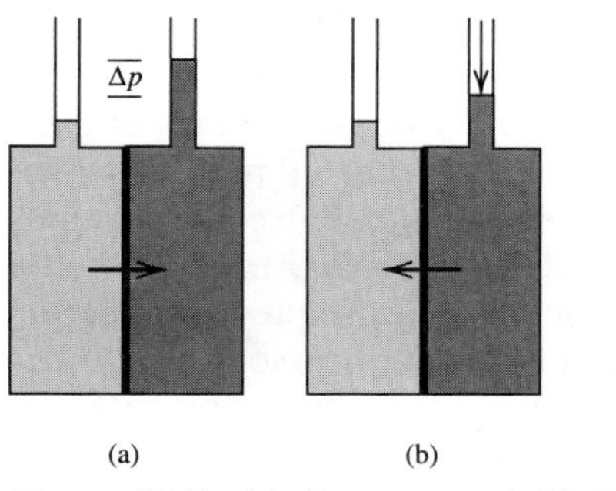

Figure 15.5 (a) Osmosis and (b) RO.

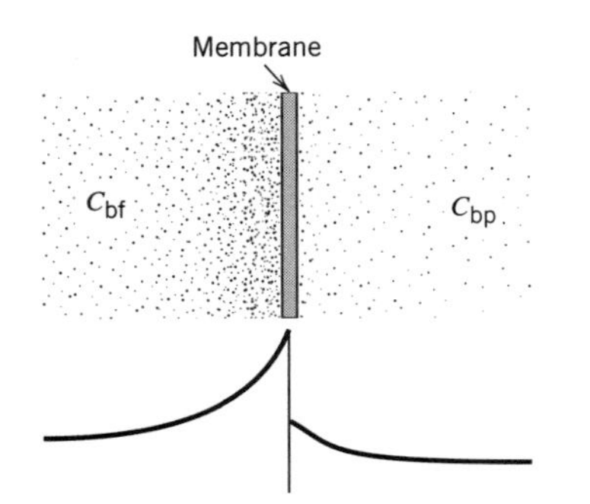

Figure 15.6 Concentration polarization.

The mass transfer of water through a membrane is proportional to the area of the membrane, the net pressure difference above the osmotic pressure, and the distance over which the pressure difference occurs.

$$\frac{\Delta m}{\Delta t} \propto \frac{A}{\Delta x}(\Delta p - \Delta p_o) \qquad \text{or} \qquad N_w = K_w(\Delta p - \Delta p_o) \tag{15.32}$$

where
- m is mass of water passed
- t is time
- A is the area of the membrane
- x is length
- N_w is the flux of water (mass/area/time)
- K_w is a resistance coefficient
- Δp is the applied pressure
- Δp_o is the osmotic pressure

The resistance coefficient is a function of membrane characteristics, solutes in the water, temperature, fouling, and other effects. Other effects include polarization phenomena, which increase the resistance. Concentration polarization (Fig. 15.6) refers to the buildup of ions on either side of the membrane. As water is removed from the solution in the right-hand cell, the concentration of ions at the membrane surface increases and movement of these ions from the surface occurs by mixing and diffusion. There will be some solute carried with the water through the membrane. These ions must diffuse away or be hydraulically transported from the left-hand membrane surface.

In electrodialysis an electrical potential is used to provide the driving force. A voltage is impressed across an anode and a cathode. Membranes that are selective to cations or anions are inserted in the water between the electrodes. The ions migrate to the electrode of opposite charge, producing a brine and desalinated product water.

Concentration polarization is one cause of membrane fouling. High concentrations of dissolved ions on the feed side of the membrane cause solubility products to be exceeded, with deposition of the precipitates on the membrane. Sequestering agents can be added to the feed water to retain precipitation species in solution. Other measures such as pretreatment or pH control can be used to prevent precipitation. Acidification to a pH of 4–6 prevents the formation of carbonate and precipitation of $CaCO_3$ (AWWA Membrane Committee, 1992).

The concentrations of solutes in the bulk phases of the feed and permeate are C_{bf} and C_{bp}, respectively, in Fig. 15.6. The osmotic pressure depends on the concentrations and the boundary layers at the membrane surface.

As noted earlier, some solute will pass through the membrane. The flux of solute depends on the concentration gradient and a resistance parameter.

$$N_s = K_s(C_f - C_p) \tag{15.33}$$

where

N_s is the flux of solute
K_s is the resistance parameter for solute passage
C_f and C_p are the concentrations of solute in the feed and permeate, respectively

Another parameter describing the performance of a membrane is the solute rejection coefficient, which is a measure of the ability of membrane to reject the passage of a species, i.

$$R_i = \frac{C_{if} - C_{ip}}{C_{if}} \tag{15.34}$$

where

R_i is the rejection coefficient
C_{if} and C_{ip} are the concentrations of species i in the feed and permeate, respectively

Two types of elements for membranes are shown in Fig. 15.7. The parallel plate arrangement is also known as tangential flow. The elements are bundled together in modules. The permeate is collected in separate channels from the retentate or concentrate. Another configuration is a spiral wound module, which consists of membranes and spacers wound around a central permeate collection tube.

The size ranges of particles removed by various membrane processes are shown in Fig. 15.8. Larger opening membranes used for microfiltration are classified by their pore size. These membranes remove particles on the order of a micron (10^{-6} m) in size. Microfiltration can remove *Giardia*, coliforms, *Cryptosporidium*, and particles that shield pathogens from disinfectants (Yoo et al., 1995). Ultrafiltration and nanofiltration membranes are usually rated according to the smallest molecular weight substances removed by the membrane. The size of the pore openings is on the order of a nanometer (10^{-9} m). Nanofiltration is used for softening of water (removal of calcium and magne-

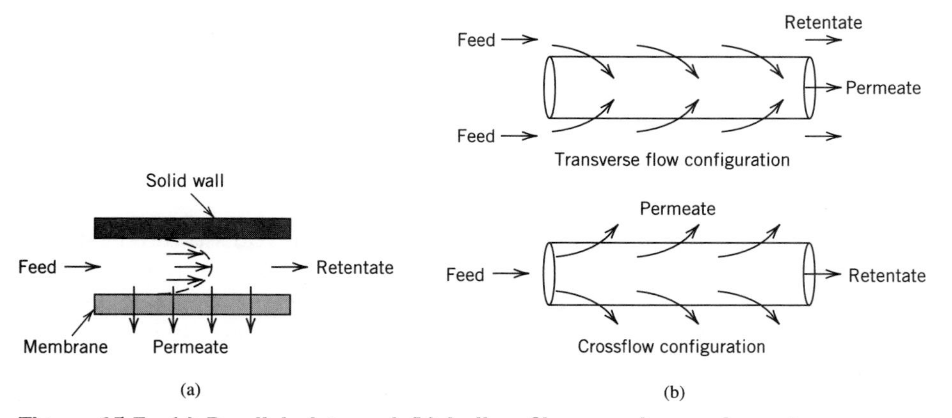

Figure 15.7 (a) Parallel plate and (b) hollow fiber membrane elements.

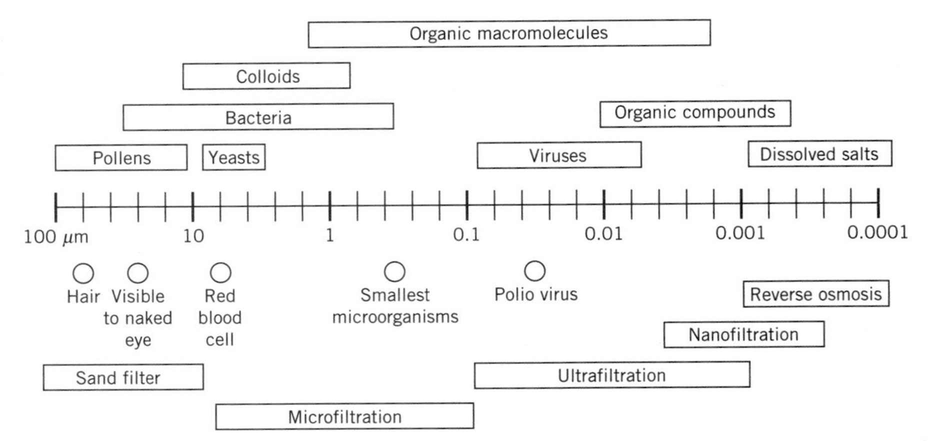

Figure 15.8 Particle size ranges removed by membrane processes. Courtesy of Degrémont Infilco.

sium). RO membranes are able to remove ions such as sodium and chloride in addition to calcium, magnesium, and other ions.

Fouling of a membrane increases resistance to flow and reduces the flux of water through a membrane. Prophylactic measures for fouling have been mentioned above. Backwashing or chemical treatment may be applied to remove foulants. Colmatage refers to flux reduction by foulants that may be reversibly removed by hydraulic or chemical treatments. Irreversible fouling of membranes is the most serious problem. Oxidizing agents such as chlorine or ozone attack membranes and change their structure. Other particles may be irreversibly sorbed to membranes. Raw water characteristics and water quality objectives must be considered along with pretreatment technologies and membrane type in the system design.

Operating pressures decrease as the pore opening size of the membrane increases. Operating pressures for newer RO systems are in the range of 1 550–3 200 kPa (225–460 psi) (Tan and Sudak, 1992). Nanofiltration systems can be operated at pressures of 500–1 000 kPa (70–140 psi). Recovery is the percentage of influent converted to product water. Typical RO and nanofiltration plants have recoveries between 50 and 85% (Wiesner and Chellam, 1992). A microfiltration plant operated at transmembrane pressures less than 101 kPa (15 psi) to achieve >3 log removal of 4–10 μm particles; *Giardia* and *Cryptosporidium* were also effectively removed (Yoo et al., 1995). Operating transmembrane pressures up to 300 kPa (45 psi) are typical for microfiltration membranes with openings around 0.2 μm.

15.9 CARBON ADSORPTION

Carbon has been known since the middle ages to be able to remove dissolved substances from liquids. Understanding of the process has advanced considerably but there is still much to learn. Carbon is able to adsorb dissolved substances onto its porous, fissured surfaces. The activated carbon is the adsorbent and substances being adsorbed are adsorbates. There are other materials that can be used as adsorbents but carbon is the choice for water treatment because it is able to remove a broad range of adsorbates.

Activated carbon is most often used to remove organic contaminants, particularly synthetic organic chemicals (SOCs), but it effectively removes many inorganic contaminants such as radon-222, mercury, and other toxic metals (Faust and Aly, 1987). Significantly, trihalomethanes are not adsorbed to a large extent by activated carbon. Dechlorination is another use of activated carbon. Chlorine and chloramines react with the carbon to form chloride and carbon dioxide products.

Activated carbon is prepared in a manner that results in large surface area (fissures) within the medium. In granular activated carbon (GAC) beds there is also some degree of filtration phenomena. An alternative to GAC is powdered activated carbon (PAC), which may be added to biological treatment or water treatment operations.

Eventually the carbon medium becomes saturated with adsorbates and the carbon must be regenerated. Regeneration can be performed by chemical or thermal means, which remove or destroy the adsorbed compounds.

Activated Carbon

Activated carbon can be prepared from almost any carbonaceous material (e.g., wood, coal, lignite, and coconut shells) by heating it with or without addition of dehydrating chemicals in the absence of air to liberate carbon from its associated atoms. This step is carbonization. Activation of the carbon occurs by passing mildly oxidative hot gases (carbon dioxide or steam) through the carbon at temperatures between 315 and 925°C. This causes the formation of tiny fissures or pores. Some noncarbonaceous materials and volatile organics will also be removed in this process. The apparent dry density of GAC is from 22 to 50 g/100 mL and the pore volume is around 0.85–0.95 mL/g (Flick, 1991). The apparent dry density of PAC is in the range of 34–74 g/100 mL and pore volumes are in the range of 2.2–2.5 mL/g.

Adsorption is a surface area phenomenon. The surface areas of commercial GACs range from 600–1 600 m^2/g (Flick, 1991; Narbaitz, 1985). Commercially available GAC has most of its particles ranging in size from 40 mesh (0.425 mm) to 8 mesh (2.36 mm), whereas 80% of PAC particles are smaller than 325 mesh (0.025 mm). Because the majority of the surface area in activated carbon is within internal pores there is essentially no difference in the surface area per unit mass, and thus adsorption capacity, for GAC and PAC (Narbaitz, 1985). The smaller particle diameters of PAC allow it to reach equilibrium more quickly than GAC.

15.9.1 Adsorption Isotherms

Many factors affect the amount of adsorption: chemical properties of the adsorbate, activated carbon properties, and liquid phase characteristics such as pH and temperature. Oxygen concentration was found to affect the adsorption capacity for chlorophenol of activated carbons made from bituminous coal and lignite coal but it did not affect adsorption for a wood base carbon (Sorial et al., 1993). The molecular size and structure and polarity are important characteristics of the adsorbate. The activated carbon adsorption properties are similar for PAC and GAC, as noted earlier. Most waters contain many substances that will adsorb and affect the adsorption capacities of each other. This makes it important to maintain laboratory conditions similar to field conditions when gathering adsorption data.

An adsorption isotherm describes the relation between the amount or concentration of adsorbate that accumulates on the adsorbent and the equilibrium concentration of dissolved adsorbate. An adsorption isotherm is an expression of the principle of microscopic reversibility, although adsorption can be irreversible. The most common method for gathering isotherm data is the bottle point technique. Accurately measured concentrations of adsorbate are placed in about 10 or more bottles, each containing different amounts of adsorbent. The bottles are mixed with shakers or other means to provide good mixing and contact of the carbon and the solution until a constant liquid phase concentration of adsorbate is achieved. The data from the bottles are then analyzed according to the following equation:

$$V(C_0 - C) = M(q_e - q_i) \tag{15.35}$$

where

V is the volume of the sample
C_0 is the initial concentration of adsorbate in solution
C is the equilibrium concentration in solution
M is the mass of carbon in the bottle
q_e is the final (equilibrium) solid phase concentration of the contaminant on the carbon, mass or moles of contaminant/mass of carbon
q_i is the initial solid phase concentration of the contaminant

The tests are usually conducted with virgin carbon ($q_i = 0$). With this condition, the above equation can be rearranged to

$$q_e = \frac{(C_0 - C)V}{M} \tag{15.36}$$

These equilibrium data are then formulated into an adsorption isotherm model. The first adsorption isotherm was derived by Langmuir (1918), who from theoretical principles developed an equation similar to the Michaelis–Menten equation. Langmuir theorized that only a single adsorption layer could exist. This is a reasonable assumption for gases because the chemical forces bonding the gas to the adsorbent probably do not extend beyond a single layer.

At equilibrium a fraction of the surface of the adsorbent will be covered with adsorbate and the rate of adsorption will equal the rate of desorption. Only molecules striking the bare surface have an opportunity to be retained because the forces of attraction do not extend over multilayers of the adsorbed component. Likewise, the rate of removal of adsorbed molecules from the surface is proportional to the surface area covered by them. Stating these principles in equation form:

Define λ as the fraction of surface area covered by adsorbate at equilibrium. Then $1 - \lambda$ is the fraction of bare surface at equilibrium. Because rate of adsorption is equal to rate of desorption,

$$kC(1 - \lambda) = k'\lambda \tag{15.37}$$

where

C is the concentration of the substance being adsorbed (if the adsorbate were a gas the C would be replaced by the partial pressure of the gas)
k and k' are the rate constants for adsorption and desorption, respectively

The fraction of covered surface area can be determined from Eq. (15.37).

$$\lambda = \frac{kC}{kC + k'} \tag{15.38}$$

Define q_e as the number of moles or mass of adsorbate per unit weight of adsorbent at concentration C of solute when equilibrium exists. The amount of adsorption is directly proportional to the covered surface area.

$$q_e \propto \lambda \qquad \text{or} \qquad q_e = k''\lambda \tag{15.39}$$

where
k'' is a proportionality constant

Using Eq. (15.39) with Eq. (15.38) and rearranging it,

$$q_e = \frac{k''kC}{kC + k'} = \frac{k''KC}{KC + 1} \tag{15.40}$$

where
$K = k/k'$

The maximum amount of solute that can be adsorbed is defined as Q_0. The maximum adsorption occurs when the surface is fully covered. From Eq. (15.39),

$$Q_0 = k''(1) = k''$$

Substituting this into Eq. (15.40),

$$q_e = \frac{Q_0 KC}{KC + 1} \tag{15.41}$$

Another commonly used adsorption isotherm is the Freundlich isotherm (Freundlich, 1926), which is an empirical equation.

$$q_e = K_F C^{1/n} \tag{15.42}$$

where
K_F and n are constants

The Brunauer, Emmett, Teller (BET) isotherm was developed from theory as an extension to the Langmuir isotherm for the case where multilayer adsorption was occurring (Brunauer et al., 1938). It is:

$$q_e = \frac{BCQ_o}{(C_s - C)\left[1 + \frac{(B-1)C}{C_s}\right]} \tag{15.43}$$

where
B and Q_0 are constants
C_s is the saturation concentration of the solute in water

The three adsorption isotherms result in significantly different functional relations. The choice of the best relation will depend on which one best describes the data. The Freundlich isotherm most often proves to be the best relation. The isotherms are equilibrium relationships that do not indicate the rate of reaction.

15.9.2 Granular Activated Carbon Adsorbers

Batch treatment can be accomplished by simply letting water remain in contact with carbon for a period of time. In addition to batch units, there are three types of continuous flow adsorbers illustrated in Fig. 15.9. In a fixed bed adsorber the flow is

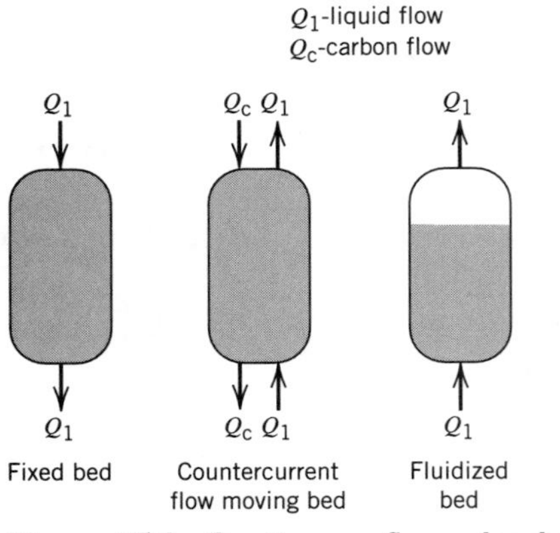

Figure 15.9 Continuous flow adsorbers.

introduced at the top or bottom of the bed. The bed is replaced upon exhaustion. In a countercurrent operation, the adsorbent material is continuously input into the unit in the opposite direction of the fluid movement. The third option is a fluidized bed unit, in which the liquid is introduced at the bottom of the unit and its velocity is sufficient to expand the medium particles. An activated carbon system incorporating onsite regeneration is shown in Fig. 15.10.

From an examination of the isotherm equations, the countercurrent flow configuration will achieve the most efficient use of the carbon because rate of adsorption depends on the degree of unsaturation of the carbon. The other two methods will result in lower effluent concentrations of adsorbate.

Design of Fixed Bed Adsorbers

As adsorbate-bearing water is introduced into the top of a fixed bed adsorber containing fresh medium, the top layer of the medium begins to remove adsorbate until

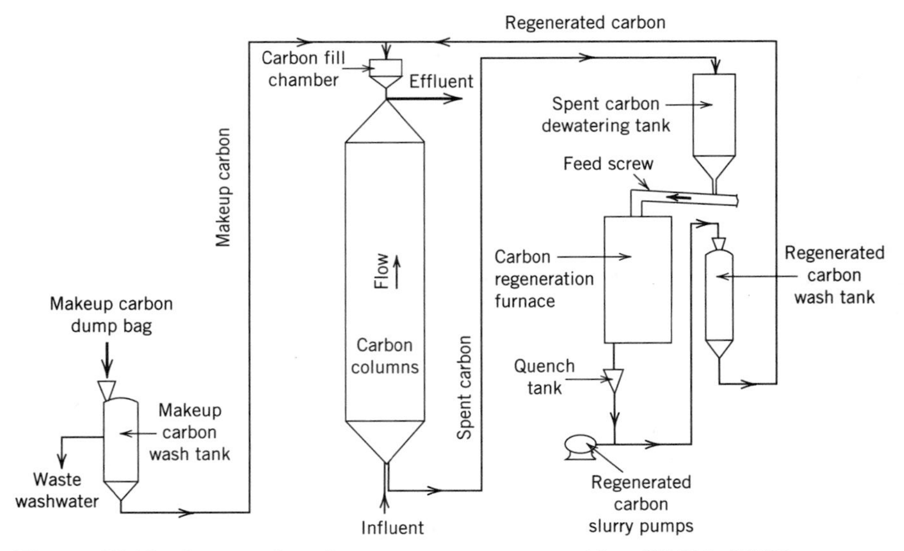

Figure 15.10 Activated carbon treatment system. After USEPA (1973).

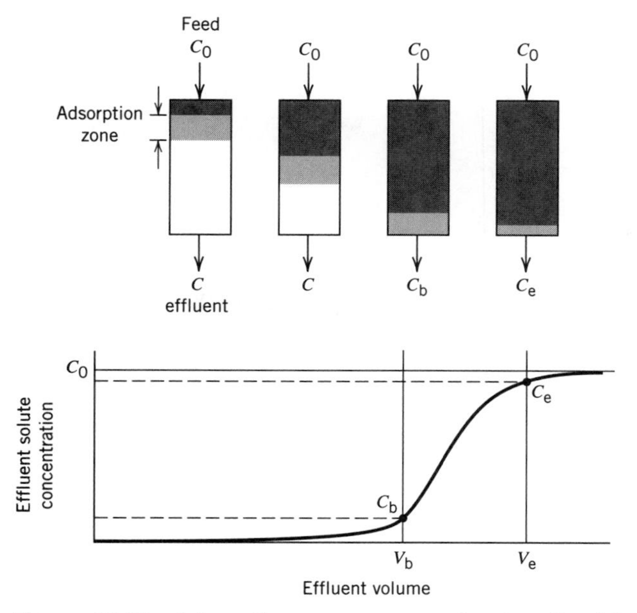

Figure 15.11 Adsorption zone progression in a fixed bed adsorber.

it becomes saturated at a level corresponding to the influent concentration. Successive layers of medium will have lower accumulations of adsorbate because of the diminishing concentration of adsorbate in the liquid. At some point the liquid adsorbate concentration will decrease to essentially zero, and medium below this point will not have any accumulation of adsorbate. The zone where the concentration in the liquid decreases from the influent value to a very small value is the adsorption zone and this zone moves downward through the column as time progresses. The situation is depicted in Fig. 15.11 along with a graph of effluent concentration as a function of volume passed through the bed.

Concentration of the adsorbate exhibits an S-shaped curve in the adsorption zone with ends asymptotically approaching zero and the influent concentration C_0. This curve is known as a breakthrough curve. The breakthrough concentration, C_b, is selected based on the effluent criterion with a safety factor. The volume of water treated at breakthrough is V_b. The volume of water passed through the bed at exhaustion is V_e. The exhaustion concentration, C_e, is 0.90–0.99 of C_0 depending on the shape of the curve. When the breakthrough curve exhibits a gradual rise to the asymptotic concentration (C_0), C_e is chosen near $0.90C_0$; when the curve does not exhibit a significant amount of tailing C_e is chosen near $0.99C_0$.

Empty bed contact time (EBCT) is a primary design variable. It is the detention time in a bed without medium. Conventional water treatment plant adsorbers have EBCTs in the range of 7–20 min.

The volumetric throughput rate influences the shape of the curve as shown in Fig. 15.12. As volumetric throughput and velocity through the bed decrease, the depth of the adsorption zone decreases because there is more time for adsorption in each layer.

Rate Formulation for Adsorption

The rate formulation for adsorption is based on an analogy to Fick's first law. The driving force for adsorption is due to a liquid adsorbate concentration that is higher

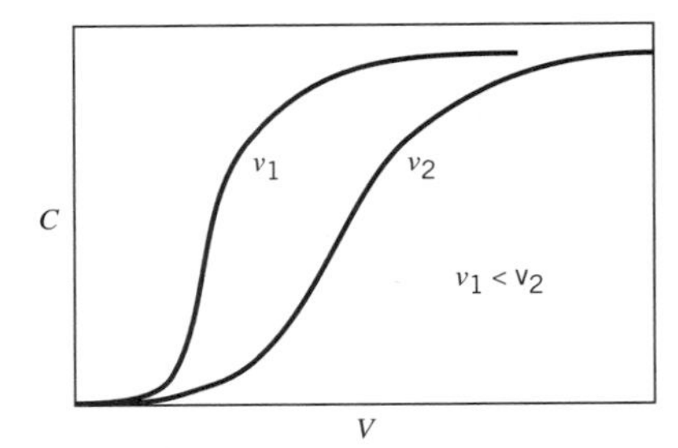

Figure 15.12 Effects of flow-through velocity on the breakthrough curve.

than the concentration that would be in equilibrium with the amount of mass adsorbed on the adsorbent.

$$N = \frac{1}{a}\frac{dm}{dt} = -k'\frac{C - C^*}{x} = k(C^* - C) \tag{15.44}$$

where

N is mass flux
m is mass of adsorbate
t is time
a is the adsorption area
C is concentration in the liquid
C^* is concentration that exists when the liquid is in equilibrium with the adsorbent
k' is the resistance to mass transfer
x is a nominal distance
$k = k'/x$

All resistances resulting from the tortuous nature of flow and diffusion are lumped into the coefficients k' or k. Rearranging Eq. (15.44) and dividing by V, the volume,

$$r = \frac{ka}{V}(C^* - C) = k\alpha(C^* - C) \tag{15.45}$$

where

r is rate of mass removal, mass/volume/time
$\alpha = a/V$

Theory of Fixed Bed Adsorber Systems

The design and theory of fixed bed adsorption systems centers on establishing the shape of the breakthrough curve and its velocity through the bed. The breakthrough curve is depicted in Fig. 15.13. Above the breakthrough curve the capacity of the

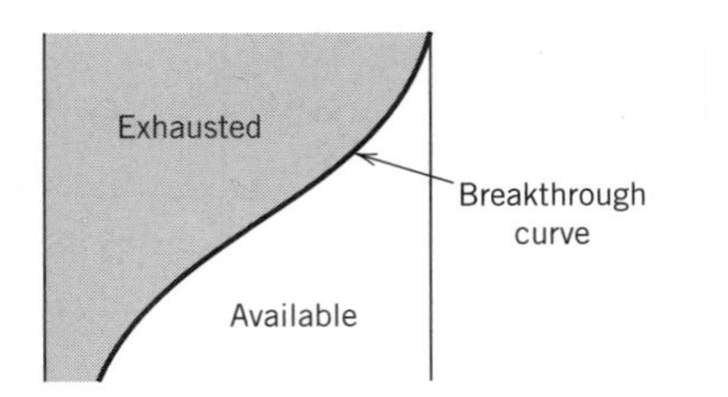

Figure 15.13 Breakthrough curve.

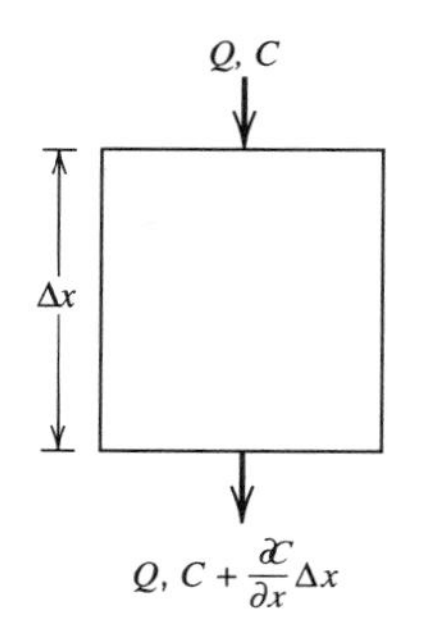

Figure 15.14 Elemental volume.

adsorbent is exhausted. Below the curve, the adsorbent is essentially virgin. The depiction of the curve is a conceptualization of the phenomenon.

The amount of mass adsorbed in the adsorption zone varies from the ultimate capacity of the adsorbent at the input concentration to no mass adsorbed at the leading edge of the adsorption zone. For practical purposes, the adsorption zone is defined as the zone where the liquid concentration changes from C_0 to C_b. The capacity of the adsorbent varies in the characteristic S manner given in Fig. 15.13. The adsorption capacity of any layer in the adsorption zone is distributed throughout the layer. Uniform flow patterns will produce a uniform adsorption capacity at each location across a layer of medium.

The medium must be replaced when the effluent concentration begins to rise, i.e., when the leading edge of the adsorption zone has reached the end of the column. The capacity of the bed above the adsorption zone is readily established based on the ultimate capacity of the adsorbent. The capacity of the adsorbent utilized in the adsorption zone must be established to be added to the capacity of the saturated adsorbent above the zone to determine the total capacity of the bed at the influent conditions and bed flow-through velocity.

Moving with the adsorption zone, taking down as the positive direction and setting up a mass balance for a point within the zone (Fig. 15.14),

$$QC - Q\left(C + \frac{\partial C}{\partial x}\Delta x\right) + r\Delta V = e\frac{\partial C}{\partial t}\Delta V$$

where

e is the porosity of the bed

The cross-sectional area of the column is A. Substituting $A\Delta x$ for ΔV and Eq. (15.45) for r and assuming steady state, this equation reduces to

$$\frac{Q}{A}\frac{dC}{dx} = v\frac{dC}{dx} = k\alpha(C^* - C) \qquad (15.46)$$

where

v is the nominal flow velocity through the bed

The development is most conveniently performed in terms of mass-related parameters. The mass flux, F_m, of liquid through the column is given by

$$F_m = \frac{\rho_l Q}{A} \tag{15.47}$$

where
ρ_l is the density of the liquid

Substituting this into Eq. (15.46) and taking the up direction as positive:

$$F_m = \frac{dC}{dx} = k\rho_l \frac{Q}{A}\frac{dC}{dx} = k_d\alpha(C - C^*) \tag{15.48}$$

where
$k_d = k\rho_l$

Defining the mass of adsorbent in the bed as m,

$$m = \rho_p V = \rho_p AD \qquad \text{and} \qquad dm = \rho_p A\, dD \tag{15.49}$$

where
ρ_p is the packed density of the adsorbent
D is the depth of the bed

Substituting Eq. (15.49) into Eq. (15.48) results in the differential equation describing adsorption in the adsorption zone.

$$\rho_p F_m A \frac{dC}{dm} = k_d\alpha(C - C^*) \tag{15.50}$$

The Capacity Utilized in the Adsorption Zone

As noted above, the adsorption zone is defined as that part of the bed where the concentration in the liquid varies from C_0 to C_b. The adsorption zone is assumed to have a constant depth, δ.

$$\delta = \frac{V_e - V_b}{A} \tag{15.51}$$

where
V_e is the volume of liquid passed through the bed at exhaustion
V_b is the volume of liquid passed at breakthrough

The time for the adsorption zone to travel the full depth of the column is

$$t_e = \frac{V_e}{Q} \tag{15.52}$$

where
t_e is the time to bed exhaustion

The time for the adsorption zone to move a distance δ, t_δ, is

$$t_\delta = \frac{\delta}{v} \tag{15.53}$$

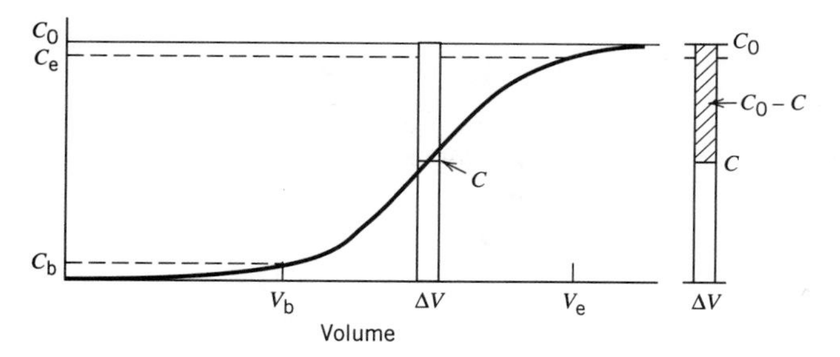

Figure 15.15 Adsorption in an incremental volume.

There will be a finite time, t_f, for the adsorption zone to form. The time to exhaustion will be composed of the time for the zone to form plus the time for the zone to travel the depth of the bed, D.

$$t_e = t_f + t_D \tag{15.54}$$

where

t_D is the time for the zone to travel the depth of the bed

Noting that $vt_D = D$ and $vt_\delta = \delta$, the ratio of t_δ and t_D is determined from Eqs. (15.53) and (15.54):

$$\frac{t_\delta}{t_D} = \frac{\delta}{D} = \frac{t_\delta}{t_e - t_f} \tag{15.55}$$

The total mass adsorbed in the adsorption zone is the sum of the masses adsorbed in each incremental layer in the zone. From Fig. 15.15, the concentration adsorbed in an incremental volume, ΔV, is $C_0 - C$. If the concentration in the liquid is C_0, there is no adsorption in the incremental volume; if it is 0, then all mass in the liquid is adsorbed. The incremental mass adsorbed is

$$\Delta m = (C_0 - C)\Delta V$$

The adsorption zone volume is $V_e - V_b$. The total mass that may be adsorbed in the adsorption zone, m_T, is

$$m_T = C_0(V_e - V_b)$$

Defining the fractional capacity of the adsorption zone as f, it is calculated from

$$f = \frac{\int_{V_b}^{V_e} (C_0 - C)\, dV}{C_0(V_e - V_b)} \tag{15.56}$$

(C is the function plotted on the breakthrough curve.)

The extreme values for f are 0 and 1. Figure 15.16 shows two adsorption zones with small and large depths. It is seen from Fig. 15.16a that in an adsorption zone with a small depth, the capacity of the zone is nearly fully utilized. In the limiting case of Fig. 15.16a, the adsorption zone has no depth and the concentration gradient

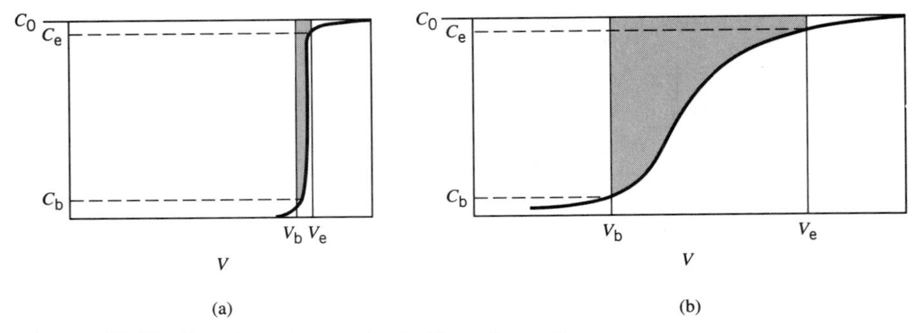

Figure 15.16 Fractional capacity in the adsorption zone.

is a vertical line. The average concentration of adsorbate in the liquid within the zone is zero and the fractional capacity of the zone is 1.

Figure 15.16b approaches the other extreme. As the zone stretches, less capacity in the zone is utilized. In the limiting case the average concentration of adsorbate in the liquid within the zone is C_0 and the fractional capacity utilized is 0. From the definition of f, when $f = 0$, the concentration of the adsorbate in the adsorption zone is 0 and the adsorption zone has a large available adsorption capacity. Conversely when f is 1, the adsorption zone has a low available capacity.

The integral in Eq. (15.56) may be rearranged to more clearly illustrate these principles.

$$f = \frac{\int_{V_b}^{V_e} (C_0 - C)\, dV}{C_0(V_e - V_b)} = \int_{V_b}^{V_e} \left(1 - \frac{C}{C_0}\right) d\left(\frac{V - V_b}{V_e - V_b}\right) \tag{15.57}$$

Substituting $C = 0$ and $C = C_0$ into Eq. (15.57) will result in $f = 1$ and 0, respectively.

When a large amount of adsorption has occurred in the adsorption zone, $f = 0$, and the liquid concentration in the zone, $C \approx C_0$. Because the adsorbent in the zone is almost saturated, the time to form the zone (t_f) will be approximately the time for the zone to travel its own depth, δ. When the zone is compact, the formation time is nearly zero. Therefore a reasonable time of formation–f relation is

$$f = 1 - \frac{t_f}{t_\delta} \quad (15.58a) \qquad \text{or} \qquad t_f = (1 - f)t_\delta \quad (15.58b)$$

Either Eq. (15.58a) or (15.58b) may be substituted into Eq. (15.55) to find

$$\frac{\delta}{D} = \frac{t_\delta}{t_e + t_\delta(f - 1)}$$

Using $t_e = \dfrac{V_e}{Q}$ and $t_\delta = \dfrac{V_e - V_b}{Q}$

$$\frac{\delta}{D} = \frac{V_e - V_b}{V_b + f(V_e - V_b)} \tag{15.59}$$

Define:

M_∞ = the amount of adsorbate that will accumulate at complete saturation

$$q_\infty = \frac{\text{amount of adsorbate at the influent concentration}}{\text{unit mass of adsorbent}}$$

$$M_\infty = \rho_p A D q_\infty \quad (15.60a) \qquad \text{and} \qquad \frac{M_\infty}{A} = \rho_p D q_\infty \quad (15.60b)$$

At the breakpoint, the primary adsorption zone is just beginning to exit the column. The fraction of the saturation mass contained in the depth, δ is $1 - f$. Defining M_b as the amount of solute that has accumulated in the column at the time of breakthrough:

$$\frac{M_b}{A} = \rho_p q_\infty (D - \delta) + \rho_p q_\infty \delta(1 - f)$$
$$= \rho_p q_\infty [(D - \delta) + \delta(1 - f)] \quad (15.61)$$

The percentage saturation of the whole column at breakthrough, S_b, is

$$S_b = \frac{M_b/A}{M_\infty/A} \times 100 = \frac{\rho_p q_\infty (D - \delta) + \rho_p q_\infty \delta(1 - f)}{\rho_p q_\infty D} \times 100 = \frac{D - \delta f}{D} \times 100 \quad (15.62)$$

With this information, the volume of the column required to remove the mass of influent solute at a mass loading rate of QC_0 over the time t_e can be determined.

To establish the shape of the breakthrough curve, the isotherm for the adsorbate (Fig. 15.17) and the kinetic expression (Eq. 15.50) must be used. From the isotherm plot in Fig. 15.17, the liquid concentration C_0 is in equilibrium with q_0. Now $q_0 = q_\infty$, the ultimate or saturation capacity for the influent characteristics.

Define the superficial mass rate of saturation, m_s, as

$$m_s = \frac{\text{mass of adsorbent saturated}}{\text{time} \cdot \text{area of the column}}$$

then

$$F_m C_0 = m_s q_\infty$$

(A check on the word definitions and dimensions of each term in this equation should be performed to verify it.)

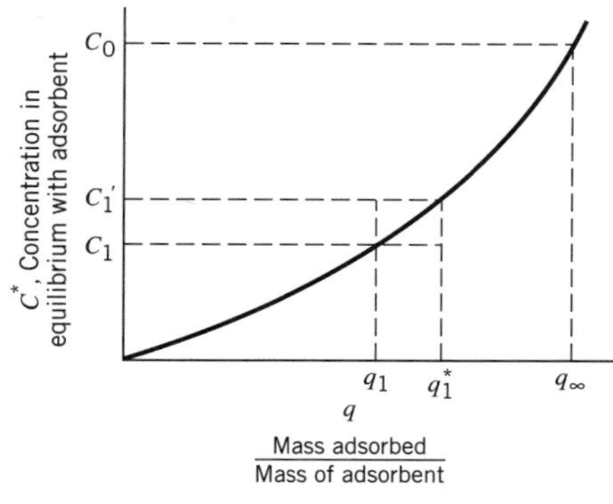

Figure 15.17 Adsorption isotherm.

Likewise, at any point C on Fig. 15.17, the corresponding equilibrium adsorbate concentration on the adsorbent (mass basis) is q, and

$$F_m C = m_s q \tag{15.63}$$

Equation (15.63) is known as the operating line. The slope of the operating line is m_s/F_m and it is constant. If virgin carbon is used and the effluent concentration is initially negligible, the operating line passes through the origin and through the point (q_∞, C_0).

As liquid moves through the adsorption zone, some adsorbate is removed in a layer, which results in an increase in q for the layer. The liquid then moves to the next layer, which has a q value that is less than the value that would be in equilibrium with the current liquid adsorbate concentration. Consider an elemental volume with a concentration C_1' that is passing a layer of adsorbent at concentration q_1. This is depicted by the C_1'–q_1 lines in Fig. 15.17. The layer would be in equilibrium with a liquid concentration of C_1. Mass transfer occurs from the liquid to the adsorbent since C_1' is greater than C_1. The driving force for mass transfer from the liquid to the adsorbent is $C_1' - C_1$. The liquid concentration decreases toward C_1 and the adsorbent concentration increases toward q_1^* that would be in equilibrium with C_1'. The equilibrium and adsorbent concentrations are between C_1' and C_1 and q_1^* and q_1, respectively. Equilibrium between the liquid and adsorbent phases is probably not reached because the elemental volume is moved on to the next layer by the following elemental volume of liquid. Equation (15.48) describes the rate of mass transfer that is given next using the definition in Eq. (15.47).

$$\rho_1 \frac{Q}{A}\frac{dC}{dx} = F_m \frac{dC}{dx} = k_d\alpha(C - C^*)$$

Rearranging this equation and integrating it across the primary adsorption zone,

$$\frac{k_d\alpha}{F_m}\int_0^\delta dx = \int_{C_b}^{C_e} \frac{dC}{C - C^*} = \frac{k_d\alpha\delta}{F_m} \tag{15.64}$$

The relation between δ and the volume of liquid passed through the bed is given by Eq. (15.51). Similarly, for a point at a distance x below the beginning of the primary adsorption zone:

$$x = \frac{V_x - V_b}{A} \tag{15.65}$$

where

V_x is the volume passed when a depth x of the primary adsorption zone has exited

The concentration of solute at the point x is C.

Taking the ratio of the above equation to Eq. (15.51),

$$\frac{x}{\delta} = \frac{V_x - V_b}{V_e - V_b} = \frac{\int_{C_b}^{C} \frac{dC}{C - C^*}}{\int_{C_b}^{C_e} \frac{dC}{C - C^*}} \tag{15.66}$$

The breakthrough curve can now be plotted by numerical or graphical integration of Eq. (15.66).

A relation between $C - C^*$ and C is required. The procedure to find this relation is as follows.

1. Determine the best fit isotherm and plot the isotherm relation.
2. Draw the operating line.
3. Referring to the accompanying sketch, pivot about the operating line to find the liquid concentration C_1 and the corresponding concentration, q_1, of the solute on

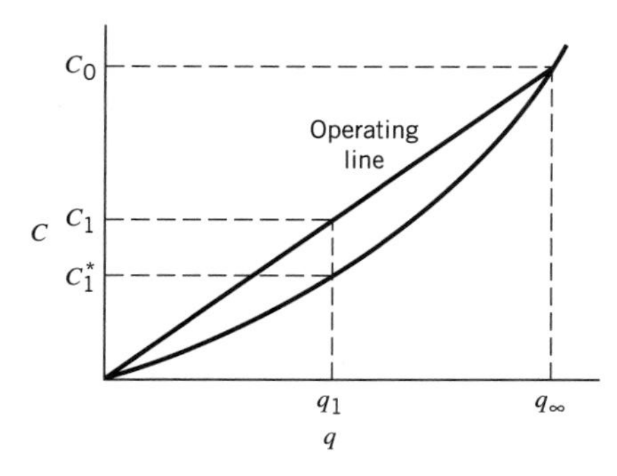

the adsorbent layer. Liquid at concentration C_1 is passing the adsorbent layer at q_1. Now use the isotherm curve to find the concentration in the liquid C_1^* that would be in equilibrium with adsorbent at q_1.
4. Repeat Step 3 and construct the curve shown in Fig. 15.18.
5. For graphical integration, discretize Eq. (15.66):

$$\int_{C_b}^{C_1} \frac{dC}{C - C^*} = \sum_{i=1}^{3} \frac{\Delta C_i}{(C - C^*)_i} = A_1 + A_2 + A_3$$

$$\int_{C_b}^{C_e} \frac{dC}{C - C^*} = \sum_{i=1}^{n} \frac{\Delta C_i}{[(C - C^*)_i]} = A$$

Now

$$\frac{x}{\delta} = \frac{V_x - V_b}{V_e - V_b} \qquad \frac{x_1}{\delta} = \frac{A_1 + A_2 + A_3}{A}$$

6. The values of C/C_0 are also found from the concentration difference plot; i.e., for the point x_1, the concentration ratio is C_1/C_0. These values are used to construct

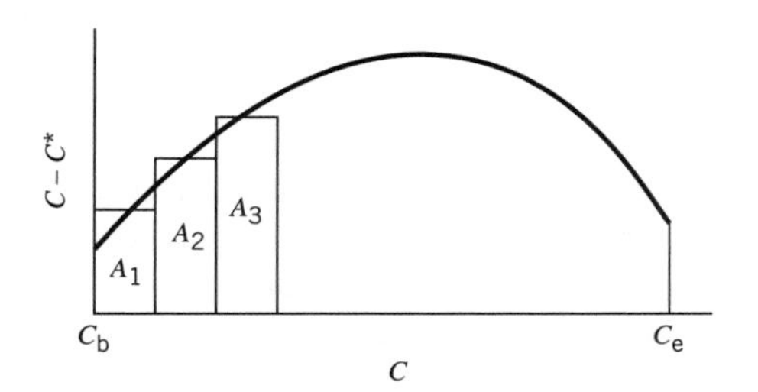

Figure 15.18 Concentration difference curve.

the breakthrough curve and find f. For the breakthrough curve, plot $\frac{C}{C_0}$ versus $\frac{V_x - V_b}{V_e - V_b}$ as shown in the accompanying sketch. This curve is numerically or graphically integrated to find f.

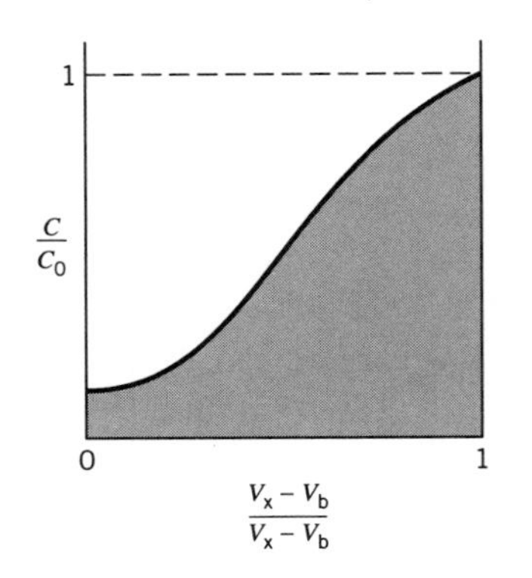

■ Example 15.4 Design of a Fixed Bed Adsorber

The data in columns 1, 2, and 3 in the following table were found for 1,1,2-trichloroethane (TCE) adsorption on activated carbon from equilibrium bottle studies (Narbaitz,

(1)	(2)	(3)	(4)
C_0	C	M/V	q_e
0.430	0.000 23	2.72	0.158
0.430	0.000 39	1.71	0.251
0.430	0.002 45	0.632	0.600
0.430	0.038 7	0.097 2	3.52
0.430	0.109	0.044 8	6.08
0.522	0.0160	0.208	2.44
0.522	0.007 40	0.343	1.50
4.21	0.009 55	2.67	1.57
4.21	0.041 2	1.091	3.82
4.21	0.119	0.602	6.80
4.21	0.206	0.405	9.89
5.78	0.140	0.666	8.46
9.09	0.242	0.766	11.6
9.09	0.374	0.596	14.2
9.09	1.182	0.310	25.5
9.09	1.778	0.233	31.4

1985). Virgin carbon was used in the bottle studies and the water was obtained from the field site that contained background concentrations of TOC. Design a fixed bed adsorber to remove 1,1,2-TCE down to 0.01 mg/L when the influent concentration is 4.0 mg/L and the flow is 1 250 m^3/d (0.330 Mgal/d). The bulk density of the carbon in the bed is 0.6 g/cm^3 (37.4 lb/ft^3) and the bed height is 3.5 m (11.5 ft). Water is to be fed to the bed at a rate of 0.15 m^3/m^2/min (3.68 gal/ft^2/min).

Laboratory studies with water from the site have shown that $k_d\alpha$ is 13.5 g/cm³/min (0.487 lb/in.³/min). The fresh carbon contains no 1,1,2-TCE or TOC and the initial effluent contains negligible TOC; therefore, it can be assumed that the operating line passes through the origin.

The breakthrough concentration will be set at 80% of the effluent criterion to provide a safety factor. The lab studies have shown that the breakthrough curve does not exhibit significant tailing, so C_e will be set at $0.98C_0 = 3.92$ mg/L.

The first step in the solution is to establish the adsorption isotherm. Equation (15.36) was used to find the q_e values in column 4 of the table.

To evaluate the fitness of the Langmuir isotherm, a regression analysis of $1/q_e$ against $1/C$ was performed. The Freundlich equation (Eq. 15.42) was linearized by taking logarithms of each side of the equation and regressing $\ln q_e$ against $\ln C$. C_s data were unavailable; therefore, the BET isotherm was not appropriate.

The results of the regression analyses were:

Langmuir

Intercept: $0.248\,7 = 1/Q_0 \Rightarrow Q_0 = 4.021$ mg/g

Slope: $0.001\,429 = 1/(Q_0K) \Rightarrow K = 173.9$ L/mg

The correlation coefficient (R^2) is $R^2 = 0.977$

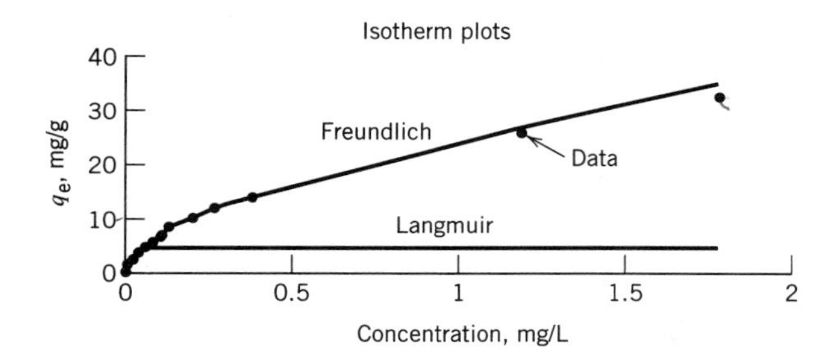

Freundlich

Intercept: $3.260 = \ln K_F \Rightarrow K_F = 24.66$

Slope: $0.591\,9 = 1/n \Rightarrow n = 1.689$

$R^2 = 0.997$

Plots of the two isotherms are shown above. From observation of the plots and the R^2 values, the Freundlich isotherm best described the data. The Langmuir isotherm describes only data near the origin well. The working equation is

$$q_e = 24.66C^{0.591\,9} \tag{i}$$

Now the isotherm relation (Eq. i) is plotted and the operating line which passes through the origin and the point (C_0, q_∞) is drawn (Eq. 15.63). The plot is shown in Fig. 15.19.

The curves in Fig. 15.19 are now used to construct a graph as shown in Fig. 15.18.

Data from the two curves in Fig. 15.19 are tabulated below for the $C - C^*$ versus C plot (Fig. 15.20). The starting value of q corresponds to $C = C_b$. The equation describing the operating line is $C = 0.067\,8q$ and Eq. (i) describes the isotherm (C^*). These relations were used to construct the table. These computations and others in

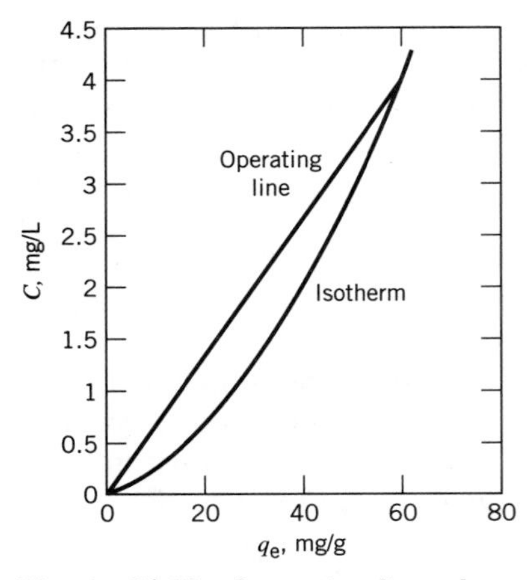

Figure 15.19 Operating line plot.

the table are readily performed with a spreadsheet. A smaller interval size leads to more accurate results.

Graphical integration of Eq. (15.66) using the curve in Fig. 15.20 was performed. The data for A_i and C/C_0 are tabulated with the C and C^* data and used to construct the plot of the breakthrough curve shown in Fig. 15.21.

The equation for A_i is

$$A_i = (C_i - C_{i-1})\frac{[(C_i - C_i^*) + (C_{i-1} - C_{i-1}^*)]}{2}$$

$$A_t = \Sigma A_i$$

The breakthrough curve does not exhibit tailing. The fractional capacity in the adsorption zone is calculated from

$$f = \sum_{j=1}^{n}\left[\left(\frac{\sum_1^j A_i}{A_t} - \frac{\sum_1^{j-1} A_i}{A_t}\right)\left(\frac{C_j - C_{j-1}}{2}\right)\right] = 0.49$$

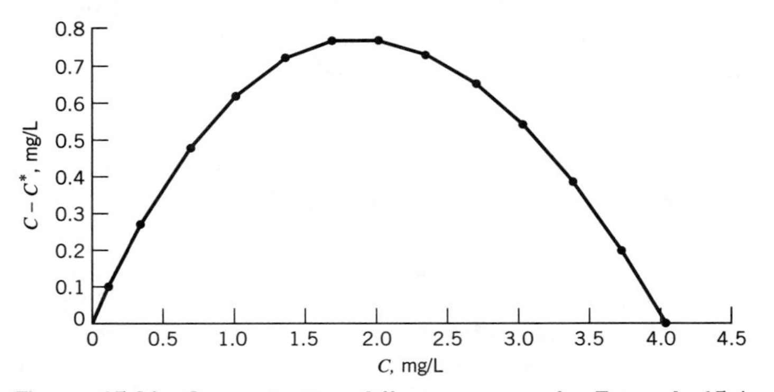

Figure 15.20 Concentration difference curve for Example 15.4.

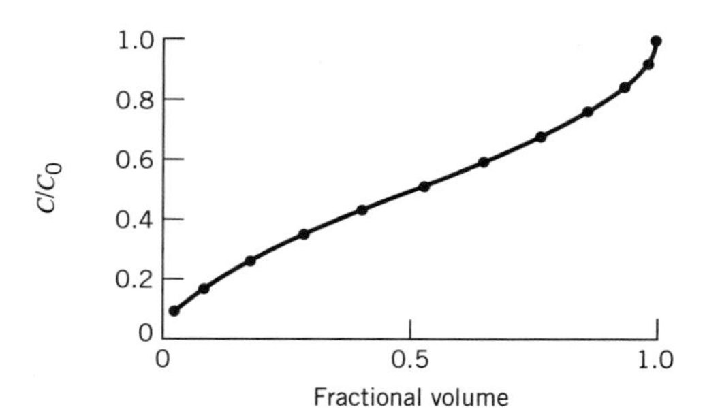

Figure 15.21 Breakthrough curve for Example 15.4.

From Fig. 15.19 or Eq. (i) the saturated concentration in the adsorbate (at C = 4.0 mg/L) is q_∞ = 59.2 mg/g. Now Eq. (15.64) must be applied. The table below can be used to find the value of the integral, which is approximated as $\Sigma\Delta C/(C - C^*)$. By starting at C_b and carrying the integration to C_e = 3.92 mg/L,

$$\frac{k_d \alpha \delta}{F_m} = \int_{C_b}^{C_e} \frac{dC}{C - C^*} = 7.53$$

The mass flow rate is

$$F_m = \rho \frac{Q}{A} = (1.00 \times 10^3 \text{ kg/m}^3)(0.15 \text{ m/min}) = 150 \text{ kg/m}^2\text{/min}$$

$$\text{In U.S. units: } F_m = \rho \frac{Q}{A} = \left(62.4 \frac{\text{lb}}{\text{ft}^3}\right)\left(3.68 \frac{\text{gal}}{\text{ft}^2\text{-min}}\right)\left(\frac{1 \text{ ft}^3}{7.48 \text{ gal}}\right) = 30.7 \text{ lb/ft}^2\text{/min}$$

q	1.72	5	10	15	20	25	30	35	40
C	0.117	0.339	0.678	1.017	1.356	1.695	2.034	2.373	2.712
C^*	0.010 1	0.061 5	0.198	0.393	0.639	0.931	1.267	1.644	2.060
$C - C^*$	0.106	0.278	0.480	0.624	0.717	0.764	0.767	0.729	0.652
A_i		0.043	0.128	0.187	0.227	0.251	0.259	0.253	0.234
$\Sigma A_i/A_t$		0.021	0.083	0.173	0.282	0.404	0.529	0.651	0.764
C/C_0		0.085	0.169	0.254	0.339	0.424	0.508	0.593	0.678
$\Delta C/(C - C^*)$		1.16	0.895	0.614	0.506	0.458	0.443	0.453	0.491

q	45	50	55	59.2	
C	3.051	3.390	3.729	4.014	
C^*	2.514	3.003	3.528	3.995	
$C - C^*$	0.537	0.387	0.201	0.019	
A_i	0.202	0.157	0.100	0.031	$A_t = 2.072$
$\Sigma A_i/A_t$	0.861	0.937	0.985	1.000	
C/C_0	0.763	0.847	0.932	1.000	
$\Delta C/(C - C^*)$	0.570	0.734	1.15	0.052[a]	$\Sigma\Delta C/(C - C^*) = 7.53$

[a] 0.052 is based on q = 57.8, corresponding to C_e = 3.92 mg/L.

δ can now be found:

$$\delta = \frac{(7.53)F_m}{k_d\alpha} = \frac{(7.53)\,(150\text{ kg/m}^2\text{/min})}{(13.5\text{ g/cm}^3\text{/min})}\left(\frac{1\,000\text{ g}}{\text{kg}}\right)\left(\frac{1\text{ m}}{100\text{ cm}}\right)^3 = 0.083\,7\text{ m} = 8.37\text{ cm}$$

$$\text{In U.S. units: }\delta = \frac{(7.53)\,(30.7\text{ lb/ft}^2\text{/min})}{(0.487\text{ lb/in.}^3\text{/min})}\left(\frac{1\text{ ft}}{12\text{ in.}}\right)^2 = 3.30\text{ in.} = 0.275\text{ ft}$$

In this example the height of the adsorption zone is very small compared to the height of the bed.

The fractional saturation of the carbon at the time of breakthrough is

$$\text{Fractional saturation} = \frac{D - \delta f}{D} = \frac{3.5\text{ m} - (0.083\,7\text{ m})(0.49)}{3.5\text{ m}} = 0.99$$

$$\text{In U.S. units: Fractional saturation} = \frac{11.5\text{ ft} - (0.275\text{ ft})(0.49)}{11.5\text{ ft}} = 0.99$$

At breakthrough the carbon contains

$$(0.99)\left(\frac{0.059\,2\text{ g}}{\text{g}}\right)\left(\frac{0.6\text{ g}}{\text{cm}^3}\right)\left(\frac{1\text{ kg}}{1\,000\text{ g}}\right)\left(\frac{100\text{ cm}}{\text{m}}\right)^3(3.5\text{ m}) = 123\text{ kg TCE/m}^2$$

$$\text{In U.S. units: }(0.99)\left(\frac{0.059\,2\text{ g}}{\text{g}}\right)\left(37.4\,\frac{\text{lb}}{\text{ft}^3}\right)(11.5\text{ ft}) = 25.2\text{ lb TCE/ft}^2$$

$$F_mC_0 = \left(\frac{150\text{ kg}}{\text{m}^2\text{-min}}\right)\left(\frac{4.0\text{ mg}}{\text{L}}\right)\left(\frac{1\text{ kg}}{10^6\text{ mg}}\right)\left(\frac{1\,000\text{ L}}{\text{m}^3}\right)\left(\frac{1\text{ m}^3}{10^3\text{ kg}}\right) = 6.00\times10^{-4}\text{ kg/m}^2\text{/min}$$

$$\text{In U.S. units: }F_mC_0 = \left(\frac{30.7\text{ lb}}{\text{ft}^2\text{-min}}\right)\left(\frac{4.0\text{ mg}}{\text{L}}\right)\left(\frac{1\text{ g}}{10^3\text{ mg}}\right)\left(\frac{1\text{ lb}}{454\text{ g}}\right)\left(\frac{3.79\text{ L}}{\text{gal}}\right)\left(\frac{7.48\text{ gal}}{\text{ft}^3}\right)\left(\frac{1\text{ ft}^3}{62.4\text{ lb}}\right)$$

$$= 1.22\times10^{-4}\text{ lb/ft}^2\text{/min}$$

The run time is

$$t = \frac{123\text{ kg TCE/m}^2}{6.00\times10^{-4}\text{ kg/m}^2\text{/min}} = 2.05\times10^5\text{ min} = 142\text{ d}$$

$$\text{In U.S. units: }t = \frac{25.2\text{ lb TCE/ft}^2}{1.22\times10^{-4}\text{ lb/ft}^2\text{/min}} = 2.06\times10^5\text{ min} = 142\text{ d}$$

The volume of water passed through the column at breakthrough is

$$V_b = \left(\frac{150\text{ kg}}{\text{m}^2\text{-min}}\right)(2.05\times10^5\text{ min})\left(\frac{1\text{ m}^3}{10^3\text{ kg}}\right) = 3.08\times10^4\text{ m}^3\text{/m}^2$$

$$\text{In U.S. units: }V_b = \left(\frac{30.7\text{ lb}}{\text{ft}^2\text{-min}}\right)(2.06\times10^5\text{ min})\left(\frac{7.48\text{ gal}}{1\text{ ft}^3}\right)\left(\frac{1\text{ ft}^3}{62.4\text{ lb}}\right) = 7.58\times10^5\text{ gal/ft}^2$$

Bed Depth Service Time Method

The design methodology presented in the previous section assumes that the breakthrough curve does not change its shape as the adsorption wave moves through the column and it is assumed that equilibrium is established between the feed and exhausted carbon. Significant biological activity and complex adsorption isotherms will negate these conditions. A practical generalized method for design of continuous flow carbon adsorption systems based on laboratory or pilot plant data is the bed-depth-

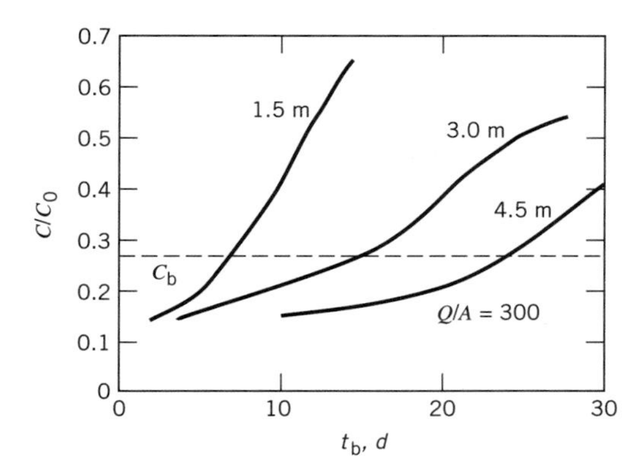

Figure 15.22 Breakthrough curves at different bed depths.

service time (BDST) method (Hutchins, 1973). The method is not restricted to favorable isotherms or nonbiodegradable wastewaters (Benedek, 1974). Breakthrough curves are obtained at several different column lengths that span the lengths to be used in the eventual design.

The design must meet two constraints: the minimum contact time must be provided and carbon must be supplied at a rate equal to the exhaustion rate. The actual minimum exhaustion rate and theoretical minimum exhaustion rates may be quite different. Data from the BDST technique will provide the information to determine actual minimum exhaustion rates and give the designer information to optimize contact time and exhaustion rate combinations to give the minimum cost of the operation.

Wastewater of a relatively constant concentration is applied to columns of different depth at the same loading rate for each column and the effluent concentrations are monitored over time. The fractional effluent concentration data are plotted as shown in Fig. 15.22. These curves are the breakthrough curves for the columns. If the curves do not begin at $C/C_0 = 0$, some of the material is nonadsorbable. The curves in Fig. 15.22 indicate that the fraction of nonadsorbable matter is approximately 0.10 to 0.15.

A horizontal line is drawn at the specified breakthrough concentration as indicated on the plot. The intersection of the breakthrough curves and the effluent limiting concentration (chosen with a suitable safety factor) defines the time between regenerations for a bed, t_b. The service time is less than the regeneration time because it includes the filling time of the bed, which is given here expressed in terms of the available volume of the bed.

$$t_s = t_b - \frac{eAD}{Q} \tag{15.67}$$

where

t_s is the service time of the bed (similar to t_D in Eq. 15.54)

Normally t_s will be equal to t_b but it should be checked with the above equation.

The overall rate of carbon exhaustion for a bed depth in Fig. 15.22 is calculated from

$$R_e = \frac{\rho_p AD}{t_b} \tag{15.68}$$

where

R_e is the rate of carbon exhaustion

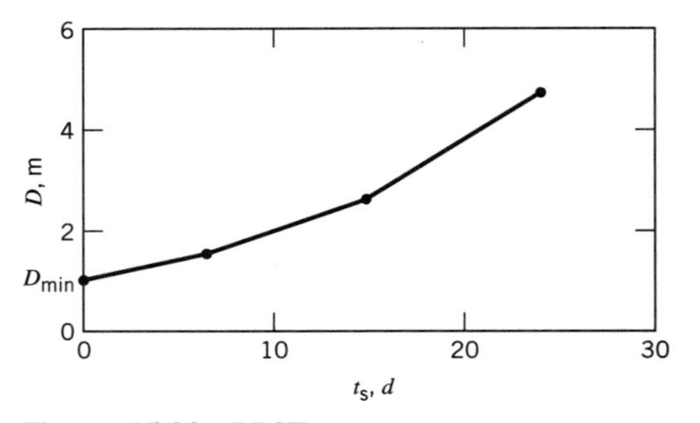

Figure 15.23 BDST curve.

Equation (15.68) is applied to each bed.

The EBCT is also calculated for each bed.

$$t_d = \frac{V_B}{Q} = \frac{D}{Q/A} \tag{15.69}$$

where

t_d is EBCT

V_B is the volume of the bed (carbon + void volume)

The BDST curve is obtained by plotting bed depth versus service time (Fig. 15.23). A regression is performed to find the curve of best fit through the points. The point where the curve passes the ordinate defines the minimum bed depth, D_{min} (or detention time), required to achieve the effluent criterion.

The minimum contact time is determined from Fig. 15.23.

$$t_{d,min} = \frac{D_{min}}{Q/A} \tag{15.70}$$

where

$t_{d,min}$ is the minimum EBCT

D_{min} is the minimum depth determined by the intersection of the curve with the ordinate on Fig. 15.23

The carbon exhaustion rate data are plotted against the EBCT data to form an operating line (Fig. 15.24) for the process (Erskine and Schuliger, 1971). The designer

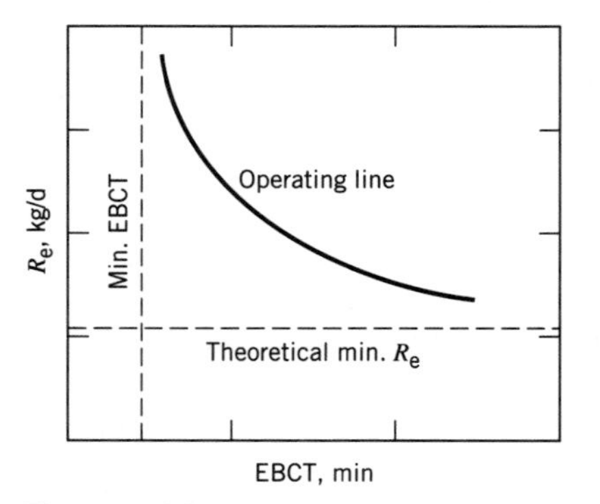

Figure 15.24 Operating line and theoretical minimum EBCT and exhaustion rate when only adsorption is significant.

can choose any point on the operating line for a proper design. An economic evaluation of the costs associated with larger bed depths and rates of carbon regeneration at each depth can be performed to find the least cost design. A more comprehensive analysis can be performed by applying different surface loading rates and performing the economic analysis for each operating line.

The theoretical minimum exhaustion rate is based on the equilibrium isotherm for the carbon and the wastewater. The rate at which carbon is being exhausted is equal to the rate at which adsorbate is being removed.

$$R_{e,min} q_\infty = QC_0 \Rightarrow R_{e,min} = \frac{QC_0}{q_\infty} \tag{15.71a}$$

If there is a concentration, C_n, of nonadsorbable matter then

$$R_{e,min} = \frac{Q(C_0 - C_n)}{q_\infty} \tag{15.71b}$$

The operating line asymptotically approaches the minimum EBCT and exhaustion rate lines. The theoretical minimum exhaustion rate assumes that only adsorption is significant and that the adsorption zone is a square wave moving through the column. The actual exhaustion rate will be less than the theoretical minimum rate. If biological activity or other phenomena contribute to the removal, the actual rate of exhaustion may be lower than the theoretical minimum rate depending on the extent of phenomena other than adsorption (see the following example).

■ Example 15.5 Carbon Adsorber Design with the BDST Method

The breakthrough curves for COD shown in Fig. 15.22 were obtained for three columns with depths of 1.5, 3.0, and 4.5 m. The loading rate applied to each column was 300 m/d. The influent flow rate is 960 m^3/d. The influent contains 150 mg/L COD, of which 15 mg/L is nonadsorbable. The breakthrough concentration was set at $0.28C_0$, which is indicated on Fig. 15.22.

The carbon has a porosity of 0.50 and packed density of 0.50 g/cm^3. From isotherms ran with the wastewater and carbon, q_∞ is 0.31 kg COD/kg carbon. Find the operating line.

The required area of the column is

$$A = \frac{Q}{Q/A} = \frac{960 \text{ m}^3/\text{d}}{300 \text{ m/d}} = 3.20 \text{ m}^2$$

The time to breakthrough in each column was found from Fig. 15.22 and entered in the following table.

Depth m	t_b d	t_s d	R_e kg/d	EBCT min
1.5	6.5	6.5	369	7.2
3.0	15.0	15.0	320	14.4
4.5	24.0	24.0	300	21.6

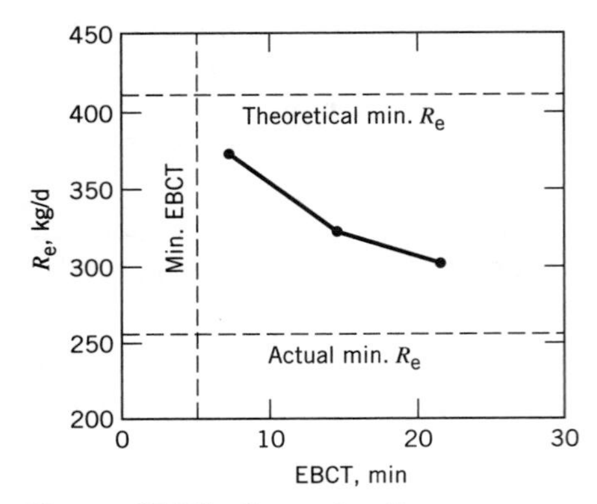

Figure 15.25 Operating line.

Equation (15.67) was applied to find the service time. For the 1.5-m bed,

$$t_s = t_b - \frac{eAD}{Q} = 6.5\text{ d} - \frac{(0.50)(3.20\text{ m}^2)(1.5\text{ m})}{960\text{ m}^3/\text{d}} = 6.5\text{ d} - 0.002\,5\text{ d} = 6.5\text{ d}$$

The other t_s values were calculated similarly and are given in the table. These data were plotted in Fig. 15.23. The curve in Fig. 15.23 was regressed using the model $D = Ke^{-mt_s}$ to find the intercept value, D_{min}. The regression equation is

$$D = 1.054e^{-0.063t_s} \quad R^2 = 0.973$$

The regression indicated that $D_{min} = 1.05$ m.

The overall carbon exhaustion rates were calculated from Eq. (15.68). For the 1.5 m bed,

$$R_e = \frac{\rho_p AD}{t_b} = \frac{(0.50\text{ g/cm}^3)(3.20\text{ m}^2)(1.5\text{ m})}{6.5\text{ d}}\left(\frac{1\text{ kg}}{1\,000\text{ g}}\right)\left(\frac{10^6\text{ cm}^3}{\text{m}^3}\right) = 369\text{ kg/d}$$

The other values were calculated similarly and entered into the table.

EBCTs for the bed were calculated from Eq. (15.69). For the 1.5-m depth column,

$$t_d = \frac{V_B}{Q} = \frac{D}{Q/A} = \frac{1.5\text{ m}}{300\text{ m/d}} = 0.005\text{ d} = (0.005\text{ d})(1\,440\text{ min/d}) = 7.2\text{ min}$$

The other values were calculated similarly and entered into the table. The R_e and EBCT data were plotted in Fig. 15.25 to obtain the operating line. From the limited amount of data it appears that $R_{e,min} = 250$ kg/d. A regression using a power law indicates this is reasonable. The equation was

$$R_e = 537t_d^{-0.191} \quad R^2 = 0.996$$

The theoretical minimum carbon exhaustion rate (Eq. 15.71b) is

$$R_{e,min} = \frac{Q(C_0 - C_n)}{q_\infty} = \frac{(960\text{ m}^3/\text{d})(150\text{ mg/L} - 15\text{ mg/L})}{0.31\text{ kg/kg}}\left(\frac{1\text{ kg}}{10^6\text{ mg}}\right)\left(\frac{1\,000\text{ L}}{\text{m}^3}\right)$$
$$= 418\text{ kg/d}$$

The minimum contact time is

$$t_{d,min} = \frac{D_{min}}{Q/A} = \frac{1.05\ \text{m}}{300\ \text{m/d}}\left(\frac{1\,440\ \text{min}}{\text{d}}\right) = 5.0\ \text{min}$$

The minimum contact time, theoretical minimum exhaustion rate, and actual exhaustion rates were drawn with dashed lines on Fig. 15.25. The operating line asymptotically approaches the minimum contact time and actual exhaustion rate lines. Any point on the operating line is a valid design. Costs of greater depth of the bed must be compared against carbon regeneration rates to establish the optimum point.

Comparing the actual and theoretical exhaustion rate lines it is apparent that biological activity is very significant in these columns. A power law does not yield a definite minimum exhaustion rate. More data would allow better definition of the line but the data clearly indicate that the actual minimum carbon exhaustion rate is well below the theoretical minimum.

QUESTIONS AND PROBLEMS

1. What are the total and calcium hardnesses as $CaCO_3$ of a water that contains ions at the concentrations given in the following table?

Ion	Ca^{2+}	Mg^{2+}	Fe^{3+}	Na^{+}	K^{+}	CO_2
Concentration, mg/L	48	20	0.3	1.2	0.7	2

2. If a significant amount of alkalinity in a water came from species other than inorganic carbon species or OH^- ions, would this portion of alkalinity contribute to carbonate hardness and assist softening?
3. Calculate the K_{sp} for $CaCO_3$ based on the practical limits of $CaCO_3$ removal in a treatment process (0.6 meq/L of $CaCO_3$). Using the K_{sp} based on the practical removal limit, calculate how much $CaCO_3$ would precipitate from initial concentrations of Ca^{2+} and CO_3^{2-} of 1.90 and 2.20 meq/L, respectively.
4. The following compositions of major constituents of four natural waters are given in Samuel D. Faust and Osman M. Aly (1981). Draw meq/L bar graphs for waters 2 and 4. List the concentrations of hypothetical compounds in the waters.

Constituent[a]	1	2	3	4
Ca	40	94	140	126
Mg	22	40	43	43
Na	0.4	17	[b]	13
K	1.2	2.2	[b]	2.1
HCO_3^-	213	471	241	440
SO_4^{2-}	4.9	49	303	139
Cl^-	2.0	9.0	38	8.0

[a]All concentrations are in mg/L of the constituent.
[b]Na and K concentrations were not specified.

5. What pH is attained by adding 0.7 meq/L of lime to pure water? What pH is attained by adding the same amount of lime to water with a pH of 7.00 that has an alkalinity of 100 mg/L as $CaCO_3$?

6. (a) For hardness removal in Example 15.1(a), prepare bar graphs describing the state of the water at stages given below. It will be assumed that reactions occur in a definite sequence.
   (i) The initial state of the water
   (ii) After addition of lime
   (iii) After reaction of lime with CO_2 and bicarbonate
   (iv) After precipitation of $CaCO_3$
   (b) Prepare bar graphs as indicated in part (a) for Example 15.1(b) In addition prepare the following bar graphs.
   (v) Showing the state of the water after $Mg(OH)_2$ has precipitated
   (vi) After Na_2CO_3 has been added
   (vii) The final state of the water

7. For each water in problem 4 what are the stoichiometric quantities of lime and soda required to remove hardness to practical limits? Ignore excess lime requirements for Mg removal; also ignore the lime requirement for CO_2 removal.

8. Calculate the lime dose, carbon dioxide dose after the first stage, and Na_2CO_3 dose for hardness removal of water No. 3 in Problem 4 using a split 2-stage recarbonation process. The effluent Mg concentration is to be 40 mg/L as $CaCO_3$ and the total hardness should not exceed 120 mg/L as $CaCO_3$. The excess lime for Mg removal is 1.00 meq/L. The carbon dioxide concentration of the water was 0.45 meq/L.

9. Calculate the lime and soda ash requirements for a split treatment process. What is the fraction of bypassed flow? The influent water contains

   Ca^{2+}, 4.80 meq/L Mg^{2+}, 2.70 meq/L CO_2, 0.60 meq/L
   Alkalinity, 6.30 meq/L

   The product water should have a magnesium concentration of 0.80 meq/L and a total hardness concentration of 2 meq/L or lower. Perform the analysis when the concentration of magnesium from the first stage process is 0. Assume the excess lime required to precipitate $Mg(OH)_2$ is 1.0 meq/L and an excess of 0.6 meq/L of $CaCO_3$ is required for Ca^{2+} removal. What are the concentrations of all softening related species after each stage?

10. Determine the total hardness, alkalinity, Cl^-, and HCO_3^- concentrations for the following water, analyzed at 25°C. The pH of the water is 7.9.

Constituent	Concentration mg/L
CO_2	3.0
Ca^{2+}	45.5
K^+	37.5
Mg^{2+}	21.0
Cl^-	?
HCO_3^-	?
So_4^{2-}	46.5

11. Algae in stabilization ponds or reservoirs may cause hardness removal during daylight hours on hot summer days when their metabolic activities are high. Explain how this may happen.

12. Calculate the saturation concentration of Ca^{2+} in equilibrium with an atmospheric CO_2 concentration of $10^{-3.5}$ atm at pH values of 5, 6, 7, 8, and 9. Ignore ionic strength effects and use K_H = 1 450 mg/L/atm (Eq. 3.20), $K_0 = 10^{-2.8}$ for Eq. (3.21), and $K_1' = 10^{-3.5}$ for Eq. (3.22a). Other equilibrium constants are given in Tables 1.5 and 3.2.

13. What is the error in estimation of the saturation pH of a water that has an alkalinity concentration of 190 mg/L and Ca^{2+} concentration of 120 mg/L, both measured as $CaCO_3$? Perform the calculations for water samples with pHs of 7.00 and 8.50 at a temperature of 25°C. Assume $S = 0.145$. (Hint: Eq. 15.15 assumes that all alkalinity is HCO_3^- ion.)

14. (a) What is the saturation pH for a water at 10°C that contains Ca^{2+} at 100 mg/L as $CaCO_3$, an alkalinity of 150 mg/L as $CaCO_3$, and TDS of 400 mg/L? (b) Perform the same exercise for this water at a temperature of 25°C. What is the SI of these waters if the pH of the water is 7.80?

15. What is the error in estimating the saturation pH of the water in Problem 14 if the salinity correction is ignored?

16. How much strong acid or $Ca(OH)_2$ must be added to the water in Problem 14 to achieve a saturation index of 0.10? Only consider activity corrections in calculating pH_s.

17. (a) Although the oxidation of iron with permanganate is favored by a low pH, why is the overall reaction for the removal of iron favored by a high pH? (b) Considering that each oxygen atom can acquire two electrons, what are the overall equations for the removal of iron and manganese using ozone as the oxidant?

18. Describe the methods for iron and manganese removal, including chemical equations.

19. (a) In the case of iron and aluminum, why is significantly more metal coagulant required beyond the stoichiometric amount to reduce soluble orthophosphate concentrations from 0.5 to 1 mg P/L? (b) What is the largest factor regulating the dose of calcium to remove phosphorus?

20. A metal plating industry produces effluent containing Cu^{2+}, Fe^{2+}, and Ni^{2+} at concentrations of 25, 25, and 35 mg/L, respectively. The wastewater flow rate averages 290 m^3/d. An ion exchanger with a capacity of 3.00 meq/g is to be used to remove these metals before discharge of the waste to the sewer. The exchange capacity of the resin on a volumetric basis is 2.0 meq/L. The regenerant solution will be H_2SO_4 at a concentration of 50 g/L. What is the volume of the ion exchanger and mass of resin in it to remove these ions if the resin is to be regenerated once per week? What is the volume of regenerant used on a weekly basis?

21. On which side of a solute removing membrane is precipitation a problem? Why?

22. Explain why and how an increase in flow-through velocity affects the breakthrough curve. Draw the breakthrough curve for a nominal flow-through velocity near zero.

23. How can steady state be assumed in deriving Eq. (15.46) when the adsorption zone is moving through the bed, causing the accumulation on any layer of carbon within the adsorption zone to change with time?

24. Would recycle of the effluent improve the efficiency of a continuous flow fixed bed carbon adsorption unit? Why or why not?

25. Design a fixed bed carbon adsorber to remove a contaminant down to 0.005 mg/L with a breakthrough criterion of 0.004 mg/L. Plot the breakthrough curve and find the run time. The influent concentration is 0.5 mg/L in a flow of 850 m^3/d (0.224 Mgal/d). Virgin carbon with a density of 0.52 g/cm^3 (32.4 lb/ft^3) is to be used in the unit. The column will have a depth of 4.0 m (13.1 ft) and the surface loading rate is 0.25 $m^3/m^2/min$ (8 840 gal/ft^2/d). $k_d\alpha = 6.0$ $g/cm^3/min$ (374 $lb/ft^3/min$). Use C_e of 0.46 mg/L. The following table provides the adsorption data for the compound.

C	mg/L	0.000 20	0.000 42	0.000 59	0.001 0	0.002 5	0.004 2	0.008 8
q_e	mg/g	0.016 5	0.028 8	0.041 5	0.062 6	0.121	0.170	0.322
C	mg/L	0.019	0.068	0.11	0.36	0.95	1.33	
q_e	mg/g	0.667	1.61	2.58	5.55	12.4	17.1	

26. A painter is cleaning her paintbrush with solvent. The mass of the brush fibers is 250 g and the fibers have 50 g of paint on them. The brush will be immersed in solvent and agitated until equilibrium conditions are established. A Freundlich isotherm applies with $K_F = 0.90$ and $n = 3.14$ for q in mg/g and C in mg/L.
(a) Find the mass of paint remaining on the brush if it is immersed in a solvent volume of 1.0 L.
(b) Find the total volume of solvent required to achieve the same degree of cleansing as in (a) if two equal volumes of solvent will be used in succession. The brush is immersed in the first volume and agitated until equilibrium is established, removed and the exercise is repeated with the second volume. Assume that the solvent remaining in the brush after it is removed from the first batch is insignificant.

27. Solve Example 15.5 in U.S. units for the operating line with the following changes. Assume the curves in Fig. 15.22 apply but they were obtained for column depths of 4.0, 8.0, and 12.0 ft at a nominal velocity (Q/A) of 6 000 gal/ft^2/d. The influent flow rate is 0.30 Mgal/d. Influent COD is 150 mg/L of which 15 mg/L is nonabsorbable. The breakthrough concentration is $0.28C_0$. The carbon porosity is 0.48 and its packed density is 35 lb/ft^3. The value of q_∞ is 0.30 lb COD/lb carbon.

KEY REFERENCE

CLIFFORD, D. A. (1990), "Ion Exchange and Inorganic Adsorption," in *Water Quality and Treatment,* F. W. Pontius, ed., American Water Works Association, Denver, CO, pp. 561–639.

REFERENCES

ASCE and AWWA (1990), *Water Treatment Plant Design,* 2nd ed., McGraw-Hill, Toronto.

AWWA Joint Task Group (1990), "Suggested Methods for Calculating and Interpreting Calcium Carbonate Saturation Indexes," *J. American Water Works Association,* 82, 7, pp. 71–77.

AWWA Membrane Committee (1992), "Committee Report: Membrane Processes in Potable Water Treatment," *J. American Water Works Association,* 84, 1, pp. 59–67.

BENEDEK, A. (1974), "Carbon Evaluation and Process Design," Proceedings of the Physical–Chemical Treatment Activated Carbon Adsorption in Pollution Control, Environment Canada, Ottawa, ON.

BRINCK, W. L., R. J. SCHLIEKELMAN, D. L. BENNET, C. R. BELL, AND I. MARKWOOD (1978), "Radium Removal Efficiencies in Water Treatment Processes," *J. American Water Works Association,* 70, 1, pp. 31–35.

BRUNAUER, S., P. H. EMMETT, AND E. TELLER (1938), "Adsorption of Gases in Multimolecular Layers," *J. American Chemistry Society,* 60, pp. 309–319.

BULUSU, K. R., B. B. SUNDARESAN, B. N. PATHAK, W. G. NAWLAKHE, D. N. KULKARNI, AND V. P. THERGAONKAR (1979), "Fluorides in Water, Defluoridation Methods and Their Limitations," *J. of the Institution of Engineers (India),* 60 1, pp. 1–25.

CALDWELL, D. H. AND W. B. LAWRENCE (1953), "Water Softening and Conditioning Problems," *Industrial and Engineering Chemistry,* 45, 3, pp. 535–547.

CASTRO, K. AND A. K. ZANDER (1995), "Membrane Air-Stripping: Effects of Pretreatment," *J. American Water Works Association,* 87, 3, pp. 50–61.

CLEASBY, J. L. AND J. H. DILLINGHAM (1966), "Rational Aspects of Split Treatment," *J. of the Sanitary Engineering Division, ASCE,* 92, SA2, pp. 1–7.

COLE, L. D. (1976), "Surface-Water Treatment in Cold Weather," *J. American Water Works Association,* 68, 1, pp. 22–25.

COMSTOCK, G. W. (1979), "Water Hardness and Cardiovascular Disease," *American J. of Epidemiology,* 110, pp. 375–400.

CULP, R. L. AND H. A. STOLTENBERG (1958), "Fluoride Reduction at LaCrosse, Kan.," *J. American Water Works Association,* 50, 3, pp. 423–431.

Dept. of National Health and Welfare (1980), *Guidelines for Canadian Drinking Water Quality 1978, Supporting Documentation*, Supply and Services Canada, Ottawa, ON.

ECKENFELDER, W. W., JR. AND Y. ARGAMAN (1991), "Principles of Biological and Physical/Chemical Nitrogen Removal," in *Phosphorus and Nitrogen Removal from Municipal Wastewater,* 2nd ed., R. Sedlak, ed., Lewis Publishers, Chelsea, MI, pp. 3–42.

ERSKINE, D. B. AND W. G. SCHULIGER (1971), "Graphical Method to Determine the Performance of Activated Carbon Processes for Liquids," *American Institute of Chemical Engineers Symposium Series,* 68, 124, pp. 185–190.

FAUST, S. D. AND O. M. ALY (1981), *Chemistry of Natural Waters,* Ann Arbor Science, Ann Arbor, MI.

FAUST, S. D. AND O. M. ALY (1987), *Adsorption Processes for Water Treatment,* Butterworth Publishers, Toronto.

FLICK, E. W. (1991), *Water Treatment Chemicals: An Industrial Guide,* Noyes Publications, Park Ridge, NJ.

FREUNDLICH, H. (1926), *Colloid and Capillary Chemistry,* Methuen & Co., London.

GLAZE, W. H. (1990), "Chemical Oxidation" in *Water Quality and Treatment,* F. W. Pontius, ed., American Water Works Association, Denver, CO, pp. 747–779.

HUTCHINS, R. A. (1973), "Optimum Sizing of Activated Carbon Systems," *Industrial Water Engineering,* 10, 3, pp. 40–43.

JENKINS, D. AND S. W. HERMANOWICZ (1991), "Principles of Chemical Phosphate Removal," in *Phosphorus and Nitrogen Removal from Municipal Wastewater,* 2nd ed., R. I. Sedlak, ed., Lewis Publishers, Chelsea, MI, pp. 91–110.

KUCHENRITHER, R. D., G. K. ELMUND, AND C. P. HOUCK (1992), "Sludge Quality Benefits Realized from Drinking Water Stabilization," *Water and Environment Research,* 64, 2, pp. 150–153.

LANGELIER, W. F. (1936), "The Analytical Control of Anti-Corrosion Water Treatment," *J. American Water Works Association,* 28, pp. 1500–1521.

LANGMUIR, I. (1918), "The Adsorption of Gases on Plane Surfaces of Glass, Mica and Platinum," *J. American Chemistry Society,* 40, pp. 1361–1402.

LARSON, T. E., R. W. LANE, AND C. H. NEFF (1959), "Stabilization of Magnesium Hydroxide in the Solids-Contact Process," *J. American Water Works Association,* 51, 12, pp. 1551–1558.

MCCLEAF, P. R. AND E. D. SCHROEDER (1995), "Denitrification Using a Membrane-Immobilized Biofilm," *J. American Water Works Association,* 87, 3, pp. 77–86.

MERRILL, D. T. (1978), "Chemical Conditioning for Water Softening and Corrosion Control," in *Water Treatment Plant Design,* R. L. Sanks, ed., Ann Arbor Science, Ann Arbor, MI, pp. 497–565.

MONTGOMERY, J. M., Consulting Engineers, Inc. (1985), *Water Treatment Principles and Design,* John Wiley & Sons, Ltd., Toronto.

MOUCHET, P. (1992), "From Conventional to Biological Removal of Iron and Manganese in France," *J. American Water Works Association,* 84, 4, pp. 158–167.

NARBAITZ, R. M. (1985), *Modelling the Competitive Adsorption of 1,1,2-Trichloroethane with Naturally Occurring Background Organics onto Activated Carbon,* Ph.D. Thesis, McMaster University, Hamilton, ON.

NAWLAKHE, W. G., D. N. KULKARNI, B. N. PATHAK, AND W. L. BULUSU (1975), "Defluoridation of Water by Nalgonda Technique," *Indian J. of Environmental Health,* 17, 1, pp. 26–65.

NERI, L. C. AND H. L. JOHANSEN (1978), "Water Hardness and Cardiovascular Disease," *Annals of the New York Academy of Sciences,* 304, pp. 203–219.

PLUMMER, L. N. AND E. BUSENBERG (1982), "The Solubilities of Calcite, Aragonite, and Veterite in CO_2-H_2O Solutions between 0 and 90°C and an Evaluation of the Aqueous Model for the System $CaCO_3$-CO_2-H_2O," *Geochimica et Cosmochimic Acta,* 46, 6, pp. 1011–1040.

ROBINSON, R. B., G. D. REED, AND B. FRAZIER (1992), "Iron and Manganese Sequestration Facilities Using Sodium Silicate," *J. American Water Works Association,* 84, 2, pp. 77–82.

ROSSUM, J. R. AND D. T. MERRILL (1983), "An Evaluation of the Calcium Carbonate Saturation Indexes," *J. American Water Works Association,* 75, 2, pp. 95–100.

RUBEL, F., JR. AND R. D. WOOSELY (1979), "The Removal of Fluoride from Drinking Water by Activated Alumina," *J. American Water Works Association,* 71, 1, pp. 45–48.

SANKS, R. L (1978), "Ion Exchange," in *Water Treatment Plant Design,* R. L. Sanks, ed., Ann Arbor Science Publishers, Ann Arbor, MI, pp. 597–622.

SAYELL, K. M. AND R. R. DAVIS (1975), "Removal of Iron and Manganese from Raw Water Supplies Using Manganese Greensand Zeolite," *Industrial Water Engineering,* 12, 5, pp. 20–23.

SENGUPTA, A. K. AND B. SHI (1992), "Selective Alum Recovery from Clarifier Sludge," *J. American Water Works Association,* 84, 1, pp. 96–103.

SIMS, R. C. AND E. HINDIN (1978), "Use of Clinoptilolite for Removal of Trace Levels of Ammonia in Reuse Water," in *Chemistry of Wastewater Technology,* A. J. Rubin, ed., Ann Arbor Science Publishers, Ann Arbor, MI, pp. 305–323.

SNOEYINK, V. L. AND D. JENKINS (1980), *Water Chemistry*, John Wiley & Sons, Toronto.

SORIAL, G. A., M. T. SUIDAN, R. D. VIDIC, AND R. C. BRENNER (1993), "Effect of GAC Characteristics on Adsorption of Organic Pollutants," *Water Environment Research,* 65, 1, pp. 53–57.

TAN, L. AND R. G. SUDAK (1992), "Removing Color from a Groundwater Source," *J. American Water Works Association,* 84, 1, pp. 79–87.

TRUSSELL, R. R., A. TRUSSELL, AND P. KREFT (1980), *Selenium Removal from Groundwater Using Activated Alumina,* USEPA Report No. 600/2-80-153, Cincinnati, OH.

USEPA (1973), *Process Design Manual for Carbon Adsorption,* Center for Environmental Research Information, Cincinnati, OH.

USEPA (1987a), *Dewatering Municipal Wastewater Sludges,* Design Manual No. EPA/625/1-87/014, Center for Environmental Research Information, Cincinnati, OH.

USEPA (1987b), *Phosphorus Removal,* Design Manual No. EPA/625/1-87/001, Center for Environmental Research Information, Cincinnati, OH.

USEPA (1987c), *Retrofitting POTWS for Phosphorus Removal in the Chesapeake Bay Drainage Basin,* Handbook No. EPA/625/6-87/017, Center for Environmental Research Information, Cincinnati, OH.

WEBER, W. J., JR. (1972), *Physicochemical Processes for Water Quality Control,* Wiley-Interscience, Toronto.

WIESNER, M. R. AND S. CHELLAM (1992), "Mass Transport Considerations for Pressure-Driven Membrane Processes," *J. American Water Works Association,* 84, 1, pp. 88–95.

YOO, R. S., D. R. BROWN, R. J. PARDINI, AND G. D. BENTSON (1995), "Microfiltration: A Case Study," *J. American Water Works Association,* 87, 3, pp. 38–39.

CHAPTER 16

DISINFECTION

Disinfection is the destruction of pathogenic microorganisms in a water. The water is not necessarily sterilized. The protection of the public health from waterborne disease transmission by disinfection of water has been recognized since the turn of the century. The eradication of waterborne pathogens is the most important treatment of water. Disinfection is also applied to wastewater effluents to reduce the risk of disease for recreational users of surface waters and to minimize the contamination to downstream drawers of water.

Pathogen kill efficiency is not the only consideration in selecting a disinfectant. The characteristics of a good disinfectant are:

1. Effective kill of pathogenic microorganisms.
2. Nontoxic to humans or domestic animals.
3. Nontoxic to fish and other aquatic species.
4. Easy and safe to store, transport, and dispense.
5. Low cost.
6. Easy and reliable analysis in water.
7. Provides residual protection in drinking water.

Boiling water for 15 to 20 min eradicates pathogenic microorganisms but it is too energy intensive to be used for water or wastewater disinfection. It was used for a short time in only one municipality, Pathernay, France around the turn of the century (Baker, 1981). However, when municipal water treatment works fail to produce water of suitable microbiological quality, boil water orders are immediately issued to the public.

Currently, disease outbreaks in surface water systems occur primarily because of inadequate or interrupted disinfection, especially in systems that provide disinfection as the only treatment (Craun, 1988). There are many agents that effect disinfection, including chemical oxidants, irradiation, thermal treatment, and electrochemical treatment (Patermarakis and Fountoukidis, 1990). But the history of water and wastewater disinfection is the history of chlorination (White, 1992, 1978). The first continuous application of chlorine for disinfection of a municipal water supply in the United States was in 1908 for the city of Jersey City, NJ (Craun, 1988).

Chlorine satisfies the characteristics of a good disinfectant to a large degree. Developments in disinfection technology have produced many alternatives to chlorination, including other oxidizing agents such as chloramines, chlorine dioxide, permanganate, and ozone. Bromine, bromine chloride, and iodine are feasible alternatives. Ultraviolet (UV) and gamma irradiation may also be used. Silver is also a bactericide.

Chlorine is the least cost disinfection agent but recent research has identified byproducts of chlorination, some of which have been proved to be carcinogenic. In particular there has been much research and concern over trihalomethanes (THMs) which are produced with chlorine. The hazards associated with these and other byproducts are small at the concentrations produced in a typical operation and modern chlorination practice has been modified to further minimize their formation. Even without the improvements in chlorination practice, it has been stated by many that the benefits of chlorination have far outweighed the risks.

The issue is not solely confined to weighing the benefits and disadvantages of chlorine. Other disinfection agents may produce less risk, although at an elevated cost. Using the Ames test (Section 8.2.2) for comparison among disinfectants for the amount of mutagenic (mutagenic compounds are often carcinogenic) material produced, the ranking of disinfectants generally follows the order of ozone $<$ chlorine dioxide $<$ chloramine $<$ chlorine for the primary disinfectant for a given treatment system (Noot et al., 1989). Ozone can sometimes produce risk values as high as those for chlorine. There is a considerable amount of controversy and ongoing research being devoted to quantifying the risks associated with all disinfection technologies.

Once mutagenic compounds have been formed they can be partially removed by coagulation–flocculation–filtration processes but granular activated carbon treatment is very effective.

In a review of epidemiological studies (Craun, 1988), it was noted that there was an increased risk of certain types of cancer in populations using chlorinated water; however, no direct evidence was obtained to show that THMs were responsible. Other chlorination byproducts may contribute to the risk. The problem of assessing causal relationships between chlorination and cancer is confounded by many factors. Preliminary data from animals also indicate that there may be an association between chlorination and cardiovascular disease (Craun, 1988).

An intense research effort is underway to identify and quantify the risks of chlorination. These must be weighed against the well-established benefits of chlorination. Data used in most studies were obtained for chlorination practices that are not current. Modern chlorination practice reduces the formation of chlorine byproducts. There is even less information available on the risks associated with the use of alternative disinfectants compared to chlorine, which further clouds the decision on the most appropriate disinfectant. For instance, one study has shown that subchronic toxicity of chlorine dioxide $>$ monochloramine $>$ chlorine when administered at the same doses (mg/L basis) to rats for 90 days (Daniel et al., 1990). More studies of this nature are required.

At the moment, the general feeling is that chlorination does not elevate risk levels significantly to eliminate it as a primary alternative. However, increasingly severe regulations on THMs will favor the change to other disinfectants, particularly ozone.

The focus of this chapter is on the disinfectants and practices for pathogen removal in waters. Beyond the disinfectants discussed in this chapter there are many synthetic biocides used for general control of microorganism growth in pipelines, cooling towers, and other industrial applications. These agents will be present in wastewaters with potential to affect biological treatment processes and to contribute toxicity in the effluent.

16.1 KINETICS OF DISINFECTION

The kinetics of disinfection is described by a first-order law from studies by Chick (1908).

$$\frac{dN}{dt} = -kN \tag{16.1}$$

where
    N is the number of microorganisms
    t is time
    k is the dieoff coefficient

The dieoff coefficient is a function of disinfectant dose, type of microorganism, and conditions in the water. Watson (1908) refined the rate coefficient to explicitly include the concentration of disinfectant and another term related to disinfection power of the disinfectant.

$$k = \alpha C^n \tag{16.2}$$

where
    C is the concentration of disinfectant
    n is termed a constant of dilution
    α is an inactivation constant

In Eq. (16.2) the exponent n is commonly assumed to be 1, although this should be experimentally verified. Combining the above two equations and integrating, the Chick–Watson disinfection model (Eq. 16.3) is obtained.

$$\ln \frac{N}{N_0} = -\alpha C^n t \tag{16.3}$$

where
    N_0 is the actual number of microorganisms

The inactivation constant is specific to the microorganism and disinfectant being used and is also sensitive to environmental conditions in the water. Furthermore, disinfectants undergo chemical reactions that diminish their disinfection power but do not destroy it. Therefore, deviations from the Chick–Watson law are common. The general principle embodied in the law is that as concentration or contact time (Ct) is increased, inactivation of microorganisms is increased. Practically, $C \times t$ plots are prepared for in situ conditions to arrive at the dose and contact times required to attain the desired removal.

The bactericidal action of a disinfectant is, of course, not limited to pathogens. Fortunately waterborne disease-causing microorganisms are removed to levels generally considered safe at lower doses of disinfectant than required for complete sterilization of the water. Protozoans and viruses require higher doses of disinfectant than bacteria. Another common use for disinfecting agents is to control microbial growth in treatment units, pipes, and other appurtenances in the operation. Slime growth, which is primarily bacteria, can result in the addition of taste- and odor-causing compounds to a water. Microbial buildup increases headloss in conveyance systems, where the additional energy costs must be weighed against the costs of disinfectant addition. Many disinfectants increase the corrosiveness of a water, which is another cost consideration.

16.2 CHLORINATION

Chlorine participates in a number of reactions that affect its disinfecting capability.

16.2.1 Chemistry of Chlorine

Chlorine is a gas and dissolved chlorine has a tendency to escape to the atmosphere. Henry's law describes the equilibrium relation.

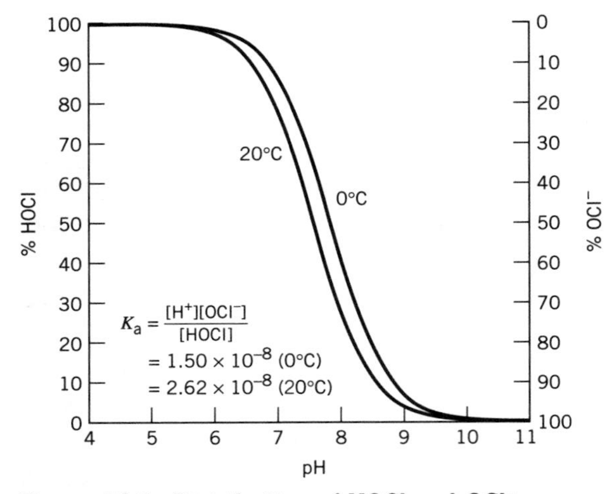

Figure 16.1 Distribution of HOCl and OCl^-.

$$Cl_2(aq) \rightleftharpoons Cl_2(g) \tag{16.4}$$

However the loss of chlorine by volatilization is minimal because chlorine rapidly hydrolyzes in water according to the following reactions. The reaction in Eq. (16.5) occurs in a fraction of a second at 20°C and takes only a few seconds at 0°C (White, 1992).

$$Cl_2 + H_2O \rightleftharpoons HOCl + H^+ + Cl^- \tag{16.5}$$

$$HOCl \rightleftharpoons H^+ + OCl^- \tag{16.6}$$

Hypochlorous acid (HOCl) is also volatile but it is on the order of 1.28×10^5 less volatile than Cl_2 (i.e., its Henry's law constant is 1.28×10^5 larger than that for Cl_2) (Blatchley et al., 1992).

The equilibrium expressions for reactions (16.5) and (16.6) are

$$K = \frac{[H^+][Cl^-][HOCl]}{[Cl_2]} = 4 \times 10^{-4} \text{ (at 25°C)} \tag{16.7}$$

$$K_a = \frac{[H^+][OCl^-]}{[HOCl]} \tag{16.8}$$

From Eq. (16.7), it can be seen that Cl_2 is much less than 1% of the total moles of chlorine species (Cl_2, HOCl, and OCL^-) in the pH range 6–9 of treated waters. Free available chlorine is chlorine in the form of Cl_2, HOCl, or OCl^- (hypochlorite ion). The dissociation of HOCl (Eq. 16.8) is also temperature and pH dependent as shown in Fig. 16.1 and Table 16.1. The variation of pK_a in Eq. (16.8) as a function of temperature was studied by Morris (1966), who found the following equation:

$$pK_a = \frac{3\,000.00}{T} - 10.068\,6 + (0.025\,3)T \tag{16.9}$$

where

T is in °K

TABLE 16.1 Dissociation of Hypochlorous Acid with pH and Temperature

	% HOCl			% HOCl	
pH	0°C	20°C	pH	0°C	20°C
4	100	100	8	40.1	27.6
5	99.9	99.7	8.5	17.4	10.8
6	98.5	97.4	9	6.3	3.7
7	86.9	79.2	10	0.7	0.4
7.5	67.9	54.7	11	0.07	0.04

The following observations are made from Fig. 16.1 and Table 16.1.

At pH 5.0 and below, almost all chlorine is in the form of HOCl.

At pH 10.0 and above, almost all chlorine is in the form of OCl^-.

HOCl is a very strong disinfectant, about 80–200 times as strong as OCl^-; therefore, pH exerts a strong influence on the effectiveness of chlorine. HOCl reacts with the enzymes essential to the metabolic processes of living cells. If hypochlorite salts are used, the following reactions occur.

$$\begin{aligned} &Ca(OCl)_2 \rightarrow Ca^{2+} + 2OCl^- \\ &H^+ + OCl^- \rightleftharpoons HOCl \end{aligned} \tag{16.10}$$

Chlorine reacts with the following substances:

a. Reducing agents such as S^{2-}, Fe^{2+}, Mn^{2+}, and NO_2^-
Using Table 1.3, the overall reaction between chlorine and a reducing agent may be formed. An example reaction is

$$H_2S + 4Cl_2 + 4H_2O \rightarrow H_2SO_4 + 8HCl$$

b. Organic matter
Chlorine is able to react in a variety of ways with functional groups and other reaction sites on organic molecules. As an example, the electrophilic addition of Cl_2 to a double bond of an alkene is shown below.

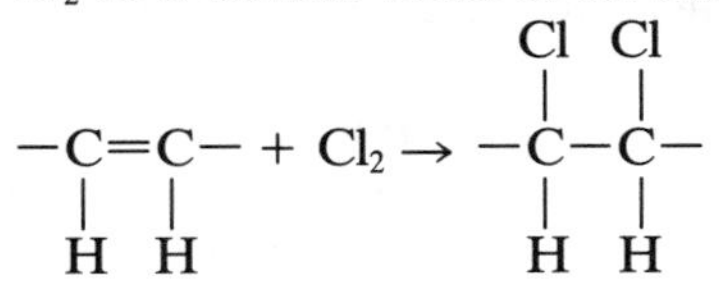

c. Ammonia
Hypochlorous acid reacts with ammonia to produce chloramines.

$$NH_3 + HOCl \rightarrow NH_2Cl + H_2O \quad \text{monochloramine} \tag{16.11}$$

$$NH_3 + 2HOCl \rightarrow NHCl_2 + 2H_2O \quad \text{dichloramine} \tag{16.12}$$

$$NH_3 + 3HOCl \rightarrow NCl_3 + 3H_2O \quad \text{trichloramine} \tag{16.13}$$

The distribution of the three types of chloramines is a function of pH. Chloramines are combined chlorine residuals. The electron uptake capability of a chloramine is directly dependent on the number of hypochlorous acid molecules used to form it. Ammonia and chloride are released by the reaction of a chloramine with a

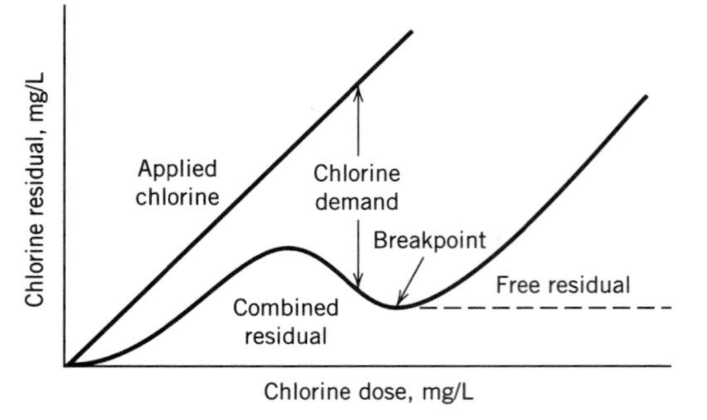

Figure 16.2 Chlorine demand curve.

reducing agent. For instance, the balanced half-reaction for dichloramine is

$$NHCl_2 + 2H^+ + 4e^- = NH_3 + 2Cl^-$$

There were two HOCl molecules required to form $NHCl_2$ and each HOCl molecule can take up two electrons.

Dichloramine produces a disagreeable odor and taste whereas monochloramine does not. Trichloramine is not stable and it breaks down to N_2 with a loss of oxidation power. Ammonia and chlorine are both biocides; chloramines are disinfectants but their strength is significantly less than HOCl or OCl^-. However, chloramines are much longer lasting in the water and provide a degree of residual protection. It was found in one study that the combined disinfection effect of free chlorine and monochloramine was greater than the additive effects of each agent; i.e., there was synergism for inactivation of *E. coli* (Kouame and Haas, 1991).

As a result of ammonia and other impurities in a water, there is a chlorine demand. The chlorine demand is assessed by adding known amounts of chlorine to a water and measuring the residual chlorine concentrations after a specified contact time. A typical chlorine demand curve is shown in Fig. 16.2.

The chlorine demand curve shape results from the reactions of chlorine. The curve changes its location as the contact time at which residuals are measured is changed. The first amount of chlorine is consumed by inorganic reducing substances that convert the chlorine into chloride, which has no residual oxidizing power. Excess chlorine after this point is converted into chloramines and chloroorganics but the former usually outweighs the latter. The dominant chloramine formed depends on the molar ratio of chlorine added to ammonia nitrogen present. When the molar ratio exceeds 1, reaction (16.12) begins to dominate; when the ratio exceeds 2, reaction (16.13) begins to dominate. There is some formation of NCl_3 even when the ratio of chlorine to nitrogen is less than 1, even though monochloramine is by far predominant.

The unstable nitrogen trichloride is formed when the molar ratio of chlorine to nitrogen exceeds 2:1 to 3:1 and a large amount of it begins to break down to N_2. There are also many other inorganic forms of nitrogen with various oxidation states such as nitrate that may be formed from oxidation of ammonia nitrogen as more chlorine is added. Formation of these inorganic species reduces the combined residual. The reactions that occur after the hump in the curve are not well understood. Monochloramines dominate the combined residual up to the hump in the curve. After the dip in the curve, the ammonia reactions are finished and further addition of chlorine will result in a free chlorine residual (HOCl and OCl^-). The dip in the curve is known as the breakpoint and addition of chlorine beyond this point is called breakpoint chlorination. There will be a relatively small residual amount of chloramines (all three of them) remaining after the breakpoint that contribute to the total chlorine residual.

Reactions of chlorine with organic matter will occur coincidentally with the chloramine and other reactions. Some organic compounds will completely oxidize chlorine but other chloroorganics formed will have some oxidizing power. Reactions of chlorine with organic matter for the formation of combined chlorine residuals are usually not significant compared to the ammonia reactions. The chlorine demand curve will shift as time progresses and the more slowly occurring reactions consume or convert chlorine to other forms.

It is not desirable to be in the region immediately after the hump in the curve. The rapid change in the curve is an instable region. Dichloramines exist in this region, which are malodorous as noted above, and although they are more powerful disinfectants than monochloramine, they are more unstable. Also NCl_3 is being formed, which is the most unstable chloramine.

In drinking water supplies the dose of chlorine is determined by the amount of chlorine demand and the desired residual concentration, which may be up to 0.5–1.0 mg/L. Significant concentrations of chloramines in water are toxic to pet fish and measures must be taken to remove the chloramines before using the water in aquariums.

■ Example 16.1 Chlorine Demand

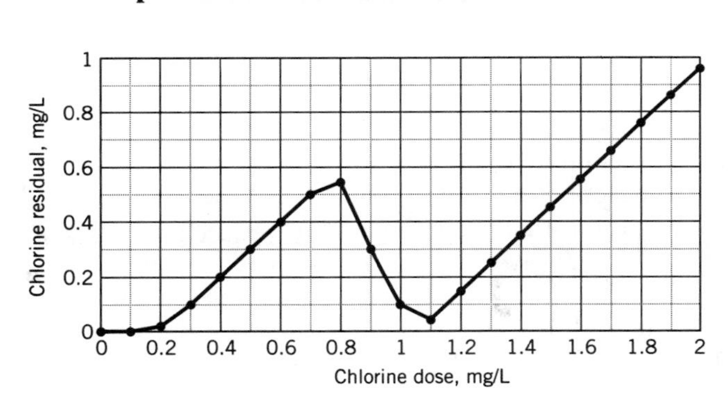

The chlorine demand curve on the graph was obtained for a drinking water for a 1-h contact time. Determine the daily amount of NaOCl to be applied to this water to produce a combined residual of 0.4 mg/L and a free residual of 0.5 mg/L after a contact time of 1 h in a flow of 24 000 m^3/d (6.34 Mgal/d).

From the graph the chlorine dose to achieve a combined residual of 0.4 mg/L is 0.60 mg/L.

The combined residual remaining after the breakpoint is approximately 0.08 mg/L. The chlorine dose to reach the breakpoint is 1.1 mg/L. After the breakpoint, the curve shows a linear increase in chlorine residual with chlorine dose. The dose of chlorine required to obtain a free residual of 0.5 mg/L is 1.1 mg/L + 0.50 mg/L = 1.60 mg/L. The total residual is 0.58 mg/L at this dose.

NaOCl is a salt that dissociates to

$$NaOCl \rightarrow Na^+ + OCl^-$$

The MW of NaOCl is 23 + 16 + 35.5 = 74.5 g. One mole of OCl^- is equivalent to 1 mole of Cl_2 from reactions (16.5) and (16.6). The amount of NaOCl that must be added each day for a combined residual of 0.4 mg/L after a 1-h contact time is

$$QC = \left(24\,000\,\frac{\text{m}^3}{\text{d}}\right)\left(0.60\,\frac{\text{mg Cl}_2}{\text{L}}\right)\left(\frac{74.5\text{ mg NaOCl}}{71\text{ mg Cl}_2}\right)\left(\frac{1\,000\text{ L}}{\text{m}^3}\right)\left(\frac{1\text{ kg}}{10^6\text{ mg}}\right) = 15.1\text{ kg/d}$$

In U.S. units:

$$QC = \left(6.34 \times 10^6\,\frac{\text{gal}}{\text{d}}\right)\left(0.60\,\frac{\text{mg Cl}_2}{\text{L}}\right)\left(\frac{74.5\text{ mg NaOCl}}{71\text{ mg Cl}_2}\right)\left(\frac{3.79\text{ L}}{\text{gal}}\right)\left(\frac{1\text{ lb}}{454 \times 10^3\text{ mg}}\right)$$
$$= 33.3\text{ lb/d}$$

Similarly, the daily amount of NaOCl to be added to achieve a free residual of 0.5 mg/L is

$$QC = \left(\frac{1.60\text{ mg/L}}{0.60\text{ mg/L}}\right)(15.1\text{ kg/d}) = 40.3\text{ kg/d}$$

$$\text{U.S. units: } QC = \left(\frac{1.60\text{ mg/L}}{0.60\text{ mg/L}}\right)(33.3\text{ lb/d}) = 88.8\text{ lb/d}$$

The chlorine demand of wastewaters will be much higher than the demand for water treated for consumption. The major factor controlling chlorine demand in wastewaters is the concentration of ammonia. Secondary biological treatment processes are often not designed to remove ammonia from the effluent. The amount of chlorine required to achieve a free residual in this case will be excessively high.

16.2.2 Chlorine Decay

Chlorine demand is not exerted immediately upon addition of chlorine to the water. First-order kinetics are assumed to describe the rate of reaction between chlorine and chlorine demand substances. The nature of chlorine demanding substances and flow patterns affects the amount of chlorine decay between the influent and the effluent in a reactor to which chlorine has been added. Teefy and Singer (1990) found first-order rate constants for free chlorine of 0.162 and 0.062 4 h^{-1} at two water treatment plants. Lawler and Singer (1993) suggest that 0.2 h^{-1} is a reasonable value for water treatment plants based on other studies.

Abdel-Gawad and Bewtra (1988) studied the decay of total residual chlorine from chlorination of physical–chemical treatment effluent in natural river water. They found a first-order decay model to be suitable with the rate constant influenced by turbulence, evaporation, photolysis, and temperature. The rate constant was formulated as

$$k_T = F_{\text{TB}}(k_{\text{ev}} + k_{\text{S}} + k_{\text{ox}})\theta^{(T-20)} \tag{16.14}$$

where

k_T is overall decay coefficient at temperature T
k_{ev} is rate constant for evaporation
k_S is rate constant for photooxidation
k_{ox} is rate of free radical oxidation by chlorine
F_{TB} is a turbulence factor
θ is the Arrhenius constant
T is temperature (°C)

The values of constants in the above equation were: $F_{TB} = 2.05$ for turbulent conditions, $F_{TB} = 1.0$ for quiescent conditions; $k_{ev} = 0.013/H\ d^{-1}$ (H is depth of the flow in m); $k_S = 0.03\ d^{-1}$; $k_{ox} = 0.065\ d^{-1}$; and $\theta = 1.08$.

Nowell and Hoigné (1992) also found that photodegradation of chlorine followed a first-order reaction. Sunlight in the wavelength range 320–340 nm controls photolysis. The hypochlorite ion was more sensitive to sunlight than hypochlorous acid. Covering basins minimizes the dosage of chlorine.

16.2.3 Determination of Chlorine Residuals

Free available chlorine in water is usually measured by either the DPD (*N,N*-diethyl-*p*-phenylenediamine) method or amperometric titration. DPD is a colorimetric redox indicator. In the former method, DPD is mixed along with other agents into a solution containing chlorine. Chlorine oxidizes DPD and the oxidized DPD is measured by titration or spectrophotometrically. Amperometric titration involves titration of chlorine or other halogens with a reducing agent of known normality (usually 0.005 64 *N* phenylarsine oxide) at a constant applied voltage. These methods are both subject to positive errors, especially in samples high in organic nitrogen and at high temperature, but these methods are currently the best analytical techniques available (Jensen and Johnson, 1989).

Two iodometric methods are also specified in *Standard Methods* (1992). In iodometric method I, iodide is added to a sample and oxidized to iodine, which is then titrated with thiosulfate reducing agent. In iodometric method II, which is suitable only for wastewaters, reducing agent is added to the sample, which is then titrated with iodine or iodate to determine the amount of unreacted reducing agent.

Gordon et al. (1988) have summarized and critiqued the various methods of measuring disinfectant residuals in water. A comparison of recommended techniques for chlorine residuals in wastewaters was performed by Derrigan et al. (1993).

16.3 DISINFECTION PRACTICE

Both drinking water and wastewater effluents are disinfected.

16.3.1 Water Treatment

Current United States practice specified in the Surface Water Treatment Rule[1] (SWTR) for disinfection requires water treatment systems to inactivate 99.9% of *Giardia* cysts and 99.99% of enteric viruses (3 and 4 log reductions, respectively). These organisms were chosen as standards because of their resistance to disinfection.

The SWTR is based on Eq. (16.3), i.e., the *Ct* concept. The appropriate *Ct* value must be satisfied at peak flow conditions through the plant. The concentration of disinfectant is its value measured in the effluent from a reactor. This provides a safety factor considering the decay of chlorine and has been found to be reasonable in a thorough analysis of the effects of mixing (flow pattern) and chlorine decay in reactors by Lawler and Singer (1993).

[1]National Primary Drinking Water Regulations: Filtration, Disinfection, Turbidity, *Giardia lamblia*, Viruses, *Legionella* and Heterotrophic Bacteria. Federal Regulation 54:124:27486 (June 29, 1989).

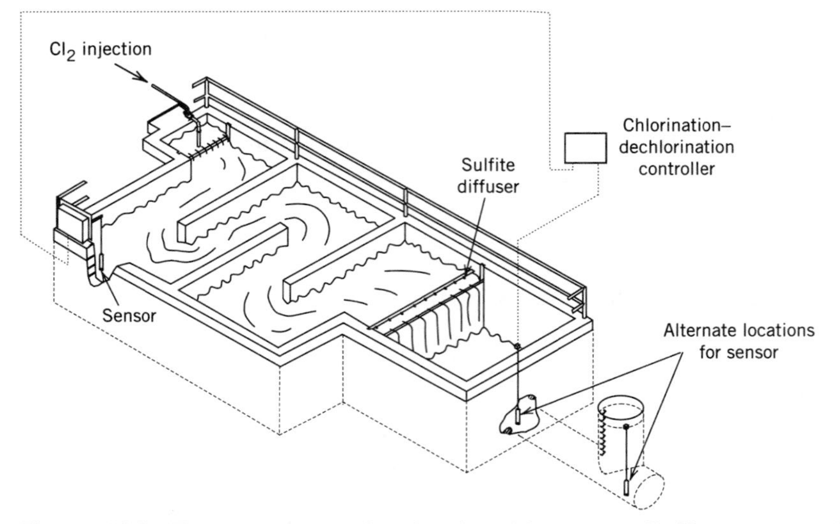

Figure 16.3 Typical contact chamber for chlorination. Baffles are provided to promote plug flow. When chlorine has been applied at elevated concentrations, sulfite is added to reduce chlorine to levels that will not cause consumer reaction to chlorine taste and odor. Sketch courtesy of Stranco.

The best disinfection kinetics are realized in a plug flow (PF) basin (Fig. 16.3). The characteristic time specified in the SWTR is the time for 10% of the water that enters the reactor at one instant to leave the reactor, t_{10}. (For the step feed input in Example 10.3, this corresponds to the time at which $C = 0.10C_I$.) In the case of a PF reactor, t_{10} and t_d are the same; however, for complete mixed (CM) reactors t_{10} is approximately $0.105t_d$. For an intermediate mixed reactor t_{10} is between these values. The use of t_{10} does not properly account for the effects of mixing on the rate of disinfection, which can be determined from application of principles developed in Chapter 10 and other advanced theory (Lawler and Singer, 1993). The SWTR approach normally tends to underestimate the amount of disinfection.

Table 16.2 gives *Ct* values for chlorine and other disinfectants to meet the SWTR standard. Free chlorine hydrolyzes into hypochlorous acid, which then dissociates

TABLE 16.2 SWTR *Ct*[a] Values for Achieving 99.9% Reduction of *Giardia lamblia*

		Temperature, °C					
Disinfectant	pH	≤1	5	10	15	20	25
Free chlorine[b]	6	165	116	87	58	44	29
(2 mg/L)	7	236	165	124	83	62	41
	8	346	243	182	122	91	61
	9	500	353	265	177	132	88
Ozone	6–9	2.9	1.9	1.4	0.95	0.72	0.48
Chlorine dioxide	6–9	63	26	23	19	15	11
Chloramines (preformed)	6–9	3 800	2 200	1 850	1 500	1 100	750

[a]*C* is in mg/L and *t* is in minutes.
[b]*Ct* values depend on the concentration of free chlorine (see text).

TABLE 16.3 SWTR *Ct*[a] Values for Achieving 90% Reduction of *Giardia lamblia*

		Temperature, °C					
Disinfectant	pH	≤1	5	10	15	20	25
Free chlorine[b]	6	55	39	29	19	15	10
(2 mg/L)	7	79	55	41	28	21	14
	8	115	81	61	41	30	20
	9	167	118	88	59	44	29
Ozone	6–9	0.97	0.63	0.48	0.32	0.24	0.16
Chlorine dioxide	6–9	21	8.7	7.7	6.3	5	3.7
Chloramines (preformed)	6–9	1 270	735	615	500	370	250

[a]*C* is in mg/L and *t* is in minutes.
[b]*Ct* values depend on the concentration of free chlorine (see text).

(Eqs. 16.5 and 16.6). Because the disinfection powers of hypochlorous acid and hypochlorite ion are different, *Ct* values for free chlorine in Table 16.2 are variable with pH; furthermore, for the same reason, *Ct* values for free chlorine are dependent on the total concentration of free chlorine. The SWTR provides tables of *Ct* values for different free chlorine concentrations. For chlorine, the value of n in Eqs. (16.2) and (16.3) is taken as 0.82 by the USEPA.

If filtration of a water is included in its treatment, the SWTR allows a credit of 2 log reductions of *Giardia* cysts. *Ct* values for disinfectants are decreased to values given in Table 16.3.

Total chlorine dosages at treatment plants vary from 0.2 to 40 mg/L (White, 1992). Maximum doses rarely exceed 15 mg/L. Water is commonly chlorinated to produce finished water free chlorine residuals near 1.0 mg/L. Prechlorination doses vary considerably depending on raw water quality and treatment objectives.

Concern over disinfection byproducts has substantially modified application of chlorine over recent years. THMs, which are a primary byproduct of chlorination, are regulated in both Canada and the United States (Section 8.6). In addition to chlorination as the final treatment, chlorine may be added near the beginning of the treatment plant (prechlorination) to reduce microbial populations throughout the plant. There may be other purposes for prechlorination such as oxidation of problematic metals to cause their precipitation and removal in sedimentation basins and filters.

The alterations in practice generally reduce the amount of chlorine applied. Elimination of prechlorination or reducing the prechlorination dosage is common (Disinfection Committee, 1992). Organic matter contains the precursors for THM formation. Organic matter concentrations are highest in the raw water and decrease throughout the plant. The potential for THM formation increases as the concentrations of chlorine or organic matter rise; thus, the application of chlorine in the earlier stages of treatment is discouraged. Conventional coagulation and sedimentation processes remove approximately 50% of the total organic carbon and precursors of THMs or total organic halides (TOX) (Singer and Chang, 1989).

There are many measures that can be taken to reduce THM formation. Other disinfectant–oxidants can be substituted for chlorine. Chloramines do not form THMs in significant amounts. Preformed chloramines can be added for disinfection purposes or ammonia can be added along with chlorine to form chloramines. Coagulation can be improved by adjusting pH. Activated carbon can be used to remove THM precursors at different stages of treatment.

Current North American practice is to maintain a residual concentration of disinfectant throughout the distribution system to provide protection against leaks and other breakdowns in system integrity. Residuals also retard the growth and regrowth of heterotrophic bacteria in the distribution system. Chloramines are the most commonly chosen agent. Chloramines are more stable and more effective than chlorine for maintaining residual protection throughout the distribution system (Neden et al., 1992). The average chloramine residual in water utilities serving more than 50 000 people in the United States in 1991 was 2.5 mg/L (Kirmeyer et al., 1991).

■ Example 16.2 Disinfection Design According to the *Ct* Concept

Calculate the quantity of chlorine consumed on a daily basis and the detention time required for 99.9% reduction of *G. lamblia* according to the SWTR rule. The minimum temperature of the water is 5°C, the free chlorine residual in the effluent from the basins is to be 2 mg/L, the decay rate for chlorine is assumed to be 0.2 h^{-1}, and the flow rate is 1.50×10^4 m^3/d (3.96 Mgal/d). The pH of the water varies between 7.2 and 8.00. Make the calculations for both a CM and a PF basin.

The highest pH requires the largest *Ct* value from Table 16.2. At a pH of 8.00 and temperature of 5°C, the *Ct* value is 243. The effective contact time, $t = 243/2 = 122$ min at these conditions. If the basin is PF, the detention time in the basin is 122 min because $t_d = t_{10}$. Using a first-order decay model with a rate constant of 0.2 h^{-1} (see Eq. 10.4f),

$$\frac{dC}{dt} = -kt \Rightarrow C = C_0 e^{-kt_d} \qquad \text{or} \qquad C_0 = Ce^{kt_d} = (2.00\ \text{mg/L})e^{(0.2\,\text{h}^{-1})(122\,\text{min})(1\,\text{h}/60\,\text{min})}$$

$$= 3.00\ \text{mg/L}$$

The quantity of chlorine consumed is

$$QC = \left(1.50 \times 10^4 \frac{\text{m}^3}{\text{d}}\right)\left(1\,000 \frac{\text{L}}{\text{m}^3}\right)\left(3.00 \frac{\text{mg}}{\text{L}}\right)\left(\frac{1\ \text{kg}}{10^6\ \text{mg}}\right) = 45\ \text{kg/d}$$

$$\text{In U.S. units: } QC = \left(3.96 \times 10^6 \frac{\text{gal}}{\text{d}}\right)\left(3.00 \frac{\text{mg}}{\text{L}}\right)\left(\frac{3.79\ \text{L}}{\text{gal}}\right)\left(\frac{1\ \text{lb}}{454 \times 10^3\ \text{mg}}\right) = 99.2\ \text{lb/d}$$

If the basin is CM, the detention time in the basin must be

$$t_d = t_{10}/0.105 = (122\ \text{min})/0.105 = 1\,162\ \text{min} = 19.4\ \text{h}$$

Equation (10.21) was derived for first-order decay in a CM basin.

$$C = \frac{C_0}{1 + kt_d} \Rightarrow C_0 = C(1 + kt_d) = (3.00\ \text{mg/L})[1 + (0.2\,\text{h}^{-1})(19.4\ \text{h})] = 14.6\ \text{mg/L}$$

The quantity of chlorine consumed in this case is

$$QC = \left(1.50 \times 10^4 \frac{\text{m}^3}{\text{d}}\right)\left(1\,000 \frac{\text{L}}{\text{m}^3}\right)\left(14.6 \frac{\text{mg}}{\text{L}}\right)\left(\frac{1\ \text{kg}}{10^6\ \text{mg}}\right) = 219\ \text{kg/d}$$

$$\text{In U.S. units: } QC = \left(3.96 \times 10^6 \frac{\text{gal}}{\text{d}}\right)\left(14.6 \frac{\text{mg}}{\text{L}}\right)\left(\frac{3.79\ \text{L}}{\text{gal}}\right)\left(\frac{1\ \text{lb}}{454 \times 10^3\ \text{mg}}\right) = 482\ \text{lb/d}$$

The benefits of a PF regime are apparent from both the size of the basins and the quantity of chlorine required.

16.3.2 Wastewater Treatment

Effluents from sewage treatment plants must often be disinfected before discharge to receiving waters, particularly during warm weather when water recreation activity is high. Chlorine dosages vary depending on treatment objectives and the degree of purification of the wastewater. Chlorination of raw sewage can be used to control odors, slime growth, and corrosion through oxidation of sulfides. Chlorine is applied to biological processes as a curative measure for poor sludge settling, foaming, and nuisance larvae. Breakpoint chlorination can be used to remove ammonia. Grease may also be removed by chlorination.

The concentration and variety of substances able to react with chlorine in wastewater are much higher than in raw waters to be treated for water consumption. Chlorine will reduce BOD. The Chick–Watson disinfection model is still reasonable but parameters become significantly different from those for a raw water. The high concentration of organics can result in variable formation of partially oxidized compounds that may have some germicidal properties. These reactions may be slow and produce deviations from the Chick–Watson law as reactions progress and disinfection power decreases.

Chloroorganics are formed in greater amounts from the chlorination of sewage. Aquatic toxicity must also be considered against the benefit of indicator microorganism reduction. It may be necessary to dechlorinate the effluent before discharge to the receiving water.

Ammonia concentrations are high in the effluent from facilities where nitrification is not occurring. In these plants, ammonia will dominate the chlorine demand and breakpoint chlorination will consume a large amount of chlorine. The complexity of disinfection efficiency and chlorine demand in wastewaters is exemplified by ammonia behavior. Dhaliwal and Baker (1983) found that the chlorine demand of wastewater supplemented with various amounts of ammonia decreased for the wastewater when NH_3-N concentrations rose above 2 mg/L. Chlorine dose was 7 mg/L and contact time was 1 h in the study. Slower reaction of chloramines with chlorine-demanding substances compared to the reaction rate of free chlorine with these substances was hypothesized as the reason. The amount of disinfection was equivalent or even slightly better for the same chlorine dose but higher ammonia concentrations. Dhaliwal and Baker (1983) also found that the amounts of ammonia and nitrite in the effluent could influence the chlorine demand and germicidal effectiveness of chlorination for similar reasons. Nitrite will react with free chlorine at a much higher rate than with combined chlorine. The destruction of free chlorine increases its consumption and reduces disinfection ability.

The 15-min chlorine demands of biotreatment effluent vary from 2 to 10 mg/L (White, 1992). Primary effluent has higher demands, typically in the range of 10–16 mg/L. Longer contact times will increase the chlorine demand. Disinfection basins are often baffled serpentine flow basins that promote PF. The variability of chlorine demand and disinfection illustrated for wastewaters indicates the necessity for determinations at each location.

16.3.3 Disinfection as the Sole Treatment of Surface Water

Surface waters are always subject to contamination from wildlife and surface wash. For this reason, disinfection must be applied to surface waters and as raw water quality deteriorates, additional treatments beginning with filtration must be applied. The

TABLE 16.4 Raw Water Quality Not Requiring Filtration[a,b]

Raw water criterion	Monitoring requirement
Total coliforms	<100/100 mL (90% of samples over 6 months)
Fecal coliforms	<20/100 mL (90% of samples over 6 months)
Turbidity	<5 NTU over any 12 consecutive months with two periods >5 NTU permitted per year, provided a boil water order is issued

[a]From USEPA (1988).
[b]Bacteriological sampling frequency increases from 1/week to 5/week as the population served increases from <501 to >25 000.

USEPA has set the criteria in Table 16.4 to distinguish when a raw water would not require filtration in addition to disinfection.

Considering the more important phenomena that influence the efficacy of chlorination, Geldreich et al. (1990) arrived at the following essential criteria for a raw surface water not to require filtration in addition to disinfection: fecal coliforms, 20 organisms/100 mL; turbidity, 1.0 NTU; color, 15 acu (apparent color units); and chlorine demand, 2 mg/L. The study examined data from 34 raw water sources. Turbidity levels above 1.0 NTU are associated with significant increases in total coliform densities. High color values are an indication of elevated humic acids and other dissolved organics that increase chlorine demand and result in higher trihalomethane formation potential.

16.3.4 Other Applications of Chlorine

Chlorine is used for a variety of purposes beyond disinfection. It has already been pointed out that it can be used to retard microbial growth in pipes and treatment units. As an oxidant, chlorine is used for iron and manganese removal and sulfide oxidation. Taste and odor control is another common application of chlorination. The oxidizing power of chlorine destroys many taste- and odor-causing compounds. Not all taste- and odor-causing compounds are affected by chlorine. Indeed, chlorine oxidizes phenols to more intensive disagreeable odor and taste compounds. Chlorine and chloramines themselves produce odors and tastes that are disagreeable to many people. A trial and error approach is used to determine the effectiveness of chlorine in controlling tastes and odors.

Monochloramines were not as effective as free chlorine in controlling tastes and odors caused by dimethylsulfides (Krasner et al., 1989). The threshold odor concentration for dimethyltrisulfide is 10 ng/L whereas the other dimethylsulfides require concentrations in the μg/L range to produce significant odor. Chloramination was adopted instead of prechlorination to meet trihalomethane regulations. In instances where these compounds are a problem other oxidants may have to be considered.

The use of chlorine as an oxidant of inorganic substances has been discussed in other sections of the text. Oxidation of sulfides that are a taste and odor problem and oxidation of metals to higher, more insoluble oxidation states are examples of this application.

The Asiatic clam, *Corbicula fluminea*, was introduced to the United States from southeast Asia in 1938 and now inhabits every major river system south of 40° latitude (Cameron et al., 1989). The mollusk causes clogging and taste and odor problems in cities with source waters with high organics content and long transmission lines. Chlorine has been the control agent of choice for this organism. The clam is more sensitive to all biocides at elevated temperatures; however, the spring and fall larval release

Figure 16.4 Zebra mussel. Courtesy of Stranco.

periods do not necessarily correspond to the highest water temperatures (Belanger et al., 1991).

Monochloramines and chlorine dioxide were effective biocides for this organism and they did not produce trihalomethanes (Cameron et al., 1989). Monochloramine effectiveness is increased in the presence of a small excess of un-ionized ammonia (Belanger et al., 1991). Copper sulfate is another biocide for this species.

The zebra mussel (*Dreissena polymorpha;* Fig. 16.4) is another recent addition to the North American ecosystem. It is also causing clogging problems at intakes. The control of zebra mussels and the Asiatic clams is a chronic problem. A monitoring study of zebra mussels at the intake of a plant in the Niagara River found that veligers (a larval stage) of this organism were present in the river from May to as late as December (Brady et al., 1993). Optimal larval development is at 20–25°C.

Hypochlorite was the most effective oxidant controlling zebra mussels although permanganate was nearly as effective (Klerks and Fraleigh, 1991).

Chlorine is applied in wastewater treatment facilities for some reasons given earlier. White (1992) suggests the application capacities below at points in the facility. These are guidelines that should be evaluated in each case.

1. Prechlorination: fresh domestic sewage, 10–12 mg/L; septic domestic sewage, 30–40 mg/L
2. Secondary sedimentation tanks, 10 mg/L
3. Return line for control of activated sludge bulking, 5 mg/L
4. Influent to biofilters or trickling filters, 10–15 mg/L
5. Sludge thickener line (odor control), 50 mg/L

16.4 OZONE

Ozone is a more powerful oxidizing agent than other disinfectants discussed in this chapter and a very effective biocide. Ozone has been used for drinking water treatment on a municipal scale since 1906, when it was installed in treatment facilities in Nice, France (Singer, 1990). More than 2 000 water treatment works, primarily in France and other European countries, now use ozone for disinfection and taste and odor control (Tate, 1991). There are currently more than 50 plants in Canada and more than 40 plants in the United States using ozone.

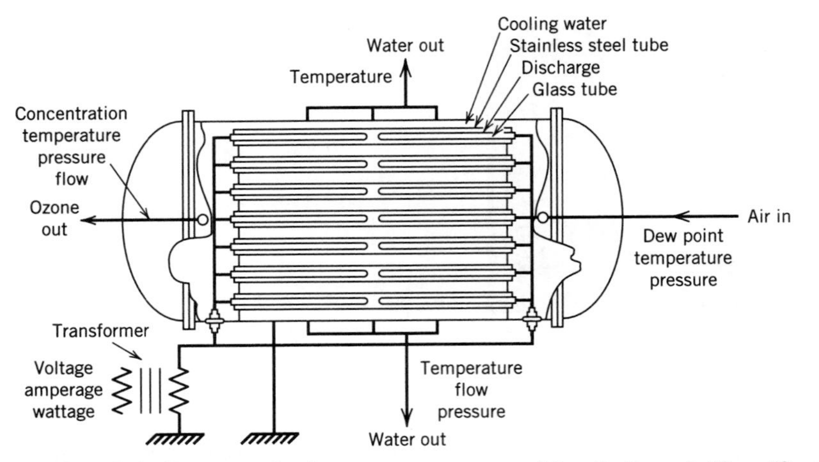

Figure 16.5 Horizontal tube ozone generator. After R. Gerval, "Specifications and Performance Control for Ozone Generators," in *Ozonization Manual for Water and Wastewater Treatment,* W. J. Masschelein, ed., copyright 1982, John Wiley & Sons. Reprinted by permission of John Wiley & Sons, Ltd.

Ozone is created by an electrical discharge (corona discharge is most efficient) in a gas containing oxygen. Ultraviolet irradiation (wavelengths <200 nm) of a gas containing oxygen is an alternative method. The equation for ozone formation is

$$3O_2 \rightarrow 2O_3 \tag{16.15}$$

The rate of ozone production is a function of the oxygen concentration and impurities such as dust and water vapor in the gas.

A common horizontal tube ozone generator is shown in Fig. 16.5. A high alternating current is applied to produce voltages from 6 to 20 kV across glass dielectric tubes that are internally metal coated to conduct the current. The influent gas is dried. The quantity of energy required lies between 15 and 20 W-h/g of ozone produced from air (Masschelein, 1982a). Other types of ozonators are discussed in Masschelein (1982b). The ozone is held in contact with the water through a variety of technologies but the most common contact system in North America is a bubble diffusion system with water depths in the range of 4.5–7.3 m (14.8–24 ft) (Ferguson et al., 1991). Other contactors are over and under or around the end baffled basins, packed contactors, and deep U-tubes. Ozone in the offgas from contactors must be destroyed. Thermal systems or more commonly thermal-catalytic systems are used in North America for destroying ozone (Tate, 1991).

The mass transfer coefficient (K_La) in a gas-sparged CM reactor was found to lie in the range of 0.25–0.45 min^{-1} except at very low gas flow rates (Grasso et al., 1990). The rate of ozone decomposition has been found to follow an autocatalytic model (Grasso and Weber, 1989):

$$\frac{d[O_3]}{dt} = -k_1[O_3] - k_2[O_3]^2 \tag{16.16}$$

The rate constants in the above equation are functions of $[OH^-]$ other alkalinity ions, the amount of UV radiation, and hydrogen peroxide concentration, which all have been found to catalyze ozone decomposition. If the solution pH is not well buffered k_2 will vary with rate constants associated with a number of intermediate reactions in ozone decomposition (Grasso and Weber, 1989). The rate constants should be empirically assessed for each situation. The decomposition of ozone to oxygen is quite rapid and it is impossible to maintain free ozone residuals in a water for any significant length of time. The half-life of ozone is near 20 min at typical conditions.

Ozone reacts with most organic matter. Ozone attacks organic matter directly or free radical species (such as the hydroxyl radical, OH) formed by ozone decomposition oxidize organic matter. Byproducts formed from ozone attack with human health significance in water treatment are organic peroxides, unsaturated aldehydes, and epoxides (Singer, 1990). On the other hand, ozonation destroys the precursors of most halogenated byproducts, e.g., THMs, haloacetic acids, and haloacetonitriles, that can be formed by subsequent chlorination. A UV–ozone oxidation treatment unit is shown in Fig. 16.6.

Partial oxidation of organic matter by ozone produces more readily biodegradable compounds. As a consequence, using ozone as a final disinfectant can lead to extensive biological regrowth in the distribution system. If the ozone treatment is followed by a porous media treatment such as filtration or granular activated carbon treatment, biological growth in the bed is enhanced (Section 14.11) and a biologically stable water, not subject to regrowth, will be produced (Singer, 1990).

Ozone is effective in oxidizing inorganic substances. The presence of significant carbon dioxide–associated alkalinity improves ozone's ability to destroy organics and as a disinfectant (Singer, 1990) and results in more rapid consumption of ozone as noted above. Ozonation was found to be effective for removing musty, earthy, fishy, and muddy tastes and odors from water that had been chlorinated, coagulated, and sand filtered; astringent and plastic tastes were not significantly removed. However, ozonation was responsible for the development of fruity odors with high intensities

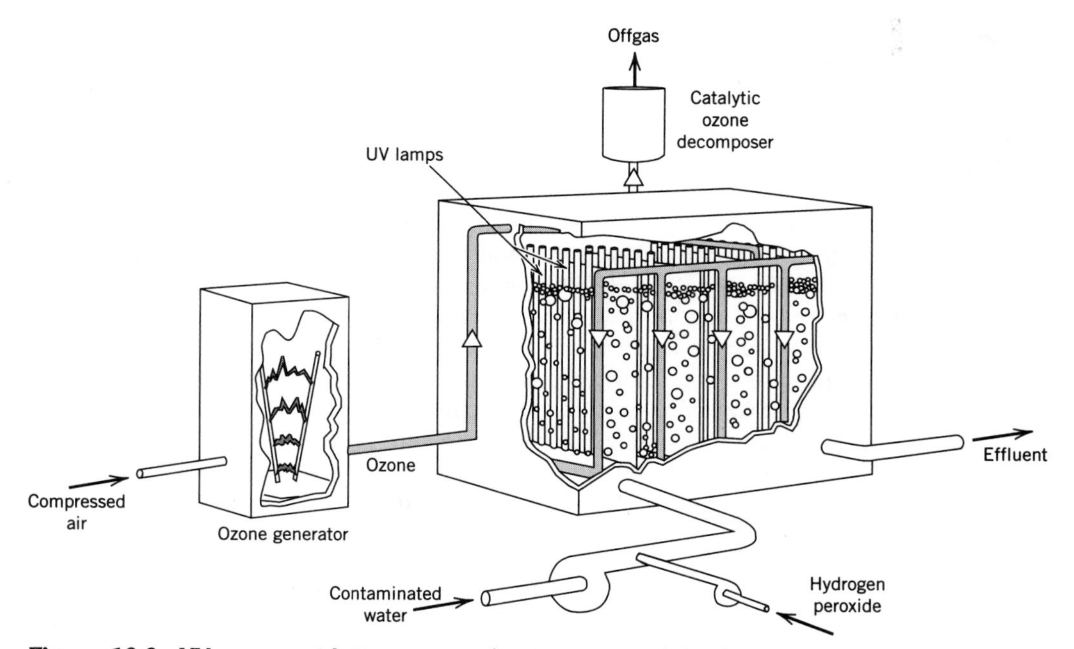

Figure 16.6 UV-ozone oxidation system for treatment of drinking water or wastewater. Courtesy of Zimpro Environmental, Inc.

(Anselme et al., 1989). In general ozone is highly effective as an oxidant of off-flavors compared to chlorine derivatives or permanganate oxidants (Ferguson et al., 1991). Ozone is also very effective for color removal.

Ultraviolet irradiation greatly enhances the ability of ozone to decompose humic acids and other organic compounds. In one study the destruction of dissolved organic substances with UV and ozone proceeded at least 10 times faster than in the presence of ozone alone (Kusakabe et al., 1990). Major final products of decomposition were acetic acid, formic acid, and oxalic acid, which are refractory to ozone attack and are not precursors of THMs. This is an example of an advanced oxidation process using ozone that yields free radical oxidizing species upon decomposition. The process is accelerated by addition of hydroxide ion, hydrogen peroxide, or UV irradiation. Advanced oxidation processes can be used to reduce TOC levels.

16.5 ULTRAVIOLET AND IONIZING RADIATION

Ultraviolet radiation, which is electromagnetic radiation in the shorter wavelength range of the spectrum from 5 to 400 nm, causes dieoff of microorganisms and leaves no residual radiation (and thus no residual disinfecting power) in a water. Ultraviolet radiation has been used since the turn of this century to disinfect water. In the past 20 years there has been an increase in the application of UV radiation to wastewater treatment plant effluents.

Ultraviolet radiation is not effective against *Giardia* cysts. Current United States' regulations (SWTR) for capability of inactivating *Giardia* cysts preclude UV radiation as a primary disinfectant for surface waters where potential for *Giardia* contamination exists (USEPA, 1990). Ultraviolet radiation is appropriate as a primary disinfectant for groundwaters that do not contain *Giardia*.

There is no chemical consumption for UV disinfection and no harmful byproducts are formed. Ultraviolet radiation was observed to have effects equivalent to a residual by Lund and Hongve (1994). The growth of heterotrophic bacteria was inhibited in water that had been irradiated. The effect lasted for 1 week or more and was explained by the production of oxidizing reagents such as hydroxyl radicals formed by UV irradiation of humic substances.

The penetrating power of UV radiation is not as great as ionizing radiation. Ultraviolet rays cause damage to nucleic acids; some cells may be able to recover after exposure to UV radiation. Water and constituents in water affect the transmission and absorption of UV rays. The Beer–Lambert law (Section 5.6) applies. Suspended solids not only reduce the transmission of UV radiation but also shield bacteria from exposure, particularly bacteria within suspended particulates. The correlations between transmittance and turbidity or suspended solids in clarified secondary effluent may be weak at these low concentrations (less than 20 mg/L) (Darby et al., 1993) and the dose of UV radiation may have to be empirically established.

Some bacteria have the ability to repair UV induced damage to DNA. The process is thought to occur in the following manner (Lindenauer and Darby, 1994). Ultraviolet light absorption produces photoproducts, the most important of which are pyrimidine dimers formed from adjacent pyrimidine molecules on the same strand of DNA. With light, a photoreactivating enzyme can split the dimers, causing the repair of the DNA. Lindenauer and Darby (1994) found that the percentage of photoreactivation of total coliforms in wastewater inversely related to UV dose. The overall extent of recovery was less than 1% for all UV doses.

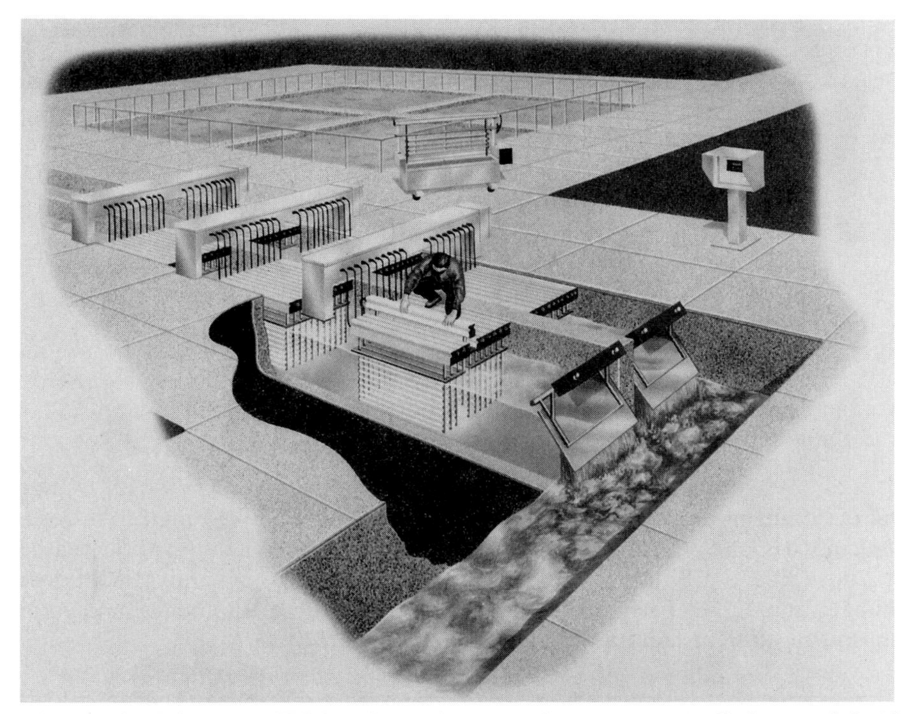

Figure 16.7 A typical UV disinfection system. Lamps are installed in modules for ease of lamp cleaning and replacement. Courtesy of Trojan Technologies Inc.

Ultraviolet radiation is generated by current flow between electrodes in ionized mercury vapor. The maximum energy output of low-pressure mercury arc lamps is at a wavelength of 253.7 nm, which is in the middle of the UV range. Depending on the intensity of the UV radiation and the degree of dispersion in the flow, the contact time necessary to inactivate microorganisms ranges from a few seconds to a few minutes. Long tubular lamps are usually used. Water flows through a bank of lamps to achieve the desired contact time (Fig. 16.7).

The output intensity of lamps is somewhat unstable during the first 100 h of operation and decays exponentially during this period. The 100% output is the output after 100 h of operation. The germicidal effectiveness of the lamps deteriorates with time, which must be considered when designing the system. Manufacturers supply curves that provide the correction factor for lamp output as a function of age of the lamp as shown in Fig. 16.8. Ultraviolet lamp output is also variable with temperature. Manufacturer's rating curves should be consulted.

The rate of inactivation of microorganisms with UV exposure is assumed to follow a first-order reaction:

$$N = N_0 e^{-kEt} = N_0 e^{-kD} \tag{16.17}$$

where

N and N_0 are the number or concentration of bacteria at time t and initially, respectively

k is an inactivation coefficient

E is average UV intensity in the reactor

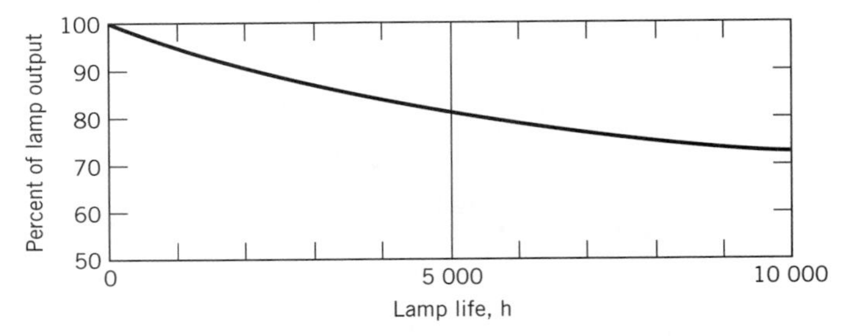

Figure 16.8 Typical UV lamp output. (Time 0 is after 100 h of operation.)

t is average detention time in the irradiated portion of the reactor
D is radiation dose ($D = Et$)

Darby et al. (1993) found a fairly well-defined relation between UV dose (D) and log survival of coliforms in clarified secondary effluent with TSS <15 mg/L. At doses of 30 mW-s/cm^2 (0.054 mWh/in.^{2}), log survival was near -3.7; log survival decreased to -4.6 at a dose of 150 mW-s/cm^2 (0.16 mWh/in.^{2}). The change in survival with dose was not linear. Table 16.5 lists inactivation coefficients for a number of indicator microorganisms in effluent from a biological treatment process.

■ Example 16.3 UV Disinfection

Assume Fig. 16.8 describes the output of a UV lamp system and find the detention time in a UV system to inactivate 99.9% of *E. coli* after 10 000 h of lamp operation. The system is designed to provide an intensity of 2 mW/cm^2 (12.9 mW/in.^{2}) with new lamps (100-h operation). The rate constants in Table 16.5 apply. For this system, what removal of *E. coli* would occur with new lamps?

From Eq. (16.17), $\ln (N/N_0) = -kD$; from Table 16.5, $k = 0.013$ m^2/J (504 ft^2/Wh).

$$D = -\frac{\ln (N/N_0)}{k} = -\frac{\ln (1/1\,000)}{0.013 \text{ m}^2/\text{J}} = 531 \text{ J/m}^2$$

$$\text{In U.S. units: } D = -\frac{\ln (1/1\,000)}{504 \text{ ft}^2/\text{Wh}} = 0.013\,7 \text{ Wh/ft}^2$$

The intensity of the lamps after 10 000 h (1.14 yr) of operation is approximately 72% of their intensity when new.

$E_{1.14} = 0.72(2 \text{ mW/cm}^2) = 1.44 \text{ mW/cm}^2$
In U.S. units: $E_{1.14} = 0.72(12.9 \text{ mW/in.}^2) = 9.29 \text{ mW/in.}^2$

$$Et = D \qquad t = \frac{D}{E} = \left(\frac{531 \text{ J/m}^2}{1.44 \text{ mW/cm}^2}\right)\left(\frac{1\,000 \text{ mW}}{\text{J/s}}\right)\left(\frac{1 \text{ m}}{100 \text{ cm}}\right)^2 = 36.9 \text{ s}$$

$$\text{In U.S. units: } t = \left(\frac{0.013\,7 \text{ Wh/ft}^2}{9.29 \text{ mW/in.}^2}\right)\left(\frac{10^3 \text{ mW}}{1 \text{ W}}\right)\left(\frac{1 \text{ ft}}{12 \text{ in.}}\right)^2\left(\frac{3\,600 \text{ s}}{\text{h}}\right) = 36.9 \text{ s}$$

When the lamps are new,

$D = Et = (2.0 \text{ mW/cm}^2)(36.9 \text{ s}) = 73.8 \text{ mW-s/cm}^2$
In U.S. units: $D = Et = (12.9 \text{ mW/in.}^2)(36.9 \text{ s}) = 476 \text{ mW-s/in.}^2$

$$N = N_0 e^{-kD} \qquad \frac{N}{N_0} = \exp\left[-\left(0.013\,\frac{\text{m}^2}{\text{J}}\right)\left(73.8\,\frac{\text{mW-s}}{\text{cm}^2}\right)\left(\frac{1\text{ J/s}}{1\,000\text{ mW}}\right)\left(\frac{100\text{ cm}}{1\text{ m}}\right)^2\right]$$

$$= 6.81 \times 10^{-5}$$

$$\text{In U.S. units: } \frac{N}{N_0} = \exp\left[-\left(504\,\frac{\text{ft}^2}{\text{Wh}}\right)\left(476\,\frac{\text{mW-s}}{\text{in.}^2}\right)\left(\frac{1\text{ W}}{1\,000\text{ mW}}\right)\left(\frac{1\text{ h}}{3\,600\text{ s}}\right)\left(\frac{12\text{ in.}}{1\text{ ft}}\right)^2\right]$$

$$= 6.80 \times 10^{-5}$$

The survival is 0.006 8% when the lamps are new.

Ionizing radiation such as gamma radiation and X rays has wavelengths in the range of 0.001 to 100 nm. Electron beams are also ionizing. Ionizing radiation is a powerful disinfectant, causing irreversible damage to a cell. There is no residual radiation left in the water. Gamma radiation appears to produce slightly better disinfection than high-energy electron beams (Farooq et al., 1993). The application of ionizing radiation as a disinfectant in the water or wastewater industry has been limited, although gamma radiation has been used to inactivate microorganisms in food and medical products for a number of years.

Gamma irradiation requires a radioactive source, whereas electron beam technology does not, which is a major advantage for electron beam technology. Both processes avoid problems associated with use of chemical oxidants (Farooq et al., 1993).

High-energy electron beams have been used for other applications, including aromatic hydrocarbon destruction. The principal reactive species responsible for attack on organics is the hydroxyl radical (Nickelsen et al., 1994). Toluene and benzene were decayed according to a first-order reaction with respect to adsorbed radiation dose:

$$\frac{dR}{dD} = -kR \Rightarrow R_D = R_0 e^{-kD}$$

TABLE 16.5 Inactivation Rate Constants for Indicator and Other Microorganisms for UV Radiation[a]

Organism	k (m^2/J)	k (ft^2/Wh)
Escherichia coli	0.013	504
Fecal streptococci	0.006 7	260
Sulfite-reducing bacteria spores	0.001 5	58
Somatic coliphages	0.016	620
F-specific bacteriophages	0.005 3	205
MS2 bacteriophages	0.011	426
Reoviruses	0.005 5	213

[a]Reprinted from *Water Research,* 25, T. J. Nieuwstad, A. H. Havelaar, and M. van Olphen, "Hydraulic and Microbiological Characterization of Reactors for Ultraviolet Disinfection of Secondary Wastewater Effluent," pp. 775–783, Copyright 1991, with kind permission from Elsevier Science Ltd., The Boulevard, Langford Lane, Kidlington, 0X5 1GB, UK.

where

R_D is the solute concentration at a dose D
R_0 is the initial concentration of the solute
k is a constant

The rate constant k is a function of pH and the amount of organic matter in solution. Organic matter is a scavenger for the radicals. Phenols are initial byproducts, along with a variety of hydroxylated and other highly oxidized byproducts.

16.6 OTHER DISINFECTANTS

Other disinfectants are being examined as suitable replacements for the common disinfectants. All of the halogens have disinfecting power.

16.6.1 Chlorine Dioxide and the Other Halogens

Chlorine dioxide (ClO_2, with the bonding structure O=Cl=O) is a powerful oxidizing agent with nearly 2.5 times the oxidizing power of chlorine. Historically, chlorine dioxide has been primarily used for the removal of taste and odor caused by phenolic compounds (White, 1992). Research has shown that chlorine dioxide does not produce THMs and that the production of precursors for THMs is reduced. This has sparked new interest in chlorine dioxide as a disinfectant. The disinfection efficiency of chlorine dioxide does not vary with pH, in contrast to chlorine, which is converted to pH-sensitive HOCl, and chlorine dioxide does not oxidize bromide (Hoigné and Bader, 1994). Chlorine dioxide is also superior to chlorine for the oxidation of iron and manganese. For the removal of iron and manganese, chlorine dioxide is added as a treatment preceding flocculation and sedimentation.

The benefits of chlorine dioxide for disinfection with THM reduction are potentially offset by its production of disinfection byproducts. As chlorine dioxide exerts its oxidizing power, it degrades into chlorite (ClO_2^-) and to a lesser extent chlorate (ClO_3^-) and chloride ions. Because the concentration of chlorine added with chlorine dioxide is smaller than the amount of chlorine that would be required for a given amount of oxidizing power, the potential for formation of chloroorganics is reduced (see Problem 19). But chlorite ion can oxidize hemoglobin and cause methemoglobinemia. Infants are more susceptible to this disease because they have a reduced capability of reducing methemoglobin. Furthermore, hemolytic anemia, which is a decrease in the blood concentration of hemoglobin, is induced at low concentrations of chlorite ion. Dialysis patients may be particularly susceptible to this condition. There have been studies, with chlorite ion doses well above those found in water disinfection applications, that have found no adverse effects on the exposed population. White (1992) has discussed various studies. Clearly, there is need for more research to quantify the risks and benefits for chlorine dioxide against other disinfectants.

There is no evidence that chlorine dioxide is mutagenic or carcinogenic. However, Eckhardt et al. (1982) found that chlorate was mutagenic to a strain of *Salmonella typhimurium* and to *Drosophila*. But Meier et al. (1985) found that neither chlorite nor chlorate were mutagenic to mouse micronuclei, mouse sperm head, or mouse bone marrow chromosomes.

Chlorine dioxide application in water treatment plants has been practiced for many years at installations throughout the world. There have been many installations that have adopted chlorine dioxide as a disinfectant during the past 25 years.

Chlorine dioxide is generated on site from the sodium salt of chlorite or chlorate ion. The processes are generally named after the company that developed them (White, 1992). The reactions for the primary processes used in water treatment plants are as follows.

The Mathieson or SO_2 Process

$$2NaClO_3 + H_2SO_4 + SO_2 \rightarrow 2ClO_2 + H_2SO_4 + Na_2SO_4 \quad (16.18)$$

A side reaction for this process is

$$2NaClO_3 + 5SO_2 + 4H_2O \rightarrow Cl_2 + 3H_2SO_4 + Na_2SO_4 \quad (16.19)$$

The Solvay or Methanol Process

$$2NaClO_3 + CH_3OH + H_2SO_4 \rightarrow 2ClO_2 + HCHO + Na_2SO_4 + 2H_2O \quad (16.20)$$

A side reaction in this process is

$$2NaClO_3 + CH_3OH + 2H_2SO_4 \rightarrow ClO_2 + \tfrac{1}{2}Cl_2 + CO_2 + 2NaHSO_4 + 3H_2O \quad (16.21)$$

The Chloride Reduction Process

$$2NaClO_3 + 2NaCl + 2H_2SO_4 \rightarrow 2ClO_2 + Cl_2 + 2Na_2SO_4 + 2H_2O \quad (16.22)$$

The Jaszka–CIP Process

This process is a feasible alternative for the production of low amounts of chlorine dioxide.

$$2NaClO_3 + 2SO_2 \rightarrow 2ClO_2 + Na_2SO_4 \quad (16.23)$$

Chlorine dioxide does not react with ammonia and therefore there is no demand associated with ammonia. Also there are no chloramine residuals produced. The reaction of chlorine dioxide with a number of inorganic and organic compounds has been studied and found to be first-order with respect to chlorine dioxide concentration and the concentration of the compound, i.e., a second-order reaction overall (Hoigné and Bader, 1994).

$$\frac{d[ClO_2]}{dt} = -k[ClO_2][C] \quad (16.24)$$

where

C is any compound

Rate constants for various reactions of ClO_2 have been provided for a number of compounds by Hoigné and Bader (1994) and Tratnyek and Hoigné (1994).

In anticipation of more stringent rules on the byproducts of chlorite and chlorate ions there has been research devoted to their destruction. Granular activated carbon can be used but the life of the carbon is short and it is not considered economically feasible (Dixon and Lee, 1991). Chlorate ion does not react with activated carbon but it is reversibly sorbed (Gonce and Voudrais, 1994). The reducing agents of sulfur dioxide, thiosulfate, and sulfite have also not been efficacious because the chlorite ion is converted into chlorate by these reductants (Dixon and Lee, 1991; Griese et al., 1991). Ferrous iron was extremely effective for removing undesirable ClO_2 and ClO_2^- residuals but chlorate was unaffected by ferrous iron (Griese et al., 1991; Iatrou and Knocke, 1992).

Iodine and Bromine

Bromine and iodine, like other halogens, have disinfecting power. Their use as disinfectants has been limited for various reasons but cost is a primary reason. Historically, bromine and iodine have been about 3 and 40 times more expensive than chlorine on a weight basis (Mills, 1977). Both bromine and iodine require 2 to 3 times as much energy to produce as chlorine; but ozone requires between 4 and 16 times as much energy to produce as chlorine. Research on the byproducts of reactions of these compounds is limited.

Bromine is relatively soluble in water, whereas elemental iodine is only slightly soluble in water. Both are widely distributed in nature in small quantities; commercial sources of bromine and iodine are more highly concentrated brines or saline deposits. Mills (1977) describes processes for production of Br_2 and I_2 from the naturally occurring forms of bromide and iodate.

Bromine undergoes reactions similar to chlorine, forming hypobromous acid and bromamines as well as reacting with organics. The equilibrium constants for some chlorine and bromine compounds are given in Table 16.6. The higher pK for hypobromous acid compared to hypochlorous acid results in more undissociated acid over the pH range of treated waters. The undissociated acid is a more effective disinfecting agent than the ion. Bromine is also volatile; Blatchley et al. (1992) have studied its Henry's law constant, which is on the order of that for chlorine.

In a study comparing bromine and chlorine disinfection of sewage, bromine was more effective than chlorine above pH of 4 for equal doses on a weight basis (Johnson and Sun, 1977). Bromamines are also effective disinfection agents.

Brominated drinking water would not significantly increase the amount of bromine taken internally and the body is quite capable of eliminating the excess bromide (Mills, 1977).

Iodine is also an effective disinfecting agent and apparently more effective than chlorine or bromine against cysts and spores (Chang, 1958; Stringer et al., 1977). Iodine doses of 2.0 mg/L were more effective than 1.0 mg/L doses of chlorine in removing fecal coliforms from turbid waters above a pH of 7.4 (Ellis and van Vree, 1989). Similar results were found by Ellis et al. (1993).

Both undissociated iodine, I_2, and hypoiodous acid, HOI, have disinfecting power. Iodine reacts only slowly with organic matter and does not react with ammonia to form iodamines (Ellis and van Vree, 1989).

Possible physiological effects of iodine after long-term ingestion have remained a barrier to using iodine for water disinfection, although there are studies that have not shown any harmful effects (Black et al., 1965). Iodine deficiencies cause many disorders, of which goiter is the most well known. Iodine diet supplements, particularly iodized salt, are used to provide adequate amounts of iodine. The dual purpose of

TABLE 16.6 Distribution of Chlorine and Bromine Species at pH 8.0 in Water at 0°C

Halogen Species	pK[a]	%HOX	$\%OX^-$	$\%NH_2X$
HOCl	7.58	40	60	—
HOBr	8.7	83	17	—
NH_2Cl	≈1.0	—	—	100
NH_2Br	≈6.5	82	8	<10

[a]pK data for all species except HOCl taken from Mills (1977).

controlling iodine deficiency and providing disinfection can be achieved with iodination of water supplies. There have been several attempts to control iodine deficiency (see Squatrito et al., 1986, for example). Studies have indicated that iodination for both purposes is feasible (Cook, 1977; Maberly et al., 1981; Thomas et al., 1979). Cook (1977) indicates that iodine levels from 0.5 to 1 mg/L are safe for long-term human consumption. Iodination is much more expensive than chlorination, which is the major reason for lack of iodination. Iodination can be an alternative for disinfection and diet supplement in particularly remote areas of developing countries where both problems exist.

Bromine and iodine have been used for the disinfection of swimming pool waters. Free iodine concentrations from 0.2 to 0.6 mg/L were used successfully for disinfection of swimming pools (Marshall et al., 1960). A more common use of iodine is to disinfect small quantities of water for drinking. Iodine tablets to be added to a small container are supplied for this purpose. Iodine has also found use as a disinfectant in other applications such as the milk industry (Salvato, 1982).

Halogens can be combined to form disinfectants. The most commonly studied agent is bromine chloride, formed with equivalent amounts of bromine and chlorine.

$$Br_2 + Cl_2 \rightleftharpoons 2BrCl$$

Bromine chloride exists in equilibrium with bromine and chlorine in both liquid and gas phases (Mills, 1977). Investigations have shown that the germicidal properties of bromine chloride are superior to either agent used alone (Mills, 1977).

16.6.2 Sunlight and Silver

Sunlight is a natural alternative to synthetic disinfectants. Silver has been used in limited circumstances for disinfection.

Sunlight

The germicidal action of UV radiation is well known as described earlier. Ultraviolet radiation (shorter wavelengths between 5 and 400 nm) is a component of solar radiation. Studies have shown that it is not only UV rays that are harmful to bacteria but that other portions of the sunlight spectrum contribute to bacteria mortality. There is an exponential decrease in germicidal effectiveness as radiation wavelength increases from the shorter UV range through the visible to infrared ranges, where the kill efficiency is nil (Acra et al., 1987).

Acra et al. (1984, 1987) have performed studies on solar disinfection of water supplies. These studies provide an indication that solar disinfection is feasible but there are no known applications of the technology. The belt between latitudes of 15° and 35° on either side of the equator receives about 2 500 h of sunlight per year and has potential for solar disinfection of water supplies. The zone around the equator has high sunlight intensity, but more frequent cloud cover, and radiation scattering is high.

Solar disinfection of sewage treatment plant effluent has also been performed with successful results at a pilot-scale level in Israel (Acher et al., 1994). A sensitiver, methylene blue, was added to the wastewater to enhance the adsorption and transfer of sunlight to dissolved oxygen, which causes the formation of active oxidative species such as H_2O_2, OH, and O_2^-. The dye was added at 0.6–0.8 mg/L. Shallow units (water depth varied from 18 to 23 cm or 7 to 9 in.) were operated at a detention time around

35 min. Log_{10} reductions of total coliforms, fecal coliforms, fecal streptococci, and poliovirus were 3.2 ± 0.3, 3.12 ± 0.2, 3.9 ± 0.3, and 1.9 ± 0.25, respectively, under these conditions.

Silver

Silver also has bactericidal powers. Dosages of 1 μg to 0.5 mg/L have been reported to sterilize water (McKee and Wolfe, 1963). Woodward (1963) reviewed literature on silver disinfection and found that studies showed that Ag^+ doses from 30 to 90 μg/L reduced *E. coli* microorganisms by 99.9% in one-half hour to 14 h in normal pH ranges of water. There are no toxic effects of silver to humans at these dosages. Silver oxide has even been used to purify a water supply for Warsaw, Poland. The costs of silver for large-scale disinfection use are too high to make it a practical alternative. Colloidal silver may cause permanent discoloration of the skin (argyria), eyes, and mucous membranes but the precise concentration to cause these effects is unknown (McKee and Wolfe, 1963; Salvato, 1982).

A more recent use of silver has been to impregnate the filters in point-of-use tap water purification devices to retard the growth of microorganisms in them. In a review of studies on the efficacy of silver in these devices, Bell (1991) found that silver had some bactericidal action against total coliforms but the bactericidal effects of silver on heterotrophic bacteria (heterotrophic plate count) were negligible.

QUESTIONS AND PROBLEMS

1. What are the characteristics of a good disinfectant?
2. What is the oxidation number of Cl in Cl_2, HOCl, and OCL^-?
3. What are the initial concentrations of Cl_2, HOCl, and OCl^- in a water with a temperature of 20°C after dissolution and hydrolysis of 5 000 mg/L of Cl_2? The pH of the water is buffered at 6.00.
4. Give some reasons that substances which are able to take up the same number of electrons may differ in their ability to disinfect.
5. List the uses of chlorine.
6. In the iodometric analysis for chlorine residuals, chlorine or its derivatives oxidize iodide to iodine, which is then titrated with thiosulfate. Write the chemical equation for the oxidation of iodide by $NHCl_2$. (Hint: to determine the products, 1 mole of chlorine produces 1 mole of iodine and 1 mole of chlorine is equivalent to 1 mole of HOCl.)
7. What are the concentrations of HOCl and OCL^- if 40 mg/L of $Ca(OCl)_2$ are added to a water with a final pH of 6.35 at a temperature of 20°C?
8. Calculate the chlorine dose equivalent to 0.8 mg/L of Cl_2 at a pH of 7.00 if HOCl is 80 times as effective as OCl^- for disinfection and the pH of the water is 6.00. Perform the same calculation if HOCl is 200 times as effective as OCL^-. Use a temperature of 25°C.
9. What are the amounts of chlorine and anhydrous ammonia that must be added to a flow of 22 000 m^3/d (5.81 Mgal/d) to achieve a combined residual of 1.0 mg/L as Cl_2? Assume there are no side reactions of either substance.
10. How much $Ca(OCl)_2$ must be added to a flow of 1 250 m^3/d (0.330 Mgal/d) to produce a total combined chlorine residual of 0.70 mg/L as Cl_2 if the water contains 0.09 mg/L of S^{2-} and 4.0 mg/L of NH_4^+? There are no other significant chlorine demanding substances; assume reaction stoichiometries are strictly obeyed.
11. A water contains the following concentration of chlorine related species: HOCl-0.41 mg/L;

OCl^--0.35 mg/L; NH_2 Cl-0.12 mg/L; and NH_2 Cl-0.05 mg/L. What are the free, combined, and total residuals as Cl_2?

12. Draw a typical chlorine demand curve and explain each phase of the curve.

13. Using the following figure, what is the chlorine dose required to attain a combined residual of 0.5 mg/L and a free residual of 0.5 mg/L? What is the combined residual present when a free residual of 0.5 mg/L exists?

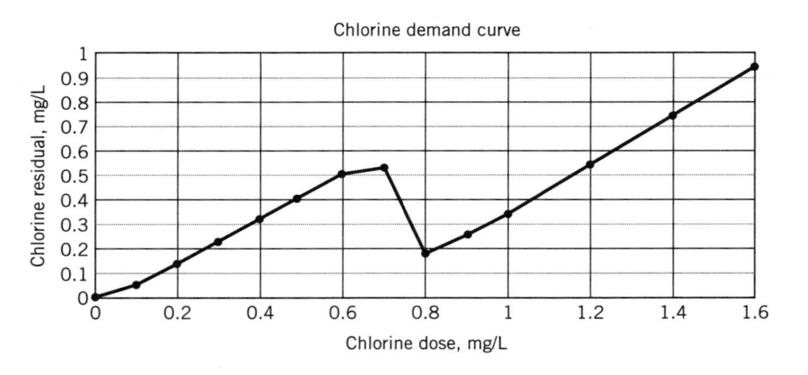

14. What mass of sulfite (SO_3^{2-}) must be added on a daily basis to a flow of 3 450 m^3/d to dechlorinate the water from a total residual concentration of 3.05 mg/L as Cl_2 to a total residual concentration of 1.00 mg/L as Cl_2?

15. For Abdel-Gawad and Bewtra's model for chlorine decay (Eq. 16.14), which factor is the most important influence on chlorine decay? Use the values they found for the coefficients, a flow depth of 2 m, and temperature of 10°C. Would a temperature increase from 10 to 15°C have a larger effect on the decay coefficient than a 10% change in any of the other factors?

16. It is found that a combined chlorine residual of 1.2 mg/L results in a reduction of indicator microorganisms by 99.0% for a contact time of 30 min in stabilization pond effluent. (a) Assuming the Chick–Watson model applies with $n = 1$, calculate the change in (i) combined chlorine residual or (ii) contact time required to increase removals to 99.9%. Assume that the contact chamber is PF. (b) Perform the same exercise as in (a) but use a value of $n = 0.8$.

17. A plant must be in compliance with SWTR for 99.9% removal of *G. lamblia*. The plant disinfects to an effluent concentration of 2.00 mg/L free chlorine from its disinfection basins, which exhibit PF behavior. What is the daily use of chlorine at the plant with flow variation between 1.85×10^4 and 3.15×10^4 m^3/d (4.88–8.32 Mgal/d), pH fluctuation between 7.0 and 8.0, and temperature extremes of 5 and 20°C? The low temperature applies to the low flow and the high temperature applies to the high flow but pH can be at either extreme. Assume that the detention time in the disinfection basins is able to be adjusted to the appropriate value but effluent free chlorine concentration is always 2.00 mg/L. Use a chlorine decay rate of 0.15 h^{-1}.

18. What are the required detention times in a plug flow basin at a temperature of 5°C and pH of 7.0 to meet the SWTR for 90% reduction of *G. lamblia* for (a) free chlorine at 2.0 mg/L and (b) chlorine dioxide at 2.0 mg/L?

19. (a) Write the half-reactions for the reduction of ClO_2 to Cl^- and Cl_2 to Cl^-. How many atoms of chlorine are added in chlorine dioxide compared to chlorine gas to obtain 1 eq of oxidizing power?

(b) The answer to (a) represents the minimum ratio of $ClO_2 : Cl_2$ per electron equivalent. Why is the actual ratio higher than this value?

20. What is the equivalent weight of chlorine dioxide if 60, 25, and 15% of the chlorine in chlorine dioxide is converted into chlorite, chlorate, and chloride ions, respectively?

21. What are the effects of pH on the disinfection power of Cl_2 and ClO_2?

22. What average intensity must be provided in a UV system to achieve 99.99% removal of indicator microorganisms with an inactivation rate constant of 0.010 m^2/J and detention time of 2 min in the UV zone?

23. If the rate constants in the autocatalytic model for ozone decomposition are $k_1 = 0.095$ min^{-1} and $k_2 = 1.20$ L/mole/min at pH 6.00 of a water, how long would it take for 99% of the ozone to be removed if the initial concentration is 0.5 mg/L?

KEY REFERENCES

HAAS, C. N. (1990), "Disinfection," in *Water Quality and Treatment,* 4th ed., F. W. Pontius, ed., McGraw-Hill, Toronto, pp. 877–932.

WHITE, G. C. (1978), *Disinfection of Wastewater and Water for Reuse,* Van Nostrand Reinhold, Toronto.

WHITE, G. C. (1992), *Handbook of Chlorination and Alternative Disinfectants,* 3rd ed., Van Nostrand Reinhold, Toronto.

REFERENCES

ABDEL-GAWAD, S. T. AND J. K. BEWTRA (1988), "Decay of Chlorine in Diluted Municipal Effluents," *Canadian J. of Civil Engineering,* 15, 6, pp. 948–954.

ACHER, A. J., E. FISCHER, AND Y. MANOR (1994), "Sunlight Disinfection of Domestic Effluents for Agricultural Use," *Water Research,* 28, 5, pp. 1153–1160.

ACRA, A., Z. RAFFOUL, AND Y. KARAHAGOPIAN (1984), *Solar Disinfection of Drinking Water and Oral Rehydration Solutions: Guidelines for Household Application in Developing Countries,* UNICEF, New York.

ACRA, A., M. JURDI, H. MU'ALLEM, Y, KARAHAGOPIAN, AND Z. RAFFOUL (1987), *Solar Ultraviolet Radiation: Assessment and Application for Drinking-Water Disinfection,* report for International Development Research Centre, Ottawa, Canada.

ANSELME, C., I. H. SUFFET, AND J. MALLEVIALLE (1989), "Effects of Ozonation on Tastes and Odors," *J. American Water Works Association,* 80, 10, pp. 45–51.

BAKER, M. N. (1981), *The Quest for Pure Water,* vol. 1, 2nd ed., American Water Works Association, Denver, CO.

BELANGER, S. E., D. S. CHERRY, J. L. FARRIS, K. G. SAPPINGTON, AND J. CAIRNS, JR. (1991), "Sensitivity of the Asiatic Clam to Various Biocidal Control Agents," *J. American Water Works Association,* 83, 10, pp. 79–87.

BELL, F. A., JR. (1991), "Review of Effects of Silver-Impregnated Carbon Filters on Microbial Water Quality," *J. American Water Works Association,* 83, 8, pp. 74–76.

BLACK, A. P., R. N. KINMAN, W. C. THOMAS, G. FREUND, AND E. D. BIRD (1965), "Use of Iodine for Disinfection," *J. American Water Works Association,* 57, 11, pp. 1401–1421.

BLATCHLEY, E. R., III, R. W. JOHNSON, J. E. ALLEMAN, AND W. F. MCCOY (1992), "Effective Henry's Law Constants for Free Chlorine and Free Bromine," *Water Research,* 26, 1, pp. 99–106.

BRADY, T. J., J. E. VAN BENSCHOTEN, J. N. JENSEN, D. P. LEWIS, AND J. SFERRAZZA (1993), "Sampling and Enumeration of Zebra Mussel Veligers: Implications for Control," *J. American Water Works Association,* 85, 6, pp. 100–103.

CAMERON, G. N., J. M. SYMONS, S. R. SPENCER, AND J. Y. MA (1989), "Minimizing THM Formation during Control of the Asiatic Clam: A Comparison of Biocides," *J. American Water Works Association,* 81, 10, pp. 53–62.

CHANG, S. L. (1958), "The Use of Active Iodine as a Water Disinfectant," *J. American Pharmaceutical Association,* 47, pp. 417–423.

CHICK, H. (1908), "An Investigation of the Laws of Disinfection," *J. Hygiene,* 8, pp. 92–158.

COOK, B. (1977), "Using Iodine to Disinfect Water Supplies," Proc. EPA/National Individual On-Site Wastewater Conference, USEPA, Washington, DC, pp. 217–226.

CRAUN, G. F. (1988), "Surface Water Supplies and Health," *J. American Water Works Association,* 80, 2, pp. 40–52.

DANIEL, F. B., L. W. CONDIE, M. ROBINSON, J. A. STOBER, R. G. YORK, G. R. OLSON, AND S. WANG. (1990), "Comparative Subchronic Toxicity Studies of Three Disinfectants," *J. American Water Works Association,* 82, 10, pp. 61–69.

DARBY, J. L., K. E. SNIDER, AND G. TCHOBANOGLOUS (1993), "Ultraviolet Disinfection for Wastewater Reclamation and Reuse Subject to Restrictive Standards," *Water Environment Research,* 65, 2, pp. 169–180.

DERRIGAN, J., L. Y. LIN, AND J. N. JENSEN (1993), "Comparison of Free and Total Chlorine Measurements in Municipal Wastewaters," *Water Environment Research,* 65, 3, pp. 205–211.

DHALIWAL, B. S. AND R. A. BAKER (1983), "Role of Ammonia-N in Secondary Effluent Chlorination," *J. Water Pollution Control Federation,* 55, 5, pp. 454–456.

Disinfection Committee (1992), "Survey of Water Utility Disinfection Practices," *J. American Water Works Association,* 84, 9, pp. 121–128.

DIXON, K. L. AND R. G. LEE (1991), "The Effect of Sulfur-Based Reducing Agents and GAC Filtration on Chlorine Dioxide By-products," *J. American Water Works Association,* 83, 5, pp. 48–55.

ECKHARDT, K., E. GOCKE, M. T. KING, AND D. WILD (1982), "Mutagenic Activity of Chlorate, Bromate and Iodate," *Mutation Research,* 97, p. 185.

ELLIS, K. V. AND H. B. R. J. VAN VREE (1989), "Iodine Used as a Water-Disinfectant in Turbid Waters," *Water Research,* 23, 6, pp. 671–676.

ELLIS, K. V., A. P. COTTON, AND M. A. KHOWAJA (1993), "Iodine Disinfection of Poor Quality Waters," *Water Research,* 27, 3, pp. 369–375.

FAROOQ, S., C. N. KURUCZ, T. D. WAITE, AND W. J. COOPER (1993), "Disinfection of Wastewaters: High-Energy Electron vs Gamma Radiation," *Water Research,* 27, 7, pp. 1177–1184.

FERGUSON, D. W., J. T. GRAMITH, AND M. J. MCGUIRE (1991), "Applying Ozone for Organics Control and Disinfection: A Utility Perspective," *J. American Water Works Association,* 83, 5, pp. 32–39.

GELDREICH, E. E., J. A. GOODRICH, AND R. M. CLARK (1990), "Characterizing Surface Waters That May Not Require Filtration," *J. American Water Works Association,* 82, 12, pp. 40–50.

GERVAL, R. (1982), "Specifications and Performance Control for Ozone Generators," in *Ozonization Manual for Water and Wastewater Treatment,* W. J. Masschelein, ed., John Wiley & Sons, Toronto, pp. 24–27.

GONCE, N. AND E. A. VOUDRAIS (1994), "Removal of Chlorite and Chlorate Ions from Water Using Granular Activated Carbon," *Water Research,* 28, 5, pp. 1059–1069.

GORDON, G., W. J. COOPER, R. G. RICE, AND G. E. PACEY (1988), "Methods of Measuring Disinfectant Residuals," *J. American Water Works Association,* 80, 9, pp. 94–108.

GRASSO, D. AND W. J. WEBER, JR. (1989), "Mathematical Interpretation of Aqueous-Phase Ozone Decomposition Rates," *J. of Environmental Engineering, ASCE,* 115, EE3, pp. 541–559.

GRASSO, D., E. FUJIKAWA, AND W. J. WEBER, JR. (1990), "Ozone Mass Transfer in a Gas-Sparged Turbine Reactor," *Research J. Water Pollution Control Federation,* 62, 3, pp. 246–253.

GRIESE, M. H., K. HAUSER, M. BERKEMEIER, AND G. GORDON (1991), "Using Reducing Agents to Eliminate Chlorine Dioxide and Chlorite Ion Residuals in Drinking Water," *J. American Water Works Association,* 83, 5, pp. 56–61.

HOIGNÉ, J. AND H. Bader (1994), "Kinetics of Reactions of Chlorine Dioxide (OClO) in Water—I. Rate Constants for Inorganic and Organic Compounds," *Water Research,* 28, 1, pp. 45–55.

IATROU, A. AND W. R. KNOCKE (1992), "Removing Chlorite by the Addition of Ferrous Iron," *J. American Water Works Association,* 84, 11, pp. 63–68.

JENSEN, J. N. AND J. D. JOHNSON (1989), "Specificity of the DPD and Amperometric Titration Methods for Free Available Chlorine: A Review," *J. American Water Works Association,* 81, 12, pp. 59–64.

JOHNSON, D. J. AND W. SUN (1977), "Bromine Disinfection of Wastewater," in *Disinfection: Water and Wastewater,* J. D. Johnson, ed., Ann Arbor Science, Ann Arbor, MI, pp. 179–191.

KIRMEYER, G., G. FOUST, AND M. LECHEVALLIER (1991), "Optimization of Chloramination for Distribution System Water Quality Control," Proceedings of the 1991 AWWA Annual Conference, Philadelphia, PA.

KLERKS, P. L. AND P. C. FRALEIGH (1991), "Controlling Zebra Mussels with Oxidants," *J. American Water Works Association,* 83, 12, pp. 92–100.

KOUAME, Y. AND C. N. HAAS (1991), "Inactivation of *E. coli* by Combined Action of Free Chlorine and Monochloramine," *Water Research,* 25, 9, pp. 1027–1032.

KRASNER, S. W., S. E. BARRETT, M. S. DALE, AND C. J. HWANG (1989), "Free Chlorine Versus Monochloramine for Controlling Off-Tastes and Off-Odors," *J. American Water Works Association,* 81, 2, pp. 86–93.

KUSAKABE, K., S. ASO, J. HAYASHI, K. ISOMURA, AND S. MOROOKA (1990), "Decomposition of Humic Acid and Reduction of Trihalomethane Formation Potential in Water by Ozone with UV Irradiation," *Water Research,* 24, 6, pp. 781–785.

LAWLER, D. F. AND P. C. SINGER (1993), "Analyzing Disinfection Kinetics and Reactor Design: A Conceptual Approach Versus the SWTR," *J. American Water Works Association,* 85, 11, pp. 67–76.

LINDENAUER, K. G. AND J. L. DARBY (1994), "Ultraviolet Disinfection of Wastewater: Effect of Dose on Subsequent Photoreactivation," *Water Research,* 28, 4, pp. 805–817.

LUND, V. AND D. HONGVE (1994), "Ultraviolet Irradiated Water Containing Humic Substances Inhibits Bacterial Metabolism," *Water Research,* 28, 5, pp. 1111–1116.

MABERLY, G. F., C. J. EASTMAN, AND J. M. CORCORAN (1981), "Effect of Iodization of a Village Water Supply on Goitre Size and Thyroid Function, *The Lancet,* London.

MARSHALL, J. D., J. D. MCLAUGHLIN, AND E. W. CARSCALLEN (1960), "Iodine Disinfection of a Cooperative Pool," *The Sanitarian,* 22, 4, pp. 199–203.

MASSCHELEIN, W. J. (1982a), "Thermodynamic Aspects of the Formation of Ozone and Secondary Products of Electrical Discharge," in *Ozonization Manual for Water and Wastewater Treatment,* W. J. Masschelein, ed., John Wiley & Sons, Toronto, pp. 9–12.

MASSCHELEIN, W. J., ed. (1982b), *Ozonization Manual for Water and Wastewater Treatment,* John Wiley & Sons, Toronto.

MCKEE, J. E. AND H. W. WOLFE (1963), *Water Quality Criteria,* California State Water Resources Control Board, Sacramento, CA.

MEIER, J. R., R. J. BULL, J. A. STOBER, AND M. C. CIMINO (1985), "Evaluation of Chemicals Used for Drinking Water Disinfection for Production of Chromosomal Damage and Sperm-Head Abnormalities in Mice," *Environmental Mutagenesis,* 7, 2, pp. 201–211.

MILLS, J. F. (1977), "Interhalogens and Halogen Mixtures as Disinfectants," in *Disinfection: Water and Wastewater,* J. D. Johnson, ed., Ann Arbor Science, Ann Arbor, MI, pp. 113–143.

MORRIS, J. C. (1966), "The Acid Ionization Constant of HOCl," *J. Physical Chemistry,* 70, 12, pp. 3798–3805.

NEDEN, D. G., R. J. JONES, J. R. SMITH, G. J. KIRMEYER, AND G. W. FOUST (1992), "Comparing Chlorination and Chloramination for Controlling Bacterial Regrowth," *J. American Water Works Association,* 84, 7, pp. 80–88.

NICKELSEN, M. G., W. J. COOPER, K. LIN, C. N. KURUCZ, AND T. D. WAITE (1994), "High Energy Electron Beam Generation of Oxidants for the Treatment of Benzene and Toluene in the Presence of Radical Scavengers," *Water Research,* 28, 5, pp. 1227–1237.

NIEUWSTAD, T. J., A. H. HAVELAAR, AND M. VAN OLPHEN (1991), "Hydraulic and Microbiological Characterization of Reactors for Ultraviolet Disinfection of Secondary Wastewater Effluent," *Water Research,* 25, 7, pp. 775–783.

NOOT, D. K., W. B. ANDERSON, S. A. DAIGNAULT, D. T. WILLIAMS, AND P. M. HUCK (1989), "Evaluating Treatment Processes with the Ames Mutagenicity Assay," *J. American Water Works Association,* 81, 9, pp. 87–102.

NOWELL, L. H. AND J. HOIGNÉ (1992), "Photolysis of Aqueous Chlorine at Sunlight and Ultraviolet Wavelengths—I. Degradation Rates," *Water Research,* 26, 5, pp. 593–598.

PATERMARAKIS, G. AND E. FOUNTOUKIDIS (1990), "Disinfection of Water by Electrochemical Treatment," *Water Research,* 24, 12, pp. 1491–1496.

SALVATO, J. A. (1983), *Environmental Engineering and Sanitation*, 3rd ed., John Wiley & Sons, Toronto.

SINGER, P. C. (1990), "Assessing Ozone Research Needs in Water Treatment," *J. American Water Works Association,* 82, 10, pp. 78–88.

SINGER, P. C. AND S. D. CHANG (1989), "Correlations between Trihalomethanes and Total Organic Halides Formed during Water Treatment," *J. American Water Works Association,* 81, 8, pp. 61–65.

SQUATRITO, S., R. VIGNERI, F. RUNELLO, A. M. EVANS, R. D. POLLEY, AND S. H. INGBAR (1986), "Prevention and Treatment of Endemic Iodine Deficiency Goitre by Iodization of a Municipal Water Supply," *J. Clinical Endocrinology and Metabolism,* 63, 2, pp. 368–375.

Standard Methods for the Examination of Water and Wastewater (1992), 18th ed., American Public Health Association, Washington, DC.

STRINGER, R. P., W. N. CRAMER, AND C. W. KRUSÉ (1977), "Comparison of Bromine, Chlorine and Iodine as Disinfectants for Amoebic Cysts," in *Disinfection: Water and Wastewater,* J. D. Johnson, ed., Ann Arbor Science, Ann Arbor, MI, pp. 193–209.

TATE, C. (1991), "Survey of Ozone Installations in North America," *J. American Water Works Association,* 83, 5, pp. 40–47.

TEEFY, S. M. AND P. C. SINGER (1990), "Performance and Analysis of Tracer Tests to Determine Compliance of a Disinfection Scheme with the SWTR," *J. American Water Works Association,* 82, 12, pp. 88–98.

THOMAS, W. C., M. H. MALAGODI, T. W. OATES, AND J. P. MCCOURT (1979), "Effects of an Iodinated Water Supply," *Transcripts of the American Clinical and Climatological Association,* 90, pp. 153–162.

TRATNYEK, P. G. AND J. HOIGNÉ (1994), "Kinetics of Reactions of Chlorine Dioxide (OClO) in Water—II. Quantitative Structure–Activity Relationships for Phenolic Compounds," *Water Research,* 28, 1, pp. 57–66.

USEPA (1988), National Primary Drinking Water Regulations, *Federal Register,* Mar. 14, 1988.

USEPA (1990), *Technologies for Upgrading Existing or Designing New Drinking Water Treatment Facilities,* Center for Environmental Research Information, Cincinnati, OH.

WATSON, H. E. (1908), "A Note on the Variation of the Rate of Disinfection with Change in Concentration of the Disinfectant," *J. Hygiene,* 8, p. 536.

WOODWARD, R. L. (1963), "Review of the Bactericidal Effectiveness of Silver," *J. American Water Works Association,* 55, 7, pp. 881–886.

SECTION V

BIOLOGICAL WASTEWATER TREATMENT

CHAPTER 17

AEROBIC BIOLOGICAL TREATMENT

Biological wastewater treatment is primarily used to remove dissolved and colloidal organic matter in a wastewater. Some suspended organics will also be metabolized and because of the natural flocculation and settling characteristics of the biomass formed in biological treatment, the biomass along with other suspended matter can be removed in a sedimentation basin.

Biological treatment is a "natural" process. Organic matter in water will naturally decay as a result of the presence of microorganisms in receiving bodies of water. High organic loads in a wastewater will upset the biocenosis of receiving bodies of water and cause other undesirable effects. Biological treatment is engineered to accelerate natural decay processes and neutralize the waste before it is finally discharged to receiving waters.

Aerobic biological treatment is the focus of this chapter; however, many concepts developed in this chapter apply to any biological treatment process. Further information on other biological treatment processes is given in Chapters 18, 19, and 20.

17.1 MICROORGANISMS IN AEROBIC BIOLOGICAL TREATMENT

Chapter 6 describes general characteristics of microorganisms. Biomass in a reactor is the sum total of all living organisms in the process. Other organic matter present, suspended or dissolved, may have been associated with living biomass at some time but this matter is not part of the biological engine driving the process.

Bacteria are the primary agents of treatment in any biological treatment process. Taken as a whole, their diverse characteristics and minimal growth requirements allow them to proliferate in a wastewater environment. Aerobic processes are usually operated at low dissolved oxygen (DO) concentrations and the microorganisms are subjected to varying periods of time when no DO is present. Therefore, many facultative microorganisms will be found in an aerobic process.

Viruses are present in a biological treatment process but they have no significance in removal of organic compounds.

Protozoans will also be found in large numbers in an aerobic treatment process; however, they will still only account for a small percentage of the biomass. There are some protozoans that are saprophytic, i.e., primary feeders on the raw organics present in the influent. Others are predators of bacteria and other eukaryotes. The presence of ciliated protozoans is usually indicative of good treatment. These protozoans occur in the presence of higher amounts of DO and lower amounts of dissolved organic matter.

Yeast and fungi are not common in wastewater treatment processes. As discussed in Chapter 6, their lower nitrogen requirements and ability to survive at lower pHs can enhance their role in treatment of certain industrial wastewaters. The occurrence of fungi, in particular, is not beneficial for most normally operated biological treatment processes because of their poor settleability, which makes it difficult to separate them from the wastewater and deteriorates final effluent quality as well as causes other operational problems.

Rotifers and crustaceans are the next higher stage of life beyond protozoans. Their role in providing treatment is minimal but their presence is indicative that the treatment process is healthy. Sludge worms are not significant in biological treatment of a wastewater but they will be found.

17.2 THE ACTIVATED SLUDGE PROCESS

Activated sludge is defined as a suspension of microorganisms, both living and dead, in a wastewater. The microorganisms are activated by an input of air (oxygen). The influent to the process is usually settled. The process involves two distinct operations usually performed in two separate basins: aeration and settling.

The aeration basin is the first basin in the process. Microorganisms are mixed with the sewage and oxygen is supplied by aeration. Here the organics in the waste are metabolized to produce end products and new biomass. Mixing must be adequate to prevent the sedimentation of microorganisms and to mix oxygen, sewage, microorganisms, and nutrients. The contents of the aeration basin are known as the mixed liquor (ML).

The second operation is the separation of the biomass and other suspended solids from the wastewater. This is accomplished in a clarifier designed according to type III sedimentation principles (Section 11.5). The clarified effluent is relatively devoid of any suspended particles compared to the clarifier influent. A portion of the sludge from the clarifier underflow is usually returned to the aeration basin, whereas the remainder is discarded or sent for further processing.

The process was developed in the early 1900s by Ardern and Lockett (1914a, 1914b, 1915; Ardern, 1917) who used fill and draw reactors (a batch process) to successfully treat wastes in a short period of time. Continuous flow reactors were designed shortly thereafter and regularly used because of problems in controlling a number of batch reactors throughout fill–react–settle–draw cycles with variable influent flow rates. Although batch treatment was ignored for over 50 years, interestingly, batch treatment systems have been reestablished as a viable treatment alternative in modern times (Irvine and Busch, 1979).

Some of the substrate will be completely oxidized to harmless end products of CO_2, H_2O, and other inorganic substances to provide energy for growth of the microorganisms. Oxygen, which is usually supplied by aeration, must be input continuously or semicontinuously. This is the major energy consuming operation in the process (Owen, 1982). Energy requirements for return sludge pumping from the clarifier to the aeration basin and operation of the clarifier itself are relatively insignificant in the energy balance.

A portion of the substrate is used for synthesis of biomass. The aerobic mode of metabolism is the most efficient in terms of energy recovered by the biomass per unit of substrate processed. This results in a relatively large quantity of sludge production, which is the other primary characteristic of this process. Sludge processing and disposal is also a major operating expense (Owen, 1982). It may be possible to use the sludge as a soil conditioner. Whether the sludge can ultimately be applied to crops grown for human consumption or only to land restricted for aesthetic or recreational use or no access at all depends on the sludge's accumulation of heavy metals and other toxicants. The occurrence of metals or other toxicants in the sewage depends on the presence of certain industries discharging untreated or partially treated wastes to the sewer system.

17.3 SUBSTRATE REMOVAL AND GROWTH OF MICROORGANISMS

Substrate removal and growth of microorganisms are closely tied together. Even though starved conditions exist in biological treatment processes there is net growth of microorganisms.

17.3.1 Substrate Removal

The growth cycle of microorganisms and equations to describe substrate uptake have been given in Chapter 6. The Monod (Monod, 1949) equation (Eqs. 17.1a and 17.1b) has been found to suitably describe substrate removal in a wide variety of biological treatment processes. As noted before, the equation is empirical and its constants should be developed for each situation. A wide variety of substrates and microorganisms are involved in a complex dynamic environment when wastewater is being treated. The equation describes the interaction of this large number of variables.

$$r_S = -\frac{kX_VS}{K+S} \tag{17.1a}$$

$$r_S = -\frac{kS}{K+S} \tag{17.1b}$$

where

r_S is the rate of substrate removal ($ML^{-3}T^{-1}$)
X_V is the concentration of volatile suspended solids (VSS)
k and K are the maximum and half-velocity constants (see Section 4.12)
S is concentration of substrate

The maximum velocity constant, k, and the half-velocity constant, K, are functions of environmental variables such as DO concentration, pH, temperature, inhibitory substances, and nutrients as well as the degradability of the substrate.

First-order relations (Eqs. 17.2a and 17.2b) or other expressions such as a retardant model have also been found to provide good correlation of results. The Monod equation reduces to a first-order formulation at low substrate concentrations. The applicability of a first-order model is not surprising because wastewater treatment processes produce an effluent that is low in degradable substrate concentration. The constant k does not have the same value in Eqs. (17.1) and (17.2).

$$r_S = -kX_VS \quad (17.2a) \qquad r_S = -kS \tag{17.2b}$$

The constants in the substrate removal expressions are developed over the range of operating conditions of the process.

The biomass concentration is included in Eqs. (17.1a) and (17.2a). Precisely, the active biomass, X_a, should be used as opposed to VSS, which is only a rough approximation of the concentration of active microorganisms. In addition to viable microorganisms there will be a significant amount of organic debris in the wastewater. Some of this material will come in the influent but a significant amount of suspended organic material will be dead biomass or associated with the decay of biomass (endogenous VSS) because treatment processes are normally operated with a considerable excess of microorganisms to ensure effective treatment. VSS is chosen as a measure of biomass as opposed to total suspended solids (TSS) because biological solids will be primarily composed of organic matter, although there will be an associated inorganic component. Inorganic suspended solids may also be present in the influent.

The measurement of VSS is convenient and practical compared to the assessment of the active biomass (X_a), which may consist not only of viable cells but also of enzymes and other catalytic agents in solution or suspension. As long as conditions in the prototype are similar to conditions in the model the rate constants remain valid. Adenosine triphosphate (ATP) and dehydrogenase activity measurements are the best measures of viable (active) biomass in biological treatment processes (Jørgensen et al., 1992; Weddle and Jenkins, 1971).

Although it has been common practice to incorporate VSS into the substrate removal expression, there is research to demonstrate that it is not always the best relation (Droste et al., 1993; Goodman and Englande, 1974). Substrate removal depends on the active mass present and a given amount of substrate can support only a given amount of active mass when the process is operated under starved conditions for the biomass. This is the common practice to maintain high removal rates and also to maintain the sludge in a state where it flocculates and settles well in the secondary clarifier.

When substrate limited conditions exist, advanced models and studies show that the active mass concentration remains approximately constant but it will decrease as a percentage of the total biomass as the amount of solids in the system is increased (Goodman and Englande, 1974; McKinney, 1962; Weddle and Jenkins, 1971). The direct relation between substrate removal and VSS expressed by Eqs. (17.1a) and (17.2a) becomes doubtful. Therefore, Eqs. (17.1b) and (17.2b) have also been found to give valid results.

The treatment models to be developed in this text will use one of Eqs. (17.1a), (17.1b), (17.2a), or (17.2b) as the descriptor of substrate removal kinetics. Regardless of which equation is used, any of the other models could be substituted and carried through each development. The important point is to apply the models consistently. In practice, the model of choice will be the one that best describes operational data. Regardless of the choice to incorporate VSS into the substrate removal expressions, the models are coarse and, although they have been used satisfactorily time and again, they do not account for many phenomena that are evident. The generation of secondary metabolites through primary substrate metabolism and generation of secondary substrate from decay of the biomass are examples of secondary phenomena that are incorporated into advanced models, which are beyond the scope of this presentation.

Temperature Dependence of Rate Coefficients

The Arrhenius equation is used to describe the temperature dependence of the maximum velocity coefficient k in Eqs. (17.1a) and (17.1b) or the velocity coefficient k in Eqs. (17.2a) and (17.2b).

$$k_T = k_{20}\theta^{(T-20)} \tag{17.3}$$

where

θ is a constant

k_T and k_{20} are the rate coefficients at temperatures of T and 20°C, respectively

Metcalf and Eddy (1991) report a range for θ of from 1.0 to 1.8, with a typical value of 1.04 for activated sludge systems. The temperature dependence of K in Eqs. (17.1a) and (17.1b) has been described by a variety of empirical correlations.

BOD, COD, and TOC Removal

Substrate is usually expressed in terms of BOD, COD, or TOC. Sometimes removal rates of specific compounds are examined such as nitrate or components that are toxic. The removal rates of the nonspecific measures of BOD, COD, and TOC will be different. Organics become more oxidized as biological treatment progresses but there is an accumulation of byproducts of microbial growth and metabolism that are difficult to degrade. This is reflected in the ratios of BOD and COD to TOC shown in Table 17.1, which also shows the effect of primary sedimentation on these ratios. The data in the table are for only a limited number of plants but the variation of the ratios is typical for treatment of domestic wastewater.

17.3.2 Growth of Microorganisms and Biological Sludge Production

Sludge production is another major characteristic of the process. The removal and metabolism of substrate result in the growth of new biomass. This production of biomass can be described by:

$$r_{X\text{p}} = -Yr_{\text{S}}$$

where

$r_{X\text{p}}$ is the production of VSS (biomass) from substrate removal ($\text{ML}^{-3}\text{T}^{-1}$ or mg/L/d)

Y (cf. Eq. 6.3) is a yield factor (mass of microorganisms produced per mass of substrate removed)

Because the process is operated in substrate limited conditions, the decay of biomass through starvation, death, predation, and autooxidation becomes significant. These phenomena, collectively known as endogenous decay, can be modeled by a first-order expression.

$$r_{X\text{e}} = -k_{\text{e}}X_{\text{V}}$$

where

$r_{X\text{e}}$ is the rate of decrease of VSS caused by endogenous decay ($\text{ML}^{-3}\text{T}^{-1}$ or mg/L/d)

k_{e} is a rate constant (mass removed through endogenous decay/mass present/d, T^{-1})

The net growth rate of microorganisms is the summation of the above two phenomena.

$$r_X = -Yr_{\text{S}} - k_{\text{e}}X_{\text{V}} \tag{17.4}$$

TABLE 17.1 Organics Variation in Treatment of Municipal Wastewater[a,b]

	BOD_5 mg/L		COD mg/L		TOC mg/L		BOD_5/TOC		COD/TOC	
	Ave.	Range	Ave.	Range	Ave.	Range	Ave.	Range	Ave.	Range
Raw	86	72–105	236	136–304	56	41–70	1.50	1.31–1.88	4.16	3.32–4.68
Primary effluent	58	46–68	204	146–299	52	44–61	1.11	1.00–1.33	3.90	3.19–5.85
Final effluent	15	11–20	84	77–95	35	33–40	0.44	0.20–0.69	2.40	2.02–2.58
Ave. removal (%)	83		64		38					

[a]From Eckenfelder and Ford (1970).
[b]Data are for eight plants. All plants did not necessarily report all three measures.

TABLE 17.2 Yield and Endogenous Decay Coefficients for the Activated Sludge Process[a]

Coefficient	Basis	Range	Typical
Y	g VSS/g BOD_5	0.4–0.84	0.6
Y	g VSS/g COD	0.24–0.4	0.4
k_e	d^{-1}	0.004–0.10	0.06

[a]Derived from Lawrence and McCarty (1970), Metcalf and Eddy (1991), and WEF and ASCE (1992a).

where
r_X is the net growth rate of microorganisms ($ML^{-3}T^{-1}$)

Equation (17.4) accommodates the two phenomena of growth from substrate removal and endogenous decay separately; it is more versatile than simply using a net yield factor times the rate of substrate removal, which is the observed yield (Eq. 17.5). The latter approach does not correlate the data as well as Eq. (17.4).

$$r_X = -Y_{obs} r_S \tag{17.5}$$

where
Y_{obs} is the observed yield factor

Typical values for Y and k_e are given in Table 17.2. Equation (17.4) describes the net yield of solids from substrate removal. The total production of solids depends on influent solids that are not degradable. These include refractory organics as well as inorganic solids. There is also a fixed (inorganic) component of biological solids, discussed in Sections 17.4.1 and 17.9.

Equation (17.4) shows how sludge production is related to the substrate removal rate and concentration of biomass in the system. In a later section, the biomass concentration will be related to the residence time of sludge in the system or sludge age. Some treatment installations, particularly smaller operations, may not have a primary clarifier. In this case there is a significant deviation from the solids production predicted by Eq. (17.4) because of the input of high solids concentrations. Some of the influent solids will be transformed into biological solids. Schulz et al. (1982) surveyed a number of installations that did not settle sewage entering the activated sludge process. Different modifications of the activated sludge process were included in the survey. The average rate of sludge production was 0.86 kg TSS/kg BOD_5 removed (range: 0.60–1.22 kg TSS/kg BOD_5 removed), with considerable variation for a single plant and among different plants. Good correlations did not exist between solids production and other process parameters. Solids handling facilities should be designed for 150% of design values.

Sludge Composition and Nutrient Requirements

The major nutrient requirements are nitrogen and phosphorus. The amounts of these nutrients required for activated sludge or any biological process depend on the net amount of biomass formed and removed from the process. Typically, sludge has a chemical formula near $C_5H_7NO_2$, determined by Hoover and Porges (1952) in one of the earliest studies of this nature in the field of environmental engineering, and this formula is almost universally used in textbooks and research papers.

Dairy waste was the substrate in the study of Hoover and Porges; there can be significant variations from their formula at different operating conditions and with different waste sources (Burkehead and Waddell, 1969). Incorporating typical biomass phosphorus content, the formulation becomes $C_5H_7NO_2P_{0.074}$. The phosphorus requirement is one sixth of the nitrogen requirement on a mass basis when this formula applies. The rule of thumb that applies to activated sludge processes is that the ratio of influent degradable matter expressed on an ultimate BOD or COD basis to nitrogen and phosphorus should be COD:N:P = 100:5:1 on a mass basis. When nutrient concentrations are less than this value, chemical nutrient supplements of ammonium or phosphate salts are added to the influent.

It is commonly assumed that effluent from an activated sludge process contains a dissolved phosphorus concentration of 1 mg/L. The total phosphorus requirement depends on soluble effluent phosphorus and the amount of sludge produced. Other nutrients are required in trace amounts but they are normally present in excess in most wastewaters.

Writing the chemical equation for the oxidation of biomass, the formula $C_5H_7NO_2$ results in 1.42 mg COD/mg VSS. Different chemical formulas will change the value of the oxygen demand of biomass. It is not necessary to know the chemical composition of biomass to determine its COD. Simply measuring the COD on filtered and unfiltered aliquots from a sample and determining the VSS of the sample is sufficient to determine the COD of the VSS. If this simple exercise were performed more regularly, COD balances for biological processes would exhibit improvement over making the usual assumption that the COD of VSS is 1.42 mg/mg. The value of 1.42 will be commonly used in this textbook but other values will also be used to reflect reality. When no data are available, 1.42 is probably the most reasonable value to use.

17.4 ACTIVATED SLUDGE CONFIGURATIONS

The primary flow and mixing regimes in activated sludge processes are shown in Fig. 17.1. As discussed in Chapter 10, plug flow (PF) and complete mixed (CM) are the extremes in mixing regimes. Many alternative influent addition and effluent withdrawal schemes in the reactor have been used that result in intermediate mixing regimes. The mixing regime is a fundamental process embodiment that influences substrate removal kinetics, settleability of the sludge produced in the process, and conditions in the reactor. Before these various processes are analyzed the symbolic notation scheme is described.

17.4.1 Definition of Symbols for the Activated Sludge Process

The symbols Q, S, and X refer to volumetric flow rate, substrate (measured as BOD_u, BOD_5, COD, or TOC) concentration, and solids concentration, respectively. For Q and S, subscripts indicate the location of the quantity. For X, there are usually two subscripts: the first subscript indicates the type of solids (e.g., TSS, VSS, or inert) and the second subscript indicates the location. X never appears unsubscripted—it must have at least one subscript indicating the type of solids.

Influent

The influent refers solely to the external flow entering the system. Recycled flow (an internal phenomenon) is excluded.

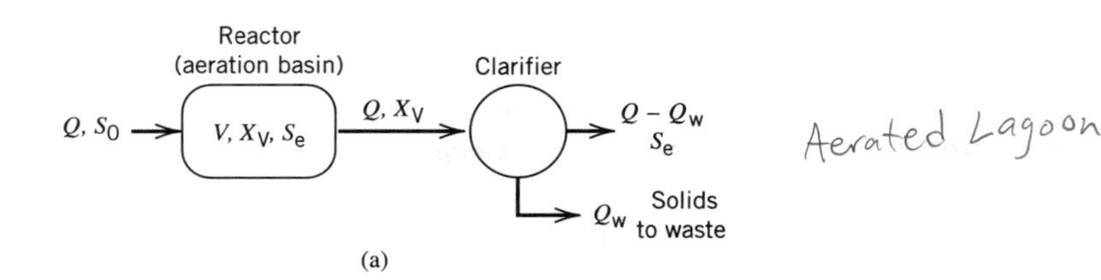

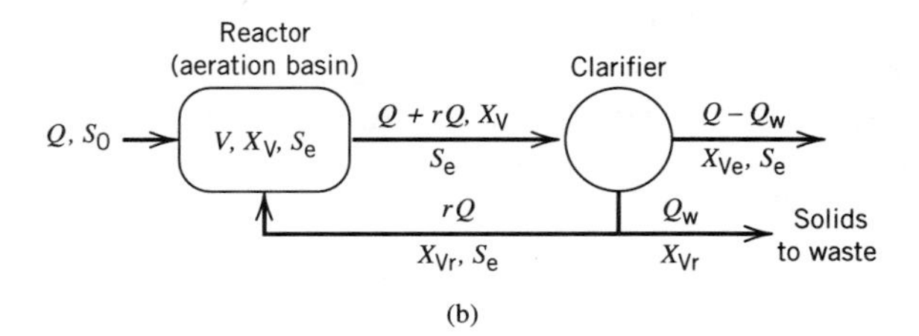

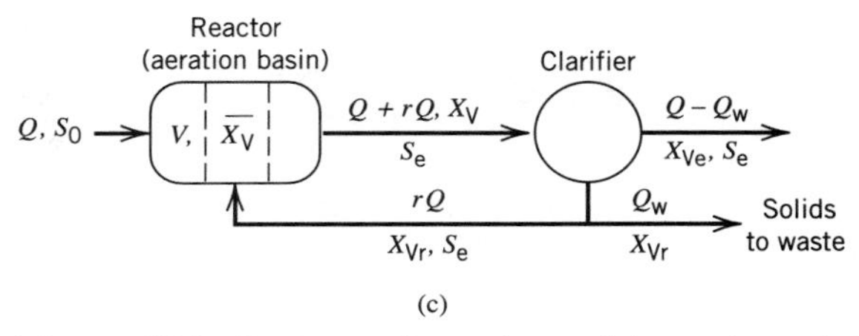

Figure 17.1 Basic configurations of the activated sludge process. (a) Complete mixed, no recycle; (b) complete mixed, biological solids recycle; (c) plug flow, biological solids recycle.

Q = volumetric flow rate

S_0 = substrate concentration (0 refers to influent)

X_{V0} = VSS (V refers to VSS)

X_{Ti0} = inert (inorganic) solids concentration (ISS) (Ti refers to the inorganic portion of TSS)

X_{T0} = TSS

Influent TSS is the sum of influent VSS and influent inert solids defined above:

$$X_{T0} = X_{Ti0} + X_{V0}$$

Reactor

Symbols that designate quantities in the reactor usually have no secondary subscripts associated with them.

S = soluble substrate concentration

This is the major exception to the rule given at the beginning of this section. In a CM reactor, the substrate concentration in the effluent is the same as the substrate concentration in the reactor (see Sections 10.1.1 and 10.1.3); therefore S_e (see the next subsection) is commonly used. In a PF reactor, S (unsubscripted) must be used because substrate concentration is a continuously varying function of time or distance in the reactor. S_e cannot be used because it refers to substrate concentration at a specific location.

X_V = VSS

V = volume

X_T = TSS

X_a = active mass concentration

X_{Ti} = inert (inorganic SS)(ISS) (Note this is in accord with influent notation and Ti, when together, is taken as a primary symbol group.)

There are two components of inorganic SS. Inorganic SS that are present in the influent to the reactor will pass through the process. Biosolids also contain inorganic solids; typically, biomass is composed of 0.75–0.80 VSS and 0.20–0.25 ISS. From a determination of TSS–VSS in the reactor, it is impossible to distinguish between ISS originating in the influent and biomass ISS.

Reactor Effluent

Note that for any reactor without recycle, the reactor effluent flow rate is the same as the influent flow rate.

For a CM reactor, effluent and reactor concentrations of any substance are the same. For a PF reactor, the solids concentration varies along the length of the reactor as shown later. However, for most processes the variation in solids concentration is small. Therefore, the average solids (of any type) concentration in the reactor is approximately the same as the reactor effluent concentration. In PF reactors, often a bar is placed over the symbol to indicate "average" concentration.

S_e = soluble substrate concentration

X_V = VSS

X_T = TSS

X_{Ti} = ISS

System Effluent

No biological treatment is assumed to occur in the clarifier and therefore only solids concentrations change from the influent to the various outflows from the clarifier.

Substrate concentration in clarified effluent and underflow from the clarifier is S_e (There is no soluble substrate removal in the clarifier according to the assumption).

X_{Ve} = VSS in clarified effluent

X_{Te} = TSS in clarified effluent

X_{Tie} = ISS in clarified effluent

Recycled and Waste Flow

The underflow from the clarifier is split into two flow paths. One is the waste flow and the other is the recycled flow. For solids concentrations, the subscripts "w" or "r" are appropriate but "r" is usually employed to describe the solids concentrations.

Q_w = waste sludge flow rate

X_{Tr} = TSS in waste sludge or recycled underflow

X_{Vr} = VSS in waste sludge or recycled underflow

X_{Tir} = ISS in waste sludge or recycled underflow

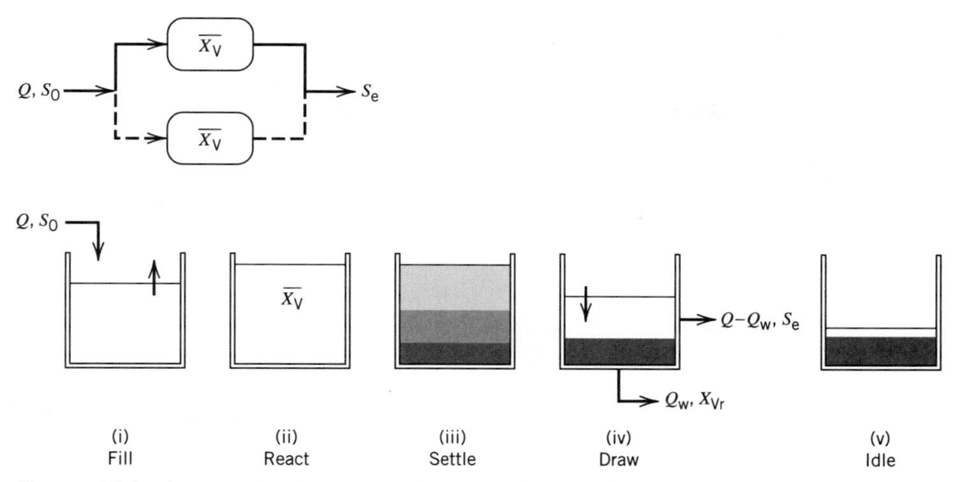

Figure 17.2 Sequencing batch reactor operating cycle.

r = recycle factor (This is the ratio of the recycled flow, Q_r, to the influent flow = Q_r/Q.)

Clarifier

X_{Tc} = average TSS in secondary clarifier

X_{Vc} = average VSS in secondary clarifier

X_{Tic} = average ISS in secondary clarifier

Sequencing batch reactors (Fig. 17.2) are another operational mode. One reactor may be operated in a fill and draw mode. Usually two or more reactors are used and operated in staggered sequences. In a multireactor sequencing batch treatment operation, each reactor cycles through the periods indicated in Fig. 17.2. Depending on influent flow quantity and quality variation, it may not be necessary to have an idle period in the cycle. In a continuous flow situation, it is always necessary to have one reactor filling while the other reactors are staggered throughout other periods of the operating cycle.

17.5 PROCESS ANALYSIS

Inert solids are subject to minimal biological transformation and dissolution is normally not significant. As a preliminary mass balance exercise, it is useful to consider the accumulation of inert solids [such as $Al(OH)_3(s)$] in a system with and without recycle from the secondary clarifier. It has been shown in Section 10.3.2 that recycle does not change the average residence time of liquid in the aeration basin.

The system without recycle is shown in Fig. 17.3a. Steady state conditions are assumed to exist. Because the clarifier is downstream of the aeration basin and there is no feedback from the clarifier to the aeration basin, the mass balance will be made around the aeration basin only. Initially it will be assumed that the solids concentrations are different in the influent, reactor, and reactor effluent. A mass balance on inert

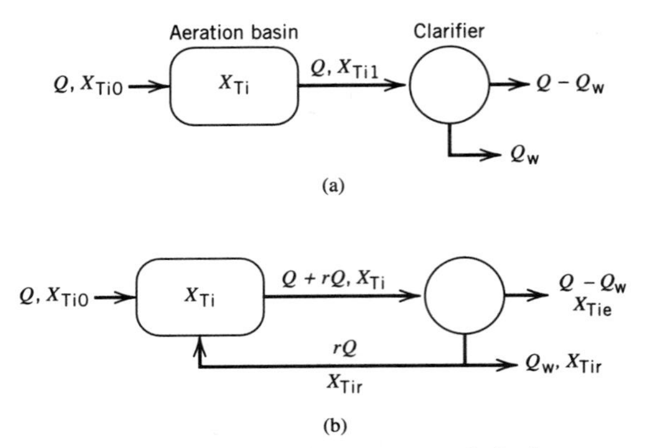

Figure 17.3 Inert solids in activated sludge systems. (a) Nonrecycle system; (b) recycle system.

solids is

$$\text{IN} - \text{OUT} + \text{GENERATION} = \text{ACCUMULATION}$$
$$QX_{\text{Ti0}} - QX_{\text{Ti1}} + 0 = \frac{dX_{\text{Ti}}}{dt}V = 0 \tag{17.6}$$

where
X_{Ti1} is the reactor effluent inert solids concentration

The equation simply shows that influent and effluent concentrations of inert solids from the reactor are the same and confirms the obvious. Therefore, the concentration of inert solids in the reactor is also equal to the influent concentration.

$$X_{\text{Ti1}} = X_{\text{Ti}} = X_{\text{Ti0}}$$

The mixing regime in the reactor does not change this result because inert solids (originating in the influent) concentrations are not changed by any reaction in the aeration basin. If the concentrations of inert solids in the clarified effluent and the underflow from the sedimentation basin are X_{Tie} and X_{Tiw}, respectively, the mass balance around the clarifier is

$$QX_{\text{Ti}} = (Q - Q_{\text{w}})X_{\text{Tie}} + Q_{\text{w}}X_{\text{Tiw}}$$
$$X_{\text{Ti}} = X_{\text{Ti0}} = \frac{(Q - Q_{\text{w}})X_{\text{Tie}} + Q_{\text{w}}X_{\text{Tiw}}}{Q} \tag{17.7}$$

The influence of solids recycle from the secondary clarifier to the aeration basin on the accumulation of inert solids in the reactor will now be examined (Fig. 17.3b). It is now possible to isolate the system as a whole, the aeration basin, or the clarifier, and make mass balances for each case.

It is assumed that the solids in the return line are well mixed with the influent flow. From an exercise similar to the previous one it is again obvious that the concentrations of inert solids in the reactor and reactor effluent are the same because there is no degradation of inert solids in the aeration basin. The word statement of the equation is dropped and the other assumptions noted above apply.

1. Aeration Basin

$$QX_{Ti0} + rQX_{Tir} - (1 + r)QX_{Ti} = \frac{dX_{Ti}}{dt} V = 0 \qquad (17.8a)$$

2. Clarifier

$$(1 + r)QX_{Ti} - rQX_{Tir} - (Q - Q_w)X_{Tie} - Q_w X_{Tir} = \frac{dX_{Tic}}{dt} V_c = 0 \qquad (17.8b)$$

where

V_c is the volume of the clarifier

3. System

$$QX_{Ti0} - (Q - Q_w)X_{Tie} - Q_w X_{Tir} = \frac{dX_{Ti}}{dt} V + \frac{dX_{Tic}}{dt} V_c = 0 \qquad (17.8c)$$

Equation (17.8a) can be used to relate the inert solids concentration in the aeration basin to the recycle line inert solids concentration and the recycle flow rate, r.

$$X_{Ti} = \frac{X_{Ti0} + rX_{Tir}}{1 + r} \qquad (17.9)$$

The solids concentration in the return line, which is drawn from the clarifier underflow, is much higher than the concentrations of inert solids in the effluent from the aeration basin and the effluent from the system. Consequently, Q_w is small compared to Q. Using the system balance,

$$X_{Ti0} = \frac{(Q - Q_w)}{Q} X_{Tie} + \frac{Q_w}{Q} X_{Tir}$$

To simplify manipulation of the equations it will be assumed that the clarifier is ideal, which means that no solids escape in the clarified overflow and all inert solids leave the system through the waste solids line from the underflow of the clarifier. In this case the previous equation reduces to

$$X_{Ti0} = \frac{Q_w}{Q} X_{Tir}$$

Substituting this result into Eq. (17.9),

$$X_{Ti} = \frac{X_{Ti0} + r\left(\frac{Q}{Q_w}\right) X_{Ti0}}{1 + r} = \frac{X_{Ti0}}{1 + r}\left(1 + \frac{rQ}{Q_w}\right) \qquad (17.10)$$

It is seen that the concentration of inert solids in the aeration basin is a function of the compactibility of the sludge in the clarifier; Q/Q_w (or X_{Tir}/X_{Ti0}) depends on how well the sludge compacts. It is also seen that X_{Ti} is greater than X_{Ti0}. Given that Q_w is small compared to Q, a rather large increase in inert solids concentration in the aeration basin compared to the influent flow will occur because of recycling solids from the underflow of the secondary clarifier. Other types of solids are also concentrated in the reactor by recycle.

17.6 FOOD TO MICROORGANISM RATIO AND SLUDGE AGE

Two important parameters for operating the process are the food: microorganism (F:M) ratio and the sludge age. These parameters originate from mass balances for the systems. The F:M ratio, U, is

$$U = \frac{Q(S_0 - S_e)}{VX_V} = \frac{(S_0 - S_e)}{X_V \theta_d} \tag{17.11a}$$

where

θ_d is the hydraulic retention time (HRT)

The F:M ratio describes the degree of starvation of the microorganisms. Because biological treatment processes should remove nearly all of the influent substrate, the F:M ratio is often expressed as

$$U = \frac{S_0}{X_V \theta_d} \tag{17.11b}$$

Equation (17.11b) also expresses the potential food availability to the microbial population.

The sludge age, θ_X, describes the residence time of the sludge in the system. The sludge or biomass requires a certain amount of time to assimilate the substrate and reproduce. If the sludge is not able to reproduce itself before being washed out of the system, failure will result. Also, the sludge age is related to the F:M ratio describing the relative state of starvation of the microorganisms. Higher sludge ages cause the sludge to undergo more endogenous decay. This has an effect on the settleability of the sludge as well as on the total amount of sludge produced in the system.

Merely storing the sludge under conditions that have a minimal effect on its activity and maintain it in a dormant state will not increase the sludge age. Because the sludge is in an aerobic state in the aeration basin and the small amount of DO in the aeration basin effluent is rapidly exhausted in the clarifier, the sludge is in an anoxic or anaerobic condition in the clarifier. Also, there is no supply of exogenous substrate in the clarifier if the process is operating efficiently. The change in DO and lack of substrate mildly shock the sludge, putting it in an essentially dormant state. Therefore, residence time of the sludge in the clarifier does not contribute to the effective sludge age. It is only in the aeration basin, where a fresh supply of oxygen and substrate is maintained, that the sludge metabolic activity is significant.

Under these conditions, the sludge age or sludge residence time (SRT) is the average amount of time the sludge spends in the aeration basin. The SRT is completely analogous to the HRT, which is the average residence time of a particle of water in the aeration basin, although the two times are not necessarily equal.

$$\begin{aligned} \theta_X &= \frac{\text{mass of solids in aeration basin}}{\text{solids removal rate from the system}} \\ &= \frac{VX_V}{\text{solids removal rate from the system}} \end{aligned} \tag{17.12}$$

The specific expression of the solids removal rate is given below for each system. The concepts of F:M and SRT also become clearer after the mass balance relations are examined.

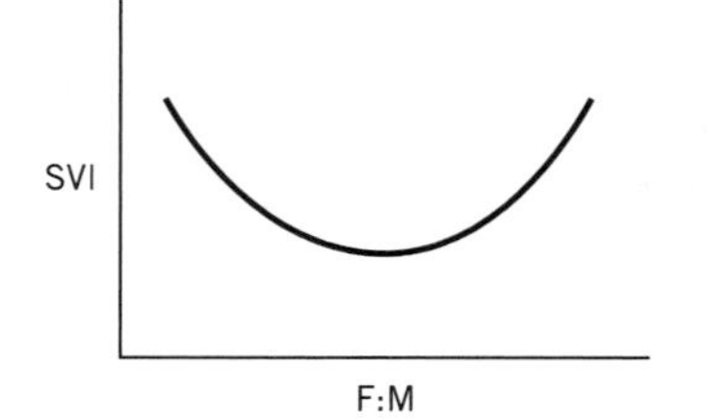

Figure 17.4 SVI as a function of F:M ratio.

17.6.1 Sludge Volume Index

A measure of the settleability and compactibility of sludge is made from a laboratory column settling test. The procedure is outlined in *Standard Methods* (1992). Mixed liquor with a known TSS content (X_T) is mixed and placed in a 1- or 2-L cylinder. The larger cylinder is desirable to minimize bridging of the sludge floc and wall effects. Gentle stirring during the test is also recommended to obtain the most efficient settling. The mixed liquor is allowed to settle for a period of time ranging from 30 min to 1 or 2 h. One-half hour is the more common settling time.

At the end of the settling period the volume of sludge is read from the cylinder. The sludge volume index (SVI) is defined as the volume in milliliters occupied by 1 g of sludge after it has settled for a specified period of time. If a 1-L cylinder is used and the sludge occupies a volume of y mL at the end of the settling period,

$$\text{SVI(mL/g)} = \frac{y}{X_T}(1\,000 \text{ mg/g}) \tag{17.13}$$

A low SVI is indicative of a sludge that settles well. The SVI can be used to estimate the concentrations of VSS and TSS in the recycle line if the ratio of VSS to TSS in the mixed liquor is known. Typically, the ratio of VSS to TSS in the mixed liquor is in the range of 0.75 to 0.80.

$$X_{Tr} = \frac{10^6}{\text{SVI}} \quad (17.14a) \qquad X_{Vr} = \left(\frac{X_V}{X_T}\right) X_{Tr} \quad (17.14b)$$

0.75

where

X_{Tr} and X_{Vr} are in mg/L

The F:M ratio and the SRT (which is directly related to F:M as shown below) influence the settleability and compactibility of the sludge. When the biomass is in a state of endogenous decay, it tends to form polymers that result in natural flocculation under quiescent conditions. In a CM reactor at low sludge ages the sludge tends to become populated with filamentous organisms that exhibit poor settleability and the sludge does not flocculate well. At the other extreme of highly starved conditions or a very high SRT, the sludge forms pinpoint floc (like the head of a needle) and does not flocculate as well as in intermediate ranges. A typical plot of SVI versus F:M ratio is shown in Fig. 17.4. Using the relations developed below, the F:M ratio can be replaced with the sludge age. Other factors discussed in Section 17.12 affect SVI.

17.6.2 CM Reactor without Recycle

The simplest system to examine is a CM reactor that does not receive recycled sludge from the clarifier. The clarifier is included for the sake of completeness but its operation and efficiency, which are related to conditions in the aeration basin, do not have any influence on phenomena in the aeration basin.

Figure 17.1a is a schematic of the process. Steady state conditions will be assumed. Also, the influent to the reactor will be assumed to not contain any VSS. As noted before, the influent is usually settled; therefore, influent suspended solids tend to be colloidal in nature. These colloidal particles normally contain degradable organic matter and many of the particles may be microorganisms themselves. But these microorganisms will not likely represent a significant contribution to active microorganisms in the aeration basin. The microorganisms that populate the aeration basin are largely adventitious in origin. They survive and reproduce in the conditions existent in the aeration basin. Our analysis will focus on the removal of dissolved organic matter for the moment.

Substrate Balance

$$\text{In} - \text{Out} + \text{Generation} = \text{Accumulation}$$

$$QS_0 - QS_e + r_S V = \frac{dS_e}{dt} V = 0$$

Even though generation (substrate transformation) is, in fact, negative in the case of wastewater treatment, the mass balance should not be incorrectly biased at this stage by putting a negative sign in front of the generation term. The kinetic formulation with the correct sign will be substituted later. Solving for r_S in terms of the physical conditions of the process,

$$r_S = -\frac{Q(S_0 - S_e)}{V} = -\frac{S_0 - S_e}{\theta_d} = -\rho \qquad (17.15)$$

where

ρ is the rate of wastewater treatment

The rate of wastewater treatment, ρ, is simply a statement of the substrate mass balance without any kinetic formulation. A kinetic relation will be substituted for r_S and functionally it is related to ρ by Eq. (17.15). It avoids confusion to use ρ and r_S separately.

If a Monod model is substituted for r_S,

$$-\frac{kS_e}{K + S_e} = -\frac{S_0 - S_e}{\theta_d} \qquad (17.16)$$

Solving for S_e,

$$S_e^2 + kS_e\theta_d - S_0S_e + KS_e - KS_0 = 0 \qquad (17.17)$$

Defining $b = k\theta_d - S_0 + K$ and $c = -KS_0$ $(a = 1)$,

$$S_e = \frac{-b + \sqrt{b^2 - 4ac}}{2a} \qquad (17.18)$$

The negative root will be physically meaningless.

When the form of the kinetic expression is known (i.e., first-order, second-order, Monod, or other model) along with the kinetic coefficients, substitution for r_S in Eq. (17.15) and solving for S_e is fairly routine. This approach is used to predict system behavior under various operating conditions. The other problem is to analyze data to determine the kinetic expression. The rate of wastewater treatment can be determined for any operating condition by measuring the readily determined parameters of θ_d, S_0, and S_e. However, the kinetic expression and the values of the kinetic coefficients should be determined from analysis of a number of operating conditions (see Example 17.1).

It is obvious from Eq. (17.17) that in the case of the two-parameter Monod model, data from at least two operating conditions must be available to solve for k and K. Given the multitude of substrates and microorganisms present in any wastewater, variability is inevitable and more than two operating conditions should be analyzed with regression techniques to establish the best model and its coefficient values.

The efficiency of treatment, η, is defined on the basis of degradable substrate removal. If substrate concentrations are defined on a BOD basis, the efficiency of treatment is

$$\eta = \frac{S_0 - S_e}{S_0} \tag{17.19a}$$

If substrate concentration is defined on a COD basis, there is likely to be a component of the COD that is not removable. The nondegradable component (S_n) is subtracted from the influent COD concentration. The effluent COD consists of the nondegradable component and COD that has not been removed in the process. The equations are based on removable COD only and the nondegradable component is ignored in the formulations for kinetic design of the process.

Defining S_{ST0} as the total soluble influent COD to the process and S_{STe} as the total soluble effluent COD from the process, the efficiency of treatment on a COD basis is defined as

$$\eta = \frac{(S_{ST0} - S_n) - (S_{STe} - S_n)}{S_{ST0} - S_n} \times 100 \tag{17.19b}$$

where

$S_0 = S_{ST0} - S_n$

$S_e = S_{STe} - S_n$

It is more usual to compare total influent COD or BOD to soluble effluent COD or BOD, which is discussed later. For now we will work with the definition given in Eq. (17.19b).

The soluble effluent BOD_5 depends on the ease of biodegradation of soluble effluent organics. The organic matter remaining in solution after a biological treatment process is not likely to be as readily degraded as the influent organic matter to the process; therefore, the BOD rate constant will be lower for the effluent compared to the influent. An in situ determination must be performed.

The interpretation of S_n can be difficult and elaborate studies may be required to determine the true concentration of nondegradable components in the influent as opposed to soluble nondegradable components that are actually generated in the reactor. Practically, "correcting" S_0 and S_e with some value S_n simply provides the best correlation for the Monod equation in some circumstances (see for example, Selna and Schroeder, 1978). The lowest value of effluent soluble COD obtained over a wide range of operating conditions would be the maximum value for S_n.

Biomass Balance

The biomass balance applies to biological solids active in the process.

$$QX_{V0} - QX_V + r_X V = \frac{dX}{dt} V = 0$$

Influent biomass VSS is assumed to be negligible.

$$QX_V = r_X V \tag{17.20}$$

If the influent VSS is significant, whether it is incorporated into the biomass balance or not depends on whether the influent VSS are biological solids that will be active in the aeration basin.

The production rate of solids is equal to the net generation rate of solids. Substituting the kinetic relation for r_X (Eq. 17.4),

$$QX_V = (-Yr_S - k_e X_V)V \tag{17.21}$$

The sludge age or SRT for this system is

$$\theta_X = \frac{X_V V}{QX_V} = \theta_d \tag{17.22}$$

The sludge age and HRT are equal in this system but they are not necessarily equal in other systems.

Solving Eq. (17.21) for the sludge age and noting that $-\rho$ or $-UX_V$ can be substituted for r_S (see Eqs. 17.11 and 17.15),

$$\frac{1}{\theta_X} = -\frac{Yr_S}{X_V} - k_e = \frac{Y\rho}{X_V} - k_e = YU - k_e \tag{17.23}$$

Equation (17.23) relates the F:M ratio (also known as the process loading factor) to the sludge age.

Equation (17.23) can be further manipulated by substituting relations for r_S and ρ. First, substituting for ρ, the concentration of VSS in the aeration basin may be determined.

$$\frac{1}{\theta_X} = \frac{Y(S_0 - S_e)}{\theta_d X_V} - k_e \Rightarrow X_V = \frac{Y(S_0 - S_e)}{1 + k_e \theta_X} \tag{17.24}$$

The minimum sludge age, θ_X^m (Lawrence and McCarty, 1969, 1970), is a useful concept in the design of activated sludge systems and other biological treatment processes. The minimum sludge age is determined by the maximum rate at which the sludge can grow in the system. The sludge cannot be washed out of the system at a rate faster than the minimum sludge age or the microorganisms will not be able to sustain themselves in the system unless they are being supplied from an outside source. The maximum rate at which the microorganisms can grow is dependent on the maximum rate of substrate removal. The highest possible substrate concentration in the aeration basin is S_0. Choosing Eq. (17.1b) for the substrate removal expression and substituting it into Eq. (17.23), and replacing S_e with S_0 (to provide the highest rate of substrate removal),

$$\frac{1}{\theta_X^m} = \frac{YkS_0}{X_V(K + S_0)} - k_e$$

Furthermore, if $S_0 \gg K$,

$$\theta_X^m = \frac{X_V}{Yk - k_e X_V} \tag{17.25}$$

Similar expressions can be derived when other substrate removal relations apply.

Total Effluent COD from the Process

The total effluent COD from the process consists of the soluble effluent COD and the suspended or particulate COD. The former depends on the efficiency of biological treatment in the aeration basin, whereas the latter depends on the efficiency of solids removal in the secondary clarifier. There are no mechanistic models that can be developed to predict the concentrations of solids in the clarified effluent or in the clarifier underflow. The mixed liquor VSS (MLVSS) influent to the clarifier may contain anywhere from 700 to 5 000 mg/L of VSS and the clarified effluent can contain less than 20 mg/L to higher values depending on the operating conditions of the process. Likewise, the compactibility of the sludge can vary over a wide range. Observations can be used to correlate each of these parameters empirically with system operating conditions (HRT, SRT, temperature, pH, and other variables) and other parameters such as time of year or sewage strength.

Defining the total COD in the clarified effluent as S_{STe},

$$S_{STe} = S_e + fX_{Ve} \tag{17.26}$$

where

f is a factor for the COD content of VSS

If the composition of VSS in the system is $C_5H_7NO_2$, the COD content of VSS (f) is 1.42 mg COD/mg VSS (see Section 17.3.2).

Removal of Influent Suspended Organic Matter

Effluent from primary clarifiers normally still contains significant amounts of suspended organic matter. Influent suspended organic matter is removed by two mechanisms in a biological treatment process:

1. Biomass metabolizes some of these organic solids. As for soluble organics, for the portion of suspended organics that is metabolized, there is a fraction that is oxidized, and the other fraction is transformed (synthesized) into biomass.
2. Suspended organics are flocculated with the biomass and settle in the secondary clarifier.

The efficiency of treatment is usually gauged by comparing the total influent COD or BOD to the soluble effluent BOD or COD; S_0 in Eqs. (17.19a) or (17.19b) would be measured on an unfiltered sample of influent. It is reasonable to assess the treatment on this basis because the objective of treatment in the biological reactor is to reduce the soluble degradable organics to a minimum. The separation of suspended solids is a function of the efficiency of the secondary clarifier. Separation of suspended matter from the clarified effluent can be further improved by filtration.

Laboratory studies conducted for research often use a totally soluble waste to avoid complications in modeling the removal rate of influent suspended organics. Modeling the rate of influent suspended degradable organics results in a more complex model. Influent VSS that is not microorganisms which will become part of the biological engine are actually part of the substrate. Substrate removal expressions remain based on the amount of soluble organics remaining in the water but, in fact, substrate removal incorporates hydrolysis and metabolism of suspended organics along with soluble organics. VSS concentrations less accurately reflect the amount of acclimated biomass in the aeration basin as influent VSS rises. It should be noted that some of the influent

VSS is microorganisms that have acclimatized to the waste as the wastewater flows through the sewer system and preliminary treatment operations.

It is possible to use the total influent BOD or COD concentration for S_0 in any of the equations given in this chapter if the kinetic coefficients and other parameters (such as the yield factor or ρ) have been developed on a basis of total influent organics. Regardless of which approach is used, it is necessary to develop treatment parameters for each waste and apply the model consistently.

17.6.3 CM Reactor with Recycle

The assumptions made in the previous section apply to a CM reactor with recycle also. In addition, it will be assumed that the storage of solids in the clarifier is insignificant. As discussed previously, the sludge is essentially dormant in the clarifier and this is equivalent to ignoring its presence (in a biological sense) in the clarifier. Studies have shown that for typical holding times in secondary clarifiers, the sludge is able to recover its activity when returned to the aeration basin (Ford and Eckenfelder, 1967).

Making a mass balance on soluble substrate around the system leads to the same results as given by Eqs. (17.15) and (17.16). Because the sludge is dormant in the clarifier, substrate removal occurs only in the aeration basin and the active volume in the system is V, the volume of the aeration basin. Equation (17.26) is applicable for the total effluent COD from the process.

Biomass Balance

Making the balance around the system,

$$QX_{V0} - Q_w X_{Vr} - (Q - Q_w)X_{Ve} + r_X V = \frac{dX_V}{dt} V = 0$$

There are two exit points for sludge: the clarified effluent and the underflow waste sludge line from the clarifier. Again it is noted that the sludge is only active in the aeration basin.

Because influent biomass VSS is assumed to be negligible,

$$Q_w X_{Vr} + (Q - Q_w)X_{Ve} = r_X V \tag{17.27}$$

The sludge age for this system is given by

$$\theta_X = \frac{X_V V}{Q_w X_{Vr} + (Q - Q_w)X_{Ve}} \tag{17.28}$$

Solving Eq. (17.27) for the inverse of the sludge age, the same equation (Eq. 17.23) obtained for the nonrecycle case is obtained.

Note in this case that the actual residence time of the sludge in the entire system (θ_T) is given by

$$\theta_T = \frac{X_V V + X_{Vc} V_c}{Q_w X_{Vr} + (Q - Q_w)X_{Ve}}$$

where

V_c is the volume of the clarifier

The effective sludge age or SRT is different from θ_T for reasons discussed in Section 17.6.

Substituting the expression for ρ (Eq. 17.15) in place of r_S in Eq. (17.23) allows X_V to be determined.

$$\frac{1}{\theta_X} = \frac{Y(S_0 - S_e)}{X_V \theta_d} - k_e$$

which can be rearranged to

$$X_V = \frac{\theta_X}{\theta_d}\left[\frac{Y(S_0 - S_e)}{1 + k_e \theta_X}\right] \tag{17.29}$$

The difference between Eqs. (17.29) and (17.24) is the θ_X/θ_d term, which is a physical concentration factor resulting from recycling the concentrated sludge. In the nonrecycle case this term equals 1 and so it does not explicitly appear in Eq. (17.24). Because the sludge age is often more than an order of magnitude greater than the HRT, a high degree of concentration can be attained by recycle.

Following the same procedures used for the nonrecycle case, the same expression for the minimum sludge age (Eq. 17.25) can be developed for the recycle case. The same expressions for treatment efficiency defined in Eqs. (17.19a) and (17.19b) also apply.

If influent VSS is significant but does not contain biomass adapted to the process, it is not included in the biomass balance. If the influent VSS is biomass acclimated to the process then it is included in the biomass balance and effectively reduces the sludge age or reproduction time for microorganisms. This is not problematic. Supply of microorganisms from an outside source reduces growth time required for microorganisms in the process.

If a model is developed ignoring influent VSS when it is significant, the model will be limited. More rigorous analysis of the data can be used to develop a more refined and accurate model that will have more validity if the proper data are obtained.

17.6.4 The Rate of Recycle

The design engineer has control over the HRT and the SRT. The operator is able to vary each of these within the design limitations of the system. The recycle ratio r is related to these and other parameters. To derive the relation, consider a mass balance for VSS around the clarifier (Fig. 17.1b).

$$Q(1 + r)X_V = (Q - Q_w)X_{Ve} + rQX_{Vr} + Q_w X_{Vr}$$

It is recognized that the expression for sludge age (Eq. 17.28) can be used to simplify the equation by removing the terms involving Q_w. Q_w is not a primary control variable. It is a function of the sludge age, which influences sludge settleability and the concentration of solids in the recycle line. Rearranging Eq. (17.28),

$$\frac{VX_V}{\theta_X} = (Q - Q_w)X_{Ve} + Q_w X_{Vr} \tag{17.30}$$

Substituting this into the mass balance,

$$(1 + r)X_V = \frac{VX_V}{Q\theta_X} + rX_{Vr}$$

The following relation is obtained:

$$\frac{1}{\theta_X} = \frac{Q}{V}\left(1 + r - r\frac{X_{Vr}}{X_V}\right) \tag{17.31}$$

The term X_{Vr}/X_V is a measure of the sludge compactibility. A measure of X_{Vr} may be obtained from the SVI test. Solving Eq. (17.31) for r,

$$r = \frac{1 - \dfrac{\theta_d}{\theta_X}}{\dfrac{X_{Vr}}{X_V} - 1} \tag{17.32}$$

■ Example 17.1 Determination of Parameters for an Activated Sludge Process

The data in the following table have been obtained from bench-scale studies of a continuous flow activated sludge process treating a synthetic waste. The data are the average values determined after steady state conditions have been achieved. Determine the kinetic coefficients and yield factor for this system. Substrate concentration was measured as COD.

Run	θ_X, d	θ_d, d	S_0, mg/L	S_e, mg/L	X_V, mg/L
1	12.0	0.25	310	35	2 850
2	8.8	0.25	300	62	2 004
3	6.1	0.25	315	59	1 688
4	12.3	0.33	284	47	1 914
5	8.9	0.33	302	53	1 625
6	6.0	0.33	300	66	997

Regardless of whether recycle has been used in the system, Eq. (17.15) applies. Substituting Eq. (17.1a) for r_S and rearranging Eq. (17.15) in the linear Lineweaver–Burke form (see Section 4.12):

$$\frac{X_V}{\rho} = \frac{X_V \theta_d}{(S_0 - S_e)} = \frac{K}{k}\frac{1}{S_e} + \frac{1}{k}$$

The data were plotted according to this equation and a regression was performed to determine the coefficients. The R^2 value of the regression was 0.60. The values of the intercept and slope were 0.573 and 78, respectively. From this

$$k = 1/0.573 = 1.75\ \text{d}^{-1}$$
$$K = k(78) = (1.75)(78) = 136\ \text{mg/L}$$

The fitted curve is also plotted on the figure below. Equation (17.1b) was also checked as the kinetic model but the R^2 value was much lower than 0.60. Equation (17.1a) was chosen as the kinetic model.

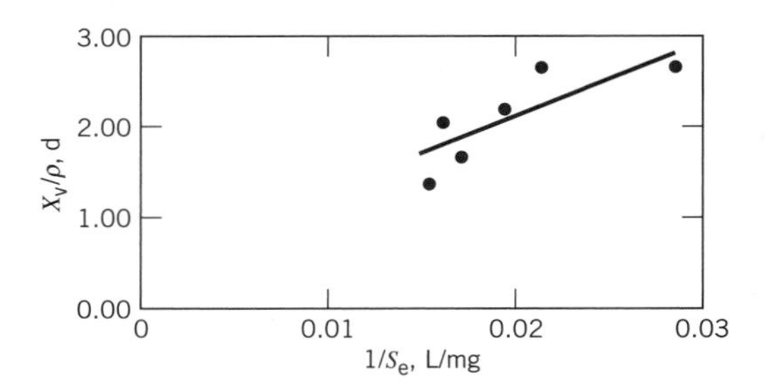

Estimating kinetic parameters of the Monod equation by linearization procedures applied to nonlinear equations is subject to instability and error. For instance, if S_e had been measured as 45 mg/L (this could easily occur in a biological treatment process) instead of 35 mg/L for run 1, the Lineweaver–Burke formulation would lead to a negative K value. Nonlinear parameter estimation techniques are the best choice when possible.

The yield factor and endogenous decay coefficient were determined from Eq. (17.23), which applies to either the recycle or the nonrecycle case (sludge age must be correctly calculated depending on recycle).

From a regression of $1/\theta_X$ against U (r_S/X_V), the slope and intercept were

$$-k_e = -0.019 \text{ d}^{-1} \qquad Y = 0.28 \quad (R^2 = 0.94)$$

The equation and fitted curve are plotted in the accompanying figure.

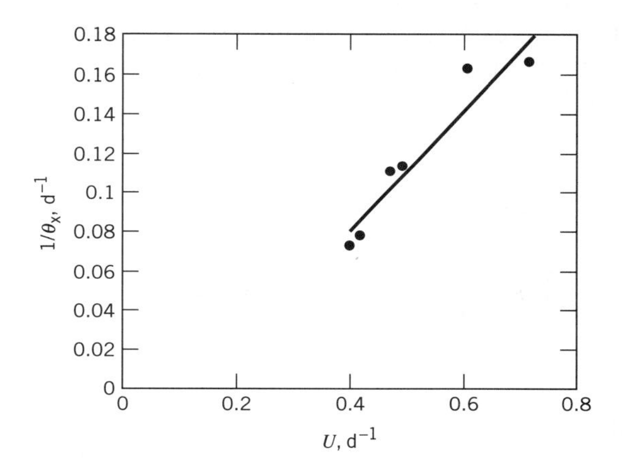

Similar procedures would be applied to data generated for other variations of the activated sludge process.

If Eq. (17.1a) is used as the substrate removal model, a biomass balance dictates that effluent quality is solely a function of sludge age. MLVSS will vary inversely to the detention time in the system. If Eq. (17.1b) is used as the kinetic model, a substrate balance will show that effluent quality is solely dependent on the detention time in the aeration basin and MLVSS will vary directly with the sludge age. In the former case a minimum detention time is required for mass transfer of substrates and other substances into and out of the microorganisms and metabolism regardless of the concentration of microorganisms. This is illustrated by effluent quality variation with detention time in a contact stabilization basin (discussed in Section 17.8.4).

In the second case, the time (detention time) for organics uptake does not ensure that there is enough time for the microorganisms to reproduce. The sludge age cannot be arbitrarily varied to extremes. The coefficients developed will apply to the model selected and be reasonably constant within a range of operating conditions around the conditions used to generate data for coefficient determination. The models developed here are simple representations of a very complex process and the mathematical constructs must be applied judiciously.

■ Example 17.2 CM Activated Sludge Process Effluent

A CM activated sludge process with recycle is being operated at the following conditions:

$$Q = 1\,650 \text{ m}^3/\text{d } (0.436 \text{ Mgal/d}) \qquad r = 0.60 \qquad S_0 = 200 \text{ mg COD/L}$$

Lab studies have shown that Eq. (17.1a) is the best kinetic model. The rate constants associated with the process are as follows:

$$k = 0.95 \text{ mg COD/mg VSS/d} \qquad K = 85 \text{ mg COD/L}$$
$$Y = 0.48 \text{ mg VSS/mg COD} \qquad k_e = 0.07 \text{ d}^{-1}$$

The average VSS concentration in the clarified effluent from the secondary clarifier has been found to be 15 mg/L; VSS in the influent is negligible.

Determine the effluent soluble substrate concentration, total effluent COD concentration, waste flow rate from the underflow of the secondary clarifier, sludge age, and MLVSS if X_{Vr} is 5 000 mg/L and θ_d is 8 h.

The volume of the basin is

$$V = Q\theta_d = \left(1\,650 \frac{\text{m}^3}{\text{d}}\right)(8 \text{ h})\left(\frac{1 \text{ d}}{24 \text{ h}}\right) = 550 \text{ m}^3$$

$$\text{In U.S. units: } V = \left(0.436 \times 10^6 \frac{\text{gal}}{\text{d}}\right)(8 \text{ h})\left(\frac{1 \text{ d}}{24 \text{ h}}\right)\left(\frac{1 \text{ ft}^3}{7.48 \text{ gal}}\right) = 19\,430 \text{ ft}^3$$

The substrate balance results in

$$-\frac{kX_V S_e}{K + S_e} = -\frac{S_0 - S_e}{\theta_d}$$

There are two unknowns in this equation, S_e and X_V. Because r and X_{Vr} are known, a biomass balance around the aeration basin will also provide information on X_V.

$$\text{In} - \text{Out} + \text{Generation} = 0$$

$$rQX_{Vr} - (1 + r)QX_V + \left(Y\frac{kX_V S_e}{K + S_e} - k_e X_V\right)V = 0$$

This expression and the substrate balance expression have the same two unknowns, X_V and S_e. This equation may be solved for X_V, which can then be substituted into the substrate balance expression.

$$X_V = \frac{rQX_{Vr}}{(1 + r)Q + k_e V - \dfrac{YkS_e}{K + S_e}V}$$

Making the substitution,

$$\frac{S_0 - S_e}{\theta_d} = \left(\frac{kS_e}{K + S_e}\right)\left[\frac{rQX_{Vr}}{(1 + r)Q + k_e V - \dfrac{YkS_e}{K + S_e}V}\right]$$

This equation may be converted into a polynomial or solved by trial and error for S_e. The polynomial expression is

$$[-(1+r)Q + YkV - k_eV]S_e^2 + [(1+r)Q(S_0 - K) - YkVS_0 - k_eV(K - S_0) + krQX_{Vr}\theta_d]S_e + (1+r)QKS_0 + k_eVKS_0 = 0$$

Substituting the values for the parameters and solving the quadratic (only the positive root is meaningful) results in S_e = 34 mg/L.

The total effluent COD, S_{Te}, assuming VSS has a composition of $C_5H_7NO_2$ is found from Eq. (17.26),

$$S_{Te} = 34 + 1.42(15) = 55 \text{ mg/L}$$

X_V may now be determined from the substrate balance.

$$X_V = \frac{(S_0 - S_e)(K + S_e)}{\theta_d k S_e} = \frac{\left(200\,\frac{\text{mg}}{\text{L}} - 34\,\frac{\text{mg}}{\text{L}}\right)\left(85\,\frac{\text{mg}}{\text{L}} + 34\,\frac{\text{mg}}{\text{L}}\right)}{(0.333\text{ d})(0.95\text{ d}^{-1})\left(34\,\frac{\text{mg}}{\text{L}}\right)}$$

$$= 1\,835 \text{ mg/L}$$

The sludge age is determined from the relation

$$\frac{1}{\theta_X} = YU - k_e = \frac{\frac{YkX_VS_e}{K + S_e}}{X_V} - k_e = \frac{(0.48)(0.95\text{ d}^{-1})\left(34\,\frac{\text{mg}}{\text{L}}\right)}{(85 + 34)\,\frac{\text{mg}}{\text{L}}} - 0.07\text{ d}^{-1} = 0.060\text{ d}^{-1}$$

$$\theta_X = 16.6\text{ d}$$

The waste flow rate is determined from a mass balance around the system, which shows that the solids wasted are equal to the solids produced.

$$r_XV = \left(\frac{YkX_VS_e}{K + S_e} - k_eX_V\right)V = (Q - Q_w)X_{Ve} + Q_wX_{Vr}$$

Solving this equation for Q_w,

$$Q_w = \frac{r_XV - QX_{Ve}}{X_{Vr} - X_{Ve}}$$

First r_XV will be determined.

$$r_XV = \left[\frac{0.48(0.95\text{ d}^{-1})\left(1\,835\,\frac{\text{mg}}{\text{L}}\right)\left(34\,\frac{\text{mg}}{\text{L}}\right)}{85\,\frac{\text{mg}}{\text{L}} + 34\,\frac{\text{mg}}{\text{L}}} - (0.07\text{ d}^{-1})\left(1\,835\,\frac{\text{mg}}{\text{L}}\right)\right] \times (550\text{ m}^3)\left(\frac{1\,000\text{ L}}{\text{m}^3}\right)\left(\frac{1\text{ kg}}{10^6\text{ mg}}\right) = 60.8\text{ kg/d}$$

In U.S. units:

$$r_X V = \left[\frac{0.48(0.95\ \text{d}^{-1})\left(1\,835\ \dfrac{\text{mg}}{\text{L}}\right)\left(34\ \dfrac{\text{mg}}{\text{L}}\right)}{85\ \dfrac{\text{mg}}{\text{L}} + 34\ \dfrac{\text{mg}}{\text{L}}} - (0.07\ \text{d}^{-1})\left(1\,835\ \frac{\text{mg}}{\text{L}}\right) \right]$$

$$\times (19\,430\ \text{ft}^3)\left(\frac{28.3\ \text{L}}{\text{ft}^3}\right)\left(\frac{1\ \text{lb}}{454 \times 10^3\ \text{mg}}\right) = 134\ \text{lb/d}$$

Substituting $r_X V$ in the equation for Q_w,

$$Q_w = \frac{r_X V - Q X_{Ve}}{X_{Vr} - X_{Ve}}$$

$$= \frac{\left(60.8\ \dfrac{\text{kg}}{\text{d}}\right)\left(\dfrac{10^6\ \text{mg}}{\text{kg}}\right)\left(\dfrac{1\ \text{m}^3}{1\,000\ \text{L}}\right) - \left(15\ \dfrac{\text{mg}}{\text{L}}\right)\left(1\,650\ \dfrac{\text{m}^3}{\text{d}}\right)}{5\,000\ \dfrac{\text{mg}}{\text{L}} - 15\ \dfrac{\text{mg}}{\text{L}}}$$

$$= 7.2\ \text{m}^3/\text{d}$$

In U.S. units:

$$Q_w = \frac{\left(134\ \dfrac{\text{lb}}{\text{d}}\right)\left(\dfrac{454 \times 10^3\ \text{mg}}{\text{lb}}\right)\left(\dfrac{1\ \text{gal}}{3.79\ \text{L}}\right) - \left(15\ \dfrac{\text{mg}}{\text{L}}\right)\left(0.436 \times 10^6\ \dfrac{\text{gal}}{\text{d}}\right)}{5\,000\ \dfrac{\text{mg}}{\text{L}} - 15\ \dfrac{\text{mg}}{\text{L}}}$$

$$= 1\,908\ \text{gal/d}$$

Q_r is only 0.4% of Q.

17.7 PLUG FLOW ACTIVATED SLUDGE TREATMENT

If the same substrate removal expression applies in a PF and a CM reactor, the PF system will be more efficient. However, effluent quality may be similar between these two systems because of production of slowly degraded compounds (Sykes, 1983) and the safety factor applied in the design. Byproducts are formed as a result of substrate removal and growth and products of microbial maintenance and cell lysis (these phenomena are endogenous decay phenomena). The balance between byproduct formation between these two routes depends on the sludge age and the HRT in the process. If the exogenous substrate contained in the influent is largely being removed, then the effluent quality will mostly consist of relatively nondegradable byproducts; the rate of production of soluble organics is primarily controlled by biomass level, which is controlled by sludge age. The hydraulic flow regime and raw substrate removal kinetics would have only a secondary influence on effluent quality. Various models have been proposed (Hao and Lau, 1988; Sykes, 1982) for byproduct formation but there is not a large amount of agreement at present.

Achievement of flow regimes near a PF condition is difficult because of the intense aeration and mixing in aerobic processes. From tracer tests Murphy and Boyko (1970) found that many aeration basins designed to be PF were actually nearer a CM flow regime. Baffles or separation of basin compartments are required to achieve PF conditions. PF conditions promote the growth of biomass that has good settling qualities (Chudoba et al., 1973).

A PF system provides the opportunity to tailor the oxygen input into the aeration basin to correspond to the demand as the sewage travels through the aeration basin. The oxygen demand will be higher at the influent end of the basin as readily degradable substrates are metabolized. At the effluent end of the basin the rate of oxygen uptake is lower because of the lower rate of endogenous metabolism and absence of readily degraded substrates. Operation of a system where the oxygen input is decreased along the aeration basin is a process variation known as tapered aeration.

Operation of conventional activated sludge systems in the tapered aeration mode has a concomitant benefit of improving the quality of the floc exiting the aeration basin and lowering the TSS concentration in clarified effluent (Das et al., 1993). This result is not surprising considering the performance of flocculation systems (Chapter 13), wherein decreasing velocity gradients produce the most highly flocculated effluent.

The PF configuration is shown in Fig. 17.1c. A PF basin will have recycled flow from the underflow of the secondary clarifier returned to the head of the basin. The flow rate into the basin will be

$$Q' = Q + rQ$$

where

Q is the flow rate into the system
Q' is the flow rate into the aeration basin

The substrate concentration at the beginning of the basin (S_i) depends on the recycle flow rate, which contains substrate at S_e and the influent flow rate and substrate concentration S_0. A simple mixing equation describes S_i. It is based on the mass balance at the confluence point of the influent and recycle lines.

$$S_i = \frac{QS_0 + rQS_e}{Q + rQ} = \frac{S_0 + rS_e}{1 + r} \tag{17.33}$$

The time required to achieve a given amount of substrate removal is determined by setting up a mass balance on an elemental volume in a PF reactor as outlined in Section 10.1.2. In a steady state process, for the elemental volume given in Fig. 10.3,

$$\text{In} - \text{Out} + \text{Generation} = \text{Accumulation}$$

$$Q'C - Q'\left(C + \frac{dC}{dx}\Delta x\right) + r_S \Delta V = 0 \tag{17.34}$$

Substituting $A\Delta x$ for ΔV and simplifying the mass balance,

$$\frac{Q'}{A}\frac{dS}{dx} = r_S$$

The appropriate substrate removal kinetic expression (Eqs. 17.1a–17.2b) is substituted for r_S and the equation is integrated. A change of variable from x to time t is made by replacing Q'/A (velocity) with dx/dt and applying the chain rule. Using Eq. (17.1a) as the kinetic model:

$$\frac{Q'}{A}\frac{dS}{dx} = \frac{dx}{dt}\frac{dS}{dx} = \frac{dS}{dt} = -\frac{k\overline{X}_V S}{K + S}$$

In PF reactors, for kinetic expressions that incorporate X_V (Eqs. 17.1a or 17.2a), the average VSS concentration in the aeration basin, $\overline{X}_V$, is used instead of X_V. The concentration of biomass increases at the front of the basin as substrate is metabolized. In the latter portion of the reactor, where substrate concentration is near exhaustion, VSS concentration will decrease because of endogenous decay. But the variation in concentration of VSS throughout the basin will not be sufficient to cause erroneous results in a typical activated sludge process (SRT >5 HRT) (Lawrence and McCarty, 1970) and the integration can be performed analytically.

Separating the variables in the above expression and integrating it,

$$K\int_{S_i}^{S}\frac{dS}{S}+\int_{S_i}^{S}dS=-k\overline{X}_V\int_0^{\theta_p}dt$$
$$K\ln\frac{S_e}{S_i}+S_e-S_i=-k\overline{X}_V\theta_p \tag{17.35}$$

where

S_i is the substrate concentration at the beginning of the basin
S_e is the substrate concentration at the effluent end of the basin
θ_p is the time for liquid to pass through the basin ($\theta_p = V/Q'$)

Note that the time of treatment is different from the liquid HRT which is $\theta_d = V/Q$ (as shown in Section 10.3). By substituting the relations for S_i and Q' into Eq. (17.35) the treatment can be related to the HRT and influent concentration, S_0.

The equations for sludge age and rate of recycle (Eqs. 17.28 and 17.32, respectively) are the same as developed for the CM process with recycle except that $\overline{X}_V$ is substituted for X_V. Also Eq. (17.29) applied to a PF reactor will yield $\overline{X}_V$ instead of X_V. The substrate balance expression cannot be replaced with the kinetic expression ($-r_S$) because of the variable substrate concentration in the aeration basin (see Problem 24).

17.8 VARIATIONS OF THE ACTIVATED SLUDGE PROCESS

Variation in flow regime is a fundamental process embodiment. Many other modifications may be incorporated into an activated sludge process.

17.8.1 Sequencing Batch Reactors

The operation of sequencing batch reactor (SBR) systems is described in Fig. 17.2. SBR systems are hybrid systems with some characteristics of continuous flow PF and CM systems but they have other characteristics that are truly unique. During the fill phase the reactor contents are mixed with the continuously incoming wastewater. Aeration may be delayed or not be applied at all during the fill period to improve sludge settleability by favoring the growth of microorganisms that form flocs (Section 17.12). During the react phase, although the reactor contents are mixed, the whole reactor contents remain in the tank for the specified duration of this phase. This is an ideal PF condition for this phase. At the end of the react phase, aeration and mixing are terminated. Ideal quiescent settling, not subject to flow currents or other irregularities, occurs.

In a SBR system the fill time (t_f) for an individual reactor depends on the available volume in the reactor. The available volume is the total volume of the reactor less the portion occupied by the settled sludge remaining in the reactor after the previous cycle. In equation form:

$$V_{AB} = \alpha V_B \tag{17.36}$$

$$t_f = \frac{V_{AB}}{Q} = \alpha \frac{V_B}{Q} \tag{17.37}$$

where

Q is the influent flow rate during the fill period

α is the unoccupied fraction of the total volume of the batch reactor at the beginning of the fill period

V_B is the volume of a batch reactor when it is empty

V_{AB} is the available volume at the beginning of the fill period

One complete cycle is composed of the fill, react, settle, draw, and idle periods.

$$t_c = t_f + t_r + t_s + t_d + t_i \tag{17.38}$$

where

the subscripts c, f, r, s, d, and i indicate cycle, fill, react, settle, draw, and idle, respectively

In a steady flow situation, each of the reactors will pass through the same cycle period and a quasi-steady state condition will be achieved for the system.

In a two-tank SBR system, while one reactor fills, the other reactor passes through the other periods of the cycle; in a three-tank system, after a reactor fills, it has two fill times to pass through the remaining periods of a cycle. The equation expressing the general relation is

$$(n - 1)t_f = t_r + t_s + t_d + t_i \tag{17.39}$$

where

n is the number of reactors in the SBR system

The cycle time and the durations of each period, in particular the fill time, and the available volume are the physical operating constraints on the system. Although there is flexibility in changing the duration of any phase, there are tradeoffs in increasing the duration of one phase at the expense of the others. Minimum settling times must be observed or sludge will wash out of the system and deteriorate effluent quality as well as reduce the inventory of sludge in the reactor.

The substrate concentration remaining at the end of the fill period is a variable function of the volume and the kinetic expression for substrate removal that applies during the fill period. The mass balance during the fill period is

$$\text{In} - \text{Out} + \text{Generation} = \text{Accumulation}$$

$$QS_0 - 0 + r_{Sf}V = \frac{d}{dt}(VS) \tag{17.40}$$

where

r_{Sf} is the rate of substrate removal during the fill period

Using Eq. (17.2b) for r_{Sf}, Eq. (17.40) becomes

$$QS_0 - kSV = S\frac{dV}{dt} + V\frac{dS}{dt}$$

The equation can be further simplified by recognizing that $dV/dt = Q$. Using this, dividing by V, and rearranging the equation,

$$\frac{dS}{dt} + \frac{Q}{V}S + kS = \frac{Q}{V}S_0 \tag{17.41}$$

Q/V is $1/t$. At the beginning of the fill period, the reactor contains a volume V_I and $t = V_I/Q$; at the end of the fill period $t = V_B/Q$. The differential equation to be solved is

$$\frac{dS}{dt} + \frac{S}{t} + kS = \frac{S_0}{t} \tag{17.42}$$

The equation is solved using an integrating factor. The solution of the equation is

$$S = e^{-\int\left(\frac{1}{t}+k\right)dt}\int \frac{S_0}{t} e^{\int\left(\frac{1}{t}+k\right)dt}\,dt + Ce^{-\int\left(\frac{1}{t}+k\right)dt}$$

where

C is an integrating constant

$$S = S_0\frac{e^{-kt}}{t}\int e^{kt}\,dt + \frac{C}{t}e^{-kt}$$

Performing the integration,

$$S = \frac{S_0}{kt} + \frac{C}{t}e^{-kt} \tag{17.43}$$

Using the initial condition, $t = V_I/Q$ and $S = S_e$:

$$S_e = \frac{S_0Q}{kV_I} + \frac{CQ}{V_I}e^{-k(V_I/Q)} \Rightarrow C = \left(\frac{V_I}{Q}S_e - \frac{S_0}{k}\right)e^{k(V_I/Q)}$$

Making the substitution for C, the final equation becomes

$$S = \frac{S_0}{kt} + \left(\frac{V_I}{Q}S_e - \frac{S_0}{k}\right)\frac{1}{t}e^{k\left(\frac{V_I}{Q}-t\right)} \tag{17.44}$$

The substrate concentration at the end of the fill period can be found by using

$t_f = V_B/Q \qquad S = S_f$

$$S_f = \frac{S_0Q}{kV_B} + \left(\frac{V_I}{Q}S_e - \frac{S_0}{k}\right)\frac{Q}{V_B}e^{k\left(\frac{V_I}{Q}-\frac{V_B}{Q}\right)} = \frac{S_0Q}{kV_B} + \left(\frac{V_I}{V_B}S_e - \frac{QS_0}{V_Bk}\right)e^{k\left(-\frac{V_{AB}}{Q}\right)} \tag{17.45}$$

For most of the substrate removal expressions a numerical solution is required to solve the differential equation describing substrate removal during the fill period. Droste (1990) has evaluated physical and reaction constraints on SBR systems for various models and compared them to other activated sludge processes.

Calculating the amount of removal during the react period is straightforward. The equation for substrate removal is integrated directly as in the case for PF. Using Eq. (17.2b),

$$\frac{dS}{dt} = -kS \Rightarrow \int_{S_f}^{S_e}\frac{dS}{S} = -\int_0^{t_r}dt \qquad \text{or} \qquad S_e = S_fe^{-kt_r}$$

where
S_f is substrate concentration at the end of the fill period
S_e is the substrate concentration at the end of the react period

The amount of reaction time is found directly from Eq. (17.39) once the durations of the other periods are established. The concentration of substrate at the end of the react period is the effluent concentration. Because the feast–famine feed cycle in a SBR process promotes the growth of microorganisms that settle well and there are no currents from inflow or outflow during the settle phase, the amount of time for settling is likely to be substantially less than detention times used in conventional continuous flow secondary clarifiers.

The draw time depends on the design of the drawoff intake device and the disturbance caused as the water leaves the reactor. Many intake devices float. The solids do not need to be completely settled to begin drawoff of the clarified supernatant with this type of intake.

The sludge age in a SBR system is based on the principles discussed earlier. The effective sludge age depends on the time during which the biomass is active. Biomass activity occurs during the fill and react periods when food is available and other conditions favorable for metabolism exist. During the settle and draw phases, substrate should be exhausted and oxygen supply is negligible.

Sludge age is calculated based on the average amount of solids in the SBR. The actual sludge residence time in the system is

$$\theta_T = \frac{V_B \overline{X_V}}{(QX_V)_w} \tag{17.46}$$

where
$\overline{X_V}$ is the average concentration of sludge in the reactor. $\overline{X_V}$ is calculated by taking the average of the amount of sludge present at the beginning of the fill period divided by the total volume of the reactor and the concentration of sludge present at the end of the react period.
$(QX_V)_w$ is the amount of sludge wasted (in the clarified effluent and the concentrated sludge drawn off during the draw or idle period) on a daily basis

The average amount of solids will be slightly different from the value obtained using this procedure because of the varying amount of sludge in the reactor during the react phase resulting from growth and endogenous decay and, also, solids loss in the clarified effluent during the draw phase will cause the amount of solids in the reactor to vary during this phase. However, these variations will be small and the procedure just given is a practical approximation to the average amount of biomass in the system.

The effective sludge age depends on the fraction of the cycle time in the fill and draw periods. The actual sludge residence time is reduced by this fraction to obtain the effective sludge age.

$$\theta_X = \frac{t_f + t_r}{t_c}\,\theta_T \tag{17.47}$$

SBR systems are sometimes operated with an anoxic or anaerobic fill period. However, substrate is being supplied to the biomass during this period and the substrate will at least undergo the initial stages of metabolism, where it is adsorbed and absorbed by the biomass and some extracellular enzyme activity begins. Therefore, the fill period should be included in the effective sludge age when the process is operated in this manner.

Consistent definitions of sludge age have not been used in the literature. The reader is cautioned to check the definitions of sludge age used when making comparisons among different studies.

17.8.2 Fixed Film Activated Sludge Processes

Incorporating fixed medium into the reactor provides a surface on which microorganisms can attach and grow. The microorganisms on the medium surface are not washed out of the system with the liquid and higher sludge ages and biomass concentrations are obtained in the reactor.

Submerged fixed media will be used with aeration supplied by diffused aerators. Mass balances for substrate and biomass are made in the normal manner. Hamoda (1989) studied a bench-scale system fed a synthetic, sugar-based waste. Biomass washout was determined by plotting HRT versus SRT at different influent COD concentrations. SRT depends on attached and suspended biomass in the system. The slope of the best fit straight line for an influent COD concentration is the washout factor, F.

$$F = \frac{\Delta \text{HRT}}{\Delta \text{SRT}} \tag{17.48}$$

The washout factor was then plotted against influent COD concentration and a linear relation ($F = 2.2 \times 10^{-4} S_0$) was found.

17.8.3 Extended Aeration

Activated sludge processes with long HRTs (on the order of 24 h) are referred to as "extended aeration" processes. A high sludge age is also maintained in this process. Extended aeration processes are typically designed to handle domestic waste from small communities. Often the operator will not be present on a continuous basis.

The amount of sludge generated from low flow rates does not generate enough sludge to be wasted on a regular basis. Also, the high sludge age gives rise to more endogenous decay of sludge. Sludge is allowed to accumulate in the system until the sludge blanket in the clarifier rises to the point where excessive solids begin to wash out with the clarified effluent. The time between sludge removals may be 6 months or more.

The process is a continuous flow process where the contents of the aeration basin are CM. The long aeration time ensures a high removal of degradable organics; however, the high sludge age promotes the formation of pinpoint floc, which does not settle as well as zoogleal floc formed at lower sludge ages (see Fig. 17.4). Consequently, effluent quality deteriorates somewhat because of the presence of unsettled fines. The secondary clarifier may be a separate unit from the aeration basin or an aeration–clarifier unit (Fig. 17.5) may be supplied. The clarifier is separated from the aeration basin by a solid baffle wall as shown in the figure. The mixing pattern in the reactor causes the recycle of solids from the clarifier.

The process operation provides a practical solution for wastewater treatment for small communities, schools, isolated residences (such as hospitals or retirement complexes), or other small operations where a full-time attendant is not warranted.

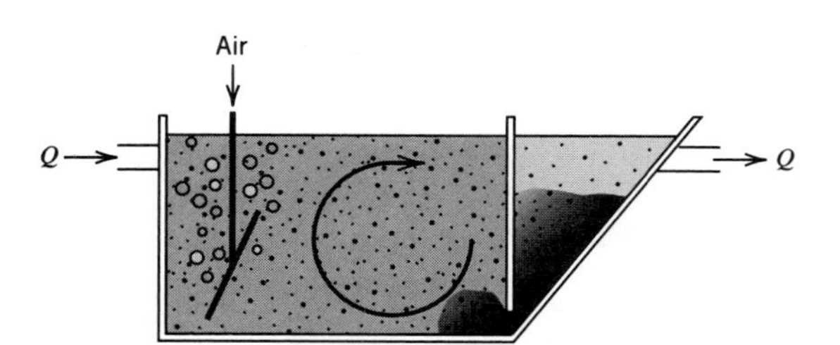

Figure 17.5 Extended aeration system with internal clarifier.

A high sludge age and a long average hydraulic time enable the process to cope with more highly variable or intermittent flows from small operations and to maintain stable treatment. In many extended aeration plants the influent to the aeration basin is not settled in a primary clarifier. Many commercial manufacturers supply extended aeration package plants.

17.8.4 Other Process Variations

Other process variations have been utilized for various situations. The step aeration process depicted in Fig. 17.6 is a process modification that changes the environment in a PF reactor to an environment that would be observed in a CM reactor. By introducing the influent along several points in the reactor, more uniform environmental conditions are made to occur in the reactor. In particular, loading rates and oxygen demand rates are, respectively, approximately the same at any point in the reactor. The flow regime remains PF in the ideal case.

If a conventional process is operated with shorter detention times and higher F : M ratios it is referred to as a modified aeration process. Removal efficiencies in this process are lower. However, if high mixed liquor suspended solids (MLSS) concentrations are maintained with low detention times, process performance improves. This variation is known as a high rate process. F : M ratios are intermediate between the conventional and modified aeration processes.

The contact stabilization process (Fig. 17.7a) economizes the volume required for treatment. The influent is introduced into a small reactor or contact tank that also receives biomass in a starved condition. Soluble substrate is readily adsorbed and absorbed by the starved biomass. The mixed liquor leaving the contact reactor is settled and the biomass is concentrated. The biomass is then sent to an aeration basin,

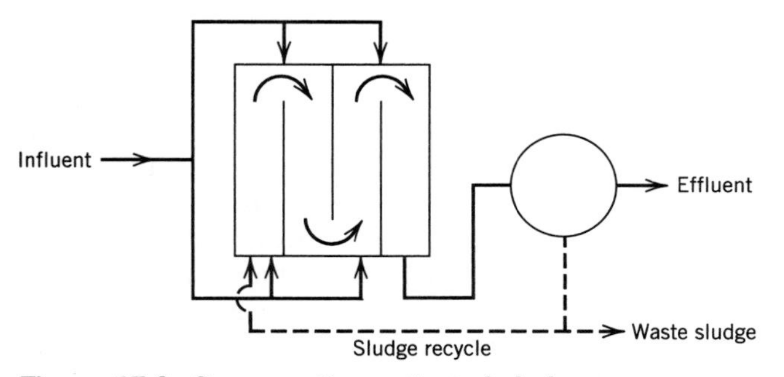

Figure 17.6 Step aeration activated sludge process.

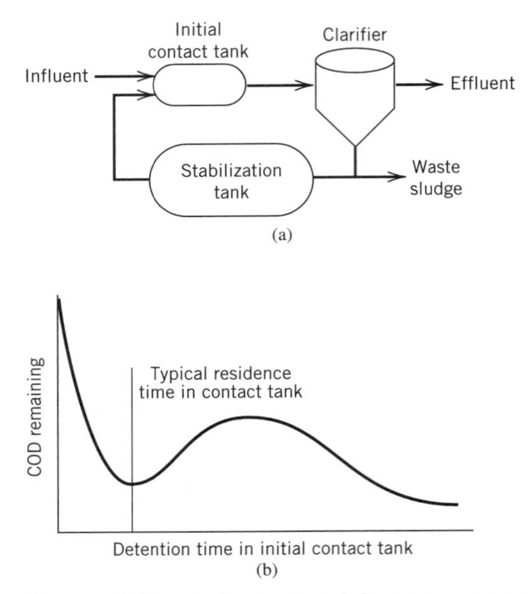

Figure 17.7 (a) Contact stabilization process. (b) COD uptake in the contact tank.

where it resides for a much longer time and metabolism and stabilization of the captured substrate occurs.

The key to a successful operation is the residence time in the contact chamber. After the substrate is taken up by the microorganisms, metabolism initiates the release of secondary byproducts and results in a typical COD variation with time as shown in Fig. 17.7b. Over- or underdesign of the contact time will result in effluent quality deterioration. Contact basin retention times are in the range of 0.5–1 h (Metcalf and Eddy, 1991). Aeration times for the sludge in the stabilization basin are typically in the range of 3–6 h.

Effluent quality from this process will not be as high as that achieved in conventional processes because of the nature of the interplay between uptake and release of COD in the contact phase and release of small amounts of COD from the biomass while it is in the clarifier. However the volume requirements of the process will be approximately 50% of a conventional process (Metcalf and Eddy, 1991).

The deep shaft activated sludge process (Fig. 17.8) is another variation that is used where costs of land are high. There are a number of installations in Japan. These reactors may extend 150 m (500 ft) into the ground (Sandford and Chisolm, 1977). A mixture of sewage and air travels down in the downcomer and upward in a riser. The higher pressures achieved in the shaft improve oxygen transfer rates and more oxygen can be dissolved with beneficial effects on substrate stabilization. DO levels can range from 25 to 60 mg/L in the shaft. Microorganism metabolic rates are not affected by the high pressures. The nature of flow in the reactor is PF (Sandford and Chisolm, 1977). Temperature is nearly constant throughout the year in deep shaft reactors.

Separation of the biomass from the mixed liquor is amenable to flotation (see Section 20.6.2). The effluent from the deep shaft reactor is supersaturated with air when exposed to atmospheric pressure. The release of dissolved air from the mixed liquor forms bubbles that float suspended solids to the surface.

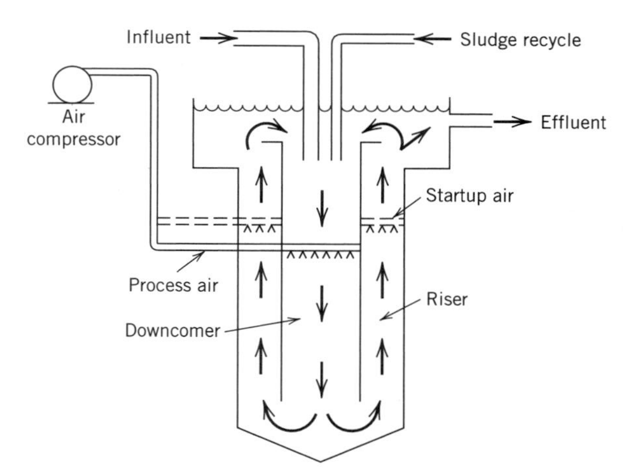

Figure 17.8 Deep shaft activated sludge process.

Pure Oxygen Activated Sludge Process

Pure oxygen activated sludge processes are conventional activated sludge configurations but relatively pure oxygen as opposed to air is supplied to the aeration basin. Larger plants (>40 000 m^3/d or 10.5 Mgal/d) generate the oxygen cryogenically whereas smaller plants use a selective sorption process, pressure swing adsorption. Because the gas flow rates are significantly reduced, supplemental mixing must be supplied in the mixed liquor to keep solids in suspension. The aeration basins are also covered to minimize the escape of the oxygen-rich gas above the basin; however, some gas must be exhausted to remove carbon dioxide produced in the process. Nevertheless, there will be an accumulation of CO_2 in the mixed liquor, which results in a decrease of pH. Buffering may be required with alkaline agents, particularly if nitrification is desired.

Higher DO concentrations will be maintained in the aeration basin compared to an air activated sludge process. The generally accepted minimum DO concentration for an activated sludge process is 2.0 mg/L (Parker and Merrill, 1976). At this concentration mass transfer of oxygen to the flocs will not be a limiting factor. Higher DO concentrations in a pure oxygen activated sludge process will have some influence on the flora that are established.

Powdered Activated Carbon Activated Sludge Process

Carbon may be added to a conventional activated sludge process to enhance treatment, particularly with respect to removal of recalcitrant and toxic substances. The process is known as the PAC (or PACT) process because powdered activated carbon is used (Fig. 17.9). In addition to adsorbing various constituents, the carbon provides a support surface for microorganisms. The adsorbed compounds are exposed to the biomass on the carbon surface for the sludge age as opposed to the HRT. Sludge production is higher in a PACT process because of carbon additives.

Design Parameters and Operating Conditions for Activated Sludge Processes

Typical design parameters and operating conditions for various modifications of activated sludge processes are summarized in Table 17.3. The ranges given are suitable

TABLE 17.3 Design Parameters and Operating Conditions for Activated Sludge Processes[a]

Process modification	θ_X, d	F:M kg BOD_5/kg MLVSS-d	Loading rate kg BOD_5/m^3-d	θ_d, h	MLTSS mg/L	r, [0]
Conventional	5–15	0.2–0.4	0.3–0.6	4–8	1 500–3 000	0.25–0.5
Tapered aeration	5–15	0.2–0.4	0.3–0.6	4–8	1 500–3 000	0.25–0.5
Step aeration	5–15	0.2–0.4	0.6–1.0	3–58	2 000–3 500	0.25–0.75
Modified aeration	0.2–0.5	1.5–5.0	1.2–2.4	1.5–3	200–500	0.05–0.15
Contact stabilization	5–15	0.2–0.6	1.0–1.2	0.5–1.0[b] 3–6[c]	1 000–3 000[b] 4 000–9 000[c]	0.5–1.5
Extended aeration	20–30	0.05–0.15	0.1–0.4	18–36	1 500–5 000	0.5–1.5
High rate aeration	5–10	0.4–1.5	1.6–16	2–4	3 000–6 000	1–5
Pure oxygen	3–10	0.25–1.0	1.6–3.3	1–3	3 000–8 000	0.25–0.5

[a]From Metcalf and Eddy (1991), *Wastewater Engineering: Treatment, Disposal, Reuse,* 3rd ed., G. Tchobanoglous and F. L. Burton, eds., McGraw-Hill, Toronto, used with permission of McGraw-Hill, Inc.

[b]Contact unit.

[c]Solids stabilization unit.

Figure 17.9 Bench scale CM (left) and PAC (right) activated sludge systems.

for a wide variety of wastewaters; however, some industrial wastewaters may require longer HRTs.

17.9 SLUDGE PRODUCTION IN ACTIVATED SLUDGE SYSTEMS

The sludge produced in an activated sludge process or any biological treatment process is a function of the substrate characteristics, sludge age, and other environmental conditions, particularly temperature of the mixed liquor. The rate of production of biological solids is given by Eqs. (17.4) or (17.27). Note that these equations describe the total production of biological solids, which appear in the effluent from the secondary clarifier and in the waste sludge line from the underflow of the secondary clarifier. An observed yield factor (a net yield factor as discussed in Section 17.3.2) can be formulated for a process at given operating conditions.

$$P_{\mathrm{Xb}} = (-Yr_{\mathrm{S}} - k_{\mathrm{e}}X_{\mathrm{V}})V = -Y_{\mathrm{obs,b}}r_{\mathrm{S}}V \tag{17.49}$$

where

P_{Xb} is rate of biological sludge production
$Y_{\mathrm{obs,b}}$ is the observed yield factor for biological solids

Sludge age affects the concentration of MLVSS and temperature and other environmental factors affect both the yield factor and the endogenous decay coefficient.

The rate of total solids production from a biological treatment process will be higher than the rate calculated from Eq. (17.49) because this equation only describes biological VSS production and there is normally a significant amount of solids in the influent to the aeration basin. Degradation of influent solids has been discussed in Section 17.6.2. The rate of production of solids from a biological process is

$$P_{\mathrm{Xt}} = -\frac{Yr_{\mathrm{S}} + k_{\mathrm{e}}X_{\mathrm{V}}}{f_{\mathrm{b}}}V + Qf_{\mathrm{t}}X_{\mathrm{T0}} = -Y_{\mathrm{obs,t}}r_{\mathrm{S}}V \tag{17.50}$$

where

P_{Xt} is rate of sludge production
f_b is the ratio of VSS:TSS for biological solids
f_t is the fraction of influent TSS that was not degraded
$Y_{obs,t}$ is the observed yield factor for total solids

Studies are needed to determine the factor f_t in Eq. (17.50). It is a function of the degradable portion of the influent VSS and the VSS:TSS ratio for the influent. An example will illustrate the factors on which f_t depends.

■ Example 17.3 Sludge Production in an Activated Sludge Process

An activated sludge process is treating primary settled effluent that has a BOD_5 of 150 mg/L and TSS concentration of 90 mg/L in a flow of 25 000 m^3/d (6.61 Mgal/d). The detention time in the aeration basins is 5 h. MLVSS is 2 100 mg/L and the effluent contains 20 mg/L of BOD_5. The yield factor is 0.68 mg VSS/mg BOD_5 and the endogenous decay coefficient is 0.05 d^{-1}. The VSS:TSS ratios for biological solids and influent SS are 0.80 and 0.50, respectively. Find the observed biological yield factor and the range of the observed yield factor for total solids. Also find the possible range for rate of total solids production.

The rate of substrate removal (Eq. 17.15) is

$$r_S = -\frac{(S_0 - S_e)}{\theta_d} = -\frac{(150 - 20)\ \text{mg/L}}{5\ \text{h}}\left(\frac{24\ \text{h}}{\text{d}}\right) = -624\ \text{mg/L/d}$$

The volume of the aeration basins is

$$V = Q\,\theta_d = \left(25\,000\,\frac{\text{m}^3}{\text{d}}\right)(5\ \text{h})\left(\frac{1\ \text{d}}{24\ \text{h}}\right) = 5\,208\ \text{m}^3 = 5.21 \times 10^6\ \text{L}$$

Applying Eq. (17.49),

$$Y_{obs,b} = \frac{Yr_S + k_e X_V}{r_S} = \frac{0.68(-624\ \text{mg/L/d}) + 0.05\ \text{d}^{-1}\,(2\,100\ \text{mg/L})}{-624\ \text{mg/L/d}}$$
$$= 0.512\ \text{mg VSS/mg BOD}_5$$

The rate of biological solids production is

$$P_{Xb} = -Y_{obs,b}r_S V = -\left(0.512\,\frac{\text{mg VSS}}{\text{mg BOD}_5}\right)\left(-624\,\frac{\text{mg BOD}_5\text{/L}}{\text{d}}\right)(5.21 \times 10^6\ \text{L})\left(\frac{1\ \text{kg}}{10^6\ \text{mg}}\right)$$
$$= 1.67 \times 10^3\ \text{kg/d}$$

If none of the influent VSS were degradable then $f_t = 1$ in Eq. (17.50) and the maximum theoretical solids production would occur.

$$Y_{obs,t} = \frac{Yr_S + k_e X_V}{f_b r_S} - \frac{Qf_t X_{T0}}{r_S V} = \frac{0.512\ \text{mg VSS/mg BOD}_5}{0.80\ \text{mg VSS/mg TSS}} - \frac{(1)(90\ \text{mg TSS/L})}{(5\ \text{h})(-624\ \text{mg/L/d})}\left(\frac{24\ \text{h}}{\text{d}}\right)$$
$$= 0.640\ \text{mg TSS/mg BOD}_5 + 0.692\ \text{mg TSS/mg BOD}_5 = 1.332\ \text{mg TSS/mg BOD}_5$$

$$P_{Xt} = -Y_{obs,t}r_S V = -\left(1.332\,\frac{\text{mg TSS}}{\text{mg BOD}_5}\right)\left(-624\,\frac{\text{mg BOD}_5\text{/L}}{\text{d}}\right)(5.21 \times 10^6\ \text{L})\left(\frac{1\ \text{kg}}{10^6\ \text{mg}}\right)$$
$$= 4.33 \times 10^3\ \text{kg/d}$$

The other extreme for degradation of the influent VSS is that all influent were degradable suspended substrate that were transformed into biological solids. To solve the equations it is also necessary to know the contribution of these degradable VSS to the influent BOD_5. Commensurate with the assumption that all VSS were metabolized in the process it will be assumed that the influent VSS fully contributed to the influent BOD_5.

The influent VSS concentration is 0.50(90 mg/L) = 45 mg/L.

It is also necessary to know the VSS : TSS ratio of the degradable VSS. It will not be 0.50 and a value of 0.75 will be used (the ratio of degradable VSS to TSS in the influent is not necessarily the same as the ratio of VSS : TSS for biological solids). Now the degradable VSS is associated with (45 mg/L)/0.75 = 60 mg TSS/L. Besides the 15 mg/L of suspended inorganic solids associated with the biomass there would be 30 mg/L ISS in the influent. Some of the influent ISS could be solubilized in the process but it will be assumed that no ISS has been solubilized. The factor f_t in Eq. (17.50) can now be determined as

$$f_t = 30/90 = 0.333$$

The VSS component of influent TSS is implicitly contained in the influent BOD_5. The inorganic solids associated with the influent degradable VSS are assumed to be released with metabolism but the formation of biomass from metabolism of the degradable VSS will reassimilate inorganics. The observed yield factor and rate of solids production are

$$Y_{obs,t} = \frac{Yr_S + k_e X_V}{f_b r_S} - \frac{Qf_t X_{T0}}{r_S V}$$

$$= 0.640 \text{ mg TSS/mg BOD}_5 - \frac{(0.333)(90 \text{ mg TSS/L})}{(5 \text{ h})(-624 \text{ mg/L/d})}\left(\frac{24 \text{ h}}{\text{d}}\right)$$

$$= 0.640 \text{ mg TSS/mg BOD}_5 + 0.231 \text{ mg TSS/mg BOD}_5 = 0.871 \text{ mg TSS/mg BOD}_5$$

$$P_{Xt} = -Y_{obs,t} r_S V = -\left(0.871 \frac{\text{mg TSS}}{\text{mg BOD}_5}\right)\left(-624 \frac{\text{mg BOD}_5/\text{L}}{\text{d}}\right)(5.21 \times 10^6 \text{ L})\left(\frac{1 \text{ kg}}{10^6 \text{ mg}}\right)$$

$$= 2.83 \times 10^3 \text{ kg/d}$$

The actual solids production rate will lie between 2.83×10^3 and 4.33×10^3 kg/d. There are numerous factors to be determined in arriving at the actual rate of solids production. Observed yield factors can be calculated from pilot studies and with sufficient measurements the values of the factors can be assessed. Their values are of academic interest because the overall yield factor is known. But this example illustrates the phenomena that are involved.

Observed yield factors for domestic wastewater with primary sedimentation range from 0.33 to 0.88 mg TSS/mg BOD_5; without primary sedimentation the range is 0.62–1.18 mg TSS/mg BOD_5 for sludge ages up to 30 d and temperatures between 10 and 30°C (WEF and ASCE, 1992a). Processes that operate at high sludge ages such as extended aeration have lower observed yield factors than processes operated at lower sludge ages. Higher temperatures result in lower observed yield factors than lower temperatures.

17.10 DESIGN OF ACTIVATED SLUDGE PROCESSES FOR NITROGEN AND PHOSPHORUS REMOVAL

Specialized treatment processes can be designed to provide higher removals of nitrogen and phosphorus. These processes are more complex than activated sludge systems designed to remove carbonaceous BOD. They incorporate anaerobic–anoxic–aerobic sequences that favor the growth and metabolism of organisms responsible for nitrogen removal and phosphorus uptake.

17.10.1 Enhanced Nitrogen Removal

At higher sludge ages and longer detention times a typical activated sludge process will produce a nitrified effluent. Denitrification occurs when certain microorganisms (e.g., *Pseudomonas denitrificans*) in an environment devoid of oxygen use nitrate as an electron acceptor to oxidize organic matter. Nitrate is reduced to nitrogen gas. The reactions describing nitrogen transformations in sewage treatment are as follows:

$$\text{Organic matter} + O_2 \rightarrow NH_3 + \text{microorganisms} + CO_2 \tag{17.51a}$$

$$NH_3 + 2O_2 \rightarrow NO_3^- + H^+ + H_2O + \text{microorganisms} \tag{17.51b}$$

$$\text{Organic matter} + NO_3^- \rightarrow N_2 + \text{microorganisms} \tag{17.51c}$$

Ammonia is produced from the decomposition of organic matter. If the sludge age is high enough and other environmental conditions are suitable, nitrifying bacteria establish themselves and convert the ammonia to nitrates (also see Section 5.9.3). Oxygen requirements for the process rise because of nitrification.

The autotrophic nitrifiers have a lower growth rate than heterotrophic bacteria responsible for the removal of carbonaceous BOD. Nitrifier growth rates are strongly dependent on temperature and other environmental variables. The interaction of carbonaceous BOD removal and ammonia production followed by conversion to nitrate is complex and various expressions and values for constants appear in the literature. The nitrifiers are more fastidious in their environmental requirements than other organisms in an activated sludge process. The USEPA (1993) discusses the theory and design approaches in detail. The minimum sludge age used to ensure nitrification at average conditions is 7 d at 10°C (USEPA, 1993). Nitrification–denitrification has been demonstrated to occur at temperatures as low as 2°C (Oleskiewicz and Berquist, 1988). Rates for nitrification and denitrification at 2°C were about one fifth and one fourth, respectively, of rates observed at 15°C.

The minimum DO concentration is 0.5 mg/L, although DO concentrations below 2.5 mg/L may limit the rate of nitrification. Nitrification consumes alkalinity, as shown in Eq. (17.51b), and sufficient buffering capacity should be present to maintain pH in the range of 6.5–8.0. There is a rather abrupt drop in rate of nitrification outside of this range.

Nitrification may be coupled with carbonaceous BOD removal in the same aeration basin or an additional basin with its own clarifier may be used for nitrification as illustrated in Fig. 17.10. Providing means for fixed film growth in CM activated sludge reactors significantly improves settling properties of the sludge and makes nitrification efficiency independent of the SRT or suspended biomass (Wanner et al., 1988).

A separate reactor is used to accomplish denitrification. When a denitrifying reactor follows a conventional reactor, the influent to the basin contains a high amount of nitrate and very little degradable carbon. Therefore, addition of organic matter is

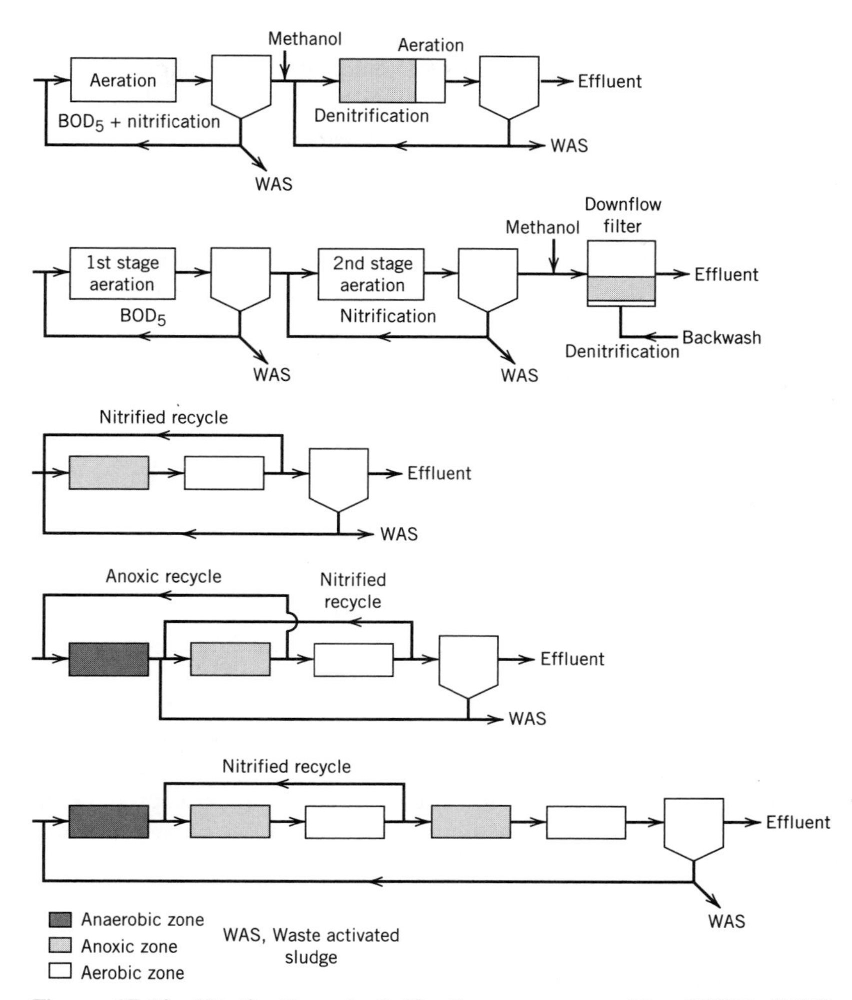

Figure 17.10 Nitrification–denitrification processes. After USEPA (1993).

required for denitrification. The organic matter is metabolized by the denitrifiers with no impairment of effluent soluble organics unless an excess is added. Any readily degradable carbon source can serve as substrate for the denitrifiers. Ease of degradation and expense are the primary considerations of supplemental carbon sources. Organics can be added by directing some influent to the denitrification basin. Methanol is a common substrate; other carbon sources that have been used are ethanol, molasses, and acetate.

The amount of carbon consumed by oxidation with nitrite or nitrate can be calculated by using half-reactions 26 and 27 in Table 1.3 and the half-reaction for the carbon source. Example 17.4 illustrates the procedure. This does not yield the total carbon required for the process, as discussed after the example, but it does provide a baseline for the amount of carbon to be added.

No oxygen is added to the reactor, which causes microorganisms to shift to metabolic pathways using nitrate as an electron acceptor. The water is in an anoxic state. If oxygen or nitrite is present in the influent to the denitrification reactor, then the amount of carbon source must be increased to reduce these components. The carbon requirements for oxidation of these components are calculated in a manner similar to the procedure in Example 17.4 using the appropriate half-reactions.

■ Example 17.4 Methanol Demand for Oxidation with Nitrate

How much methanol is consumed by oxidation with 30 mg/L of NO_3^-?

From the principles in Section 1.5, the half-reaction for methanol can be formulated as

$$\tfrac{1}{6}CO_2 + H^+ + e^- = \tfrac{1}{6}CH_3OH + \tfrac{1}{6}H_2O$$

which is combined with half-reaction 27 from Table 1.3 to find the overall reaction as

$$\tfrac{1}{6}CH_3OH + \tfrac{1}{5}NO_3^- + \tfrac{1}{5}H^+ \rightarrow \tfrac{1}{10}N_2 + \tfrac{1}{6}CO_2 + \tfrac{13}{30}H_2O$$

The molecular weight of nitrate is 48 and the molecular weight of methanol is 32. Therefore, 5.33 mg of CH_3OH are required for 9.60 mg of NO_3^-. The amount of methanol consumed is

$$M = \left(\frac{5.33}{9.60}\right) 30 \text{ mg/L} = 16.7 \text{ mg/L}$$

In addition to the carbon consumed by microbial oxidation with nitrate, nitrite, or oxygen, carbon must be supplied for growth of the microorganisms; i.e., a portion of the carbon source is oxidized to provide energy for the microorganisms to incorporate the remainder of the carbon source into new cells. The commonly accepted overall stoichiometries of reactions for methanol involving the three oxidation agents are as follows (McCarty et al., 1969):

$$1.08CH_3OH + NO_2^- + 0.24H_2CO_3 \rightarrow 0.056C_5H_7NO_2 + 0.47N_2 + HCO_3^- + 1.68H_2O \quad (17.52a)$$

$$0.67CH_3OH + NO_3^- + 0.53H_2CO_3 \rightarrow 0.04C_5H_7NO_2 + 0.48N_2 + HCO_3^- + 1.23H_2O \quad (17.52b)$$

$$0.93CH_3OH + O_2 + 0.056NO_3^- \rightarrow 0.056C_5H_7NO_2 + 0.48N_2 + 0.59H_2CO_3 + 1.04H_2O \quad (17.52c)$$

The overall amount of methanol required (M) can be calculated from Eqs. (17.52a–c) as

$$M = 2.47(NO_3^-\text{-N}) + 1.53(NO_2^-\text{-N}) + 0.87O_2 \quad (17.53)$$

For carbon sources other than methanol, the amount of carbon source required can be estimated by multiplying the amount of methanol by the ratio of the COD of the carbon source to the COD of methanol.

The redox reaction for denitrification with methanol can be rearranged to

$$\tfrac{5}{6}CH_3OH + NO_3^- + \tfrac{1}{6}H_2CO_3 \rightarrow \tfrac{1}{2}N_2 + HCO_3^- + \tfrac{4}{3}H_2O$$

which clearly illustrates that alkalinity is produced. Most of the carbon source will be consumed by denitrification as opposed to biomass synthesis. Denitrification in general results in a net production of alkalinity. The optimal pH range for denitrification is 6.0–8.0 (USEPA, 1993).

A 2 to 3-h retention time is typical in a suspended growth denitrification reactor. The sludge age in a denitrifying reactor must be beyond the minimum age of 1–2.5 d required to produce a flocculating sludge (USEPA, 1993). An alternative to a suspended growth process is the proprietary downflow packed-bed filter reactor of Tetra Technologies. The filter is similar to a gravity filter and contains coarse, round, high-density medium to which denitrifying bacteria attach and grow. The medium also filters out solids and eliminates the need for a secondary clarifier. The filter is backwashed for a few seconds every 4–8 h to remove accumulated nitrogen gas. The filter is backwashed for a longer period of time every 1–5 d to remove solids and excess biomass growth without totally cleaning the media.

Loading rates to the filter are generally in the range of 58–117 $m^3/m^2/d$ (1 400–2 900 $gal/ft^2/d$), with an empty bed contact time of 30 min or greater (USEPA, 1993). The fixed medium retains biomass well in excess of minimum sludge ages required for denitrification.

Effluent from a separate suspended growth denitrification reactor is usually freshened by aeration for a short period of time from 20 to 60 min (USEPA, 1993). Aeration promotes removal of excess carbon source added for denitrification and improves settleability of the sludge by stripping nitrogen gas.

Oxygen does not repress denitrifying enzymes in activated sludge processes (Simpkin and Boyle, 1988). The versatility of microorganisms is illustrated by *Nitrobacter* species, which are capable of growing effectively in the absence of oxygen, and some strains are able to use nitrate as an electron acceptor and soluble organic substances as a carbon source provided that nitrite concentrations remain below 23 mg/L (Bock et al., 1988). Therefore, single sludge nitrification–denitrification systems are another alternative (see the last three processes in Fig. 17.10). In these systems nitrified effluent from a later stage in the process is recycled to an anoxic stage near the beginning of the process. Carbonaceous BOD in the influent serves as the carbon source for denitrification in the reactor. There are a variety of single sludge systems in addition to those shown in Fig. 17.10. The design and operation of these systems relies on the principles just given. The USEPA (1993) gives a thorough discussion of these systems. SBR systems can also be operated to produce nitrification–denitrification.

17.10.2 Enhanced Phosphorus Uptake

There are bacteria with the ability to accumulate phosphorus in the form of polyphosphates well in excess of the phosphorus requirements for growth of microorganisms. Conventional activated sludge biomass typically contains 1–2% phosphorus on a dry weight basis, whereas biomass in an enhanced phosphate removal process is capable of accumulating phosphorus in excess of 3%; in some cases phosphorus contents up to 18% have been obtained with artificial, tailored substrates (Appeldoorn et al., 1992). The highest phosphorus concentration found in the biomass with domestic sewage as a substrate is near 7%. The microbiological and chemical processes that lead to enhanced phosphorus uptake are not clear (Bark et al., 1992) but the operational features of an enhanced phosphorus uptake process are known.

The essential features of the process are an anaerobic phase followed by an aerobic phase. It is generally thought that microorganisms are responsible for the phenomenon (bio-P microorganisms) but chemical precipitation of phosphorus may be a significant factor (Bark et al., 1992). The most commonly implicated species are from the genus *Acinetobacter* but other related species may be involved. Operation of the process with anaerobic–aerobic sequencing provides favorable conditions for enrichment of

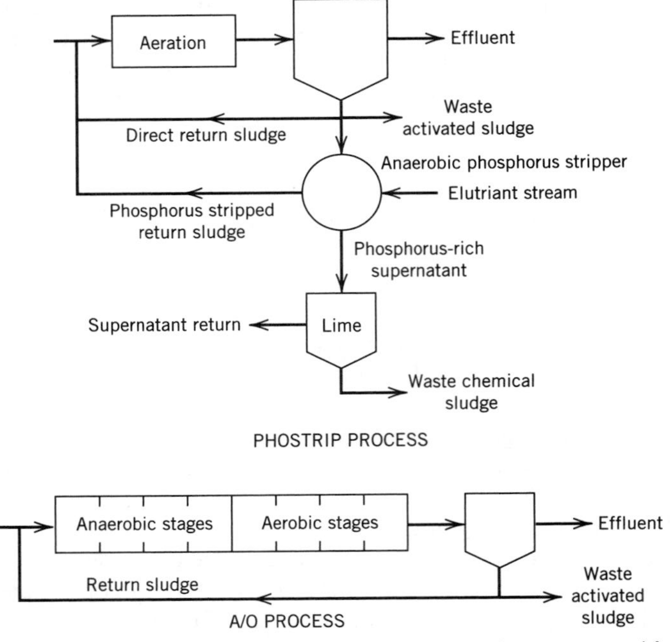

Figure 17.11 Enhanced phosphorus removal processes. After USEPA (1987).

the sludge with bio-P microorganisms. It is hypothesized that in the presence of short chain fatty acids under aerobic conditions the bio-P microorganisms are able to store polyphosphates as a phosphorus source and for energy generation. The initial anaerobic phase is required to produce short chain acids. Phosphorus is released from the sludge during the anaerobic phase but the released phosphorus is taken up later in the process. These acids are able to be utilized by the bio-P microorganisms with concomitant phosphorus removal in a subsequent aerobic reactor. The phosphorus-rich sludge formed is settled and removed from the wastewater.

Detention times of 0.5–2 h under anaerobic conditions are typical. The aerobic stage is designed conventionally. Two commercial processes designed for BOD removal and enhanced phosphorus uptake are shown in Fig. 17.11. In the A/O (anaerobic/oxic) process the reactor is compartmentalized into a number of equal sized anaerobic and aerobic stages. Each compartment is CM but compartmentalization promotes PF. The A/O process directly applies the principle of an anaerobic phase followed by an aerobic phase.

The A/O process can be modified for denitrification by incorporating an anoxic stage into the middle of the process and recycling effluent from the end of the aerobic (nitrification) stage to the anoxic stage. The anoxic stage is also divided into equal sized compartments.

The Phostrip process incorporates both biological and chemical removal of phosphorus. In the anaerobic stripper, sludge thickening occurs under anaerobic conditions. Phosphorus release occurs in this tank. The underflow from the stripper may be recycled to the influent to the stripper or an elutriation stream may be passed through the stripper. The elutriation stream may be primary effluent, secondary effluent, or supernatant from the lime precipitation reactor. The supernatant from the stripper is sent to a lime precipitation clarifier, where a chemical sludge is formed. A portion of the phosphorus is removed in the waste activated sludge and the remainder is removed in the sludge from the lime clarifier. Biological removal occurs through an anaerobic–aerobic sequence.

The modified Bardenpho process (the last process pictured in Fig. 17.10) is another process that accomplishes both enhanced phosphorus uptake and nitrification–denitrification. Volatile acids are produced in the initial anaerobic reactor. Some BOD removal and denitrification occurs in the second reactor. Phosphorus uptake occurs in the third reactor along with BOD removal and nitrification. In the second anoxic reactor additional denitrification occurs through endogenous decay of the biomass with nitrate as an electron acceptor. The final aerobic stage strips nitrogen gas and supplies oxygen to the biomass to minimize losses of phosphorus in the clarifier.

Operating conditions for the three enhanced phosphorus removal processes are given in Table 17.4. SBR systems can be operated to achieve enhanced phosphorus uptake.

As noted above, activated sludge that has aerobically accumulated phosphorus in an enhanced phosphorus uptake system will begin to release phosphate when an oxygen deficiency occurs (Schön et al., 1993). Under anaerobic conditions Rasmussen et al. (1994) found that most of the release was accomplished in the short time of 4 h. Designing clarifiers to minimize residence time of the sludge within them is essential to proper operation of an enhanced phosphorus removal process.

TABLE 17.4 Typical Operating Conditions for Enhanced Bio-P Processes[a]

	Process			
Parameter	Phostrip	A/O	A/O with nitrification	Modified Bardenpho
F:M, kg TBOD[b]/kg MLVSS/d	[c]	0.2–0.7	0.15–0.25	0.1–0.2
SRT,[d] d	[c]	2–6	4–8	10–30
MLSS, mg/L	600–5 000	2 000–4 000	3 000–5 000	2 000–4 000
HRT,[e] h	1–10			
Anaerobic		0.5–1.5	0.5–1.5	1–2
Anoxic 1		[f]	0.5–1.0	2–4
Aerobic 1		1–3	3.5–6.0	4–12
Anoxic 2		[f]	[f]	2–4
Aerobic 2		[f]	[f]	0.5–1.0
Return sludge flow, % of influent		25–40	20–50	100
Internal recycle, % of influent			100–300	400
HRT in stripper, h	5–20			
Elutriation flow, % of stripper feed	50–100			
Stripper feed, % of influent	20–30			
Stripper underflow, % of influent	10–20			
Lime clarifier loading, $m^3/m^2/d$	48			
($gal/ft^2/d$)	(1 180)			
Lime dosage in lime clarifier, mg/L	100–300			

[a]From USEPA (1987).
[b]TBOD is total (particulate + soluble) BOD.
[c]Based on activated sludge system design.
[d]Average mass of solids in the system divided by average solids waste rate.
[e]HRT is based on volume divided by influent flow rate.
[f]Not present.

17.11 OXYGEN UPTAKE IN ACTIVATED SLUDGE PROCESSES

Oxygen is the terminal electron acceptor in aerobic metabolic processes. As noted above, DO concentrations of 2.0 mg/L are generally considered adequate to maintain process efficiency. Oxygen transfer is discussed in Chapter 12 and aeration devices are described in Sections 12.4 and 12.5.

The model developed in Section 17.3.2 distinguishes between the metabolic processes of substrate removal and endogenous decay. In addition to metabolic processes, the waste may contain substances that exert a significant chemical oxygen demand independently of microbial metabolism. (Do not confuse this with the standard COD, which is a measure of the total oxygen demand of the waste.) The equation for oxygen consumption is

$$r_{O2} = -ar_S + bX_V + k_C \tag{17.54}$$

where

r_{O2} is the rate of oxygen removal ($ML^{-3}T^{-1}$)
a and b are constants
k_C is a constant dependent on substances initially present in the waste that exert an immediate oxygen demand

The constant a represents the fraction of substrate removed that has been used for energy and thus oxidized. The constant b is the fraction of VSS that is oxidized when VSS is expressed on an oxygen equivalence basis.

The immediate chemical oxygen demand can be determined by supplying a known quantity of oxygen to the unseeded raw waste and measuring the amount of oxygen consumed in a short period of time. The determination can be made in a similar manner to the BOD test by mixing the waste with water with a known oxygen concentration and measuring the decrease in DO concentration. The coefficient k_C depends on the rate at which these substances enter the reactor.

Equation (17.54) can be rearranged to

$$\frac{r_{O2}}{X_V} = -\frac{ar_S}{X_V} + b + \frac{k_C}{X_V} = aU + b + \frac{k_C}{X_V} \tag{17.55}$$

If the rate of oxygen consumption is known, a plot of r_{O2}/X_V (adjusted for the initial chemical oxygen demand, if it is significant) versus U will yield the values of the constants.

Oxygen uptake rate (OUR) is a useful process control parameter. Changes in OUR can signal the need for a change in operation. *Standard Methods* (1992) outlines a procedure for OUR. It consists of taking an ML sample, transferring it to a BOD bottle (which is completely filled), and measuring the decrease in DO over time. The sample must be quickly transferred from the aeration basin to the BOD bottle for accurate results. If DO in the sample is too low for accurate DO determination, the sample must be initially aerated before DO decrease is monitored. This may introduce significant error in OUR determination and other techniques provide more accurate results (Chiesa et al., 1990).

17.12 BULKING AND FOAMING SLUDGE

Foam formation and poorly settling sludge are two of the most common problems of the activated sludge process. A sludge that exhibits poor settling characteristics is referred to as a bulking sludge. Filamentous microorganisms are usually responsible for a bulked sludge. Large surface area to volume ratios for these microorganisms retard their settling velocities. The causes and combative measures for filamentous bulking are many (Chiesa and Irvine, 1985; Chudoba, 1985).

Fungi are the most familiar filamentous microorganisms. The vegetative structure of most fungi is composed of filaments, which actually contain a number of nuclei. Fungi are not commonly significant in wastewater treatment.

Some bacteria grow in filamentous sheaths that contain distinct cells or there may be no cross walls in the filaments, which can be considered as long multinucleate cells (Brock, 1970). Type 021N bacteria are among the most frequently reported filamentous organisms in surveys of bulking sludge. Utilization of a wide variety of carbon and nitrogen sources and achievement of maximum growth rate at low substrate concentration (low half-velocity constant) support the potential for this organism to compete effectively in activated sludge systems characterized by low or variable nutrient concentrations (Williams and Unz, 1989). Characteristics of this microorganism suggested that it can be appropriately placed within the genus *Thiotrix* or *Leucothrix*. Williams and Unz (1989) have studied nutritional characteristics of filamentous sulfur bacteria (*Thiotrix, Beggiatoa,* and *Leucothrix*). Organic acids are an important class of carbon sources for growth of the filamentous sulfur bacteria. These bacteria do not develop well at low pH values.

The same types of microorganisms are often found in foam and bulked sludge. Foaming was originally thought to be caused primarily by *Nocardia amarae* but recent work has shown many filamentous types to be present in foams. In a survey of plants with foam in Australia (Seviour et al., 1990), *Microthrix parvicella, N. amarae*-like organisms, *N. pinensis*-like organisms, and type 0092 were most commonly found. *Nocardia* growth is supported by high sludge ages, low F : M ratios, and higher rather than lower waste temperatures. The most successful strategies for control of this organism based on a nationwide survey in the United States were reduction of the sludge age to less than 6 d and chlorination of return activated sludge (Pitt and Jenkins, 1990). When the system sludge age is greater than 6 d an aerobic selector (selectors are described later) does not appear to provide any control of *Nocardia* spp. but an anoxic selector was effective in a nitrifying system with a sludge age of 12 d (Cha et al., 1992).

Nocardia filaments are concentrated in foam compared to the mixed liquor. Foam removal is a logical and beneficial control measure of these and other species that cause foaming (Cha et al., 1992).

CM systems are more prone to give rise to a poorly settling sludge. A feast–famine cycle has been shown to favor the growth of well-settling microorganisms. Batch and PF systems naturally progress through a feast–famine metabolic regime. For CM systems, a feast–famine cycle can be achieved by adding a much smaller CM reactor in front of the main CM reactor. The smaller CM reactor is known as a selector. High–substrate loading rates exist within the selector and this condition promotes the growth of floc-forming bacteria. Substrates assimilated in the selector are metabolized in the main CM reactor, resulting in starvation conditions.

The loading rates in the selector should be higher than 12 kg BOD_5/kg MLVSS/d or 20 kg COD/kg MLVSS/d (Chiesa and Linne, 1990). This loading is somewhat higher than results from simulations and experience in Europe of Kappeler and Gujer (1994). Using a CM selector with an HRT of 13 min (based on combined influent and recycled flow of approximately 35%) in front of an aeration operated at an HRT of either 3 or 6 h improved sludge settleability compared to the control (Linne et al.,

1989). The setup with the aeration basin at 6 h HRT was superior to the 3 h aeration basin HRT. Nitrification was improved in the selector systems compared to the control.

In addition to aerobic selectors, anoxic and anaerobic selectors have also been able to control bulking (Daigger and Nicholson, 1990; Nowak and Brown, 1990). HRTs were about 90 min in the anaerobic and anoxic selectors compared to HRTs of 11 and 16 min in aerobic selectors. Anaerobic conditions were attained by supplying mixing but no aeration. Anoxic conditions were attained by recycling nitrified mixed liquor to the selector, which was not aerated.

Other studies have shown similar results. Anoxic conditions have been shown to suppress growth of some filamentous organisms (Wanner et al., 1987a). Anaerobic conditions are also effective but dissimilative sulfate reduction may occur during anaerobiosis, which leads to collapse of biological phosphorus removal and results in excessive growth of *Thiotrix* (Wanner et al., 1987b).

Chemical removal of phosphorus in the primary clarifier enhanced total removals in the plant along with savings in chemical, energy, and sludge handling costs (Ericsson and Eriksson, 1988). However, addition of Fe^{3+} at extreme dosages of 20 mg/L or more to the primary clarifier deteriorated settling characteristics of sludge formed in the bioreactor and favored growth of filamentous organisms because of phosphorus depletion.

The most common sludge bulking problem is caused by excessive growth of filamentous microorganisms. Zoogloea-like microorganisms can cause a rare form of nonfilamentous bulking. These microorganisms form exocellular slime (capsule) that has a high water content, forming voluminous flocs that do not compact (Novák et al., 1993). When these microorganisms are present in large amounts, their jellylike consistency can cause foaming and scum formation as well as poor settling. The remedies for this type of bulking are unclear, although a high content of oleic acid in the substrate has been thought to stimulate their growth and capsule formation.

Although much progress has been made in the past few years in identifying and characterizing microorganisms responsible for bulking and foaming, the occurrence of these problems often defies easy explanation and remedy. Remedies are not quickly made because there has to be a significant change in the flora inhabiting the process. Changes in process operating conditions may have to be in place for some time before the success or failure of the modification is apparent because of the time required for new microorganisms to reproduce to significant numbers and current undesirable species to be removed as their competitive advantage subsides. In addition to measures suggested above, classical control measures are adjustment of the F:M ratio, raising or lowering DO, or applying a disinfectant. The latter measure will probably only have a temporary effect without other long-term adjustments. The common agent is chlorine, which will indiscriminately act on all species, although to various degrees, but there is no research to demonstrate that filamentous types are more or less innately susceptible to chlorine or any other biocide. Because filaments extend beyond floc agglomerates, the filamentous organisms have greater exposure to biocidal agents than organisms in the center of flocs. With significant dieoff of the existing population and change in operating conditions that favors more desirable species the recovery process can be accelerated.

Carbohydrate-rich wastes are more prone to give rise to filamentous populations. Selectors, batch, or PF systems should be considered for treating these wastes. Nutrient deficiencies also cause growth of filamentous microorganisms. One of the first checks made on a bulking system is to measure nutrient concentrations to determine whether they are sufficient for the amount of sludge being produced. This problem is readily solved by addition of nutrients to the waste.

Another problem that can be encountered in secondary clarifiers is rising sludge caused by denitrification in the clarifier. Denitrification results in the formation of nitrogen gas bubbles, which buoy settled sludge, deteriorating the clarified effluent quality. In nitrifying activated sludge reactors there is always potential for denitrification to occur in the secondary clarifier. High temperatures (above 20°C) accelerate the rate of denitrification. When influent nitrate concentrations exceed 6–8 mg N/L and the temperature exceeds 20°C Henze et al. (1993) found that enough nitrogen gas would be produced in detention times of 1 h or less to cause rising sludge. DO levels in the influent would delay the onset of denitrification but the delay was insignificant.

Plants in warm climates will be susceptible to this problem. The only practical solution is to denitrify the effluent from the activated sludge process.

17.13 PROCESS OPERATING CHARACTERISTICS AND KINETIC COEFFICIENT VALUES

Kinetic coefficients and process operating conditions are influenced by many factors such as the concentration and components of dissolved and suspended matter in the influent and temperature. Table 17.5 lists values of coefficients for various wastes obtained from a number of studies.

Experience has shown that effluent from CM and PF systems has approximately the same quality (see for example, Chudoba et al., 1991; Toerber et al., 1974) notwithstanding the kinetic developments in Chapter 10. There are some reasons for this discrepancy. The most satisfying theoretical explanation is that substrate removal kinetics are zero-order, which would cause no difference between PF and CM performance as advanced by Argaman (1991) who quotes studies that demonstrate that removal of individual wastewater components each follows a zero-order law. The overall rate of removal of all components together, described with a nonspecific measure such as BOD or COD, can exhibit pseudo-Monod or first-order kinetics.

Another explanation that has experimental validity is that the effluent organics concentration is primarily composed of microbial products of metabolism and decay, as discussed in Sections 17.6.2 and 17.7. There are many shortcomings of the basic models that only describe gross phenomena and indirectly account for phenomena such as microbial secondary or decay products. It is impossible to handle the changing nature of organics and microorganisms in a dynamic situation with one (or even a few) substrate formulations incorporated into a model of the process. Much more detailed models are required, accounting for consumption and production of classes of compounds and growth of different groups of microorganisms. Because of the coarse nature of basic models, the effects of flow regimes may be masked in the models.

17.14 FIXED FILM PROCESSES

The alternative to suspended growth processes is to place fixed media in the reactor. Microorganisms attach to the medium and grow in place. They are not washed out of the reactor with the liquid, so very high sludge ages and concentrations of microorganisms can be achieved in these reactors. The two most common aerobic fixed film processes are the trickling filter and rotating biological disk reactor.

TABLE 17.5 Kinetic Coefficients[a] of Activated Sludge Systems

Waste	HRT, h	SRT, d	T, °C	Y, g/g	k_e, d^{-1}	k, d^{-1}	K, mg/L	Reference
Domestic sewage	3–5	6.0–16.6	≈15	0.70[b]	0.025			Keyes and Asano (1975)
Domestic sewage[c]			20	0.46–0.69[d]		4.3–28.7[d]	10–180[d]	Henze et al. (1987)
Domestic sewage				0.5[b]	0.055			Lawrence and McCarty (1970)[c]
Domestic sewage			20–21	0.67[b]	0.048			Lawrence and McCarty (1970)[c]
Domestic sewage				0.67[d]	0.07	5.6[d]	22[d]	Lawrence and McCarty (1970)[c]
Domestic sewage				0.5[b]	0.06	26.4[b]	120[b]	Mynhier and Grady (1975)[c]
Glucose	8	0.9–4.1	25	0.59[c]	0.14	21.2–25.2[d]	115–123[d]	Srinivasaraghavan and Gaudy (1975)
Glucose				0.42[b]	0.087	3.0[b]	355[b]	Lawrence and McCarty (1970)[c]
Glucose			10	0.59[b]		3.3[b]		Lawrence and McCarty (1970)[c]
Glucose				0.40[e]		20.5[e]	8[e]	Lawrence and McCarty (1970)[c]
Glucose–peptone			20	0.49[b]		10.3[b]		Lawrence and McCarty (1970)[c]
Industrial			53	0.60[d]	0.52	5.6[d]	890[d]	Jackson (1982)
Peptone	6.6	5.5	22				37[d]	Selna and Schroeder (1978)
Peptone	6.0–6.9	10	21.5				47[d]	Selna and Schroeder (1978)
Peptone			30	0.43[b]		14.5[b]	65[b]	Lawrence and McCarty (1970)[c]
Poultry processing				1.32[b]	0.72	54.5[b]	500[b]	Mynhier and Grady (1975)[c]
Shrimp processing				0.50[b]	1.6	37.0[b]	85.5[b]	Mynhier and Grady (1975)[c]
Skim milk				0.48[b]	0.045	5.1[b]	100[b]	Lawrence and McCarty (1970)[c]
Slaughterhouse	6–30	0.25–1.25	52	0.30[d]	0.32	19.8[d]	30[d]	Couillard et al. (1989)
Slaughterhouse	6–30	0.25–1.25	58	0.32[d]	0.78	31.1[d]	992[d]	Couillard et al. (1989)
Slaughterhouse			22	0.34[d]	0.03	1.7[d]	362[d]	Lovett et al. (1984)
Slaughterhouse			19–23	0.41[d]	0.04	0.67[d]	150[d]	Lovett et al. (1983)
Soybean				0.74[b]	0.14	16.2[b]	355[b]	Mynhier and Grady (1975)[c]
Synthetic waste				0.65[b]	0.18			Lawrence and McCarty (1970)[c]

[a]These coefficients apply to Eq. (17.1a).
[b]BOD_5 basis.
[c]Based on a review.
[d]COD basis.
[e]Weight basis.

17.14.1 Trickling Filters

Trickling filters (Fig. 17.12) are not filters, although this name is historically used to describe them. Filtration is not a significant removal mechanism. Media in a trickling filter may consist of rock or plastic pieces. Wastewater is uniformly dosed over the medium and trickles downward through it. Mass transport processes move the substrate to and through the biological film attached to the medium. A highly varied microflora populates a trickling filter, although bacteria are still the main group of microorganisms responsible for substrate removal. Air (oxygen) supply is by natural convection currents that are caused by temperature differences between the air and sewage.

The film attached to the medium grows until lack of substrate and/or the bulk weight of the film is not able to withstand the shear force of liquid flowing over the film. The film then sloughs off. Because of the clumpy nature of sloughed biofilm,

(a)

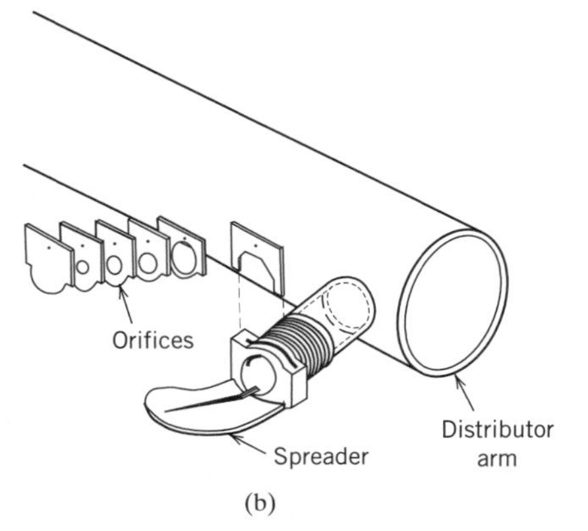

(b)

Figure 17.12 (a) Trickling filter with mast type rotary distributor and synthetic media. (b) Typical spreader assembly. Courtesy of Eimco.

settling problems are not generally encountered in trickling filter operations. However the clarified effluent may have more floating organic particulates that lower effluent quality compared to activated sludge processes.

Large surface area and a high void volume both promote more efficient treatment. Synthetic media may be optimized for maximum surface area to volume ratio (specific surface) and high void volume but there is a cost premium for synthetic media. Much higher loading rates can be obtained for synthetic media filters. Characteristics for synthetic, rock, and other media are given in Table 17.6. Figure 18.7 shows some media. Loading rates, depth, and other characteristics of different types of trickling filters are given in Table 17.7.

A variety of liquid and sludge recirculation schemes are employed to achieve satisfactory hydraulic and organic loading rates (Fig. 17.13). One or two filters are used depending on the strength of the wastewater and the desired effluent quality. Two-stage trickling filter systems provide the best effluent quality. Sometimes intermediate clarifiers are incorporated into two-stage processes. The WEF and ASCE (1992a) provide additional performance comparison information on the systems in Fig. 17.13.

The development of equations describing trickling filter performance has progressed along the following lines. A correlation between depth of the filter and substrate removal is obvious; Velz (1948) assumed a first-order reaction. The amount of biomass in the filter is difficult to estimate and biomass concentration is not normally incorporated into the reaction rate expression.

$$\frac{dS}{dD} = -kS \tag{17.56}$$

where

D is the depth of the filter
k is a rate constant

which results in

$$S_e = S_i e^{-kD} \tag{17.57}$$

The influent substrate concentration depends on the amount of recirculation (r), similar to the PF activated sludge process.

$$S_i = \frac{QS_0 + rQS_e}{Q + rQ} = \frac{S_0 + rS_e}{1 + r} \tag{17.58}$$

The amount of treatment depends on the length of time the sewage is in contact with the film. The nominal contact time, t, to flow through the filter can be calculated from the surface velocity, v.

$$t = \frac{D}{v} = \frac{D}{Q_t/A} \tag{17.59}$$

where

$Q_t = Q + rQ$
A is the surface area of the filter

Flow in the filter is tortuous and a function of the media geometry and packing characteristics. Therefore the actual contact time, t_c, is calculated from a modified form of Eq. (17.59).

TABLE 17.6 Media Characteristics for Fixed Film Processes[a]

Medium	Size		Unit Weight		Specific surface[b]		Void volume, %
	cm	in.	kg/m^3	lb/ft^3	m^2/m^3	ft^2/ft^3	
Rock	2.5–6.5	1–2.5	1 250–1 442	78–90	56–69	17–21	40–50
	10–12.5	4–5	800–993	50–62	39–164	12–50	50–60
Slag	5–7.5	2–3	897–1 200	56–75	56–69	17–21	40–50
	7.5–12.5	3–5	800–993	50–62	46–59	14–18	50–60
Redwood	122 × 122 × 51	48 × 48 × 20	32–96	2–6	39–49	12–15	70–80
Crossflow	61 × 61 × 122	24 × 24 × 48	24–63	1.5–3.9	98–226	30–69	95–>95
Random pack	1.5–8.9	0.6–3.5	53–112	3.3–7.0	102–330	31–101	88–95
Flexible	na[b]	na	98–223	30–68	na	na	na
Hanging sheets							

[a]Compiled from Metcalf and Eddy (1991), USEPA (1993), WEF and ASCE (1992a).
[b]na, not applicable.

TABLE 17.7 Typical Design Information for Trickling Filters[a]

Item	Filter classification			
	Low rate	Intermediate rate	High rate[b]	Roughing
Hydraulic loading, $m^3/m^2/d$ ($gal/ft^2/d$)	1.2–3.7 (30–86)	3.7–9.4 (86–230)	9.4–38 (230–922)	47–190 (1 150–4 600)
BOD_5 loading, $kg/m^3/d$ ($lb/1\,000\ ft^3/d$)	0.08–0.40 (5–25)	0.24–0.48 (15–30)	0.48–0.96 (30–60)	1.6–8.0 (100–500)
Depth, m (ft)	1.8–2.4 (6–8)	1.8–2.4 (6–8)	0.9–1.8 (3–6)	4.5–12 (15–40)
Recirculation ratio	0	0–1	1–2	1–4
Filter media	Rock, slag, etc.	Rock, slag, etc.	Rock, slag, synthetic	Synthetic, redwood
Power requirements $kW/10^3\ m^3$ ($kW/10^3\ ft^3$)	2–4 (70–140)	2–8 (70–280)	6–10 (210–350)	10–20 (350–700)
Filter flies	Many	Intermediate	Few	Few or none
Sloughing	Intermittent	Intermittent	Continuous	Continuous
Dosing intervals	Not more than 5 min	15–60 s	Not more than 15 s	Continuous
BOD_5 removal efficiency	80–90	50–70	65–85	40–65
Effluent	Usually fully nitrified	Partially nitrified	Nitrified at low loadings	Nitrified at low loadings

[a]Compiled from Metcalf and Eddy (1991, 1979), WPCF (1988).

[b]Synthetic media filters may have hydraulic and organic loading rates up to double those listed for high-rate filters.

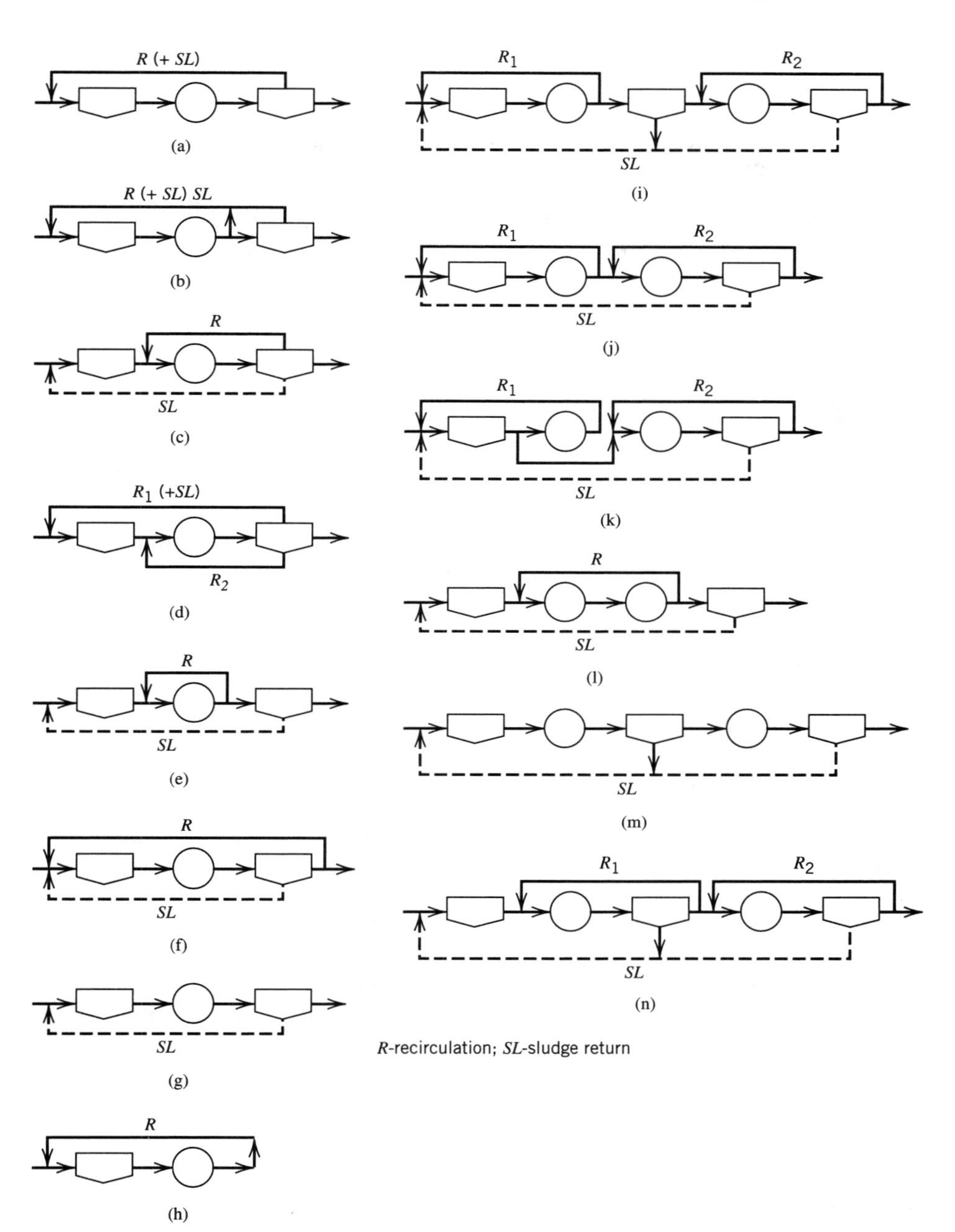

Figure 17.13 Trickling filter configurations. From WEF and ASCE, (1992a), *Design of Municipal Wastewater Treatment Plants,* vol. 1, WEF, © WEF 1992.

$$t_c = \frac{CD}{(Q_t/A)^n} \tag{17.60}$$

where

C and n are constants dependent on the medium

A first-order relation is used to describe removal with respect to contact time.

$$\frac{dS}{dt} = -k'S \tag{17.61}$$

where

k' is the rate constant

Integrating Eq. (17.61) and substituting Eq. (17.60) for the time limit,

$$S_e = S_i \exp\left[-\frac{k'CD}{(Q_t/A)^n}\right] = S_i \exp\left[-\frac{kD}{(Q_t/A)^n}\right] \tag{17.62}$$

where

$k = k'C$

Equation (17.62) is known as the Schultz–Germain formula (Germain, 1966; Schultz, 1960). Effectively,

$$k_{\text{Schultz-Germain}} = k_{\text{Velz}}(Q_t/A)^n$$

Eckenfelder (1963) proposed the following three-parameter formula, which is similar to those just developed:

$$S_e = S_i \exp\left[-kDS_a^m\left(\frac{A}{Q_t}\right)^n\right] \tag{17.63}$$

where

m and n are constants characteristic of the medium

S_a is the specific surface of the medium (m^2/m^3 or ft^2/ft^3)

Once data are available to determine the coefficients in Eqs. (17.62) and (17.63), these equations are solved to provide an acceptable effluent quality with recirculation, area, and a depth that fall within the appropriate hydraulic and organic loading rates specified in Table 17.7. The design is checked at limiting conditions, which may be when the maximum load is delivered to the filter or at cool temperature conditions.

The Arrhenius equation (Eq. 17.3) is used to describe the temperature variation of the rate coefficient. The WEF and ASCE (1992a) report a range of θ from 1.015 to 1.045 for carbonaceous BOD removal and 1.035 is suggested for a conservative design.

Another group of equations that has been used to characterize trickling filter performance is the National Research Council (NRC) equations (National Research Council, 1946).

$$E_1 = \frac{100}{1 + 0.443\sqrt{\dfrac{w_1}{VF}}} \qquad E_2 = \frac{100}{1 + \dfrac{0.443}{1 - E_1}\sqrt{\dfrac{w_2}{VF}}} \qquad F = \frac{1 + r}{(1 + 0.1r)^2}$$

where

w is quantity of organics supplied to the filter in kg BOD_5/d

F is a recirculation factor

V is the volume of the filter in m^3

subscripts 1 and 2 refer to the first and second filter, respectively

The efficiency, E, in these equations is

$$E_1 = \frac{S_0 - S_1}{S_0} \qquad E_2 = \frac{S_1 - S_e}{S_1}$$

The NRC equations were developed for filters containing rock media from data taken at a number of military bases and are more empirical in nature than the principled developments given earlier, so they should be used with caution, if at all. The equation for E_2 only applies when an intermediate settling basin is incorporated between the filters.

■ Example 17.5 Calculation of Constants in the Schultz–Germain Formula from Lab Data

From the laboratory data on a trickling filter in the following table obtained at 20°C, find the constants in the Schultz–Germain formula.

	% BOD_5 remaining			
	Hydraulic loading, $m^3/m^2/d$ ($gal/ft^2/d$)			
Depth, m (ft)	0.50 (12.3)	1.00 (24.5)	1.50 (36.8)	2.00 (49.1)
1.50 (4.92)	81	84	87	93
3.00 (9.84)	65	73	79	83
4.50 (14.8)	51	62	69	75
6.00 (19.7)	42	54	62	69

The Schultz–Germain formula is

$$S_e = S_i \exp\left[-\frac{kD}{(Q_t/A)^n}\right] \tag{1}$$

To find the exponent n, and the rate constant k, transform the formula by taking the logarithm of each side of the equation.

$$\ln\left(\frac{S_e}{S_i}\right) = -kD\left(\frac{A}{Q_t}\right)^n \tag{2}$$

A plot of S_e/S_i versus D on a semilog graph paper (Fig. 17.14) will yield a family of curves, each with a slope, s.

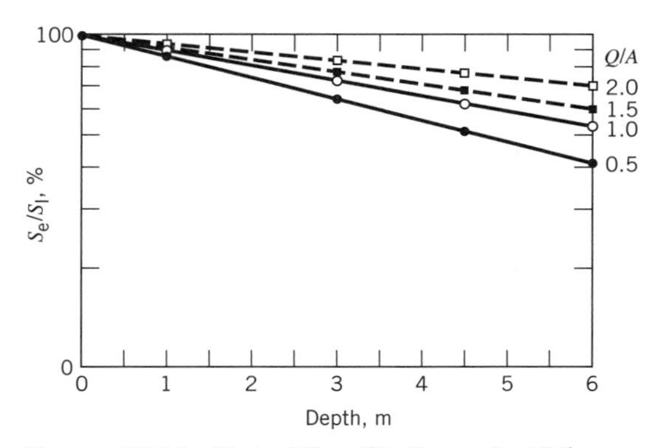

Figure 17.14 Plot of Eq. (2), Example 17.5.

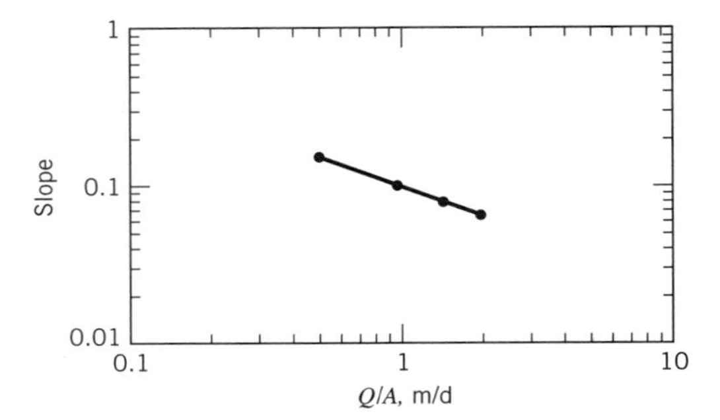

Figure 17.15 Plot to determine n.

$$s = -k\left(\frac{A}{Q_t}\right)^n \tag{3}$$

Taking the logarithm of Eq. (3),

$$\ln s = \ln(-k) - n \ln\left(\frac{Q_t}{A}\right)$$

To find n, the slope–Q_t/A data are plotted in Fig. 17.15. A regression of these data yields the slope of the line, which is n.

$$n = -\frac{\ln 0.05 - \ln 0.20}{\ln 0.29 - \ln 3.0} = 0.59$$

The slopes of the lines in Fig. 17.15 are as follows:

Hydraulic Loading, $m^3/m^2/d$ (gal/ft^2/d)	Slope, m^{-1} (ft^{-1})
0.50	$-(\ln 100 - \ln 70)/2.50 = -0.143$
(12.3)	(−0.043 5)
1.00	$-0.357/3.40 = -0.105$
(24.5)	(−0.032 0)
1.50	$-0.357/4.32 = -0.082\,6$
(36.8)	(−0.025 2)
2.00	$-0.357/6.75 = -0.052\,8$
(49.1)	(−0.016 1)

Although k may be found from the line in Fig. 17.15, the following procedure provides a more accurate determination of k. Because n is known, Eq. (2) may now be used to find k as the slope of a line given by the following equation.

$$\ln\left(\frac{S_e}{S_i}\right) = -k\left[D\left(\frac{A}{Q_t}\right)^n\right]$$

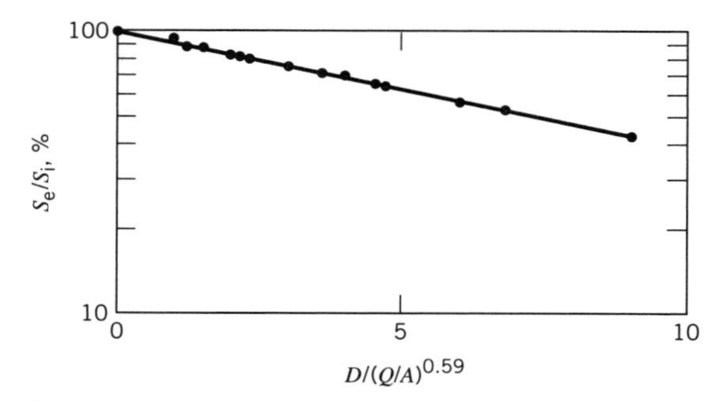

Figure 17.16 Plot to determine k.

The data for the above equation are given in the following table and plotted in Fig. 17.16.

SI units				U.S. units			
D	Q_t/A	$D/(Q_t/A)^{0.59}$	S_e/S_i	D	Q_t/A	$D/(Q_t/A)^{0.59}$	S_e/S_i
1.50	0.50	2.26	81	4.92	12.3	1.12	81
3.00	0.50	4.51	65	9.84	12.3	2.25	65
4.50	0.50	6.77	51	14.8	12.3	3.37	51
6.00	0.50	9.02	42	19.7	12.3	4.49	42
1.50	1.00	1.50	84	4.92	24.5	0.747	84
3.00	1.00	3.00	73	9.84	24.5	1.50	73
4.50	1.00	4.50	62	14.8	24.5	2.24	62
6.00	1.00	6.00	54	19.7	24.5	2.99	54
1.50	1.50	1.18	87	4.92	36.8	0.589	87
3.00	1.50	2.36	79	9.84	36.8	1.18	79
4.50	1.50	3.54	69	14.8	36.8	1.77	69
6.00	1.50	4.73	62	19.7	36.8	2.36	62
1.50	2.00	1.00	93	4.92	49.1	0.497	93
3.00	2.00	2.00	83	9.84	49.1	0.994	83
4.50	2.00	2.99	75	14.8	49.1	1.49	75
6.00	2.00	3.99	69	19.7	49.1	1.99	69

From Fig. 17.16, the slope, k, of the line is found to be:

$$k = \frac{\ln 100 - \ln 60}{5.14} = 0.099$$

In U.S. units:

$$k = \frac{\ln 100 - \ln 60}{2.54} = 0.201$$

The design equation is

$$S_e = S_i \exp\left[-0.099D\left(\frac{A}{Q_t}\right)^{0.59}\right]$$

where

D is in m

Q_t/A is in $m^3/m^2/d$

In U.S. units, the design equation is

$$S_e = S_i \exp\left[-0.20D\left(\frac{A}{Q_t}\right)^{0.59}\right]$$

where

D is in ft

Q_t/A is in $gal/ft^2/d$

Note: Q_t/A includes the recycle flow into the filter and S_i is the influent substrate concentration diluted by the recycled flow.

Sludge Production from Trickling Filters

The sludge age in trickling filters is high compared to suspended growth systems. Sludge ages over 100 d can be easily attained. The equation of net biological growth applies to all biological processes but the actual biomass content in a trickling filter is difficult to measure. The long residence time of sludge in the filter promotes the endogenous decay of the biomass, lowering the production of biomass and producing a more highly mineralized sludge even though less efficient oxygen transfer to the slime growth lowers the rates of endogenous decay. Overall observed yields from trickling filter processes are at the lower range of observed yield factors for activated sludge processes and on the order of 60–80% of the observed yield factors for typical activated sludge processes. High-rate trickling filter operations produce more sludge than low-rate operations.

Air Supply in Trickling Filters

Air supply in trickling results from natural draft currents induced by the temperature difference between the wastewater and the ambient air. When the wastewater is cooler than the ambient air, heat transfer from the air to the wastewater results in an increase in the density of air and airflow is downward through the filter. Warm wastewater relative to ambient air transfers heat to the air in the filter, decreasing its density and causing air upflow through the filter. There will be times when temperature differences will not result in airflow. When temperatures are below freezing, increased airflow will cause the filter to freeze. Airflow should be restricted under these conditions.

Forced ventilation of filters is not usually designed. There are a number of recommended design practices to maintain adequate natural ventilation (WPCF, 1977). Air enters or exits through the underdrains and collecting channels. These passageways should be designed to flow no more than half full. Underdrain blocks should have openings at the top that are at least 15% of the surface area of the filter. Ventilating manholes and vent stacks with open gratings should be installed at ends of the main collection channel and at the filter periphery to provide at least 1 m^2 of ventilating area for each 250 m^2 (1 ft^2 per 250 ft^2) of filter surface area.

Operation of Trickling Filters

Sustained high organic loading rates beyond the design capacity of the filters can result in their plugging. Performance deteriorates as part of the volume of the filter is taken out of service and ponding occurs on the surface preventing the flow of air. Odor problems are also intensified. Chlorine can be used to dislodge biological growth on a short-term basis. If the problem persists, increased recirculation or more porous media may be required. If these measures are not adequate, more filter capacity through an enlarged filter or another filter should be added.

Cold weather operation may lead to freezing of the filter. The microbiological performance of the filter deteriorates to low to negligible rates. The filter can be covered. Increasing the rate of recirculation in an uncovered filter actually leads to a decrease in the average temperature within a filter. As sewage trickles down a filter, the air current cools it. Recirculating the cooled filter effluent contributes to a further lowering of the temperature.

Filters also provide a suitable environment for breeding of small *Psychoda* flies, or "filter" flies as they are commonly known. These flies are able to pass through the openings in window screens. In warm weather the breeding cycle of these flies can be less than 1 week. Periodically spraying the filter with insecticide (at one time DDT was suggested as the insecticide!) or submerging the filter (which drowns the larvae) will minimize this problem.

17.14.2 Hydraulic Design of Distributors for Trickling Filters

The distribution system (Fig. 17.12) for trickling filters is designed to apply the sewage evenly over the filter at a predetermined rate. Revolving distributors are preferred in North America. These distributors usually have two or four arms that revolve about a central post. Guy wires connected to the central post support the arms. The arms may be self-propelled or motor driven. Speeds vary from $\frac{1}{2}$ to 2 rpm for small and large distributors, respectively. Using data and formulas in Metcalf and Eddy (1991), the peripheral speed of a two-arm distributor would range from 0.5–3.7 m/min (1.5–12 ft/min). The rotational speed is reduced directly proportional to the number of arms.

The arms range from 4.5 to 70 m (15–230 ft) in diameter and discharge ports on the arms are typically 95 mm (3.7 in.) in diameter. The arms are a minimum of 15 cm (6 in.) above the filter media and higher if the possibility of ice formation exists. The velocity in the arms should be above 0.3–0.6 m/s (1–2 ft/s) to prevent deposition of particulates (WEF and ASCE, 1992a).

The basic approach to designing the distributor arms is described by Ordon (1966). The area serviced by the arms increases as the square of the distance from the center of the filter. Maintaining an even distribution of sewage over the filter requires that

$$\frac{Q}{A} = \frac{Q_0}{A_t} \quad \text{or} \quad Q = Q_0 \frac{\ell^2}{L^2} \tag{17.64}$$

where

Q is the flow discharged over an area A, bounded at a distance ℓ from the center
A_t is the total surface area
Q_0 is the flow rate into the arm
L is the length of the arm

The flow in the distributor pipe at any distance, ℓ is

$$Q_\ell = Q_0 - Q = Q_0 - Q_0 \frac{\ell^2}{L^2} \tag{17.65}$$

The Darcy–Weisbach equation can be used to calculate the hydraulic gradient in the pipe.

$$S = \frac{dh}{d\ell} = \frac{f}{D}\frac{v^2}{2g} = \frac{f}{2r}\frac{Q_\ell^2}{2gA_p^2} = \frac{f}{4}\frac{Q_\ell^2}{\pi^2 g r^5} \tag{17.66a}$$

or

$$-\frac{dh}{d\ell} = kQ_\ell^2 \tag{17.66b}$$

where

- S is the slope of the energy gradient
- h is the total energy head
- A_p is the pipe area
- D is the pipe diameter
- r is the radius of the pipe
- f is the friction coefficient (function of roughness and Reynold's number)
- $k = \dfrac{f}{4\pi^2 r^5 g}$

The Reynold's number varies because the velocity of flow in the pipe varies. A single f value can be chosen to simplify calculations after examination of the Reynold's numbers in each reach of the pipe.

Substituting Eq. (17.65) into Eq. (17.66b) and integrating,

$$h_L = -\int_{h_0}^{h_f} dh = kQ_0^2\left[\int_0^L d\ell - \frac{2}{L^2}\int_0^L \ell^2\, d\ell + \frac{1}{L^4}\int_0^L \ell^4\, d\ell\right] = 0.59kLQ_0^2 \tag{17.67}$$

where

- h_0 and h_f are the initial and final total energy heads, respectively
- h_L is the headloss over the length L

The flow, Q_ℓ, in the pipe varies not continuously but in a stepwise fashion according to the spacing of the ports. Divergence of the streamlines as the flow expands to fully occupy the pipe cross-section at each port will cause additional turbulence and losses.

The headloss from a point ℓ to the end of the arm can be approximated by discretizing the headloss relation (Eq. 17.66b).

$$\Delta h_{\ell - L} = zkQ_\ell^2(L - \ell)$$

From Eq. (17.67), $z \approx 0.6$ and a suitable relation for the headloss from the center to a point ℓ is

$$h_\ell = 0.6k[LQ_0^2 - (L - \ell)Q_\ell^2] \tag{17.68}$$

The flow through a port is related to the static pressure head at the port by the orifice equation.

$$q = C_d a\sqrt{2gh_{\ell s}} \tag{17.69}$$

where

q is the discharge
C_d is the discharge coefficient
a is the area of the port
$h_{\ell s}$ is the static pressure head at location ℓ

An energy balance for the distributor must contain terms for the headloss, kinetic energy head resulting from velocity within the pipe, and kinetic energy head developed from the rotational motion of the distributor in addition to the static pressure head. The velocity at any point ℓ is calculated from Eq. (17.65) by dividing it by the cross-sectional area of the pipe, A_p.

$$v_\ell = v_0 - v_0 \frac{\ell^2}{L^2} \quad (17.70)$$

The head caused by rotational motion of the distributor is

$$h_c = \frac{v_c^2}{2g} = \frac{\omega^2 \ell^2}{2g} = \frac{\left(\frac{2\pi N}{60}\right)^2 \ell^2}{2g} = 5.59 \times 10^{-4} N^2 \ell^2 \quad (17.71)$$

where

h_c is the head caused by centrifugal forces
ω is the angular velocity
N is rpm of the arm

Note that centrifugal head is gained along the length of the arm.

The static head at any location is determined from

$$h_{\ell s} = h_0 - 0.6kLQ_0^2 + 0.6k(L - \ell)Q_\ell^2 + 5.59 \times 10^{-4} N^2 \ell^2 - \frac{v_\ell^2}{2g} \quad (17.72)$$

or the following equation may be used:

$$h_{\ell s} = h_0 - \sum h_{li} + 5.59 \times 10^{-4} N^2 \ell^2 - \frac{v_\ell^2}{2g}$$

where

Σh_{li} is the sum of head losses between ports from the beginning of the arm up to a distance ℓ

The spacing of the ports is determined by plotting Q versus ℓ according to Eq. (17.65) and then drawing a step curve to represent the actual flow in the pipe. The steps occur at each port and are equal to the discharge through the port. The area of the port can be sized to give the predetermined discharge.

It is more common to use a constant port size. In this case, the method just outlined is used to obtain the static pressure distribution in the pipe with Eq. (17.72). Using Eq. (17.69), the discharge through each port is calculated and the spacing of the orifices is determined that corresponds to the step discharges that have been

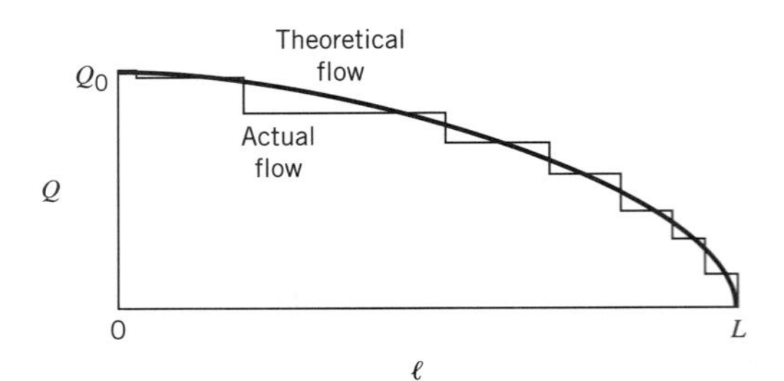

Figure 17.17 Ideal and actual flow distribution in the arm.

calculated with Eq. (17.69) (Fig. 17.17). With the values for q and the new spacing, repeat the exercise beginning with the static head calculation. Usually only two or three iterations are required to obtain the port spacing. There is a possibility of offsetting the port spacing on different arms to maximize coverage of the filter. Also, the distribution arms can be tapered, which will result in a more favorable hydraulic gradient for uniform flow distribution.

The distributor arms are supplied from a pump or dosing siphon. Dual siphons are used to maintain continuous flow. The distribution system must be designed to accommodate the variable flows from the siphons. Daily variation of flows should also be considered in the design. Manufacturers supply specifications for siphons and distribution systems. Metcalf and Eddy (1935) provide a detailed description of hydraulic computations for a trickling filter distribution system.

The jets issuing from the port will cause a rotary motion of the distributor. From a momentum balance in the horizontal plane for the ith orifice,

$$T_i = \ell_i F_i = \rho q_i v_i \ell_i$$

where

- T is the torque
- F is force
- ρ is the density of water
- v is the velocity of the jet

$$T = \Sigma T_i = \Sigma \rho q_i v_i \ell_i \tag{17.73}$$

If the resulting torque is too low to overcome frictional resistance and drive the arm at the required speed then a mechanical assist will have to be provided or a greater entrance head (h_0) must be provided. The minimum head for self-propulsion is 30–36 cm (12–14 in.). If the speed is too high, then the arms may be rotated so that the jets do not discharge horizontally.

Calculations must be performed for the minimum and maximum flows. If the expected lowest flows are too low to drive the rotary distributors, dosing tanks are provided. In the event of low flows, four-arm distributors are often operated with only two arms.

17.14.3 Rotating Biological Contact Units

Rotating biological contact (RBC) units contain circular disks made of styrofoam, high-density plastic, or other lightweight material, which are rotated at 1–2 rpm. Oxygen is transferred from the atmosphere to the exposed film. An RBC unit is shown

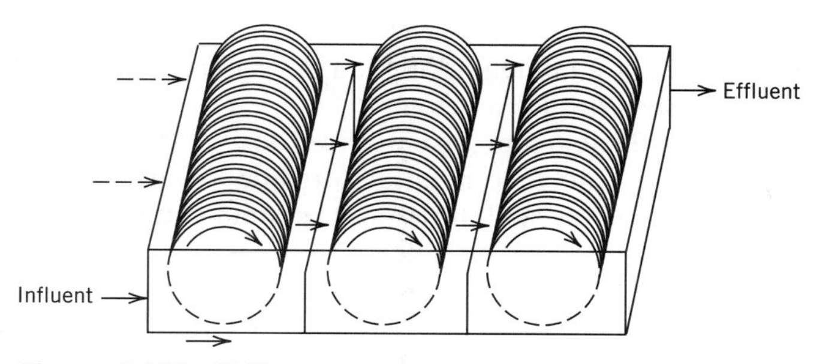

Figure 17.18 RBC process.

in Fig. 17.18. Diffused aeration may be used to enhance performance. The process has similarities to trickling filters and the activated sludge process but the biofilm performance is the main feature of the process.

The WEF and ASCE (1992a) note that RBC units have been prone to operational problems with the shafts and medium. Excessive biomass buildup may occur on the medium. When aeration is used shaft rotation may be uneven. However, there are a number of RBC installations and manufacturers of the units.

The advantages of an RBC process are the low power input to supply oxygen to the biomass and the high biomass concentration and sludge age attained with the fixed medium. The power to rotate a shaft ranges from 2 to 4 kW per shaft, or a reasonable value is 32 kW/1 000 m^2 (3 kW/1 000 ft^2) of medium surface (Owen, 1982). Antonie (1976) calculated that RBC system would require one third to one half of the energy of an activated sludge system. There is no need for recycle in an RBC system and maintenance is minimal. The exposure of the disks to the atmosphere promotes cooling of the biomass; therefore the reactors must be covered in cold climates.

Mechanical or compressed air drives are used to rotate the media. Disks have diameters ranging from 2 to 3.2 m (6.5–10.5 ft). Disks are approximately 1.3 cm (0.5 in.) thick and are spaced on 3.4-cm (1.3-in.) centers (Antonie, 1976).

Influent to an RBC should be settled or at least screened. The medium in an RBC unit is typically about 40% submerged. After a ripening period of 1–2 weeks the medium will be covered with biomass. As biomass builds up on the medium the film is subjected to increasing shear in the liquid. Increased weight of the film and shear cause the film to slough off. The sloughed biomass and other suspended solids exit with the effluent, which is sent to a final clarifier. Flow through an RBC unit is more commonly parallel to the media (Grady and Lim, 1980) resulting in more uniform conditions in a module. Including a number of modules in series provides the best treatment.

Modeling RBC systems is difficult. Mass transfer of substrate, nutrients, and oxygen significantly affect the rate of substrate removal and biomass growth in the films. Suspended biomass in the liquid also participates in substrate removal. As the thickness of a biofilm increases, mass transfer limitations prevent substrate or oxygen from being transported to the base layers of the film. A limiting biofilm thickness is reached beyond which there is no improvement in the rate of substrate removal. Anaerobic activity in the base layers of the film leads to production of H_2S and deteriorates process performance. Various measures can be used to control film thickness. The rotational speed of the disks can be increased to increase the level of shear; aeration can be periodically used to increase the level of shear. The direction of rotation can be reversed. Chemicals such as chlorine can be applied to kill biomass and cause it to slough. A step feed, wherein influent is applied at various points along the reactor, results in more uniform growth on the disks.

Modeling complexity has led to development of empirical design criteria. The most important feature in design of the units is to maintain organic loading rates at levels that will not exceed the oxygen transfer capabilities of the system. Supplemental aeration can be used if necessary to improve performance.

An equation similar to the Schultz–Germain formula for trickling filters is recommended for modeling RBC performance (Eq. 17.74), although a variety of other semiempirical formulations have been used (Spengel and Dzombak, 1992).

$$S_e = S_i e^{k(V/Q)^{0.5}} \tag{17.74}$$

where

V is in m^3

Q is in m^3/s

A reasonably conservative range for k is 90–112 at a temperature of 13°C for the units just given on V and Q (WEF and ASCE, 1992a). This value was arrived at by considering the performance variation of a number of installations. If the temperature drops to 5°C the value of k should be decreased to 60–74 pending studies at lower temperatures.

Medium surface organic loadings to the first stage of an RBC should not exceed 3.1 kg total $BOD_5/100\ m^2/d$ (0.6 $lb/100\ ft^2/d$) or 1.2 kg soluble $BOD_5/100\ m^2/d$ (0.25 $lb/100\ ft^2/d$) (WEF and ASCE, 1992a). Loading rates from 0.76–1.2 kg total $BOD_5/100\ m^2/d$ (0.15–0.25 $lb/100\ ft^2/d$) achieved nitrification in various studies when the temperature was above 4.4–12.4°C (Antonie, 1976). WEF and ASCE (1992a) provide a design example for an RBC based on typical commercially available modules.

The scaleup of bench-scale treatability studies to full-scale systems for RBC systems presents more difficulties compared to other biological treatment processes (Spengel and Dzombak, 1992). Oxygen transfer is one of the more problematic variables. If high rotation rates are used in small-scale systems to achieve the same disk tip speed as in a large system, shear levels and turbulence in the liquid will be different. Film thickness affects all mass transfer phenomena. The biomass will be exposed to the atmosphere and wastewater at a higher frequency in the small-scale system. These factors will cause differences in the two systems. Preserving conditions of similitude will be unable to resolve these complex interactions.

Sludge production from RBC units will be similar to observed yields from trickling filter processes. The sludge age is high in an RBC process because of the adherence of biological solids to the disks.

In a four-stage RBC plant without aeration, mean DO levels were less than 0.75 mg/L at all organic loading rates (Surampalli and Baumann, 1989). Supplemental aeration was applied to increase the mean DO level to 1.5–1.8 mg/L in the first stage and up to a mean value of 5.2 mg/L in the fourth stage at lower organic loadings. Soluble COD removal increased by 4 and 20%, respectively, at low and high organic loadings at the higher DO levels. Ammonia nitrification improved by a factor from 3 to 4 depending on the loading rate.

QUESTIONS AND PROBLEMS

1. In a system with recycle, the following data on inert (nonbiodegradable inorganics that originated in the influent) solids concentrations were obtained:

$$X_{Ti0} = 20\text{ mg/L} \qquad X_{Tie} = 0\text{ mg/L} \qquad X_{Tir} = 100\text{ mg/L}$$

The system is at steady state and the clarifier is ideal. What are the values for Q_w and X_{Ti}, if Q is 0.35 m^3/s (7.99 Mgal/d) and the recycle ratio, r, is (i) 0.5 (ii) 1.0? Does the flow regime (i.e., CM or PF) in the reactor make any difference? If it does, perform the calculations for each regime.

2. If the inert solids concentration escaping in the clarified effluent (X_{Tie}) is not negligible, what is the equation describing the relation between the reactor inert solids concentration and influent clarified effluent inert solids concentrations (see Eq. 17.10)? Solve Problem 1 with $X_{Tie} = 3$ mg/L.

3. What are the units and word descriptions (identify the parameter specifically, e.g., not only mass but mass of liquid, biomass, substrate, or other) for the following parameters in a CM activated sludge system with recycle: S_0, S_e, k, k_e, X_V, Y, r_S, r, and V. The substrate kinetic model being applied is

$$r_S = -kX_VS_e$$

Example:

$$Q{:}\frac{\text{volume (sewage)}}{\text{time}}, L^3T^{-1}$$

4. From the following list of parameters, which are primary (independent) variables and which are dependent (designer's choice) variables?

$$S_0,\ S_e,\ \theta_X,\ \theta_d,\ Y,\ k,\ K,\ X_V$$

5. Discuss why using Eq. (17.4) as opposed to using a net yield factor multiplied by the rate of substrate removal better describes the rate of biomass production in an activated sludge process.

6. To more precisely control the sludge age, laboratory studies are often conducted with recycle systems incorporating CM or PF reactors but the sludge is wasted directly from the aeration basin as opposed to wasting sludge from the clarifier underflow line. Sketch a recycle system similar to Fig. 17.1b except that Q_w (the waste solids flow) is a separate flow directly withdrawn from the aeration basin. Underflow from the clarifier is recycled to the aeration basin. Write steady state solids balance equations for ISS and VSS for the system, aeration basin, and clarifier. What is the sludge age in this system if 10% of the volume of the aeration basin is wasted each day and VSS concentrations in the influent and clarified effluent are negligible?

7. Why does recycle of a portion of the settled effluent from an activated sludge reactor result in an increase in the average residence time of sludge in the reactor but does not result in an increase of the average residence time of water in the reactor?

8. If studies conducted with a wastewater indicated that the biomass VSS formed had an associated inorganic component of 0.30 g/g VSS, what is the mixed liquor concentration of inert solids that originated from the influent if the mixed liquor TSS and VSS measured 2 000 and 1 480 mg/L, respectively? Assume that there was no solubilization of influent inert suspended solids.

9. How do operating conditions of an activated sludge basin generally influence the settleability of sludge in the secondary clarifier?

10. If the nutrient requirement for nitrogen and phosphorus is 100:5:1 (mass basis) when degradable organic matter is expressed on a COD basis, what is the ratio if BOD_5 is used to express degradable organic matter and BOD_5 is assumed to be two thirds of ultimate BOD?

11. The Reynold's and Froude numbers are dimensionless groups governing the behavior of fluid flow and used to scale fluid flow processes. What are the parameters that would be used to scale up an activated sludge process?

12. Discuss why the sludge age used in calculations is not the actual residence time of sludge in the system (aeration basin and clarifier)?

13. If the sludge concentration in the return sludge line from the secondary clarifier were the same as the MLVSS concentration (X_V), would recycle of return sludge increase the sludge age?

14. If an activated sludge process operator decided to reduce Q_w to arbitrary low rates, describe the effects on the system. What would happen to substrate removal and effluent quality?

15. Derive the equation for the minimum sludge age if Eq. (17.1a) is the substrate removal expression.

16. (a) What is the value of the rate constant in the first-order substrate removal expression (Eq. 17.2a) equivalent to the Monod expression (Eq. 17.1a) that has a k of 1.10 d^{-1} and K of 100 mg/L for S values of 200 and 20 mg/L? Assume that X_V is unchanged.
(b) Perform the same exercise as in (a) to relate Eq. (17.2b) to Eq. (17.1a) if k and K are 1.10 d^{-1} and 100 mg/L, respectively, in the latter equation and X_V is 2 500 mg/L.

17. Why does f equal 1.42 mg COD/mg VSS in Eq. (17.26) if microorganisms have a composition of $C_5H_7NO_2$?

18. What is the composition of VSS ($C_xH_yO_zN_a$) if the COD of VSS is 1.55 mg COD/mg VSS and the nitrogen content is 0.10 mg N/mg VSS? (Hint: nitrogen is not oxidized in a COD determination. Set $x = 1$ to facilitate solution of the problem.)

19. The yield factor (Y) for a wastewater was determined to be 0.65 mg VSS/mg BOD_5. What is the yield factor on a COD basis (mg VSS/mg COD) if the rate constant for BOD exertion was 0.090 d^{-1} (base 10) for this wastewater?

20. What is the observed yield factor when Y is 0.57 g VSS/g BOD_5 and k_e is 0.065 d^{-1} at sludge ages of (a) 4 d and (b) 6 d?

21. How would Fig. 17.4 change if sludge age were plotted along the abscissa?

22. Laboratory studies on a domestic sewage have given the following information: k = 100 mg/L/h, K = 40 mg COD/L, k_e = 0.07 d^{-1}, Y = 0.45 mg VSS/mg COD, and VSS = 0.75 TSS. A settling column test has shown that the SVI is 125 mL/g. Design a steady state CM activated sludge basin with recycle for these data. The design flow is 1 000 m^3/d with a COD of 220 mg/L. Use a sludge age of 5 d. There is a nonbiodegradable component of influent COD of 20 mg/L. The efficiency of treatment is 90%. Assume that the clarified effluent contains a negligible amount of solids. Lab studies have shown that the rate of substrate removal is essentially independent of the VSS at a concentration around 1 000 mg VSS/L or higher in the aeration basin. The substrate kinetic model under these conditions is

$$r_S = -\frac{kS_e}{K + S_e}$$

The solids concentration coming into the process is negligible and also assume that the solids in the clarified effluent are negligible.
(a) Find: ρ, S_e, θ_d, r, X_V, X_{Vr}, V, Q_w, and U.
(b) Using the basin designed in (a) with the same rate constants, S_0, θ_X, and SVI, what must happen to X_V, S_e, Q_w, r, and ρ if Q doubles? Assume that the same substrate kinetic model applies.

23. Design a CM recycle activated sludge system to treat a waste flow of 5 400 m^3/d (1.43 Mgal/d) with a COD of 375 mg/L. Influent suspended solids are negligible. The nondegradable portion of influent COD is 15 mg/L. The process must attain a removal of 95% of the degradable COD. A detention time of 6 h will be used. VSS in the return sludge line will have a concentration of 9 000 mg/L. The following parameters have been determined based on the following substrate removal model:

$$r_S = -\frac{kX_VS_e}{K + S_e}$$

Y = 0.45 mg VSS/mg COD $\quad$ k = 4.4 mg COD/mg VSS/d $\quad$ K = 170 mg/L
k_e = 0.05 d^{-1}

What are the sludge age, MLVSS, waste sludge flow rate, and recycle ratio in this process? Assume there are no solids in the clarifier overflow.

24. Repeat Problem 22 for a PF activated sludge system.

25. Why should recycled sludge be input at the front end of a PF basin?

26. For an activated sludge system with recycle assume that the flow regime for solids in all channels and pipes in the system is PF. Also assume that the solids movement from entrance to underflow in the clarifier is PF (first in, first out). Solids are wasted from the underflow of the clarifier and the sludge age is maintained at θ_X.
(a) If the bioreactor flow regime is ideal CM, would the SRT of all solids be near θ_X or would there be an age distribution for the solids? Explain your answer.
(b) Answer the same question as (a) for an ideal PF bioreactor flow regime.

27. (a) What is the SVI of mixed liquor with a TSS concentration of 3 150 mg/L that settles to a volume of 160 mL is a 1-L cylinder?
(b) What is the maximum SVI that can be obtained from MLTSS concentrations of 1 000, 3 000, and 6 000 mg/L?

28. Find the expression for $\rho \left[\frac{Q(S_0 - S_e)}{V} \right]$ in terms of k, r, θ_d, $\overline{X}_V$, S_0, and S_e for a PF activated sludge aeration basin when the kinetic expression is

$$r_s = -k\overline{X}_V S$$

29. For an SBR process, the sludge takes 45 min to settle and the depth of liquid in the reactor is 40% of the depth of the reactor after drawoff to prevent disturbing the settled sludge. The drawoff period takes an additional 20 min after the settle phase. An idle time of 10 min is to be allowed. What is the required volume of each of the tanks in 2-tank and 3-tank SBR systems to handle a flow of 55 000 m^3/d to provide a react time of 60 min?

30. A sequencing batch reactor system is being designed to treat a wastewater with a flow of 40 000 m^3/d and influent COD concentration of 450 mg/L. The system will have two tanks. From laboratory studies, the substrate removal model that applies is a first-order expression:

$$r_S = -kS$$

where
S is the soluble COD concentration
$k = 8.4\ d^{-1}$

Aeration will be supplied during the fill and react periods and the model applies during each of these phases. Aeration is suspended during the settle and draw periods and there is no substrate removal during these periods.

The overall net yield of biomass is 0.31 g VSS/g COD removed at an effective sludge age of 5 d. What is the MLVSS concentration that exists at the end of the react period? The SVI is 95. The ratio of VSS:TSS is 0.77.

Size the volume of the tanks to treat the waste to a soluble effluent COD concentration of 50 mg/L. Find the soluble substrate concentration at the end of the fill period. Each tank will pass through fill, react, settle, and draw periods. The sludge will settle in a time of 35 min and the draw time will be 20 min. There is no idle time. Sludge will be wasted at the end of the draw period. Assume $\alpha = 0.65$. Calculate the sludge waste rate assuming there are no solids in the clarified effluent from the reactor.

31. Explain how a spike in influent concentration of a substance is changed by flow regime without other transformations in (a) an ideal CM reactor, (b) an ideal PF reactor, and (c) an SBR reactor.

32. If a wastewater is denitrified, is there any net oxygen requirement for nitrogen in the wastewater treatment process?

33. What influent concentration of phosphorus is able to be treated by an A/O enhanced phosphorus uptake process with nitrification where the biomass accumulates phosphorus at 3.5% of its weight? The process is removing 300 mg/L of COD and the observed yield factor is 0.30. Assume that the effluent contains 1.0 mg/L of phosphorus.

34. An economist will tell you that economies are generally achieved as the scale of an operation increases. If the use of water in a community increases with a concomitant increase in quantities of wastewater but the amount of organics discharged into the wastewater does not change will economies of scale apply to treatment of the greater quantities of more dilute wastewater?

35. What is the rate of oxygen consumption (kg/d) for an activated sludge process with operational conditions as follows?

$Q = 8\,000\ \text{m}^3/\text{d}$ $S_0 = 240\ \text{mg COD/L}$ $S_e = 30\ \text{mg COD/L}$

$\theta_X = 4.0\ \text{d}$ $\theta_d = 5.5\ \text{h}$ $X_V = 1\,500\ \text{mg/L}$

The COD of VSS was measured at 1.58 mg COD/mg VSS.

36. What are some operational problems and their solutions for trickling filters?

37. The laboratory data in the following table were obtained at 20°C.

Trickling Filter Lab Data

	% BOD_5 remaining			
	Hydraulic loading, $m^3/m^2/d$			
Depth, m	60	100	140	180
1	86	90	92	94
2	68	75	80	84
3	52	63	70	76
4	43	55	63	70
5	35	45	54	63
6	26	38	47	57

(a) Using these data find k and n in the Schultz–Germain formula for this waste and medium.

(b) The lowest expected temperature is 10°C. For a depth of 5.5 m and recirculation ratio of 2.0, what is the design filter diameter for a flow of 4 000 m^3/d for an influent containing a BOD_5 concentration of 220 mg/L? The desired BOD_5 removal is 90%. Assume $\theta = 1.030$.

38. Using information and rate constants developed in Problem 37, perform a sensitivity analysis on the effects on effluent quality of a +10% change in each of the following variables: recycle ratio, filter depth, and filter surface area. Vary each parameter independently. Use a temperature of 20°C.

39. Design a four-arm distributor for a trickling filter with a diameter of 28 m (91.8 ft), receiving a flow (included recycle) of 6 000 m^3/d (1.59 Mgal/d). The center area of the filter with a diameter of 1.0 m (3.30 ft) houses the distribution apparatus and does not contain medium. The arms will discharge continuously. The equivalent roughness height of the arms is 0.000 046 m (0.001 81 in.) and the design temperature is 15°C. Each arm will contain 20 evenly spaced ports with varying areas. The discharge coefficient for the ports is 0.72. The minimum velocity in the constant diameter arm is 0.30 m/s (1 ft/s). The peripheral speed of an arm is 0.60 m/s (2.0 ft/s) and the head provided by the dosing tank is 2.20 m (7.22 ft). Use an average friction factor in your analysis.

40. How much surface area is available on an RBC with a 9-m shaft supporting 2.5-m diameter disks spaced at 3.5 cm (center to center) with a thickness of 1.4 cm? What flow rate could be handled for an influent BOD_5 concentration of 150 mg/L to achieve a maximum loading of 1.2 kg soluble $BOD_5/100\ m^2/d$? Size the tank assuming 40% submergence of the media.

KEY REFERENCES

Metcalf and Eddy (1991), *Wastewater Engineering: Treatment Disposal, Reuse,* 3rd ed., G. Tchobanoglous and F. L. Burton, eds., McGraw-Hill, Toronto.

WEF and ASCE (1992a), *Design of Municipal Wastewater Treatment Plants,* vol. I, Water Environment Federation, Alexandria, VA.

WEF and ASCE (1992b), *Design of Municipal Wastewater Treatment Plants,* vol. II, Water Environment Federation, Alexandria, VA.

REFERENCES

Antonie, R. L. (1976), *Fixed Biological Surfaces—Wastewater Treatment,* CRC Press, Cleveland, OH.

Appeldoorn, K. J., G. J. J. Kortstee, and A. B. Zehnder (1992), "Biological Phosphate Removal by Activated Sludge under Defined Conditions," *Water Research,* 26, 4, pp. 453–460.

Ardern, E. (1917), "A Resume of the Present Position of the Activated Sludge Process of Sewage Purification," *J. Society of Chemical Industries,* 36, pp. 822–883.

Ardern, E. and W. T. Lockett (1914a), "Experiments on the Oxidation of Sewage without the Aid of Filters," *J. Society of Chemical Industries,* 33, pp. 523–539.

Ardern, E. and W. T. Lockett (1914b), "The Oxidation of Sewage without the Aid of Filters II," *J. Society of Chemical Industries,* 33, pp. 1122–1124.

Ardern, E. and W. T. Lockett (1915), "The Oxidation of Sewage without the Aid of Filters III," *J. Society of Chemical Industries,* 34, pp. 937–943.

Argaman, Y. (1991), "Chemical Reaction Engineeering and Activated Sludge," *Water Research,* 25, 12, pp. 1583–1586.

Bark, K., A. Sponner, P. Kämpfer, S. Grund, and W. Dott (1992), "Differences in Polyphosphate Adsorption by *Acinetobacter* Isolates from Wastewater Producing Polyphosphate: AMP Phosphotransferase," *Water Research,* 26, 10, pp. 1379–1388.

Bock, E., P. A. Wilderer, and A. Freitag (1988), "Growth of Nitrobacter in the Absence of Dissolved Oxygen," *Water Research,* 22, 2, pp. 245–249.

Brock, T. D. (1970), *Biology of Microorganisms,* Prentice-Hall, Englewood Cliffs, NJ.

Burkehead, C. E. and S. L. Waddell (1969), "Composition Studies of Activated Sludge," presented at the 24th Industrial Waste Conference, Purdue University, Lafayette, IN.

Cha, D. K., D. Jenkins, W. P. Lewis, and W. H. Kido (1992), "Process Control Factors Influencing *Nocardia* Populations in Activated Sludge," *Water Environment Research,* 64, 1, pp. 37–43.

Chiesa, S. C. and Irvine, R. L. (1985), "Growth and Control of Filamentous Microbes in Activated Sludge: an Integrated Hypothesis," *Water Research,* 19, 4, pp. 471–479.

Chiesa, S. C. and S. R. Linne (1990), "Discussion of: The Effects of Sludge Age and Selector Configuration on the Control of Filamentous Bulking in the Activated Sludge Process," *Research J. Water Pollution Control Federation,* 62, 6, pp. 828–829.

Chiesa, S. C., M. G. Rieth, and T. E. K. L. Ching (1990), "Evaluation of Activated Sludge Oxygen Uptake Rate Test Procedures," *J. of Environmental Engineering, ASCE,* 116, EE3, pp. 472–486.

Chudoba, J. (1985), "Control of Activated Sludge Filamentous Bulking—VI. Formulation of Basic Principles," *Water Research,* 19, 8, pp. 1017–1022.

CHUDOBA, J., V. OTTOVA, AND V. MADERA (1973), "Control of Activated Sludge Filamentous Bulking—I. Effect of the Hydraulic Regime or Degree of Mixing in an Aeration Tank," *Water Research,* 7, 8, pp. 1163–1182.

CHUDOBA, J., P. STRAKOVÁ, AND M. KONDO (1991), "Compartmentalized versus Completely-Mixed Biological Wastewater Treatment Systems," *Water Research,* 25, 8, pp. 973–978.

COUILLARD, D., S. GARIÉPY, AND F. T. TRAN (1989), "Slaughterhouse Effluent Treatment by Thermophilic Aerobic Process," *Water Research,* 23, 5, pp. 573–579.

DAIGGER, G. T. AND G. A. NICHOLSON (1990), "Performance of Four Full-Scale Nitrifying Wastewater Treatment Plants Incorporating Selectors," *Research J. Water Pollution Control Federation,* 62, 5, pp. 676–683.

DAS, D., T. M. KEINATH, D. S. PARKER, AND E. J. WAHLBERG (1993), "Floc Breakup in Activated Sludge Plants," *Water Environment Research,* 65, 2, pp. 138–145.

DROSTE, R. L. (1990), "Kinetic Comparison of Continuous Flow and Sequencing Batch Reactors for Biological Wastewater Treatment," in *Proceedings Annual CSCE Conference,* in vol. I-2, Canadian Society of Civil Engineering, pp. 536–556.

DROSTE, R. L., L. FERNANDES, AND X. SUN (1993), "Substrate Utilization and VSS Relations in Activated Sludge Processes," Third International Conference on Waste Management in the Chemical and Petrochemical Industries, IAWQ, Salvador, Brazil.

ECKENFELDER, W. W., JR. (1963), "Trickling Filtration Design and Performance," *Transactions American Society of Civil Engineers,* 128, Part III, pp. 371–384.

ECKENFELDER, W. W., JR. AND D. L. FORD (1970), *Water Pollution Control: Experimental Procedures for Process Design,* Pemberton Press, New York.

ERICSSON, B. AND L. ERIKSSON (1988), "Activated Sludge Characteristics in a Phosphorous Depleted Environment," *Water Research,* 22, 2, pp. 151–162.

FORD, D. L. AND W. W. ECKENFELDER, JR. (1967), "Effect of Process Variables on Sludge Floc Formation and Settling Characteristics," *J. Water Pollution Control Federation,* 39, 11, pp. 1850–1859.

GERMAIN, J. E. (1966), "Economic Treatment of Domestic Waste by Plastic-Medium Trickling Filter," *J. Water Pollution Control Federation,* 38, 2, pp. 192–203.

GOODMAN, B. L. AND A. J. ENGLANDE, JR. (1974), "A Unified Model of the Activated Sludge Process," *J. Water Pollution Control Federation,* 46, 2, pp. 312–332.

GRADY, C. P. L., JR. AND H. C. LIM (1980), *Biological Wastewater Treatment: Theory and Applications,* Marcel Dekker, New York.

HAMODA, M. F. (1989), "Kinetic Analysis of Aerated Submerged Fixed-Film (ASFF) Bioreactors," *Water Research,* 23, 9, pp. 1147–1154.

HAO, O. J. (1988), "Kinetics of Microbial By-Product Formation in Chemostat Pure Cultures," *J. of Environmental Engineering, ASCE,* 114, EE5, pp. 1097–1115.

HENZE, M., R. DUPONT, P. GRAU, AND A. SOTA (1993), "Rising Sludge in Secondary Settlers due to Denitrification," *Water Research,* 27, 2, pp. 231–236.

HENZE, M., C. P. L. GRADY, JR., W. GUJER, G. V. R. MARAIS, AND T. MATSUO (1987), "A General Model for Single-Sludge Wastewater Treatment Systems," *Water Research,* 21, 5, pp. 505–515.

HOOVER, S. R. AND N. PORGES (1952), "Assimilation of Dairy Waste by Activated Sludge," *Sewage and Industrial Waste,* 24, pp. 306–312.

IRVINE, R. L., AND A. W. BUSCH (1979), "Sequencing Batch Reactors—An Overview," *J. Water Pollution Control Federation,* 51, 2, pp. 235–243.

JACKSON, M. L. (1982), "Thermophilic Treatment of a High Biological Demand Wastewater: Laboratory, Pilot Plant and Design," *Proc. 37th Industrial Waste Conference,* Purdue University, Lafayette, IN pp. 753–763.

JØRGENSEN, P. E., T. ERIKSEN, AND B. K. JENSEN (1992), "Estimation of Viable Biomass in

Wastewater and Activated Sludge by Determination of ATP, Oxygen Utilization Rate and FDA Hydrolysis," *Water Research,* 26, 11, pp. 1495–1501.

KAPPELER, J. AND W. GUJER (1994), "Verification and Applications of a Mathematical Model for 'Aerobic Bulking,'" *Water Research,* 28, 2, pp. 311–322.

KEYES, T. W. AND T. ASANO (1975), "Application of Kinetic Models to the Control of Activated Sludge Processes," *J. Water Pollution Control Federation,* 47, 11, pp. 2574–2585.

LAWRENCE, A. L. AND P. L. MCCARTY (1969), "Kinetics of Methane Fermentation in Anaerobic Treatment," *J. Water Pollution Control Federation,* 41, 2, pp. R1–R17.

LAWRENCE, A. L. AND P. L. MCCARTY (1970), "Unified Basis for Biological Treatment Design and Operation," *J. Sanitary Engineering Division, ASCE,* 96, SA3, pp. 757–778.

LINNE, S. R., S. C. CHIESA, M. G. RIETH, AND R. C. POLTA (1989), "The Impact of Selector Operation on Activated Sludge Settleability and Nitrification: Pilot-Scale Results," *J. Water Pollution Control Federation,* 61, 1, pp. 66–72.

LOVETT, D. A., B. V. KAVANAGH, AND L. S. HERBERT (1983), "Effect of Sludge Age and Substrate Composition on the Settling and Dewatering Characteristics of Activated Sludge," *Water Research,* 17, 11, pp. 1511–1515.

LOVETT, D. A., S. M. TRAVERS, AND K. R. DAVEY (1984), "Activated Sludge Treatment of Abattoir Wastewater—I. Influence of Sludge Age and Feeding Pattern," *Water Research,* 18, 4, pp. 429–434.

MCCARTY, P. L., L. BECK, AND P. ST. AMANT (1969), "Biological Denitrification of Wastewaters by Addition of Organic Materials," *Proc. of the 24th Industrial Waste Conference,* Purdue University, Lafayette, IN, pp. 1271–1285.

MCKINNEY, R. E. (1962), "Mathematics of Complete-Mixing Activated Sludge," *J. Sanitary Engineering Division, ASCE,* 88, SA3, pp. 87–113.

METCALF, L. AND H. P. EDDY (1935), *American Sewerage Practice,* vol. III, McGraw-Hill, Toronto.

Metcalf and Eddy (1979), *Wastewater Engineering: Treatment Disposal, Reuse,* 2nd ed., G. Tchobanoglous, ed., McGraw-Hill, Toronto.

MONOD, J. (1949), "The Growth of Bacterial Cultures," *Annual Reviews of Microbiology,* 3, pp. 371–394.

MURPHY, K. L. AND B. BOYKO (1970), "Longitudinal Mixing in Spiral Flow Aeration Tanks," *J. Sanitary Engineering Division, ASCE,* 96, SA2, pp. 211–221.

MYNHIER, M. D. AND C. P. L. GRADY, JR. (1975), "Design Graphs for the Activated Sludge Process," *J. of the Environmental Engineering Division, ASCE,* 101, EE5, pp. 829–846.

National Research Council (1946), "Sewage Treatment at Military Installations," *Sewage Works J.,* 18, 5, pp. 791–1028.

NOVÁK, L., L. LARREA, J. WANNER, AND J. L. GARCI-HERAS (1993), "Non-Filamentous Activated Sludge Bulking in a Laboratory Scale System," *Water Research,* 27, 8, pp. 1339–1346.

NOWAK, G. AND G. D. BROWN (1990), "Characteristics of *Nostocoida limicola* and Its Activity in Activated Sludge Suspension," *J. Water Pollution Control Federation,* 62, 2, pp. 137–142.

OLESZKIEWICZ, J. A. AND S. A. BERQUIST (1988), "Low Temperature Nitrogen Removal in Sequencing Batch Reactors," *Water Research,* 22, 9, pp. 1163–1171.

ORDON, C. J. (1969), "Manifolds, Rotating and Stationary," *J. Sanitary Engineering Division, ASCE,* 92, SA1, pp. 269–280.

OWEN, W. F. (1982), *Energy in Wastewater Treatment,* Prentice-Hall, Toronto.

PARKER, D. S. AND M. S. MERRILL (1976), "Oxygen and Air Activated Sludge: Another View," *J. Water Pollution Control Federation,* 48, 11, pp. 2511–2528.

PITT, P. AND D. JENKINS (1990), "Causes and Control of *Nocardia* in Activated Sludge," *J. Water Pollution Control Federation,* 62, 2, pp. 143–150.

RASMUSSEN, H., J. H. BRUUS, K. KEIDING, AND P. H. NIELSEN (1994), "Observations of Dewaterability and Physical, Chemical and Microbiological Changes in Anaerobically Stored Activated Sludge from a Nutrient Removal Plant," *Water Research,* 28, 2, pp. 417–425.

Sandford, D. S. and K. A. Chisholm (1977), "The Treatment of Municipal Wastewater Using the ICI Deep Shaft Process," presented at the 29th Western Canada Water & Sewage Treatment Conference, Edmonton, Alberta, Canada.

Schön, G., S. Geywitz, and F. Mertens (1993), "Influence of Dissolved Oxygen and Oxidation–Reduction Potential on Phosphate Release and Uptake by Activated Sludge from Sewage Plants with Enhanced Biological Phosphorus Removal," *Water Research,* 27, 3, pp. 349–354.

Schulz, J. R., B. A. Hegg, and K. L. Rakness (1982), "Realistic Sludge Production for Activated Sludge Plants without Primary Clarifiers," *J. Water Pollution Control Federation,* 54, 10, pp. 1355–1359.

Schultz, K. L. (1960), "Load and Efficiency of Trickling Filters," *J. Water Pollution Control Federation,* 33, 3, pp. 245–260.

Selna, M. W. and E. D. Schroeder (1978), "Response of Activated Sludge Processes to Organic Transients—Kinetics," *J. Water Pollution Control Federation,* 50, 5, pp. 944–956.

Seviour, E. M., C. J. Williams, R. J. Seviour, J. A. Soddell, and K. C. Lindrea (1990), "A Survey of Filamentous Bacteria Populations from Foaming Activated Sludge Plants in Australia," *Water Research,* 24, 4, pp. 493–498.

Simpkin, T. J. and W. C. Boyle (1988), "The Lack of Repression by Oxygen of the Denitrifying Enzymes in Activated Sludge," *Water Research,* 22, 2, pp. 201–206.

Spengel, D. B. and D. A. Dzombak (1992), "Biokinetic Modeling and Scale-Up Considerations for Biological Contractors," *Water Environment Research,* 64, 3, pp. 223–235.

Srinivasaraghavan, R. and A. F. Gaudy, Jr. (1975), "Operational Performance of an Activated Sludge Process with Constant Sludge Feedback," *J. Water Pollution Control Federation,* 47, 7, pp. 1946–1960.

Standard Methods for the Examination of Water and Wastewater (1992), 18th ed., American Public Health Association, Washington, DC.

Surampalli, R. Y. and E. R. Baumann (1989), "Supplemental Aeration Enhanced Nitrification in a Secondary RBC Plant," *J. Water Pollution Control Federation,* 61, 2, pp. 200–207.

Sykes, R. M. (1982), "Hydraulic Regime and Activated Sludge Performance," *J. Environmental Engineering Division, ASCE,* 108, EE2, pp. 286–296.

Sykes, R. M. (1983), "Author's Closure to Discussion of Hydraulic Regime and Activated Sludge Performance," *J. of Environmental Engineering, ASCE,* 109, EE2, pp. 527–529.

Toerber, E. D., W. L. Paulson, and H. S. Smith (1974), "Comparison of Completely Mixed and Plug Flow Biological Systems," *J. Water Pollution Control Federation,* 46, 8, pp. 1995–2014.

USEPA (1987), *Phosphorus Removal,* Design Manual No. EPA/625/1-87/001, Center for Environmental Research Information, Cincinnati, OH.

USEPA (1993), *Nitrogen Control,* Manual No. EPA/625/R-93/010, Center for Environmental Research Information, Cincinnati, OH.

Velz, C. J. (1948), "A Basic Law for the Performance of Biological Beds," *Sewage Works J.,* 20, 4, pp. 607–617.

Wanner, J., J. Cudoba, K. Kucman, and L. Proske (1987a), "Control of Activated Sludge Filamentous Bulking—VII. Effect of Anoxic Conditions," *Water Research,* 21, 12, pp. 1447–1451.

Wanner, J., K. Kucman, V. Ottova, and P. Grau (1987b), "Effect of Anaerobic Conditions on Activated Sludge Filamentous Bulking in Laboratory Systems," *Water Research,* 21, 12, pp. 1541–1546.

Wanner, J., K. Kucman, and P. Grau (1988), "Activated Sludge Process Combined with Biofilm Cultivation," *Water Research,* 22, 2, pp. 207–215.

WEDDLE, C. L. AND D. JENKINS (1971), "The Viability and Activity of Activated Sludge," *Water Research,* 5, 8, pp. 621–640.

WILLIAMS, T. M. AND R. F. UNZ (1989), "The Nutrition of Thiotrix, Type 021N, Beggiatoa and Leucothrix Strains," *Water Research,* 23, 1, pp. 15–22.

WPCF (1977), *Wastewater Treatment Plant Design,* Manual of Practice No. 8, Water Pollution Control Federation, Washington, DC.

WPCF (1988), *Operation and Maintenance of Trickling Filters, RBCs, and Related Processes,* Manual of Practice No. OM-10, Water Pollution Control Federation, Washington, DC.

CHAPTER 18

ANAEROBIC WASTEWATER TREATMENT

Anaerobic treatment, being a biological treatment, bears many similarities to aerobic treatment. This chapter develops the concepts of anaerobic treatment in light of previous discussions for aerobic treatment given in Chapter 17, pointing out the distinguishing features of anaerobic treatment in comparison to aerobic treatment.

Until recently "biological treatment" has meant an aerobic process in which oxygen is supplied by aeration to allow aerobic bacteria to break down and assimilate the wastes. The supply of air is a major operational expense of this process. Large quantities of sludge are also produced, which may or may not have resale value. If no resale opportunity exists, sludge disposal is another major expense associated with the process.

Anaerobic treatment, in contrast to aerobic treatment, does not require air input and generates considerably smaller quantities of sludge. The sludge produced has approximately the same fertilizer value per unit weight as sludge produced in an aerobic process. Anaerobic treatment may require substantial quantities of heat energy input but this energy penalty may be completely offset by the methane produced. There are also other methods or circumstances that reduce the heat energy requirement. The biogas (methane) feature of anaerobic treatment has been promoted as a solution to energy problems in various articles and, although claims may be exaggerated in some circumstances, many industrial wastes provide opportunities for realization of these benefits. The food and beverage industry, for example, produces wastes that are often rich in organic content and have temperatures above the ambient temperature.

In recent decades, several developments have occurred that have greatly increased energy efficiency and attractiveness of anaerobic waste treatment. Research groups throughout the world have developed anaerobic reactors to treat wastes more quickly, more reliably, and with a greater net production of methane gas. Full-scale implementations of these developments have met with success and competitive installations will continue to take advantage of the new technologies.

History

The production of biogas, consisting primarily of methane and carbon dioxide, was discovered in the seventeenth century after scientists observed "marsh gas" burning on the surface of swamps. Marsh gas is methane, a product of anaerobic biological degradation of organic materials. Mass transfer of oxygen from the atmosphere is unable to maintain measurable concentrations of oxygen under stagnant water conditions and high concentrations of organic matter. Under anaerobic conditions algae, which produce oxygen, are unable to survive.

Wastewaters, unless applied to a broad surface area or supplied from artificial aerators, naturally give rise to anaerobic treatment. Not necessarily by design, anaerobic treatment occurs in any holding tank for wastewater, including cesspools and septic tanks, which are among the first "treatments" applied to wastewater. Solids are metabolized and solubilized in anaerobic treatment. Solids are easily collected to form a highly concentrated waste and solids reduction was the primary application of anaerobic treatment until modern times. Research and implementation of engineered anaerobic treatment processes for solids reduction have been in progress for over a century.

Ironically, this oldest form of wastewater treatment was not developed and was applied only circumstantially in ponds for high-strength wastewaters. Reasons for this will become clear as concepts in this chapter are developed. But domestic wastewater, particularly in North America, is dilute and anaerobic treatment of low-strength wastewaters remains a challenge today.

Except for anaerobic ponds, the first application of anaerobic treatment to raw wastewaters was in the 1950s, when the anaerobic contact process, which is a configuration similar to a recycle activated sludge process, was developed. Industrial wastewaters often have organics concentrations high enough to make anaerobic treatment feasible and this process proved to be successful. Since then at least five major options have been developed for anaerobic waste treatment and most of these have lowered the organic strength threshold at which a waste can be treated anaerobically.

18.1 ANAEROBIC METABOLISM

There is no oxygen present or consumed in an anaerobic process. In the fermentation that occurs, a portion of the energy-rich organic compounds are oxidized but other organic compounds or inorganic carbon dioxide are used as the electron (hydrogen) acceptors. The transfer of electrons does release relatively small amounts of chemically bound energy (reduction in free energy) which is used for growth by the anaerobes. Because the energy yield of anaerobic fermentation is only about one seventh of the yield for aerobes (Zoetemeyer, 1982), the growth of anaerobes is slow. However, this does not mean that their rate of processing substrate is low.

Because there is no oxygen consumed in the process there is no reduction in COD in an anaerobic process. The removal of COD is accomplished by the conversion of organics into methane, which is a relatively insoluble gas (see Henry's law constants in Table 1.4). There will also be a significant production of CO_2. The production of hydrogen and other gases also occurs but it is small enough to be negligible. Anaerobic treatment is realized in the final conversion of metabolic intermediates to methane. If the process is stopped short of this step, the effluent contains soluble products from intermediate stages of metabolism with the same oxygen demand as the initial material.

Anaerobic digestion involves a complex consortium of microorganisms. Toerien et al. (1970) suggested that the biochemical processes as well as the microbial species involved could be divided into three categories. A schematic of the reaction steps is outlined in Fig. 18.1.

1. Hydrolysis

Large molecules and suspended solids cannot be directly metabolized by anaerobes. When their concentrations are significant, hydrolysis reactions become an important first stage of anaerobic metabolism. Hydrolysis is the breakdown of large, complex soluble and insoluble molecules into smaller molecules that can

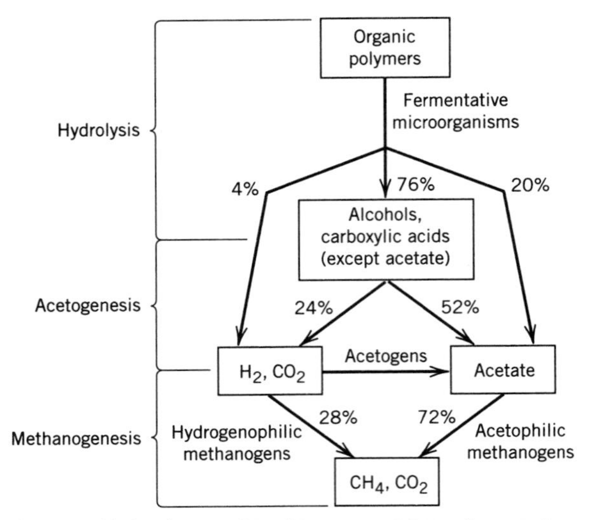

Figure 18.1 Anaerobic decomposition of organic matter. From Zehnder et al. (1982).

be transported into the cells and metabolized. Extracellular enzymes associated with the primary fermentative microorganisms are used to accomplish the task. There is an expenditure of energy in hydrolysis reactions. The energy for hydrolysis and synthesis is obtained from the catabolism of the smaller molecules resulting from hydrolysis. The fermentative microorganisms responsible for this step do not form methane.

2. Acetogenesis and Acid Formation
The same microorganisms that perform hydrolysis reactions carry the fermentation through this stage. The end products of hydrolysis are fermented into organic acids, other low molecular weight compounds, hydrogen, and carbon dioxide. The primary product of this fermentation is acetic acid. As shown in Fig. 18.1, microbial degradation of hydrolysis products is accompanied by a significant amount of hydrogen production. Bacteria that produce acetic acid and hydrogen are acetogenic bacteria. Other acid forming bacteria form butyric and propionic acid as well as other low molecular weight compounds. These microorganisms are relatively hardy and can tolerate a wide range of environmental conditions. The acid formers have an optimal pH between 5–6. Digesters are normally operated at a pH near 7 but their metabolic rates at this pH are still favorable compared to the methane formers responsible for the final conversion of organics into methane. Without proper operation of the process, particularly pH control, the acid formers can create highly unfavorable conditions for the methane formers.

3. Methanogenesis
The formation of methane, which is the ultimate product of anaerobic digestion, occurs by two major routes. The primary route is the fermentation of the major product of the acid forming phase, acetic acid, to methane and carbon dioxide. Bacteria that utilize acetic acid are acetoclastic (or acetophilic) bacteria. The overall reaction is

$$CH_3COOH \rightarrow CH_4 + CO_2$$

Based on thermodynamic considerations and from experimental data, Zeikus (1975) proposed the following reaction for acetate conversion to methane:

$$CH_3COOH + 4H_2 \rightarrow 2CH_4 + 2H_2O$$

The most common acetoclastic methanogens in reactors treating wastes with a high volatile fatty acid content are from the genera *Methanosarcina* and *Methanosaeta* (formerly *Methanothrix*[1]). *Methanosarcina* spp. are coccoid bacteria with doubling times near 1.5 d, and *Methanosaeta* spp. are sheathed rods, sometimes growing as long filaments with doubling times near 4 d (Zinder, 1988). These doubling times occur at optimal conditions for the methane formers. Even though *Methanosaeta* spp. grow more slowly, they are most frequently the dominant genus.

Some methanogens are also able to use hydrogen to reduce carbon dioxide to methane (hydrogenophilic methanogens) with an overall reaction of

$$4H_2 + CO_2 \rightarrow CH_4 + 2H_2O$$

There is a synergistic relation between the hydrogen producers and the hydrogen scavengers. Subtle changes in hydrogen conditions can change the end products of the acid forming phase. Furthermore, as the hydrogen partial pressure rises, hydrogen oxidation becomes more thermodynamically favorable than acetate degradation and acetate production is increased (Harper and Pohland, 1987). Alcohol degradation is also inhibited by high hydrogen levels. As noted earlier the net hydrogen production from the two phases is very small but it is an important intermediate in the metabolic processes.

It is recommended that overall hydrogen partial pressures be below 10^{-4} atm (corresponding to approximately a 10^{-8} *M* solution) for stability and good performance in anaerobic systems. At these hydrogen levels continuous production of acetic acid from influent and intermediate organics is assured and acetate utilization capacity is not inhibited (Harper and Pohland, 1987). These recommendations are based on thermodynamic considerations and the assumption of equilibrium between the liquid and the gas phase for hydrogen. Equilibrium between liquid and gas phases is not achieved, even in active reactors (Pauss and Guiot, 1993). Hydrogen concentrations in the liquid are higher than the equilibrium value because of mass transfer limitations. More research is required to understand hydrogen dynamics.

The methane formers are much more fastidious in their environmental requirements than the acid formers. Their rates of metabolism are also lower than the rates of the acid formers and therefore methane production is generally the rate-limiting step in anaerobic digestion. The optimal pH for methane formers is around 7.0 and their activity drops to very low values when the pH falls outside of the range of 6.0–8.0.

Inhibition of the methanogens can exacerbate deteriorating conditions. The methane formers fail to remove the organic acids produced in the initial stages of metabolism while the acid formers continue acid production. This adds more environmental stress on the methane formers. The process "pickles" itself, with no removal of COD. However, there are measures that can be taken to guard against this situation.

18.2 PROCESS FUNDAMENTALS

The design and operation of anaerobic processes must be cognizant of the characteristics of microorganisms and features of their metabolic pathways outlined in Section

[1]The original culture used to define the type species *Methanothrix sohengenii* was based on a mixed culture (Patel and Sprott, 1990).

18.1. The most important considerations for operation of anaerobic processes are discussed next.

18.2.1 Environmental Variables

Both physical and chemical variables influence the habitat of the microorganisms in the reactor. Oxidation–reduction potentials in anaerobic reactors are usually below −350 mV.

Temperature

As temperature increases the rate of reaction generally increases. For biological systems the rate increases are usually not as great as for chemical reactions.

There are two optimal ranges for process operation to produce methane: 30–40°C (the mesophilic range is from 15 to 40°C) and 50–60°C (the thermophilic range is for temperatures above 40°C). The psychrophilic range is temperatures below 15–20°C. Methane has been produced at temperatures down to 10°C or lower, but for reasonable rates of methane production, temperatures should be maintained above 20°C. Rates of methane production approximately double for each 10°C temperature change in the mesophilic range. Loading rates must decrease as temperature decreases to maintain the same extent of treatment. Operation in the thermophilic range is not generally practical because of the high heating energy requirement. A stable temperature is more conducive to stable operation than any specific temperature.

pH

The most important process control parameter is pH. The optimum pH range for all methanogenic bacteria studied by Zehnder et al. (1982) was between 6 and 8 but the optimum pH for the group as a whole is near 7.0. The lower growth rates of the methanogens require that the process be run at conditions most favorable to them. Numerous references report that the pH required in anaerobic systems for good performance and stability is in the range of 6.5–7.5, although stable operation has been observed outside this range. The system must contain adequate buffering capacity to accommodate the production of volatile acids and carbon dioxide that will dissolve at the operating pressure. Excess alkalinity or ability to control pH must be present to guard against the accumulation of excess volatile acids.

Anaerobic processes can operate over a wide range of volatile acids concentrations (from less than 100 mg/L to over 5 000 mg/L) if proper pH control is practiced. A constant pH lends stability to the process. Automatic pH control is often the most economical means of pH control because less chemicals are consumed. The alkalinity requirement varies with the waste, system operation, and type of process.

The three major chemical sources of alkalinity are lime, sodium bicarbonate, and sodium carbonate. Sodium hydroxide, which is often used in vegetable and fruit peeling operations, is also an important source of alkalinity. Other substances that produce alkalinity are soaps or other salts of organic acids.

Mixing

Septic tanks, Imhoff tanks, and sludge digesters were the first anaerobic treatment operations. All of these processes employed single-tank anaerobic reactors. In septic tanks incoming solids are settled and digested in the same compartment. An additional compartment may be included for polishing of the effluent but the first compartment

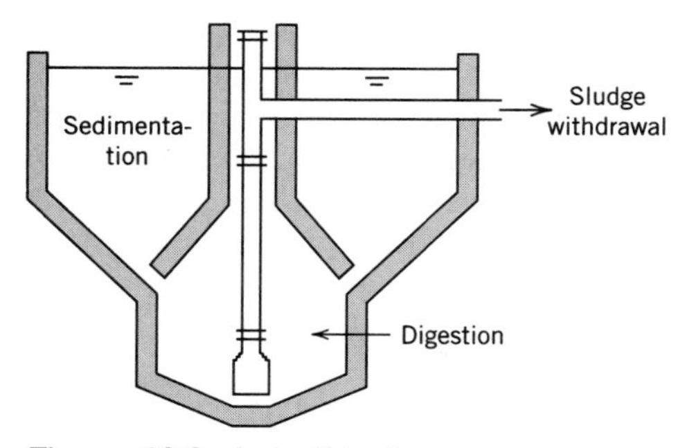

Figure 18.2 Imhoff tank.

is the essential digestion chamber. The Imhoff tank (Fig. 18.2) which is a two-story treatment system, consists of a sedimentation basin with a compartment below for digestion of the settled solids. This system is primitive by modern standards. Inadequate mixing is one reason for poor performance. Mixing of the digestion chamber is accomplished only by the evolution of gas.

Early single-tank anaerobic sludge digesters may have been mixed mechanically or with other methods. The importance of mixing for improved anaerobic process performance was recognized in the 1940s (Babbitt and Baumann, 1958). The separation of digestion from other processes and the application of mixing were the first major advancements in anaerobic treatment.

Mixing is an important factor in pH control and maintenance of uniform environmental conditions. Without adequate mixing, unfavorable microenvironments can develop. Mixing distributes buffering agents throughout the reactor and prevents local buildup of high concentrations of intermediate metabolic products that can be inhibitory to methanogens.

Ammonia and Sulfide Control

Free ammonia can inhibit anaerobic metabolism at high concentrations. Anaerobes can acclimatize to high ammonia concentrations but large fluctuations can be deleterious to the process. Free ammonia, which is much more toxic than the ammonium ion, is more prevalent at high pH. Both elevated pH and NH_3 levels contribute to process failure but the situation can be controlled by addition of acid.

Wastes high in protein content will produce significant amounts of ammonia, which increases alkalinity. It is often desirable to recycle effluent from a reactor receiving proteinaceous waste to add alkalinity to the influent. The protein content of most wastes is usually not high enough to cause ammonia toxicity problems. However, wastes containing blood can produce enough ammonium bicarbonate to elevate pH beyond the optimal range without acid addition.

Sulfate can be used as an electron acceptor under anaerobic conditions, resulting in sulfide production. Sulfides are inhibitory to methanogens and sulfate reducers themselves. Wastes high in sulfate can be prone to sulfide toxicity. Metals will precipitate sulfides; iron can be added to eliminate sulfide toxicity when sulfide concentrations are inhibitory (Gupta et al., 1994).

Nutrient Requirements

The low growth yields of anaerobes (both groups) from a given amount of substrate result in lower nutrient requirements compared to aerobes. The normal composition of microorganisms (both aerobic and anaerobic) is usually assumed to be $C_5H_7NO_2$. Phosphorus content is about one fifth of nitrogen content on a weight basis. The quantity of sludge or volatile suspended solids (VSS) produced depends on process operating conditions and is also related to influent waste strength measured on a COD basis (assuming all other growth requirements are in excess). For a typical activated sludge process, this results in a COD:N:P requirement of 100:5:1 on a mass basis. Anaerobic systems produce 20% or less of the amount of sludge produced in aerobic systems for the same substrate and the N and P requirements should decrease proportionately. The COD:N ratio has been observed to be as high as 700:5. A value of 250:5 is reasonable for highly loaded processes (0.8–1.2 kg COD/kg VSS/d); for processes operating at a lower loading rate the ratio can conservatively increase from 250:5 by multiplying it by a factor equal to the loading rate divided by 1.2 kg COD/kg VSS/d (Henze and Harremoës, 1983).

There are a number of trace elements required for successful anaerobic digestion. Nickel and cobalt have been shown to promote methanogenesis (Murray and van den Berg, 1981). For typical wastes, these substances will normally be present in excess.

18.2.2 Solids Yield and Retention Time

The most important advancement in the field of anaerobic treatment has been recognition of the central role of sludge age or solids retention time (SRT) in controlling the process. The SRT is the average time that a solid particle, particularly a biological particle, stays in the reactor (Sections 17.6 and 17.8.1). In suspended growth processes SRT is the same for all solid particles, whether they are of biological origin or not, but in other processes this is not the case. Conventional anaerobic sludge digestion reactors with artificial mixing are suspended growth reactors.

The importance of increasing SRT by solids recycle in activated sludge systems was recognized from the development of the process, but similar approaches for anaerobic systems were not implemented until the 1950s, when the anaerobic contact process (a suspended growth sludge recycle process) was developed. In a suspended growth reactor without recycle, the SRT and hydraulic retention time (HRT) are the same. This places a severe constraint on an anaerobic process because of the slow growth rates of the anaerobes. Very large reactors are required.

The new anaerobic process modifications either control SRT independently of HRT or by process design promote occurrence of very high SRTs, which is desirable. An increase in SRT will produce an increase in sludge (microorganism) concentration in the reactor. A minimum SRT must be maintained to allow the working microorganisms to reproduce themselves before being washed out and wasted from the system. The equation that defines SRT has already been given in Chapter 17. It is

$$\theta_X = \frac{\text{Mass of sludge in the reactor}}{\text{Mass removal rate of sludge from the reactor}} \tag{18.1}$$

The sludge of interest in an anaerobic reactor is specifically the anaerobic biomass. The concentration and ease of substrate degradation, concentration of microorganisms, and HRT interact to determine the amount of substrate removal. In a suspended growth reactor, the parameters in the equation for sludge age are readily formulated. The development is exactly the same as the development for activated sludge processes in Chapter 17 (Eqs. 17.21–17.23 and 17.27–17.28).

$$\frac{1}{\theta_X} = \frac{Q_W X_W}{V X_V} \tag{18.2}$$

where

θ_X is SRT

Q_w is volumetric flow rate of waste solids from the system

X_w is VSS concentration in Q_w

V is the volume of the reactor

X_V is the average concentration of VSS in the reactor

The waste flow, Q_w, is usually equal to the influent flow rate, Q, except in systems that employ recycle of biological solids. If there is more than one solids removal stream from the suspended growth system, then the solids removed in all streams must be taken into account in the materials balance. The application of Eq. (18.1) to other systems is not straightforward because of the difficulty of measuring biomass in the reactor in these systems. A brief discussion of some ramifications of changes in SRT is given here. The discussion overlaps with material in Chapter 17, which the reader is advised to review.

Below the minimum SRT it is impossible to obtain any treatment; beyond this point, an increase in SRT will produce an increase in removal of substrate. Each incremental increase in SRT beyond the minimum will produce a diminishing improvement in treatment. The amount and concentration of sludge in the reactor are approximately directly proportional to the SRT. Higher SRT will result in an excess of sludge in the reactor. The excess sludge inventory aids in keeping the process stable because this reserve biomass allows the system to cope with changes in influent quality and quantity. If a toxicant appears that inactivates a portion of the biomass, the reserve biomass is able to maintain reasonable treatment.

From laboratory studies of many wastes the minimum reproduction time (SRT) for methane formers is 3–5 days at 35°C. All biological systems require a safety factor of from 3 to 20 times the minimum SRT for successful operation (Lawrence and McCarty, 1969).

There are four methods of maintaining and increasing the amount of biological solids in an anaerobic reactor:

1. Separation of solids from the reactor effluent and recycle of these solids to the reactor
2. Provision of fixed surfaces on which bacteria grow and are retained in the system
3. Development of a dense sludge blanket, which results in holdup of solids in the system
4. Operation of the system at long hydraulic detention times

The fourth method is used in conventional sludge digestion units and is not suitable for lower strength wastes because of the large reactor volumes required and the difficulty of concentrating soluble wastes. Regarding methods 1 and 3, it is not possible to separate biological solids from inert organic or inorganic solids in the liquid. Increasing the SRT by these methods will increase the concentration of all types of particles and inhibitory levels of inert solids may be reached that ultimately limit the maximum SRT which can be maintained.

Bacteria attach to and grow on the surfaces mentioned in method 2. The tenacity with which the bacteria adhere to the surfaces in these types of reactors can allow very high SRTs to be attained. The first three methods have been implemented with notable success in the more recent reactor designs that are discussed in detail in Section 18.5. Methods 1 and 2 are used in familiar aerobic processes, activated sludge units and trickling filters, respectively. Method 3 is unique to anaerobic systems. Under proper conditions dense granules containing biomass form in this system. In methods 2 and 3 the density of biomass in the reactor is variable.

18.2.3 Biogas Potential

Assays analogous to the BOD test have been devised to assess the potential for a waste to be treated under anaerobic conditions.

Biochemical Methane Potential and Anaerobic Toxicity Assay

The methane potential of a waste is related to the concentration of organics (COD) in it and the efficiency of treatment. The 5-d biochemical oxygen demand (BOD_5) is also a common parameter used to measure waste strength. A BOD_5 value can be conservatively converted to a COD value by multiplying BOD_5 by 1.5. The maximum theoretical yield of methane, M, is 0.35 m^3 CH_4/kg COD (5.6 ft^3/lb) removed as shown in Example 18.1.

The maximum methane potential of a waste may not be realized in a treatment process for reasons such as toxicity or the refractory nature of some of the organics. In 1979 a procedure named the biochemical methane potential (BMP) test, analogous to the BOD test, was defined to assess the methane potential of a waste. The procedure is readily modified to a toxicity assay (Owen et al., 1979). Although these procedures have not been incorporated into *Standard Methods* (1992), they are widely used in the field.

In the BMP test a sample is inoculated with an active culture and supplemented with growth requirements to provide optimal conditions for anaerobic metabolism. Stock solutions containing minerals, nutrients, vitamins, and other growth factors are prepared. The appropriate sample volume is anaerobically transferred to a serum bottle. The prepared solutions are combined, inoculum is added, and the mixture is transferred to the serum bottle. Anaerobic transfers prevent oxygen toxicity. The tubing and flasks used to transfer the sample, medium, and inoculum are flushed with 70:30 nitrogen:carbon dioxide gas before a transfer is made. One of the prepared solutions contains sodium sulfide to provide a reducing environment and another solution contains resazurin, an indicator that turns pink when it is oxidized, which indicates the presence of oxygen.

The serum bottle is capped after all solutions are added and incubated at the desired temperature, which is usually 35°C. Gas production and composition are monitored over time. The incubation period is typically 30 d or the time for gas production to cease. Gas volume produced is monitored with a glass syringe that is allowed to equilibrate with atmospheric pressure after the needle is inserted into the serum bottle. Samples for analysis of gas content are also taken with a syringe.

The anaerobic toxicity assay uses the procedure for BMP except that an acetate–propionate spike is added to the serum bottle to provide a readily degradable substrate. Methane production from various sample sizes (and dilutions in the serum bottle) is compared to gas production from a control to assess the toxicity of the sample.

Methane Production in Anaerobic Treatment

The normal composition of biogas from anaerobic processes ranges from 60 to 70% methane (CH_4) and a balance of 30–40% carbon dioxide. Small amounts of hydrogen and traces of hydrogen sulfide (H_2S), ammonia, water vapor, and other gases are also present. The energy content is entirely associated with methane, which has an energy content of 37 MJ/m^3 (994 Btu/ft^3). The presence of carbon dioxide in biogas reduces its energy content to the range of 22–26 MJ/m^3 (591–698 Btu/ft^3).

Because of its corrosive nature, H_2S is an undesirable component of biogas. It can be removed by passing the gas over iron filings, which are easily regenerated by exposure to air.

Water vapor in biogas may also present a problem. Gas leaving the digester is saturated with water vapor that may condense in pipe lines, causing blockage. Moisture may be removed by using condenser traps that are periodically drained.

The BMP procedure indicates the maximum potential methane production from a wastewater. The rate of methane production is related to the flow rate and substrate removal by Eq. (18.3).

$$Q_m = Q(S_{T0} - S_{Te})M = QEMS_{T0} \tag{18.3}$$

where

Q_m is the quantity of methane per unit time
Q is influent flow rate
S_{T0} is the total influent COD (suspended + soluble)
S_{Te} is the total effluent COD (suspended + soluble)
E is an efficiency factor (dimensionless, ranging from 0 to 1)
M is the volume of CH_4 produced per unit of COD removed

Environmental conditions in a treatment process and losses will decrease the yield of methane from the theoretical or BMP values. A conservative value for M is 0.20 m^3/kg (3.2 ft^3/lb). Observed M values in the literature range from 0.10 to 0.35 m^3/kg (1.6 to 5.6 ft^3/lb) COD removed. Part of the deviation from theoretical values may result from gas leakage or the conversion of some substances to compounds that are not oxidized under conditions of the COD test. Some of the COD removed is converted to biomass. The COD balance must compare the total COD input to the process against the total COD, both soluble and suspended, accumulating in the reactor and exiting from the process. If only effluent soluble COD is accounted for and the net accumulation of biomass in the reactor is ignored, the calculated M values will definitely be less than the theoretical maximum. In some processes it is difficult to monitor the accumulation of solids in the reactor over short periods of time (which may be a few months).

The efficiency factor is related to overall treatment efficiency (in terms of COD removed from the wastewater) and the concentration of nonbiodegradable components in the influent. It is typically in the range of 0.6–0.9. The influent COD originating from components such as seeds and skins is practically nonremovable. In general, soluble components in the influent are readily removed and solid components are removed to an intermediate degree. Laboratory treatability studies are needed to define E.

There is a practical minimum limit of 1 000 mg/L on the influent COD concentration needed to obtain successful anaerobic treatment, although some studies have successfully treated waste at lower COD concentrations (Droste et al., 1988). The low amount of synthesis of anaerobes makes control of solids losses from the reactor critical at low substrate concentrations. As influent substrate concentration increases, the efficiency of any biological process improves. Also, as influent concentration increases, reactor loadings can be increased within limits while maintaining suitable HRTs. In anaerobic treatment this means a larger output of methane per unit volume of reactor per unit time.

■ **Example 18.1 Methane Yield**

Prove that the maximum yield of CH_4 is 0.35 m^3 CH_4/kg COD (5.6 ft^3/lb). Also calculate the daily volume of methane produced from a waste containing a COD of 3 000 mg/L if 80% of the waste is degraded and the wastewater flow rate is 675 m^3/d (0.178 Mgal/d).

Because all COD removed in an anaerobic process is converted to methane, it is necessary to determine the COD equivalence of methane. This is done, as outlined in Chapter 5, by calculating the amount of oxygen required to completely oxidize 1 mole of CH_4 at STP. The balanced reaction is

$$\underset{16}{CH_4} + \underset{64}{2O_2} \rightarrow \underset{44}{CO_2} + \underset{36}{2H_2O}$$

The COD of methane is 64 g O_2/16 g CH_4 or 4.00 g/g. The complete metabolism of 1.00 kg of COD will produce 0.25 kg of CH_4. The number of moles of CH_4 produced will be 250 g/16 g = 15.6 moles. The volume of 1 mole of gas is 22.4 L. The total volume of gas produced is

$$V = 22.4\,\frac{\text{L}}{\text{mole}} \times 15.6\text{ moles} = 349\text{ L} = 0.35\text{ m}^3$$

Equation (18.3) is used to determine the methane volume produced from a waste containing 3 000 mg/L COD at a flow rate of 675 m^3/d with 80% COD removal.

$$Q_m = QEMS_{T0} = \left(675\,\frac{\text{m}^3}{\text{d}}\right)(0.80)\left(0.35\,\frac{\text{m}^3\text{ CH}_4}{\text{kg}}\right)\left(3\,000\,\frac{\text{mg}}{\text{L}}\right)\left(\frac{1\text{ kg}}{10^6\text{ mg}}\right)\left(\frac{1\,000\text{ L}}{1\text{ m}^3}\right)$$

$$= 567\text{ m}^3/\text{d}$$

In U.S. units:

$$Q_m = \left(0.178 \times 10^6\,\frac{\text{gal}}{\text{d}}\right)(0.80)\left(0.35\,\frac{\text{m}^3\text{ CH}_4}{\text{kg}}\right)\left(3\,000\,\frac{\text{mg}}{\text{L}}\right)\left(\frac{1\text{ kg}}{10^6\text{ mg}}\right)\left(\frac{3.79\text{ L}}{\text{gal}}\right)\left(\frac{35.3\text{ ft}^3}{\text{m}^3}\right)$$

$$= 20\,000\text{ ft}^3/\text{d}$$

Note: The total volume of gas produced will depend on the volume of CO_2 produced. If the CH_4:CO_2 ratio is 2:1 on a volumetric basis, the total volume of gas produced will be 851 m^3/d (30 030 ft^3/d).

18.3 PROCESS ANALYSIS

The complexity of the anaerobic process makes it difficult to model. Models developed here are restricted to suspended growth processes because they are most easily modeled. An overall model is developed for the process before a detailed model is developed that takes into consideration each of the major phases of anaerobic decomposition. A model should be applied consistently with the assumptions and conditions for which it has been developed.

The developments follow the principles given in Chapter 17, which the reader is advised to consult.

18.3.1 General Model for an Anaerobic Process

A model of the overall process considers only a single biomass and lumps substrates together. There is validity to the overall model because methane formation is the slowest step in anaerobic fermentation. In many situations this governs the rate of process performance. Coefficients developed for the overall model will largely reflect the coefficients of the methane formers.

The reactor considered will be a CM suspended growth reactor without recycle. The substrate balance, considering a Monod removal rate expression, is

$$QS_0 - QS_e - \frac{kX_VS_e}{K + S_e}V = \frac{dS_e}{dt}V \tag{18.4}$$

where

- S_0 is influent COD
- S_e is effluent COD
- X_V is anaerobic VSS in the reactor
- k is an overall rate coefficient
- K is the half-velocity constant
- t is time

The biomass balance (equation of net biological growth) is

$$-QX_V + Y\frac{kX_VS_e}{K + S_e}V - k_eX_VV = \frac{dX_V}{dt}V \tag{18.5}$$

where

- Y is the overall yield factor
- k_e is the overall endogenous decay factor

Equations (18.4) and (18.5) are the same as equations for suspended growth activated sludge systems. Other equations developed in Chapter 17 for a nonrecycle reactor also apply and are not repeated here. If an anaerobic reactor is followed by a clarifier or other solids separation device with a portion of the separated solids recycled to the reactor, the equations developed for a recycle activated sludge system apply. The above equations apply when the influent substrate concentration is mostly soluble organics.

The equation for methane production depends on the rate of substrate removal and the conversion factor for COD to methane (0.25 g CH_4/g COD).

$$Q_m = 0.25\frac{kX_VS_e}{K + S_e}V \tag{18.6}$$

where

- Q_m is on a mass per unit time basis

Equations (18.4) and (18.5) should be applied with caution; the model is highly simplified. Operating data and coefficients for a given waste and process are less likely to be applicable for different reactors or wastewaters.

Definition of terms in the substrate and biomass solids balances is straightforward when the substrate is largely soluble. However, as discussed in later sections, a CM nonrecycle reactor is a poor choice for treatment of a waste with this characteristic. However, a nonrecycle CM reactor is a suitable choice for a waste that has a high degree of suspended solids such as in the case of an anaerobic digester for solids generated in an aerobic biological treatment process. Recycle suspended growth processes are also suitable for wastes that have an intermediate degree of suspended solids.

The maximum velocity constant k is related to the concentration of active biomass in the reactor. When the influent contains significant amounts of suspended solids, the VSS in the reactor will consist of a mixture of biomass and partially degraded influent VSS. As noted earlier, the influent VSS is generally not acclimated anaerobic biomass (either acid formers or methanogens), and using total VSS in the reactor in kinetic expressions is erroneous. From ATP studies and modeling of anaerobic digestion of a high solids waste in a suspended growth reactor, only 40–50% of biomass was viable, which represented only 5–10% of the total VSS in the reactor (Chung and Neethling, 1990).

Differentiation of anaerobic VSS from other organic particulates cannot be readily performed. But experiments can be performed with variable influent VSS and variable operating conditions to fit a model that realistically describes the process performance over all conditions.

The second difficulty in modeling an influent with a high VSS content is the definition of the substrate. The substrate consists of the influent organic suspended and soluble organics. Clearly the degradation of suspended solids and soluble solids will be different, which is one of the weaknesses of the general model; moreover, similar to aerobic systems, the problem of distinguishing nonanaerobic VSS from anaerobic biomass reappears here. The equations for a system in which a significant portion of the influent COD is particulate are as follows.

$$X_{V0} = X_{Van0} + X_{Vb0} + X_{Vn0} \tag{18.7}$$

where

X_{V0} is total influent VSS
X_{Van0} is anaerobic VSS in the influent. This is normally negligible.
X_{Vb0} is other biodegradable VSS in the influent
X_{Vn0} is nonbiodegradable VSS in the influent

The nonbiodegradable VSS in the influent will not change its concentration through the process. Therefore $X_{Vn0} = X_{Vn}$.

Based on this discussion the influent organics concentration is divided into the following fractions:

$$S_{T0} = S_0 + fX_{V0} \tag{18.8}$$

where

S_{T0} is the total influent COD concentration
S_0 is the influent soluble (degradable) COD concentration
f is the COD equivalence of VSS (f = 1.42 mg COD/mg VSS if VSS has a composition of $C_5H_7NO_2$)

The degradable influent COD concentration is

$$S_{d0} = S_0 + fX_{Vb0} \tag{18.9}$$

where

S_{d0} is the degradable substrate concentration in the influent

The VSS in the reactor and the effluent follows Eq. (18.8).

$$X_V = X_{Van} + X_{Vb} + X_{Vn} \tag{18.10}$$

Likewise, the reactor and effluent substrate concentrations are differentiated in the following manner:

$$S_{Te} = S_e + fX_V = S_e + f(X_{Vb} + X_{Van} + X_{Vn}) \tag{18.11}$$

where

S_{Te} is the total effluent soluble COD concentration

$$S_{de} = S_e + fX_{Vb} \tag{18.12}$$

where

S_{de} is the degradable soluble substrate concentration in the effluent

The substrate and biomass balances are reformulated as

$$QS_{d0} - QS_{de} - \frac{kX_{Van}S_{de}}{K + S_{de}}V = \frac{dS_{de}}{dt}V \tag{18.13}$$

$$-QX_{Van} + Y\frac{kX_{Van}S_{de}}{K + S_{de}}V - k_eX_{Van}V = \frac{dX_{Van}}{dt}V \tag{18.14}$$

From this series of equations other parameters such as the process loading factor and the sludge age can be developed in a manner similar to developments in Chapter 17. Coefficients that apply to the general model are given in Table 18.1. Example 18.2 illustrates the use of a model based on anaerobic VSS.

The model applied most commonly in textbooks and other references is a model based on total VSS in the reactor. The model is developed for a CM suspended growth anaerobic digester that is receiving mainly suspended solids. Sludge age can be formulated in terms of the solids production rate, P_X.

TABLE 18.1 Rate Coefficients[a] for Anaerobic Cultures at Temperatures in the Range of 30–40°C

Coefficient[b]	Range	Typical	Comment
Acid Formers			
k_A, g COD/g VSS/d	9.5–176	13	For the Monod model (Eq. 18.19b)
K_A, g/L	0.023–31	0.2	
Y_a, g VSS/g COD	0.12–0.54	0.15	
k_{ea}, d^{-1}	0.08–6.1	0.5	
Methane Formers			
k_m, g COD/g VSS/d	1.8–17	13	For the Monod model (Eq. 18.21b)
K_m, g/L	0.002–3.9	0.05	
Y_m, g VSS/g COD	0.02–0.28	0.03	
k_{em}, d^{-1}	0.01–0.04	0.02	
Overall Process			
k, g COD/g VSS/d	—	2.2	For a Monod kinetic model
K, g/L	0.2–4	1	
Y, g VSS/g COD	0.024–0.21	0.18	
k_e, d^{-1}	0.01–0.04	0.03	

[a]Adapted from Henze and Harremoës (1983), "Anaerobic Treatment of Wastewater in Fixed film Reactors—A Literature Review," *Water Science and Technology,* 15, 8/9, pp. 1–101. Used with permission from the publishers, Pergamon Press, and the copyright holders, IAWQ.

[b]The coefficients are generally high for field conditions because many of the studies were conducted with readily degradable wastes under highly favorable laboratory conditions.

$$\theta_X = \frac{VX_V}{P_X} \qquad P_X = \frac{VX_V}{\theta_X}$$

Equation (17.29) can be substituted for X_V.

$$P_X = \frac{V}{\theta_X}\left[\left(\frac{\theta_X}{\theta_d}\right)\frac{Y(S_{T0} - S_{Te})}{1 + k_e\theta_X}\right] = \frac{YQ(S_{T0} - S_{Te})}{1 + k_e\theta_X}$$

where

θ_d is the HRT

S_{T0} and S_{Te} should be the total COD or total BOD (suspended plus soluble) entering and exiting the process, respectively. Although the equation is based on the equation of net biological growth, it is used to estimate total solids production, which includes undigested solids. Obviously the yield factor based on production of anaerobic biomass from COD removal will not apply. Therefore, the right-hand side of the equation is multiplied by a so-called efficiency factor to adjust it to the observed solids exiting the reactor. This empirical adjustment is required because of the flawed premises of the model. Coefficients developed for this model will not be directly applicable in an anaerobic VSS-based model.

■ Example 18.2 Design of Complete Mixed Anaerobic Sludge Digester

This example illustrates the application of the mass balance equations and other considerations that must be applied to treatment in anaerobic reactors.

Design a nonrecycle CM anaerobic reactor to produce an effluent with a concentration of 40% of the influent VSS (this is a net removal of 60%) for a mixture of primary sludge and waste activated sludge that has been thickened to a concentration of 65 g/L. Soluble COD in the influent is negligible. The nonbiodegradable component of the influent VSS is 15%. The sludge contains 77% VSS and the flow rate is 80 m^3/d. The process will be operated at a temperature of 25°C. Assume that influent and reactor VSS compositions are each $C_5H_7NO_2$. Steady state conditions are assumed and effluent soluble COD is expected to be 1 000 mg/L.

Use the model for the overall process with coefficient values given below for the limiting conditions. Size the reactor and characterize the solids within the reactor. Find the solids production and gas production from the process.

The coefficients are as follows:

$k = 1.1$ g COD/g VSS/d $\quad K = 2$ g/L $\quad Y = 0.18$ g VSS/g COD $\quad k_e = 0.03$ d^{-1}

The influent VSS concentration is 0.77(65 g/L) = 50 g/L.

$$X_{Vn} = 0.15(50 \text{ g/L}) = 7.5 \text{ g/L}$$

Effluent VSS is

$$X_V = X_{Van} + X_{Vb} + X_{Vn} = 0.40(50 \text{ g/L}) = 20 \text{ g/L}$$

Because the VSS composition is $C_5H_7NO_2$, $f = 1.42$ mg COD/mg VSS. The influent COD concentration is

$$S_{T0} = (1.42 \text{ mg COD/mg VSS})(50 \text{ g VSS/L}) = 71 \text{ g/L}$$

The influent degradable COD concentration is

$$S_{d0} = S_0 + fX_{Vb0} = 0 + 1.42(50 - 7.5) = 60.35 \text{ g/L}$$

The effluent COD concentration is

$$1.42(20 \text{ g/L}) + 1 \text{ g/L} = 29.4 \text{ g/L}$$

The COD removal is

$$\frac{S_{T0} - S_{Te}}{S_{T0}} \times 100 = \frac{71 - 29.4}{71}(100) = 58.6\%$$

$$X_{Van} + X_{Vb} = 20 \text{ g/L} - 7.5 \text{ g/L} = 12.5 \text{ g/L}$$

$$X_{Vb} = 12.5 - X_{Van}$$

$$-QX_{Van} + Y\frac{kX_{Van}S_{de}}{K + S_{de}}V - k_e X_{Van} V = 0$$

$$S_{de} = S_e + fX_{Vb} \qquad \text{and} \qquad X_{Vb} = 12.5 - X_{Van}$$

$$S_{de} = S_e + f(12.5 - X_{Van})$$

Substituting this into the biomass balance, rearranging, and dividing by X_{Van},

$$\frac{Q}{V} = Y\frac{k[S_e + f(12.5 - X_{Van})]}{K + S_e + f(12.5 - X_{Van})} - k_e$$

The substrate balance can now be used to find another equation in terms of θ_d and X_{Van}.

$$QS_{d0} - QS_{de} - \frac{kX_{Van}S_{de}}{K + S_{de}}V = 0$$

$$\frac{Q}{V} = \frac{kX_{Van}S_{de}}{K + S_{de}}\left(\frac{1}{S_{d0} - S_{de}}\right) = \frac{kX_{Van}[S_e + f(12.5 - X_{Van})]}{K + S_e + f(12.5 - X_{Van})}\left\{\frac{1}{S_{d0} - [S_e + f(12.5 - X_{Van})]}\right\}$$

The two equations for Q/V may now be equated to solve for X_{Van}.

$$\frac{kX_{Van}[S_e + f(12.5 - X_{Van})]}{K + S_e + f(12.5 - X_{Van})}\left\{\frac{1}{S_{d0} - [S_e + f(12.5 - X_{Van})]}\right\} = Y\frac{k[S_e + f(12.5 - X_{Van})]}{K + S_e + f(12.5 - X_{Van})} - k_e$$

$$\frac{kX_{Van}[S_e + f(12.5 - X_{Van})]}{S_{d0} - [S_e + f(12.5 - X_{Van})]} = Yk[S_e + f(12.5 - X_{Van})] - k_e[K + S_e + f(12.5 - X_{Van})]$$

The equation is cumbersome to manipulate; therefore, values will be substituted to simplify it.

$$\frac{1.1X_{Van}[1.0 + 1.42(12.5 - X_{Van})]}{60.4 - [1.0 + 1.42(12.5 - X_{Van})]} = (0.18)(1.1)[1.0 + 1.42(12.5 - X_{Van})] - 0.03[2.0 + 1.0 + 1.42(12.5 - X_{Van})]$$

$$\frac{20.6X_{Van} - 1.56X_{Van}^2}{41.7 + 1.42X_{Van}} = 3.09 - 0.239X_{Van}$$

$$1.22X_{Van}^2 - 26.2X_{Van} + 1.29 = 0$$

The quadratic yields $X_{Van} = 13.8$ and 7.67 g/L as solutions. The only feasible solution is

$$X_{Van} = 7.67\ g/L = 7.7\ g/L$$
$$X_{Vb} = 12.5 - X_{Van} = 12.5 - 7.7 = 4.8\ g/L$$
$$S_{de} = S_e + f(12.5 - X_{Van}) = 1.0 + 1.42(12.5 - 7.7) = 7.8\ g/L$$
$$\frac{Q}{V} = Y\frac{kS_{de}}{K + S_{de}} - k_e = 0.18\frac{1.1(7.8)}{2 + 7.8} - 0.03 = 0.128\ d^{-1}$$

The required detention time or sludge age in this suspended growth system is

$$\theta_d = V/Q = 1/0.128\ d^{-1} = 7.8\ d$$

The volume of the reactor is

$$V = Q\theta_d = (80\ m^3/d)(7.8\ d) = 624\ m^3$$

The solids wasted (or produced) from this system is

$$P_X = QX_V = \left(80\frac{m^3}{d}\right)\left(20\frac{g\ VSS}{L}\right)\left(\frac{1\,000\ L}{m^3}\right)\left(\frac{1\ kg}{1\,000\ g}\right) = 1\,600\ kg/d$$

Using a factor of 0.25 g CH_4/g COD removed, the mass rate of methane production (M_{CH4}) is

$$M_{CH4} = (0.25)Q(S_{T0} - S_{Te})$$
$$= \left(0.25\frac{g\ CH_4}{g\ COD}\right)\left(80\frac{m^3}{d}\right)\left(71\frac{g\ COD}{L} - 29.4\frac{g\ COD}{L}\right)\left(\frac{1\,000\ L}{m^3}\right)\left(\frac{1\ kg}{1\,000\ g}\right)$$
$$= 832\ kg/d$$

At STP, 1 mole of gas occupies 22.4 L or 0.022 4 m^3. The volumetric flow rate of methane is

$$Q_{CH4} = \left(832\frac{kg}{d}\right)\left(\frac{1\ mole}{0.016\ kg}\right)\left(\frac{0.022\,4\ m^3}{mole}\right) = 1\,164\ m^3/d$$

If the volumetric ratio of $CO_2 : CH_4$ is 1 : 2, the gas flow rate from the reactor (Q_g) is

$$Q_g = 1.5(1\,164\ m^3/d) = 1\,746\ m^3/d$$

The results obtained from these calculations are in agreement with theoretical values discussed later. It is usual practice to apply a safety factor in the actual design of a system. The rate constants used in this example tend to provide a higher rate of substrate removal than is observed for municipal sludges.

18.3.2 Advanced Model for an Anaerobic Process

Phase separation is possible in anaerobic treatment and a number of studies have been performed examining each of the major metabolic steps. Systems designed for phase separation have two reactors, where hydrolysis and acid production are accomplished in the first reactor and methane production occurs primarily in the second reactor. These systems, along with pure culture studies, allow the measurement of parameters specific to each group of anaerobes.

Substrate Removal

In the general case, the influent will contain some hydrolyzable substrate, volatile acids, and simple compounds that are amenable to metabolism by the acid formers.

$$S_{T0} = S_{h0} + S_{A0} + S_{s0} \tag{18.15}$$

where
S_{T0} is total influent COD concentration
S_{h0} is the influent COD concentration of hydrolyzable substrate
S_{A0} is the influent COD concentration of volatile acids
S_{s0} is the influent COD concentration of simple substrate compounds

The concentration of soluble refractory substances in the influent is ignored in the development. Refractory compounds will simply pass through the reactor and appear in the effluent at the same concentration as in the influent. The mass balances are more easily manipulated when all forms of substrate are expressed as COD. If a substrate is expressed as a particular entity, then the COD equivalence of the substrate must be found and appropriate conversion factors must be included in the equations. For instance, volatile acids can be expressed as acetic acid and simple substrate can be expressed as glucose. The COD conversion factors for these two compounds are both 1.07 g COD/g substrate.

Hydrolyzable substrate will be represented as suspended solids, although large dissolved organic molecules will also have to be hydrolyzed.

The anaerobic biomass in the reactor is comprised of the two main groups of anaerobes: acid formers and the methanogens.

$$X_{Van} = X_{Va} + X_{Vm} \tag{18.16}$$

where
X_{Van} is VSS of total anaerobic biomass
X_{Va} is VSS of acid formers
X_{Vm} is VSS of methanogens

As for the activated sludge process the kinetic expressions for substrate removal can be formulated as first-order or zero-order with respect to biomass. In the following development, the biomass of the appropriate group will be included in the kinetic expressions.

For wastes containing significant amounts of solids or large complex molecules, hydrolysis is an important, time-consuming step. Eastman and Ferguson (1981) have demonstrated that hydrolysis of particulate organic matter is the rate-limiting step in a separate acid producing reactor. Hydrolysis is generally modeled as a first-order reaction with respect to substrate. In fact, any kinetic expression that fits the data can be used for the rate of hydrolysis or any of the following kinetic expressions. The equations will be formulated for a CM suspended growth reactor.

$$r_{Sh} = -k_h X_{Va} S_{he} \tag{18.17}$$

where
r_{Sh} is rate of hydrolysis
k_h is the hydrolysis rate constant
S_{he} is the COD concentration of hydrolyzable substrate

The mass balance for hydrolyzable substrate for the reactor is

$$QS_{h0} - QS_{he} - k_h X_{Va} S_{he} V = \frac{dS_{he}}{dt} V \tag{18.18}$$

The products of the hydrolysis reaction are simple substrates. The products of hydrolysis and simple substrates in the influent are acted on by the acid forming bacteria to produce volatile acids. The rate of volatile acids production can be formulated as a first-order or Monod-type reaction.

$$r_A = -k_A X_{Va} S_{se} \quad (18.19a) \qquad r_A = -\frac{k_A X_{Va} S_{se}}{K_A + S_{se}} \quad (18.19b)$$

where

r_A is the rate of volatile acids production
k_A and K_A are kinetic constants
S_{se} is the effluent concentration of simple substrates

Note that the value of k_A does not have the same numerical value in Eqs. (18.19a) and (18.19b).

In the mass balance for simple substrates, the rate of hydrolysis is an input term in addition to the simple substrate present in the influent. The first-order kinetic formulation will be used in the mass balance for simple substrates. It will be assumed that the same rate constant applies to the transformation of simple substrate and the products of hydrolysis.

$$QS_{s0} - QS_{se} + k_h X_{Va} S_{he} V - k_A X_{Va} S_{se} V = \frac{dS_{se}}{dt} V \tag{18.20}$$

The rate of methane formation can also be formulated as a first-order or Monod reaction.

$$r_m = -k_m X_{Vm} S_{Ae} \quad (18.21a) \qquad r_m = -\frac{k_m X_{Vm} S_{Ae}}{K_m + S_{Ae}} \quad (18.21b)$$

where

S_{Ae} is the effluent concentration of volatile acids

Again, note that k_m does not have the same numerical value in Eqs. (18.21a) and (18.21b).

The mass balance for volatile acids depends on the influent volatile acids and their rate of production as given by Eq. (18.20) as well as their destruction by the methanogens given by either Eq. (18.21a) or Eq. (18.21b). Also, endogenous decay of biomass will result in the loss of anaerobic VSS and digestion of this biomass. Equation (18.21a) will be used in the mass balance.

$$QS_{A0} - QS_{Ae} + (1 - 1.42Y_a)k_A X_{Va} S_{se} V + 1.42(k_{ea} X_{Va} + k_{em} X_{Vm})V - k_m X_{Vm} S_{Ae} V = \frac{dS_{Ae}}{dt} V \tag{18.22}$$

where

$1.42Y_a$ is the yield of acid formers on a COD basis (g biomass COD formed/g substrate COD removed, assuming VSS has a composition of $C_5H_7NO_2$)
k_{ea} is the endogenous decay coefficient of the acid formers
k_{em} is the endogenous decay coefficient of the methanogens

The production term for volatile acids states that 1 g of primary substrate COD removed produces $1 - 1.42Y_a$ g of volatile acid COD. $1.42Y_a$ g of primary substrate COD has been converted to acid former biomass COD, which will be accounted for in the biomass balance for acid formers (Eq. 18.26).

The term $1.42(k_{ea}X_{Va} + k_{em}X_{Vm})V$ is the rate of endogenous decay (converted to a COD basis) of the two groups of anaerobes. This term could have been added to Eq. (18.18) instead of including it in Eq. (18.22). The coefficients developed from experimental data will reflect the choice. Arguments could be advanced for including these terms in either equation but they are topics for discussion in more advanced courses. In any case the COD released by endogenous decay must be accounted for in the model.

The equations for the rate of methane production corresponding to Eqs (18.21a) and (18.21b) are

$$Q_m = k_m X_{Vm} S_{Ae} V \quad (18.23a) \qquad Q_m = \frac{k_m X_{Vm} S_{Ae}}{K_m + S_{Ae}} V \quad (18.23b)$$

where

Q_m is on a COD basis

Although the solution of the advanced model appears imposing with the number of equations involved, one key to simplifying the solution is to recognize that the rate of COD removal is the rate of methane production.

Biomass Growth and Solids Balances

The total VSS in an anaerobic system consists of the anaerobic biomass and influent VSS that are not metabolized to soluble compounds or methane. In general, it is not likely that influent VSS will contain significant amounts of either acid formers or methanogens that are acclimatized to the operating conditions in the anaerobic reactor. Therefore, in the following development, it will be assumed that there are no acid formers or methanogens present in the influent. The balances will be based on a CM suspended growth reactor, as earlier.

In addition to the biomass and degradable VSS there may be a significant amount of refractory (inert) VSS in the influent to the reactor. The refractory VSS will not be treated and in a CM system they will appear in the effluent at the same concentration as in the influent. The total VSS in the reactor is

$$X_V = X_{Vb} + X_{Va} + X_{Vm} + X_{Vn} \quad (18.24)$$

where

X_V is total VSS in the reactor
X_{Vb} is unmetabolized influent VSS
X_{Vn} is inert (nonbiodegradable) VSS

The concentration of X_{Vb} depends on the rate of hydrolysis of VSS (Eq. 18.17). The mass balance for X_{Vb} is Eq. (18.18) expressed on a VSS basis. If the COD equivalence of influent VSS is f (g COD/g VSS) the mass balance is

$$QX_{V0} - QX_{Vb} - k_h X_{Va} \frac{S_{he}}{f} V = \frac{dX_{Vb}}{dt} V \quad (18.25)$$

where

X_{V0} is the influent degradable VSS concentration

Note that multiplying Eq. (18.25) by f results in Eq. (18.18).

The net growth rate of acid formers depends on their rates of processing substrate and endogenous decay. Because there is a net expenditure of energy in the hydrolysis reaction there is no growth of acid formers resulting from this step. Growth of acid formers occurs as a result of capturing free energy released in the conversion of substrate to acids and other byproducts. Either Eq. (18.19a) or Eq. (18.19b) describes the energy producing reaction. Recalling that there are no acid formers in the influent, the mass balance for the acid formers is

$$-QX_{\mathrm{Va}} + Y_{\mathrm{a}}k_{\mathrm{A}}X_{\mathrm{Va}}S_{\mathrm{se}}V - k_{\mathrm{ea}}X_{\mathrm{Va}}V = \frac{dX_{\mathrm{Va}}}{dt}V \tag{18.26}$$

where

Y_{a} is the yield factor for acid formers on a VSS basis (g VSS formed/g COD removed)

The first-order substrate removal expression has been used in Eq. (18.26) to agree with Eq. (18.20). The same kinetic expression, whether first-order, Monod, or other formulations, must be used in the corresponding substrate removal mass balance and the biomass balance.

The mass balance for the methanogens is similar to the mass balance for the acid formers.

$$-QX_{\mathrm{Vm}} + Y_{\mathrm{m}}k_{\mathrm{m}}X_{\mathrm{Vm}}S_{\mathrm{VAe}}V - k_{\mathrm{em}}X_{\mathrm{Vm}}V = \frac{dX_{\mathrm{Vm}}}{dt}V \tag{18.27}$$

where

Y_{m} is the yield factor for methanogens

The yield factors and endogenous decay rate coefficients for the acid formers and the methanogens are not, in general, equal.

Equations (18.15)–(18.27) define the more significant phenomena occurring in an anaerobic process. The equations illustrate the complexity of the process; moreover, there have been some simplifications in the development. For instance, hydrogen generation and consumption have been ignored.

Information on a considerable number of parameters over variable conditions is required to fully define all constants and coefficients in the preceding model. For steady state conditions, the time variable terms are set to 0 and the equations can be solved. For variable conditions, the equations must be solved simultaneously using a standard numerical procedure, e.g., the Runge–Kutta technique.

Other Models of Anaerobic Processes

In some cases the advanced model equations can be simplified. If the influent contains insignificant concentrations of volatile acids, influent volatile acids are eliminated from the appropriate equations. If the waste contains only soluble substrate or the suspended solids concentration is small, hydrolysis will not generally be significant. Equations (18.17), (18.18), and (18.25) would not be considered and Eq. (18.20) would be modified to

$$QS_{\mathrm{s0}} - QS_{\mathrm{se}} - k_{\mathrm{A}}X_{\mathrm{Va}}S_{\mathrm{se}}V = \frac{dS_{\mathrm{se}}}{dt}V \tag{18.28}$$

Detailed modeling of the process becomes more difficult as the reactor configuration changes from a CM suspended growth system. Some of the more recent anaerobic reactors introduce inert media into the reactor to enhance the retention of biomass. Whether the medium is fixed or fluidized, measurement of the biomass growth on the medium becomes more difficult. Droste and Kennedy (1988) developed a detailed model for a downflow fixed film reactor and verified it with many runs under variable conditions. The values of coefficients and constants in their model were in good agreement with coefficients from other studies of the individual phases of anaerobic digestion.

Without comprehensive study of a process, the most reasonable approach to designing a process is to take advantage of operational results from processes that are treating wastes similar to the waste under consideration.

Table 18.1 lists values for the coefficients in the models presented so far. A comparison of the maximum substrate removal constants between the acid formers and the methane formers shows that the former is about 6 times larger than the latter. The yield factor, Y, is the parameter most responsible for differences in sludge production between aerobic and anaerobic processes. The yield factor for aerobic microorganisms may be 3 to 10 times greater than the yield factor for anaerobic microorganisms for the same substrate. Typical yield factors are 0.15 and 0.50 for anaerobic and aerobic processes, respectively. When the influent contains VSS, studies on determination of the yield factor must be cognizant of the differentiation between undigested and refractory VSS in the reactor and anaerobic biomass formed in the reactor. Typical endogenous decay coefficients are 0.06 d^{-1} and 0.03 d^{-1} (i.e., 6 and 3% of total bacterial biomass are decayed per day, respectively) for aerobic and anaerobic processes, respectively.

Eastman and Ferguson (1981) found that the hydrolysis rate constant was 3 d^{-1} in a separate acid phase reactor. This high of a rate constant would not have a significant slowing effect on the rate of treatment in a digester treating sludge. Gujer and Zehnder (1983) evaluated results from a number of studies and found that the first-order hydrolysis constant ranged from 0.04 to 0.30 d^{-1} over a temperature range from 10 to 35°C.

The effect of temperature is significant in anaerobic processes as for all biological processes. In Eq. (18.29), Henze and Harremoës (1983) found a θ value of 1.105 to apply over the temperature range from 10 to 30°C from examination of results from a number of studies.

$$r_{S,T_1} = r_{S,T_2}\,\theta^{(T_2-T_1)} \tag{18.29}$$

where

$r_{S,T}$ is overall substrate removal rate at a temperature, T
θ is a constant

Above temperatures of 40°C the rate of substrate removal rapidly declines for anaerobic cultures acclimated to mesophilic temperatures; however, processes can be acclimatized to higher temperatures and operated under thermophilic conditions. Loading rates can be increased significantly while maintaining good process performance at thermophilic conditions; Harris and Dague (1993) were able to double organic loads to anaerobic filters at 56°C and achieve performance equal to or better than filters at 35°C.

Many difficulties in modeling anaerobic processes are illustrated by these developments. A large amount of data on various parameters is required to fully characterize the process. Modeling is easier and more clearly defined when most of the influent organics are soluble. The coefficients in Table 18.1 are largely based on studies in which soluble substrates were used. Distinguishing between anaerobic biomass and other VSS that have originated in the influent is difficult. The designer and modeler must be cognizant of the basis on which the coefficients have been determined and apply the model consistently. The applicability of the model is diminished when simplifications such as using the total VSS when influent VSS is significant in the substrate removal rate expression are performed.

Researchers do gather sufficient data to elucidate detailed models, which have advanced understanding of the processes. However, because of the onerous tasks of gathering data, more empirical approaches are often used to design systems.

18.4 MISCONCEPTIONS AND BARRIERS ABOUT ANAEROBIC TREATMENT

To a considerable extent, use of anaerobic digestion processes has been limited by popular misconceptions. Speece (1981) has given a good discussion of many of the issues covered in this section. These barriers need discussion to allow an objective appraisal of anaerobic treatment. Many notions on problems associated with anaerobic treatment have stemmed from lack of understanding of process principles and from improper designs or operations that have resulted in failure. Older pre-1950-designed, low-rate, conventional systems contain design flaws that do not promote stable operation.

A fundamental mistake has been the comparison of efficiencies of anaerobic sludge digestion to aerobic treatment of wastewaters. Aerobic biological treatment efficiency is based primarily on the removal of soluble organics initially present in the waste. Naturally, anaerobic digestion of biological solids produced in aerobic treatment and other solids separated in a primary clarifier is more difficult than digestion or treatment of the initial dissolved substrate. This gives rise to the false conception that anaerobic processes are inefficient. Aerobic sludge digestion is also inefficient compared to aerobic wastewater treatment. Anaerobic processes can be designed to remove degradable soluble substrates from wastewaters as rapidly or more rapidly than aerobic processes as shown by typical loading rates given later.

The thermodynamic limitations on the low amount of energy that may be obtained from reduced organics by microorganisms under anaerobic conditions also leads to the erroneous notion that anaerobic systems have low removal rates. Aerobic microorganisms are able to extract much larger amounts of energy from a given substrate than that obtained anaerobically. From sugars, aerobes can obtain more than 14 times as much energy as anaerobes. In this sense anaerobes are less efficient than aerobes. However, this does not affect the kinetics of actual treatment systems. From a treatment perspective, there are two major consequences: aerobes, being more efficient in this respect, produce more sludge than anaerobes per unit of substrate processed; and second, the chemical energy not captured by the anaerobes results in the formation of methane. Because sludge is usually a disposal problem, its lower production is a desirable feature of anaerobic processes and, furthermore, methane is a fuel. Wastewater becomes a low-energy resource through anaerobic treatment.

Some substances in a wastewater will be susceptible to biological treatment, whereas others will not be affected. The latter are known as refractory or nonbiodegradable materials. Relative amounts of biologically removable and nonremovable substances determine the overall biodegradability of the waste. From a general review of characteristics of domestic and other wastes in aerobic and anaerobic processes, there is little difference in biodegradability for a given waste under either treatment regime. In other words, removals attainable are approximately the same regardless of the type of biological treatment. Effluent quality from an anaerobic process is usually not as good as that from an aerobic process, but in terms of overall removal the difference is marginal, particularly when high-strength wastes are involved.

Another misconception is that anaerobic treatment is unstable. Inadequate mixing in older conventional reactors is a common design flaw. Older conventional reactors are suspended growth systems with mixing supplied by the microorganisms through gas production or by supplemental mechanical power input. Effective mixing is essential in the conventional process to achieve high loading rates and corresponding low liquid detention times.

Poorly mixed or dead (unmixed) zones decrease detention times below design values. This leads to a series of phenomena that at the minimum cause reduced treatment and instability and can ultimately result in complete failure. Field evaluations (Monteith and Stephenson, 1981) have revealed that more than 50% of the reactor volume in a number of installations with different types of mixing systems was dead space. Furthermore, significant short circuiting, which is flow that passes through the reactor too quickly for treatment, was found.

Although microorganisms involved in anaerobic fermentation are fairly sensitive to their environment, proper design and operation can overcome major problems. It is obvious from this discussion that actual detention times in the conventional process of the past were often significantly below design detention times because of inadequate mixing, a problem for which there are a number of solutions.

Besides instability problems associated with insufficient retention time, low concentrations of the microorganisms in the process also contribute to instability. The new designs inherently promote stability by maintaining large inventories of microorganisms and many of these designs eliminate the need for additional mixing systems.

A common misconception is that anaerobic processes are more sensitive than aerobic processes to toxic substances. This misconception stems from anaerobic digestion's history as a municipal sludge reduction treatment. Sludges generated in sewage treatment concentrate various toxicants, which makes them more difficult to treat by any biological means. A comparison of LC_{50}s for a wide range of chemical classes between aerobic heterotrophs and methanogens showed that there was no significant difference between the sensitivities of the two groups (Blum and Speece, 1991). There was one exception: the methanogens were more sensitive to chlorinated hydrocarbons and alcohols. Anaerobic processes may handle some toxicants better than aerobic processes. Anaerobic processes generate sulfides that complex or precipitate heavy metals, which remove them as a toxicant.

It is true that the relationship among microbial species in the anaerobic process is more complex than in an aerobic process. There exists more potential for one group, the acid formers, to cause conditions that are unsuitable for the other major group, the methane formers, which have stricter environmental requirements. However, with proper buffering, SRT, and pH control, stable operation can be achieved.

The last fallacy concerns the temperature at which the process must be operated. The methane-forming anaerobes have two optimal temperature ranges depending on the species: one in the range of 33–45°C and the other between 65 and 70°C. Because most digesters have been built for operation near 35°C, there is a general feeling that this temperature must always be used. On the other hand, aerobic systems are almost always designed for ambient temperatures, even though these temperatures are not optimal for aerobes. Aerobic systems are designed to operate at lower temperatures because this is more economical, not because the bacteria perform better at these temperatures. Massé and Droste (1993) have found excellent treatment in SBR anaerobic digesters treating pig manure at a temperature of 20°C.

Maintaining an elevated temperature in an anaerobic process generally requires consumption of the excess energy produced by the process and often requires an external heat source in addition. High temperatures are not always necessary. Many studies have shown that the total amount of methane that can be produced, and thus the treatability of the waste, do not vary with temperature. Reactor loading rates and rate of methane production per unit volume of reactor do vary, but as explained in later sections, loading rates are still equivalent to, or higher than, loading rates for aerobic systems.

Startup is also cited as a problem for anaerobic systems. Lower rates of growth for anaerobic bacteria result in startup times longer than those for aerobic systems. However, startup should be required only once and there are practical steps that can be taken to minimize startup times (Kennedy, 1985). The problem of startup is compensated for by the relatively quick recovery time of an anaerobic system that has been shut down for a significant time. It is difficult for aerobic cultures to be maintained over long periods of dormancy.

The above misconceptions and a lack of familiarity with the new anaerobic technologies have presented barriers to their implementation. Scientifically, the new technologies present sound alternatives for wastewater treatment over a wide range of circumstances.

18.5 NEW ANAEROBIC TREATMENT PROCESSES

The characteristics of recently developed anaerobic treatment processes are examined in this section. Any biological treatment process requires laboratory investigations and pilot plant studies before implementation. The processes overlap in their ability to treat various wastes and laboratory studies, along with costs, will determine the best system design. Figure 18.3 shows all the processes to be discussed.

Only the major onsite processes are discussed. There are a number of variations of these processes and other experimental designs that have yet to be proved.

18.5.1 Conventional Anaerobic Treatment

Before the advent of improved anaerobic treatment technologies, anaerobic treatment referred to "anaerobic digestion" of solids generated in aerobic biological wastewater treatment operations. Conventional treatment consists of a well-mixed reactor without solids recycle. All solids are in suspension. The SRT is equal to the hydraulic retention or detention time (HRT) in a suspended solids reactor without recycle. The reactor must provide a minimum HRT of about 10 d at 35°C; depending on the waste, 10–30 d is a reasonable HRT range. Before the 1950s anaerobic digesters were frequently not supplied with mixing systems or not operated at elevated temperatures. Modern conventional systems incorporate these features and are often referred to as high-rate digesters compared to the older (conventional) systems. These conventional high-rate systems should not be confused with newer developments in anaerobic treatment discussed in the following sections, which can accommodate even higher loading rates. A high-rate digester is shown in Fig. 18.4.

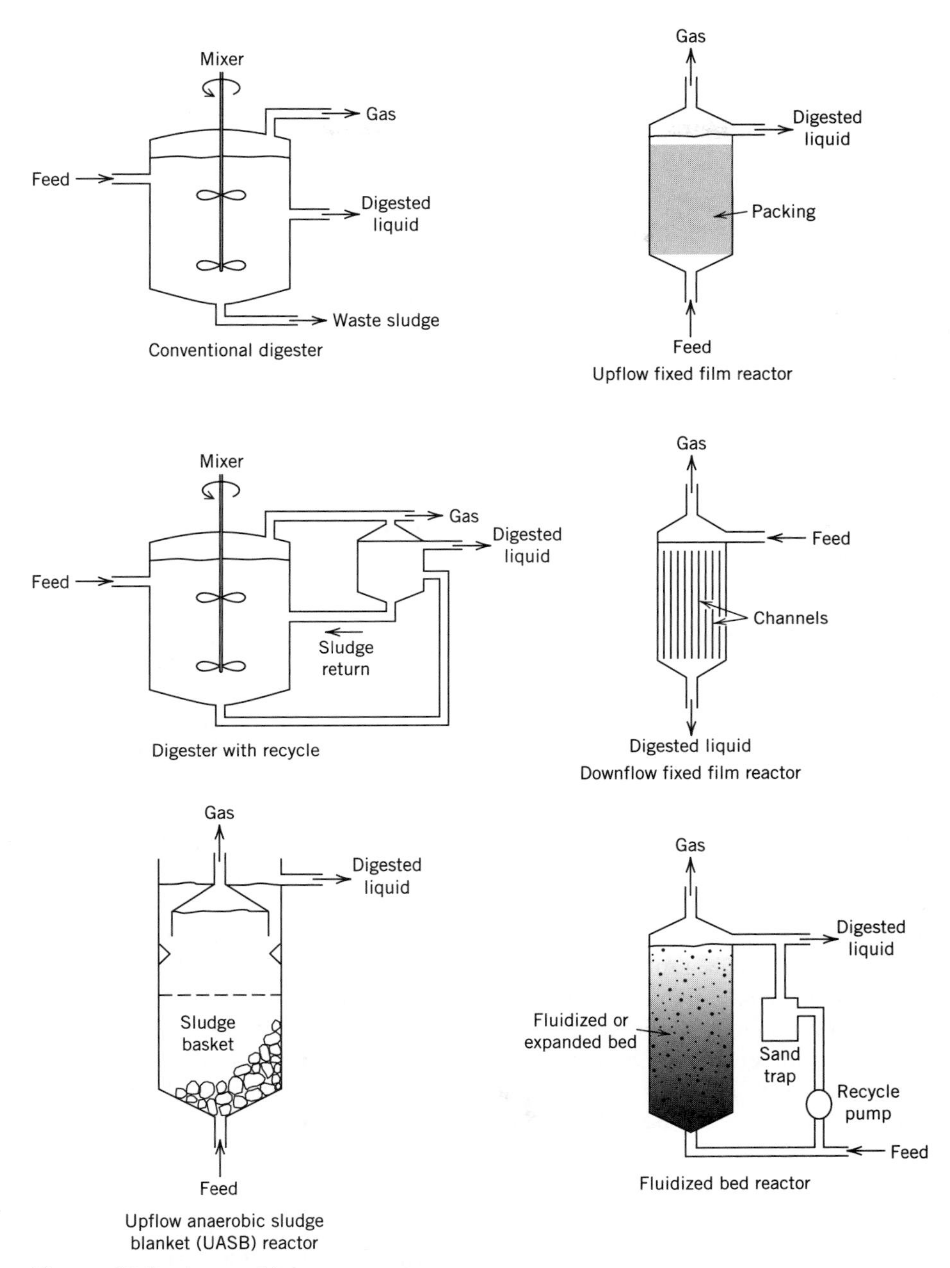

Figure 18.3 Anaerobic treatment processes.

Conventional systems designed for sludge digestion usually contain a second covered vessel following the reactor. There is no mixing provided in this vessel and it may be heated or unheated. The two stages are operated in series. The primary purpose of the second stage is to separate the solids from the influent and thicken them. The solids in the second unit provide a reserve of biomass in the event of reactor failure.

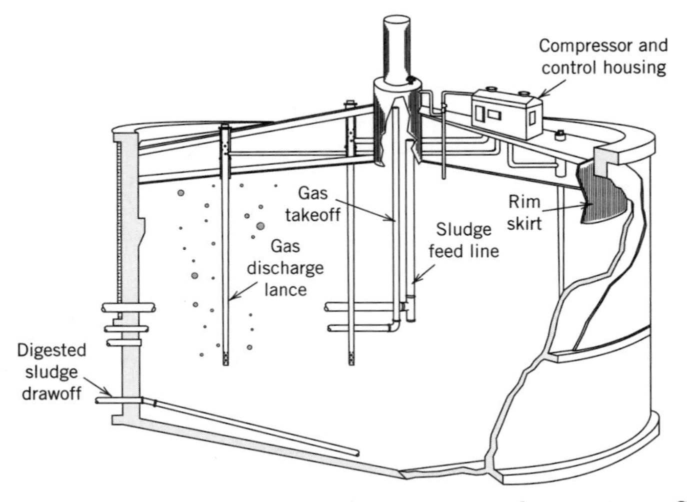

Figure 18.4 High-rate anaerobic digester with gas mixing. Courtesy of Envirex.

Suggested SRTs (HRTs) versus temperature for sewage sludge digestion are given in Table 18.2. The SRTs in Table 18.2 apply to the first stage in a two-stage system. Because of longer liquid residence times and lack of small void spaces that would be subject to plugging, the conventional process remains the most suitable choice for sludges or other wastes high in suspended solids. Wastes high in soluble COD content may be treated in conventional reactors but other options are usually better choices.

Once the reactor is installed, the only method to control SRT is to limit influent flow rates to levels that do not decrease HRT below the design value. Wastes must have a consistently high COD to ensure that enough active biomass is produced to keep minimum biomass concentrations in the reactor.

A high degree of mixing is necessary for successful performance of these reactors. It may be accomplished in a number of ways, including recirculation of biogas or by using mechanical mixers (EPA, 1987). The latter must be made of materials able to withstand the corrosive environment in the reactor. Typical values of the velocity gradient used in conventional systems vary between 50 and 80 s^{-1} (USEPA, 1979) but there is little research relating velocity gradients and liquid rheological behavior in these reactors (EPA, 1987). The EPA (1987) provides a good review of mixing practice and systems in anaerobic reactors.

TABLE 18.2 Suggested SRTs for Sludge Digestion at Various Operating Temperatures

Temperature °C	Minimum SRT d	SRT suggested for design d
18	11	28
24	8	20
30	6	14
35	4	10
40	4	10

Loading rates for these reactors are expressed in terms of VSS, which are the organic portion of total suspended solids (TSS). Commonly, VSS is near 75% of TSS. The range of loading rates is from 0.5 to 6.0 kg VSS/m^3/d (0.03–0.4 lb/ft^3/d). The lower loading rates from 0.5 to 1.0 kg VSS/m^3/d (0.03–0.06 lb/ft^3/d) are for lower temperatures. Assuming the COD of VSS to be 1.5 mg COD/mg VSS (although this can range from 1.0 to greater than 2.0), influent COD concentrations could vary from 15 000 to 180 000 mg/L or more if a significant soluble COD concentration is present. Typical solids reductions achieved in sewage sludge digesters range from 40 to 60% (WPCF, 1987).

Loading below design rates is more likely to result in process upset in conventional compared to other systems. In a reactor without recycle, insufficient influent COD will have an immediate effect of reducing the inventory of bacteria. Upon return to normal loading rates, the acidogens will recover more quickly than the methanogens, which can result in souring of the digester. Other anaerobic processes have means to retard the loss of bacteria. Continuous feeding is most desirable for conventional reactors. These systems, compared to other anaerobic processes, are less tolerant of changes in feed and environmental conditions.

There is no other major system component required for conventional reactors. A sludge concentration device may be beneficial for dilute sludges because it will reduce the reactor volume and other associated costs such as heating the influent if it is required. Sometimes a second unmixed, unheated tank is added for sedimentation and thickening of effluent from the first tank. The second tank may be open or closed. When closed it allows time for additional digestion and the sludge in it is a ready reserve for the first tank in the event of process failure.

18.5.2 Contact Process

Contact processes use well-mixed reactors but incorporate solids recycle from a solids separation device to increase SRT. These systems are similar to conventional activated sludge systems. Because SRT can be controlled separately from HRT, HRT can be considerably reduced. Besides maintaining the SRT necessary for the biomass, recycling solids retains influent solids in the system for the SRT, which is significantly longer than the HRT, thus providing more treatment time for these solids.

If an overall model is used, the equations describing the process are similar to those developed in Chapter 17 for a CM reactor aerobic process with solids recycle.

The contact process is best suited for treatment of wastes that have intermediate suspended solids concentrations ranging from 10 000 to 20 000 mg/L. Influent COD can range from 2 000 to 100 000 mg/L. Separation of solids from the reactor effluent requires a clarifier or other device. Achieving consistent solids concentration for recycle can be problematic as influent quantity and quality to the clarifier vary. Therefore, as suspended solids concentration decreases relative to total influent COD concentration to the reactor, fixed film processes become more attractive because they do not rely on solids separation and recycle as primary process control operations. As solids concentrations increase beyond the intermediate range, the conventional process becomes more desirable for the same reason. However, in choosing between contact and other processes a close examination of system costs, waste characteristics, and treatment efficiencies is needed. For a soluble substrate, studies have demonstrated that the removal efficiency of a contact process is similar to that of a fixed film reactor.

Suitable HRTs range from 1 to 7 d depending on waste characteristics, but in any case, a significant reduction in reactor volume is achieved compared to the conventional

Figure 18.5 Bench scale sequencing batch reactors at the end of a settle phase.

process. The minimum SRT remains the same as for conventional reactors (Table 18.2). The most effective loading range for contact reactors is 2–10 kg COD/m^3/d (0.12–0.6 lb/ft^3/d). Higher loadings have been used with no process upsets, but removal efficiency drops.

The smaller volume required for a contact vessel compared to a conventional vessel represents a cost saving and more effective mixing may be obtained at lower power input into the smaller volume. Solids may be separated and recycled to a degree largely independent of variations in influent flow and load. This provides more effective control of SRT.

Settling devices for anaerobic processes may be subject to flotation problems because of gas formation and the subsequent attachment of gas bubbles to biological flocs. The use of a flotation separator/thickener is therefore an alternative. If conventional gravity sedimentation is used, temporary inhibition of the anaerobes should be incorporated to halt gas production. Oxygen may be added to the reactor effluent or its temperature may be rapidly decreased by 5–10°C. Each of these has only a temporary effect on the sludge and does not impair its activity in the reactor. Vacuum degasification before sedimentation does not stop gas production. Lamella or tube settlers are the most efficient gravity sedimentation devices; however, they are susceptible to clogging when treating effluent from a biological treatment process.

Sequencing batch reactors may be used in anaerobic treatment. The operating cycle of a sequencing batch process has been described in Chapter 17. The process is a contact process wherein the sludge is recycled internally. During fill and react phases it may be necessary to apply mixing. Figure 18.5 shows a bank of sequencing anaerobic reactors at the end of the settle phase.

18.5.3 Upflow Anaerobic Sludge Blanket Reactors

The upflow anaerobic sludge blanket (UASB) process was developed in the Netherlands (Lettinga et al., 1980). The UASB system is the most commonly used high rate anaerobic treatment process. The reactor (Fig. 18.3) relies on development of a dense, active sludge mass in the lower portion of the reactor. Distinct sludge granules near the size of peas are usually formed (Fig. 18.6), but in some cases the sludge blanket is flocculent. Whether the sludge bed is granular or flocculent does not affect the operation or performance of UASB systems (Lettinga and Hulshoff Pol, 1991). VSS concentrations up to 100 g/L may occur in this zone. The waste is introduced at the bottom of the reactor, where upon contact with the sludge bed, it is degraded to CH_4 and CO_2. Gas formation and evolution supply sufficient mixing in the bed. Some solids are buoyed up by rising gas bubbles but a quiescent settling zone is provided for their separation and return into the lower portions of the reactor. This internal recycling of solids removes the need for external solids recycle.

Design of the gas separator and solids settling zone is important to the success of the reactor and, if incorrect, significant solids losses may occur, which could result in failure. The clarifier is normally incorporated into the top of the reactor to conserve space. New designs sometimes include flexirings or other media in the upper portion of the reactor to trap solids and encourage fixed film growth. This results in a hybrid sludge blanket–fixed film reactor.

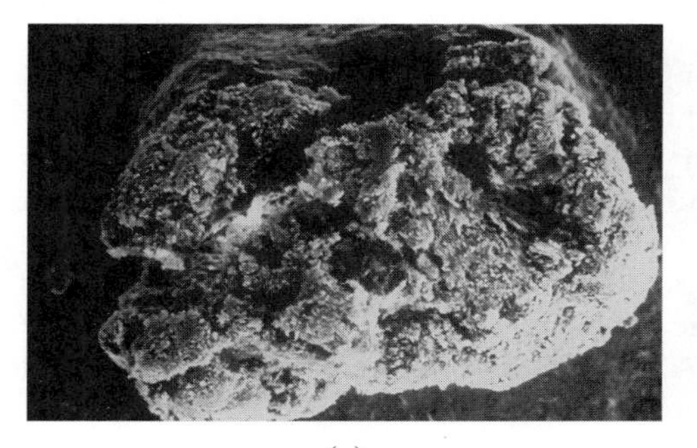

(a)

(b)

(c) (d)

Figure 18.6 Granulation in UASB reactors. (*a*) Electron micrograph of a granule. (*b*) A granule split in half, showing the hollow core which usually develops. Note the evolution of a gas bubble. (*c*) Closeup of granules. (*d*) UASB reactors with a packed bed of granules.

Mechanisms leading to formation of the granular sludge, which is the most significant aspect of these reactors, are not currently well understood but dilute soluble wastes with TSS concentrations less than 1 000–2 000 mg/L give rise to better sludge blankets. The UASB is reported to be more sensitive than other processes to waste composition. Startup is more difficult and requires special considerations to develop the sludge blanket. It has also been reported that wastes high in NH_4^+ content or low in divalent cation content are not suited for treatment in a UASB.

Upflow velocities in typical UASB reactors range up to 1–2 m/h (3.3–6.6 ft/h) although Lettinga and Hulshoff Pol (1991) and van Haandel and Lettinga (1994) recommend that the average daily upflow velocity should not exceed 1 m/h (3.3 ft/h). Expanded bed UASB reactors have also been studied (van Haandel and Lettinga (1994). Upflow velocities are in the range of 6–12 m/h (20–40 ft/h) to expand the granules to gain some of the advantages of fluidized bed reactors (discussed in the next section). The potential application of expanded bed UASB reactors is treatment of low-strength wastewaters even at low temperatures. More surface area of the granules is available for microorganism–wastewater contact and the additional mixing intensity from fluid flow compensates for the lower mixing intensity resulting from reduced biogas production of the low strength waste. Influent suspended solids will largely be carried through and out of the expanded bed reactor; therefore, the waste should be pre-settled.

The UASB process can be used for strong wastewaters provided that the VSS:COD ratio is less than 1.0 or that the VSS are highly biodegradable if the ratio is higher. The reactors have been found capable of treating dilute wastes from 500–600 mg COD/L up to 20 000 mg/L. There are relatively few studies on anaerobic reactors of any type utilizing low-strength wastes (less than 1 000 mg COD/L) and, as noted above, under practical operating conditions, 1 000 mg COD/L is a realistic minimum cutoff point, although unsettled domestic sewage (BOD_5, 357 mg/L; COD, 627 mg/L) has been successfully treated in UASB reactors operated with a 4-h HRT at temperatures ranging from 18 to 28°C (Barbosa and Sant'Anna, 1989).

Loading rates for UASB reactors range on a COD basis from 0.5 to 40 kg/m^3/d (0.03–2.5 lb/ft^3/d). The HRT can be 1 d or less. The high density of solids in the sludge zone is responsible for retaining the biomass well in excess of minimum SRT for methanogens.

A UASB reactor was found to outperform an anaerobic fixed film reactor and a continuous flow aerobic system (Latkar and Chakrabarti, 1994). Maximum loading rates for 100% conversion of urea were 7.09, 2.73, and 1.3 kg urea/m^3/d (0.44, 0.17, and 0.081 lb/ft^3/d), respectively. A UASB system can operate safely with an NH_4^+-N loading in the range of 2.4–3.4 kg NH_4^+-N/m^3/d (0.15–0.21 lb/ft^3/d).

18.5.4 Fixed Film Reactors

Fixed film reactors employ a fixed medium that provides a surface on which bacteria attach and grow. The medium immobilizes the bacteria; consequently, these reactors can achieve SRT in excess of 100 d. As a result, these processes adjust well to changes in both flow rate and quality of influent and are the most stable of the high-rate reactors. Nonattached biomass also plays a significant role in substrate conversion in fixed film reactors.

As temperature drops, biomass concentrations must rise to maintain the same rate of treatment. These reactors have been demonstrated to perform well at low temperatures with dilute wastes. Fixed film reactors have been shown capable of treating domestic wastewater at ambient temperature (25°C) on a lab scale and, along with the UASB, hold the most promise for full-scale domestic sewage treatment.

Startup of a fixed film process is, in general, a more lengthy process than startup for a conventional suspended growth process. On the other hand, the UASB, which is an unconventional suspended growth process, is even slower than the fixed film process in startup. Size of inoculum has less effect for fixed film processes because it does not directly attach to the medium. It may take 6 mo. or more for a biofilm to develop for a low-strength waste treated at a low temperature.

This disadvantage is compensated by the reactor's lowered sensitivity to hydraulic and load variations and its ability to withstand discontinuous operation. A 5-d feeding schedule is adequate and the reactor can recover in 1 or 2 weeks from a period of dormancy of 6 mo. or more. The reactor will retain active biomass when shut down with liquid remaining in it. This feature makes it particularly attractive for seasonal industries.

There are three variations of fixed film reactors: the upflow, downflow, and fluidized bed reactors. Each has particular strengths.

Upflow Fixed Film Reactors

Upflow fixed film reactors are often called "biofilters." This is a misnomer, because filtration does not play a significant role in their performance. They are actually "packed biological reactors," filled with rocks or plastic modules that provide a multitude of random channels and a large surface area. Wastewater is introduced at the bottom and allowed to overflow at the top. Bacteria are present in clumps in channels near the filter bottom as well as being attached to media surfaces throughout the reactor. Flow rates are low and very little shear is exerted on the bacterial slime growth.

Upflow reactors keep some solids in suspension in the pores and treatment derives from both suspended and fixed film growth. Because of the presence of active suspended solids, the specific surface area of medium in upflow fixed film reactors is less important than it is for downflow reactors. Suitable media that have been used are raschig rings, flexirings, pall rings, rock or plastic balls, and crossflow and tubular media (Fig. 18.7). Using rock or plastic balls of 20 mm (0.8 in.) diameter will result in a void or pore volume of about 40%. Flexirings provide a void volume of about 95% and are therefore more desirable, but they are also more costly. Media used in upflow reactors generally have diameters ranging from 20 to 170 mm (0.8–6.7 in.).

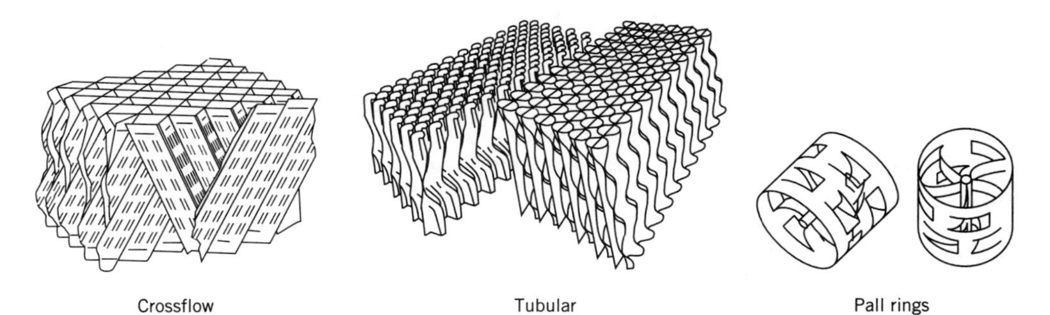

Figure 18.7 Commercial media used in upflow fixed film reactors.

TABLE 18.3 Performance and Features of Full-Scale Upflow Anaerobic Reactors in North America[a,b]

Characteristic	Range
Height, m (ft)	3–12.2 (10–40)
Temperature, °C	32–37
Influent COD, mg/L	2 500–24 000
Loading, kg COD/m^3/d (lb/ft^3/d)	1.5–15 (0.09–0.9)
Recycled flow : influent flow	0–10 : 1
HRT, h	20–96
COD removal efficiency, %	61–90+

[a]Adapted from Young and Yang (1989).
[b]Data on two atypical reactors were not included.

Media characteristics affect the operation and performance of the reactors. Ideal media have more surface area and high porosity. Higher porosity medium allows the accumulation of more biomass. Sintered glass, raschig rings, or granular activated carbon in addition to flexirings are examples of more desirable media. Organic loading rates can be increased, which decreases required reactor volume and HRT and also increases the upflow velocity through the reactor. Higher upflow velocities retard clogging but the structure of highly porous media prevents significant biomass loss at higher velocities. A reactor with sintered glass media was able to withstand upflow velocities of 20 m/d (66 ft/d) without biomass washout at organic loading rates up to 13 kg COD/m^3/d (0.8 lb/ft^3/d) (Anderson et al., 1994); a reactor with PVC rings could only deliver equivalent removals at loading rates less than 6 kg COD/m^3/d (0.4 lb/ft^3/d) and upflow velocities of 10 m/d (33 ft/d) or lower.

The specific surface of commercial media used in full-scale reactors averages about 100 m^2/m^3 regardless of the type of medium (Young and Yang, 1989). No instances of plugging or reactors using crossflow or tubular media were found by Young and Yang (1989) in their survey.

COD loadings that can be accommodated by upflow fixed film reactors range from 5 to 30 kg/m^3/d. The minimum influent COD concentration is 1 000 and 30 000 mg/L is the upper limit. The reactors function best when influent suspended solids amount to less than 500 mg/L and the ratio of soluble to insoluble COD is greater than 1. Higher suspended solids input may lead to clogging. An HRT of 1 d is usually sufficient.

Information on full-scale upflow anaerobic reactors is contained in Table 18.3. As noted below the table, data from two reactors were not included. One of these was treating domestic sewage with an influent BOD_5 between 100 and 150 mg/L and operating at a loading rate of 0.1–1.2 kg COD/m^3/d (0.006–0.075 lb/ft^3/d). Removals were 60–70% on a BOD_5 basis. The other was treating landfill leachate and operating at HRTs between 30–40 d. Over 90% of the BOD_5 was being removed.

More recent designs of upflow reactors incorporate a zone that does not have medium at the bottom of the reactor. Suspended growth can accumulate in this zone, resulting in a hybrid reactor.

Downflow Fixed Film Reactors

The downflow stationary fixed film (DSFF) reactor was developed at Canada's National Research Council laboratories. In this reactor bacteria are grown on vertically oriented surfaces. Influent is introduced at the top of the reactor and effluent is withdrawn from the bottom of the reactor.

The DSFF reactor is similar in loadings and HRT to the upflow reactor but is less subject to plugging because larger channels are provided in the medium. In a fully developed reactor, gas production from the film provides a considerable degree of mixing. This mixing helps keep the solids in suspension for some time rather than settling directly to the bottom. Suspended solids concentrations up to 3% can be accommodated by downflow fixed film reactors. At higher influent suspended solids concentrations it may be necessary to incorporate recycle to prevent them from settling and plugging the bottom of the reactor.

Various types of vertically oriented supports may be used in DSFF reactors but porous, rough media such as clay or fibrous polyester materials have been found to give the best results. Most reactors have been designed with channels with sides on the order of 1–2.5 cm (0.4–1 in.) and specific surfaces in the range of 100–150 m^2/m^3 (30–46 ft^2/ft^3). The void volume is 60–90% of the total reactor volume (Kennedy and Droste, 1991).

Fluidized Bed Reactors

The fluidized or expanded bed reactor is the most recent innovation in anaerobic treatment technology. In these reactors, bacteria are grown on particles of a medium such as sand, and the liquid is pumped through the reactor at a high rate, thereby suspending the medium particles and obtaining high rates of mass transfer. Fluid shear developed by high liquid velocities limits the growth of bacterial films to small thicknesses. Smaller film thickness provides less diffusional resistance to mass transfer of substrate and nutrients than thicker films. Bacterial end products are also removed more quickly from thinner films.

As liquid velocity in the bed is increased, media particles are forced apart and supported by the fluid as opposed to resting on each other. Also there is more movement of the particles as higher velocities are used. Authors distinguish between expanded and fluidized beds: expanded beds are expanded to a lesser degree than fluidized beds and there is little particle movement in an expanded bed. In this text, both expanded and fluidized beds are referred to as fluidized beds.

Approaches described in Chapter 14 for backwashing filters can be used to calculate fluid velocities and bed expansion. Most of the media materials used in fluidized reactors have been natural materials such as crushed rock and sand (Iza, 1991). Recommended media sizes range between 0.1 and 0.7 mm (0.004–0.028 in.) nominal diameter. Granular activated carbon medium promotes more rapid accumulation of biomass during startup than anthracite or sand media (Fox et al., 1990).

In any case, fluidized bed reactors are all high-shear reactors. Flow rates that achieve fluidization require more pumping energy than that needed for other anaerobic processes but the requirement can be lessened by using media with specific gravities near that of water. A media trap, particularly in the case of low-density media, may be needed to prevent media from being carried out with the effluent.

Bed expansion achieves many desirable features. Plugging potential is reduced because growth of thick films is limited. Mass transfer improvements have been mentioned above. However, in a fluidized bed reactor, sufficiently large masses of dense solids still form to provide SRTs in excess of the minimum desirable range. Expansion of the medium provides more surface area upon which bacteria can grow, resulting in more treatment capability per unit volume of reactor. In addition, flows through the reactor are more uniform.

Fluidized beds have been demonstrated on a laboratory scale to treat domestic sewage at ambient temperatures. The range of acceptable influent COD concentrations is from 1 000 to 30 000 mg/L. The loading rate range is from 1 to 30 kg COD/m^3/d (0.06–1.9 lb/ft^3/d). HRTs of 9 h to 1 d are normal. Wastes should have soluble to insoluble COD ratios greater than 1 and should not contain solids in excess of 1 000 mg/L.

18.5.5 Two-Phase Anaerobic Digestion

Two-phase anaerobic digestion is adapted to the two-stage nature of anaerobic metabolism. Separate reactors are designed for acidogenesis and methanogenesis. The advantage of this configuration is that conditions in each reactor can be optimized for each group of microorganisms. Low pH and low SRT will limit the growth of methanogens in the acid forming reactor. As noted before the optimal pH for the acid formers is around 5.6, whereas the methane formers have an optimum pH around 7.0. If a low pH in the range of 5.0–6.0 is maintained in the acid forming reactor, then any of the process modifications can be used and acid forming bacteria will be dominant with negligible methanogenic activity. For instance, Fongastitkul et al. (1994) used UASB reactors for each phase; Sutton and Li (1983) used fluidized bed reactors for each phase.

Hydrolysis and formation of volatile fatty acids occurs in the first reactor (phase) and their conversion to methane occurs in the second reactor (phase). Any type of reactor may be designed for the second reactor. Although studies have found that the overall treatment efficiency may be improved in two-phase systems (Ghosh, 1990; Sutton and Li, 1983) there are not a large number of installations. One disadvantage of the process is the additional reactor.

18.6 PROCESS STABILITY AND MONITORING

Anaerobic metabolism is inherently more unstable than aerobic metabolism and, in general, anaerobic treatment is more unstable than aerobic treatment. However, aerobic treatment is not without problems and some of the new anaerobic processes, particularly the fixed film options, are virtually as stable as conventional aerobic treatment. The UASB and conventional processes appear to be the least stable anaerobic processes. This does not mean any of these processes are not viable alternatives.

Anaerobic processes will generally require more time to recover from poisoning by toxic wastes than aerobic processes because of the lower synthesis rate of the anaerobes. Recovery time also depends on the inventory of biomass in the system. Fixed film processes are better able to handle toxicants because of their relatively larger SRT and larger biomass inventory. Wastes from the food and beverage industries, which are often amenable to anaerobic treatment because of their high strength, will not normally have toxicity problems.

More parameters must be monitored in anaerobic processes but the workload increase is not substantial. Besides the routine parameters of pH, COD, VSS, TSS, and alkalinity, an anaerobic process usually requires measurement of volatile acids concentrations and gas composition. A gas chromatograph is needed for each of the latter parameters. These are available for less than $5 000. Methane and carbon dioxide are the two components monitored in the gas. Hydrogen content of the gas requires a specialized instrument but it is another control parameter.

Alkalinity, pH, and gas composition are the primary indicators of process stability. A change in any one or two of these parameters may not be significant but adverse changes in all of them signal upset. The usual pH range for anaerobic treatment is from 6.5 to 7.5. Volatile acids and alkalinity affect the pH. Processes can be operated over a wide range of volatile acids concentrations. The ratio of volatile acids to alkalinity is more important than the values of either parameter. Increases of the ratio above 0.3–0.4 indicate upset and the necessity for corrective action (Hartwig, 1981); when the ratio exceeds 0.8, pH depression and inhibition of methane production occurs, resulting in process failure. The methane content of gas is typically in the range of 65–75%, with carbon dioxide comprising the bulk of the remainder.

Volatile acids concentrations are significant in any anaerobic digestion process. Both bicarbonate ion and volatile acids will contribute to the alkalinity determined according to the standard titration endpoint of pH 4.3–4.6 (Section 3.7.1). Therefore, the total alkalinity result is corrected with the following equation (Jenkins et al., 1983) to find the bicarbonate alkalinity.

$$TBA_{4.3} = ALK_{4.3} - 0.85VA \tag{18.30}$$

where

$TBA_{4.3}$ is the total bicarbonate alkalinity determined by titration to pH of 4.3
$ALK_{4.3}$ is the total alkalinity determined by titration to pH of 4.3
VA is the concentration of volatile acids (expressed as $CaCO_3$)

If individual volatile acids have been determined the correct bicarbonate alkalinity can be calculated using equilibria relationships. The standard alkalinity procedure is inexact when significant volatile acids concentrations exist in the sample. Equation (18.30) assumes that 85% of the volatile acids have been titrated, which may not be the case.

In an alkalinity determination, volatile acids will not begin to consume significant amounts of titrant until lower pH levels are attained because of their lower pK_as (acetic and propionic acids will be dominant; see Table 3.2). Jenkins et al. (1983, 1985) have proposed the following useful change to the procedure of determining the usable (i.e., bicarbonate) alkalinity in anaerobic reactors. Titrating to a pH of 5.75 will convert approximately 80% of the bicarbonate ion to carbon dioxide (see Eq. 3.22b) and at this pH less than 20% of the volatile acids will have contributed to alkalinity. The formula for usable alkalinity is

$$TBA_{5.75} = 1.25 \times ALK_{5.75} \tag{18.31}$$

where

subscript 5.75 indicates the endpoint pH

Using Eq. (18.31), the error contributed by volatile acids will be less than 20%. The modification in the procedure reduces the effort and produces useful results.

Noting the disparity between hydrogen concentrations in the gas and liquid phases discussed earlier, hydrogen concentrations in the gas phase are not a good indicator of liquid phase concentrations. But in a process running under relatively stable conditions, changes in gas phase hydrogen concentration could be an indicator of process upset.

Sulfate reduction to sulfide will occur in anaerobic reactors as discussed before. Fixed film systems are able to withstand higher concentration of H_2S than CM systems (Maillacheruvu et al., 1993). Undissociated H_2S at 60–75 mg S/L caused stress in CM systems at organic loading rates of 0.25–0.50 kg COD/m^3/d. For anaerobic filters at organic loading rates of 1–5 kg COD/m^3/d, undissociated H_2S in excess of 200 mg

S/L could be tolerated. However, when influent sulfate concentrations are above 200 mg/L, there is potential for sulfide toxicity (Parkin and Owen, 1986).

18.6.1 Thermophilic Digestion

Operation of reactors in the thermophilic range has been studied but seldom applied in full-scale systems. Most of the studies have been performed on conventional systems treating sludge generated from domestic wastewater. At temperatures in the thermophilic range (50–60°C), digestion rates are double or higher than those at mesophilic conditions. There is not any clear evidence that the overall degradability of sludge is increased at higher temperatures. Most studies have found that sludge dewaterability is improved with digestion at thermophilic temperatures (Parkin and Owen, 1986).

The major reason discouraging the application of thermophilic digestion is the energy usually required to heat the influent to reactor temperature. Other disadvantages cited for the process are poor supernatant quality and less stable operation (Parkin and Owen, 1986).

Temperature fluctuation contributes to instability of anaerobic digestion. A constant uniform temperature provides the best treatment.

18.6.2 Sludge Production

Lowered sludge production and costs associated with its disposal are a major advantage of anaerobic treatment. Aerobic treatment always produces more sludge than anaerobic treatment.

Sludge processing for final disposal is energy intensive. It is usually more economical to dewater sludge before transporting it to an approved disposal site, which is often located at a significant distance from the plant site. Gravity or flotation thickening prior to sludge dewatering may be cost effective. Drying beds, vacuum and pressure filters, and various types of centrifuges are used for dewatering. Except for drying beds, which require large land areas, these dewatering devices consume large amounts of energy. Also, chemical conditioning may be required for effective dewatering.

Nutrient content of aerobic and anaerobic sludges is similar. If an opportunity for sludge resale exists, then the associated income will be greater for an aerobic process because of its larger sludge production. This income will be reduced by costs of processing the larger amounts of sludge before it can be sold. Nutrients not contained in the sludge from an anaerobic process, for the most part, leave in the effluent in soluble form.

18.6.3 Anaerobic Treatment of Low-Strength Wastes

It has been noted that a practical minimum influent COD concentration of 1 000 mg/L is required for successful anaerobic treatment. There have been a number of laboratory studies on anaerobic treatment of domestic sewage. Full-scale treatment data for anaerobic treatment of domestic sewage are scarce.

A fluidized bed reactor operating at a hydraulic retention time of 1.5 h and temperature of 10°C was able to achieve total COD removals of 70% (Sanz and Fdz-Polanco, 1990) on an unsettled sewage (TSS of 285 mg/L) with soluble and total CODs of 390 and 760 mg/L, respectively. Gradually decreasing the temperature to 5°C did not have a large effect on effluent quality.

A full-scale upflow anaerobic filter operating at ambient temperatures between 15 and 25°C was removing 50–71% of either BOD or COD in domestic sewage (Young and Yang, 1989). The HRT in the system was 12–18 h and influent BOD was in the range of 100–150 mg/L.

18.7 COMPARISON OF ANAEROBIC AND AEROBIC TREATMENT PROCESSES

The HRT and loading ranges for processes discussed above are summarized in Table 18.4. They are compared only among themselves but usually any anaerobic process compares well against an aerobic process in terms of overall efficiency and loading rates. Loading rates for aerobic processes range from 0.15 to 5 kg COD/m^3/d. Operating temperature ranges given in this table for the conventional and contact processes reflect more commonly used temperatures but it is feasible to operate these processes at temperatures down to 20°C.

In general, comparison between aerobic and anaerobic treatment processes should be undertaken with caution because each individual case has peculiarities that may make only certain processes feasible. In many cases, both anaerobic and aerobic processes should be used together for optimal treatment.

The primary objective of all processes is pollutant removal. For an anaerobic process it takes a highly concentrated biodegradable waste to produce quantities of methane that will make the process yield a net energy income, as discussed later. However, the waste concentration at which anaerobic treatment becomes more economical than aerobic treatment is much lower. In an overall comparison, besides pollutant removal, the following items should be considered:

1. Number and size of operations in each process
2. Energy and chemical inputs into the process
3. Process stability

TABLE 18.4 Characteristics and Energy Use of Anaerobic Processes

Parameter	Conventional	Contact	Filter[a]	UASB	Fluidized bed
HRT, d	15	5	1	1	0.5
Loading rate, kg COD/m^3/d (lb/ft^3/d)	0.5–6.0[b] (0.03–0.4)	2–10 (0.12–0.6)	5–30 (0.3–1.9)	0.5–40 (0.03–2.5)	1–30 (0.06–1.9)
Heat energy consumption, MJ/m^3 (kWh/ft^3)	105 (0.83)	95 (0.75)	93 (0.73)	93 (0.73)	93 (0.73)
Mixing and pumping energy consumption, MJ/m^3 (kWh/ft^3)	88 (0.69)	26 (0.20)	0.1 (0.000 8)	0.1 (0.000 8)	1–29[c] (0.008–0.23)
COD for energy self-sufficiency, kg/m^3 (lb/ft^3)	26 (1.6)	17 (1.06)	14 (0.87)	14 (0.87)	15–19 (0.94–1.19)

[a]Includes DSFF reactors.
[b]On a kg VSS/m^3/d basis.
[c]Depends on organic loading and media.

4. Sludge production
5. Laboratory facilities and monitoring involved

There are other considerations that may be important such as startup time and ability of a process to operate well for seasonal industries. Whether methane can be used within the plant makes a large difference in the energy credit that may be obtained from an anaerobic process.

Pollutant Removal Efficiency

Removal of organics is related to the amount of biodegradable organics in the waste. As discussed earlier, the total amount of biodegradable substances is nearly the same whether the waste is treated under aerobic or under anaerobic conditions. Although aerobic treatment systems can usually produce a better quality effluent than anaerobic systems, total removals will be more similar as the wastewater becomes more highly concentrated. Removals attained in an anaerobic process alone should be well within effluent quality limits prescribed by Canada's Environmental Protection Service for biodegradable wastes.

Number and Size of Operations

Conventional aerobic treatment processes normally require primary sedimentation. This has two disadvantages: an extra vessel and appurtenances are required and the sludge that settles in this vessel is not reduced in volume or putrescibility. Many anaerobic processes do not normally require primary sedimentation. However, grit removal is mandatory for some biological reactors (particularly those with high SRT) and desirable for all to prevent accumulation of inert sludge which will displace biomass. Grit removal should precede both activated sludge and high rate anaerobic treatment.

The requirement for equalization is about the same between anaerobic and aerobic processes. When load variations exceed 1 : 4 it is good practice to provide equalization. It is more difficult to control SRT in suspended growth processes whether they are anaerobic or aerobic. In contrast, fixed film processes are less susceptible to problems resulting from load variation.

For industrial wastes, reactors will usually be larger for an aerobic process compared to an anaerobic process. This can be deduced from an examination of the loading rates reported in Table 18.4 for the respective processes. Upper loading ranges reported for anaerobic processes are much higher than upper loading ranges for aerobic processes (about 6 $kg/m^3/d$ or 0.4 $lb/ft^3/d$).

Generally, all aerobic and anaerobic processes will require a secondary clarifier or other solids separation device. There are few available data on settleability of anaerobic reactor effluents to be compared against aerobic reactor effluents. However, anaerobic processes may require a vacuum degasifier or means to inhibit gas production before conventional sedimentation. Flotation separation may be used but this also requires an energy input beyond gravity clarifiers.

Energy and Chemical Inputs

The lowest temperature for operation of anaerobic reactors is near 10°C, but 20°C is a more conservative practical minimum. The size of a reactor will need to be increased by a factor of about 2 for each 10°C decrease in operating temperature. For a reactor operating at 35°C, degradable wastes should easily result in a production of 1 m^3 CH_4/m^3 reactor/d. Methane derived from digestion will usually satisfy heating requirements of the waste. Heating the influent, when required, is usually the major energy input into an anaerobic reactor. Conventional and contact reactors also require mixing energy.

Aerobic treatment, except in the case of trickling filters, always requires energy input to pump air or oxygen into the system. Trickling filters use more energy for pumping but, in general, require less energy input than suspended growth aerobic processes. In the Canadian and northern United States climate, trickling filters can provide little treatment during winter if they are not enclosed and heated.

The need for recycle, which requires pumping energy, is variable depending on the aerobic or anaerobic process. Fluidized bed reactors definitely require recycle and incur a more significant energy penalty to keep the bed fluidized. UASB and conventional reactors do not require recycle. It may prove more economical to recycle effluent, as opposed to adding chemicals, in upflow anaerobic filters to reduce alkalinity requirements. All aerobic processes require recycle of sludge or liquid.

The other major energy input into a treatment process is for solids processing. Less energy is consumed for solids processing in anaerobic systems compared to aerobic systems because of lower sludge production in anaerobic systems.

The usual chemical inputs for any biological process are alkalinity or acidity for pH adjustment and buffer capacity and nutrient addition to satisfy synthesis needs of the microorganisms. Alkalinity is the most common agent needed for pH control. Alkalinity addition is quite often the most expensive operation cost of anaerobic systems. In anaerobic digestion the volatile acids:alkalinity ratio should be in the range of 0.3–0.4. The most economical way of minimizing alkalinity additions may be to install a pH control system on the reactor itself as opposed to adding excess alkalinity to the influent. Anaerobic processes treating industrial wastes or sludges commonly produce 200–500 mg/L of volatile acids, although the range is greater. For the range given, 570–1 430 mg/L of alkalinity as $CaCO_3$ would be required. Aerobic processes do not normally require pH adjustment unless they are designed to convert ammonia to nitrate (nitrification).

As discussed in Section 18.2.1, the nutrient requirement for a given amount of substrate is greater for an aerobic process than for an anaerobic process. Many wastes will contain an adequate amount of nutrients for treatment in either the anaerobic or aerobic mode. The ratio of carbohydrates and fats to protein is the major controlling factor. The latter contain significant amounts of nitrogen and phosphorus.

18.8 ENERGY ASSESSMENT OF ANAEROBIC AND AEROBIC TREATMENT

Energy comparisons are important because energy comprises one of the major operating costs of any treatment process.

18.8.1 Anaerobic Treatment

The net energy available from an anaerobic process is equal to the chemical energy (methane) produced by the process minus the thermal energy required by the process. The methane yield depends on a number of factors, including waste composition, temperature, HRT, and SRT, which determine organic (COD) and solids (VSS) loadings.

Thermal energy required by the process is the sum of energy required to heat the wastewater to digester operating temperature and energy needed to replace reactor heat losses to the environment. These heat losses account for only a small percentage, normally less than 10%, of total energy requirements.

Operating conditions and other parameter values that have been assumed for the anaerobic processes to calculate energy consumption and the influent COD concentrations needed to make a process energy self-sufficient are listed in Table 18.4. The data are for an influent temperature of 15°C and a reactor operating temperature of 30°C. It was assumed that the methane yield has been 0.25 m^3/kg COD (4 ft^3/lb) removed and that COD removals of 80% have been attained for each process. Use of heat exchangers was not considered for heating energy consumption. They could significantly reduce energy consumption in some circumstances.

It is observed that energy consumption of anaerobic treatment is very dependent on the type of process used and therefore on the type of waste treated. Mixing and pumping energy is the most variable item. Heating energy is directly related to the temperature difference between the reactor and the influent. In addition to the energy consumption factors given, there will be energy costs associated with sludge processing (dewatering) and transport to the ultimate disposal site.

18.8.2 Anaerobic versus Aerobic Treatment

A generalized plot of net specific energy consumption (energy consumed per unit volume of influent flow) versus influent substrate concentration for aerobic and anaerobic processes (Droste et al., 1983) is shown in Fig. 18.8. Energy consumption for sludge dewatering was incorporated into the data used to plot these lines. Use of a heat exchanger was not considered. Installation of a heat exchanger will favor an anaerobic process operated at temperatures above ambient temperature.

An important observation to be made from this figure is that aerobic processes always involve net energy consumption. In an anaerobic process, the energy required to heat the waste to the reactor temperature is the most significant energy demand; therefore, the temperature difference (DT) between the influent and reactor contents is used as a parameter on the plot. Many industries, particularly the food and beverage industries, produce wastes that are warm.

The intersection between the anaerobic process and the aerobic process lines determines the substrate concentration at which net energy consumption is equal for an aerobic and an anaerobic process. At substrate concentrations above this value, anaerobic processes are more favorable for a given temperature difference.

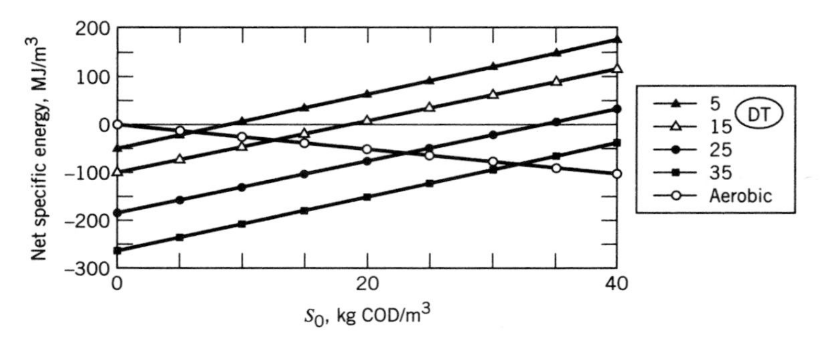

Figure 18.8 Energy comparison between aerobic and anaerobic processes.

18.8.3 Calculation of the Energy Potential of a Waste

A rough procedure is outlined here to calculate the net energy potential of a waste. Equation (18.3) can be used to estimate the quantity of methane, Q_m (m^3/d), that will be produced from the waste given its flow rate, Q (m^3/d), and COD concentration, S_0 (kg/m^3). If the 5-d BOD of the waste is known, multiply it by 1.5 to obtain COD. In Eq. (18.3), typical values for E and M are 0.75 and 0.25, respectively. Multiplying Q_m by 37 MJ/m^3 will result in the energy generation potential per day from wastewater treatment.

The heating energy required, Q_H (MJ/d), without using heat exchangers can be determined by

$$Q_H = Q \times \text{DT} \times 4.18 \times 1.1/B \tag{18.32}$$

where

DT is the temperature difference between the reactor and the influent (°C)

B is the boiler efficiency factor (dimensionless) which accounts for losses in converting the methane to heat energy. A typical value for B is 0.70.

The dimensionless factor 1.1 accounts for energy losses through the reactor walls.

The energy per volume of flow required for pumping and mixing, E_{MP}, can be obtained from Table 18.4 and the mixing and pumping energy per day, Q_{MP}, can be found from

$$Q_{MP} = E_{MP} \times Q \tag{18.33}$$

The net energy yield (or consumption) per day, E_T, for the process is

$$E_T = 37Q_m - (Q_H + Q_{MP}) \tag{18.34}$$

18.9 PATHOGEN REDUCTION IN ANAEROBIC PROCESSES

Sludge generated in waste treatment processes can be a valuable soil amendment, completing a natural cycle wherein the produce from the land is returned to the land. Pathogens in the sludge pose a health risk. The environment in an anaerobic digester will significantly reduce pathogen viability. Fecal coliforms (FC) are the best indicator of pathogens. In the United States, recent legislation (discussed in Section 20.8) for land application of sludge specifies that one alternative for sludge to meet class B status for land application is for the sludge to contain less than 2 000 000 FC/g total solids (TS) (6.3 log FC/TS).[2]

Reduction in FC densities for mostly conventional anaerobic digesters of sewage sludge are given in Table 18.5 along with other plant operating information. Analysis of data from the plants showed that a $\log_{10}$ reduction of 2.0 can be expected for typical processes operating in the mesophilic range (Stukenberg et al., 1994). The removal of FC was greater than reduction in FC densities because solids reductions of 50% or greater were being achieved by more than 75% of the plants.

Operation of reactors in the thermophilic range (50–60°C) achieves greater FC reductions (Parkin and Owen, 1986). Garber (1977) found in pilot studies that FC $\log_{10}$/g TS reductions averaged 1.79 and 4.91 for processes operated in the mesophilic and thermophilic temperature ranges, respectively.

[2] 40 CFR Part 503 regulations, *Federal Register,* Washington, DC.

TABLE 18.5 FC Reduction in Anaerobic Digestion Processes[a,b]

	Rated capacity m^3/s (ft^3/s)	Primary digester HRT[c] d	VS load $kg/m^3/d$ ($lb/ft^3/d$)	Raw solids FC log FC/g TS	Digested biosolids FC log FC/g TS	Log_{10} reduction
Range	0.11–18.40 (3.9–650)	6–57	0.59–4.1 (0.037–0.26)	5.43–10.55	3.10–8.11	0.36–4.22

[a]Adapted from Stukenberg et al. (1994).
[b]Survey of 54 plants, all operated within temperatures of 34–37°C.
[c]Approximately half of the plants were two-stage systems equipped with primary and secondary digesters. Second stage digesters were usually unheated and unmixed. Two of the plants were two-phase systems with separate acid and methane digesters.

Higher reductions were observed in a two-phase facility, which is probably caused by the harsh acidic environment in the acid reactor and routing the flow through two separate reactors. The reductions in FC for conventional anaerobic sewage sludge digesters will not necessarily be obtained in the other anaerobic processes described here. HRTs in these processes are usually significantly lower than for a conventional process.

QUESTIONS AND PROBLEMS

1. Describe the main steps of metabolism in anaerobic digestion. What is the rate-limiting step in the process?
2. Distinguish between acetogenic, acetoclastic, and hydrogenophilic bacteria.
3. What is the methane generation potential of a waste with a BOD_5 of 1 600 mg/L that is treated to a BOD_5 removal of 70%? The BOD rate constant was 0.08 d^{-1} (base 10).
4. What is the energy content of a gas that contains 50% methane and 50% carbon dioxide?
5. How is COD removed in an anaerobic process? Write the chemical equations for the conversion of glucose ($C_6H_{12}O_6$) to acetic acid and for the conversion of acetic acid to methane assuming that there is 100% conversion (i.e., there is no substrate used for microbial growth). How much acetic acid could be produced from 100 g of glucose? How much CH_4, and CO_2 will be produced from 100 g of glucose?
6. Discuss nutrient (N and P) removal in an anaerobic process compared to an activated sludge process.
7. In the closed environment of a BOD bottle, discuss the ultimate COD remaining in the bottle.
8. If 1 kg of soluble COD removed in an anaerobic treatment process produced a net yield (see Eq. 17.5) of 0.15 kg of anaerobic biomass, what is the theoretical volume of methane produced?
9. What are the volumetric loading rates for solids (VSS basis) and COD and the F : M ratio for a CM anaerobic digester that receives sludge containing 13 000 mg VSS/L and a COD of 18 700 mg/L. The detention time in the digester is 12 d and a 50% reduction of VSS is achieved.
10. A waste with a COD of 4 200 mg/L and flow of 1 850 m^3/d (0.489 Mgal/d) at steady state operation under one set of treatment conditions produced an effluent containing total volatile acids (as acetic acid) of 250 mg/L and soluble COD of other substances of 300 mg/L. The average concentration of VSS in the effluent was 460 mg/L. What was the rate of methane production (weight and volume) from this wastewater?

11. Explain why anaerobic organisms produce less sludge per unit of substrate metabolized compared to aerobes.

12. (a) If the same rate of substrate processing existed for both acid and methane formers, what would be the ratio of the acid formers to the methane formers in a CM reactor if the following parameters on a COD basis applied to the Monod models for substrate removal?

k_A-25 d^{-1} K_A-1 g/L k_m-10 d^{-1} K_m-0.05 g/L

The concentrations of primary substrate and volatile acids are each 500 mg COD/L.

(b) In a steady state process is it necessary that the acid formers and methane formers process substrate at the same rate?

13. Design a conventional CM anaerobic sludge digester to treat a sludge flow of 1 150 m^3/d containing VSS at 18 000 mg/L. Soluble influent COD is negligible. Use the overall model with coefficients given below. The temperature will be 20°C. Assuming that the effluent contains a VSS concentration that is 35% of the influent VSS concentration, what are the methane production rates, volume of the reactor, and total VSS and anaerobic VSS in the effluent from the reactor? Assume that VSS has a composition of $C_5H_7NO_2$. The soluble effluent COD is 850 mg/L. Use an M value of 0.35 m^3 CH_4/kg COD. The nonremovable VSS is estimated to be 20% of influent VSS.

k = 0.85 g COD/g VSS/d K = 3 g/L Y = 0.16 g VSS/g COD k_e = 0.03 d^{-1}

14. Solve Problem 13 for a waste that is totally soluble with a nonbiodegradable component of 20% of influent COD. The influent COD is 25 500 mg/L and a removal of 65% based on the effluent soluble COD is desired. Effluent solids will be 100% removed in a clarifier. Would you recommend the installation of this process for this waste?

15. Design an anaerobic contact process to treat the waste in Problem 14. Assume that 20% of the influent COD is nonbiodegradable. Rate coefficients are given in Problem 13. The removal is 65% based on the total influent COD to the process and the soluble effluent COD. What are the volume of the contact tank, concentration of VSS in the contact tank, recycle rate, solids waste rate, methane production rate, and sludge age? The sludge age will be 5 times the HRT. The clarifier is an ideal clarifier with no solids overflow and sludge concentrates to 14.5 g/L in the underflow from the clarifier. Would you recommend the installation of this process for this waste?

16. (a) Derive the equation: $P_X = \frac{V}{\theta_X}\left[\left(\frac{\theta_X}{\theta_d}\right)\frac{Y(S_{TO} - S_{Te})}{1 + k_e\theta_X}\right] = \frac{YQ(S_{TO} - S_{Te})}{1 + k_e\theta_X}$

(b) Discuss the applicability of this equation to an anaerobic process receiving a waste (i) with a high suspended solids content; (ii) that contains mostly soluble COD.

17. Give brief descriptions of the six major anaerobic reactor configurations noting the types of waste for which they are most suited.

18. Based on the minimum sludge ages given in Table 18.2 what is the value of θ in the Arrhenius equation (Eq. 18.29) for an anaerobic sludge digestion process? How does this compare to values for other processes?

19. Size a UASB reactor to treat a waste with a COD of 10 000 mg/L at maximum flow. The maximum and average flow rates are 2 450 and 1 200 m^3/d, respectively. Calculate the loadings at maximum and average flows.

20. What conditions could be used in the acid formation reactor of a two-phase digestion system to minimize the growth of methanogens?

21. Calculate the energy required or consumed on a daily basis for a waste with a COD content of 12 000 mg/L that is to be treated in an upflow anaerobic filter operating at a temperature of 35°C. The waste flow rate is 4 500 m^3/d (1.19 Mgal/d) and its temperature is 10°C. Use a mixing and pumping energy requirement of 0.1 MJ/m^3 (0.001 05 hp-h/ft^3). Assume that the treatment efficiency is 75% and that M = 0.35 m^3/kg (5.60 ft^3/lb).

22. Calculate the energy required to heat a waste flow of 2 400 m^3/d from 10°C to 20, 35, and 55°C. Calculate the minimum COD concentration required in the influent to provide the energy for heating the waste to the reactor temperature. Assume a removal efficiency of 80% and M = 0.35 m^3/kg and ignore all losses and other energy requirements. If the detention time in the reactor is 2 d, what is the methane production per unit volume of reactor for each of the operating temperatures?

23. A steady state anaerobic contact process with an influent COD of 12 000 mg/L and flow rate of 1 480 m^3/d (0.391 Mgal/d) is operated at a detention time of 3.0 d. All solids are assumed to settle in the clarifier and are wasted or recycled to the contact tank. The following coefficients apply:

$k_A = 30$ g COD/g VSS/d	$K_A = 6$ g/L	$Y_a = 0.15$ g VSS/g COD	$k_{ea} = 1.0\ d^{-1}$
$k_m = 6$ g COD/g VSS/d	$K_m = 2$ g/L	$Y_m = 0.05$ g VSS/g COD	$k_{em} = 0.02\ d^{-1}$
$f = 1.42$ mg COD/mg VSS			

All COD in the influent is degradable. What are the effluent COD, X_{Vm}, X_{Va}, size of the reactor, sludge age, and rate of methane production for this system? Also, find the rate of recycle in the system. Assume that hydrolysis is rapid and can be ignored. The Monod models Eqs. (18.19b and 18.21b) apply. There are no volatile acids in the influent and the VSS in the clarifier concentrates to 8 000 mg/L.

KEY REFERENCES

HENZE, M. AND P. HARREMOËS (1983), "Anaerobic Treatment of Wastewater in Fixed Film Reactors—A Literature Review," *Water Science and Technology,* 15, 8/9, pp. 1–101.

PARKIN, G. F. AND W. F. OWEN (1986), "Fundamentals of Anaerobic Digestion of Wastewater Sludges," *J. of Environmental Engineering, ASCE,* 112, EE5, pp. 867–920.

SPEECE, R. E. (1996), *Anaerobic Biotechnology for Industrial Wastewaters,* Archae Press, Nashville, TN.

VAN HAANDEL, A. C. AND G. LETTINGA (1994), *Anaerobic Sewage Treatment,* John Wiley & Sons, Ltd., Toronto.

REFERENCES

ANDERSON, G. K., B. KASAPGIL, AND O. INCE (1994), "Comparison of Porous and Non-Porous Media in Upflow Anaerobic Filters When Treating Dairy Wastewater," *Water Research,* 28, 7, pp. 1619–1624.

BABBITT, H. E. AND E. R. BAUMANN (1958), *Sewerage and Sewage Treatment,* John Wiley & Sons, Toronto.

BARBOSA, R. A. AND G. L. SANT'ANNA, JR. (1989), "Treatment of Raw Domestic Sewage in an UASB Reactor," *Water Research,* 23, 12, pp. 1483–1490.

BLUM, D. J. W. AND R. E. SPEECE (1991), "A Database of Chemical Toxicity to Environmental Bacteria and Its Use in Interspecies Comparisons and Correlations," *J. Water Pollution Control Federation,* 63, 3, pp. 198–207.

CHUNG, Y-C. AND J. B. NEETHLING (1990), "Viability of Anaerobic Digester Sludge," *J. of Environmental Engineering, ASCE,* 116, EE2, pp. 330–342.

DROSTE, R. L. AND K. J. KENNEDY (1988), "Dynamic Model of Downflow Fixed Film Reactor," *J. of Environmental Engineering, ASCE,* 114, EE3, pp. 606–620.

DROSTE, R. L. AND W. A. SANCHEZ (1983), "Comparison of Wastewater and Process Parameters on Energy Consumption in Aerobic and Alternative Anaerobic Treatment," Proceedings:

International Association on Water Pollution Research Conference on Energy Savings in Water Pollution Control, Paris.

DROSTE, R. L., A. SIMPSON, AND W. A. SANCHEZ (1983), "Potential for Energy Conservation in the Food and Beverage Industries through Anaerobic Digestion of Wastes to Methane," report, Conservation and Renewable Energy Branch, Energy, Mines & Resources, Canada, Ottawa.

DROSTE, R. L., S. GUIOT, S. GORUR, AND K. J. KENNEDY (1988), "Anaerobic Treatment of Dilute Wastewaters with a USB Reactor," *Water Pollution Research J. of Canada,* 22, 3, pp. 474–490.

EASTMAN, J. A. AND J. F. FERGUSON (1981), "Solubilization of Particulate Organic Carbon during the Acid Phase of Anaerobic Digestion," *J. Water Pollution Control Federation,* 53, 3, pp. 352–366.

EPA (1987), "Anaerobic Digester Mixing Systems," *J. Water Pollution Control Federation,* 59, 3, pp. 162–170.

FOX, P., M. T. SUIDAN, AND J. T. BANDY (1990), "A Comparison of Media Types in Acetate Fed Expanded-Bed Anaerobic Reactors," *Water Research,* 24, 7, pp. 827–835.

FONGASTITKUL, P., D. S. MAVINIC, AND K. V. LO (1994), "A Two-Phased Anaerobic Digestion Process: Concept, Process Failure and Maximum System Loading Rate," *Water Environment Research,* 66, 3, pp. 243–254.

GARBER, W. F. (1977), "Certain Aspects of Anaerobic Digestion of Wastewater Solids in the Thermophilic Range at the Hyperion Treatment Plant," *Progress in Water Technology,* 8, 6, pp. 401–406.

GHOSH, S. (1990), "Closure Discussion on 'Improved Sludge Gasification by Two-Phase Anaerobic Digestion'," *J. of Environmental Engineering, ASCE,* 116, EE4, pp. 786–791.

GUJER, W. AND J. B. ZEHNDER (1983), "Conversion Processes in Anaerobic Digestion," *Water Science and Technology,* 15, 8/9, pp. 127–167.

GUPTA, A., J. R. V. FLORA, M. GUPTA, G. D. SAYLES, AND M. T. SUIDAN (1994), "Methanogenesis and Sulfate Reduction in Chemostats—I. Kinetic Studies and Experiments," *Water Research,* 28, 4, pp. 781–793.

HARPER, S. R. AND F. G. POHLAND (1987), "Enhancement of Anaerobic Treatment Efficiency Through Process Modification," *J. Water Pollution Control Federation,* 59, 3, pp. 152–161.

HARRIS, W. L. AND R. R. DAGUE (1993), "Comparative Performance of Anaerobic Filters at Mesophilic and Thermophilic Temperatures," *Water Environment Research,* 65, 6, pp. 764–771.

HARTWIG, T. L. (1981), "Anaerobic Sludge Digestion of Municipal Wastewater Sludge: Operation and Maintenance," *Proceedings: Anaerobic Wastewater and Energy Recovery Seminar,* Duncan Lagnese and Associates, Inc., 1315 Babcock Blvd., Pittsburgh, PA.

IZA, J. (1991), "Fluidized Bed Reactors for Anaerobic Wastewater Treatment," *Water Science and Technology,* 24, 8, pp. 109–132.

JENKINS, S. R., J. M. MORGAN, AND C. L. SAWYER (1983), "Measuring Anaerobic Sludge Digestion and Growth by a Simple Alkalimetric Titration," *J. Water Pollution Control Federation,* 55, 5, pp. 448–453.

JENKINS, S. R., J. M. MORGAN, AND X. ZHANG (1985), "Measuring the Usable Carbonate Alkalinity of Operating Anaerobic Digesters," *J. Water Pollution Control Federation,* 63, 1, pp. 28–34.

KENNEDY, K. J. (1985), *Startup and Steady State Kinetics of Anaerobic Downflow Stationary Fixed Film Reactors,* Ph.D. Thesis, Dept. of Civil Engineering, University of Ottawa, Ottawa.

KENNEDY, K. J. AND R. L. DROSTE (1991), "Anaerobic Wastewater Treatment in Downflow Stationary Fixed Film Reactors," *Water Science and Technology,* 24, 8, pp. 157–177.

LAKTAR, M. AND T. CHAKRABARTI (1994), "Performance of Upflow Anaerobic Sludge Blanket

Reactor Carrying Out Biological Hydrolysis of Urea," *Water Environment Research,* 66, 1, pp. 12–15.

LAWRENCE, A. L. AND P. L. MCCARTY (1969), "Kinetics of Methane Fermentation in Anaerobic Treatment," *J. Water Pollution Control Federation,* 41, 2, pp. R1–R17.

LETTINGA, G., A. F. VAN VELSEN, S. W. HOBMA, W. DE ZEEUW, AND A. KLAPWIJK (1980), "Use of the Upflow Sludge Blanket (USB) Reactor Concept for Biological Wastewater Treatment, Especially for Anaerobic Treatment," *Biotechnology and Bioengineering,* 22, 4, pp. 699–734.

LETTINGA, G. AND L. W. HULSHOFF POL (1991), "UASB-Process Design for Various Types of Wastewaters," *Water Science and Technology,* 24, 8, pp. 87–107.

MAILLACHERUVU, K. Y., G. F. PARKIN, C. Y. PENG, W. C. KUO, Z. I. OONGE, AND V. LEBDUSCHKA (1993), "Sulfide Toxicity in Anaerobic Systems Fed Sulfate and Various Organics," *Water Environment Research,* 65, 2, pp. 100–109.

MASSÉ, D. AND R. L. DROSTE (1993), "Psychrophilic Anaerobic Digestion of Swine Manure Slurry in Sequencing Batch Reactors," *Third International Conference on Waste Management in the Chemical and Petrochemical Industries,* IAWQ, Salvador, Brazil, Oct., 1993.

MONTEITH, H. D. AND J. R. STEPHENSON (1981), "Mixing Efficiencies in Full-Scale Anaerobic Digesters by Tracer Methods," *J. Water Pollution Control Federation,* 53, 1, pp. 78–84.

MURRAY, W. D. AND L. VAN DEN BERG (1981), "Effects of Nickel, Cobalt and Molybdenum on Performance of Methanogenic Fixed-Film Reactors," *Applied and Environmental Microbiology,* 42, pp. 502–505.

OWEN, W. F., D. C. STUCKEY, J. B. HEALY, L. Y. YOUNG, AND P. L. MCCARTY (1979), "Bioassay for Monitoring Biochemical Methane Potential and Anaerobic Toxicity," *Water Research,* 13, 6, pp. 485–492.

PATEL, G. B AND G. D. SPROTT (1990), "*Methanosaeta concilii* New Genus, New Species ("*Methanothrix concilii*") and *Methanosaeta thermoacetophila* New Combination, Revived Name," *International J. of Systematic Bacteriology,* 40, 1, pp. 79–82.

PAUSS, A. AND S. GUIOT (1993), "Hydrogen Monitoring in Anaerobic Sludge Bed Reactors at Various Hydraulic Regimes and Loading Rates," *Water Environment Research,* 65, 3, pp. 276–280.

SANZ, I. AND F. FDZ-POLANCO (1990), "Low Temperature Treatment of Municipal Sewage in Anaerobic Fluidized Bed Reactors," *Water Research,* 24, 4, pp. 463–469.

SPEECE, R. E. (1981), "Fundamentals of the Anaerobic Digestion of Municipal Sludges and Industrial Wastewaters," *Proceedings: Anaerobic Wastewater and Energy Recovery Seminar,* Duncan Lagnese and Associates, Inc., 1315 Babcock Blvd., Pittsburgh, PA.

Standard Methods for the Examination of Water and Wastewater (1992), 18th ed., American Public Health Association, Washington, DC.

STUKENBERG, J. R., G. SHRIMP, SR., J. SANDINO, J. H. CLARK AND J. T. CROSSE (1994), "Compliance Outlook: Meeting 40 CFR Part 503, Class B Pathogen Reduction Criteria with Anaerobic Digestion," *Water Environment Research,* 66, 3, pp. 255–263.

SUTTON, P. M. AND A. LI (1983), "Single Phase and Two Phase Anaerobic Stabilization in Fluidized Bed Reactors," *Water Science and Technology,* 15, pp. 333–344.

TOERIEN, D. F., P. G. THIELE, AND W. A. PRETORIUS (1970), "Substrate Flow in Anaerobic Digestion," 5th International Conference on Water Pollution Research, San Francisco, CA.

USEPA (1979), *Process Design Manual for Sludge Treatment and Disposal,* Design Manual No. EPA-625/1-70-011, Center for Environmental Research Information, Cincinnati, OH.

WPCF (1987), *Anaerobic Sludge Digestion,* Manual of Practice, No. 16, 2nd ed., Water Pollution Control Federation, Alexandria, VA.

YOUNG, J. C. AND B. S. YANG (1989), "Design Considerations For Full-Scale Anaerobic Filters," *J. Water Pollution Control Federation,* 61, 9–10, pp. 1576–1587.

ZEHNDER, A. J., K. INGVORSEN, AND T. MARTI (1982), "Microbiology of Methanogen Bacteria," in *Anaerobic Digestion,* D. E. Hughes, D. A. Stafford, B. I. Wheatley, W. Baader, G. Lettinga, E. J. Nyns, and W. Verstraeten eds., Elsevier, Amsterdam, pp. 45–68.

ZINDER, S. H. (1988), "Conversion of Acetic Acid to Methane by Thermophiles," in *Anaerobic Digestion 1988,* Proc. of 5th International Symposium on Anaerobic Digestion, Bologna, Italy. Pergamon, New York.

ZOETEMEYER, R. J. (1982), *Acidogenesis of Soluble Carbohydrate Containing Wastewaters,* Ph.D. Thesis, University of Amsterdam, The Netherlands.

CHAPTER 19

TREATMENT IN PONDS, LAND SYSTEMS, AND WETLANDS

Ponds are historically the oldest form of wastewater treatment. Much more attention is currently being paid to land treatment of wastewater and treatment of wastewater in wetlands. Ecological and economical benefits may be realized from these options. Treatment in all of these systems is complex because there are a diversity of natural phenomena that influence their performance. Proper design and controlled operation of these systems improves their performance over naturally occurring systems.

19.1 OVERVIEW OF STABILIZATION PONDS

Stabilization ponds, often referred to as lagoons or oxidation ponds, are holding basins where natural processes stabilize the waste and pathogen dieoff occurs. Ponds may or may not be artificially aerated. They are suited for treatment of municipal and many industrial wastes.

The ecosystem in a pond is complex. As in any biological treatment process, microorganisms attack and convert organic matter into new microorganisms; end products of CO_2, H_2O, and other inorganic substances; and byproducts. Algae are present in significant amounts in many ponds and they use CO_2 and sunlight (photosynthesis) to produce new cells and O_2. The oxygen balance depends on activities of algae and aerobic microorganisms as well as oxygen transfer from the atmosphere. Settleable organics and microorganisms settle to the pond bottom (benthos) and undergo anaerobic decay. Complete anaerobic decay results in the conversion of the organic matter into CO_2 and CH_4. Intermediate products of anaerobic decay are soluble low molecular weight organic acids and other compounds that are released into the upper layers in the pond and become available for attack by the suspended aerobic microorganisms. These processes are illustrated in Fig. 19.1. Depending on the loading rate, pond geometry, and temperature all of the processes depicted in Fig. 19.1 may not be present in a single pond.

Light penetration into ponds is important for photosynthetic activity and also inactivation of pathogens. The Beer–Lambert law (Section 5.6) governs light transmission. Photosynthetically active radiation falls between 400 and 700 nm (Curtis et al., 1994). The attenuation coefficient of light in this region depends on the algae concentration and the wavelength of light. A Secchi disk is an inexpensive useful tool for determining the attenuation coefficient of photosynthetically active radiation. Curtis et al. (1994) found:

$$1/k_{pa} = 0.022\,2 + 0.261\ (\text{Secchi depth}) \tag{19.1}$$

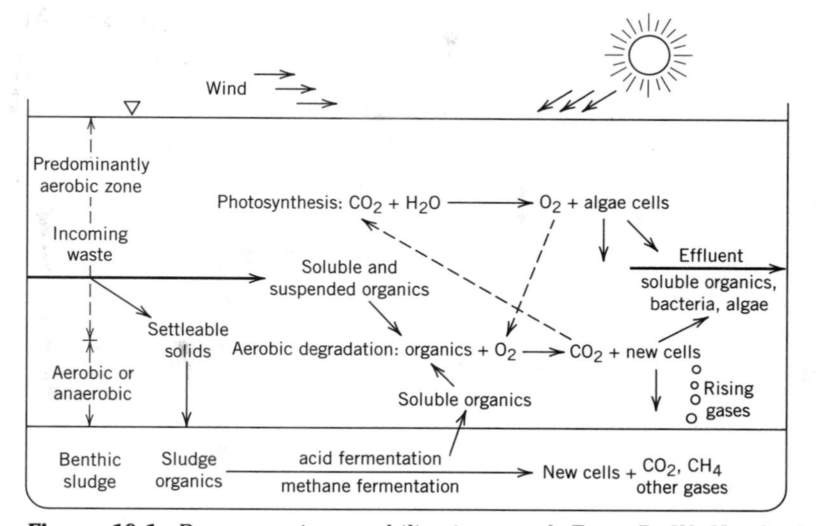

Figure 19.1 Processes in a stabilization pond. From D. W. Hendricks, and W. D. Pote (1974), "Thermodynamic Analysis of a Primary Oxidation Pond," *J. Water Pollution Control Federation,* 46, 2, pp. 333–351, © WEF 1974.

where

k_{pa} is the overall coefficient (base e) of attenuation of photosynthetically active radiation (m^{-1}).

Secchi depth is the depth at which the white disk becomes invisible

Effluent from a pond consists of soluble organic matter and suspended microorganisms and organic matter. Algae, when present in a pond, are a primary component of effluent suspended solids (SS) because of their natural ability to remain in suspension. Green algae and diatoms grow most readily in pond systems; bluegreen algae are not able to compete with green algae except under adverse conditions such as a nutrient limitation (McKinney, 1976). *Chlorella* and *Scenedesmus* are common algal species found in ponds.

Simple design models for pond systems are described in Section 19.3 but the complex environments in stabilization ponds limit the ability of simple models to describe pond performance. Extensive data collection and calibration runs are necessary to obtain an operational model (see, for instance, the model of Fritz et al., 1979, as updated by Colomer and Rico, 1993). Most of the design criteria for pond systems are empirically formulated.

Pond Operation

The operation of a stabilization pond system is relatively simple and inexpensive. Preliminary treatment is not usually applied to the influent to pond systems. Except in aerated lagoons there are no other mechanical devices used in pond systems other than pumps where required. Flow through a pond system is generally by gravity. The lack of a controlled environment slows reaction rates in ponds and much longer retention times are required to attain effective treatment. Ambient temperatures are the primary influence on rate of reaction. The coolest sustained temperatures are used for system design. The time required for treatment ranges from a few days in warm climates to 6 mo. or more in the north. The relatively long detention time in pond systems provides equalization of flows and quality. There is also flexibility in discharging effluent depending on receiving water conditions. A hydrologically controlled release is governed by the flow in the receiving waters as well as the quality of effluent from the pond system to more fully take advantage of the assimilative capacity of the receiving water.

Sludge will slowly accumulate in ponds. In ponds designed for small communities in eastern Ontario, desludging is needed every 15–20 yr. A more frequent maintenance task is ridding ponds of macrophytes, which are aquatic plants with large root systems. Macrophytes can have suspended root systems or roots that are attached to the bottom. Cattails are an example of the latter. Besides occupying a portion of the pond's volume, these growths create dead volumes and short circuiting through a pond.

Stabilization ponds are often used for small communities or industries located in remote areas. Land, which is the major capital expense of ponds, is usually less expensive in remote areas. The warm climates in tropical countries and the minimal operational demands of a pond system make them a primary choice for treating sewage in developing countries. Ponds are often the method of choice in northern communities because of low costs for design and operation and simplicity of operation (Smith and Knoll, 1986). In these severe climates their functional lifetime is also longer than mechanical treatment plants.

Pond Effluent Quality

Assessment of pond effluent quality is also a complex issue. Effluent from conventional pond systems is often not treated for SS removal and is directly discharged to the receiving water. Suspended solids concentrations in the effluent will be higher than in the effluent from mechanical biological treatment plants and likely will be in the range of 70–100 mg/L. Examination of the reactions in Fig. 19.1 shows that a portion of the influent organics is ultimately transformed into algae that are removed with the effluent. Studies have shown that the majority of effluent SS are algae and not sewage solids or bacteria.

Soluble and total BOD_5 concentrations in pond effluents from properly designed systems will be less than 25–30 mg/L. The high concentrations of algae in the effluent will have a small impact on a laboratory BOD_5 determination as discussed in Section 5.9.2. The total COD of pond effluents will reflect the presence of the algae and other soluble and suspended organics. Dead algae cells similar to bacteria have an ultimate oxygen demand in the range of 1.0–1.5 times the weight of the cells (Oswald and Ramani, 1976; Wittmann, 1982). One chemical formula for the composition of algae cells is $C_{106}H_{263}O_{110}N_{16}P$. The discharge of algae will also contribute nitrogen and phosphorus to receiving waters.

The nature of algal solids is different from bacteria or other organics. The primary differentiating factor is that algae produce oxygen through photosynthesis. The characteristic reaction is

$$6CO_2 + 6H_2O \xrightarrow{\text{sunlight}} 6C_6H_{12}O_6 + 6O_2 \qquad (19.2)$$

Glucose in the above equation is generally representative of algae biomolecules because it can be synthesized into cell constituents as required.

Algae are relatively hardy; introducing pond algae into the different environment of the receiving water will probably cause some algal cells to die but many will readjust to the new environment and remain viable. Algae also will play a role in the food chain in the receiving waters. Gloyna and Tischler (1981) note that a number of studies have shown algae laden pond effluents do not produce a measurable impact on dissolved oxygen resources of receiving waters. This would apply to moving streams. Gloyna and Tischler (1982) also note that nuisance conditions or environmental problems will occur when stream flows are small with little dilution capacity for the pond effluent.

The fate of algae discharged into long river systems is complex. The algae discharged will enrich the trophic level of the system. Retardation of eutrophication is one of the primary objectives of wastewater treatment. Algae that do not adjust to the river environment will ultimately die and contribute to the oxygen demand load in the stream. This is offset by the production of oxygen from viable algae.

Discharge of pond effluent to lakes or streams that flow into lakes will definitely contribute to eutrophication of the lake. Measures to remove the algae and other SS would be demanded in this case because algae growth is a major problem. Rivers that discharge to nearby estuaries would probably not experience problems. Issues revolving around the quality and impacts of effluent from pond systems are discussed by Gloyna and Tischler (1981), Wittmann (1982), and Gloyna and Tischler (1982).

The removal of algae from pond effluents significantly increases the costs of pond systems and also adds to operation costs and demands of the system, which are primary advantages of pond systems.

19.2 POND TYPES

Ponds are classified according to the metabolic regime present in the pond; the organic loading rate is the primary determinant of the metabolic regime. The following designations are used:

1. Anaerobic
2. Facultative
3. Aerobic or maturation ponds
4. Aerated ponds

Anaerobic ponds are devoid of oxygen throughout their depth. Anaerobic metabolism has been described in detail in Chapter 18 and is summarized here. Anaerobic decay is a two-stage process. In the first stage organics are broken down into small molecules that are fermented into acids. There is no reduction in COD in the first stage fermentation. Methane-producing bacteria then convert the organic acids into methane and carbon dioxide. Anaerobes do not reproduce as quickly as aerobes. The methanogens are particularly sensitive to environmental conditions and cool temperatures (less than 10°C) retard their metabolic rates to low to negligible values.

When wastewater strength is high, anaerobic ponds can be used as the first treatment in a series of ponds to reduce organic strength by 50% or more. High organic loading rates and deep ponds result in an anaerobic pond. Loading rates higher than those discussed later for facultative ponds will induce anaerobic conditions. In tropical Asia, McGarry and Pescod (1970) surveyed a number of pond systems and found the following relation to describe the maximum BOD_5 loading that could be applied to a facultative pond without the pond becoming completely anaerobic.

$$L_{\max} = k_c(1.054)^{(1.8T+32)} \tag{19.3}$$

where

L_{max} is the maximum BOD_5 loading rate, kg/ha/d (lb/acre/d)
k_c is a conversion factor: $k_c = 11.0$ (SI); $k_c = 9.81$ (U.S.)
T is temperature in °C

Oswald (1968) surveyed a number of pond systems in the fall in temperate climates and found that ponds would become completely anaerobic when BOD loadings exceeded 85 kg BOD_5/ha/d (76 lb/acre/d). Table 19.1 provides operating and performance data on anaerobic ponds. In warm climates anaerobic ponds can achieve BOD_5 removals of 60–80%.

Municipal wastewaters are not normally treated with an anaerobic pond but high-strength industrial or agricultural wastes are amenable to preliminary treatment in an

TABLE 19.1 Anaerobic Pond Data

S_0[a] mg/L	L_a[a] kg/ha/d (lb/acre/d)	Depth m (ft)	t_d d	T °C	Removal[a] %	Comment	Reference
374	1 960 (1 750)	2.2 (7.2)	4.2	29.5	61	Domestic, 2-mo. data	Espino de la O and Martinez (1976)
226	2 280 (2 034)	4.5 (14.8)	4.5	23.5	52	Domestic, 3-mo. data	
893	—	—	1.0	27	31	Slaughterhouse, 1 wk data	
1 304	636 (567)				87	Cannery, 4-yr ave. of summer data	Parker (1976)
415	107 (95)				73	Cannery, 4-yr ave. of summer data	
708	155 (138)				83	Milk prod., 3-yr ave. of spring peak data	
	1 120–3 360 (1 000–3 000)	2.4–3.0 (7.9–9.8)	2–7		70–80	General data on ponds treating a variety of industrial wastes in India	Sastry and Mohanrao (1976)
	404 (360)	0.9 (3.0)	60		70	Chemical waste, Texas	Orgeron (1976)
	439 (392)	1.8 (5.9)	15		51	Canning	Ford and Tischler (1976)[b]
	1 411 (1 258)	2.2 (7.2)	16		80	Meat and poultry	
	60 (54)	1.1 (3.6)	65		89	Chemical	
	388 (346)	1.8 (5.9)	18.4		50	Paper	
	1 604 (1 430)	1.8 (5.9)	3.5		44	Textile	
	27 (24)	2.1 (6.9)	50		61	Sugar	
	179 (160)	1.8 (5.9)	245		37	Rendering	
	3 360 (3 000)	1.3 (4.3)	6.2		68	Leather	

[a] BOD_5 basis.
[b] Ave. data.

anaerobic pond. Depths of the ponds range from 2.5 to 5 m (8–16 ft). The pond contents are black, and gas bubbles containing primarily methane and carbon dioxide will be observed exiting at the pond surface. Algae are unable to exist in an anaerobic pond. Odors are a problem with anaerobic ponds. Hydrogen sulfide is one of the common odoriferous substances produced under anaerobic conditions and it also causes aquatic toxicity.

Facultative ponds are the most common type of pond. Intermediate organic loading rates are applied to these ponds and they are not as deep as anaerobic ponds. Depths of these ponds range from 1.2 to 2.5 m (4–8 ft). The larger surface area and lower organic loading rate result in a measurable dissolved oxygen concentration in the upper layers of the pond. The lower depths of the pond will be anaerobic. Some of the solids that settle to the bottom are anaerobically decayed to methane. Partial anaerobic decay of other settled solids converts them to soluble acids and other soluble byproducts that are be returned to the upper layers. All of the processes shown in Fig. 19.1 occur to a significant extent in these ponds. A number of design equations have been developed for facultative ponds.

Aerobic or maturation ponds are very shallow ponds 30–45 cm (1–1.5 ft) deep that receive effluent from facultative ponds. The organic loading rate is low and aerobic conditions exist throughout the pond depth. There are always some algae present in a maturation pond. These ponds are designed to polish the final effluent by settling SS and stabilizing the low concentration of influent soluble organics.

Aerated ponds rely on mechanical aeration in addition to algae and atmospheric aeration to supply oxygen. The depth of these ponds is larger than that for nonaerated facultative ponds, ranging from 2 to 6 m (6.5–20 ft). These ponds are essentially activated sludge systems, although mixing efficiency may be considerably less than in a regular activated sludge process depending on the number of aerators installed and the pond geometry and volume. Mechanically aerated ponds are referred to as lagoons in this text and are discussed in Section 19.5.

Aquaculture (fish production) can be incorporated into stabilization pond systems. Aquaculture in sewage fed systems is not common in North America but it is widely practiced throughout other continents in the world (Wrigley et al., 1988). Organic loading rates at the facultative pond level or lower are required for fish survival and growth. Hardy species such as *Talipia* (*Sarotheroden mossambicus* and *S. niloticus*) and the carp family (*Hypophthalmichthys molitrix, Amstichthys nobilis, Cyprinus carpio, Cerrhina molitcrella,* and *Ctenopharyngodon idella*) survive best in the conditions in stabilization ponds.

19.3 DESIGN OF POND SYSTEMS

Empirical design loading rates for facultative pond systems in the United States were surveyed by Canter and Englande (1970). Pond systems are usually designed with more than one cell. The loading rates are based on the total number of cells in the system. They found the following average design loading rate criteria: states in the north above a latitude of 42° had design loading rates that averaged 29 kg BOD_5/ha/d (26 lb/acre/d) (range of 19–45 kg BOD_5/ha/d or 17–40 lb/acre/d); states in the southern region below a latitude of 37° had an average design loading rate of 49 kg BOD_5/ha/d (44 lb/acre/d) (range of 33–56 kg BOD_5/ha/d or 29–50 lb/acre/d); and states in the central region had an average design loading rate criterion of 37 kg BOD_5/ha/d (33 lb/acre/d) (range of 20–90 kg BOD_5/ha/d or 18–80 lb/acre/d). The average deten-

TABLE 19.2 Design Loading Rates for Facultative Ponds[a]

Average winter air temperature, °C	Organic loading rate	
	kg BOD_5/ha/d	lb/acre/d
<0	11–22	9.8–20
0–15	22–45	20–40
>15	45–90	40–80

[a]After Reed et al. (1988).

tion time criteria for states in the north, central, and southern regions were 117, 82, and 31 d, respectively. USEPA (1983) and Reed et al. (1988) reported that these design criteria can result in repeated violation of effluent BOD_5 criteria.

Based on many years of experience, Reed et al. (1988) recommend the design loading rates in Table 19.2 based on average winter air temperatures.

These loading rates are similar to those found by Canter and Englande (1970), although the recommended loading rate for northern states is lower than the average loading rate found in the survey. The BOD loading rate in the first cell is usually limited to 40 kg/ha/d (36 lb/acre/d) or less and the total hydraulic detention time in the system is 120–180 d when the temperature drops below 0°C. In Canada similar design criteria exist. There is no discharge during the winter months. In climates where the average winter air temperature is above 15°C the loadings in the primary cell can be as high as 100 kg BOD_5/ha/d (Reed et al., 1988).

In the far north, system detention times up to a year are used because of the long winter ice cover. The criteria for pond systems in the Northwest Territories (Table 19.3) have been developed from empirical observation on the performance of systems. In pond systems for the Northwest Territories and cooler regions of Canada, a two-

TABLE 19.3 Design Criteria for Pond Systems in the Northwest Territories[a]

Parameter	Design criteria
Short-Detention Ponds	
Number of cells (flow <400 m^3/d)	2 minimum
(flow >400 m^3/d)	4
Hydraulic detention time, d/cell	3 minimum
Depth, m (ft)	3 (9.8) minimum
Sludge removal frequency, yr	5–10
Sludge accumulation rate, L/cap-d (ft^3/cap-d)	0.35 (0.012)
Long-Detention Treatment Ponds	
Number of cells	2 minimum
Summer treatment detention time, d	60
Depth, m (ft)	1.5–2.0 (4.9–9.8)
Storage Ponds	
Number of cells	2 minimum
Storage time, d	365
Depth, m (ft)	1.5–2.0 (4.9–9.8)

[a]After Heinke et al. (1991).

stage system is recommended (Heinke et al., 1988, 1991). The first stage is a short-detention pond with a minimum hydraulic residence time of 3 d, where SS and associated BOD will settle. These ponds are essentially primary clarifiers. Settled sludge digestion is slow and the depth in these cells should be 3.0 m (9.8 ft) with an additional allowance made for sludge accumulation based on removal every 5 yr. The suggested sludge accumulation rate is 0.35 L/cap/d (0.092 gal/cap/d). Whitley and Thirumurthi (1992) monitored a short-detention lagoon system in Whitehorse, Yukon Territories over 1 yr. The system had four cells with average detention times of 2.8 d/cell and performance was better than the guidelines in Table 19.3 probably because of the larger number of cells.

Effluent from the short-detention cells is discharged into long-detention ponds. There are two types of long-detention ponds: continuous flow ponds with a minimum hydraulic residence time of 60 d and intermittent discharge and storage ponds with storage provided for 1 yr. Continuous discharge ponds will be ineffective during winter conditions. Effluent from intermittent discharge storage ponds is released during the time of year when dilution is high and temperatures are warmer. Even in southern Canada the trend is to design 365-d storage ponds and this storage period is recommended for ponds in the Northwest Territories (Heinke et al., 1991). The major factor influencing effluent quality from intermittent discharge ponds is the length of time that wastewater is held after spring breakup. Secondary treatment objectives are generally achieved within 40 d after breakup.

In all types of ponds, a minimum of two cells is required to provide flexibility in operation and maintenance.

Models for Facultative Ponds

Many researchers have proposed models for design of facultative ponds. There is considerable variability in the predictions of the models (predictions can vary by a factor of 10 or more). Models are usually based on BOD_5 as opposed to COD. Data used for models often compare total influent BOD to filtered or settled effluent samples, which does not describe the total organic load being discharged from the pond. Application of a model based on filtered BOD to unsettled wastewater with a high concentration of algae will naturally lead to erroneous results.

The USEPA (1983) compared the most common design approaches but did not make any recommendations on the preferred method. Empirical correlations have been developed and models based on flow regimes of plug flow (PF) through partially mixed flow to complete mixed (CM) flow have been used. The reaction rate coefficient in the substrate removal expression is usually not referred to the concentration of volatile suspended solids (VSS) in the pond. The fluctuation of VSS concentrations and variable makeup (algae, bacteria, and unsettled influent particulates) of VSS disallow useful correlations between VSS concentration and substrate removal rate.

The standard PF model (see Section 10.1.2) is

$$S_e = S_0 e^{-k\theta_d} \tag{19.4}$$

where

S_e is effluent BOD_5
S_0 is influent BOD_5
k is the reaction rate constant
θ_d is the detention time in the pond

TABLE 19.4 Plug Flow Reaction Rate Constant Variation with Organic Loading Rate[a,b]

Organic loading rate kg/ha/d (lb/acre/d)	k d^{-1}
22 (20)	0.024
45 (40)	0.044
67 (60)	0.059
90 (80)	0.077
112 (100)	0.101

[a]From Neel et al. (1961).
[b]Computed from data in the article based on average total BOD_5 removals over a 1-yr period.

Based on a study of four facultative ponds at various locations in the United States, the USEPA (1983) found that k at an average temperature of 14°C was 0.015 8 d^{-1} and 0.027 9 d^{-1} on a total and soluble BOD basis, respectively. Using the Arrhenius equation formulated as

$$k_2 = k_1 \theta^{(T_2 - T_1)} \tag{19.5}$$

where
k_1 and k_2 are the rate coefficients at temperatures of T_1 and T_2 °C, respectively

θ was found to be 0.962 on a total or soluble BOD_5 basis. Note that a θ value less than 1 predicts that k decreases with an increase in temperature. There was significant scatter in the data. Using data in Neel et al. (1961) the reaction rate constant varied with organic loading rate as shown in Table 19.4. The rate coefficients in Table 19.4 predict the same amount of BOD_5 removal for loading rates at 67 kg/ha/d (60 lb/acre/d) up to 112 kg/ha/d (100 lb/acre/d). There was slightly improved performance below a loading of 67 kg/ha/d (60 lb/acre/d).

The CM model has also been frequently used to model pond systems. Moreno (1990) found that hydraulic mixing regimes were CM in five facultative ponds. Marais and Shaw (1961) have used an equation based on a CM model. When ponds are modeled, they are usually treated as CM reactors. Daily fluctuations in temperature can cause the pond contents to mix as water in the upper layers warms during the day and then cools at night and sinks to the bottom. Assuming that a pond is CM provides a safety factor in estimating its performance. The CM model (see Section 10.1.3) is

$$\frac{S_e}{S_0} = \frac{1}{1 + k\theta_d} \tag{19.6}$$

or for a series of CM ponds,

$$\frac{S_e}{S_0} = \frac{1}{(1 + k\theta_d)^n} \tag{19.7}$$

where
n is the number of ponds in series

For CM ponds at an average temperature of 14°C, the USEPA (1983) found a k value of 0.055 4 d^{-1} from regression of data for four pond systems in the United States. Using Eq. (19.5), θ was found to be 0.987. Mara (1975) suggests that k for CM ponds be estimated according to

$$k = 0.30(1.05)^{(T-20)} \tag{19.8}$$

where

k has units of d^{-1}

Equation (19.8) is valid above temperatures of 15°C.

The flow in ponds has been found to be intermediate between PF and CM mixing regimes (Thirumurthi, 1974). A dispersion model is required to calculate the degree of treatment. Thirumurthi (1974) has supplied a graph to facilitate the calculation of effluent quality when dispersed flow conditions exist. But effective use of the graph requires a number of tracer studies to establish the critical (most highly mixed) flow condition. Intermediate mixing regimes in a single pond can also be simulated by considering the pond to be comprised of a number of individual CM cells. For a multipond system each pond would be considered to be made up of a number of CM cells that provide the correct amount of dispersion in each pond. The limiting dispersed flow condition is a CM regime. Considering a pond to be CM provides a safety factor and is the most practical approach to designing a pond.

By providing a series of ponds, other flow irregularities, particularly short circuiting, are minimized. As shown by Eq. (19.7) the level of treatment is also improved as the total system volume is divided into a larger number of cells. In a study of four pond systems, the pond system that had a design where the water moved sequentially through a seven-cell system had the best performance compared to three other three-cell systems (USEPA, 1983). In a CM system, the pond geometry theoretically does not have any influence on flow through the pond and a square pond can be used. However, rectangular ponds with a length to width ratio of 3:1 to 4:1 tend to promote a PF regime but more importantly they provide more distance between the influent and effluent points and minimize short circuiting. Short-circuited flow does not receive any treatment.

The following empirical equation based on loading rate has been used by Gloyna (1976) for estimation of the pond volume required to achieve BOD removal of 80–90%. BOD removal efficiency is based on unfiltered influent samples and filtered effluent samples.

$$V = f_a f_S (3.5 \times 10^{-2}) Q L_a \theta^{(35-T)} \tag{19.9}$$

where

V is the pond volume, m^3

f_a is an algal toxicity factor

f_S is sulfide oxygen demand

Q is influent flow in m^3/d

L_a is influent ultimate BOD or COD concentration (mg/L)

T is in °C

The pond depth should be taken to be 1.0 m (3.3 ft) when Eq. (19.9) is applied to find the pond surface area. But the actual pond depth is suggested to be 1.5 m (4.9 ft) when seasonal variations in temperature or daily fluctuations in flow are significant. If the influent contains high amounts of slowly degradable organics where detention time becomes the controlling factor, depths up to 2 m (6.6 ft) can be used. Regardless of the pond depth, the required surface area should be based on a depth of 1.0 m when applying Eq. (19.9). The temperature correction coefficient, θ, was found to be 1.085 for Eq. (19.9). The algal toxicity factor, f_a, is assumed to be 1.0 for domestic wastes and many industrial wastes and the sulfide oxygen demand, f_S, is also 1.0 when equivalent sulfate concentrations are less than 500 mg/L.

The application of the above equations and rate coefficients can result in widely different designs (see Problem 9). The best procedure for ultimate assessment of coefficients is to evaluate local ponds receiving wastes similar to the waste under consideration.

Suspended Solids Removal

Influent SS are primarily removed in the first cell in a stabilization pond system. There will be a buildup of solids near the inlet of the first pond. Maturation ponds are designed for settling SS but their performance is variable, particularly with respect to algae.

■ Example 19.1 Modelling Stabilization Pond Systems

A pond system consists of a large pond with a detention time of 6 d, followed by two ponds in series that each have detention times of 3 d. The BOD_5 reduction through the system is 88%. What is the rate constant describing BOD removal if the ponds are considered to be CM?

For a CM pond,

$$\frac{S_e}{S_0} = \frac{1}{1 + k\theta_d}$$

For the first pond

$$\frac{S_{e1}}{S_0} = \frac{1}{1 + k(6\text{ d})}$$

for the second pond

$$\frac{S_{e2}}{S_{e1}} = \frac{1}{1 + k(3\text{ d})}$$

for the third pond

$$\frac{S_{e3}}{S_{e2}} = \frac{1}{1 + k(3\text{ d})}$$

$$\frac{S_{e1}}{S_{e0}} \times \frac{S_{e2}}{S_{e1}} \times \frac{S_{e3}}{S_{e2}} = \frac{S_{e3}}{S_0} = \frac{1 - 0.88}{1} = 0.12 = \left[\frac{1}{1 + k(6\text{ d})}\right]\left[\frac{1}{1 + k(3\text{ d})}\right]^2$$

Solving the cubic equation, $k = 0.265\text{ d}^{-1}$

Nitrogen and Phosphorus Removal

Nitrogen is removed by a number of processes in a stabilization pond. Ammonia and nitrate are assimilated into algal, plant, and bacteria biomass. Biological solids in the effluent from the pond are one source of nitrogen loss. Ammonia can be converted into nitrate in the aerobic zone of the pond and biological denitrification can occur in the anoxic zone. Pond pHs are generally in the alkaline range and during intense algae activity can rise above 9.0. Free ammonia concentrations increase as pH rises and favors the stripping of gaseous ammonia. Total nitrogen removals ranged from 46 to 95% on an annual basis in pond systems (USEPA, 1983). The bulk of the removal occurred in the primary cells where ammonia stripping was a significant factor.

Phosphorus dynamics and transformations are complex in a pond system. Primary reactions that affect the amount of phosphorus in the pond effluent include biological uptake by algae and bacteria and reactions with cations that form settleable precipitates. Secondary acid–base and complexation reactions influence the primary reactions. Biological uptake of phosphorus is a cyclical process. Suspended aerobic microorganisms remove phosphorus from solution. Those that settle and die are anaerobically digested in the pond benthos, which releases some of their phosphorus into solution. Anaerobic microorganisms remove phosphorus through growth but release phosphorus through self-digestion. The quasi-equilibria established among all processes will not result in a significant amount of phosphorus removal.

Large plants that grow in ponds will also take up nitrogen and phosphorus. Certain plants may provide suitable habitats for nitrifying and denitrifying bacteria. Regular harvest of plants can have a significant impact on phosphorus removal. The water hyacinth (*Eichhornia crassipes*) is a plant that will grow in pond environments in warm climates. This plant is seeded into ponds and grows rapidly. Regular harvest improves nutrient removal and has other beneficial effects on metal removal.

If stringent effluent discharge limits exist for phosphorus then chemical control will be necessary. Addition of alum to the pond and settling is required. Ponds are usually discharged on an annual or semiannual basis in Ontario. Batch treatment of the pond with liquid alum is applied 1- or 2-weeks before discharge is begun. The alum is dosed by discharging it from a boat to mix it with the pond contents. For stabilization pond systems receiving domestic waste at loadings from 4 to 11 kg BOD_5/ha/d (3.6–9.8 lb/acre/d), alum concentrations in the pond of 80–100 mg/L reduced effluent total phosphorus concentrations below 1 mg/L (Droste et al., 1992). Concomitant removal of dissolved phosphorus and SS occurs as a result of flocculation and precipitation with alum. Higher pond loading rates will require higher alum dosages to produce equivalent results (Section 15.5). Sludge accumulation will not be significantly affected by batch treatment with alum at these rates.

Heat Balance for Ponds

The temperature for design of a pond is usually taken as the mean temperature during the coolest month. In Alaska, similar to guidelines for the Northwest Territories, the general guideline is to provide a 12-mo. retention time in the pond system before discharge (Smith and Knoll, 1986). If this volume of storage is not feasible, the minimum amount of storage required is usually specified to be 1 mo. longer than the normal ice cover expected in the region.

Pond temperatures can be calculated from heat balances. Heat energy transfer occurs through all surfaces of a pond. Heat is lost through the pond surface exposed to the atmosphere by evaporation, convection, and radiation. Wind velocity and water vapor pressure influence the rate of heat transfer. Heat is gained from solar radiation. Semiempirical models have been developed for these phenomena (see, for example, Velz, 1970) but a more practical approach for the heat balance has been developed by Mancini and Barnhart (1968).

If the pond is CM, a heat balance is made and the rate of heat transfer is modeled similarly to mass transfer. The following equation applies.

$$QT_0 - QT_p - \phi A(T_p - T_a) = \frac{dT_p}{dt}V \tag{19.10}$$

where

T_0, T_p, and T_a are influent, pond, and ambient temperatures, respectively, in °C
ϕ is a heat transfer coefficient
A is pond surface area

Note the similarity of the heat transfer expression to Eq. (12.7); V is eliminated in the third term on the left-hand side in Eq. (19.10) because V multiplies the rate of heat transfer and it is included in the denominator of the heat transfer expression. If steady state conditions apply, the pond temperature in terms of the other variables is

$$T_p = \frac{A\phi T_a + QT_0}{A\phi + Q} \tag{19.11}$$

All of the heat transfer phenomena are lumped into the rate coefficient, ϕ. Mancini and Barnhart (1968) statistically evaluated ϕ values for mechanically aerated lagoons in various regions of the United States (Table 19.5). In addition to the information in Table 19.5 they recommended a ϕ value of 0.80 m/d (2.60 ft/s) for the Gulf Coast area of Texas based on limited information.

A higher ϕ value will result in a pond temperature that is nearer the ambient temperature. Mancini and Barnhart (1968) recommend using the 90th percentile values for ϕ. Because sewage is more likely to be cooler than the ambient temperatures, lower values of ϕ would result in a more conservative design because the pond temperature would be lower and reaction rates would also decrease.

The rate of heat transfer would be higher in aerated lagoons compared to stabilization ponds because of the continuous input of air. Equivalent data on ϕ values for stabilization ponds have not been reported.

19.4 INDICATOR MICROORGANISM DIEOFF IN PONDS

One of the most important functions of stabilization pond systems is removal of pathogenic organisms. The amount of dieoff is a function of the residence time, temperature, and solar radiation intensity. First-order kinetics are generally suitable to describe dieoff of indicator organisms in ponds. The dieoff rate coefficient varies throughout a 24-h period as sunlight intensity changes. Each of these effects is separately incorporated into the overall dieoff coefficient.

TABLE 19.5 Heat Transfer Coefficient for Aerated Lagoons[a]

	ϕ, m/d (ft/d)		
Region	90th percentile	50th percentile	10th percentile
Eastern U.S.	0.50 (1.6)	0.14 (0.46)	0.06 (0.20)
Midwest U.S.	2.6 (8.5)	1.6 (5.2)	0.92 (3.0)

[a]After Mancini and Barnhart (1968).

$$k = k_d + k_l \quad (19.12)$$

where

k is the overall dieoff coefficient, d^{-1}
k_d is the dieoff coefficient during darkness
k_l is the average rate of dieoff from solar radiation

The latter term is a function of the average solar radiation intensity in the pond, which is in turn a function of the pond depth and incident solar radiation according to the Beer–Lambert law (Section 5.6).

$$k_l = k_s\bar{I} = \frac{1}{H}\int_0^H I_0 e^{-Kh}\,dh = \frac{k_s I_0}{KH}(1 - e^{-KH}) \quad (19.13)$$

where

k_s is the specific solar dieoff coefficient, $cm^2/cal/d$
I_0 is the average daily intensity of solar radiation at the pond surface, cal/cm^2
K is the light attenuation coefficient, m^{-1}
H is the pond depth, m

The e^{-KH} term can be neglected when the pond depth is greater than 0.9 m (3.0 ft) (Mayo, 1989). Using this modification and substituting Eq. (19.13) into Eq. (19.12), the overall dieoff rate coefficient is

$$k = k_d + \frac{k_s I_0}{K H} \quad (19.14)$$

For fecal coliforms, the value of k_s/K has been found to vary from 5.79×10^{-4} to 6.72×10^{-4} m-cm^2/cal/d for facultative and maturation ponds in various locations in Africa and Asia (Mayo, 1989). In Dar es Salaam, Tanzania, the value of k_d was 0.108 d^{-1} and I_0 was 550 cal/cm^2 on a daily basis (Mayo, 1989). To ensure a low pathogen content in effluent from stabilization pond systems in the tropics, total retention times of 20 d are recommended. In the recent pandemic of cholera that occurred in Brazil, all *Vibrio cholerae* were removed by the sixth pond in a waste stabilization pond system that was comprised of a 1-d anaerobic pond followed by a 2-d facultative pond and eight 2-d maturation ponds (i.e., a total detention time of 11 d was sufficient) (Oragui et al., 1993). Temperature in the pond system was 25–27°C.

19.5 AERATED LAGOONS

An aerated lagoon incorporates features of activated sludge and stabilization pond operations. The essential feature of an aerated lagoon is mechanical aeration to supply oxygen to maintain aerobic conditions and improve mixing in the pond. Mixing improves contact between bacteria and dissolved organics. The maintenance of measurable dissolved oxygen concentrations also reduces the time required to stabilize organic matter in aerobic lagoons compared to stabilization ponds. Sludge is not normally recycled in aerobic lagoon systems, which lowers the amount of solids in suspension, and there is less operator control, which decreases the treatment efficiency compared to activated sludge processes.

The degree of mixing provided in aerated lagoons is variable. Some systems have a degree of aeration sufficient to keep all solids in suspension, whereas others referred to as partial-mixed ponds, have less power input. There are no rational methods to

TABLE 19.6 Typical Low-Speed Surface Aerator Performance Data[a]

Power kW	Depth m (ft)	Zone of complete mixing m (ft)	Zone of oxygen dispersion m (ft)
2.2	1.8 (6)	15 (50)	46 (150)
3.7	1.8 (6)	21 (70)	64 (210)
7.5	2.4 (8)	27 (90)	79 (260)
14.9	3.0 (10)	35 (115)	101 (330)
18.6	3.0 (10)	40 (130)	114 (375)

[a]After Benefield and Randall (1980).

assess the power requirements to keep solids in suspension. Malina et al. (1972), from studies of lagoon systems, found that power inputs above 5.9 kW/1 000 m^3 (0.17 kW/1 000 ft^3) were sufficient to maintain all solids in suspension. Benefield and Randall (1980) and the USEPA (1983) quote data from other sources which indicate that the power requirement ranges from 3 to 10.8 kW/1 000 m^3 (0.085–0.30 kW/1 000 ft^3) for lagoon depths ranging from 2.4 to 5.5 m (8–18 ft). A minimum freeboard depth of 0.6 m (2 ft) should be added to the depth.

The zone of influence of aerators where all solids are kept in suspension is smaller than the zone to which oxygen will be delivered. Table 19.6 should be consulted for typical data. Aerators should be placed such that there is some overlap in their zones of influence to assure the desired distribution of solids and oxygen.

When ponds are well mixed, the CM model (Eqs. 19.6 and 19.7) is the appropriate model. The choice of the rate constant is the critical factor and temperature effects should be evaluated from pilot studies or data taken from a similar pond. The lowest temperature period should be used for design of the system.

When data are unavailable, the USEPA (1983) recommends a design value for k of 2.5 d^{-1} at 20°C. However the Great Lakes Upper Mississippi River Board (GLUMRB) (1978) recommends a design value of 0.276 d^{-1} at 20°C. The value of θ to be used in Eq. (19.5) for temperature correction is 1.085 according to the USEPA (1983) and 1.036 according to GLUMRB (1978). Although the USEPA recommendations result in more typical designs, pilot studies or data from nearby lagoons operating under conditions similar to the proposed lagoons are required to resolve these wide discrepancies in rate coefficients. Equation (19.11) can be used to calculate temperature variation in the lagoon.

Partially mixed ponds should be studied with tracers to determine whether the PF or CM flow regime is more nearly approached in the pond. As noted above, the CM model will be the most conservative approach to sizing the lagoons.

19.6 TREATMENT OF WASTEWATER IN NATURAL SYSTEMS

Natural systems for wastewater treatment and disposal include application of wastewater to land and treating wastewater through natural or artificial wetlands. Both alternatives are land intensive but when land is available at low cost these alternatives are an economical option. These methods of treatment are satisfying approaches because they render wastewater as a resource rather than as a problem to be rectified. The accelerated disappearance of wetlands over the past few decades has serious ecological ramifications. Many states now protect wetlands and foster reestablishment of wetlands.

Municipal wastewaters are good candidates for treatment in natural systems. Not all wastewaters will be suitable for treatment in these systems because they contain hazardous substances. Conventional treatments produce small quantities of sludge that can be disposed in small controlled areas. Natural treatment applies the wastewater to large areas and disperses contaminants over these areas. Present and future use of the area must be considered. Pretreatment may render a wastewater suitable for natural treatment. Natural systems are also used for polishing effluents from conventional treatment processes.

High BOD and TSS removals can be achieved in these systems. Nitrogen removal depends on many factors, including soil characteristics as well as biological, chemical, and mass transfer phenomena. Phosphorus is much less mobile than nitrogen and its removal ultimately depends on the soil capabilities to bind it. Soil may be forming in the process. Plant growth is a factor in nutrient transformations and may be significant for their removal if plants are harvested.

19.6.1 Land Treatment of Wastewater

There are many factors involved in the design of a land application system (Fig. 19.2). Land application systems fall into one of three categories depending on land availability and use and treatment objectives (USEPA, 1981). Treatment objectives often extend beyond the sole objective of removal of objectionable constituents in the wastewater and may include crop growth, irrigation, and groundwater recharge. The latter is an important objective in arid countries, particularly the Middle East. The USEPA (1981) has detailed design information on land treatment systems and presents a number of design examples.

Slow rate (SR) systems rely on plants and soil percolation to provide treatment. These systems have also been referred to as irrigation systems. Rapid infiltration (RI) systems have higher hydraulic loading rates than SR systems. Higher hydraulic loading

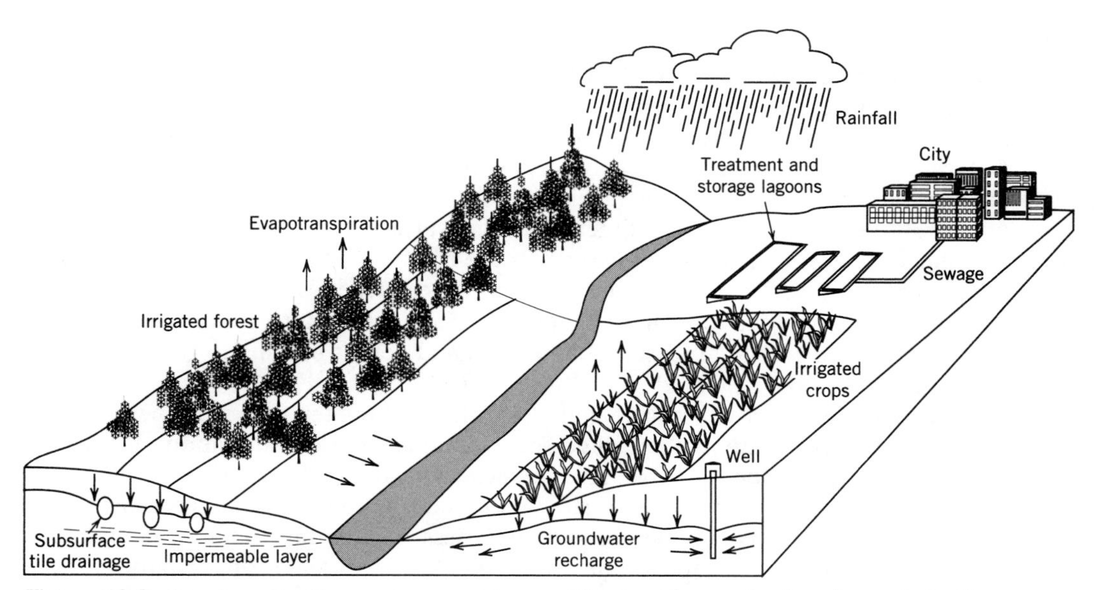

Figure 19.2 Land application systems can be used for forest or agricultural crop irrigation. After USEPA (1979).

TABLE 19.7 Characteristics of Land Treatment Systems[a]

Characteristic	System		
	Slow rate	Rapid infiltration	Overland flow
Hydraulic application rate, m/yr (ft/yr)	0.5–6 (1.6–20)	6–125 (20–410)	3–20 (10–66)
Soil permeability, cm/h (in./h)	>0.15 (0.06)	>1(0.4)	<0.5 (0.2)
Minimum pretreatment	Primary	Primary	Fine screening
BOD loading, kg/ha/yr (lb/acre/yr)	370–1 830 (330–1 633)	8 000–46 000 (7 140–41 000)	2 000–7 500 (1 780–6 690)
Application method	Sprinkler or surface	Surface	Sprinkler or surface
Land slope suitability			
0–12%	High	High	High
12–20%	Low, high[b]	Low	Moderate
>20%	Very low, moderate[b]	Do not use	Do not use
Land use suitability			
Open or cropland	High, moderate[b]	High	High
Partially forested	Moderate, moderately high	Moderate	Moderate
Heavily forested	Low, high[b]	Low	Low
Developed (residential, commercial, or industrial)	Low, very low[b]	Very low	Very low

[a]After Reed et al. (1988) and USEPA (1981).
[b]The first factor applies to agricultural systems; the second factor applies to forest land systems.

rates are obtained by applying more pretreatment to the wastewater and using soils or artificially prepared beds with higher hydraulic conductivities. The third system is overland flow (OF) treatment, which is used when surface soil permeabilities are low. Vegetation is used in OF systems and the effluent is collected as runoff and conveyed to the receiving water. Characteristics of each of the systems are given in Table 19.7.

Application of wastewater to land is more subject to climatological factors than conventional treatment. Seasonal change in rainfall and temperature affect the rate at which wastewater can be applied to the soil. The design must take into account wastewater generation rates against variable application rates throughout the year.

It is necessary to survey the soil profile at any site for a land application system. Before studying the conveyance properties of the soil at a site, it is necessary to conduct site reconnaissance by digging small pits over the site to observe if any undesirable soil features exist. Fractures, near surface rock, lenses of open layers, mottling of the profile, and other anomalies are characteristics that should be avoided in application sites. The soil profile to be evaluated depends on the type of system: for OF systems only the top 1 m (3.3 ft) needs examination; for SR systems the depth examined should extend at least to 1.5 m (5 ft) and depths of 3 m (10 ft) or more are recommended to be evaluated (USEPA, 1981).

Hydraulic conductivity of the soil is another parameter that must be assessed. Low hydraulic conductivities preclude some options from consideration, necessitate changes in pretreatment, may require replacing topsoil, or have other ramifications on the design. RI systems are ultimately governed by the hydraulic conductivity of the soil. Other systems may be designed with less reliable information on hydraulic conductivity but a more extensive investigation of the hydraulic conductivity over the site for any of the processes will enable the designer to make the best decisions and avoid problems.

Measurement of Hydraulic Conductivity

There are various techniques to measure the hydraulic conductivity of soil. A good description of many techniques is given by the USEPA (1981). There are practical field techniques designed to directly measure the infiltration capacity of a soil. Infiltration capacity can vary widely over an area; therefore, as many tests as possible should be made over a site. Also, larger test areas give more accurate results than smaller test areas. The two best techniques for infiltration capacity are the flooding basin technique and sprinkling infiltrometer, which are direct measurement techniques.

In the flooding basin technique, a circular flashing with a diameter of at least 3 m (10 ft) is placed in the ground to a depth of 15 cm (6 in.) below the surface and height of 20 cm (8 in.) above the surface. Tensiometers are placed within a concentric circle within the flashing. The soil is saturated by flooding the basin a few times. Then water is continuously applied to the basin. The test is conducted for 4–12 h and the cumulative rate of water uptake or infiltration rate variation with time is plotted on log–log paper. After flooding is terminated, drainage of the soil profile is gauged from the tensiometers. Plotting the average rate versus time on arithmetic paper will yield a curve that asymptotically approaches a limiting value that is chosen for design. At least two tests should be conducted in an area to determine whether they are in agreement. If results are different, more tests are indicated. In RI installations the soil must be excavated to the depth that will be at the bottom of the infiltration basins.

Sprinkling infiltrometers are used to determine the limiting application rate when sprinklers will be used to apply wastewater. A grid of rain gauges is placed over the area to be wetted. A revolving sprinkler is used with a shield with an opening that allows water to be sprayed over about one eighth of the plan area of a circle defined by the range of the sprinkler. The shield blocks the remainder of the water sprayed in the traverse of the sprinkler and allows it to be collected and returned to the supply reservoir. The test is conducted for about 1 h. The area selected for measurement of the application rate is determined where the applied water just disappears from the soil as the sprinkler jet returns to the spot. The amount of water collected in the gauges is measured and the infiltration rate is calculated. USEPA (1981) gives more detailed information on these techniques and a description of other techniques.

19.6.2 Slow Rate Land Application Systems

SR systems are the oldest and most common application of land treatment. Crop water requirements, soil characteristics, and climate factors interact to determine the wastewater application rate. SR systems are designed either to maximize the hydraulic loading rate, and thus minimize the land area required, or to produce a cash crop. The former is the more common objective and each objective has different design constraints.

Hydraulic loading rates are the limiting loading rate constraint in all SR systems when the wastewater has been screened and settled. BOD and SS removals will be excellent after the water has percolated through 1.5 m (5 ft) of soil. USEPA (1981) shows data from a number of sites where BODs were less than 2 mg/L at shallow sampling depths. Clogging of the soil will not be a problem.

Crop selection is commensurate with the treatment objectives. If revenue generation is considered, the market for various crops must be determined. Agricultural or forest crops may be used. Grasses are most beneficial for maximizing the hydraulic loading rate. Table 19.8 provides nutrient uptake and other characteristics of various crops. All characteristics given in Table 19.8 are subject to a number of conditions, including rate of harvesting, plant age when harvested, supplemental nutrients, salinity, toxicity, and soil characteristics, as well as application rate of wastewater. Wastewaters will generally be deficient in potassium and it will be necessary to add potassium to wastewaters applied for agricultural crops. More detailed discussions of the impact of various factors on crop growth and selection is given by the USEPA (1981) and the Food and Agriculture Organization (FAO) (1977).

The hydraulic loading rate is determined from a mass balance on the receiving soil. Runoff of applied wastewater is ignored; wastewater should not be applied when the potential for overland wastewater flow is high. Wastewater will not be applied when it is raining.

$$L_{w(H)} = ET - P_r + F_c F_A K \tag{19.16}$$

where

$L_{w(H)}$ is the hydraulic loading rate (volumetric flow per unit area)
F_c is a factor accounting for soil clogging
F_A is a factor accounting for periods of nonapplication
ET is the rate of evapotranspiration
P_r is the rate of precipitation
K is the soil permeability (LT^{-1})

Equation (19.16) is first applied on an annual basis to determine the average loading rate. This hydraulic loading rate is then compared with the allowable loading rate for nitrogen. The lower loading rate is the design value. If the hydraulic loading rate is limiting, Eq. (19.16) is then applied on a monthly basis considering monthly variation in climatological terms.

The permeability value (K) to be used in Eq. (19.16) is the lowest measured permeability at the site. It is reduced by a factor F_c, which accounts for the long-term acceptance rate at the site. The USEPA (1981) suggests that F_c be taken at 0.04–0.10, whereas Metcalf and Eddy (1991) give a range of 0.02–0.06 for F_c. Some experience is needed to choose the proper value of F_c; higher values of F_c are used when more uniform conditions exist at the site and a large amount of reliable data are available.

The factor F_A accounts for periods of time without application of wastewater. Freezing and crop management will suspend operation. Local data must be obtained to find the number of days during winter months when the temperature is less than 0°C (or at a minimum, −4°C). For forest crops, wastewater can often be applied during subfreezing conditions. No wastewater is applied during planting, cultivation, and harvesting of plants. For these months F_A is reduced by the proportion of days when these conditions or activities occur. The expression $F_c F_A K$ is the design hydraulic permeability.

The maximum allowable hydraulic loading rate is given by the above development. If cash crops are being grown the allowable hydraulic loading rate is dictated by their moisture requirements and moisture tolerances. Wastewater may have to be applied intermittently. More time will probably be allotted for cultivation and harvesting. Agricultural agencies should be consulted for specific requirements beyond those given in Table 19.8.

Wastewater treatment objectives may include protection of potable groundwater aquifers in which case nitrogen must be removed by the system to a potential nitrate concentration of no more than 10 mg N/L, which is the general standard throughout the world.

TABLE 19.8 Nutrient Uptake Rates and Water Characteristics for Selected Crops[a]

	Nutrient uptake, kg/ha/yr (lb/acre/yr)			Water characteristics	
	Nitrogen	Phosphorus	Potassium	Potential as water user[b]	Moisture tolerance[c]
Forage Crops					
Alfalfa	225–540 (200–480)	22–35 (20–30)	175–225 (155–200)	High	Low
Bentgrass	—	—	—	High	High
Bromegrass	130–225 (425–200)	40–55 (35–50)	245 (220)	High	High
Coastal Bermudagrass	400–675 (360–600)	35–45 (30–40)	225 (200)	High	High
Kentucky bluegrass	200–270 (180–240)	45 (40)	200 (180)	High	Moderate
Quackgrass	235–280 (210–250)	30–45 (30–40)	275 (245)	—	—
Reed canarygrass	335–450 (300–400)	40–45 (35–40)	315 (280)	High	High
Ryegrass	200–280 (180–250)	60–85 (55–75)	270–325 (240–290)	—	—
Sweet clover[d]	175 (155)	20 (18)	100 (90)	High	Moderate-high
Tall fescue	150–325 (135–290)	30 (27)	300 (270)	High	High
Orchardgrass	250–350 (225–315)	20–50 (18–45)	225–315 (200–280)	High	Moderate
Field Crops					
Barley	125 (110)	15 (13)	20 (18)	Moderate	Low
Corn	175–200 (155–180)	20–30 (18–27)	110 (100)	Moderate	Moderate
Cotton	75–110 (67–100)	15 (13)	40 (35)	Moderate	Low
Grain, sorghum	135 (120)	15 (13)	70 (63)	Low	Moderate
Oats	—	—	—	Moderate	Low
Potatoes	230 (205)	20 (18)	245–325 (220–290)	—	—
Rice	—	—	—	High	High
Safflower	—	—	—	Moderate	Moderate
Soybeans[d]	250 (225)	10–20 (9–18)	30–55 (27–50)	Moderate	Moderate
Wheat	160 (145)	15 (13)	20–45 (18–40)	Moderate	Low
Hardwoods	110–340 (100–305)			High	High
Forest Crops[e]					
Pine	110–320 (110–285)			High	Moderate-low
Douglas fir	150–250 (135–225)			High	Moderate

[a]From USEPA (1981).
[b]Water use definitions expressed as a fraction of alfalfa consumptive use: high, 0.8–1.0; moderate, 0.6–0.79; low, <0.6.
[c]Moisture tolerance ratings: high, withstands prolonged soil saturation >3 d; moderate, withstands soil saturation 2–3 d; low, withstands no soil saturation.
[d]Legumes also fix nitrogen from the atmosphere.
[e]Values are for overstory and understory vegetation in vigorously growing forests. There is significant regional variation in the rates. If all trees are harvested then the values in the table apply. If only salable stems are harvested, nitrogen removal rates will be less than 30% of these rates.

Nitrogen losses occur through crop uptake and denitrification. After losses from the incoming nitrogen are subtracted, the remaining amount of nitrogen percolates into the soil and is diluted by excess precipitation. Dilution of the percolating water with native groundwater is also ignored to provide conservative results. Rearranging the mass balance and solving for the annual allowable loading rate on a nitrogen basis:

$$L_{w(N)} = \frac{C_p(P_r - ET) + k_c N}{(1 - f)C_N - C_p} \tag{19.17}$$

where

$L_{w(N)}$ is the allowable annual hydraulic loading rate based on nitrogen limits, cm/yr (in./yr)
P_r and ET are on an annual basis, cm/yr (in./yr)
C_p is the nitrogen concentration in percolating water, mg N/L
N is the nitrogen uptake by crops, kg/ha/yr (lb/acre/yr)
C_N is the nitrogen concentration in the applied wastewater, mg N/L
f is the fraction of applied nitrogen removed by denitrification and volatilization
k_c is a conversion factor, $k_c = 10$ (SI); $k_c = 4.42$ (U.S.)

In Eq. (19.17), C_p is usually taken as 10 mg/L. Nitrogen uptake by crops is given in Table 19.8. The annual nitrogen uptake is proportioned to each month based on the ratio of ET for the month to annual ET, which reflects the amount of growth during the month. More detailed information on the growth and nutrient uptake of crops throughout their season may be available from agricultural agencies and can be used to more accurately estimate monthly uptake rates. Denitrification and volatilization rates can range from 3 to 70% of nitrogen in the applied wastewater but an f value from 0.15 to 0.25 is used for a conservative design (USEPA, 1981).

Equation (19.17) is first solved on an annual basis and compared with the maximum hydraulic loading rate. The lower value governs the design. If $L_{w(N)}$ is lower than $L_{w(H)}$, then monthly application rates are calculated by applying Eq. (19.17) on a monthly basis.

The land area required is calculated from

$$A = V/L_w \tag{19.18}$$

where

A is the field area
V is the annual volume of wastewater to be applied
L_w is the design annual loading rate expressed as a depth

Storage will be required in most cases to balance monthly wastewater generation and allowable application rates. The problem is similar to the design of an equalization basin (Section 10.4). The water balance equation including storage is

$$Q_m t_m - L_{w,m} A t_m = \Delta S \tag{19.19}$$

where

Q_m is the flow during a month
t_m is the duration of the month in days
$L_{w,m}$ is the allowable application rate per unit area during the month (LT^{-1})
S is storage expressed as depth over the seepage area

To determine the required storage with Eq. (19.19), list the months in order with their allowable application rates and wastewater flow rates normalized per unit area of application field. Calculate the difference between application rate and normalized flow rate for each month. The difference is the change in storage. Beginning with the first month for which normalized flow rate exceeds application rate, add the successive differences until application rate exceeds normalized wastewater flow rate. The differences that have been added together form the cumulative storage volume required.

Finally, the storage volume is refined by calculating the volume of water lost through evaporation and seepage and the volume gained by precipitation based on the area of the storage pond. The mass balance equation is

$$\Delta S = Q_m t_m - L_{w(d)} t_m - ET_m t_m - S_{pm} t_m + P_{rm} t_m \qquad (19.20)$$

where

ET_m is the monthly rate of evaporation
$L_{w(d)}$ is the design hydraulic loading rate
S_{pm} is the monthly rate of seepage
P_{rm} is the monthly rate of precipitation

Local data should be used for evaporation. There will be an accumulation of solids on the bottom of the storage pond, which will retard the rate of seepage. A conservative assumption for seepage is to ignore it.

A buffer zone is usually required around the application field. Depending on the degree of pretreatment, public access to the site may be curtailed or free if, for example, a golf course is being irrigated.

■ Example 19.2 Slow Rate Infiltration System Design

The evapotranspiration and precipitation (5-yr return period) monthly data in the second and third columns, respectively, in the following table were obtained for an area. The evapotranspiration data are for an annual cover of a grass crop. Use these data to design an SR system for a community of 5 000 residents that generates wastewater at an average rate of 325 L/cap/d (86 gal/cap/d) except during the months of June–August when the rate rises to 375 L/cap/d (99 gal/cap/d).

Field tests indicate that the soil permeability is 1.5–3.0 cm/h (0.6–1.2 in./h) and that the soil is fairly uniform over the application area. The total nitrogen content of screened and settled wastewater is 30 mg N/L. The crop chosen for the application field has an annual nitrogen uptake rate of 350 kg/ha/yr (312 lb/acre/yr).

The percolation rate to be used will be taken as 8% of the lower measured value because the soil is fairly uniform. The potential monthly percolation rate ($P_{w,m}$) not considering days when operation is not feasible, (i.e., F_A is taken as 1) is

$$P_{w,m} = 0.08\left(1.5\,\frac{\text{cm}}{\text{h}}\right)\left(24\,\frac{\text{h}}{\text{d}}\right)\left(30\,\frac{\text{d}}{\text{mo.}}\right) = 86.3\text{ cm/mo.}$$

$$\text{In U.S. units: } P_{w,m} = 0.08\left(0.6\,\frac{\text{in.}}{\text{h}}\right)\left(24\,\frac{\text{h}}{\text{d}}\right)\left(30\,\frac{\text{d}}{\text{mo.}}\right) = 34.6\text{ in./mo.}$$

If a nitrogen limitation exists for the percolating water, Eq. (19.17) is used.

$$L_{w(N)} = \frac{C_p(P_r - ET) + N(10)}{(1-f)C_N - C_p} = \frac{\left(10\,\frac{\text{mg}}{\text{L}}\right)\left(-10.4\,\frac{\text{cm}}{\text{yr}}\right) + \left(350\,\frac{\text{kg}}{\text{ha-yr}}\right)\left(10\,\frac{\text{mg-cm-ha}}{\text{L-kg}}\right)}{(1-0.20)\left(30\,\frac{\text{mg}}{\text{L}}\right) - 10\,\frac{\text{mg}}{\text{L}}}$$

$$= 242.6\text{ cm/yr}$$

Data and Loading Rate Calculations for Example 19.2

(1)	(2)	(3)	(4)	(5)	(6)	(7)	(8)	(9)
Month	ET cm (in.)	P_r cm (in.)	ET − P_r cm (in.)	No. of inoperative days d	$P_{w,m}$[a] cm (in.)	$L_{w(H)}$ cm (in.)	N kg/ha (lb/acre)	$L_{w(N)}$ cm (in.)
January	3.05	5.30	−2.25	19	31.6	29.4	11.0	9.5
	(1.20)	(2.09)	(−0.93)		(12.4)	(11.6)	(9.8)	(3.7)
February	4.88	5.35	−0.47	12	51.8	51.3	17.5	12.8
	(1.92)	(2.11)	(−0.19)		(20.4)	(20.2)	(6.9)	(5.1)
March	6.36	6.85	−0.49	3	77.7	77.2	22.9	16.7
April	10.29	6.90	3.39	4	74.8	78.0	37.0	24.0
May	12.20	7.21	4.99	0	86.3	91.0	43.8	27.7
June	12.95	7.27	5.68	0	86.3	92.0	46.5	29.2
July	12.95	8.66	4.29	0	86.3	90.6	46.5	30.2
August	11.40	8.64	2.76	0	86.3	89.1	41.0	27.3
September	9.15	8.09	1.06	0	86.3	87.4	32.9	22.7
October	6.60	6.23	0.37	0	86.3	86.7	23.7	16.7
November	4.55	8.45	−3.90	2	80.5	76.6	16.4	14.5
December	3.00	8.05	−5.05	8	63.3	58.3	10.8	11.3
Annual	97.38	87.00	10.38	48	897.5	907.8	350.0	242.6
	(38.3)	(34.3)	(4.09)		(353.3)	(357.4)	(312.3)	(95.5)

[a]Based on a mean monthly percolation rate of 86.3 cm (34.6 in.).

In U.S. units:

$$L_{w(N)} = \frac{\left(10\,\frac{\text{mg}}{\text{L}}\right)\left(-4.10\,\frac{\text{in.}}{\text{yr}}\right) + \left(312\,\frac{\text{lb}}{\text{acre-yr}}\right)\left(4.42\,\frac{\text{mg-acre-in.}}{\text{lb-L}}\right)}{(1-0.20)\left(30\,\frac{\text{mg}}{\text{L}}\right) - 10\,\frac{\text{mg}}{\text{L}}} = 95.6\ \text{in./yr}$$

The annual loading rate for a nitrogen limitation is significantly less than the hydraulic loading rate. The values for N are calculated based on the ratio of monthly ET to annual ET and are given in col. 8 of the table. Monthly $L_{w(N)}$ rates (col. 9) are calculated by applying Eq. (19.17) to monthly data. In this example, $L_{w(N)}$ values are always less than $L_{w(H)}$ for each month. If $L_{w(H)}$ was less than $L_{w(N)}$ in any month then $L_{w(H)}$ would have been the design value for that month.

The volume of wastewater generated for the year is

$$V = (30\ \text{d})(9)(5\,000)(0.325\ \text{m}^3/\text{d}) + (30\ \text{d})(3)(5\,000)(0.375\ \text{m}^3/\text{d}) = 607\,500\ \text{m}^3$$

$$V = \left[(30\ \text{d})(9)(5\,000)\left(\frac{86\ \text{gal}}{\text{d}}\right) + (30\ \text{d})(3)(5\,000)\left(\frac{99\ \text{gal}}{\text{d}}\right)\right]\left(\frac{1\ \text{ft}^3}{7.48\ \text{gal}}\right) = 21.5 \times 10^6\ \text{ft}^3$$

The land area required is

$$A = \frac{V}{L_w} = \frac{607\,500\ \text{m}^3}{242.6\ \text{cm}}\left(\frac{100\ \text{cm}}{\text{m}}\right) = 250\,412\ \text{m}^2 = 25.0\ \text{ha}$$

$$\text{In U.S. units: } A = \frac{V}{L_w} = \frac{21.5 \times 10^6\ \text{ft}^3}{95.5\ \text{in.}}\left(\frac{12\ \text{in.}}{\text{ft}}\right)\left(\frac{1\ \text{acre}}{43\,560\ \text{ft}^2}\right) = 62.0\ \text{acre}$$

The normalized wastewater generation rates during the two periods (V_{m1} and V_{m2}, respectively) are

$$V_{m1} = \frac{Qt}{A} = \frac{(0.325\ \text{m}^3/\text{d})(5\,000)(30\ \text{d})}{250\,412\ \text{m}^2}\left(\frac{100\ \text{cm}}{\text{m}}\right) = 19.4\ \text{cm}$$

$$V_{m2} = \frac{0.375}{0.325}(19.4\ \text{cm}) = 22.4\ \text{cm}$$

In U.S. units:

$$V_{m1} = \frac{Qt}{A} = \frac{(86\ \text{gal/d})(5\,000)(30\ \text{d})}{62.0\ \text{acre}}\left(\frac{1\ \text{acre}}{43\,560\ \text{ft}^2}\right)\left(\frac{1\ \text{ft}^3}{7.48\ \text{gal}}\right)\left(\frac{12\ \text{in.}}{\text{ft}}\right) = 7.7\ \text{in.}$$

$$V_{m2} = \frac{99}{86}(7.7\ \text{in.}) = 8.9\ \text{in.}$$

Month	$L_{w,m}$ cm (in.)	V_m cm (in.)	$V_m - L_{w,m}$ cm (in.)	Storage cm (in.)
January	9.5	19.4	9.9	15.7
	(3.7)	(7.7)	(3.9)	(6.2)
February	12.8	19.4	6.6	25.6
	(5.1)	(7.7)	(2.6)	(10.1)
March	16.7	19.4	2.7	32.2
		(7.7)	(1.1)	(12.7)
April	24.0	19.4	−4.6	34.9
		(7.7)	(−1.8)	(13.7)
May	27.7	19.4	−8.3	30.3
June	29.2	22.4	−6.8	22.0
July	30.2	22.4	−7.8	15.2
August	27.3	22.4	−4.9	7.4
September	22.7	19.4	−3.3	2.5
October	16.7	19.4	2.7	0[a]
November	14.5	19.4	4.9	2.7
December	11.3	19.4	8.1	7.6
Annual	242.6	241.8	−0.8[b]	
	(95.5)	(95.2)	(−0.3[b])	

[a]Adjusted to 0.

[b]The value is not 0 because of roundoff error.

The calculations to determine the storage requirement are given in the preceding table. The minimum monthly loadings are taken from this table. The first month in the sequence of months where the wastewater generation is greater than the application rate is October. After this month there will be an accumulation of 2.7 cm (1.1 in.) of wastewater. The additional accumulations are added in sequence.

The maximum storage is

$$S = (34.9\ \text{cm})(25.0\ \text{ha})\left(\frac{1\ \text{m}}{100\ \text{cm}}\right)\left(\frac{10\,000\ \text{m}^2}{\text{ha}}\right) = 87\,250\ \text{m}^3$$

In U.S. units: $S = (13.7\ \text{in.})(62.0\ \text{acre})\left(\frac{1\ \text{ft}}{12\ \text{in.}}\right)\left(\frac{43\,560\ \text{ft}^2}{\text{acre}}\right) = 3.08 \times 10^6\ \text{ft}^3$

An analysis of seepage, evaporation, and precipitation will decrease the volume required for storage.

19.6.3 Rapid Infiltration Systems

RI systems are used to renovate wastewater with a common objective of replenishing groundwaters. Crop growth is not a consideration. Frequently RI systems are designed for final polishing of a wastewater that has been given primary and secondary treatment. Nitrification or nitrification–denitrification will be achieved along with phosphorus removal.

Wastewater is applied intermittently in all RI systems, which improves the treatment performance of these systems and maximizes long-term infiltration rates. The cycle time for a site consists of the period during which wastewater is applied and the drying period. From a survey of existing RI installations (USEPA, 1981), application periods ranged from less than a day to 2 weeks. The drying period was at least as long as the application period, ranging up to 20 times the duration of the application period. Suggested loading cycles for different treatment objectives are given in Table 19.9.

The maximum allowable application rate is the rate that can be sustained over the application period without causing significant ponding after application of the wastewater has ceased. The rate of infiltration should be conservatively taken near the lower values of infiltration tests at the site. Soil may be excavated and replaced with high conductivity soil such as sand to achieve desired infiltration rates.

The annual application rate and field area requirements are readily established once the infiltration rate and cycle pattern are determined. The field is divided into the requisite number of sections or basins that accommodate the loading cycle. Equalization requirements are determined as illustrated above.

The basic design equations begin with determination of the hydraulic loading rate. The design hydraulic loading rate applies a safety factor of 0.02 to 0.15 (depending on the technique used to measure conductivity) to the lowest infiltration rate observed for the site (USEPA, 1981).

TABLE 19.9 Suggested Loading Cycles for RI Systems[a]

Treatment objective	Pretreatment	Season	Application period d	Drying period d
Maximize infiltration rates	Primary	Summer	1–2	5–7
		Winter	1–2	7–12
	Secondary	Summer	1–3	4–5
		Winter	1–3	5–10
Maximize nitrogen removal	Primary	Summer	1–2	10–14
		Winter	1–2	12–16
	Secondary	Summer	7–9	10–15
		Winter	9–12	12–16
Maximize nitrification	Primary	Summer	1–2	5–7
		Winter	1–2	7–12
	Secondary	Summer	1–3	4–5
		Winter	1–3	5–10

[a]From USEPA (1981).

$$L_{w(d)} = F_s K \tag{19.21}$$

where

$L_{w(d)}$ is expressed on a yearly basis (m/yr or ft/yr)
F_s is a safety factor (0.10–0.15 for basin infiltration test; 0.02–0.04 for cylinder infiltrometer; 0.04–0.10 for vertical hydraulic conductivity measurement)

The cycle time and number of annual cycles are

$$T_c = T_A + T_d \tag{19.22}$$

$$N_c = 365 \text{ d}/T_c \tag{19.23}$$

where

T_c is the time for one cycle
T_A is the time period for wastewater application
T_d is the drying time
N_c is the number of cycles per year

The loading rate per cycle and the application rate in a cycle can now be calculated.

$$L_{wc} = L_{w(d)}/N_c \tag{19.24}$$

$$r_A = L_{wc}/T_A \tag{19.25}$$

where

L_{wc} is the loading rate per cycle (m/cycle or ft/cycle)
r_A is the application rate applied over T_A (m/d or ft/d)

Add the maximum monthly precipitation expected once every 10 years to r_A. Then check if ponding is significant. The criterion is (USEPA, 1981)

$$(r_A + P - K)T_A < 0.30 \text{ to } 0.46 \text{ m (1 to 1.5 ft)} \tag{19.26}$$

where

P is the maximum monthly precipitation once in 10 yr expressed on a daily basis (m/d or ft/d)

If the criterion is exceeded, reduce the application rate. The wastewater flow is delivered to infiltration zones sequentially. The area receiving flow during one application period and the total infiltration area are

$$A_c = \frac{Q}{r_A - P}$$

$$A = \frac{T_c}{T_A} A_c = \frac{T_c}{T_A}\left(\frac{Q}{r_A - P}\right) \tag{19.27}$$

where

A_c is the average area that receives flow once per cycle
A is the total infiltration area

It is necessary to prevent renovated water from mixing with native groundwater. The depth fluctuation of the water table should be determined throughout the year and over a number of years to find the limiting depth available for infiltration. Direction of renovated and native groundwater flows must be determined. Drainage is necessary to maintain treatment and infiltration rates. Topography may allow natural drainage of infiltrated waste to collecting channels or streams. Renovated water may be collected in underdrains or wells and pumped to desired streams.

19.6.4 Overland Flow Systems

OF systems are used when soil permeabilities are too low to support the application rates for SR systems. Infiltration of the wastewater will be minimal, generating wastewater "runoff" that must be collected and conveyed to the receiving body of water.

Vegetation plays an important role in OF systems. The vegetative cover should be a mixture of grasses to promote the most advantageous treatment. A mixture will be more likely to contain plants that are adapted to the particular conditions of the system. The crop will have to be mowed periodically and the clippings can be sold as hay if a market exists. For a normal mowing frequency of 2–3 times per year, the first cuttings may be left in place.

There are various empirical formulas used to describe treatment performance in OF systems. The application rate and the slope of the land influence the contact time between the wastewater and soil with vegetation. Land slopes in the range of 2–8% do not affect performance (USEPA, 1981). Design guidelines for OF systems are given in Table 19.10. Hydraulic loading rates and application rates increase with the degree of treatment given to the wastewater. Lower application rates within the ranges are used at lower temperatures.

The hydraulic loading is based on the number of operative days of the system.

$$L_{w(d)} = \frac{Q(365\text{ d}) + \Delta V_s}{TA} \tag{19.28}$$

where

Q is the average daily flow rate
ΔV_s is the change in volume of wastewater in a storage basin
T is the number of days during the year when wastewater will be applied

Storage may be designed to store wastewater during winter periods when lower application rates are required; this would decrease land area requirements. Seepage, evaporation, and precipitation will change the volume of wastewater to be applied when storage exists.

TABLE 19.10 Overland Flow Design Guidelines[a]

Pretreatment	Hydraulic loading rate cm/d (in./d)	Application rate[b] m^3/m/h (gal/ft/h)	Application period[c] h/d	Side length of sloped area m (ft)
Screening	0.9–3 (0.4–1.2)	0.07–0.12 (5.6–9.7)	8–12	36–45 (120–150)
Primary sedimentation	1.4–4 (0.6–1.6)	0.08–0.12 (6.4–9.7)	8–12	30–36 (100–150)
Stabilization pond	1.3–3.3 (0.50–1.3)	0.03–0.10 (2.4–8.0)	8–18	45 (150)
Secondary biological	2.8–6.7 (1.1–2.6)	0.10–0.20 (8.0–16)	8–12	30–36 (100–150)

[a]From USEPA (1981).
[b]Application rate is per m (or ft) of slope length.
[c]Application frequencies of 5–7 d/wk are recommended.

The area, A, is the area of the sloped surface to which wastewater is applied. Side length ranges for the sloped area are given in Table 19.10. Rectangular strips are designed. Water is distributed by sprays, slotted pipes, or sprinklers.

$$A = l_D l_s \tag{19.29}$$

where
l_D is the length of the distributor (width of the sloped area)
l_s is the length of the sloped surface

The application rate is

$$r_A = \frac{Q(365 \text{ d}) + \Delta V_s}{l_D T D} \tag{19.30}$$

where
D is the number of hours per day of operation

Less area will be required with seven-day-a-week operation. Choosing any three of the four variables from Table 19.10 and the above equations will define the system. Additional area is required for storage and a buffer zone around the site.

19.7 WETLANDS FOR WASTEWATER TREATMENT

Wetlands are natural wet ecosystems with diverse and complex roles in nature. Definitions of wetlands vary (Mitsch and Gosselink, 1993) but fundamentally all wetlands are at least intermittently flooded with water. Although historically many cultures lived in harmony with wetlands, modern society has until recently regarded wetlands as nuisances to be drained for development; thus, they have disappeared at a high rate in North America. A variety of wildlife from fish to waterfowl is found in wetlands and dependent on them for survival. Peat, foods, and timber may be harvested from wetlands. They are an important sink for nutrients or other substances and perform other ecological functions besides being aesthetic. Over the past 30 yr the importance of wetlands has been recognized and formalized into government support programs and legislation for wetlands preservation and reestablishment of wetlands beyond the traditional support of conservation groups and other interested parties. Wastewater is an amenity to establish and support wetlands.

Although the definition of wetlands systems varies, estimates of wetlands area in Canada are 14% of Canada or 127 million ha (Zoltai, 1988). Dahl and Johnson (1991) estimate the wetland area in the conterminous United States at 41.8 million ha (103 million acres). The Government of Canada has adopted a policy that includes goals of preventing net loss of wetlands, enhancement and rehabilitation of wetlands in critical areas, and recognition of wetlands functions in resource planning (Government of Canada, 1991). Similarly the U.S. Government supports a variety of wetland protection programs. Increased emphasis on wastewater treatment and government wetlands support programs converged near the mid-1980s, resulting in a significant rise in the number of wetlands systems designed for wastewater treatment with a concomitant rise in study of these systems.

Natural wetlands throughout the world are populated by emergent vegetation. Cattail (*Typha* spp.), reed (*Phragmites* spp.), sedges (*Carex* spp.), bulrushes (*Scirpus* spp.), rushes (*Juncus* spp.), and grasses are the more familiar species (Mitsch and Gosselink, 1992; USEPA, 1988). Floating aquatic plants most extensively used in wastewater treatment systems are the water hyacinth (*Eichhornia crassipes*) and duckweeds (*Lemna* spp.). Vegetation in natural wetlands is a function of the type of wetlands and its location. Mitsch and Gosselink (1993) provide a more extensive list of plant species that are found naturally and have been planted in constructed wetlands.

Wetlands are designed to remove conventional pollutants of BOD, SS, and nutrients. Heavy metals can also be removed to a significant extent. Settled sewage, effluent from secondary treatment, stormwater runoff, leachate, acid mine wastes, and industrial wastewaters have been treated in wetlands; the most common application of wetlands is for polishing secondary effluent. Removal of BOD, SS, N, P, and heavy metals from secondary effluent does not occur to a greater extent in wetlands than in other forms of land treatment (Richardson and Nichols, 1985).

When water is present, anaerobic conditions develop in the upper layers of the soil that influence the biology and chemistry functions of wetlands systems. Plants must be able to survive in anoxic or anaerobic soil conditions over extended periods of time. Denitrification, which is important for wastewater treatment, and swamp gas (methane) production are some consequences of anaerobic conditions. The capability of wetlands to remove nitrogen through denitrification is well established.

The accumulation of nutrients and organics in a wetlands follows from growth of vegetation, death or dieback, decay, and finally soil accretion. The times for 50% of a plant to dieback and fall to the ground (half-life) were surveyed by Kadlec (1989a) for a number of common plants found in wetlands. Half-lives of above ground plants ranged from 220 to 850 d; roots and woods half-lives vary from 14 to 70 yr. From analysis of plant growth and decay times, Kadlec (1989a) concluded that probable startup times for wetlands are 5–10 yr and that data available are overly optimistic since the wetlands are still in a saturation phase.

Biomass typically contains 2 and 0.4% nitrogen and phosphorus, respectively. Once stationary conditions have been reached only biomass accumulation results in removal of nutrients. Kadlec (1989a) calculated that a wetland with 10-d detention time, hydraulic loading rate of 200 m^3/ha/d (21 400 gal/acre/d), and influent nitrogen and phosphorus concentrations of 10 mg/L would only remove 1.4 and 0.7% of the applied nitrogen and phosphorus, respectively, with a soil accumulation rate of 2.5 mm/yr (0.1 in./yr) with soil having the same nutrient content as living biomass. This soil accumulation rate was observed for a natural wetland 800 years old, but wetlands receiving wastewater may exhibit significantly different rates.

Mitsch and Gosselink (1993) note that there is no consensus on whether wetlands serve as net sources or sinks for nitrogen and phosphorus. Extensive data acquisition is required over long periods of time to establish average mass balances in the complex, dynamic environment of a wetland. They note with caution that freshwater marshes have been more often evaluated as sinks for nitrogen and phosphorus, particularly when sewage is applied to the wetlands. There are many contradictory studies and phosphorus is among the least predictable elements.

Harvesting vegetation is generally not practical because of the labor involved and the small portion of nutrients that may be removed by harvesting (USEPA, 1988). In one operation, weed eaters, machetes, and rakes were used for manual harvesting. Harvesting rates were approximately 181 m^2/person/d (1 950 ft^2/person/d) (Gearhart et al., 1989). Harvested cells had slightly higher effluent SS concentrations than unharvested cells; BOD removal was not affected by harvesting. A survey of the literature by Guntenspergen et al. (1989) showed that emergent vegetation only incorporated a small percentage of added nutrients into biomass, with the exception of water hyacinths and other free floating wetland plants, which sequester significant amounts of nutrients and heavy metals.

Constructed wetlands for wastewater treatment at Listowel, ON. Courtesy of Gore & Storrie Limited.

Two types of wetlands systems are designed. Surface flow systems are similar to natural wetlands and a free water surface is maintained. The other alternative is a subsurface flow system (Fig. 19.3) where the water flows through a permeable medium. Emergent vegetation is supported in these systems. Treatment is generally better in subsurface flow systems and these systems do not have mosquito problems.

The water budget for a wetlands is described by the following equation:

$$P - A + Q_i + G_i - Q_o - G_o - \mathrm{ET} = \frac{dV}{dt} \tag{19.31}$$

where

- P is precipitation
- A is abstractions from precipitation
- Q_i is surface inflow
- G_i is groundwater inflow
- Q_o is surface outflow
- G_o is groundwater outflow
- ET is evapotranspiration
- V is volume of water storage in the wetlands
- t is time

A comprehensive design of a wetlands system would consider variation in all of these variables over a long-term period to ensure that minimum water depths are maintained during periods of drought and that detention times are sufficient during wet years. All terms must be considered when sewage is applied to natural wetlands. Constructed wetlands are usually designed with compacted soil where suitable or an impermeable liner to prevent groundwater contamination (Reed et al., 1988; USEPA, 1988). Organic soil layers in natural wetlands reduce infiltration rates to very low values. Stormwater runoff is normally diverted from wetlands for wastewater treatment. Field measurements are required to establish parameters used in the water balance equation.

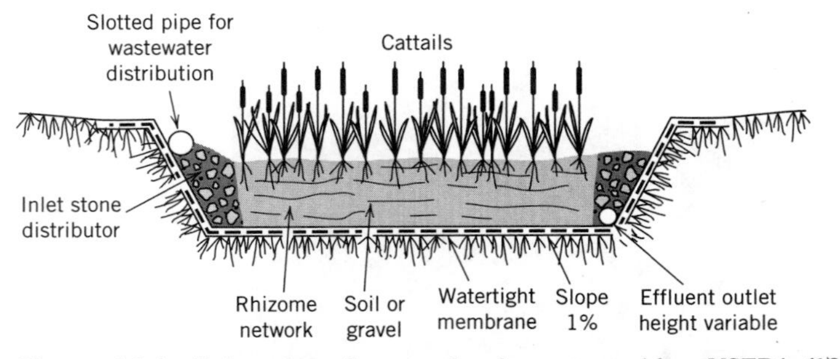

Figure 19.3 Subsurface flow wetlands system. After USEPA (1988).

Kadlec (1989b) describes some of the techniques used and gives equations for terms in the water balance equation.

Commonly a simpler approach is used for wetlands design. Hydraulic loading rate is the primary control variable for wetland systems. Variable recommendations exist in the literature. Based on a database of 44 surface flow constructed wetlands receiving effluent from biological treatment plants, hydraulic loadings of 500 m^3/ha/d (53 500 gal/acre/d) or less will achieve effective reductions (>60%) of BOD and TSS (Knight, 1993). Performance data for a number of systems are listed in Watson et al. (1989) and WPCF (1990). From these data it appears that subsurface systems can be loaded at higher surface loading rates compared to surface systems and achieve the same removals. Performance data comparisons must consider differences in operating conditions among systems. Some systems receive high concentrations of BOD and SS from only screened or settled wastewater and removals tend to be higher. A system receiving 20 mg/L of BOD_5 would only remove 75% of the BOD_5 if the effluent concentration were 5 mg/L. Effluent concentrations at lower values would be difficult to achieve. Preliminary design guidelines listed in Table 19.11 were suggested by Watson et al. (1989).

Loadings must be reduced to 100–150 m^3/ha/d (10 700–16 000 gal/acre/d) or less to achieve nitrification and nutrient removal of 60% or more. Richardson and Nichols (1985) suggested typical hydraulic loadings of 10–40 m^3/ha/d (1 050–4 300 gal/acre/d) to achieve greater than 50% removal of N and P. Using per capita loading rates of

TABLE 19.11 Suggested Hydraulic Loading Rates of Wetlands[a]

	Hydraulic loading, m^3/ha/d (gal/acre/d)	
Treatment	Surface flow	Subsurface flow
Basic treatment	[b]	230–620 (24 600–66 300)
Secondary treatment	120–470 (12 800–50 200)	470–1 870 (50 200–200 000)
Polishing treatment	190–940 (20 300–100 500)	470–1 870 (50 200–200 000)

[a]Reprinted with permission from J. T. Watson, S. C. Reed, R. H. Kadlec, R. L. Knight, and A. E. Whitehouse (1989), "Performance Expectations and Loading Rates for Constructed Wetlands," in *Constructed Wetlands for Wastewater Treatment,* D. A. Hammer, ed., Chelsea, MI. Copyright 1989 Lewis Publishers, an imprint of CRC Press, Boca Raton, FL.

[b]Insufficient data.

2.2 g P/d and 10.8 g N/d, Nichols (1983) estimated that 1 ha (2.5 acres) of wetland would remove 50% of these nutrients for 60 people. Area requirements would increase to 1 ha for 20 persons to achieve 75% removals. These estimates were based on performance of 9 wetlands systems receiving secondary effluent; the wetlands had been in operation for 3–69 years.

Depths in surface flow wetlands are usually less than 0.5 m (1.6 ft) (WPCF, 1990). As depths increase to 0.7–1.0 m (2.3–3.3 ft), floating plants replace emergent vegetation. At a depth of 0.5 m (1.6 ft) and hydraulic loading rate of 500 m^3/ha/d (53 500 gal/acre/d) the nominal detention time in a wetlands would be 10 d. The actual detention time would be less than this because of the volume occupied by vegetation. Many sources agree that the minimum detention time should be around 5 d, particularly if high nitrogen removals are desired.

A PF first-order removal expression is commonly used as the basic performance model in surface flow wetlands.

$$C = C_0 e^{-kAD/Q} \tag{19.32}$$

where

C and C_0 are the effluent and influent BOD_5 concentrations, respectively
D is depth of submergence of the bed
A is the surface area of the bed
Q is the influent flow rate

Different variations of this model have been reported by Reed et al. (1988) and Kadlec (1989b) and all models incorporate some empirical coefficients. These models and other equations below have not been tested extensively; more study will be required. Constructed wetlands are often configured as a series of cells. Flow through a series of cells with vegetation will be close to a PF regime. Single-cell and more open configurations will exhibit variable flow.

The rate constant is subject to temperature. In northern climates ice cover will not prevent water movement under the ice but treatment efficiency is reduced and some storage must be provided (Kadlec, 1989b). Performances reported above are for nonfreezing conditions. Pilot plant studies are required to determine the rate constant variation in different seasons.

There is much more agreement on the equations used to design subsurface flow systems (Cooper and Hobson, 1989; Reed et al., 1988; Watson et al., 1989). Flow through subsurface systems will be PF and Eq. (19.32) is applied. Reed et al. (1988) propose a relationship for k that they note is only tentative:

$$k_{20} = k_0(37.31e^{4.172}) \tag{19.33}$$

where

k_0 is the optimum rate constant for a medium with a fully developed root zone; 1.839 d^{-1} for typical municipal wastewater, 0.198 d^{-1} for industrial wastewater with high COD
k_{20} is the rate constant at 20°C, d^{-1}
e is total porosity of the medium

Typical values of porosity for sand are given in Table 14.3. Porosity will vary from 0.18 to 0.35 for coarse to fine gravel (Reed et al., 1988). Equation (19.5) is used to describe the temperature variation of k; θ has a value of 1.1.

Europeans have much more experience with subsurface flow systems. For Q in m^3/d and A in m^2, k in Eq. (19.32) is taken in the range of 7–12 d^{-1} according to the typical designs in the United Kingdom for reed bed systems (Cooper and Hobson, 1989). The design strength of domestic sewage in the United Kingdom is 330 mg BOD_5/L.

The cross-sectional area of a bed is calculated from Darcy's law:

$$Q = KA_c \frac{\Delta h}{\Delta L} \quad (19.34)$$

where

K is the hydraulic conductivity of the fully developed bed in m/s
A_c is the cross-sectional area of the bed in m^2
$\Delta h / \Delta L$ is the slope of the bed

In a shallow, saturated bed the hydraulic gradient and the slope will practically be the same. Reed et al. (1988) suggest that 0.001 or a lower value should be used for $\Delta h / \Delta L$ when the bed is flat and the gradient is controlled with an overflow weir.

Cooper and Hobson (1989) note that using hydraulic conductivities less than 10^{-4} m/s (3.9×10^{-3} in./s) results in short, unacceptably wide beds and most United Kingdom beds have been designed using values between 10^{-4} and 3×10^{-4} m/s (3.9×10^{-3}–0.12 in./s) but these values will also tend to result in wide beds. These values are higher than conductivities experienced in typical systems and some overland flow is expected. If a gravel medium is used, conductivities may reach 10^{-3} m/s (0.039 in./s). The typical depth of beds is 0.6 m (2 ft) because root systems tend to weaken below this depth and to prevent beds from freezing. Slopes depend on the hydraulic conductivity but values up to 4–5% or higher may be used. In young reed beds generally receiving screened domestic sewage and monitored over a 2-yr period, BOD and SS performance was good with BOD removals of 65% or better and SS removals generally above 65%. Nitrogen and phosphorus removals were low to negligible.

There are only two published quantitative design criteria for wetlands for acid mine drainage (Wieder et al., 1989). The U.S. Bureau of Mines from a survey proposed the empirical design rule of 4.9 m^2 for each L/min of flow (200 ft^2 per gal/min) (Girts and Kleinmann, 1986). The survey excluded sites where outflow metals exceeded inflow metals. The guideline was developed for flows of 19–38 L/min (5–10 gal/min), pH above 4.0, and Fe and Mn concentrations less than 50 and 20 mg/L, respectively. The Tennessee Valley Authority (TVA) developed the following empirical design criteria to achieve effluent Fe and Mn concentrations less than 3 and 2 mg/L, respectively (Brodie et al., 1988).

For Fe:

pH <5.5, 2 m^2 (22 ft^2) is required per mg Fe/min in the influent
pH >5.5, 0.75 m^2 (8 ft^2) is required per mg Fe/min in the influent

For Mn:

pH <5.5, 7 m^2 (75 ft^2) is required per mg Mn/min in the influent
pH >5.5, 2 m^2 (22 ft^2) is required per mg Mn/min in the influent

Mosquito control is an important issue associated with wetlands. Measures recommended for mosquito control include keeping BOD loading rates under 110 kg/ha/d (98 lb/acre/d) to avoid stagnant zones where anaerobic conditions may occur. Natural predators of mosquito larvae such as dragonflies and water beetles require aerobic conditions. Uniform distribution or application of the influent throughout the wetlands should be practiced (Stowell et al., 1985). Control measures are not well understood.

Weed growth can be a problem; weeds can shade the more desirable species and retard their growth. In subsurface flow systems weeds were best controlled by flooding the bed (Cooper and Hobson, 1989). A level surface enhances flooding of the bed.

Pathogen Removal in Wetlands

In studies on surface flow and subsurface flow wetlands in California, Gersberg et al. (1989) found bacterial and viral indicators to be removed in the range of 90–99%. From their review of the literature they concluded that constructed wetlands could reduce total coliforms to 10^3/100 mL or lower if secondary wastewaters were being treated. Raw or primary wastewaters would not experience removals to this level in wetlands without disinfection.

QUESTIONS AND PROBLEMS

1. Give some reasons that an anaerobic pond should be deep.
2. Describe the characteristics of each type of stabilization pond.
3. What is the COD of algae if their composition is $C_{106}H_{263}O_{110}N_{16}P$? Assume that P is oxidized to PO_4^{3-}.
4. Using the McGarry–Pescod relation (Eq. 19.3) determine the maximum loading rate that may be applied to the primary facultative pond at a temperature of 20°C. If the waste flow rate is 10 000 m^3/d with a BOD_5 concentration of 400 mg/L, what is the area of the pond?
5. Review and summarize the articles by Gloyna and Tischler (1981), Wittmann (1982), and Gloyna and Tischler (1982), which expand on some of the arguments presented in Section 19.1. Draw your conclusions on stabilization pond systems as a viable treatment.
6. Why does assuming the mixing regime of a pond to be CM provide a safety factor in estimating its performance?
7. Why are correlations between VSS and substrate removal rate weak for stabilization ponds?
8. Design a facultative pond system with four cells to treat a flow of 3 000 m^3/d with a BOD_5 concentration of 275 mg/L. Use a depth of 1.5 m for each cell. Assume that the average winter temperature is in the range of 0–15°C and design the pond surface areas using the average value of the loading rates given in Table 19.2. Then find the predicted effluent BOD_5 at temperatures of 5 and 25°C assuming that rate coefficients are valid at 20°C. (a) Use the CM model with $k = 0.055\,4\ d^{-1}$ and $\theta = 0.987$; and (b) use a PF model with $k = 0.015\,8\ d^{-1}$ and $\theta = 1$.
9. Compare the surface area required in a facultative pond system at 20°C using PF, CM, and Gloyna (Eq. 19.9) models. Assume $f_a = f_S = 1$ in the Gloyna model. There will be three ponds in the system, each with a depth of 1.5 m (4.92 ft). The influent flow rate is 890 m^3/d (0.235 Mgal/d). The influent organics concentration is 150 mg BOD_5/L (unfiltered sample) and the system will achieve a BOD removal of 85%. What is the surface loading rate in each case?
10. Design a stabilization pond system that will contain an anaerobic pond followed by three facultative ponds and two maturation ponds. The maturation ponds will be operated in parallel but the other ponds are operated in series. The design temperature is 12°C. The influent BOD_5 is 520 mg/L and the flow rate is 1 700 m^3/d. Assume that a BOD_5 removal of 40% is achieved in the anaerobic pond. Base other design details of the system on typical criteria given in the text. The residence time in a maturation pond is 24 h. Generally use reasonable loading rate criteria.
11. How is nitrogen removed in stabilization pond systems?
12. What is the lagoon temperature in a 2.7-m-deep aerated lagoon with a surface area of 1.1

ha in the Gulf Coast area of Texas? The lagoon has a detention time of 3 d and influent and ambient temperatures are 15 and 25°C, respectively.

13. What is the percentage reduction of indicator microorganisms in a 1.3 m deep facultative stabilization pond in Dar es Salaam if k_S/K is 6.00×10^{-4} m-cm²/cal/d? The retention time in the pond is 3 d.

14. Size a two cell (equal sized cells) aerated lagoon system according to the USEPA recommendations for a temperature of 10°C, flow of 2 550 m³/d, and influent BOD_5 concentration of 280 mg/L. Removals of 90% or better should be attained.

15. If an aerated lagoon has a depth of 2.4 m (7.87 ft), a surface area of 9 000 m² (96 800 ft²), and a length : width ratio of 2.5 : 1 to minimize short circuiting, what would be the number of low-speed surface aerators required based on the following specifications: (a) to achieve oxygen dispersion where the circles of dispersion just touch each other; (b) to achieve oxygen dispersion throughout the lagoon; (c) to achieve complete mixing where the circles of influence of each aerator just touch each other; and (d) to achieve complete mixing throughout the pond? What is the power consumption per unit volume of lagoon in each case?

16. Use the mass balance expressions for nitrogen and water to derive Eq. (19.17).

17. For the rainfall distribution data, evapotranspiration rates, and inoperative days data that are given in the accompanying table, design an SR system to treat a wastewater flow of 5 600 m³/d. Assume the wastewater flows are the same for each month. The design soil hydraulic conductivity is 1.0 cm/h and F_c will be taken as 0.05. The wastewater nitrogen concentration is 28 mg/L. The crop selected will have a nitrogen uptake of 250 kg/ha/yr. Determine the storage required ignoring evaporation, seepage, and precipitation in the storage pond.

Monthly Rainfall, Evapotranspiration, and Inoperative Days Data

Month	Jan	Feb	Mar	Apr	May	Jun	Jul	Aug	Sep	Oct	Nov	Dec
P_r, mm	74	84	107	82	110	95	63	130	137	166	121	99
ET, mm	20	40	55	81	98	100	102	98	73	50	25	18
Inoperative days	25	23	10	3						1	7	18

18. For a wastewater flow of 1 000 m³/d and an application rate (r_A) of 0.15 m/d determine the area of each zone receiving flow during an application (A_c), the total infiltration area (A), and the hydraulic loading based on the total area (Q/A) for an RI system with the following T_A and T_d periods (given in days): (a) 1, 13; (b) 2, 12; (c) 1, 6; and (d) 2, 5.

19. Justify and derive the equation immediately before Eq. (19.27).

20. Find the infiltration area of an RI system for a flow of 1 900 m³/d on land with an infiltration rate of 4.0 cm/h. The permeability has been measured with a basin infiltration test; use $F_s = 0.10$. The wastewater has been treated in a secondary system. For the warm half of the year, wastewater will be applied for 3 d with a drying period of 5 d; during the cool half of the year, wastewater will be applied over 2 d with a drying period of 12 d. The maximum one in 10 yr monthly precipitation is 15 and 24 cm/mo for the warm and cool seasons, respectively.

21. An OF system is to be designed for a community with an average wastewater flow of 500 m³/d. The design hydraulic loading rate is 1.5 cm/d. The wastewater is being discharged from a stabilization pond system; therefore no additional storage is required.
(a) Determine r_A, A, and l_D for an application schedule of 5 d/wk and 8 h/d.
(b) If a more conservative r_A of 0.06 m³/m/h was specified, what change is required in

the application period each day to achieve this loading rate? Other parameters are given in (a) and in the problem introduction.

(c) Find the required area of the application site if a 7 d/wk application schedule is used with the other conditions in (a) and the problem introduction.

22. Design a subsurface flow constructed wetlands to treat a wastewater flow from a community of 680 residents where the average per capita wastewater flow is 285 L/cap/d (75.2 gal/cap/d). The wastewater is to be treated to 85% BOD removal. The slope of the bed will be 0.015. Use the recommended depth of 0.6 m (2.0 ft) and gravel with a hydraulic conductivity of 5×10^{-4} m/s (1.64×10^{-3} ft/s). The porosity of the medium is 0.25 and the low temperature is 5°C. There are no unusual characteristics of the wastewater from this community. Find the length and width of the wetlands.

KEY REFERENCES

GLOYNA, E. F., J. F. MALINA, JR., AND E. M. DAVIS, eds. (1976), *Ponds as a Waste Treatment Alternative, Water Resources Symposium No. 9,* University of Texas, Austin, TX.

HAMMER, D. A., ed. (1989), *Constructed Wetlands for Wastewater Treatment,* Lewis Publishers, Chelsea, MI.

REED, S. C., E. J. MIDDLEBROOKS, AND R. W. CRITES (1988), *Natural Systems for Waste Management and Treatment,* McGraw-Hill, Toronto.

USEPA (1981), *Land Treatment of Municipal Wastewater: Process Design Manual,* No. EPA-625/1–81-013, Center for Environmental Research Information, Cincinnati, OH.

WPCF (1990), *Natural Systems for Wastewater Treatment,* Manual of Practice No. FD-16, Water Environment Federation, Alexandria, VA.

REFERENCES

BENEFIELD, L. D. AND C. W. RANDALL (1980), *Biological Process Design for Wastewater Treatment,* Prentice-Hall, Englewood Cliffs, NJ.

BRODIE, G. A., D. A. HAMMER, AND D. A. TOMLJANOVICH (1988), "Constructed Wetlands for Acid Drainage Control in the Tennessee Valley," in *Mine Drainage and Surface Mine Reclamation,* U.S. Bureau of Mines Information Circular 9183, pp. 325–331.

CANTER, L. W. AND A. J. ENGLANDE (1970), "States' Design Criteria for Waste Stabilization Ponds," *J. Water Pollution Control Federation,* 42, 10, pp. 1840–1847.

COLOMER, F. L. AND D. P. RICO (1993), "Mechanistic Model for Facultative Stabilization Ponds," *Water Environment Research,* 65, 5, pp. 679–685.

COOPER, P. F. AND J. A. HOBSON (1989), "Sewage Treatment by Reed Bed Systems: The Present Situation in the United Kingdom," in *Constructed Wetlands for Wastewater Treatment,* D. A. Hammer, ed., Lewis Publishers, Chelsea, MI, pp. 153–171.

CURTIS, T. O., D. D. MARA, N. G. H. DIXO, AND S. A. SILVA (1994), "Light Penetration in Waste Stabilization Ponds," *Water Research,* 28, 5, pp. 1031–1038.

DAHL, T. E. AND C. E. JOHNSON (1991), *Wetlands Status and Trends in the Conterminous United States Mid-1970s to Mid-1980s,* U.S. Department of Interior, Fish and Wildlife Service, Washington, DC.

DROSTE, R. L., M. E. HULLEY, G. DAGG-FOSTER, L. M. JULIEN, AND A. C. ROWNEY (1992), *Wastewater Allocation Study for the South Nation River,* vol. 1, Gore & Storrie Limited, Ottawa, ON.

ESPINO DE LA O, E. AND J. A. MARTINEZ (1976), "Evaluation of Waste Stabilization Pond Performance in Mexico," in *Ponds as a Waste Treatment Alternative, Water Resources Symposium No. 9,* E. F. Gloyna, J. F. Malina, Jr., and E. M. Davis, eds., University of Texas, Austin, TX, pp. 276–283.

Food and Agriculture Organization (1977), *Guidelines for Predicting Crop Water Requirements,* Food and Agriculture Organization of the United Nations, Rome.

FORD, D. L. AND L. F. TISCHLER (1976), "Process and Economic Considerations of Ponds for the Treatment of Industrial Wastewaters," in *Ponds as a Waste Treatment Alternative, Water Resources Symposium No. 9,* E. F. Gloyna, J. F. Malina, Jr., and E. M. Davis, eds., University of Texas, Austin, TX, pp. 236–251.

FRITZ, J. J., A. C. MIDDLETON, AND D. D. MEREDITH (1979), "Dynamic Process Modeling of Wastewater Stabilization Ponds," *J. Water Pollution Control Federation,* 51, 11, pp. 2724–2743.

GEARHART, R. A., F. KLOPP, AND G. ALLEN (1989), "Constructed Free Surface Wetlands to Treat and Receive Wastewater: Pilot Project to Full Scale," in *Constructed Wetlands for Wastewater Treatment,* D. A. Hammer, ed., Lewis Publishers, Chelsea, MI, pp. 121–137.

GERSBERG, R. M., R. A. GEARHART, AND M. IVES (1989), "Pathogen Removal in Constructed Wetlands," in *Constructed Wetlands for Wastewater Treatment,* D. A. Hammer, ed., Lewis Publishers, Chelsea, MI, pp. 431–445.

GIRTS, M. A. AND R. L. P. KLEINMANN (1986), "Constructed Wetlands for the Treatment of Acid Mine Drainage: A Preliminary Review," *National Symposium on Surface Mining, Hydrology, Sedimentology, and Reclamation,* University of Kentucky Press, Lexington, pp. 165–171.

GLOYNA, E. F. (1976), "Facultative Waste Stabilization Pond Design," in *Ponds as a Waste Treatment Alternative, Water Resources Symposium No. 9,* E. F. Gloyna, J. F. Malina, Jr., and E. M. Davis, eds., University of Texas, Austin, TX, pp. 143–157.

GLOYNA, E. F. AND L. F. TISCHLER (1981), "Recommendations for Regulatory Modifications: The Use of Waste Stabilization Ponds," *J. Water Pollution Control Federation,* 53, 11, pp. 1559–1563.

GLOYNA, E. F. AND L. F. TISCHLER (1982), "Author's Response," *J. Water Pollution Control Federation,* 54, 10, pp. 1429–1431.

GLUMRB (1978), *Ten States Recommended Standards for Sewage Treatment Works,* A Report of the Great Lakes Upper Mississippi River Board of State Sanitary Engineers, Albany, NY.

Government of Canada (1991), *The Federal Policy on Wetland Conservation,* Ottawa, ON.

GUNTENSPERGEN, G. R., F. STEARNS, AND J. A. KADLEC (1989), "Wetland Vegetation," in *Constructed Wetlands for Wastewater Treatment,* D. A. Hammer, ed., Lewis Publishers, Chelsea, MI, pp. 73–88.

HEINKE, G. W., D. W. SMITH, AND G. R. FINCH (1988), *Guidelines for the Planning, Design, Operation and Maintenance of Wastewater Lagoon Systems in the Northwest Territories,* vols. 1 and 2, Department of Municipal and Community Affairs, Government of the Northwest Territories, Yellowknife, NWT.

HEINKE, G. W., D. W. SMITH, AND G. R. FINCH (1991), "Guidelines for the Planning and Design of Wastewater Lagoon Systems in Cold Climates," *Canadian J. of Civil Engineering,* 18, 4, pp. 556–567.

HENDRICKS, D. W. AND W. D. POTE (1974), "Thermodynamic Analysis of a Primary Oxidation Pond," *J. Water Pollution Control Federation,* 46, 2, pp. 333–351.

KADLEC, R. H. (1989a), "Decomposition in Wastewater Wetlands," in *Constructed Wetlands for Wastewater Treatment,* D. A. Hammer, ed., Lewis Publishers, Chelsea, MI, pp. 459–468.

KADLEC, R. H. (1989b), "Hydrologic Factors in Wetland Water Treatment," in *Constructed Wetlands for Wastewater Treatment,* D. A. Hammer, ed., Lewis Publishers, Chelsea, MI, pp. 21–40.

KNIGHT, R. L. (1993), "Operating Experience with Constructed Wetlands for Wastewater Treatment," *Tappi J.,* 76, 1, pp. 109–112.

MALINA, J. F., JR., R. KAYSER, W. W. ECKENFELDER, JR., E. F. GLOYNA, AND W. R. DRYNAN

(1972), *Design Guidelines for Biological Wastewater Treatment Processes,* Center for Research in Water Resources Report, CRWR-76, University of Texas, Austin, TX.

MANCINI, J. L. AND E. L. BARNHART (1968), "Industrial Waste Treatment in Aerated Lagoons," in *Advances in Water Quality Improvement, Water Resources Symposium No. 1,* E. F. Gloyna, and W. W. Eckenfelder, Jr., eds., University of Texas, Austin, TX, pp. 313–324.

MARA, D. D. (1975), "Author's Reply," *Water Research,* 9, pp. 596–597.

MARAIS, G. V. R. AND V. A. SHAW (1961), "A Rational Theory for the Design of Sewage Stabilization Ponds," *Transactions South African Institution of Civil Engineers,* 3, 11, pp. 205–227.

MAYO, A. (1989), "Effect of Pond Depth on Bacterial Mortality Rate," *J. of Environmental Engineering, ASCE,* 115, EE5, pp. 964–977.

McGARRY, M. G. AND M. B. PESCOD (1970), "Stabilization Pond Design Criteria for Tropical Asia," *Proc. of Second International Symposium for Waste Treatment Lagoons,* Kansas City, MO.

McKINNEY, R. E. (1976), "Functional Characteristics Unique to Ponds," in *Ponds as a Waste Treatment Alternative, Water Resources Symposium No. 9,* E. F. Gloyna, J. F. Malina, Jr., and E. M. Davis, eds., University of Texas, Austin, TX, pp. 317–325.

Metcalf and Eddy, Inc. (1991), *Wastewater Engineering: Treatment, Disposal, and Reuse,* 3rd ed., G. Tchobanoglous and F. L. Burton, eds., McGraw-Hill, Toronto.

MITSCH, W. J. AND J. G. GOSSELINK (1993), *Wetlands,* 2nd ed., Van Nostrand Reinhold, New York.

MORENO, M. D. (1990), "A Tracer Study of the Hydraulics of Facultative Stabilization Ponds," *Water Research,* 24, 8, pp. 1025–1030.

NICHOLS, D. S. (1983), "Capacity of Natural Wetlands to Remove Nutrients from Wastewater," *J. Water Pollution Control Federation,* 55, 5, pp. 495–505.

NEEL, J. K., J. H. McDERMOTT, AND C. A. MONDAY, JR. (1961), "Experimental Lagooning of Raw Sewage at Fayette, Missouri," *J. Water Pollution Control Federation,* 33, 6, pp. 603–641.

ORAGUI, J. I., H. ARRIDGE, D. D. MARA, H. W. PEARSON, AND S. A. SILVA (1993), "*Vibrio Chloerae* O1 (El Tor) Removal in Waste Stabilization Ponds in Northeast Brazil," *Water Research,* 27, 4, pp. 727–728.

ORGERON, D. J. (1976), "Biological Waste Stabilization Ponds at E.I. DuPont de Nemours & Company Sabiune River Works, Orange, Texas," in *Ponds as a Waste Treatment Alternative, Water Resources Symposium No. 9,* E. F. Gloyna, J. F. Malina, Jr., and E. M. Davis, eds., University of Texas, Austin, TX, pp. 221–235.

OSWALD, W. J. (1968), "Advances in Anaerobic Pond Systems Design," in *Advances in Water Quality Improvement, Water Resources Symposium No. 1,* E. F. Gloyna and W. W. Eckenfelder, Jr., eds., University of Texas, Austin, TX, pp. 409–426.

OSWALD, W. J. AND R. RAMANI (1976), "The Fate of Algae in Receiving Waters," in *Ponds as a Waste Treatment Alternative, Water Resources Symposium No. 9,* E. F. Gloyna, J. F. Malina, Jr., and E. M. Davis, eds., University of Texas, Austin, TX, pp. 111–121.

PARKER, C. D. (1976), "Pond Design for Industrial Use in Australia with Reference to Food Wastes," in *Ponds as a Waste Treatment Alternative, Water Resources Symposium No. 9,* E. F. Gloyna, J. F. Malina, Jr., and E. M. Davis, eds., University of Texas, Austin, TX, pp. 285–297.

RICHARDSON, C. J. AND D. S. NICHOLS (1985), "Ecological Analysis of Wastewater Management Criteria in Wetland Ecosystems," in *Ecological Considerations in Wetlands Treatment of Municipal Wastewaters,* P. J. Godfrey, E. R. Kaynor, and S. Pelezarski, eds., Van Nostrand Reinhold, New York, pp. 351–391.

SASTRY, C. A. AND G. J. MOHANRAO (1976), "Waste Stabilization Pond Design and Experiences in India," in *Ponds as a Waste Treatment Alternative,* Water Resources Symposium No. 9,

E. F. GLOYNA, J. F. MALINA, JR., AND E. M. DAVIS, eds., University of Texas, Austin, TX, pp. 299–313.

SMITH, D. W. AND H. KNOLL, eds., (1986), *Cold Climate Utilities Manual,* Canadian Society for Civil Engineering, Montreal, PQ.

STOWELL, R., S. WEBER, G. TCHOBANOGLOUS, B. A. WILSON, AND K. R. TOWNZEN (1985), "Mosquito Considerations in the Design of Wetland Systems for the Treatment of Wastewater," in *Ecological Considerations in Wetlands Treatment of Municipal Wastewaters,* P. J. Godfrey, E. R. Kaynor, and S. Pelezarski, eds., Van Nostrand Reinhold, New York, pp. 38–47.

THIRUMURTHI, D. (1974), "Design Criteria for Waste Stabilization Ponds," *J. Water Pollution Control Federation,* 46, 9, pp. 2094–2106.

USEPA (1979), *Environmental Pollution Control Alternatives: Municipal Wastewater,* No. EPA-625/5–79-012, Center for Environmental Research Information, Cincinnati, OH.

USEPA (1983), *Municipal Wastewater Stabilization Ponds: Design Manual,* No. EPA-625/1–83-015, Center for Environmental Research Information, Cincinnati, OH.

USEPA (1988), *Constructed Wetlands and Aquatic Plant Systems for Municipal Wastewater Treatment: Process Design Manual,* No. EPA-625/1–88-022, Center for Environmental Research Information, Cincinnati, OH.

VELZ, C. J. (1970), *Applied Stream Sanitation,* Wiley–Interscience, Toronto.

WATSON, J. T., S. C. REED, R. H. KADLEC, R. L. KNIGHT, AND A. E. WHITEHOUSE (1989), "Performance Expectations and Loading Rates for Constructed Wetlands," in *Constructed Wetlands for Wastewater Treatment,* D. A. Hammer, ed., Lewis Publishers, Chelsea, MI, pp. 319–351.

WHITLEY, G. AND D. THIRUMURTHI (1992), "Field Monitoring and Performance Evaluation of the Whitehorse Sewage Lagoon," *Canadian J. of Civil Engineering,* 19, 5, pp. 751–759.

WITTMANN, J. W. (1982), "Discussion of Recommendations for Regulatory Modifications: The Use of Waste Stabilization Ponds," *J. Water Pollution Control Federation,* 54, 10, pp. 1428–1429.

WIEDER, R. K., G. TCHOBANOGLOUS, AND R. W. TUTTLE (1989), "Preliminary Considerations Regarding Constructed Wetlands for Wastewater Treatment," in *Constructed Wetlands for Wastewater Treatment,* D. A. Hammer, ed., Lewis Publishers, Chelsea, MI, pp. 297–305.

WRIGLEY, T. J., D. F. TOERIEN, AND I. G. GAIGHER (1988), "Fish Production in Small Oxidation Ponds", *Water Research,* 22, 10, pp. 1279–1285.

ZOLTAI, S. C. (1988), "Wetland Environments and Classification," in *Wetlands of Canada, Ecological Land Classification Series,* No. 24, National Wetlands Working Group, ed., Environment Canada, Ottawa, pp. 1–26.

CHAPTER 20

SLUDGE PROCESSING AND LAND APPLICATION

Sludge disposal is one of the major operating expenses of any wastewater treatment process. Recent increasingly restrictive regulations on sludge disposal have contributed to the need for and expense of sludge processing. There are many biological and physical–chemical options to reduce the volume and quantity of sludge. Concentrated sludges are also more easily controlled at their ultimate disposal site. Anaerobic digestion, covered in Chapter 18, is another often used option in addition to those processes covered here.

It is not the intent of this chapter to provide all details on the operation of sludge processing operations. There are many subtleties to fine tuning individual processes and the overall sludge processing train to provide the least cost sludge for disposal. More details are available in Metcalf and Eddy (1991), USEPA (1979, 1987), and from manufacturers of processes. In addition to typical information it is essential to pilot processes with trial and error adjustment.

Sludge handling begins with the processes that generate sludge. Chemical agents and blending in the primary sludge generation processes affect quantities of sludge to be processed and efficiency of sludge concentration processes.

Sludge is a rich source of nutrients and soil conditioning substances; therefore, land application is a natural and desirable disposal option.

20.1 SLUDGE CHARACTERISTICS AND CONDITIONING

The high concentration of solids in sludge slurries affects the density and viscosity of the suspension. Dewatered sludges can have solids concentrations ranging up to 40% on a weight basis.

Sludge Density

The mass of solids in a slurry is normally related to the volatile suspended solids (VSS) and fixed suspended solids (FSS) components in the slurry. The specific gravity (s.g.) of a slurry, S_s, is

$$S_s = \frac{\text{mass of slurry/volume of slurry}}{1.0} = \frac{m_w + m_V + m_F}{V_s} \tag{20.1}$$

where

m_w, m_V, and m_F are the masses of water, VSS, and FSS, respectively
V_s is the volume of the sludge slurry

Define S_w, S_V, and S_F as the s.g.s of water, VSS, and FSS, respectively. It is assumed that each of these entities is distinct and the volumes contributed by each of them can therefore be added. This is not the case for biological solids, where the inorganic (fixed) solids are intimately associated with the organic (volatile) solids. However, suitable working relations can be developed for particular situations. Expressing the volume relation:

$$V_s = V_w + V_V + V_F$$

where

V_s, V_w, V_V, and V_F are the volumes of sludge, water, VSS, and FSS, respectively

Using the s.g.s,

$$\frac{m_w}{S_w} + \frac{m_V}{S_V} + \frac{m_F}{S_F} = \frac{m_s}{S_s}$$

where

m_s is the total mass in the slurry

Dividing by m_s,

$$\frac{m_w/m_s}{S_w} + \frac{m_V/m_s}{S_V} + \frac{m_F/m_s}{S_F} = \frac{1}{S_s} = \sum_i \frac{f_{mi}}{S_i} \tag{20.2}$$

where

f_{mi} is the mass fraction of the ith component in the slurry. The mass fractions of the solids are on a dry basis.

The water (moisture) or total solids content of a sludge expressed on a percentage basis is often used to express its concentration. The volume of a sludge related to its total solids content is

$$V_s = \frac{M}{(p_M/100)S_s} = \frac{M}{[(100 - p_w)/100]S_s} \tag{20.3}$$

where

M is the mass of dry solids in the slurry; $M = m_V + m_F$

p_M and p_w are the percent solids and water content (both on a mass basis), respectively

The s.g. of organic matter is near the s.g. of water, i.e., near 1.00; the density range of activated sludge solids is 1.01–1.10 g/L (Dammel and Schroeder, 1991). Mineral solids have much higher s.g.s. A value of 2.5 is commonly used for the s.g. of FSS but solids in a chemically conditioned water may have s.g.s ranging from 1.5 to 2.5.

■ Example 20.1 Sludge Volume and Specific Gravity

(a) Calculate the s.g. of a biological sludge that contains 75% VSS and 25% FSS when the sludge has a concentration of 10 000 mg/L (≈1% by weight).

(b) Calculate the volume of 1 kg of this sludge when it is concentrated to 5, 10, and 20% solids concentration by weight.

(a) At 10 000 mg/L, 1 L of the sludge contains 10 g of solids that consist of 7.5 g VSS and 2.5 g FSS. The first inclination might be to assume that a liter of sludge had a mass of 1 010 g. But because the s.g.s of VSS and FSS are ≥1.00, the water content of 1 L of sludge will be very near 990 g. Assuming the s.g.s of VSS and FSS are 1.00 and 2.50, respectively, from Eq. (20.2):

$$\frac{1}{S_s} = \frac{990/1\,000}{1.0} + \frac{7.5/1\,000}{1.0} + \frac{2.5/1\,000}{2.5} = 0.998\,5 \qquad S_s = \frac{1}{0.998\,5} = 1.001\,5$$

It is seen that the assumption about the water content is correct. Further iterations are needless to refine the estimate of the mass of water in the sludge.

(b) At a solids content of 5% by weight, the mass of solids in 1 kg of sludge is 50 g. Using 50 g, the s.g. of the sludge (by trial and error solution of Eq. 20.2) is

$$\frac{1}{S_s} = \frac{958/1\,008}{1.0} + \frac{37.5/1\,008}{1.0} + \frac{12.5/1\,008}{2.5} = 0.992\,6 \qquad S_s = \frac{1}{0.992\,6} = 1.008$$

From this result, the solids content is 50 g/1 008 g = 0.049 6 or 4.96%. The actual solids content of 1 L should be increased by a factor of approximately 5.00/4.96 = 1.008 and Eq. (20.2) should be solved again using 37.8 g and 25.2 g for VSS and FSS, respectively. The error is negligible in this case but at higher solids concentrations it would be necessary to perform the correction.

Applying Eq. (20.3),

$$V_s = \frac{50\text{ g}}{(5/100)\left(1.008\,\frac{\text{g}}{\text{cm}^3}\right)\left(\frac{1\,000\text{ cm}^3}{\text{L}}\right)} = 0.992\text{ L}$$

By similar calculations, the s.g.s of sludges at 10 and 20% solids content are 1.014 and 1.031, respectively. The volumes of 1 kg of sludge are 0.985 and 0.970 L, respectively.

As Example 20.1 shows, the volume of sludge to be removed is essentially indirectly proportional to the solids content of the sludge. Figures 20.1a and 20.1b show the variation in the volume of sludge as a function of FSS : VSS ratio, mass percentage of

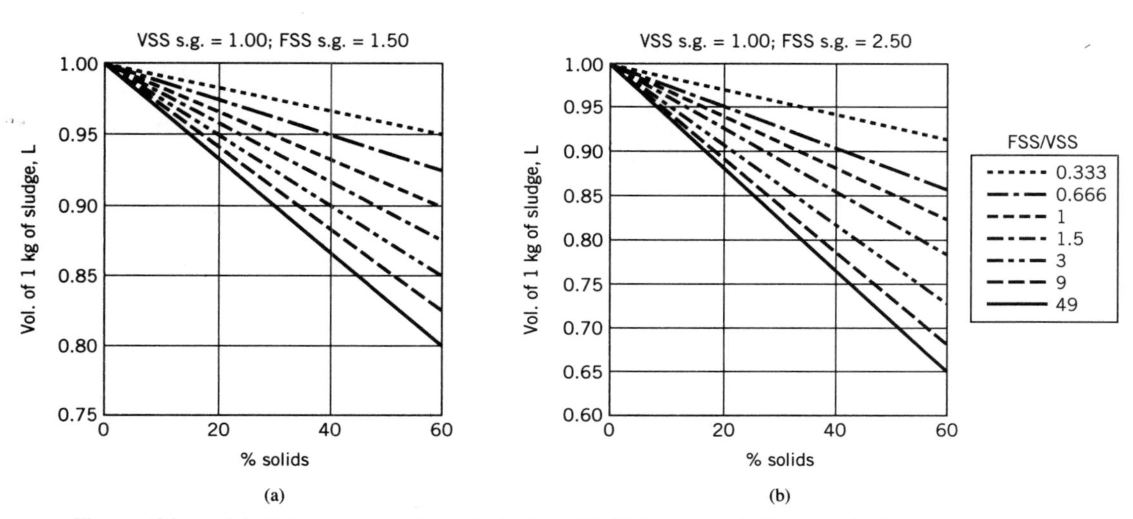

Figure 20.1 (a) Volume variation of sludge. (b) Volume variation of sludge.

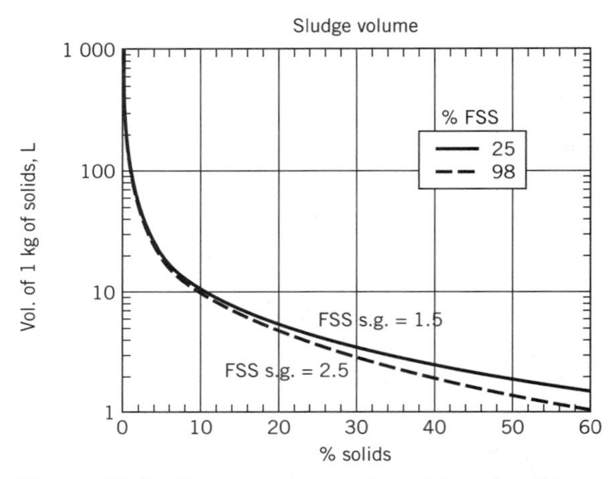

Figure 20.2 Volume occupied by 1 kg of solids.

solids in the sludge, and s.g. of FSS. These graphs were prepared by solving Eq. (20.2) and then applying Eq. (20.3). Biological sludges have FSS : VSS ratios near 1 : 3, which results in the lowest overall solids density in sludges. Mineral sludges from water treatment plants do not contain significant amounts of VSS.

Figure 20.2 shows the extremes in volume variation that can occur for sludges found in water and wastewater treatment plants as a function of mass concentration. Solids content has to be above 10% to significantly change the volume of 1 kg of sludge from 1 liter. Sludge concentration processes can have a dramatic effect on the volume of sludge produced. Influent water to a water treatment plant will contain 1 kg of solids in 10^4–10^5 L (1 lb per 10^3–10^4 gal) of water. Municipal raw wastewaters will contain 1 kg of solids in 10^3–10^4 L (1 lb per 10^2–10^3 gal) of water.

Sludge Viscosity

Vesilind (1979) notes that pumping sludge is more of an art than a science. The capricious nature of sludge viscosity is one of the reasons for this statement. Sludge varies from a Newtonian fluid, where shear is proportional to the velocity gradient, to a plastic fluid, where a threshold shear must be reached before the sludge will move. Most wastewater sludges are pseudoplastic; this behavior can be described by the following viscosity law (Vesilind, 1979):

$$\tau = \eta \left(\frac{dv}{dy} \right)^n \tag{20.4}$$

where

τ is the shear stress
η is the plastic viscosity
n is a constant
dv/dy is the velocity gradient

Vesilind (1979) notes that it is impractical to apply Eq. (20.4) because there are no typical sludges and the rheological characteristics of a given sludge vary with concentration of solids and other factors. Practical approaches to estimating headloss for sludge pumping use empirical correlations of a friction factor as a function of a Reynold's number and apply the Darcy–Weisbach equation. Otherwise empirical plots of headloss as a function of velocity are utilized. Solids concentration is a parameter on the plots.

20.2 SLUDGE GENERATION AND TREATMENT PROCESSES

Solids generated in water treatment operations consist of essentially all suspended solids in the influent plus all chemical agents added that produce precipitates or conditioning agents that are incorporated into the sludge in sludge processing operations. Lime and alum recovery are sometimes used but all other solids, regardless of recycling within the water treatment plant, ultimately are present in the sludge streams.

Solids generation in physical–chemical wastewater treatment plants is similar to water treatment plants. Solids generation in biological wastewater treatment plants is more difficult to estimate. All suspended solids in the influent do not appear in the sludge. Some are biologically metabolized to soluble or gaseous end products in the biological treatment process or sludge digestion process. Some soluble wastewater components will be transformed into biological solids that can be reduced in digestion. Solids generation in a biological process is a function of the type of process and the operation of the process. Final effluent from a wastewater treatment plant may contain a solids concentration that has a significant effect on the solids remaining for sludge processing.

Solids formed from addition of precipitation agents can be estimated from the stoichiometry of the reaction (for example, the reactions in Table 13.2); however, the stoichiometry of precipitate formation can vary from the chemical equations because of side reactions. Solids generation (or destruction) from biological treatment is discussed in Chapters 17 and 18, which should be reviewed.

Water treatment plants may dispose of their sludge streams in sewers for processing at the wastewater treatment plant. This economizes sludge handling at one central location. Chemical agents added at water treatment plants normally are not harmful to biological treatment processes and they often contribute to enhanced solids separation at wastewater treatment plants. Return of sludge generated in water treatment to surface waters is discouraged or prohibited.

Onsite sludge treatment refers in general to processes used to concentrate sludge. The exceptions in water treatment are processes to recover lime and metal coagulants (alum and iron salts) from $CaCO_3$ and metal coagulant sludges (Sections 13.1 and 15.2), respectively. In wastewater treatment anaerobic and aerobic sludge digestion reduce quantities and therefore volumes of sludge. Sludge quantities may be reduced by thermal processes (e.g., incineration, pyrolysis, or wet-air oxidation) before ultimate disposal. Landfill and land application are the most commonly used means of ultimate disposal. Some wastewater sludges may be composted for land application. Ocean dumping of sludge is discouraged or prohibited.

Except for land application of sludge this chapter focuses on sludge concentration processes. There are a variety of standard and proprietary devices for processing sludge. Some are described in more detail in following sections. Performance of processes varies widely and recommendations for different processes from various sources are even contradictory. Careful study of the options and pilot-scale testing is recommended to make the best choice.

TABLE 20.1 Water Treatment Sludge Concentrations

Sludge type	Solids concentration %
Sedimentation basin underflow	0.5–2
Slurry from upflow clarifier	2–5
Filter backwash water	50–1 000[a]
Settled solids from lime–soda softening	2–15
Alum–lime coagulation softening settled sludge	Up to 10
Iron–lime coagulation softening settled sludge	10–25
Gravity thickener	
Coagulation settlings and backwash water	2–20 (2–4 is more typical)
Filter backwash water	Up to 4
Lime sludge	15–30
Vacuum filter	
Coagulant sludge	10–20
Lime softening (>85% $CaCO_3$ content)	50–70
Lime softening [high $Mg(OH)_2$ content]	20–25
Centrifuge	
Coagulant sludge	10–20
Lime and alum sludge	15–40
Lime sludge	30–70
Pressure filter	
Coagulant sludge	30–45
Lime sludge	55–70
Belt filter	
Coagulant sludge	10–15
Sand drying bed	
Coagulant sludge	20–25
Lime sludge	50
Storage lagoon	
Coagulant sludge	7–15
Lime sludge	50–60

[a]Given in mg/L.

Options for sludge concentration or treatment for water treatment and wastewater sludges are given in Figs. 20.3a and 20.3b in the general sequence in which they occur. All of the processing phases are not necessarily used. Chemical conditioning agents may be added before each of the processes. Chemicals must be added before some of the processes.

Performance ranges for water treatment plant sludge processes are given in Table 20.1. Such tables can only be used as guides. The sources of sludge are the sedimentation basins and backwash water from the filters. Backwash water is commonly returned ahead of the coagulation basin or it may be separately thickened and combined with other sludges for dewatering. Softening sludges with a high $CaCO_3$ content are able to achieve the highest concentrations.

Primary sedimentation solids and biological solids generated in wastewater treatment processes are different in nature and offer more choices for blending and processing before dewatering. Waste solids from the secondary clarifier may be recycled to the primary clarifier if there is a net gain in underflow solids concentration. Table 20.2 gives characteristics of primary solids and solids generated in different biological treatment processes. Solids removed in primary clarifiers concentrate to a higher

(a)

Thickening	Solids Reduction	Dewatering
Dissolved air flotation Gravity	$CaCO_3$ recovery Alum recovery	Centrifugation Drying bed Horizontal belt filter Lagoon Pressure filter Vacuum filter

(b)

Thickening	Solids Reduction	Physical–Chemical Stabilization	Dewatering
Centrifugation			Centrifugation
Dissolved air flotation			Drying bed
Gravity belt	Aerobic sludge digestion	Heat Treatment	Horizontal belt filter
Gravity	Anaerobic sludge digestion	Lime stabilization	Lagoon
Rotary drum			Pressure filter
			Vacuum filter

Figure 20.3 (a) Water treatment sludge operations. (b) Wastewater sludge processing operations.

degree than biological solids and if recycle of the biological solids to the primary clarifier is not practiced, biological solids may be separately thickened before blending with primary clarifier solids. A greater variety of processes are used to thicken sludge for digestion or dewatering. Typical performance ranges of thickening processes are given in Table 20.3. Performance of dewatering processes are given in Table 20.4. Variation in performance data is evident from the tables and the values should only be taken as guides.

TABLE 20.2 Wastewater Treatment Sludge Concentrations[a]

Sludge type	Sludge concentration %
Primary sludge	5–8
Waste activated sludge	0.5–2.0
Fixed film waste sludge	3–10
Primary and waste activated sludge	2.5–4
Primary and fixed film sludge	3–5
Aerobically digested sludge (thickened)	1–2
Anaerobically digested sludge (thickened)	6–12

[a]Adapted from Metcalf and Eddy (1991) and USEPA (1979, 1987).

TABLE 20.3 Thickeners Performance for Wastewater Sludges[a]

	Solids concentration, %			
Sludge type	Dissolved air flotation	Gravity thickener	Belt thickener (with polymer)	Centrifuge
Primary clarifier		8–10	9–12	9–12
Waste activated sludge	3–5	2–2.5	4–6	4–6
Fixed film waste sludge	3–5	2.5–3	5–7	5–7
Primary + waste activated sludge	4–6	4–5	5–7	5–7
Primary + fixed film sludge	4–6	5–6	5–10	6–10
	Solids capture, %			
Without chemicals	80–95	80–92	na[b]	80–90
With chemicals	90–98	na[b]	—	90–98

[a]Adapted from Metcalf and Eddy (1991) and USEPA (1987).
[b]na-not applicable.

Higher solids concentration in the influent to a sludge concentration process yield higher concentrations of solids in the product. Sludge processing is both energy and labor intensive. Gravity sludge thickeners are the most common primary sludge concentration processes in both water and wastewater treatment because the process requires minimal energy input. Sludge thickeners should not be used for storage of sludge or their performance deteriorates. General considerations for selection of sludge volume reduction processes are as follows:

1. Sludge characteristics.
2. Volume reduction attainable.
3. Energy inputs and other operation and maintenance costs. Skills required to operate the process must also be considered.
4. Chemical inputs.
5. Other operations in the sludge processing train.
6. Ultimate disposal. It may be desirable to have sludge in a liquid form for spreading on land but a sludge with minimal moisture content is desirable for landfill. Sludge may be pumped to a disposal location or hauled by truck or railway which have opposing moisture requirements.

TABLE 20.4 Performance of Dewatering Units for Wastewater Sludges

	Sludge concentration, %		Solids capture, %	
Process	Without chemicals	With chemicals	Without chemicals	With chemicals
Centrifuge	10–30	10–35	55–90	85–98
Drying beds	30–60	na[a]		
Horizontal belt press	na	15–30	na	85–98
Lagoon	15–40	na		
Pressure filter	na	20–50	na	90–98
Vacuum filter	na	15–30		90–98

[a]na, not applicable.

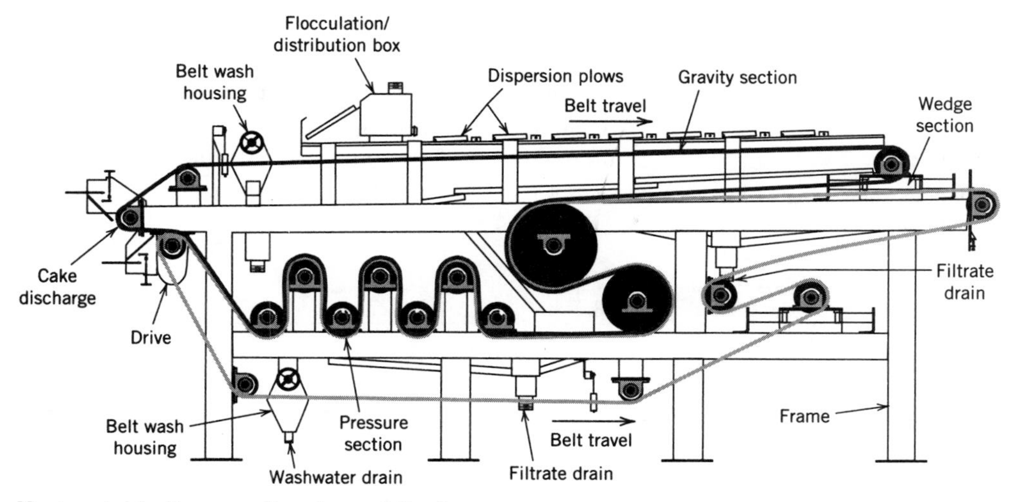

Horizontal belt press. Courtesy of Envirex.

7. Concentration of solids in the reject liquid.
8. Space requirements. Mechanical processes require much smaller areas than land dewatering.

More details on the design and operation of the most common concentration operations are given in later sections.

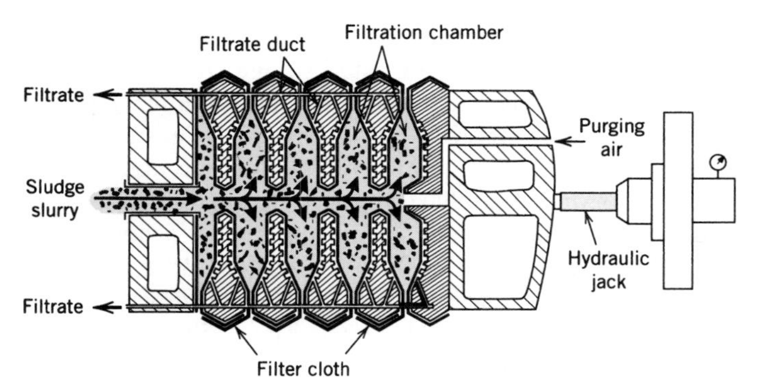

Filter press. Courtesy of Degrémont Infilco.

20.3 SLUDGE CONDITIONING

Sludges are conditioned to improve their dewatering and cake forming properties in processes described in the above section and below. The tables in Section 20.2 show that both product cake concentrations and percentage solids capture generally improve with chemical addition. Common coagulating agents such as ferric chloride and lime in addition to a variety of synthetic chemicals are good conditioning agents. Inorganic chemical conditioning is associated principally with vacuum and pressure filtration processes with lime and ferric chloride being the most often used agents (USEPA,

TABLE 20.5 Effect of Freeze–Thaw Conditioning on Dewatered Cake Solids Concentration of Chemical Sludges[a,b]

Sludge type	Dewatered solids concentration, % dry weight			
	Gravity thickening	Vacuum filtration	Centrifuge	Pressure filter
Polymer coagulating agent (unconditioned)	1.2–2.7	15–17	5–11	15–21
Polymer coagulating agent (freeze–thaw conditioned)	20–25	45	28–33	37–40
Iron coagulating agent (unconditioned)	8.1–8.9	21–27	17–31	18–30
Iron coagulating agent (freeze–thaw conditioned)	7–16	47–54	38–44	48–53

[a]From W. R. Knocke, C. M. Dishman, and G. F. Miller (1993), "Measurement of Chemical Sludge Floc Density and Implications Related to Sludge Dewatering," *Water Environment Research,* 65, 6, pp. 735–743, © WEF 1993.

[b]All sludges originated from water treatment plants treating surface waters.

1987). Dosage rates for iron salts are in the range of 20–62 kg/tonne (40–124 lb/ton) of dry solids in the sludge feed. Lime dosage is in the range of 75–277 kg/tonne (150–554 lb/ton) of dry solids in the sludge feed. Organic polymer dosages range from 0.25 to 5 kg/tonne (0.5–10 lb/ton) of dry solids. Lower G values are used to mix chemicals with sludges so as not to destroy the solids matrix as noted in Chapter 13.

Natural conditioning of sludge can be accomplished by freeze–thawing. Ice crystal formation during freezing separates solid and liquid fractions. The frozen sludge is thawed and thickened by gravity or mechanically dewatered. Martel (1989a) found that anaerobically digested sludge could be drained to a solids content of 20% or greater after being frozen and thawed. Aerobically digested sludge could only attain total solids contents of from 6 to 18% after freezing, thawing, and draining. Improvement from freeze–thaw conditioning in dewaterability of chemical sludges formed at water treatment plants is shown in Table 20.5. Equations along with an approach to design freeze-thaw beds are given by Martel (1989b).

Heat treatment or thermal conditioning of sludge is an option that accomplishes conditioning and inactivation of microorganisms within the sludge. Both heat and pressure are applied to the sludge for 15–40 min (USEPA, 1987). Temperatures in the range of 177–204°C and pressures from 1 720 to 2 750 kPa (250–400 psi) are achieved in the units that break down floc structures to allow easier water release.

Storage of activated sludge changes the dewatering properties of the sludge. Storage under anaerobic conditions reduces its dewaterability and also increases the resistance to flow of water from the sludge (Novak et al., 1988; Parker et al., 1972; Rasmussen et al., 1994). The most dramatic changes in dewaterability occur within 3–4 d. Phosphorus release from anaerobic storage of sludge from an enriched phosphorus removal process occurs rapidly within the first few hours of storage (Rasmussen et al., 1994).

■ Example 20.2 Sludge Volume Reduction Through Dewatering

A water treatment plant is processing a flow of 16 000 m^3/d (4.22 Mgal/d). The raw water contains 15 mg/L TSS and alum [$Al_2(SO_4)_3 \cdot 14H_2O$] is added at 30 mg/L as a coagulant. Partial softening is performed by adding 50 mg/L of $Ca(OH)_2$ (only calcium will be removed). The sedimentation basin sludge and backwash water are thickened in a gravity thickener and sent to a vacuum filter. Calculate the volume of sludge produced on a daily basis from the thickener and the vacuum filter. Assume that the processes perform at average values given in Table 20.1 and that natural alkalinity in the water is sufficient for both alum and lime softening additions.

The mass of solids generated in the process will be calculated first. From Table 13.2, 1 mole of $Al_2(SO_4)_3$ produces 2 moles of $Al(OH)_3$. The mass of precipitate produced per unit of metal coagulant is 156/342.3 = 0.456 mg/mg. The hydrated alum contains 342.3/594.3 = 0.576 mg of aluminum sulfate per mg of coagulant. The amount of precipitate (C_p) produced from 30 mg/L of $Al_2(SO_4)_3 \cdot 14H_2O$ is

$$C_p = (0.456 \text{ mg/mg})(0.576 \text{ mg/mg})(30 \text{ mg/L}) = 7.9 \text{ mg/L}$$

From Eq. (15.2), 2 moles of $CaCO_3$ will be produced for each mole of lime added; however, 0.6 meq/L or 30 mg/L of $CaCO_3$ will not precipitate. The mass of $CaCO_3$ precipitate per mass of lime added is 2(100)/74 = 2.70 mg/mg. The amount of precipitate (C_L) produced from 50 mg/L of lime is

$$C_L = (2.70 \text{ mg/mg})(50 \text{ mg/L}) - 30 \text{ mg/L} = 105 \text{ mg/L}$$

The total solids generated (C_T) in the treatment consists of the solids in the influent in addition to the precipitates formed from chemical addition.

$$C_T = (15 + 7.9 + 105) \text{ mg/L} = 128 \text{ mg/L}$$

The sludge is a highly mineralized sludge. The influent solids will be assumed to contain a VSS : FSS ratio of 3 : 1, which probably overestimates the content of organics. This ratio is typical of organic matter but any silt content of influent solids would decrease this ratio. The ratio of FSS : VSS in the sludge solids is

$$\text{FSS}:\text{VSS} = \frac{7.9 + 105 + (0.25)(15)}{(0.75)(15)} = 116.7:11.3 = 10.4:1$$

The s.g.s of VSS and FSS will be taken as 1.00 and 2.50, respectively.

The total quantity of solids generated on a daily basis is

$$QC_T = \left(128 \frac{\text{mg}}{\text{L}}\right)\left(16\,000 \frac{\text{m}^3}{\text{d}}\right)\left(\frac{1\,000 \text{ L}}{\text{m}^3}\right)\left(\frac{1 \text{ kg}}{10^6 \text{ mg}}\right) = 2\,050 \text{ kg/d}$$

$$\text{In U.S. units: } QC_T = \left(128 \frac{\text{mg}}{\text{L}}\right)\left(4.22 \times 10^6 \frac{\text{gal}}{\text{d}}\right)\left(\frac{3.79 \text{ L}}{\text{gal}}\right)\left(\frac{1 \text{ lb}}{454 \times 10^3 \text{ mg}}\right) = 4\,510 \text{ lb/d}$$

From Table 20.1 the concentration range for a gravity thickened sludge for coagulation settlings and backwash water is 2–4%. Using the middle value of 3%, trial and error solution of Eq. (20.2) shows that the s.g. of the sludge is

$$\frac{1}{S_s} = \frac{m_w/m_s}{S_w} + \frac{m_V/m_s}{S_V} + \frac{m_F/m_s}{S_F} = \frac{986/1\,016.9}{1.00} + \frac{2.7/1\,016.9}{1.00} + \frac{28.2/1\,016.9}{2.5} = 0.983\,8$$

$$S_s = 1/0.983\,3 = 1.017$$

From Eq. (20.3), the volume of the 2 050 kg of sludge after gravity thickening is

$$V_s = \frac{M}{(p_M/100)S_s} = \frac{2\,050 \text{ kg}}{(3/100)\left(1.017 \frac{\text{g}}{\text{cm}^3}\right)\left(\frac{1\,000 \text{ cm}^3}{\text{L}}\right)\left(\frac{1\,000 \text{ L}}{\text{m}^3}\right)\left(\frac{1 \text{ kg}}{1\,000 \text{ g}}\right)} = 67 \text{ m}^3$$

In U.S. units: $V_s = \dfrac{4\,510 \text{ lb}}{(3/100)\left(1.017 \dfrac{\text{g}}{\text{cm}^3}\right)\left(\dfrac{1\,000 \text{ cm}^3}{\text{L}}\right)\left(\dfrac{3.79 \text{ L}}{\text{gal}}\right)\left(\dfrac{1 \text{ lb}}{454 \text{ g}}\right)} = 17\,700 \text{ gal}$

For vacuum filtration of a water treatment coagulant sludge, the middle value for product solids concentration is 15%. Trial and error solution of Eq. (20.3) for s.g. of the slurry yields

$$\frac{1}{S_s} = \frac{926/1\,086}{1.00} + \frac{14.1/1\,086}{1.00} + \frac{148.9/1\,086}{2.5} = 0.920\,5$$

$$S_s = 1/0.920\,5 = 1.086$$

From Eq. (20.3), the volume of 2 050 kg (4 510 lb) of sludge is

$$V_s = \frac{M}{(p_M/100)S_s} = \frac{2\,050 \text{ kg}}{(15/100)\left(1.086 \dfrac{\text{g}}{\text{cm}^3}\right)\left(\dfrac{1\,000 \text{ cm}^3}{\text{L}}\right)\left(\dfrac{1\,000 \text{ L}}{\text{m}^3}\right)\left(\dfrac{1 \text{ kg}}{1\,000 \text{ g}}\right)}$$

$$= 12.6 \text{ m}^3 \text{ (3 320 gal)}$$

The volume of sludge is 0.08% of the influent flow.

20.4 AEROBIC SLUDGE DIGESTION

Aerobic sludge digestion is a frequently chosen sludge reduction process for small installations. Sludge produced from the primary clarifiers and biological treatment is normally concentrated and sent to an aerated digester, where endogenous respiration (decay) of the solids occurs. The process is stable and relatively easy to operate. The long residence time of sludge in an aerobic digester allows nitrifiers to become established; soluble nitrogen in the effluent is in the form of nitrate (Section 5.9.3).

In a batch digestion process the reduction of VSS proceeds according to the curve in Fig. 20.4. It is seen that the concentration of VSS becomes essentially constant after a period of time. All of the influent volatile solids are not degradable and, furthermore, some organic substances are transformed into refractory compounds by metabolism. The residual VSS concentration is subtracted from the initial VSS concentration to obtain the biodegradable VSS.

In order for solids to be metabolized they must first undergo hydrolysis to break down large molecules to smaller molecules that may be taken up by viable microorganisms. The starved conditions in the digester cause death and lysis of cells. Therefore, some solubilization of VSS occurs during digestion.

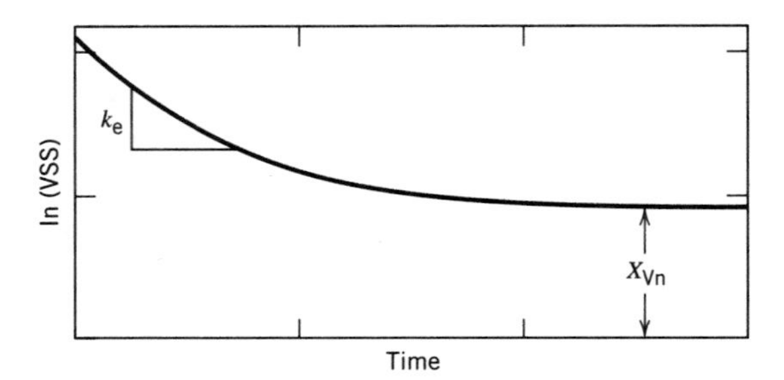

Figure 20.4 VSS variation in a batch digestion.

20.5 MODEL FOR AEROBIC SLUDGE DIGESTION

As for any biological process the equation of net biological growth (Eq. 17.4) applies and is restated here:

$$r_X = -Yr_S - k_e X_V \tag{20.5}$$

where

X_V is VSS concentration
Y is a yield factor
r_X is rate of sludge production
S is substrate concentration
k_e is the endogenous decay rate coefficient

If a biological sludge generated from an activated sludge or trickling filter process is being digested, the first term on the right-hand side of the equation is considered to be zero and the sludge reduction is described by a first-order decay reaction.

For sludge collected in a primary clarifier, the organic solids are first synthesized into biological solids, which then undergo endogenous decay. The overall phenomenon is still found to be adequately described by a first-order relation as given in Eq. (20.6).

$$r_{Xe} = -k_e X_V \tag{20.6}$$

where

r_{Xe} is the rate of endogenous decay

Krishnamoorthy and Loehr (1989) found that either a first-order or retardant first-order model adequately described aerobic sludge digestion of primary sludge and waste activated sludge. The retardant model was hypothesized to offer a better description of the process because active solids decrease during the process and there is probably a buildup of refractory substances, but it was found to provide no better correlations than a normal first-order model.

Only degradable solids are subject to decay. Biological stabilization will be accomplished through hydrolysis of the solids and metabolism of the soluble products; there may be some solubilization of inorganics and other suspended organic matter, which also results in solids reduction. The solubilization stage is not specifically accounted for in the model given by Eq. (20.6) but it would be reflected in endogenous decay rate constants determined from total suspended solids (TSS) or VSS data. The net amount of solids reduction through solubilization can be assessed by measuring suspended and soluble COD in both the influent and effluent. Solubilization has often not been measured in aerobic digestion studies but it can be significant (Droste and Sanchez, 1986). The degree of solubilization is less predictable than the rate of endogenous decay.

The model presented below is the commonly applied model of Adams et al. (1974) refined with incorporation of solubilization. Let the following terms be defined for a complete mixed aerobic digestion reactor.

X_{Vd0} = influent degradable VSS concentration
X_{Vd} = reactor degradable VSS concentration

The definition sketch in Fig. 20.5 applies.

Performing a mass balance around the digester, which is assumed to be complete mixed:

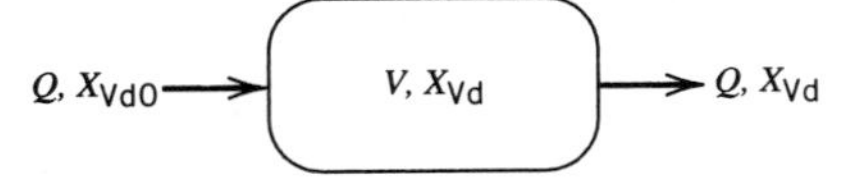

Figure 20.5 Aerobic digester schematic.

$$\text{IN} - \text{OUT} + \text{GENERATION} = \text{ACCUMULATION}$$
$$QX_{\text{Vd0}} - QX_{\text{Vd}} + r_{\text{d}}V + r_{\text{sl}}V = \frac{dX_{\text{Vd}}}{dt}V \tag{20.7}$$

where
- r_{d} is the kinetic expression for decay
- r_{sl} is the rate of solubilization
- Q is volumetric flow rate
- V is volume of the reactor

In this model formulation, solubilization is the excess rate of solubilization over the rate of metabolism. The rate of decay is also influenced by the rate of hydrolysis. Another approach would be to consider solids stabilization as a two-stage process consisting of solubilization (hydrolysis) followed by a second stage of metabolism similar to the development for anaerobes in Chapter 18.

For the steady state situation, $dX_{\text{Vd}}/dt = 0$. The decay of VSS is adequately modeled with first-order kinetics. Solubilization has not been modeled to any significant extent but a first-order model will be used because, in fact, solubilization has been implicitly incorporated into the first-order rate constants reported for the process. Also, hydrolysis has been modeled as a first-order process in anaerobic metabolism.

$$r_{\text{d}} = -k_{\text{e}}X_{\text{Vd}}$$
$$r_{\text{sl}} = -k_{\text{sl}}X_{\text{Vd}}$$

where
- k_{sl} is the rate coefficient for solubilization

If solubilization were not determined the observed decay coefficient would be

$$k_{\text{eobs}} = k_{\text{e}} + k_{\text{sl}}$$

where
- k_{eobs} is the observed rate coefficient for VSS removal

Substituting this rate model into Eq. (20.7) and solving it for the required hydraulic retention time leads to

$$\theta_{\text{d}} = \frac{X_{\text{Vd0}} - X_{\text{Vd}}}{(k_{\text{e}} + k_{\text{sl}})X_{\text{Vd}}} \tag{20.8}$$

where
- θ_{d} is the hydraulic retention time

There is a nonbiodegradable component of the VSS that does not change as it passes through the reactor. Define the following terms:

$$X_{\text{V0}} = X_{\text{Vd0}} + X_{\text{Vn0}} \tag{20.9}$$
$$X_{\text{V}} = X_{\text{Vd}} + X_{\text{Vn}} \tag{20.10}$$

where
X_{V0} is influent VSS concentration
X_V is the VSS concentration in the reactor
X_{Vd0} is the degradable component of influent VSS
X_{Vn0} is the nonbiodegradable component of influent VSS
X_{Vn} is the nonbiodegradable component of X_V

Because there is no change in the nonbiodegradable component,

$$X_{Vn0} = X_{Vn} \tag{20.11}$$

After substituting Eqs. (20.9)–(20.11) into Eq. (20.8), the following expression for θ_d is obtained.

$$\theta_d = \frac{X_{V0} - X_V}{(k_e + k_{sl})(X_V - X_{Vn})} \tag{20.12}$$

Nondegradable VSS may also be subject to solubilization. A mass balance for total VSS in the system at steady state is

$$QX_{V0} - QX_V + r_d V + r_{slt} V = 0 \tag{20.13}$$

where
r_{slt} is the rate of solubilization of total VSS
r_d has the same formulation as above

$$r_{slt} = -k_{sl}(X_{Vd} + X_{Vn}) \tag{20.14}$$

This model assumes that the rate of solubilization of degradable and nondegradable VSS is the same, which may not be the case. The model is readily modified to incorporate separate terms for solubilization of each type of VSS if necessary. Rate constants reported are usually the overall observed rate constants for the process.

The other equation that applies from a mass balance around the system describes the production of soluble COD:

$$k_{sl} f(X_{Vd} + X_{Vn})V = Q(S_e - S_0) \tag{20.15}$$

where
f is the COD of VSS
S_e is soluble effluent COD
S_0 is soluble influent COD

The issue of solubilization is not important from a water quality perspective because the supernatant from any solids separation process following aerobic digestion will be returned to the primary biological treatment unit. However, solubilization does affect solids removal and the detention time needed to achieve the desired amount of solids reduction.

It must also be remembered that there is no digestion of fixed (inert) solids, which will pass through the digestion basin, although some of these may be solubilized as noted above.

■ Example 20.3 Aerobic Sludge Digestion and Solubilization

Waste solids from a wastewater treatment plant are concentrated to 9 000 mg/L and sent to a CM aerobic digester operated at a detention time of 16 d. Studies have shown that the nondegradable portion of the influent VSS is 20%. Effluent VSS is 3 000 mg/L. The COD of VSS in the system has been determined to be 1.39 mg/mg. Soluble COD in the influent is 100 mg/L and soluble COD in the effluent is 900 mg/L.

(a) Derive the observed endogenous decay coefficient based solely on VSS reduction.

(b) What is the range for the true value of the endogenous decay coefficient?

(a) Ignoring solubilization, the degradable and nondegradable influent and effluent VSS values are

$$X_{Vn} = 0.20X_{V0} = 0.20(9\,000 \text{ mg/L}) = 1\,620 \text{ mg/L}$$
$$X_{Vd0} = X_{V0} - X_{Vn} = 9\,000 \text{ mg/L} - 1\,620 \text{ mg/L} = 7\,380 \text{ mg/L}$$
$$X_{Vd} = X_{V} - X_{Vn} = 3\,000 \text{ mg/L} - 1\,620 \text{ mg/L} = 1\,480 \text{ mg/L}$$

Rearranging Eq. (20.8),

$$k_e = \frac{X_{Vd0} - X_{Vd}}{\theta_d X_{Vd}} = \frac{7\,380 \text{ mg/L} - 1\,480 \text{ mg/L}}{(15 \text{ d})(1\,480 \text{ mg/L})} = 0.27 \text{ d}^{-1}$$

(b) There is insufficient information to determine the portions of degradable and nondegradable VSS that have been solubilized.

The amount of VSS that has been solubilized (X_{Vs}) is

$$X_{Vs} = (900 \text{ mg COD/L} - 100 \text{ mg COD/L})/(1.39 \text{ mg COD/mgVSS}) = 576 \text{ mg/L}$$

X_{Vd0} is unchanged from the above value. If only degradable VSS has been solubilized, then X_{Vn} also remains at 1 620 mg/L. The amount of VSS that was actually endogenously decayed is

$$7\,380 \text{ mg/L} - 1\,480 \text{ mg/L} - 576 \text{ mg/L} = 5\,324 \text{ mg/L}$$

The endogenous decay coefficient is

$$k_e = \frac{5\,324 \text{ mg/L}}{(15 \text{ d})(1\,480 \text{ mg/L})} = 0.24 \text{ d}^{-1}$$

If only nondegradable VSS was solubilized then X_{Vn} is

$$X_{Vn} = 1\,620 \text{ mg/L} - 576 \text{ mg/L} = 1\,044 \text{ mg/L}$$
$$X_{Vd} = 3\,000 \text{ mg/L} - 576 \text{ mg/L} = 2\,424 \text{ mg/L}$$
$$k_e = \frac{7\,380 \text{ mg/L} - 2\,424 \text{ mg/L}}{(15 \text{ d})(2\,424 \text{ mg/L})} = 0.14 \text{ d}^{-1}$$

The true value of k_e lies between 0.14 and 0.24 d^{-1}. If this process was operated at different detention times, the models would predict significantly different amounts of endogenous decay.

20.5.1 Rate Constants and Sludge Degradability

Primary sludge is generally more refractory than biologically generated solids. Krishnamoorthy and Loehr (1989) found for batch digestion of primary sludge alone that k_e values (based on degradable solids) were in the range of 0.028–0.056 d^{-1}. The initial biologically degradable VSS to total VSS ratio was in the range of 74–91%. The first-order model was reasonable for a lengthy stabilization period but the first stage of primary sludge digestion proceeded at a lower rate than predicted by the model.

For waste activated sludge, k_e values were found to be in the range of 0.016–0.426 d^{-1} based on degradable VSS concentrations (Krishnamoorthy and Loehr, 1989). Biodegradable VSS were 44–67% of total VSS. Lower k_e values were a result of high sludge ages in the activated sludge process. Goodman and Englande (1974) also found that a high sludge age in the activated sludge process decreased k_e values in subsequent aerobic digestion.

Several studies have found that an inverse relation exists between initial concentration of VSS and k_e (Ganczarczyk et al., 1980; Krishnamoorthy and Loehr, 1989; Reynolds, 1973). The reason for this effect is not well understood, although oxygen transfer at high VSS concentrations may be limiting.

The Arrhenius equation describes the effect of temperature on the rate constant.

$$k_{e,T} = k_{e,20}\theta^{(T-20)} \tag{20.16}$$

where

$k_{e,T}$ is the digestion decay constant at any temperature
$k_{e,20}$ is the digestion decay constant at 20°C
T is temperature in °C

The endogenous decay coefficient is different between batch and semicontinuous flow reactors. A batch reactor will undergo a change in flora as digestion proceeds, whereas a semicontinuous flow reactor has a more stable environment. Droste and Sanchez (1986) found the k_e values in Table 20.6 for batch and semicontinuous flow digestion of activated sludge at temperatures in the range of 10–30°C. An equation similar to Eq. (10.19) would apply to batch digestion, where concentration is based on degradable VSS.

Krishnamoorthy and Loehr (1989) found a θ value of 1.097 and that value applied for temperatures up to 30°C. Mavinic and Koers (1977) found that the degree-day product (P_T in Eq. 20.17) is an important design parameter. In laboratory and field scale studies between temperatures of 5 and 20°C, the amount of digestion achieved attained a maximum at P_T equal to 250°-d. Above this value there was little increase in the amount of stabilization.

$$P_T = \theta_d \times T \tag{20.17}$$

TABLE 20.6 Endogenous Decay Coefficients in Batch and Semicontinuous Flow Reactors[a]

		k_e,[b] d^{-1}	
Temperature, °C	θ_d, d	Semicontinuous[c]	Batch
10	10	0.22	
	20	0.24	0.056
	30	0.27	
20	10	0.25	
	20	0.29	0.151
	30	0.63	
30	10	0.19	
	20	0.16	0.12
	30	0.16	

[a]From Droste and Sanchez (1986).
[b]Based on degradable VSS in waste activated sludge.
[c]Reactors were fed once per day, approximating a continuous flow reactor.

where

θ_d is in d
T is in °C

There will be a natural decrease in pH as digestion proceeds because of nitrification. The decrease in pH will not stop digestion but it does appear to have a minor inhibitory effect on the rate of digestion. Maintaining a pH in the range of 7–8 provides the maximum rate of aerobic digestion (Anderson and Mavinic, 1987; Bhargava and Datar, 1988; Krishnamoorthy and Loehr, 1989). The lowering of pH does not affect the degradability of VSS. In the work of Krishnamoorthy and Loehr (1989), adding phosphate buffer to sludge that had been digesting 42 d, to raise the pH to 7.0, did not result in further decrease in the volatile solids in the reactor. Waste composition, temperature, pH, and other environmental conditions result in some variability of decay rates among different wastes and dictate a need for field studies.

Thermophilic Aerobic Digestion

Increasing the temperature of digestion accelerates the rate of degradation considerably but at the cost of energy required to heat the sludge and maintain the reactor at temperatures greater than 40°C up to the 50–60°C range. At these temperatures the time required to achieve digestion can decrease to 7 d or less.

Autothermal aerobic digestion, where the temperature of the process is maintained by the heat released from the metabolism of the organic matter is a feasible alternative. Reactors must be covered and insulated. Kelly et al. (1993) found that no additional heat source was required to maintain temperatures in the range of 55–70°C in digesters located in the province of British Columbia, Canada. Recommendations from their study of three full-scale systems are given in Table 20.7. They found that θ_d less than 4–5 d resulted in washouts of the biomass. The sludge feed rate has to be adjusted to maintain the desired P_T value based on the rate of VS destruction and ambient temperatures.

There were no odor problems with a properly operated process. Sludge dewaterability appeared to be about the same as in mesophilic processes. Except for fecal streptococci, which exhibited an increase, indicator and pathogen dieoff in the process was also excellent when there were two reactors in series, with a temperature of at least 55°C being consistently achieved in the final reactor. Existing aerobic digestion processes can be retrofitted to autothermal digestion by covering and insulating the reactors.

TABLE 20.7 Autothermal Aerobic Digestion Design Parameters[a]

Parameter	Value
θ_d, d[b,c]	7
P_T, °-d[b]	400
Mixing energy, G, s^{-1}	450
Air flow, vol. air/vol. reactor/h	0.5–2.0
Organic loading, kg VS/m^3/d	4–20
(lb/ft^3/d)	(0.25–1.25)

[a]From Kelly et al. (1993).
[b]To achieve VS reductions of 38%.
[c]Minimum value for Eq. (20.17).

TABLE 20.8 Empirical Reaction Rate Coefficients[a] for Indicator Microorganisms and Enteroviruses[b]

Detention time d	Temperature °C	Total coliforms	Fecal coliforms	Fecal streptococci	Enteroviruses
20	8.0	0.03	0.03	0.03	0.04
20	14.6	0.06	0.05	0.05	0.05
20	17.5	0.08	0.07	0.06	0.05
20	21.7	0.03[c]	0.06	0.05	0.04
20	23.7	0.11	0.09	0.07	0.05
20	25.6	0.07	0.03[c]	0.04[c]	0.06
15	29.0	0.10	0.09	0.05	0.07
10	31.1	0.08	0.10	0.06	0.11
10	37.5	0.09	0.11	0.08	0.24
20	38.2	0.13	0.12	0.08	0.16
15	39.8	0.15	0.11	0.08	0.16

[a]The rate coefficient is given by Eq. (20.18).
[b]Reprinted from J. H. Martin, H. E. Bostian, and G. Stern, "Reductions of Enteric Microorganisms during Aerobic Sludge Digestion," *Water Research,* 24, pp. 1377–1385, Copyright 1990, with kind permission from Elsevier Science Ltd., The Boulevard, Langford Lane, Kidlington, 0X5 1GB, UK.
[c]Statistical analysis showed these coefficients to be outliers.

20.5.2 Indicator Microorganism Reduction in Aerobic Digestion

Martin et al. (1990) studied indicator microorganism dieoff in a fill and draw operated, complete mixed aerobic digester. The digester received waste activated sludge from an activated sludge process that was not preceded with primary clarification. Four indicator groups were studied: total coliforms, fecal coliforms, fecal streptococci, and enteroviruses. Fecal coliforms and fecal streptococci in the influent to the digester were strongly correlated (correlation coefficient of 0.925), as expected, but total coliforms and fecal coliforms or fecal streptococci were not highly correlated, probably because of the presence of non-fecal coliforms in the total coliform group.

Zero-order or first-order kinetics were not found to describe well the dieoff of any indicator group. An empirical model (Eq. 20.18) gave good correlations for all groups.

$$k = \frac{\log_{10} C_0 - \log_{10} C_e}{\theta_d} \tag{20.18}$$

where

C_0 and C_e are influent and effluent indicator densities, respectively, in colony forming units (CFU) or plaque forming units (PFU) per 100 mL
k is the empirical rate coefficient
θ_d is the detention time in days

The Arrhenius equation modeled the empirical rate coefficient as a function of temperature. Their data are summarized in Table 20.8.

20.6 SLUDGE THICKENING

Sludge removed from clarifiers in water and wastewater treatment operations is often further concentrated by gravity or air flotation before further processing or disposal.

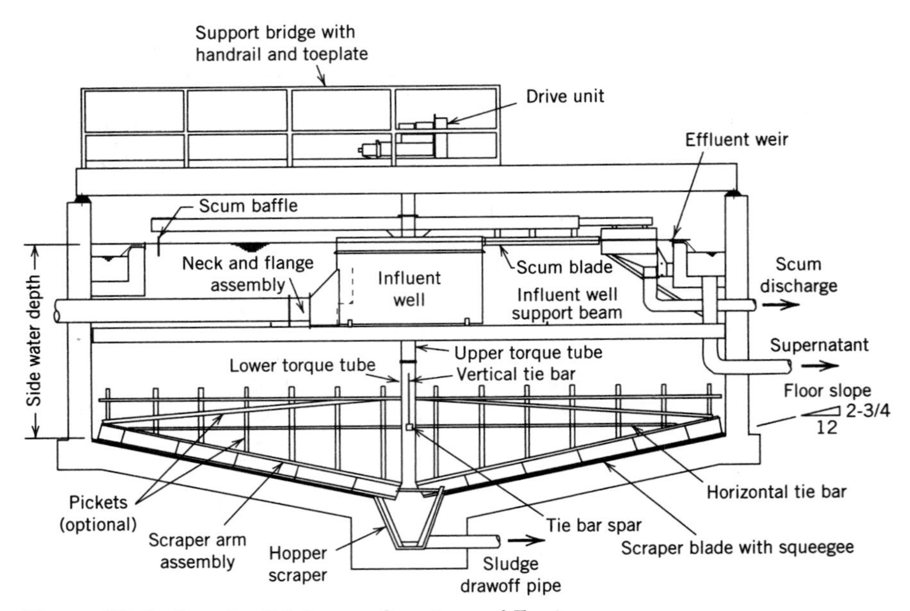

Figure 20.6 Gravity thickener. Courtesy of Envirex.

20.6.1 Gravity Thickening

Sludge in the thickening zone of a secondary clarifier slowly squeezes water out of the suspension as it compacts, as shown in Section 11.5. This compaction settling is sometimes classified as type IV settling. Gravity thickeners are sedimentation units (Fig. 20.6) designed to achieve higher concentrations of sludge that has been collected in primary and secondary clarifiers. The thickened sludge removed from the bottom of these units is sent to further processing or a landfill. The supernatant removed from these units contains a substantial quantity of solids and it is returned to the primary clarifier.

As the sludge compacts, the water has smaller channels for the water to escape, which increases resistance to compaction according to a first-order expression.

$$-\frac{dH}{dt} = i(H - H_\infty) \tag{20.19}$$

where

H is height of the sludge
H_∞ is the ultimate height or compaction
i is a constant specific to the sludge

Rearranging the above expression and integrating,

$$H_t - H_\infty = (H_0 - H_\infty)e^{-i(t-t_0)} \tag{20.20}$$

where

H_t is the height of the sludge at any time t
subscript 0 refers to the initial condition

The ultimate compaction is determined from laboratory or field studies after thickening a suspension for about 24 h or until the sludge does not change concentration. Measuring the height of the sludge in the laboratory provides the data to find i in Eq. (20.20).

If waste activated sludge is to be thickened in a gravity thickener, the sludge should not reside in the thickener for more than 18 h to reduce gas production and other undesirable effects (e.g., odors) resulting from biological activity (WEF and ASCE, 1992). The most common design for gravity thickeners is a circular basin with a side depth of 3–4 m (9.8–13.1 ft) and floor sloping at 2:12–3:12. Overflow rates between 4 and 8 $m^3/m^2/d$ (100–200 $gal/ft^2/d$) are suitable for secondary sludges but maximum overflow rates of 16–32 $m^3/m^2/d$ (390–785 $gal/ft^2/d$) can be used for primary sludges.

20.6.2 Flotation Thickening

Flotation thickening is an alternative to gravity thickening of sludge. It is also commonly used for water treatment in Europe instead of sedimentation (Edzwald et al., 1992). Figure 20.7 shows a schematic of the process. Air is dissolved in water at pressures above atmospheric pressure in enclosed units. This water, saturated in air with respect to atmospheric pressure, is then fed with wastewater into a vessel at atmospheric pressure. Bubble formation occurs at the pressure reducing valve, which may be a needle valve or nozzle, and dissolved air is also released from solution in the flotation vessel. Bubbles are formed which are entrapped in flocs or adhere to flocs and buoy them to the surface. Polyelectrolyte addition can improve the adhesion of bubbles to flocs. A skimmer collects the floating solids and a portion of the effluent is recycled through the air dissolution tank.

Dissolving air directly into the sludge feed is not recommended because high shear forces will occur on the sludge particles, causing them to break. The tanks should be covered to prevent rain and wind from breaking up the surface sludge layer.

Bubble sizes produced should be less than 100–120 μm in diameter for optimal performance (de Rijk et al., 1994). Typical bubble sizes have median nominal diameters from 60 to 80 μm. Smaller bubbles are advantageous for a number of reasons:

- Contact time and the possibility of collision increase as rise velocity decreases.

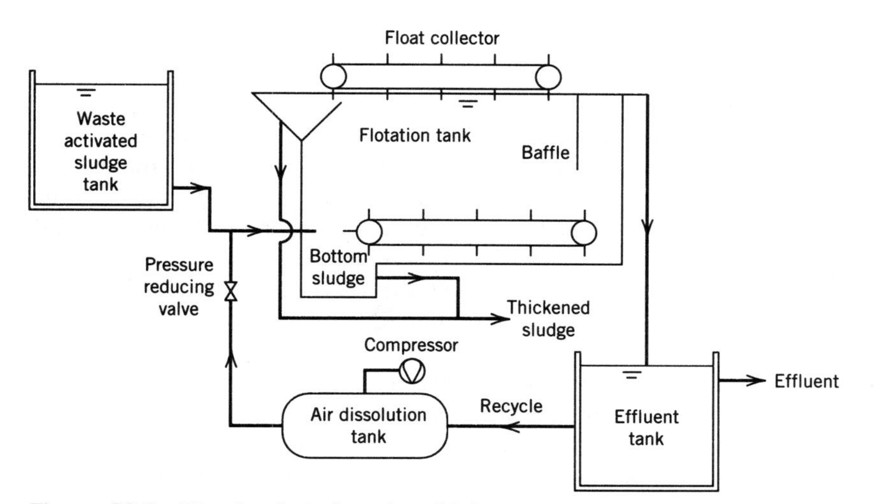

Figure 20.7 Dissolved air flotation thickener.

TABLE 20.9 Henry's Law Constants for Air[a]

Temperature, °C	0	10	20	30	40	50	60
K_H, mg/L/atm	37.3	29.3	24.3	20.9	18.5	17.0	16.0

[a]At 1 atm and 0°C the density of dry air is 1.292 8 g/L.

- Smaller bubbles are more readily incorporated into the floc.
- More bubble and more contact opportunities are produced with a larger number of smaller bubbles.
- Adhesion forces (arising from surface tension) between the bubble and solid particle are stronger with smaller bubbles.
- Larger bubbles rise more rapidly and increase shear forces on floc particles.

Henry's law (Section 1.10) governs the dissolution and release of air.

$$C_s = K_H p \tag{20.21}$$

where
C_s is the saturation concentration
K_H is Henry's constant
p is pressure

Table 20.9 gives the values of Henry's law constant for air at different temperatures.

In a nozzle or needle valve, pressure is decreased in accordance with Bernoulli's equation as the water passes through a small opening.

$$p_1 + \frac{1}{2}\rho v_1^2 = p_2 + \frac{1}{2}\rho v_2^2 + \rho g h_L$$

where
subscripts 1 and 2 refer to the upstream and valve channels, respectively
h_L is headloss

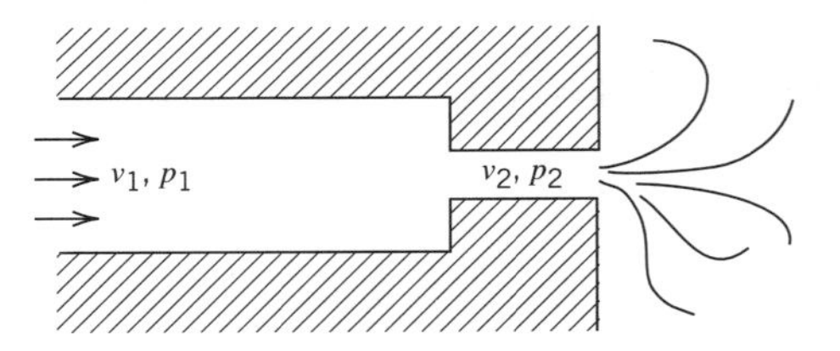

Bubbles are probably formed via cavitation, which is the growth of a bubble because of a pressure drop, followed by implosion and bursting of the bubble into many small bubbles. Placing a tube after the pressure-reducing valve before injecting the air-saturated water into the wastewater shifted the bubble sizes to larger values (de Rijk et al., 1994). There was no change in the size of the bubbles when the saturation pressure was greater than 5 atm (73.5 psi). Below saturation pressures of 5 atm (75 psi) higher flow rates of the air-laden water stimulated formation of smaller bubbles from the valve.

The air : solids ratio is an important design parameter. As more air is dissolved in the influent, higher removals and higher float solids concentrations will be achieved. However performance levels off rapidly at air : solids ratios somewhere above a value of 0.02–0.04 on a mass basis. The mass of air released in the flotation tank is

$$M_a = Q_r(C_{r0} - C_e) \tag{20.22}$$

where
M_a is the mass of air released
Q_r is the flow rate in the recycle line
C_{r0} is the concentration of air in the flow leaving the air dissolution tank
C_e is the concentration of air in the effluent tank

Equation (20.22) assumes that the influent waste is saturated with air.

Applying Henry's law, the mass of air released can be expressed in terms of pressure.

$$\frac{C_{r0}}{C_e} = \frac{p_{r0}}{p_{atm}} = p_{r0} \tag{20.23}$$

where
p_{r0} is the air pressure in the air dissolution tank
p_{atm} is atmospheric pressure which is assumed to be 1 atm (14.7 psi)

Equation (20.23) further assumes that equilibrium is achieved between the liquid in the flotation tank and the atmosphere. Substituting Eq. (20.23) into Eq. (20.22), the mass of air released can be calculated from

$$M_a = Q_r C_e (p_{r0} - 1) \tag{20.24}$$

Complete saturation of the liquid in the air dissolution tank will not be achieved. It is a function of the air–water interface, detention time, and mixing in the tank. Detention times can range from a few seconds to a few minutes. A correction factor, f, is applied in Eq. (20.24) to account for the efficiency of air dissolution.

$$M_a = Q_r C_e (f p_{r0} - 1) \tag{20.25}$$

Current practice in the United States is to achieve 85–90% efficiency (USEPA, 1979). In typical practice, f is 70% for unpacked saturators and 90% for packed saturators (Edzwald et al., 1992).

Typical design values for dissolved air flotation systems for thickening waste activated sludge are given in Table 20.10. The removal efficiencies and float concentration given in this table are on the conservative side for many situations. Float concentra-

TABLE 20.10 Design Parameters for Dissolved Air Flotation of Waste Activated Sludge with Addition of Polyelectrolyte Aids[a]

Parameter	Typical design value
Solids loading, kg/m³/d (lb/ft²/d)	
Without chemical addition	40–120 (8.2–25)
With chemical addition	60–240 (12–50)
Float concentration, %	4
Removal efficiency, %	90–95
Polyelectrolyte addition, kg/tonne (lb/ton) of dry solids	1.5–10 (3.0–20)
Air : solids ratio, kg air/kg solids (lb air/lb solids)	0.02–0.06 (0.02–0.06)
Effluent recycle ratio, % of influent	40–70
Hydraulic loading, m³/m²/d (gal/ft²/d)	
With chemical addition	25–120 (615–2 950)
Without chemical addition	42 (1 030) maximum

[a]Adapted from Viessman and Hammer (1985) and WEF and ASCE (1992).

tions exceeding 4% have been obtained without the addition of polyelectrolytes. Air pressures up to 6 atm have been used. Experimental variation of the air pressure in the air dissolution tank should be performed to find conditions that produce the smallest bubbles with the valve in the system.

20.7 MECHANICAL SLUDGE DEWATERING

Sludge dewatering is used to remove as much water as possible from a sludge to produce a highly concentrated cake. There is a variety of dewatering devices on the market. The capital and operating expenses of these devices are usually significantly less than the costs of hauling large quantities of sludge long distances to a landfill or other disposal site.

A sludge suspension contains bulk water, which is not bound to the sludge particles, and bound water, of which there are three different types (Vesilind, 1994): interstitial water, vicinal water, and water of hydration. Interstitial water is water captured in the interstitial spaces within flocs and within cells. Vicinal water is bound to the surfaces of solids. Water of hydration is chemically bonded to the sludge. Dewatering devices primarily remove bulk water; some interstitial water can be removed but it is likely that the major fraction of bound water is vicinal water that cannot be removed mechanically (Vesilind, 1994). Freeze–thaw conditioning physically disrupts floc and cell structure and produces the greatest degree of bound water release for dewatering (Robinson and Knocke, 1992).

20.7.1 Centrifugation

Centrifuges are analogous to sedimentation tanks except that the suspended particles are accelerated by a centrifugal force that is higher than the gravity force. A schematic of a centrifuge is shown in Fig. 20.8. In this centrifuge, slurry is input into a central feed pipe. The bowl is rotated typically between 200 and 8 000 rpm, creating the centrifugal force that moves the particles to the wall of the bowl. An extraction screw rotated at a different speed from the bowl scrapes solids from the bowl wall and moves them to the solids discharge end. Liquid (centrate) is discharged at the opposite end of the centrifuge. Centrifuges may also be designed for unidirectional flow and tangential feed.

The centrate contains a significant amount of solids and it is returned to the

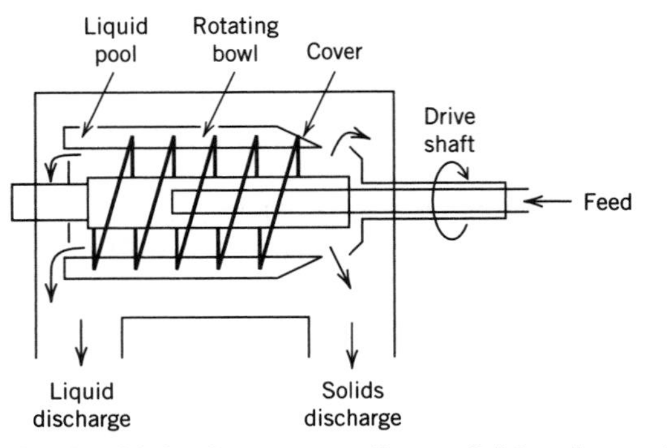

Figure 20.8 Continuous flow solid bowl centrifuge.

treatment plant. In a wastewater treatment plant it is returned ahead of the primary clarifier. Centrifuges are not as commonly used in water treatment plants because the sludge is in a liquid state. Centrate is returned to the head of the water treatment plant.

The major forces acting on a particle are the centrifugal force (Eq. 20.26) and the drag force (Eq. 20.27). The force balance is similar to the force balance on a particle settling in a fluid. The centrifugal force is

$$F_c = m_e\omega^2 r = (\rho_s - \rho)V_p\omega^2 r \tag{20.26}$$

where
- F_c is the centrifugal force
- m_e is effective mass of the particle
- ω is the angular velocity of the fluid
- r is the distance of the particle from the center
- ρ_s is the density of the particle
- ρ is the density of the fluid
- V_p is the volume of the particle

and the drag force is

$$F_D = \tfrac{1}{2}\rho C_D A v^2 \tag{20.27}$$

where
- F_D is the drag force
- C_D is the drag coefficient
- A is the cross-sectional area of the particle
- v is the velocity of the particle

The drag coefficient is a function of the Reynold's number as discussed in Section 11.3.

Ignoring the gravity force which is small compared to the other forces and making a force balance on a particle:

$$\rho_s V_p \frac{dv}{dt} = (\rho_s - \rho)V_p\omega^2 r - \tfrac{1}{2}\rho C_D A v^2$$

Making the substitution of *dr/dt* for v and using the nominal diameter, *d*, for the particle, the equation becomes

$$\rho_s \frac{\pi d^3}{6}\frac{d^2 r}{dt^2} = (\rho_s - \rho)\frac{\pi d^3}{6}\omega^2 r - \tfrac{1}{2}\rho C_D \frac{\pi d^2}{4}\left[\frac{dr}{dt}\right]^2$$

or

$$\frac{d^2 r}{dt^2} + C_D \frac{\rho}{\rho_s}\frac{3}{4d}\left[\frac{dr}{dt}\right]^2 - \frac{(\rho_s - \rho)}{\rho_s}\omega^2 r = 0 \tag{20.28}$$

If the particle is in a laminar flow regime, the drag coefficient is 24/Re. Substituting this into Eq. (20.28) the differential equation becomes

$$\frac{d^2 r}{dt^2} + \frac{18\mu}{\rho_s d^2}\frac{dr}{dt} - \frac{(\rho_s - \rho)}{\rho_s}\omega^2 r = 0$$

where
- μ is the viscosity of the fluid

The solution of this differential equation is

$$r = C_1 \exp\left(\frac{-b + \sqrt{b^2 + 4c}}{2}\right)t + C_2 \exp\left(\frac{-b - \sqrt{b^2 + 4c}}{2}\right)t$$

where

$b = \dfrac{18\mu}{\rho_s d^2}$, $c = \dfrac{(\rho_s - \rho)}{\rho_s}\,\omega^2$, and for all values $b^2 + 4c > 0$ (it is impossible that the roots be equal).

The critical initial conditions are

$$t = 0: \qquad r = 0: \qquad \frac{dr}{dt} = 0$$

because the particle entering at the center of the unit has the farthest distance to travel. The velocity dr/dt must be zero at the center because of symmetry. However a particle that enters exactly at the center of the unit will not be subjected to any net centrifugal force; it will simply spin around and be moved longitudinally until it is moved from the centerline, at which time it will begin to accelerate toward the wall. The equation has no solution using these initial conditions, i.e., $r = 0$ at all times.

A slight perturbation is all that is required to move the particle from the centerline. Irregularities in the particle shape are sufficient to cause a force imbalance on the particle and initiate its movement radially. Also gravity has been ignored in the preceding development. Normal fluctuation in flow streamlines will move the particle away from the centerline or a particle may be displaced by another particle. The moment that the particle is moved from the centerline, it is subjected to the centrifugal force and travels toward the perimeter. The differential equation can be solved if an initial condition of a small displacement, δ, from the centerline is specified. Using this condition the solution of the differential equation is

$$r = \frac{\delta}{\sqrt{b^2 + 4c}}\left(\alpha_1 e^{\alpha_2 t} - \alpha_2 e^{\alpha_1 t}\right) \tag{20.29}$$

where

$$\alpha_1 = \frac{-b + \sqrt{b^2 + 4c}}{2}, \qquad \alpha_2 = \frac{-b - \sqrt{b^2 + 4c}}{2}$$

Equation (20.29) can be plotted to find the time for a particle to move a given radial distance.

The acceleration of the particle is commonly ignored (or assumed to be constant). Assuming that the change in acceleration of the particle with respect to radial distance is negligible, Eq. (20.28) integrates to

$$t = \frac{18\mu}{d^2\omega^2} \ln \frac{r_o}{r_i} \tag{20.30}$$

where

r_i and r_o are the initial and final radial distances of the particle from the centerline

The high concentration of solids in a slurry will affect the movement of the particles. This problem and the simplifying assumptions used to derive Eqs. (20.29) and (20.30) result in a more empirical approach to describe thickening in centrifuges.

Vesilind (1979) describes the empirical modifications commonly used to model centrifuge performance.

20.7.2 Vacuum Dewatering

Vacuum filtration is the large-scale continuous flow version of a laboratory solids determination. A sludge slurry is brought into contact with a filter and a vacuum is applied to draw the water from the sludge and through the filter leaving a sludge cake deposit on the filter. The continuous flow vacuum filter (Fig. 20.9) consists of a horizontal cylinder that is 20–35% submerged in a vat of previously conditioned sludge. The filter medium is wrapped around the drum. As the drum rotates the medium is rolled on and off the drum. The cake is cleaned from the medium by roll discharge (the cake crumbles from the medium when it is passed over a small roller) or by a scraping blade. The medium is washed before it is returned to the drum. As for centrifuges, the filtrate is returned to the water or wastewater treatment plant for reprocessing.

There are two types of media (USEPA, 1987). Coil spring medium consists of a permanent layer of springs on the drum, which is overlaid with a second layer of coil springs that travels off and on to the drum. The open area of the coil springs is 7–14%. The feed sludge concentration must be high with sufficient fiber content to prevent the loss of solids through the large open area. The other medium is cloth, which must be used when filtering unthickened secondary sludge.

Solids capture ranges from 85 to 99.5%. The product sludge cake moisture normally falls between 60 and 90%, depending on the feed type, solids concentration, chemical conditioning agents, and operation. Typical loading rates are 34–75 $kg/m^2/h$ (7–15 $lb/ft^2/h$) for primary sludges and 17–24.5 $kg/m^2/h$ (3.5–5 $lb/ft^2/h$) for mixed digested sludges. Lime, polyelectrolytes, ferric chloride, and other chemicals are used as conditioning agents. Precoating a filter with diatomaceous earth may considerably improve its performance.

Flow through the cake and the filter medium is flow through porous media. The theory of vacuum filter operation is based on an adaptation of the Carman–Kozeny equation or Darcy's law (Coackley and Jones, 1956).

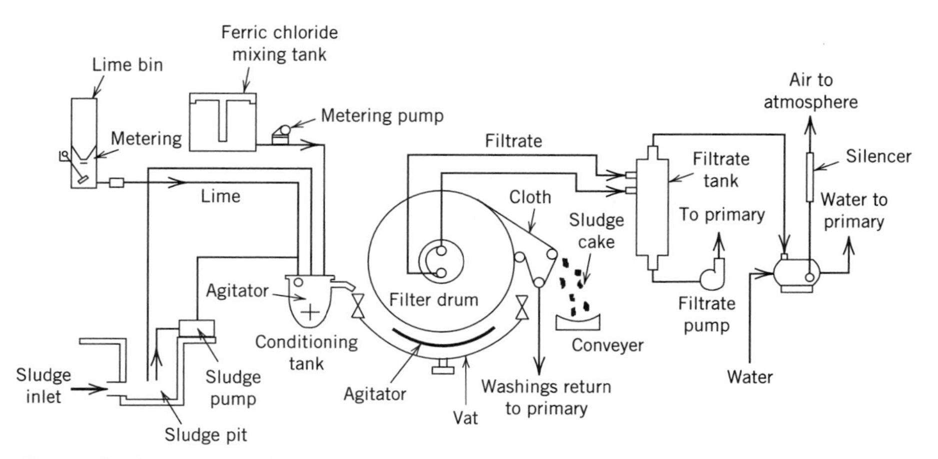

Figure 20.9 Vacuum filter system. After USEPA (1979).

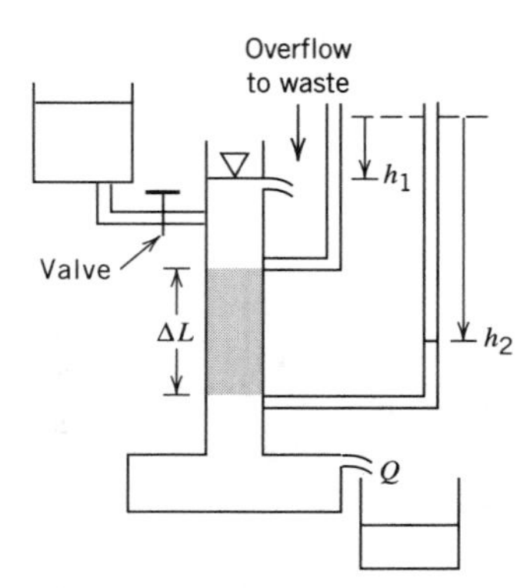

Figure 20.10 Darcy's law apparatus.

Darcy's law (see the illustration in Fig. 20.10) is

$$Q = \frac{dV}{dt} = KA\frac{\Delta h}{\Delta L} \tag{20.31}$$

where
- Q is volumetric flow rate of the filtrate
- V is volume of water
- A is area of flow
- K is conductivity
- h is head
- L is distance

The resistance parameter depends on the Reynold's number, porosity of the media, distribution of grains, and other characteristics of the media. From dimensional analysis or analogy to the Hagen–Poiseuelle law for pipe flow:

$$K = kg/\nu \tag{20.32}$$

where
- k is intrinsic permeability
- ν is the kinematic viscosity

and

$$\frac{dV}{dt} = \frac{kA}{\mu}\frac{\Delta P}{\Delta L} \tag{20.33}$$

where
- ΔP is the positive pressure differential
- ΔL is the depth of the medium
- μ is the dynamic viscosity of the filtrate

The intrinsic resistance, r can be defined as $r = 1/k$. The resistance from the cake, R_c, is

$$R_c = r_c\Delta L \tag{20.34}$$

where
- r_c is the intrinsic resistance of the cake

The resistances contributed by the cake and the filter medium are considered to be independent and therefore able to be summed. The total resistance, R, is

$$R = R_c + R_f \tag{20.35}$$

where

R_f is the resistance of the filter medium

Now the volume of the cake formed, V_c, is

$$V_c = A\Delta L$$

Defining the specific deposit, σ, as the volume of cake formed per volume of filtrate obtained, the above equation becomes

$$\sigma V = A\Delta L \tag{20.36}$$

Considering the total resistance and Eqs. (20.34) and (20.36), Eq. (20.33) becomes

$$\frac{dV}{dt} = \frac{\Delta P A^2}{\mu(r_c \sigma V + R_f A)} \tag{20.37}$$

The intrinsic resistance, r_c, is on a volume basis. The resistance can also be formulated in terms of the mass of dry cake solids formed per unit volume of filtrate, w. The specific resistance, r_{wc}, is related to r_c by

$$r_{wc} w = r_c \sigma \tag{20.38}$$

Substituting this into Eq. (20.37),

$$\frac{dV}{dt} = \frac{\Delta P A^2}{\mu(r_{wc} w V + R_f A)} \tag{20.39}$$

For a constant pressure difference over time Eq. (20.39) integrates to

$$\int_0^t dt = \mu \int_0^V \left(\frac{w r_{wc} V}{\Delta P A^2} + \frac{R_f}{\Delta P A} \right) dV$$

$$t = \frac{\mu w r_{wc} V^2}{2\Delta P A^2} + \frac{\mu R_f V}{\Delta P A} \quad \text{or} \quad \frac{t}{V} = \frac{\mu w r_{wc} V}{2\Delta P A^2} + \frac{\mu R_f}{\Delta P A} \tag{20.40}$$

The terms in Eq. (20.40) can be determined from a plot of t/V versus V, which will yield a straight line. The specific resistance, r_{wc}, can be calculated from the slope, m, of the line.

$$r_{wc} = \frac{2\Delta P A^2}{\mu w} m \tag{20.41}$$

A laboratory setup to determine the specific resistance of sludge cakes is shown in Fig. 20.11. The procedure (USEPA, 1987) for the determination is to add 50–200 mL of sludge (with any conditioning agent added and mixed with the sludge) to the filter pad in the Büchner funnel. The pinch clamp is closed and the sludge is allowed to drain by gravity for 2 min. The vacuum pump, supplying a vacuum of 37 cm (14.6 in.) Hg is turned on and the clamp is opened. At 15-s intervals the volume of filtrate collected in the graduated cylinder is noted. Observations are made until the vacuum breaks or the filtrate volume does not change significantly between readings. The theory developed in this section ignores changes in resistance as the cake forms; porosity and tortuosity, among other cake characteristics, change. However, the theory

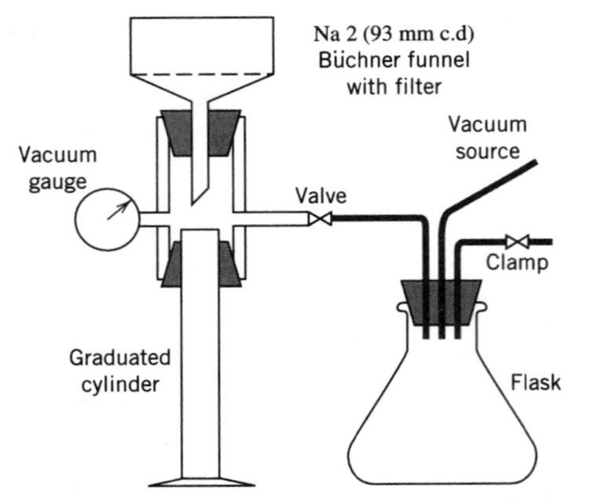

Figure 20.11 Laboratory setup to determine cake resistance.

is adequate to describe vacuum filtration and the test procedure. A plot of t/V versus V yields the specific resistance. The test is useful to evaluate different doses and combinations of sludges and conditioning agents on the specific resistance and quality of the cake. A plot of resistance versus dose and type of conditioning agent can be used to determine the optimal operating condition. Observations of performance of the full-scale unit can be related to results obtained from the laboratory setup to further improve the utility of the lab test.

The mass (weight) of solids deposited can be approximated from the solids concentration in kg/m^3 in the feed slurry.

Ranges for specific resistances of various sludges are given in Table 20.11.

20.8 LAND APPLICATION OF SLUDGE

Application of sludge to land, where feasible, is the most desirable alternative because it uses sludge in a natural cycle. Ideally, sludges produced from wastewater treatment processes contain all foreign matter introduced into water through domestic and industrial use that has not been removed or transformed into neutral substances by the wastewater treatment process. Except for chemical additions, the residues accumulated in wastewater treatment processes largely consist of organics and minerals ultimately derived from soil. Industrial processes and home use of synthetic agents may introduce metals above ambient levels and toxic compounds into sludge. Pathogens or

TABLE 20.11 Specific Resistances of Sludges[a]

Sludge type	Specific resistance m/kg
Raw	$10\text{–}30 \times 10^{13}$
Raw, coagulated	$3\text{–}10 \times 10^{11}$
Digested	$3\text{–}30 \times 10^{12}$
Digested, coagulated	$2\text{–}20 \times 10^{11}$
Activated	$4\text{–}12 \times 10^{13}$

[a]From USEPA (1987).

parasites occur in sludge as a result of sewage being a carriage vehicle for human excrement.

Per capita sludge production rates in Ontario are near 80 and 115 g/d for primary and secondary treatment, respectively, without sludge digestion (Schmidtke, 1981). Adding iron or aluminum salts for phosphorus removal will increase sludge production by about 40% for primary treatment or 25% for secondary treatment. Sludge digestion typically reduces the sludge mass by 40%. The final sludge concentration depends on the dewatering treatment applied.

Sludge may be applied in liquid, semisolid (wet cake), or dried form. Subsurface application devices may be used. Sludge transportation requirements and method of application will dictate the degree of dewatering. Liquid sludge application is preferred where feasible. Dewatering sludge removes soluble components with the liquid. About one third of the nitrogen in digested sludge is in the liquid fraction (Environment Canada, 1984); thickening and dewatering will reduce the soluble nitrogen level significantly.

Metals that are toxic and biocumulative and nutrients are constituents of interest in sludge. Composition of sludge varies widely depending on industrial activity and the treatment applied. Some typical values are given in Table 20.12. Environment Canada (1984) and USEPA (1983) contain more information on these and other constituents typically found in sludges. If the sludge contains high concentrations of toxic organics, it will be a hazardous waste subject to more restrictive disposal measures.

Sludge may be applied to agricultural lands, forests, other lands with vegetation (golf courses, turf farms, parks, and other recreational areas), or for land reclamation (surface mines or marginal lands). There are also dedicated disposal sites for sludge application. Controlling factors on the application rate are the utilization rate of nutrients by the crop or vegetation; potential of plants for uptake of toxic components, particularly metals, by the plants; accumulation of salts and metals in the soil; and aesthetics. Public health risk through direct exposure or consumption of edible crops and groundwater contamination are always considered in applying sludge to land.

Federal regulations in the United States for sludge disposal on land have been recently revised after extensive study, public consultation, and debate (Bastian et al., 1992). The regulations (Section 503 of the Clean Water Act) were promulgated in 1993 (Water Environment Federation Residuals Management Committee, 1993). Many factors influence contaminant movement through food chains. Plant uptake of toxic contaminants usually reaches a plateau with increasing sludge application rates (Chaney and Ryan, 1992). Loading rates specified in the regulations considered various food-chain pathways and the most exposed individual which could be a person, animal, or plant (Chaney and Ryan, 1992). Based on toxicity studies and bioaccumulation within individual species and through food chains the standards in Table 20.13 were determined. States or localities may adopt more severe regulations than the federal regulations.

For general land application, sludge quality and loading rates must meet the minimum quality limits given in Table 20.13. There are some restrictions on the use of minimum quality sludge in agricultural or other applications. Clean sludge can be distributed or sold to the public for virtually any use if pathogen and vector attraction limits are also attained.

Pathogen and vector attraction reduction regulations are specified in addition to metals. There are two classifications for pathogen reduction: Class A and Class B. There are a number of criteria or treatment alternatives specified in the regulations to meet either classification. Class A sludge has either fecal coliform densities under

TABLE 20.12 Typical Composition of Sludges

	Raw primary sludge[a]		All sludges[b]		Digested sludge[c]		Septage[d]	
Constituent	Range	Typical	Range	Typical	Range	Typical	Range	Typical
Total solids (%)	2.0–7.0	4.0			2.0–6.0	3.5	0.1–13	34
Volatile solids (% of TS)	60–80	65			35–65	51		63
pH	5–8	6			7.2–7.8	7.5		6.9
Alkalinity (mg/L as $CaCO_3$)	500–1 500	600			200–7 600	4 800	522–4 190	970
Dry weight basis								
Total N (g/kg)	15–40	25	<1–176	33	1.6–4.0[e]	2.7[e]		0.69[e]
Al (g/kg)			1–135	4	4.1–61	9.6		
As (mg/kg)			1.1–230	10				
Ca (g/kg)			1–250	39	44	26–67		
Cd (mg/kg)			3–3 410	16	5–260	10		
Cl (g/kg)					1.7–190	7.1		
Co (mg/kg)			1–18	4.0	1–42	9.0		
Cr (mg/kg)			10–99 000	500	200–1 280	375		
Cu (mg/kg)			84–10 400	850	280–2 570	970		
Fe (g/kg)	20–40	25	<1–153	11	14–110	51		
Hg (μg/kg)			0.2–10 600	5	0.43–4.7	2.1		
K (g/kg)	0–8.3	4	0.2–26.4	3	0.04–0.16	0.09		
Mg (g/kg)			0.3–<19.7	4.5	3.1–11	6.8		
Mn (mg/kg)			18–7 100	260	170–2 090	320		
Mo (mg/kg)			5–39	30	7.0–97	12		
Na (g/kg)			0.1–30.7	2.4	0.07–0.42	0.16		
Ni (mg/kg)			2–3 520	82	23–410	120		
P (g/kg)	3.5–12.2	7	<1–143	23	14–57	24	20–760[f]	210[f]
Pb (mg/kg)			13–19 700	500	200–1 280	375		
Sn (mg/kg)			2.6–329	14				
Zn (mg/kg)			101–27 800	500	400–5 130	1 600		

[a]From Metcalf and Eddy (1972).
[b]From Chaney (1983) and Sommers (1977); includes data on aerobically and anaerobically digested sludge, lagooned, primary, tertiary, and unspecified sludges.
[c]MOE (1977).
[d]Adapted from USEPA (1984) for United States data.
[e]NH_3 + Kjeldahl N in g/L.
[f]In mg/L.

1 000 MPN/g dry solids or *Salmonella* sp. densities under 3 MPN/g dry solids. Treatment alternatives that may be used in lieu of meeting the microorganism density criteria consist of various heat and pH treatments. The pH must be raised above 12 when the latter is used as a treatment. There are also virus reduction criteria.

For Class B sludge, the fecal coliform density requirement is relaxed to under 2×10^6 MPN/g dry solids. There are many treatment processes that can be used to meet Class B status, including aerobic and anaerobic digestion. Minimum treatment performance or other process operating constraints are specified for each treatment. Harvesting times for food crops are also specified.

Vector attraction reduction refers to reducing the potential for the application site to attract agents of disease such as flies, rodents, and birds. There are eight methods that may be used to meet vector reduction requirements for sludge that will be sold or distributed for domestic use. In addition to these methods there are two other techniques that may be used for general application of sludge to agricultural and forest lands or other sites. The treatment performance requirements or operating constraints are specified to achieve a significant reduction in septicity of the sludge. For instance,

TABLE 20.13 Land Application Standards and Guidelines

	United States[a]				Canada[b,c]	
	Minimum quality sludge		Clean sludge			
Pollutant	Ceiling concentration limit mg/kg	Cumulative pollutant loading[d] kg/ha	Pollutant concentration limit mg/kg	Annual pollutant loading rate kg/ha/yr	Maximum acceptable concentration mg/kg	Cumulative acceptable loading kg/ha
Arsenic	75	41	41	2.0	75	15
Cadmium	85	39	39	1.9	20	4
Chromium	3 000	3 000	1 200	150		
Cobalt					150	30
Copper	4 300	1 500	1 500	75		
Lead	840	300	300	15	500	100
Mercury	57	17	17	0.85	5	1
Molybdenum	75	18	18	0.90	20	4
Nickel	420	420	420	21	180	36
Selenium	100	100	36	5.0	14	2.8
Zinc	7 500	2 800	2 800	140	1 850	370

[a]Waste Environment Federation Residuals Management Committee (1993).
[b]Environment Canada (1984); see Bradley et al. (1992) for criteria for individual provinces.
[c]The limits are for agricultural applications.
[d]The time to achieve the cumulative loading criteria depends on the application rate and the net accumulation of the metals in the soil.

one acceptable measure is to reduce the mass of volatile solids in the sludge by a minimum of 38%. Any process can be used.

Canadian federal guidelines are also given in Table 20.13. Provinces are not required to adopt the federal guidelines. Ontario guidelines are generally near federal guidelines (MAF and MOE, 1992). At the metals concentrations in sludge specified in Table 20.13, the minimum time period to reach the cumulative pollutant loadings in soils will be 25 yr or more if the sludge is being applied to uncontaminated soils (MAF and MOE, 1992).

Land requirements are dependent on the application rate. Typical application rates for various land disposal options are given in Table 20.14. These data were generated under less severe regulations or guidelines than currently exist in North America.

Site selection and characterization are the initial steps in a design for land sludge application. Public participation is essential in the final selection. Background concentrations of metals in the soil will dictate the timespan over which sludge may be applied. Maximum possible application rates of the sludge are calculated by

$$A_T = \frac{K_c L}{C} \tag{20.42}$$

where

- A_T is the application rate of sludge in tonne/ha
- L is the metal limitation in kg/ha
- K_c is a conversion factor (1 000 for SI; 446 for U.S.)
- C is the concentration of the metal in the sludge in mg/kg (ppm)

Equation (20.42) is applied for each metal. The permissible application rate is dictated by the metal constituent that yields the lowest application rate.

TABLE 20.14 General Guidelines for Land Application of Sludge[a]

Disposal option	Time period of application	Application rates (dry weight basis),[b] tonne/ha/yr (ton/acre/yr)	
		Range	Typical
Agricultural	Annual	2–70 (1–30)	11 (5)
Forest	One time or at 3–5 yr intervals	10–220 (4–1 008)	44 (20)
Land reclamation	One time	7–450 (3–200)	112 (50)
Dedicated disposal site	Annual	220–900 (100–400)	340 (150)

[a]From USEPA (1983).
[b]Rates are only for the application area and do not include area for the buffer zone, sludge storage, or other project area requirements.

The fertilizer value of sludge is one of the benefits of land application. The nutrient requirements of the crop to be grown or vegetation present on the site must be considered. Optimum application times and nutrient needs differ for various crops, growing conditions, and frequency of harvest. Table 20.15 provides information on

TABLE 20.15 Nutrients Removed in Grain and Straw[a]

Crop	Yield tonne/ha (ton/acre)	Nutrients removed, kg/ha (lb/acre)			
		N	P	K	S
Alfalfa	9.0[b] (4.0)	278 (248)	19 (17)	160 (143)	22 (20)
Barley	3.2 (1.4)	101 (90)	12 (11)	52 (46)	9 (8)
Corn	13.4 (6.0)	156 (139)	30 (27)	118 (105)	19 (17)
Flax	1.1 (0.50)	61 (54)	8 (7)	41 (37)	3.4 (3)
Grass	6.7 (3.0)	96 (86)	17 (15)	139 (124)	9 (8)
Oats	3.0 (1.3)	112 (100)	15 (13)	72 (64)	15 (13)
Potatoes	34 (15)	252 (225)	29 (26)	303 (270)	24 (21)
Rapeseed	1.7 (0.8)	133 (119)	18 (16)	67 (60)	7.8 (7)
Sugar beets	34 (15)	179 (160)	17 (15)	120 (107)	21 (19)
Wheat	2.7 (1.2)	95 (85)	13 (12)	52 (47)	9 (8)

[a]From Alberta Farm Guide (1976).
[b]Legumes obtain much of their nitrogen from the atmosphere.

TABLE 20.16 Mineralization Rates for Organic Nitrogen in Sludges[a]

	Mineralization rate,[b] %		
Time after sludge application, yr	Raw sludge or septage	Digested anaerobic sludge	Composted
1	40	20	10
2	20	10	5
3	10	5	3
4	5	3	3

[a]From USEPA (1983) and Water Environment Federation Residuals Management Committee (1993).
[b]Mineralization rates are 3% for years 5 through 10.

nutrient uptake of some crops (also see Table 19.8). Farmers must be advised to adjust commercial fertilizer application rates in accordance with sludge application rates.

Nitrogen is the most abundant nutrient in sludge and high application rates can pose a nitrate contamination risk to groundwaters and accelerate fertilization of surface waters. An agronomic application rate of sludge is adjusted to the nutrient demands of the crop being grown and minimizes the amount of nitrogen percolating with runoff below the root zone of the plants and the amount of nitrogen in surface runoff. (In the United States clean sludge is not required by the 503 regulations to be applied at the agronomic rate, although local permitting authorities may impose this restriction.)

Ammonia nitrogen in anaerobically digested sewage sludge is subject to losses through volatilization. If the sludge is not immediately incorporated into the soil, ammonia losses through volatilization can reach 50% (MAF and MOE, 1992). This must be incorporated into the nitrogen balance for an agronomic application rate.

For nitrogen to be available to plants it must be in a mineralized form. The USEPA in the 503 regulation (Water Environment Federation Residuals Management Committee, 1993) adopted mineralization rates[1] that were used in an earlier manual (USEPA, 1983). Through mineralization, depending on the sludge, varying amounts of organic nitrogen in the sludge is assumed to be available in the year of application and the second through the fourth years after application. Thereafter, 3% of the original organic nitrogen becomes available in each successive year until the nitrogen is exhausted. The mineralization factors during the first 4 yr are given in Table 20.16.

An equation describing the available nitrogen in the first year after application is

$$N_{a1} = k_c[(NO_3^-) + k_v(NH_3) + f_m N_o] \tag{20.43}$$

where

N_{a1} is the available nitrogen in the first year after application in kg/tonne (lb/ton) dry solids

(NO_3^-) is the mass fraction (g of NO_3^--N/g of solids or lb of NO_3^--N/lb of solids) of nitrate-N in the sludge

k_c is a conversion factor (1 000 kg/tonne for SI; 2 000 lb/ton for U.S.)

k_v is a volatilization factor

0.50 for surface applied sludge

1.0 for subsurface applied liquid sludge (Reed et al., 1988)

(NH_3) is the mass fraction of ammonia-N in the sludge (g of NH_3-N/g of solids or lb of N/lb of solids)

[1]There are some discrepancies for values of the mineralization factors in Water Environment Federation Residuals Management Committee (1993) (Rule 503) and USEPA (1983), but they are minor.

f_m is the mineralized fraction of the organic nitrogen based on the organic N in the freshly applied sludge (see Table 20.16)
N_o is the mass fraction of organic nitrogen in the sludge at application (g of organic N/g of solids or lb of organic N/lb of solids)

In the years following the year of application the available nitrogen is

$$N_{an} = k_c f_{mn}(N_{n-1} - f_{mn-1}N_{n-1}) \tag{20.44}$$

where
N_{an} is the available nitrogen in year n after sludge application in kg/tonne (lb/ton) dry solids
f_{mn}, f_{mn-1} are the availability factors for years n and $n - 1$, respectively, given in Table 20.16

Equation (20.44) is applied recursively for each year.

The limiting application rate based on nitrogen requirements is calculated from

$$A_N = \frac{U_N}{N_{a1} + \Sigma N_{ai}} \tag{20.45}$$

where
A_N is nitrogen limiting application rate in tonne/ha/yr (ton/acre/yr)
U_N is the amount of nitrogen needed by the crop in kg/ha/yr (lb/acre/yr)
ΣN_{ai} is the sum of mineralized nitrogen available from applications in previous years in kg/tonne (lb/ton), $i = 2$ to n

The loading rates are determined based on the dry solids content of the sludge. Then, using the concentration of solids in the sludge, the application rate is converted to a volumetric basis.

In the United States, septage, which is the solids accumulated in septic tanks, is not regulated as sewage sludge. Sampling and testing septage, which is usually collected by small operations (one to three or four trucks), is impractical. These septage disposal firms are widely distributed throughout rural areas. The pollutant limiting septage application is nitrogen. The maximum application rate on a volumetric basis for septage based on data from nitrogen concentrations in septage and using the above principles is

$$A_{NV} = KU_N \tag{20.46}$$

where
A_{NV} is volumetric application rate of liquid sludge, m^3/ha/yr (Mgal/acre/yr)
$K = 3.2 \times 10^6$ (SI); $K = 385$ (U.S.)
U_N is in kg/ha/yr (lb/acre/yr)

There are other regulations on the land application of septage.

MAF and MOE (1992) suggest that 40% of the acid-soluble phosphorus in the sludge will be available to plants as fertilizer. Metcalf and Eddy (1991) suggest that 50% of the total phosphorus in the sludge is normally assumed to be available to plants. If phosphorus control is specified, an equation similar to Eq. (20.45) is used to find the maximum application rate.

$$A_P = \frac{U_P}{C_P P} \tag{20.47}$$

where
A_P is phosphorus limiting application rate in tonne/ha/yr (ton/acre/yr)

U_P is the amount of phosphorus needed by the crop in kg/ha/yr (lb/acre/yr)
C_P is the availability of phosphorus in the sludge in kg/kg (lb/lb)
P is the total phosphorus content of the sludge in kg/tonne (lb/ton) dry solids

The lowest application rate calculated for any constituent specified in the local regulations is the design application rate.

■ Example 20.4 Land Application of Sludge

Determine the allowable application rate based on the nitrogen constraint of raw sludge with a nitrogen content as follows: NO_3^--N, 10 g/kg; NH_3, 0; and N_o, 20 g/kg to agricultural land where the crops will take up 100 kg/ha of nitrogen during each year over a 3-yr period. The sludge will be applied below the ground surface. What is the volumetric application rate in the third year if the sludge concentration is 10%?

$$NO_3^-\text{-N} = 10 \text{ g}/1\,000 \text{ g} = 0.010 \qquad N_o = 20 \text{ g}/1\,000 \text{ g} = 0.020$$

The nitrogen available during the first year from application of sludge is

$$N_{a1} = k_c[(NO_3^-) + k_v(NH_3) + f_m N_o] = (1\,000)[0.010 + 0.40(0.020)] = 18 \text{ kg/tonne}$$
$$U_N = 100 \text{ kg/ha}$$

During the first year:

$$A_{N1} = \frac{U_N}{N_{a1} + \Sigma N_{ai}} = \frac{100 \text{ kg/ha}}{18 \text{ kg/tonne}} = 5.56 \text{ tonne/ha}$$

The organic nitrogen remaining after the first year is

$$N_{o1} = N_o - f_1 N_o = (1\,000)[0.020 - 0.40(0.020)] = 12 \text{ kg/tonne}$$

The nitrogen available during the second year from the first application is

$$N_{a2} = f_2 N_{o1} = 0.20(12 \text{ kg/tonne}) = 2.4 \text{ kg/tonne}$$

The allowable application rate during the second year is

$$A_{N2} = \frac{U_N}{N_{a1} + \Sigma N_{ai}} = \frac{100}{18 + 2.4} = 4.90 \text{ tonne/ha}$$

The nitrogen remaining from the first sludge application is now

$$N_{o2} = N_1 - f_2 N_1 = (1\,000)[0.012 - 0.20(0.012)] = 9.60 \text{ kg/tonne}$$

The nitrogen released from this sludge is

$$N_{a3} = f_3 N_{o2} = 0.10(9.60 \text{ kg/tonne}) = 0.96 \text{ kg/tonne}$$

The sludge was applied at a lower rate during the second year and the nitrogen release rate from it must be reduced by the ratio of the application rates to determine an effective release rate.

$$N_{a2} = (4.90/5.56)(2.4 \text{ kg/tonne}) = 2.12 \text{ kg/tonne}$$
$$A_{N3} = \frac{100}{18 + 2.12 + 0.96} = 4.74 \text{ tonne/ha}$$

The solids concentration in the sludge is 10%. From Fig. 20.2, 1 kg of solids will occupy 9.7 L. The volume of sludge applied during the third year is

$$A_{NV} = \left(\frac{4.74 \text{ tonne}}{\text{ha-yr}}\right)\left(\frac{1\,000 \text{ kg}}{\text{tonne}}\right)\left(\frac{9.7 \text{ L}}{\text{kg}}\right)\left(\frac{\text{m}^3}{1\,000 \text{ L}}\right) = 46.0 \text{ m}^3\text{/ha/yr}$$

QUESTIONS AND PROBLEMS

1. What are the s.g. and volume of 1 kg of a sludge that has a solids content of 3.5% and contains 70% VSS and 30% FSS? Assume that FSS have an s.g. of 1.75.
2. (a) If lime dosage at 150 kg/tonne (300 lb/ton) of dry solids produces a dewatered cake with a solids concentration of 20% for a sludge flow rate of 380 m³/d (0.100 Mgal/d) with a solids concentration of 2.5%, what is the volume of cake to be disposed on a daily basis? Assume that the lime produces 2.0 mg of precipitate per mg of lime added.
   (b) Answer the same question for a polymer dose of 1 kg/tonne (2.0 lb/ton) that produces a cake with a solids concentration of 25% and compare the result to (a). Assume that the polymer produces 1.50 mg of precipitate per mg of polymer added.
3. What are the s.g. and volume of 100 kg of a 40% solids biological sludge with a VSS : FSS ratio of 80 : 20? Assume that the s.g. of FSS is 2.5. How many truckloads per month will be required for a plant that generates 120 kg/d of this sludge dewatered to a solids content of 40% if the capacity of the truck is 4.0 m³?
4. List wastewater operations that can be used to increase the concentration of sludge.
5. (a) What are the hydraulic retention times required to reduce the biodegradable content of a sludge by 90% using aerobic sludge digestion when the k_e values are 0.15 and 0.30 d^{-1}? A continuous flow CM basin will be used.
   (b) What are the times required to achieve the same amount of reduction in a batch process if the batch k_e rate constants are one half of the values given in part (a)?
6. Use the data in Table 20.6 and assume that the degradable VSS is 55% of influent VSS. Ignore solubilization.
   (a) What is the ratio of effluent VSS to influent VSS in a continuous flow reactor (assume that semicontinuous flow and continuous flow are the same) operating at a sludge age of 20 d for temperatures of 10 and 30°C?
   (b) Answer the same question for a batch system at the same sludge age and temperatures.
7. In Fig. P1 a plot of VSS concentration remaining over time from a batch aerobic digestion laboratory study is shown. Assuming that the batch study kinetics are applicable to a continuous flow CM reactor, size a continuous flow CM reactor to remove 80% of the degradable VSS when the influent VSS is 10 000 mg/L and the flow rate is 500 m³/d (0.132 Mgal/d). Base your analysis on the overall observed "endogenous" decay.

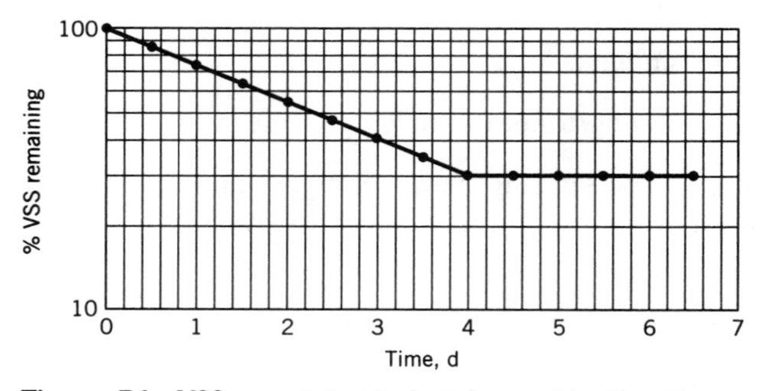

Figure P1 VSS remaining in batch aerobic digestion.

8. Using the P_T concept, what volume is required for an aerobic digester treating a waste solids flow rate of 750 m^3/d (0.198 Mgal/d) at a minimum temperature of 12°C?

9. What is the equivalent value of θ in Eq. (20.17) for the P_T concept? Assume that the P_T concept results in 90% removal.

10. In an aerobic sludge digestion process, assume that the rate of solubilization is first-order and that nondegradable and degradable VSS are equally subject to solubilization (i.e., the same rate constant applies to each component). If the degradable VSS is 68% of influent VSS and an analyst measured an influent and effluent VSS of 12 400 and 4 900 mg/L, respectively, and a soluble effluent COD of 720 mg/L at a detention time of 16 d in a CM process, what are the endogenous decay and solubilization constants? The soluble influent COD was 80 mg/L and the COD content of VSS is 1.42 mg/mg.

11. Using the data in Table 20.8, estimate the reduction of fecal coliforms and enteroviruses in an aerobic digester operated at a temperature of 25°C and detention time of 18 d.

12. Determine θ values according to Eq. (20.18) for total coliform and fecal streptococci dieoff coefficients in aerobic digestion using the data in Table 20.8.

13. If a sludge concentrates from an initial concentration of 6% to a concentration of 18% in a cylinder in 15 h, what is the value of i in Eq. (20.19)? The ultimate compaction obtainable was 20%. The initial height of the sludge in the cylinder used was 45 cm.

14. What is the operating pressure in an air dissolution tank to achieve an air : solids ratio of 0.03 for the following conditions? The influent waste activated sludge flow rate is 600 m^3/d (0.219 Mgal/d) containing solids at 4 500 mg/L. The recycle flow will be 70% of the influent flow. Assume f is 0.65 and that the influent to the process and effluent from the flotation tank is saturated with air. The temperature is 10°C and the ambient pressure is 1 atm.

15. Would recycle of centrate water directly to the influent of a centrifuge be beneficial (Hint: a mass balance could be useful)? Perform calculations for a centrifuge receiving a sludge with a concentration of 6% by weight with a capture of 90% of the solids. The cake from the centrifuge contains 21% solids. Find the volumetric flow rates of the cake and centrate in terms of the influent flow rate without recycle. Polymer is added to the influent at 5 kg/tonne of dry solids. The polymer does not hydrate or react with any dissolved solids in the influent.

16. The following data were obtained at 20°C from a setup as shown in Fig. 20.11 for anaerobically digested sludge thickened to a concenration of 8%. Conditioning agents had been added to the thickened sludge. The VSS : FSS ratio was 3 : 1; assume that the s.g. of FSS is 2.5. The vacuum applied was 5.0 m of H_2O and the diameter of the Büchner funnel was 5.00 cm. Find the specific resistance of the sludge.

Time, min	1	2	3	4	5	6	7
Volume of filtrate, mL	1.5	2.8	3.8	5.0	6.2	6.8	7.7

17. (a) Derive the solution (or nonsolution) of Eq. (20.28) using the initial conditions of $r = 0$ and $dr/dt = 0$. Assume laminar flow conditions.

(b) Derive Eq. (20.29) using the initial condition $r = \delta$, $dr/dt = 0$. Assume laminar flow conditions.

18. If sludge contains mercury at the minimum quality ceiling (U.S.) concentration limit in Table 20.12 and it is applied at a rate of 70 tonne/ha/yr to agricultural land, how long will it take to reach the cumulative pollutant loading for minimum quality sludge? Assume that all mercury deposited on the soil remains at the site.

19. A digested sludge contains metals at the concentrations given in the following table. What is the allowable cumulative application of this sludge in Canada (Table 20.13) and the annual application rate if the sludge is to be applied continuously over 25 years?

Metal	As	Cd	Co	Pb	Hg	Mo	Ni	Se	Zn
Sludge content, mg/kg	5	10	30	150	1.8	18	42	6	880

20. (a) An anaerobically digested sludge is to be applied to agricultural land. The sludge contains a total nitrogen content of 40 g/kg (80 lb/ton); its ammonia nitrogen content is 15 g/kg (30 lb/ton). If surface application is used to apply sludge to a site over a 5 yr period of time, what is the loading rate that may be applied to the site in each of the years if crops with nitrogen demands at the level of oats are being grown?

(b) What is the constant annual application rate that would satisfy the nitrogen limitation if the nitrogen demand of the crop remained at the level of oats and the sludge composition is as given in (a)? Assume all of the organic N is eventually released.

(c) If the sludge concentration is 8% and the rate of sludge generation is 75 m^3/d (19 810 gal/d) what is the land area required to dispose of the sludge using the application rate found in (b)?

KEY REFERENCES

REED, S. C., E. J. MIDDLEBROOKS, AND R. W. CRITES (1988), *Natural Systems for Waste Management and Treatment,* McGraw-Hill, Toronto.

USEPA (1987), *Dewatering Municipal Wastewater Sludges,* Design Manual No. EPA/625/1–87/014, Center for Environmental Research Information, Cincinnati, OH.

VESILIND, P. A. (1979), *Treatment and Disposal of Wastewater Sludges,* Ann Arbor Science Publishers, Ann Arbor, MI.

REFERENCES

ADAMS, C. E., W. W. ECKENFELDER, JR., AND R. M. STEIN (1974), "Modifications to Aerobic Digester Design," *Water Research,* 8, 4, pp. 213–218.

Alberta Farm Guide (1976), Alberta Dept. of Agriculture, Edmonton, AL.

ANDERSON, B. C. AND D. S. MAVINIC (1987), "Improvement in Aerobic Digestion through pH Control: Initial Assessment of Pilot-Scale Studies," *Canadian J. of Civil Engineering,* 14, 4, pp. 477–484.

BASTIAN, R., J. B. FARRELL, T. C. GRANATO, C. LUE-HING, R. I. PIETZ, K. C. RAO, AND R. M. SOUTHWORTH (1992), "Regulatory Issues," in *Municipal Sewage Management: Processing, Utilization and Disposal,* C. Lue-Hing, D. R. Zenz and R. Kuchenrither, eds., vol. 4 in Water Quality Management Library, Technomic, Lancaster, PA, pp. 3–68.

BHARGAVA, D. S. AND M. T. DATAR (1988), "Progress and Kinetics of Aerobic Digestion of Secondary Sludges," *Water Research,* 22, 1, pp. 37–47.

BRADLEY, J. W., S. KYOSAI, P. MATTHEWS, K. SATO, AND M. WEBBER (1992), "Worldwide Sludge Management Practices," in *Municipal Sewage Management: Processing, Utilization and Disposal,* C. Lue-Hing, D. R. Zenz and R. Kuchenrither, eds., vol. 4 in Water Quality Management Library, Technomic, Lancaster, PA, pp. 537–657.

CHANEY, R. L. (1983), "Potential Effects of Waste Constituents on the Food Chain," in *Land Treatment of Hazardous Wastes,* J. F. Parr, P. B. Marsh, and J. M. Kla, eds., Noyes Data Corp., Park Ridge, NJ, pp. 52–140.

CHANEY, R. L. AND J. A. RYAN (1992), "Regulating Residuals Management Practice," *Water Environment & Technology,* 4, 4, pp. 36–41.

COACKLEY, P. AND B. R. S. JONES (1956), "Vacuum Sludge Filtration," *Sewage and Industrial Wastes,* 28, 8, pp. 963–976.

DAMMEL, E. E. AND E. D. SCHROEDER (1991), "Density of Activated Sludge Solids," *Water Research,* 25, 7, pp. 841–846.

de Rijk, S. E., J. H. J. M. van der Graaf, and J. G. den Blanken (1994), "Bubble Size in Flotation Thickening," *Water Research,* 28, 2, pp. 465–473.

Droste, R. L. and W. A. Sanchez (1986), "Modeling Active Mass in Aerobic Sludge Digestion," *Biotechnology and Bioengineering,* 28, 11, pp. 1699–1706.

Edzwald, J. K., J. P. Walsh, G. S. Kaminski, and H. J. Dunn (1992), "Flocculation and Air Requirements for Dissolved Air Flotation," *J. American Water Works Association,* 84, 3, pp. 92–100.

Environment Canada (1984), *Manual for Land Application of Treated Municipal Wastewater and Sludge,* Environmental Protection Service Manual No. EPS 6-EP-84-1, Ottawa.

Ganczarczyk, J., M. F. Hamoda, and H. L. Wong (1980), "Performance of Aerobic Digestion at Different Sludge Solid Levels and Operating Patterns," *Water Research,* 14, 11, pp. 627–633.

Goodman, B. L. and A. J. Englande, Jr. (1974), "A Unified Model of the Activated Sludge Process," *J. Water Pollution Control Federation,* 46, 2, pp. 312–332.

Kelly, H. G., H. Melcer, and D. S. Mavinic (1993), "Autothermal Thermophilic Aerobic Digestion of Municipal Sludges: A One-Year, Full-Scale Demonstration Project," *Water Environment Research*, 65, 7, pp. 849–861.

Knocke, W. R., C. M. Dishman, and G. F. Miller (1993), "Measurement of Chemical Sludge Floc Density and Implications Related to Sludge Dewatering," *Water Environment Research,* 65, 6, pp. 735–743.

Krishnamoorthy, R. and R. C. Loehr (1989), "Aerobic Sludge Stabilization—Factors Affecting Kinetics," *J. of Environmental Engineering, ASCE,* 115, EE2, pp. 283–301.

MAF and MOE (1992), *Guidelines for Sewage Sludge Utilization on Agricultural Lands,* Ontario Ministry of Agriculture and Food and Ontario Ministry of the Environment, Toronto.

Martel, C. J. (1989a), "Dewaterability of Freeze-Thaw Conditioned Sludges," *J. Water Pollution Control Federation,* 61, 2, pp. 237–241.

Martel, C. J. (1989b), "Development and Design of Sludge Freezing Beds," *J. of Environmental Engineering, ASCE,* 115, EE4, pp. 799–808.

Martin, J. H., H. E. Bostian, and G. Stern (1990), "Reductions of Enteric Microorganisms during Aerobic Sludge Digestion," *Water Research,* 24, 11, pp. 1377–1385.

Mavinic, D. S. and D. A. Koers (1977), "Aerobic Sludge Digestion at Cold Temperatures," *Canadian J. of Civil Engineering,* 4, 4, pp. 445–454.

Metcalf and Eddy (1972), *Wastewater Engineering: Collection, Treatment, Disposal,* McGraw-Hill, Toronto.

Metcalf and Eddy (1991), *Wastewater Engineering: Treatment, Disposal and Reuse,* 3rd ed., G. Tchobanoglous and F.L. Burton, eds., McGraw-Hill, Toronto.

Ministry of the Environment MOE (1977), *Plant Operating Summary, Water Pollution Control Projects,* Ontario, Toronto.

Novak, J. T., G. L. Goodman, A. Pariroo, and J. C. Huang (1988), "The Binding of Sludges during Filtration," *J. Water Pollution Control Federation,* 60, 2, pp. 206–214.

Parker, D. G., C. W. Randall, and P. H. King (1972), "Biological Conditioning for Improved Sludge Filterability," *J. Water Pollution Control Federation,* 44, 11, pp. 2066–2077.

Rasmussen, H., J. H. Bruus, K. Keiding, and P. H. Nielsen (1994), "Observations of Dewaterability and Physical, Chemical and Microbiological Changes in Anaerobically Stored Activated Sludge from a Nutrient Removal Plant," *Water Research,* 28, 2, pp. 417–425.

Reynolds, T. D. (1973), "Aerobic Digestion of Thickened Waste Activated Sludge," *Proc. of the 28th Industrial Waste Conference,* Purdue University, Lafayette, IN, pp. 12–37.

Robinson, J. and W. R. Knocke (1992), "Use of Dilatometric and Drying Techniques for Assessing Sludge Dewatering Characteristics," *Water Environment Research,* 64, 1, pp. 60–68.

Schmidtke, N. W. (1981), "Sludge Generation, Handling and Disposal at Phosphorus Control Facilities in Ontario," in *Characterization, Treatment and Use of Sewage Sludge,* Proceedings Second European Symposium, Commission of the European Communities, Vienna, pp. 190–225.

Sommers, L. E. (1977), "Chemical Composition of Sewage Sludges and Analysis of Their Potential as Fertilizers," *J. of Environmental Quality,* 6, 2, pp. 225–232.

USEPA (1979), *Sludge Treatment and Disposal,* Design Manual No. EPA 625/1–79-011, Center for Environmental Research Information, Cincinnati, OH.

USEPA (1983), *Land Application of Municipal Sludge,* Design Manual No. EPA 625/1–83-016, Center for Environmental Research Information, Cincinnati, OH.

USEPA (1984), *Septage Treatment and Disposal,* Design Manual No. EPA 625/1–84-009, Center for Environmental Research Information, Cincinnati, OH.

Vesilind, P. A. (1994), "The Role of Water in Sludge Dewatering," *Water Environment Research*, 66, 1, pp. 4–11.

Viessman, W., Jr. and M. J. Hammer (1985), *Water Supply and Pollution Control,* 4th ed., Harper & Row, San Francisco.

Water Environment Federation Residuals Management Committee (1993), "Biosolids and the 503 Standards, Final Rule and Phased-in Submission of Sewage Sludge Permit Application, Final Rule," WEF, Washington, DC.

WEF and ASCE (1992), *Design of Municipal Wastewater Treatment Plants,* vol. I, Water Environment Federation, Washington, DC.

CHAPTER 21

EFFLUENT DISPOSAL IN NATURAL WATERS

Effluents from sewage treatment plants may be disposed in streams, lakes, oceans, underground, or on land. The most common choice for disposing of the effluent is to return it to the source from which it has been withdrawn. The disposal point should be downstream of any location where water is to be withdrawn for human consumption.

21.1 POLLUTANTS IN NATURAL WATERS

Standards, characteristics of the receiving body of water, and the uses of water from the receiving body of water influence the degree of treatment required. The impact of a pollutant has a characteristic time and concomitant space scale (Fig. 21.1) depending on processes that transform the pollutant to its ultimate fate or transport the pollutant away from the point of entry to the receiving water.

The pristine state of natural waters is variable and difficult to assess because almost all natural waters are influenced to a significant extent by anthropogenic activity. A number of indices have been developed to gauge the degree of pollution in water bodies (Ott, 1978). These indices apply different weights to various substances found in natural waters. The water quality classification scheme given in Table 21.1 was chosen by Prati et al. (1971) after reviewing classification systems in countries around the world.

Fish Survival and Temperature

Warm effluents from power plants and other sources also modify ecosystems. Power plants discharge large quantities of water used for cooling. Table 21.2 lists temperatures required for fish survival. The USEPA (1976b) used the formula

$$\begin{matrix}\text{Maximum weekly}\\ \text{average}\\ \text{temperature}\end{matrix} = \begin{matrix}\text{optimum}\\ \text{temperature}\end{matrix} + \frac{1}{3}\left(\begin{matrix}\text{ultimate upper}\\ \text{incipient lethal}\\ \text{temperature}\end{matrix} - \begin{matrix}\text{optimum}\\ \text{temperature}\end{matrix}\right)$$

and data from another source to calculate the temperatures in Table 21.2. Note that the more desirable fish species generally need lower temperatures for spawning and growth.

Nutrient Loadings to Lakes

The concentration and loading of nutrients in a lake directly influence the trophic state (abundance and activity of organisms in the lake). As the overall nutrient level rises, a lake passes from oligotrophic (paucity of nutrients) to mesotrophic to eutrophic

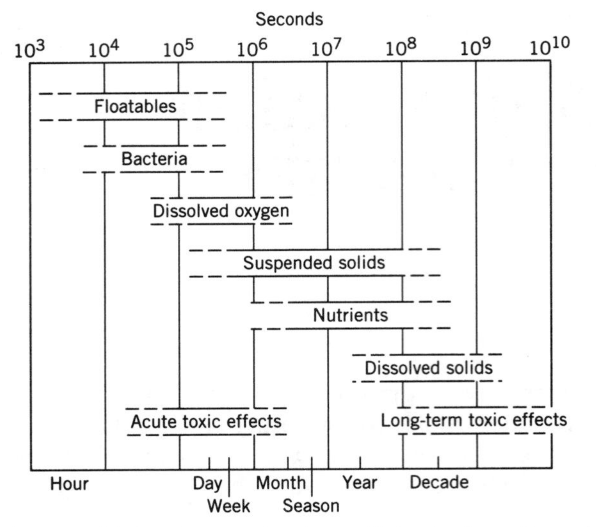

Figure 21.1 Time scales for pollutant effect. From USEPA (1976a).

(abundance of nutrients) conditions. The total biomass of virtually all groups of organisms increases along this gradient. However, the types and diversity of species change with the trophic condition. The diversity of species is highest in moderately oligotrophic systems and lower in very low or in higher primary production systems. As a lake becomes nutrient rich, less desirable types of life are present. Therefore, one indicator of the trophic level of a lake is the species types and diversity index.

The number of objectionable algae blooms increases as a lake becomes more eutrophic. Total algal biomass increases in proportion to nutrient levels, which affect water transparency as well as dissolved oxygen (DO) concentrations. Cyanobacteria (bluegreen algae) comprise an increasing proportion of plankton biomass at higher

TABLE 21.1 Surface Water Quality Classification[a]

	Condition of water				
Parameter	Excellent	Acceptable	Slightly polluted	Polluted	Grossly polluted
pH	6.5–8.0	6.0–8.4	5.0–9.0	3.9–10.1	<3.9–>10.1
Dissolved oxygen, %	88–112	75–125	50–150	20–200	<20–>200
BOD_5 mg/L	1.5	3.0	6.0	12.0	>12.0
COD, mg/L	10	20	40	80	>80
Suspended solids, mg/L	20	40	100	278	>278
Ammonia, mg/L	0.1	0.3	0.9	2.7	>2.7
Nitrate, mg/L	4	12	36	108	>108
Chloride, mg/L	50	150	300	620	>620
Iron, mg/L	0.1	0.3	0.9	2.7	>2.7
Manganese, mg/L	0.05	0.17	0.5	1.0	>1.0

[a]Reprinted from L. Prati, R. Pavanello, and J. T. Haney, "Assessment of Surface Water Quality by a Single Index of Pollution," *Water Research,* 5, pp. 741–751, Copyright 1971, with kind permission from Elsevier Science Ltd., The Boulevard, Langford Lane, Kidlington 0X5 1GB UK.

TABLE 21.2 Maximum Average Weekly Temperatures for Growth of Fish[a]

Temperature °C	Fish species tolerance
32	Bluegill, channel catfish, largemouth bass
30	Emerald shiner
29	Smallmouth bass, yellow perch
28	Northern pike, white crappie, white sucker
27	Black crappie
25	Sauger
20	Atlantic salmon
19	Brook trout, rainbow trout
18	Coho salmon, sockeye salmon
17	Lake herring (Cisco)

[a]From USEPA (1976b).

nutrient levels. Cyanobacteria form surface blooms, produce undesirable tastes and odors in the water, and can produce neurotoxins as noted in Section 6.2.1.

Acceptable levels of nutrient loadings causing fertilization in lakes have been arrived at from an extensive study of lakes in the northern climates throughout the world by Vollenweider (1970). Table 21.3 lists the permissible loadings for N and P, which are the maximum possible that will allow the lake to remain oligotrophic in the long run. Dangerous loadings are twice the permissible loadings and are sufficient to cause a lake to become eutrophic.

Discharge of nutrient laden waters to the ocean can also result in ecosystem upset. Bell (1992) reported that runoff and sewage effluents are implicated in eutrophication and damage to coral reefs off the coast of Australia. The concentrations of N and P associated with the onset of eutrophication in coral reef communities are less well defined than for freshwater systems but inorganic nitrogen and phosphorus levels of approximately 1 μM and 0.1–0.2 μM, respectively, were suggested. These levels are in accord with threshold levels for eutrophication in sensitive freshwater systems.

TABLE 21.3 Criteria for Allowable Loadings of N and P for Lakes[a]

	Permissible loadings			
	g/m^2 surface area/yr (g/ft^2 surface area/yr)		g/m^3/yr (g/ft^3/yr)	
Mean depth m (ft)	N	P	N	P
5 (16)	1.0 (0.09)	0.07 (0.006 5)	0.20 (0.005 7)	0.014 0 (0.000 40)
10 (33)	1.5 (0.014)	0.10 (0.009 3)	0.15 (0.004 3)	0.010 0 (0.000 28)
50 (164)	4.0 (0.37)	0.25 (0.023)	0.08 (0.002 3)	0.005 0 (0.000 14)
100 (328)	6.0 (0.56)	0.40 (0.037)	0.06 (0.001 7)	0.004 0 (0.000 11)
150 (492)	7.5 (0.70)	0.50 (0.046)	0.05 (0.001 4)	0.003 3 (0.000 094)
200 (656)	9.0 (0.84)	0.60 (0.056)	0.05 (0.001 4)	0.003 0 (0.000 085)

[a]From R. A. Vollenweider, *Scientific Fundamentals of the Eutrophication of Lakes and Flowing Waters with Particular Reference to Nitrogen and Phosphorus as Factors in Eutrophication,* copyright Organization for Economic Co-Operation and Development 1970.

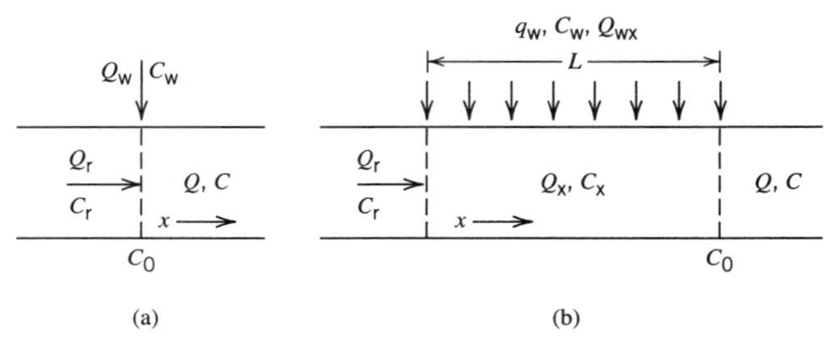

Figure 21.2 (a) Point and (b) distributed pollutant discharges.

21.2 LOADING EQUATIONS FOR STREAMS

Discharge of pollutants into a stream may originate from a point source or a distributed source. Point sources include sewer outfalls from municipal or industrial sewage treatment plants or other distinct locations where liquid wastes are discharged such as meltoff from a snow dump site. Distributed sources, such as runoff or irrigation, discharge over a wide area. The distinction between point and nonpoint (distributed) sources depends on the time and space scales of the pollutant. Whether the source is point or nonpoint becomes irrelevant for substances with long space-time scales. Definition sketches for the two types of sources are given in Fig. 21.2, (a) and (b).

In Fig. 21.2 the subscript r refers to the stream condition immediately above the point where the pollution source enters the stream; subscript w refers to the waste source. Subscript 0 will be used in the equations below to refer to the location immediately after the waste discharge mixes with the stream. It is assumed that a point source discharge mixes immediately in the cross-section with the receiving stream water. For a distributed source, the discharge occurs over a distance L and the concentration and stream flow rate vary in this reach primarily as a result of the discharge of the waste. A uniform discharge per unit length of stream, q_w, is assumed and the total volumetric flow from the distributed source is Q_w. The relations among discharges are

$$q_w = \frac{Q_w}{L} \quad (21.1a) \qquad q_w x = Q_{wx} \quad (21.1b)$$

where

x is distance downstream of the first point of entry of pollution source

Q_{wx} is cumulative flow discharged from the distributed source over the reach from 0 to x

21.2.1 Pollutant Decay in Streams

The concentration of a pollutant in the stream is influenced by dilution and various other physical, chemical, and biological processes. The setup of mass balances for this type of problem has been discussed in Chapter 10. The equations developed in the following sections are used to quantify allowable loadings that can be delivered to a stream.

Steady state conditions will be assumed to exist in the developments that follow. The flow regime in the stream is assumed to be plug flow (PF). Actual flow regimes in streams will exhibit some dispersion but over typical space–time scales for most substances, PF will be more realistic than complete mixed conditions. Tracer tests will provide the definitive definition of the flow regime. There can be instances where mixing in the longitudinal direction may be significant.

21.2.2 Conservative Substance

Conservative substances are not affected by any process other than dilution. Chlorides are generally a good example of a conservative substance. They are soluble and participate in insignificant amounts only in complexation reactions and are taken up in minimal amounts by microorganisms and higher life forms. However, if the concentration of chlorides is high in the waste stream, a density difference exists between the waste stream and the receiving stream which impairs the mixing and dispersion of the waste stream in the receiving body of water.

The equations for a conservative substance are simple.

Point Source

The flow of water immediately increases at the point of discharge.

$$Q = Q_r + Q_w \tag{21.2a}$$

The concentration in the receiving water at the discharge point is determined from a mixing equation that results from a mass balance at the point of confluence.

$$C_0 = C = \frac{C_r Q_r + C_w Q_w}{Q} \tag{21.2b}$$

Distributed Source

The corresponding equations for flow and concentration in the reach receiving the discharge and downstream of the discharge are as follows.

$0 \rightarrow L$:

$$Q_x = Q_r + q_w x = Q_r + Q_{wx} \tag{21.3a}$$

$$C_x = \frac{Q_r C_r + C_w q_w x}{Q_x} = \frac{Q_r C_r + C_w Q_{wx}}{Q_x} \tag{21.3b}$$

after L:

$$Q = Q_r + q_w L = Q_r + Q_w \tag{21.3c}$$

$$C_0 = \frac{Q_r C_r + C_w q_w L}{Q} = \frac{Q_r C_r + C_w Q_w}{Q} \tag{21.3d}$$

21.2.3 Substances That Are Transformed by One Reaction

Many substances are transformed by one or more natural phenomena. In some instances only one mechanism is dominant. In other cases many mechanisms are significant, but the overall process is too complex to separate it into its individual components; therefore, all phenomena are lumped into a single reaction model. The reaction may incorporate physical, chemical, and biological phenomena. Indicator bacteria are a good example. Temperature, sedimentation, sunlight, predation, and nutrient availability are some phenomena that influence the survival of bacteria, but generally indicator bacteria disappearance is modeled according to a first-order model, in which the rate constant is a function of temperature (Section 7.3.7).

All rate constants are normally assumed to follow the Arrhenius equation.

$$k_T = k_{20}\,\theta^{(T-20)} \tag{21.4}$$

where

k_{20} is the value at 20°C
k_T is the value at temperature T in °C
θ is a constant

Point Source

For a substance that decays according to first-order kinetics, the equations have been developed in Chapter 10 for a point source and PF situation. The reaction model is

$$r = -kC \tag{21.5a}$$

where
  r is the rate of reaction
  k is the rate constant

As developed in Section 10.1.2, the mass balance results in:

$$\frac{dC}{dx} = -\frac{k}{v}C \qquad \text{or} \qquad \frac{dC}{dt} = -kC \tag{21.5b}$$

where
  v is the average flow velocity in the stream

Time and distance are synonymous, although irregular stream geometries cause variable stream velocities. In reaches where the channel geometry is irregular, the distance can be divided into reaches where the velocity is approximately constant and the first form of Eq. (21.5b) can be applied over each reach. Otherwise, a distance–velocity relation must be determined before this equation can be integrated (see Section 10.1.2). Regardless of velocity variation, if the travel time between two points is known, the second form of Eq. (21.5b) can be used. The solution of the second form for Eq. (21.5b) is

$$C = C_0 e^{-kt} = \left(\frac{C_r Q_r + C_w Q_w}{Q}\right) e^{-kt} \tag{21.6}$$

The starting concentration is given by the mixing equation, Eq. (21.2b).

Distributed Source

In the reach receiving waste, concentration varies from dilution as well as according to the reaction model which will again be assumed to be a first-order decay.

Setting up a mass balance for an elemental volume (Fig. 21.3):

$$Q_x C_x + q_w C_w \Delta x - (Q_x + \Delta Q_x)(C_x + \Delta C_x) - kC_x \Delta V = \frac{\partial C_x}{\partial t} \Delta V = 0 \tag{21.7}$$

where
  ΔV is volume ($a\Delta x$) of the element

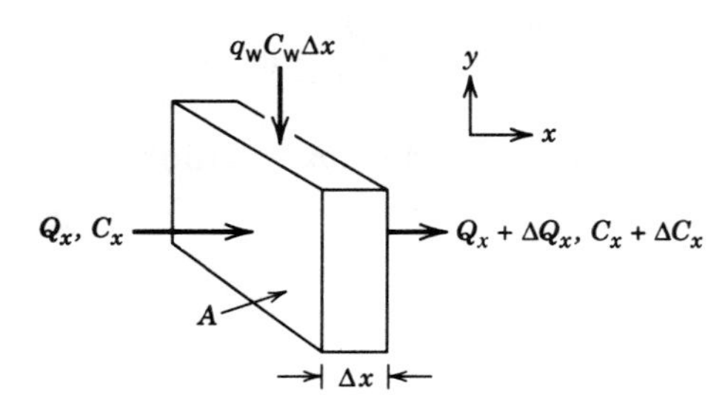

Figure 21.3 Elemental volume in the reach receiving distributed flow.

It is further assumed that the velocity of flow, v (Q_x/A), is constant in the reach. Expanding and simplifying the mass balance and dividing by A and Δx results in:

$$\frac{q_w}{A}C_w - \frac{Q_x}{A}\frac{dC_x}{dx} - \frac{C_x}{A}\frac{dQ_x}{dx} = kC_x$$

In addition to $v = Q_x/A$, it is noted that $dQ_x/dx = q_w$.

Using these relations the equation may be further simplified to

$$\frac{q_w}{A}C_w - v\frac{dC_x}{dx} - \frac{q_w C_x}{A} = kC_x$$

Finally, the equation may be divided by v and rearranged.

$$\frac{dC_x}{dx} = -\frac{k}{v}C_x - \frac{q_w}{Av}C_x + \frac{q_w}{Av}C_w \tag{21.8}$$

The differential equation (21.8) is readily integrated. Defining two new lumped constants, K_1 and K_2, as follows:

$$K_1 = -\frac{k}{v} - \frac{q_w}{Av} = \frac{1}{v}\left(-k - \frac{q_w}{A}\right) \qquad \text{and} \qquad K_2 = \frac{q_w}{Av}C_w$$

The differential equation is now:

$$\frac{dC_x}{dx} = K_1 C_x + K_2$$

$$\int_{C_r}^{C_x} \frac{dC_x}{K_1 C_x + K_2} = \int_0^x dx$$

The solution of the integral is

$$C_x = \left(C_r + \frac{K_2}{K_1}\right)e^{K_1 x} - \frac{K_2}{K_1}$$

$$= C_r e^{-\left(k+\frac{q_w}{A}\right)\frac{x}{v}} - \frac{q_w C_w}{A\left(k + \frac{q_w}{A}\right)}\left[e^{-\left(k+\frac{q_w}{A}\right)\frac{x}{v}} - 1\right]$$

Substituting t for x/v,

$$C_x = C_r e^{-\left(k+\frac{q_w}{A}\right)t} - \frac{q_w C_w}{A\left(k + \frac{q_w}{A}\right)}\left[e^{-\left(k+\frac{q_w}{A}\right)t} - 1\right] \tag{21.9}$$

By solving Eq. (21.9) with $x = L$ or $t = L/v$, the concentration, C_0 (in the stream immediately after the reach receiving the distributed flow), will be found. Equation (21.6) may then be used to find the concentration, C, at any distance downstream of the reach receiving the discharge.

■ Example 21.1 Indicator Bacteria Decay from Point and Distributed Sources

A stream is receiving waste containing indicator microorganisms from a point source and from a distributed source located 1.0 km (0.62 mi) downstream of the point source (refer to the accompanying sketch). The subscripts p and d refer to the point source and the distributed source, respectively. The distributed source discharges over a distance of 0.8 km (0.50 mi).

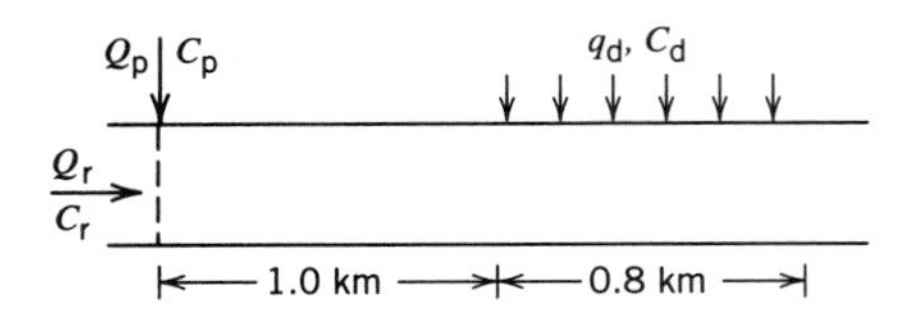

Other quantities indicated in the sketch have the following values:

$Q_r = 9\,500$ m³/d (3.88 ft³/s) $\qquad C_r = 22/100$ mL
$Q_p = 1\,050$ m³/d (0.429 ft³/s) $\qquad C_p = 1\,400/100$ mL
$q_d = 500$ m³/km/d (28 390 ft³/mi/d) $\qquad C_d = 2\,700/100$ mL

The velocity of the stream is constant at 3 km/d (1.86 mi/d). A first-order model describes microorganism decay and the rate constant (base 10) is 0.072 h^{-1}. Find the concentration of indicator microorganisms 2.0 km (1.24 mi) below the distributed source. Assume that the point and distributed source discharges mix well with the stream.

The stream concentration immediately after mixing with the point source discharge is calculated from Eq. (21.2b):

$$C_0 = C = \frac{C_r Q_r + C_p Q_p}{Q_r + Q_p}$$

$$= \frac{\left(9\,500\,\frac{\text{m}^3}{\text{d}}\right)(22/100\text{ mL}) + \left(1\,050\,\frac{\text{m}^3}{\text{d}}\right)(1\,400/100\text{ mL})}{9\,500\,\frac{\text{m}^3}{\text{d}} + 1\,050\,\frac{\text{m}^3}{\text{d}}}$$

$$= 159/100\text{ mL}$$

In U.S. units:

$$C_0 = C = \frac{\left(3.88\,\frac{\text{ft}^3}{\text{s}}\right)(22/100\text{ mL}) + \left(0.429\,\frac{\text{ft}^3}{\text{s}}\right)(1\,400/100\text{ mL})}{3.88\,\frac{\text{ft}^3}{\text{s}} + 0.429\,\frac{\text{ft}^3}{\text{s}}}$$

$$= 159/100\text{ mL}$$

The time to travel from the point source down to the beginning of the distributed source (separated by a distance, d_1) is

$t = d_1/v = (1.0\text{ km})/(3\text{ km/d}) = 0.333\text{ d} = 8.00\text{ h}$
In U.S. units: $t = (0.62\text{ mi})/(1.86\text{ mi/d}) = 0.333\text{ d}$

Using Eq. (21.6), the concentration of indicator microorganisms in the stream at the beginning of the distributed source is

$$C = C_0 10^{-kt} = \frac{159}{100\text{ mL}}\,10^{-(0.072\,\text{h}^{-1})(8.00\,\text{h})} = 42.2/100\text{ mL}$$

Equation (21.9) applies to the reach to which the distributed source discharges. The area of flow is required to solve the equation. The area is calculated based on the average flow in the reach. At the beginning and end of the reach the flows are Q_1 and Q_2, respectively.

$Q_1 = 9\,500\ \text{m}^3/\text{d} + 1\,050\ \text{m}^3/\text{d} = 10\,550\ \text{m}^3/\text{d}$

In U.S. units: $Q_1 = 3.88\ \text{ft}^3/\text{s} + 0.429\ \text{ft}^3/\text{s} = 4.31\ \text{ft}^3/\text{s}$

$Q_2 = Q_1 + q_d d_2 = 10\,550\ \text{m}^3/\text{d} + (500\ \text{m}^3/\text{km/d})(0.800\ \text{km}) = 10\,950\ \text{m}^3/\text{d}$

In U.S. units: $Q_2 = 4.31\ \text{ft}^3/\text{s} + (28\,390\ \text{ft}^3/\text{mi/d})(0.50\ \text{mi})(1\ \text{d}/86\,400\ \text{s}) = 4.47\ \text{ft}^3/\text{s}$

The average flow in the reach is

$$\frac{Q_1 + Q_2}{2} = \frac{10\,550 + 10\,950}{2} = 10\,750\ \text{m}^3/\text{d}\ (4.39\ \text{ft}^3/\text{s})$$

Now,

$$v = Q/A \text{ and } A = Q/v = (10\,750\ \text{m}^3/\text{d})/(3\,000\ \text{m/d}) = 3.58\ \text{m}^2$$

$$\text{In U.S. units: } A = \frac{\left(4.39\,\frac{\text{ft}^3}{\text{s}}\right)\left(\frac{86\,400\ \text{s}}{\text{d}}\right)}{\left(1.86\,\frac{\text{mi}}{\text{d}}\right)\left(\frac{5\,280\ \text{ft}}{\text{mi}}\right)} = 38.6\ \text{ft}^2$$

The term

$$\frac{q_d}{A} = \frac{500\ \text{m}^3/\text{d}}{(1\,000\ \text{m})(3.58\ \text{m}^2)} = 0.140\ \text{d}^{-1}$$

$$\text{In U.S. units: } \frac{q_d}{A} = \frac{28\,390\ \text{ft}^3/\text{d}}{(1\ \text{mi})(38.6\ \text{ft}^2)(5\,280\ \text{ft/mi})} = 0.140\ \text{d}^{-1}$$

Converting the dieoff rate coefficient to base e:

$$k_e = 2.3k_{10} = 2.3(0.072\ \text{h}^{-1}) = 0.166\ \text{h}^{-1} = 3.97\ \text{d}^{-1}$$

The time to travel 0.8 km is

$$t = x/v = (0.8\ \text{km})/(3\ \text{km/d}) = 0.267\ \text{d}$$

Substituting these quantities into Eq. (21.9), the concentration of indicator micro-organisms at the end of the reach is

$$C_2 = \frac{42.2}{100\ \text{mL}} e^{-(0.140+3.97)(0.267)} + \frac{(0.140\ \text{d}^{-1})(2.7 \times 10^3/100\ \text{mL})}{(0.140 + 3.97)\text{d}^{-1}}[1 - e^{-(0.140+3.97)(0.267)}]$$
$$= 75.2/100\ \text{mL}$$

Finally, applying Eq. (21.6) again for a distance of 2.0 km (1.24 mi) downstream of the distributed source:

$$C = C_0 10^{-kt} = \frac{75.2}{100\ \text{mL}} 10^{-(0.072\ \text{h}^{-1})\left(\frac{2.00\ \text{km}}{3\ \text{km/d}}\right)\left(24\frac{\text{h}}{\text{d}}\right)} = 5.3/100\ \text{mL}$$

$$\text{In U.S. units: } C = \frac{75.2}{100\ \text{mL}} 10^{-(0.072\ \text{h}^{-1})\left(\frac{1.24\ \text{mi}}{1.86\ \text{mi/d}}\right)\left(24\frac{\text{h}}{\text{d}}\right)} = 5.3/100\ \text{mL}$$

21.3 DISSOLVED OXYGEN VARIATION IN A STREAM

The DO concentration variation in a stream depends on many factors. The classical development of Streeter and Phelps (1925) considers the two main influences: oxygen decrease resulting from the exertion of BOD and oxygen replenishment by natural reaeration from the atmosphere. The approach is similar to the developments presented in Section 21.2 except that the reaction is a coupled phenomena. Flow in the stream is assumed to be in the PF regime and steady state conditions are also assumed to exist.

The consumption of BOD (which is conveniently expressed in terms of oxygen) is expressed by

$$r_L = -k_1 L \tag{21.10}$$

where

r_L is the rate of BOD exertion
L is BOD concentration
k_1 is the rate constant for BOD exertion

The saturation concentration of DO is governed by Henry's law. Natural turbulence in the stream enhances the mass transfer of oxygen between the stream and atmosphere. The change in DO caused by reaeration (Fig. 21.4) is expressed by

$$r_{aer} = k_2(C_s - C) \tag{21.11}$$

where

r_{aer} is the rate of reaeration
C is concentration of DO
C_s is saturation concentration of oxygen
k_2 is the rate constant for reaeration

Figure 21.4 Natural reaeration in a stream.

Similar to these developments, only one-dimensional flow and concentration variation are considered; i.e., the stream is well mixed in any cross-sectional plane. An elemental volume of the stream is shown in the accompanying sketch.

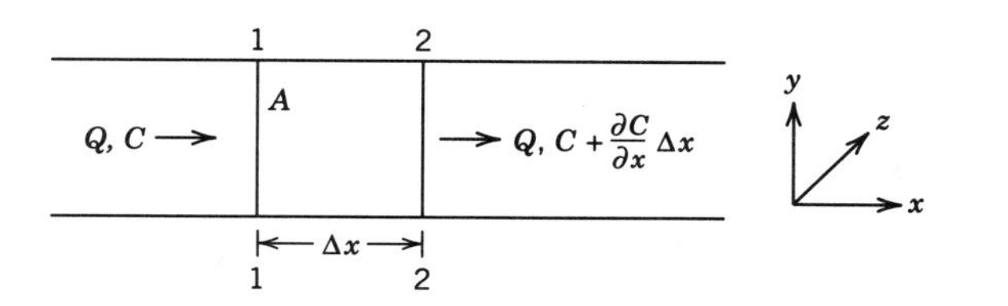

The equation for the mass balance is

$$QC - Q\left(C + \frac{\partial C}{\partial x}\Delta x\right) + r\,\Delta V = \frac{\partial C}{\partial t}\Delta V$$

where
ΔV is the volume ($A\Delta x$) of the element
r is the volumetric reaction rate

Assuming steady state ($\partial C/\partial t = 0$), simplifying the preceding equation, and substituting the relations for velocity and volume results in the usual expression with the application of the chain rule.

$$v\frac{\partial C}{\partial x} = \frac{dC}{dx}\frac{dx}{dt} = \frac{dC}{dt} = r \qquad (21.12)$$

The reaction rate is a coupled phenomena depending on BOD consumption and reaeration.

$$r = -k_1 L + k_2(C_s - C) \qquad (21.13)$$

Substituting this equation into Eq. (21.12), the resulting differential equation is

$$\frac{dC}{dt} = -k_1 L + k_2(C_s - C) \qquad (21.14)$$

Differential equations that include a constant coupled with the variable are usually simplified by substitution of a new variable as follows. Define the oxygen deficit, D, as

$$D = C_s - C \qquad (21.15)$$

and substitute it into Eq. (21.14) to yield,

$$\frac{dD}{dt} = k_1 L - k_2 D$$

Before attempting to solve this equation it must be recognized that BOD (L) is a function of time. The equation describing the relation of L and t is readily found by making a mass balance for BOD from which it will be determined that

$$\frac{dL}{dt} = -k_1 L \Rightarrow L = L_0 e^{-k_1 t}$$

where
L_0 is the initial ultimate BOD

Substituting this into the differential equation:

$$\frac{dD}{dt} = k_1 L_0 e^{-k_1 t} - k_2 D \tag{21.16}$$

Equation (21.16) must be integrated between the following limits:

$$t = 0, \quad D = D_0$$
$$t = t, \quad D = D$$

also, from Section 21.2.3,

$$t = 0, \quad L = L_0$$

D_0 and L_0 are the values after mixing of the waste flow with the stream and found by applying the mixing equation (Eq. 21.2b).

Rearranging Eq. (21.16) and applying an integrating factor:

$$e^{k_2 t}\left(\frac{dD}{dt} + k_2 D\right) = e^{k_2 t} k_1 L_0 e^{-k_1 t}$$

$$\int d(De^{k_2 t}) = \int k_1 L_0 e^{(k_2 - k_1)t}\, dt$$

The expression resulting from the integration and substitution of the initial condition is

$$D = \frac{k_1 L_0}{k_2 - k_1}(e^{-k_1 t} - e^{-k_2 t}) + D_0 e^{-k_2 t} \tag{21.17}$$

or, expressing the equation in base 10 (where the primed constants are referred to base 10):

$$D = \frac{k_1' L_0}{k_2' - k_1'}(10^{-k_1' t} - 10^{-k_2' t}) + D_0 10^{-k_2' t}$$

A plot of the deficit curve, known as a DO sag curve, is shown in Fig. 21.5. The resulting deficit curve is the summation of the two contributing phenomena.

The maximum deficit defines the critical point, which is defined by

$$\frac{dD}{dt} = 0 \qquad \frac{d^2 D}{dt^2} < 0$$

The values of time and deficit at the critical point are t_c and D_c, respectively. Equations (21.16) and (21.17) are used to solve for t_c and D_c. The equation for t_c is

$$t_c = \left(\frac{1}{k_2 - k_1}\right) \ln\left[\frac{k_2}{k_1}\left(1 - \frac{(k_2 - k_1)D_0}{k_1 L_0}\right)\right] \tag{21.18}$$

D_c is found by substituting t_c into Eq. (21.17):

$$D_c = \frac{k_1 L_0}{k_2 - k_1}(e^{-k_1 t_c} - e^{-k_2 t_c}) + D_0 e^{-k_2 t_c}$$

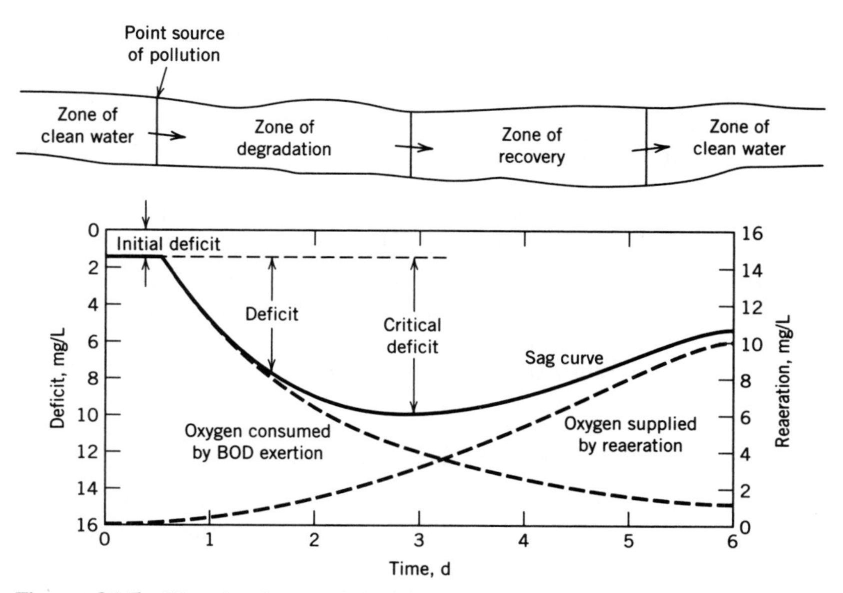

Figure 21.5 Dissolved oxygen sag curve.

or using Eq. (21.18),

$$D_c = \frac{k_1 L_0}{k_2 - k_1}\left\langle \exp\left\{-\frac{k_1}{k_2 - k_1}\ln\left[\frac{k_2}{k_1}\left(1 - \frac{(k_2 - k_1)D_0}{k_1 L_0}\right)\right]\right\} - \exp\left\{-\frac{k_2}{k_2 - k_1}\ln\left[\frac{k_2}{k_1}\left(1 - \frac{(k_2 - k_1)D_0}{k_1 L_0}\right)\right]\right\}\right\rangle + D_0 \exp\left\{-\frac{k_2}{k_2 - k_1}\ln\left[\frac{k_2}{k_1}\left(1 - \frac{(k_2 - k_1)D_0}{k_1 L_0}\right)\right]\right\} \quad (21.19)$$

21.3.1 Nitrification in Natural Waters

The oxidation of ammonia to nitrite and nitrate also exerts a significant oxygen demand (see Section 5.9.3). If sewage with a typical composition of $C_{10}H_{19}O_3N$ (Christensen and McCarty, 1975) is treated only for carbonaceous BOD removal, the potential ammonia oxygen demand is 0.32 mg O_2/mg $C_{10}H_{19}O_3N$ removed. Because some of the $C_{10}H_{19}O_3N$ removed will be incorporated into biomass in the biological treatment process, the effluent soluble ammonia-N will not be as high as the potential ammonia-N. In sewage treatment plants without nitrification, effluent ammonia-N concentrations can range up to 25 mg/L or more.

The overall ammonia conversion process (Eq. 5.32c) is normally modeled as a first-order reaction.

$$r_N = -k_N N \quad (21.20)$$

where

r_N is the rate of ammonia oxidation to nitrate
k_N is the rate coefficient
N is the concentration of ammonia

Incorporating this into a mass balance for oxygen taken over an elemental volume:

$$\frac{dC}{dt} = -k_1L - k_NN + k_2(C_s - C) \tag{21.21}$$

It is left as an exercise for the student to solve this equation for the DO deficit as a function of time or distance (see Problem 9).

Temperature and pH have a more pronounced effect on the rate of nitrification compared to other phenomena (Tetra Tech, 1978). The majority of reported values for k_N found by Tetra Tech (1978) were in the range of 0.1–0.54 d^{-1} at temperatures ranging from 20 to 28°C. QUAL2E is a river water quality modeling program widely used, for which the typical range for k_N is given as 0.1–1.0 d^{-1} (Brown and Barnwell, 1987); others specify a narrower range of 0.1–0.5 d^{-1} (Thomann and Mueller, 1987).

21.3.2 Factors Affecting the Dissolved Oxygen Sag Curve

There are many factors that influence the development and severity of the DO sag curve. Not all of them have been taken into account by the preceding development. However, those that have been ignored generally result in a conservative analysis. Camp and Meserve (1974) have discussed some of these additional phenomena and formulated equations describing their influence. Factors affecting the oxygen sag curve are as follows.

Initial Deficit. An increase or decrease in the initial deficit will lower or raise the sag curve, respectively.

Amount and Type of Load (Quality and Quantity). The concentration of organic matter in the waste stream and its volumetric flow rate influence the dilution and starting concentration in the stream. The amount of organic matter in the waste stream that is degradable and the ease with which it is degraded will also affect the severity of the critical deficit.

Stream Characteristics. The velocity of the stream creates turbulence that aids the transfer of oxygen. The depth of the stream is also an important factor influencing gas transfer. Environmental conditions in the stream (such as pH and nutrient availability) affect metabolic rates.

Fraction of Settleable Solids. Settleable organic matter will deposit on the bed of the stream and anaerobically decay there. This does not impose an oxygen burden on the stream and ignoring it lends a safety factor to the analysis.

Algae. Algae are another net source of oxygen over a 24-h cycle but their production of oxygen is not constant throughout the day. During daylight they produce oxygen. During darkness, respiration of algae consumes oxygen.

Temperature. Temperature influences all rate coefficients in the formulations. An increase of temperature produces the worst case scenario. The saturation concentration of oxygen decreases, thus decreasing the oxygen deficit in the reaeration equation. The rate constants for BOD exertion and reaeration both increase; however, the increase is usually greater for the BOD constant. Also, fish and other life forms tend to be more severely stressed at higher temperatures.

Ice Cover. Ice cover presents a barrier to oxygen transfer from the atmosphere but it occurs at low temperatures when BOD exertion and fish oxygen requirements are minimal.

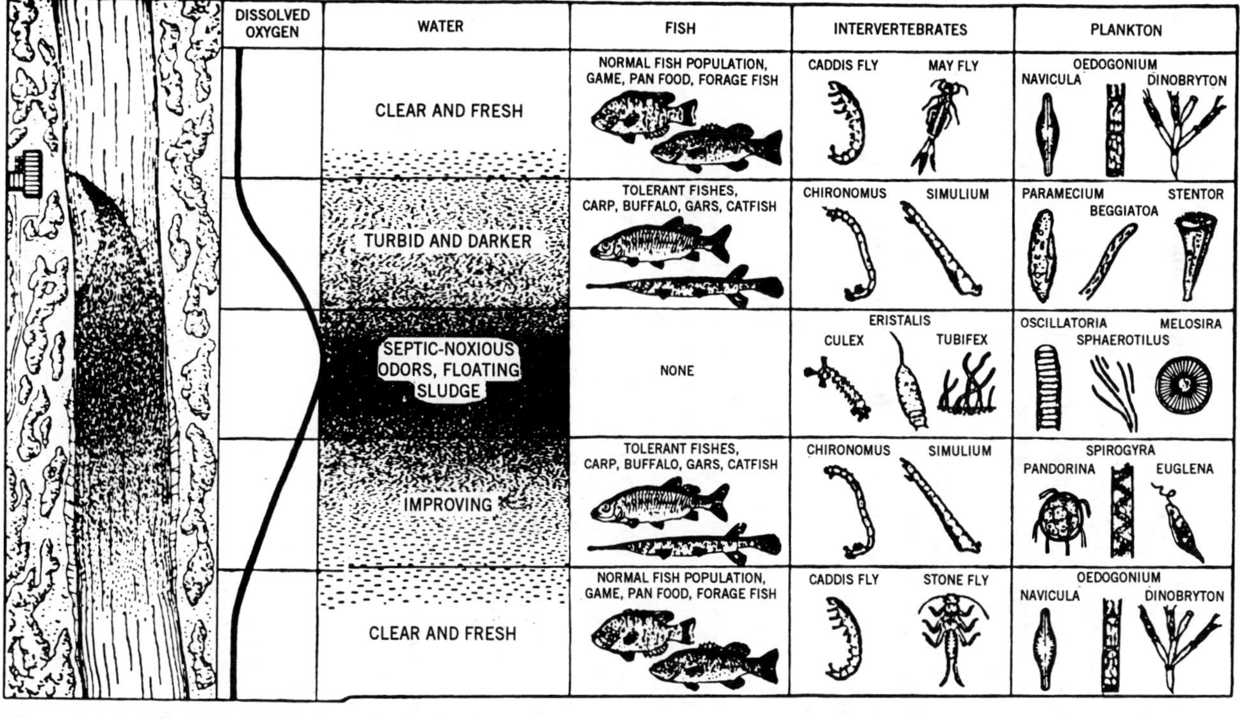

Figure 21.6 Effects of DO variation on the biotic community in a stream. From "Stream Pollution," R. Eliassen. Copyright © 1952 by Scientific American, Inc. All rights reserved.

Chlorination of Effluent. The application of chlorine or other disinfectants that maintain a residual in the effluent can delay the onset of rapid BOD exertion in the stream. Chlorine is also toxic to aquatic species.

Nitrification. Nitrification also imposes a significant oxygen demand load on the stream.

The critical deficit is assessed under extreme conditions. The volumetric flow rate in the stream is normally taken as one of the following:

1. The mean of the driest 7-d period of record
2. The flow that is exceeded 90–98% of the time.
3. The minimum average 7 consecutive day flow once in 10 or 20 years (known as 7Q10 or 7Q20 flows, respectively).

The temperature should be taken as the highest temperature likely to occur during the critical period when flows are low or when the discharge load is the highest.

DO variation has an encompassing effect on both the flora and fauna in a stream as illustrated in Fig. 21.6.

■ Example 21.2 Application of the Streeter–Phelps Equation

A community has a waste stabilization pond treatment system. The effluent is discharged annually during the spring freshet when flows in the receiving stream are highest. The volume of water to be discharged is 400 000 m^3. Analysis of streamflow data has produced the low flow hydrograph that is expected to occur once every 20 yr. From this hydrograph the minimum flows that will be sustained in the stream are 3.10, 1.65, and 0.98 m^3/d over 19, 28, and 40 d, respectively.

The simplest operating strategy is to discharge the ponds at a constant rate. The problem is to determine which discharge period will result in the highest DO being maintained in the stream. Also, find the critical DO and the time at which it occurs.

Pond effluent is assumed to have a DO of 4.00 mg/L and a BOD_5 of 30 mg/L. Background stream DO and BOD_5 concentrations have been taken at the 75th percentile values, which are 8.00 and 7.90 mg/L, respectively. The 75th percentile temperature is 12°C for this river.

The rate constants (base e) k_1 and k_2 have values of 0.35 and 1.10 d^{-1}, respectively, at 20°C. Use θ values of 1.047 and 1.024 d^{-1} for k_1 and k_2, respectively.

The saturation DO concentration is 10.8 mg/L (see Table A.2 in the Appendix). The discharge rates (Q_w) over each time period (t_p) are 400 000 m^3/t_p and corresponding initial DO deficit (D_0) and BOD_u (L_0) after mixing the waste and stream flows (Eq. 21.2b) are given in the following table.

t_p d	Q_w m^3/s	Q_r m^3/s	C_0 mg/L	D_0 mg/L	L_0 mg/L
19	0.243	3.10	7.71	3.09	13.54
28	0.165	1.65	7.64	3.16	14.11
40	0.116	0.98	7.54	3.27	14.91

The initial deficits are also given by $D_0 = C_s - C_0$.

Initial BOD_5 values ($x_{5,0}$) after mixing were adjusted to initial ultimate BOD values by

$$L_0 = \frac{x_{5,0}}{1 - e^{-5k_1}} = \frac{x_{5,0}}{1 - e^{-(0.24\,d^{-1})(5\,d)}} = 1.43x_{5,0}$$

The rate constants must be adjusted to the stream temperature according to Eq. (21.4).

$$k_{1,12} = (0.35\ d^{-1})1.047^{(12-20)} = 0.24\ d^{-1} \qquad k_{2,12} = (1.10\ d^{-1})1.024^{(12-20)} = 0.91\ d^{-1}$$

Equation (21.18) was solved for the critical time. For the first scenario:

$$t_c = \left(\frac{1}{k_2 - k_1}\right) \ln\left[\frac{k_2}{k_1}\left(1 - \frac{(k_2 - k_1)D_0}{k_1 L_0}\right)\right]$$

$$= \left[\frac{1}{(0.91 - 0.24)d^{-1}}\right] \ln\left[\left(\frac{0.91\ d^{-1}}{0.24\ d^{-1}}\right)\left(1 - \frac{(0.91\ d^{-1} - 0.24\ d^{-1})(3.09\ mg/L)}{(0.24\ d^{-1})(13.54\ mg/L)}\right)\right]$$

$$= 0.48\ d$$

The critical deficit is

$$D_c = \frac{k_1 L_0}{k_2 - k_1}(e^{-k_1 t_c} - e^{-k_2 t_c}) + D_0 e^{-k_2 t_c}$$

$$= \left[\frac{(0.24\ d^{-1})(13.54\ mg/L)}{(0.91 - 0.24)d^{-1}}\right][e^{-(0.24\,d^{-1})(0.48\,d)} - e^{-(0.91\,d^{-1})(0.48\,d)}] + (3.09\ mg/L)e^{-(0.91\,d^{-1})(0.48\,d)}$$

$$= 3.20\ mg/L$$

The corresponding critical DO concentration is

$$C_c = C_s - D_c = 10.80\ mg/L - 3.20\ mg/L = 7.60\ mg/L$$

Proceeding similarly for the other discharge scenarios, the critical times, deficits, and DO concentrations are

t_p, d	t_c, d	D_c, mg/L	C_c, mg/L
28	0.54	3.30	7.50
40	0.60	3.44	7.36

The best discharge scenario would be to discharge over the shortest period when the stream flows are the highest.

The critical deficit varies among regulating agencies. The critical deficit is often specified as 40% of the saturation DO concentration. The Ontario Ministry of the Environment uses the DO criteria given in Table 21.4, which take into account the fish community and the additional stress on aquatic species at higher temperatures.

21.3.3 The Reaeration Rate Coefficient

The reaeration coefficient is a function of temperature and follows the normal Arrhenius law (Eq. 21.4). The value of θ for k_2 in Eq. (21.4) is often reported as 1.024 but it should be checked at stream conditions.

There are many equations used to estimate k_2. Two of the most commonly used

TABLE 21.4 Minimum DO Guidelines for Ontario Surface Waters[a]

Temperature °C	Minimum DO concentration			
	Cold water biota[b]		Warm water biota[b]	
	% Saturation	mg/L	% Saturation	mg/L
0	54	8	47	7
5	54	7	47	6
10	54	6	47	5
15	54	6	47	5
20	57	6	47	4
25	63	5	48	4

[a]From MOE (1984).
[b]A cold water biota is defined as fresh water with mixed fish population including some salmonids. A warm water biota is fresh water containing a mixed fish population and no salmonids.

formulations were reported by Churchill et al. (1962) (Eq. 21.22a) and O'Connor and Dobbins (1958) (Eq. 21.22b).

$$k_2 = \frac{5.03v^{0.969}}{H^{1.673}} \qquad (21.22a) \qquad\qquad k_2 = \frac{(D_{O2}v)^{1/2}}{H^{3/2}} \qquad (21.22b)$$

where
- v is the stream velocity in m/s
- H is the steam depth in m
- D_{O2} is the diffusivity of oxygen, 2.05×10^{-9} m²/s (2.21×10^{-8} ft²/s) at 20°C
- k_2 (base e) has units of d^{-1} in Eq. (21.22a) when v is in m/s and H is in m
- k_2 (base e) is consistent with the dimensions of v and H in Eq. (21.22b)

(See Problem 13 for conversion of the Churchill et al. equation to U.S. units.)

Owens et al. (1964) studied reaeration in shallow streams and determined the correlation given in Eq. (21.22c).

$$k_2 = \frac{k_c v^{0.67}}{H^{1.85}} \qquad (21.22c)$$

where
- k_c = 5.34 (SI); 21.7 (U.S.)
- k_2 (base e) is in d^{-1} for v in m/s and H in m (SI); k_2 (base e) in d^{-1}, v in ft/s and H in ft (U.S.)

Covar (1976) examined the field data collected by the preceding three groups and plotted the formulas they developed. He found that the dividing line between the data of O'Connor and Dobbins (1958) and Churchill et al. (1962) was also the line where the equations developed by these two groups gave identical results. Covar also arbitrarily chose a line at a depth of 0.61 m (2 ft) that divided most of the data from Owens et al. (1964) from the data from the other two groups. The equation of Owens et al. gave results that were very close to results obtained by the other groups at the 0.61 m (2 ft) dividing line. The plot of the equations and dividing lines is shown in

Fig. 21.7. The equations are consistent and the work of Covar (1976) unifies field measurements and theoretical developments made by these groups.

Figure 21.7 shows that the reaeration coefficient is highest for shallow, fast moving streams, as one would expect. As the depth of the stream increases or stream velocity decreases, the reaeration coefficient decreases.

Tetra Tech (1978) has reviewed other formulations for the reaeration coefficient and tabulated results determined for various streams. A wide variation in reaeration constants was found for natural streams. It is always best to make in situ measurements for the reaeration coefficient.

21.3.4 Reaeration at Dams

Waterfalls, rapids, and dams exhibit different aeration capabilities from normal channels. Empirical formulations are used to describe reaeration. For navigable dams on large rivers with flows from 12 m^3/s (425 ft^3/s) up to 4 000 m^3/s (141 300 ft^3/s), a linear equation (Eq. 21.23) was found to provide better results than earlier formulations (Railsback et al., 1990). Some of the dams were also used for hydropower. Headloss through the dams ranged from 3.0 to 8.5 m (10–28 ft).

$$D_b = \beta D_a - \alpha \tag{21.23}$$

where

D_a and D_b are the DO deficits above and below the dam
β and α are constants

The values of β and α were in the ranges of 0.12–1.07 and −0.40–0.92, respectively. Including any of the variables of flow rate, headloss, and temperature was not generally found to improve the predictions of the model. The structural type of the dam, whether it was fixed crest, gated, or gated with submerged discharge, did not provide consistent results. Field studies are required to determine the best values of the constants for each situation and the correlation may not be applicable for dams with a headloss less than 3.0 m (10 ft).

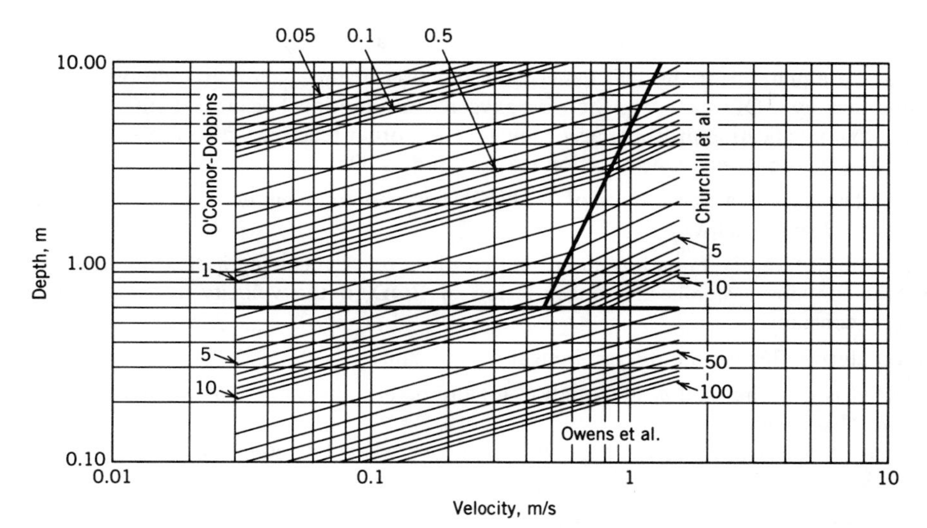

Figure 21.7 Reaeration coefficient velocity–depth relations. After Covar (1976).

QUESTIONS AND PROBLEMS

1. (a) Discuss the factors that distinguish the time scales for the various types of pollutants in Fig. 21.1. (b) Would it make any difference in the response of a body of water to a decaying pollutant if the pollutant source was point or nonpoint and the body of water was (i) complete mixed or (ii) plug flow?
2. Are suspended solids a conservative pollutant? Is DO a conservative substance?
3. Devise a proof that shows that the temperature of a stream that mixes with effluent from a waste source follows the mixing equation if the waters are well mixed at the point of discharge. Is in-stream temperature a conservative pollutant?
4. A point source is discharging waste at 10^3 m^3/d (0.364 Mgal/d) containing 10^4 indicator microorganisms/100 mL into a stream with a flow rate of 10^6 m^3/d (409 ft^3/s) and microorganism count of 20/100 mL immediately above the point source. The stream velocity is 23 km/d (0.87 ft/s). At a location 5 km (3.1 mi) downstream of the point source, a distributed source is discharging at 320 m^3/km/d (136 080 gal/mi/d) over a distance of 1 km (0.62 mi) with a microorganism concentration of 10^3/100 mL. The t_{90} time for indicator microorganisms in the stream is 22 h. What is the concentration of indicator microorganisms 2, 5.5, and 10 km (1.24, 3.42, and 6.21 mi) below the point source?
5. A city discharges 1.50×10^4 m^3/d (3.96 Mgal/d) of raw sewage into a stream with a discharge of 2.2 m^3/s (77.7 ft^3/s) and a temperature of 17°C. The sewage temperature is 15°C and its $BOD_{5,20}$ is 250 mg/L. What is the critical oxygen deficit and at what time does it occur? The stream is saturated with oxygen above the sewage discharge. Assume the stream-sewage temperature does not change after confluence. Use $k_1 = 0.15$ d^{-1}, $k_2 = 0.28$ d^{-1}, and assume that k_1 is the same for the stream and the sewage. Rate constants are given for a temperature of 20°C and referred to base 10.
6. A stream with $k_2 = 0.31$ d^{-1}, temperature of 28°C, and minimum flow of 23.0 m^3/s (812 ft^3/s) receives 3.80×10^4 m^3/d (10.0 Mgal/d) of sewage from a city. What is the maximum permissible BOD of the sewage if the DO content of the stream is never to fall below 4.0 mg/L? The river is saturated with DO above the sewage outfall. Assume $k_1 = 0.17$ d^{-1}. Rate constants are given for a temperature of 28°C in base 10.
7. (a) Discuss each of the phenomena influencing the DO sag curve. (b) Discuss the effect of a temperature change on each of the terms in the DO sag curve equation. Would a temperature increase or a temperature decrease result in an increased critical deficit? Would the time to reach the critical deficit increase or decrease?
8. Differentiate Eq. (21.17) and prove that it results in Eq. (21.16).
9. What is the equation for the DO deficit as a function of time when in addition to DO consumption by BOD exertion, there is oxygen consumption because of nitrification? Assume that nitrification proceeds according to a first-order reaction with a rate coefficient k_N and that the ammonia concentration is expressed as nitrogen.
10. Prove that waste with a composition of $C_{10}H_{19}O_3N$ has a potential ammonia oxygen demand of 0.32 mg O_2/mg $C_{10}H_{19}O_3N$.
11. What is the differential equation describing DO in a reach receiving a distributed waste flow that contains a significant concentration of BOD, if the velocity in the reach is relatively constant?
12. How do algae influence the oxygen budget of a stream?
13. (a) What is the Churchill et al. equation for the reaeration coefficient when v is in ft/s and H is in ft?
    (b) What is the equation for the dividing line between the equations of O'Connor-Dobbins and Churchill et al. on Fig. 21.7?
14. Which of the following changes will produce the largest increase in the rate of stream reaeration? (a) Decreasing the depth by 25%; (b) increasing the velocity by 25%; or (c)

increasing the temperature from 10 to 20°C. Use the Churchill et al. equation. Assume all other variables remain constant except the variable indicated in (a) and (b). For (c) assume that the DO is 0.

KEY REFERENCE

THOMANN, R. V. AND J. A. MUELLER (1987), *Principles of Surface Water Quality Modeling and Control,* Harper & Row, Washington, DC.

REFERENCES

BELL, P. R. F. (1992), "Eutrophication and Coral Reefs—Some Examples in the Great Barrier Reef Lagoon," *Water Research,* 26, 5, pp. 553–568.

BROWN L. C. AND T. O. BARNWELL, JR. (1987), *The Enhanced Stream Water Quality Models QUAL2E and QUAL2E-UNCAS: Documentation and User Manual,* Environmental Research Laboratory, USEPA, Athens, GA.

CAMP, T. R. AND R. L. MESERVE (1974), *Water and Its Impurities,* 2nd ed., Dowden, Hutchinson and Ross, Stroudsbourg, PA.

CHURCHILL, M., H. ELMORE, AND R. BUCKINGHAM (1962), "The Prediction of Stream Reaeration Rates," *J. Sanitary Engineering Division, ASCE,* 88, SA4, pp. 1–46.

CHRISTENSEN, D. R. AND P. L. MCCARTY (1975), "Multi-Purpose Biological Treatment Model," *J. Water Pollution Control Federation,* 47, 11, pp. 2652–2664.

COVAR, A. P. (1976), "Selecting the Proper Reaeration Coefficient for Use in Water Quality Models," presented at USEPA Conference on Environmental Simulation and Modeling, April 19–22.

ELIASSEN, R. (1952), "Stream Pollution," *Scientific American,* 186, 3, pp. 17–21.

MOE (1984), *Water Management: Goals, Policies, Objectives and Implementation Procedures of the Ministry of the Environment,* Ontario Ministry of the Environment, Toronto.

O'CONNOR, D. J. AND W. E. DOBBINS (1958), "Mechanism of Reaeration in Natural Streams," *Transactions of the American Society of Civil Engineers,* 123, pp. 641–666.

OTT, W. R. (1978), *Environmental Indices: Theory and Practice*, Ann Arbor Science, Ann Arbor, MI.

OWENS, M., R. W. EDWARDS, AND J. W. GIBBS (1964), "Some Reaeration Studies in Streams," *International J. Air and Water Pollution,* 8, pp. 469–486.

PRATI, L., R. PAVANELLO, AND J. T. HANEY (1971), "Assessment of Surface Water Quality by a Single Index of Pollution," *Water Research,* 5, 9, pp. 741–751.

RAILSBACK, S. F., J. M. BOWNDS, M. J. SALE, M. M. STEVENS, AND G. H. TAYLOR (1990), "Aeration at Ohio River Basin Navigation Dams," *J. of Environmental Engineering, ASCE,* 116, EE2, pp. 361–375.

STREETER, H. W. AND E. B. PHELPS (1925), "A Study of the Pollution and Purification of the Ohio River," *Public Health Bulletin,* 146, U.S. Public Health Service, Washington, DC.

TETRA TECH, INC. (1978), *Rates, Constants, and Kinetics Formulations in Surface Water Quality Modeling,* Report No. EPA-600/3–78-105, EPA, Athens, GA.

USEPA (1976a), *Areawide Assessment Procedures Manual,* vol. I, EPA-600/9–76-014, Municipal Environmental Research Laboratory, Cincinnati, OH.

USEPA (1976b), *Quality Criteria for Water,* USEPA, Washington, DC.

VOLLENWEIDER, R. A. (1970), *Scientific Fundamentals of the Eutrophication of Lakes and Flowing Waters with Particular Reference to Nitrogen and Phosphorus as Factors in Eutrophication,* Organization for Economic Co-Operation and Development, Paris, France.

APPENDIX

TABLE A.1 Most Probable Number of Coliforms per 100 mL of Sample

Number of positive tubes				Number of positive tubes				Number of positive tubes				Number of positive tubes				Number of positive tubes				Number of positive tubes			
10 mL	1 mL	0.1 mL	MPN	10 mL	1 mL	0.1 mL	MPN	10 mL	1 mL	0.1 mL	MPN	10 mL	1 mL	0.1 mL	MPN	10 mL	1 mL	0.1 mL	MPN	10 mL	1 mL	0.1 mL	MPN
0	0	0		1	0	0	2.0	2	0	0	4.5	3	0	0	7.8	4	0	0	13	5	0	0	23
0	0	1	1.8	1	0	0	4.0	2	0	1	6.8	3	0	1	11	4	0	1	17	5	0	1	35
0	0	2	3.6	1	0	2	6.0	2	0	2	9.1	3	0	2	13	4	0	2	21	5	0	2	43
0	0	3	5.4	1	0	3	8.0	2	0	3	12	3	0	3	16	4	0	3	25	5	0	3	58
0	0	4	7.2	1	0	4	10	2	0	4	14	3	0	4	20	4	0	4	30	5	0	4	76
0	0	5	9.0	1	0	5	12	2	0	5	16	3	0	5	23	4	0	5	36	5	0	5	95
0	1	0	1.8	1	1	0	4.0	2	1	0	6.8	3	1	0	11	4	1	0	17	5	1	0	33
0	1	1	3.6	1	1	0	6.1	2	1	1	9.2	3	1	1	14	4	1	1	21	5	1	1	46
0	1	2	5.5	1	1	2	8.1	2	1	2	12	3	1	2	17	4	1	2	26	5	1	2	64
0	1	3	7.3	1	1	3	10	2	1	3	14	3	1	3	20	4	1	3	31	5	1	3	84
0	1	4	9.1	1	1	4	12	2	1	4	17	3	1	4	23	4	1	4	36	5	1	4	110
0	1	5	11	1	1	5	14	2	1	5	19	3	1	5	27	4	1	5	42	5	1	5	130
0	2	0	3.7	1	2	0	6.1	2	2	0	9.3	3	2	0	14	4	2	0	22	5	2	0	49
0	2	1	5.5	1	2	0	8.2	2	2	1	12	3	2	1	17	4	2	1	26	5	2	1	70
0	2	2	7.4	1	2	2	10	2	2	2	14	3	2	2	20	4	2	2	32	5	2	2	95
0	2	3	9.2	1	2	3	12	2	2	3	17	3	2	3	24	4	2	3	38	5	2	3	120
0	2	4	11	1	2	4	15	2	2	4	19	3	2	4	27	4	2	4	44	5	2	4	150
0	2	5	13	1	2	5	17	2	2	5	22	3	2	5	31	4	2	5	50	5	2	5	180

0	3	0	5.6	1	3	0	8.3	2	3	0	12	3	3	0	17	4	3	0	27	5	3	0	79
0	3	1	7.4	1	3	0	10	2	3	1	14	3	3	1	21	4	3	1	33	5	3	1	110
0	3	2	9.3	1	3	2	13	2	3	2	17	3	3	2	24	4	3	2	39	5	3	2	140
0	3	3	11	1	3	3	15	2	3	3	20	3	3	3	28	4	3	3	45	5	3	3	180
0	3	4	13	1	3	4	17	2	3	4	22	3	3	4	31	4	3	4	52	5	3	4	210
0	3	5	15	1	3	5	19	2	3	5	25	3	3	5	35	4	3	5	59	5	3	5	250
0	4	0	7.5	1	4	0	11	2	4	0	15	3	4	0	21	4	4	0	34	5	4	0	130
0	4	1	9.4	1	4	0	13	2	4	1	17	3	4	1	24	4	4	1	40	5	4	1	170
0	4	2	11	1	4	2	15	2	4	2	20	3	4	2	28	4	4	2	47	5	4	2	220
0	4	3	13	1	4	3	17	2	4	3	23	3	4	3	32	4	4	3	54	5	4	3	280
0	4	4	15	1	4	4	19	2	4	4	25	3	4	4	36	4	4	4	62	5	4	4	350
0	4	5	17	1	4	5	22	2	4	5	28	3	4	5	40	4	4	5	69	5	4	5	430
0	5	0	9.4	1	5	0	13	2	5	0	17	3	5	0	25	4	5	0	41	5	5	0	240
0	5	1	11	1	5	0	15	2	5	1	20	3	5	1	29	4	5	1	48	5	5	1	350
0	5	2	13	1	5	2	17	2	5	2	23	3	5	2	32	4	5	2	56	5	5	2	540
0	5	3	15	1	5	3	19	2	5	3	26	3	5	3	37	4	5	3	64	5	5	3	920
0	5	4	17	1	5	4	22	2	5	4	29	3	5	4	41	4	5	4	72	5	5	4	1 600
0	5	5	19	1	5	5	24	2	5	5	32	3	5	5	45	4	5	5	81				

TABLE A.2 Dissolved Oxygen Saturation Concentrations,[a] mg/L

Temperature °C	Chloride concentration, mg/L					
	0	5 000	10 000	15 000	20 000	25 000
0	14.621	13.728	12.888	12.097	11.355	10.657
1	14.216	13.356	12.545	11.783	11.066	10.392
2	13.829	13.000	12.218	11.483	10.790	10.319
3	13.460	12.660	11.906	11.195	10.526	9.897
4	13.107	12.335	11.607	10.920	10.273	9.664
5	12.770	12.024	11.320	10.656	10.031	9.441
6	12.447	11.727	11.046	10.404	9.799	9.228
7	12.139	11.442	10.783	10.162	9.576	9.023
8	11.843	11.169	10.531	9.930	9.362	8.826
9	11.559	10.907	10.290	9.707	9.156	8.636
10	11.288	10.656	10.058	9.493	8.959	8.454
11	11.027	10.415	9.835	9.287	8.769	8.279
12	10.777	10.183	9.621	9.089	8.586	8.111
13	10.537	9.961	9.416	8.899	8.411	7.949
14	10.306	9.747	9.218	8.716	8.242	7.792
15	10.084	9.541	9.027	8.540	8.079	7.642
16	9.870	9.344	8.844	8.370	7.922	7.496
17	9.665	9.153	8.667	8.207	7.770	7.356
18	9.467	8.969	8.497	8.049	7.624	7.221
19	9.276	8.792	8.333	7.896	7.483	7.090
20	9.092	8.621	8.174	7.749	7.346	6.964
21	8.915	8.456	8.021	7.607	7.214	6.842
22	8.743	8.297	7.873	7.470	7.087	6.723
23	8.578	8.143	7.730	7.337	6.963	6.609
24	8.418	7.994	7.591	7.208	6.844	6.498
25	8.263	7.850	7.457	7.083	6.728	6.390
26	8.113	7.711	7.327	6.962	6.615	6.285
27	7.968	7.575	7.201	6.845	6.506	6.184
28	7.827	7.444	7.079	6.731	6.400	6.085
29	7.691	7.317	6.961	6.621	6.297	5.990
30	7.559	7.194	6.845	6.513	6.197	5.896
35	6.950	6.624	6.314	6.017	5.734	5.464
40	6.412	6.121	5.842	5.576	5.321	5.078
45	5.927	5.665	5.414	5.174	4.944	4.724
50	5.477	5.242	5.016	4.799	4.591	4.392

[a]From *Standard Methods for the Examination of Water and Wastewater,* 18th ed., copyright 1992 by APHA, AWWA, and WEF, American Public Health Association, Washington, DC. Reprinted with permission.

TABLE A.3 Chemicals Used in Water and Wastewater Treatment[a]

Agent	Bulk density g/100 mL	Active ingredients %
Chelating Agents		
Na_4EDTA (liq)	127–131	39
$Na_4EDTA \cdot 4H_2O$ (dry)	78	99
Nitrilotriacetic acid (dry)	89	99.5
Coagulants and Flocculants		
$Al_2(SO_4)_3 \cdot 14H_2O$ (dry)	90–115	
$Al_2(SO_4)_3 \cdot 14H_2O$ (liq)	134	49
$FeCl_3$ (liq)	137	28–47
$FeCl_2$ (liq)	118–132	18–28
$Fe_2(SO_4)_3$ (50% soln)	149	%Fe(III), 10–10.5
Polymers[b]	110–140	%TS, 24–55
Oxidants and Disinfectants		
$BrCl_2$ (mixture of Br_2, Cl_2, and $BrCl_2$)	232	%Br, 70 %Cl, 30
$Ca(OCl)_2$ (dry)	80–98	65
Cl_2 (liq)	1.47 (0°C)	>99
H_2O_2	90–129	30–70
$KMnO_4$	138–160	97–99
$NaHSO_3$ (liq)	133	40
NaOCl (liq)	122	13.1–14.9
Other		
CaO (dry)	96	80–95
$Ca(OH)_2$ (dry)	48	82–95
$CuSO_4$ (dry)	120–144	99
H_2SO_4	170–182	62–93
H_3PO_4	157–169	75–85
H_2SiF_6 (fluosilicic acid)	123	24
Greensand	136	—
KCl (dry)	103	>99
NaCl (dry)	105–121	98.6–99.9
Na_2CO_3 (dry)	253	>99
NaF (dry)	144–168	95–98
NaOH (liq)	153–170	50–73
NH_4OH (liq)	90	$\%NH_3$, 28–30

[a]From E. W. Flick (1991), *Water Treatment Chemicals: An Industrial Guide,* Noyes Publications, Park Ridge, NJ, and S. Kawamura (1991), *Integrated Design of Water Treatment Facilities,* Wiley-Interscience, Toronto.
[b]MW: 2 000–60 000.

TABLE A.4 Conversion Factors and Constants

Acceleration of gravity, g = 9.807 m/s^2 (32.174 ft/s^2)
Universal gas constant, R = 1.987 2 cal deg^{-1} mole^{-1}
= 0.082 054 L-atm deg^{-1} mole^{-1}
= 8.314 4 joules deg^{-1} mole^{-1}
Standard atmosphere = 101.325 kN/m^2 (14.696 lb$_f$/in.^{2})
= 101.325 kPa (1.013 bar)
= 10.333 m (33.899 ft) of water
1 bar = 10^5 N/m^2 (14.504 lb$_f$/in.^{2})
Density of dry air at 0°C and 1 atm = 0.001 293 g/cm^3
1 gram molecular volume of a gas at 0°C and 1 atm = 22.4 L
1 meter head of water (20°C) = 0.009 79 N/mm^2 (1.420 lb$_f$/in.^{2})
= 9.790 kN/m^2 (1.420 lb$_f$/in.^{2})

1 acre = 43 560 ft^2	1 ft^2 = 2.296 × 10^{-5} acre
1 amp = 1 coulomb/s	
1 atm = 760 mm of Hg = 29.921 3 in. of Hg	
1 Btu = 1.055 1 kJ	1 kJ = 0.947 8 Btu
1 cal (20°C) = 0.003 966 Btu	1 Btu = 252.1 cal (20°C)
1 cal = 4.184 J	1 J = 0.239 0 cal
1 centipoise = 0.01 g/cm-s	
1° Centigrade = 1.800°F	1° Fahrenheit = 0.555 6°C
°C = $\frac{5}{9}$ (°F − 32)	°F = $\frac{9}{5}$°C + 32
°K = °C + 273.16	
1 Faraday = 96 493 coulombs/equiv	
= 23 062.4 cal/volt-equiv	
1 fathom = 1.829 m	1 m = 0.546 8 fathom
1 ft^3 = 7.480 51 U.S. gal	1 U.S. gal = 0.133 680 ft^3
1 ft^3 = 28.316 L	1 L = 0.035 31 ft^3
1 ft-lb/s = 1.355 8 J/s	1 J/s = 0.737 6 ft-lb/s
1 grain = 0.064 80 g	1 g = 15.43 grain
1 Imp. gal = 4.546 L	1 L = 0.220 0 Imp. gal
1 U.S. gal = 3.785 L	1 L = 0.264 2 U.S. gal
1 U.S. gal = 0.133 68 ft^3	1 ft^3 = 7.480 5 U.S. gal
1 U.S. gal = 0.003 785 m^3	1 m^3 = 264.17 U.S. gal
1 ha = 2.471 acres	1 acre = 0.404 7 ha
= 10 000 m^2	
1 hp = 745.7 J/s	1 J/s = 0.001 341 hp
1 in. = 25.4 mm	1 mm = 0.039 4 in.
1 kg = 2.204 62 lb	1 lb = 0.453 59 kg
1 knot = 1.852 km/h	1 km/h = 0.540 0 knot
1 km = 0.621 4 mi	1 mi = 1.609 km
1 000 L = 1 m^3	1 L = 0.001 00 m^3
1 lumen (at 5 550 A) = 0.001 471 watt	1 watt = 679.8 lumen (at 5 550 A)
1 lumen/cm^2 = 1 Lambert	
1 lumen/m^2 = 0.092 9 foot-candle	1 foot-candle = 10.76 lumen/cm^2
1 m= 3.281 ft	1 ft = 0.304 8 m
1 m^3 = 35.31 ft^3	1 ft^3 = 0.028 32 m^3
1 mi = 5 280 ft	1 ft = 1.894 × 10^{-4} mi
1 N = 1 kg-m/s^2	
pi = 3.141 592 65	
1 lb = 453.6 g	1 g = 0.002 204 lb
= 16 oz	
1 stoke = 1 cm^2/s	
1 ton (U.S.) = 2 000 lb	
1 tonne (metric) = 1 000 kg	
1 tonne = 2 204.6 lb	1 lb = 4.536 × 10^{-4} tonne
1 W = 1 J/s	
1 W = 0.737 6 ft-lb/s	1 ft-lb/s = 1.355 8 W

TABLE A.4 Continued

Loading Rate and Other Conversion Factors Commonly Used in Environmental Engineering	
1 kg/ha/d = 0.892 2 lb/acre/d	1 lb/acre/d = 1.120 9 kg/ha/d
1 $kg/m^2/d$ = 0.204 8 $lb/ft^2/d$	1 $lb/ft^2/d$ = 4.882 $kg/m^2/d$
1 $kg/m^3/d$ = 0.062 43 $lb/ft^3/d$	1 $lb/ft^3/d$ = 16.01 $kg/m^3/d$
1 kg/m^3 = 0.062 43 lb/ft^3	1 lb/ft^3 = 16.019 kg/m^3
1 $m^3/m^2/d$ = 24.54 $gal/ft^2/d$	1 $gal/ft^2/d$ = 0.040 7 $m^3/m^2/d$
1 $m^3/m/d$ = 80.519 6 gal/ft/d	1 $gal/ft^2/d$ = 0.133 7 ft/d
1 m^3/kg = 16.019 ft^3/lb	1 $gal/ft^2/d$ = 0.000 001 547 ft/s
1 $L/m^2/s$ = 1.473 $gal/min/ft^2$	1 gal/ft/d = 0.012 42 $m^3/m/d$
1 kJ/kg = 0.430 3 Btu/lb	1 ft^3/lb = 0.06 243 m^3/kg
1 kg/ha = 0.892 2 lb/acre	1 $gal/min/ft^2$ = 0.679 1 $L/m^2/s$
1 kg/tonne = 2.000 lb/ton	1 Btu/lb = 2.324 kJ/kg
1 kW/m^3 = 0.038 0 hp/1 000 ft^3	1 lb/acre = 1.120 9 kg/ha
1 $m^3/ha/d$ = 106.91 gal/acre/d	1 lb/ton = 0.500 kg/tonne
1 L/mg = 0.119 8 Mgal/lb	1 hp/1 000 ft^3 = 26.33 kW/m^3
1 kPa (gauge) = 0.145 0 $lb/in.^2$ (gauge)	1 gal/acre/d = 0.009 354 $m^3/ha/d$
	1 Mgal/lb = 8.344 L/mg
	1 $lb/in.^2$ (gauge) = 6.894 8 kPa (gauge)

NORMAL DISTRIBUTION

Density function for normally distributed variable with mean, μ, and standard deviation, σ:

$$p(x) = \frac{1}{\sigma\sqrt{2\pi}} e^{-(x-\mu)^2/2\sigma^2} \qquad \text{or} \qquad f(z) = \frac{1}{\sqrt{2\pi}} e^{-z^2/2}$$

where

$$z = \frac{x - \mu}{\sigma}$$

Cumulative distribution function for normally distributed variable:

$$F(z) = \frac{1}{\sqrt{2\pi}} \int_0^z e^{-z^2/2}\, dz$$

INTEGRATING FACTOR FOR LINEAR DIFFERENTIAL EQUATIONS OF THE FIRST ORDER

A first-order linear differential equation of the form

$$y' + P(x)y = Q(x)$$

where

P and Q are continuous functions of x

can be solved by setting $Q(x) = 0$. This leads to the separable equation

$$\frac{y'}{y} = -P(x)$$

if $y \neq 0$. Integrating,

$$\ln |y| = -\int P(x)\,dx + \ln |C|$$

where

C is a constant

Raising each side to the base e,

$$y = Ce^{-\int P(x)\,dx} \qquad \text{or} \qquad ye^{\int P(x)\,dx} = C$$

Now,

$$\frac{d}{dx}\left(ye^{\int P(x)\,dx}\right) = y'e^{\int P(x)\,dx} + P(x)ye^{\int P(x)\,dx} = e^{\int P(x)\,dx}[y' + P(x)y]$$

If each side of the original differential equation is multiplied by $e^{\int P(x)\,dx}$, the equation may be written

$$\frac{d}{dx}\left(ye^{\int P(x)\,dx}\right) = Q(x)e^{\int P(x)\,dx}$$

The implicit solution of this equation is

$$ye^{\int P(x)\,dx} = \int Q(x)e^{\int P(x)\,dx}\,dx + C$$

which can be solved for an explicit solution.

The expression $e^{\int P(x)\,dx}$ is known as an integrating factor.

TABLE A.5 Physical Properties of Water (SI Units)[a]

Temperature °C	Specific weight γ kN/m^3	Density ρ kg/m^3	Dynamic viscosity $\mu \times 10^{3}$[b] N-s/m^2	Kinematic viscosity $\nu \times 10^{6}$[b] m^2/s	Surface tension σ N/m	Vapor pressure p_v kN/m^2
0	9.805	999.8	1.781	1.785	0.076 5	0.61
5	9.807	1 000.0	1.518	1.519	0.074 9	0.87
10	9.804	999.7	1.307	1.306	0.074 2	1.23
15	9.798	999.1	1.139	1.139	0.073 5	1.70
20	9.789	998.2	1.002	1.003	0.072 8	2.34
25	9.777	997.0	0.890	0.893	0.072 0	3.17
30	9.764	995.7	0.798	0.800	0.071 2	4.24
40	9.730	992.2	0.653	0.658	0.069 6	7.38
50	9.689	988.0	0.547	0.553	0.067 9	12.33
60	9.642	983.2	0.466	0.474	0.066 2	19.92
70	9.589	977.8	0.404	0.413	0.064 4	31.16
80	9.530	971.8	0.354	0.364	0.062 6	47.34
90	9.466	965.3	0.315	0.326	0.060 8	70.10
100	9.399	958.4	0.282	0.294	0.058 9	101.33

[a]From J. K. Vennard and R. L. Street, *Elementary Fluid Mechanics,* 5th ed., copyright © 1975 by John Wiley & Sons, Inc. Reprinted by permission of John Wiley & Sons, Inc.

[b]In this text, when a table heading indicates that the number is multiplied by some factor, the recorded number must be divided by this factor to obtain the correct result. For instance, the dynamic viscosity at 0°C is $z = 1.781 \times 10^{-3}$ N-s/m^2. The number recorded is $z \times (10^3) = 1.781 \times 10^{-3} \times 10^3 = 1.781$.

TABLE A.6 Physical Properties of Water (U.S. Units)[a]

Temperature °F	Specific weight γ lb/ft³	Density ρ slug/ft³	Dynamic viscosity $\mu \times 10^5$ lb-s/ft²	Kinematic viscosity $\nu \times 10^5$ ft²/s	Surface tension σ lb/ft	Vapor pressure p_v lb/ft²
32	62.42	1.940	3.746	1.931	0.005 18	0.09
40	62.43	1.940	3.229	1.664	0.006 14	0.12
50	62.41	1.940	2.735	1.410	0.005 09	0.18
60	62.37	1.938	2.359	1.217	0.005 04	0.26
70	62.30	1.936	2.050	1.059	0.004 98	0.36
80	62.22	1.934	1.799	0.930	0.004 92	0.51
90	62.11	1.931	1.595	0.826	0.004 86	0.70
100	62.00	1.927	1.424	0.739	0.004 80	0.95
120	61.71	1.918	1.168	0.609	0.004 67	1.69
140	61.38	1.908	0.981	0.514	0.004 54	2.89
160	61.00	1.896	0.838	0.442	0.004 41	4.74
180	60.58	1.883	0.726	0.385	0.004 27	7.51
200	60.12	1.868	0.637	0.341	0.004 13	11.52
212	59.83	1.860	0.593	0.319	0.004 04	14.70

[a]From J. K. Vennard and R. L. Street, *Elementary Fluid Mechanics,* 5th ed., copyright © 1975 by John Wiley & Sons, Inc. Reprinted by permission of John Wiley & Sons, Inc.

TABLE A.7 Derived Units

Quantity	SI Unit	Symbol	Expression in other units	Expression in SI units
Electric charge	coulomb	C		A s
Electric potential	volt	V	W/A	m^2 kg s^{-3} A^{-1}
Electric resistance	ohm	Ω	V/A	m^2 kg s^{-3} A^{-1}
Energy, work, heat	joule	J	N m	m^2 kg s^{-1}
Frequency	hertz	Hz		s^{-1}
Force	newton	N		m kg s^{-2}
Pressure, stress	pascal	Pa	N/m^2	kg m^{-1} s^{-2}
Power	watt	W	J/s	m^2 kg s^{-2}

FIRMS AND EQUIPMENT SUPPLIERS QUOTED IN THE TEXT

Degrémont Infilco Ltd.
160-D, Saint-Joseph Blvd.
Lachine, QC H8S 2L3

Eimco Process Equipment
669 West Second South
P.O. Box 300
Salt Lake City, UT 84110-0300

Envirex Inc.
P.O. Box 1604
Waukesha, WI 53187

Gore & Storrie Limited
Consulting Engineers
255 Consumers Road
North York, ON M2J 5B6

Lakeside Equipment Corporation
1022 E. Devon Ave.
P.O. Box 8448
Bartlett, IL 60103

Parkson
9050 Ryan
Dorval, QC H9P 2M8

Phipps & Bird
8741 Landmark Road
P.O. Box 27324
Richmond, VA 23261-7324

Stranco
P.O. Box 389
Bradley, IL 60915-0389

Supelco Canada Ltd.
1300 Aimco Blvd.
Mississauga, ON L4W 1B2

Trojan Technologies, Inc.
3020 Gore Road
London, ON N5V 4T7

Zimpro Environmental, Inc.
301 W. Military Rd.
Rothschild, WI 54474

TABLE A.8 Commercial Pipe Sizes

Metric sizes	Existing U.S. pipe sizes	
mm	in.	mm
	4	101.6
	5	127.0
150	6	152.4
200	8	203.2
250	10	254.0
300	12	304.8
	14	355.6
	15	381.0
400	16	406.4
	18	457.2
500	20	508.0
	21	533.4
600	24	609.6
700	27	685.8
800	30	762.0
900	36	914.4
1 000	42	1 066
1 200	48	1 219
1 400	54	1 371
1 600	60	1 524
	66	1 676
1 800	72	1 829
2 000	78	1 981
	84	2 134
2 200	90	2 286
2 400	96	2 438
	102	2 591
2 700	108	2 743
	114	2 896
3 000	120	3 048
3 300		
3 600	144	3 658
	180	4 572
	204	5 182
	240	6 096

TABLE A.9 Typical Wall Roughness Factors for Commercial Conduits[a]

Material (new)	Roughness height	
	m	ft
Riveted steel	0.000 09–0.009	0.003–0.03
Concrete	0.000 3–0.003	0.001–0.01
Wood stave	0.000 2–0.000 9	0.000 6–0.003
Cast iron	0.000 26	0.000 85
Galvanized iron	0.000 15	0.000 5
Asphalted cast iron	0.000 1	0.000 4
Commercial steel or wrought iron	0.000 046	0.000 15
Drawn brass or copper tubing	0.000 001 5	0.000 005
Glass and plastic	Smooth	Smooth

[a]Adapted from L. F. Moody (1944), "Friction Factors for Pipe Flow," *Transactions of American Society of Mechanical Engineers,* 66, pp. 671–684. Used with permission of the American Society of Mechanical Engineers.

MOODY DIAGRAM[1]

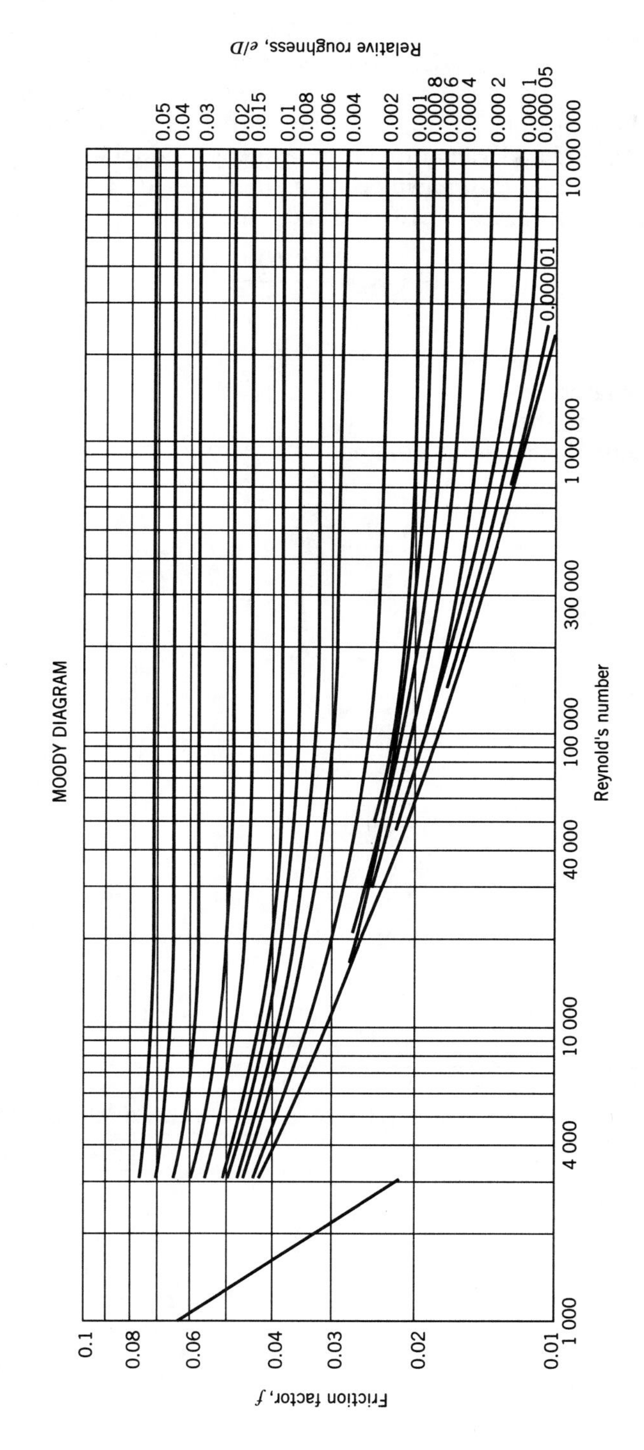

[1]L. F. Moody (1944), "Friction Factors for Pipe Flow," *Transactions American Society of Mechanical Engineers,* 66, pp. 671–684. Used with the permission of The American Society of Mechanical Engineers.

NOMOGRAPH FOR HYDROXYL ALKALINITY DETERMINATION

Adapted from *Standard Methods for the Examination of Water and Wastewater* (1992), 18th ed., APHA, AWWA, and WEF, American Public Health Association, Washington, DC and based on the regression for pK_w given by L. N. Plummer and E. Busenberg (1982), "The Solubilities of Calcite, Aragonite and Veterite in CO_2-H_2O Solutions between 0 and 90°C and an Evaluation of the Aqueous Model for the System $CaCO_3$-CO_2-H_2O," *Geochimica et Cosmochimic Acta,* 46, 6, pp. 1011–1040. Activity coefficients were calculated from Eq. (1.10) and the Güntelberg approximation:

$$-\log \gamma_i = \frac{0.5 Z_i^2 \mu^{1/2}}{1 + \mu^{1/2}}$$

where

μ is ionic strength
Z_i is the charge on the ion
γ_i is the activity coefficient of the ion.

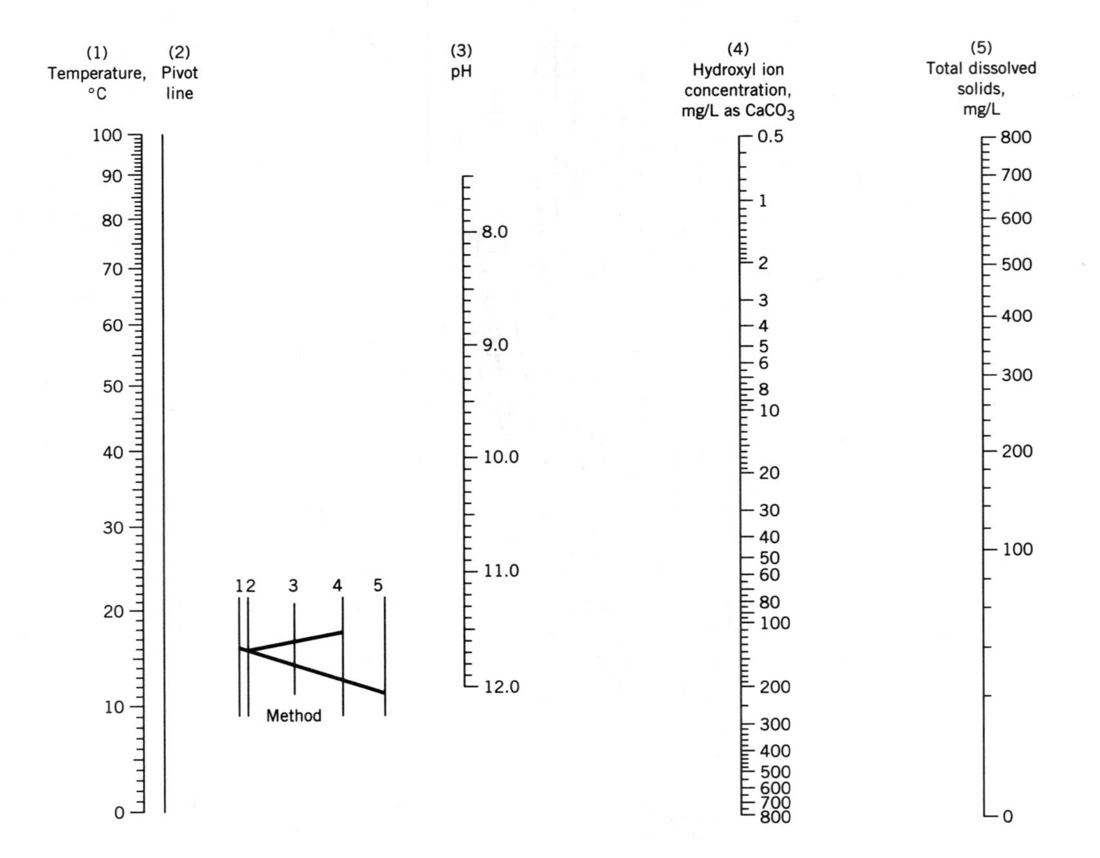

NOMOGRAPH FOR BICARBONATE ALKALINITY DETERMINATION

See notes under Nomograph for Hydroxyl Alkalinity Determination. The regression for pK_2 was used in this nomograph. Use the nomograph for hydroxyl alkalinity to find $[OH^-]$ for use in $[Alk] - [OH^-]$.

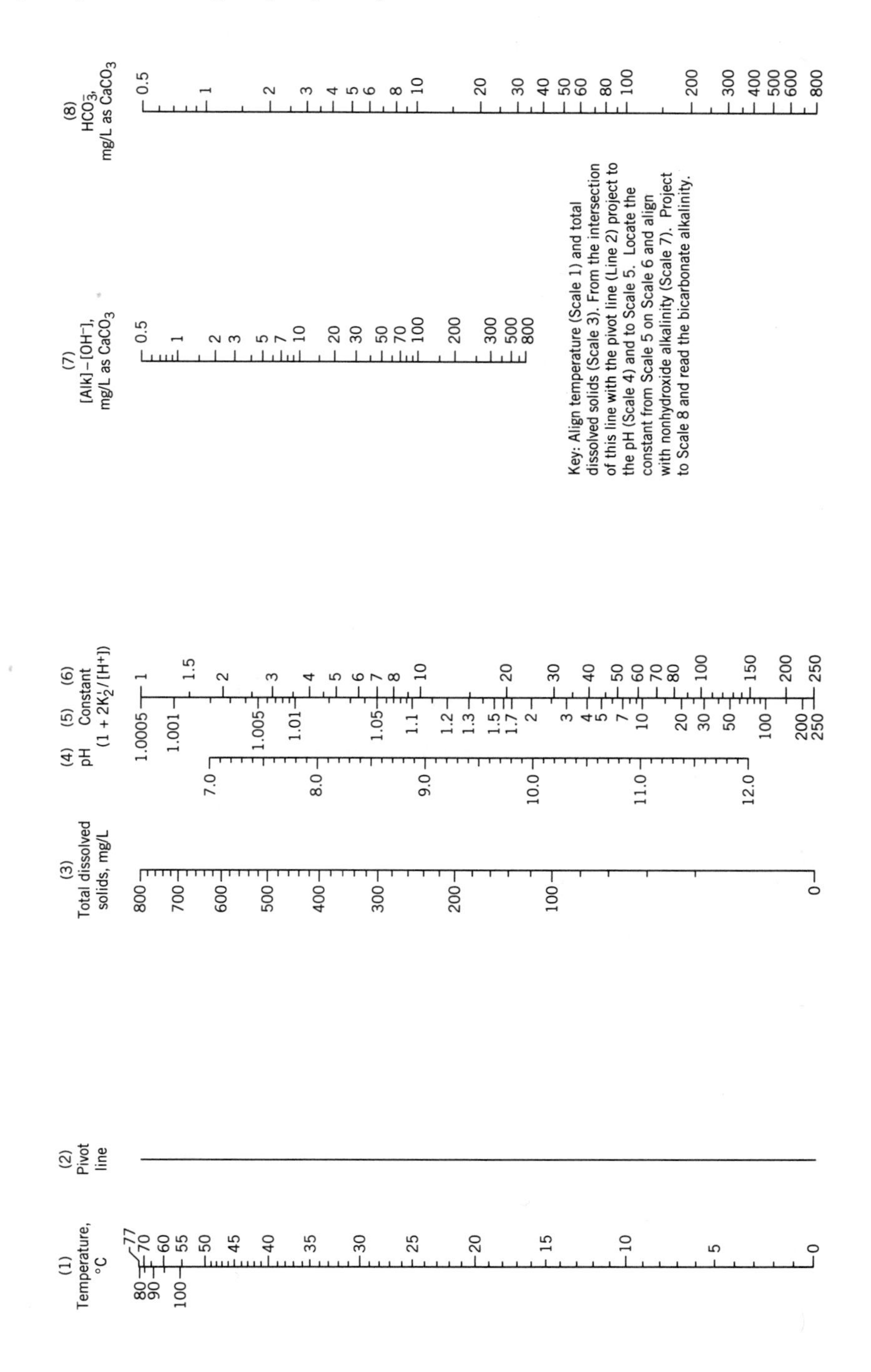

NOMOGRAPH FOR CARBONATE ALKALINITY DETERMINATION

See notes under Nomograph for Hydroxyl Alkalinity Determination. The regression for pK_2 was used in this nomograph. Use the nomograph for hydroxyl alkalinity to find $[OH^-]$ for use in $[Alk] - [OH^-]$.

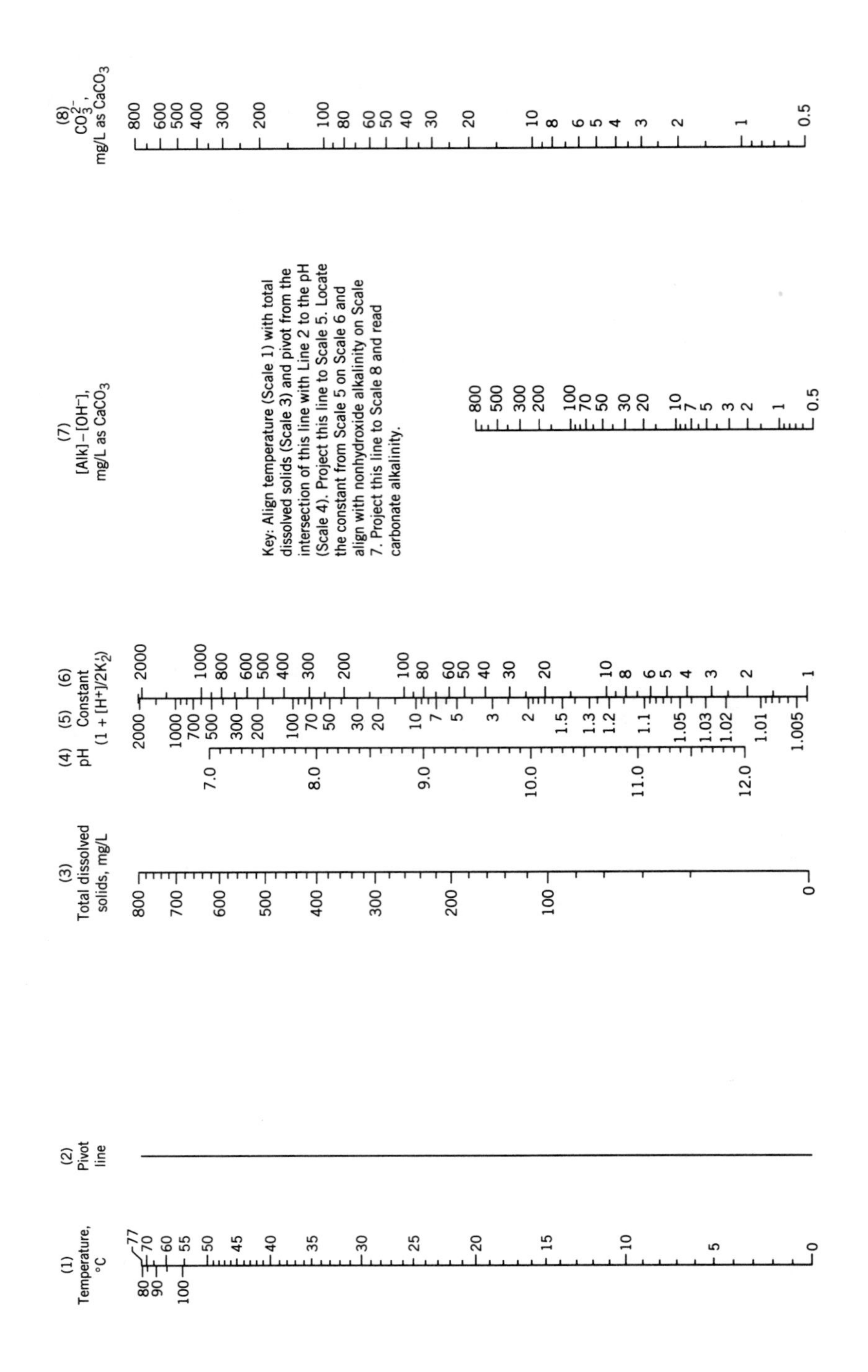

NOMOGRAPH FOR FREE CARBON DIOXIDE DETERMINATION

See notes under Nomograph for Hydroxyl Alkalinity Determination. The regression for pK_1 was used in this nomograph. Use the nomograph for bicarbonate alkalinity for $[HCO_3^-]$.

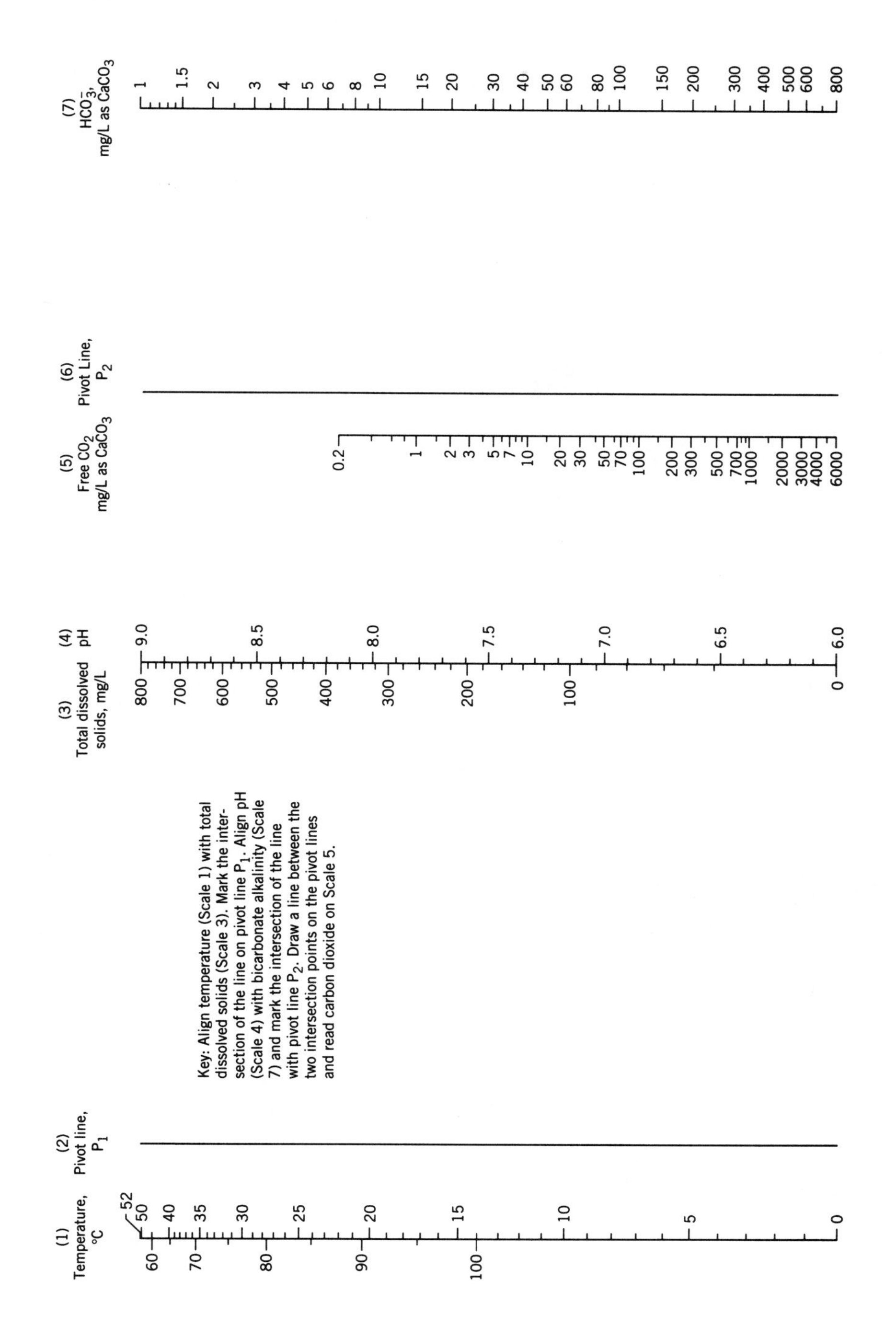

INDEX